Lecture Notes in Computer Science 6281

Advanced Research in Computing and Software Science

Subline of Lectures Notes in Computer Science

Petr Hliněný Antonín Kučera (Eds.)

Mathematical Foundations of Computer Science 2010

35th International Symposium, MFCS 2010
Brno, Czech Republic, August 23-27, 2010
Proceedings

Springer

Volume Editors

Petr Hliněný
Antonín Kučera
Masaryk University
Falculty of Informatics
Botanicka 68a
602 00 Brno, Czech Republic
E-mail: {hlineny,tony}@fi.muni.cz

Library of Congress Control Number: 2010932133

CR Subject Classification (1998): F.2, G.2, E.1, F.1, I.3.5, F.3

LNCS Sublibrary: SL 1 – Theoretical Computer Science and General Issues

ISSN 0302-9743
ISBN-10 3-642-15154-X Springer Berlin Heidelberg New York
ISBN-13 978-3-642-15154-5 Springer Berlin Heidelberg New York

Typesetting: Camera-ready by author, data conversion by Scientific Publishing Services, Chennai, India
Printed on acid-free paper 06/3180 5 4 3 2 1 0

Preface

The series of MFCS symposia, organized in rotation by Poland, Slovakia, and the Czech Republic since 1972, has a long and well-established tradition. The symposia encourage high-quality research in all branches of theoretical computer science. Their broad scope provides an opportunity to bring together researchers who do not usually meet at specialized conferences.

The 35th International Symposium on Mathematical Foundations of Computer Science (MFCS 2010) was organized in parallel with the 19th EACSL Annual Conference on Computer Science Logic (CSL 2010). The federated MFCS and CSL 2010 conference had shared plenary sessions and social events for all participants, but the scientific program and the proceedings were prepared independently for both events.

Out of 149 regular submissions to MFCS 2010, the Program Committee selected 56 papers for presentation at the conference and publication in this volume. Each paper was reviewed by at least three Program Committee members with the help of outside experts, and the actual selection was based on a subsequent electronic discussion.

In addition to the contributed papers, the scientific program of MFCS 2010 included five MFCS and CSL plenary talks delivered by David Basin (ETH Zürich), Herbert Edelsbrunner (IST Austria and Duke University), Erich Grädel (RWTH Aachen), Bojan Mohar (University of Ljubljana and Simon Fraser University), and Joseph Sifakis (CNRS), and three invited MFCS lectures by Andris Ambainis (University of Latvia), Juraj Hromkovič (ETH Zürich), and Daniel Lokshtanov (Universitetet i Bergen). We are grateful to the invited speakers for accepting our invitation and sharing their knowledge and skills with all MFCS 2010 participants.

The federated MFCS and CSL 2010 symposium was accompanied by the following satellite workshops on more specialized topics:

- *Program Extraction and Constructive Proofs / Classical Logic and Computation: Joint Workshop in Honor of Helmut Schwichtenberg*, organized by Steffen van Bakel, Stefano Berardi, Ulrich Berger, Hannes Diener, Monika Seisenberger, and Peter Schuster
- *Randomized and Quantum Computation*, organized by Rūsiņš Freivalds
- *Workshop on Fixed Points in Computer Science (FICS)*, organized by Zoltán Ésik and Luigi Santocanale
- *Young Researchers Forum*, organized by Jan Strejček
- *Theory and Algorithmic Aspects of Graph Crossing Number*, organized by Drago Bokal, Petr Hliněný, and Gelasio Salazar
- *YuriFest: Symposium on Logic in Computer Science Celebrating Yuri Gurevich's 70th Birthday*, organized by Nachum Dershowitz
- *International Workshop on Reachability Problems*, organized by Igor Potapov and Antonín Kučera

- *Games and Probabilistic Models in Formal Verification*, organized by Tomáš Brázdil and Krishnendu Chatterjee
- *Mathematical Foundations of Fuzzy Logics*, organized by Agata Ciabattoni
- *International Workshop on Categorical Logic*, organized by Jiří Rosický
- *Logic, Combinatorics and Computation*, organized by Bruno Courcelle, Petr Hliněný, and Johann A. Makowsky
- *Parametrized Complexity of Computational Reasoning*, organized by Stefan Szeider

As the editors of these proceedings, we would like to thank everyone who contributed to the success of the symposium. First of all, we thank the authors of all submitted papers for considering MFCS 2010 as an appropriate platform for presenting their work. Since the number of submissions was high, many good papers could not be accepted. We also thank the Program Committee members for their demanding and responsible work, the reviewers for careful reading of all the submissions, the EasyChair system for smooth conference submission management, and the staff at Springer for the professional support in producing this volume.

We also gratefully acknowledge financial support from the *European Research Consortium in Informatics and Mathematics (ERCIM)*, as well as the support from the Czech national research center *Institute for Theoretical Computer Science (ITI)*.

June 2010

Petr Hliněný
Antonín Kučera

Conference Organization

The 35th International Symposium on Mathematical Foundations of Computer Science (MFCS 2010) took place in Brno, Czech Republic, at the Faculty of Informatics of Masaryk University during August 23–27, 2010. MFCS 2010 was organized in parallel with the 19th EACSL Annual Conferences on Computer Science Logic (CSL 2010), and the federated conference was accompanied by 12 satellite workshops scheduled during August 21–22 and 28–29.

Program Committee

Luca Aceto *(Reykjavik)*
Jiří Adámek *(Braunschweig)*
Christel Baier *(Dresden)*
Patricia Bouyer-Decitre *(Cachan)*
Sergio Cabello *(Ljubljana)*
Witold Charatonik *(Wroclaw)*
Jurek Czyzowicz *(Quebec)*
Volker Diekert *(Stuttgart)*
Rod Downey *(Wellington)*
Fedor Fomin *(Bergen)*
Gregory Gutin *(London)*
Michel Habib *(Paris)*
Edward A. Hirsch *(St. Petersburg)*
Petr Hliněný, Co-chair *(Brno)*
Juraj Hromkovič *(Zurich)*
Juhani Karhumäki *(Turku)*
Ken-ichi Kawarabayashi *(Tokyo)*
Petr Kolman *(Prague)*
Daniel Král *(Prague)*
Rastislav Královič *(Bratislava)*
Antonín Kučera, Co-chair *(Brno)*
Luděk Kučera *(Prague)*
Madhavan Mukund *(Chennai)*
Jean-Eric Pin *(Paris)*
Alexander Rabinovich *(Tel Aviv)*
Peter Rossmanith *(Aachen)*
Davide Sangiorgi *(Bologna)*
Vladimiro Sassone *(Southampton)*
Helmut Seidl *(Munich)*
Jiří Sgall *(Prague)*
Daniel Štefankovič *(Rochester)*
Colin Stirling *(Edinburgh)*
Stefan Szeider *(Vienna)*
Andrzej Tarlecki *(Warsaw)*
Wolfgang Thomas *(Aachen)*
Pawel Urzyczyn *(Warsaw)*
Sue Whitesides *(Victoria)*
James Worrell *(Oxford)*
Shmuel Zaks *(Haifa)*

Organizing Committee

Jan Bouda, Chair
Tomáš Brázdil
Libor Caha
Matěj Čuhel
Jan Holeček
Dagmar Janoušková
Dana Komárková
Zbyněk Mayer
Matěj Pivoluska
Adam Rambousek
Tomáš Rebok
Jan Staudek
Tomáš Staudek
Šimon Suchomel
Marek Trtík

External Reviewers

Isolde Adler
Susanne Albers
Dmitry Antipov
Pedro Baltazar
Evangelos Bampas
Andrej Bauer
Christoph Behle
Andre Berger
Dietmar Berwanger
Laurent Bienvenu
Hans-Joachim Boeckenhauer
Artur Boronat
Glencora Borradaile
Yacine Boufkhad
Romain Brenguier
Václav Brožek
Binh-Minh Bui-Xuan
Sergiu Bursuc
Julia Böttcher
Arnaud Carayol
Felice Cardone
Pascal Caron
Olivier Carton
Xi Chen
Siu-Wing Cheng
Miroslav Chlebik
Jacek Chrzaszcz
Josef Cibulka
David Cohen
Thomas Colcombet
José Félix Costa
Bruno Courcelle
Robert Crowston
Deepak D'Souza
Peter Damaschke
Carsten Damm
Samir Datta
Claire David
Hans de Nivelle
Frederic Dorn
Laurent Doyen
Vida Dujmovic
Zdeněk Dvořák
Ehab ElSalamouny
Jörg Endrullis
Zoltan Esik
Louis Esperet
Piotr Faliszewski
Michael Fellows
Marcelo Fiore
Diana Fischer
Herbert Fleischner
Michal Forisek
Dimitris Fotakis
Wojciech Fraczak
Pierre Fraigniaud
Sibylle Froeschle
Christiane Frougny
Emanuele Fusco
Johannes Klaus Fichte
Peter Gacs
William Gasarch
Leszek Gasieniec
Mauro Gaspari
Tomáš Gavenčiak
Paweł Gawrychowski
Qi Ge
Archis Ghate
George Giakoupis
Amy Glen
Mordecai Golin
Petr Golovach
Jean Goubault-Larrecq
Marcus Grösser
Pierre Guillon
Christoph Haase
MohammadTaghi Hajiaghayi
Vesa Halava
Magnús Halldórsson
Kristoffer Arnsfelt Hansen
Ramesh Hariharan
Tero Harju
Ulrich Hertrampf
Alexander Herz
Tom Hirschowitz
Mika Hirvensalo
John Hitchcock
Jan Hladký

Michael Hochman
Nigel Horspool
Paul Hunter
David Ilcinkas
Toshimasa Ishii
Dmitry Itsykson
Tao Jiang
Mark Jones
Eun Jung Kim
Lukasz Kaiser
Naonori Kakimura
Michael Kaminski
Naoyuki Kamiyama
Jarkko Kari
Alexander Kartzow
Telikepalli Kavitha
Akitoshi Kawamura
Lucia Keller
Przemysław Kiciak
Stefan Kiefer
Emanuel Kieroński
Daniel Kirsten
Bjoern Kjos-Hanssen
Ralf Klasing
Pavel Klavík
Joachim Klein
Bartek Klin
Joachim Kneis
Yusuke Kobayashi
Dennis Komm
Beata Konikowska
Michal Koucký
Ioannis Koutis
Máté Kovács
Łukasz Kowalik
Tomasz Krawczyk
Stephan Kreutzer
Manfred Kufleitner
Alexander Kulikov
Dietrich Kuske
Tomi Kärki
Geoffrey LaForte
Michael Lampis
Alexander Langer
Jürn Laun
Ranko Lazic
Didier Le Botlan
Troy Lee
Wen-shin Lee
Jerome Leroux
Asaf Levin
Fei Li
Vincent Limouzy
Kamal Lodaya
Markus Lohrey
Sylvain Lombardy
John Longley
Michael Luttenberger
Carsten Lutz
Borut Lužar
Kamesh Madduri
Meena Mahajan
Sebastian Maneth
Martin Mareš
Dimitri Marinakis
Nicolas Markey
Wim Martens
Simone Martini
Luke Mathieson
John Matthews
Richard Mayr
Pierre McKenzie
Ingmar Meinecke
Wolfgang Merkle
Viola Meszaros
Alessia Milani
Joe Miller
Neeldhara Misra
Joe Mitchell
Tobias Moemke
Pat Morin
Hiroki Morizumi
Ben Moszkowski
Andrzej Murawski
Anca Muscholl
Phuong Nguyen
Cyril Nicaud
Sergey Nikolenko
Dirk Nowotka
Sergey Nurk

Alexander Okhotin
Jörg Olschewski
Sebastian Ordyniak
Joel Ouaknine
Sang-il Oum
Michele Pagani
Rasmus Pagh
Tivadar Papai
Paweł Parys
Anthony Perez
Mihaly Petreczky
Giovanni Pighizzini
Vladimir Podolskii
Chris Pollett
Igor Potapov
Sylvia Pott
Giuseppe Prencipe
Svetlana Puzynina
Karin Quaas
Venkatesh Raman
R. Ramanujam
Michael Rao
Jan Reimann
Christian Reitwießner
Andrei Romashchenko
Thomas Rothvoss
Aleksi Saarela
Nicoletta Sabadini
Joshua Sack
Antonino Salibra
Saket Saurabh
Cem Say
Aleksy Schubert
Victor Selivanov
Jean-Sebastien Sereni
Alexander Shen
Amir Shpilka
Somnath Sikdar
Christian Sommer
Issam Souilah
Slawek Staworko
Björn Steffen
Benjamin Steinberg
Monika Steinova
Svetlana Stolpner
Hisao Tamaki
Tami Tamir
Lidia Tendera
Sophie Tison
Anthony Widjaja To
Ioan Todinca
Paolo Tranquilli
Tomasz Truderung
Michael Ummels
Pim van 't Hof
Erik Jan van Leeuwen
Vincent van Oostrom
Rob van Stee
Jeffrey Vaughan
Nicholas Vining
Ivan Visconti
Mikhail Volkov
Tomasz Waleń
Todd Wareham
Pascal Weil
Ryan Williams
Thomas Wolle
Marcin Wolski
Karianto Wong
Michał Wrona
Mu Yang
Chee Yap
Grigory Yaroslavtsev
Neal Young
Norbert Zeh
Marc Zeitoun
Lijun Zhang
Jie Zheng
Martin Zimmermann
Charalampos Zinoviadis

Previous MFCS Symposia

1972 Jablonna (Poland)
1973 Štrbské Pleso (Czechoslovakia)
1974 Jadwisin (Poland)
1975 Mariánské Lázně (Czechoslovakia)
1976 Gdańsk (Poland)
1977 Tatranská Lomnica (Czechoslovakia)
1978 Zakopane (Poland)
1979 Olomouc (Czechoslovakia)
1980 Rydzyna (Poland)
1981 Štrbské Pleso (Czechoslovakia)
1984 Praha (Czechoslovakia)
1986 Bratislava (Czechoslovakia)
1988 Karlovy Vary (Czechoslovakia)
1989 Porabka-Kozubnik (Poland)
1990 Banská Bystrica (Czechoslovakia)
1991 Kazimierz Dolny (Poland)
1992 Praha (Czechoslovakia)
1993 Gdańsk (Poland)
1994 Košice (Slovak Republic)
1995 Praha (Czech Republic)
1996 Kraków (Poland)
1997 Bratislava (Slovak Republic)
1998 Brno (Czech Republic)
1999 Szklarska Poreba (Poland)
2000 Bratislava (Slovak Republic)
2001 Mariánské Lázně (Czech Republic)
2002 Warsaw-Otwock (Poland)
2003 Bratislava (Slovak Republic)
2004 Praha (Czech Republic)
2005 Gdańsk (Poland)
2006 Bratislava (Slovak Republic)
2007 Český Krumlov (Czech Republic)
2008 Toruń (Poland)
2009 Nový Smokovec (Slovak Republic)

Table of Contents

New Developments in Quantum Algorithms

Andris Ambainis*

Faculty of Computing, University of Latvia, Raina bulv. 19, Riga, LV-1586, Latvia
andris.ambainis@lu.lv

Abstract. In this talk, we describe two recent developments in quantum algorithms.

The first new development is a quantum algorithm for evaluating a Boolean formula consisting of AND and OR gates of size N in time $O(\sqrt{N})$. This provides quantum speedups for any problem that can be expressed via Boolean formulas. This result can be also extended to *span problems*, a generalization of Boolean formulas. This provides an optimal quantum algorithm for any Boolean function in the black-box query model.

The second new development is a quantum algorithm for solving systems of linear equations. In contrast with traditional algorithms that run in time $O(N^{2.37...})$ where N is the size of the system, the quantum algorithm runs in time $O(\log^c N)$. It outputs a quantum state describing the solution of the system.

1 History of Quantum Algorithms

1.1 First Quantum Algorithms

Quantum computing (and, more broadly, quantum information science) is a new area at the boundary of computer science and physics. It studies how to apply quantum mechanics to solve problems in computer science and information processing. The area of quantum computing was shaped by the discoveries of two major quantum algorithms in mid-1990s.

The first of the these two discoveries was Shor's polynomial time quantum algorithm for factoring and discrete logarithms. Factoring and discrete logarithm are very hard number theoretic problems. The difficulty of these problems has been used to design cryptosystems (such as RSA and Diffie-Helman key exchange) for secure data transmission over an insecure network (such as Internet). The security of data transmission is based on the assumption that it is hard to factor (or find discrete logarithm of) large numbers. Until recently, this assumption was not in doubt. Mathematicians had tried to devise an efficient way of factoring large numbers for centuries, with no success.

In 1994, Shor [56] discovered a fast algorithm for factoring large numbers - on a quantum mechanical computer. This shook up the foundations of cryptography.

* Supported by FP7 Marie Curie Grant PIRG02-GA-2007-224886 and ESF project 1DP/1.1.1.2.0/09/APIA/VIAA/044.

P. Hliněný and A. Kučera (Eds.): MFCS 2010, LNCS 6281, pp. 1–11, 2010.

If a quantum mechanical computer is built, today's methods for secure data transmission over the Internet will become insecure.

Another, equally strikingly discovery was made in 1996, by Lov Grover [34]. He invented a quantum algorithm for speeding up exhaustive search problems. Grover's algorithm solves a generic exhaustive search problem with N possible solutions in time $O(\sqrt{N})$. This provides a quadratic speedup for a range of search problems, from problems that are in P classically to NP-complete problems.

Since then, each of the two algorithms has been analyzed in great detail. Shor's algorithm has been generalized to solve a class of algebraic problems that can be abstracted to *Abelian hidden subgroup problem* [39]. Besides factoring and discrete logarithm, the instances of Abelian HSP include cryptanalysis of hidden linear equations [18], solving Pell's equation, principal ideal problem [35] and others.

Grover's algorithm has been generalized to the framework of *amplitude amplification* [21] and extended to solve problems like approximate counting [23,47] and collision-finding [22].

1.2 Quantum Walks and Adiabatic Algorithms

Later, two new methods for designing quantum algorithms emerged: quantum walks [4,41,9,54,60] and adiabatic algorithms [31].

Quantum walks are quantum generalizations of classical random walks. They have been used to obtain quantum speedups for a number of problems. The typical setting is as follows. Assume that we have a classical Markov chain, on a state-space in which some states are special (marked). The Markov chain starts in a uniformly random state and stops if it reaches a marked state. If the classical Markov chain reaches a marked state in expected time T, then there is a quantum algorithm which can find it in time $O(\sqrt{T})$, assuming some conditions on the Markov chain [58,44,42].

This approach gives quantum speedups for a number of problems: element distinctness [5], search on a grid [13,59], finding triangles in graphs [45], testing matrix multiplication [19] and others.

Another application of quantum walks is to the "glued trees" problem [26]. In this problem, we have a graph G with two particular vertices u, v, designed as the entrance and the exit. The problem is to find the vertex v, if we start at the vertex u. There is a special exponential size graph called "glued trees" on which any classical algorithm needs exponential time to find v but a quantum algorithm can find v in polynomial time [26].

Adiabatic computation is a physics-based paradigm for quantum algorithms. In this paradigm, we design two quantum systems:

- H_{sol} whose lowest-energy state $|\psi_{sol}\rangle$ encodes a solution to a computational problem (for example, a satisfying assignment for SAT).
- H_{start} whose lowest-energy state $|\psi_{start}\rangle$ is such that we can easily prepare $|\psi_{start}\rangle$.

We then prepare $|\psi_{start}\rangle$ and slowly transform the forces acting on the quantum system from H_{start} to H_{sol}. Adiabatic theorem of quantum mechanics guarantees that, if the transformation is slow enough, $|\psi_{start}\rangle$ is transformed into a state close to $|\psi_{sol}\rangle$ [31].

The key question here is: what is "slowly enough"? Do we need a polynomial time or an exponential time to transform H_{start} to H_{sol} (thus solving SAT by a quantum algorithm)? This is a subject of an ongoing debate [31,29,2].

Adiabatic computation has been used by D-Wave Systems [30] which claims to have built a 128-bit adiabatic quantum computer. However, the claims of D-Wave have been questioned by many prominent scientists (see e.g. [1]).

1.3 Most Recent Algorithms

Two most recent discoveries in this field are the quantum algorithms for formula evaluation [32] and solving systems of linear equations [36]. Both of those algorithms use the methods from the previous algorithms but do it in a novel, unexpected way. Formula evaluation uses quantum walks but in a form that is quite different from the previous approach (which we described above). Quantum algorithm for formula evaluation uses *eigenvalue estimation* [46] which is the key technical subroutine of Shor's factoring algorithm [56] and the related quantum algorithms. But, again, eigenvalue estimation is used in a very unexpected way.

These two algorithms are the main focus of this survey. We describe them in detail in sections 2 and 3.

2 Formula Evaluation

2.1 Overview

We consider evaluating a Boolean formula of variables $x_1, \ldots, x_N$ consisting of ANDs and ORs, with each variable occuring exactly once in the formula. Such a formula can be described by a tree, with variables x_i at the leaves and AND/OR gates at the internal nodes. This problem has many applications because Boolean formulas can be used to describe a number of different situations. The most obvious one is determining if the input data $x_1, \ldots, x_N$ satisfy certain constraints that can be expressed by AND/OR gates.

For a less obvious application, we can view formula evaluation as a black-box model for a 2-player game (such as chess) if both players play their optimal strategies. In this case, the game can be represented by a game tree consisting of possible positions. The leaves of a tree correspond to the possible end positions of the game. Each of them contains a variable x_i, with $x_i = 1$ if the 1st player wins and $x_i = 0$ otherwise. If an internal node v corresponds to a position in which the 1st player makes the next move, then v contains a value that is OR of the values of v's children. (The 1st player wins if he has a move that leads to a position from which he can win.) If v is a node for which the 2nd player makes the next move, then v contains a value that is AND of the values of v's children. (In this case, the 1st player wins if he wins for any possible move of the 2nd player.)

The question is: assuming we have no further information about the game beyond the position tree, how many of the variables x_i do we have to examine to determine whether the 1st player has a winning strategy?

Classically, the most widely studied case is the full binary tree of depth d, with $N = 2^d$ leaves. It can be evaluated by looking at $\Theta(N^{.754\ldots})$ leaves and this is optimal [54,53,57]. A natural question was whether one could achieve a better result, using quantum algorithms. This question was a well known open problem in the quantum computing community since mid-1990s. Until 2007, the only known result was that $\Omega(\sqrt{N})$ quantum steps are necessary, for any AND-OR tree [3,14].

2.2 The Model

By standard rules from Boolean logic (de Morgan's laws), we can replace both AND and OR gates by NAND gates. A NAND gate $NAND(y_1, \ldots, y_k)$ outputs 1 if $AND(y_1, \ldots, y_k) = 0$ (i.e., $y_i = 0$ for at least one $i \in \{1, \ldots, k\}$) and 0 otherwise. Then, we have a tree with $x_1, \ldots, x_N$ at the leaves and NAND gates at the internal vertices. The advantage of this transformation is that we now have to deal with just one type of logic gates (instead of two - AND and OR).

We work in the quantum query model. In the discrete-time version of this model [6,20], the input bits $x_1, \ldots, x_N$ can be accessed by queries O to a black box.

To define O, we represent basis states as $|i, z\rangle$ where $i \in \{0, 1, \ldots, N\}$. The query transformation O_x (where $x = (x_1, \ldots, x_N)$) maps $|0, z\rangle$ to $|0, z\rangle$ and $|i, z\rangle$ to $(-1)^{x_i}|i, z\rangle$ for $i \in \{1, ..., N\}$ (i.e., we change phase depending on x_i, unless $i = 0$ in which case we do nothing).

Our algorithm can involve queries O_x and arbitrary non-query transformations that do not depend on $x_1, \ldots, x_N$. The task is to solve a computational problem (e.g., to compute a value of a NAND formula) with as few queries as possible.

2.3 Results

In 2007, in a breakthrough result, Farhi et al. [32] showed that the full binary AND-OR tree can be evaluated in $O(\sqrt{N})$ quantum time in continuous-time counterpart of the query model.

Several improvements followed soon. Ambainis et al. [27,8,12] translated the algorithm of [32] to the conventional discrete time quantum query model and extended it to evaluating arbitrary Boolean formulas with $O(N^{1/2+o(1)})$ quantum queries.

Soon after, Reichardt and Špalek [52] discovered a far-reaching generalization of this result. Namely, the quantum algorithm was generalized to evaluating span programs. A span program is an algebraic model of computation, originally invented for proving lower bounds on circuit size [40].

In a span program, we have a *target* vector v in some linear space. We also have other vectors $v_1, \ldots, v_m$, each of which is associated with some condition $x_i = 0$ or $x_i = 1$. The span program evaluates to 1 on an input $x_1, \ldots, x_n$ if v is equal to a linear combination of vectors $v_{i_1}, \ldots, v_{i_k}$ which are associated with conditions that are true for the given input $x_1, \ldots, x_n$. Otherwise, the span program evaluates to 0.

Here is an example of a span program. We have a two dimensional linear space, with the following vectors:

$$v = \begin{pmatrix} 1 \\ 0 \end{pmatrix}, v_1 = \begin{pmatrix} 1 \\ a \end{pmatrix}, v_2 = \begin{pmatrix} 1 \\ b \end{pmatrix}, v_3 = \begin{pmatrix} 1 \\ c \end{pmatrix}$$

where a, b, c are any three distinct non-zero numbers. Vectors v_1, v_2, v_3 are associated with conditions $x_1 = 1$, $x_2 = 1$, $x_3 = 1$, respectively.

Given any two of v_1, v_2, v_3, we can express any vector in two dimensions (including v) as their linear combination. Thus, this span program computes the majority function $MAJ(x_1, x_2, x_3)$ which is 1 whenever at least 2 of variables x_1, x_2, x_3 are equal to 1.

Logic formulae can be embedded into span programs. That is, if we have two span programs computing functions $f_1(x_1, \ldots, x_N)$ and $f_2(x_1, \ldots, x_N)$, we can combine them into span programs for f_1 AND f_2 and f_1 OR f_2 in a fairly simple way.

Reichardt and Špalek [52] invented a complexity measure, *witness size* for span programs. This measure generalizes formula size: a logic formula of size S can be transformed into a span program with witness size S. [52] gave a quantum algorithm for evaluating a span program of witness size S with $O(\sqrt{S})$ queries. This is a very interesting result because it allows to evaluate formulas with gates other than AND and OR by designing span programs for those gates and them composing them into one big span program. The next step was even more interesting.

The next step was even more surprising. Reichardt [48,49,51] discovered that the span program approach is optimal, for any Boolean function $f(x_1, \ldots, x_N)$. That is [51], if $Q(f)$ is the minimum number of quantum queries for evaluating f (by any quantum algorithm), then there is a span program with witness size $O(Q^2(f))$. Thus, a span-program based algorithm can evaluate f with $O(Q(f))$ queries, within a constant factor of the best possible number of queries.

This fact linked two lines of research: quantum formula evaluation algorithms and "quantum adversary" lower bounds. "Quantum adversary" (invented in [3]) is a method for proving lower bounds on the number of queries to evaluate $f(x_1, \ldots, x_N)$ by a quantum algorithm. Several progressively stronger versions of "quantum adversary" have been invented [7,15,43,38], with the strongest being the "negative adversary" method of [38].

Finding the best lower bound for quantum algorithms provable via "negative adversary" method is a semidefinite program (SDP). Reichardt [48,49,51] considered the dual of this SDP and showed that the dual SDP gives the span program with the smallest witness size. Thus, the span programs are optimal (in terms of query complexity) for any Boolean function $f(x_1, \ldots, x_N)$. (Finding the optimal span program, however, requires solving a semidefinite program of size 2^N.)

As a by-product, this gave a better algorithm for formula evaluation, improving the complexity from $O(N^{1/2+o(1)})$ in [12] to $O(\sqrt{N}\log N)$ in [50].

2.4 Algorithmic Ideas

We now give a very informal sketch the simplest version of formula evaluation algorithm. We augment the formula tree with a finite "tail" of length L as shown in Figure 1. We then consider a quantum walk on this tree. At the leaves of the tree, the transformations that are performed depend on whether the leaf holds $x_i = 0$ or $x_i = 1$. (This is achieved by querying the respective x_i and then performing one of two transformations, depending on the outcome of the query.)

The starting state is an appropriately chosen quantum state $|\psi_{start}\rangle = \sum_i \alpha_i |i\rangle$ consisting of the states $|i\rangle$ in the tail. If the quantum walk is set up properly, an amazing thing happens! Whenever the formula evaluates to 0, the state $|\psi_{start}\rangle$ remains almost unchanged. Whenever the formula evaluates to 1, after $O(N^{1/2+o(1)})$ steps, the state is almost completely different from $|\psi_{start}\rangle$. This means that we can distinguish between the two cases by running the walk for $O(N^{1/2+o(1)})$ steps and measuring whether the state is still $|\psi_{start}\rangle$. Surprisingly, the behaviour of the walk only depends on the value of the formula and not on which particular variables $x_1, \ldots, x_N$ are 1.

The algorithm for evaluating span programs is essentially the same, except that the quantum walk is performed on a weighted graph that corresponds to the span program.

For more information on this topic, we refer the reader to the survey [10] and the original papers.

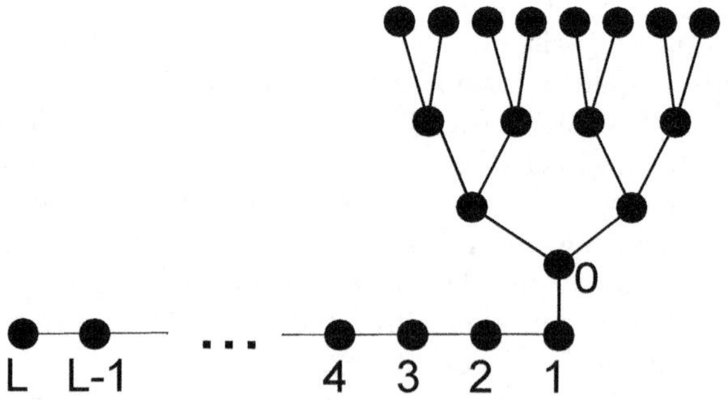

Fig. 1. A formula tree augmented with a finite "tail"

3 Linear Equations

3.1 Overview

Solving large systems of linear equations is a very common problem in scientific computing, with many applications. We consider solving a system of N linear equations with N unknowns: $Ax = b$ where

$$A = \begin{pmatrix} a_{11} & a_{12} & \dots & a_{1N} \\ a_{21} & a_{22} & \dots & a_{2N} \\ \dots & \dots & \dots & \dots \\ a_{N1} & a_{N2} & \dots & a_{NN} \end{pmatrix}, x = \begin{pmatrix} x_1 \\ x_2 \\ \dots \\ x_N \end{pmatrix}, b = \begin{pmatrix} b_1 \\ b_2 \\ \dots \\ b_N \end{pmatrix}.$$

A and b are given to us. The task is to find x.

The best classical algorithm for solving a general system $Ax = b$ runs in time $O(N^{2.37\dots})$. The reason for that was that even outputting the solution requires time $\Omega(N)$ because the solution contains values for N variables. Thus, it seemed that there was no hope for achieving more than a polynomial speedup by a quantum algorithm.

Recently, Harrow, Hassidim and Lloyd [36] discovered a surprising quantum algorithm that allows to solve systems of linear equations in time $O(\log^c N)$ - in an unconventional sense. Namely, the algorithm of [36] generates the quantum state

$$|\psi\rangle = \sum_{i=1}^{N} x_i |i\rangle$$

with the coefficients x_i being equal to the values of variables in the solution $x = (x_1, x_2, \dots, x_N)$ of the system $Ax = b$.

What can we do with this quantum state? We cannot extract all the values x_i from it. If we measured this state, we would obtain one value i, with probabilities of different i proportional to $|x_i|^2$.

We can, however, estimate some quantities that depend on all of x_i. For example, if all variables in the solution had values 1 or -1, having the quantum state $|\psi\rangle$ would enable us to estimate the fraction of variables $x_i = -1$. Moreover, similar tasks appear to be hard classically. As shown by [36], a classical $O(\log^c N)$-time algorithm for computing any quantity of this type implies a polynomial time classical algorithm for simulating any quantum computation. Thus (unless P=BQP), this quantum algorithm provides a genuine speedup over the classical algorithms.

3.2 More Details

In more detail, the running times of both classical and quantum algorithms for solving systems of linear equations actually depend on several parameters. One parameter is N, the number of equations (and variables). Another parameter is κ, the condition number of the system. κ is defined as $\frac{\mu_{max}}{\mu_{min}}$ where μ_{min} and μ_{max} are the smallest and the largest singular values of the matrix A [37, Chapter 5.8].

Intuitively, the condition number describes the closeness of the matrix A to a singular matrix. For a singular matrix, $\mu_{min} = 0$ and $\kappa = \infty$. Larger condition number means that the matrix A is closer to a singular matrix. In this case, small changes to input data A and b (or small numerical inaccuracies) can cause large changes to solution x. To compensate, if we have a matrix A with large κ, we have to perform the computation with a higher accuracy. This increases the running time of all algorithms for solving systems of linear equations but the exact increase varies greatly from algorithm to algorithm.

The main classical algorithms for solving systems of linear equations are:

1. LU-decomposition [33, Chapter 3] which runs in time $O(N^{2.376\ldots} \log^c(\kappa))$ [24]. Here, $w = 2.376\ldots$ is the constant from the running time $O(N^w)$ of the best matrix multiplication algorithm [28].
2. Conjugate gradient [33, Chapter 10], which runs in time $O(m\sqrt{\kappa})$ [55] where m is the number of non-zero entries in the matrix. If we know that each row of A contains at most s non-zero entries, this is at most $O(Ns\sqrt{\kappa})$.

The running time of the quantum algorithm [36] is $O(\kappa^2 T \log^c N)$ where T is the time necessary to implement the transformation e^{iA} on a quantum computer. T varies greatly, depending on A. For sparse A with at most s nonzero values in each row and each column, $T = O(s^2 \log N)$ [16]. Thus, in this case the running time of the quantum algorithm is $O(\kappa^2 s^4 \log^c N)$. This achieves an exponential speedup for the case when N is large and κ is relatively small (e.g., $\kappa = O(1)$ or $\kappa = O(\log N)$).

The key bottleneck is the dependence on κ which is actually worse than in the classical algorithms. We have been able to improve it to $O(\kappa^{1+o(1)} \log^c N)$ [11]. Unfortunately, further improvement is very unlikely. [36] have shown that an $O(\kappa^{1-\epsilon} \log^c N)$ time quantum algorithm would imply BQP=PSPACE.

For non-sparse A, one could use the algorithms of [25,17] to simulate e^{iA}. The dependence on N is better than $O(N^{2.376\ldots})$ in the classical LU decomposition but the speedup is only polynomial.

References

1. Aaronson, S.: D-Wave Easter Spectacular. A blog post (April 7, 2007), `http://scottaaronson.com/blog/?p=225`
2. Altshuler, B., Krovi, H., Roland, J.: Anderson localization casts clouds over adiabatic quantum optimization, arxiv:0912.0746
3. Ambainis, A.: Quantum lower bounds by quantum arguments. Journal of Computer and System Sciences 64, 750–767 (2002), quant-ph/0002066
4. Ambainis, A.: Quantum walks and their algorithmic applications. International Journal of Quantum Information 1, 507–518 (2003), quant-ph/0403120
5. Ambainis, A.: Quantum walk algorithm for element distinctness. SIAM Journal on Computing 37(1), 210–239 (2007), FOCS 2004 and quant-ph/0311001
6. Ambainis, A.: Quantum search algorithms (a survey). SIGACT News 35(2), 22–35 (2004), quant-ph/0504012
7. Ambainis, A.: Polynomial degree vs. quantum query complexity. Journal of Computer and System Sciences 72(2), 220–238 (2006), quant-ph/0305028

8. Ambainis, A.: A nearly optimal discrete query quantum algorithm for evaluating NAND formulas, arxiv:0704.3628
9. Ambainis, A.: Quantum random walks - New method for designing quantum algorithms. In: Geffert, V., Karhumäki, J., Bertoni, A., Preneel, B., Návrat, P., Bieliková, M. (eds.) SOFSEM 2008. LNCS, vol. 4910, pp. 1–4. Springer, Heidelberg (2008)
10. Ambainis, A.: Quantum algorithms for formula evaluation. In: Proceedings of the NATO Advanced Research Workshop Quantum Cryptography and Computing: Theory and Implementations (to appear)
11. Ambainis, A.: Variable time amplitude amplification and a faster quantum algorithm for systems of linear equations (in preparation, 2010)
12. Ambainis, A., Childs, A., Reichardt, B., Spalek, R., Zhang, S.: Any AND-OR formula of size N can be evaluated in time $N^{1/2+o(1)}$ on a quantum computer. In: Proceedings of FOCS 2007, pp. 363–372 (2007)
13. Ambainis, A., Kempe, J., Rivosh, A.: Coins make quantum walks faster. In: Proceedings of SODA 2005, pp. 1099–1108 (2005), quant-ph/0402107
14. Barnum, H., Saks, M.: A lower bound on the quantum complexity of read once functions. Journal of Computer and System Sciences 69, 244–258 (2004), quant-ph/0201007
15. Barnum, H., Saks, M.E., Szegedy, M.: Quantum query complexity and semi-definite programming. In: IEEE Conference on Computational Complexity 2003, pp. 179–193 (2003)
16. Berry, D.W., Ahokas, G., Cleve, R., Sanders, B.C.: Efficient quantum algorithms for simulating sparse Hamiltonians. Communications in Mathematical Physics 270(2), 359–371 (2007), quant-ph/0508139
17. Berry, D.W., Childs, A.: The quantum query complexity of implementing black-box unitary transformations, arxiv:0910.4157
18. Boneh, D., Lipton, R.: Quantum cryptanalysis of hidden linear functions (extended abstract). In: Coppersmith, D. (ed.) CRYPTO 1995. LNCS, vol. 963, pp. 424–437. Springer, Heidelberg (1995)
19. Buhrman, H., Špalek, R.: Quantum verification of matrix products. In: Proceedings of SODA 2006, pp. 880–889 (2006), quant-ph/0409035
20. Buhrman, H., de Wolf, R.: Complexity measures and decision tree complexity: a survey. Theoretical Computer Science 288, 21–43 (2002)
21. Brassard, G., Høyer, P., Mosca, M., Tapp, A.: Quantum amplitude amplification and estimation. In: Quantum Computation and Quantum Information Science. AMS Contemporary Mathematics Series, vol. 305, pp. 53–74 (2002), quant-ph/0005055
22. Brassard, G., Høyer, P., Tapp, A.: Quantum cryptanalysis of hash and claw-free functions. In: Lucchesi, C.L., Moura, A.V. (eds.) LATIN 1998. LNCS, vol. 1380, pp. 163–169. Springer, Heidelberg (1998), quant-ph/9705002
23. Brassard, G., Høyer, P., Tapp, A.: Quantum counting. In: Larsen, K.G., Skyum, S., Winskel, G. (eds.) ICALP 1998. LNCS, vol. 1443, pp. 820–831. Springer, Heidelberg (1998), quant-ph/9805082
24. Bunch, J.R., Hopcroft, J.E.: Triangular factorization and inversion by fast matrix multiplication. Mathematics of Computation 28, 231–236 (1974)
25. Childs, A.: On the relationship between continuous- and discrete-time quantum walk. Communications in Mathematical Physics 294, 581–603 (2010), arXiv:0810.0312

26. Childs, A.M., Cleve, R., Deotto, E., Farhi, E., Gutmann, S., Spielman, D.A.: Exponential algorithmic speedup by quantum walk. In: Proceedings of STOC 2003, pp. 59–68 (2003), quant-ph/0209131
27. Childs, A., Reichardt, B., Špalek, R., Zhang, S.: Every NAND formula on N variables can be evaluated in time $O(N^{1/2+\epsilon})$, quant-ph/0703015, version v1 (March 2, 2007)
28. Coppersmith, D., Winograd, S.: Matrix multiplication via arithmetic progressions. Journal of Symbolic Computation 9(3), 251–280 (1990)
29. van Dam, W., Mosca, M., Vazirani, U.: How Powerful is Adiabatic Quantum Computation? In: Proceedings of FOCS 2001, pp. 279–287 (2001)
30. D-Wave Systems, `http://www.dwavesys.com`
31. Farhi, E., Goldstone, J., Gutman, S., Sipser, M.: A quantum adiabatic algorithm applied to random instances of an NP-complete problem. Science 292, 472–476 (2001), quant-ph/0104129
32. Farhi, E., Goldstone, J., Gutman, S.: A Quantum Algorithm for the Hamiltonian NAND Tree. Theory of Computing 4, 169–190 (2008), quant-ph/0702144
33. Golub, G., van Loan, C.: Matrix Computations, 3rd edn. John Hopkins University Press, Baltimore (1996)
34. Grover, L.: A fast quantum mechanical algorithm for database search. In: Proceedings of STOC 1996, pp. 212–219 (1996), quant-ph/9605043
35. Hallgren, S.: Polynomial-time quantum algorithms for Pell's equation and the principal ideal problem. Journal of the ACM 54(1) (2007)
36. Harrow, A., Hassidim, A., Lloyd, S.: Quantum algorithm for linear systems of equations. Physical Review Letters 103, 150502 (2008), arXiv:0907.3920
37. Horn, R., Johnson, C.: Matrix Analysis. Cambridge University Press, Cambridge (1985)
38. Høyer, P., Lee, T., Špalek, R.: Negative weights make adversaries stronger. In: Proceedings of STOC 2007, pp. 526–535 (2007), quant-ph/0611054
39. Jozsa, R.: Quantum factoring, discrete logarithms, and the hidden subgroup problem. Computing in Science and Engineering 3, 34–43 (2001), quant-ph/0012084
40. Karchmer, M., Wigderson, A.: On Span Programs. In: Structure in Complexity Theory Conference 1993, pp. 102–111 (1993)
41. Kempe, J.: Quantum random walks - an introductory overview. Contemporary Physics 44(4), 307–327 (2003), quant-ph/0303081
42. Krovi, H., Magniez, F., Ozols, M., Roland, J.: Finding is as easy as detecting for quantum walks. In: Proceedings of ICALP (to appear, 2010), arxiv:1002.2419
43. Laplante, S., Magniez, F.: Lower Bounds for Randomized and Quantum Query Complexity Using Kolmogorov Arguments. SIAM Journal on Computing 38(1), 46–62 (2008), Also CCC 2004 and quant-ph/0311189
44. Magniez, F., Nayak, A., Roland, J., Santha, M.: Search via quantum walk. In: Proceedings of STOC 2007, pp. 575–584 (2007), quant-ph/0608026
45. Magniez, F., Santha, M., Szegedy, M.: Quantum algorithms for the triangle problem. SIAM Journal on Computing 37(2), 413–424 (2007), quant-ph/0310134
46. Mosca, M., Ekert, A.: The Hidden Subgroup Problem and Eigenvalue Estimation on a Quantum Computer. In: Williams, C.P. (ed.) QCQC 1998. LNCS, vol. 1509, pp. 174–188. Springer, Heidelberg (1999), quant-ph/9903071
47. Nayak, A., Wu, F.: The quantum query complexity of approximating the median and related statistics. In: Proceedings of STOC 1999, pp. 384–393 (1999), quant-ph/9804066

48. Reichardt, B.: Span programs and quantum query complexity: The general adversary bound is nearly tight for every boolean function. In: Proceedings of FOCS (2009), arXiv:0904.2759
49. Reichardt, B.: Span-program-based quantum algorithm for evaluating unbalanced formulas, arXiv:09071622
50. Reichardt, B.: Faster quantum algorithm for evaluating game trees, arXiv:0907.1623
51. Reichardt, B.: Reflections for quantum query algorithms, arxiv:1005.1601
52. Reichardt, B., Špalek, R.: Span-program-based quantum algorithm for evaluating formulas. In: Proceedings of STOC 2008, pp. 103–112 (2008), arXiv:0710.2630
53. Saks, M., Wigderson, A.: Probabilistic Boolean decision trees and the complexity of evaluating game trees. In: Proceedings of FOCS 1986, pp. 29–38 (1986)
54. Santha, M.: Quantum walk based search algorithms. In: Agrawal, M., Du, D.-Z., Duan, Z., Li, A. (eds.) TAMC 2008. LNCS, vol. 4978, pp. 31–46. Springer, Heidelberg (2008), arXiv:0808.0059
55. Shewchuk, J.: An introduction to the conjugate gradient method without the agonizing pain. Technical Report CMU-CS-94-125, School of Computer Science, Carnegie Mellon University (1994)
56. Shor, P.: Algorithms for Quantum Computation: Discrete Logarithms and Factoring. In: Proceedings of FOCS 1994, pp. 124–134 (1994), quant-ph/9508027
57. Snir, M.: Lower bounds on probabilistic linear decision trees. Theoretical Computer Science 38, 69–82 (1985)
58. Szegedy, M.: Quantum speed-up of Markov chain based algorithms. In: Proceedings of FOCS 2004, pp. 32–41 (2004)
59. Tulsi, T.A.: Faster quantum walk algorithm for the two dimensional spatial search. Physical Review A 78, 012310 (2008), arXiv:0801.0497
60. Venegas-Andrade, S.E.: Quantum Walks for Computer Scientists. Morgan and Claypool (2008)

Persistent Homology under Non-uniform Error*

Paul Bendich[1,2,3], Herbert Edelsbrunner[1,2,3,4], Michael Kerber[1,2], and Amit Patel[1,2]

[1] IST Austria (Institute of Science and Technology Austria), Klosterneuburg, Austria
[2] Dept. Comput. Sci., Duke Univ., Durham, North Carolina
[3] Dept. Mathematics, Duke Univ. Durham, North Carolina
[4] Geomagic, Research Triangle Park, North Carolina

Abstract. Using ideas from persistent homology, the robustness of a level set of a real-valued function is defined in terms of the magnitude of the perturbation necessary to kill the classes. Prior work has shown that the homology and robustness information can be read off the extended persistence diagram of the function. This paper extends these results to a non-uniform error model in which perturbations vary in their magnitude across the domain.

Keywords: Topological spaces, continuous functions, level sets, perturbations, homology, extended persistence, error models, stability, robustness.

1 Introduction

Numerical errors are an inescapable by-product of scientific and other computations, and thus they have inspired the creation of entire fields of mathematical inquiry, including numerical analysis and statistics. There are many sources, such as randomness, limited resolution, and limited computational resources. There are also many coping strategies, such as improving accuracy, or finding credible substitutes for the elusive ideal. The contributions of this paper belong to the analytic approach that gives estimates on how far the result is from the ideal. More specifically, we study real-valued functions, which model many real world applications, including medical images and satellite pictures. To extract information from a function, we consider level and sublevel sets and their topology expressed in terms of homology groups. The question thus arises to what extent the homology of a level or sublevel set is sensitive to perturbations of the function. In this paper, we study the effect of perturbations which mimic the common situation in which measurements vary in their accuracy, and we call these *non-uniform perturbations*. We assume that this variation is tied to the location, and that we have complete information on how the accuracy varies across the domain. We capture this information in a (non-uniform) error model, which will be formally defined in Section 3.

On a technical level, we extend the algebraic and measure theoretic concept of persistent homology to non-uniform error models. Specifically, we define the non-uniform

* This research is partially supported by the Defense Advanced Research Projects Agency (DARPA), under grants HR0011-05-1-0057 and HR0011-09-0065, as well as the National Science Foundation (NSF), under grant DBI-0820624.

P. Hliněný and A. Kučera (Eds.): MFCS 2010, LNCS 6281, pp. 12–23, 2010.

persistence diagram of a function under an error model, and we relate it to the conventional persistence diagram (defined for uniform error) and to non-uniform persistence diagrams of other functions under the same error model. Using results from prior work, we then extend these results to well diagrams, which characterize the robustness of level sets. The main technical concept is a transformation of functions that turns non-uniform into uniform error and thus extends much of the machinery of persistent and robust homology to non-uniform error. Under the additional assumption of the linearity of the error model, this includes one of the cornerstones of the theory of persistent homology, namely the stability of the diagram.

Outline. Section 2 reviews the background from persistent homology. Section 3 introduces error models and discusses their effect on the persistence diagrams of functions. Section 4 introduces dual error models and uses them to transform non-uniform to uniform error. Section 5 demonstrates that linear error models give rise to a richer theory than the more general error models. Finally, Section 6 concludes the paper.

2 Background

In this section, we review the background on homology and on persistence; see Munkres [13] and Hatcher [12] for standard texts in algebraic topology, and Edelsbrunner, Harer [10] for a recent book in computational topology. While the material in this section may seem dry and technical, we remind the reader of the connection to the fundamental problems of feature extraction, matching, and classification for images, shapes, and more general data. Indeed, persistent homology has been applied to a host of different shape and data analysis questions, including for natural images [3], trademark images [4], brain structure [5], sensor networks [8], and gene expression [9].

Persistence. The persistence of homology classes along a filtration of a topological space can be defined in a quite general context. For the purposes of this paper, we need only a particular type of filtration, one defined by the sublevel sets of a real-valued function on a compact topological space. Given such a space $\mathbb{X}$ and a function $f : \mathbb{X} \to \mathbb{R}$, we call a set $\mathbb{X}_r(f) = f^{-1}(-\infty, r]$ a *sublevel set*, and we consider the nested sequence of these sets. Whenever $r \leq s$, the inclusion $\mathbb{X}_r(f) \hookrightarrow \mathbb{X}_s(f)$ induces maps on the homology groups $\mathsf{H}(\mathbb{X}_r(f)) \to \mathsf{H}(\mathbb{X}_s(f))$. Here we use modulo 2 homology, that is, the coefficient group is $\mathbb{Z}/2\mathbb{Z}$. In addition, we simplify the notation by taking the direct sum of the homology groups over all dimensions. A real value r is called a *homological regular value* of f if there exists an $\varepsilon > 0$ such that the inclusion $\mathbb{X}_{r-\delta}(f) \hookrightarrow \mathbb{X}_{r+\delta}(f)$ induces an isomorphism between homology groups for all $\delta < \varepsilon$. Otherwise we call r a *homological critical value*. We say that f is *tame* if the homology groups of each sublevel set have finite rank and if there are only finitely many homological critical values. Assuming that f is tame, with ordered homological critical values $r_1 < r_2 < \ldots < r_n$, we select $n+1$ homological regular values s_i such that $s_0 < r_1 < s_1 < \ldots < r_n < s_n$, and set $\mathbb{X}_i = \mathbb{X}_{s_i}(f)$. Note that $\mathbb{X}_0 = \emptyset$ and $\mathbb{X}_n = \mathbb{X}$, by compactness. The inclusions $\mathbb{X}_i \hookrightarrow \mathbb{X}_j$ induce maps $\mathsf{f}^{i,j} : \mathsf{H}(\mathbb{X}_i) \to \mathsf{H}(\mathbb{X}_j)$ for $0 \leq i \leq j \leq n$ and give a *filtration* of the homology groups:

$$0 = \mathsf{H}(\mathbb{X}_0) \to \mathsf{H}(\mathbb{X}_1) \to \ldots \to \mathsf{H}(\mathbb{X}_n) = \mathsf{H}(\mathbb{X}). \quad (1)$$

Given a class $\alpha \in \mathsf{H}(\mathbb{X}_i)$, we say that α is *born* at $\mathbb{X}_i$ if $\alpha \notin \operatorname{im} \mathsf{f}^{i-1,i}$. A class α born at $\mathbb{X}_i$ is said to *die entering* $\mathbb{X}_j$ if $\mathsf{f}^{i,j}(\alpha) \in \operatorname{im} \mathsf{f}^{i-1,j}$ but $\mathsf{f}^{i,j-1}(\alpha) \notin \operatorname{im} \mathsf{f}^{i-1,j-1}$. We remark that if a class α is born at $\mathbb{X}_i$, then every class in the coset $[\alpha] = \alpha + \operatorname{im} \mathsf{f}^{i-1,i}$ is born at the same time. Of course, whenever such an α dies entering $\mathbb{X}_j$, the entire coset $[\alpha]$ also dies with it.

Extended persistence. The filtration in (1) begins with the zero group but ends with a potentially nonzero group. Hence, it is possible to have classes that are born but never die. We call these *essential* classes, as they represent the actual homology of the space $\mathbb{X}$. To measure the persistence of the essential classes, we follow [7] and extend (1) using relative homology groups. More precisely, we consider for each i the *superlevel set* $\mathbb{X}^i = f^{-1}[s_{n-i}, \infty)$. By compactness, we have $\mathbb{X}^0 = \emptyset$ and $\mathbb{X}^n = \mathbb{X}$ and therefore $\mathsf{H}(\mathbb{X}, \mathbb{X}^0) = \mathsf{H}(\mathbb{X})$ and $\mathsf{H}(\mathbb{X}, \mathbb{X}^n) = 0$. For $0 \leq i \leq j \leq n$, the inclusions $\mathbb{X}^i \hookrightarrow \mathbb{X}^j$ induce maps on relative homology. We then consider the extended filtration:

$$\begin{aligned} 0 = \mathsf{H}(\mathbb{X}_0) &\to \mathsf{H}(\mathbb{X}_1) \to \ldots \to \mathsf{H}(\mathbb{X}_n) = \mathsf{H}(\mathbb{X}) \\ &= \mathsf{H}(\mathbb{X}, \mathbb{X}^0) \to \mathsf{H}(\mathbb{X}, \mathbb{X}^1) \ldots \to \mathsf{H}(\mathbb{X}, \mathbb{X}^n) = 0, \end{aligned} \tag{2}$$

and extend the notions of birth and death in the obvious way. Now all classes will eventually die, as this filtration begins and ends with the zero group. The information contained within the extended filtration (2) can be compactly represented by *persistence diagrams* $\mathrm{Dgm}_p(f)$, one for each dimension p in homology. These diagrams are multisets of points drawn in three copies of the extended plane, shrunk to finite size and arranged side by side, as shown in Figure 1. For technical reasons, we always consider the diagram to contain infinitely many copies of each point on the baseline, where the birth and death coordinates coincide. By $\mathrm{Dgm}(f)$, we mean the points of all diagrams in all dimensions, overlaid as one multiset of points.

Contained within each $\mathrm{Dgm}_p(f)$ are three subdiagrams, corresponding to three different combinations of birth and death location. The *ordinary subdiagram*, $\mathrm{Ord}_p(f)$,

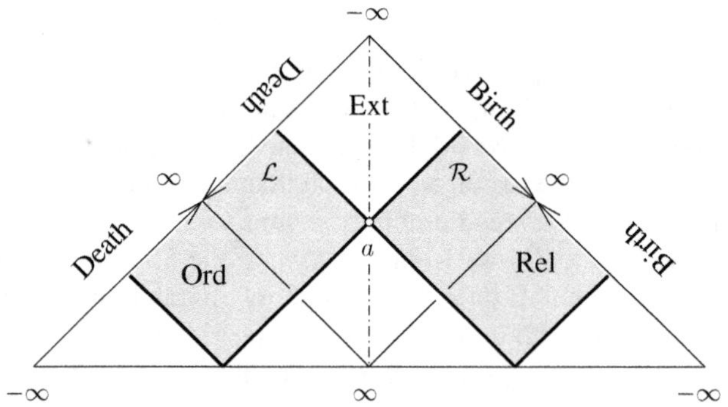

Fig. 1. The persistence diagram of a function consists of three subdiagrams, Ord, Ext, and Rel, arranged in a triangle as shown. The points in the two shaded rectangles, $\mathcal{L}$ and $\mathcal{R}$, represent the homology of the level set defined by a.

contains the point (r_i, r_j) for each coset of classes that are born at $\mathbb{X}_i$ and die entering $\mathbb{X}_j$. Here, birth and death both happen during the first half of (2). The *extended subdiagram*, $\mathrm{Ext}_p(f)$, contains (r_i, r_j) for each coset of classes that is born at $\mathbb{X}_i$ and dies entering $(\mathbb{X}, \mathbb{X}^{n-j+1})$. Finally, the *relative subdiagram*, $\mathrm{Rel}_p(f)$, contains (r_i, r_j) for each coset of classes that is born at $(\mathbb{X}, \mathbb{X}^{n-i+1})$ and dies entering $(\mathbb{X}, \mathbb{X}^{n-j+1})$. We arrange the three subdiagrams side by side, while reversing the birth axis of the extended subdiagram and both axes of the relative subdiagram. We do so to simplify the interpretation of the diagram, as will be explained later.

Stability. An essential property of the persistence diagrams is their stability under small changes of the function. To make this precise, we need to define a distance between functions and a distance between diagrams. Given two functions $f, h : \mathbb{X} \to \mathbb{R}$, and a real number $r \geq 0$, we call h an *r-perturbation* of f if $|f(x) - h(x)| \leq r$, for every $x \in \mathbb{X}$. This relation is symmetric and can be used to define a metric on the space of real-valued functions on $\mathbb{X}$, setting $\|f - h\|_\infty$ equal to the minimum r such that f and h are r-perturbations of each other. This is of course the standard L_∞-distance between functions. Given any two persistence diagrams, $\mathcal{D}$ and $\mathcal{D}'$, we define the *bottleneck distance* between them as the largest distance between matched points (in maximum norm) under the best possible matching between the diagrams. More formally,

$$W_\infty(\mathcal{D}, \mathcal{D}') = \inf_\gamma \sup_u \|u - \gamma(u)\|_\infty, \tag{3}$$

where u ranges over all points of the diagram $\mathcal{D}$, and γ ranges over all bijections from $\mathcal{D}$ to $\mathcal{D}'$. We then have:

1 (Stability Theorem [6]). *Given continuous and tame functions $f, h : \mathbb{X} \to \mathbb{R}$ on a compact topological space, we have $W_\infty(\mathrm{Dgm}_p(f), \mathrm{Dgm}_p(h)) \leq \|f - h\|_\infty$ for each homological dimension p.*

This result seems natural as we can construct a homotopy between f and h in which the values change continuously, each by at most $\|f - h\|_\infty$. However, consider that critical points may appear and disappear and global rearrangements may cause the pairing between critical values change during the homotopy.

Robustness. The persistence diagram of a real-valued function f carries a wealth of information. For example, it allows us to measure the robustness of the homology of level sets to perturbations of f. We now make this precise.

Fixing some value $a \in \mathbb{R}$ and a real number $r \geq 0$, we consider the preimage of the interval: $\mathbb{X}_{[a-r,a+r]}(f) = f^{-1}[a - r, a + r]$. For every r-perturbation h of f, the level set $h^{-1}(a)$ will be a subset of this preimage, and hence there is an induced map on homology $\mathsf{j}_h : \mathsf{H}(h^{-1}(a)) \to \mathsf{H}(\mathbb{X}_{[a-r,a+r]}(f))$. Following [11], we say that a class $\alpha \in \mathsf{H}(\mathbb{X}_{[a-r,a+r]}(f))$ is *supported* by h if it belongs to the image of j_h; in other words, if the set $h^{-1}(a)$ carries a chain representative of α. The *well group* of f and $r \geq 0$ is then defined to consist of those classes that are supported by all r-perturbations of f. It is a subgroup of $\mathsf{H}(\mathbb{X}_{[a-r,a+r]}(f))$. The sequence of well groups no longer forms a filtration but a more general zigzag module, introduced recently in [2]. These modules can still be characterized, albeit less directly, by their persistence diagrams. In the case of well groups, all births happen at the beginning, so the diagram simplifies to a multiset of points that mark deaths on the real line. We refer to this multiset as the *well diagram* of the function and the value defining the level set. It expresses what we call the *robustness* of the homology classes, that is, their resilience to perturbations of the function. In [1], the authors demonstrate a simple relationship between the persistence diagrams of f and the well diagrams of f and a, for every $a \in \mathbb{R}$. To describe this relationship, we first define for each homological dimension p two multisets of points:

$$\begin{aligned}\mathcal{L}_p[a] &= \{(x,y) \in \mathrm{Ord}_p(f) \mid x < a, y > a\} \sqcup \{(x,y) \in \mathrm{Ext}_p(f) \mid x < a, y > a\},\\ \mathcal{R}_p[a] &= \{(x,y) \in \mathrm{Ext}_p(f) \mid x > a, y < a\} \sqcup \{(x,y) \in \mathrm{Rel}_p(f) \mid x > a, y < a\},\end{aligned}$$

where x refers of course to the birth coordinate and y to the death coordinate of the point; see Figure 1. Then the p-dimensional homology of $f^{-1}(a)$ is characterized by points (x,y) in $\mathcal{L}_p[a] \cup \mathcal{R}_{p+1}[a]$. If the point belongs to $\mathcal{L}_p[a]$, then its robustness is equal to $\min\{a-x, y-a\}$, while if the points belongs to $\mathcal{R}_{p+1}[a]$, then its robustness is $\min\{x-a, a-y\}$. To get the well diagram of f and a, we then just plot the robustness value for every point in $\mathcal{L}_p[a] \cup \mathcal{R}_{p+1}[a]$ on the real line.

3 Non-uniform Error

In this section, we extend the concepts of persistence and robustness from a uniform to a non-uniform notion of error. We begin by introducing the error model as a 1-parameter family of functions.

Error model. It is convenient to substitute the extended real line, $\bar{\mathbb{R}} = \mathbb{R} \cup \{-\infty, \infty\}$, for $\mathbb{R}$ as the range of our functions, and therefore also of the error model.

2 (Definition). *An* error model *on a compact topological space* $\mathbb{X}$ *is a continuous mapping* $E : \mathbb{X} \times \bar{\mathbb{R}} \to \bar{\mathbb{R}}$ *that satisfies the following two properties:*

monotonicity: $E(x,r) < E(x,s)$ *for all* $x \in \mathbb{X}$ *and all* $r < s$*;*
normalization: $E(x,0) = 0, E(x,\infty) = \infty, E(x,-\infty) = -\infty$*, for all* $x \in \mathbb{X}$.

The error model is uniform *if* $E(x,r) = r$ *for all* $(x,r) \in \mathbb{X} \times \bar{\mathbb{R}}$*, and it is* non-uniform *otherwise.*

Fixing either a point x or a radius r, we get restricted functions $e_x : \bar{\mathbb{R}} \to \bar{\mathbb{R}}$ and $e_r : \mathbb{X} \to \bar{\mathbb{R}}$ defined by $e_x(r) = e_r(x) = E(x,r)$. Note that the two conditions above guarantee that e_x is invertible, a property we make use of later. Intuitively, one might imagine r to be a global noise parameter which leads, via the error model E, to a variable amount $e_x(r)$ of noise across $\mathbb{X}$. The continuity of E models our assumption that errors are not independent and indeed are closely related for nearby points. In Section 5, we will add the further assumption that E is linear, meaning $E(x,r) = r \cdot E(x,1)$ for all $(x,r) \in \mathbb{X} \times \bar{\mathbb{R}}$. For the moment we make no such requirement of our model.

Non-uniform filtrations and persistence. Given a function $f : \mathbb{X} \to \bar{\mathbb{R}}$ and any two values $r \leq s$, the standard notion of an *interlevel set* is the preimage of the interval: $\mathbb{X}_{[r,s]}(f) = f^{-1}[r,s]$. In applications in which $\mathbb{X}$ is 3-dimensional, this construct is often referred to as an *interval volume*. Extending it to our framework, we define the *non-uniform interlevel set* as the set of points with image between the bounds specified by the error model:

$$\mathbb{X}_{[r,s]}(f,E) = \{x \in \mathbb{X} \mid E(x,r) \leq f(x) \leq E(x,s)\}.$$

As illustrated in Figure 2, we can construct the non-uniform interlevel set by intersecting the graph of f with the strip of points between the graphs of e_r and e_s, and projecting the intersection to $\mathbb{X}$. In the special case in which $r = -\infty$, we write $\mathbb{X}_s(f,E) = \mathbb{X}_{[-\infty,s]}(f,E)$ and call it a *non-uniform sublevel set*. Similarly, if $s = \infty$, we write $\mathbb{X}^r(f,E) = \mathbb{X}_{[r,\infty]}(f,E)$ and call it a *non-uniform superlevel set*.

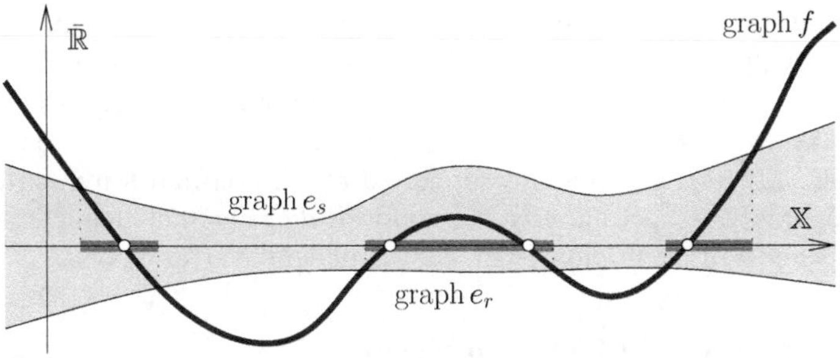

Fig. 2. The graph of a function f, the shaded strip bounded from below and above by the graphs of restrictions of the error model, and the non-uniform interlevel set obtained by projecting the intersection. For $r \leq 0 \leq s$, the non-uniform interlevel set contains the zeroset of f.

Whenever $r \leq s$, the monotonicity requirement guarantees the inclusion of $\mathbb{X}_r(f,E)$ in $\mathbb{X}_s(f,E)$, and the inclusion of $\mathbb{X}^s(f,E)$ in $\mathbb{X}^r(f,E)$. Hence, just as in Section 2, the non-uniform sublevel and superlevel sets give an extended filtration of $\mathbb{X}$. As a result, we have in each homological dimension p a *non-uniform persistence diagram*, denoted by $\mathrm{Ngm}_p(f,E)$. As before, we write $\mathrm{Ngm}(f,E)$ for the overlay of the diagrams in all dimensions. In Section 5, we will see that, under the assumption of a linear error model, these non-uniform diagrams are stable.

Non-uniform perturbations and robustness. Now suppose that we have a function $f : \mathbb{X} \to \bar{\mathbb{R}}$ as well as an error model $E : \mathbb{X} \times \bar{\mathbb{R}} \to \bar{\mathbb{R}}$. As promised, we create a theoretical language to quantify the robustness of the homology of a level set $f^{-1}(0)$ under non-uniform perturbation. Given another function $h : \mathbb{X} \to \bar{\mathbb{R}}$ and a value $r \geq 0$, we say that h is a *non-uniform r-perturbation* of f, with respect to E, if

$$E(x,-r) \leq f(x) - h(x) \leq E(x,r), \tag{4}$$

for all $x \in \mathbb{X}$. For example, every function of the form $f - e_s$, with $s \in [-r, r]$, is a non-uniform r-perturbation of f. If E is linear, or indeed if each e_x is odd, meaning $e_x(-r) = -e_x(r)$ for every r, then f will also be a non-uniform r-perturbation of h, but this need not be true in the general case. It is useful to understand the connection between non-uniform perturbations and interlevel sets.

3 (Non-uniform Perturbation Lemma). *A function $h : \mathbb{X} \to \bar{\mathbb{R}}$ is a non-uniform r-perturbation of $f : \mathbb{X} \to \bar{\mathbb{R}}$, under the error model $E : \mathbb{X} \times \bar{\mathbb{R}} \to \bar{\mathbb{R}}$, only if $h^{-1}(0) \subseteq \mathbb{X}_{[-r,r]}(f, E)$.*

Proof. Starting with the definition of a non-uniform r-perturbation, we get

$$E(x, -r) \ \leq \ f(x) \ \leq \ E(x, r),$$

for all points x with $h(x) = 0$. The two inequalities define the non-uniform interlevel set defined by $-r \leq r$, which implies the claimed containment. ☐

We note that $\mathbb{X}_{[-r,r]}(f, E)$ is the smallest interlevel set that contains the zeroset of every non-uniform r-perturbation. In other words, it is the union of all these zerosets. Compare this with the fact that $\mathbb{X}_{[-r,r]}(f, E)$ is also the union of the zerosets of the functions $f - e_s$, for all $s \in [-r, r]$. Adopting the terminology from Section 2, we can now define the *non-uniform well group* of f, E, and a value $r \geq 0$ to consist of those classes in $\mathsf{H}(\mathbb{X}_{[-r,r]}(f, E))$ that are supported by all non-uniform r-perturbations of f. Correspondingly, we get the *non-uniform well diagram* that characterizes the *non-uniform robustness* of the homology of $f^{-1}(a)$ under the error model E.

4 Transformation to Uniform Error

In this section, we show that the non-uniform persistence and well diagrams of a function f and an error model E are really just the uniform diagrams of another function. To do so, we create a dual error model, E^*, which enables us to transform non-uniform interlevel sets and perturbations into uniform interlevel sets and perturbations.

Dual error model. Given an error model $E : \mathbb{X} \times \bar{\mathbb{R}} \to \bar{\mathbb{R}}$, we recall that the function $e_x : \bar{\mathbb{R}} \to \bar{\mathbb{R}}$ defined by $e_x(r) = E(x, r)$ is invertible for every $x \in \mathbb{X}$. We thus have the following definition.

4 (Definition). *The* dual error model *of E is the unique mapping $E^* : \mathbb{X} \times \bar{\mathbb{R}} \to \bar{\mathbb{R}}$ that satisfies $E^*(x, E(x, r)) = r$ for every (x, r) in $\mathbb{X} \times \bar{\mathbb{R}}$.*

Considering the restrictions of E and E^* obtained by fixing a point x, we note that $e_x^* = e_x^{-1}$. The name of E^* is justified by the following result. For technical reasons, it needs the additional assumption that $\mathbb{X}$ be first-countable [14]. This assumption is rather mild; for example, it is satisfied whenever $\mathbb{X}$ can be embedded in finite-dimensional Euclidean space.

5 (Duality Lemma). *Given an error model $E : \mathbb{X} \times \bar{\mathbb{R}} \to \bar{\mathbb{R}}$ on a compact and first-countable topological space $\mathbb{X}$, then (i) E^* is an error model, (ii) $(E^*)^* = E$, (iii) E^* is linear iff E is linear.*

Proof. Claims (ii) and (iii) are obvious. For Claim (i), we note that monotonicity and normalization for E immediately imply the same properties for E^*. So we must only prove that E^* is a continuous function from $\mathbb{X} \times \bar{\mathbb{R}}$ to $\bar{\mathbb{R}}$. To do this, we make use of the following lemma from point-set topology: Assuming a first-countable space $\mathbb{W}$ and a compact space $\mathbb{Y}$, a mapping $H : \mathbb{W} \to \mathbb{Y}$ is continuous iff the graph of H is a closed subset of $\mathbb{W} \times \mathbb{Y}$; see eg. [14]. In our context, we have $\mathbb{W} = \mathbb{X} \times \bar{\mathbb{R}}$ and $\mathbb{Y} = \bar{\mathbb{R}}$.

By assumption, E is continuous, and thus the graph of E is a closed subset of $\mathbb{X} \times \bar{\mathbb{R}} \times \bar{\mathbb{R}}$. On the other hand, the graphs of E and E^* are homeomorphic. To see this, recall that $\text{graph}\, E = \{(x, r, a) \in \mathbb{X} \times \bar{\mathbb{R}} \times \bar{\mathbb{R}} \mid E(x, r) = a\}$. Switching the last two arguments gives a homeomorphism $\psi : \mathbb{X} \times \bar{\mathbb{R}} \times \bar{\mathbb{R}} \to \mathbb{X} \times \bar{\mathbb{R}} \times \bar{\mathbb{R}}$ defined by $\psi(x, r, a) = (x, a, r)$. Then $(x, r, a) \in \text{graph}\, E$ iff $(x, a, r) \in \text{graph}\, E^*$. In other words, the restriction of ψ to the graph of E provides a homeomorphism between $\text{graph}\, E$ and $\text{graph}\, E^*$. In particular, the graph of E^* is closed. Applying the lemma once again, we conclude that E^* is continuous. □

Transforming functions. The dual error model E^* of E allows us to associate to each function f a transformation Γ_f that turns non-uniform interlevel sets into uniform ones, as we now explain. Given such f, we create a new function $\Gamma_f(f)$ defined by the rule $\Gamma_f(f)(x) = E^*(x, f(x))$ for each $x \in \mathbb{X}$. This definition is illustrated in Figure 3. To construct $\Gamma_f(f)$ geometrically, we use the subset of the graph of E that projects to the graph of f in $\mathbb{X} \times \bar{\mathbb{R}}$. The projection of the same subset to $\mathbb{X}$ times the other copy of $\bar{\mathbb{R}}$ gives the graph of $\Gamma_f(f)$. As mentioned earlier, this transformation forms a correspondence between non-uniform and uniform interlevel sets.

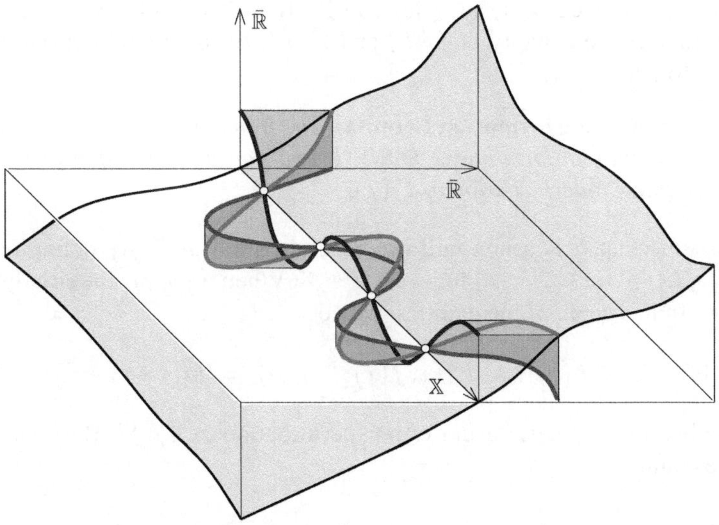

Fig. 3. The surface is the graph of the error model, E. We also see the graph of f in the vertical plane, the graph of $\Gamma_f(f)$ in the horizontal plane, and the curve in the surface that projects to both. We note that f and $\Gamma_f(f)$ have the same zeroset.

6 (Interlevel Set Transformation Lemma). *For every $r \leq s$, the non-uniform interlevel set of f and E is the uniform interlevel set of $\Gamma_f(f)$; that is,* $\mathbb{X}_{[r,s]}(f, E) = \mathbb{X}_{[r,s]}(\Gamma_f(f))$.

Proof. Assume first that $x \in \mathbb{X}_{[r,s]}(f, E)$. Applying the strictly increasing function e_x^* to the chain of inequalities $e_x(r) \leq f(x) \leq e_x(s)$ gives

$$r \leq e_x^*(f(x)) = E^*(x, f(x)) = \Gamma_f(f)(x) \leq s,$$

and so $x \in \mathbb{X}_{[r,s]}(\Gamma_f(f))$. Reversing the argument proves the claim. □

In particular, the non-uniform sublevel and superlevel sets of f and E are the uniform sublevel and superlevel sets of $\Gamma_f(f)$. This implies that the extended filtrations defined by f and E and by $\Gamma_f(f)$ are the same. Hence, they define the same sequence of homology groups and maps between them, and therefore also the same persistence diagrams.

7 (Persistence Diagram Lemma). *The transformation Γ_f turns non-uniform persistence diagrams into uniform ones; that is,* $\mathrm{Ngm}(f, E) = \mathrm{Dgm}(\Gamma_f(f))$.

Transforming perturbations. We now generalize the above construction, applying Γ_f to any other function $h : \mathbb{X} \to \mathbb{R}$ by setting

$$\Gamma_f(h)(x) = \Gamma_f(f)(x) - E^*(x, f(x) - h(x)),$$

for each $x \in \mathbb{X}$. Of course, we could perform a similar operation using Γ_h, but $\Gamma_f(h)$ and $\Gamma_h(h)$ are not necessarily the same. In Section 5, we will see that a linear error model guarantees the equality of $\Gamma_f(h)$ and $\Gamma_h(h)$. Even without this assumption, we have the following:

8 (Perturbation Transformation Lemma). *The transformation Γ_f turns non-uniform r-perturbations into uniform ones; that is, h is a non-uniform r-perturbation of f iff $\Gamma_f(h)$ is a uniform r-perturbation of $\Gamma_f(f)$.*

Proof. Assume that h is a non-uniform r-perturbation of f. By definition, we have $e_x(-r) \leq f(x) - h(x) \leq e_x(r)$ for every $x \in \mathbb{X}$. When we apply the strictly increasing function e_x^* to this chain of inequalities, we get

$$-r \leq e_x^*(f(x) - h(x)) = E^*(x, f(x) - h(x)) = \Gamma_f(f)(x) - \Gamma_f(h)(x) \leq r,$$

which implies that $\Gamma_f(h)$ is a uniform r-perturbation of $\Gamma_f(f)$. Reversing the steps proves the claim. □

The two Transformation Lemmas together imply that the non-uniform well diagram of f and E is identical to the uniform well diagram of Γ_f. In other words, the robustness of each level set to non-uniform perturbation can be read directly off $\mathrm{Ngm}(f, E) = \mathrm{Dgm}(\Gamma_f(f))$ in the manner discussed in Section 2.

5 Linear Error

In this section, we demonstrate that the imposition of linearity on our error model E leads to a richer theory of persistence. More specifically, we assume that E is linear and use this fact to define a metric on the space of all functions. Then we show that, under this metric, non-uniform persistence diagrams are stable.

Metric. Recall that an error model E is *linear* if $E(x,r) = r \cdot E(x,1)$, for all $x \in \mathbb{X}$ and all $r \in \bar{\mathbb{R}}$. Put another way, E is linear if all of the functions e_x are linear. Whenever we have such a model and two functions f and h, we may rewrite the inequalities in (4) as $|f(x) - h(x)| \leq E(x,r)$. In other words, f is a non-uniform r-perturbation of h iff h is a non-uniform r-perturbation of f. This leads us to define the following notion of *non-uniform distance* between h and f:

$$d_E(f,h) \;=\; \min\{r \mid |f(x) - h(x)| \leq E(x,r), \text{ for all } x \in \mathbb{X}\}.$$

9 (Metric Lemma). *Assuming a linear error model E, d_E is a metric on the vector space of all $\bar{\mathbb{R}}$-valued functions on $\mathbb{X}$.*

Proof. Since $E(x,0) = 0$, we have $f = h$ iff $d_E(f,h) = 0$. Furthermore, $d_E(f,h) = d_E(h,f)$, as discussed above. It remains to prove the triangle inequality: $d_E(f,h) \leq d_E(f,g) + d_E(g,h)$. Put $R = d_E(f,g)$ and $S = d_E(g,h)$ and fix a point $x \in \mathbb{X}$. Then using the triangle inequality for the standard metric in $\bar{\mathbb{R}}$, we find

$$\begin{aligned} |f(x) - h(x)| &\leq |f(x) - g(x)| + |g(x) - h(x)| \\ &\leq E(x,R) + E(x,S) \\ &\leq E(x,R+S). \end{aligned}$$

This implies $d_E(f,h) \leq R + S$ and the claim follows. □

We note that the Metric Lemma still holds under a weaker assumption on E. Namely, we need only assume that for each $x \in \mathbb{X}$, $e_x(-r) = -e_x(r)$ and $e_x(r) + e_x(s) \leq e_x(r+s)$, for all $r, s \geq 0$. In words, each e_x is odd and also convex on the non-negative half of the extended real line.

Stability. We now compare the non-uniform persistence diagrams of two functions f and h on $\mathbb{X}$, and prove that their bottleneck distance is bounded from above by the non-uniform distance between the two functions. First we need to show that for a linear error model, the transformations defined by different functions are the same.

10 (Linear Transformation Lemma). *Let $f, h : \mathbb{X} \to \bar{\mathbb{R}}$ be two functions. Assuming a linear error model E, we have $\Gamma_f(h) = \Gamma_h(h)$.*

Proof. Fix a point $x \in \mathbb{X}$. Starting with the definition, we get

$$\begin{aligned} \Gamma_f(h)(x) &= \Gamma_f(f)(x) - E^*(x, f(x) - h(x)) \\ &= E^*(x, f(x)) - E^*(x, f(x) - h(x)) \\ &= E^*(x, h(x)), \end{aligned}$$

where we use linearity to get from the second to the third line. But the third line is equal to $\Gamma_h(h)(x)$, and the claim follows. □

The Linear Transformation Lemma justifies the notation $\Gamma(h)$ to refer to the common function $\Gamma_f(h)$ for any other function f. As an immediate consequence of the Perturbation Transformation Lemma, we then find $d_E(f,h) = \|\Gamma(f) - \Gamma(h)\|_\infty$. Finally, we get the stability of the non-uniform persistence diagrams.

11 (Non-uniform Stability Theorem). *For any two $f, h : \mathbb{X} \to \bar{\mathbb{R}}$, and a linear error model E, we have: $W_\infty(\mathrm{Ngm}(f,E), \mathrm{Ngm}(h,E)) \leq d_E(f,h)$.*

Proof. First we transform f and h into $\Gamma(f)$ and $\Gamma(h)$, recalling from the Persistence Diagram Lemma that $\mathrm{Ngm}(f,E) = \mathrm{Dgm}(\Gamma(f))$ and $\mathrm{Ngm}(h,E) = \mathrm{Dgm}(\Gamma(h))$. Applying the Stability Theorem to the two uniform persistence diagrams, we find

$$W_\infty(\mathrm{Dgm}(\Gamma(f)), \mathrm{Dgm}(\Gamma(h))) \leq \|\Gamma(f) - \Gamma(h)\|_\infty = d_E(f,h),$$

and the result follows. □

6 Discussion

The main contribution of this paper is the extension of the machinery of persistent homology to non-uniform error models. This extension is not complete and many questions remain yet unanswered.

On the technical level, it would be interesting to gain a more detailed understanding on how the difference between $\mathrm{Dgm}(f)$ and $\mathrm{Ngm}(f,E)$ relates to the error model, E. Similarly, can we extend the Non-uniform Stability Theorem from linear to non-linear error models? A more challenging question is the extension of robustness under non-uniform error to a full-blown probability theory of the homology of level sets.

Acknowledgments

The authors thank Dmitriy Morozov for insightful discussions on the topic of this paper.

References

1. Bendich, P., Edelsbrunner, H., Morozov, D., Patel, A.: Robustness of level and interlevel sets. Manuscript, IST Austria, Klosterneuburg, Austria (2010)
2. Carlsson, G., de Silva, V.: Zigzag persistence. Found. Comput. Math. (2010)
3. Carlsson, G., Ishkhanov, T., de Silva, V., Zomorodian, A.: On the local behavior of spaces of local images. Internat. J. Comput. Vision 76, 1–12 (2008)
4. Cerri, A., Ferri, M., Giorgi, D.: Retrieval of trademark images by means of size functions. Graphical Models 68, 451–471 (2006)
5. Chung, M.K., Bubenik, P., Kim, P.T.: Persistence diagrams of cortical surface data. In: Prince, J.L., Pham, D.L., Myers, K.J. (eds.) Information Processing in Medical Imaging. LNCS, vol. 5636, pp. 386–397. Springer, Heidelberg (2009)

6. Cohen-Steiner, D., Edelsbrunner, H., Harer, J.: Stability of persistence diagrams. Discrete Comput. Geom. 28, 511–533 (2007)
7. Cohen-Steiner, D., Edelsbrunner, H., Harer, J.: Extending persistence using Poincaré and Lefschetz duality. Found. Comput. Math. 9, 79–103 (2009)
8. de Silva, V., Ghrist, R.: Coverage in sensor networks via persistent homology. Alg. Geom. Topology 7, 339–358 (2007)
9. Dequèant, M.-L., Ahnert, S., Edelsbrunner, H., Fink, T.M.A., Glynn, E.F., Hattem, G., Kudlicki, A., Mileyko, Y., Morton, J., Mushegian, A.R., Pachter, L., Rowicka, M., Shiu, A., Sturmfels, B., Pourquié, O.: Comparison of pattern detection methods in microarray time series of the segmentation clock. PLoS ONE 3, e2856 (2008), doi:10.1371/journal.pone.0002856
10. Edelsbrunner, H., Harer, J.L.: Computational Topology. An Introduction. Amer. Math. Soc. (2009)
11. Edelsbrunner, H., Morozov, D., Patel, A.: Quantifying transversality by measuring the robustness of intersections. Manuscript, Dept. Comput. Sci., Duke Univ., Durham, North Carolina (2009)
12. Hatcher, A.: Algebraic Topology. Cambridge Univ. Press, Cambridge (2002)
13. Munkres, J.R.: Elements of Algebraic Topology. Perseus, Cambridge, Massachusetts (1984)
14. Thron, W.J.: Topological Structures. Holt, Rinehart & Winston, New York (1966)

Information Complexity of Online Problems*

Juraj Hromkovič[1], Rastislav Královič[2], and Richard Královič[1]

[1] Department of Computer Science, ETH Zurich, Switzerland
{juraj.hromkovic,richard.kralovic}@inf.ethz.ch
[2] Department of Computer Science, Comenius University, Bratislava, Slovakia
kralovic@dcs.fmph.uniba.sk

Abstract. What is information? Frequently spoken about in many contexts, yet nobody has ever been able to define it with mathematical rigor. The best we are left with so far is the concept of entropy by Shannon, and the concept of information content of binary strings by Chaitin and Kolmogorov. While these are doubtlessly great research instruments, they are hardly helpful in measuring the amount of information contained in particular objects. In a pursuit to overcome these limitations, we propose the notion of information content of algorithmic problems. We discuss our approaches and their possible usefulness in understanding the basic concepts of informatics, namely the concept of algorithms and the concept of computational complexity.

1 Introduction

Looking at the terms "informatics" or "information processing" as reasonable alternatives to "computer science" for naming our scientific discipline, we identify two basic notions describing the fundamental stones of informatics. The first one is the notion of "algorithm", which is related to "automatic processing", and the second one is "information". The relationships of computer scientists to these two notions are different. The notion of an algorithm as a method for solving a problem was fixed in formal mathematical definitions [11], and it seems that we understand the concept of algorithms very well. The formal concept of algorithms belongs to informatics; the development of this concept is a fundamental contribution of informatics to science.

The formal definition of an algorithm can be viewed as an axiom of the formal language of mathematics; we even relate the birth of computer science to the development of this fundamental concept. The situation with the notion of "information" is completely different. This notion is not a property of informatics or a contribution of informatics to other scientific disciplines, as it is used in physics, biology, technical sciences, and many other disciplines. The crucial point is that we work intuitively with this term and yet we do not really know what "information" is. We are even very far from giving a precise mathematical definition of "information". We have the concept of entropy due to Shannon [10] and

* This work was partially supported by ETH grant TH 18 07-3, and APVV grant 0433-06.

P. Hliněný and A. Kučera (Eds.): MFCS 2010, LNCS 6281, pp. 24–36, 2010.

the concept of Kolmogorov complexity due to Kolmogorov [9] and Chaitin [4]. Both concepts are great research instruments, but at the very end they cannot be used to estimate the information content of particular objects (binary strings).

On the other hand, computer scientists frequently speak about automatic information processing. What does it mean to solve an instance of a computing problem? To compute a solution of a problem instance is nothing else than extracting a desired information about the problem instance that is somehow hidden in the problem instance description. Remember, we cannot speak about the computational complexity of a problem without fixing the representation of the problem instances. The aim of information processing in the usual sense is to express some part of information contained in the input in a prescribed way – from this point of view, computing is a transformation of one representation of information to another one. Complexity theory indicates that the cost of these transformations may be arbitrarily high, depending on the problem. The concept of NP-completeness shows that many problems can be "equally" hard from the computational point of view, because one can efficiently transform the representation of the instances of one problem into the representation of the instances of another problem.

All this suggests that in order to advance in understanding the nature of computing, we should intensify our efforts to figure out what information really is. What about introducing a concept of "information content of algorithmic problems" in order to classify the problems in this way, and then trying to learn something essential about the computational hardness of the corresponding problems?

Another reason to focus more on studying the meaning of information is the 40-years old unsuccessful effort to prove nontrivial lower bounds on the computational complexity of concrete algorithmic problems. On the one hand, we believe that NP-hard problems require exponential time to be solved, and on the other hand we are unable to prove a superlinear lower bound on the time complexity of at least one of them. Our lower bound methods are often strongly related to "information transfer" between different parts of hardware (memory) or between different parts of computations. It seems that arguments of this kind are not strong enough to reach our goals. Understanding more about the concept of information and its role in computation could be the way out.

In this paper, we introduce the notion of "information content of algorithmic problems" with the hope that the study of this information measure could be helpful in understanding the fundamentals of computing (information processing).

2 Information Content of Online Problems

Our first attempt starts with defining the information content of a problem as a measure that does not depend on the computational hardness of the problem. Because of that, we do not start with thinking about classical offline problems. In an online setting, the algorithm receives instances of the problem piece by piece. After getting a part of the input, the algorithm is required to make a corresponding decision with respect to the problem instance solution and is not allowed to revise decisions later after seeing a new part of the input. This means that the algorithm must generate a part of the solution with zero information about the missing parts of the problem instance. Thinking in terms of optimization problems, one defines the competitive ratio $\mathrm{cr}_{\mathtt{A}}(I)$ of an online algorithm A on a problem instance I as the ratio between the cost of the solution computed by A and the cost of the optimal solution that can be computed based on the complete information about the instance. An algorithm A, whose competitive ratio $\mathrm{cr}_{\mathtt{A}}(I)$ is at most r for any problem instance I, is called r-competitive.

Our first idea is to define the information content $\inf_P$ of an online problem P as the minimum number of advice bits sufficient and necessary to guarantee that it is possible to compute an optimal solution to any problem instance (to get the competitive ratio 1). The advice bits are nothing else but "information" about the not yet known part of the input. For sure, the Kolmogorov complexity of the instance is an upper bound on the number $\inf_P(I)$ of advice bits needed to solve the instance optimally in an online manner, however, this is only an upper bound on the information content of this problem instance. Often one does not need the full information about the instance to distinguish it from other ones. Moreover, one does not necessarily need to distinguish each instance from all other ones; it is enough to take the right decision, which can be common for many problem instances. And this is exactly what we are searching for, because this is what we intuitively relate to the desired notion of "information content of a problem".

One can define the information content of an online problem in the common worst-case manner, parameterized by the size of the problem instances. Hence, $\inf_P(n)$ would be the maximum of $\inf_P(I)$ over all instances I of size n. However, to get an exact definition of $\inf_P(n)$, one still has to fix the way in which the advice bits are communicated. The first paper [5] investigating the number of advice bits necessary for solving online problems optimally proposed that the online algorithm may pose a sequence of questions and it always gets the right answer as a non-empty sequence of bits. The sum of the lengths of all answers is the advice complexity. This kind of measurement is rough, because one can code also a lot of information in the lengths of the individual answers and in this way the whole communication can be viewed as a string over three symbols. In fact, the authors of [5] proved that this measurement is rough up to a multiplicative constant factor.

One possible solution to fix the problems of the model introduced in [5] has been proposed in [6]. Here, the algorithm receives an advice of fixed constant length k with each request. The length k is used as a measure of the advice complexity of the problem; for an input consisting of n requests, exactly kn bits of advice are used. This model enforces a smooth flow of the advice information, since the same amount of advice is made available to the algorithm in every request. It is not possible, however, to employ this model to analyze online algorithms that use less than n bits of advice. This is a significant drawback, because many well-known online problems are very easy to solve optimally with one bit of advice available in every request.

Another approach to define the communication of the advice bits, introduced in [2], is to relax the requirement of a smooth advice information flow and to follow a way analogous to the definition of randomized Turing machines. In the case of randomization, the Turing machine is augmented with a separate tape containing random bits. It is then possible to measure the amount of randomness used as the number of bits accessed on the random tape. Similarly, we can provide an *advice tape* to the online algorithm. The algorithm can access this tape in an arbitrary way and the advice complexity is measured as the total number of advice bits accessed over the whole computation. In this way, it is perfectly possible to analyze problems that use less than 1 advice bit per request on average. The advice complexity can be sublinear as well.

In the sequel, we provide a more formal definition of the information content of an online problem which coincides with the advice complexity as defined in [2]. At first, we define the advice complexity of an online algorithm:

Definition 1. *An* online algorithm $\mathtt{A}$ with advice *computes the output sequence* $\mathtt{A}^{\phi}(I) = (y_1, \dots, y_n)$ *such that* y_i *is computed from* $\phi, x_1, \dots, x_i$*, where* ϕ *is the content of the advice tape, i. e., an infinite binary sequence, and* $I = (x_1, \dots, x_n)$ *is an input instance of the considered online problem. Algorithm* $\mathtt{A}$ *is* c-competitive with advice complexity $s(n)$ *if, for every* n *and for each input sequence* I *of length* n*, there exists some* ϕ *such that the competitive ratio of algorithm* $\mathtt{A}$ *on* I *is at most* c*, i. e.,* $\mathrm{cr}_{\mathtt{A}}(I) \leq c$*, and no positions of* ϕ *except the first* $s(n)$ *bits are accessed during the computation of* $\mathtt{A}^{\phi}(I)$*.*

Now we are ready to formally define the information content $\inf_P(n)$ of an online problem P. More precisely, we define upper and lower bounds on the information content $\inf_P(n)$. The definition of the upper bound is straightforward:

Definition 2. *A function* $f(n)$ *is* an upper bound on $\inf_P(n)$ *if and only if there exists a* 1*-competitive online algorithm* $\mathtt{A}$ *with advice complexity* $s(n)$ *such that* $s(n) \leq f(n)$ *for all* n*.*

It is, however, more tricky to find a suitable definition of the lower bound on the information content. As a first attempt, we can make a simple modification to the definition of the upper bound:

Definition 3. *A function* $g(n)$ *is* a lower bound on $\inf_P(n)$ *if and only if, for every* 1*-competitive online algorithm* $\mathtt{A}$ *with advice complexity* $s(n)$*, it holds that* $s(n) \geq g(n)$ *for all* n*.*

Unfortunately, this definition has a serious drawback: It is possible to store a constant, but arbitrarily high, amount of information in the online algorithm itself. In certain cases, such information can be used to solve some input instances very efficiently. In such cases, no nontrivial lower bound on $\inf_P(n)$ can be given in the sense of Definition 3.

As an example of such a situation, consider the following (rather artificial) online problem:

Definition 4 (Guess the Input Length Problem). *An input instance of the* Guess the Input Length *problem (*GIL*) consists of n requests. Each of the first $n-1$ requests is the symbol* 0 *and the last request is the symbol* 1*. The online algorithm must guess the number of requests after the first request is received, i. e., it must output n as an answer to the first request. No answer is required for any other request.*

For any n, we can construct an algorithm that, given advice containing a single bit 0, outputs n. If the first bit of advice is 1, the algorithm reads the complete number from the advice tape and outputs it. In such a way, $\Theta(\log n')$ advice bits are used for all numbers $n' \neq n$, but a single bit of advice is sufficient for an input of length n. Hence, the only feasible lower bound on $\inf_{\mathrm{GIL}}(n)$ in the sense of Definition 3 is $g(n) = 1$.

One possible approach to avoid this problem is to allow the lower bound to be valid only from some minimal n_0 that depends on the online algorithm:

Definition 5. *A function $g(n)$ is* a lower bound on $\inf_P(n)$ *if and only if, for every* 1*-competitive online algorithm* A *with advice complexity $s(n)$, there exists some n_0 such that $s(n) \geq g(n)$ holds for all $n \geq n_0$.*

Easily, the Kolmogorov complexity $K(n)$ of the number n is an upper bound on $\inf_{\mathrm{GIL}}(n)$. Intuitively, $K(n)$ should be a lower bound on $\inf_{\mathrm{GIL}}(n)$ as well. Indeed, the online algorithm must output the number n without any a-priori information, the whole information about n must be contained in the used advice. To define the exact value of $K(n)$, we have to fix some "programming language" with respect to which we measure the Kolmogorov complexity. However, once this programming language is fixed, it might still be possible to create a more sophisticated online algorithm for GIL that saves some constant amount of advice for infinitely many inputs. Furthermore, we are not able to provide any bounds on such a constant. Hence, we cannot use the function $K(n)-c$ as a lower bound of $\inf_{\mathrm{GIL}}(n)$, regardless of the choice of the constant c. We may, however, relax Definition 5 in the following way:

Definition 6. *A function $g(n)$ is* a lower bound on $\inf_P(n)$ *if and only if, for every* 1*-competitive online algorithm* A *with advice complexity $s(n)$, there exists some constant Δ such that $s(n) + \Delta \geq g(n)$ holds for all n.*

It is not difficult to see that $K(n)$ is a lower bound on $\inf_{\mathrm{GIL}}(n)$ in the sense of Definition 6. A drawback of such a definition is that a lower bound can be in fact larger than an upper bound. The difference between the two, however, can be at most constant, hence it is not an issue if an asymptotic analysis is done.

Alternatively, we could use another way to relax Definition 5:

Definition 7. *A function $g(n)$ is* a lower bound on $\inf_P(n)$ *if and only if, for every* 1*-competitive online algorithm* A *with advice complexity $s(n)$, $s(n) \geq g(n)$ holds for infinitely many n.*

Such a definition of a lower bound is very weak – a lower bound can be arbitrarily larger than an upper bound on infinitely many values. Nevertheless, it is not possible that a lower bound is asymptotically larger than an upper bound. On a positive side, Definition 7 allows us to easily claim a lower bound of $\log_2(n)$ on $\inf_{\text{GIL}}(n)$ due to the existence of incompressible strings.

Another approach to avoid the problems of Definition 3 is to actually modify the definition of advice complexity (Definition 1) to require that $s(n)$ bits of advice are sufficient to solve all input instances of length *at most* n. After such modification, we can easily claim $\log_2(n)$ to be a lower bound on $\inf_{\text{GIL}}(n)$ in the sense of Definition 3.

We have presented several possible ways how to define a lower bound on the information content of an online problem. Nevertheless, there is usually little difference between them for the relevant online problems. In the rest of this paper, we stick to the strongest Definition 3, despite to its deficiencies, since all presented results hold with respect to this definition as well. As a shorthand, we write

$$\inf_P(n) \leq f(n)$$

if $f(n)$ is an upper bound on $\inf_P(n)$ and

$$\inf_P(n) \geq g(n)$$

if $g(n)$ is a lower bound on $\inf_P(n)$.

In the sequel, we show that the information content of online problems can be very different even for well-known natural problems. As an example of a problem with a very low information content, consider the SKIRENTAL problem (for a definition, see, e. g., [3]). For any input instance, the optimal solution of the SKIRENTAL problem has very simple structure: Either it consists of buying the skis at the very beginning, or it consists of renting the skis all the time. Hence, a single bit of advice is sufficient (and necessary) to solve the SKIRENTAL problem optimally for input instances of arbitrary lengths. Thus, we have a tight bound on the information content of the SKIRENTAL problem: $\inf_{\text{SKIRENTAL}}(n) = 1$.

Next, we show that the well-known PAGING problem is an example of an online problem that has linear information content, i. e., any optimal algorithm for this problem requires $\Theta(n)$ advice bits, where n is the number of requests.

Definition 8 (Paging Problem). *The input is a sequence of integers representing requests to logical pages $I = (x_1, \ldots, x_n)$, $x_i > 0$. An online algorithm* A *maintains a buffer (content of the physical memory) $B = \{b_1, \ldots, b_K\}$ of K integers, where K is a fixed constant known to* A*. Before processing the first request, the buffer gets initialized as $B = \{1, \ldots, K\}$. Upon receiving a request x_i, if $x_i \in B$, then $y_i = 0$. If $x_i \notin B$, then a* page fault *occurs, and the algorithm has to find some* victim *b_j, i. e., $B := B \setminus \{b_j\} \cup \{x_i\}$, and $y_i = b_j$. The cost of the solution $\mathcal{A} = \mathtt{A}(I)$ is the number of page faults, i. e., $C(\mathcal{A}) = |\{y_i : y_i > 0\}|$.*

In fact, the paging problem is not a single problem, but rather a collection of problems parameterized by the buffer size K. For simplicity, however, we omit this parameter from the notation of the problem, and assume that K is a fixed constant known to the online algorithm. It is not difficult to prove a linear upper bound on the information content of PAGING [2, 1]:

Theorem 1

$$\inf_{\text{PAGING}}(n) \leq n + K$$

Proof. It is sufficient to show that there is an optimal online algorithm (i. e., a 1-competitive algorithm) solving PAGING with advice complexity $n + K$.

Consider an input sequence I and an optimal offline algorithm OPT processing it. In each step of OPT, call a page currently in the buffer *active* if it will be requested again before OPT replaces it by some other page. We design A such that, in each step i, the set of OPT's active pages will be in B, and A will maintain with each page an *active* flag identifying this subset. If A gets an input x_i that causes a page fault, some *passive* (i. e., non-active) page is replaced by x_i. Moreover, A reads with each input also one bit from the advice tape telling whether x_i is active for OPT. Since the set of active pages is the same for OPT and A, it is immediate that A generates the same sequence of page faults.

Algorithm A consumes one bit of advice with every request. Moreover, it needs to receive the information about which pages that are stored in the buffer at the beginning of the computation are active. Hence, A reads $n + K$ bits of advice in total. □

Next, we provide a lower bound on $\inf_{\text{PAGING}}(n)$ that shows that the above-described upper bound is tight if the buffer size K is large [2, 1], that is, any optimal algorithm for PAGING must consume almost one bit of advice for every request:

Theorem 2. *There exists a fixed constant C that does not depend on K such that*

$$\inf_{\text{PAGING}}(n) \geq n\left(1 - \frac{\log(K-1) + C}{4(K-1)}\right) - \mathcal{O}(1)$$

holds. The constant of $\mathcal{O}(1)$ depends on K.

Proof. The main idea of the proof is, for any input length n, to construct a set of input instances that are sufficiently different, i. e., that any two instances from this set provably need a different advice if they are to be solved in an optimal way.

We now formally describe how to construct such a set of inputs $\mathcal{I}$. Let $\nu := \lfloor n/(2K-2) \rfloor$ and $Z := \binom{2K-2}{K-1}$. The set $\mathcal{I}$ consists of Z^ν inputs organized in a complete Z-ary tree of height ν, where each edge is labeled with a sequence of $2K - 2$ requests. Each leaf of the tree represents one input from $\mathcal{I}$ which is obtained by concatenating the request sequences on all edges of the path from the root to this leaf. Let N_h denote the sequence of $K - 1$ requests $(hK + 1, hK + 2, \ldots, hK + K - 1)$. Let us denote the root of the tree to be on level 1, the sons of the root to be on level 2, etc. Each edge e leading from a vertex v of level h is labeled with the sequence N_h followed by the sequence $S(e)$. The sequences $S(e)$ are defined recursively as follows. For each vertex v, consider a set of K elements $m(v)$; the intuition is that $m(v)$ is the content of the memory of some optimal algorithm processing the given input. Let us define $m(root) = \{1, \ldots, K\}$. Inductively, consider a vertex v of level h with $m(v)$ already defined.

The sequences $S(e)$ of the outgoing edges contain $(K-1)$-element subsets of the set $m(v) \cup N_h$ such that they do not contain the element $hK + K - 1$ (the ordering within the sequence is not important; to avoid ambiguity, let us assume the sequences $S(e)$ are increasing). For any edge $e = (v, v')$, let us set $m(v') := S(e) \cup \{hK + K - 1\}$. Since, by using this procedure, we keep the invariant that $m(v) \cap N_h = \emptyset$, there are exactly $Z = \binom{2K-2}{K-1}$ possible subsets, a unique one for every son of v.

In this way, we obtain a set of instances of length $\nu(2K-2)$. Each of these instances can be extended to the length n by repeating the last request $n - \nu(2K-2) = n \bmod (2K-2)$ times.

The intuition behind this construction is as follows. Every vertex of the tree represents some prefix of some input in $\mathcal{I}$. When an optimal algorithm has processed a prefix of an input instance that corresponds to a vertex v, the content of its buffer is $m(v)$. At that moment, a sequence N_h of requests arrives, where h is the level of v. All of the requested pages in N_h are new. Thus, they are not in the buffer and the algorithm must make a page fault for each of them. However, the algorithm can choose which pages in the buffer will be overwritten. In this way, after processing N_h, the algorithm can obtain any content of the buffer that is a subset of $m(v) \cup N_h$ that contains K elements and one of these elements is $hk + K - 1$ (the last request of N_h).

After processing N_h, the set of control requests $S(e)$ follows. If the algorithm made a correct choice while processing N_h, all elements of $S(e)$ are in the buffer after processing N_h, hence no page faults are generated while processing $S(e)$. On the other hand, if a wrong choice was made, at least one page fault is necessary. Easily, once a wrong choice is made, the algorithm cannot be optimal.

Assume that two inputs from $\mathcal{I}$ are given the same advice. Let v be the vertex representing the longest common prefix of these inputs, and h be the level of v. The online algorithm makes the same choice while processing N_h. Therefore, this choice is wrong in at least one of the inputs, and the online algorithm fails to achieve optimality.

Hence, the advice complexity of A has to be at least $\log(|\mathcal{I}|) = \log(Z^\nu) = \nu \log(Z)$ bits:

$$\begin{aligned} s(n) \geq \nu \log(Z) &= \left\lfloor \frac{n}{2K-2} \right\rfloor \log \binom{2K-2}{K-1} \\ &\geq \left(\frac{n}{2K-2} - 1 \right) \log \binom{2K-2}{K-1} = \frac{n}{2K-2} \log \binom{2K-2}{K-1} - \mathcal{O}(1). \end{aligned} \tag{1}$$

Using the Stirling formula, we have

$$\binom{2t}{t} = \frac{(2t)!}{(t!)^2} \geq \frac{\sqrt{2\pi 2t}\left(\frac{2t}{e}\right)^{2t}}{\left(\sqrt{2\pi t}\left(\frac{t}{e}\right)^t \left(1 + \mathcal{O}\left(\frac{1}{t}\right)\right)\right)^2} \geq 2^{2t} \cdot \frac{D}{\sqrt{t}} \tag{2}$$

for some constant D not depending on K. Substituting $t := K - 1$ into (2) and (1), we complete the proof:

$$\begin{aligned} s(n) &\geq n\left(\frac{1}{2K-2}\log\left(2^{2(K-1)}\frac{D}{\sqrt{K-1}}\right)\right) - \mathcal{O}(1) \\ &= n \cdot \frac{2(K-1) + \log(D) - \frac{1}{2}\log(K-1)}{2K-2} - \mathcal{O}(1) \\ &= n\left(1 - \frac{-2\log(D) + \log(K-1)}{4K-4}\right) - \mathcal{O}(1). \end{aligned}$$

□

The information content of some other online problems has been analyzed in [2, 1]. In particular, the problem of *Disjoint Path Allocation* (DisPathAlloc for short) and the problem of *Job Shop Scheduling for Two Jobs* (JSSchedule for short) have been considered. For definitions and more information about these problems, see, e. g., [3] and [7, 8], respectively.

The DisPathAlloc problem is an another example of an online problem with high information content: Any optimal algorithm for DisPathAlloc needs at least $n/2-1$ bits of advice to solve input instances with n requests. On the other hand, there is a very simple optimal algorithm that consumes one bit of advice for every request. Hence, we can write $n \geq \inf_{\text{DisPathAlloc}}(n) \geq n/2 - 1$.

The JSSchedule problem, on the other hand, has a lower information content. If there are n jobs to be scheduled, $2\lceil\sqrt{n}\rceil$ advice bits are sufficient to find the optimal solution. This bound is asymptotically tight, i. e., $\Omega(\sqrt{n})$ of advice bits are necessary.

The following table summarizes the known results about the information content of the discussed online problems:

Table 1. Upper and lower bounds on the information content of online problems

P	Upper bound on $\inf_P(n)$	Lower bound on $\inf_P(n)$
Paging	$n + K$	$n\left(1 - \frac{\log(K-1)+C}{4(K-1)}\right) - \mathcal{O}(1)$
DisPathAlloc	n	$\frac{n}{2} - 1$
JSSchedule	$2\lceil\sqrt{n}\rceil$	$\Omega(\sqrt{n})$

3 Relative Information Content of Online Problems

So far, we have focused on the full information content of online problems, defined as the size of advice about future requests that is needed to solve the problem optimally. The requirement of an optimal solution is, however, very strict. In the context of online problems, suboptimal solutions are often analyzed and their quality is measured by their competitive ratio. Hence, it makes sense to consider a trade-off between the competitive ratio of an online algorithm and the size of the advice needed to achieve this ratio. This leads us to the definition of *information content of an online problem P relative to the competitive ratio r*, denoted as $\inf_P^{(r)}(n)$. Again, we define upper and lower bounds on $\inf_P^{(r)}(n)$:

Definition 9. *A function $f(n)$ is* an upper bound on $\inf_P^{(r)}(n)$ *if and only if there exists an r-competitive online algorithm* A *with advice complexity $s(n)$ such that $s(n) \leq f(n)$ for all n.*

Definition 10. *A function $g(n)$ is* a lower bound on $\inf_P^{(r)}(n)$ *if and only if, for every r-competitive online algorithm* A *with advice complexity $s(n)$, it holds that $s(n) \geq g(n)$ for all n.*

The discussion about alternative definitions of the information content is valid for the definition of relative information content as well. Thus, it is possible to create variants of Definition 10 in an analogous way to Definitions 5, 6, and 7.

Several results about the relative information content of online problems are known as well [2, 1]. For the paging problem, it is known that linear advice is necessary to achieve competitive ratios close to 1. More precisely, for any $1 \leq r \leq 1.25$ and any fixed K, $\inf_{\text{PAGING}}^{(r)} \geq \Omega(n)$. The constant hidden in the asymptotic notation, however, depends on r and K. The precise bound can be formulated as follows [1, Theorem 2]:

Theorem 3. *Let r be any constant such that $1 \leq r \leq 1.25$. It holds that*

$$\inf_{\text{PAGING}}^{(r)} \geq \frac{n}{2K-2}\left[1 + \log(3-2r) - (2r-2)\log\left(\frac{1}{2r-2} - 1\right)\right] - \mathcal{O}(1).$$

The constant of $\mathcal{O}(1)$ depends on K and the parameter r.

The idea behind the proof of Theorem 3 is to consider the set of instances defined in the proof of Theorem 2 and adapt the analysis for the case of algorithms with a fixed competitive ratio. The lower bound of Theorem 3 can be complemented by the following upper bound [1, Theorem 1] which states that, for any r, it is possible to achieve an r-competitive algorithm for PAGING with linear advice. Moreover, the linearity constant tends to 0 with growing r:

Theorem 4. *For each constant $r \geq 1$, it holds that*

$$\inf_{\text{PAGING}}^{(r)} \leq n \log\left(\frac{r+1}{r^{\frac{r}{r+1}}}\right) + 3\log n + \mathcal{O}(1).$$

Theorems 3 and 4 suggest that large advice is necessary for PAGING if a constant competitive ratio is required. Now we focus on the opposite end of the competitive ratio spectrum, where the competitive ratio is not a fixed constant.

It is a well-known fact that any deterministic algorithm for paging cannot be better than K-competitive [3]. It is, however, an interesting fact that even extremely small advice can significantly improve this ratio. For example, just two bits of advice for the whole input instance are sufficient to obtain an algorithm with competitive ratio $K/2 + 7.5$, regardless of the number of input requests. In general, b bits of advice are sufficient to obtain a competitive ratio of $\frac{2(K+1)}{2^b} + 3b + 1$ [1, Theorem 5]:

Theorem 5. *Let b be any fixed integer and let*

$$r := \frac{2(K+1)}{2^b} + 3b + 1.$$

Then

$$\mathrm{inf}^{(r)}_{\textsc{Paging}} \leq b.$$

The main idea behind this result is to construct 2^b deterministic paging algorithms that are sufficiently different and that the total number of page faults they make together has a reasonable upper bound. It turns out that the idea of marking algorithms (as described in [3]) can be used to achieve this goal. Afterwards, we can argue that, for any input instance, at least one of these algorithms works well. The advice provided is then just an identification of the best possible algorithm from the constructed set.

On the other hand, it is possible to prove a lower bound corresponding to Theorem 5 [1, Theorem 6]. In fact, this lower bound shows that the upper bound of Theorem 5 is almost tight for small b:

Theorem 6. *Let b be any fixed integer and let $r := K/2^b$. Then*

$$\mathrm{inf}^{(r)}_{\textsc{Paging}} \geq b.$$

The idea behind this lower bound is not very involved. It is sufficient to consider a set of all possible input instances of length n that use only numbers 1 to $K+1$ in their requests. It is not difficult to see that the optimal algorithm makes at most n/K page faults for any such instance. On the other hand, if the advice is short, there are many different inputs with the same advice, what can be used to infer that the algorithm makes sufficiently many page faults on at least one of them.

For the DisPathAlloc problem, we have the following lower bound on the relative information content [1, Theorem 8]:

Theorem 7

$$\mathrm{inf}^{(r)}_{\textsc{DisPathAlloc}} \geq \frac{n+2}{2r} - 2$$

Furthermore, it is not difficult to provide an upper bound on $\mathrm{inf}^{(r)}_{\textsc{DisPathAlloc}}$ that is only a factor of $\log n$ away from the lower bound of Theorem 7 for large r, and that is asymptotically tight for constant r [1, Theorem 9]:

Theorem 8

$$\mathrm{inf}^{(r)}_{\textsc{DisPathAlloc}} \leq \min\left\{ \left(\frac{n}{r} + 3\right) \log n + \mathcal{O}(1), n \log \frac{r}{(r-1)^{\frac{r-1}{r}}} + 3 \log n + \mathcal{O}(1) \right\}$$

For the JSSCHEDULE problem, it is straightforward to adapt the randomized algorithm of [7] to achieve an algorithm with small (logarithmic) advice and a competitive ratio close to 1:

Theorem 9. *There is an online algorithm for* JSSCHEDULE *with advice complexity $s(n) \leq 1 + \log(n)$ that achieves competitive ratio $\mathcal{O}(1 + 1/\sqrt{n})$. Hence,*

$$\inf^{(r)}_{\text{JSSCHEDULE}} \leq 1 + \log(n),$$

for any $r > 1$.

4 Relating Information Content to Computational Complexity

As already mentioned, the first fundamental contribution of informatics to science was the development of the formal concept of an algorithm. The second most important concept discussed in informatics is the concept of computational complexity. The above proposed definitions of the information content of online problems are not related to this concept. This is the same case as for the Kolmogorov complexity that is also not related to the complexity of generating strings from their representations. But for relating the Kolmogorov complexity to the computational complexity of string generation, the concept of resource-bounded Kolmogorov complexity has been developed. It is possible to apply a similar idea in order to combine the information content of online problems with the efficiency of online algorithms.

For any function $f\colon \mathbb{N} \to \mathbb{N}$, we define the *$f$-resource-bounded information content of an online problem P relative to the competitive ratio r*, denoted as $\inf^{(f,r)}_P(n)$:

Definition 11. *Consider the function $f\colon \mathbb{N} \to \mathbb{N}$. The function $h(n)$ is* an upper bound on $\inf^{(f,r)}_P(n)$ *if and only if there exists an r-competitive online algorithm* `A` *with advice complexity $s(n)$ such that $s(n) \leq h(n)$ for all n, and the running time of* `A` *is at most $f(n)$ for any input of length n.*

Definition 12. *Consider the function $f\colon \mathbb{N} \to \mathbb{N}$. The function $g(n)$ is* a lower bound on $\inf^{(f,r)}_P(n)$ *if and only if, for every r-competitive online algorithm* `A` *with advice complexity $s(n)$ such that the running time of* `A` *is at most $f(n)$ for any input of length n, it holds that $s(n) \geq g(n)$ for all n.*

In our examples of online problems such as PAGING, DISPATHALLOC, or JSSCHEDULE, the computational complexity does not matter, because all upper bounds on the number of advice bits can be achieved by efficient online algorithms with advice. But this is not necessarily true for any online problem. If one considers the special online problem GIL (estimating the length of the input), the Kolmogorov complexity of the lengths does not need to be an upper bound on the number of advice bits. An interesting research idea would be to study this phenomenon for "natural" optimization problems. If an NP-hard optimization problem has to be solved in polynomial time in an online manner, then it is not necessarily sufficient to get a compressed version of the whole input.

References

1. Böckenhauer, H.-J., Komm, D., Královič, R., Královič, R., Mömke, T.: Online algorithms with advice. Technical Report 614, ETH Zürich (2009)
2. Böckenhauer, H.-J., Komm, D., Královič, R., Královič, R., Mömke, T.: On the advice complexity of online problems. In: Dong, Y., Du, D.-Z., Ibarra, O. (eds.) ISAAC 2009. LNCS, vol. 5878, pp. 331–340. Springer, Heidelberg (2009)
3. Borodin, A., El-Yaniv, R.: Online Computation and Competitive Analysis. Cambridge University Press, Cambridge (1998)
4. Chaitin, G.J.: On the length of programs for computing finite binary sequences. Journal of the ACM 13(4), 547–569 (1966)
5. Dobrev, S., Královič, R., Pardubská, D.: How much information about the future is needed? In: Geffert, V., Karhumäki, J., Bertoni, A., Preneel, B., Návrat, P., Bieliková, M. (eds.) SOFSEM 2008. LNCS, vol. 4910, pp. 247–258. Springer, Heidelberg (2008)
6. Emek, Y., Fraigniaud, P., Korman, A., Rosén, A.: Online computation with advice. In: Albers, S., Marchetti-Spaccamela, A., Matias, Y., Nikoletseas, S.E., Thomas, W. (eds.) ICALP 2009. LNCS, vol. 5555, pp. 427–438. Springer, Heidelberg (2009)
7. Hromkovič, J.: Design and Analysis of Randomized Algorithms: Introduction to Design Paradigms, Texts in Theoretical Computer Science. An EATCS Series. Springer, New York (2005)
8. Hromkovič, J., Mömke, T., Steinhöfel, K., Widmayer, P.: Job shop scheduling with unit length tasks: bounds and algorithms. Algorithmic Operations Research 2(1), 1–14 (2007)
9. Kolmogorov, A.N.: Three approaches to the definition of the concept "quantity of information". Problemy Peredachi Informatsii 1, 3–11 (1965)
10. Shannon, C.E.: A mathematical theory of communication. Mobile Computing and Communications Review 5(1), 3–55 (2001)
11. Turing, A.M.: On computable numbers, with an application to the Entscheidungsproblem. Proceedings of the London Mathematical Society 2(42), 230–265 (1936)

Algorithmic Lower Bounds for Problems on Decomposable Graphs

Daniel Lokshtanov

Universitetet i Bergen, Institutt for Informatikk,
Postboks 7803, 5020 Bergen, Norway
daniello@ii.uib.no

Abstract. The treewidth and cliquewidth of a graph are central notions in graph theory and graph algorithms. Many NP-hard problems become tractable when the treewidth or cliquewidth of the input graph is bounded by a constant. In this talk I will briefly survey the known algorithmic results for graphs of bounded treewidth and cliquewidth, and give an overview of a line of work that explores the limits of tractability of problems on graphs of bounded treewidth or cliquewidth. Specifically, we will consider the following questions:

- Which problems are solvable in polynomial time on graphs of bounded treewidth, but require that the degree of the polynomial grows with the treewidth?
- Which problems are harder on bounded cliquewidth graphs than on bounded treewidth graphs?
- Can the known algorithms for problems on graphs of bounded treewidth and cliquewidth be improved?

P. Hliněný and A. Kučera (Eds.): MFCS 2010, LNCS 6281, p. 37, 2010.

Do We Really Understand the Crossing Numbers?

Bojan Mohar*,**

Department of Mathematics
Simon Fraser University
Burnaby, B.C. V5A 1S6
mohar@sfu.ca

Abstract. The crossing number of a graph is the minimum number of crossings that occur in a drawing of the graph in the plane. This notion is natural and easy to understand, yet we do not know much about it apart from some basic properties. History, successes and pitfalls, some recent developments, and future directions will be presented.

Introduction

The crossing number $\text{CR}(G)$ of a graph G is the minimum number of crossings that occur in a drawing of the graph in the plane. This notion is natural and easy to understand. At least it seems so, when one encounters it for the first time. However, despite many attempts, even some of the most basic questions remain unanswered. What is the crossing number of the complete graph K_n? What about K_{13}? Are there any non-trivial graph families whose crossing number is easy to determine?

The study of crossing numbers brought some beautiful results. The Crossing Lemma turned out to be one of the fundamental mathematical tools with diverse applications in discrete geometry, graph theory, number theory and elsewhere. Yet, we still do not fully understand what the crossing number is. There are different ways how to define it, and there is very little understanding how all these definitions differ from each other, if at all.

Ups and downs in the theory of crossing number minimization include its practical applications. F. T. Leighton developed basic theory in relation to applications in VLSI design [1,11,12]. When making layout for large electrical networks, it is desirable to eliminate crossings of wires, and when we design a printed circuit, its obvious model is a planar drawing. Today it is clear that crossing number minimization is hard even for very simple graphs. One positive result may be FPT solution using theory of graph minors (Grohe [7]; Kawarabayashi and Reed [8]). However, this is still miles away from being usable in practical computations. It is

* Supported in part by the ARRS, Research Program P1-0297, by an NSERC Discovery Grant, and by the Canada Research Chair Program.

** On leave from IMFM & FMF, Department of Mathematics, University of Ljubljana, 1000 Ljubljana, Slovenia.

P. Hliněný and A. Kučera (Eds.): MFCS 2010, LNCS 6281, pp. 38–41, 2010.

only very recently that we got deeper understanding about graphs whose crossing number is one or two.

There is vast literature about this subject. We refer to [13,18,19] and to [20] for more details about diverse applications of this important notion. Some specific directions presented in the talk are outlined in the sequel.

Crossing Number Definitions

A *drawing* of a graph G is a representation of G in the Euclidean plane $\mathbb{R}^2$ (or in some other surface) where vertices are represented as distinct points and edges by simple polygonal arcs joining points that correspond to their endvertices. It is required that the interior of every arc representing an edge contains no points representing the vertices of G. If interiors of two arcs intersect, we speak about a *crossing* of the drawing.

The *crossing number* $\mathrm{CR}(G)$ of a graph G is the minimum number of crossings taken over all drawings of G. When each edge e of G has a weight $w_e \in \mathbb{N}$, the weighted crossing number $\mathrm{WCR}(\mathcal{D})$ of a drawing $\mathcal{D}$ is the sum $\sum w_e \cdot w_f$ taken over all crossings in $\mathcal{D}$ (where e, f are the edges involved in the crossing). The *weighted crossing number* $\mathrm{WCR}(G)$ of G is the minimum $\mathrm{WCR}(\mathcal{D})$ taken over all drawings G. Of course, if all edge-weights are equal to 1, then $\mathrm{WCR}(G) = \mathrm{CR}(G)$.

One can also consider the *pair crossing number* $\mathrm{PAIR\text{-}CR}(G)$, where we only consider the number of pairs of edges that cross each other (and count possible multiple crossings only once), the *odd crossing number* $\mathrm{ODD\text{-}CR}(G)$, where we only count those pairs of edges that cross an odd number of times, or the *independent odd crossing number* $\mathrm{IODD\text{-}CR}(G)$, where we only count the pairs of nonadjacent edges that cross an odd number of times. If we insist that the edges in drawings are represented by straight line segments, we obtain the notion of the *rectilinear crossing number* $\mathrm{LIN\text{-}CR}(G)$. It is obvious that the following inequalities hold:

$$\mathrm{IODD\text{-}CR}(G) \le \mathrm{ODD\text{-}CR}(G) \le \mathrm{PAIR\text{-}CR}(G) \le \mathrm{CR}(G) \le \mathrm{LIN\text{-}CR}(G).$$

It is known that the odd crossing number can be strictly smaller than the usual crossing number [16] and that the latter can be strictly smaller than the rectilinear crossing number [2].

Structural graph theory based on the Robertson and Seymour theory of graph minors gives powerful results in relation to topological realizations of graphs. However, it does not work well with crossing numbers. To overcome this deficiency, Bokal et al. [3] introduced a related notion of the *minor crossing number*, $\mathrm{MCR}(G)$, which is defined as the minimum of $\mathrm{CR}(H)$ taken over all graphs H that contain G as a minor. The study of the minor crossing numbers may be closely related to structure of some surface embeddings.

Near-planar Graphs

A nonplanar graph G is *near-planar* if it contains an edge e such that $G - e$ is planar. Such an edge e is called a *planarizing edge*. It is easy to see that near-planar graphs can have arbitrarily large crossing number. However, it seems that computing the crossing number of near-planar graphs should be much easier than in unrestricted cases. A less known, but particularly interesting result was obtained by Riskin [17], who proved that the crossing number of a 3-connected cubic near-planar graph G can be computed easily as the length of a shortest path in the geometric dual graph of the planar subgraph $G - x - y$, where $xy \in E(G)$ is the edge whose removal yields a planar graph. Riskin asked if a similar correspondence holds in more general situations, but this was disproved by Mohar [14] (see also [9]). Another relevant paper about crossing numbers of near-planar graphs was published by Hliněný and Salazar [10].

Several generalizations of Riskin's result are indeed possible. Cabello and Mohar [4,14] provided efficiently computable upper and lower bounds on the crossing number of near-planar graphs in a form of min-max relations. These relations can be extended to the non-3-connected case and even to the case of weighted edges.

On the other hand, Cabello and Mohar [5] recently proved that computing the crossing number of near-planar graphs is NP-hard. This discovery is a surprise and brings more questions than answers.

The Crossing Function

Despite the fact that our understanding of crossing numbers is limited, we suggest the study of the *crossing number function*. This function is universal for the crossing number since it contains the complete information about crossing number of *all* graphs. It has been introduced recently and can be defined as follows.

Let n be a positive integer, and let $\binom{[n]}{2}$ denote the set of all unordered pairs $\{i, j\}$, where $i, j \in [n] = \{1, \ldots, n\}$ and $i \neq j$. A *weight function* is a function $w : \binom{[n]}{2} \to \mathbb{R}$. It can be viewed as a weighting of the edges of the (labeled) complete graph K_n of order n (negative weights allowed). Let K_n^w denote that weighted complete graph, and let $\mathrm{WCR}(w) = \mathrm{WCR}(K_n^w)$. The mapping $\mathrm{WCR} : \mathbb{R}^{\binom{[n]}{2}} \to \mathbb{R}$ defined by this rule is called the *crossing function* (of order n).

Theorem 1. *The crossing function of order n is a contiuous, piecewise quadratic function with $\binom{n}{2}$ real parameters, with finitely many domains of smoothness in $\mathbb{R}^{\binom{[n]}{2}}$, and is symmetric with respect to the natural action of the symmetric group S_n on $\binom{[n]}{2}$.*

Every (labeled) graph G of order n can be viewed as a weight function w_G, whose values are 1 for those pairs that correspond to the edges of G, and 0 otherwise. Clearly, $\mathrm{WCR}(w_G) = \mathrm{CR}(G)$. This shows that the crossing function contains the complete information about crossing numbers of all graphs.

As this is a relatively new concept, not much is known about the crossing function. The author and Stephen [15] studied the behaviour of the crossing function on small random graphs. As expected, the study of the crossing function seems to be hard, but we expect that it may become a tool used for better understanding of crossing numbers over all.

References

1. Bhatt, S.N., Leighton, F.T.: A framework for solving VLSI graph layout problems. J. Comput. System Sci. 28, 300–343 (1984)
2. Bienstock, D., Dean, N.: Bounds for rectilinear crossing numbers. J. Graph Theory 17, 333–348 (1993)
3. Bokal, D., Fijavž, G., Mohar, B.: The minor crossing number. SIAM J. Discret. Math. 20(2), 344–356 (2006)
4. Cabello, C., Mohar, B.: Crossing and weighted crossing number of near-planar graphs. In: Tollis, I.G., Patrignani, M. (eds.) GD 2008. LNCS, vol. 5417, pp. 38–49. Springer, Heidelberg (2009)
5. Cabello, C., Mohar, B.: Adding one edge to planar graphs makes crossing number hard. In: SoCG 2010 (2010)
6. Garey, M.R., Johnson, D.S.: Crossing number is NP-complete. SIAM J. Alg. Discr. Meth. 4, 312–316 (1983)
7. Grohe, M.: Computing crossing numbers in quadratic time. In: Proceedings of the thirty-third annual ACM symposium on Theory of computing, Hersonissos, Greece, pp. 231–236 (July 2001), doi:10.1145/380752.380805
8. Kawarabayashi, K., Reed, B.: Computing crossing number in linear time. In: Proceedings of the thirty-ninth annual ACM symposium on Theory of computing, San Diego, California, USA, June 11-13 (2007), doi:10.1145/1250790.1250848.
9. Gutwenger, C., Mutzel, P., Weiskircher, R.: Inserting an edge into a planar graph. Algorithmica 41, 289–308 (2005)
10. Hliněný, P., Salazar, G.: On the crossing number of almost planar graphs. In: Kaufmann, M., Wagner, D. (eds.) GD 2006. LNCS, vol. 4372, pp. 162–173. Springer, Heidelberg (2007)
11. Leighton, F.T.: Complexity issues in VLSI. MIT Press, Cambridge (1983)
12. Leighton, F.T.: New lower bound techniques for VLSI. Math. Systems Theory 17, 47–70 (1984)
13. Liebers, A.: Planarizing graphs—a survey and annotated bibliography. J. Graph Algorithms Appl. 5, 74 pages (2001)
14. Mohar, B.: On the crossing number of almost planar graphs. Informatica 30, 301–303 (2006)
15. Mohar, B., Stephen, T.: The expected crossing number for random edge-weights (in preparation)
16. Pelsmajer, M.J., Schaefer, M., Štefankovič, D.: Odd Crossing Number Is Not Crossing Number. In: Healy, P., Nikolov, N.S. (eds.) GD 2005. LNCS, vol. 3843, pp. 386–396. Springer, Heidelberg (2006)
17. Riskin, A.: The crossing number of a cubic plane polyhedral map plus an edge. Studia Sci. Math. Hungar. 31, 405–413 (1996)
18. Shahrokhi, F., Sýkora, O., Székely, L.A., Vrt'o, I.: Crossing numbers: bounds and applications. In: Barany, I., Böröczky, K. (eds.) Intuitive geometry (Budapest, 1995). Bolyai Society Mathematical Studies, vol. 6, pp. 179–206. Akademia Kiado (1997)
19. Székely, L.A.: A successful concept for measuring non-planarity of graphs: The crossing number. Discrete Math. 276, 331–352 (2004)
20. Vrt'o, I.: Crossing number of graphs: A bibliography, `ftp://ftp.ifi.savba.sk/pub/imrich/crobib.pdf`

Balanced Queries: Divide and Conquer*

Dmitri Akatov[1] and Georg Gottlob[1,2]

[1] Oxford University Computing Laboratory, University of Oxford
[2] Oxford Man Institute of Quantitative Finance, University of Oxford
{dmitri.akatov,georg.gottlob}@comlab.ox.ac.uk

Abstract. We define a new hypergraph decomposition method called *Balanced Decomposition* and associate *Balanced Width* to hypergraphs and queries. We compare this new method to other well known decomposition methods, and analyze the complexity of finding balanced decompositions of bounded width and the complexity of answering queries of bounded width. To this purpose we define a new complexity class, allowing recursive divide and conquer type algorithms, as a resource-bounded class in the nondeterministic auxiliary stack automaton computation model, and show that finding decompositions of bounded balanced width is feasible in this new class, whereas answering queries of bounded balanced width is complete for it.

1 Introduction

The aim of the study of *hypergraph decompositions* is to find tractable subclasses of the *Boolean Conjunctive Query* (BCQ) evaluation problem in databases and the *Constraint Satisfaction Problem* in AI. Both these problems are equivalent and well known to be NP-complete [6,19] with the cyclicity of the hypergraphs causing the state explosion. A *hypergraph decomposition* transforms a hypergraph into an acyclic structure (a labelled tree), reducing the complexity of the associated problem. The tractability results of these problems rely on the acyclicity of the tree on the one hand and on certain properties of its labels on the other. Probably the most prominent decomposition method is the *tree decomposition* of [22], originally developed for graphs, but also applicable to hypergraphs. [10,7,16,17] provide an overview of more recent decomposition methods including *(generalized) hypertree decompositions*, spread cut decompositions and fractional hypertree decompositions. An important notion in most decompositions is their *width*, which often ensures tractability if it is independent of the hypergraph under consideration. Thus the two main complexity-theoretic problems usually considered are the following:

- **Decomposition problem**: What is the complexity of recognizing hypergraphs admitting a decomposition of fixed width?

* Work funded by EPSRC Grant EP/G055114/1 "Constraint Satisfaction for Configuration: Logical Fundamentals, Algorithms and Complexity. G. Gottlob would also like to acknowledge the Royal Society Wolfson Research Merit Award.

P. Hliněný and A. Kučera (Eds.): MFCS 2010, LNCS 6281, pp. 42–54, 2010.

- **BCQ evaluation problem**: What is the complexity of evaluating BCQs for the class of queries with a decomposition of fixed width?

A particularly nice property for queries with bounded tree and (generalized) hypertree width, thus also including acyclic queries, is that the BCQ evaluation problem is not only tractable, but also complete for the complexity class LOGCFL [12]. This complexity class lies very low within the NC-AC hierarchy (and hence within P) between NC^1 and AC^1 and hence is highly parallelizable. In [11] Gottlob et al. present a parallel algorithm for the BCQ evaluation problem[1] which is optimal under its complexity restrictions, and whose running time does not depend on the shape of the hypertree. The problem of recognizing hypergraphs of bounded hypertree width is also in LOGCFL [12],[2] however, most efficient sequential algorithms, see e.g. [15], compute the decomposition node by node in a top-down manner, following the shape of the resulting tree, which for most hypergraphs is often deep (linear in the size of the hypergraph) and narrow (branching factor of 1 for most nodes). This has negative effects on the parallelization of such hypertree computation algorithms, which is easiest when the *computation tree* is balanced, indicating a *division* of the problem into smaller independent subproblems which can be *conquered* recursively.

While looking for better, more "balanced", algorithms, we decided to analyze hypertrees which are balanced *a priori*, but are not necessarily valid (generalized) hypertree decompositions. These *balanced decompositions* constitute an entirely new hypergraph decomposition method in its own right, and hence deserve further analysis. In particular they possess beneficial properties for parallelization, they capture wider classes of hypergraphs than other known decomposition methods, and provide more insight into the structure of NP-complete problems.

In section 3 we provide the formal definition of balanced decompositions and compare them to generalized hypertree decompositions. In section 4 we characterize balanced decompositions game-theoretically by defining the *Robber and Sergeants Game* for hypergraphs. To better understand the complexity of the decomposition and the BCQ evaluation problems, we define a **new complexity class $\mathbf{DC^1}$** in section 5 by limiting resources of Nondeterministic auxiliary Stack Automata [18], and identify its lower bounds as LOGCFL and $\mathrm{GC}(\log^2 n, \mathrm{NL})$ in the Guess-and-Check model and its upper bound as $\mathrm{NTiSp}(n^{O(1)}, \log n)$, the space-bounded subclass of NP. In section 6 and section 7 we show that recognizing hypergraphs of bounded *balanced width* (BW) is feasible in DC^1, while the **BCQ evaluation problem** for the class of queries of **bounded balanced width** is **complete for $\mathbf{DC^1}$**. We conclude the paper in section 8.

Omitted proofs can be found in the full version of this paper [1].

---

[1] Actually, the algorithm was developed for acyclic BCQs, but can easily be adapted to generalized hypertree decompositions.

[2] Unfortunately recognizing hypergraphs of generalized hypertree width at least 3 is NP-complete [14].

2 Preliminaries

All sets in this paper are finite.

We assume the reader to be familiar with the standard formalizations of rooted and ordered trees. We use T to denote a tree, its node set and its "child function", we write T_p for a subtree rooted at a node p, $O(T)$ for the "root" of T, and $\sqsubseteq_T$ for the "ancestor relation", with $O(T) \sqsubseteq_T p$ for all other $p \in T$.

A *hypergraph* is a tuple $H = (V(H), E(H))$ where $V(H)$ is a set called the *vertices* of H and $E(H) \subseteq \mathcal{P}(V(H) \setminus \{\emptyset\}$ is a set called the *edges* or *hyperedges* of H. Given a hypergraph H and $R, S \subset E(H)$, we say $Q \subset R \setminus S$ is an $[S]$-component of R iff either $Q = \{e\}$ with $e \subseteq \bigcup S$, or for any two edges in Q there exists a sequence (an $[S]$*-path*) of edges in Q between them, such that every two consecutive edges share some vertex not covered by $\bigcup S$.

Given a hypergraph H, a *hypertree for* H is a triple (T, χ, λ), where T is a rooted tree, and χ and λ are labeling functions which associate to each vertex $p \in T$ two sets $\chi(p) \subseteq V(H)$ and $\lambda(p) \subseteq E(H)$.
Given $p \in T$ we define $\chi(T_p) = \bigcup\{\chi(q)|q \in T_p\}$ and $\lambda(T_p) = \bigcup\{\lambda(q)|q \in T_p\}$.
The *width* of a hypertree is $\max_{p\in T} |\lambda(p)|$.

A *hypertree decomposition* is a hypertree satisfying the following conditions:

1. For all $e \in E(H)$, there exists $p \in T$, such that $e \subseteq \chi(p)$,
2. for all $v \in V(H)$, the set $\{p \in T | v \in \chi(p)\}$ induces a connected subtree of T,
3. for each $p \in T$, $\chi(p) \subseteq \bigcup \lambda(p)$,
4. for each $p \in T$, $(\bigcup \lambda(p)) \cap \chi(T_p) \subseteq \chi(p)$.

A *generalized hypertree decomposition* is a hypertree only satisfying the first three of these conditions. The width of a (generalized) hypertree decomposition is the width of its hypertree. The *(generalized) hypertree width* (GHW / HW) of a hypergraph H is the minimal width over all its (generalized) hypertree decompositions [12].

The *monotone Robber and Marshals Game* and its equivalence with hypertree decompositions is studied in [13] and [3].

A *Database* is a relational structure over a schema (signature). A *Boolean Conjunctive Query* (BCQ) is also a relational structure (over the same schema) containing no constants. We call every tuple occurring in some relation also an *atom*, and the objects of the base set occurring in the query or an atom its *variables*. We write atoms(q) for the set of atoms of a query q and var(a) for the set of variables occurring in an object a (e.g. an atom or a query). We say that a database D *satisfies* a BCQ q ($D \models q$) iff there exists a homomorphism from q to D.[3]

The *underlying hypergraph* of a BCQ q is a the hypergraph $H(q)$, with $V(H(q))$ = $var(q)$ and $E(H(q)) = \{\text{var}(a)|a \in \text{atoms}(q)\}$. A decomposition of a BCQ is simply a decomposition of its underlying hypergraph.

[3] An alternative definition of databases and BCQs can be found e.g. in [2].

3 Balanced Decompositions

Definition 1. *Let H be a hypergraph. A* hypercut decomposition *of H is a hypertree (T, χ, λ) for H which satisfies the following conditions:*

1. *For each $e \in E(H)$ there exists $p \in T$ such that $e \in \lambda(p)$,*
2. *for each $Y \in V(H)$, the set $\{p \in T | Y \in \chi(p)\}$ contains its $\sqsubseteq_T$-meet,*
3. *for each vertex $p \in T$, $\chi(p) = \bigcup \lambda(p)$.*

A hypercut decomposition is a shallow decomposition *if additionally* $\text{depth}(T) \leq \log |E(H)|$ *holds. A hypercut decomposition is a* balanced decomposition *if additionally $|\lambda(T_q)| \leq |\lambda(T_p)|/2$ holds for all $p \in T, q \in T(p)$. The* width *of a shallow decomposition or a balanced decomposition is the width of its hypertree. The* shallow width, *resp.* balanced width, *of H is the minimum width over all its shallow, resp. balanced, decompositions. We write* $\text{SW}(H)$ *for the shallow width and* $\text{BW}(H)$ *for the balanced width of H.*

Notice that a hypercut decomposition is uniquely defined by T and λ alone, while the χ-labels are only required for the second condition. We do not define a hypercut width, since then every hypergraph would trivially have width one.

Example 1. The balanced and shallow widths of the following hypergraph are two and we also show a balanced decomposition, which is also shallow (in the decomposition we only present the λ-labels):

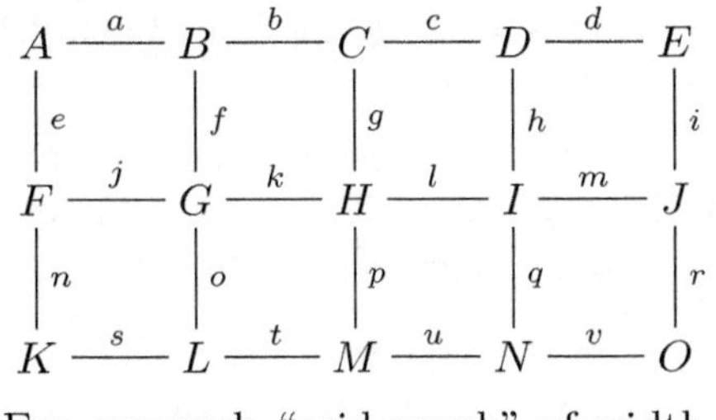

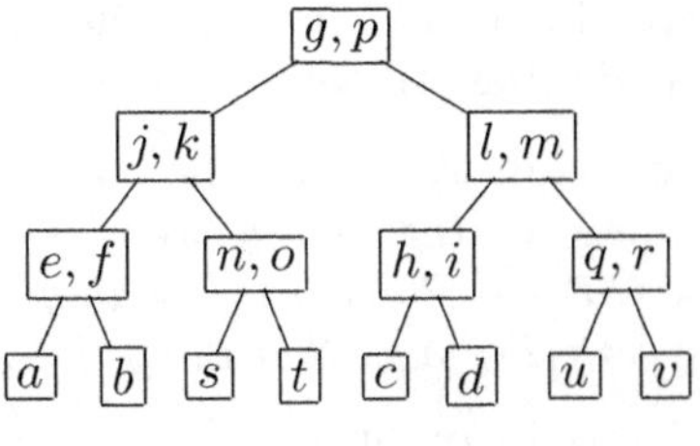

For any such "grid graph" of width k and length m, it is easy to construct a balanced decomposition of width k. The generalized hypertree width of such hypergraphs, however, is always $k+1$, and any decomposition tree will have depth at least m (linear in the size of the hypergraph), and will generally contain a long "chain" with no branching.

Proposition 1. *Let H be a hypergraph. The following holds:*

$$SW(H) \leq BW(H) \leq GHW(H) \leq SW(H) \log |E(H)| .$$

A shallow tree will have "good" branching at some point, however this branching does not have to occur at every internal node, as in a (perfectly) balanced tree. Hence also the distinction between balanced and shallow decompositions, which also capture slightly different classes of hypergraphs. For instance, every graph consisting of a single cycle has shallow width one, whereas its balanced width is always two. However, for all complexity-theoretic purposes, the distinction between balanced and shallow decompositions is not relevant, since our results apply to both of them.

4 Robber and Sergeants

As with many other decomposition methods for hypergraphs it helps to visualize a decomposition in terms of a two player game between a Robber and some Law Enforcement Entity, [24,13,16]. We shall define the *Robber & k Sergeants game* on a hypergraph H (R&S$^k(H)$). It resembles the Robber and Marshals game [13], but has two important differences: The robber is positioned on edges rather than vertices (an escape space hence becomes a set of edges rather than a set of vertices). Also, the sergeants only have to cover any edge once and it remains covered for the rest of the game (the robber can never go to that edge again). Hence the game is by definition monotone (the escape space can never increase).

Let H be a hypergraph, let k be a positive integer, and let $A \subseteq E(H)$ such that A is connected, be the *initial escape space*. The Robber and k Sergeants Game from A (R&S$^k(A)$) is played by two players - $\mathcal{R}$ (the robber) and $\mathcal{S}$ (the sergeants). Player $\mathcal{S}$ announces moves by choosing a set S of up to k edges of A. If S covers the whole of A, player $\mathcal{S}$ wins. Otherwise, player $\mathcal{R}$ chooses an $[S]$-connected component of A, say B. They then proceed to play the game R&S$^k(B)$. If the game is shallow, then player $\mathcal{R}$ wins R&S$^k(A)$ if he can sustain play for more than $\log|A|$ moves. If the game is balanced, then player $\mathcal{R}$ wins if from any escape space A and a sergeants' move S he can select an $[S]$-component B of A such that $|B| > |A|/2$. The game R&S$^k(H)$ is the game R&S$^k(E(H))$ (on the full hypergraph). Player $\mathcal{S}$ has a winning strategy, if for any possible move of player $\mathcal{R}$, he can still win the game. This leads to the formal definition of a winning strategy:

Definition 2. *Let k be a positive integer, let H be a connected hypergraph. A* winning strategy *for R&S$^k(H)$ is a tuple (T, ρ, λ), where T is a rooted tree and $\rho, \lambda : T \to \mathcal{P}(A)$ are labelling functions (escape space and sergeants' moves, respectively) such that the following conditions hold:*

1. Initial Condition: $\rho(O(T)) = E(H)$.
2. Boundedness: *For all* $t \in T$, $1 \leq |\lambda(t)| \leq k$.
3. Completeness: *For all* $s \in T$, $\rho(s) = \mu(s) \cup \bigcup_{t \in T(s)} \rho(t)$.
4. Separation: *For all* $s \in T, t \neq u \in T(s)$, $\rho(t) \cap \rho(u) = \emptyset$.
5. Connectedness: *For all* $s \in T, t \in T(s), e \in \rho(s), f \in \rho(t)$, *$e$ is $[\lambda(s)]$-connected to f in $\rho(s)$ iff $e \in \rho(t)$.*

A winning strategy in the shallow R&S^k game additionally satisfies $\text{depth}(T) \leq \log|E(H)|$. *A winning strategy in the balanced R&S^k game additionally satisfies $|\rho(t)| \leq |\rho(s)|/2$, for all $s \in T, t \in T(s)$.*

The separation and connectedness conditions say that escape space labels of the children of a node s are distinct $[\lambda(s)]$-components of $\rho(s)$. The completeness condition says that we include all such components, and also that for each node s we have $\lambda(s) \subseteq \rho(s)$.

Lemma 1. *Let H be a hypergraph. There exists a k-width shallow/balanced decomposition of H iff there exists a winning strategy in the shallow/balanced R&S^k game on H.*

5 The DC Hierarchy

An *auxiliary pushdown automaton* (AuxPDA) is a generalization of both the Turing Machine (TM) and the Pusdown Automaton (PDA) — it possesses both a tape and a pushdown. Adding a pushdown makes these machines more powerful than Turing Machines, since they admit *recursive algorithms*, which push "temporary variables" before a recursive call and pop them after the call returns. For instance, a nondeterministic TM using simultaneously logarithmic space and polynomial time (in $\mathrm{NTiSp}(n^{O(1)}, \log n)$, see e.g. [20]) can solve problems precisely in NL. A nondeterministic AuxPDA (NauxPDA) with the same time and space bound on the worktape can precisely solve problems in LOGCFL, which contains the latter complexity class. We do not usually limit the space on the pushdown (the *maximal pushdown height*), however, Ruzzo showed that problems in LOGCFL only require $O(\log^2 n)$ space on the pushdown. We write $\mathrm{NTiSpPh}(T(n), S(n), H(n))$ for the class of problems solvable by a NauxPDA which is simultaneously bounded by time $O(T(n))$, worktape space $O(S(n))$ and maximal pushdown height $O(H(n))$.

A *stack* acts like a pushdown for writing (pushing and popping), but like a tape for reading (any cell can be read). Thus, *Stack Automata* (SA), introduced by Ginsburg et al. [8], are more powerful than PDAs. Analogously to extending TMs with a pushdown, Ibarra proposed to do the same with SAs [18], yielding the model of the *auxiliary stack automaton* (AuxSA). AuxSAs allow recursive algorithms, just like AuxPDAs, but these algorithms additionally have access to all previously computed temporary variables (the accumulated temporary variables) and are thus more powerful. We write $\mathrm{NTiSpPh}(T(n), S(n), H(n))$ for the class of problems solvable by a nondeterministic SA (NauxSA) which is simultaneously bounded by time $O(T(n))$, worktape space $O(S(n))$ and maximal stack height $O(H(n))$.

Since a pushdown can be simulated by a stack, and a stack can in turn be simulated by a worktape we have the following relationship of complexity classes:

$$\begin{aligned}&\mathrm{NTiSpPh}(T(n), S(n), H(n))\\ \subseteq&\mathrm{NTiSpSh}(T(n), S(n), H(n))\\ \subseteq&\mathrm{NTiSp}(T(n), \max(S(n), H(n)))\,.\end{aligned}$$

Definition 3. *For integers $k \geq 0$, let $\mathrm{DC}^k = \mathrm{NTiSpSh}(n^{O(1)}, \log n, \log^{k+1} n)$.*

This new hierarchy of complexity classes lies between $\mathrm{NL}=\mathrm{DC}^0$ and NP. The name suggests on the one hand the way an algorithm in such a particular class might work (Divide and Conquer through recursion), and on the other hand the maximal depth of recursive calls ($O(\log^k n)$ for DC^k, each time storing $O(\log n)$ cells on the stack). A single function call then is an NL algorithm which additionally can access previously computed temporary variables. In this paper we will only consider the class DC^1, allowing a logarithmic number of recursive calls. In particular, $\log n$ recursive calls exactly allow at each call to divide an input of length n into at least two parts each at most half as big until the parts have constant size. We have $\mathrm{LOGCFL} \subseteq \mathrm{DC}^1 \subseteq \mathrm{NTiSp}(n^{O(1)}, \log^2 n)$.

We can use SAs to simulate the Guess-and-Check model of [5], by nondeterministically guessing an "advice string" and placing it on the stack. In particular, we have $\mathrm{GC}(\log^{k+1} n, \mathrm{NL}) \subseteq \mathrm{DC}^k$, where the former class is the class of languages for which an advice string of length $O(\log^{k+1} n)$ can be guessed such that the original input plus the advice string can be decided in NL. Note that $\mathrm{GC}(\log^2 n, \mathrm{NL})$ is **not known or believed to be contained in** P, which strongly indicates that neither is DC^1.

We get the following inclusion diagram of complexity classes:

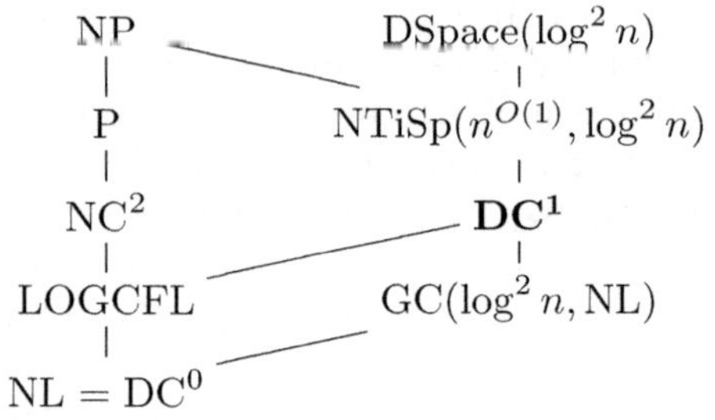

6 Membership in DC^1

Winning strategies in the R&S^{k}game give us an easy way to find balanced decompositions. Consider the algorithm k-robber-sergeants:

Algorithm 1. k-robber-sergeants

```
input Hypergraph H
fixed parameter k: Integer
check-win(firstEdge(H), |V(H)|)
accept

procedure check-win(Edge r, Integer size)
   guess Edge[1...k] sergs
   for each Edge f ∈ E(H) do
      if connected(r,f) then
         Integer n := count-connected-edges(f, sergs)
         if n > size/2 then reject
         else if n > 0 then check-win(f, sergs)
```

Here the function *count-connected-edges(e, sergs)* counts the number of edges $[S]$-connected to e, where S is the set of all sergeant hyperedges already on the stack plus *sergs*. This can obviously be done in NL, and even in L by using the undirected st-connectivity algorithm of [21].

Theorem 1. *Let k be a positive integer, and H a hypergraph. The algorithm k-robber-sergeants accepts iff a balanced decomposition of H of width at most k exists. Moreover this algorithm is in* DC^1.

k-robber-sergeants repeats a lot of work, however, this affects neither its correctness nor its complexity bounds. A deterministic algorithm would of course trade space for time, avoiding any redundancy. It is an easy exercise to adapt k-robber-sergeants to check for shallow decompositions instead.

As for the bounded BW BCQ evaluation problem, consider the algorithm k-hd-bcq:

Algorithm 2. k-hd-bcq

```
fixed parameter k: Integer
input Database d
input Query q with a k-width hypercut decomposition
satisfiable(root(q))
accept

procedure satisfiable(Node u)
  for each Atom a ∈ λ(u) do                    // at most k repetitions
    guess Tuple t ∈ table(d, name(a))
    for each (Atom, Tuple) (b, s) on stack do
      if not compatible((a, t), (b, s)) then reject
    push (a, t)
  for each Node v ∈ children(u) do satisfiable(v)
  pop all (a, t) pairs which were pushed during current function call
```

We assume that the tree of the hypercut decomposition is an ordered tree, and that we can access its root using the function *root*, and that given a node u, we can iterate through *children(u)* using a single pointer. Given an atom a the function *name(a)* returns the "schema" s of that atom, and we can use *table(d, s)* to access the appropriate table in d. When we call satisfiable recursively, we assume that the current temporary variables (in this case only u and v) are placed on the stack, and popped after the recursive call returns. We push and pop the atom a and tuple t explicitly at every iteration within one call, because we need them on the stack before satisfiable is called recursively. The function *compatible*$((a,t),(b,u))$ checks whether tuples t and u are compatible under the schemas of a and b, respectively, i.e. whether all shared variables of a and b have the same values in t and u.

Theorem 2. *Fix a positive integer k. Given a database D, a boolean conjunctive query Q with associated hypergraph H and a hypercut decomposition (T,χ,λ) of H of width at most k, the algorithm k-hd-bcq accepts iff $D \models Q$. Moreover, the algorithm operates in* $\mathrm{NTiSpSh}(n^{O(1)}, \log n, d\log n)$, *where n is the size of the whole input and $d = \mathrm{depth}(T)$.*

Corollary 1. *For queries of fixed (bounded) shallow or balanced width the BCQ evaluation problem is in* DC^1.

7 DC¹-Completeness

To show hardness for DC^1 of the BCQ evaluation problem for queries of fixed balanced width we will need some preliminary results about NauxSAs. For any AuxSA, AuxPDA, PDA or SA we define the *push-pop tree* to be an ordered tree representing the sequence of the non-read stack operations of the machine. The root of the tree represents an empty pushdown/stack, any one node represents some distinct state of the pushdown/stack, and a new child is added to that node whenever the machine pushes while the pushdown/stack is in that state. For example, if the sequence is Push, Pop, Push, Push, Pop, Push, Pop, Push, Pop, Pop, then the push-pop tree looks like this:

```
      *
     / \
    *   *
       /|\
      * * *
```

Definition 4. *We call a NauxSA M* regular *if it has the following properties:*

1. *The push-pop tree of M is a full binary tree.*
2. *After every push the entire contents of the stack is read.*
3. *Whenever M doesn't push, it acts deterministically (the only nondeterministic steps are the pushes).*

Lemma 2. *Let M be a NauxSA with stack size $O(\log^2 n)$, tape size $O(\log n)$ running in polynomial time and deciding the language L. Then there exists a regular NauxSA M' with same stack and tape sizes running in polynomial time deciding L. Moreover, there exists a* log-space *reduction from M to M'.*

Theorem 3. *Let M be a regular NauxSA running in polynomial time, logarithmic space and with a push-pop tree of height $k = O(\log n)$ and x a string. Then there exists a database B and a Boolean Conjunctive Query Q with a Balanced Decomposition of width* 8 *such that $B \models Q$ iff M accepts x. Moreover there exists a* log-*space reduction from (M, x) to (B, Q).*

Proof. The database B has the tables $I(C)$, $A(C)$, $D(C_1, C_2)$ and for each i, $1 \leq i \leq k$, the tables $U_i(C_1, S, C_2)$, $R_i(C_1, S, C_2)$ and $O_i(C_1, C_2)$.[4] The table I contains the initial configuration of M as the only tuple. The table A contains all accepting configurations of M. A tuple (c, d) is in D iff M starting in configuration c (deterministically) reaches configuration d without performing any stack operations, and the next operation of M would be a stack operation. A tuple (c, d) is in O_i iff M starting in configuration c would next perform a pop and end up in configuration d. A tuple (c, s, d) is in U_i iff M starting in configuration c would next push s into the ith cell and end up in configuration d. Similarly a

[4] For a schema, $T(A, B, C)$ here indicates that we have a table called T which has three *attributes* (with "types" A, B and C), such that every tuple in that table has exactly three elements.

tuple (c, s, d) is in R_i iff M starting in configuration c would next read s from the ith cell and end up in configuration d. All these tables can be computed in log-space.

Now we build the query Q which corresponds to the run of M, contracting multiple consecutive deterministic steps into one atom: Let T be a full binary tree of depth k. For a node N let $p^j(N)$, $l(N)$ and $r(N)$ denote the j-th ancestor, left child and right child of N, respectively. For each $N \in T \setminus O(T)$, define the query Q_N^R in the following way: if $\mathrm{depth}_T(N) = 1$, then $Q_N^R = R_1(X_N^2, S_N, X_N^3)$, otherwise

$$\begin{aligned} Q_N^R =& R_{d(N)}(X_N^2, S_N, Y_N^{d(N)}) \wedge D(Y_N^{d(N)}, Z_N^{d(N)}) \\ &\wedge R_{d(N)-1}(Z_N^{d(N)}, S_{p(N)}, Y_N^{d(N)-1}) \wedge D(Y_N^{d(N)-1}, Z_N^{d(N)-1}) \\ &\wedge R_{d(N)-2}(Z_N^{d(N)-1}, S_{p^2(N)}, Y_N^{d(N)-2}) \\ &\dots \\ &\wedge D(Y_N^2, Z_N^2) \wedge R_1(Z_N^2, S_{p^{d(N)-1}(N)}, X_N^3)\,. \end{aligned}$$

Q_N^R now encodes the actions of the machine which read and process the contents of the stack, after it reached the node N in the push-pop tree. The variables S_N represent the strings already pushed to the stack, and the variables Y_N^i represent the intermediate configurations between successive reads. Note how this query will be "attached" into the "simulation" of the machine through its first and last variables (corresponding to the starting and finishing configurations of the "stack processing").

For each $N \in T$, define a query Q_N in the following way:
If N is the root, then

$$\begin{aligned} Q_N =& D(X_N^1, X_N^2) \wedge U_1(X_N^2, S_{l(N)}, X_{l(N)}^1) \wedge D(X_{l(N)}^7, X_N^3) \wedge \\ & U_1(X_N^3, S_{r(N)}, X_{r(N)}^1) \wedge D(X_{r(N)}^7, X_N^4)\,, \end{aligned}$$

if N is a leaf, then

$$Q_N = D(X_N^1, X_N^2) \wedge Q_N^R \wedge D(X_N^3, X_N^4) \wedge O_{d(N)}(X_N^4, X_N^7)\,,$$

otherwise

$$\begin{aligned} Q_N =& D(X_N^1, X_N^2) \wedge Q_N^R \wedge \\ & D(X_N^3, X_N^4) \wedge U_{d(N)+1}(X_N^4, S_{l(N)}, X_{l(N)}^1) \wedge \\ & D(X_{l(N)}^7, X_N^5) \wedge U_{d(N)+1}(X_N^5, S_{r(N)}, X_{r(N)}^1) \wedge \\ & D(X_{r(N)}^7, X_N^6) \wedge O_{d(N)}(X_N^6, X_N^7)\,. \end{aligned}$$

Q_N now encodes all the actions of the machine after it reached the node N in the push-pop tree. The variables X_N^i represent the configurations of M while it has S_N pushed in the $d(N)$th stack cell. First it reads and processes the stack (unless N is the root and the stack is empty), then it pushes to the stack and "passes control" to the first child, then it pushes to the stack again and passes control to the second child (unless N is a leaf and it does not have children), and finally pops the stack and passes control to its parent (unless N is the root). This passing of control is achieved through sharing of variables, which correspond to the according configurations of the machine at any such point in the computation.

Between all steps accessing the stack we also introduce a deterministic step. In case the machine does not need any deterministic steps, this can be encoded in the table D by having a tuple with two equal elements.

Finally define

$$Q = I(X^1_{O(T)}) \wedge (\bigwedge_{N \in T} Q_N) \wedge A(X^4_{O(T)}) .$$

Here we glue all partial queries together, and we also require the first configuration to be the initial configuration of M, and the last configuration to be an accepting configuration. A valid instantiation of the variables in Q corresponds to a successful run of M. Hence $B \models Q$ iff M accepts x.

The hypergraph corresponding to Q has a hyperedge for every atom in the query, since they are all different. In particular, every subquery Q^R_N will correspond to $2\text{depth}_T(N) - 1$ hyperedges. Additionally we have 5 hyperedges for the root, 3 hyperedges for each leaf, 7 hyperedges for every internal node, and 2 more hyperedges for the overall query. Altogether we get $4(k+1)2^k - 1$ hyperedges. We can build a balanced decomposition in the following way: For every Q^R_N it is easy to build an incomplete binary tree such that each node is labelled with exactly one atom (D or R) and that the tree is a balanced decomposition of Q^R_N of width 1, since Q^R_N is acyclic (ignoring the variables S_N which will be present in the final tree already). Call each of these trees R_N. Now let T' be a tree which is like T, and label every $N \in T$ with Q_N without Q^R_N. Now "merge" each R_N with the corresponding node N of T' by adding the label of the root of R_N to the label of N and attaching the rest of the subtree to N. Also, add two more nodes as children of the root, one labelled with the I-atom and the other with the A-atom, to produce the final (T', λ). There will be at most 8 atoms in every label.

It is obvious that (T', λ) is balanced, since every node has two or four children and is built absolutely symmetrically. □

Corollary 2. *The BCQ evaluation problem for queries of bounded balanced width is complete for* DC^1.

8 Determinization, Parallelization and Future Work

A standard technique to make a nondeterministic algorithm deterministic is a brute-force search of the computation tree, trying out all nondeterministic choices. In many cases this increases the time requirement, however not always the space requirement. For the class DC^1 this is also the case, in particular since it is a subclass of DSpace($\log^2 n$). The deterministic algorithm works its way along the push-pop tree, backtracking its steps whenever some "nondeterministic" choice leads to rejection. Notice, that once a node in the push-pop tree has several descendants, we can split the work to different processors, since the results for the subtasks do not depend on each other. The only downside is the amount of work that needs to be done at every such node, since there are $O(n^{O(1)\log n})$ possibilities for the contents of the stack (at the deepest level). Hence the algorithm, even with parallelization, remains superpolynomial. It is however still *quasipolynomial.*

Future Work includes a better analysis of relations between resource-bounded (N)AuxSAs and other models of computation, in order to relate the complexity classes in the DC hierarchy to other known complexity classes, in particular those presented in [23], [9] and [4]. Another direction of work is to establish whether the problem of recognizing hypergraphs of bounded BW is complete for DC^1 or whether it belongs to a lower complexity class. Finally, an important aspect of future work is the implementation and testing of the parallelization methods described above in practice.

References

1. `http://benner.dbai.tuwien.ac.at/staff/gottlob/DAGG-MFCS10.pdf`
2. Abiteboul, S., Hull, R., Vianu, V.: Foundations of Databases. Addison Wesley, Reading (November 1994)
3. Adler, I.: Marshals, monotone marshals, and hypertree-width. Journal of Graph Theory 47(4), 275–296 (2004)
4. Beigel, R., Fu, B.: Molecular computing, bounded nondeterminism, and efficient recursion. In: Proceedings of the 24th International Colloquium on Automata, Languages, and Programming, vol. 25, pp. 816–826 (1998)
5. Cai, L., Chen, J.: On the amount of nondeterminism and the power of verifying. SIAM Journal on Computing 26, 311–320 (1997)
6. Chandra, A.K., Merlin, P.M.: Optimal implementation of conjunctive queries in relational data bases. In: STOC 1977: Proceedings of the ninth annual ACM symposium on Theory of computing, pp. 77–90. ACM, New York (1977)
7. Cohen, D., Jeavons, P., Gyssens, M.: A unified theory of structural tractability for constraint satisfaction and spread cut decomposition. In: IJCAI 2005: Proceedings of the 19th international joint conference on Artificial intelligence, pp. 72–77. Morgan Kaufmann Publishers Inc., San Francisco (2005)
8. Ginsburg, S., Greibach, S.A., Harrison, M.A.: Stack automata and compiling. J. ACM 14(1), 172–201 (1967)
9. Goldsmith, J., Levy, M.A., Mundhenk, M.: Limited nondeterminism (1996)
10. Gottlob, G., Leone, N., Scarcello, F.: A comparison of structural csp decomposition methods. Artificial Intelligence 124(2), 243–282 (2000)
11. Gottlob, G., Leone, N., Scarcello, F.: The complexity of acyclic conjunctive queries. J. ACM 48(3), 431–498 (2001)
12. Gottlob, G., Leone, N., Scarcello, F.: Hypertree decompositions and tractable queries. Journal of Computer and System Sciences 64(3), 579–627 (2002)
13. Gottlob, G., Leone, N., Scarcello, F.: Robbers, marshals, and guards: Game theoretic and logical characterizations of hypertree width. J. Comput. Syst. Sci. 66(4), 775–808 (2003)

14. Gottlob, G., Miklos, Z., Schwentick, T.: Generalized hypertree decompositions: Np-hardness and tractable variants. In: PODS 2007: Proceedings of the twenty-sixth ACM SIGMOD-SIGACT-SIGART symposium on Principles of database systems, pp. 13–22. ACM Press, New York (2007)
15. Gottlob, G., Samer, M.: A backtracking-based algorithm for hypertree decomposition. J. Exp. Algorithmics 13 (2009)
16. Grohe, M., Marx, D.: Constraint solving via fractional edge covers. In: SODA 2006: Proceedings of the seventeenth annual ACM-SIAM symposium on Discrete algorithm, pp. 289–298. ACM, New York (2006)
17. Hlineny, P., Oum, S.-i., Seese, D., Gottlob, G.: Width parameters beyond tree-width and their applications. The Computer Journal 51(3), 326–362 (2007)
18. Ibarra, O.H.: Characterizations of some tape and time complexity classes of turing machines in terms of multihead and auxiliary stack automata. Journal of Computer and System Sciences 5(2), 88–117 (1971)
19. Mackworth, A.: Consistency in networks of relations. Artificial Intelligence 8(1), 99–118 (1977)
20. Monien, B., Sudborough, I.H.: Bandwidth constrained np-complete problems. Theoretical Computer Science 41, 141–167 (1985)
21. Reingold, O.: Undirected st-connectivity in log-space. In: STOC 2005: Proceedings of the thirty-seventh annual ACM symposium on Theory of computing, pp. 376–385. ACM, New York (2005)
22. Robertson, N., Seymour, P.: Graph minors. ii. algorithmic aspects of tree-width. Journal of Algorithms 7(3), 309–322 (1986)
23. Ruzzo, W.L.: Tree-size bounded alternation. Journal of Computer and System Sciences 21(2), 218–235 (1980)
24. Seymour, P., Thomas, R.: Graph searching and a min-max theorem for tree-width. Journal of Combinatorial Theory, Series B 58(1), 22–33 (1993)

Slowly Synchronizing Automata and Digraphs*

Dmitry Ananichev, Vladimir Gusev, and Mikhail Volkov

Department of Mathematics and Mechanics,
Ural State University, 620083 Ekaterinburg, Russia
Dmitry.Ananichev@usu.ru, vl.gusev@gmail.com, Mikhail.Volkov@usu.ru

Abstract. We present several infinite series of synchronizing automata for which the minimum length of reset words is close to the square of the number of states. These automata are closely related to primitive digraphs with large exponent.

1 Background and Overview

A *complete deterministic finite automaton* (DFA) is a triple $\mathscr{A} = \langle Q, \Sigma, \delta \rangle$, where Q and Σ are finite sets called the *state set* and the *input alphabet* respectively, and $\delta : Q \times \Sigma \to Q$ is a totally defined function called the *transition function*. Let Σ^* stand for the collection of all finite words over the alphabet Σ, including the empty word. The function δ extends to a function $Q \times \Sigma^* \to Q$ (still denoted by δ) in the following natural way: for every $q \in Q$ and $w \in \Sigma^*$, we set $\delta(q, w) = q$ if w is empty and $\delta(q, w) = \delta(\delta(q, v), a)$ if $w = va$ for some $v \in \Sigma^*$ and $a \in \Sigma$. Thus, via δ, every word $w \in \Sigma^*$ acts on the set Q.

A DFA $\mathscr{A} = \langle Q, \Sigma, \delta \rangle$ is called *synchronizing* if the action of some word $w \in \Sigma^*$ resets $\mathscr{A}$, that is, leaves the automaton in one particular state no matter at which state in Q it is applied: $\delta(q, w) = \delta(q', w)$ for all $q, q' \in Q$. Any such word w is said to be a *reset word* for the DFA. The minimum length of reset words for $\mathscr{A}$ is called the *reset length* of $\mathscr{A}$.

Synchronizing automata serve as transparent and natural models of error-resistant systems in many applications (coding theory, robotics, testing of reactive systems) and also reveal interesting connections with symbolic dynamics and other parts of mathematics. For a brief introduction to the theory of synchronizing automata we refer the reader to the recent surveys [15,22]. Here we focus on the so-called Černý conjecture that constitutes a major open problem in this area.

In 1964 Černý [5] constructed for each $n > 1$ a synchronizing automaton $\mathscr{C}_n$ with n states whose reset length is $(n-1)^2$. Soon after that he conjectured that these automata represent the worst possible case, that is, every synchronizing automaton with n states can be reset by a word of length $(n-1)^2$. This simply looking conjecture resists researchers' efforts for more than 40 years. Even though the conjecture has been confirmed for various restricted classes of synchronizing

* Supported by the Russian Foundation for Basic Research, grants 09-01-12142 and 10-01-00524, and by the Federal Education Agency of Russia, grant 2.1.1/3537.

P. Hliněný and A. Kučera (Eds.): MFCS 2010, LNCS 6281, pp. 55–65, 2010.

automata (cf., e.g., [9,6,11,19,20,2,23]), no upper bound of magnitude $O(n^2)$ for the reset length of n-state synchronizing automata is known in general. The best upper bound achieved so far is $\frac{n^3-n}{6}$, see [13].

One of the difficulties that one encounters when approaching the Černý conjecture is that there are only very few *extreme* automata, that is, n-state synchronizing automata with reset length $(n-1)^2$. In fact, the Černý series $\mathscr{C}_n$ is the only known infinite series of extreme automata. Besides that, only a few isolated examples of such automata have been found, see [22] for a complete list. Moreover, even *slowly* synchronizing automata, that is, automata with reset length close to the Černý bound are very rare. This empirical observation is supported also by probabilistic arguments. For instance, the probability that a composition of $2n$ random self-maps of a set of size n is a constant map tends to 1 as n goes to infinity [10]. In terms of automata, this result means that the reset length of a random automaton with n states and at least $2n$ input letters does not exceed $2n$. For further results of the same flavor see [16]. Thus, there is no hope to find new examples of slowly synchronizing automata by a lucky chance or via a random sampling experiment.

We therefore have designed and performed a set of exhaustive search experiments. Our experiments are briefly described in Section 5 while the main body of the paper is devoted to a theoretical analysis of their outcome. We concentrate on two principal issues. In Section 3 we discuss a similarity between the distribution of reset lengths of synchronizing automata and the distribution of exponents of primitive digraphs. Section 4 presents a few series of slowly synchronizing automata. Most of these series have been expanded from new examples discovered in the course of our experiments. In our opinion, the proof technique is also of interest; in fact, we provide a transparent and uniform approach to all sufficiently large slowly synchronizing automata with 2 input letters, both new and already known ones.

2 Preliminaries

We start with recalling two elementary and well-known number-theoretic results.

Lemma 1 ([14, Theorem 1.0.1]). *If $k_1, \ldots, k_m$ are positive integers whose greatest common divisor is equal to 1, then there exists an integer N such that every integer larger than N is expressible as a non-negative integer combination of $k_1, \ldots, k_m$.*

The question of how the least N with the property stated in Lemma 1 depends on the integers $k_1, \ldots, k_m$ is known as the *diophantine Frobenius problem* and in general is highly non-trivial, see [14]. There is, however, a simple special case which we will need in Section 4.

Lemma 2 ([14, Theorem 2.1.1]). *If k_1, k_2 are relatively prime positive integers, then $k_1k_2 - k_1 - k_2$ is the largest integer that is not expressible as a non-negative integer combination of k_1 and k_2.*

A *directed graph* (digraph) is a pair $D = \langle V, E\rangle$ where V is a finite set and $E \subseteq V \times V$. We refer to elements of V and E as *vertices* and *edges*. Observe that our definition allows loops but excludes multiple edges. If $v, v' \in V$ and $e = (v, v') \in E$, the edge e is said to be *outgoing* for v. We assume the reader's acquaintance with basic notions of the theory of directed graphs such as (directed) path, cycle, isomorphism etc.

Given a DFA $\mathscr{A} = \langle Q, \Sigma, \delta\rangle$, its *underlying digraph* $D(\mathscr{A})$ has Q as the vertex set and $(q, q') \in Q \times Q$ is an edge of $D(\mathscr{A})$ if and only if $q' = \delta(q, a)$ for some $a \in \Sigma$. It is easy to see that a digraph D is isomorphic to the underlying digraph of some DFA if and only if each vertex of D has at least one outgoing edge. In the sequel, we always consider only digraphs satisfying this property. Every DFA $\mathscr{A}$ such that $D \cong D(\mathscr{A})$ is called a *coloring* of D. Thus, every coloring of D is defined by assigning non-empty sets of labels (colors) from some alphabet Σ to edges of D such that the label sets assigned to the outgoing edges of each vertex form a partition of Σ. Fig. 1 shows a digraph and two of its colorings by $\Sigma = \{a, b\}$.

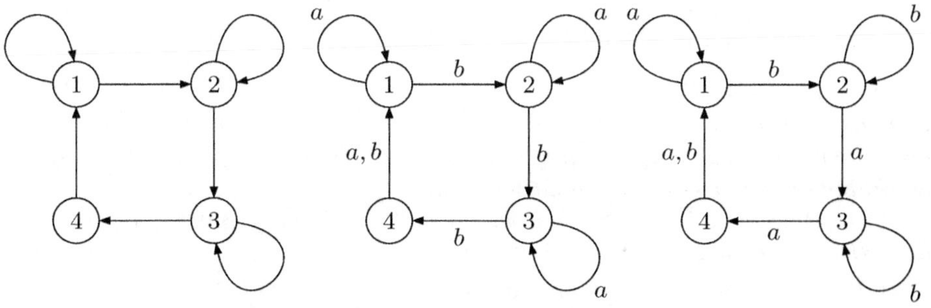

Fig. 1. A digraph and two of its colorings

The *matrix* of a digraph $D = \langle V, E\rangle$ is just the incidence matrix of the edge relation, that is, a $V \times V$-matrix whose entry in the row v and the column v' is 1 if $(v, v') \in E$ and 0 otherwise. For instance, the matrix of the digraph in Fig. 1 (with respect to the chosen numbering of its vertices) is $\begin{pmatrix} 1&1&0&0\\ 0&1&1&0\\ 0&0&1&1\\ 1&0&0&0 \end{pmatrix}$. Conversely, given an $n \times n$-matrix $P = (p_{ij})$ with non-negative real entries, we assign to it a digraph $D(P)$ on the set $\{1, 2, \ldots, n\}$ as follows: (i, j) is an edge of $D(P)$ if and only if $p_{ij} > 0$. This "two-way" correspondence allows us to formulate in terms of digraphs several important for the sequel notions and results which originated in the classical Perron–Frobenius theory of non-negative matrices.

Recall that a digraph $D = \langle V, E\rangle$ is said to be *strongly connected* if for every pair $(v, v') \in V \times V$, there exists a path from v to v'. By the tth *power* of D we mean the digraph D^t with the same vertex set V, such that $(v, v') \in V \times V$ is an edge of D^t if and only if there is a path in D from v to v' of length precisely t. If M is the matrix of D, then the digraph D^t can be equivalently defined as $D(M^t)$, where M^t is the usual t^{th} power of M.

A strongly connected digraph D is called *primitive* if the greatest common divisor of the lengths of all cycles in D is equal to 1. (In the literature such graphs are sometimes called *aperiodic*.) Lemma 1 readily implies that if D is a primitive digraph, then in some power D^t of D every pair of vertices constitutes an edge, i.e., D^t is a complete digraph with loops. (This is equivalent to saying that every entry of the matrix M^t, where M is the matrix of D, is positive.) The least t with this property is called the *exponent* of the digraph D and is denoted by $\gamma(D)$. We need some results on exponents of digraphs summarized as follows.

Theorem 1. (a) (Wielandt's theorem, see [24,7], [8, Theorem 1]) *If a primitive graph D has n vertices, then $\gamma(D) \leq (n-1)^2+1$.*

(b) [8, Theorem 6 and Corollary 4] *Up to isomorphism, there is exactly one primitive digraph D on $n > 2$ vertices with $\gamma(D) = (n-1)^2+1$, and exactly one with $\gamma(D) = (n-1)^2$. The matrices of the digraphs are*

$$\begin{pmatrix} 0\,1\,0 \ldots 0\,0 \\ 0\,0\,1 \ldots 0\,0 \\ \ldots\ldots\ldots\ldots \\ 0\,0\,0 \ldots 0\,1 \\ 1\,1\,0 \ldots 0\,0 \end{pmatrix} \text{ and } \begin{pmatrix} 0\,1\,0 \ldots 0\,0 \\ 0\,0\,1 \ldots 0\,0 \\ \ldots\ldots\ldots\ldots \\ 1\,0\,0 \ldots 0\,1 \\ 1\,1\,0 \ldots 0\,0 \end{pmatrix} \text{ respectively.} \tag{1}$$

(c) [8, Theorem 7] *If $n > 4$ is even, then there is no primitive digraph D on n vertices such that $n^2-4n+6 < \gamma(D) < (n-1)^2$, and, up to isomorphism, there are either 3 or 4 primitive digraphs D on n vertices with $\gamma(D) = n^2-4n+6$, according as n is or is not a multiple of 3.*

(d) [8, Theorem 8] *If $n > 3$ is odd, then there is no primitive digraph D on n vertices such that $n^2-3n+4 < \gamma(D) < (n-1)^2$, and, up to isomorphism, there is exactly one primitive digraph D on n vertices with $\gamma(D) = n^2-3n+4$, exactly one with $\gamma(D) = n^2-3n+3$, and exactly two with $\gamma(D) = n^2-3n+2$. The matrices of these digraphs are:*

$$\begin{pmatrix} 0\,1\,0 \ldots 0\,0 \\ 0\,0\,1 \ldots 0\,0 \\ \ldots\ldots\ldots\ldots \\ 0\,0\,0 \ldots 1\,0 \\ 0\,0\,0 \ldots 0\,1 \\ 1\,0\,1 \ldots 0\,0 \end{pmatrix}, \begin{pmatrix} 0\,1\,0 \ldots 0\,0 \\ 0\,0\,1 \ldots 0\,0 \\ \ldots\ldots\ldots\ldots \\ 0\,0\,0 \ldots 1\,0 \\ 0\,1\,0 \ldots 0\,1 \\ 1\,0\,1 \ldots 0\,0 \end{pmatrix}, \begin{pmatrix} 0\,1\,0 \ldots 0\,0 \\ 0\,0\,1 \ldots 0\,0 \\ \ldots\ldots\ldots\ldots \\ 1\,0\,0 \ldots 1\,0 \\ 0\,1\,0 \ldots 0\,1 \\ 1\,0\,1 \ldots 0\,0 \end{pmatrix}, \begin{pmatrix} 0\,1\,0 \ldots 0\,0 \\ 0\,0\,1 \ldots 0\,0 \\ \ldots\ldots\ldots\ldots \\ 1\,0\,0 \ldots 1\,0 \\ 0\,0\,0 \ldots 0\,1 \\ 1\,0\,1 \ldots 0\,0 \end{pmatrix}. \tag{2}$$

(e) [8, Theorem 8] *If $n > 3$ is odd, then there is no primitive digraph D on n vertices such that $n^2-4n+6 < \gamma(D) < n^2-3n+2$, and, up to isomorphism, there are either 3 or 4 primitive digraphs D on n vertices with $\gamma(D) = n^2-4n+6$, according as n is or is not a multiple of 3.*

3 Exponents of Digraphs vs. Lengths of Reset Words

As mentioned in Section 1, this paper has grown from certain observations made when we analyzed experimental results. One such observation has been a similarity between the "upper parts" of two sequences: the sequence of possible reset lengths of 2-letter synchronizing automata with n states and the sequence of possible exponents of primitive digraphs with n vertices. As it is clear from Theorem 1, the upper part of the latter sequence has certain gaps whose sizes and positions depend on the parity of n; our experiments have revealed a similar pattern of gaps in the upper part of the former sequence. Table 1 illustrates this observation for $n = 9$.

Table 1. Exponents of primitive digraphs with 9 vertices vs lengths of shortest reset words for 2-letter synchronizing automata with 9 states

N	65	64	63	62	61	60	59	58	57	56	55	54	53	52	51
Number of non-isomorphic primitive digraphs with exponent N	1	1	0	0	0	0	0	1	1	2	0	0	0	0	4
Number of non-isomorphic 2-letter synchronizing automata with reset length N	0	1	0	0	0	0	0	1	2	3	0	0	0	4	4

The data in the second row of Table 1 are calculated from Theorem 1, while the data in the third row come from our experiments.

Concerning gaps in the upper part of the sequence of possible reset lengths of 2-letter synchronizing automata with a given number of states, we notice that the first gap was registered in earlier experiments. (Namely, according to [17,18], for $n = 7, 8, 9, 10$ there exists no 2-letter synchronizing automata with n states with reset lengths between $n^2 - 2n$ and $n^2 - 3n + 5$.) However, to the best of our knowledge, the second gap as seen in Table 1 has not been reported in the literature up to now.

We strongly believe that the observed similarity is more than a coincidence. Clearly, there are deep connections between primitive digraphs and synchronizing automata. Indeed, it is well known (see [1]) that if the underlying digraph of a synchronizing automaton is strongly connected that the digraph must be primitive; on the other hand, as follows from Trahtman's proof [21] of the so-called Road Coloring conjecture by Adler, Goodwyn, and Weiss [1], every primitive digraph admits a synchronizing coloring. This, however, does not suffice to explain similarities such as in Table 1 because many of slowly synchronizing automata "responsible" for non-zero entries in the third row cannot be obtained as colorings of primitive digraphs with large exponents corresponding to non-zero entries in the second row. In the next section we demonstrate some new connections between primitive digraphs with large exponents and slowly synchronizing automata with two input letters. In this way, we derive all known series of such automata and construct many new ones.

4 Some Series of Slowly Synchronizing Automata

Due to space limitations, we present here only a part of our results on slowly synchronizing automata. Namely, we restrict ourselves to series derived from three of the primitive digraphs whose matrices are listed in Theorem 1. These series, in particular, ensure that the "island" of reset lengths between n^2-3n+2 and n^2-3n+4 exists for each n.

We start with the digraph W_n corresponding to the first matrix in (1). The digraph (more precisely, its matrix) first appeared in Wielandt's seminal paper [24]. It has n vertices $1, 2, \ldots, n$, say, and the following $n+1$ edges: $(i, i+1)$ for $i = 1, \ldots, n-1$, $(n, 1)$, and $(n, 2)$.

It is easy to see that, up to isomorphism and renaming of letters, there exists a unique coloring of the digraph W_n by two letters. Let $\mathscr{W}_n$ denote this coloring. Fig. 2 shows the digraph W_n and the DFA $\mathscr{W}_n$.

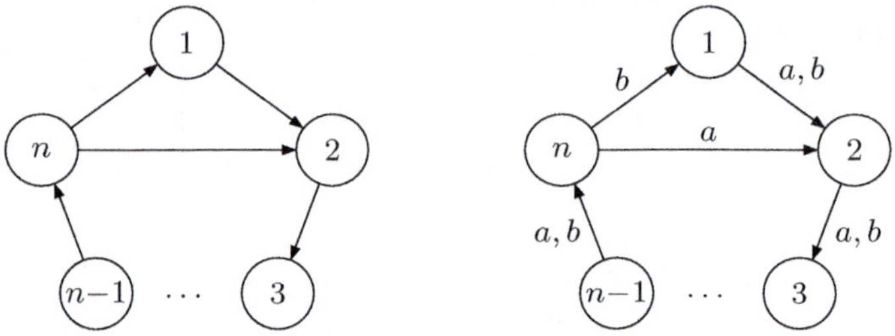

Fig. 2. The digraph W_n and its unique coloring $\mathscr{W}_n$

Theorem 2. *The automaton $\mathscr{W}_n$ is synchronizing and its reset length is $n^2 - 3n + 3$.*

Proof. It is routine to verify that the word $(ab^{n-2})^{n-2}a$, whose length is $(n-1)(n-2)+1 = n^2-3n+3$, is a reset word for $\mathscr{W}_n$.

Now let w be a reset word for $\mathscr{W}_n$ and assume that the length of w (denoted $|w|$) is minimal. Let $j \in Q = \{1, 2, \ldots, n\}$ be the state to which the action of w brings $\mathscr{W}_n$. Then from every state in Q there is a path to j labelled w. It is clear that for each $j \neq 2$ all paths ending at j share the last edge. Therefore, if $j \neq 2$, removing the last letter from the word w produces a word that still would be a reset word for $\mathscr{W}_n$. We conclude that $j = 2$ because $|w|$ is minimal.

If $u \in \{a, b\}^*$, the word uw also is a reset word and it also brings the automaton to the state 2. Hence, for every $\ell \geq |w|$, there is a path of length ℓ in W_n from any given vertex i to 2. In particular, setting $i = 2$, we conclude that for every $\ell \geq |w|$ there is a cycle of length ℓ in W_n. The digraph W_n has only two simple cycles: one of length n and one of length $n-1$. Each cycle of W_n must consist of these two cycles traversed several times whence each number $\ell \geq |w|$ must be expressible as a non-negative integer combination of n and $n-1$. Here we invoke Lemma 2 which implies that $|w| > n(n-1) - n - (n-1) = n^2 - 3n + 1$. Suppose that $|w| = n^2 - 3n + 2$. Then there should be a path of this length from the vertex 1 to the vertex 2. The only outgoing edge of 1 is $(1, 2)$, and thus, in the path it must be followed by a cycle of length $n^2 - 3n + 1$. No cycle of such length may exist by Lemma 2. Hence $|w| \geq n^2 - 3n + 3$.

The series $\mathscr{W}_n$ was discovered by the first author in 2008 (unpublished). His rather involved proof of Theorem 2 used a technique developed in [4].

As mentioned in Section 3, Trahtman's recent result [21] implies that every primitive digraph admits a synchronizing coloring. This gives rise to the following natural question: given a primitive digraph on n vertices, what is the minimum length of reset words for its synchronizing colorings? Observe that in general underlying digraphs of slowly synchronizing automata may admit colorings with rather short reset words. Fig. 1 illustrates this phenomenon: the first coloring of the 4-vertex digraph in Fig. 1 is the Černý automaton $\mathscr{C}_4$ with shortest reset word of length 9 while the second coloring can be reset of the word a^3 of length 3. Wielandt's digraphs W_n, however, can be colored in an essentially unique way, whence Theorem 2 gives the lower bound $n^2 - 3n + 3$ for the value in question. We strongly believe that this lower bound is in fact tight, in other words, we suggest a conjecture that is in a sense parallel to the Černý one.

Conjecture 1. Every primitive digraph on n vertices admits a synchronizing coloring that can be reset by a word of length $n^2 - 3n + 3$.

We observe that while there is a clear analogy between Conjecture 1 and the Černý conjecture, the validity of none of them immediately implies the validity of the other.

Now we discuss a less straightforward way to get a slowly synchronizing series from Wielandt's digraphs W_n. Namely, we aim to show that the Černý automata $\mathscr{C}_n$ are closely related to these digraphs. First, recall the definition of $\mathscr{C}_n$. We may assume that the state set of $\mathscr{C}_n$ is $Q = \{1, 2, \dots, n\}$ and the letters a and b act on Q as follows:

$$\delta(i,a) = \begin{cases} i & \text{if } i < n, \\ 1 & \text{if } i = n; \end{cases} \qquad \delta(i,b) = \begin{cases} i+1 & \text{if } i < n, \\ 1 & \text{if } i = n. \end{cases}$$

The automaton $\mathscr{C}_n$ is shown in Fig. 3 on the left.

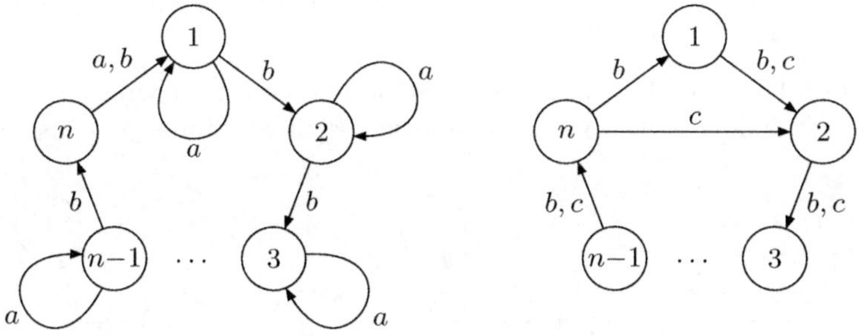

Fig. 3. The automaton $\mathscr{C}_n$ and the automaton induced by the actions of b and $c = ab$

Now we present a new simple proof for the following classic result.

Theorem 3 ([5, Lemma 1]). *The automaton $\mathscr{C}_n$ is synchronizing and its reset length is $(n-1)^2$.*

Proof. It is easy to see that the word $(ab^{n-1})^{n-2}a$ of length $n(n-2)+1 = (n-1)^2$ is a reset word for $\mathscr{C}_n$.

Now let w be a reset word of minimum length for $\mathscr{C}_n$. Since the letter b acts on Q as a cyclic permutation, the word w cannot end with b. (Otherwise removing the last letter gives a shorter reset word.) Thus, we can write w as $w = w'a$ for some $w' \in \{a,b\}^*$ such that the image of Q under the action of w' is precisely the set $\{1,n\}$.

Since the letter a fixes each state in its image $\{1,2,\ldots,n-1\}$, every occurrence of a in w except the last one is followed by an occurrence of b. (Otherwise a^2 occurs in w as a factor and reducing this factor to just a results in a shorter reset word.) Therefore, if we let $c = ab$, then the word w' can be rewritten into a word v over the alphabet $\{b,c\}$. The actions of b and c induce a new automaton on the state set Q; this induced automaton (shown in Fig. 3 on the right) is obviously isomorphic to the automaton $\mathscr{W}_n$. Since w' and v act on Q in the same way, the word vc is a reset word for the induced automaton. By Theorem 2 the length of vc (as a word over $\{b,c\}$) is at least n^2-3n+3. Since the action of b on any set S of states cannot change the cardinality of S and the action of c can decrease the cardinality by 1 at most, the word vc must contain at least $n-1$ occurrences of c. Hence the length of v over $\{b,c\}$ is at least n^2-3n+2 and v contain at least $n-2$ occurrences of c. Since each occurrence of c in v corresponds to an occurrence of the factor ab in w', we conclude that the length of w' over $\{a,b\}$ is at least $n^2-3n+2+n-2 = n^2-2n$. Thus, $|w| = |w'a| \geq n^2-2n+1 = (n-1)^2$.

We have found two more series of slowly synchronizing automata related to Wielandt's digraphs W_n: a series with reset length n^2-3n+2 and another one with reset length n^2-4n+6. These two series will be presented in an extended version of the paper.

Now we discuss a few series related to the digraph D_n defined by the second matrix in (1). The digraph is obtained from W_n by adding the edge $(n-1,1)$. Fig. 4 shows the digraph D_n and its colorings $\mathscr{D}'_n$ and $\mathscr{D}''_n$.

Theorem 4. *The automata $\mathscr{D}'_n$ and $\mathscr{D}''_n$ are synchronizing with reset lengths n^2-3n+4 and n^2-3n+2 respectively.*

The proof of Theorem 4 is similar to that of Theorem 2. It will be presented in an extended version of this paper. The series $\mathscr{D}'_n$ is of interest because for $n > 6$

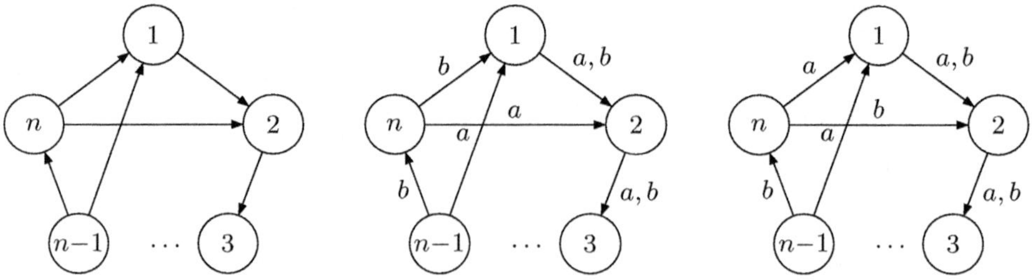

Fig. 4. The digraph D_n and its colorings $\mathscr{D}'_n$ and $\mathscr{D}''_n$

it yields the maximum known value of reset length beyond the Černý series $\mathscr{C}_n$ and also the maximum known value of reset length for synchronizing automata without loops. The series $\mathscr{D}''_n$ also enjoys an extremal property: it provides the maximum known value of reset length for synchronizing automata in which no letter acts as a permutation.

One more series of slowly synchronizing automata related to the digraphs D_n has reset length $n^2 - 4n + 6$. It will be presented in an extended version of this paper.

Except the Černý series $\mathscr{C}_n$, the only infinite series of 2-letter slowly synchronizing automata published so far was the series $\mathscr{B}_n$ ($n > 3$ is odd) constructed in [4]. The automaton $\mathscr{B}_n$ has $Q = \{1, 2, \ldots, n\}$ as its state set, and its input letters a and b act on Q as follows:

$$\delta(i,a) = \begin{cases} i & \text{if } i < n-1, \\ 1 & \text{if } i = n-1, \\ 2 & \text{if } i = n; \end{cases} \qquad \delta(i,b) = \begin{cases} i+1 & \text{if } i < n, \\ 1 & \text{if } i = n. \end{cases}$$

The automaton $\mathscr{B}_n$ is shown in Fig. 5 on the left.

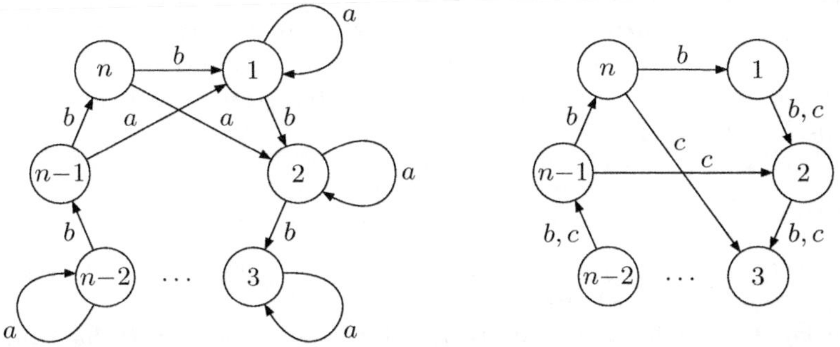

Fig. 5. The automaton $\mathscr{B}_n$ and the automaton induced by the actions of b and $c = ab$

Theorem 5 ([4, Theorem 1.1]). *If $n > 3$ is odd, then the automaton $\mathscr{B}_n$ is synchronizing and its reset length is $n^2 - 3n + 2$.*

The proof of Theorem 5 in [4] is quite involved. Now we can easily prove this result using an argument similar to that in our proof of Theorem 3. The key observation is that the automaton induced by the actions of b and $c = ab$ on the set Q as shown in Fig. 5 on the right is nothing but a coloring of one of the digraphs with exponent $n^2 - 3n + 2$, namely, of the digraph defined by the second matrix in (2). The details of the proof will appear in an extended version of this paper.

5 Experiments

Here we briefly describe the settings of our experiments. Recall that a DFA $\mathscr{A} = \langle Q, \Sigma, \delta \rangle$ is said to be *initially-connected* if there exists a state $q_0 \in Q$ from which every state $q \in Q$ is reachable, that is, $q = \delta(q_0, w)$ for some $w \in \Sigma^*$. In general a synchronizing automaton need not be initially-connected. However, it is well known that when studying the Černý conjecture, we may restrict ourselves to DFA whose underlying digraphs are strongly connected because the validity of the conjecture can be easily reduced to this case (see [12] for example). Clearly, DFA with strongly connected underlying digraphs are initially-connected.

We used a convenient string representation of initially-connected DFA (ICDFA) developed in [3] to generate all such DFA with up to 9 states and 2 input letters. Each ICDFA was tested for synchronizability and then for each synchronizing automaton its reset length was calculated. For these tasks, we implemented standard algorithms (see [15,22]) in C.

The main difficulty that had to be overcome is that the number of ICDFA dramatically grows with the number of states. (For 9 states, there are about 700 billions ICDFA with 2 input letters.) The problem, however, can be efficiently parallelized. For this, a dedicated processor was programmed to generate ICDFA in portions (slices in terminology of [3]) that were fed to other processors for synchronization tests etc. The management program was written in C with MPI. Calculations organized this way took less than a day of running a small size computer grid based on a number of AMD Opteron 2.6 GHz processors.

All slowly synchronizing automata that we found were double-checked by running on them the package TESTAS developed by Trahtman [17].

Our experiments have also produced some interesting statistical results that will be discussed elsewhere.

References

1. Adler, R.L., Goodwyn, L.W., Weiss, B.: Equivalence of topological Markov shifts. Israel J. Math. 27, 49–63 (1977)
2. Almeida, J., Steinberg, B.: Matrix mortality and the Černý–Pin conjecture. In: Diekert, V., Nowotka, D. (eds.) Developments in Language Theory. LNCS, vol. 5583, pp. 67–80. Springer, Heidelberg (2009)
3. Almeida, M., Moreira, N., Reis, R.: Enumeration and generation with a string automata representation. Theor. Comput. Sci. 387, 93–102 (2007)
4. Ananichev, D.S., Volkov, M.V., Zaks, Y.I.: Synchronizing automata with a letter of deficiency 2. Theor. Comput. Sci. 376, 30–41 (2007)
5. Černý, J.: Poznámka k homogénnym eksperimentom s konečnými automatami. Matematicko-fyzikalny Časopis Slovensk. Akad. Vied 14(3), 208–216 (1964) (in Slovak)
6. Dubuc, L.: Sur les automates circulaires et la conjecture de Černý. RAIRO Inform. Théor. Appl. 32, 21–34 (1998) (in French)
7. Dulmage, A.L., Mendelsohn, N.S.: The exponent of a primitive matrix. Can. Math. Bull. 5, 241–244 (1962)
8. Dulmage, A.L., Mendelsohn, N.S.: Gaps in the exponent set of primitive matrices. Ill. J. Math. 8, 642–656 (1964)
9. Eppstein, D.: Reset sequences for monotonic automata. SIAM J. Comput. 19, 500–510 (1990)
10. Higgins, P.M.: The range order of a product of i transformations from a finite full transformation semigroup. Semigroup Forum 37, 31–36 (1988)

11. Kari, J.: Synchronizing finite automata on Eulerian digraphs. Theoret. Comput. Sci. 295, 223–232 (2003)
12. Pin, J.-E.: Le problème de la synchronization et la conjecture de Černý. Thèse de 3ème cycle. Université de Paris 6 (1978) (in French)
13. Pin, J.-E.: On two combinatorial problems arising from automata theory. Ann. Discrete Math. 17, 535–548 (1983)
14. Ramírez Alfonsín, J.L.: The diophantine Frobenius problem. Oxford University Press, Oxford (2005)
15. Sandberg, S.: Homing and synchronizing sequences. In: Broy, M., et al. (eds.) Model-Based Testing of Reactive Systems. LNCS, vol. 3472, pp. 5–33. Springer, Heidelberg (2005)
16. Skvortsov, E., Yu, Z.: Synchronizing random automata. In: Rigo, M. (ed.) AutoMathA 2009, Université de Liège (2009) (submitted)
17. Trahtman, A.N.: An efficient algorithm finds noticeable trends and examples concerning the Černý conjecture. In: Královič, R., Urzyczyn, P. (eds.) MFCS 2006. LNCS, vol. 4162, pp. 789–800. Springer, Heidelberg (2006)
18. Trahtman, A.N.: Notable trends concerning the synchronization of graphs and automata. Electr. Notes Discrete Math. 25, 173–175 (2006)
19. Trahtman, A.N.: The Černý conjecture for aperiodic automata. Discrete Math. Theor. Comput. Sci. 9(2), 3–10 (2007)
20. Trahtman, A.N.: Some aspects of synchronization of DFA. J. Comput. Sci. Technol. 23, 719–727 (2008)
21. Trahtman, A.N.: The Road Coloring Problem. Israel J. Math. 172, 51–60 (2009)
22. Volkov, M.V.: Synchronizing automata and the Černý conjecture. In: Martín-Vide, C., Otto, F., Fernau, H. (eds.) LATA 2008. LNCS, vol. 5196, pp. 11–27. Springer, Heidelberg (2008)
23. Volkov, M.V.: Synchronizing automata preserving a chain of partial orders. Theoret. Comput. Sci. 410, 2992–2998 (2009)
24. Wielandt, H.: Unzerlegbare, nicht negative Matrizen. Math. Z. 52, 642–648 (1950) (in German)

Weights of Exact Threshold Functions

László Babai[1], Kristoffer Arnsfelt Hansen[2,⋆],
Vladimir V. Podolskii[3,⋆⋆], and Xiaoming Sun[4,⋆⋆⋆]

[1] The University of Chicago
laci@cs.uchicago.edu
[2] Aarhus University
arnsfelt@cs.au.dk
[3] Steklov Mathematical Institute
podolskii@mi.ras.ru
[4] ITCS, Tsinghua University
xiaomings@tsinghua.edu.cn

Abstract. We consider Boolean exact threshold functions defined by linear equations, and in general degree d polynomials. We give upper and lower bounds on the maximum magnitude (absolute value) of the coefficients required to represent such functions. These bounds are very close and in the linear case in particular they are almost matching. The quantity is the same as the maximum magnitude of integer coefficients of linear equations required to express every possible intersection of a hyperplane in $\mathbf{R}^n$ and the Boolean cube $\{0,1\}^n$, or in the general case intersections of hypersurfaces of degree d in $\mathbf{R}^n$ and the Boolean cube $\{0,1\}^n$. In the process we construct new families of ill-conditioned matrices. We further stratify the problem (in the linear case) in terms of the dimension k of the affine subspace spanned by the solutions, and give upper and lower bounds in this case as well. Our bounds here in terms of k leave a substantial gap, a challenge for future work.

1 Introduction

A (linear) exact threshold function is a Boolean function that decides whether a real valued linear equality in its Boolean inputs holds. The related class of (linear) threshold functions consist of those Boolean functions that decide whether a real valued linear inequality in their Boolean inputs holds. To be more precise, an exact threshold function on n Boolean inputs $x_1, \dots, x_n$ is a Boolean function

⋆ Work done at Aarhus University supported by a postdoc fellowship from the Carlsberg Foundation and at The University of Chicago, supported by a Villum Kann Rasmussen postdoc fellowship.

⋆⋆ Partially supported by grant 09-01-00709-a from Russian Foundation for Basic Research Fund.

⋆⋆⋆ Supported in part by the National Natural Science Foundation of China Grant 60553001, 60603005, 60621062, the National Basic Research Program of China Grant 2007CB807900,2007CB807901 and Tsinghua University Initiative Scientific Research Program 2009THZ02120.

P. Hliněný and A. Kučera (Eds.): MFCS 2010, LNCS 6281, pp. 66–77, 2010.

that decides whether $w_1x_1 + \cdots + w_nx_n = t$, where $w_1, \ldots, w_n$ are real valued *weights*, and t is a real valued *threshold*. A threshold function on n Boolean inputs $x_1, \ldots, x_n$ is then a Boolean function that decides whether $w_1x_1 + \cdots + w_nx_n \geq t$.

Threshold functions have been studied extensively in many areas of computer science (cf. [12,15,14]). Less attention has been given to exact threshold functions, but they have been considered in Boolean circuit complexity [4,6,7,8,9] and in structural complexity theory [1,10]. We believe that studying exact threshold functions in itself is natural and interesting. However, an important reason for this is that such a study may bring additional insight to the study of threshold functions. For the results of [4,6,7,8] proved for exact threshold functions, it is not currently known whether they also hold for threshold functions. A long standing open question for threshold functions is to prove good lower bounds for depth two circuits. Recent work [9] shows that circuit classes defined using exact threshold functions seamlessly interleave in the usual hierarchy of threshold circuit classes. In particular the class of depth two exact threshold circuits is shown to be a subclass of depth two threshold circuits. For this class no good lower bounds are known as well.

One can readily extend the notions of exact threshold functions as well as threshold functions to higher degree. A polynomial exact threshold function of degree d is a Boolean function that decides whether a real valued polynomial of degree d vanishes when evaluated on the Boolean input. Polynomial threshold functions are defined as an analogous generalization of threshold functions.

In this work we are interested in exact threshold functions from the fundamental perspective of representations of Boolean functions. More precisely, we are interested in the magnitude (absolute value) of integer weights needed to represent any possible exact threshold function. It is not hard to see that without loss of generality one may assume that the real valued weights and threshold defining an exact threshold function are in fact integers (as is also the case with threshold functions), thereby making the question we study well-defined.

The analogous question of the magnitude of weights required for threshold functions has a long history of research. An upper bound on the magnitude of integer weights required to represent any threshold function was obtained by Muroga, Toda and Takasu [13] (cf. [14]). They showed that weights of magnitude $\leq (n+1)^{(n+1)/2}/2^n$ are sufficient. Also, several examples of explicit functions that require weights of magnitude $2^{\Omega(n)}$ are known [14,15]. The existence of such functions may also be established by a counting argument, since there are at least $2^{n(n-1)/2}$ threshold functions on n variables [19,18].

An almost optimal lower bound was obtained by Håstad [11]. Let n be a power of 2. Then Håstad constructed a threshold function requiring an integer weight of magnitude at least $(1/n)e^{-4n^\beta}n^{n/2}/2^n$, where $\beta = \log(3/2)$. Thus in the case when n is a power of 2 the upper bound and this lower bound differ only by a subexponential factor. Generalizing this work, for any constant d, [17] constructed a polynomial threshold function of degree d that requires an integer weight of magnitude $n^{\Omega(n^d)}$.

Alon and Vũ [2], building on the techniques of Håstad, gave a new construction of ill-conditioned matrices. For a non-singular $n \times n$ matrix A, let $B = A^{-1} = (b_{ij})$ and define $\chi(A) = \max_{i,j} |b_{ij}|$. Define further $\chi_1(n)$ as the maximum of $\chi(A)$ over all non-singular $n \times n$ $(0,1)$ matrices A. Define $\chi_2(n)$ to be the analogous quantity where $(-1,1)$ matrices are considered instead.

With these definitions, Alon and Vũ provide for every n an explicit $n \times n$ $(0,1)$ matrix A_1 and an explicit $n \times n$ $(-1,1)$ matrix A_2 such that $\chi(A_i) \geq n^{n/2}/2^{n(2-o(1))}$ for $i = 1,2$. When n is a power of 2 these lower bounds may be improved to $n^{n/2}/2^{n(1-o(1))}$. Upper bounds for $\chi_i(n)$ are derived from the Hadamard inequality.

Theorem 1 (Alon and Vũ). $n^{\frac{n}{2}}/2^{n(2-o(1))} \leq \chi_i(n)$ *for* $i = 1,2$. $\chi_1(n) \leq n^{\frac{n}{2}}/2^{n-1}$, *and* $\chi_2(n) \leq (n-1)^{\frac{n-1}{2}}/2^{n-1}$.

Alon and Vũ are able to apply their construction of ill-conditioned matrices to answer questions about flat simplices, weights of threshold functions, coin weighing, and indecomposable hypergraphs. In particular, they construct a threshold function on n variables that requires a weight of magnitude $\geq n^{n/2}/2^{n(2-o(1))}$.

Let $\text{maxw}^T(n)$ denote the minimal W such that every possible possible threshold function on n variables can be realized using integer weights of magnitude $\leq W$. We summarize the above discussion in the following theorem:

Theorem 2 (Muroga et al.; Håstad; Alon and Vũ). $n^{\frac{n}{2}}/2^{n(2-o(1))} \leq \text{maxw}^T(n) \leq (n+1)^{\frac{n+1}{2}}/2^n$.

We are now in position to state our first theorem. We define $\text{maxw}^E(n)$ for exact threshold functions as the analogous quantity of $\text{maxw}^T(n)$. For this quantity we obtain the following upper and lower bounds:

Theorem 3. $n^{\frac{n}{2}}/2^{n(2-o(1))} \leq \text{maxw}^E(n) \leq n^{\frac{n}{2}+1}$.

As is evident from Theorems 1, 2 and 3, the quantities $\chi_i(n)$, $\text{maxw}^T(n)$ and $\text{maxw}^E(n)$ are very close, in fact they are equal up to an exponential factor. Furthermore, when n is a power of 2, the bounds for $\chi_i(n)$ and $\text{maxw}^T(n)$ differ only by a subexponential factor. We do not know if the same holds for $\text{maxw}^E(n)$.

While some of our methods are related to the methods employed in the study of threshold functions, we do not see an explicit relationship between the quantities $\text{maxw}^T(n)$ and $\text{maxw}^E(n)$. The proofs of Theorem 2 and Theorem 3 in fact show that $\chi_2(n) \leq \text{maxw}^T(n)$ and $\chi_1(n-1) \leq \text{maxw}^E(n)$, but we do not know whether it is possible to turn a threshold function that requires certain magnitude into an exact threshold function requiring a similar magnitude or conversely.

A property that seems to be unique to (linear) exact threshold functions is that we can speak of the *dimension* of such functions. For a given exact threshold function f defined by a linear equation $w_1x_1 + \cdots + w_nx_n = t$, consider the real affine space V generated by the points of the Boolean cube that satisfy the equation. We can then define the dimension of f to be the dimension of V.

Let $\text{maxw}^E(n,k)$ denote the minimal W such that every exact threshold function of dimension k of n Boolean variables can be realized using integer weights of magnitude $\leq W$. Our second result gives the following for this quantity:

Theorem 4. *For all n and all $1 \leq k \leq n$, $(\lfloor \frac{n}{k} \rfloor^k - 1)/2k \leq \text{maxw}^E(n,k) \leq n^{2^k}$.*

For polynomial exact threshold functions, let $\text{maxw}_d^E(n)$ denote the magnitude of weights required to represent every possible exact degree d polynomial threshold function on n variables. Our final result is the following generalized bounds:

Theorem 5. $n^{\frac{1}{2}n^d}/2^{2n^d+o(n^d)+d} \leq \text{maxw}_d^E(2dn) \leq n^{\frac{dn^d}{2}+d}$.

This result is analogous to results for threshold functions, see [17]. For the lower bounds in this theorem we prove a specific generalization of Theorem 1.

The remainder of the paper is organized as follows. In Section 2 we state precisely the definitions we use and present some simple observations. In Section 3.1 we provide an example of a function of an exact threshold function that requires exponential magnitude of weights by an elementary argument. We prove Theorem 3 in Section 3.2. The proof of Theorem 4 is given in Sections 3.3 and 3.4 respectively. Theorem 5 is proved in Section 3.5. We conclude with open problems in Section 4. Due to page limitations several proofs as well as additional explanations are omitted. These will appear in the full version of this paper.

2 Preliminaries

We consider here a Boolean function f to be a function $f : \{0,1\}^n \to \{0,1\}$. We say that a Boolean function f on n variables is an exact threshold function if there exist real numbers $w_1, \ldots, w_n$ (the weights) and a real number t (the threshold value), such that $f(x) = 1$ if and only if $\sum_{i=1}^n w_i x_i = t$, for all $x \in \{0,1\}^n$. We may say that the list of weight $w_1, \ldots, w_n$ and the threshold t as well as the expression $w_1 x_1 + \cdots + w_n x_n = t$ are a *realization* of the function f. Similarly, a Boolean function f on n variables is a threshold function if there exist real numbers $w_1, \ldots, w_n$ and a real number t, such that $f(x) = 1$ if and only if $\sum_{i=1}^n w_i x_i \geq t$ for any $x \in \{0,1\}^n$.

One may observe that without loss of generality it can be assumed that the real valued weights as well as the real valued threshold are in fact integers. Often when considering threshold functions one considers the Boolean cube $\{-1,1\}^n$ instead of $\{0,1\}^n$ as we do in this work. This is of no consequence to the possible weights of exact threshold and threshold functions, as may easily be observed. Furthermore, since, in this work we are only interested in the weights of exact threshold functions, we may without loss of generality restrict our attention to exact threshold functions f for which $f(0,\ldots,0) = 1$.

Identifying a Boolean function f with the subset $f^{-1}(1)$, an exact threshold function corresponds to the intersection of a hyperplane in $\mathbf{R}^n$ with the Boolean n-cube $\{0,1\}^n$, and in the higher degree case an intersection with a degree d hypersurface. In the linear case, from the remark above, for studying weights of exact threshold functions we may restrict our attention to affine spaces that are in fact subspaces of $\mathbf{R}^n$ and are spanned by vectors of the Boolean cube $\{0,1\}^n$. Because of this fact we will mainly phrase our theorems in vector space terminology.

3 Weights of an Exact Threshold Function

3.1 Example

An interesting example of an exact threshold function is the (sequence) equality function. Define the equality function EQ on $2n$ variables $x_1, \dots, x_n$ and $y_1, \dots, y_n$ by $\mathrm{EQ}(x, y) = 1$ if and only if $x_i = y_i$ for all i.

It turns out that the set of weights that realize this exact threshold function corresponds *precisely* to solutions to the well known problem of finding n positive integers $a_1 < \dots < a_n$ such that all sums of the form $\sum_{i \in I} a_i$ are distinct. Using this insight it is easy to see that weights of exponential magnitude are needed to realize the EQ function.

3.2 Upper and Lower Bounds for the Linear Case

Before proving the upper bound, we state without proof the following three lemmas:

Lemma 1 (Faddeev and Sominskii [5]). *Let A be a $n \times n$ matrix with all entries 0 or 1. Then $|\det(A)| \le (n+1)^{\frac{n+1}{2}}/2^n$.*

Lemma 2. *Let $v_1, \dots, v_{n-1} \in \{0,1\}^n$ be linearly independent. Then there exist integers $w_1, \dots, w_n$ such that the equation $w_1x_1 + \dots + w_nx_n = 0$, defines the linear subspace $\mathrm{span}(\{v_1, \dots, v_{n-1}\})$, and satisfy $|w_i| \le n^{\frac{n}{2}}/2^{n-1}$.*

Lemma 3. *Let V be a vector space and let $k = n - \dim(\mathrm{span}(V \cap \{0,1\}^n))$. Then there exist vector spaces $V_1, \dots, V_k$ such that $\dim(\mathrm{span}(V_i \cap \{0,1\}^n)) = n-1$ for all i and $V \cap \{0,1\}^n = \left(\bigcap_{i=1}^k V_i\right) \cap \{0,1\}^n$.*

Theorem 6. *Let V be a vector space in $\mathbf{R}^n$. Then there exist integers $w_1, \dots, w_n$ such that for all $x \in \{0,1\}^n$ we have $x \in V$ if and only if $w_1x_1 + \dots + w_nx_n = 0$ and furthermore satisfy $|w_i| \le n^{\frac{n}{2}+1}$ for all i.*

Thus every exact threshold function f on n variables can be realized using integer weights of absolute value at most $n^{\frac{n}{2}+1}$ as well.

Proof. Let $k = n - \dim(\mathrm{span}(V \cap \{0,1\}^n))$. Then by Lemma 3 there exist vector spaces $V_1, \dots, V_k$ of dimension $n-1$ spanned by vectors from $\{0,1\}^n$ such that $V \cap \{0,1\}^n = \left(\bigcap_{i=1}^k V_i\right) \cap \{0,1\}^n$. For each V_i, by Lemma 2 there exist integers w_{ij} such that for all $x \in \{0,1\}^n$ we have $x \in V_i$ if and only if the equation $w_{i1}x_1 + \dots + w_{in}x_n = 0$ is satisfied, and furthermore $|w_{ij}| \le n^{\frac{n}{2}}/2^{n-1}$. We conclude the proof by the probabilistic method.

For $i = 1, \ldots, k$, pick $c_i \in \{-2^{n-1}, \ldots, 2^{n-1}\}$ uniformly at random, and consider the combined equation $\sum_{i=1}^{k} \sum_{j=1}^{n} c_i w_{ij} x_j = 0$. Clearly, every $x \in V \cap \{0,1\}^n$ must satisfy this equation, since every such x satisfies the individiual equations for all i.

Now consider $x \in \{0,1\} \setminus V$. There must be some i such that $x \notin V_i$, and for this i the chosen x does not satisfy the corresponding equation. This implies that the chosen x satisfies the combined equation with probability $< 2^{-n}$. Since $|\{0,1\}^n \setminus V| \leq 2^n$, the probability that some $x \in \{0,1\}^n \setminus V$ satisfies the combined equation is < 1, and thus there is a fixed choice $\widehat{c}$ for c with this property. Hence, $w_j = \sum_{i=1}^{k} \widehat{c}_i w_{ij}$ satisfies the requirements, since each $|\widehat{c}_i w_{ij}| \leq n^{\frac{n}{2}}$. □

Ziegler [20], using the results of Theorem 1, gave a lower bound on the maximal coefficient of an inequality defining a facet of a full-dimensional (0, 1) polytope in $\mathbf{R}^n$. Since the polytope is of full dimension, the hyperplane given by the facet is uniquely determined by points from $\{0,1\}^n$ and hence corresponds to a unique exact threshold function, and the lower bound applies to our setting also. We state the lower bound below, and additionally point out that the construction provides a lower bound on the magnitude of *all* the coefficients.

Theorem 7. *For any n, there exists an exact threshold function f on n variables such that any realization of f requires an integer weight of magnitude $n^{\frac{n}{2}}/2^{n(2-o(1))}$.*

Observation 1. Alon and Vũ show when $n-1$ is a power of 2, that the inverse of the $(n-1) \times (n-1)$ matrix A they construct actually has a column (in fact many columns) where *all* entries are of magnitude $n^{\frac{n}{2}}/2^{\Theta(n)}$. This means that in the construction above, when $n-1$ is a power of 2, one can obtain that all the first $n-1$ coefficients are of this magnitude.

In other words, for every n, not only is it such that one may be required to use $\Omega(n \log n)$ bits to store the largest weight, but to store all the weights one may be required to use $\Omega(n^2 \log n)$ bits.

3.3 Small Dimension Upper Bound

Theorem 8. *Let V be a vector space in $\mathbf{R}^n$ and let $k = \dim(\mathrm{span}(V \cap \{0,1\}^n))$. Then there exist integers $w_1, \ldots, w_n$ such that for all $x \in \{0,1\}^n$ we have $x \in V$ if and only if $w_1 x_1 + \cdots + w_n x_n = 0$ and furthermore satisfy $|w_i| \leq n^{2^k}$ for all i. Thus every exact threshold function f on n variables of dimension at most k can be realized using integer weights of absolute value at most n^{2^k} as well.*

Proof. Let $v_1, \ldots, v_k \in \{0,1\}^n$ be a basis of $\mathrm{span}(V \cap \{0,1\}^n)$. For $\alpha \in \{0,1\}^k$ define the set $S_\alpha = \{i \in \{1, \ldots, n\} \mid \forall j \in \{1, \ldots, k\} : (v_j)_i = \alpha_j\}$. Number the nonempty such sets $S_1, \ldots, S_K$ for $K \leq 2^k$. For $i \in \{1, \ldots, K\}$, let $n_i = |S_i|$, and assume $S_i = \{a_{i1}, \ldots, a_{in_i}\}$. Further, define $\chi_i \in \{0,1\}^n$ to be the characteristic vector of the set S_i.

Now, consider the vector space W in $\mathbf{R}^K$ defined by $y \in W$ if and only if $\sum_{i=1}^{K} y_i \chi_i \in V$. Note that, if $\sum_{i=1}^{K} y_i \chi_i \in \{0,1\}^n$, we must have $y \in \{0,1\}^K$

by construction. By Theorem 6 we have integer weights $\hat{w}_1, \ldots, \hat{w}_K$ such that for all $y \in \{0,1\}^K$ we have $y \in W$ if and only if $\hat{w}_1 y_1 + \cdots + \hat{w}_K y_K = 0$ and furthermore satisfy $|\hat{w}_i| \leq K^{\frac{K}{2}+1}$. Thus, we also have for all $y \in \{0,1\}^K$ it holds that $\sum_{i=1}^K y_i \chi_i \in V$ if and only if $\hat{w}_1 y_1 + \cdots + \hat{w}_K y_K = 0$.

We now define integer weights $w_1, \ldots, w_n$ as follows. Let $N = \prod_{l=1}^K n_l$. If $n_i = 1$ we simply define $w_{a_{i1}} = N\hat{w}_i$. Otherwise we give the first element of each set S_i weight as $w_{a_{i1}} = -(n_i - 1)\prod_{l=1}^{i-1} n_l + N\hat{w}_i$, and the remaining elements are given weights as $w_{a_{ij}} = \prod_{l=1}^{i-1} n_l$, for $j = 2, \ldots, n_i$. By construction we obtain the property that if $w_1 x_1 + \cdots + w_n x_n = 0$ for $x \in \{0,1\}^n$ we must have that for all i we have that all coordinates $x_{a_{ij}}$ have the same value, 0 or 1. Thus x must be of the form $\sum_{i=1}^K y_i \chi_i \in W$ for $y \in \{0,1\}^K$. The converse trivially holds. Now, finally note that for all i we have $|w_i| \leq K^{\frac{K}{2}+1} \prod_{j=1}^K n_j \leq K^{\frac{K}{2}+1} \left(\frac{n}{K}\right)^K = \frac{n^K}{K^{K/2-1}} \leq n^K \leq n^{2^k}$. □

3.4 Small Dimension Lower Bound

Suppose $k \leq n/2$, let $d = \lfloor \frac{n}{k} \rfloor$, and define a k dimensional vector space V by

$$V = \mathrm{span}\{\overbrace{1\ldots1}^{d}\overbrace{0\ldots0}^{n-d}, \overbrace{0\ldots0}^{d}\overbrace{1\ldots1}^{d}\overbrace{0\ldots0}^{n-2d}, \ldots, \overbrace{0\ldots0}^{(k-1)d}\overbrace{1\ldots1}^{d}\overbrace{0\ldots0}^{n-kd}\}.$$

Theorem 9. *Suppose $w_1, \ldots, w_n$ are integers satisfying $x \in V$ if and only if $\sum_{i=1}^n w_i x_i = 0$ for all $x \in \{0,1\}^n$. Then we must have $\max_i |w_i| \geq \frac{d^k-1}{2k} \sim \frac{n^k}{2k^{k+1}}$. Thus there exist an exact threshold function on n variables of dimension k that requires an integer weight of absolute value at least $(d^k-1)/2k$ as well.*

Proof. First we relabel the weights, $w_1, \ldots, w_{kd}$ by $w_{1,1}, \ldots, w_{1,d}$; $w_{2,1}, \ldots, w_{2,d}$; $\ldots$;$w_{k,1}, \ldots, w_{k,d}$, where $w_{i,j} = w_{(i-1)d+j}$ ($i \in [k]$, $j \in [d]$). For each $i \in [k]$ define the (multi)-set S_i by $S_i = \{w_{i,1}, w_{i,2}, \ldots, w_{i,d}\}$. For a (multi-)set $S \subset \mathbf{Z}$ define $\mathrm{sum}(S) = \sum_{y \in S} y$. By assumption and by the definition of V we have $\mathrm{sum}(S_i) = 0$, for all i. We claim that if $\max_i |w_i| < \frac{d^k-1}{2k}$, then there exist subsets $\tilde{S}_i \subseteq S_i$ for each $i \in [k]$, such that $\sum_{i=1}^k \mathrm{sum}(\tilde{S}_i) = 0$, and at least one $\tilde{S}_i$ satisfies $\tilde{S}_i \neq \emptyset$, $\tilde{S}_i \neq S_i$. In other words there would exist $x \in \{0,1\}^n$ satisfying $\sum_{i=1}^n w_i x_i = 0$ and $x \notin V$, leading to a contradiction. Hence we can conclude that $\max_i |w_i| \geq \frac{d^k-1}{2k}$. We next prove this claim, thereby completing the proof of the theorem.

Let $M = \max_i |w_i|$. By assumption we have $M < \frac{d^k-1}{2k}$. Since $\mathrm{sum}(S_i) = 0$, i.e. $w_{i,1} + w_{i,2} + \cdots + w_{i,d} = 0$, we can arrange $\{w_{i,1}, \ldots, w_{i,d}\}$ in an order $\{\tilde{w}_{i,1}, \ldots, \tilde{w}_{i,d}\}$ such that: (1) $\tilde{w}_{i,1} \geq 0$; (2) for each $2 \leq j \leq d$, if $\sum_{l=1}^{j-1} \tilde{w}_{i,l} \geq 0$, then $\tilde{w}_{i,j} \leq 0$, otherwise $\tilde{w}_{i,j} > 0$. It is easy to see that for any $i \in [k]$, $j \in [d]$ we have, $-M \leq \tilde{w}_{i,1} + \tilde{w}_{i,2} + \cdots + \tilde{w}_{i,j} \leq M$.

Now, for each k-tuple $(l_1, \ldots, l_k) \in [d]^k$, consider the double summation $\sum_{i=1}^k \sum_{j=1}^{l_i} \tilde{w}_{i,j}$. By the previous equation we have $-kM \leq \sum_{i=1}^k \sum_{j=1}^{l_i} \tilde{w}_{i,j} \leq kM$. But in total there are d^k different k-tuples $(l_1, \ldots, l_k)$, and

$2kM + 1 < 2k \cdot \frac{d^k-1}{2k} + 1 = d^k$. Therefore there exist two k-tuples $(l_1, \ldots, l_k) \neq (l'_1, \ldots, l'_k)$, such that $\sum_{i=1}^{k} \sum_{j=1}^{l_i} \tilde{w}_{i,j} = \sum_{i=1}^{k} \sum_{j=1}^{l'_i} \tilde{w}_{i,j}$.

For each $i \in [k]$, we define a set $\tilde{S}_i$. If $l_i \geq l'_i$, let $\tilde{S}_i = \{\tilde{w}_{i,l'_i+1}, \ldots, \tilde{w}_{i,l_i}\}$. Otherwise, let $\tilde{S}_i = S_i \setminus \{\tilde{w}_{i,l_i+1}, \ldots, \tilde{w}_{i,l'_i}\}$. Combining the previous equation with the fact that $\mathrm{sum}(S_i) = 0$, we obtain $\sum_{i=1}^{k} \mathrm{sum}(\tilde{S}_i) = 0$.

It is clear that $\tilde{S}_i \neq S_i$ for all i, and since $(l_1, \ldots, l_k) \neq (l'_1, \ldots, l'_k)$, there must exist i such that $\tilde{S}_i \neq \emptyset$. □

3.5 Higher Degree Bounds

The upper bound for polynomial exact threshold functions follows easily from the upper bound for the linear case. We omit the proof.

As to the lower bound we proceed similarly to the case of degree 1. We start by a specific generalization of Alon and Vũ's lower bound on $\chi(A)$ [2]. This could be done by translating the analysis of [17] to matrix terminology analogous of Alon and Vũ's translated analysis of Håstad's result [11]. But instead we shall prefer to present a technically simpler proof using matrix terminology based on the results of [2]. We note, however, that although this proof differs from the proof in [17], the underlying ideas are the same.

In what follows we use parameters n and d, where d denotes the degree of the polynomial and nd is the number of input variables. We denote input variables by x_{ij} for $i = 1, \ldots, d$, $j = 1, \ldots, n$ and suppose they range over $\{0, 1\}$. Let $\overline{x}^1 = (x_{1,1}, \ldots, x_{1,n}), \ldots, \overline{x}^d = (x_{d,1}, \ldots, x_{d,n})$. Let $x = (\overline{x}^1, \ldots, \overline{x}^d)$. We denote by M_d the set of monomials $\{x_{1,j_1} x_{2,j_2} \cdots x_{d,j_d} | j_1, \ldots, j_d \in \{1, \ldots, n\}\}$ (that is, we take exactly one variable from each $\overline{x}^i$). For the matrix A we denote by A_{ij} its submatrix obtained by deleting the ith row and the jth column.

To state our generalization we first need the following definition:

Definition 1. *A matrix $A \in \{0, 1\}^{m \times m}$ is called* d-generable *if we can label each column of A with a unique monomial from M_d and label each row of A by an assignment to the variables (an input) in such a way that each entry a_{ij} is exactly the value of the monomial corresponding to the column j when evaluated on the input corresponding to the row i.*

With this we can provide our generalization of Alon and Vũ's lower bound.

Theorem 10. *For any d and any n there exist an (explicit) matrix $A^{(d)} \in \{0, 1\}^{n^d \times n^d}$ such that $A^{(d)}$ is d-generable, $|\det A^{(d)}_{1,n^d} / \det A^{(d)}| \geq n^{\frac{1}{2} n^d} / 2^{2n^d + o(n^d)}$ and $A^{(d)}$ has the form $\begin{bmatrix} e_1^T \\ B \end{bmatrix}$, where e_1^T is a row $(1, 0, \ldots, 0)$ of length n^d and B is a $(n^d - 1) \times n^d$ matrix. (Furthermore, we note that the function hidden in the o-notation above depends on n but does not depend on d.)*

Proof (sketch). The proof goes by induction on d. In the base case, $d = 1$, the matrix $A^{(1)}$ can be easily obtained from the matrix constructed in [2]. For the inductive step we need the following notation:

For a matrix $A = \{a_{ij}\}_{i,j=1}^n$, let $\overline{A}$ denote the matrix obtained by reflecting A in the vertical median, i.e., $\overline{A} = \{a_{i,n+1-j}\}_{i,j=1}^n$. Now we define $A^{(d)}$ by

$$A^{(d)} = \left[\begin{array}{c|c|c|c} A^{(d-1)} & 0 & 0 & \cdots \\ \hline \begin{matrix} e_n^T \\ \end{matrix} & \overline{A}^{(d-1)} & 0 & \cdots \\ \hline 0 & \begin{matrix} e_1^T \\ \end{matrix} & A^{(d-1)} & \ddots \\ \hline \vdots & \ddots & \ddots & \ddots \end{array}\right]$$

This matrix has $n \times n$ blocks. In the diagonal blocks we have matrices $A^{(d-1)}$, $\overline{A}^{(d-1)}$, $A^{(d-1)}$, ... In the blocks right below the diagonal only the first row is nonzero and these rows are $e_n^T, e_1^T, e_n^T, \ldots$. All other blocks consists exclusively of zeros.

It is now just a matter of checking that $A^{(d)}$ satisfies all required properties. To verify the claimed inequality, we repeatedly apply Lemma 2.3.1 from [2]. We omit the details. □

Now we prove the main result of this section.

Theorem 11. $\mathrm{maxw}_d^E(2dn) \geq n^{\frac{1}{2}n^d}/2^{2n^d+o(n^d)+d}$.

Proof. We will now have variables $x_{ij}, y_{ij} \in \{0,1\}$ for $i = 1, \ldots, d$ and $j = 1, \ldots, n$ (that is, we have twice the number of variables as above), and consider integer polynomials $p(x, y)$ of degree d. In fact it will be more convenient for us to consider polynomials $q(x-y, x+y)$ instead (recall that x and y are vectors of variables). It is easy to check that we can always convert a polynomial $p(x, y)$ into an equivalent polynomial $q(x-y, x+y)$ and vice versa. Moreover, if $q(x-y, x+y)$ has a large coefficient then the corresponding polynomial $p(x, y)$ will also have a large coefficient.

Observation 2. Suppose all coefficients of a degree-d integer polynomial $p(x, y)$ are of absolute value at most s. Let $q(u, v) = p((v+u)/2, (v-u)/2)$. Then $p(x, y) = q(x-y, x+y)$, and $2^d q(u, v)$ have integer coefficients of absolute value at most $2^d s$. □

We will now construct a polynomial exact threshold function with the desired properties. We start with the matrix $A^{(d)}$ given by Theorem 10. First of all we switch the labeling of the matrix: in each monomial in the labeling of $A^{(d)}$ we substitute each variable x_{ij} by the expression $x_{ij} - y_{ij}$. Thus now our columns correspond to monomials in variables $x - y$. Let us denote this new set of monomials corresponding to columns of $A^{(d)}$ by M_d'. To complete the labeling we must change the row labels correspondingly. This is possible since we can find values of x_{ij} and y_{ij} in such a way that the old value of x_{ij} is equal to the new value of $x_{ij} - y_{ij}$.

Now, let us consider the matrix $B = \left[A^{(d)}\ e_1\right]$. That is, we add one column to the matrix $A^{(d)}$. Let $z \in \{0,1\}$ be a new variable and let the new column correspond to the monomial $\theta = (x_{11} - y_{11})(x_{21} - y_{21})\dots(x_{d1} - y_{d1})z$. We next need to extend the assignments labeling the rows to include values of z. For the first row we let $z = 1$ and for the others rows we let $z = 0$. Now it is easy to see that each entry of B is equal to the value of monomial corresponding to the column on the assignment corresponding to the row.

Next, let us consider an exact polynomial threshold gate over the set of monomials $M'_d \cup \{\theta\}$ where the coefficient vector w is a nonzero integer solution of the system $Bw = 0$. Note that the dimension of the solution space of this system is 1 since $A^{(d)}$ is nonsingular and thus w is uniquely determined up to a multiplicative factor. We denote this polynomial by $p(x-y, z)$ and the corresponding exact threshold function by $f(x,y,z)$.

Now we will prove that any integer representation w' of the same exact threshold function over the set of all monomials over the variables $x-y$, $x+y$ and over the monomial θ must have a large coordinate. First, note that such representation should satisfy the system

$$[B\ B'] \begin{pmatrix} u \\ v \end{pmatrix} = 0, \tag{1}$$

where the columns B' corresponds to monomials in variables $x-y$ and $x+y$ which are not in M'_d and $\begin{pmatrix} u \\ v \end{pmatrix}$ is w' where u corresponds to the B-part and v corresponds to the B'-part. Entries of the matrix B' are as usual equal to the value of the monomial corresponding to the column on the input corresponding to the row. So, B' is $\{0, \pm 1\}$-matrix.

Proposition 1. $Bu = 0$.

We omit the proof of this proposition which uses a symmetry argument.

Now if we detach the last column of the matrix and move it to the right-hand side we will get system with the square matrix:

$$A^{(d)} \begin{pmatrix} u_1 \\ u_2 \\ \vdots \\ u_{n^d} \end{pmatrix} = \begin{pmatrix} u_{n^d+1} \\ 0 \\ \vdots \\ 0 \end{pmatrix}.$$

Note that u_{n^d+1} should be nonzero since otherwise all u_i are zero (recall that $A^{(d)}$ is nonsingular). Now by Cramer's rule we have that $|u_{n^d}| = \left|\frac{u_{n^d+1} \det A^{(d)}_{1,n^d}}{\det A^{(d)}}\right| \geq \left|\frac{\det A^{(d)}_{1,n^d}}{\det A^{(d)}}\right|$. Now we get rid of z. Define the function $g(x,y) = f(x,y,0)$. It is obviously a polynomial exact threshold function. Moreover any degree-d integer polynomial representing $g(x, y)$ can be transformed into a degree-d integer polynomial representing $f(x, y, z)$. Indeed, let the polynomial for g be given by the vector of coefficients u. Add to it an integer coefficient for the monomial θ in such a way that resulting vector u' satisfy the first row of the system (1). Note also that it automatically satisfies all other rows of this system (since they refer to inputs with $z = 0$). Thus u' is a solution of the system (1) and hence $|u_{n^d}| \geq \left|\frac{\det A^{(d)}_{1,n^d}}{\det A^{(d)}}\right| \geq \frac{n^{\frac{1}{2}n^d}}{2^{2n^d+o(n^d)}}$.

Now we have proved the lower bound on the coefficient of an integer polynomial of degree at most d in variables $x - y$ and $x + y$ which represents $g(x, y)$. And by Lemma 2 we have desired lower bound for polynomials in variables x and y (here a factor 2^d appears in the denominator). □

Remark 1. Note that the place where we needed a new variable z is Proposition 1. If we should try instead to write a system without the variable z and with nontrivial right-hand side we would not be able to perform the symmetry argument.

Remark 2. We can reprove the lower bound on $\mathrm{maxw}_d^T(n)$ (which is a generalization of $\mathrm{maxw}^T(n)$ to degree-d threshold functions) from [17] similarly generalizing proof of the lower bound on $\mathrm{maxw}^T(n)$ from [2]. We omit the details.

4 Conclusion

We have obtained upper and lower bounds for the magnitude of integer weights required to represent exact threshold functions, for linear exact threshold functions as well as polynomial exact threshold functions in general. For the linear case, we also gave bounds for the interesting special case of small dimension functions. In the small dimension case there seems to be ample room for further improvement of the bounds. In the other cases our bounds are very close, especially for the linear case, leaving little room for improvement. However our proof raises an interesting question: Is it possible that exact threshold functions on n variables of dimension less than $n - 1$ can require larger weights than those of dimension $n-1$, or is the worse upper bound an artifact of our proof. For obtaining the lower bounds for polynomial exact threshold functions we constructed ill-conditioned matrices with special properties – these may have additional applications.

Another question arises from comparing results known for threshold functions and for exact threshold functions. Beigel [3] give a techinique for showing that a simple linear threshold function require exponential weights even for representations of degree $n^{1/2-\epsilon}$. This technique was generalized in [16] to the case of higher degree polynomial threshold functions. Beigel's approach and its extensions do not seem to be applicable to exact threshold functions. Obtaining analogous results for exact threshold functions is therefore an open problem.

References

1. Agrawal, M., Arvind, V.: Geometric sets of low information content. Theoretical Computer Science 158(1-2), 193–219 (1996)
2. Alon, N., Vũ, V.H.: Anti-hadamard matrices, coin weighing, threshold gates, and indecomposable hypergraphs. Journal of Combinatorial Theory, Series A 79(1), 133–160 (1997)
3. Beigel, R.: Perceptrons, PP, and the polynomial hierarchy. Computational Complexity 4(4), 339–349 (1994)
4. Beigel, R., Tarui, J., Toda, S.: On probabilistic ACC circuits with an exact-threshold output gate. In: Ibaraki, T., Iwama, K., Yamashita, M., Inagaki, Y., Nishizeki, T. (eds.) ISAAC 1992. LNCS, vol. 650, pp. 420–429. Springer, Heidelberg (1992)
5. Faddeev, D.K., Sominskii, I.S.: Problems in Higher Algebra. W.H. Freeman, New York (1965)
6. Green, F.: A complex-number fourier technique for lower bounds on the mod-m degree. Computational Complexity 9(1), 16–38 (2000)
7. Hansen, K.A.: Computing symmetric boolean functions by circuits with few exact threshold gates. In: Lin, G. (ed.) COCOON 2007. LNCS, vol. 4598, pp. 448–458. Springer, Heidelberg (2007)
8. Hansen, K.A.: Depth reduction for circuits with a single layer of modular counting gates. In: Frid, A., Morozov, A., Rybalchenko, A., Wagner, K.W. (eds.) Computer Science - Theory and Applications. LNCS, vol. 5675, pp. 117–128. Springer, Heidelberg (2009)
9. Hansen, K.A., Podolskii, V.V.: Exact threshold circuits. In: Proceedings of the 25th Annual IEEE Conference on Computational Complexity, pp. 270–279. IEEE Computer Society, Los Alamitos (2010)
10. Harkins, R.C., Hitchcock, J.M.: Dimension, halfspaces, and the density of hard sets. In: Lin, G. (ed.) COCOON 2007. LNCS, vol. 4598, pp. 129–139. Springer, Heidelberg (2007)
11. Håstad, J.: On the size of weights for threshold gates. SIAM Journal on Discrete Mathematics 7(3), 484–492 (1994)
12. Hertz, J., Krogh, A., Palmer, R.G.: Introduction to the Theory of Neural Computation. Addison-Wesley Publishing Company, Reading (1991)
13. Muroga, S., Toda, I., Takasu, S.: Theory of majority decision elements. Journal of the Franklin Institute 271, 376–418 (1961)
14. Muroga, S.: Threshold Logic and its Applications. John Wiley & Sons, Inc., Chichester (1971)
15. Parberry, I.: Circuit Complexity and Neural Networks. MIT Press, Cambridge (1994)
16. Podolskii, V.V.: A uniform lower bound on weights of perceptrons. In: Hirsch, E.A., Razborov, A.A., Semenov, A., Slissenko, A. (eds.) Computer Science – Theory and Applications. LNCS, vol. 5010, pp. 261–272. Springer, Heidelberg (2008)
17. Podolskii, V.V.: Perceptrons of large weight. Problems of Information Transmission 45(1), 46–53 (2009)
18. Smith, D.R.: Bounds on the number of threshold functions. IEEE Transactions on Electronic Computers EC 15(6), 368–369 (1966)
19. Yajima, S., Ibaraki, T.: A lower bound on the number of threshold functions. IEEE Transactions on Electronic Computers EC 14(6), 926–929 (1965)
20. Ziegler, G.M.: Lectures on 0/1-polytopes. In: Kalai, G., Ziegler, G.M. (eds.) Polytopes - Combinatorics and Computation, DMV Seminar, vol. 29, pp. 1–43. Birkhäuser, Basel (2000)

Proof Systems and Transformation Games

Yoram Bachrach[1], Michael Zuckerman[2], Michael Wooldridge[3], and Jeffrey S. Rosenschein[2]

[1] Microsoft Research, Cambridge, UK
[2] The Hebrew University of Jerusalem, Israel
[3] University of Liverpool, UK

Abstract. We introduce *Transformation Games* (TGs), a form of coalitional game in which players are endowed with sets of initial resources, and have capabilities allowing them to derive certain output resources, given certain input resources. The aim of a TG is to generate a particular target resource; players achieve this by forming a coalition capable of performing a sequence of transformations from its combined set of initial resources to the target resource. After presenting the TG model, and discussing its interpretation, we consider possible restrictions on the transformation chain, resulting in different coalitional games. After presenting the basic model, we consider the computational complexity of several problems in TGs, such as testing whether a coalition wins, checking if a player is a dummy or a veto player, computing the core of the game, computing power indices, and checking the effects of possible restrictions on the coalition. Finally, we consider extensions to the model in which transformations have associated costs.

1 Introduction

We consider a new model of cooperative activity among self-interested players. In a *Transformation Game* (TG), players must cooperate to generate a certain target resource. In order to generate the resource, each player is endowed with a certain set of initial resources, and in addition, each player is assumed to be capable of *transformations*, allowing it to generate a certain resource, given the availability of a certain input set of resources required for the transformation. Coalitions may thus form *transformation chains* to generate various resources. A coalition of players is successful if it manages to form a transformation chain that eventually generates the target resource. Forming such chains is typically complicated, as there are usually constraints on the structure of the chain. One example is time restrictions, in the form of deadlines. Even when there is no deadline, short chains are typically preferred, since we might expect that the more transformations a chain has, the higher the probability of some transformation failing.

We model restrictions on these chains, and consider game theoretic notions and the complexity of computing them under these different restrictions. We consider three types of domains: unrestricted domains, where there is no restriction on the chain; makespan domains, where each transformation requires a certain amount of time and the coalition must generate the target resource before a certain deadline; and limited transformation domains, where the coalition must generate the target resource without

P. Hliněný and A. Kučera (Eds.): MFCS 2010, LNCS 6281, pp. 78–89, 2010.

performing more than a certain number of transformations. We also consider two types of transformations: *simple* transformations, where a transformation simply allows building an output resource from *one* input resource, and *complex* transformations, where a transformation may require a *set* of input resources to generate a certain output resource.

TG can be viewed as a strategic, game-theoretic formulation of *proof systems*. In a formal proof system, the goal is to derive some logical statement from some logical premises by applying logical inference rules. When modelled as a TG, premises and proof rules are distributed across a collection of agents, and proof becomes a cooperative process, with different agents contributing their domain expertise (premises) and capabilities (proof rules). Game theoretic solution concepts such as the Banzhaf index provide a measure of the relevant significance of agents (and hence premises and proof rules) in the proof process. Viewed in this way, TGs provide a formal foundation for cooperative theorem proving systems such as those described in [8,10], as well as cooperative problem solving systems in general [12]. (We also believe that TGs can provide a first step towards providing a cooperative game-theoretic treatment of supply chains, although we do not explore this issue further within the present paper.)

2 Preliminaries

We briefly discuss basic game theoretic concepts that are later applied in the context of TGs (see, e.g., [13] for a detailed introduction). A *transferable utility coalitional game* is composed of a set I of n players and a characteristic function mapping any subset (coalition) of the players to a real value $v : 2^I \rightarrow \mathbb{R}$, indicating the total utility these players can obtain together. The coalition I of all the players is called the *grand coalition*. Often such games are *increasing*, i.e., for all coalitions $C' \subseteq C$ we have $v(C') \leq v(C)$. In *simple* games, v only gets values of 0 or 1 (i.e., $v : 2^I \rightarrow \{0,1\}$), and in this case we say $C \subseteq I$ *wins* if $v(C) = 1$ and *loses* otherwise. We say player i is a *critical* in a winning coalition C if the removal of i from that coalition would make it a losing coalition: $v(C) = 1$ and $v(C \setminus \{i\}) = 0$.

The characteristic function defines the value a coalition can obtain, but does not indicate how to distribute these gains to the players within the coalition. An *imputation* $(p_1, \ldots, p_n)$ is a division of the gains of the grand coalition among all players, where $p_i \in \mathbb{R}$, such that $\sum_{i=1}^{n} p_i = v(I)$. We call p_i the payoff of player a_i, and denote the payoff of a coalition C as $p(C) = \sum_{i \in \{i | a_i \in C\}} p_i$.

Game theory offers solution concepts, defining imputations that are likely to occur. A minimal requirement of an imputation is *individual-rationality* (IR): for every player $a_i \in C$, we have $p_i \geq v(\{a_i\})$. Extending IR to coalitions, we say a coalition B *blocks* the imputation $(p_1, \ldots, p_n)$ if $p(B) < v(B)$. If a blocked imputation is chosen, the grand coalition is *unstable*, since the blocking coalition can do better by working without the other players. The prominent solution concept focusing on stability is the core. The core of a game is the set of all imputations $(p_1, \ldots, p_n)$ that are not blocked by any coalition, so for any coalition C we have $p(C) \geq v(C)$.

In general, the core can contain multiple imputations, and can also be empty. Another solution, which defines a *unique* imputation, is the Shapley value.The Shapley value of a player depends on his marginal contribution over all possible coalition permutations.

We denote by π a permutation (ordering) of the players, so $\pi : \{1,\dots,n\} \to \{1,\dots,n\}$ and π is reversible, and by Π the set of all possible such permutations. Denote by $S_\pi(i)$ the predecessors of i in π, so $S_\pi(i) = \{j \mid \pi(j) < \pi(i)\}$. The Shapley value is given by the imputation $sh(v) = (sh_1(v),\dots,sh_n(v))$ where $sh_i(v) = \frac{1}{n!}\sum_{\pi\in\Pi}[v(S_\pi(i)\cup\{i\}) - v(S_\pi(i))]$.

An important application of the Shapley value is that of power indices, which try to measure a player's ability to change the outcome of a game, and are used for example to measure political power. Another game theoretic concept that is also used to measure power is the Banzhaf power index, which depends on the number of coalitions in which a player is critical, out of all the possible coalitions. The Banzhaf power index is given by $\beta(v) = (\beta_1(v),\dots,\beta_n(v))$ where $\beta_i(v) = \frac{1}{2^{n-1}}\sum_{S\subseteq I|a_i\in S}[v(S) - v(S\setminus\{i\})]$.

3 Transformation Games

Transformation games (TGs) involve a set of players, $I = \{a_1,\dots,a_n\}$, a set of resources $R = \{r_1,\dots,r_k\}$, and a certain goal resource $r_g \in R$. In these domains, each player a_i is endowed with a set of resources $R_i \subseteq R$. Players have capabilities that allow them to generate a target resource when they have certain input resources. We model these abilities via *transformations*. A transformation is a pair $\langle B, r\rangle$ where B is a subset $B \subseteq R$, indicating the resources required for the transformation, and $r \in R$ is the resource generated by the transformation. The set of all such possible transformations (over R) is D. The capabilities of each player a_i are given by a set $D_i \subseteq D$. We say a transformation $d = \langle B, r\rangle$ is *simple* if $|B| = 1$ (i.e., it generates a target resource given a *single* input resource), and *complex* if $|B| > 1$. Some caveats are worth highlighting:

First, our model of TGs has *no* notion of resource *quantity*. For example, the TG framework cannot explicitly express constraints such as 4 nails and 5 pieces of wood are required to build a table. Second, we do not model resource *consumption*: thus when a player generates a resource from base resources, the player ends up with *both* the base resources *and* the generated resource. This may at first sight seem a strange modeling choice, but it is very natural in many settings. For example, consider that derivations as corresponding to *logical proofs*. In classical logic proofs, when we derive a lemma ϕ from premises Δ, we do not "consume" Δ: both ϕ and premises Δ that were used to derive it can be used as often as required in the subsequent proof.

Formally, then, a TG Γ is a structure $\Gamma = \langle I, R, R_1,\dots,R_n, D_1,\dots,D_n, r_g\rangle$ where: I is a set of players; R is a set of resources; for each $a_i \in I$, R_i is the set of resources with which that player a_i is initially endowed; for each $a_i \in I$, $D_i \subseteq D$ is the set of transformations that player a_i can carry out; and $r_g \in R$ is a resource representing the goal of the game. We sometimes consider transformations that require a certain amount of time. In such settings, let a_i be a player with capability $d \in D_i$. We denote the time player a_i needs in order to perform the transformation as $t_i(d) \in \mathbb{N}$.

Given a TG, we can define the set of resources a coalition $C \subseteq I$ can derive. We say a coalition C is endowed with a resource r, and denote this as $has(C, r)$, if there exists a player $a_i \in C$ such that $r \in R_i$. We denote the set of resources a coalition is endowed with as $R_C = \{r \in R \mid has(C, r)\}$. We now define an infix relation $\Rightarrow\ \subseteq 2^I \times R$, with the intended interpretation that $C \Rightarrow r$ means that coalition C can produce resource r. We inductively define the relation $\Rightarrow$ as follows. We have $C \Rightarrow r$ iff either:

- $has(C,r)$ (i.e., the coalition C is directly endowed with resource r); or else
- for some $\{r_{b_1}, r_{b_2}, \dots, r_{b_m}\} \subseteq R$ we have $C \Rightarrow r_{b_1}, C \Rightarrow r_{b_2}, \dots, C \Rightarrow r_{b_m}$ and for some player $a_i \in C$ we have $\langle\{r_{b_1}, r_{b_2}, \dots, r_{b_m}\}, r\rangle \in D_i$.

Definition 1. *Unrestricted-TG: An unrestricted TG (UTG) with the goal resource r_g is the game where a coalition C wins if it can derive r_g and loses otherwise: $v(C) = 1$ if $C \Rightarrow r_g$ and $v(c) = 0$ otherwise.*

We now take into account the total number of transformations used to generate resources, and the time required to generate a resource. We denote the fact that a coalition C can generate a resource r, using at most k transformations, by $C \Rightarrow_k r$. Consider a sequence of resource subsets $S = \langle R_1, R_2, \dots R_k\rangle$, such that each R_i contains one additional resource over the previous R_{i-1} (so $R_i = R_{i-1} \cup \{r'_i\}$). We say C *allows* the sequence S if for any index i, C can generate r'_i (the additional item for the next resource subset in the sequence) given base resources in R_{i-1} (so C is capable of a transformation $d = \langle A, r'_i\rangle$, where $A \subseteq R_{i-1}$). A sequence $S = \langle R_1, R_2, \dots R_k\rangle$ (with k subsets) that C allows is called a $k-1$-*transformation sequence for resource r by coalition C* if $r \in R_k$ and the first subset in the sequence is the subset of resources the coalition C is endowed with, $R_1 = R_C$ (since C requires $k-1$ transformations to obtain r this way). If there exists such a sequence, we denote this by $C \Rightarrow_k r$. We denote the minimal number of transformations that C needs to derive r as $d(C,r) = \min\{b \mid C \Rightarrow_b r\}$, and if C cannot derive r we denote $d(C,r) = \infty$.

Definition 2. *DTG: A transformation restricted TG (DTG) with the goal resource r_g and with the transformation bound k is the game where a coalition C wins if it can derive r_g* using at most k transformations *and loses otherwise: $v(C) = 1$ if both $C \Rightarrow r_g$ and $d(C, r_g) \leq k$, and otherwise $v(c) = 0$.*

Similarly, we consider the makespan domain, where each transformation requires a certain amount of time. The main difference between the makespan domain and the DTG domain is that transformations may be done *simultaneously*.[1] We denote the fact that a coalition C can generate a resource r in time of at most t by $C \Rightarrow^t r$. We define the notion recursively. If a coalition is endowed with a resource, it can generate this resource instantaneously (with time limit of 0), i.e., if $has(C,r)$ then $C \Rightarrow^0 r$. Now consider a coalition C such that $C \Rightarrow^{t_1} r_{b_1}, C \Rightarrow^{t_2} r_{b_2}, \dots, C \Rightarrow^{t_m} r_{b_m}$, and player $a_i \in C$ who is capable of the transformation $d = \langle\{r_{b_1}, r_{b_2}, \dots, r_{b_m}\}, r\rangle$ (so $d \in D_i$), requiring a transformation time t, so $t_i(d) = t$. Given a coalition C, we denote the time in which a coalition can perform a transformation as $t_C(d) = \min_{a_i \in C} t_i(d)$, the minimal time in which the transformation can be performed, across all players in the coalition. We denote the time in which the coalition can obtain *all* of the base resources $r_{b_1}, \dots, r_{b_m}$ as $s = \max t_i$. The final transformation (which generates r) requires a time of t, so $C \Rightarrow^{s+t} r$. Again, different ways of obtaining the target resource result in different time bounds, and we consider the optimal way of obtaining the target resource (the *minimal* time a coalition C requires to derive r). If $C \Rightarrow r$ we denote the minimal transformation

[1] For example, if it takes 5 hours to convert oil to gasoline and 4 hours to convert oil to plastic, if we have oil we can obtain both gasoline and plastic in 5 hours, using parallel transformations.

time that C needs to derive r as $t(C,r) = \min\{b \mid C \Rightarrow^b r\}$, and if C cannot derive r we denote $t(C,r) = \infty$. Similarly to DTGs, we define makespan (time limited) TGs:

Definition 3. *TTG: A time limited TG (TTG) with goal resource r_g and time limit t is the game where a coalition C wins if it can derive r_g* with time of at most t *and loses otherwise: $v(C) = 1$ if both $C \Rightarrow r_g$ and $t(C, r_g) \leq t$, and otherwise $v(c) = 0$.*

3.1 Transformation Games and Logical Proofs

Structurally, TGs are similar to logical proof systems (see, e.g., [11, p. 48]). In a proof system in formal logic, we have a set of formulae of some logic, known as the *premises*, and a collection of *inference rules*, the role of which is to allow us to derive new formulae from existing formulae. Formally, if $\mathbb{L}$ is the set of formulae of the logic, then an inference rule ρ can be understood as a relation $\rho \subseteq 2^{\mathbb{L}} \times \mathbb{L}$. Given a set of premises $\Delta \subseteq \mathbb{L}$ and a set of inference rules $\rho_1 \ldots, \rho_k$, a *proof* is a finite sequence of formulae $\phi_1, \ldots, \phi_l$, such that for all i, $1 \leq i \leq l$, either $\phi_i \in \Delta$ (i.e., ϕ_i is a premise) or if there exists some subset $\Delta' \subseteq \{\phi_1, \ldots, \phi_{i-1}\}$ and some $\rho_j \in \{\rho_1, \ldots, \rho_k\}$ such that $(\Delta', \phi_i) \in \rho_j$ (i.e., ϕ_i can be derived from the formulae preceding ϕ_i by some inference rule). Typical notation is that $\Delta \vdash_{\rho_1,\ldots,\rho_k} \phi$ means that ϕ can be derived from premises Δ using rules $\rho_1, \ldots, \rho_k$. Such proofs can be modeled in our framework as follows. Resources R are logical formulae $\mathbb{L}$, and the initial allocation of resources $R_1, \ldots, R_n$ equates to the premises; capabilities $D_1, \ldots, D_n$ equate to inference rules. Notice that the assumption that resources are not "consumed" during the transformation process is very natural when considered in this setting: in classical logic proofs, premises and lemmas can be reused as often as required. Clearly the relationship between TGs and proofs is very natural: such formal proof systems can be directly modeled within our framework. There are two main differences, however, as follows.

First, in proof systems inference rules are usually given a succinct specification, as a "pattern" to be matched against premises. The classical proof rule modus ponens, for example, is usually specified as the following pattern: $\frac{\phi; \quad \phi \rightarrow \psi}{\psi}$, which says that if we have derived ϕ, and we have derived that $\phi \rightarrow \psi$, then we can derive ψ. Here, ϕ and ψ are variables, which can be instantiated with any formula. The second is that we take a *strategic* view: a proof modeled within our system is obtained through a cooperative process. TGs can be understood as a formulation both of cooperative theorem proving systems [8,10], as well as cooperative problem solving systems in general [12]. In such systems, agents have different areas of expertise (= resources) as well as different capabilities (= transformations). Game theoretic concepts such as the Banzhaf index provide a measure of how important different premises and inference rules are with respect to being able to prove a theorem.

4 Problems and Algorithms

Given a TG $\Gamma = \langle I, R, R_1, \ldots, R_n, D_1, \ldots, D_n, r_g \rangle$, the following are natural problems regarding the game. COALITION-VALUE (CV): given a coalition $C \subseteq I$, compute $v_\Gamma(C)$ (i.e., test whether a coalition is successful or not). VETO (VET): given a player a_i, check if it is a veto player, so for any winning coalition C, we have $a_i \in C$. DUMMY:

given a player a_i, check if it is a dummy player, so for any coalition C, we have $v_\Gamma(C \cup \{a_i\}) = v_\Gamma(C)$. CORE: compute the set of payoff vectors that are in the core, and return a representation of all payoff vectors in it. SHAPLEY: compute a_i's Shapley value $sh_i(v_\Gamma)$. BANZHAF: compute a_i's Banzhaf index $\beta_i(v_\Gamma)$.

We now summarize the results of the present paper, and prove them in the remainder of the paper. We provide polynomial algorithms for testing whether a coalition wins or loses (CV) for UTGs, DTGs, and TTGs with simple transformations, and for UTGs and TTGs with complex transformations, but show that the problem is NP-hard for DTGs with complex transformations. We provide polynomial algorithms for testing for veto players and computing the core in all domains where CV is computable in polynomial time, but show the problem is co-NP-hard in DTGs with complex transformations. We show that testing for dummy players and computing the Shapley value are co-NP-hard in all the TG domains defined, and provide a stronger result for the Banzhaf power index, showing that it is #P-hard in all these domains.[2] The following table summarizes our results regarding TGs with *simple* transformations.

Table 1. Complexity of TG problems. If the results differ for simple and complex transformations, the results for complex transformations are given in parentheses. Key: P = polynomial algorithm; co-NPC = co-NP-complete; co-NPH = co-NP-hard.

	UTG	DTG	TTG
CV	P	P (NPH)	P
VETO	P	P (co-NPH)	P
DUMMY	co-NPC	co-NPC (co-NPH)	co-NPC
CORE	P	P (co-NPH)	P
SHAPLEY	co-NPH	co-NPH	co-NPH
BANZHAF	#P-Hard	#P-Hard	#P-Hard

Theorem 1. *CV is in P, for all the following types of TGs with simple transformations: UTG, DTG, TTG. CV is in P for UTGs and TTGs with complex transformations.*

Proof. First consider UTG. Denote the set S of resources with which C is endowed, $S = \{r \mid has(C, r)\}$. Denote the set of transformations of the players in C as $D_C = \cup_{a_i \in C} D_i$. We say that a set of resources S matches a transformation $d = \langle B, r \rangle \in D$ if $B \subseteq S$. If S matches d then using the resources in S the coalition C can also produce r through transformation d. Consider a basic step of iterating through all transformations in D. When we find a transformation $d = \langle B, r \rangle$ that S matches, we add r to S. A test to see whether a transformation d matches S can be done in time at most $|R|^2$ (where R is the set of all resources), so the basic step takes at most $|D_C| \cdot |R|^2$ time. If after performing a basic step no transformation in D_C matches S, S holds all the resources that C can generate, and we stop performing basic steps. If S has changed during a basic step, at

[2] The complexity class #P expresses the hardness of problems that "count solutions". Informally NP deals with whether a solution to a combinatorial problem exists, while #P deals with calculating the *number* of solutions. Counting solutions generalizes the checking their existence, so we usually regard #P-hardness as a more negative result than NP-hardness.

least one resource is added to it. Thus, we perform at most $|R|$ basic steps to compute the set of all resources C can generate, so S can be computed in polynomial time. We can then check whether S contains r_g. We note that the suggested algorithm works for simple as well as complex transformations. Now consider TTGs with simple transformations. We build a directed graph representing the transformations as follows. For each resource r the graph has a vertex v_r, and for each transformation $d = \langle r_x, r_y \rangle$ the graph has an edge e_d from v_{r_x} to v_{r_y}. Given a coalition C we consider G_C, the subgraph induced by C. $G_C = \langle V, E_C \rangle$ contains only the edges of the transformations available to C, so $E_C = \{\langle v_{r_x}, v_{r_y} \rangle \mid \langle r_x, r_y \rangle \in D_C\}$. The graph G_C is weighted, and the weight of each edge $e = \langle r_x, r_y \rangle$ is $w(e) = \min_{a_i \subset C} t_i(\langle r_x, r_y \rangle)$, the minimal time to derive r_y from r_x across all players in the coalition. Denote the weight of the minimal path from r_a to r_g in G_C as $w_C(r_a, r_g)$. The coalition C is endowed with all the resources in R_C and can generate all of them instantly. The minimal time in which C can generate r_g is $\min_{r_a \in R_C} w_C(r_a, r_g)$. For each resource $r_a \in R_C$, we can compute $w_C(r_a, r_g)$ in polynomial time, so we can compute in polynomial time the minimal time in which C can generate r_g, and test whether this time exceeds the required deadline.For simple transformations, we can simulate a DTG domain as a TTG domain, by having each transformation require 1 time unit (and setting the threshold to be the threshold number of transformations[3]).

Finally, we show how to adapt the algorithm used for UTGs (with either simple or complex transformations) to be used for TTGs with complex transformations. For the TTG CV algorithm for a coalition C, for each resource r we maintain $m(r)$, a bound from above on the minimal time required to produce r. All the $m(r)$ of resources endowed by some player in the coalition C are initialized to 0, and the rest are initialized to ∞. Our basic step remains iterating through all the transformations in D. When we find a transformation $d = \langle B, r \rangle$ which S matches, where the transformation requires $t(d)$, we compute the time in which the transformation can be completed, $c(d) = max_{b \in B} m(b) + t_C(d)$ (if S does not match a transformation d, we denote $c(d) = \infty$). During each basic step, we compute the possible completion times for all the matching transformations, and apply the smallest one, $argmin_{d \in D} c(d)$. To apply a transformation $d = \langle B, r \rangle$, we simply add r to S, and update $m(r)$ to be $c(d)$. During each basic step we only apply *one* transformation (although we scan all the possible transformations). A simple induction shows that after each basic step, for any resource r such that $m(r) \neq \infty$ the value $m(r)$ is indeed the minimal time required to generate r. Again, the algorithm ends if no transformations were applied during a basic step. As before, a basic step requires time of $|D_C| \cdot |R|^2$ time, and we perform at most $|R|$ basic steps, so the algorithm requires polynomial time. We can then check whether S contains r_g, and whether $m(r_g)$ is smaller than the required time threshold.

Corollary 1. *VETO is in P, for all the following types of TGs with simple transformations: UTG, DTG, TTG, and for UTGs and TTGs with complex transformations.*

Proof. A veto player a_i is present in all winning coalitions: TGs are trivially seen to be increasing, so simply check whether $v(I_{-a_i}) = 0$.

[3] With complex transformations, this is no longer possible, since if a transformation requires several base resources, the shortest time to produce each of them may be different.

Now consider the problem of computing the core in TGs with simple transformations. In simple (0,1-valued) games, a well-known folk theorem tells us that the core of a game is non-empty iff the game has a veto player. Thus, in simple games, the core can be represented as a list of the veto players in the game. This gives the following:

Corollary 2. *CORE is in P, for all the following types of TGs with simple transformations: UTG, DTG, TTG, and for UTGs and TTGs with complex transformations.*

Theorem 2. *DUMMY is co-NP-complete, for all the following types of TGs with simple transformations: UTGs, DTGs, TTGs, and for UTGs and TTGs with complex transformations. For DTGs with complex transformations, DUMMY is co-NP-hard.*

Proof. Due to Theorem 1, we can verify in polynomial time whether a_i is beneficial to C by testing if $v(C \cup \{a_i\}) - v(C) > 0$. Thus DUMMY is in co-NP for UTGs, DTGs, TTGs with simple transformations, and for UTGs and TTGs with complex transformations. We reduce SAT to testing if a player in a UTG with simple transformations is not a dummy (TG-NON-DUMMY). Showing DUMMY is co-NP-hard in UTGs is enough to show it is co-NP-hard for DTGs and TTGs, since it is possible to set the threshold (of the maximal allowed transformations or allowed time) so high that the TG is effectively unrestricted. Hardness results also apply to complex transformations as well, since the restricted case of simple transformations is hard. Let the SAT instance be $\phi = c_1 \wedge c_2 \wedge \cdots \wedge c_m$ over propositions $x_1, \ldots, x_n$, where $c_i = l_{i_1} \vee \cdots \vee l_{i_k}$, where each such l_j is a positive or negative literal, either x_k or $\neg x_k$ for some proposition x_k. The TG-NON-DUMMY query is regarding the player a_y. For each literal (either x_i or $\neg x_i$) we construct a player (a_{x_i} and $a_{\neg x_i}$). These players are called the literal players. The generated TG game has a resource r_y, and only a_y is endowed with that resource. The game also has the resource r_z, with which all the literal players are endowed. For each proposition x_i we also have a resource r_{x_i}. For each clause c_j in the formula ϕ we have a resource r_{c_j}. The goal resource is the resource r_g. For each positive literal x_i we have transformation $d_{x_i} = \langle r_z, r_{x_i} \rangle$. For each negative literal we have transformation $d_{\neg x_i} = \langle r_{x_i}, r_g \rangle$. For each clause c_j we have transformation $d_{c_j} = \langle r_{c_j}, r_{c_{j+1}} \rangle$, where for the last clause c_m we have a transformation $d_{c_m} = \langle r_{c_m}, r_g \rangle$. Player a_y is only capable of $d_0 = \langle r_y, r_{c_1} \rangle$. Player a_{x_i} is capable of d_{x_i}, and player $a_{\neg x_i}$ is capable of $d_{\neg x_i}$. If x_i occurs in its positive form in c_j (i.e., $c_j = x_i \vee l_{i_2} \vee \cdots$) then a_{x_i} is capable of d_{c_j}. If x_i occurs in its negative form in c_j (i.e., $c_j = \neg x_i \vee l_{i_2} \vee \cdots$) then $a_{\neg x_i}$ is capable of the d_{c_j}.

We identify an assignment with a coalition, and identify a coalition with an assignment candidate (which possibly contains both a positive and a negative assignment to a variable, or which possibly does not assign anything to a variable). Let A be an assignment to the variables in ϕ. We denote the coalition that A represents as $C_A = \{a_{x_i} \mid A(x_i) = T\} \cup \{a_{\neg x_i} \mid A(x_i) = F\}$. There are only two resources with which players are endowed: r_y and r_z. It is possible to generate r_g either through a transformation chain starting with r_z, going through r_{x_i} (for some variable x_i) and ending with r_g, or through a transformation chain starting with r_y, going through r_{c_1}, through r_{c_2}, and so on, until r_{c_m}, and finally deriving r_g from r_{c_m} (no other chains generate r_g).

Given a valid assignment A, C_A does not allow converting r_z to r_g, since to do so C_A needs to be able to generate r_{x_i} from r_z (for some variable x_i) and needs to be able to generate r_g from r_{x_i}. However, the only player who can generate r_{x_i} from r_z is a_{x_i}, and the only player who can generate r_g from r_{x_i} is $a_{\neg x_i}$, and C_A can never contain both a_{x_i} and $a_{\neg x_i}$ (for any x_i) by definition of C_A. Suppose A is a satisfying assignment for ϕ. Let c_j be some clause in ϕ. A satisfies ϕ, so it satisfies c_j through at least one variable x_i. If x_i occurs positively in ϕ, $A(x_i) = T$ so $a_{x_i} \in C_A$, and if x_i occurs negatively in ϕ, $A(x_i) = F$ so $a_{\neg x_i} \in C_A$, so we have a player $a \in C$ capable of d_{c_j}. Thus, C_A can convert r_{c_1} to r_{c_2}, can convert r_{c_2} to r_{c_3}, and so on. Thus, given r_{c_1}, C_A can generate r_g. Player a_y is endowed with r_y, and can generate r_{c_1} from r_y, so $C_A \cup \{a_y\}$ wins. However, $a_y \notin C_A$, and C_A cannot generate r_{c_1}. Since A is a valid assignment, C_A cannot generate r_g through a chain starting with r_z, so C_A is a losing coalition. Thus, a_y is not a dummy, as $v(C_A \cup \{a_y\}) - v(C_A) = 1$. On the other hand, suppose a_y is not a dummy, and is beneficial to coalition C, so C is losing but $C \cup \{a_y\}$ is winning. Since C loses and cannot contain both a_{x_i} and $a_{\neg x_i}$ (for any x_i), as this would allow it to generate r_{x_i} from r_z and to generate r_g from r_{x_i} (and C would win without a_y). Consider the assignment A: if C contains a_{x_i} we set $A(x_i) = T$, and if C contains $a_{\neg x_i}$ we set $A(x_i) = F$ (if C contains neither a_{x_i} nor $a_{\neg x_i}$ we can set $A(x_i) = T$). Since $C \cup \{a_y\}$ wins, but cannot generate r_g through a chain starting with r_z, it must generate r_g through the chain starting with r_y and going through the r_{c_j}'s. Thus, for any clause c_j, C contains a player capable of transformation $d_{c_j} = \langle r_{c_j}, r_{c_{j+1}} \rangle$. That player can only be a_{x_i} or $a_{\neg x_i}$ for some proposition x_i. If that player is $a_{x_i} \in C$ then c_j has the literal x_j (in positive form) and $A(x_i) = T$, so A satisfies c_j, and if it is $a_{\neg x_i} \in C$ then c_j has the literal $\neg x_j$ (negative form) and $A(x_i) = F$, so again A satisfies c_j. Thus A satisfies all the clauses in ϕ.

Theorem 3. *For DTGs with complex transformations, CV is NP-hard even for TGs with a single player, and VETO is co-NP-hard.*

Proof. We reduce VERTEX COVER to DTG CV. We are given a graph $G = \langle V, E \rangle$ with $V = \{v_1, \ldots, v_n\}$, $E = \{e_1, \ldots, e_m\}$ such that e_i is from $v_{i,a}$ to $v_{i,b}$ and a target cover size of k. We construct the following DTG. We have a resource r_t and goal resource r_g, a resource r_{e_i} for each edge e_i, and a resource r_{v_i} for each vertex. We have a transformation from r_t to each vertex resource r_{v_i}. If e_i is from $v_{i,a}$ to $v_{i,b}$ we have two transformations: from $r_{v_{i,a}}$ to r_{e_i}, and from $r_{v_{i,b}}$ to r_{e_i}. We have a complex transformation from $\{r_{e_1}, \ldots, r_{e_m}\}$ to r_g. A single player has r_t and all the above transformations. The target maximal number of transformations for the DTG is $k + m + 1$. Now, $G = \langle V, E \rangle$ has a vertex cover of size k iff the player wins in the game so defined.

Corollary 3. *Testing whether the Shapley value or Banzhaf index of a player in TGs exceeds a certain threshold is co-NP-hard for all the following types of TGs: UTG, DTG, TTG, with simple or complex transformations.*

Proof. Theorem 2 shows DUMMY is co-NP-hard in these domains. However, the Shapley value or Banzhaf index of a player can only be 0 if the player is a dummy player. Thus, computing these indices in these domains (or the decision problem of testing whether they are greater than some value) is co-NP-hard.

Definition 4. *#SET-COVER (#SC): We are given a collection $C = \{S_1, \ldots, S_n\}$ of subsets. We denote $\cup_{S_i \in C} S_i = S$. A set cover is a subset $C' \subseteq C$ such that $\cup_{S_i \in C'} = S$. We* *are asked to compute the number of covers of S. #SC is a #P-hard problem. Counting the number of vertex covers, #VERTEX-COVER, is a restricted form of #SC.*[4]

Theorem 4. *Computing the Banzhaf index in UTGs, DTGs, and TTGs (with simple or complex transformations) is #P-hard.*

Proof. We reduce a #SC instance to checking the Banzhaf index in a UTG. Consider the #SC instance with $C = \{S_1, \ldots, S_n\}$, so that $\cup_{S_i \in C} S_i = S$. Denote the items in S as $S = \{t_1, t_2, \ldots, t_k\}$. Denote the items in S_i as $S_i = \{t_{(S_i,1)}, t_{(S_i,2)}, \ldots, t_{(S_i,k_i)}\}$. For each subset S_i of the #SC instance, the reduced UTG has a player a_{S_i}. For each item $t_i \in S$ the UTG instance has a resource r_{t_i}. The reduced instance also has a player a_{pow}, the resources r_0, r_{pow} and the goal resource r_g. For each item $t_i \in S$ there is a transformation $d_i = \langle \{r_{t_{i-1}}\}, r_{t_i} \rangle$. Another transformation is $d_{pow} = \langle \{r_{t_n}\}, r_g \rangle$, of which only a_{pow} is capable. All players have resource r_0. Each player is capable of the transformation in her subset—for the subset $S_i = \{t_{i_1}, t_{i_2}, \ldots, t_{i_k}\}$, the player a_i is capable of $d_{i_1}, d_{i_2}, \ldots, d_{i_k}$. The query regarding the power index is for player a_{pow}. Note that a coalition $C = \{a_{i_1}, a_{i_2}, \ldots, a_{i_k}\}$ wins iff it contains both a_{pow} and players who are capable of all $d_1, d_2, \ldots, d_n$. However, to be capable of d_i the coalition must contain some a_j such that $t_i \in S_j$. Consider a winning coalition $C = \{a_{pow}\} \cup \{a_{i_1}, a_{i_2}, \ldots, a_{i_k}\}$, and denote $S_C = \{S_{i_1}, S_{i_2}, \ldots, S_{i_k}\}$. A coalition C wins iff $a_{pow} \in C$ and S_C is a set cover of S. The Banzhaf index in the reduced game is $\frac{q}{2^{n-1}}$, where n is the number of players and q is the number of winning coalitions that contain a_{pow} that lose when a_{pow} is removed from the coalition. No coalition can win without a_{pow}, so q is the number of all winning coalitions, which is the number of set covers of the #SC instance. Thus we reduced #SC to BANZHAF in a UTG with simple transformations (a restricted case of complex transformations). We can do the same with DTGs and TTGs with a high enough threshold. Thus, BANZHAF is #P-hard in all considered TG domains.

5 TGs with Costs

In many domains, transformations have costs. Suppose we wish to derive a resource r_g from base resources R, and can do this either using a powerful but expensive computer or using a slower but cheaper one. Such tradeoffs are ubiquitous in real-world problem-solving. We model TGs with costs as follows. Every transformation t has cost $c(t) \in \mathbb{R}^+$. Given a coalition C and a resource r, we denote by $h(C, r)$ the minimum cost needed to obtain r from R_C, which is the sum of transformation costs in the minimal sequence of transformations from R_C to r. If r cannot be obtained from R_C, we set $h(C, r) = \infty$. The goal resource r_g has the value $v(r_g) \in \mathbb{R}^+$.

Definition 5. *CTG: A TG with costs (CTG) with the goal resource r_g and the cost function $c : D \rightarrow \mathbb{R}^+$ is the game where the value of a coalition C is the value of the goal resource r_g minus the minimum cost needed to obtain r_g from R_C—if this latter difference is positive, and 0 otherwise. Thus, $v(C) = \max(0, v(r_g) - h(C, r_g))$.*

[4] [5,3] consider a related domain (Coalitional Skill Games and Connectivity Games), and also use #SC to show that computing the Banzhaf index in that domain is #P-complete.

Algorithm 1 computes coalition values in a CTG. We define for every resource $r \in R$ a vertex in a hypergraph, v_r. We identify with every transformation $t = \langle\{r_1, \ldots, r_l\}, r\rangle$ an hyperedge $e_t = \langle\{v_{r_1}, \ldots, v_{r_l}\}, v_r\rangle$. We denote: R – resources, C – coalition, r_g – target resource, D_C – C's transformations. Subprocedure Total-Cost computes the transformations in the path from R_C to r, summing their costs to get the total path cost.

Algorithm 1. Compute Coalitional Value

Procedure Compute-Coalitional-Value (R, C, r_g, D_C):

1. For all $r \in R_C$ do $\lambda(v_r) \leftarrow 0$
2. For all $r \in R \setminus R_C$ do $\lambda(v_r) \leftarrow \infty$
3. For all $r \in R$ do $S(v_r) \leftarrow \emptyset$
4. $T \leftarrow D_C$ (T initially contains all the transformations coalition C has)
5. while $T \neq \emptyset$:
   (a) $t = \langle\{r_1, \ldots, r_l\}, r\rangle \leftarrow \arg\min_{t \in T} Total - Cost(t).first)$
   (b) $tc \leftarrow Total - Cost(t).first$, $S \leftarrow Total - Cost(t).second$
   (c) if $tc == \infty$ then (remaining transformations unreachable from R_C)
      i. return $\max(0, v(r_g) - \lambda(v_{r_g})$
   (d) if $tc < \lambda(v_r)$ then $\lambda(v_r) \leftarrow tc$, $S(v_r) \leftarrow S$
   (e) $t \leftarrow T \setminus \{t\}$
6. return $\max(0, v(r_g) - \lambda(v_{r_g}))$

Procedure Total-Cost $(t = \langle\{r_1, \ldots, r_l\}, r\rangle$

1. if $\sum_{i=1}^{l} \lambda(v_{r_i}) == \infty$ then return $pair(\infty, \emptyset)$
2. $S \leftarrow \bigcup_{i=1}^{l} S(v_{r_i}) \cup \{t\}$
3. $tc \leftarrow \sum_{t_i \in S} c(t_i)$
4. return $pair(tc, S)$

Theorem 5. *Algorithm 1 calculates the coalitional value of a coalition C in a CTG. The proof is omitted for lack of space.*

Proposition 1. *The DUMMY problem is co-NP-Complete for CTG. SH is co-NP-Hard, and BZ is #P-Hard for CTG.*

Proof. DUMMY $\in$ co-NP for CTG, since given a coalition C and a player a_i, due to Theorem 5, it is easy to test whether $v(C) < v(C \cup \{a_i\})$ (i.e., that a_i is not a dummy player). UTG is a private case of CTG (set for all the transformations t, $c(t) = 0$, and set $v(r_g) = 1$). And so all the hardness results for UTG hold for CTG as well.

6 Related Work and Conclusions

This work is somewhat reminiscent of previous work on multi-agent supply chains. Although some attention was given to auctions or procurement in such domains, (for example for forming supply chains [1] or procurement tasks [6]), previous work gave little attention to coalitional aspects. One exception is [14], which studies stability in supply chains, but focuses on pair coalitions and situations without side payments.

Previous research considered bounded resources through threshold games, in which a coalition wins if the sum of their combined resources or maximal flow exceed a stated threshold [7,9,4]. In one sense such games are simpler than TGs, as they consider a single resource; in another sense they are richer, as different quantities of resource are considered. Coalitional Resource Games (CRGs) [15] are also related to our work. In CRGs, players seek to achieve individual goals, and cooperate in order to pool scarce resources in order to achieve mutually satisfying sets of goals. The main differences are that in CRGs, players have *individual* goals to achieve, which require different quantities of resources; in addition, CRGs do not consider anything like transformation chains to achieve goals. It would be interesting to combine the models presented in this paper with those of [15]. TGs can also be considered as descended from Coalitional Skill Games [3] or related to connectivity and flow games [5,4]; the main difference is that this previous work does not consider transformation chains.

Finally, despite our hardness results, power indices can be tractably *approximated* [2] and used to determine the criticality of facts and rules in collaborative inference.

References

1. Babaioff, M., Walsh, W.E.: Incentive-compatible, budget-balanced, yet highly efficient auctions for supply chain formation. Decision Support Systems (2005)
2. Bachrach, Y., Markakis, E., Resnick, E., Procaccia, A.D., Rosenschein, J.S., Saberi, A.: Approximating power indices: theoretical and empirical analysis. The Journal of Autonomous Agents and Multi-Agent Systems 20(2), 105–122 (2010)
3. Bachrach, Y., Rosenschein, J.S.: Coalitional skill games. In: AAMAS 2008 (2008)
4. Bachrach, Y., Rosenschein, J.S.: Power in threshold network flow games. The Journal of Autonomous Agents and Multi-Agent Systems 18(1), 106–132 (2009)
5. Bachrach, Y., Rosenschein, J.S., Porat, E.: Power and stability in connectivity games. In: AAMAS 2008 (2008)
6. Chen, R., Roundy, R., Zhang, R., Janakiraman, G.: Efficient auction mechanisms for supply chain procurement. Manage. Sci. 51(3), 467–482 (2005)
7. Deng, X., Papadimitriou, C.H.: On the complexity of cooperative solution concepts. Mathematics of Operations Research 19(2), 257–266 (1994)
8. Denzinger, J., Kronenburg, M.: Planning for distributed theorem proving. In: Görz, G., Hölldobler, S. (eds.) KI 1996. LNCS (LNAI), vol. 1137, pp. 43–56. Springer, Heidelberg (1996)
9. Elkind, E., Goldberg, L., Goldberg, P., Wooldridge, M.: Computational complexity of weighted threshold games. In: AAAI 2007 (2007)
10. Fisher, M., Wooldridge, M.: Distributed problem-solving as concurrent theorem proving. In: Boman, M., Van de Velde, W. (eds.) MAAMAW 1997. LNCS, vol. 1237. Springer, Heidelberg (1997)
11. Genesereth, M.R., Nilsson, N.: Logical Foundations of Artificial Intelligence. Morgan Kaufmann Publishers, San Mateo (1987)
12. Lenat, D.B.: BEINGS: Knowledge as interacting experts. In: IJCAI 1975, pp. 126–133 (1975)
13. Osborne, M.J., Rubinstein, A.: A Course in Game Theory. MIT Press, Cambridge (1994)
14. Ostrovsky, M.: Stability in supply chain networks. American Economic Review 98(3), 897–923 (2008)
15. Wooldridge, M., Dunne, P.E.: On the computational complexity of coalitional resource games. Artificial Intelligence 170(10), 853–871 (2006)

Scheduling Real-Time Mixed-Criticality Jobs

Sanjoy K. Baruah[1,*], Vincenzo Bonifaci[2], Gianlorenzo D'Angelo[3], Haohan Li[1], Alberto Marchetti-Spaccamela[4], Nicole Megow[2], and Leen Stougie[5]

[1] The University of North Carolina, Chapel Hill, NC, USA
{baruah,lihaohan}@cs.unc.edu
[2] Max-Planck-Institut für Informatik, Saarbrücken, Germany
{bonifaci,nmegow}@mpi-inf.mpg.de
[3] Università degli Studi dell'Aquila, Montoluco di Roio, Italy
gianlorenzo.dangelo@univaq.it
[4] Sapienza University of Rome, Rome, Italy
alberto@dis.uniroma1.it
[5] Vrije Universiteit, and CWI, Amsterdam, The Netherlands
lstougie@feweb.vu.nl

Abstract. Many safety-critical embedded systems are subject to certification requirements; some systems may be required to meet multiple sets of certification requirements, from different certification authorities. Certification requirements in such "mixed-criticality" systems give rise to interesting scheduling problems, that cannot be satisfactorily addressed using techniques from conventional scheduling theory. In this paper, we study a formal model for representing such mixed-criticality workloads. We demonstrate first the intractability of determining whether a system specified in this model can be scheduled to meet all its certification requirements, even for systems subject to two sets of certification requirements. Then we quantify, via the metric of processor speedup factor, the effectiveness of two techniques, reservation-based scheduling and priority-based scheduling, that are widely used in scheduling such mixed-criticality systems, showing that the latter of the two is superior to the former. We also show that the speedup factors are tight for these two techniques.

1 Introduction

Due to cost and increased chip computational power, there is an increasing trend in embedded systems towards implementing multiple functionalities upon a single shared computing platform. It is typically the case that not all these functionalities are equally critical for the overall successful performance of the system. The analysis of such *mixed criticality* systems has been identified as one of the core foundational focal areas in the emerging discipline of Cyber Physical Systems. Coming up with procedures that will allow for the cost-effective certification of such mixed-criticality systems has been recognized as a unique,

* Supported in part by AT&T, IBM, and Sun Corps.; NSF grants CNS 0834270 and CNS 0834132; ARO grant W911NF-09-1-0535; and AFOSR grant FA9550-09-1-0549.

P. Hliněný and A. Kučera (Eds.): MFCS 2010, LNCS 6281, pp. 90–101, 2010.

particularly challenging, collection of problems [3]. Recognizing these challenges, several US government R&D organizations including the Air Force Research Laboratory, the National Science Foundation, the National Security Agency, the National Aeronautics and Space Administration, etc., have led initiatives such as the Mixed Criticality Architecture Requirements (MCAR) program aimed at streamlining the certification process for safety-critical embedded systems; these initiatives have brought together participants from industry, academia, and standards bodies to seek out more advanced, efficient, and cost-effective certification processes. Within this setting, new interesting scheduling problems arise that will be the focus of this paper.

We illustrate this by an example from the domain of unmanned aerial vehicles (UAV's), used for defense reconnaissance and surveillance. The functionalities on board such UAV's are classified into two levels of criticality:

- Level 1: the *mission-critical* functionalities, concerning reconnaissance and surveillance objectives, like capturing images from the ground, transmitting these images to the base station, etc.
- Level 2: the *flight-critical* functionalities: to be performed by the aircraft to ensure its safe operation.

For permission to operate such UAV's over civilian airspace (e.g., for border surveillance), it is mandatory that its flight-critical functionalities be certified by civilian Certification Authorities (CA's), which tend to be very conservative concerning the safety requirements. These CA's are not interested in the mission-critical functionalities, which must be validated by the clients and the vendor-manufacturer. The latter are also interested in the level 2 functionalities, but typically to standards that are less rigorous than the ones used by the civilian CA's. As such, we may consider the level 2 functionalities as a subset of the level 1 functionalities.

The difference in certification requirements is expressed by different *Worst-Case Execution Times* (WCET) for the execution of any real-time code depending on the considered critical level. In fact, each CA has its own rules, tools, etc., for determining the value of the WCET. With reference to the previous example, the WCET of the same piece of code of a flight critical functionality has two values: one lower value that express the WCET if we are considering all mission critical functions and one higher value if we restrict to all flight-critical functions. On the other side, a level 1 functionality has only one WCET.

We refer to [5] for further explications and motivations for modeling this certification requirement process. We restrict here to giving an example.

Example 1. Consider a system comprised of two jobs: J_1 is flight-critical while J_2 is only mission-critical. Both jobs arrive at time-instant 0, and have their deadlines at time-instant 10. J_1 is characterized by two WCETs: at level 1 its WCET is $P_1(1)$ and at level 2 its WCET is $P_1(2)$ (where $P_1(1) \leq P_1(2)$); J_2 is characterized by only one WCET $P_2(1)$.

Suppose that $P_1(1) = 3$, $P_1(2) = 5$ and $P_2(1) = 6$. Consider the schedule that first executes J_1 and then J_2.

- The CA responsible for safety-critical certification would determine that J_1 completes latest by time-instant 5 and meets its deadline; note that if the execution time of J_1 is 5 then in the worst case it is not possible to complete J_2 by its deadline; however, this CA is not interested in J_2; hence the system passes certification.
- The CA responsible for mission-critical certification determines that J_1 completes latest by time-instant 3, and J_2 by time-instant 9. Thus they both complete by their deadlines, and the system passes certification.

Note that by scheduling first J_2 and then J_1 we do not meet the requirements of the flight critical functionalities. In fact in this case the execution of J_1 could start at time 6 and therefore the job does not complete by its deadline if we assume its WCET of 5.

We thus see that the system is certified as being correct by both the flight-critical and the mission-critical CA's, despite the fact that the sum of the WCET's at their own criticality level (6 and 5) exceeds the length of the scheduling window over which they are to execute.

On the other side, suppose the deadline of J_2 would change to 8, then neither scheduling J_1 before J_2 nor scheduling J_2 before J_1 can be certified. In this case, scheduling J_1 before J_2 can result in a completion time of J_2 at time 9 greater than J_2's deadline. □

In Section 2, we present the model for representing mixed-criticality real-time systems, which has been proposed in [4,5]. This mixed-criticality (MC) scheduling model extends the conventional model of a real-time job by allowing for the specification of different WCET's for a job at different criticality levels.

In previous papers [4,5], the problem to decide schedulability of a given MC system was conjectured to be NP-hard, but a proof was never given. We do so here in Section 3. However, the exact complexity of the problem remains open, since it is not clear if the problem is actually in NP. We prove that it is, if the number of criticality levels is a constant. Otherwise, we can only show that it is in PSPACE.

In the same section we present an algorithm that decides MC-schedulability efficiently for a special case.

In Section 4 we study the two techniques that are most widely used in designing mixed-criticality systems for certifiability; we quantify the sub-optimality of both techniques via the metric of *processor speedup factor* (cf. resource augmentation in performance analysis of approximation algorithms, as initiated in [7]). The results here extend the results in [5], who considered the techniques for dual-criticality systems, i.e., in which there are only two different criticality levels. Our results improve the results in [4], where also the techniques for L criticality levels are studied. Moreover, we prove here that our results are tight.

2 Model and Definitions

We consider a mixed-criticality (MC) system with L criticality levels, for some L. A *job* in an MC system is characterized by a 4-tuple of parameters: $J_j = (r_j, d_j, \chi_j, P_j)$, where

- $r_j \in \mathbb{Q}_+$ is the release time;
- $d_j \in \mathbb{Q}_+$ is the deadline, $d_j \geq r_j$;
- $\chi_j \in \mathbb{N}_+$ is the criticality of the job;
- $P_j \in \mathbb{Q}_+^L$ is a vector, the ℓ-th coordinate of which specifies the worst-case execution time (WCET) estimate of job J_j at criticality level ℓ. In a job-specification we usually represent it by $(P_j(1), \ldots, P_j(L))$.

It is natural to assume $P_j(\ell)$ to be monotonically nondecreasing for increasing ℓ. This we will do throughout, and mention if the assumption can be dropped where possible. At any moment, we call a job *available* if its release time has passed and the job has not yet completed execution.

An instance I of the MC-schedulability problem consists of a set of n jobs. In this paper we assume that there is only one machine (processor) to execute the jobs. We allow jobs to be *preempted* by the machine.

To define MC-schedulability we define the notion of a *scenario*. Each job J_j requires an amount of execution time p_j within its *time window* $[r_j, d_j]$. The value of p_j is not known from the specification of J_j, but is only discovered by actually executing the job until it *signals* that it has completed execution. This characterizes the uncertainty of the problem. We call a collection of realized values $(p_1, p_2, \ldots, p_n)$ a *scenario* of instance I.

We define the *criticality level*, or shortly criticality, of a scenario $(p_1, p_2, \ldots, p_n)$ of I as the smallest integer ℓ such that $p_j \leq P_j(\ell)$ for all $j = 1, \ldots, n$. (If there is no such ℓ, we define that scenario to be *erroneous*.)

Definition 1. *A schedule for a scenario* $(p_1, \ldots, p_n)$ *of criticality* ℓ *is* feasible *if every job* J_j *with* $\chi_j \geq \ell$ *receives execution time* p_j *during its time window* $[r_j, d_j]$.

A *clairvoyant* scheduling policy knows the scenario of I, i.e., $(p_1, \ldots, p_n)$, prior to determining a schedule for I.

Definition 2. *An instance* I *is* clairvoyantly-schedulable *if for each scenario of* I *there exists a feasible schedule.*

By contrast, an *on-line* scheduling policy discovers the value of p_j only by executing J_j until it signals completion. In particular, the criticality level of the scenario becomes known only by executing jobs. At each time instant, scheduling decisions can be based only on the partial information revealed thus far.

Definition 3. *An on-line scheduling policy is* correct *for instance* I *if for any non-erroneous scenario of instance* I *the policy generates a feasible schedule.*

Definition 4. *An instance* I *is* MC-schedulable *if it admits a correct on-line scheduling policy.*

The MC-SCHEDULABILITY problem is to determine whether a given instance I is MC-schedulable or not. A little thought should make it clear that for deciding MC-schedulability one only needs to consider scenarios in which for each i, $p_i = P_i(\ell)$ for some ℓ. The following is obvious.

Proposition 1. *If an instance I is MC-schedulable on a given processor, then I is clairvoyantly-schedulable on the same processor.*

Example 2. Consider an instance of a *dual-criticality* system: a system with $L = 2$. Consider an instance I comprised of 4 jobs. Job J_2 has criticality level 1 (which is the lower criticality level), and the other 3 jobs have the higher criticality level 2.

$$\begin{aligned} J_1 &= (0, 3, 2, (1, 2)) \\ J_2 &= (0, 3, 1, (2, 2)) \\ J_3 &= (0, 5, 2, (1, 1)) \\ J_4 &= (3, 5, 2, (1, 2)) \end{aligned}$$

For this example instance, any scenario in which p_1, p_2, p_3, and p_4 are no larger than 1, 2, 1, and 1, respectively, has criticality 1; while any scenario not of criticality 1 in which p_1, p_2, p_3, and p_4 are no larger than 2, 2, 1, and 2, respectively, has criticality 2. All remaining scenarios are, by definition, erroneous. It is easy to verify that this instance is clairvoyantly-schedulable.

Policy S_0, described below, is an example of an on-line scheduling policy for instance I:

S_0: Execute J_1 over [0,1]. If J_1 has remaining execution (i.e., p_1 is revealed to be greater than 1), then continue with scheduling policy S_1 below; else, continue with executing scheduling policy S_2 below.
S_1: Execute J_1 over (1,2], J_3 over (2,3], and J_4 over (3,5].
S_2: Execute J_2 over (1,3], J_3 over (3,4], and J_4 over (4,5].

Scheduling policy S_0 is however not correct for I, as can be seen by considering the schedule that is generated on the scenario $(1, 2, 1, 2)$. This particular scenario has criticality 2, since $p_4 = 2 > P_4(1) = 1$. Hence, a correct schedule would need to complete jobs J_1, J_3 and J_4 by their deadlines. However, the schedule generated by S_0 has executed J_4 for only one unit before its deadline. In fact, it turns out that instance I is not MC-schedulable.

3 Complexity of MC-Schedulability

In this section we investigate the complexity of the MC-SCHEDULABILITY problem. We show that it is NP-hard in the strong sense. However, a little thought should make it clear that it is not trivial to decide if the problem belongs to NP or not. We prove that it actually belongs to NP if the number of criticality levels is bounded by a fixed constant. For the general case, in which the number of criticality levels is part of the input, we show that it belongs to the class PSPACE, leaving membership to NP as an open question.

A preliminary observation is that determining clairvoyant-schedulability has the same complexity as the ordinary scheduling problem with only 1 criticality level: verify for each criticality level $\ell = 1, \ldots, L$ if the jobs of that criticality level or higher can be scheduled to complete before their deadlines if each such job j has execution time $P_j(\ell)$. In particular this means that clairvoyant-schedulability of any instance on a fully preemptive processor platform can be verified in polynomial time. This also holds if $P_j(\ell)$ is not monotonic in ℓ.

We show that it is strongly NP-hard to determine whether a given clairvoyantly-schedulable system is also MC-schedulable upon a fully preemptive single-processor platform.

Theorem 1. MC-SCHEDULABILITY *is* NP-*hard in the strong sense, even when all release times are identical and there are only two criticality levels.*

Proof. The proof is by reduction from the strongly NP-complete problem 3-PARTITION [6]. In an instance I_{3P} of 3-PARTITION, we are given a set S of $3m$ positive integers $s_0, s_1, \ldots, s_{3m-1}$ and a positive integer B such that $B/4 < s_i < B/2$ for each i and $\sum_{i=0}^{3m-1} s_i = mB$. The problem is to decide whether S can be partitioned into m disjoint sets $S_0, S_1, \ldots, S_{m-1}$ such that, for $0 \leq k < m$, $\sum_{s_i \in S_k} s_i = B$.

We give here just the polynomial transformation and defer the rest of the proof to a full version of the paper. From a given instance I_{3P} we construct an MC-SCHEDU-LABILITY instance I_{MC} consisting of $4m$ jobs with release time 0, which in the 4-tuple notation are:

- *3P-jobs:* For each i, $0 \leq i < 3m$, job $J_i = (0, 2mB, 2, (s_i, 2s_i))$;
- *Blocking jobs:* For each k, $0 \leq k < m$, job $J_{3m+k} = (0, 2(k+1)B, 1, (B, B))$.

□

The question remains if MC-SCHEDULABILITY actually belongs to the complexity class NP. In case the number of criticality levels L is a constant, we answer this question affirmatively. The proof is based on a polynomial-time checkable characterization of an online scheduling policy.

Theorem 2. MC-SCHEDULABILITY *for L criticality levels is in* NP *for any fixed L, and in* PSPACE *when L is part of the input.*

Equal deadlines. Theorem 1 above shows that the problem is in general NP-hard even if release times are identical. On the other hand, we show here that the special case in which all jobs have equal deadlines ($d_j = D$, $j = 1, \ldots, n$) can be solved in polynomial time. We first derive a necessary condition for such an instance I to be MC-schedulable. Consider the criticality level ℓ scenario of I in which each job J_j needs exactly $p_j = P_j(\ell)$ execution time.

Necessary condition: If I is MC-schedulable then for each ℓ, a scheduling policy exists that allocates to each job J_j with $\chi_j \geq \ell$ at least $P_j(\ell)$ execution time within time window $[r_j, D]$, i.e., the *makespan* of the scenario is at most D.

This condition is easily checked: Let $I_\ell = \{J_j \in I \mid \chi_j \geq \ell\}$ and $|I_\ell| = n_\ell$. Let (after renumbering) $J_1, J_2, \ldots, J_{n_\ell}$ denote the jobs in I_ℓ in order of non-decreasing release times: $r_1 \leq r_2 \leq \ldots \leq r_{n_\ell}$. Clearly, the makespan of I_ℓ is given by

$$C_{\max}^{\ell} := \max_{j=1,\ldots,n_\ell} r_j + \sum_{i=j}^{n_\ell} P_j(\ell). \tag{1}$$

The necessary condition is then verified by checking if

$$\max_{\ell=1,\ldots,L} C_{\max}^{\ell} \leq D. \tag{2}$$

Consider the *criticality-monotonic* (CM) on-line scheduling policy, which schedules at each time instant an available job of highest criticality.

Theorem 3. *CM is correct for all for MC-schedulable instances in which all jobs have the same deadline.*

Proof. We prove this by showing that the necessary condition is also sufficient. Consider any scenario of I that has criticality level ℓ. In a CM-schedule, the scheduling of jobs of criticality ℓ or higher is not effected by the presence of lower-criticality jobs, since their execution is postponed as soon as jobs in I_ℓ become available. Hence, a CM-schedule can be thought of as a schedule that minimizes the makespan of the jobs in I_ℓ. By the necessary condition, this does not exceed the common deadline D. □

Notice that this theorem also holds when $P_j(\ell)$ is not monotonic in ℓ. Some other well-solved sub-problems will be presented in the full version of the paper.

4 MC-Schedulability Testing Using Resource Augmentation

Since MC-SCHEDULABILITY is intractable even for dual-criticality instances, we concentrate here on sufficient MC-schedulability conditions that can be verified in polynomial time. We study two such scheduling policies that yield such sufficient conditions and compare their strength under augmenting the speed of the machine or server. Taking the required speed to give a necessary condition for MC-schedulability as a measure of performance quality, the second policy we present outperforms the former one.

We make here the assumption that for each job J_j, $P_j(\ell) = P_j(\chi_j)$ for all $\ell \geq \chi_j$. That is, no job executes longer than the WCET at its own specified criticality. This is without loss of generality for any correct scheduling policy: any such policy will immediately interrupt (and no longer schedule) a job J_j if its execution time p_j exceeds $P_j(\chi_j)$, since this makes the scenario of higher criticality level than χ_j, and therefore the completion of J_j becomes irrelevant for the scenario.

As stated in Section 1, one straightforward approach is to map each MC job J_j into a "traditional" (i.e., non-MC) job with the same arrival time r_j and deadline d_j and processing time $p_j = P_j(\chi_j) = \max_\ell P_j(\ell)$ (by monotonicity), and determine whether the resulting collection of traditional jobs is schedulable using some preemptive single machine scheduling algorithm such as the *Earliest Deadline First* (EDF) rule. This test can clearly be done in polynomial time. We will refer to mixed-criticality instances that are MC-schedulable by this test as *worst-case reservations schedulable* (WCR-schedulable) instances. The speed-up factor in the following theorem has been proved in [4]. We complement it by proving tightness.

Theorem 4. *If an instance is WCR-schedulable on a processor, then it is MC-schedulable on the same processor. Conversely, if an instance I with L criticality levels is MC-schedulable on a given processor, then I is WCR-schedulable on a processor that is L times as fast, and this factor is tight.*

We now present another schedulability condition that can also be tested in polynomial time, but offers a performance guarantee (as measured by the processor speedup factor) that is superior to the performance guarantee offered by the WCR-approach.

In this algorithm we determine off-line, before knowing the actual execution times, a total ordering of the jobs in a priority list and for each scenario execute at each moment in time the available job with the highest priority.

The priority list is constructed recursively using the approach commonly referred to in the real-time scheduling literature as the "Audsley approach" [1,2]; it is also related to a technique introduced by Lawler [8]. First determine the lowest priority job: Job J_i has lowest priority if there is at least $P_i(\chi_i)$ time between its release time and its deadline available when every other job J_j is executed before J_i for $P_j(\chi_i)$ time units (the WCET of job J_j according to the criticality level of job i). The procedure is repeatedly applied to the set of jobs excluding the lowest priority job, until all jobs are ordered, or at some iteration a lowest priority job does not exist. If job J_i has higher priority than job J_j we write $J_i \triangleright J_j$.

Because the priority of a job is based only on its own criticality level, the instance I is called *Own Criticality Based Priority (OCBP)-schedulable* if we find a complete ordering of the jobs.

If at some recursion in the algorithm no lowest priority job exists, we say the instance is not OCBP-schedulable. We can simply argue that this does not mean that the instance is not MC-schedulable: Suppose that scheduling according to the fixed priority list J_1, J_2, J_3 with $\chi_2 = 1$ and $\chi_1 = \chi_3 = 2$, proves the instance to be schedulable. It may not be OCBP-schedulable since this does not take into account that J_2 does not need to be executed at all if J_1 receives execution time $p_1 > P_1(1)$.

Clearly, if a priority list exists, it can be determined in polynomial time.

It turns out that the OCBP-test is more powerful than the WCR-test according to the speedup criterion.

Theorem 5. *If an instance is OCBP-schedulable on a given processor, then it is MC-schedulable on the same processor. Conversely, if instance I with L criticality levels is MC-schedulable on a given processor, then I is OCBP-schedulable on a processor that is s_L times as fast, with s_L equal to the root of the equation $x^L = (1+x)^{L-1}$, and this factor is tight. It holds $s_L = \Theta(L/\ln L)$.*

Proof. We prove here the critical part of the claim: that a speedup of s_L is sufficient. The proof that OCBP-schedulability implies MC-schedulability has been given in [4].

Notice that $s_1 = 1$, and that (as one can verify using elementary calculus) $s_{L'} \geq s_L$ if $L' > L$. Let I be an instance with at most L criticality levels that is MC-schedulable on a speed-1 processor, but not OCBP-schedulable on a speed-s processor for some $s \geq s_L$, and amongst such instances let it be minimal with respect to L and the number of jobs. Suppose I has n jobs. Minimality of I implies that there is no time-instant t such $t \notin \cup_{j=1}^{n}[r_j, d_j]$, otherwise either the jobs with deadline before t or the jobs with release time after t would comprise a smaller instance with the same property.

Claim. Any job in I with the latest deadline must be of criticality L.

Proof. Suppose that a job J_i with $\chi_i = h < L$ has latest deadline. Create from I an instance I_h with level h by "truncating" all jobs with criticality level greater than h to their worst-case level-h scenarios:

$$J_j = (r_j, d_j, \chi_j, (P_j(1), \ldots, P_j(L))) \in I \rightarrow$$
$$J'_j = (r_j, d_j, \min(\chi_j, h), (P_j(1), \ldots, P_j(h))) \in I_h.$$

Clearly, I_h being a restricted instance of I, is MC-schedulable as well, and, by minimality of I, I_h is OCBP-schedulable on a speed-s_h processor.

That J_i has latest deadline in I but cannot be assigned lowest priority on a speed-s processor implies that the scenario with $p_j = P_j(h)$ cannot be feasibly scheduled on a speed-s processor; thus I_h is not clairvoyantly schedulable on a speed-s processor. But I_h not being clairvoyantly schedulable implies I_h not being OCBP-schedulable, and because $s \geq s_L \geq s_h$, this contradicts the OCBP-schedulability of I_h on a speed-s_h processor. □

For each $\ell \in \{1, \ldots, L\}$, let $d(\ell)$ denote the latest deadline of any criticality-ℓ job in I: $d(\ell) = \max_{J_j|\chi_j=\ell} d_j$. A *work-conserving* schedule on a processor is a schedule that never leaves the processor idle if there is a job available. Consider any such a work-conserving schedule on a unit-speed processor of all jobs in I of the scenario in which $p_j = P_j(\ell)$ for all j. We define Λ_ℓ as the set of time intervals on which the processor is idle before $d(\ell)$, and λ_ℓ as the total length of this set of intervals.

Claim. For each ℓ and each $J_j \in I$ with $\chi_j \leq \ell$ we have $[r_j, d_j] \cap \Lambda_\ell = \emptyset$.

Proof. Observe that since $s \geq s_L \geq 1$, all idle intervals of Λ_ℓ are also idle intervals in any work-conserving schedule of I on a speed-s processor. Hence, any job J_j

with $\chi_j \leq \ell$ with $[r_j, d_j] \cap \Lambda_\ell \neq \emptyset$ would meet its deadline in such a schedule if it were assigned lowest priority. Since I is assumed to be non-OCBP schedulable on a speed-s processor, this implies that $(I \setminus \{J_i\})$ is non-OCBP schedulable on a speed-s processor, contradicting the minimality of I. □

As a corollary, $\Lambda_L = \emptyset$ and $\lambda_L = 0$.

For each $h = 1, \ldots, L$ and $\ell = 1, \ldots, L$, let

$$c_h(\ell) = \sum_{J_j | \chi_j = h} P_j(\ell)$$

Notice that by assumption

$$\forall \ \ell \ \forall h \leq \ell : c_h(\ell) = c_h(h). \tag{3}$$

Since instance I is clairvoyantly schedulable on a unit-speed processor, clearly we must have

$$\forall \ \ell : c_\ell(\ell) \leq d(\ell) - \lambda_\ell. \tag{4}$$

But also, due to clairvoyant schedulability, the criticality-ℓ scenario, in which each job J_j with criticality $\geq \ell$ receives exactly $P_j(\ell)$ units of execution, completes by the latest deadline $d(L)$:

$$\forall \ell : \sum_{i=\ell}^{L} c_i(\ell) \leq d(L) - \lambda_\ell. \tag{5}$$

Instance I is not OCBP-schedulable on a speed-s processor, which translated in terms of the introduced notation is:

$$\forall \ell : \sum_{i=1}^{L} c_i(\ell) > s(d(\ell) - \lambda_\ell). \tag{6}$$

Hence, for each ℓ,

$$\begin{aligned}
s(d(\ell) - \lambda_\ell) &< \sum_{i=1}^{\ell-1} c_i(\ell) + \sum_{i=\ell}^{L} c_i(\ell) \\
&= \sum_{i=1}^{\ell-1} c_i(i) + \sum_{i=\ell}^{L} c_i(\ell) \quad \text{(by (3))} \\
&\leq \sum_{i=1}^{\ell-1} (d(i) - \lambda_i) + (d(L) - \lambda_\ell) \quad \text{(by (4) and (5))} \\
&\leq \sum_{i=1}^{\ell-1} (d(i) - \lambda_i) + d(L).
\end{aligned}$$

Therefore, for all $\ell = 1, \ldots, L$,

$$s < \frac{d(L) + \sum_{i=1}^{\ell-1}(d(i) - \lambda_i)}{d(\ell) - \lambda_\ell}$$

Using notation $\delta_\ell = d(\ell) - \lambda_\ell$ (hence $\delta_L = d(L)$ since $\lambda_L = 0$) this yields

$$s < \min_{\ell=1,\ldots,L} \frac{\delta_L + \sum_{i=1}^{\ell-1} \delta_i}{\delta_\ell} \tag{7}$$

The minimum is maximized if all L terms are equal. Let x be this maximum value. Then for all $\ell = 1, \ldots, L$,

$$x = \frac{\delta_L + \delta_1 + \delta_2 + \cdots + \delta_{\ell-1}}{\delta_\ell} = \frac{x\delta_{\ell-1} + \delta_{\ell-1}}{\delta_\ell} = \left(\frac{1+x}{\delta_\ell}\right)\delta_{\ell-1}.$$

Hence,

$$\delta_\ell = \left(\frac{1+x}{x}\right)\delta_{\ell-1} \; \forall \ell = 1, \ldots, L \quad \text{which implies} \quad \delta_L = \left(\frac{1+x}{x}\right)^{L-1} \delta_1 .$$

Since, in particular, $x = \frac{\delta_L}{\delta_1}$, we have

$$x = \left(\frac{1+x}{x}\right)^{L-1},$$

which concludes the proof. □

We note that for $L = 2$ in the above theorem, $s_2 = (1 + \sqrt{5})/2$, the golden ratio; thus the result is a true generalization of earlier results in [5]. In general, $s_L = \Theta(L/\ln L)$; hence, this priority-based scheduling approach asymptotically improves on the reservations-based approach by a factor of $\Theta(\ln L)$ from the perspective of processor speedup factors.

Notice that the proof of the speedup bound for OCBP-schedulability in Theorem 5 only uses the clairvoyant-schedulability of the instance, which is a weaker condition than MC-schedulability (recall Proposition 1). The following claim shows that it is not possible to get an improved test if the proof of its speedup bound is based on clairvoyant-schedulability alone.

Proposition 2. *There are dual-criticality instances that are clairvoyantly schedulable on a given processor, but that are not MC-schedulable on a processor that is less than $(1 + \sqrt{5})/2$ times as fast.*

Proof. Consider the following instance

- $J_1 = (0, 1, 1, (1, 1))$;
- $J_2 = (0, \sigma, 2, (\sigma - 1, \sigma))$.

This system is clairvoyantly-schedulable. To analyze its MC-schedulability, consider the possible policies on a higher speed-s processor. The first one starts with J_2 and runs it till $P_2(1) = (\sigma - 1)/s$, and if it signals completion, schedule J_1 which then finishes latest by $(\sigma - 1)/s + 1/s = \sigma/s$. This is feasible only if $\sigma/s \leq 1$, that is, $s \geq \sigma$. The other policy is simply to first schedule J_1 and then J_2, which may require a total execution time $1/s + \sigma/s$, which is feasible only if $(1 + \sigma)/s \leq \sigma$, that is, $s \geq (\sigma + 1)/\sigma$. Hence, if the processor has speed $s < \min\{\sigma, (\sigma+1)/\sigma\}$, neither of the possible scheduling policies is correct. Taking $\sigma = (\sigma + 1)/\sigma$, that is, $\sigma = (1 + \sqrt{5})/2$, implies $s \geq \sigma$. □

Nevertheless, it remains the question if a test other than OCBP can test MC-schedulability within a smaller speedup bound. We do not give a full answer to this question. However, we can rule out *fixed-priority policies*, that is, policies which execute the jobs in some ordering fixed before execution time. This ordering is not adapted during execution, except that we do not execute jobs of criticality level $i < h$ after a scenario was revealed to be a level-h scenario. Such a policy admits a simple representation as a sequence of jobs. The following result shows that OCBP is best possible among fixed-priority policies.

Theorem 6. *There exist MC-instances with L criticality levels that are MC-schedulable, but that are not Π-schedulable for any fixed priority policy Π on a processor that is s_L times as fast, with s_L being the root of the equation $x^L = (1 + x)^{L-1}$.*

References

1. Audsley, N.C.: Optimal priority assignment and feasibility of static priority tasks with arbitrary start times. Technical Report YCS 164, Department of Computer Science, University of York, England (1991)
2. Audsley, N.C.: Flexible Scheduling in Hard-Real-Time Systems. PhD thesis, Department of Computer Science, University of York (1993)
3. Barhorst, J., Belote, T., Binns, P., Hoffman, J., Paunicka, J., Sarathy, P., Stanfill, J.S.P., Stuart, D., Urzi, R.: White paper: A research agenda for mixed-criticality systems (April 2009), `http://www.cse.wustl.edu/~cdgill/CPSWEEK09_MCAR/`
4. Baruah, S., Li, H., Stougie, L.: Mixed-criticality scheduling: improved resource-augmentation results. In: Proc. of the 25th ISCA International Conference on Computers and their Applications (to appear, 2010)
5. Baruah, S., Li, H., Stougie, L.: Towards the design of certifiable mixed-criticality systems. In: Proc. of the 16th IEEE Real-Time Technology and Applications Symposium, Stockholm, Sweden (2010)
6. Garey, M., Johnson, D.: Computers and Intractability: a Guide to the Theory of NP-Completeness. W. H. Freeman and company, New York (1979)
7. Kalyanasundaram, B., Pruhs, K.: Speed is as powerful as clairvoyance. Journal of the ACM 47(4), 617–643 (2000)
8. Lawler, E.: Optimal sequencing of a single machine subject to precedence constraints. Management Science 19(5), 544–546 (1973)

A DEXPTIME-Complete Dolev-Yao Theory with Distributive Encryption*

A. Baskar[1,**], R. Ramanujam[2,***], and S.P. Suresh[1]

[1] Chennai Mathematical Institute, Chennai, India
{abaskar,spsuresh}@cmi.ac.in
[2] Institute of Mathematical Sciences Chennai, India
jam@imsc.res.in

Abstract. In the context of modelling cryptographic tools like blind signatures and homomorphic encryption, the Dolev-Yao model is typically extended with an operator over which encryption is distributive. We consider one such theory which lacks any obvious locality property and show that its derivability problem is hard: in fact, it is DEXPTIME-complete. The result holds also when blind pairing is associative. The lower bound contrasts with PTIME decidability for restricted theories of blind signatures, and the upper bound with non-elementary decidability for abelian group operators with distributive encryption.

1 Introduction

Dolev-Yao style term models [DY83] for cryptographic protocols (the so-called "symbolic models") use a term algebra containing operations like pairing, encryption, signatures, hash functions, and nonces to build terms that are sent as messages in the protocol. The adversary against a protocol is modeled as a powerful network, which is only restricted in the way in which messages may be derived from the ones sent by the "honest" principals. Since these models are used for algorithmic analysis, the following *term derivability problem* is of basic interest: given a finite set of terms X and a term t, is there a way for the adversary to derive t from X?

In this paper, we study a security problem for a set of cryptographic primitives in an extension of the Dolev-Yao model which includes a **blind pairing** that commutes over encryption. That is, we can "push" an encryption by key k inside $[t, t']$ and get $[\{t\}_k, \{t'\}_k]$. We can also form a blind pair $[t, t']$ from t and t', and extract t' or t from $[t, t']$, provided we have the other part of the blind pair. We show that the existence of a passive attack (that is, by an attacker who cannot forge messages) is decidable in exponential time.

Though the blind pairing constructor finds natural use in the Dolev-Yao modelling of electronic voting protocols [FOO92], more restricted uses of blind pairing may well

* We thank the anonymous referees for many helpful comments that helped improve the presentation. Please see http://www.cmi.ac.in/~spsuresh/pdffiles/mfcs2010-tr.pdf for a technical report version with detailed proofs.

** Supported by the Council of Scientific and Industrial Research (CSIR), India.

*** I thank NIAS (http://www.nias.knaw.nl) for a Lorentz Fellowship from February to June 2010.

P. Hliněný and A. Kučera (Eds.): MFCS 2010, LNCS 6281, pp. 102–113, 2010.

suffice in many applications. What then can be interesting about such a result, in a framework with a fixed set of primitives, a weak attacker model and offering an algorithm with such high complexity? Perhaps the fact that the algorithm is presented as an automaton construction; but then it should be noted that the original Dolev-Yao paper used an automaton construction (indeed, a deterministic one) to solve the secrecy problem for a class of protocols called ping-pong protocols.

Indeed the result is of a technical nature and relates to the theoretician's toolkit in the study of Dolev-Yao models. The standard strategy to prove the derivability problem decidable is to prove a so-called **locality property** [RT03, CS03], that if t is derivable from X, then there is a special kind of derivation (a **normal derivation**) π such that every term occurring in π comes from $S(X \cup \{t\})$, where S is a function mapping a finite set of terms to *another finite set of terms*. Typically S is the **subterm function** *st*, but in many cases it is a minor variant. The locality property is used to provide a decision procedure for the derivability problem (which is typically a PTIME algorithm).

As we will show later, our system does not have an obvious locality property, and so we cannot follow the standard route to decidability. In fact, we can construct a set of terms X and a term t such that the set of terms occurring in any derivation of t from X is *exponential* in the size of $X \cup \{t\}$. This suggests that it would be difficult to define a function S of the kind mentioned above such that any term occurring in a normal derivation of t from X comes from $S(X \cup \{t\})$.

The first technical contribution of this paper is to show one way of working around this difficulty. We prove a *weak locality property*: we define a function S which maps every finite set of terms X to an *infinite* set of terms $S(X)$. We then prove that all terms occurring in a normal derivation of t from X are from $S(X \cup \{t\})$, and that the set of terms in $S(X \cup \{t\})$ that are derivable from X is regular. This facilitates an automaton construction and yields a decision procedure for checking whether t is derivable from X.

The second technical contribution is to settle the complexity of the derivability problem by proving a DEXPTIME-hardness result by reduction from the backwards reachability problem for alternating pushdown systems. While many lower bound results for the *active* intruder deduction problem exist in the literature, under various settings, this is one of the few lower bound results for the *passive* intruder deduction problem.

The third technical contribution of the paper is the use (in our decision procedure) of the alternating automaton saturation technique in itself (similar to the one in [BEM97]). In fact, the lower bound reduction shows the close connections to alternating pushdown systems, and so it is no surprise that automaton saturation, one of the standard tools for analysis of pushdown systems, is used for our upper bound proofs. This should also be viewed in the context of the use of tree automata for protocol verification, specifically the idea of representing (an over-approximation of) the set of deducible terms using tree automata. This has been explored in a number of papers [Mon99, Gou00, GK99]. Applications of two-way alternating tree automata to security protocol verification has been touched upon in [CDG+07]. The saturation technique that we use offers yet another tool that may be of use in other contexts.

Where does the high complexity of this problem originate from? It arises from the fact that blind pairing is distributive over encryption. This can be seen in the light of results on closely related constructors.

In [DKR09, BC06, CRZ05], a different way of modelling blind signatures is considered. Two operators, blind and unblind are used, with the following rules:

$$\mathsf{unblind}(\mathsf{blind}(m, r), r) = m$$
$$\mathsf{unblind}(\mathsf{sign}(\mathsf{blind}(m, r), k), r) = \mathsf{sign}(m, k)$$

The restriction (as compared to distributive encryption) here is that the r in the above equations is an atomic term, typically a random number, and whenever a blind pair is signed, the signature gets pushed only to the first component and not the second. Because of this, the system enjoys a locality property, and the basic derivability problem is decidable in PTIME.

In earlier work in [BRS07], we proposed essentially the same system described in this paper, but we imposed a restriction that the second component of blind pairs are always of the form n or $\{n\}_k$ where n is an atomic term (or nonce). And the only rule that involves pushing an encryption inside a blind pair is the derivation of $[\{t\}_k, n]$ from $[t, \{n\}_{inv(k)}]$ and k. This restricted system also satisfies a locality property.

At the other end of the spectrum, a much more powerful system is considered in [LLT07]. They study an abelian group operator + such that $\{t_1 + \cdots + t_n\}_k = \{t_1\}_k + \cdots + \{t_n\}_k$, i.e. encryption is homomorphic over +. They employ a very involved argument and prove the derivability problem in the general case to be decidable with a non-elementary upper bound. They also give a DEXPTIME algorithm in the case when the operator is xor, and a PTIME algorithm in the so-called binary case. The blind pair operator we consider has very different characteristics than xor, and the arguments in [LLT07] do not apply here.

2 Extension of the Dolev-Yao Model with Blind Pairs

Assume a set of basic terms $\mathscr{N}$, which includes the set of keys $\mathscr{K}$. Let $inv(k)$ be a function on $\mathscr{K}$ such that $inv(inv(k)) = k$. The set of **terms** $\mathscr{T}$ is defined to be:

$$\mathscr{T} ::= m \mid (t_1, t_2) \mid [t_1, t_2] \mid \{t\}_k$$

where $m \in \mathscr{N}$, $k \in \mathscr{K}$, and t, t_1, and t_2 range over $\mathscr{T}$.

The set of **subterms** of t, $st(t)$, is the smallest $X \subseteq \mathscr{T}$ such that 1) $t \in X$, 2) if $(t, t') \in X$ or $[t, t'] \in X$, then $\{t, t'\} \subseteq X$, and 3) if $\{t\}_k \in X$ then $\{t, k\} \subseteq X$. $st(X)$ is defined to be $\bigcup_{t \in X} st(t)$. A **keyword** is an element of $\mathscr{K}^*$. Given a term t and a keyword $x = k_1 \cdots k_n$, $\{t\}_x = \{\cdots\{t\}_{k_1} \cdots\}_{k_n}$. If $x = \varepsilon$, $\{t\}_x$ is t itself.

For simplicity, we assume henceforth that all terms are **normal**. These are terms which do not contain a subterm of the form $\{[t_1, t_2]\}_k$. For a term t, we get its normal form $t\!\downarrow$ by "pushing encryptions over blind pairs, all the way inside." Formally, it is defined as follows: $m\!\downarrow = m$ for $m \in \mathscr{N}$; $(t_1, t_2)\!\downarrow = (t_1\!\downarrow, t_2\!\downarrow)$; $[t_1, t_2]\!\downarrow = [t_1\!\downarrow, t_2\!\downarrow]$; and

$$\{t\}_k\!\downarrow = \begin{cases} [\{t_1\}_k\!\downarrow, \{t_2\}_k\!\downarrow] & \text{if } t = [t_1, t_2] \\ \{t\!\downarrow\}_k & \text{otherwise} \end{cases}$$

Definition 1. *A* **derivation** *or a* **proof** π *of t from assumptions X is a tree whose nodes are labelled by terms, whose root is labelled t, whose leaves are instances of the Ax* *rule and labelled by terms from X, and whose internal nodes are instances of one of the analz-rules or synth-rules in Figure 1. We use* $X \vdash t$ *to also denote that there is a proof of t from X. For a set of terms X,* $clos(X) = \{t \mid X \vdash t\}$ *is the* **closure** *of X.*

analz-rules		$\dfrac{\{t\}_k\!\downarrow \quad inv(k)}{t}\,decrypt$	$\dfrac{(t_0, t_1)}{t_i}\,split_i$	$\dfrac{[t_0, t_1]\!\downarrow \quad t_i\!\downarrow}{t_{1-i}}\,blindsplit_i$
synth-rules	$\dfrac{}{t}\,Ax\ (t \in X)$	$\dfrac{t \quad k}{\{t\}_k\!\downarrow}\,encrypt$	$\dfrac{t_1 \quad t_2}{(t_1, t_2)}\,pair$	$\dfrac{t_1 \quad t_2}{[t_1, t_2]}\,blindpair$

Fig. 1. Proof system for normal terms (with assumptions from $X \subseteq \mathcal{T}$). In the *decrypt* rule, $\{t\}_k\!\downarrow$ is the **major premise** and k is the **minor premise**. In the $blindsplit_i$ rule, $[t_0, t_1]\!\downarrow$ is the **major premise** and t_i is the **minor premise**.

It is significant that the main premise of the *decrypt* rule is $\{t\}_k\!\downarrow$. This allows us to derive $[t, t']$ from $[\{t\}_k, \{t'\}_k]$ and $inv(k)$, for instance.

Definition 2. *The* **derivability problem** *(also called the* **passive intruder deduction problem***) is the following: given a finite set* $X \subseteq \mathcal{T}$ *and* $t \in \mathcal{T}$*, determine whether* $X \vdash t$*.*

As we mentioned in the introduction, the standard strategy to prove this problem decidable is to define a notion of **normal proofs**, show that every proof can be transformed to a normal proof, and prove a so-called **locality property**, that every term occurring in a normal proof of $X \vdash t$ comes from $S(X \cup \{t\})$, where $S : 2^{\mathcal{T}} \to 2^{\mathcal{T}}$ is a function mapping a finite set of terms to another finite set of terms. Typically S is the subterm function *st*, but in many cases it is a minor variant. This typically yields a PTIME algorithm for the derivability problem.

But there is no obvious locality property for the proof system considered here. For instance, to derive the term $\{a\}_k$ from $[a, b]$, $\{b\}_k$ and k, we necessarily need to go via the term $[\{a\}_k, \{b\}_k]$, which is not a subterm of either the premises or the conclusion. In fact, the structure of terms occurring in a proof of $X \vdash t$ can get very complex.

For example, one can code up some kind of a counter – a set X of $O(n)$ terms and another term t, each of size $O(1)$, with $X \vdash t$, but such that every proof of t from X has at least 2^n terms occurring in it. The reader can refer to [BRS10] for details.

3 Normal Proofs

Even though our proof system lacks an obvious locality property, we can prove a weak locality property, which will help us derive a decision procedure for the derivability problem. This section is devoted to a proof of the weak locality property (or **weak subterm property**).

We first define the notion of a normal proof. These are proofs got by applying the transformations of Figure 2 repeatedly. Any subproof that matches a pattern on the left column is meant to be replaced by the proof on the right column in the same row. The idea behind normalization is to perform applications of the *encrypt* and *decrypt* rules as early as possible in the proof.

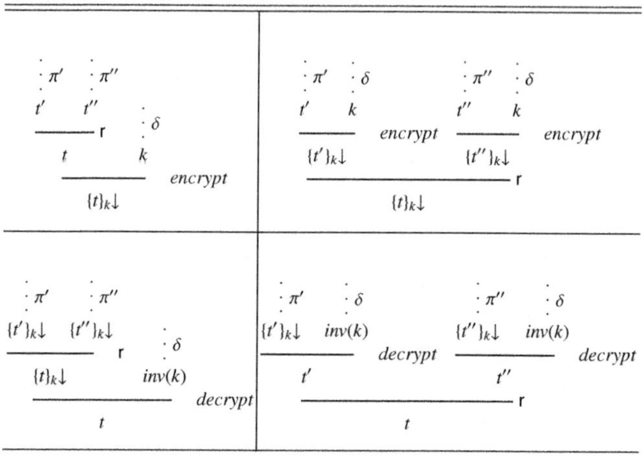

Fig. 2. The normalization rules. Rule r is meant to be either *blindpair* (in which case $t = [t', t'']$), or $blindsplit_0$ (in which case $t' = [t'', t]$), or $blindsplit_1$ (in which case $t' = [t, t'']$).

Definition 3. *A proof π of t from assumptions X is a* **minimal proof** *if t occurs only in the root of the proof.*

A proof π is a **normal proof** *if the following two conditions hold:*

1. *every subproof of π is minimal, and*
2. *the transformations in Figure 2 cannot be applied to π.*

Lemma 1. *Whenever $X \vdash t$, there is a normal proof of t from X.*

We now state the weak locality property for normal proofs. The standard locality property can be viewed as giving a bound on the "width" and encryption depth of terms occurring in a proof of $X \vdash t$. We prove a weaker property, where only the width of terms is bounded. So the set of terms occurring in any normal proof of $X \vdash t$ is got by encrypting terms (perhaps repeatedly) from a "core" set, using keys derivable from X. The core, it turns out, is $st(X \cup \{t\})$. For every $p \in st(X \cup \{t\})$, define $\mathscr{L}_p$ to be $\{x \in (st(X \cup \{t\}) \cap \mathscr{K})^* \mid X \vdash \{p\}_x\}$. We shall show in the next section that $\mathscr{L}_p$ is regular for each p.

We introduce a bit of notation first that will help us conveniently state the weak locality lemma. We say that a proof π of $X \vdash t$ is **purely synthetic** if:

- it ends in an application of the *Ax* or *blindpair* or *pair* rules, or
- it ends in an application of the *encrypt* rule and $t\downarrow$ is not a blind pair.

Lemma 2 (Weak locality property). *Let π be a normal proof of t from X, and let δ be a subproof of π with root labelled r. Then the following hold:*

1. *For every u occurring in δ, there is a term $p \in st(X \cup \{t\})$ and a keyword x such that $u = \{p\}_x$. Moreover, if δ is not a purely synthetic proof then $p \in st(X)$.*
2. *If the last rule of δ is decrypt or split with major premise r_1, then $r_1 \in st(X)$.*

The main difficulty is in coming up with the right statement. The proof itself is a standard induction on derivations, with an exhaustive case analysis, and is presented in full detail in [BRS10].

4 The Automaton Construction

We recall here some definitions relating to alternating pushdown systems and alternating automata (with ε-moves). The former will be needed for the lower bound argument in the next section, and the latter for the decision procedure to be presented here.

An **alternating pushdown system** (APDS) is a triple $\mathscr{P} = (P, \Gamma, \hookrightarrow)$, where P is a *finite set of control locations*, Γ is a *finite stack alphabet*, and $\hookrightarrow \subseteq P \times \Gamma^* \times 2^{(P \times \Gamma^*)}$ is a *finite set of transition rules*. We write transitions as $(a, x) \hookrightarrow \{(b_1, x_1), \ldots, (b_n, x_n)\}$. A *configuration* is a pair (a, x) where $a \in P$ and $x \in \Gamma^*$. Given a set of configurations C, a configuration (a, x), and $i \geq 0$, we say that $(a, x) \Rightarrow_{\mathscr{P},i} C$ iff:

- $(a, x) \in C$ and $i = 0$, or
- there is a transition $(a, y) \hookrightarrow \{(b_1, y_1), \ldots, (b_n, y_n)\}$ of $\mathscr{P}$, $z \in \Gamma^*$, and $i_1, \ldots, i_1 \geq 0$ such that $i = i_1 + \cdots + i_n + 1$ and $x = yz$ and for all $j \in \{1, \ldots, n\}$, $(b_j, y_j z) \Rightarrow_{\mathscr{P},i_j} C$.

We say that $(a, x) \Rightarrow_{\mathscr{P}} C$ iff $(a, x) \Rightarrow_{\mathscr{P},i} C$ for some $i \geq 0$.

An **alternating automaton** is an APDS $\mathscr{P} = (Q, \Sigma, \hookrightarrow)$ such that $\hookrightarrow \subseteq Q \times (\Sigma \cup \{\varepsilon\}) \times 2^{(Q \times \{\varepsilon\})}$. For $q \in Q$, $a \in \Sigma \cup \{\varepsilon\}$, and $C \subseteq Q$, we use $q \overset{a}{\hookrightarrow} C$ to denote the fact that $(q, a, C \times \{\varepsilon\}) \in \hookrightarrow$. For ease of notation, we will also write $q \overset{a}{\hookrightarrow} q'$ to mean $q \overset{a}{\hookrightarrow} \{q'\}$. Given $C \subseteq Q$, and $x \in \Sigma^*$, we use the notation $q \overset{x}{\Rightarrow}_{\mathscr{P},i} C$ to mean that $(q, x) \Rightarrow_{\mathscr{P},i} C \times \{\varepsilon\}$. For $C = \{q_1, \ldots, q_m\}$ and $C' \subseteq Q$, we use the notation $C \overset{x}{\Rightarrow}_{\mathscr{P},i} C'$ to mean that for all $j \leq m$, there exists i_j such that $q_j \overset{x}{\Rightarrow}_{\mathscr{P},i_j} C'$, and $i = i_1 + \cdots + i_m$. We also say $q \overset{x}{\Rightarrow}_{\mathscr{P}} C$ and $C \overset{x}{\Rightarrow}_{\mathscr{P}} C'$ to mean that there is some i such that $q \overset{x}{\Rightarrow}_{\mathscr{P},i} C$ and $C \overset{x}{\Rightarrow}_{\mathscr{P},i} C'$, respectively.

We typically drop the superscript $\mathscr{P}$ if it is clear from the context which APDS is referred to.

Fix a finite set of terms X_0 and a term t_0. We let Y_0 denote $st(X_0 \cup \{t_0\})$ and $K_0 = Y_0 \cap \mathscr{K}$. In this section, we address the question of whether there exists a normal proof of t_0 from X_0. Lemma 2 provides a key to the solution – every term occurring in such a proof is of the form $\{p\}_x$ for $p \in Y_0$ and $x \in K_0^*$. Therefore it is easy to see that the different $\mathscr{L}_p$ (for $p \in Y_0$) satisfy the following equations (among others):

$$\begin{array}{l}
kx \in \mathscr{L}_p \text{ iff } x \in \mathscr{L}_{\{p\}_k} \\
\text{if } x \in \mathscr{L}_p \text{ and } x \in \mathscr{L}_{p'} \text{ then } x \in \mathscr{L}_{[p,p']} \\
\text{if } x \in \mathscr{L}_p \text{ and } x \in \mathscr{L}_{[p,p']} \text{ then } x \in \mathscr{L}_{p'} \\
\text{if } x \in \mathscr{L}_{p'} \text{ and } x \in \mathscr{L}_{[p,p']} \text{ then } x \in \mathscr{L}_p \\
\text{if } x \in \mathscr{L}_p \text{ and } \varepsilon \in \mathscr{L}_k \text{ then } xk \in \mathscr{L}_p \\
\text{if } \varepsilon \in \mathscr{L}_{\{p\}_k} \text{ and } \varepsilon \in \mathscr{L}_{inv(k)} \text{ then } \varepsilon \in \mathscr{L}_p
\end{array}$$

This immediately suggests the construction of an alternating automaton $\mathscr{A}$ such that for every $t \in Y_0$ and keyword x, $x \in \mathscr{L}_t$ if and only if there is a run of $\mathscr{A}$ on the word x from the state t to a designated "final state" f. Then checking whether $X \vdash t_0$ (or in other words, $\varepsilon \in \mathscr{L}_{t_0}$) is simply a matter of checking if there is a run of $\mathscr{A}$ on ε from the state t_0 to the state f.

The states of the automaton are terms from Y_0 and the transitions are a direct transcription of the above equations. For instance there is an edge $t \overset{k}{\hookrightarrow} \{t\}_k$, and there is an edge $t \overset{\varepsilon}{\hookrightarrow} \{[t,t'],t'\}$. In the construction, we wish every $x \in \mathscr{L}_t$ to be witnessed by a run $t \overset{x}{\Rightarrow} \{f\}$ (f is a designated final state). This forces us to apply a saturation construction. For instance, suppose that $kx \in \mathscr{L}_t$ and this fact is witnessed by a run $t \overset{kx}{\Rightarrow} \{f\}$ (at some stage of the automaton construction). It is also the case that $x \in \mathscr{L}_{\{t\}_k}$, and this ought to be witnessed by a run $\{t\}_k \overset{x}{\Rightarrow} \{f\}$. To achieve this, we introduce a new transition $\{t\}_k \overset{\varepsilon}{\hookrightarrow} C$ whenever $t \overset{k}{\Rightarrow} C$. In fact, it does not suffice to stop after revising the automaton once. The procedure has to be repeated till no more new transitions can be added.

Thus we define a sequence of alternating automata $\mathscr{A}_0, \mathscr{A}_1, \ldots, \mathscr{A}_i, \ldots$, each of which adds transitions to the previous one, as given by the definition in Figure 3.

For each $i \geq 0$, $\mathscr{A}_i$ is given by $(Q, \Sigma, \hookrightarrow_i)$ where $Q = Y_0 \cup \{f\}$ ($f \notin Y_0$) and $\Sigma = K_0$. We define $\hookrightarrow_i$ by induction.

- $\hookrightarrow_0$ is the smallest subset of $Q \times (\Sigma \cup \{\varepsilon\}) \times 2^Q$ that satisfies the following:
  1. if $t \in Y_0, k \in K_0$ such that $\{t\}_k{\downarrow} \in Y_0$, then $t \overset{k}{\hookrightarrow}_0 \{\{t\}_k{\downarrow}\}$.
  2. if $t, t', t'' \in Y_0$ such that t is the conclusion of an instance of the *blindpair* or *blindsplit*$_i$ rules with premises t' and t'', then $t \overset{\varepsilon}{\hookrightarrow}_0 \{t', t''\}$.
- $\hookrightarrow_{i+1}$ is the smallest subset of $Q \times (\Sigma \cup \{\varepsilon\}) \times 2^Q$ such that:
  1. if $q \overset{a}{\Rightarrow}_i C$, then $q \overset{a}{\hookrightarrow}_{i+1} C$.
  2. if $\{t\}_k{\downarrow} \in Y_0$ and $t \overset{k}{\Rightarrow}_i C$, then $\{t\}_k{\downarrow} \overset{\varepsilon}{\hookrightarrow}_{i+1} C$.
  3. if $k \in K_0$ and $k \overset{\varepsilon}{\Rightarrow}_i \{f\}$, then $f \overset{k}{\hookrightarrow}_{i+1} \{f\}$.
  4. if $\Gamma \subseteq Y_0, t \in Y_0$, and if there is an instance r of one of the rules of Figure 1 (nullary, unary or binary) whose set of premises is (exactly) Γ and conclusion is t—note that *Ax* is a nullary rule, and hence this clause covers all $t \in X_0$—the following holds:

$$\text{if } u \overset{\varepsilon}{\Rightarrow}_i \{f\} \text{ for every } u \in \Gamma, \text{ then } t \overset{\varepsilon}{\hookrightarrow}_{i+1} \{f\}.$$

Fig. 3. The sequence of automata for analyzing $X_0 \vdash t_0$, with $Y_0 = st(X_0 \cup \{t_0\})$ and $K_0 = Y_0 \cap \mathscr{K}$. We use $\hookrightarrow_i$ for $\hookrightarrow_{\mathscr{A}_i}$ and $\Rightarrow_i$ for $\Rightarrow_{\mathscr{A}_i}$.

We would like to emphasize that saturating an alternating automaton fits in very naturally with our problem. For example, $X \vdash m$ where $X = \{[\{t\}_k, m], t, k\}$. To detect this, we need to test if $m \overset{\varepsilon}{\hookrightarrow}_i \{f\}$ for some i. This test turns out to be true for $i = 4$, as witnessed by the following sequence of edges and paths. Other constructions like two-way automata do not seem immediately applicable to this situation.

$m \overset{\varepsilon}{\hookrightarrow}_0 \{[\{t\}_k, m], \{t\}_k\}$.
$t \overset{\varepsilon}{\hookrightarrow}_1 \{f\}$, $k \overset{\varepsilon}{\hookrightarrow}_1 \{f\}$, $[\{t\}_k, m] \overset{\varepsilon}{\hookrightarrow}_1 \{f\}$.
$f \overset{k}{\hookrightarrow}_2 \{f\}$, $t \overset{k}{\Rightarrow}_2 \{f\}$.
$\{t\}_k \overset{\varepsilon}{\hookrightarrow}_3 \{f\}$ (this is a crucial use of saturation), $m \overset{\varepsilon}{\Rightarrow}_3 \{f\}$.
$m \overset{\varepsilon}{\hookrightarrow}_4 \{f\}$.

The following lemma essentially shows that the saturation procedure terminates in exponential time.

Lemma 3. *1. For all $i \geq 0$ and all $a \in \Sigma \cup \{\varepsilon\}$, the relation $\overset{a}{\Rightarrow}_i$ is constructible from $\hookrightarrow_i$ in time $2^{O(d)}$, where $d = |Q|$.*
2. For all $i \geq 0$ and all $a \in \Sigma$, the relation $\overset{a}{\hookrightarrow}_{i+1}$ is constructible from $\Rightarrow_i$ in time $2^{O(d)}$.
3. There exists $d' \leq d^2 \cdot 2^d$ such that for all $i \geq d'$, $q \in Q$, $a \in \Sigma \cup \{\varepsilon\}$, and $C \subseteq Q$, $q \overset{a}{\hookrightarrow}_i C$ if and only if $q \overset{a}{\hookrightarrow}_{d'} C$.

We now present theorems that assert the correctness of the above construction. It is **sound**, i.e. none of the automata accept an x starting from r where $\{r\}_x$ is not derivable from X_0; and that it is **complete**, i.e. whenever $\{r\}_x$ is derived from X_0, one of the $\mathscr{A}_i$'s has an accepting run over x starting from r. To simplify the statement and proof in the rest of this section, we first introduce the following notations:

- for $X \subseteq \mathscr{T}$ and keyword x, we use $X \vdash x$ to mean that $X \vdash k$ for every k occurring in x.
- for $C \subseteq Y_0$ and keyword y, $\{C\}_y = \{\{t\}_y\downarrow \mid t \in C\}$.
- for $q \in Q, C \subseteq Q$, $q \overset{x}{\Rightarrow}_{i,d} C$ iff $q \overset{x}{\Rightarrow}_{\mathscr{A}_i,d} C$.
- for $C, C' \subseteq Q$, $C \overset{x}{\Rightarrow}_{i,d} C'$ iff $C \overset{x}{\Rightarrow}_{\mathscr{A}_i,d} C'$.

Theorem 1 (Soundness). *For any i, any $t \in Y_0$, and any keyword x, if $t \overset{x}{\Rightarrow}_i \{f\}$, then $X_0 \vdash \{t\}_x\downarrow$.*

Soundness is an immediate consequence of the following lemma, taking $C = \{f\}$ and $y = \varepsilon$.

Lemma 4. *Suppose $i, d \geq 0$, $t \in Y_0$, $x, y \in K_0^*$, and $C \subseteq Q$ (with $D = C \cap Y_0$). Suppose the following also hold: 1) $t \overset{x}{\Rightarrow}_{i,d} C$, and 2) $C \subseteq Y_0$ or $X_0 \vdash y$. Then $X_0 \cup \{D\}_y \vdash \{t\}_{xy}$.*

As one may expect, the proof is by induction on the size of the run labelled x from t to C, but the difficulty with the proof is that in a run over x from t to C, each branch may hit f after reading a different prefix of x. Hence the inductive statement is subtle and this is why the statement of the Lemma is complex. In fact, formulating Lemma 4 precisely turned out to be the trickiest part of the upper bound proof. A detailed proof is presented in the technical report [BRS10].

Theorem 2 (Completeness). *For any $t \in Y_0$ and any keyword x, if $X_0 \vdash \{t\}_x\downarrow$, then there exists $i \geq 0$ such that $t \overset{x}{\Rightarrow}_i \{f\}$.*

The proof is by induction on derivations, and is reasonably straightforward.

Theorem 3. *Given $X_0 \subseteq \mathcal{T}$ and $t_0 \in \mathcal{T}$, it is decidable in* DEXPTIME *whether $X_0 \vdash t_0$.*

Proof. Let X_0 and t_0 be given, and let $Y_0 = st(X_0 \cup \{t_0\})$.

By Lemma 3, there is d' such that for all $q \in Q$, $a \in \Sigma \cup \{\varepsilon\}$, and $C \subseteq Q$, and any $i \geq 0$,

$$\text{if } q \overset{a}{\hookrightarrow}_i C \text{ then } q \overset{a}{\hookrightarrow}_{d'} C.$$

Further $\hookrightarrow_{d'}$ is computable in time $2^{O(d)}$, where $d = |Y_0|$.

By the soundness theorem (Theorem 1), for all i, any $t \in Y_0$ and any keyword x, if $t \overset{x}{\Rightarrow}_i \{f\}$, then $X_0 \vdash \{t\}_x\downarrow$. In particular, this holds for $i = d'$. On the other hand, by the completeness theorem (Theorem 2), whenever $X_0 \vdash \{t\}_x\downarrow$ for $t \in Y_0$ and keyword x, there is an i such that $t \overset{x}{\Rightarrow}_i \{f\}$, and hence $t \overset{x}{\Rightarrow}_{d'} \{f\}$. Thus to check whether $X_0 \vdash t_0$, it suffices to check if $t_0 \overset{\varepsilon}{\Rightarrow}_{d'} \{f\}$. But by construction, if $t_0 \overset{\varepsilon}{\Rightarrow}_{d'} \{f\}$, then $t_0 \overset{\varepsilon}{\hookrightarrow}_{d'+1} \{f\}$, but this means that $t_0 \overset{\varepsilon}{\hookrightarrow}_{d'} \{f\}$.

Thus one only needs to check—in constant time—whether $t_0 \overset{\varepsilon}{\hookrightarrow}_{d'} \{f\}$. Thus the derivability problem is solvable in DEXPTIME. □

5 A DEXPTIME Lower Bound for the Derivability Problem

We recall the following fact about alternating pushdown systems.

Fact 4. *The* **backwards-reachability problem for alternating pushdown systems,** *which asks, given an APDS $\mathcal{P}$ and two configurations (s, x_s) and (f, x_f), whether $(s, x_s) \Rightarrow_{\mathcal{P}} (f, x_f)$, is* DEXPTIME-*complete [SSE06].*

We reduce this problem to the problem of checking whether $X \vdash t$ in our proof system, given $X \subseteq \mathcal{T}$ and $t \in \mathcal{T}$.

Assume that we are given an APDS $\mathcal{P} = (P, \Gamma, \hookrightarrow)$, and two configurations (s, x_s) and (f, x_f). Let us assume that the rules in $\hookrightarrow$ are numbered 1 to ℓ.

We will take $M = P \cup \{\mathsf{c}_m \mid 1 \leq m \leq \ell\}$ to be a set of atomic terms, and $K = \Gamma \cup \{d, e\}$ to be a set of *non-symmetric keys* (such that none of them is the inverse of another, and such that $d, e \notin \Gamma$).

We translate each rule to a term as follows. Suppose the mth rule is:

$$(a, x) \hookrightarrow \{(b_1, x_1), \ldots, (b_n, x_n)\}.$$

This gets translated to the following term r_m:

$$\mathsf{r}_m = [[\cdots[[\mathsf{r}'_m, \{b_1\}_{x_1}], \{b_2\}_{x_2}], \cdots, \{b_{n-1}\}_{x_{n-1}}], \{b_n\}_{x_n}], \text{ where}$$
$$\mathsf{r}'_m = [[\cdots[[\{\mathsf{c}_m\}_d, \{a\}_x], \{b_1\}_{x_1}], \cdots, \{b_{n-1}\}_{x_{n-1}}], \{b_n\}_{x_n}].$$

We take X to be the set $\{\mathsf{r}_m \mid 1 \leq m \leq \ell\} \cup \{\{f\}_{x_f e}\} \cup \{\{\mathsf{c}_m\}_d \mid 1 \leq m \leq \ell\} \cup \Gamma \cup \{e\}$.

The reduction is almost a straight transcription of the APDS rules. But we need to take some care because given a blind pair $[t, t']$, we can split it using either t or t'. Further, we have to avoid an accidental split of r_m using a part of r_n, for distinct $m, n \leq \ell$. This explains the need for the "tags" c_m $(m \leq \ell)$.

We claim that $(s, x_s) \Rightarrow_{\mathscr{P}} (f, x_f)$ iff $X \vdash \{s\}_{x_s e}$. A detailed proof for both directions is presented in [BRS10]. Here we just present a high-level sketch of the proof. We prove the harder direction, that if there is a normal proof of $X \vdash \{a\}_{xe}$ then $(a, x) \Rightarrow_{\mathscr{P}} (f, x_f)$. The overall strategy is to prove that whenever a term of the form $\{a\}_{xe}$ is proved, there has to be a rule of $\mathscr{P}$ of which (a, y) is the left side, $x = yz$, and there is a shorter proof of $\{b\}_{y_i z}$, for every (b_i, y_i) on the right side of that rule. This requires to do a careful analysis of the proof of $X \vdash \{a\}_{xe}$. Here it is crucial to consider normal proofs, since the weak locality property (Lemma 2) imposes some structure on the terms occurring in such proofs. For instance, throughout the following we will use the fact that the *pair* rule will never be used in normal derivations that we encounter in the following proof.

We now introduce the following bit of notation, for conveniently presenting the argument. For any term t whose normal form is $[t_1, \ldots, t_n]$, we define *comps*(t) to be the set $\{t_1, \ldots, t_n\}$. If $t \in st(X)$ such that $\{\mathsf{c}_m\}_d \in st(t)$, then *residues*$(t)$ is defined by the following:

- *residues*$(\mathsf{r}_m) = \emptyset$
- if $t \neq \mathsf{r}_m$, then *residues*(t) = *residues*$([t, t']) \cup \{t'\}$, where t' is the unique term such that $[t, t'] \in st(\mathsf{r}_m)$.

Lemma 5. *For any configuration (a, x), if there is a normal proof of $X \vdash \{a\}_{xe}$, then*

$$(a, x) \Rightarrow_{\mathscr{P}} (f, x_f)$$

The lemma follows easily, by induction on the size of normal proofs, from the next assertion.

Lemma 6. *If there is a normal proof π of $X \vdash \{a\}_{xe}$, then either $(a, x) = (f, x_f)$ or there is a rule of $\mathscr{P}$, $(a, y) \hookrightarrow \{(b_1, y_1), \ldots, (b_n, y_n)\}$, and $z \in \Gamma^*$ such that $x = yz$, and for each $j \leq n$, a* subproof *π_j of π with conclusion $X \vdash \{b_j\}_{y_j z e}$.*

Proof. The observation that drives the proof of this lemma is the following.

For any normal proof π of $X \vdash \{a\}_{xe}$ and any subproof δ of π with conclusion $\{p\}_{we}$, and any $m \leq \ell$:

1. if the last rule of δ is an application of *blindpair*, and if $\{\mathsf{c}_m\}_d \in st(p)$, then $X \vdash \{r\}_{we}$ is the conclusion of some subproof of δ, for every $\{r\}_{we} \in$ *comps*$(\{p\}_{we})$.
2. if the last rule of δ is an application of *blindsplit*, and if $\{\mathsf{c}_m\}_d \in st(p)$, then $X \vdash \{r\}_{we}$ is the conclusion of some subproof of δ, for every $r \in$ *residues*(p).

Let π be a normal proof of $X \vdash \{a\}_{xe}$ and suppose that $(a, x) \neq (f, x_f)$. Then it is clear that for all prefixes y of xe, $\{a\}_y \notin X$. Thus π does not end in an application of *encrypt* (an easy consequence of the structure of X). It obviously cannot end in an application of *blindpair*. So it is clear that the last rule is an application of *blindsplit*, with major premise t and minor premise t'. Now t is a blind pair, and hence there is a unique

$p \in st(X)$ and $z \in \Gamma^*$ such that $t = \{p\}_{ze}$ (again a consequence of the structure of X). It can be seen that $\{\mathsf{c}_m\}_d \in st(p)$ for some $m \leq \ell$. If t is obtained as the result of an application of *encrypt*, then it can be seen that $p = \mathsf{r}_m$ and thus p has no residues, and hence it is vacuously true that $\{r\}_{ze}$ occurs in δ for all $r \in residues(p)$. Otherwise, t is the result of a blind split, and hence, by the observation at the start of the proof, $\{r\}_{ze}$ occurs in δ for all $r \in residues(p)$.

Now if $p \in st(\mathsf{r}'_m)$, then among the residues of p will be found $\{b_j\}_{y_j}$ for every (b_j, y_j) on the right hand side of the rule numbered m. So by what has been proved above, there is a subproof π_j of π whose conclusion is $X \vdash \{b_j\}_{y_jze}$, and we are done.

Suppose $p \notin st(\mathsf{r}'_m)$. Then, it can be seen that $t' = \{p'\}_{ze}$ for some $p' \in st(X)$ such that $\mathsf{r}'_m \in st(p')$. Now clearly $p' \notin X$ (since it is a proper subterm of r_m, missing a component of the form $\{a\}_w$ as it does) and hence t' is not the result of an application of *encrypt* (again an easy consequence of the structure of X). It cannot also be the result of an application of *blindsplit*, since then one of the premises has to be $\{a\}_{xe}$, contradicting minimality. Thus t' is the result of an application of *blindpair*, but the observation at the beginning of this proof tells us that $X \vdash \{r\}_{ze}$ for all $\{r\}_{ze} \in comps(\{p'\}_{ze})$. But notice that $\mathsf{r}'_m \in st(p')$, and hence we can conclude that among $comps(p')$ will be found $\{b_j\}_{y_j}$ for every (b_j, y_j) on the right hand side of the rule numbered m. So by the observation at the start of the proof, we can conclude that for each j, there is a subproof π_j of π whose conclusion is $X \vdash \{b_j\}_{y_jze}$, and we are done. □

And the following theorem is the end result.

Theorem 4. *The passive intruder deduction problem is* DEXPTIME-*hard.*

6 Discussion

We can think of a number of extensions of our system by considering more algebraic properties of the blind pair operator, like associativity, commutativity, unitariness, etc. It then becomes more convenient to treat an extension of the Dolev-Yao model with a polyadic + operator, over which encryption distributes. In this framework, a very powerful system is studied in [LLT07], where + is treated as an abelian group operator.

The decidability results in [LLT07] are driven by a set of normalization rules whose effect is drastically different from ours. Our rules ensure that the "width" of terms occurring in a normal proof of $X \vdash t$ is bounded by $X \cup \{t\}$. But their normalization rules ensure that the encryption depth of terms occurring in a normal proof of $X \vdash t$ is bounded by $X \cup \{t\}$. On the other hand, the width of terms, represented by coefficients in the +-terms, can grow unboundedly. The rest of their decidability proof is an involved argument using algebraic methods.

The techniques of our paper do not seem to extend to the system with an abelian group operator, nor for slightly weaker systems where + is associative and commutative, or when + is a (not necessarily commutative) group operator and the term syntax allows terms of the form $-t$. But the techniques for our upper bound proofs extends to the case when + is just an associative operator (not necessarily commutative, or has inverses). Another extension that is usually considered is encryption with constructed keys rather than atomic keys. The upper bound results go through for this system as well, with much of the hard work lying in extending the weak locality theorem. A sketch of the proofs is presented in the technical report [BRS10], and will be developed further in a companion paper.

References

[BC06] Bernat, V., Comon-Lundh, H.: Normal proofs in intruder theories. In: Okada, M., Satoh, I. (eds.) ASIAN 2006. LNCS, vol. 4435, pp. 151–166. Springer, Heidelberg (2008)

[BEM97] Bouajjani, A., Esparza, J., Maler, O.: Reachability analysis of pushdown automata: Application to model-checking. In: Mazurkiewicz, A., Winkowski, J. (eds.) CONCUR 1997. LNCS, vol. 1243, pp. 135–150. Springer, Heidelberg (1997)

[BRS07] Baskar, A., Ramanujam, R., Suresh, S.P.: Knowledge-based modelling of voting protocols. In: Proc. of TARK XI, pp. 62–71 (2007)

[BRS10] Baskar, A., Ramanujam, R., Suresh, S.P.: A DEXPTIME-complete Dolev-Yao theory with distributive encryption. Technical report (May 2010), http://www.cmi.ac.in/~spsuresh/pdffiles/mfcs2010-tr.pdf

[CDG+07] Comon, H., Dauchet, M., Gilleron, R., Löding, C., Jacquemard, F., Lugiez, D., Tison, S., Tommasi, M.: Tree Automata Techniques and Applications (2007), http://www.grappa.univ-lille3.fr/tata

[CS03] Comon-Lundh, H., Shmatikov, V.: Intruder Deductions, Constraint Solving and Insecurity Decisions in Presence of Exclusive or. In: Proc. LICS 18, June 2003, pp. 271–280 (2003)

[CRZ05] Cortier, V., Rusinowitch, M., Zalinescu, E.: A resolution strategy for verifying cryptographic protocols with cbc encryption and blind signatures. In: PPDP, pp. 12–22 (2005)

[DKR09] Delaune, S., Kremer, S., Ryan, M.D.: Verifying privacy-type properties of electronic voting protocols. Journal of Computer Security 17(4), 435–487 (2009)

[DY83] Dolev, D., Yao, A.: On the Security of public-key protocols. IEEE Transactions on Information Theory 29, 198–208 (1983)

[FOO92] Fujioka, A., Okamoto, T., Ohta, K.: A practical secret voting scheme for large scale elections. In: ASIACRYPT, pp. 244–251 (1992)

[GK99] Genet, T., Klay, F.: Rewriting for cryptographic protocol verification. Technical report, CNET-France Telecom (1999)

[Gou00] Goubault-Larrecq, J.: A method for automatic cryptographic protocol verification. In: Rolim, J.D.P. (ed.) IPDPS-WS 2000. LNCS, vol. 1800, pp. 977–984. Springer, Heidelberg (2000)

[LLT07] Lafourcade, P., Lugiez, D., Treinen, R.: Intruder deduction for the equational theory of abelian groups with distributive encryption. Information and Computation 205(4), 581–623 (2007)

[Mon99] Monniaux, D.: Abstracting cryptographic protocols with tree automata. In: Cortesi, A., Filé, G. (eds.) SAS 1999. LNCS, vol. 1694, pp. 149–163. Springer, Heidelberg (1999)

[RT03] Rusinowitch, M., Turuani, M.: Protocol Insecurity with Finite Number of Sessions and Composed Keys is NP-complete. TCS 299, 451–475 (2003)

[SSE06] Suwimonteerabuth, D., Schwoon, S., Esparza, J.: Efficient algorithms for alternating pushdown systems with an application to the computation of certificate chains. In: Graf, S., Zhang, W. (eds.) ATVA 2006. LNCS, vol. 4218, pp. 141–153. Springer, Heidelberg (2006)

On Problem Kernels for Possible Winner Determination under the k-Approval Protocol

Nadja Betzler*

Institut für Informatik, Friedrich-Schiller-Universität Jena
Ernst-Abbe-Platz 2, D-07743 Jena, Germany
nadja.betzler@uni-jena.de

Abstract. The POSSIBLE WINNER problem asks whether some distinguished candidate may become the winner of an election when the given incomplete votes (partial orders) are extended into complete ones (linear orders) in a favorable way. Under the k-approval protocol, for every voter, the best k candidates of his or her preference order get one point. A candidate with maximum total number of points wins. The POSSIBLE WINNER problem for k-approval is NP-complete even if there are only two votes (and k is part of the input). In addition, it is NP-complete for every fixed $k \in \{2, \ldots, m-2\}$ with m denoting the number of candidates if the number of votes is unbounded. We investigate the parameterized complexity with respect to the combined parameter k and "number of incomplete votes" t, and with respect to the combined parameter $k' := m - k$ and t. For both cases, we use kernelization to show fixed-parameter tractability. However, we show that whereas there is a polynomial-size problem kernel with respect to (t, k'), it is very unlikely that there is a polynomial-size kernel for (t, k). We provide additional fixed-parameter algorithms for some special cases.

1 Introduction

Voting situations arise in political elections, multi-agent systems, human resource departments, etc. This includes scenarios in which one is interested in finding a small group of winners (or losers), such as awarding a small number of grants, picking out a limited number of students for a graduate school, or voting for a committee with few members. Such situations are naturally reflected by a variant of approval voting, the *k-approval* voting system, where every voter gives one point to each of the k alternatives/candidates which he or she likes best and the candidates having the most points in total win. On the one side, k-approval extends *plurality* where a voter gives one point to one candidate, that is $k = 1$, and, on the other side, it extends *veto* where a voter gives one point to all but one candidate, that is, $k' = 1$ for $k' := m - k$ and m candidates.

At a certain point in the decision making process one might face the situation that the voters have made up their minds "partially". For example, for the decision about the Nobel prize for peace in 2009, a committee member might have already known that he (or she) prefers Obama and Bono to Berlusconi, but might have not decided on the order of Obama and Bono yet. This immediately leads to the question whether, given a

* Supported by the DFG, research project PAWS, NI 369/10.

P. Hliněný and A. Kučera (Eds.): MFCS 2010, LNCS 6281, pp. 114–125, 2010.

set of "partial preferences", a certain candidate may still win. The formalization of this question leads to the POSSIBLE WINNER problem.

The POSSIBLE WINNER problem has been introduced by Konczak and Lang [15] and since then its computational complexity has been studied for several voting systems [2,3,5,17,18]. Even for the comparatively simple k-approval voting, it turned out that POSSIBLE WINNER is NP-complete except for the special cases of plurality and veto [3], that is, for any k greater than one and smaller than the number of candidates minus one. A multivariate complexity study showed that it is NP-complete if there are only two voters when k is part of the input but fixed-parameter tractable with respect to the "number of candidates" [5]. In contrast, for the approval voting variant where each voter can assign a point to *up to* k candidates, it can easily be seen that POSSIBLE WINNER can be solved in polynomial-time. A prominent special case of POSSIBLE WINNER is the MANIPULATION problem, where the input consists of a set of linear orders and a set of completely unspecified votes. For k-approval, it is easy to see that MANIPULATION is solvable in polynomial time for unweighted votes but for weighted votes it is NP-complete for all fixed $k \neq 1$ [14].

The above described hardness results motivate a multivariate analysis with respect to the combined parameter "number of voters" *and* "number of candidates to which a voter gives one/zero points" for k-approval. Can we efficiently solve POSSIBLE WINNER in the case that these parameters are both small? Directly related questions are whether we can ignore or delete candidates which are not relevant for the decision process and how to identify such candidates. In this context, parameterized algorithmics [10,16] provides the concept of kernelization by means of polynomial-time data reduction rules that "preprocess" an instance such that the size of the "reduced" instance only depends on the parameters [6,13].

In this work, we use kernelization to show the fixed-parameter tractability of POSSIBLE WINNER for k-approval in two "symmetric" scenarios. First, we consider the combined parameter "number of incomplete votes" t *and* "number of candidates to which every voter gives zero points" $k' := m - k$ for m candidates (directly extending the veto voting system with $k' = 1$). Making use of a simple observation we show that POSSIBLE WINNER admits a polynomial-size problem kernel with respect to (t, k') and provide two algorithms: one with exponential running time factor $2^{O(k')}$ in case of constant t and one with exponential running time factor $2^{O(t)}$ in case of constant k'. Second, we consider the combined parameter t and k, where k denotes the "number of candidates to which a voter gives a point". We observe that here one cannot argue symmetrically to the first scenario. Using other arguments, we give a superexponential-size problem kernel showing the fixed-parameter tractability of POSSIBLE WINNER with respect to (t, k). For the special case of 2-approval, we give an improved polynomial-size kernel with $O(t^2)$ candidates. Using a methodology due to Bodlaender et al. [7], our main technical result shows that POSSIBLE WINNER is very unlikely to admit a polynomial-size problem kernel with respect to (t, k).

Preliminaries. A *linear vote* is a transitive, antisymmetric, and total relation on a set C of candidates and *partial vote* a transitive and antisymmetric relation on a set C of candidates. We use $>$ to denote the relation between candidates in a linear vote and $\succ$

to denote the relation between candidates in a partial vote. We often specify a subset $D \subseteq C$ of candidates instead of single candidates in a partial vote; for a candidate $e \in C \setminus D$ and $D = \{d_1, \ldots, d_s\}$, the meaning of "$e \succ D$" is "$\{e \succ d_1, e \succ d_2, \ldots, e \succ d_s\}$". A linear vote v^l *extends* a partial vote v^p if $v^p \subseteq v^l$, that is, for every $i, j \leq m$, from $c_i \succ c_j$ in v^p it follows that $c_i > \cdots > c_j$ in v^l. An extension E of a set of partial votes $V^p = \{v_1^p, \ldots, v_n^p\}$ is a mapping from V^p to a set of linear votes $V^l := \{v_1^l, \ldots, v_n^l\}$ such that v_i^l extends v_i^p for every i. Given a set of partial votes V^p on C, a candidate $c \in C$ is a *possible winner* if there exists a *winning extension* E, that is, c wins in E with respect to a considered voting system. For any voting system R, the underlying decision problem is defined as follows.

POSSIBLE WINNER

Given: A set of candidates C, a set of partial votes V on C, and a distinguished candidate $c \in C$.

Question: Is there an extension E of V such that c wins with respect to R in E?

We focus on the voting system k-*approval* where, given a set V of linear votes on a set C of candidates, the first k candidates within a vote get one point and the remaining candidates get zero points. For every candidate $c' \in C$, one sums up the points over all votes from V to obtain its *score* $s(c')$ and the candidates with maximum score win. We call the first k positions of a vote *one-positions* and the remaining positions *zero-positions*. All results are given for the *unique winner* case, that is, looking for a single candidate with maximum score, but can be adapted easily to hold for the "co-winner" case where several candidates may get the maximum score and all of them win.

A parameterized problem L is a subset of $\Sigma^* \times \Sigma^*$ for some finite alphabet Σ [10,16]. An instance of a parameterized problem consists of (x, p) where p is called the parameter. We mainly consider "combined" parameters which are tuples of positive integers. A parameterized problem is *fixed-parameter tractable* if it can be solved in time $f(|p|) \cdot \text{poly}(|x|)$ for a computable function f. A kernelization algorithm consists of a set of *(data) reduction rules* working as follows [6,13,16]. Given an instance $(x, p) \in \Sigma^* \times \Sigma^*$, they output in time polynomial in $|x| + |p|$ an instance $(x', p') \in \Sigma^* \times \Sigma^*$ such that the following two conditions hold. First, (x, p) is a yes-instance if and only if (x', p') is a yes-instance (termed *soundness*). Second, $|x'| + |p'| \leq g(|p|)$ where g is a computable function. If g is a polynomial function, then we say that the parameterized problem admits a *polynomial kernel*.

Some of the reduction rules given in this work will not directly decrease the instance size by removing candidates or votes but instead only decrease the number of possible extensions of a vote, for example, by "fixing" candidates. To *fix* a candidate at a certain position means to specify its relation to all other candidates. Clearly, a candidate may not be fixed at every position in a specific partial vote. To take this into account, an important concept is the notation of *shifting* a candidate. More precisely, we say a candidate c' can shift a candidate c'' to the left (right) in a partial vote v if $c'' \succ c'$ ($c' \succ c''$) in v, that is, setting c' to a one-position (zero-position) implies setting c'' to a one-position (zero-position) as well. For every candidate $c' \in C$ and a partial vote $v \in V$, let $L(v, c') := \{c'' \in C \mid c'' \succ c' \text{ in } v\}$ and $R(v, c') := \{c'' \in C \mid c' \succ c'' \text{ in } v\}$.

Then, fixing a candidate $c' \in C$ *as good as possible* means to add $L(v, c') \succ c' \succ C \setminus (L(v, c') \cup \{c'\})$ to v. Analogously, fixing a candidate *as bad as possible* is realized by adding $C \setminus (R(v, c') \cup \{c'\}) \succ c' \succ R(v, c')$ to v. If a candidate $c' \in C$ is fixed in all partial votes, this implies that also its score $s(c')$ is fixed.

The votes of an input instance of POSSIBLE WINNER can be partitioned into a (possibly empty) set of linear votes, called V^l, and a set of proper (non-linear) partial votes, called V^p. We state all our results for the parameter $t := |V^p|$. All positive results also hold for the parameter number of total votes $n := |V^l| + |V^p|$. Due to the space restrictions, several (parts of) proofs are deferred to a full version of this work.

2 Fixed Number of Zero-Positions

For $(m - k')$-approval with $k' < m$, k' denotes the number of zero-positions. We give a polynomial kernel with respect to (t, k') for POSSIBLE WINNER where t is the number of partial votes. In addition, we provide two parameterized algorithms for special cases.

2.1 Problem Kernel

Consider a POSSIBLE WINNER instance with candidate set C, vote set $V = V^l \cup V^p$, and distinguished candidate $c \in C$ for $(m - k')$-approval. We start with a simple reduction rule that is a crucial first step for all kernelization results in this work.

Rule 1. *For every vote $v_i \in V^p$, if $|L(v_i, c)| < m - k'$, fix c as good as possible in v_i.*

The soundness and polynomial-time running time of Rule 1 is easy to verify. The condition $|L(v_i, c)| < m - k'$ is crucial since otherwise c might shift a candidate c' to a one-position whereas c is assigned to a zero position and this could cause c' to beat c. After applying Rule 1, the score of c is fixed at the maximum possible value since it makes one point in all votes in which this is possible. Now, for every candidate $c' \in C \setminus \{c\}$, by counting the points that c' makes within the linear votes V^l, compute the number of zero positions that c' must assume within the partial votes V^p such that it is beaten by c. Let this number be $z(c')$ and $Z_+ := \{c' \in C \setminus \{c\} \mid z(c') > 0\}$. Since there are only tk' zero positions in V^p, one can observe the following.

Observation 1 *In a yes-instance, $\sum_{c' \in C \setminus \{c\}} z(c') \leq tk'$ and $|Z_+| \leq tk'$.*

Observation 1 provides a simple upper bound for the number of candidates in Z_+. By formulating a data reduction rule that bounds the number of remaining candidates and replacing the linear votes V^l by a bounded number of "equivalent votes" we can show the following theorem. The basic idea is that since a remaining candidate from $C \setminus (Z_+ \cup \{c\})$ can be set arbitrarily in every vote without beating c, it is possible to replace the set of all remaining candidates by tk'^2 "representative candidates".

Theorem 1. *For $(m - k')$-approval,* POSSIBLE WINNER *with t partial votes admits a polynomial kernel with at most $tk'^2 + tk' + 1$ candidates.*

Initialization:
For every $D' \in \mathcal{D} \setminus \{(d_1, \ldots, d_p)\}$, set $T(0, D') = 0$.
Set $T(0, (d_1, \ldots, d_p)) = 1$.

Update:
For $0 \leq i \leq t-1$,
for every $D' = (d'_1, \ldots, d'_p) \in \mathcal{D}$,
$T(i+1, D') = 1$ if there are two candidates z_g, z_h that can take the zero-positions in v_{i+1}
and $T(i, D'') = 1$ with $D'' := \{d''_1, \ldots, d''_p\}$ and
$d''_j = d'_j$ for $j \in \{1, \ldots, q\} \setminus \{g, h\}$, $d''_g \leq d'_g + 1$, and $d''_h \leq d'_h + 1$.

Output:
"yes" if $T(t, (0, \ldots, 0)) = 1$, "no" otherwise

Fig. 1. Dynamic programming algorithm for $(m-2)$-approval

2.2 Parameterized Algorithms

We give algorithms running in $2^{O(p)} \cdot \text{poly}(n, m)$ time with p denoting either k' or t where the other parameter is of constant value. Note that the kernelization from the previous subsection does not imply such running times.

Constant number of partial votes. For two partial votes, there can be at most $2k'$ candidates that must take a zero-position in a yes-instance (see Observation 1). Branching into the two possibilities of taking the zero-position in the first or in the second vote for every such candidate, results in a search tree of size $2^{2k'} = 4^{k'}$. For every "leaf" of the search tree it is easy to check if there is a corresponding extension. Using similar arguments, one arrives at the following.

Proposition 1 *For a constant number t of partial votes,* Possible Winner *for $(m-k')$-approval can be solved in $2^{t^2 k'} \cdot \text{poly}(n, m)$ time.*

Constant number of zero-positions. For constant k' the existence of an algorithm with running time $2^{O(t)} \cdot \text{poly}(n, m)$ seems to be less obvious than for the case of constant t. We start by giving a dynamic programming algorithm for $(m-2)$-approval. Employing an idea used in [4, Lemma 2], we show that it runs in $4^t \cdot \text{poly}(n, m)$ time and space.

As in the previous subsection, fix c according to Rule 1 such that it makes the maximum possible score and let $Z_+ := \{z_1, \ldots, z_p\}$ denote the set of candidates that take at least one zero-position in a winning extension. Let $d_1, \ldots, d_p$ denote the corresponding number of zero-positions that must be assumed and let $\mathcal{D} := \{(d'_1, \ldots, d'_p) \mid 0 \leq d'_j \leq d_j \text{ for } 0 \leq j \leq p\}$. Then, the dynamic programming table T is defined by $T(i, D')$ for $1 \leq i \leq t$ and $D' = (d'_1, \ldots, d'_p) \in \mathcal{D}$. Herein, $T(i, D') = 1$ if the partial votes from $\{v_1, \ldots, v_i\}$ can be extended such that candidate z_j takes at least $d_j - d'_j$ zero-positions for $1 \leq j \leq p$; otherwise $T(i, D') = 0$. Intuitively, d'_j stands for the number of zero-positions which z_j must still take in the remaining votes $\{v_{i+1}, \ldots, v_t\}$. Clearly, if $T(t, (0, \ldots, 0)) = 1$ for an instance, then it is a yes-instance. The dynamic programming algorithm is given in Figure 1. By further extending it to work for any constant k' we can show the following.

Theorem 2. *For* $(m - 2)$*-approval with* t *partial votes,* POSSIBLE WINNER *can be solved in* $4^t \cdot \text{poly}(n, m)$ *time and* $O(t \cdot 4^t)$ *space. For* $(m - k')$*-approval with* t *partial votes,* POSSIBLE WINNER *can be solved in* $2^{O(t)} \cdot \text{poly}(n, m)$ *time for constant* k'*.*

3 Fixed Number of One-Positions

We study POSSIBLE WINNER for k-approval with respect to the combined parameter k and number t of partial votes. The problem can be considered as "filling" tk one-positions such that no candidate beats c. In the previous section, we exploited that the number of candidates that must take a zero-position is already bounded by the combined parameter t and "number of zero-positions" in a yes-instance (Observation 1). Here, we cannot argue analogously: Our combined parameter (t, k) only bounds the number of one-positions but there can be an unbounded number of candidates that may take a one-position in different winning extensions of the partial votes. Hence, we argue that if there are too many candidates that can take a one-position, then there must be several choices that lead to a valid extension. We show that it is sufficient to keep a set of "representative candidates" that can take the required one-positions if and only if this is possible for the whole set of candidates. This results in a problem kernel of super-exponential size showing fixed-parameter tractability with respect to (t, k). We complement this result by showing that it is very unlikely that there is a kernel of polynomial size. In addition, we give a polynomial kernel with $O(t^2)$ candidates for 2-approval.

3.1 Problem Kernels

We first describe a kernelization approach for POSSIBLE WINNER for k-approval in general and then show how to obtain a better bound on the kernel size for 2-approval.

Problem kernel for k-approval. In order to describe more complicated reduction rules, we assume that a considered instance is exhaustively reduced with respect to some simple rules. To this end, we fix the distinguished candidate c as good as possible by Rule 1 (using that $m - k' = k$). Afterwards, we apply a simple reduction rule to get rid of "irrelevant" candidates and check whether an instance is a trivial no-instance:

Rule 2. *First, for every candidate* $c' \in C \setminus \{c\}$*, if making one point in the partial votes causes* c' *not to be beaten by* c*, then fix* c' *as bad as possible in every vote. Second, compute the set* D *of candidates that can be deleted: For every candidate* $c' \in C \setminus \{c\}$ *with* $|L(v, c')| > k$ *for all* $v \in V^p$*, if the score* $s(c')$ *is at least* $s(c)$*, then output "no solution", otherwise add* c' *to* D*. Delete* D *and replace* V^l *by an equivalent set.*

The soundness of Rule 2 is easy to see: Every candidate fixed by the first part cannot be assigned to a one-position in any winning extension. For the second part, every winning extension of an unreduced instance can easily be transformed into a winning extension for the reduced one by deleting the candidates specified by Rule 2 and *vice versa.* A set of equivalent linear votes can be found according to [3, Lemma 1][1].

[1] Herein, it might be necessary to add one new candidate. However, this will not affect the following analysis and will be discussed in more detail in the full version of this work.

$$v: a \succ b \succ d \succ x,$$
$$e \succ f,$$
$$g \succ f \succ y \succ c,$$
$$a \succ h$$

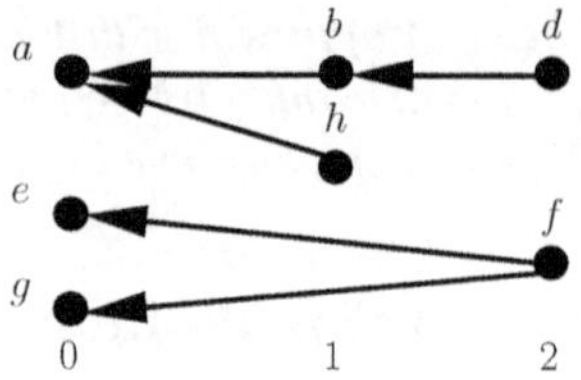

Fig. 2. Example for 3-approval: Partial vote v (left-hand side) and corresponding digraph with levels 0, 1, and 2. Arcs following by transitivity are omitted. Note that x, y, and c do not appear in the digraph since they are irrelevant for v.

In the following, we assume that Rule 2 has been applied, that is, all remaining candidates can make at least one point in an extension without beating c. To state further reduction rules, a partial vote v is represented as a digraph with vertex set $\{c' \mid c' \in C \setminus \{c\} \text{ and } |L(v, c')| < k\}$. All other candidates are considered as "irrelevant" for this vote since they cannot take a one-position. The vertices are organized into k levels. For $0 \leq j \leq k-1$, let $L_j(v) := \{c' \mid c' \in C \setminus \{c\} \text{ and } |L(v, c')| = j\}$ containing all candidates that shift exactly j candidates to a one-position if they are assigned to the best possible position. There is a directed arc from c' to c'' if and only if $c'' \in L(v, c')$. Figure 2 displays an example for the representation of a partial vote for 3-approval.

In general, the number of candidates per level is unbounded. However, for some cases it is easy to see that one can "delete" all but some representative candidates. The following reduction rule provides such an example using the fact that in any vote a candidate from the first level can be set to an arbitrary one-position without shifting any other candidate.

Rule 3. *For $v \in V^p$ with $|L_0(v)| \geq tk$, consider any subset $L' \subseteq L_0(v)$ with $|L'| = tk$. Add $L' \succ C \setminus L'$ to v.*

To see the soundness of Rule 3 consider a winning extension E for a non-reduced instance and a vote $v \in V^p$ with $|L_0(v)| \geq tk$. Since there are tk one-positions in the partial votes, there must be at least k candidates from L' not having assumed a one-position within the other $t-1$ votes. Setting these k candidates to the one-positions in v leads to a winning extension of the reduced instance. The other direction is obvious.

If Rule 3 applies to all partial votes, then in a reduced instance at most t^2k candidates are not fixed at zero-positions in V^p and the remaining candidates can be deleted by Rule 2. Hence, we consider the situation that there is a partial vote v with $|L_0(v)| < tk$. Then, we cannot ignore the candidates from the other levels but replace them by a bounded number of representatives. We first discuss how to find a set of representatives for 2-approval and then extend the underlying idea to work for general k.

For 2-approval, for a vote v with $|L_0(v)| < 2t$, it remains to bound the size of $L_1(v)$. This is achieved by the following reduction rule: Fix all but $2t$ in-neighbors of every candidate from $L_0(v)$ at zero-positions. To see the soundness, we show, given a winning extension E for the non-reduced instance, how to obtain a winning extension E' for v after the reduction (the other direction is obvious). Clearly, in $E(v)$ the first position must be assigned to a candidate c' from $L_0(v)$ and c' can also be assigned to the first position in $E'(v)$. If there is another candidate from $L_0(v)$ that takes the second position in $E(v)$, one can do the same in $E'(v)$. Otherwise, distinguish two cases. First, c' has less than $2t$ in-neighbors, then the reduction rule has not fixed any candidate that shifts c' to the first position and thus v can be extended in the same way as in E. Second, c' has at least $2t$ in-neighbors. Since there are only $2t$ one-positions and $2t$ non-fixed in-neighbors, the second position of v can be assigned to a candidate that does not take a one-position in any other vote of E.

Altogether, for 2-approval, one ends up with up to $4t^2$ non-fixed candidates per vote and hence with $O(t^3)$ non-reduced candidates in total. For general k, extend this approach iteratively by bounding the number of candidates for every level:

Rule 4. *Consider a partial vote $v \in V^p$ with $|L_0(v)| < tk$. Start with $i = 1$ and repeat until $i = k$.*
- For every candidate $c' \in L_i(v)$, if there are more than tk candidates in $L_i(v)$ which have the same neighborhood as c' in $L_0(v) \cup L_1(v) \cup \cdots \cup L_{i-1}(v)$, fix all but tk of them as bad as possible.
- Set $i := i + 1$.

Using Rule 4 one can show the following.

Theorem 3. *For k-approval,* POSSIBLE WINNER *admits a problem kernel with size bounded by a computable function in k and the number of partial votes t.*

Improved problem kernel for 2-approval. As discussed above, the kernelization as stated for k-approval in general leads to a polynomial kernel with $O(t^3)$ candidates for 2-approval. To give a kernel with $O(t^2)$ candidates, we use some properties of bipartite graphs. For a bipartite graph $(G \cup H, E)$ with vertex set $G \cup H$ and edge set $E \subseteq \{\{g, h\} \mid g \in G \text{ and } h \in H\}$, a *matching* denotes a subset $M \subseteq E$ such that for all $e, e' \in M$, $e \cap e' = \emptyset$. A vertex contained in e for an $e \in M$ is called *matching vertex* and, for $\{g, h\} \in M$, g and h are *matching neighbors*. A *maximum matching* is a matching with maximum cardinality. The *open neighborhood* of a vertex $g \in G$ is denoted by $N(g) := \{h \mid \{g, h\} \in E\}$ and, for $G' \subseteq G$, $N(G') := \bigcup_{g \in G'} N(g)$.

Lemma 1. *For a bipartite graph $(G \cup H, E)$ with maximum matching M, there is a partition of G into $G_1 \uplus G_2$, such that the following holds. First, all neighbors of G_1 are part of M. Second, every vertex from G_2 has a matching neighbor outside $N(G_1)$.*

Now, we employ Lemma 1 to design a reduction rule. Note that similar arguments are used in several works, see [8,16]. In the following, we assume that Rule 1 and Rule 2 have been applied. We define a bipartite graph $(G \cup H, E)$ as follows. For a partial profile with partial votes V^p and candidate set C, let $V' := \{v' \in V^p \mid |L_0(v')| < 2t\}$. For every $v'_i \in V'$, for $1 \leq j \leq |L_0(v'_i)|$, add a vertex g_i^j to G. Intuitively, for every candidate that can take a first position in v'_i there is a corresponding vertex in G. If a candidate can take the first position in several votes, then there are several vertices corresponding to this candidate. The vertex set H contains one vertex for every candidate from $(\bigcup_{v' \in V'} L_1(v')) \setminus (\bigcup_{v' \in V'} L_0(v'))$. There is an edge between $g_i^j \in G$ and $h \in H$ if setting the candidate corresponding to h to the second position in v'_i shifts the candidate corresponding to g_i^j to the first position. Now, we can state the following.

Rule 5. *Compute a maximum matching M in $(G \cup H, E)$. Fix every candidate corresponding to a non-matched vertex in H as bad as possible in every vote from V'.*

Lemma 2. *Rule 5 is sound and can be carried out in $O(|E| \cdot |G \cup H| + |V| \cdot |C|)$ time.*

Proof. A winning extension for an instance reduced with respect to Rule 5 is also a winning extension for an unreduced instance. Now, we show the other direction. Given a winning extension E for an unreduced instance, we construct a winning extension E_r for a reduced instance. Since Rule 5 does not fix any candidate which can take the first position in at least one vote, the first positions in E_r can be assumed by the same candidates as in E. It remains to fix the second positions without beating c. For every vote v_i, let g_i^e denote the candidate that takes the first position in v_i in E. For the corresponding vertex g_i^e one can distinguish two cases: First, $g_i^e \in G_1$. In this case, none of the neighbors of g_i^e have been fixed and, thus, the candidate which takes the second position in v_i in E can also take the second position E_r. Second, $g_i^e \in G_2$. In this case, set the candidate corresponding to the matching neighbor from g_i^e to the second position. Now, it is not to hard to see that c wins in E_r: The only candidates that possibly make more points in E_r than in E are the candidates corresponding to the matching neighbors of vertices from G_2. Due to the matching property, every such candidate makes at most one point in V'. By definition, G only contains vertices that can make at least one point and for all votes from $V^p \setminus V'$ one can easily find a winning extension which does not assign the "matching-candidates" to one-positions (see Rule 2). It follows that c also wins in the extension E_r. The claimed running time follows since a maximum bipartite matching can be found in $O(|E| \cdot |G \cup H|)$ time. □

Bounding the size of candidates in level 0 by Rule 3 and the (remaining) candidates in level 1 by Rule 5 one arrives at the following.

Theorem 4. *For 2-approval with t partial votes,* POSSIBLE WINNER *admits a polynomial kernel with less than $4t^2$ candidates.*

3.2 Kernel Lower Bound

We use a method introduced by Bodlaender et al. [7] and Fortnow and Santhanam [12] to show that, for k-approval, POSSIBLE WINNER cannot have a polynomial kernel with respect to (t, k). They provide a general scheme to show the non-existence of polynomial kernels under some reasonable assumptions from classical complexity theory.

Definition 1. *[7] A composition algorithm for a parameterized problem $L \subseteq \Sigma^* \times \mathbb{N}$ is an algorithm that receives as input a sequence $((x_1, p), \ldots, (x_q, p))$ with $(x_i, p) \in \Sigma^* \times \mathbb{N}$ for each $1 \le i \le q$, uses time polynomial in $\sum_{i=1}^{q} |x_i| + p$, and outputs $(y, p') \in \Sigma^* \times \mathbb{N}$ with (a) $(y, p') \in L \Leftrightarrow (x_i, p) \in L$ for some $1 \le i \le q$ and (b) p' is polynomial in p. A parameterized problem is compositional if there is a composition algorithm for it.*

Theorem 5. *[7,12] Let L be a compositional parameterized problem whose unparameterized version is NP-complete. Then, unless* coNP $\subseteq$ NP / poly, *there is no polynomial kernel for L.*

Dom et al. [9] provide a framework to build composition algorithms by using "identifiers". One of the necessary conditions is the existence of an algorithm running in $2^{p^\gamma} \cdot$ poly time for the considered parameter p and a fixed constant γ. Considering the combined parameter "number of ones" k and "number of partial votes" t for POSSIBLE WINNER under k-approval, there is no known algorithm running in $2^{(tk)^\gamma} \cdot \text{poly}(|X|)$ time. Hence, we apply the following overall strategy (which might be also useful for other problems).

Overall strategy. We employ a proof by contradiction. Assume that there is a polynomial kernel with respect to (t, k). Then, since for POSSIBLE WINNER there is an obvious brute-force algorithm running in $m^{tk} \cdot \text{poly}(n, m)$ time for m candidates and n votes, there must be an algorithm A with running time $\text{poly}(t, k)^{tk} \cdot \text{poly}(n, m) < 2^{(tk)^\gamma} \cdot \text{poly}(n, m)$ for an appropriate constant γ. In the next paragraph, we use the existence of algorithm A to design a composition algorithm for the combined parameter (t, k). Since it is easy to verify that the unparameterized version[2] of POSSIBLE WINNER is NP-complete, it follows from Theorem 5 that there is no polynomial kernel with respect to (t, k), a contradiction unless coNP $\subseteq$ NP / poly. Altogether, it remains to give a composition algorithm.

Composition algorithm. Consider a sequence $((x_1, (t, k)), \ldots, (x_q, (t, k)))$ of q POSSIBLE WINNER instances for k-approval. To simplify the construction, we make two assumptions. First, we assume that there is no "obvious no-instance", that is, an instance in which a candidate c' beats c even if c' makes zero points in all of the partial votes. This does not constitute any restriction since such instances can be found and removed in time polynomial in $\sum_{i=1}^{q} |x_i|$. Second, we assume that for $x_j, 1 \leq j \leq q$, within the partial votes the distinguished candidate makes zero points in every extension. Since it follows from known constructions [3,5] that the unparameterized version of the problem remains NP-complete for this case, this assumption leads to a non-existence result for this special case and thus also for the general case.

Now, we give the composition algorithm. If $q > 2^{(tk)^\gamma}$ for γ as given by algorithm A, the composition algorithm applies A to every instance. This can be done within the running time bound required by Definition 1 and, in the following, we can assume that the number of instance is at most $2^{(tk)^\gamma}$. This can be used to assign an "identifier" of sufficiently small size to every instance. Basically, the identifiers, which will be realized by specific sets of candidates, rely on the binary representation of the numbers from $\{1, \ldots, q\}$. The size of an identifier will be linear in $s := \lceil \log q \rceil$ which is polynomial in the combined parameter (t, k) since $q \leq 2^{(tk)^\gamma}$.

Compose the sequence of instances to one big instance $(X, (3s + 4, 2t))$ with $X = (C, V^l \cup V^p, c)$ as follows. For $1 \leq i \leq q$, let x_i be $(C_i, V_i^l \cup V_i^p, c_i)$. Then,

$$C := \biguplus_{1 \leq i \leq q} (C_i \setminus \{c_i\}) \uplus \{c\} \uplus D \uplus Z \uplus A \uplus B$$

with $D := \{d_0^0, \ldots, d_s^0\} \cup \{d_0^1, \ldots, d_s^1\}$, $Z := \bigcup_{1 \leq j \leq t} Z_j$ with $Z_j := \{z_{h,j}^0 \mid 0 \leq h \leq s\} \cup \{z_{h,j}^1 \mid 0 \leq h \leq s\}$, $A := \{a_1, \ldots, a_q\}$, and a set B with $|B| := 2s + 3 - k$. The candidates from D and Z will be used as identifiers for the different instances: Every

[2] See [7] for a formal definition.

$$\begin{array}{llll} V_1^p : Z_{w,1} > \overline{D_w} > \overline{Z_{w,t}} & > a_w > C \setminus (Z_{w,1} \cup \overline{D_w} \cup \overline{Z_{w,t}}) & \\ \quad Z_{w,j} > \overline{D_w} > \overline{Z_{w,j-1}} & > a_w > C \setminus (Z_{w,j} \cup \overline{D_w} \cup \overline{Z_{w,j-1}}) & \text{for } 2 \leq j \leq t \\ V_2^p : B \quad > D_w > w_j & > C \setminus (B \cup D_w \cup (C_w \setminus \{c_w\})) & \text{for } 1 \leq j \leq t \end{array}$$

Fig. 3. Extension for X in which c wins. For a winning extension $E(x_w) = w'_1, \ldots, w'_t$ of x_w, let w_j denotes the linear order given by w'_j restricted to the candidates from $C_w \setminus \{c\}$.

instance x_i is uniquely identified by the binary code of the integer $i = b_0 \cdot 2^0 + b_1 \cdot 2^1 + \cdots + b_s \cdot 2^s$ with $b_h \in \{0, 1\}$. Then, a subset $D_i \subset D$ identifies x_i when $d_h^1 \in D_i$ if and only if $b_h = 1$ and $d_h^0 \in D_i$ if and only if $b_h = 0$. Let $\overline{D_i} := D \setminus D_i$. Similarly, for every $1 \leq j \leq t$, the set $Z_{i,j}$ denotes the candidates from Z_j that identify i, that is,

$$Z_{i,j} := \{z_{h,j}^0 \mid h \in \{0, \ldots, s\} \text{ and } b_h = 0\} \cup \{z_{h,j}^1 \mid h \in \{0, \ldots, s\} \text{ and } b_h = 1\}.$$

Let $\overline{Z_{i,j}} := Z_j \setminus Z_{i,j}$ denote the remaining candidates from Z_j.

The set of partial votes V^p consists of two subsets V_1^p and V_2^p, both containing t partial votes. The basic idea is that a winning extension of V_1^p "selects" an (arbitrary) instance x_i and there is a winning extension for x_i if and only if V_2^p can be extended such that c wins. The set V_1^p contains the vote

$$\{Z_{i,1} \cup \overline{D_i} \cup \overline{Z_{i,t}} \succ a_i \mid 1 \leq i \leq q\},\ D \cup Z \cup A \succ C \setminus (D \cup Z \cup A),$$

meaning that the vote contains the condition $Z_{i,1} \cup \overline{D_i} \cup \overline{Z_{i,t}} \succ a_i$ for every i. Furthermore, for every $j \in \{2, \ldots, t\}$, the set V_1^p contains the vote

$$\{Z_{i,j} \cup \overline{D_i} \cup \overline{Z_{i,j-1}} \succ a_i \mid 1 \leq i \leq q\},\ D \cup Z \cup A \succ C \setminus (D \cup Z \cup A).$$

The set V_2^p consists of the partial votes $v_1, \ldots, v_t$. Every vote $v_j \in V_2^p$ "composes" the votes v_i^j for $1 \leq i \leq q$ where v_i^j denotes the jth vote from instance x_i after deleting c_i. Then, for $1 \leq j \leq t$, the vote v_j is

$$B \succ (C \backslash B),\ \{v_i^j \mid 1 \leq i \leq q\},\ \{D_i \succ C_i \backslash \{c_i\} \mid 1 \leq i \leq q\},\ C \backslash (A \cup Z \cup \{c\}) \succ A \cup Z \cup \{c\}.$$

Using [3, Lemma 1], one can construct a set V_l of linear votes polynomial in $|C|$ and $|V^p|$ such that in every winning extension, within V^p,
(a) for $i \in \{1, \ldots, q\}$, the number of points a candidate $c' \in C_i \setminus \{c_i\}$ makes is at most the maximum number of points which c' makes in a winning extension within the partial votes from x_i,
(b) every candidate from $A \cup D \cup B$ makes at most t points, and
(c) every candidate from Z makes at most one point.

Fig. 3 sketches a winning extension of the composed instance X making use of a winning extension of an instance x_w. We omit to show that the constructed instanced X is a yes-instance for $(3s+4)$-approval if and only if there is an $i \in \{1, \ldots, q\}$ such that x_i is a yes-instance for k-approval.

Theorem 6. *For k-approval,* POSSIBLE WINNER *with t partial votes does not admit a polynomial problem kernel with respect to (t, k) unless* NP $\subseteq$ coNP / poly.

4 Outlook

Can similar results as in this paper be obtained for "more general" problems such as SWAP BRIBERY [11] or the counting version of POSSIBLE WINNER[1]? This comprises the development of reduction rules preserving all winning extensions. Another interesting scenario might be as follows. Given a number s of winners in the input, for example, the size of a committee, one is interested in the s candidates such that each of them has more points than the remaining candidates. For this scenario, the negative results for POSSIBLE WINNER for k-approval as given in this work and related work [3,5] can be adapted by adding $s - 1$ fixed candidates that always win, but as to the algorithmic results, it is open whether they extend to this scenario.

References

1. Bachrach, Y., Betzler, N., Faliszewski, P.: Probabilistic possible winner determination. In: Proc. of 24th AAAI (to appear 2010)
2. Baumeister, D., Rothe, J.: Taking the final step to a full dichotomy of the Possible Winner problem in pure scoring rules. In: Proc. of 19th ECAI (2010) (short paper)
3. Betzler, N., Dorn, B.: Towards a complexity dichotomy of finding possible winners in elections based on scoring rules. In: Královič, R., Niwiński, D. (eds.) MFCS 2009. LNCS, vol. 5734, pp. 124–136. Springer, Heidelberg (2009)
4. Betzler, N., Guo, J., Niedermeier, R.: Parameterized computational complexity of Dodgson and Young elections. Inform. Comput. 208(2), 165–177 (2010)
5. Betzler, N., Hemmann, S., Niedermeier, R.: A multivariate complexity analysis of determining possible winners given incomplete votes. In: Proc. of 21st IJCAI, pp. 53–58 (2009)
6. Bodlaender, H.L.: Kernelization: New upper and lower bound techniques. In: Chen, J., Fomin, F.V. (eds.) IWPEC 2009. LNCS, vol. 5917, pp. 17–37. Springer, Heidelberg (2009)
7. Bodlaender, H.L., Downey, R.G., Fellows, M.R., Hermelin, D.: On problems without polynomial kernels. J. Comput. Syst. Sci. 75(8), 423–434 (2009)
8. Chor, B., Fellows, M., Juedes, D.W.: Linear kernels in linear time, or how to save k colors in $o(n^2)$ steps. In: Hromkovič, J., Nagl, M., Westfechtel, B. (eds.) WG 2004. LNCS, vol. 3353, pp. 257–269. Springer, Heidelberg (2004)
9. Dom, M., Lokshtanov, D., Saurabh, S.: Incompressibility through colors and IDs. In: Albers, S., Marchetti-Spaccamela, A., Matias, Y., Nikoletseas, S., Thomas, W. (eds.) ICALP 2009. LNCS, vol. 5555, pp. 378–389. Springer, Heidelberg (2009)
10. Downey, R.G., Fellows, M.R.: Parameterized Complexity. Springer, Heidelberg (1999)
11. Elkind, E., Faliszewski, P., Slinko, A.: Swap bribery. In: Mavronicolas, M., Papadopoulou, V.G. (eds.) Algorithmic Game Theory. LNCS, vol. 5814, pp. 299–310. Springer, Heidelberg (2009)
12. Fortnow, L., Santhanam, R.: Infeasibility of instance compression and succinct PCPs for NP. In: Proc. of 40th STOC, pp. 133–142. ACM, New York (2008)
13. Guo, J., Niedermeier, R.: Invitation to data reduction and problem kernelization. ACM SIGACT News 38(1), 31–45 (2007)
14. Hemaspaandra, E., Hemaspaandra, L.A.: Dichotomy for voting systems. J. Comput. Syst. Sci. 73(1), 73–83 (2007)
15. Konczak, K., Lang, J.: Voting procedures with incomplete preferences. In: Proc. of IJCAI 2005 Multidisciplinary Workshop on Advances in Preference Handling (2005)
16. Niedermeier, R.: Invitation to Fixed-Parameter Algorithms. Oxford University Press, Oxford (2006)
17. Pini, M.S., Rossi, F., Venable, K.B., Walsh, T.: Incompleteness and incomparability in preference aggregation. In: Proc. of 20th IJCAI, pp. 1464–1469 (2007)
18. Xia, L., Conitzer, V.: Determining possible and necessary winners under common voting rules given partial orders. In: Proc. of 23rd AAAI, pp. 196–201. AAAI Press, Menlo Park (2008)

Counting Minimum (s,t)-Cuts in Weighted Planar Graphs in Polynomial Time

Ivona Bezáková[1] and Adam J. Friedlander[2]

[1] Rochester Institute of Technology, Rochester, NY, USA
ib@cs.rit.edu
[2] IBM, Poughkeepsie, NY, USA
afriedl@us.ibm.com

Abstract. We give an $O(nd+n\log n)$ algorithm computing the number of minimum (s,t)-cuts in weighted planar graphs, where n is the number of vertices and d is the length of the shortest s-t path in the corresponding unweighted graph. Previously, Ball and Provan gave a polynomial-time algorithm for unweighted graphs with both s and t lying on the outer face. Our results hold for all locations of s and t and weighted graphs, and have direct applications in image segmentation and other computer vision problems.

1 Introduction

Graph cuts play an important role in a number of computer vision algorithms. For example, in image segmentation, see e. g. [4,5,9], an image is represented by a graph with pixels as the vertices and an edge connects two pixels if they are neighboring and considered similar; the edge weights capture the similarity measure between the pixels. Naturally, the underlying graph tends to be planar, typically grid-like. One of the problems of image segmentation is to separate an object from the background. Many segmentation algorithms rely on finding a minimum cut between two positions, one from the object and one from the background (often provided as input from the user). The weight of the minimum cut corresponds to the least energy contour between the positions, viewing the edge weights as the strength of the connection between the respective pixels. Notice that the two chosen locations are very unlikely to be physically close to each other, hence our assumptions on planarity and arbitrary positions of the locations are well aligned with the image segmentation applications.

Counting problems are closely related to random sampling from the same universe [14]. In image segmentation the current algorithms might suffer from finding an "atypical" cut that does not represent the contour well. Ideally the user would get the opportunity to choose the best contour out of all possible minimum-cut-based segmentations. However, this might be infeasible because the number of minimum cuts between two chosen positions might be exponential. Having the ability to sample from several minimum cuts provides the user with the option to choose the best of these segmentations while keeping the running time reasonably small.

P. Hliněný and A. Kučera (Eds.): MFCS 2010, LNCS 6281, pp. 126–137, 2010.

Minimum cuts are also related to network reliability problems where the vertices are individual computers in a network, edges are connections between computers, and the edge weight captures the probability of connection failure. The number of minimum cuts between two end-points is useful in estimating the probability of disconnecting the network, see e. g., [1]. Ball and Provan [1] showed that, in case of unweighted (multi)graphs, the problem of counting all minimum (s,t)-cuts is polynomially reducible to the problem of counting all antichains in a poset. (An (s,t)-cut can be visualized as a set of edges that, if removed, disconnect the vertices s and t, see Section 2 for the formal definition.) Both problems are known to be #P-complete, as shown by the same authors in [18]. Nevertheless, they were able to devise a polynomial-time algorithm for counting all minimum (s,t)-cuts in planar graphs, under the assumption that both vertices s and t lie on the outer face. Different variants of the network reliability problem and its connection to minimum cuts were studied in a number of previous works, for example [2,19,17,15,6].

Our main contribution is an efficient polynomial-time algorithm for computing the number of minimum (s,t)-cuts in all weighted planar graphs and for all pairs of s and t (i. e., we do not impose assumptions on the locations of s and t). Recall that this is exactly the scenario of many vision applications. (For directed graphs we work under the natural and commonly assumed condition that all vertices are reachable from s and lead to t, see, e. g., [7]. Otherwise, the typical definition of cuts leads to pathological cases, as discussed in Section 4.) We extend the result of Ball and Provan to the case of weighted graphs, showing that this case also polynomially reduces to the problem of counting (unweighted) antichains in a poset. Our main result, summarized in Theorem 1 below, uses the reduction to devise a polynomial-time algorithm for counting minimum (s,t)-cuts for weighted planar graphs, for all possible pairs of s and t.

Theorem 1. *Let $G = (V, E, w)$ be any (directed) planar graph with edge weights $w : E \to \mathbf{R}^+$. Let $s, t \in V$, $s \neq t$ and assume that all vertices are reachable from s and lead to t. Then, there is an $O(nd + n \log n)$ algorithm for counting all minimum (s,t)-cuts in G, where $n = |V|$ and d is the smallest number of edges forming a path from s to t in G.*

When both s and t lie on the outer face, it is possible to connect them by an edge, splitting the outer face into two faces. The idea of [1] relies on the fact that the antichain problem can then be solved by counting the number of paths between the two new faces in the dual (directed) planar graph. However, planarity does not allow to add the (s,t) edge for arbitrary locations of s and t. We overcome this problem by showing that we can utilize one of the paths from s to t. Our proof of correctness is significantly more elaborate than the outer face case, yet the underlying algorithm is still relatively simple, as summarized in Algorithm 1.

For completeness, we review results studying the problem of finding one of the minimum (s,t)-cuts in a given weighted planar graph. Building on the work of Itai and Shiloach [12], Reif [20] developed an $O(n \log^2 n)$ divide-and-conquer algorithm for undirected graphs. Janiga and Koubek [13] designed an $O(n \log^2 n \log \log n)$ algorithm for directed planar graphs. The result of

Algorithm 1. Counting minimum-(s,t)-cuts in a weighted planar graph G

1: Compute a maximum s-t flow such that the directed flow edges do not form a cycle.
2: Construct $\hat{G}$ by contracting every strongly connected component of the residual graph, let $\hat{s}$ and $\hat{t}$ be the vertices of $\hat{G}$ corresponding to s and t, respectively.
3: Let p be a $\hat{t}$-$\hat{s}$ path in $\hat{G}$. Duplicate all edges of p, the new edges are on the left of p when traveling from $\hat{t}$ to $\hat{s}$.
4: Follow the new edges from $\hat{t}$ to $\hat{s}$. If the current edge shares the same face on the left as the previous edge, merge the edges into one edge by bypassing the middle vertex. This process results in an $\hat{t}$-$\hat{s}$ path p'.
5: Let G' be the graph $\hat{G}$ with the path p'. Construct a (directed) unweighted dual planar graph G'_d of G', omit edges that cross the edges of p'.
6: For every pair of vertices a, b in G'_d that correspond to faces that share an edge in p', compute the number of all a-b paths in G'_d, using an algorithm for directed acyclic graphs.
7: Return the sum of all numbers computed in step 6.

Borradaile and Klein [3] yields an $O(n \log n)$ algorithm for all planar graphs. The dual graph plays a central role in all these works.

This paper is organized as follows. We present preliminaries, graph terminology, and notation, in Section 2. We state the reduction result in Section 3 and we prove the main result, Theorem 1, in Section 4. Section 5 contains the proofs of the results from Section 3.

2 Preliminaries

We denote by $\mathbf{R}$, $\mathbf{R}^+$, and $\mathbf{R}_0^+$ the sets of all real numbers, positive real numbers, and nonnegative real numbers, respectively.

We work with directed graphs throughout the paper. The usual conversion of undirected graphs into directed graphs (for every undirected edge include two directed edges) provides corresponding algorithms for undirected graphs.

Let $G = (V, E, w)$ be a directed graph with positive edge weights $w : E \to \mathbf{R}^+$. Let $s, t \in V, s \neq t$ be two vertices. An *(s,t)-cut of G* is a set of vertices $S \subseteq V$ that contains s but not t. The *value of the cut S* is the sum of the edge weights of the edges going out of the set S, i.e., $\sum_{(u,v):u\in S, v\notin S} w(u,v)$. A *minimum (s,t)-cut* has the smallest possible value of all (s,t)-cuts.

Our objective is to *count* the number of all possible minimum (s,t)-cuts of an input graph G.

Minimum cuts are related to network flows. A *flow network* is a directed graph $G = (V, E, c)$ where $c : E \to \mathbf{R}^+$ defines non-negative *edge capacities*. Let $s, t \in V, s \neq t$ be two vertices called *source* and *sink*, respectively. A *flow* from s to t is a function $f : E \to \mathbf{R}_0^+$ satisfying the following properties:

- *capacity constraint:* $f(e) \leq c(e)$ for every $e \in E$, and
- *flow conservation:* $\sum_{u:(u,v)\in E} f(u,v) = \sum_{w:(v,w)\in E} f(v,w)$ for every $v \in V - \{s,t\}$.

The *value of the flow* f is the sum of the values of flow edges out of s minus the sum of the flow edges into s, i. e., $\sum_{w:(s,w)\in E} f(s,w) - \sum_{w:(w,s)\in E} f(w,s)$. A flow is said to be *maximum* if it has the largest possible value among all flows from s to t (we also refer to such flows as s-t-flows).

The *residual graph of the flow* f, denoted $G_f = (V, E_f, w_f)$, is a weighted directed graph where E_f contains the following two types of edges:

- for every $e \in E$ with $f(e) < c(e)$, the set E_f contains a *forward edge* e with weight $w_f(e) = c(u,v) - f(u,v)$, and
- for every $e = (u,v) \in E$ with $f(e) > 0$, the set E_f contains a *backward edge* $e' = (v,u)$ with weight $w_f(e') = f(e)$.

An *augmenting path* in a residual graph G_f is any path from s to t.

The following is a well-known Maximum-flow Minimum-cut Theorem by Ford and Fulkerson [8].

Theorem 2. *Let $G = (V, E, c)$ be a directed graph with positive edge weights and let $s, t \in V$. Then, the value of the minimum (s,t)-cut in G equals the maximum s-t-flow value in the flow network G.*

For more information about network flows, see, e. g., [16]. Most of our terminology and notation follows this reference.

3 Reduction to Forward-Cuts

We give a polynomial reduction from the problem of counting minimum (s,t)-cuts in a positively weighted graph to the problem of counting antichains in a poset. A poset can be represented by a directed acyclic graph and an antichain is a set of pairwise unrelated vertices (i. e., no vertex has a predecessor in the set).

Instead of proving our results for antichains, we define a closely related notion that we call forward-cuts. A forward-cut contains the antichain elements and all their predecessors. Moreover, a forward-(s,t)-cut contains the vertex s but not t. The formal definition is summarized below.

Definition 1. *Let $G = (V, E)$ be a directed acyclic (multi)graph, and let $s \in V$ be a vertex in G of indegree 0 and $t \in V$ be a vertex in G of outdegree 0. Let S be a subset of the vertices V such that $s \in S$ and $t \notin S$. We say that S is a forward-(s,t)-cut of G if there is no edge $(u,v) \in E$ such that $v \in S$ and $u \notin S$.*

The reduction result is stated in the following theorem.

Theorem 3. *Let $G = (V, E, c)$ be a (directed) flow network with edge capacities $c : E \to \mathbf{R}^+$. Let $s \in V$ be the source and $t \in V$ be the sink. There exists a directed acyclic graph $\tilde{G} = (\tilde{V}, \tilde{E})$ and vertices $\tilde{s}, \tilde{t} \in \tilde{V}$ such that the number of minimum (s,t)-cuts in G is equal to the number of forward-$(\tilde{t}, \tilde{s})$-cuts in $\tilde{G}$. Moreover, $|\tilde{V}| \le |V|$, $|\tilde{E}| \le 2|E|$, and it is possible to construct $\tilde{G}$ in time $O(|V|^3 + |E|^2)$. Also, if G is planar, then $\tilde{G}$ is planar as well and it can be constructed in time $O(|V| \log |V|)$.*

The proof of the theorem is included in Section 5. In the next section we will deal with graphs where every vertex is reachable from s and leads to t. The following corollary will be used in the proof of Theorem 1.

Corollary 1. *Suppose that there exists a path from s to every vertex of G and a path from every vertex of G to t. Then, $\tilde{t}$ is the only vertex of indegree 0 and $\tilde{s}$ is the only vertex of outdegree 0 in $\tilde{G}$.*

4 Minimum Cuts in Planar Graphs

The following theorem states that we can count the number of minimum (s,t)-cuts in weighted planar graphs in polynomial-time. We impose a natural condition on the input graphs: we can get to every vertex from s and we can get to t from every vertex. Without this condition, vertices that do not influence connectivity of s and t may artificially increase the number of minimum (s,t)-cuts. For example for a graph on vertices $\{s,t,a,b,c\}$ with arcs $\{(s,a),(a,t),(a,b),(a,b)\}$ we have (s,t)-cuts $\{s\}$, $\{s,a\}$, $\{s,a,b\}$, $\{s,a,c\}$, and $\{s,a,b,c\}$. However, the "true" (s,t)-cuts are only $\{s\}$ and $\{s,a\}$.

Theorem 1. *Let $G=(V,E,w)$ be any (directed) planar graph with edge weights $w: E \to \mathbf{R}^+$. Let $s,t \in V, s \neq t$, be two of its vertices and assume that all vertices are reachable from s and t is reachable from every vertex. Then, there is an $O(nd + n\log n)$ algorithm for counting all minimum (s,t)-cuts in G, where $n = |V|$ and d is the smallest number of edges forming a path from s to t in G.*

Before we prove the theorem, it will be useful to observe that a simple dynamic programming idea can be used to count the number of all paths between two end-points in a given directed acyclic graph.

Proposition 1. *Let D be a directed acyclic graph and let $a,b \in V(D)$ be two of its vertices. The number of paths from a to b can be counted in linear time. Moreover, if D is a weighted graph where a path from a to b gets the weight of the product of its edge weights, we can compute the sum of the weights of all paths from a to b in linear time.*

Now we are ready to prove the main theorem of this section.

Proof (of Theorem 1). Let $\tilde{G}$, $\tilde{s}$, and $\tilde{t}$ be the graph, the source, and the sink from Theorem 3 applied to graph G. The theorem states that we need to count the number of forward-$(\tilde{t},\tilde{s})$-cuts in $\tilde{G}$. Suppose $\tilde{G}$ is already embedded in the plane (this can be done in linear time, see, e. g., [11]).

Let $p = (\tilde{t} = v_0, v_1, \ldots, v_k = \tilde{s})$ be a (directed) path from $\tilde{t}$ to $\tilde{s}$ in $\tilde{G}$. To simplify our language, let us redraw $\tilde{G}$ so that the path p goes horizontally from left to right. We duplicate every edge of the path, drawing the duplicate edges just above the original edges, thus the i-th duplicate edge e_i (drawn between vertices v_{i-1} and v_i) lies inside a face f_i. (Notice that we are allowing multi-edges.) We replace every consecutive block of the duplicate edges that lie inside the same face by a single edge. Formally, if $e_i, e_{i+1}, \ldots, e_j$ lie inside the same face $f_i = f_{i+1} = \cdots = f_j$ (and if either edge e_{i-1} or edge e_{j+1} exist, neither lies in the face f_i), then the edges get replaced by a single edge from v_{i-1} to v_j that lies inside f_i. After this process we obtain a new path $p' = (\tilde{t} = v'_0, v'_1, \ldots, v'_{k'} = \tilde{s})$ where no two consecutive edges lie inside the same original face. We will refer to this new graph by G'. Notice that G' is a planar directed acyclic graph with $\tilde{t}$ and $\tilde{s}$ being the only vertices of in- and out-degree 0, respectively.

Thus, every original face bordered by an edge in p on the south got split into two or more faces in G'. More precisely, there the face got split into one or more "south" faces and exactly one "north" face. The "south" faces are in bijection with the e'_i edges and we refer to the "south" face corresponding to edge e'_i by f_i^{south}. There could be several e'_i edges bordering the same "north" face. We refer to the "north" face above the edge e'_i by f_i^{north}.

Next we construct a dual graph G'_d and its planar embedding as follows. The faces of G' become the vertices of G'_d and the edges will connect neighboring faces (with one exception, see below). For two neighboring faces f'_1 and f'_2 that share an edge $e' = (u'_1, u'_2)$, there is an edge from f'_1 to f'_2 in G'_d, drawn starting in f'_1, cutting across e', and ending in f'_2, if both of the following conditions are satisfied:

- edge e' is not on the path p',
- if G' is redrawn so that e' is vertical with u'_1 being the bottom end-point, then f'_1 is on the left of e' and f'_2 is on the right.

Notice that G'_d is a planar directed graph that allows multiple edges in case when two faces of G' neighbor in more than one edge.

We claim that G'_d is also acyclic. Suppose that G'_d contains a cycle going through faces $x_1, x_2, \ldots, x_z$ (for convenience let $x_{z+1} = x_1$). The edges in G'_d defining this cycle form a face in G'_d, let us call it X. By definition of edges in G'_d, for every pair of faces x_i, x_{i+1} (in G') there is an edge in G' shared by both faces that leads from inside of X to outside of X, and there are no other edges in G' crossing through the border of X. Let y_i be the end-point of the edge shared by x_i, x_{i+1} that lies inside X. Since G' is acyclic, following the predecessors of the y_i's, we must get to a vertex of indegree 0. Thus, $\tilde{t}$, the only vertex of indegree 0, lies inside X. Following the successors of the y_i's, we must get to $\tilde{s}$; therefore, $\tilde{s}$ lies outside of X. Then, the path p' needs to cut through the border of X and must go through one of the y_i's. But, by definition of G'_d, we did not include dual edges crossing through the edges of the path p', a contradiction.

Next we claim that every path in G'_d that starts at one of the "south" faces f_i^{south} and ends at the corresponding "north" face f_i^{north} uniquely corresponds to a forward-$(\tilde{t}, \tilde{s})$-cut in $\tilde{G}$, and, vice versa, every forward-$(\tilde{t}, \tilde{s})$-cut has a corresponding path from a "south" to the corresponding "north" face in G'_d.

Let q be a path from f_i^{south} to f_i^{north} in G'_d. Let us connect the end-points of q, forming a cycle q'. Notice that the edge connecting the end-points cuts through the path p'. The cycle q' splits the plane into two regions. Let T be the set of vertices of $\tilde{G}$ that lie in the same region as $\tilde{t}$. The path q cuts through at least one edge of $\tilde{G}$, the edge bordering the face f_i^{south} on the south. The set T includes the left end-point of this edge and the complement of T includes the right end-point. Since $\tilde{s}$ is the only vertex with outdegree 0 and there is a vertex outside T, $\tilde{s}$ must lie outside of T as well. Thus, T is an $(\tilde{t}, \tilde{s})$-cut. By the definition of edges in G'_d, all edges cutting through q start in T and end outside T. By planarity, no other edges connect T with its complement, therefore, T is a forward-$(\tilde{t}, \tilde{s})$-cut. Thus, every "south-north" path defines a forward-$(\tilde{t}, \tilde{s})$-cut.

Vice versa, let T be a forward-$(\tilde{t}, \tilde{s})$-cut. Let (u, v) be such that $u \in T$ and $v \notin T$. We claim that on a face f adjacent to (u, v) there must be exactly one other edge (u', v') such that $u' \in T$ and $v' \notin T$.

First we show that there must be an edge (u', v') such that $u' \in T$ and $v' \notin T$. Let us follow the vertices on the face f, starting with v, going to u, etc. The existence of (u', v') follows by parity: we start in $v \notin T$, then visit $u \in T$, then there might be other vertices in T, but eventually we come back to v, a vertex outside of T. Thus, there must be a pair of consecutive vertices $u' \in T$ and $v' \notin T$. Since T is a forward-cut, the edge between u' and v' must go from u' to v'.

Next we show that (u, v) and (u', v') are the only two edges on the face f crossing the boundary of T. By contradiction, suppose there is another edge (u'', v'') leading out of T. Notice that the vertices u, u', u'' are not necessarily distinct but at least two of them are different (since every vertex has only two adjacent edges bordering the face and u, u', u'' are adjacent to three distinct edges (u, v), (u', v'), and (u'', v'')). Similarly, vertices v, v', v'' are not all the same vertex. Suppose we follow the edges on the face f in the cyclic order given by following its boundary (either clockwise or counterclockwise). By parity, we encounter an (u, v)-edge forward, followed by a (u, v) edge backward, followed by a (u, v) edge forward (followed, by parity, by a (u, v) edge backward, but we do not need this edge for our argument). Without loss of generality, assume the edges are encountered in order $\overrightarrow{(u, v)}$, $\overleftarrow{(v', u')}$, and $\overrightarrow{(u'', v'')}$. We know that $\tilde{t}$ leads to every vertex, in particular also u and u'. Consider the region defined by a $\tilde{t} - u$ path, a $\tilde{t} - u'$ path, and the $u, v, \ldots, v', u'$ part of the face f's boundary. We also know that every vertex, including v and v'' leads to $\tilde{s}$. Notice that $\tilde{s}$ does not lie strictly inside f since f is a face. Also notice that $\tilde{s}$ cannot lie on the paths defining the region since $\tilde{s}$ has outdegree 0. The last possibility to consider is if $\tilde{s}$ lies inside (but not on the paths) or outside (but not on the paths) of the region defined by the paths. Then, either v or v'' cannot get to $\tilde{s}$ since exactly one of v and v'' is inside the region. We obtained a contradiction with the existence of the third edge (u'', v'').

Therefore, we know that every face containing an edge cutting through T contains exactly two edges leading out of T. Such faces are neighboring by exactly the edges leading out of T, therefore, in the dual graph G'_d with edges between the "north" and "south" faces included, there is a cycle leading through all of the faces cutting through T. Since $\tilde{t}$ and $\tilde{s}$ are separated by the cycle, the path p' must cut through the cycle. Therefore, there is a "north" and "south" face pair on the cycle and the cycle can be represented by a "south" to "north" path in G'_d.

Therefore, we need to count the number of all paths starting at a "south" face and ending at the corresponding "north" face in G'_d. This can be done, by Proposition 1 with $a = f_i^{\text{south}}$ and $b = f_i^{\text{north}}$, in linear time. (Notice that G'_d could be a multi-graph. We can replace multiple edges by a single edge with edge weight equal to the duplicity of the original edges. The weighted path count corresponds to the path count in G'_d.) We need to count the paths for every "south" face f_i^{south} (this happens as many times as is the length of p', i. e., at most n times) and sum the returned values. The overall running time is $O(|p'|(|E| + |V|)) = O(dn)$ since for planar graphs $|E| = O(|V|)$. This running time includes the construction of the paths p, p', and the dual graph G'_d. Accounting for the running time from Theorem 3, we get an overall running time of $O(dn + n \log n)$. □

5 Proof of Theorem 3

We mentioned the connection between minimum cuts and maximum flows. We utilize the connection in our proofs where we work with network flows that are acyclic, as defined below.

Definition 2. *Let $G = (V, E, c)$ be a (directed) flow network with edge capacities $c : E \to \mathbf{R}^+$. Let $s \in V$ be the source and $t \in V$ be the sink. We say that an s-t flow $f : E \to \mathbf{R}_0^+$ is acyclic if the (directed) graph $F_f = (V, D_f)$, where D_f consists of edges in E that carry positive f-values (formally, $D_f = \{e \in E \mid f(e) > 0\}$), is acyclic. We call the graph F_f the flow graph of the flow f.*

Notice that the flow graph consists of the backward edges in the corresponding residual graph, reversed. The next claim, stated without proof, observes that there exists an acyclic maximum flow in every flow network.

Proposition 2. *There exists an acyclic maximum s-t flow in any flow network $G = (V, E, c)$. The acyclic flow can be found in time $O(T(G) + |E|^2)$, where $T(G)$ is the fastest maximum flow algorithm for G. For example, the push-relabel algorithm [10] yields $T(G) = O(|V|^3)$. The term $O(|E|^2)$ comes from sequential elimination of cycles in the flow. For planar graphs Borradaile and Klein's [3] maximum flow algorithm returns an acyclic maximum s-t flow in time $O(|V| \log |V|)$.*

In subsequent proofs we rely on the fact that every maximum flow can be obtained from a sequence of augmenting paths that do not use backward edges, as spelled out by the following lemma, stated without proof.

Lemma 1. *Let $f : E \to \mathbf{R}_0^+$ be an acyclic s-t flow in a flow network $G = (V, E, c)$ with source s and sink t. Then, f can be decomposed into augmenting paths $p_1, \ldots, p_d$, where $d \le |E|$.*

Next we state that if a vertex lies in a minimum cut set, then the cut set must contain all of the vertex's predecessors.

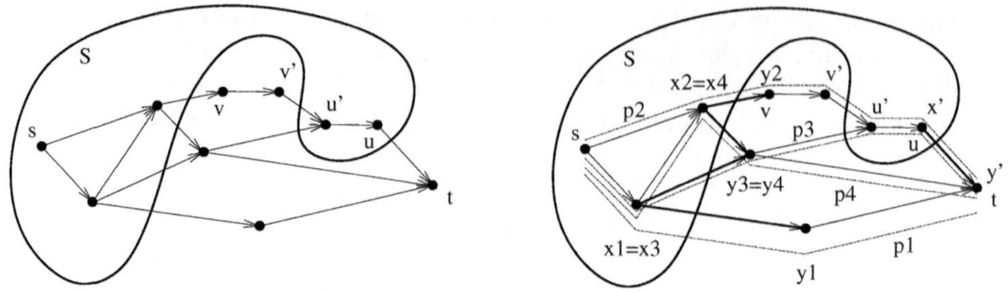

Fig. 1. Proof of Lemma 2: The left figure shows a graph F_f and an (s, t) cut S such that there are vertices $u \in S$ and $v \notin S$, and there is a path from v to u. The right figure shows the graph F_f decomposed into paths p_1, p_2, p_3, p_4 and the cut-edges (x_i, y_i) and (x', y') (these edges are highlighted).

Lemma 2. *Let $f : E \to \mathbf{R}_0^+$ be an acyclic maximum s-t flow in a flow network $G = (V, E, c)$ with source s and sink t. Let S be a minimum (s, t)-cut in G. Then, if a vertex u is in S then every vertex v that precedes u in the flow graph F_f must be in S as well.*

Proof. By contradiction, assume that there are two vertices u, v such that $u \in S$, $v \notin S$ and there exists a path from v to u in F_f, see Figure 1. It follows that there is an edge (v', u') in F_f such that $u' \in S$ and $v' \notin S$. By Lemma 1, the flow f can be decomposed into d augmenting paths $p_1, \ldots, p_d$ where the path p_i carries flow of value $\phi_i > 0$. At least one of the paths contains the edge (v', u'), let p_j be such a path. For every $i \in \{1, \ldots, d\}$ there exists an edge (x_i, y_i) on the path p_i such that $x_i \in S$, $y_i \notin S$. Moreover, for the path p_j there exists another edge (x', y') besides (x_j, y_j) such that $x' \in S$ and $y' \notin S$.

The cut value is defined as the sum of the capacities of the edges (x, y) where $x \in S$ and $y \notin S$. Therefore,

$$\text{value}(S) = \sum_{(x,y)\in E, x\in S, y\notin S} c(x, y) \geq \sum_{(x,y)\in E: \exists i: x=x_i, y=y_i \text{ or } x=x', y=y'} c(x, y)$$

$$\geq \sum_{i=1}^{d} \phi_i + \phi_j = \text{value}(f) + \phi_j.$$

Since $\phi_j > 0$, the value of the cut S is strictly greater than the value of the flow f. Therefore, by the Max-flow-min-cut Theorem (Theorem 2), S cannot be a minimum (s, t)-cut of G. □

Next we claim that no minimum cut cuts through strongly connected components of a residual graph corresponding to a maximum flow.

Lemma 3. *Let $f : E \to \mathbf{R}_0^+$ be an acyclic maximum s-t flow in a flow network $G = (V, E, c)$ with source s and sink t. Let G_f be the corresponding residual graph. If S is a minimum (s, t)-cut in G, then for each strongly connected component C in G_f, either $C \subset S$ or $C \cap S = \emptyset$.*

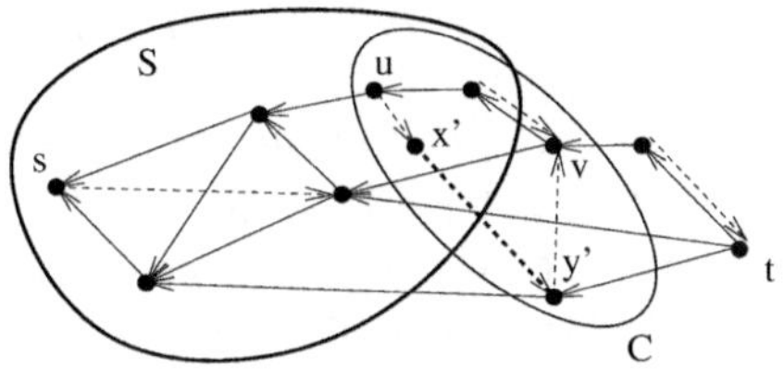

Fig. 2. Proof of Lemma 3: The figure shows a residual graph G_f and an (s,t)-cut S cutting through a strongly connected component C. Backward edges are depicted by straight arrows while the forward edges are dashed. The edge (x',y') is highlighted. For clarity we do not include residual capacities.

Proof. By contradiction, suppose that there exists a strongly connected component C and a minimum (s,t)-cut S such that there exist vertices $u,v \in C$ such that $u \in S$ and $v \notin S$, see Figure 2. Since u,v belong to the same strongly connected component, there exists a path from u to v in G_f, as well as a path from v to u in G_f. By the definition of a residual graph, G_f contains two types of edges: forward and backward edges where the backward edges are simply the edges of f reversed. We will show that there must exist a forward edge (x',y') in G_f such that $x' \in S$ and $y' \notin S$. This will imply that S cannot be a minimum (s,t)-cut.

Consider a path from u to v in G_f. Since $u \in S$ and $v \notin S$, there must exist an edge (x',y') on this path such that $x' \in S$ and $y' \notin S$. If the edge (x',y') is a backward edge then its reverse (y',x') is a flow edge and, by Lemma 2, if S is a minimum (s,t)-cut such that $x' \in S$, then y' must also be in S. Therefore, (x',y') is a forward edge.

Now let us look at the value of the cut S. By the same argument as in the proof of Lemma 2, we get

$$\text{value}(S) = \sum_{(x,y)\in E, x\in S, y\notin S} c(x,y) \geq \sum_{(x,y)\in E: \exists i: x=x_i, y=y_i \text{ or } x=x', y=y'} c(x,y)$$

$$\geq \sum_{i=1}^{d} \phi_i + \phi' = \text{value}(f) + \phi',$$

where $\phi' > 0$ is the residual capacity of the (forward) edge (x',y') in G_f. Notice that the edge (x',y') is distinct from every (x_i,y_i) since it is a forward edge. Therefore, the value of the cut S is strictly bigger than the value of the maximum flow f, a contradiction with the assumption that S is a minimum cut. □

Therefore, we define a contraction graph that contracts every strongly connected component. We obtain the following corollary of the last lemma.

Definition 3. *Let $G=(V,E)$ be a directed graph. We define the SCC-contraction graph of G, denoted $\widehat{G}$, to be the graph obtained from G by contracting each strongly connected component into a single vertex.*

Corollary 2. *Let $\widehat{G_f}$ be the SCC-contraction graph of the residual graph corresponding to an acyclic maximum s-t flow f in a flow network $G = (V, E, c)$. Suppose $\hat{s} \in V(\widehat{G_f})$ is the vertex obtained by contracting the strongly connected component containing s and $\hat{t} \in V(\widehat{G_f})$ is the vertex obtained by contracting the strongly connected component containing t. Moreover, we define a function α from the set of $(\hat{t}, \hat{s})$-cuts of $\widehat{G_f}$ to the set of (s,t)-cuts of G: for a $(\hat{t}, \hat{s})$-cut $\hat{T}$ in $\widehat{G_f}$, let $\alpha(\hat{T})$ contain all vertices v that belong to a strongly connected component $\hat{v} \notin \hat{T}$. Then,*

(i) the function α is injective, and
(ii) for every minimum (s,t)-cut S in G there exists a $(\hat{t}, \hat{s})$-cut $\hat{T}$ in $\widehat{G_f}$ such that $\alpha(\hat{T}) = S$.

Proof. To prove (i), let us first look at the (s,t)-cuts S for which there exists a $(\hat{t}, \hat{s})$-cut $\hat{T}$ such that $S = \alpha(\hat{T})$. By the definition of α it follows that any such S must satisfy the property that for every strongly connected component C of G_f, either $C \subset S$ or $C \cap S = \emptyset$.

Suppose that we have an S satisfying this property. Then we can uniquely construct $\hat{T}$ such that $S = \alpha(\hat{T})$: for every strongly connected component C of G_f, if $C \cap S = \emptyset$ then the vertex corresponding to C in $\widehat{G_f}$ is in $\hat{T}$ (and otherwise this vertex is not in $\hat{T}$). Since we can reconstruct $\hat{T}$ uniquely, α is injective.

To show part (ii), by Lemma 3, every minimum (s,t)-cut S satisfies the property that for every strongly connected component C, either $C \subset S$, or $C \cap S = \emptyset$. Therefore, there exists $\hat{T}$ (containing all strongly connected components not in S) such that $S = \alpha(\hat{T})$. □

The next two lemmas show that there is a bijection between minimum (s,t)-cuts in G and forward $(\hat{t}, \hat{s})$-cuts in the SCC-contraction graph. The bijection is given by the function α defined in the above corollary. The proofs of the lemmas, omitted due to space constraints, are similar to the proofs of Lemmas 2 and 3.

Lemma 4. *Under the assumptions of Corollary 2, suppose S is a minimum (s,t)-cut in G. Then $\alpha^{-1}(S)$ is a forward $(\hat{t}, \hat{s})$-cut in $\widehat{G_f}$.*

Lemma 5. *Under the assumptions of Corollary 2, let $\hat{T}$ be any forward-$(\hat{t}, \hat{s})$-cut in $\widehat{G_f}$. Then the (s,t)-cut $S = \alpha(\hat{T})$ is a minimum (s,t)-cut in G.*

The proof of Theorem 3, the main reduction theorem, follows from Proposition 2 and Lemmas 4 and 5.

Acknowledgments. We would like to thank the anonymous referees for many useful and insightful comments.

References

1. Ball, M.O., Provan, J.S.: Calculating Bounds on Reachability and Connectnedness in Stochastic Networks. Networks 13, 253–278 (1983)
2. Ball, M.O., Provan, J.S.: Computing Network Reliability in Time Polynomial in the Number of Cuts. Operations Research 32(3), 516–526 (1984)

3. Borradaile, G., Klein, P.: An $O(n \log n)$ algorithm for maximum st-flow in a directed planar graph. Journal of the ACM 56(2) (2009)
4. Boykov, Y., Kolmogorov, V.: An Experimental Comparison of Min-Cut/Max-Flow Algorithms for Energy Minimization in Vision. IEEE Trans. Pattern Anal. Mach. Intell. 26(9), 1124–1137 (2004)
5. Boykov, Y., Veksler, O., Zabih, R.: Fast approximate energy minimisation via graph cuts. IEEE Transactions on Pattern Analysis and Machine Intelligence 29, 1222–1239 (2001)
6. Colbourn, C.J.: Combinatorial aspects of network reliability. Annals of Operations Research 33(1), 1–15 (2005)
7. Cormen, T.H., Leiserson, C.E., Rivest, R.L., Stein, C.: Introduction to Algorithms. The MIT Press, Cambridge (2001)
8. Ford, L.R., Fulkerson, D.R.: Maximal flow through a network. Canadian Journal of Mathematics 8, 399–404 (1956)
9. Geman, D., Geman, S.: Stochastic relaxation, Gibbs distributions and the Bayesian restoration of images. IEEE Trans. Pattern Anal. Mach. Intell. 6, 721–741 (1984)
10. Goldberg, A.V., Tarjan, R.E.: A new approach to the maximum flow problem. Journal of the ACM (JACM) 35(4), 921–940 (1988)
11. Hopcroft, J., Tarjan, R.E.: Efficient planarity testing. Journal of the Association for Computing Machinery 21(4), 549–568 (1974)
12. Itai, A., Shiloach, Y.: Maximum Flow in Planar Networks. SIAM J. Comput. 8(2), 135–150 (1979)
13. Janiga, L., Koubek, V.: Minimum Cut in Directed Planar Networks. Kybernetika 28(1), 37–49 (1992)
14. Jerrum, M.R., Valiant, L.G., Vazirani, V.V.: Random generation of combinatorial structures from a uniform distribution. Theoretical Computer Science 43(2-3), 169–188 (1986)
15. Karger, D.R.: A Randomized Fully Polynomial Time Approximation Scheme for the All-Terminal Network Reliability Problem. SIAM J. Comput. 29(2), 492–514 (1999)
16. Kleinberg, J., Tardos, É.: Algorithm Design. Addison Wesley, Reading (2005)
17. Nagamochi, H., Sun, Z., Ibaraki, T.: Counting the number of minimum cuts in undirected multigraphs. IEEE Transactions on Reliability 40(5), 610–614 (1991)
18. Provan, J.S., Ball, M.O.: The complexity of counting cuts and of computing probability that a graph is connected. SIAM J. Comput. 12(4), 777–788 (1983)
19. Ramanathan, A., Colbourn, C.J.: Counting almost minimum cutsets with reliability applications. Mathematical Programming 39(3), 253–261 (1987)
20. Reif, J.H.: Minimum s-t cut of a planar undirected network in $O(n \log^2 n)$ time. SIAM Journal on Computing 12, 71–81 (1983)

Finding Best Swap Edges Minimizing the Routing Cost of a Spanning Tree*

Davide Bilò[1], Luciano Gualà[2], and Guido Proietti[1,3]

[1] Dipartimento di Informatica, Università di L'Aquila, 67010 L'Aquila, Italy
[2] Dipartimento di Matematica, Università di Tor Vergata, 00133 Roma, Italy
[3] Istituto di Analisi dei Sistemi ed Informatica, CNR, 00185 Roma, Italy

Abstract. Given an n-node, undirected and 2-edge-connected graph $G = (V, E)$ with positive real weights on its m edges, given a set of k *source* nodes $S \subseteq V$, and given a spanning tree T of G, the *routing cost* of T w.r.t. S is the sum of the distances in T from every source $s \in S$ to all the other nodes of G. If an edge e of T undergoes a *transient* failure and connectivity needs to be promptly reestablished, then to reduce set-up and rerouting costs it makes sense to temporarily replace e by means of a *swap edge*, i.e., an edge in G reconnecting the two subtrees of T induced by the removal of e. Then, a *best swap edge* for e is a swap edge which minimizes the routing cost of the tree obtained after the swapping. As a natural extension, the *all-best swap edges* problem is that of finding a best swap edge for *every* edge of T. Such a problem has been recently solved in $O(mn)$ time and linear space for arbitrary k, and in $O(n^2 + m \log n)$ time and $O(n^2)$ space for the special case $k = 2$. In this paper, we are interested to the prominent cases $k = O(1)$ and $k = n$, which model realistic communication paradigms. For these cases, we present a linear space and $\widetilde{O}(m)$ time algorithm, and thus we improve both the above running times (but for quite dense graphs in the case $k = 2$, for which however it is noticeable we make use of only linear space). Moreover, we provide an accurate analysis showing that when $k = n$, the obtained swap tree is effective in terms of routing cost.

1 Introduction

Let V be a set of n *sites* that must be interconnected, let E be a set of m potential *links* between the sites, and let $w(e)$ be some positive real *weight* associated with link e. Let $G = (V, E)$ be the corresponding weighted, undirected graph, and assume that G is 2-edge-connected (i.e., to disconnect G we have to remove at least 2 edges). Let $H = (V, E' \subseteq E)$ be a connected spanning subgraph of G, allowing all the nodes to communicate reciprocally. We call H an *all-to-all communication network*, or simply a network, in G. A network is generally built with the aim of minimizing some *cost*, which is computed using some criteria

* This work was partially supported by the PRIN 2008 research project COGENT (COmputational and GamE-theoretic aspects of uncoordinated NeTworks), funded by the Italian Ministry of Education, University, and Research.

P. Hliněný and A. Kučera (Eds.): MFCS 2010, LNCS 6281, pp. 138–149, 2010.

defined over H according to its edge weights. Thus, to maintain costs as low as possible, a network often results in a *spanning tree* of the underlying graph. There is however another crucial aspect that must be taken into consideration when designing a network: its *reliability*, or its ability to survive to possible component failures. Cost and reliability are clearly two conflicting parameters, and the main drawback of a spanning tree is that it will not even survive to a single edge malfunctioning. Then, in this undesired case, it arises the problem of promptly recovering the network connectivity after a failure.

If we restrict ourselves to temporary failures, a commonly adopted strategy to pull down the transition costs to the new solution is that of reconnecting the two subtrees generated by the failure throughout an *ad-hoc* swap edge, selected among those available in the underlying graph. Quite naturally, such an edge must be chosen according to the objective function addressed by the original tree, and it is called a *best swap edge* w.r.t. such a function, while the resulting tree is named the *swap tree*. Since the likelihood of having overlapping edge failures is quite small, it makes sense to deal with the failure of each and every single edge in the network, since we can expect that sooner or later each of them will fail. This is called an *all-best swap edges* (ABSE) problem.

Previous work. From an algorithmic point of view, the ABSE problem has been studied for the most popular tree structures, namely the *minimum spanning tree* (MST), the *single-source shortest paths tree* (SPT), the *minimum diameter spanning tree* (MDST), and the *minimum-stretch tree spanner* (MSTS).

For the MST, the ABSE problem is solvable in $O(m \log \alpha(m,n))$ by means of a *sensitivity analysis* of the MST[13]. Concerning the SPT, a best swap edge is a swap edge minimizing the total distance from the source to the disconnected nodes (this is also known as the *routing cost* from the source). The corresponding ABSE problem has been studied in [11], where an $O(n^2)$ time and space algorithm is given.[1] This result has been improved for dense graphs in [3], where an $O(m \log^2 n)$ time and $O(m)$ space algorithm is given, and extended to distributed systems in [4]. Regarding the MDST, a best swap edge is a swap edge minimizing the diameter of the swap tree, and the corresponding ABSE was studied in [10], where an $O(n\sqrt{m})$ time and $O(m)$ space algorithm is given. This was later improved for sparse graphs to $O(m \log n)$ time in [5], and then extended to distributed systems in [6]. Finally, concerning the MSTS, whose computation in known to be NP-hard, in [2] the authors present two efficient algorithms, for weighted and unweighted graphs, respectively.

Our results. The ABSE problem for the SPT has been recently generalized in [15], where the authors focused on the following problem: Given *any* input spanning tree, and given a set $S \subseteq V$ of k source nodes, find the ABSE w.r.t. the routing cost from the source nodes, i.e., the total distances between all the nodes and the sources. In [15] the authors provide an $O(m \log n + n^2)$ time and $O(n^2)$ space algorithm for the case $k = 2$, and an $O(mn)$ time and linear

[1] In the same paper, the authors define also several other swap criteria.

space algorithm for the general case of more than two sources. In this paper, we focus our attention on two significant cases concerning the size of k, namely $k = O(1)$ and $k = n$. The former models the practical situation in which only a constant number of sources is considered in the tree-based network, like in a classic client-server communication paradigm. On the other hand, the latter models the practical situation in which all the nodes are regarded as sources, and then an *all-to-all* communication protocol takes place on the tree, like in peer-to-peering. For these cases, we design an $O\left(m\, k\, 2^{O\left(\alpha(2m)^{\lceil k/2 \rceil}\right)} \log^2 n\right)$ and $O(m 2^{O(\alpha(2m))} \log^2 n)$ time algorithm, respectively, both using $O(m)$ space, where $\alpha(2m) := \alpha(2m, 2m)$ is from now on the inverse of the Ackermann function originally defined in [1]. Thus, for $k = n$ and constant $k > 2$ we significantly improve the general result given in [15],[2] while for the special but yet important case in which $k = 2$, we improve the corresponding result given in [15] for all but quite dense graphs, more precisely for $m = o(n^2/(2^{\alpha(n^2)} \log^2 n))$.

Afterwards, in an effort of establishing the effectiveness of the swapping, we concentrate on the case $k = n$, and we analyze the quality of the swap tree by measuring the approximation ratio between its routing cost and that of a *minimum (all-to-all) routing-cost spanning tree* (MRCST) of $G - e = (V, E \setminus \{e\})$ (whose computation is known to be NP-hard [9], though). In such a comparison process, we show that such ratio essentially depends on how much distances in the tree between the nodes and a *tree centroid* are *stretched* w.r.t. the corresponding distances in the underlying graph. More precisely, we first show an example in which starting from a tree that has a routing cost arbitrarily close to that of a MRCST, but with stretching unbounded, the approximation ratio is unbounded. On the positive side, we instead show that if the original tree T has a constant stretching, then the approximation ratio remains within a constant factor ρ. Notably, this is the case for the most representative elements in the set of feasible routing trees, and more precisely: (1) if T is a MRCST of G, then $\rho = 110$; (2) if T has a routing cost at most $(1 + \epsilon)$ of the minimum, as obtained by means of the PTAS for the problem provided in [16], then $\rho = 110 + 39\epsilon$ (notice that this result requires a non-trivial modification of the solution returned by the PTAS); (3) finally, if T is an SPT rooted in the median of G, which is known to provide a 2-approximation for the MRCST problem [14], then $\rho = 18$.

Due to space limitations, some proofs are omitted, for which we refer the reader to the full version of the paper.

2 Preliminaries

Let $G = (V, E)$ be an edge-weighted undirected graph with weight function $w : E \to \mathbb{R}^+$. As usual, we denote by n and m the number of nodes and edges of G, respectively. Let $S \subseteq V$ be a set of k source nodes and let H be a spanning subgraph of G. For any two given nodes a, b in G, we denote by $d_H(a, b)$ the

[2] Notice that, to be more precise, our improvement holds for $k\, 2^{O\left(\alpha(2m)^{\lceil k/2 \rceil}\right)} = o(n/\log^2 n)$, which includes $k = O(1)$.

distance (i.e., the length of a shortest path) in H between a and b. Hence, the *routing cost* of H from S is defined as $R_S(H) = \sum_{s\in S}\sum_{v\in V} d_H(s,v)$. We write $R(H)$ instead of $R_V(H)$. Then, the *minimum routing-cost spanning tree* (MRCST) problem asks for computing a spanning tree of G such that its routing cost is minimum.

Let T be a spanning tree of a graph G. For an edge e of T, let $C_T(e)$ be the set of all non-tree edges crossing the cut in $G - e$ defined by the removal of e from T. In the following, we will assume that G is 2-edge-connected, and so for any $e \in E$, $C_T(e)$ is not empty. Moreover, for any $f \in C_T(e)$, let $T_{e/f}$ denote the spanning tree obtained by replacing e with f, and let $\phi(f)$ be a function which expresses some quality measure of $T_{e/f}$. Then, we say that f is a *best swap edge for e w.r.t. φ* if f minimizes $\phi(f)$ over $C_T(e)$. In the following, the tree obtained by swapping e with its best swap edge w.r.t. the routing cost from S will be called the *swap tree* for e.

In this paper, we are interested in the following problem: *Given a weighted graph $G = (V, E)$, a set of $S \subseteq V$ of k source vertices, and a spanning tree T of G, we want to compute a best swap edge for every edge e of T w.r.t. the routing cost from S.* In the next section, we devise an efficient algorithm for solving it for constant k and for $k = n$, which is based on a geometric approach originally devised in [3] for the ABSE problem for the SPT.

3 The Algorithm

High-level description of the algorithm. We start by recalling the definition of *lower envelope* of a set of functions:

Definition 1. *Let $\mathcal{F}$ be a set of functions, where $\mathcal{F} \ni \phi_h : z \in D_h \subseteq \mathbb{R} \to \mathbb{R}$. The function*

$$L_{\mathcal{F}} : z \in D_{\mathcal{F}} = \bigcup_{\phi_h \in \mathcal{F}} D_h \mapsto \min\{\phi_h(z) : \phi_h \in \mathcal{F} \wedge z \in D_h\} \in \mathbb{R}$$

is named the lower envelope *of $\mathcal{F}$.*

Let $e_1, \ldots, e_{n-1}$ be an arbitrary order of the tree edges. For each non-tree edge f, let $\phi_f : D_f \subseteq \{1, \ldots, n-1\} \to \mathbb{R}$ be the *swap function* associated with f, which for $1 \leq i < n$, is defined as follows:

$$\phi_f(i) = \begin{cases} R(T_{e_i/f}) & \text{if } f \text{ is a swap edge for } e_i; \\ \infty & \text{otherwise.} \end{cases}$$

Then solving our problem reduces to finding the lower envelope of $\mathcal{F} = \{\phi_f : f \text{ is a non-tree edge}\}$. In [8] it was presented an efficient comparison-based algorithm to compute the lower envelope of a set of continuous real functions whose domains are real intervals. Such algorithm relies on the concept of *intersection* between functions, which can be extended to our case by defining the notion of *inversion* between pairs of swap functions.

First of all, for the sake of simplicity, we restrict the domain of each swap function ϕ_f to $[a_f, b_f] \subseteq \{1, \ldots, n-1\}$, where e_{a_f} and e_{b_f} are the leftmost and rightmost edge (w.r.t. the fixed ordering) for which f is a swap edge, respectively. Then, given two swap functions ϕ_f, ϕ_g, assume that $[a, b] = D_f \cap D_g \neq \emptyset$, and let $i \in (a, b]$. W.l.o.g., let $\phi_g(i) < \phi_f(i)$; then, i is an *inversion point* of ϕ_f, ϕ_g iff there exists a $j \in D_f, j < i$, such that $\phi_f(j) < \phi_g(j)$ or $j \notin D_g$, and for every $j < k < i$, $\phi_f(k) \leq \phi_g(k)$.

The algorithm provided in [8] is based on the efficient answering to a set of three queries that, for our problem, can be rephrased as follows:

- $\mathtt{Q}_1(f)$: Given a non-tree edge f, return the boundary of D_f;
- $\mathtt{Q}_2(f, g, e_i)$: Given a pair of non-tree edges f, g and a tree edge e_i such that $i \in D_f \cap D_g$, compare $\phi_f(i)$ and $\phi_g(i)$;
- $\mathtt{Q}_3(f, g, e_i)$: Given a pair of non-tree edges f, g and a tree edge e_i such that $i \in D_f \cap D_g$, return (if any) the next inversion point j of ϕ_f, ϕ_g with $j > i$.

The time and space complexity of the algorithm given in [8] depends on the complexity of a *Davenport-Schinzel sequence* (which in turn is related with the maximum number of inversion points between functions when compared pair-wise), as well as on the time needed for answering the queries. In the following section, we exactly address these issues as far as our problem is concerned.

Useful properties. For a given subgraph H of G and a given set $U \subseteq V$ of vertices, let us define the *broadcast cost* of H from a given node $r \in V$ w.r.t. U as $B_H(r, U) = \sum_{v \in U} d_H(r, v)$. We will write $B_H(r)$ instead of $B_H(r, U)$ when $V(H) = U$. Let $e = (x, y)$ be an edge of T, and let T_x, T_y be the trees in $T - e$ containing x and y, respectively. Let $X = S \cap V(T_x)$ be the source vertices in T_x, and let $Y = S \cap V(T_y)$ be the source vertices in T_y. We have that:

$$\begin{aligned} R_S(T) &= R_X(T_x) + R_Y(T_y) \\ &+ |T_y| B_{T_x}(x, X) + |X||T_y| w(e) + |X| B_{T_y}(y) \\ &+ |T_x| B_{T_y}(y, Y) + |Y||T_x| w(e) + |Y| B_{T_x}(x). \end{aligned} \tag{1}$$

Let $f = (x_1, y_1), g = (x_2, y_2)$ be two swap edges for e with $x_1, x_2 \in V(T_x)$. Let $\Lambda(f, g)$ denote the set of tree edges for which both f and g are swap edges, and let $\bar{x}, \bar{y}$ be the two end vertices of $\Lambda(f, g)$, with $\bar{x} \in V(T_x)$. For $i = 1, 2$, it holds that

$$\begin{cases} B_T(x_i) = B_{T_x}(x_i) + |T_y| d_T(x_i, y) + B_{T_y}(y) \\ B_T(y_i) = B_{T_y}(y_i) + |T_x| d_T(y_i, x) + B_{T_x}(x) \\ B_T(x_i, S) = B_{T_x}(x_i, X) + |Y| d_T(x_i, y) + B_{T_y}(y, Y) \\ B_T(y_i, S) = B_{T_y}(y_i, Y) + |X| d_T(y_i, x) + B_{T_x}(x, X) \end{cases} \tag{2}$$

Using (1) and (2) we have that:

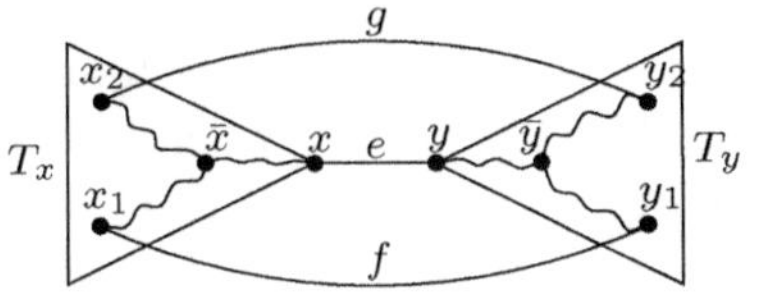

Fig. 1. A tree T and two non-tree edges $f = (x_1, y_1)$ and $g = (x_2, y_2)$

$$\begin{aligned}
\Delta^S_{f,g}(e) :=\ & R_S(T_{e/f}) - R_S(T_{e/g}) = \big(|X||T_y| + |T_x||Y|\big)(w(f) - w(g)) \\
&+ |X|\big(B_{T_y}(y_1) - B_{T_y}(y_2)\big) + |T_x|\big(B_{T_y}(y_1, Y) - B_{T_y}(y_2, Y)\big) \\
&+ |Y|\big(B_{T_x}(x_1) - B_{T_x}(x_2)\big) + |T_y|\big(B_{T_x}(x_1, X) - B_{T_x}(x_2, X)\big) \\
=\ & \big(|X||T_y| + |T_x||Y|\big)(w(f) - w(g)) \\
&+ |X|\big(B_T(y_1) - B_T(y_2)\big) + |T_x|\big(B_T(y_1, S) - B_T(y_2, S)\big) \\
&+ |Y|\big(B_T(x_1) - B_T(x_2)\big) + |T_y|\big(B_T(x_1, S) - B_T(x_2, S)\big) \\
&+ 2|X||T_x|\big(d_G(x_2, y) - d_G(x_1, y)\big) + 2|Y||T_y|\big(d_G(y_2, x) - d_G(y_1, x)\big) \\
=\ & \big((k - |Y|)|T_y| + (n - |T_y|)|Y|\big)(w(f) - w(g)) \\
&+ (k - |Y|)\big(B_T(y_1) - B_T(y_2)\big) + |Y|\big(B_T(x_1) - B_T(x_2)\big) \\
&+ (n - |T_y|)\big(B_T(y_1, S) - B_T(y_2, S)\big) + |T_y|\big(B_T(x_1, S) - B_T(x_2, S)\big) \\
&+ 2(k - |Y|)(n - |T_y|)\big(d_G(\bar{y}, y_2) - d_G(\bar{y}, y_1)\big) \\
&+ |Y||T_y|\big(d_G(\bar{x}, x_2) - d_G(\bar{x}, x_1)\big),
\end{aligned} \tag{3}$$

as $d_T(x, y_2) - d_T(x, y_1) = d_T(\bar{y}, y_2) - d_T(\bar{y}, y_1)$, $d_T(y, x_2) - d_T(y, x_1) = d_T(\bar{x}, x_2) - d_T(\bar{x}, x_1)$, $|T_x| + |T_y| = n$, and $|X| + |Y| = k$ (see Figure 1). Notice that the only two terms that depend on e are $|T_y|$ and $|Y|$. From this, we have the following:

Lemma 1. *Let us consider the edges of $\Lambda(f, g)$ in order from $\bar{y}$ to $\bar{x}$. Then, the sign of $\Delta^S_{f,g}(e)$ changes at most $k + 1$ times.*

Proof. The claim follows by observing that: (i) both $|T_y|$ and $|Y|$ do not decrease as e moves from $\bar{y}$ towards $\bar{x}$; (ii) $\Delta^S_{f,g}(e)$ is a linear function in $|T_y|$ if $|Y|$ is fixed, and (iii) $|Y|$ is an integer value in the interval $[0, k]$. □

Corollary 1. *Let us consider the edges of $\Lambda(f, g)$ in order from $\bar{y}$ to $\bar{x}$. Then, the sign of $\Delta^V_{f,g}(e)$ changes at most twice.*

Proof. The claim follows from point (i) of the previous lemma, by observing that $|T_y| = |Y|$, and from the fact that $\Delta^V_{f,g}(e)$ is quadratic in $|T_y|$. □

Implementation and analysis of the algorithm. First, we root T at an arbitrary node r, and we number the tree edges according to an arbitrary post-order visit of T, say $e_1, \ldots, e_{n-1}$. Now, we define a new set F of non-tree edges on which we will compute the lower envelope of the corresponding swap functions.

As we will prove later, this modification allows us to keep the number of inversion points low thus leading to a time-efficient algorithm.

We define F as follows. Let $F = E \setminus E(T)$ be a copy of the non-tree edges. For each $f = (y_1, y_2) \in F$, let $nca(y_1, y_2)$ denote the *nearest common ancestor* of y_1, y_2 in T. If $nca(y_1, y_2) \neq y_1, y_2$, we replace f with two auxiliary edges $f_1 = (nca(y_1, y_2), y_1), f_2 = (nca(y_1, y_2), y_2)$, and we impose $R(T_{e/f_i}) = R(T_{e/f})$ for every e in the path in T between y_i and $nca(y_1, y_2)$, $i = 1, 2$. Moreover, for each edge $f \in F$, we keep track of the corresponding original edge in $E \setminus E(T)$, which will be denoted by $\chi(f)$. Observe that for every $f, g \in F$, $\Lambda(f, g)$ is a subpath of $\Lambda(\chi(f), \chi(g))$. We now provide a bound for the number of inversion points between two swap functions.

Lemma 2. *Let f, g be two edges in F. Then, the number of inversion points between ϕ_f and ϕ_g is at most $k + 2$.*

Proof. Let $f = (x_1, y_1), g = (x_2, y_2)$, where x_i is closer to r than y_i, $i = 1, 2$, and let $D_f = [a_f, b_f], D_g = [a_g, b_g]$. W.l.o.g., let us assume that $a_f \leq a_g$. Recall that an inversion point must belong to the left-open intersection of the domains $\tilde{D} = D_f \cap D_g$. We divide $\tilde{D}$ into two disjoint intervals as follows. Let e_h be the tree edge just above $nca(y_1, y_2)$, if any. Then, $\tilde{D} = D_1 \cup D_2$, where $D_1 = (a_g, \min\{b_f, b_g, h\})$; if e_h does not exist we set $D_1 = \tilde{D}$ and consequently $D_2 = \emptyset$.

Observe that, for every $e_k \in D_1$, we have $\phi_f(k) = \infty$. As a consequence, $\phi_g(k) \leq \phi_f(k)$, for every $k \in D_1$. Moreover, $\phi_g(a_g) \neq \infty$, and $\phi_f(a_g) = \infty$. Therefore, there is no inversion point in D_1.

Now, we bound the number of inversion points in D_2. First of all, observe that for every $k \in D_2$ we have that $\phi_f(k) = \infty$ if and only if $\phi_g(k) = \infty$. Moreover, for every $e_k \in \Lambda(f, g)$, we have that $k \in D_2$, and $\phi_f(k), \phi_g(k) \neq \infty$ if and only if $e_k \in \Lambda(f, g)$. Since $e_h \in \Lambda(f, g)$, h might be an inversion point (this happens when $\phi_f(k) < \phi_g(k)$ and $a_g \neq h$). Furthermore, from Lemma 1 and because of the post-order arrangement of the tree edges, we have at most $k + 1$ additional inversion points, and the claim follows. □

Using Corollary 1 instead of Lemma 1 in the proof of the previous lemma we can prove the following

Corollary 2. *Let $S = V$ and let f, g be two edges in F. Then, the number of inversion points between ϕ_f and ϕ_g is at most 3.* □

In Figure 2, we show an example where the bound given in Corollary 2 is tight. To complete the description of the algorithm, it remains to show how to answer to the queries $\mathsf{Q}_1, \mathsf{Q}_2$, and Q_3. Query Q_1 can be answered in constant time as follows. Let $f = (x, y) \in F$ with x closer to r than y, and let e_a, e_b be the first and last edge in the path in T from y to x, respectively. Then $D_f = [a, b]$. Therefore, for every $f \in F$, we can precompute e_a and e_b in $O(m)$ time and space by means of a depth-first traversal of T.

Query Q_2 can be answered in constant time after a linear time and space preprocessing. Let $f = (u_1, y_1), g = (u_2, y_2)$ be two non-tree edges in F with u_i

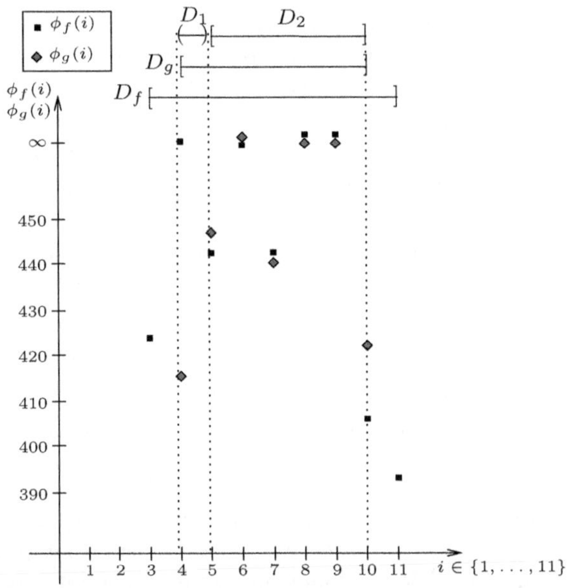

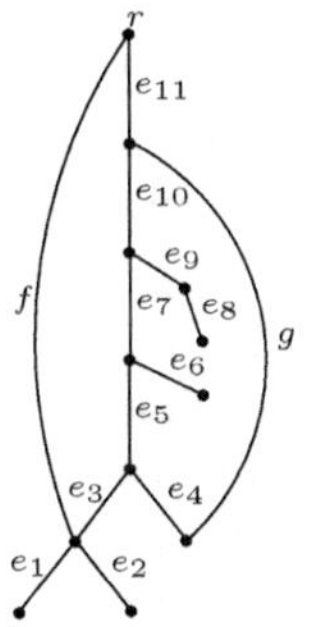

i	$\phi_f(i)$	$\phi_g(i)$
1,2	undef.	undef.
3	424	undef.
4	∞	416
5	442	446
6	∞	∞
7	442	440
8,9	∞	∞
10	406	422
11	394	undef.

Fig. 2. In this picture, it is shown an example of the mapping between non-tree edges and swap functions for the case $S = V$. On the right side, a tree T rooted at r and two non-tree edges f and g are shown. Tree edges are numbered according to a post-order traversal of T. All edges have cost 1. On the left side, a drawing of the two functions ϕ_f, ϕ_g is shown. $D_f = [3, 11]$, while $D_g = [4, 10]$. Notice that $D_1 = (4, 5) = \emptyset$, while $D_2 = [5, 10]$. Observe that 5, 7, and 10 are the three inversion points.

closer to r than y_i, $i = 1, 2$, and let $D_f = [a_f, b_f], D_g = [a_g, b_g]$. W.l.o.g., let us assume that $a_f \leq a_b$. Let $e_k = (x, y)$ be a tree edge with x closer to r than y, such that $k \in D_f \cap D_g$. First, we have to check whether e_k belongs to the paths in T between u_i and y_i, $i = 1, 2$. This can be done in $O(1)$ time by checking whether $y = nca(y_i, y)$ and $u_i = nca(u_i, x)$ [7]. Therefore, it remains to show how to compare $\phi_f(k)$ and $\phi_g(k)$ when $e_k \in \Lambda(f, g)$. In this case, the comparison of the two swap functions for point k reduces to evaluating the function $\Delta_{\chi(f),\chi(g)}(e_k)$. This can be done in constant time because all the terms of Equation (3) are available in $O(1)$ time after a linear time and space preprocessing. Indeed, if $\chi(f) = (x_1, y_1), \chi_g = (x_2, y_2)$, then $\bar{y} = nca(y_1, y_2)$, while $\bar{x}$ is the farthest vertex from r among $nca(x_1, x_2), nca(x_1, y_1)$, and $nca(x_2, y_2)$. Moreover, for every $v_1, v_2 \in V$, $d_T(v_1, v_2) = d_T(r, v_1) + d_T(r, v_2) - 2d_T(r, nca(v_1, v_2))$. Finally, for every tree edge $e' = (x', y')$ and for every $U \subseteq V$, we have $B_T(x', U) = B_T(y', U) + (|U_{y'}| - |U_{x'}|)w(e')$.

Finally, we show how to answer Q_3 in $O(k \log n)$ time after a linear time and space preprocessing. Similar arguments are enough to show that we can answer Q_3 in $O(\log n)$ time for the case $S = V$. Let $f = (x_1, y_1), g = (x_2, y_2)$ be two non-tree edges in F, with x_i closer to r than y_i, $i = 1, 2$, and let $D_f = [a_f, b_f], D_g = [a_g, b_g]$. W.l.o.g., let us assume that $a_f \leq a_g$. Let $e_k = (x, y)$ be a tree edge with

x closer to r than y such that $k \in D_f \cap D_g$. As we have shown in the proof of Lemma 2, there are two kinds of potential inversion points that have to be checked. The first one is associated with the edge just above $nca(y_1, y_2)$. The second kind of potential inversion points are related to the $k+1$ zeros $z_0, \ldots, z_k$, where z_i is the zero of $\Delta^S_{\chi(f),\chi(g)}$ when $|Y| = i$, seen as a linear function on the variable $|T_y|$ with domain $\mathbb{R}$.[3] More precisely, finding these $k+1$ potential inversion points reduces to finding the first edge $e_{h_i} = (\bar{x}_i, \bar{y}_i)$ (with $\bar{x}_i$ closer to r than $\bar{y}_i$) in the path from $nca(y_1, y_2)$ to r such that $|T_{\bar{y}_i}| > z_i$ and $|Y| = i$, for every $i = 0, \ldots, k$. As shown in [3], after a linear time and space preprocessing, we can accomplish the above task for each i in $O(\log n)$ time. We have that

Theorem 1. *Given a 2-edge-connected, undirected and positively real weighted graph $G = (V, E)$ with n vertices and m edges, given a set of k sources $S \subseteq V$, and given a spanning tree T of G, a best swap edge for every edge of T w.r.t. the routing cost from S can be computed in $O\left(m\, k\, 2^{O\left(\alpha(2m)^{\lceil k/2 \rceil}\right)} \log^2 n\right)$ time and $O(m)$ space.*[4]

Proof. By construction, we have at most $2m$ functions, which from Lemma 2 have at most $k+2$ inversion points when compared pairwise. The time complexity of the lower envelope algorithm in [8] for h functions with $k+2$ intersection points when compared pairwise is $O\left(\mathtt{Time}_{\mathcal{Q}} \cdot h \cdot \lambda_{k+3}(h) \cdot \log h\right)$, where $\mathtt{Time}_{\mathcal{Q}}$ is the time needed to answer to the three queries, and $\lambda_{k+3}(h)$ is the complexity of a Davenport-Schinzel sequence of order $k+3$. Since $\lambda_{k+3}(h) = 2^{O\left(\alpha(h)^{\lceil k/2 \rceil}\right)}$ [12], and since in our case we have shown that $\mathtt{Time}_{\mathcal{Q}} = O(k \log n)$ after a linear time and space preprocessing, it follows that the running time of our algorithm is $O\left(m\, k\, 2^{O\left(\alpha(2m)^{\lceil k/2 \rceil}\right)} \log^2 n\right)$.

To prove that the space complexity of our algorithm is linear, it is enough to show that the space complexity of the lower envelope algorithm in [8] applied to our set of functions is linear. The space complexity of the lower envelope algorithm in [8] for a set of h functions is $O(h + \mathtt{Space}_{\mathcal{L}})$, where $\mathtt{Space}_{\mathcal{L}}$ is the space needed to represent the lower envelope. The lower envelope of a set of h functions $\phi_1, \ldots, \phi_h$ with domains $[a_1, a_2], \ldots, [a_{2h}, a_{2h+1}]$ is represented as a list L of elements having pairwise disjoint intervals on the real line as keys. Each element takes $O(1)$ space. Moreover, for every key $[a, a']$ in L, we have that $[a, a'] \subseteq \bigcup_{i=1}^{h} [a_{2i}, a_{2i+1}]$ as well as $a, a' \in X$, where X is the set containing all the a_i's and all the intersection/inversion points between every pair of functions ϕ_j and ϕ_ℓ. Therefore, $\mathtt{Space}_{\mathcal{L}} = O(|X|)$. Since in our case $X \subseteq \{1, \ldots, n-1\}$, then the space complexity of the lower envelope algorithm in [8] when applied to our set of $O(m)$ functions is $O(n + m) = O(m)$. This completes the proof. □

Using Corollary 2 instead of Lemma 2 in the proof of the previous theorem, we can prove the following

[3] For the case $S = V$, the second kind of potential inversion points are related to the two zeros of the function $\Delta^V_{\chi(f),\chi(g)}$ seen as a quadratic function on the variable $|T_Y|$ with domain $\mathbb{R}$.

[4] Observe that for k constant, the time complexity is $\widetilde{O}(m)$.

Corollary 3. *Given a 2-edge-connected, undirected and positively real weighted graph* $G = (V, E)$ *with* n *vertices and* m *edges, and given a spanning tree* T *of* G*, a best swap edge for every edge of* T *w.r.t. the routing cost from* V *can be computed in* $O(m2^{O(\alpha(2m))} \log^2 n)$ *time and* $O(m)$ *space.* □

4 On the Quality of the Swap Tree When $S = V$

In this section we perform a comparison between the routing cost of the swap tree and that of a *minimum (all-to-all) routing-cost spanning tree* (MRCST) of $G - e$, namely a spanning tree minimizing the routing cost in $G - e$ w.r.t. V.

Recall that for any subgraph H of G, the *median* of H is a node w.r.t. which the broadcast cost of H is minimized. Moreover, for any spanning tree T of G, we say that a node $c \in V$ is a *centroid* of T if every subtree of T induced by the removal of c has at most $n/2$ nodes.[5] In the rest of this section we use the following notation. The failing edge will be denoted by e, while we will use f to denote the best swap edge for e w.r.t. to the routing cost. The MRCST of G and $G - e$ will be denoted by T^*_G and T^*_{G-e}, respectively.

Definition 2. *Let* T *be a spanning tree of* G*, and let* $\gamma \geq 1, \delta \geq 0$ *be reals. Let* $r \in V$*, and consider* T *as rooted at* r*. Then, we say that* T *is* (γ, δ)-stretched from r *if for any node* v *and for any descendent* t *of* v*, it holds that* $d_T(v,t) \leq \gamma\, d_G(v,t) + \delta$.

Using the results in [14] and [11] we can prove the following

Theorem 2. *For a given* $\rho \geq 1$*, let* T *be a* ρ*-approximation of* T^*_G*, i.e.,* $R(T) \leq \rho\, R(T^*_G)$*, and let* c *be a centroid of* T*. If* T *is* (γ, δ)*-stretched from* c*, then we have that* $R(T_{e/f}) \leq \left(2\rho + 4\rho\gamma^2 + 6\gamma^2 + 6\gamma + \frac{4n^2\delta}{R(G)}\right) R(T^*_{G-e})$. □

In Figure 3 we discuss a situation where, even if the original spanning tree has a routing cost arbitrarily close to the optimum one, it exhibits a bad swapping behavior, due to the fact that the tree is too (γ, δ)-stretched from its centroid(s). Now we present an algorithm that, taken a spanning tree T of G and a value $\xi \geq 0$, returns a spanning tree T' of G such that: (i) $R(T') \leq R(T)$; and (ii) T' is $(3, 2\xi/n)$-stretched from its centroid(s). For any $u, v \in V$, and any e belonging to the (unique) path in T between u and v, we define the concept of (u, v, e)-*move.* Such a move consists of modifying T in order to obtain a new tree, say T', by swapping e with $\hat{f}$, where $\hat{f} = (u, v)$ is the edge of the metric closure of G having weight $w(\hat{f}) = d_G(u, v)$. Notice that such an edge may not exist in G. Moreover, we say that an (u, v, e)-move is ξ*-improving* for T if $R(T) - R(T') \geq \xi$. The pseudo-code of the algorithm is given below, where we use the conversion algorithm given in [16] that, given a spanning tree of the metric closure of G, converts it into a spanning tree of G without increasing its routing cost.

Notice that, if T is within a factor ρ from the optimum, by choosing $\xi = R(G)/n^\beta$ for some $\beta > 0$, we have that Steps 1–5 are executed at most $2\rho n^\beta$ times, since in [14] it is shown that $R(G) \leq R(T) \leq 2\rho R(G)$.

---

[5] Recall that a tree has at most 2 centroids.

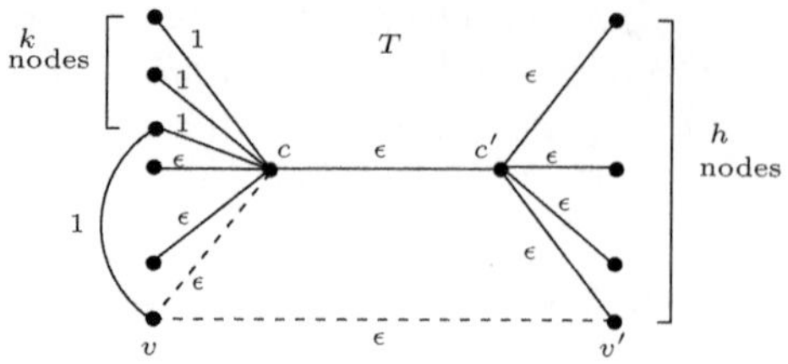

Fig. 3. A situation exhibiting a bad swapping behavior for tree T (whose edges are solid). The number of nodes is $n = 2h + 2$. There are $k + 1$ edges of weight 1, while all the other edges have a tiny weight $\epsilon > 0$. The two centroids of T are c and c'. We have that $R(T^*_G) \geq k(2h + 1)$, while $R(T) \leq k(2h + 1) + 4h + n^3\epsilon$. If edge $e = (c, c')$ fails, then the swap tree for e is $T_{e/f=(v,v')}$. Therefore, $R(T^*_{G-e}) \leq k(2h + 1) + n^3\epsilon$, while $R(T_{e/f}) \geq (k-1)(2h+1) + h(h+2) + (h-1)(h+3)$. Then, if $k = \omega(1)$ and $k = o(n)$, we have that $R(T)/R(T^*_G)$ goes to 1, while $R(T_{e/f})/R(T^*_{G-e}) = \Omega(n/k)$.

Algorithm 1

Input: A spanning tree T of G, and a parameter $\xi \geq 0$

1: **while** ($\exists$ an ξ-improving (u, v, e)-move for T) **do**
2: Let T' be the tree obtained by applying the (u, v, e)-move to T
3: Let T be the spanning tree of G obtained by the conversion algorithm in [16]
4: **end while**
5: **return** T.

Lemma 3. *For any $\xi \geq 0$, let T be the spanning tree returned by Algorithm 1. Then, T is $(3, 2\xi/n)$-stretched from its centroid(s).* □

We can finally give the following theorems

Theorem 3. *Let T coincide with T^*_G. Then, we have that $R(T_{e/f}) \leq 110\, R(T^*_{G-e})$.*

Proof. Clearly, Lemma 3 implies that T^*_G is $(3, 0)$-stretched from its centroid. Thus, from Theorem 2, since $\rho = 1$, the claim follows. □

Theorem 4. *Given $\epsilon > 0$, let T be a $(1 + \epsilon)$-approximated solution of T^*_G as returned by the* PTAS *in [17] modified according to Algorithm 1. Then, we have that $R(T_{e/f}) \leq (110 + 39\epsilon)\, R(T^*_{G-e})$.*

Proof. It suffices to choose $\xi = \frac{R(G)}{8n^{\log(1/\epsilon)+1}}$. The claim follows from Lemma 3 and Theorem 2. □

When T is an SPT of G, a refined analysis of Theorem 2 gives the following

Theorem 5. *Let T be an SPT rooted at a median of G. Then, $R(T_{e/f}) \leq 18\, R(T^*_{G-e})$.* □

References

1. Ackermann, W.: Zum Hilbertschen Aufbau der reellen Zahlen. Mathematical Annals 99, 118–133 (1928)
2. Das, S., Gfeller, B., Widmayer, P.: Computing best swaps in optimal tree spanners. In: Hong, S.-H., Nagamochi, H., Fukunaga, T. (eds.) ISAAC 2008. LNCS, vol. 5369, pp. 716–727. Springer, Heidelberg (2008)
3. Di Salvo, A., Proietti, G.: Swapping a failing edge of a shortest paths tree by minimizing the average stretch factor. Theoretical Computer Science 383(1), 23–33 (2007)
4. Flocchini, P., Mesa Enriques, A., Pagli, L., Prencipe, G., Santoro, N.: Point-of-failure shortest-path rerouting: computing the optimal swap edges distributively. IEICE Transactions 89-D(2), 700–708 (2006)
5. Gfeller, B.: Faster swap edge computation in minimum diameter spanning trees. In: Halperin, D., Mehlhorn, K. (eds.) ESA 2008. LNCS, vol. 5193, pp. 454–465. Springer, Heidelberg (2008)
6. Gfeller, B., Santoro, N., Widmayer, P.: A distributed algorithm for finding all best swap edges of a minimum diameter spanning tree. In: Pelc, A. (ed.) DISC 2007. LNCS, vol. 4731, pp. 268–282. Springer, Heidelberg (2007)
7. Harel, D., Tarjan, R.E.: Fast algorithms for finding nearest common ancestors. SIAM Journal on Computing 13(2), 338–355 (1984)
8. Hershberger, J.: Finding the upper envelope of n line segments in $O(n \log n)$ time. Information Processing Letters 33(4), 169–174 (1989)
9. Johnson, D.S., Lenstra, J.K., Rinnooy Kan, A.H.G.: The complexity of the network design problem. Networks 8, 279–285 (1978)
10. Nardelli, E., Proietti, G., Widmayer, P.: Finding all the best swaps of a minimum diameter spanning tree under transient edge failures. J. Graph Algorithms and Applications 5(5), 39–57 (2001)
11. Nardelli, E., Proietti, G., Widmayer, P.: Swapping a failing edge of a single source shortest paths tree is good and fast. Algorithmica 35(1), 56–74 (2003)
12. Nivasch, G.: Improved bounds and new techniques for Davenport–Schinzel sequences and their generalizations. J. ACM 57(3), 1–44 (2010)
13. Pettie, S.: Sensitivity analysis of minimum spanning trees in sub-inverse-Ackermann time. In: Deng, X., Du, D.-Z. (eds.) ISAAC 2005. LNCS, vol. 3827, pp. 964–973. Springer, Heidelberg (2005)
14. Wong, R.: Worst-case analysis of network design problem heuristics. SIAM J. Algebric Discrete Methods 1, 51–63 (1980)
15. Wu, B.Y., Hsiao, C.-Y., Chao, K.-M.: The swap edges of a multiple-sources routing tree. Algorithmica 50(3), 299–311 (2008)
16. Wu, B.Y., Lancia, G., Bafna, V., Chao, K.-M., Ravi, R., Tang, C.Y.: A polynomial-time approximation scheme for minimum routing cost spanning trees. SIAM J. Computing 29(3), 761–778 (1999)
17. Wu, B.Y., Chao, K.-M., Tang, C.Y.: Approximation algorithms for some optimum communication spanning tree problems. Discrete Applied Mathematics 102, 245–266 (2000)

Improved Approximability and Non-approximability Results for Graph Diameter Decreasing Problems*

Davide Bilò[1], Luciano Gualà[2], and Guido Proietti[1,3]

[1] Dipartimento di Informatica, Università di L'Aquila, 67010 L'Aquila, Italy
[2] Dipartimento di Matematica, Università di Tor Vergata, 00133 Roma, Italy
[3] Istituto di Analisi dei Sistemi ed Informatica, CNR, 00185 Roma, Italy
{davide.bilo,guido.proietti}@univaq.it, guala@mat.uniroma2.it

Abstract. In this paper we study two variants of the problem of adding edges to a graph so as to reduce the resulting diameter. More precisely, given a graph $G = (V, E)$, and two positive integers D and B, the *Minimum-Cardinality Bounded-Diameter Edge Addition* (MCBD) problem is to find a minimum cardinality set F of edges to be added to G in such a way that the diameter of $G+F$ is less than or equal to D, while the *Bounded-Cardinality Minimum-Diameter Edge Addition* (BCMD) problem is to find a set F of B edges to be added to G in such a way that the diameter of $G+F$ is minimized. Both problems are well known to be NP-hard, as well as approximable within $O(\log n \log D)$ and 4 (up to an additive term of 2), respectively. In this paper, we improve these long-standing approximation ratios to $O(\log n)$ and to 2 (up to an additive term of 2), respectively. As a consequence, we close, in an asymptotic sense, the gap on the approximability of the MCBD problem, which was known to be not approximable within $c \log n$, for some constant $c > 0$, unless P = NP. Remarkably, as we further show in the paper, our approximation ratio remains asymptotically tight even if we allow for a solution whose diameter is optimal up to a multiplicative factor approaching $\frac{5}{3}$. On the other hand, on the positive side, we show that at most twice of the minimal number of additional edges suffices to get at most twice of the required diameter.

1 Introduction

In this paper, we study two basic network design problems. In the first one, we are given a communication network and a distance requirement D. The goal is to find a minimum cardinality set of links to be added such that every pair of nodes in the network is connected by a path of at most D edges. More formally

* This work was partially supported by the PRIN 2008 research project COGENT (COmputational and GamE-theoretic aspects of uncoordinated NeTworks), funded by the Italian Ministry of Education, University, and Research.

P. Hliněný and A. Kučera (Eds.): MFCS 2010, LNCS 6281, pp. 150–161, 2010.

Minimum-Cardinality Bounded-Diameter Edge Addition (MCBD)
Instance: an undirected graph G and a positive integer value $D > 0$.
Goal: find a minimum cardinality set F of edges to be added to G such that the *diameter* of $G + F$ is less than or equal to D.

Similarly, one can define the specular problem in which we are given a communication network and a *budget* B on the number of addable links, and the goal is to add such links so that the resulting network has minimum (in terms of number of links) diameter. More formally

Bounded-Cardinality Minimum-Diameter Edge Addition (BCMD)
Instance: an undirected graph G and a positive integer value $B > 0$.
Goal: find a set F of B edges to be added to G such that the diameter of $G + F$ is minimized.

These two problems arise in practical applications like telecommunication networks and airplane flights scheduling [6,10], but they also received a lot of attention in the graph theory community (see [1,7,9,12,18,20]).

In the rest of the paper we will denote by n the number of vertices of G. Notice that the two defined problems are the optimization version of the same underlying decision problem. Therefore, for the sake of unifying the exposition, we will denote by B the cardinality of an optimal solution for MCDB, and by D the value of an optimal solution for BCMD.

Having this in mind, and following standard terminology on *bicriteria* optimization problems, for $\beta, \delta \geq 1$, a (β, δ)-approximation algorithm for BCMD will denote an algorithm which can select a set F of additional edges whose size is at most β times the budget B, and returns a graph $G + F$ of diameter of at most δD (where D is the value of an optimal solution for BCMD with budget B). Symmetrically, a (δ, β)-approximation algorithm for MCBD will denote an algorithm that returns a graph $G + F$ whose diameter is at most δ times the required value D, by using at most βB edges (where B is the size of an optimal solution for MCBD with required diameter D). Observe that a (β, δ)-approximation algorithm for BCMD is a (δ, β)-approximation algorithm for MCBD, and viceversa.

Related work. For every $D \geq 2$, MCBD is not approximable within $c \log n$ for some constant $c > 0$, unless $\mathsf{P} = \mathsf{NP}$ [16,8], while it is clearly in P for $D = 1$. As a consequence, BCMD is not approximable within a factor strictly better than $(1 + 1/D)$ for every $D \geq 2$, unless $\mathsf{P} = \mathsf{NP}$. This implies the non-existence of a $(c \log n, \delta)$-approximation algorithm for BCMD for $\delta < 1 + 1/D$, unless $\mathsf{P} = \mathsf{NP}$.

On the positive side, BCMD admits a constant $(4 + \frac{2}{D})$-approximation algorithm [16]; the same algorithm guarantees a $(2 + \frac{2}{D})$-approximation for forests. Concerning positive results for MCBD, in [8] the authors provide an $O(\log n \log D)$-approximation algorithm. In the same paper, the authors provide both approximability and non-approximability results for a more general version of MCBD in which edges are associated with a cost and a length function, and B and D are redefined accordingly. Furthermore, MCBD has been studied also for forests. Here, a 2-approximation algorithm is known for even values of D [5], while an 8-approximation algorithm has been given in [13] for odd values of D. This latter result has been improved in [4], where a $(2+\epsilon)$-approximation algorithm up to an additive constant of $O(\epsilon^{-5})$ has been given. Establishing whether BCMD and MDBC restricted to trees/forests are in P is still an open problem. Finally, concerning bicriteria approximation algorithms, in [8,14] the authors provide a polynomial time $\left(O(\log n), 2+\frac{2}{D}\right)$-approximation algorithm for BCMD.

Our results. In this paper, we provide a different analysis of the algorithm given in [16], in order to show that it actually computes a $\left(2+\frac{2}{D}\right)$-approximate solution for BCMD. Moreover, when the input instance is a forest, we achieve optimality up to small constant additive terms. More precisely, we get an approximation guarantee of $(1+\frac{2}{D})$ for even values of D, and of $(1+\frac{4}{D})$ for odd values of D. Concerning approximability of MCBD, we improve the result given in [8], by providing an $O(\log n)$-approximation algorithm. Thus, we close in an asymptotic sense the approximability of the MCBD problem. Notably, our algorithm extends to directed graphs as well as to the case when we place the distance requirements $D_{u,v}$ for each pair u, v of vertices. We regard our result as a significative contribution for the comprehension of the problem since, as we further show in the paper, our approximation ratio cannot be improved asymptotically, unless $\mathsf{P} = \mathsf{NP}$, even if we allow for a solution whose diameter is optimal up to a multiplicative factor of $\frac{5}{3} - \frac{4}{D}$. Notice that this also implies a better inapproximability threshold for BCMD for any $D \geq 8$. On the other hand, on the positive side, we also show that if a doubling of the optimal diameter is tolerated, then the MCBD problem admits a $\left(2-\frac{1}{B}\right)$-approximation algorithm. Table 1 summarizes the state of the art and our improvements for the two problems (results are given in form of bicriteria ratios for BCMD, and our contributions are marked with a †).

2 Approximation Algorithms for BCMD

We begin this section by describing the $\left(4+\frac{2}{D}\right)$-approximation algorithm for BCMD given in [16] and show that it actually computes a $\left(2+\frac{2}{D}\right)$-approximate solution. We need to introduce some notation first.

For a graph G, we denote by $V(G)$ and $E(G)$ the set of vertices and the set of edges of G, respectively. We denote by $\bar{G}$ the *complement* of G, i.e., the graph on $V(G)$ that contains the edge (u,v) iff $(u,v) \notin E(G)$. For every $F \subseteq \{(u,v) \mid u,v \in V(G)\}$, we denote by $G+F$ the graph on $V(G)$ with edge set $E(G) \cup F$. We use $G+e$ instead of $G+\{e\}$. By $\alpha(G)$ we denote the cardinality of a maximum *independent set* of G.[1] For each pair of vertices $u, v \in V(G)$, we denote by $d_G(u,v)$ the *distance* between u and v in G, i.e., the length (in terms of

[1] A set $\{v_1,\ldots,v_\ell\}$ of vertices of G is an *independent set* of G iff $(v_i,v_j) \notin E(G)$ for every $i,j = 1,\ldots,\ell$. An independent set $\{v_1,\ldots,v_\ell\}$ of G is maximum if for every independent set $\{v'_1,\ldots,v'_{\ell'}\}$ of G we have that $\ell \geq \ell'$.

Table 1. Table of known and improved results for (β, δ)-approximation algorithms for BCMD. The non-approximability results hold unless $\mathsf{P} = \mathsf{NP}$.

Problem	APPROXIMABILITY	NON-APPROXIMABILITY
General instances	in P for $D = 1$ (folklore) $(1, 4 + \frac{2}{D})$ [16] $(O(\log n), 2 + \frac{2}{D})$ [8,14] $(O(\log n \log D), 1)$ [8] $(2 - \frac{1}{B}, 2)$ † $(1, 2 + \frac{2}{D})$ † $(O(\log n), 1)$ †	$(c \log n, \delta < 1 + \frac{1}{D}), \forall D \geq 2$ [16,8] $(c \log n, \delta < \frac{5}{3} - \frac{4}{D}), \forall D \geq 8$ †
Forests	$(2, 1)$ for $D = 2h$ [5] $(2 + \epsilon, 1)$ for $D = 2h + 1$ [4] $(1, 2 + \frac{2}{D})$ $(1, 1 + \frac{2}{D})$ for $D = 2h$ † $(1, 1 + \frac{4}{D})$ for $D = 2h + 1$ †	?

number of edges) of a shortest path between u and v. The *diameter* of G will be denoted by $\mathtt{diam}(G) := \max_{u,v \in V(G)} d_G(u, v)$. Let $\lambda \geq 1$ be an integer value. We denote by G^λ the graph on $V(G)$ that contains an edge (u, v) iff $d_G(u, v) \leq \lambda$. For a given positive integer $k > 0$, and a subset $V' \subseteq V(G)$, we say that $\langle V_1, \ldots, V_k \rangle$ is a *k-clustering of* V' if (i) $\forall i, V_i \subseteq V(G)$, and (ii) $V' \subseteq \bigcup_{i=1}^k V_i$. Moreover, we define $\mathtt{radius}(V_i) := \min_{u \in V(G)} \max_{v \in V_i} d_G(u, v)$ to be the *radius of* V_i. We say that $c_i \in V(G)$ is *a center of cluster* V_i if $\max_{v \in V_i} d_G(c_i, v) = \mathtt{radius}(V_i)$. The *k-center Problem* is the problem that takes an edge weighted graph G and an integer $k \geq 1$ as inputs and asks for a k clustering $\langle V_1, \ldots, V_k \rangle$ of $V(G)$ that minimizes $\max_{i=1,\ldots,k} \mathtt{radius}(V_i)$. This problem cannot be approximated within a factor better than 2, unless $\mathsf{P} = \mathsf{NP}$ [17]. Moreover, 2-approximation algorithms are given in [11].

The algorithm in [16] makes use of any 2-approximation algorithm $\mathcal{A}$ to find a $B+1$ clustering $\langle V_0, \ldots, V_B \rangle$ of $V(G)$ for the $(B+1)$-center problem on input G. Then, it computes a center c_i for every cluster V_i and outputs the set $F = \{(c_0, c_i) \mid i = 1, \ldots, B\}$. We prove that this algorithm computes a $(2 + \frac{2}{D})$-approximate solution when $\mathcal{A}$ is the 2-approximation algorithm of Gonzalez [11]. The subsequent lemmas are the key of our proof.

Lemma 1 (Gonzalez 1985 [11]). *Let $\lambda \geq 1$ be an integer value. If $\alpha(G^\lambda) \leq k$, then Gonzalez'algorithm finds a k-clustering $\langle V_1, \ldots, V_k \rangle$ of $V(G)$ such that* $\mathtt{radius}(V_i) \leq \lambda, \forall i = 1 \ldots, k$. □

Lemma 2. $\alpha(G^D) \leq B + 1$.

Proof. Let $v_1, \ldots, v_\ell$ be an independent set in G^D. Let F^* be an optimal solution for the problem and let $G_1 = G + F^*$. Clearly, $\alpha(G_1^D) = 1$. Observe that to prove the claim, it is enough to show that for every edge $e \in E(\bar{G}), \alpha(G_e^D) \geq \ell - 1$, where $G_e = G + e$. For the sake of contradiction, assume there exists an edge $e = (u,v) \in E(\bar{G})$ such that $\alpha(G_e^D) < \ell - 1$, where $G_e = G + e$. This implies that (i) there exist four distinct indexes $i, i', j, j' \leq \ell$ such that $(v_i, v_j), (v_{i'}, v_{j'}) \in E(G_e^D)$ or (ii) there exist three distinct indexes $i, i', i'' \leq \ell$ such that $(v_i, v_{i'}), (v_{i'}, v_{i''}), (v_{i''}, v_i) \in E(G_e^D)$.

We deal with case (i) first. Since $d_G(v_i, v_j), d_G(v_{i'}, v_{j'}) > D$ and since $d_{G+e}(v_i, v_j), d_{G+e}(v_{i'}, v_{j'}) \leq D$, then, both the shortest path from v_i and v_j in G_e and the shortest path from $v_{i'}$ and $v_{j'}$ in G_e pass through edge e. Therefore, w.l.o.g., we have that

$$\begin{cases} d_G(v_i, u) + 1 + d_G(v_j, v) \leq D \\ d_G(v_{i'}, u) + 1 + d_G(v_{j'}, v) \leq D \\ D < d_G(v_i, u) + d_G(v_{i'}, u) \\ D < d_G(v_j, v) + d_G(v_{j'}, v). \end{cases}$$

If we sum up all the inequalities we get $2 < 0$, a contradiction.

Now, we deal with case (ii). Since $d_G(v_i, v_{i'}), d_G(v_{i'}, v_{i''}), d_G(v_{i''}, v_i) > D$ and since $d_{G+e}(v_i, v_{i'}), d_{G+e}(v_{i'}, v_{i''}), d_{G+e}(v_{i''}, v_i) \leq D$, then, for the pigeon principle, two of the three vertices $v_i, v_{i'}, v_{i''}$ are closer to one endvertex of e than to the other one. W.l.o.g., assume that $d_G(v_{i'}, v) \leq d_G(v_{i'}, u)$, and $d_G(v_{i''}, v) \leq d_G(v_{i''}, u)$. As a consequence,

$$\begin{aligned} d_G(v_{i'}, v_{i''}) &\leq d_G(v_{i'}, v) + d_G(v_{i''}, v) \\ &< \min\{d_G(v_{i'}, v) + 1 + d_G(v_{i''}, u), d_G(v_{i'}, u) + 1 + d_G(v_{i''}, v)\} \\ &= d_{G+e}(v_{i'}, v_{i''}) \leq D, \end{aligned}$$

and thus $(v_{i'}, v_{i''}) \in E(G^D)$, a contradiction. □

Theorem 1. *There exists a* $\left(2 + \frac{2}{D}\right)$*-approximation algorithm for BCMD.*

Proof. From Lemma 2 we have that $\alpha(G^D) \leq B+1$. As a consequence, Lemma 1 implies that Gonzalez'algorithm computes a $\langle V_0, \ldots, V_B \rangle$ clustering of $V(G)$ such that $\texttt{radius}(V_i) \leq D, \forall i = 0, \ldots, B$. Let c_i be a center of cluster V_i and let $F = \{(c_0, c_i) \mid i = 1, \ldots, B\}$ be the solution returned by the algorithm. Clearly, $|F| = B$. Moreover, every vertex $v \in V(G)$ is at distance at most $D + 1$ from c_0 in $G + F$. Therefore, $\texttt{diam}(G + F) \leq 2D + 2$. The claim follows. □

Theorem 2. *For the BCMD restricted to forests, there exists a* $\left(1+\frac{2}{D}\right)$*-approximation algorithm for even values of* D *and a* $\left(1 + \frac{4}{D}\right)$*-approximation algorithm for odd values of* D.

Proof. Let G be a forest and let $H_1, \ldots, H_\ell$ be a minimum partition of G^D in cliques.[2] As forests are perfect graphs[3] and because G^D is still a perfect graph

[2] A *partition of* G *in cliques* is a collection $\{H_1, \ldots, H_\ell\}$ of vertex-disjoint subgraphs of G such that $\bigcup_{i=1}^{\ell} V(H_i) = V(G)$ and each H_i is a clique. A partition $\{H_1, \ldots, H_\ell\}$ of a graph G in cliques is minimum iff for every other partition $\{H'_1, \ldots, H'_{\ell'}\}$ of G in cliques we have that $\ell \leq \ell'$.

[3] A graph G is *perfect* iff the minimum number of cliques necessary to cover G equals the size of the largest independent set of G [2].

(see [2]), we have that $\alpha(G^D) = \ell$. Thus, from Lemma 2, $\ell \leq B+1$. Each H_i can be viewed as a subtree T_i of G of diameter less than or equal to D. As a tree of diameter D has radius less than or equal to $\lceil D/2 \rceil$, it follows that $\langle V(T_1), \ldots, V(T_\ell) \rangle$ is an ℓ clustering of $V(G)$ such that $\mathtt{radius}(V(T_i)) \leq \lceil D/2 \rceil$, for every $i = 1, \ldots, \ell$. From Lemma 2, we have that $\alpha(G^D) = \ell \leq B+1$. As a consequence, an optimal solution for the $(B+1)$-center problem is a clustering $\langle V_0, \ldots, V_B \rangle$ such that $\mathtt{radius}(V_i) \leq \lceil D/2 \rceil$, for every $i = 0, \ldots, B$. Let c_i be a center of V_i. Let $F = \{(c_0, c_i) \mid i = 1, \ldots, B\}$. Clearly, $|F| = B$. Moreover, every vertex $v \in V(G)$ is at distance at most $\lceil D/2 \rceil + 1$ from c_0 in $G+F$. Therefore, $\mathtt{diam}(G+F) \leq 2\lceil D/2 \rceil + 2$. As the k-center problem on forests can be solved in linear time [3,15], the claim follows. □

3 An $O(\log n)$-Approximation Algorithm for MCBD

In this section we describe an $O(\log n)$-approximation algorithm for MCBD. W.l.o.g., we can restrict ourselves to the class of connected graphs. Indeed, if $\ell + 1$ denotes the number of connected components, then we can first add ℓ edges to G to let it become connected and then run our algorithm. Since the problem is clearly in P for the case $B < \ell$, we are still guaranteed to compute an $O(\log n)$-approximate solution.

In the rest of the section, for any graph H on $V(G)$, let $\mathcal{I}(H) := \{\{u,v\} \mid u, v \in V(H), d_H(u,v) > D\}$. The algorithm uses a greedy approach and consists of two phases. In the first phase, the algorithm fixes a vertex $s \in V(G)$ and computes a set F_1 of edges such that $\forall \{u,v\} \in \mathcal{I}(G), d_{G_1}(s,u) + d_{G_1}(s,v) \leq D+1$, where $G_1 = G + F_1$. Observe that this immediately implies that $\mathtt{diam}(G_1) \leq D+1$. In the second phase, the algorithm computes a set F_2 of edges such that $\mathtt{diam}(G_1 + F_2) \leq D$. More precisely, for every $\{u,v\} \in \mathcal{I}(G_1)$, there exists an edge in F_2 whose addition to G_1 let the distance between u and v decrease by at least 1. We will prove that $|F_1|, |F_2| = O(B \log n)$.

For the rest of the section, let $s \in V(G)$ be fixed. We now describe the first phase of the algorithm. Let H be a graph on $V(G)$ and let

$$\Gamma(H) := \big\{\{u,v\} \mid \{u,v\} \in \mathcal{I}(H), d_H(s,u) + d_H(s,v) > D+1\big\}.$$

For each $u, v \in V(H)$ we define

$$\mu_H(u,v) := d_H(s,u) + d_H(s,v) - (D+1),$$

and

$$\mathtt{cost}(H) := \sum_{\{u,v\} \in \Gamma(H)} \mu_H(u,v).$$

Let $F_s = \{(s,v) \mid v \in V(G)\}$. For a given edge e, define $\mathtt{gain}(e,H) := \mathtt{cost}(H) - \mathtt{cost}(H+e)$ to be the *gain of edge e w.r.t. H*. The algorithm first sets $F_1 := \emptyset$ and $H_0 := G$ and then proceeds in steps. During step $i \geq 1$, the algorithm selects an edge $e = (s,v) \in F_s$ that maximizes $\mathtt{gain}(e, H_{i-1})$, adds e to F_1, and sets $H_i := H_{i-1} + e$. The first phase of the algorithm ends at the end of step i if $\mathtt{cost}(H_i) = 0$. We prove the following

Lemma 3. *At the end of the first phase,* $|F_1| = O(B \log n)$*. Moreover, for every* $\{u, v\} \in \Gamma(G)$*, we have that* $d_{G+F_1}(s, u) + d_{G+F_1}(s, v) \leq D + 1$.

Proof. The first phase of the algorithm ends when $\texttt{cost}(H_i) = 0$, i.e., when $\mu_{H_i}(u, v) \leq 0$, for every $\{u, v\} \in \Gamma(G)$. As $H_i = G + F_1$, we have that $\mu_{G+F_1}(u, v) \leq 0$ implies $d_{G+F_1}(s, u) + d_{G+F_1}(s, v) \leq D + 1$.

Let F^* be an optimal solution for MCBD on inputs G and D and let $U = \{v_1, \ldots, v_\ell\}$ be the set of endvertices of the edges in F^*. Clearly, $\ell \leq 2B$.Let $\hat{F} = \{e_j^* \mid e_j^* = (s, v_j), j = 1 \ldots, \ell\}$ and observe that $d_{G+\hat{F}}(s, u) + d_{G+\hat{F}}(s, v) \leq D + 1$, for every $\{u, v\} \in \Gamma(G)$. Moreover, $\hat{F} \subseteq F_s$. We have that

Proposition 1. *For every* i, $\sum_{j=1}^{\ell} \texttt{gain}(e_j^*, H_i) \geq \texttt{cost}(H_i)$.

Proof. Let $\langle V_1, \ldots, V_\ell \rangle$ be an ℓ-clustering of V defined as follows. The cluster V_j contains vertex v_j and all the vertices $v \in V(G) \setminus \bigcup_{i=1}^{j-1} V_i$ such that $d_G(v, v_j) \leq d_G(v, v_k), \forall k = j + 1, \ldots, \ell$. Let $\{u, v\} \in \Gamma(H_i)$ and let $e \notin E(H_i)$. We define

$$\texttt{gain}_{u,v}(e, H_i) := \begin{cases} \mu_{H_i}(u, v) - \mu_{H_i+e}(u, v) & \text{if } \{u, v\} \in \Gamma(H_i + e); \\ \mu_{H_i}(u, v) & \text{otherwise.} \end{cases}$$

Let $\tau : V(G) \to \{1, \ldots, \ell\}$ be such that $\tau(x) = i$ if $x \in V_i$. By construction of the clusters and because edges in $\hat{F}$ are all adjacent to s, we have that $d_{H_i+e^*_{\tau(u)}}(s, u) = d_{H_i+\hat{F}}(s, u)$ and $d_{H_i+e^*_{\tau(v)}}(s, v) = d_{H_i+\hat{F}}(s, v)$. This implies that $\tau(u) \neq \tau(v)$, otherwise, if $\tau(u) = \tau(v) = \tau'$, then $D + 1 \geq d_{H_i+\hat{F}}(s, u) + d_{H_i+\hat{F}}(s, v) = d_{H_i+e^*_{\tau'}}(s, u) + d_{H_i+e^*_{\tau'}}(s, v) \geq 2 + d_{H_i}(v_{\tau'}, u) + d_{H_i}(v_{\tau'}, v) \geq 2 + d_{H_i}(u, v)$, and thus $d_{H_i}(u, v) \leq D \Rightarrow \{u, v\} \notin \Gamma(H_i)$. Clearly, if $\{u, v\} \notin \Gamma(H_i + e^*_{\tau(u)})$ or $\{u, v\} \notin \Gamma(H_i + e^*_{\tau(v)})$, then $\texttt{gain}_{u,v}(e^*_{\tau(u)}, H_i) + \texttt{gain}_{u,v}(e^*_{\tau(v)}, H_i) \geq \mu_{H_i}(u, v)$. But also in the other case we have that

$$\begin{aligned} \texttt{gain}_{u,v}(e^*_{\tau(u)}, H_i) + \texttt{gain}_{u,v}(e^*_{\tau(v)}, H_i) &= 2\big(d_{H_i}(s, u) + d_{H_i}(s, v)\big) \\ &\quad - d_{H_i+e^*_{\tau(u)}}(s, u) - d_{H_i+e^*_{\tau(u)}}(s, v) \\ &\quad - d_{H_i+e^*_{\tau(v)}}(s, u) - d_{H_i+e^*_{\tau(v)}}(s, v) \\ &\geq d_{H_i}(s, u) + d_{H_i}(s, v) \\ &\quad - d_{H_i+\hat{F}}(s, u) - d_{H_i+\hat{F}}(s, v) \\ &\geq \mu_{H_i}(u, v). \end{aligned}$$

Therefore,

$$\begin{aligned} \sum_{j=1}^{\ell} \texttt{gain}(e_j^*, H_i) &= \sum_{j=1}^{\ell} \sum_{\{u,v\} \in \Gamma(H_i)} \texttt{gain}_{u,v}(e_j^*, H_i) \\ &\geq \sum_{\{u,v\} \in \Gamma(H_i)} \Big(\texttt{gain}_{u,v}(e^*_{\tau(u)}, H_i) + \texttt{gain}_{u,v}(e^*_{\tau(v)}, H_i)\Big) \\ &\geq \sum_{\{u,v\} \in \Gamma(H_i)} \mu_{H_i}(u, v) = \texttt{cost}(H_i), \end{aligned}$$

where the first inequality follows from the fact that $\tau(u) \neq \tau(v)$. □

As a consequence of the above proposition, for every $i \geq 1$, there exists an edge $e \in F_s$ such that $\texttt{gain}(e, H_{i-1}) \geq \max\left\{1, \frac{\texttt{cost}(H_{i-1})}{\ell}\right\} \geq \max\left\{1, \frac{\texttt{cost}(H_{i-1})}{2B}\right\}$. This implies that $\texttt{cost}(H_i) \leq \texttt{cost}(H_0)\left(1 - \frac{1}{2B}\right)^{i-1} = \texttt{cost}(G)\left(1 - \frac{1}{2B}\right)^{i-1}$ for every i. Moreover, at the beginning of the last step of the algorithm, say $\eta + 1$, we have that $\texttt{cost}(H_\eta) \geq 1$. Therefore, $1 \leq \texttt{cost}(H_\eta) \leq \texttt{cost}(G)\left(1 - \frac{1}{2B}\right)^{\eta-1}$. Taking the natural logarithm and simplifying, we finally get

$$\eta \leq 2B \ln \texttt{cost}(G) = O(B \log n),$$

where the last equality comes from the fact that $\texttt{cost}(G) = O(n^3)$ as G is connected and because $|\Gamma(G)| = O(n^2)$. □

We now describe the second phase of the algorithm. Let $G_1 = G + F_1$, where F_1 is the set of edges computed by the algorithm in the first phase. We make a reduction to the *Set Cover Problem*. The *Set Cover Problem* takes as input a set of objects Z and a set $\mathcal{S}$ of subsets of Z and asks for the minimum-cardinality subset $\mathcal{S}' \subseteq \mathcal{S}$ that *covers* Z, i.e., $\bigcup_{S \in \mathcal{S}'} S = Z$. The Set Cover Problem is well-known to be approximable within $O(\log |Z|)$. We build an instance of the Set Cover Problem as follows. The objects in Z are the unordered pairs in $\mathcal{I}(G_1)$. There is a set S_e for every edge e in $\bar{G}$ which is defined as follows

$$S_e := \Big\{\{u, v\} \mid \{u, v\} \in Z, d_{G_1+e}(u, v) \leq D\Big\}.$$

Let X be a solution computed by the $O(\log |Z|)$-approximation algorithm for Set Cover Problem and let $F_2 = \{e \mid S_e \in X\}$. We can prove the following

Lemma 4. *At the end of the second phase, we have that* $|F_2| = O(B \log n)$. *Moreover,* $\texttt{diam}(G_1 + F_2) \leq D$.

Proof. Let $\mathcal{S}'$ be any solution for the covering problem defined above. Let $F = \{e \mid S_e \in \mathcal{S}'\}$ and let $\{u, v\} \in \mathcal{I}(G_1)$. By construction, object $\{u, v\} \in Z$ and there exists $S_e \in \mathcal{S}'$ such that $d_{G_1+e}(u, v) \leq D$. As $e \in F$, we have that $d_{G_1+F}(u, v) \leq d_{G+e}(u, v) \leq D$. Therefore, $\texttt{diam}(G_1 + F) \leq D$. This also implies that $\texttt{diam}(G_1 + F_2) \leq D$.

Let F^* be an optimal solution for MCBD on inputs G and D and let U be the set of endvertices of the edges in F^*. We claim that

$$\mathcal{S}^* = \{S_{(s,u)} \mid u \in U\} \cup \{S_e \mid e \in F^*\}$$

is a feasible solution for the set cover instance. Observe that this is enough to prove the claim, as $|\mathcal{S}^*| \leq 3B$ and $|Z| = |\mathcal{I}(G_1)| \leq n^2$. Let $\{u, v\} \in \mathcal{I}(G_1)$ and let P be a shortest path from u to v in $G + F^*$. As $d_{G_1}(u, v) > D$ whilst $d_{G_1+F^*}(u, v) \leq d_{G+F^*}(u, v) \leq D$, then P contains some edge of F^*. We traverse P from u to v. Let u' be the first vertex of P which is also a vertex of U, and let v' be the last vertex of P which is also a vertex of U.

If P contains exactly one edge of F^*, i.e., $(u', v') \in F^*$, then object $\{u, v\}$ is in set $S_{(u',v')}$ by construction. Moreover, $S_{(u',v')} \in \mathcal{S}^*$. If P contains two or more edges of F^*, then $d_{G_1}(u,u')+d_{G_1}(v',v) \leq D-2$. Moreover, as Lemma 3 implies that $d_{G_1}(s,u) + d_{G_1}(s,v) = D+1$, then $d_{G_1}(u,u') + d_{G_1}(v',v) \leq d_{G_1}(s,u) + d_{G_1}(s,v) - 3$. Therefore, $d_{G_1}(u,u') \leq d_{G_1}(s,u) - 2$ or $d_{G_1}(v',v) \leq d_{G_1}(s,v) - 2$. W.l.o.g., let us assume that $d_{G_1}(u,u') \leq d_{G_1}(s,u) - 2$. Thus, $d_{G_1+(s,u')}(u,v) \leq d_{G_1+(s,u')}(s,u) + d_{G_1+(s,u')}(s,v) \leq d_{G_1}(u,u') + 1 + d_{G_1}(s,v) \leq d_{G_1}(s,u) + d_{G_1}(s,v) - 1 = D$. As a consequence, the object $\{u,v\}$ is in the set $S_{(s,u')}$. Moreover, $S_{(s,u')} \in \mathcal{S}^*$. Therefore, $\mathcal{S}^*$ is a feasible solution for the set cover instance. This completes the proof. □

As a direct consequence of Lemma 3 and Lemma 4 we have that

Theorem 3. *There exists an $O(\log n)$-approximation algorithm for MCBD.* □

Remark 1. We already pointed out that our algorithm extends to directed graphs. It also extends to the case when we place the distance requirements D_{u_i,v_i} for ℓ pairs $\{u_1,v_2\},\ldots,\{u_\ell,v_\ell\}$ of vertices of G (in particular the resulting graph need not be connected). The approximation ratio becomes $O(\log \ell + \log D_{\max})$, where $D_{\max} = \max_{i=1,\ldots,\ell} D_{u_i,v_i}$.

4 On the Existence of Bicriteria Approximation Algorithms

In this section we first prove the existence of a $(2-\frac{1}{B},2)$-approximation algorithm for BCMD and then we show that for every $D \geq 6$ there is no polynomial time algorithm with an approximation guarantee of $(c\log n, \delta)$, for some constant $c>0$, and for every $\delta < \frac{5}{3} - \frac{4}{D}$, unless $\mathsf{P} = \mathsf{NP}$.

We slightly modify the algorithm described in Section 2 to show the existence of a $(2-\frac{1}{B},2)$-approximation algorithm. The correctness proof follows from the subsequent two key lemmas.

Lemma 5 (Gonzalez 1985 [11]). *Let $V' \subseteq V(G)$. Let $\langle U_1,\ldots,U_k\rangle$ be a k-clustering of V' and let $R = \max_{i=1,\ldots,k} \mathtt{radius}(U_i)$. Gonzalez'algorithm finds a k-clustering $\langle V_1,\ldots,V_k\rangle$ of V' such that $\mathtt{radius}(V_i) \leq 2R, \forall i = 1\ldots,k$.* □

Lemma 6. *There exists a $2B$-clustering $\langle U_1,\ldots,U_{2B}\rangle$ of $V(G)$ such that $\mathtt{radius}(U_1) \leq D$ and $\mathtt{radius}(U_i) \leq \frac{D-1}{2}$, for every $i = 2,\ldots,2B$.*

Proof. Let F^* be an optimal solution for BCMD on input G and B, and let $V' = \{v_1,\ldots,v_\ell\}$ be the set of endvertices of the edges in F^*. Clearly, $\ell \leq 2B$. Let $\langle U_1,\ldots,U_\ell\rangle$ be an ℓ-clustering of $V(G)$ defined as follows. The cluster U_i contains vertex v_i and all the vertices $v \in V(G) \setminus \bigcup_{k=1}^{i-1} U_k$ such that $d_G(v,v_i) \leq d_G(v,v_j), \forall j = i+1,\ldots,\ell$. Clearly, $\mathtt{radius}(U_i) \leq \max_{v\in U_i} d_G(v,v_i)$. As a consequence, we have to prove the claim for the case in which there exists a $v \in U_i$ such that $d_G(v,v_i) > \frac{D-1}{2}$. W.l.o.g., let us assume there exists $v^* \in U_1$ such that $d_G(v^*,v_1) > \frac{D-1}{2}$.

Let v, v' be two vertices of U_i. Every path P from v to v' in G+F^* passing through some edge of F^* has a length greater than or equal to $d_G(v,v_i)+1+d_G(v',v_i)$, while a shortest path in G from v to v' has a length of at most $d_G(v, v_i) + d_G(v', v_i)$. As a consequence, $d_G(v, v') = d_{G+F^*}(v, v') \leq D$.

We modify the ℓ clusters by moving vertices from $U_2, \ldots, U_\ell$ to U_1 as follows. As long as there exists a vertex $v \in U_i$, $i = 2, \ldots, \ell$, such that $d_G(v, v_i) > \frac{D-1}{2}$, we remove v from U_i and we add v to U_1. Any path in $G + F^*$ from v^* to v passing through any edge in F^* has a cost greater than or equal to $d_G(v^*, v_1) + 1 + d_G(v, v_i) > D$. Therefore, $d_G(v^*, v) = d_{G+F^*}(v^*, v) \leq D$.

As a consequence, $\texttt{radius}(U_1) \leq \max_{v \in U_1} d_G(v^*, v) \leq D$. Moreover, for every $i = 2, \ldots, \ell$, $\texttt{radius}(U_i) \leq \max_{v \in U_i} d_G(v_i, v) \leq \frac{D-1}{2}$ by construction. The claim follows. □

Let $\langle U_1, \ldots, U_{2B} \rangle$ be a $2B$-clustering of $V(G)$ such that $\texttt{radius}(U_1) \leq D$ and $\texttt{radius}(U_i) \leq \frac{D-1}{2}$, for every $i = 2, \ldots, 2B$. From Lemma 6, such a $2B$-clustering of $V(G)$ always exists. Let v_1 be a center of U_1, and let $D' = \texttt{radius}(U_1)$. Our algorithm first guesses v_1 and D'. Then, it computes a cluster $V_1 = \{v \in V(G) \mid d_G(v_1, v) \leq D'\}$ and uses Gonzalez'algorithm to find a $(2B-1)$-clustering $\langle V_2, \ldots, V_{2B} \rangle$ of $V(G) \setminus V_1$. As $U_1 \subseteq V_1$, then $\langle U_2, \ldots, U_{2B} \rangle$ is a $(2B-1)$-clustering of $V(G) \setminus V_1$. Therefore, Lemma 5 implies that $\texttt{radius}(V_i) \leq D - 1$, for every $i = 2, \ldots, 2B$. Let v_i be a center of cluster V_i. The algorithm outputs $F = \{(v_1, v_i) \mid i = 2, \ldots, 2B\}$. Observe that every vertex v is at distance of at most D from v_1 in $G + F$. Therefore, $\texttt{diam}(G + F) \leq 2D$. As $|F| \leq 2B - 1$, we have proved the following

Theorem 4. *There exists a* $\left(2 - \frac{1}{B}, 2\right)$*-approximation algorithm for BCMD.* □

Now we prove the non-existence of a $(c \log n, f(D))$-approximation algorithms, unless $\mathsf{P} = \mathsf{NP}$, where $f(D)$ goes to $\frac{5}{3}$ for non-constant values of D.

Theorem 5. *For every* $D \geq 6$*, there exists no* $(c \log n, \delta)$*-approximation algorithm for BCMD, for some constant* $c > 0$ *and for every* $\delta < \frac{5}{3} - \frac{4}{D}$*, unless* $\mathsf{P} = \mathsf{NP}$.

Proof. The reduction is from the *Minimum Dominating Set Problem* (MDS for short), i.e., the problem of finding a minimum-cardinality set of vertices U of a given graph G' on $\hat{n}$ vertices such that every vertex of G' is in U or it is a neighbor of some vertex in U. The MDS is not approximable within $c' \log \hat{n}$, for some constant $c' > 0$, unless $\mathsf{P} = \mathsf{NP}$ [19].

Let G' be a graph with $\hat{n}$ vertices and let k^* be the size of a minimum dominating set in G'. We transform the instance of MDS to an instance of BCMD with n vertices and claim that the existence of a $(c \log n, \delta)$-approximation algorithm for BCMD, with $\delta < \frac{5}{3} - \frac{4}{D}$, implies the existence of a $(c' \log \hat{n})$-approximation algorithm for MDS, for some $c' \leq \eta c$, where $\eta > 0$ is a constant. This would immediately lead to a contradiction by choosing c small enough.

In the rest of the proof we will prove a slightly stronger non-approximability result for all the even values of $D = 2(\rho + \lfloor \rho/2 \rfloor + 1)$, where $\rho \geq 1$ is an integer value and sketch the proof for all the other values of $D \geq 6$. More precisely, for every even value $D = 2(\rho + \lfloor \rho/2 \rfloor + 1)$, we will show the non existence of a

$(c \log n, \delta)$-approximation algorithm for BCMD, for some constant $c > 0$ and for every $\delta < \frac{5}{3} - \frac{7}{3D}$, unless $\mathsf{P} = \mathsf{NP}$.

Let $\rho \geq 1$ be a fixed integer value. We build the input graph G in the following way. G contains 2 copies G_1, G_2 of G' plus a singleton vertex s such that $s \notin V(G')$. For every $u \in V(G')$, denote by u_i the copy of u in G_i. Replace each edge $(u, v) \in E(G_i)$ with a path $P_{u,v}$ from u to v of length ρ by adding $\rho - 1$ new vertices and ρ new edges. For every vertex $u \in V(G')$, and for every $i = 1, 2$, append a path P_u^i to u_i of length $\lfloor \rho/2 \rfloor$ by adding $\lfloor \rho/2 \rfloor$ new vertices and $\lfloor \rho/2 \rfloor$ new edges. For every $i = 1, 2$, denote by ν_u^i the endvertex of P_u^i different from u_i (if $\lfloor \rho/2 \rfloor = 0$ then $\nu_u^i = u_i$). Set $B = 2k^*$. Observe that $n \leq \rho \hat{n}^2$.

Let U^* be a minimum cardinality dominating set in G'. By augmenting G with the $B = 2k^*$ edges from s to both the copies of each vertex in U^*, we obtain that $D \leq 2(\rho + \lfloor \rho/2 \rfloor + 1)$. Indeed, every vertex in $P_{u,v}$ is at distance of at most $\lfloor \rho/2 \rfloor$ from either u or v. Furthermore, every vertex in P_u^i is at distance of at most $\lfloor \rho/2 \rfloor$ from u_i. Finally, every copy of u is at distance of at most ρ from a copy of some vertex in U^*, as U^* is a dominating set in G'. As a consequence, every vertex is at distance at most $\rho + \lfloor \rho/2 \rfloor + 1$ from s.

Now, let F be the set of edges computed by any $(c \log n, \delta)$-approximation algorithm for BCMD, with $\delta < \frac{5}{3} - \frac{7}{3D}$. Let X be the set of the endvertices of the edges in F. We have that $|X| \leq 2cB \log n$. Let Y be equal to X. We modify Y as follows. As long as there is an $x \in Y$ which is an internal vertex of $P_{u,v}$, then we remove x from Y and we add u and v to Y. Next, as long as there is an $x \in Y$ which is a vertex of $V(P_u^i) \setminus \{u_i\}$, then we remove x from Y and we add u_i to Y. Clearly, $|Y| \leq 2|X| \leq 4cB \log n$. Let U be the set of vertices in G defined as follows: U contains a vertex u of G' iff $u_1 \in Y$ or $u_2 \in Y$. We have that $|U| \leq |Y| \leq 4cB \log n \leq 8ck^* \log(\rho \hat{n}^2) \leq 24ck^* \log \hat{n}$. To complete the proof, it is enough to show that U is a dominating set in G. Let u be any vertex in $V(G')$ and consider the two vertices ν_u^1 and ν_u^2; their distance in G is $+\infty$, while their distance in $G + F$ is upper bounded by

$$\gamma D < \frac{5}{3}D - \frac{7}{3} \leq \frac{10}{3}(\rho + \lfloor \rho/2 \rfloor + 1) - \frac{7}{3} \leq 4\rho + 2\lfloor \rho/2 \rfloor + 1.$$

As a consequence, $\gamma D \leq 4\rho + 2\lfloor \rho/2 \rfloor$. As there is no edge between $V(G_1)$ and $V(G_2)$ in G, then there exists a vertex $x \in X$ such that $d_G(x, \nu_u^1) < 2\rho + \lfloor \rho/2 \rfloor$ or $d_G(x, \nu_u^2) < 2\rho + \lfloor \rho/2 \rfloor$. Therefore, by construction, there exists a vertex v in Y such that $d_G(v, \nu_u^1) < 2\rho + \lfloor \rho/2 \rfloor$ or $d_G(v, \nu_u^2) < 2\rho + \lfloor \rho/2 \rfloor$. Since each of the vertices in Y is a copy of some vertex of G' and because $d_G(u_i, v_i) \geq 2\rho$ for every $u, v \in V(G'), (u, v) \notin E(G')$, it follows that U is a dominating set in G'.

To extend the proof for every $D \geq 7$ we can do the following. For the remaining even values of $D \geq 4$ it is enough to increase the length of every P_u^i by one while for odd values of $D \geq 7$, we build the reduction for the case $D - 1$ and we append to s two paths both of length $\lceil D/2 \rceil$ thus forcing the bound on the diameter to be an odd value. However, in both cases, we can prove the non existence of any $(c \log n, \gamma)$-approximation algorithm for some $c > 0$ and for every $\gamma < \frac{5}{3} - \frac{4}{D}$. □

References

1. Alon, N., Gyárfás, A., Ruszinkó, M.: Decreasing the diameter of bounded degree graphs. Journal of Graph Theory 35(3), 161–172 (2000)
2. Brandstädt, A., Le, V.B., Spinrad, J.: Graph classes: a survey. SIAM Monographs on Discrete Mathematics and Applications (1999)
3. Chandrasekaran, R., Daughety, A.: Location on tree networks: p-centre and n-dispersion problems. Mathematics of Operations Research 6(1), 50–57 (1981)
4. Chepoi, V., Estellon, B., Nouioua, K., Vaxès, Y.: Mixed covering of trees and the augmentation problem with odd diameter constraints. Algorithmica 45(2), 209–226 (2006)
5. Chepoi, V., Vaxès, Y.: Augmenting trees to meet biconnectivity and diameter constraints. Algorithmica 33(2), 243–262 (2002)
6. Chung, F.: Diameters of graph: old problems and new results. Congr. Numer. 60, 295–317 (1987)
7. Chung, F., Garey, M.: Diameter bounds for altered graphs. Journal of Graph Theory 8(4), 511–534 (1984)
8. Dodis, Y., Khanna, S.: Designing networks with bounded pairwise distance. In: STOC, pp. 750–759 (1999)
9. Erdös, P., Gyárfás, A., Ruszinkó, M.: How to decrease the diameter of triangle-free graphs. Combinatorica 18(4), 493–501 (1998)
10. Erdös, P., Rényi, A.: On a problem in the theory of graphs. Publ. Math. Inst. Hung. Acad. Sci. B(7), 623–639 (1963)
11. Gonzalez, T.F.: Clustering to minimize the maximum intercluster distance. Theor. Comput. Sci. 38, 293–306 (1985)
12. Grigorescu, E.: Decreasing the diameter of cycles. J. Graph Theory 43(4), 299–303 (2003)
13. Ishii, T., Yamamoto, S., Nagamochi, H.: Augmenting forests to meet odd diameter requirements. In: Ibaraki, T., Katoh, N., Ono, H. (eds.) ISAAC 2003. LNCS, vol. 2906, pp. 434–443. Springer, Heidelberg (2003)
14. Kapoor, S., Sarwat, M.: Bounded-diameter minimum-cost graph problems. Theory Comput. Syst. 41(4), 779–794 (2007)
15. Kariv, O., Hakimi, S.: An algorithmic approach to network location problems. SIAM Journal on Applied Mathematics 37(3), 513–538 (1979)
16. Li, C.-L., McCormick, S.T., Simchi-Levi, D.: On the minimum-cardinality-bounded-diameter and the bounded-cardinality-minimum-diameter edge addition problem. Operations Research Letters 11, 303–308 (1992)
17. Plesník, J.: On the computational complexity of centers locating in a graph. Aplikace Mat. 25
18. Plesník, J.: The complexity of designing a network with minimum diameter. Networks 11(1), 77–85 (1981)
19. Raz, R., Safra, S.: A sub-constant error-probability low-degree test, and a sub-constant error-probability pcp characterization of np. In: STOC, pp. 475–484 (1997)
20. Schoone, A.A., Bodlaender, H.L., van Leeuwen, J.: Diameter increase caused by edge deletion. Journal of Graph Theory 11(3), 409–427 (1987)

Distance Constraint Satisfaction Problems

Manuel Bodirsky[1], Victor Dalmau[2],
Barnaby Martin[3,⋆], and Michael Pinsker[4,⋆⋆]

[1] CNRS / LIX UMR 7161, École Polytechnique, Palaiseau, France
[2] Universitat Pompeu Fabra, Barcelona, Spain
[3] Engineering and Computing Sciences, Durham University, U.K.
[4] Équipe de Logique Mathématique, Université Paris 7, France

Abstract. We study the complexity of constraint satisfaction problems for templates Γ that are first-order definable in $(\mathbb{Z}; succ)$, the integers with the successor relation. Assuming a widely believed conjecture from finite domain constraint satisfaction (we require the *tractability conjecture* by Bulatov, Jeavons and Krokhin in the special case of *transitive* finite templates), we provide a full classification for the case that Γ is locally finite (i.e., the Gaifman graph of Γ has finite degree). We show that one of the following is true: The structure Γ is homomorphically equivalent to a structure with a certain majority polymorphism (which we call *modular median*) and CSP(Γ) can be solved in polynomial time, or Γ is homomorphically equivalent to a finite transitive structure, or CSP(Γ) is NP-complete.

1 Introduction

Constraint satisfaction problems appear naturally in many areas of theoretical computer science, for example in artificial intelligence, optimization, computer algebra, computational biology, computational linguistics, and type systems for programming languages. Such problems are typically NP-hard, but sometimes they are polynomial-time tractable. The question as to which CSPs are in P and which are hard has stimulated a lot of research in the past 10 years. For pointers to the literature, there is a recent collection of survey articles [10].

The *constraint satisfaction problem CSP* for a fixed (not necessarily finite) structure Γ with a finite relational signature τ is the problem to decide whether a given primitive positive sentence is true in Γ. A formula is *primitive positive* if it is of the form $\exists x_1, \ldots, x_n.\ \psi_1 \wedge \cdots \wedge \psi_m$ where ψ_i is an atomic formula over Γ, i.e., a formula of the form $R(y_1, \ldots, y_j)$ for a relation symbol R for a relation from Γ. The structure Γ is also called the *template* of the CSP.

The class of problems that can be formulated as a CSP for a fixed structure Γ is very large. It can be shown that for every computational problem there

⋆ Supported by EPSRC grant EP/G020604/1.
⋆⋆ Supported through Erwin-Schrödinger-Fellowship J2742-N18 and project P21209 of the Austrian Science Fund (FWF), as well as through an APART-fellowship of the Austrian Academy of Sciences.

P. Hliněný and A. Kučera (Eds.): MFCS 2010, LNCS 6281, pp. 162–173, 2010.

is a structure Γ such that the CSP for Γ is equivalent to this problem under polynomial-time Turing reductions [4]. This makes it very unlikely that we can give good descriptions of all those Γ where the CSP for Γ is in P. In contrast, the class of CSPs for a *finite* structure Γ is quite restricted, and indeed it has been conjectured that the CSP for Γ is either in P or NP-complete in this case [12]. So it appears to be natural to study the CSP for classes of infinite structures Γ that share good properties with finite structures.

In graph theory and combinatorics, there are two major concepts of *finiteness* for infinite structures. The first is ω-categoricity: a countable structure is ω-categorical if and only if its automorphism group has for all n only finitely many orbits in its natural action on n-tuples [9,19,16]. This property has been exploited to transfer techniques that were known to analyze the computational complexity of CSPs with finite domains to infinite domains [7,6]; see also the introduction of [2].

The second concept of finiteness is the property of an infinite graph or structure to be *locally finite* (see Section 8 in [11]). A graph is called locally finite if every vertex is contained in a finite number of edges; a relational structure is called locally finite if its Gaifman graph (definition given in Section 2) is locally finite. Many conjectures that are open for general infinite graphs become true for locally finite graphs, and many results that are difficult become easy for locally finite graphs.

In this paper, we initiate the study of CSPs with locally finite templates by studying locally finite templates that have a first-order definition in $(\mathbb{Z}; succ)$, where $succ = \{(x, y) \mid x = y + 1\}$ is the successor relation on the integers. As an example, consider the directed graph with vertex set $\mathbb{Z}$ which has an edge between x and y if the difference between x and y is either 1 or 3. This graph can be viewed as the structure $(\mathbb{Z}; R_{\{1,3\}})$ where $R_{\{1,3\}} = \{(x, y) \mid x - y \in \{1, 3\}\}$, which has a first-order definition over $(\mathbb{Z}; succ)$ since $R_{\{1,3\}}(x, y)$ iff

$$succ(x, y) \vee \exists u, v.\ succ(x, u) \wedge succ(u, v) \wedge succ(v, y).$$

Structures with a first-order definition in $(\mathbb{Z}; succ)$ are particularly well-behaved from a model-theoretic perspective: all such structures are strongly minimal [19,16], and therefore uncountably categorical. Uncountable models of their first-order theory will be saturated; for implications of those properties for the study of the CSP, see [5]. In some sense, $(\mathbb{Z}; succ)$ constitutes one of the simplest infinite structures that is not ω-categorical.

The corresponding class of CSPs has the flavor of assignment problems where we have to assign integers to variables such that various given constraints on differences and distances (and Boolean combinations thereof) between variables are satisfied. We therefore call the class of CSPs whose template is locally finite and definable over $(\mathbb{Z}; succ)$ *distance CSPs*. For instance, the CSP for the structure $(\mathbb{Z}; R_{\{1,3\}})$ is the computational problem to label the vertices of a given directed graph G such that if (x, y) is an arc in G, then the difference between the label for x and the label for y is one or three.

Assuming a widely accepted conjecture about finite domain CSPs, we completely classify the computational complexity of distance CSPs, and show that these problems are either NP-complete or in P. The conjecture we assume is the following special case of that of Feder and Vardi [12]: that the CSP for templates with a transitive automorphism group is either in P or NP-complete (see Section 7 for details). If Γ has a finite core then the statement is implied by this conjecture about finite domain CSPs. Otherwise, if Γ has an infinite core, we prove that either Γ is homomorphically equivalent to a structure Δ_1 over which we can define primitive positively a structure Δ_2 with a finite core Δ_3 of size larger than 2 where the inequality relation $\neq$ is primitive positive definable, and the CSP for Γ is NP-hard, or Γ has a certain majority polymorphism, which we call *modular median*, and the CSP for Γ can be solved by local consistency techniques. Polynomial-time tractability results based on local consistency were previously only known for finite or ω-categorical templates; we use the assumption that templates for distance CSPs are locally finite to extend the technique to non-ω-categorical templates.

On the way to our classification result we derive several facts about structures first-order definable in $(\mathbb{Z}; succ)$, and automorphisms and endomorphisms of these structures, which might be of independent interest in model theory, universal algebra, and combinatorics. For example, we show that every injective endomorphism of a connected locally finite structure Γ with a first-order definition in $(\mathbb{Z}; succ)$ is either of the form $x \mapsto -x + c$ or of the form $x \mapsto x + c$ for some $c \in \mathbb{Z}$.

Owing to reasons of space, many proofs are omitted – please see [3].

2 Preliminaries

A finite relational signature τ is a finite set of relation symbols R_i, each of which has an associated arity k_i. A (relational) structure Γ consists of a set D (the domain) together with a relation $R_i^\Gamma \subseteq D^{k_i}$ for each relation symbol R_i from τ. We consider only finite signature structures in this paper.

For $x, y \in \mathbb{Z}$, let $d(x, y)$ be the distance between x and y, that is, $|x - y|$. The relation $\{(x, y) \mid y = x+1\}$ is denoted by *succ*, and the relation $\{(x, y) \mid d(x, y) = 1\}$ is denoted by *sym-succ*. A k-ary relation R is said to be *first-order (fo) definable* in the τ-structure Γ if there is an fo-τ-formula $\phi(x_1, \ldots, x_k)$ such that $R = \{(x_1, \ldots, x_k) : \Gamma \models \phi(x_1, \ldots, x_k)\}$. A structure Δ is said to be fo-definable in Γ if each of its relations is fo-definable in Γ. For example, $(\mathbb{Z}; sym\text{-}succ)$ is fo-definable in $(\mathbb{Z}; succ)$ (though the converse is false).

The structure induced by a subset S of the domain of Γ is denoted by $\Gamma[S]$. We say that a structure is *connected* if it cannot be written as the disjoint union of two other structures. The *Gaifman graph* of a relational structure Γ with domain D is the following undirected graph: the vertex set is D, and there is an edge between distinct elements $x, y \in D$ when there is a tuple in one of the relations of Γ that has both x and y as entries. A structure Γ is readily seen to be connected if and only if its Gaifman graph is connected. The *degree* of a structure Γ is defined to be the degree of the Gaifman graph of Γ. The degree of a relation $R \subseteq \mathbb{Z}^k$ is defined to be the degree of the structure $(\mathbb{Z}; R)$. Throughout the paper, Γ will be a finite-degree relational structure with an fo definition in $(\mathbb{Z}; succ)$. By (Γ, R) we denote the expansion of Γ with the new relation R.

An fo-formula Θ is *primitive positive* (pp) if it is of the form $\exists x_1, \ldots, x_i.\theta(x_1, \ldots, x_i, x_{i+1}, \ldots, x_j)$ where θ is a conjunction of atoms. Note that we consider the boolean false $\bot$ to be a pp-formula, and we always allow equalities in pp-formulas. A pp-sentence is a pp-formula with no free variables. For convenience we will consider pp-formulas always to be in prenex normal form.

For a structure Γ over a finite signature, CSP(Γ) is the computational problem to decide whether a given pp-sentence is true in Γ. It is not hard to see that, for any Γ and Δ with the same domain, such that each of the relations of Δ are pp-definable in Γ, that we have CSP(Δ) $\leq_{\mathrm{P}}$ CSP(Γ) (see [18]), where $\leq_{\mathrm{P}}$ indicates polynomial-time many-to-one reduction (in fact, logspace reductions may be used, though this is harder to see and requires the result of [20]).

Let Γ and Δ be τ-structures. A *homomorphism* from Γ to Δ is a function from the domain of Γ to the domain of Δ such that, for each k-ary relation symbol R in τ and each k-tuple $(a_1, \ldots, a_k)$ from Γ, if $(a_1, \ldots, a_k) \in R^\Gamma$, then $(f(a_1), \ldots, f(a_k)) \in R^\Delta$. In this case we say that the map f *preserves* the relation R. Injective homomorphisms that also preserve the complement of each relation are called *embeddings*. Surjective embeddings are called isomorphisms; homomorphisms and isomorphisms from Γ to itself are called *endomorphisms* and *automorphisms*, respectively. The set of automorphisms of a structure Γ forms a group under composition. A (k-ary) *polymorphism* of a structure Γ over domain D is a function $f : D^k \to D$ such that, for all m-ary relations R of Γ, if $(a^i_1, \ldots, a^i_m) \in R^\Gamma$, for all $i \leq k$, then $(f(a^1_1, \ldots, a^k_1), \ldots, f(a^1_m, \ldots, a^k_m)) \in R^\Gamma$.

A unary function g (over domain D) is in the *local closure* of a set of unary functions F (over domain D) if, for every finite $D' \subseteq D$ there is a function $f' \in F$ such that g and f' agree on all elements in D'. We say the F *generates* f if we may produce f from the members of F by repeated applications of composition and local closure.

If there exist homomorphisms $f : \Gamma \to \Delta$ and $g : \Delta \to \Gamma$ then Γ and Δ are said to be *homomorphically equivalent*. It is a basic observation that CSP(Γ) = CSP(Δ) if Γ and Δ are homomorphically equivalent. A structure is a *core* if all of its endomorphisms are embeddings [1] – a *core* Δ *of a structure* Γ is an induced substructure that is itself a core and is homomorphically equivalent to Γ. It is well-known that, if a structure has a finite core, then that core is unique up to isomorphism (the same is not true for infinite cores).

We have defined the CSP to be over relational structures, but in the case of $(\mathbb{Z}; succ)$, one could equally consider the *succ* as the unary function s, instead of a binary relation. With functional signature, $(\mathbb{Z}; s)$ admits *quantifier elimination*; that is, for every fo-formula $\phi(\overline{x})$ there is a quantifier-free (qf) $\phi'(\overline{x})$ such that $(\mathbb{Z}; s) \models \forall \overline{x}.\phi(\overline{x}) \leftrightarrow \phi'(\overline{x})$ (this is easy to prove, and can be found explicitly in [13]). Thus we may have terms in ϕ' of the form $y = s^j(x)$, where s^j is the successor function composed on itself j times. Let Γ be a finite signature structure, fo-definable in $(\mathbb{Z}; succ)$, i.e. qf-definable in its functional variant $(\mathbb{Z}; s)$. Let m be the largest number such that $y = s^m(x)$ appears as a term in the qf definition of a relation of Γ. Consider now CSP(Γ), the problem to evaluate $\Phi := \exists x_1, \ldots, x_k.\phi(x_1, \ldots, x_k)$, where ϕ is a conjunction of atoms, on Γ. Let $S := \{1, \ldots, k\cdot(m+1)\}$. It is not hard to see that $(\mathbb{Z}; succ) \models \Phi$ iff $(\mathbb{Z}; succ)[S] \models \Phi$. It follows that CSP($\Gamma$) will always be in NP.

3 Endomorphisms

The main result of this section is the following theorem.

Theorem 1. *Let Γ be a relational structure with a first-order definition in $(\mathbb{Z}; succ)$ which has finite degree and which is connected. Then:*

- *The automorphism group of Γ equals either the automorphism group of $(\mathbb{Z}; succ)$, or the automorphism group of $(\mathbb{Z}; sym\text{-}succ)$.*
- *Either Γ has a finite range endomorphism, or it has an endomorphism which maps Γ onto a subset of $\mathbb{Z}$ isomorphic to a structure fo-definable in $(\mathbb{Z}; succ)$ all of whose endomorphisms are automorphisms.*

The proof of this theorem is omitted, but makes use of the following series of lemmata. We assume henceforth that Γ is a relational structure with a first-order definition in $(\mathbb{Z}; succ)$ which has finite degree and which is connected.

Before beginning the proof, we remark that although it is tempting to believe that the automorphism group of Γ equals the automorphism group of $(\mathbb{Z}; sym\text{-}succ)$ iff Γ is fo-definable in $(\mathbb{Z}; sym\text{-}succ)$, this is not true: Let

$$R := \{(x, y, u, v) \in \mathbb{Z}^4 : (y = succ(x) \wedge v = succ(u)) \vee (u = succ(v) \wedge x = succ(y))\},$$

and set $\Gamma := (\mathbb{Z}; R)$. Clearly, Γ satisfies the hypotheses of Theorem 1. The function which sends every $x \in \mathbb{Z}$ to $-x$ is an automorphism of Γ, so the automorphism group of Γ equals that of $(\mathbb{Z}; sym\text{-}succ)$, by Theorem 1. However, R is not fo-definable in $(\mathbb{Z}; sym\text{-}succ)$.

Denote by E the edge-relation of the Gaifman graph of Γ. It is clear that every endomorphism of Γ preserves E. We claim that there are $0 < d_1 < \cdots < d_n$ such that $E(x, y)$ holds iff $d(x, y) \in \{d_1, \ldots, d_n\}$. To see this, observe that if $x, y \in \mathbb{Z}$ are connected by E and $u, v \in Z$ are so that $d(x, y) = d(u, v)$, then also u, v are connected by E: This is because there is an automorphism of $(\mathbb{Z}; succ)$ (and hence of Γ) which sends $\{x, y\}$ to $\{u, v\}$ and because this automorphism also preserves E. Hence, the relation E is determined by distances. Moreover, there are only finitely many distances since Γ is assumed to have finite degree. Since Γ is connected, the greatest common divisor of $d_1, \ldots, d_n$ is 1. We will refer to the distances defining the Gaifman graph of Γ as $d_1, \ldots, d_n$. We also write D for the largest distance d_n.

Lemma 1. *Suppose that Γ is connected and of finite degree. Then there exists a constant $c = c(\Gamma)$ such that for all endomorphims e of Γ we have $d(e(x), e(y)) \leq d(x, y) + c$ for all $x, y \in \mathbb{Z}$.*

Observe that a constant $c(\Gamma)$ does not only exist, but can actually be calculated given the distances $d_1, \ldots, d_n$. In the following, we will keep the symbol c reserved for the minimal constant guaranteed by the preceding lemma.

Lemma 2. *Let e be an endomorphism of Γ. If for all $k > c+1$ there exist x, y with $d(x,y) = k$ and $d(e(x), e(y)) < k$, then e generates a finite range operation whose range has size at most $2c + 1$.*

Proof. Let $A \subseteq \mathbb{Z}$ be finite. We claim that e generates a function which maps A into a set of diameter at most $2c+1$. The lemma then follows by a standard local closure argument involving König's tree lemma.

Enumerate the pairs $(x, y) \in A^2$ with $x < y$ by $(x_1, y_1), \ldots, (x_r, y_r)$. Now the hypothesis implies that there exists t_1 generated by e such that $d(t_1(x_1), t_1(y_1)) \leq c+1$. Similarly, there exists t_2 generated by e such that $d(t_2t_1(x_2), t_2t_1(y_2)) \leq c+1$. Continuing like this we arrive at a function t_r generated by e such that $d(t_rt_{r-1}\cdots t_1(x_r), t_rt_{r-1}\cdots t_1(y_r)) \leq c+1$. Now consider $t := t_r \circ \cdots \circ t_1$. Set $f_j := t_r \circ \cdots \circ t_{j+1}$ and $g_j := t_j \circ \cdots \circ t_1$, for all $1 \leq j \leq r$; so $t = f_j \circ g_j$. Then, since by construction $d(g_j(x_j), g_j(y_j)) \leq c+1$, we have that $d(t(x_j), t(y_j)) = d(f_j(g_j(x_j)), f_j(g_j(y_j)) \leq d(g_j(x_j), g_j(y_j)) + c \leq 2c+1$ for all $1 \leq j \leq r$, and our claim follows.

Lemma 3. *If the hypothesis of the preceding lemma does not hold, i.e. if there exists $k > c+1$ such that $d(e(x), e(y)) \geq k$ for all x, y with $d(x, y) = k$, then either $e(s+D) = e(s) + D$ for all $s \in \mathbb{Z}$ or $e(s+D) = e(s) - D$ for all $s \in \mathbb{Z}$. In particular, e does not generate a finite range operation.*

The following lemma summarizes the preceding two lemmas.

Lemma 4. *The following are equivalent for an endomorphism e of Γ:*

(i) There exists $k > c+1$ such that $d(e(x), e(y)) \geq k$ for all $x, y \in \mathbb{Z}$ with $d(x, y) = k$.
(ii) e does not generate a finite range operation.
(iii) e satisfies either $e(v+D) = e(v) + D$ or $e(v+D) = e(v) - D$.

Proof. Lemma 3 shows that (i) implies (ii) and (iii). It follows from from Lemma 2 that (ii) implies (i). Finally, it is clear that (iii) implies (ii).

We know now that there are two types of endomorphisms of Γ: Those which are periodic with period D, and those which generate a finite range operation. We will now provide examples showing that both types really appear.

Example 1. Let $R := \{(x, y) : d(x, y) = 1 \vee d(x, y) = 3\}$, and set $\Gamma := (\mathbb{Z}; R)$. Set $e(3k) := 3k$, $e(3k+1) := 3k+1$, and $e(3k+2) := 3k$, for all $k \in \mathbb{Z}$. Then e is an endomorphism of Γ. It does not generate any finite range operations since it satisfies $e(v+3) = e(v) + 3$ for all $v \in \mathbb{Z}$.

Observe that in the previous example, we checked that e is of the non-finite-range type by virtue of the easily verifiable Item (iii) of Lemma 4 and without calculating $c(\Gamma)$, which is more complicated.

Example 2. For the same structure Γ, let e be the function which maps every $x \in \mathbb{Z}$ to its value modulo 4. Then e is an endomorphism which has finite range.

Example 3. Set $R := \{(x, y) : d(x, y) \in \{1, 3, 6\}\}$ and $S := \{(x, y) : d(x, y) = 3\}$. Then $\Gamma := (\mathbb{Z}; R, S)$ has the endomorphism from Example 1. However, it does not have any finite range endomorphism. To see this, consider the set $\mathbb{Z}_3 := \{3m : m \in \mathbb{Z}\}$. If e were a finite range endomorphism, it would have to map this set onto a finite set. Assume wlog that $e(0) = 0$ and $e(3) > 0$. Then $e(3) = 3$ as e preserves S. We claim $e(s) = s$ for all $s \in \mathbb{Z}_3$. Suppose to the contrary that s is the minimal positive counterexample (the negative case is similar). We have $e(s-3) = s-3$ and hence, as e preserves S, $e(s) \in \{s-6, s\}$. If we had $e(s) = s-6$, then $e(s-6) = s-6$ and $(s-6, s) \in R$ yields a contradiction.

Example 4. Let $\Gamma = (\mathbb{Z}; \textit{sym-succ})$, and let e be the function that maps every x to its absolute value. Then e does not have finite range, but does generate a function with finite range (of size 2).

The proof of Lemma 3 generalizes canonically to a more general situation, whose proof is as that of Lemma 3, with D replaced by q.

Lemma 5. *Let e be an endomorphism of Γ satisfying the various statements of Lemma 4. Let q be so that $d(x, y) = q$ implies that $d(e(x), e(y)) \leq q$. Then e satisfies either $e(v + q) = e(v) + q$ for all $v \in \mathbb{Z}$, or $e(v + q) = e(v) - q$ for all $v \in \mathbb{Z}$.*

Given an endomorphism e of Γ, we call all positive natural numbers q with the property that $e(v+q) = e(v)+q$ for all $v \in \mathbb{Z}$ or $e(v+q) = e(v)-q$ for all $v \in \mathbb{Z}$ *stable for e*. Observe that if e satisfies the various statements of Lemma 4, then D is stable for e. Note also that if p, q are stable for e, then they must have the same "direction": We cannot have $e(v + p) = e(v) + p$ and $e(v + q) = e(v) - q$ for all $v \in \mathbb{Z}$.

Lemma 6. *Let e satisfy the various statements of Lemma 4, and let q be the minimal stable number for e. Then the stable numbers for e are precisely the multiples of q. In particular, q divides D.*

Lemma 7. *Let e be an endomorphism of Γ satisfying the hypotheses of Lemma 4. Let q be its minimal stable number. Then e can be composed with automorphisms of $(\mathbb{Z}; \textit{succ})$ to obtain an endomorphism t with the following properties:*

- *t satisfies either $t(v + q) = t(v) + q$ or $t(v + q) = t(v) - q$*
- *$t(0) = 0$*
- *$t[\mathbb{Z}] = \{q \cdot s : s \in \mathbb{Z}\}$.*

Lemma 8. *Let e be an endomorphism of Γ which is not an automorphism of $(\mathbb{Z}; \textit{sym-succ})$. Then e is not injective.*

Proof. If e generates a finite range operation then the lemma follows immediately, so assume this is not the case. Then e has a minimal stable number q. Since e is not an automorphism of $(\mathbb{Z}; \textit{sym-succ})$, we have $q > 1$. But then the statement follows from the preceding lemma, since the function t is not injective.

Lemma 9. *Let e be an endomorphism of Γ which is not an automorphism of $(\mathbb{Z}; sym\text{-}succ)$ and which does not generate a finite range operation. Then e is not surjective.*

Proof. This is a direct consequence of Lemma 7, since being surjective is preserved under composition.

4 Definability of Successor

In this section we show how to reduce the complexity classification for distance constraint satisfaction problems with template Γ to the case where either Γ has a finite core, or the relation *succ* is pp-definable in Γ. We make essential use of the results of the previous section; but note that in this section we do *not* assume that Γ is connected.

Theorem 2. *Every finite degree relational structure Γ with a first-order definition in $(\mathbb{Z}; succ)$ is either homomorphically equivalent to a finite structure, or to a connected finite-degree structure Δ with a first order definition in $(\mathbb{Z}; succ)$ which satisfies one of two possibilities: CSP(Δ) (and, hence, CSP(Γ)) is NP-hard, or succ is definable in Δ.*

The proof of Theorem 2 involves the succeeding lemmata and is omitted. The following lemma demonstrates how the not necessarily connected case can be reduced to the connected case.

Lemma 10. *Every finite degree relational structure Γ with a first-order definition in $(\mathbb{Z}; succ)$ is homomorphically equivalent to a connected finite-degree structure Δ with a first order definition in $(\mathbb{Z}; succ)$.*

Lemma 11. *Let $(a_1, \ldots, a_k), (b_1, \ldots, b_k) \in \mathbb{Z}^k$. Then there is an automorphism α of $(\mathbb{Z}; succ)$ with $\alpha(a_i) = b_i$ for all $i \leq k$ if and only if $a_i - a_j = b_i - b_j$ for all $1 \leq i, j \leq k$.*

Lemma 12. *Suppose that Γ is connected. Then there is an n_0 such that $\Gamma[\{1, \ldots, n\}]$ is connected for all $n \geq n_0$.*

Lemma 13. *Suppose that Γ is connected and of finite degree. Then there are n_0 and c such that for all $n \geq n_0$ and any homomorphism f from $\Gamma[\{1, \ldots, n\}]$ to Γ we have that $d(f(x), f(y)) \leq c + d(x, y)$ for all $x, y \in \{1, \ldots, n\}$.*

Proof. Let n_0 be the number from Lemma 12. Then for all $n \geq n_0$, the structure $\Gamma[\{1, \ldots, n\}]$ is connected. Now, proceed as in Lemma 1.

Proposition 1. *Let Γ be a connected finite-degree structure with a first-order definition in $(\mathbb{Z}; succ)$. Assume that every endomorphism of Γ is an automorphism of $(\mathbb{Z}; sym\text{-}succ)$. Then for all $a_1, a_2 \in \mathbb{Z}$ there is a finite $S \subseteq \mathbb{Z}$ that contains $\{a_1, a_2\}$ such that for all homomorphisms f from $\Gamma[S]$ to Γ we have $d(f(a_1), f(a_2)) = d(a_1, a_2)$.*

Corollary 1. *Suppose that Γ is a connected finite-degree structure with a first-order definition in $(\mathbb{Z}; succ)$, and suppose that all endomorphisms of Γ are automorphisms of Γ. Then either the relation sym-succ$^k = \{(x,y) \mid d(x,y) = k\}$ is pp-definable in Γ for every $k \geq 1$, or the relation succ$^k = \{(x,y) \mid x - y = k\}$ is pp-definable in Γ for every $k \geq 1$.*

Proposition 2. *Suppose that for all k the relation sym-succ$^k = \{(x,y) \in \mathbb{Z}^2 \mid d(x,y) = k\}$ is pp-definable in Γ. Then CSP(Γ) is NP-hard.*

Proof. Observe that the primitive positive formula $\exists y. d(x,y) = 1 \wedge d(y,z) = 5$ defines the relation $R = \{(x,z) \mid d(x,z) \in \{4,6\}\}$. The structure $(\mathbb{Z}; R)$ decomposes into two copies of the structure $(\mathbb{Z}; S)$ where $S = \{(x,y) \mid d(x,y) \in \{2,3\}\}$. This structure has the endomorphism $x \mapsto x \mod 5$, and the image induced by this endomorphism is a cycle of length 5, which has a hard CSP (this is well-known; for a much stronger result on undirected graphs, see Hell and Nešetřil [14]).

5 The Power of Consistency

All tractable distance constraint satisfaction problems for templates without a finite core can be solved by an algorithmic technique known as *local consistency.* We prove these tractability results in this section.

A *majority operation* on a set X is a mapping $f : X^3 \to X$ satisfying

$$f(x,x,y) = f(x,y,x) = f(y,x,x) = x \ .$$

An n-ary relation R on a set X is 2-*decomposable* if R contains all n-tuples $(t_1, \ldots, t_n)$ such that for every 2-element subset I of $\{1, \ldots, n\}$ there is a tuple $s \in R$ such that $t_i = s_i$ for all $i \in I$.

We need the following concept to prove the algorithmic results in this paper. Let Δ be a structure with a (not necessarily finite) relational signature τ, and let ϕ be a conjunction of atomic τ-formulas with variables V. For $k > 0$, we say that ϕ is *k-consistent* (with respect to Δ) if for every assignment α of $k-1$ variables $x_1, \ldots, x_{k-1} \in V$ to elements from Δ and for every variable $x_k \in V$ the assignment α can be extended to x_k such that all conjuncts of ϕ that involve no other variables than $x_1, \ldots, x_k$ are satisfied over Δ by the extension of α. We say that ϕ is *strongly k-consistent* if ϕ is j-consistent for all j with $2 \leq j \leq k$. We say that ϕ is *globally consistent* if ϕ is k-consistent for all $k > 0$.

The following has been shown in [17] (with an explicit comment in Section 4.4 that the result also holds on infinite domains).

Theorem 3 (Special case of Theorem 3.5 of [17]). *Let Γ be a structure with a majority polymorphism. Then every relation R of Γ is 2-decomposable. Moreover, every strongly 3-consistent conjunction of atomic formulas is also globally consistent with respect to Γ.*

In the proof of the following theorem, and in the next section, it will be convenient to represent binary relations $R \subseteq \mathbb{Z}^2$ with a first-order definition in $(\mathbb{Z}; succ)$ by sets of integers as follows: the set S represents the binary relation $R_S := \{(x, x+k) \mid k \in S\}$. Conversely, when R is a binary relation with a first-order definition in $(\mathbb{Z}; succ)$, let $S(R)$ be the set such that $R_{S(R)} = R$. It is easy to see that every binary relation of finite degree and with a first-order definition in $(\mathbb{Z}; succ)$ is of the form R_S for some finite S.

Theorem 4. *Let Γ be a finite degree structure with a first-order definition in $(\mathbb{Z}; succ)$. If Γ has a majority polymorphism, then CSP(Γ) is in P.*

Definition 1. *The* d-modular median *is the ternary operation $m_d : \mathbb{Z}^3 \to \mathbb{Z}$ defined as follows:*

- *If x, y, z are congruent modulo d, then $m_d(x, y, z)$ equals the median of x, y, z.*
- *If precisely two arguments from x, y, z are congruent modulo d then $m_d(x, y, z)$ equals the first of those arguments in the ordered sequence (x, y, z).*
- *Otherwise, $m_d(x, y, z) = x$.*

Clearly, d-modular median operations are majority operations.

Corollary 2. *Let Γ be a finite-degree structure with a first-order definition in $(\mathbb{Z}; succ)$ and a finite relational signature, and suppose that Γ has a modular median polymorphism. Then CSP(Γ) is in P.*

6 Classification

In this section we finish the complexity classification for those Γ that do not have a finite core. The main result of Section 4 shows that, unless Γ has a finite core, for the complexity classification of CSP(Γ) we can assume that the structure Γ contains the relation *succ*. In the following we therefore assume that the structure Γ contains the relation *succ*; moreover, we freely use expressions of the form $x - y = d$, for fixed d, in primitive positive definitions since such expressions have themselves pp-definitions from *succ* and therefore from Γ. Our main result will be the following, whose proof is omitted.

Theorem 5. *Let Γ be a first-order expansion of $(\mathbb{Z}; succ)$. Then Γ is preserved by a modular median and CSP(Γ) is in P, or CSP(Γ) is NP-hard.*

A *d-progression* is a subset of $\mathbb{Z}$ of the form $\{k, k+d, \ldots, k+ld\}$, for some $k, l \in \mathbb{Z}$. We shall denote $\{k, k+d, \ldots, k+ld\}$ by $[k, k+ld]_d$.

Proposition 3. *Let $R \subseteq \mathbb{Z}^2$ be a finite-degree binary relation with a first-order definition in $(\mathbb{Z}; succ)$. Then the following are equivalent.*

1. *R is preserved by the d-modular median m_d;*
2. *$R = R_S$ for a d-progression S.*

Proposition 4. *Let a, b be two odd numbers such that $a < b$. Then CSP($\mathbb{Z}$; succ, $R_{\{0,a,b,a+b\}}$) is NP-hard.*

Proposition 5. *Let $a, b, c \in \mathbb{Z}$ with $b \neq c$. Then CSP($\mathbb{Z}$; succ, $R_{\{a,b\}}, R_{\{a,c\}}$) is NP-hard.*

Lemma 14. *Let S be a finite set of integers with $|S| > 1$ and let d be the greatest common divisor of all $a - a'$ with $a, a' \in S$. For any d-progression T, the relation R_T is pp-definable in ($\mathbb{Z}$; succ, R_S).*

Proposition 6. *Let Γ be a structure with only binary relations of finite degree with a first-order definition in ($\mathbb{Z}$; succ). Then either Γ is preserved by a modular median, or CSP($\mathbb{Z}$; succ) is NP-hard.*

Corollary 3. *Γ has a finite core, or CSP(Γ) is in P or NP-complete.*

Proof. Suppose that Γ does not have a finite core. Let Δ be the substructure of Γ as described in Theorem 2. Clearly, CSP(Γ) and CSP(Δ) are the same problem. Unless CSP(Γ) is NP-hard, the relation *succ* is pp-definable in Δ. By the fundamental theorem of pp definability, the CSP of the expansion of Δ by the successor relation has the same complexity as CSP(Δ). Now the claim follows from Theorem 5.

7 Concluding Remarks

Structures with a first-order definition in ($\mathbb{Z}$; *succ*) have a *transitive* automorphism group, i.e., for every $x, y \in \mathbb{Z}$ there is an automorphism of Γ that maps x to y. We call such structures Γ *transitive* as well. It is well-known and easy to prove (see e.g. [15]) that a finite core of a transitive structure is again transitive.

The complexity of the CSP for finite transitive templates has not yet been classified. The following is known.

Theorem 6 (of [8]). *Let Γ be a finite core. If there is a primitive positive interpretation of the structure $\Delta := (\{0,1\}; \{(0,0,1),(0,1,0),(1,0,0)\})$ in Γ, then CSP(Γ) is NP-complete.*

The following conjecture is widely believed in the area.

Conjecture 1 (of [8]). Let Γ be a finite core. If there is no primitive positive interpretation of the structure $\Delta := (\{0,1\}; \{(0,0,1),(0,1,0),(1,0,0)\})$ in Γ, then CSP(Γ) is in P.

We believe that this conjecture might be easier to show for *transitive* finite cores only. Note that by transitivity, the polymorphism algebra of Γ has no proper subalgebras. Since Γ is a core, all polymorphisms are surjective. It follows from known results [8] that, unless CSP(Γ) admits a primitive positive interpretation of Δ, all minimal factors of the polymorphism algebra contain an affine operation.

References

1. Bodirsky, M.: Cores of countably categorical structures. Logical Methods in Computer Science (LMCS) 3(1), 1–16 (2007), doi:10.2168/LMCS-3(1:2)
2. Bodirsky, M., Chen, H., Pinsker, M.: The reducts of equality up to primitive positive interdefinability. To appear in the Journal of Symbolic Logic (2010)
3. Bodirsky, M., Dalmau, V., Martin, B., Pinsker, M.: Distance constraint satisfaction problems. CoRR, abs/1004.3842 (2010)
4. Bodirsky, M., Grohe, M.: Non-dichotomies in constraint satisfaction complexity. In: Aceto, L., Damgård, I., Goldberg, L.A., Halldórsson, M.M., Ingólfsdóttir, A., Walukiewicz, I. (eds.) ICALP 2008, Part II. LNCS, vol. 5126, pp. 184–196. Springer, Heidelberg (2008)
5. Bodirsky, M., Hils, M., Martin, B.: On the scope of the universal-algebraic approach to constraint satisfaction. In: Proceedings of LICS 2010 (2010)
6. Bodirsky, M., Kára, J.: The complexity of temporal constraint satisfaction problems complexity of temporal constraint satisfaction problems. Journal of the ACM 57(2) (2009); An extended abstract appeared in the proceedings of STOC 2008
7. Bodirsky, M., Nešetřil, J.: Constraint satisfaction with countable homogeneous templates. Journal of Logic and Computation 16(3), 359–373 (2006)
8. Bulatov, A., Krokhin, A., Jeavons, P.G.: Classifying the complexity of constraints using finite algebras. SIAM Journal on Computing 34, 720–742 (2005)
9. Cameron, P.J.: Oligomorphic Permutation Groups. Cambridge Univ. Press, Cambridge (1990)
10. Creignou, N., Kolaitis, P.G., Vollmer, H. (eds.): Complexity of Constraints - An Overview of Current Research Themes [Result of a Dagstuhl Seminar]. LNCS, vol. 5250. Springer, Heidelberg (2008)
11. Diestel, R.: Graph Theory, 3rd edn. Springer, New York (2005)
12. Feder, T., Vardi, M.: The computational structure of monotone monadic SNP and constraint satisfaction: A study through Datalog and group theory. SIAM Journal on Computing 28, 57–104 (1999)
13. Hedman, S.: A First Course in Logic: An Introduction to Model Theory, Proof Theory, Computability, and Complexity (Oxford Texts in Logic). Oxford University Press, Inc., New York (2004)
14. Hell, P., Nešetřil, J.: On the complexity of H-coloring. Journal of Combinatorial Theory, Series B 48, 92–110 (1990)
15. Hell, P., Nešetřil, J.: Graphs and Homomorphisms. Oxford University Press, Oxford (2004)
16. Hodges, W.: A shorter model theory. Cambridge University Press, Cambridge (1997)
17. Jeavons, P., Cohen, D., Cooper, M.: Constraints, consistency and closure. AI 101(1-2), 251–265 (1998)
18. Jeavons, P., Cohen, D., Gyssens, M.: Closure properties of constraints. JACM 44(4), 527–548 (1997)
19. Marker, D.: Model Theory: An Introduction. Springer, New York (2002)
20. Reingold, O.: Undirected st-connectivity in log-space. In: STOC, pp. 376–385 (2005)

Faster Algorithms on Branch and Clique Decompositions

Hans L. Bodlaender[1], Erik Jan van Leeuwen[2], Johan M.M. van Rooij[1], and Martin Vatshelle[2]

[1] Department of Information and Computing Sciences, Utrecht University
P.O. Box 80.089, NL-3508 TB Utrecht, The Netherlands
{hansb,jmmrooij}@cs.uu.nl
[2] Department of Informatics, University of Bergen
P.O. Box 7803, N-5020 Bergen, Norway
{E.J.van.Leeuwen,martin.vatshelle}@ii.uib.no

Abstract. We combine two techniques recently introduced to obtain faster dynamic programming algorithms for optimization problems on graph decompositions. The unification of generalized fast subset convolution and fast matrix multiplication yields significant improvements to the running time of previous algorithms for several optimization problems. As an example, we give an $O^*(3^{\frac{\omega}{2}k})$ time algorithm for Minimum Dominating Set on graphs of branchwidth k, improving on the previous $O^*(4^k)$ algorithm. Here ω is the exponent in the running time of the best matrix multiplication algorithm (currently $\omega < 2.376$). For graphs of cliquewidth k, we improve from $O^*(8^k)$ to $O^*(4^k)$. We also obtain an algorithm for counting the number of perfect matchings of a graph, given a branch decomposition of width k, that runs in time $O^*(2^{\frac{\omega}{2}k})$. Generalizing these approaches, we obtain faster algorithms for all so-called $[\rho, \sigma]$-domination problems on branch decompositions if ρ and σ are finite or cofinite. The algorithms presented in this paper either attain or are very close to natural lower bounds for these problems.

1 Introduction

Graph decompositions have over the last few years shown their worth in attacking NP-hard graph optimization problems. Most of this success is due to tree decompositions, which form the basis of results in many areas, from approximation algorithms to exact algorithms and have become part of any algorithmic toolbox. This success has motivated researchers to define and study other types of graph decompositions that are 'better' than tree decompositions. In this paper, we investigate algorithms for optimization problems on two such decompositions, namely branch decompositions and clique decompositions. By for the first time combining two recent techniques used in designing algorithms on graph decompositions, generalized fast subset convolution and fast matrix multiplication, in conjunction with the use of asymmetric vertex states, we obtain significant improvements on previous results.

P. Hliněný and A. Kučera (Eds.): MFCS 2010, LNCS 6281, pp. 174–185, 2010.

Algorithmic Techniques. Fast subset convolutions were introduced by Björklund et al. [1] to improve the running time of algorithms for optimization problems admitting a convolution-like recursive definition. These ideas were recently applied and generalized to show that a whole range of problems has faster algorithms on tree decompositions [16]. In particular, Van Rooij et al. [16] showed that Minimum Dominating Set has an $O(n\,\mathrm{tw}^2(G)\,3^{\mathrm{tw}(G)})$ time algorithm. Using a generalized form of fast subset convolutions, they were able to obtain the fastest algorithms for a large class of so-called $[\rho, \sigma]$-domination problems.

Matrix multiplication has been used for a much longer time as a basic tool for solving combinatorial problems. The best possible exponent in the running time of an algorithm performing multiplication of two $n \times n$ matrices is denoted by ω, i.e. the running time is $O(n^\omega)$. Currently we know that $\omega < 2.376$, due to an algorithm by Coppersmith and Winograd [4], but it is frequently hypothesized that $\omega = 2$. Dorn [8] recently showed that matrix multiplication can also be used as a tool in solving many optimization problems on branch decompositions.

One of the main results of this paper is that fast subset convolutions and fast matrix multiplication can be combined to obtain faster algorithms on branch decompositions for many optimization problems.

Graph Decompositions. The notion of a branch decomposition was proposed by Robertson and Seymour as part of their graph minors project [12]. All of the recent results aimed at obtaining faster exact or fixed-parameter algorithms for Minimum Dominating Set on planar graphs and graphs excluding a fixed minor rely on branch decompositions [9,10,8]. The branchwidth and the treewidth of a graph are very closely related; the branchwidth of a graph is always less than its treewidth, but never by more than a factor 2/3.

The notion of cliquewidth was first studied by Courcelle et al. [5]. Whereas the treewidth of the n-vertex clique is equal to $n-1$, its cliquewidth is equal to 2. Moreover, the cliquewidth of a graph is always bounded by a function of its treewidth [6]. This makes cliquewidth an interesting graph parameter to consider on graphs where the tree- or branchwidth is too high for efficient algorithms.

Our Results. In this paper, we improve on the currently best algorithms for Minimum Dominating Set on branch and clique decompositions. Dorn [8] showed that Minimum Dominating Set has an $O(m\,4^k)$ time algorithm on branch decompositions of width k. By combining fast subset convolution and fast matrix multiplication, we improve on this algorithm and obtain an $O(mk^2\,3^{\frac{\omega}{2}k})$ time algorithm. A further innovation is the use of *asymmetric vertex states*. When combining two tables in this kind of dynamic programming algorithms, the set of vertex states used by these tables is always the same. We however use different states to obtain further speed-ups.

This result extends to counting the number of dominating sets of each size. Another counting problem where we can apply this technique is counting the number of perfect matchings of a graph. This problem generalizes the problem of computing the permanent of a matrix and is a well-known #P-hard problem [15].

We give an algorithm for this problem on branch decompositions, running in time $O(mk^2 2^{\frac{\omega}{2}k} i_\times(n))$, where $i_\times(n)$ is the time to multiply two n-bit numbers.

Using the ideas of these algorithms, we solve existence, minimization, maximization and counting variations of all $[\rho,\sigma]$-domination problems with finite or cofinite σ and ρ in $O^*(s^{\frac{\omega}{2}k})$ time, where s is the number of vertex states used. Examples of such problems are Strong Stable Set, Independent Dominating Set, Perfect Code, Induced Bounded Degree Subgraph, and p-Dominating Set.

On clique decompositions, we report an even bigger improvement for Minimum Dominating Set. We present an $O^*(4^{\mathrm{cw}(G)})$ time algorithm improving the current best $O^*(8^{\mathrm{cw}(G)})$ time algorithm obtained in [3].

2 Preliminaries

A *branch decomposition* (T,l) of a graph G is a ternary tree T and a bijection l between the edges of G and the leaves of T. Associated with every edge $e \in E(T)$ is the *middle set* X_e of e, defined as the set of vertices in $V(G)$ which have incident edges e_1, e_2 such that the leaves $l(e_1)$ and $l(e_2)$ are in different components of $T-e$. The *width* of a branch decomposition is the size of the largest set X_e. The *branchwidth* $\mathrm{bw}(G)$ is the minimum width of a branch decomposition of G.

Create a root of T as follows. Choose an edge $e=(t,t') \in E(T)$, subdivide it, and add a new vertex r to T adjacent to the vertex created in the subdivision. The middle set of each edge of the subdivision is set to X_e and the middle set of the edge incident to r is set to $\emptyset$. Root T at r. Given any edge $(t,t')=e \in E(T)$, we can now speak of T_e, which is the subtree of T induced by t, t', and their descendants. The root of T_e is t or t', whichever is closer to r in T.

Consider an arbitrary internal vertex $v \in V(T)$ and let e,f,g be its incident edges, such that e connects to the parent of v (and f,g to the two children of v). We call f the left child of e and g the right child. Consider the middle sets X_e, X_f, X_g. Note that any vertex in at least one of these sets is in at least two of them. We can then partition $X_e \cup X_f \cup X_g$ into four sets:

$$I = X_e \cap X_f \cap X_g \qquad L = (X_e \cap X_f) - I$$
$$R = (X_e \cap X_g) - I \qquad F = (X_f \cap X_g) - I$$

Observe that I, L, and R partition X_e, I, L, and F partition X_f, and I, R, and F partition X_g. We can now prove the following lemma.

Lemma 1. $|X_e \cup X_f \cup X_g| = |I|+|L|+|R|+|F| \leq \frac{3}{2}\cdot \mathrm{bw}(G)$.

The notion of cliquewidth is defined as follows. A *k-expression* combines any number of the following four operations:

- create a new labelled graph with one vertex labelled $i \in \{1,\ldots,k\}$,
- relabel all vertices with label i to j $(i \neq j)$,
- connect all vertices with label i to all vertices labelled j $(i \neq j)$,
- take the disjoint union of two labelled graphs.

The *cliquewidth* $\mathrm{cw}(G)$ of a graph G is the minimum k for which there is a k-expression that evaluates to a graph isomorphic to G. This definition can also be turned into a decomposition based on a rooted tree of degree at most three, labeling leaves of the tree by vertices of G and internal vertices by one of the above operations. We call this a *clique decomposition*. However, the definition of k-expressions is more useful in this paper.

Given sets $\rho, \sigma \subseteq \mathbb{N}$, a $[\rho, \sigma]$-*dominating set* is a subset $D \subseteq V$ such that $|N(v) \cap D| \in \rho$ for every $v \in V \setminus D$ and $|N(v) \cap D| \in \sigma$ for every $v \in D$. The $[\rho, \sigma]$-domination problems were introduced by Telle in [13,14] and form a large class of graph covering problems.

Let $i_\times(n)$ denote the time required to multiply two n-bit integers. Currently, $i_\times(n) = O(n(\log n)\, 2^{\log^* n})$ [11]. Addition and subtraction take $O(n)$ time each.

3 Dominating Set

This section gives a faster algorithm to compute a minimum dominating set of a graph using a branch decomposition. We start by describing an algorithm that is actually slower than the $O^*(4^k)$ time algorithm of Dorn [8], but which serves to simplify the presentation of our algorithm. Then we show how to improve it.

The main ingredient of the algorithm are *vertex states*. For example, the following states can be useful in the dominating set problem:

- 1 the vertex is in the dominating set
- 0_1 the vertex is not in the dominating set and it is dominated
- 0_0 the vertex is not in the dominating set and it is not dominated
- $0_?$ the vertex is not in the dominating set and it might be dominated

An assignment of states to the vertices of some middle set X_e is called a *coloring* of this set. Depending on which vertex states we are allowed to use, a dominating set can induce several colorings on a set of vertices X_e. Initially we will only use vertex states 1, 0_0, and 0_1. We will later see other combinations in use.

Let (T, l) be a branch decomposition rooted at a vertex r as described before. For each edge $(t, t') = e \in E(T)$, we will compute a function $A_e : \{1, 0_1, 0_0\}^{X_e} \times [0 \ldots n] \to \mathbb{N}$. Given a coloring c of the vertices in X_e and a number $i \in [0 \ldots n]$, $A_e(c, i)$ equals the number of dominating sets D of $G[X(T_e)]$ such that $|D| = i$ and such that D induces the coloring c on X_e. Here $X(T_e) = \bigcup_{f \in T_e} X_f$. Notice that although we speak of a dominating set, it may be so that some vertices in $G[X(T_e)]$ remain undominated. These vertices then have state 0_0 in c (or $0_?$ if we would have used a different set of states).

We can compute the functions A_e bottom-up over the tree T. We start at edges e incident with leaves of T that are not the root. These leaves correspond to an edge of G. Hence $G[X(T_e)]$ is a two-vertex graph. In this case, we simply construct the function A_e by initializing all values to zero and then enumerating all subsets D of these two vertices. For each of these sets D we add 1 to each entry of A_e in which coloring c and size i corresponds to the set D.

For any edge e not incident with a leaf, let f and g be its left and right child. Recall the definition of the sets I, L, R, F induced by X_e, X_f, and X_g. Given a coloring c, let $c(I)$ denote the coloring of the vertices of I induced by c. We define $c(L)$, $c(R)$, and $c(F)$ similarly. Given a coloring c_e of X_e, a coloring c_f of X_f, and a coloring c_g of X_g, we say that these colorings *match* if

- For any $v \in I$: either $c_e(v) = c_f(v) = c_g(v) \in \{1, 0_0\}$, or $c_e(v) = 0_1$ while $c_f(v), c_g(v) \in \{0_0, 0_1\}$ and not $c_f(v) = c_g(v) = 0_0$. (5 possibilities)
- For any $v \in F$: either $c_f(v) = c_g(v) = 1$, or $c_f(v), c_g(v) \in \{0_0, 0_1\}$ while not $c_f(v) = c_g(v) = 0_0$. (4 possibilities)
- For any $v \in L$: $c_e(v) = c_f(v) \in \{1, 0_1, 0_0\}$. (3 possibilities)
- For any $v \in R$: $c_e(v) = c_g(v) \in \{1, 0_1, 0_0\}$. (3 possibilities)

Let $i_m = \#_1(c_e(I)) + \#_1(c_f(F))$ for any matching c_e, c_f, c_g, where $\#_1(c)$ is the number of vertices that are assigned state 1 by c. Then we can count all dominating sets of $G[X(T_e)]$ per corresponding coloring c_e on X_e using:

$$A_e(c_e, i_e) = \sum_{c_e, c_f, c_g \text{ match}} \; \sum_{i_e = i_f + i_g + i_m} A_f(c_f, i_f + i_m) \cdot A_g(c_g, i_g + i_m),$$

Proposition 1. *Given a branch decomposition of width k of a graph G, one can count the number of dominating sets of each size in $O(mn^2\, 6^k i_\times(n))$ time.*

Proof. We compute the functions A_e as described above. Let e_r denote the edge incident with the root of the branch decomposition. From the definition of A_{e_r}, $A_{e_r}(\emptyset, i)$ contains the number of dominating sets of size i in G.

For all leaf edges e of T, we can compute A_e in $O(n)$ time. For all other edges, we have to compute $O(n\, 3^k)$ values for A_e, each of which requires $O(n)$ terms of the above sum per set of matchings states. Since each vertex in I has 5 possible matching states, each vertex in F has 4 possible matching states, and each vertex in L or R has 3 possible matching states, this leads to a running time of $O(n^2\, 5^{|I|} 4^{|F|} 3^{|L|+|R|} i_\times(n))$ for computing A_e.

Under the constraint that $|I| + |L| + |R|, |I| + |L| + |F|, |I| + |R| + |F| \leq k$, the running time is maximal if $|I| = 0$, $|L| = |R| = |F| = \frac{1}{2}k$. As T has $O(m)$ edges, this leads to $O(mn^2 4^{\frac{1}{2}k} 3^k i_\times(n)) = O(mn^2\, 6^k i_\times(n))$. □

3.1 Using State Changes to Improve the Algorithm

We start improving the above algorithm by using the state changes introduced in [16], based on the covering product from [1]. The algorithm of Proposition 1 uses vertex states 1, 0_1, and 0_0. On tree decompositions, it has been shown that it is more efficient to transform the problem to one using states 1, 0_0, and $0_?$ [16]. We show that we can use similar transformations on branch decompositions.

There is a big difference however between dynamic programming on both types of decompositions. On tree decompositions one can transform a tree decomposition such that forget vertices (the vertices in set F) can be dealt with separately. This is not possible on branch decompositions, which makes the situation more complicated. On branch decompositions vertices in F must be dealt with simultaneously while computing A_e, by merging the results from the two functions A_f and A_g from e's children. We will overcome this problem by using different sets of states simultaneously: the set of states used depends on whether a vertex is in L, R, I or F. Moreover, we do this asymmetrically as different states can be used on the vertices in X_e, X_f and X_g.

In Proposition 1, we used functions $A_e : \{1, 0_1, 0_0\}^{X_e} \times [0 \dots n] \to \mathbb{N}$ counting the number of solutions of a given type. We again consider constructing A_e from A_f and A_g as before. Given A_f and A_g, we can compute in $O(n\,3^k)$ time a function $A'_f : \{1, 0_1, 0_0\}^L \times \{1, 0_?, 0_0\}^I \times \{1, 0_1, 0_0\}^F \times [0 \dots n] \to \mathbb{N}$ and a function $A'_g : \{1, 0_1, 0_0\}^R \times \{1, 0_?, 0_0\}^I \times \{1, 0_1, 0_?\}^F \times [0 \dots n] \to \mathbb{N}$. The new functions again count the number of solutions that satisfy the coloring c, but in this case a single dominating set D can be counted in multiply colorings. Notice the difference in the states used for A'_f and A'_g on F.

Transforming A_f into A'_f can be done by first copying A_f into A'_f without changing the states in the domain and then coordinate-wise, for each vertex $v \in I$, iteratively transforming the states using the following formula:

$$A'_f(c_{v=0_?}, i) = A'_f(c_{v=0_1}, i) + A'_f(c_{v=0_0}, i).$$

Here $c_{v=x}$ is the coloring c with the state of v set to x. This transformation is invertible as we can also coordinate-wise apply the following similar formula: $A'_f(c_{v=0_1}, i) = A'_f(c_{v=0_?}, i) - A'_f(c_{v=0_0}, i)$. We are now ready to give the first improvement of Proposition 1.

Proposition 2. *Given a branch decomposition of width k of G, one can count the number of dominating sets of each size in $O(mn^2 3^{\frac{3}{2}k} i_\times(n))$ time.*

Proof. The algorithm works similar to the one in Proposition 1, only with a different way of computing A_e from A_f and A_g.

Given A_f and A_g, we transform these to A'_f and A'_g as shown above. Here A'_f uses the states $\{1, 0_?, 0_0\}$ on I and the states $\{1, 0_1, 0_0\}$ on L and F and A'_g uses the same states on I and the same states on R as $'A_f$ on L. However, A'_g uses different states on F, namely $\{1, 0_1, 0_?\}$. Now two colorings match if:

- For any $v \in I$: $c_e(v) = c_f(v) = c_g(v) \in \{1, 0_?, 0_0\}$. (3 possibilities)
- For any $v \in F$: either $c_f(v) = c_g(v) = 1$, or $c_f(v) = 0_0$ and $c_g(v) = 0_1$, or $c_f(v) = 0_1$ and $c_g(v) = 0_?$. (3 possibilities)
- For any $v \in L$: $c_e(v) = c_f(v) \in \{1, 0_1, 0_0\}$. (3 possibilities)
- For any $v \in R$: $c_e(v) = c_g(v) \in \{1, 0_1, 0_0\}$. (3 possibilities)

Notice the difference on I compared to Proposition 1. A vertex is not dominated (0_0) if and only if it is not so in both partial solutions we are combining, and the same for the states $0_?$ and 1, as these do not care about domination. On F we only use combinations which allow the vertices to be ignored in the rest of the process: vertices that are either in the dominating set, or are dominated. Moreover, by using different states for A'_f and A'_g, every combination of partial solutions is counted exactly once. To see this, consider each of the three combinations on F used in Proposition 1. The combination with $c_f(v) = 0_0$ and $c_g(v) = 0_1$ is counted using the same combination, while the other two combinations ($c_f(v) = 0_1$ and $c_g(v) = 0_0$ or $c_g(v) = 0_1$) are counted when combining 0_1 with $0_?$.

In this way, we can compute the function $A'_e : \{1, 0_1, 0_0\}^L \times \{1, 0_?, 0_0\}^I \times \{1, 0_1, 0_0\}^R \times [0 \ldots n] \to \mathbb{N}$ using the following formula.

$$A'_e(c_e, i_e) = \sum_{c_e, c_f, c_g \text{ match}} \; \sum_{i_e = i_f + i_g + i_m} A'_f(c_f, i_f + i_m) \cdot A'_g(c_g, i_g + i_m). \quad (1)$$

Using the new states, $O(n^2\, 3^{|I|+|L|+|R|+|F|} i_\times(n))$ time is required to evaluate the formula. After its evaluation, we reconstruct A_e from A'_e by transforming it back to using states 1, 0_1 and 0_0 as described before. As each transformation requires $O(n\, 3^k)$ additions, the improved algorithm has a running time of $O(mn^2\, 3^{|I|+|L|+|R|+|F|} i_\times(n)) = O(mn^2\, 3^{\frac{3}{2}k} i_\times(n))$ by Lemma 1. □

3.2 Using Fast Matrix Multiplication

We now apply fast matrix multiplication, as first proposed by Dorn [8]. This yields the fastest algorithm for dominating set on branch decompositions.

For multiplying a $(n \times p)$ matrix A and $(p \times n)$ matrix B, we differentiate between $p \leq n$ and $p > n$. Under the condition that $\omega = 2.376$ (the best known bound [4]), an $O(n^{1.85} p^{0.54})$ time algorithm is known if $p \leq n$ [4]. Otherwise, the matrices can be multiplied in $O(\frac{p}{n} n^\omega) = O(pn^{\omega-1})$ time by matrix splitting: split the matrices A and B into $\frac{p}{n}$ many $n \times n$ matrices $A_1, \ldots A_{\frac{p}{n}}$ and $B_1, \ldots B_{\frac{p}{n}}$, multiply each of the $\frac{p}{n}$ pairs $A_i \times B_i$, and sum up the results.

Theorem 1. *Given a branch decomposition of width k of a graph G, one can count the number of dominating sets of each size in $O(mn^2\, 3^{\frac{\omega}{2}k} i_\times(n))$ time.*

Proof. Given a coloring of I and numbers i_e and i_f, we show how to evaluate Equation 1 for all possible colorings on L, R, and F simultaneously using fast matrix multiplication. Construct a $3^{|L|} \times 3^{|F|}$ matrix M_f where each row corresponds to a coloring of L and each column corresponds to a coloring of F. Let the entries of M_f be the values of $A'_f(c_f, i_f + i_m)$ for all c_f corresponding to the colorings of L and F of the row and column, and corresponding to the fixed coloring on I and the number i_f. Construct a similar $3^{|F|} \times 3^{|R|}$ matrix M_g for A'_g such that its rows correspond to different colorings of F and its columns of M_g correspond to different colorings of R. The entries of M_g are $A'_g(c_g, i_e - i_f)$. We permute the rows of M_g such that column i of M_f and row i of M_g correspond to matching colorings on F and thus the value of i_m matches as well.

Now we can compute Equation 1 by computing $M_e = M_f \cdot M_g$. The rows of M_e correspond to colorings of L and the columns correspond to colorings of R. An entry of M_e in row i and column j contains the value of $A'_e(c_e, i_e)$ as specified by Equation 1, where c_e is the coloring induced by the coloring corresponding to row i and to column j and the coloring for I we fixed. Thus we can compute the value of $A'_e(c_e, i_e)$ by a series of n matrix multiplications. Hence we can compute the function A'_e by $n^2\, 3^{|I|}$ matrix multiplications.

The time required to compute A'_e in this way depends on $|I|$, $|L|$, $|R|$, $|F|$. Under the constraint that $|I| + |L| + |F|, |I| + |R| + |F|, |I| + |L| + |R| \leq k$

and using the matrix multiplication algorithms described above this theorem, the worst case arises when $|I| = 0$ and $|L| = |R| = |F| = \frac{k}{2}$. In this case, we compute each A'_e in $O(n^2(3^{\frac{k}{2}})^\omega i_\times(n))$ time. This proves the theorem. □

Using that dominating set has *finite integer index* [2] and by only storing the size of a minimum dominating set instead of counting, we prove the following.

Corollary 1. *Given a width k branch decomposition of a graph G, the size of a minimum dominating set of G can be found in $O(mk^2\, 3^{\frac{\omega}{2}k})$.*

4 Counting Perfect Matchings

We give a fast algorithm counting the number of perfect matchings of a graph.

Theorem 2. *Given a branch decomposition of width k of a graph G, one can compute the number of perfect matchings of G in $O(mk^2\, 2^{\frac{\omega}{2}k} i_\times(n))$ time.*

Proof. Given a branch decomposition (T, l) of a graph G, we again compute a table A_e for each $e \in E(T)$. Each vertex in X_e uses states $\{0, 1\}$, where a vertex is in state 1 if and only if it is matched to some vertex in $X(T_e)$ by an edge in a leaf below e in T. Then for any coloring c of X_e, $A_e(c)$ is the number of perfect matchings of the graph $H = (V, E)$, where $V = X(T_e) - 0(c)$ and E equals the set of edges in leaves below e in T. Here $0(c)$ is the set of vertices that are assigned state 0 by c. The number of perfect matchings then is $A_{e_r}(\emptyset)$.

To join the tables A_f and A_g of two middle sets X_f and X_g to compute table A_e for X_e, we compute a table $A_f(c, i)$ such that $A_f(c, i) = A_f(c)$ if $\#_1(c(I)) = i$ and 0 otherwise. A table $A_g(c, i)$ is computed similarly. Applying state changes as in [16] and matrix multiplication in a similar way as before, we can compute A_e in $O(mk^2 2^{\frac{\omega}{2}k} i_\times(n))$ time. The parameter i is used to track matched vertices in spite of the state changes and ensures that no vertex is matched twice. This is similar to applying fast subset convolutions [1]. Details in the full paper. □

5 $[\rho, \sigma]$-Domination Problems

We have shown how to solve two fundamental graph problems in $O^*(s^{\frac{\omega}{2}k})$ time on branch decompositions of width k, where s is the number of states used. Below, we generalize our results and show that one can solve all $[\rho, \sigma]$-domination problems with finite or cofinite ρ and σ in $O^*(s^{\frac{\omega}{2}k})$ time. This includes the existence, minimization, maximization, and counting variants of the problems.

For $[\rho, \sigma]$-domination problems, we use states ρ_j and σ_j, where ρ_j and σ_j represent that a vertex is respectively in or not in the $[\sigma, \rho]$-dominating set D and has j neighbors in D. For finite ρ, σ, we use states $\{\rho_0, \ldots, \rho_p, \sigma_0, \ldots, \sigma_q\}$. If ρ or σ is cofinite, we replace the last state by $\rho_{\geq p}$ or $\sigma_{\geq q}$, respectively. To simplify the exposition of the algorithm, we restrict our description to finite ρ and σ. Note that the number of states used equals $s = p + q + 2$.

We can essentially use the approach of [16] for the vertices in I. We thus do not repeat the details here and treat it more or less as a black box below.

There are two things we need to know however about this black box. First, the approach ensures that colorings c_e, c_f, c_g match on I if they are the same on I. Therefore, if for any fixed coloring on I we can perform a join operation on L, R, and F in $O^*(\alpha^{|L|}\beta^{|R|}\gamma^{|F|})$ time, then we can construct A_e in $O^*(\alpha^{|L|}\beta^{|R|}\gamma^{|F|}s^{|I|})$ time. Secondly, we need to know how this black box keeps track of the number of neighbors in D of each vertex in the partial solutions that we are combining. This works in the following way: if we combine two partial solutions with states ρ_i and ρ_j on a vertex $v \in I$, then the resulting combination will have state ρ_{i+j} on v. This is important because this can lead to errors in the following sense. If v has a neighbor $u \in D$ in both partial solutions, then this neighbor is counted in both i and j, and combining these partial solutions to one with state ρ_{i+j} is incorrect as u can now be counted twice.

Theorem 3. *Let $\rho, \sigma \subseteq \mathbb{N}$ be finite or cofinite, and let s be the number of states used in the given representation of the associated $[\rho, \sigma]$-domination problem. Then, given a branch decomposition of width k of a graph G, one can compute the number of $[\rho, \sigma]$-dominating sets of G of each size in $O^*(s^{\frac{\omega}{2}k})$ time.*

Proof. In this proof, we use the approach of [16] as a black box to combine solutions on I. We use states ρ_j and σ_j for A_e, A_f, and A_g such that the subscripts j represent the number of neighbors in each $[\sigma, \rho]$-dominating set D outside X_e, X_f and X_g, respectively. Because the black box uses the principle that ρ_i and ρ_j are combined to ρ_{i+j} on I, we first perform some preprocessing.

We construct functions A'_f and A'_g identical to A_f and A_g with states ρ'_i and σ'_i such that the subscripts of the states used in A'_f and A'_g count the number of neighbors in D with the following properties:

- States used in A'_f on vertices in L or I count neighbors in $(X(T_f) - X_f) \cup F$.
- States used in A'_f on vertices in F count neighbors in $(X(T_f) - X_f) \cup L \cup I$.
- States used in A'_g on vertices in R count neighbors in $(X(T_g) - X_g) \cup F$.
- States used in A'_g on vertices in F count neighbors in $(X(T_g) - X_g) \cup R$.

If we then combine partial solutions with, for a vertex in I, a state ρ'_i in A'_f and a state ρ'_j in A'_g, then the new state ρ'_{i+j} in A'_e correctly counts the number of neighbors on $(X(T_e) - X_e)$. Also, states for vertices in L and R in A_e count their neighbors in $F \cap D$. And, if we combine solutions with for a vertex in F a state ρ'_i in A'_f and a state ρ'_j in A'_g, then this vertex will have exactly $i + j$ neighbors in D. Hence the resulting A'_e equals the function A_e we want to compute.

We construct A'_f from A_f by setting $A'_f(c_f, i)$ to $A_f(c'_f, i)$, where c'_f is obtained from c_f as specified above, or to 0 if no such c'_f exists. Finding c'_f from c_f is straightforward. For example, if in a coloring c_f a vertex $v \in F$ has state ρ_j, then in c'_f this vertex must have state ρ_{j-d}, where d is the number of neighbors of v in $L \cup I$ that have state σ_i for some i. We construct A'_g from A_g in the same way, keeping in mind the difference in states used.

Using the combination on I as a black box, what remains is the combination of A'_f and A'_g on L, R, and F. For a fixed coloring of I, we combine all possible colorings on L, R, and F simultaneously using fast matrix multiplication. To do so, we permute the rows and columns of the matrices such that the part of a coloring c_f for A'_f on F is combined with a unique coloring c_g for A'_g on F. To this end, we transform the states on A'_g to different states. We thus obtain asymmetric states, similar to the algorithm for dominating set.

For colorings c_g for A'_g on F, we transform state set $\{\rho'_0, \ldots, \rho'_p, \sigma'_0, \ldots, \sigma'_q\}$ to the set $\{\rho'_{\bar{0}}, \ldots, \rho'_{\bar{p}}, \sigma'_{\bar{0}}, \ldots, \sigma'_{\bar{q}}\}$. Here $\rho'_{\bar{j}}$ is used for a vertex that is not in D and has i neighbors in D such that $j + i \in \rho$, and $\sigma'_{\bar{j}}$ is used for a vertex that is in D and has i neighbors in D such that $j + i \in \sigma$.

These transformations take $O^*(s^{|F|+1})$ time by applying the following formulas coordinate-wise to a copy A^*_g of the function A'_g.

$$A^*_g(c_{v=\rho'_{\bar{j}}}) = \sum_{i+j \in \rho} A^*_g(c_{v=\rho'_i}) \qquad\qquad A^*_g(c_{v=\sigma'_{\bar{j}}}) = \sum_{i+j \in \sigma} A^*_g(c_{v=\sigma'_i}).$$

Notice that if we combine a state ρ'_j from A'_f with a state $\rho'_{\bar{j}}$ from A^*_g, then valid combinations in which this vertex is not in the set D are counted. The same goes with σ'_j from A'_f with a state $\sigma'_{\bar{j}}$ from A^*_g for vertices in D.

We can now, for a fixed coloring on I, construct two matrices as before. Computing A_e this way requires $O^*(s^{|I|})$ matrix multiplications of a $s^{|L|} \times s^{|F|}$ matrix and a $s^{|F|} \times s^{|R|}$ matrix. This gives a running time of $O^*(s^{\frac{\omega}{2}k})$. □

6 Cliquewidth

On graphs of bounded cliquewidth, we show how to improve from $O^*(8^k)$ time obtained previously [3] to $O^*(4^k)$ time for computing Minimum Dominating Set.

Theorem 4. *Given a k-expression for a graph G, one can compute the number of dominating sets of G of each size in $O(n^3 (k^2 + i_\times(n)) 4^k)$ time.*

Proof. An operation in a k-expression applies a procedure on labelled graph H', transforming it to a graph H. In the case of the union-operation, the operation takes two labelled graphs H_1 and H_2 as an input and combines them to a graph H. For a labelled graph H, we use V_i to denote the set of vertices in H with label i. In contrast to the algorithm on branch decompositions, we do not assign states to individual vertices, but assign them to the sets $V_1, \ldots, V_k$ instead.

To each V_i, we assign two attributes, inclusion and domination. The first determines whether at least one vertex of V_i is included in a partial solution. We use states T, F, ? to indicate whether this is true, false, or optional respectively. The second attribute determines whether all vertices of V_i are dominated in a partial solution. We use states T, F, ? to indicate whether this is true, false, or optional respectively. The states we use are a combination of both attributes.

For any graph constructed in the k-expression, we will compute a table $A(c, \kappa)$ which stores the number of partial solutions to the dominating set problem of size κ that induce the coloring c.

Consider the first operation, creating a labelled graph H with one vertex with label i. Assume w.l.o.g. that $i = k$ and let c be the coloring of $V_1, \ldots, V_{i-1}$ such that the first attribute is F and the second attribute is T for all labels. Then $A(c \times (T,T), 1) = A(c \times (F,F), 0) = 1$. All other entries of A are 0.

The second operation relabels vertices with label j to label i. If $V_i = \emptyset$ or $V_j = \emptyset$, this is trivial. Otherwise (w.l.o.g. $i = k$ and $j = k-1$),

$$A(c \times (r, r') \times (s, s'), \kappa) = \sum_{t \vee u = s} \sum_{t' \wedge u' = s'} A(c \times (t, t') \times (u, u'), \kappa)$$

if $\neg r \wedge r'$, and 0 otherwise, where c is any coloring of $V_1, \ldots, V_{j-1}$. Note that for V_j, the first attribute is F and the second is T in any valid partial solution. If V_i must have a vertex in the dominating set, then it is in V_i or V_j originally. If all vertices in V_i must be dominated, all vertices in V_i and V_j must be dominated.

The third operation connects all vertices with label i to all with label j. If V_i or V_j is $\emptyset$, this is trivial. Otherwise (w.l.o.g. $i = k$ and $j = k-1$),

$$A(c \times (r, r') \times (s, s'), \kappa) = \sum_{t' \vee s = r'} \sum_{u' \vee r = s'} A(c \times (r, t') \times (s, u'), \kappa)$$

where c is any coloring of $V_1, \ldots, V_{j-1}$. If V_i and V_j are connected and V_i (resp. V_j) contains a vertex in the dominating set, then V_j (resp. V_i) is dominated.

The fourth operation joins two labelled graphs H_1 and H_2 with tables A_1 and A_2 to a labelled graph H with table A. To do this efficiently, we apply state changes. We use states F and ? for the first attribute and states T and ? for the second attribute. Changing A_1 and A_2 to tables A_1^* and A_2^* that use these states can be done in a similar manner as before. Then A^* can be computed using: $A^*(c, \kappa) = \sum_{i=0}^{\kappa} A_1^*(c, i) \cdot A_2^*(c, \kappa - i)$. Using the inverse state change formulae, we obtain A. Each of the $O(n)$ union operations takes $O(n^2\, 4^k\, i_\times(n))$ time, each of the $O(nk^2)$ other operations takes $O(n^2\, 4^k)$ time. □

Although Minimum Dominating Set has finite integer index, it is not clear what this implies for k-expressions. Using other arguments, we can still prove that for computing the number of minimum dominating sets, the table only needs to store sets of size at most $2k$ larger than the minimum set.

Theorem 5. *Given a k-expression for a graph G, one can compute (the size of) a minimum dominating set of G in $O(nk^2\, 4^k)$ time.*

7 Conclusions

We presented a novel combination of existing techniques to improve on algorithms for graph optimization problems using graph decompositions. In particular, we gave an $O^*(3^{\frac{\omega}{2} \cdot \mathrm{bw}(G)})$ and an $O^*(4^{\mathrm{cw}(G)})$ algorithm for Minimum Dominating Set. These algorithms extend immediately to their counting variant. We also showed an $O^*(2^{\frac{\omega}{2} \cdot \mathrm{bw}(G)})$ algorithm for counting perfect matchings.

On planar graphs, an $O^*(3^{\frac{\omega}{2} \cdot \mathrm{bw}(G)})$ algorithm was presented before [8], using so-called sphere cut branch decompositions [7]. This paper shows that this special branch decomposition is no longer necessary to obtain the fastest algorithm for Minimum Dominating Set on planar graphs. In particular, we get an $O^*(2^{3.99 \cdot \sqrt{n}})$ time algorithm for Planar Dominating Set and an $O^*(2^{11.98 \cdot \sqrt{k}})$ algorithm for its parameterized version, using results of Fomin and Thilikos [9,10].

The results of this paper attain, or are close to, what seem natural limits to techniques commonly used for these problems, namely the amount of space used by any dynamic programming algorithm for these problems on graph decompositions. This currently is $O(3^k)$ for Minimum Dominating Set on branch decompositions and $O(4^k)$ on clique decompositions. Also, the number of states used seems the best possible base of the exponent in the running time. Our result for Minimum Dominating Set on clique decompositions attains this bound. On branch decompositions, we are very close (the base of the exponent is 3.688) and under the hypothesis that $\omega = 2$, we attain this bound. Can we get an $O^*(3^{\mathrm{bw}(G)})$ time algorithm for Minimum Dominating Set in another way?

References

1. Björklund, A., Husfeldt, T., Kaski, P., Koivisto, M.: Fourier meets Möbius: Fast Subset Convolution. In: STOC 2007, pp. 67–74. ACM, New York (2007)
2. Bodlaender, H.L., van Antwerpen-de Fluiter, B.: Reduction algorithms for graphs of small treewidth. Inform. Comput. 167, 86–119 (2001)
3. Bui-Xuan, B.-M., Telle, J.A., Vatshelle, M.: Boolean-width of graphs. In: Chen, J., Fomin, F.V. (eds.) IWPEC 2009. LNCS, vol. 5917, pp. 61–74. Springer, Heidelberg (2009)
4. Coppersmith, D., Winograd, S.: Matrix Multiplication via Arithmetic Progressions. J. Symbolic Comput. 9, 251–280 (1990)
5. Courcelle, B., Engelfriet, J., Rozenberg, G.: Handle-rewriting hypergraph grammars. J. Comput. System. Sci. 46(2), 218–270 (1993)
6. Courcelle, B., Olariu, S.: Upper bounds on the clique-width of graphs. Discrete Appl. Math. 101, 77–114 (2000)
7. Dorn, F., Penninkx, E., Bodlaender, H.L., Fomin, F.V.: Efficient exact algorithms on planar graphs: Exploiting sphere cut branch decompositions. In: Brodal, G.S., Leonardi, S. (eds.) ESA 2005. LNCS, vol. 3669, pp. 95–106. Springer, Heidelberg (2005)
8. Dorn, F.: Dynamic Programming and Fast Matrix Multiplication. In: Azar, Y., Erlebach, T. (eds.) ESA 2006. LNCS, vol. 4168, pp. 280–291. Springer, Heidelberg (2006)
9. Fomin, F.V., Thilikos, D.M.: Dominating Sets in Planar Graphs: Branch-width and Exponential Speed-up. In: SODA 2003, pp. 386–397. ACM, New York (2003)
10. Fomin, F.V., Thilikos, D.M.: A simple and fast approach for solving problems on planar graphs. In: Diekert, V., Habib, M. (eds.) STACS 2004. LNCS, vol. 2996, pp. 56–67. Springer, Heidelberg (2004)
11. Fürer, M.: Faster Integer Multiplication. In: STOC 2007, pp. 57–66. ACM, New York (2007)
12. Robertson, N., Seymour, P.D.: Graph minors. X. Obstructions to tree-decomposition. J. Combin. Theory B 52(2), 153–190 (1991)
13. Telle, J.A.: Complexity of Domination-Type Problems in Graphs. Nordic J. Comput. 1, 157–171 (1994)
14. Telle, J.A., Proskurowski, A.: Algorithms for Vertex Partitioning Problems on Partial k-Trees. SIAM J. Discrete Math. 10, 529–550 (1997)
15. Valiant, L.G.: The Complexity of Computing the Permanent. Theor. Comput. Sci. 8, 189–201 (1979)
16. van Rooij, J.M.M., Bodlaender, H.L., Rossmanith, P.: Dynamic Programming on Tree Decompositions Using Generalised Fast Subset Convolution. In: Fiat, A., Sanders, P. (eds.) ESA 2009. LNCS, vol. 5757, pp. 566–577. Springer, Heidelberg (2009)

Exponential Space Complexity for Symbolic Maximum Flow Algorithms in 0-1 Networks

Beate Bollig

LS2 Informatik, TU Dortmund,
44221 Dortmund, Germany

Abstract. The maximum flow problem is a central problem in graph algorithms and optimization and OBDDs are one of the most common dynamic data structures for Boolean functions. Since in some applications graphs become larger and larger, a research branch has emerged which is concerned with the theoretical design and analysis of symbolic algorithms for classical graph problems on OBDD-represented graph instances. The algorithm for the maximum flow problem in 0-1 networks by Hachtel and Somenzi (1997) has been one of the first of these symbolic algorithms. Typically problems get harder when their input is represented symbolically, nevertheless not many concrete non-trivial lower bounds are known. Here, answering an open question posed by Sawitzki (2006) the first exponential lower bound on the space complexity of OBDD-based algorithms for the maximum flow problem in 0-1 networks is presented.

1 Introduction

Some modern applications require huge graphs so that explicit representations by adjacency matrices or adjacency lists are not anymore applicable. Since time and space do not suffice to consider individual vertices, one way out seems to be to deal with sets of vertices and edges represented by their characteristic functions. Since ordered binary decision diagrams, denoted OBDDs, introduced by Bryant in 1986 [4], are well suited for the representation and manipulation of Boolean functions, a research branch has emerged which is concerned with the theoretical design and analysis of so-called symbolic algorithms for classical graph problems on OBDD-represented graph instances (see, e.g., [9,10], [15,16], and [21]). Problems on implicitly given graph instances have to be solved by efficient functional operations offered by the OBDD data structure. At the beginning the OBDD-based algorithms have been justified by analyzing the number of executed OBDD operations (see, e.g., [9,10]). Since the running time for one OBDD operation depends on the sizes of the OBDDs on which the operation is performed, newer research tries to analyze the overall running time of symbolic methods including the analysis of all OBDD sizes occurring during such an algorithm (see, e.g., [21]).

Since the maximum flow problem is a central problem in graph algorithms and optimization, the maximum flow problem in 0-1 networks has been one of the first classical graph problems for which a symbolic algorithm has been

P. Hliněný and A. Kučera (Eds.): MFCS 2010, LNCS 6281, pp. 186–197, 2010.

presented [12]. It has not been the main purpose to beat explicit algorithms on graphs which can be represented explicitly but to solve the problem for very large structured graphs in reasonable time and space. Hachtel and Somenzi were able to compute a maximum flow for a graph with more than 10^{27} vertices and 10^{36} edges in less than one CPU minute. Sawitzki [15] has used a well-known method called *iterative squaring* to improve the maximum flow algorithm and has shown that his approach only uses a polylogarithmic number of OBDD operations on grid graphs. Recently, in [11] symbolic algorithms for maximum flow have been presented which use a more general model than OBDDs, called algebraic decision diagrams. The running time is not analyzed but the algorithms are compared experimentally with classical explicit ones.

Representing graphs with regularities by means of data structures smaller than adjacency matrices or adjacency lists seems to be a natural idea. But problems typically get harder when their input is represented implicitly. For circuit representations this has been shown in [1,8,14]. These results do not directly carry over to problems on OBDD-represented inputs since there are Boolean functions like some output bits of integer multiplication whose OBDD complexity is exponentially larger than its circuit size [2,5]. In [7] it has been shown that the problem of deciding whether two vertices s and t are connected in a directed graph G, the so-called graph accessibility problem GAP, is PSPACE-complete on OBDD-represented graphs. Nevertheless, OBDD-based algorithms are successful in many applications and despite the hardness results there are not many non-trivial lower bounds known for the complexity of problems on OBDD-represented graph instances. The challenge seems to be to prove small upper bounds on the OBDD size of input graphs and simultaneously large lower bounds on the size of OBDDs occuring during the computation. In [18] exponential lower bounds on OBDD-based algorithms for the single-source shortest paths problem, the maximum flow problem, and a restricted class of algorithms for the reachability problem have been presented. Only recently, an exponential lower bound on the space complexity of all OBDD-based algorithms for reachability has been shown in [3]. Note, that we look at search problems not at decision problems. The results are not very astonishing but the proofs present worst-case examples which could be helpful to realize which cases are difficult to process and why on the other hand OBDD-based algorithms are successful in many applications. Here, one aim is to obtain worst-case examples which are as simple as possible.

Our result can be summarized as follows.

Theorem 1. *Symbolic* OBDD*-based algorithms for the maximum flow problem in* 0-1 *networks need exponential space with respect to the size of the implicit representation of the input graph.*

Note that we use Sawitzki's assumption that a separation of working space and output size is not reasonable in the symbolic setting [18].

In [15] and [21] symbolic algorithms for maximum flow in 0-1 networks and topological sorting have been presented which have polylogarithmic running time with respect to the number of vertices of a given grid graph. These results rely on restrictions on the maximum number of nodes labeled by the same variable (the width) of occuring OBDDs during the computation. Since one aim is to find advantageous properties of real-world instances that cause an essentially better behavior than in the worst-case, these results have motivated the investigation of the parameterized complexity of graph problems when structured properties of input and/or output OBDDs are considered as fixed parameters. (See [6] for a comprehensive introduction into the field of parameterized complexity.) In [19] it has been shown that basic graph problems are fixed parameter intractable with respect to a fixed input OBDD width (unless P= PSPACE). The proof of Theorem 1 also demonstrates that constant input OBDD width does not suffice to guarantee polynomial space complexity for the maximum flow problem in 0-1 networks.

The organization of the paper is the following. In Section 2 we define some notation and present some basics concerning OBDDs, symbolic graph representations, and the maximum flow problem. Section 3 contains the main result of the paper. Answering an open question posed by Sawitzki in [17] and strengthening his result in [19], we prove that algorithms for the maximum flow problem in 0-1 networks need exponential space on OBDD-represented graph instances. Sawitzki's worst-case example for the general maxflow problem in [18] relies on the possibility to encode the difficulty of the problem into the edge weights. For 0-1 networks we have to use a different approach.

2 Preliminaries

In order to make the paper self-contained we briefly recall the main notions we are dealing with in this paper.

2.1 Ordered Binary Decision Diagrams

When working with Boolean functions as in circuit verification, synthesis, and model checking, ordered binary decision diagrams, denoted OBDDs, are one of the most often used data structures supporting all fundamental operations on Boolean functions efficiently.

Definition 1. *Let $X_n = \{x_1, \ldots, x_n\}$ be a set of Boolean variables. A variable ordering π on X_n is a permutation on $\{1, \ldots, n\}$ leading to the ordered list $x_{\pi(1)}, \ldots, x_{\pi(n)}$ of the variables.*

In the following a variable ordering π is sometimes identified with the corresponding ordering $x_{\pi(1)}, \ldots, x_{\pi(n)}$ of the variables, if the meaning is clear from the context.

Definition 2. *A π-OBDD on X_n is a directed acyclic graph $G = (V, E)$ whose sinks are labeled by Boolean constants and whose non-sink (or decision) nodes are labeled by Boolean variables from X_n. Each decision node has two outgoing edges one labeled by 0 and the other by 1. The edges between decision nodes have* *to respect the variable ordering* π*, i.e., if an edge leads from an* x_i*-node to an* x_j*-node, then* $\pi^{-1}(i) \leq \pi^{-1}(j)$ *(*x_i *precedes* x_j *in* $x_{\pi(1)}, \ldots, x_{\pi(n)}$*). Each node* v *represents a Boolean function* $f_v \in B_n$*, i.e.,* $f_v : \{0,1\}^n \rightarrow \{0,1\}$*, defined in the following way. In order to evaluate* $f_v(b)$*,* $b \in \{0,1\}^n$*, start at* v*. After reaching an* x_i*-node choose the outgoing edge with label* b_i *until a sink is reached. The label of this sink defines* $f_v(b)$*. The width of an* OBDD *is the maximum number of nodes labeled by the same variable. The size of a* π-OBDD G *is equal to the number of its nodes and the* π-OBDD *size of a function* f*, denoted by* π-OBDD(f)*, is the size of the minimal* π-OBDD *representing* f*. The* π-OBDD *of minimal size for a given function is unique up to isomorphism. A* π-OBDD *for a function* f *is called reduced, if it is the minimal* π-OBDD *for* f*.*

Let f be a Boolean function on the variables $x_1, \ldots, x_n$. The subfunction $f_{|x_i=c}$, $1 \leq i \leq n$ and $c \in \{0,1\}$, is defined as $f(x_1, \ldots, x_{i-1}, c, x_{i+1}, \ldots, x_n)$. The size of the reduced π-OBDD representing f is described by the following structure theorem [20].

Theorem 2. *The number of* $x_{\pi(i)}$*-nodes of the minimal* π-OBDD *for* f *is the number* s_i *of different subfunctions* $f_{|x_{\pi(1)}=a_1,\ldots,x_{\pi(i-1)}=a_{i-1}}$, $a_1, \ldots, a_{i-1} \in \{0,1\}$*, that essentially depend on* $x_{\pi(i)}$ *(a function* g *essentially depends on a Boolean variable* z*, if* $g_{|z=0} \neq g_{|z=1}$*).*

Theorem 2 implies the following simple observation which is helpful in order to prove lower bounds. Given an arbitrary variable ordering π the number of nodes labeled by a variable x in the reduced π-OBDD representing a given function f is not smaller than the number of x-nodes in a reduced π-OBDD representing any subfunction of f.

It is well known that the size of an OBDD representing a function f, that is defined on n Boolean variables and depends essentially on all of them, depends on the chosen variable ordering and may vary between linear and exponential size.

Definition 3. *The* OBDD *size or* OBDD *complexity of* f *is the minimum of all* π-OBDD(f)*.*

2.2 Symbolic OBDD-Based Graph Representations

In the following for $z = (z_{n-1}, \ldots, z_0) \in \{0,1\}^n$ let $|z| := \sum_{i=0}^{n-1} z_i 2^i$. Let $G = (V, E)$ be a graph with N vertices $v_0, \ldots v_{N-1}$. The edge set E can be represented by an OBDD for its characteristic function, where $\mathcal{X}_E(x, y) = 1 \Leftrightarrow (|x|, |y| < N) \wedge (v_{|x|}, v_{|y|}) \in E$, $x, y \in \{0,1\}^n$ and $n = \lceil \log N \rceil$. Undirected edges are represented by symmetric directed ones. In the rest of the paper we assume that N is a power of 2 since it has no bearing on the essence of our results.

2.3 The Maximum Flow Problem

The maximum flow problem is a maximization problem on directed graphs $G = (V, E)$ with a source $s \in V$, a vertex with indegree 0, and a sink $t \in V$, a vertex with outdegree 0. Given some *capacity constraint* function $c : E \rightarrow \mathbb{N}$, the aim is to compute a flow $f : E \rightarrow \mathbb{N}$ such that f respects the capacity constraint, i.e., $0 \leq f(e) \leq c(e)$ for all $e \in E$, f respects the *flow constraint* at all vertices $v \in V \setminus \{s, t\}$, i.e.,

$$\sum_{u|(u,v)\in E} f(u,v) = \sum_{w|(v,w)\in E} f(v,w),$$

and f has among all admissible flows the largest value defined by

$$val(f) = \sum_{v|(s,v)\in E} f(s,v).$$

Moreover, we assume that the graph G is antisymmetric, i.e., $(y,x) \notin E$, if $(x,y) \in E$. In the symbolic setting the source is described by a function $s(x)$ which only takes the value 1, if x is the encoding of the source and the sink by the function $t(x)$. If the capacity constraint function c is equal to the constant function 1 we look at the special case of the maximum flow problem in 0-1 networks. Because all edges have capacity 1 there is a maximum flow $f_{\max}$ such that for all edges e we know that $f_{\max}(e) \in \{0,1\}$. Therefore, we can identify $f_{\max}$ as the set of edges carrying positive flow. In the symbolic setting we are looking for an OBDD representation for the characteristic function of this edge set.

3 OBDD-Based Maximum Flow Algorithms in 0-1 Networks Need Exponential Space

In this section we prove Theorem 1 and show that OBDD-based algorithms for the maximum flow problem in 0-1 networks need exponential space. The proof structure is the following one. First, we define a pathological graph instance G_n as input graph for the 0-1 maximum flow problem. We show that the size of the corresponding OBDD representation for the characteristic function of its edge set is polynomial with respect to the number of Boolean variables. Afterwards we prove that there exists a maximum flow represented by its edge set for which the corresponding characteristic function has exponential OBDD complexity. Since in G_n a maximum flow is not unique, our proof does not rule out the possibility that there exists another maximum flow whose OBDD size is polynomial. Therefore, we refine the result by modifying the input graph such that the maximum flow is unique. As a consequence we can conclude that every OBDD-based algorithm which solves the maximum flow problem in 0-1 networks needs exponential space with respect to its input size. Now, we make our ideas more precise.

1) The definition of the input graph G_n:

The graph G_n consists of 2^{n^2} vertices $v_{i_1,\ldots,i_n}$, $i_j \in \{0,\ldots,2^n-1\}$ and $j \in \{1,\ldots,n\}$. The Boolean encoding of the indices $i_1,\ldots,i_n$ corresponds to a $n \times n$ Boolean matrix. The source s is equal to the vertex $v_{1,\ldots,1}$ and the sink t to

$v_{0,\ldots,0}$, in other words the Boolean encoding of s corresponds to the constant 1-matrix, the sink to the constant 0-matrix. There exists an edge from s to a vertex $v_{i_1,\ldots,i_n}$ in E, iff the number of ones in the Boolean encoding of $i_1,\ldots,i_n$ is odd but not n^2 (if n is odd). There exists an edge from a vertex $v_{i_1,\ldots,i_n}$ to the sink t in E, iff the number of ones is even but not 0 (nor n^2 if n is even). Furthermore, there exists an edge from a vertex $v_{i_1,\ldots,i_n}$ to $v_{j_1,\ldots,j_n}$, iff there is an index i_k, $k \in \{1,\ldots,n\}$, that is equal to $2^n - 1$ (which means that there exists a row only consisting of 1-entries in the Boolean encoding of $i_1,\ldots,i_n$), the number of ones in the Boolean encoding of the indices $i_1,\ldots,i_n$ is odd (but not n^2), the number of ones in the Boolean encoding of $j_1,\ldots,j_n$ is even (but neither 0 nor n^2) and there exists a column in the Boolean encoding of the indices that consists only of ones. Figure 1 shows the structure of the input graph G_n.

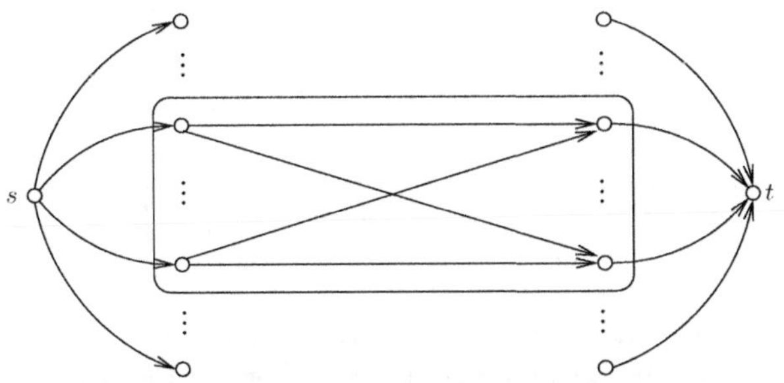

Fig. 1. The input graph G_n

2) The polynomial upper bound on the OBDD size for the characteristic function of G_n's edge set:

The characteristic function of the edge set $\mathcal{X}_E$ depends on $2n^2$ Boolean variables. Our aim is to prove that $\mathcal{X}_E$ can be represented by OBDDs of size $O(n^2)$ and constant width (maximal number of nodes labeled by the same variable) according to the variable ordering $x_{11},\ldots,x_{1n},x_{21},\ldots,x_{nn},y_{11},\ldots,y_{n1},y_{12},\ldots,y_{nn}$, where $x_{k1},\ldots,x_{kn}$ is the Boolean encoding of the index i_k and x_{kn} the least significant bit.

Applying Theorem 2 it is sufficient to prove that for every $i \in \{1,\ldots,2n^2\}$ there is only a constant number of different subfunctions obtained by replacements of the first i variables with respect to the considered variable ordering. W.l.o.g. we assume in the following that n is even. Loosely speaking, different useful information lead to different subfunctions. Therefore, we have to consider which information we have to know in order to decide whether the function value is 1. The x-variables are tested in a rowwise manner, i.e., variables that belong to the same row are tested one after another. It is checked whether all x-variables are set to 1 or whether the number of ones altogether is even (but not 0) or odd and in the latter case whether there exists a row that consists only of ones. Figure 2 shows a part of the OBDD, where the x-variables of the first row have already been tested.

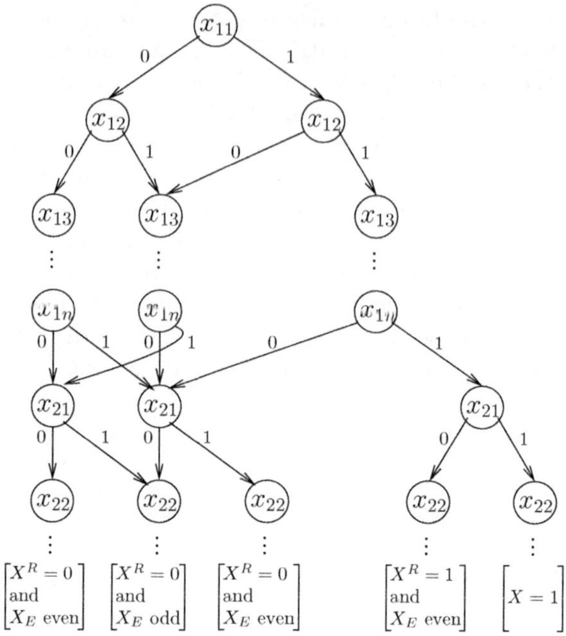

Fig. 2. A part of an OBDD that checks whether there exists a row that consists only of 1-entries and the number of 1s altogether is odd or whether there are only 1-entries. The Boolean variable X^R is 1, iff there exists a row that consists only of 1-entries in the input matrix. The Boolean variable X is 1, if all variables seen so far are set to 1. The variable X_E counts the number of 1-entries.

The OBDD width of this part is 7, since we only have to distinguish 7 different useful information about the partial assignments to the x-variables:

- whether all variables are set to 1,
- whether there has been a row for which all variables are set to 1 and the number of ones altogether is odd respectively even, and
- whether there has not yet been a row for which all variables are set to 1 but the current row contains only 1-entries and the number of ones altogether is odd respectively even,
- whether there has not yet been a row for which all variables are set to 1, the current row contains at least one 0-entry, and the number of ones altogether is odd respectively even.

In the second part of the OBDD the y-variables are tested in a columnwise manner, i.e., variables that belong to the same column are tested one after another. There are three y_{11}-variables, one is only reached by the partial assignment where all x-variables are set to 1 (that corresponds to the Boolean encoding of s), another one is reached by all assignments, where there exists an index $k \in \{1, \ldots, n\}$ with $x_{k1} = \cdots = x_{kn} = 1$ and the number of 1-entries altogether is odd, and a third one by all assignments for which the number of ones altogether is even. In the first case the function value is 1, iff the number of ones altogether is odd. This can be checked by an OBDD of width 2. In the second part it is tested whether there exists a column that consists only of ones and whether the number of ones altogether is even. Since the y-variables are tested in a columnwise manner this can be checked by an OBDD of width 6. In the latter case it is only checked whether all y-variables are set to 0 (that corresponds to the Boolean encoding of t). The three parts may share nodes, therefore, we can summarize that the width of the lower part of the OBDD for $\mathcal{X}_E$ is at most 9.

3) The exponential lower bound on the OBDD size for the characteristic function of a maximum flow in G_n:

It remains to show that there exists a maximum flow in G_n defined by the characteristic function of its edge set whose OBDD complexity is exponential. A vertex in G_n has the property $\mathcal{R}$, iff its Boolean encoding contains a row that consists only of 1-entries and the number of ones altogether is odd. A vertex in G_n has the property $\mathcal{C}$, iff its Boolean encoding contains a column that consists only of 1-entries and the number of ones altogether is even.

Each vertex except for the source s and the sink t has indegree or outdegree 1. A maximum flow contains only edges which are adjacent to vertices whose in- and outdegree is at least 1. These are the vertices that have property $\mathcal{R}$ or $\mathcal{C}$. Each maximum flow consists of edges from the source s to the vertices with property $\mathcal{R}$ and edges from vertices with property $\mathcal{C}$ to the sink t. Furthermore a maximum flow contains in a certain sense a bijective mapping from the vertices with property $\mathcal{R}$ to the vertices with property $\mathcal{C}$.

In the rest of this section we consider the maximum flow in G_n that can be identified by the following bijective mapping from the vertices with property $\mathcal{R}$ to the vertices with property $\mathcal{C}$: a vertex $v_{i_1,\ldots,i_n}$ with property $\mathcal{R}$ has an edge to a vertex $v_{j_1,\ldots,j_n}$ with property $\mathcal{C}$, iff the Boolean matrix X that encodes $i_1,\ldots,i_n$ is almost equal to the Boolean matrix Y^T, where Y is the Boolean encoding of $j_1,\ldots,j_n$. Almost equal means that, if $i_1 = 2^n - 1$ (the first row in the Boolean matrix that encodes $i_1,\ldots,i_n$ contains only 1-entries): $x_{n1} \neq y_{1n}$ and $x_{ij} = y_{ji}$, for $i \neq n$ or $j \neq 1$. Otherwise ($i_1 \neq 2^n - 1$): $x_{11} \neq y_{11}$ and $x_{ij} = y_{ji}$, for $i \neq 1$ or $j \neq 1$. Obviously, the considered edges exist in G_n and the mapping defined above is bijective. Now, we have a complete definition of a maximum flow in G_n. Let $\mathcal{X}_M$ be the characteristic function of this edge set. Next, we prove that the OBDD complexity of $\mathcal{X}_M$ is exponential. In the rest of this section let π be an arbitrary but fixed variable ordering. Applying Theorem 2 it is sufficient to prove that there is an exponential number of different subfunctions obtained by replacements by constants of the first variables with respect to π. Using the variable ordering π we define a partition of the input variables in the following way. The first part Z_U contains the first variables according to π until for the first time $n/2 - 1$ rows or columns have a tested x- or y-variable. The second part Z_L contains the remaining variables.

Case 1: There are $n/2 - 1$ columns that have an x-variable in Z_U.

We set the y-variables to 0 (that corresponds to the Boolean encoding of t). Now, we consider the following assignments to the x-variables in Z_U. We set the x-variables tested first in the first $n/2-1$ columns with respect to π to all possible assignments. Variables that belong to the same column are set to the same constant. Next, we prove that two different assignments b and b' of these $2^{n/2-1}$ partial assignments lead to different subfunctions. For this reason we consider the following assignment b_r to the remaining x-variables in Z_L. Let ℓ be a column for which the x-variables in b are set to 1 and in b' to 0 (or vice versa). W.l.o.g. we assume that the $x_{\cdot\ell}$-variables in b are set to 1 and in b' to 0. The last x-variable with respect to π which does not belong to the column ℓ is called a free x-variable. Let $x_{\cdot\ell'}$ be the free x-variable. In b_r the $x_{\cdot\ell}$ -variables are set to 1, the remaining x-variables except for the free variable $x_{\cdot\ell'}$ are set to 0. The free x-variable is set to 1, iff the number of x-variables set to 1 in b and b_r together is odd. The column ℓ' does not contain only 1-entries because there are more than one $x_{\cdot\ell'}$-variables in Z_L and at least one of them is set to 0. There exist $x_{\cdot\ell}$-variables in Z_L because otherwise there are n rows that have a variable in Z_U. Using the same argument we know that the free x-variable belongs to Z_L. The function value of the considered subfunction obtained by b respectively b' for b_r is 1 respectively 0, therefore the induced subfunctions are different and we are done.

Case 2: There are $n/2-1$ rows that have a y-variable in Z_U.

This case is similar to the first one. Here, we set the x-variables to 1 (that corresponds to the Boolean encoding of s). Changing the roles of the rows and columns and of the x- and y-variables we are done.

Case 3: There are $n/2-1$ rows that have an x-variable in Z_U.

There exist indices $r, r', c, c' \in \{1, \ldots, n\}$ for which all variables $x_{r\cdot}$, $x_{r'\cdot}$, $y_{\cdot r}$, $y_{\cdot r'}$, $x_{\cdot c}$, $x_{\cdot c'}$, $y_{c\cdot}$, and $y_{c'\cdot}$ are not in Z_U but in Z_L. The reason is that because of our case inspection there are at most $n/2-2$ columns that have an x-variable, at most $n/2-2$ columns that have a y-variable, and at most $n/2-2$ rows that have a y-variable in Z_U. W.l.o.g. let $r \neq 1$ and $c \neq 1$. In the rest of the proof we call x_{rc} the free x-variable. Now, we consider the subfunction of $\mathcal{X}_M$ where we set x_{11} to 0 and y_{11} to 1. Using Theorem 2 it is sufficient to prove that there are at least $2^{n/2-2}$ different subfunctions obtained by replacements of the variables in Z_U. We investigate all possible assignments to the x-variables tested first in the first $n/2-1$ rows with respect to π without the first row. All x-variables in Z_U that belong to the same row are set to the same constant. Variables y_{ji} in Z_U for which the variable x_{ij} is also in Z_U are set to the same constant as x_{ij}. The same is done for y_{ji}-variables for which there exists a variable $x_{i\cdot}$ in Z_U, y_{ji} is set to the same constant as the $x_{i\cdot}$-variables. The variables y_{ji} for which there are no $x_{i\cdot}$-variables in Z_U are set to 0. Altogether we obtain at least $2^{n/2-2}$ different partial assignments. Next, we show that two arbitrary of these assignments b and b' induce different subfunctions. Let ℓ be a row for which the x-variables in Z_U are set differently in b and b'. W.l.o.g. the $x_{\ell\cdot}$-variables in b are set to 1. Now, we consider the following assignment b_r to the remaining variables in Z_L except for the free x-variable. The variables $x_{\ell \cdot}$ are set to 1, all other x-variables without the free x-variable are set to 0. The free x-variable is set to 1, iff the number of x-variables set so far to 1 in b and b_r together is odd. The y_{ji}-variables in Z_L are equal to the x_{ij}-variables. The function value of the considered subfunction obtained by b respectively b' for b_r is 1 respectively 0 and we are done.

Case 4: There are $n/2 - 1$ columns that have a y-variable in Z_U.

This case is similar to the third one. Changing the roles of the rows and columns and of the x- and y-variables we are done.

Altogether, we have shown that the OBDD size for the characteristic function of the considered maximum flow in G_n is at least $2^{n/2-2}$.

In the following our aim is to strengthen the intuition why the considered graph instance is difficult for symbolic OBDD-based algorithms for the maximum flow problem. The crucial properties of our graph instance are the following ones. Our input graph can be partitioned into three parts, one part consists of directed edges from the source s, another one of directed edges into the sink t, and the third one of a complete directed bipartite subgraph. The first two parts can be represented by OBDDs of small size according to arbitrary variable orderings, the third one can be represented in small size, if the variable ordering is chosen carefully, but not, if we change the directed edges into undirected ones. The third part is in some sense *hidden* in the whole graph, such that the OBDD representation of the characteristic function of all three parts together is small (see Figure 1). In a maximum flow each vertex in the subgraph of the third part has incoming and outgoing edges and the corresponding OBDD size becomes large. We conjecture that there exists a maximum flow in G_n whose OBDD representation has polynomial size, iff there exists a perfect (undirected) matching between the vertices with property $\mathcal{R}$ and the vertices with property $\mathcal{C}$ whose OBDD representation has polynomial size.

An OBDD of polynomial size for the characteristic function of a maximum flow $\mathcal{X}_{MF}(x, y)$ in G_n with respect to a variable ordering π only exists, if the following requirements are fulfilled:

1. the π-OBDD size for the subfunction $\mathcal{X}_{MF'}(y)$, where the x-variables are set to the encoding of s, is polynomial,
2. the π-OBDD size for the subfunction $\mathcal{X}_{MF''}(x)$, where the y-variables are set to the encoding of t, is polynomial,
3. the π-OBDD size for $\mathcal{X}_{BM}(x, y)$ which is defined as $\mathcal{X}_{MF}(x, y) \wedge V_{\overline{s}}(x) \wedge V_{\overline{t}}(y)$, where $V_{\overline{s}}(x)$ respectively $V_{\overline{t}}(y)$ is the characteristic function for all vertices in G_n without s respectively t, is polynomial.

Since it is well-known that the size of the output OBDD of a synthesis operation on input OBDDs G_1 and G_2 according to the same variable ordering is at most the product of the sizes of G_1 and G_2, and $V_{\overline{s}}(x)$ and $V_{\overline{t}}(y)$ can be represented by OBDDs of constant width even with respect to an arbitrary variable ordering, the $\wedge$-synthesis of $\mathcal{X}_{MF}(x, y)$, $V_{\overline{s}}(x)$, and $V_{\overline{t}}(y)$, has an OBDD-representation of small size, if $\mathcal{X}_{MF}(x, y)$ has one. $\mathcal{X}_{BM}(x, y)$ represents a bijective mapping between the vertices with property $\mathcal{R}$ in G_n and the vertices with property $\mathcal{C}$. The intuition is that there exists no variable ordering π fulfilling all three properties mentioned above, because a variable ordering that leads to OBDDs of small size for $\mathcal{X}_{MF'}(y)$ has to test the y-variables more or less in a rowwise manner and a variable ordering for $\mathcal{X}_{MF''}(x)$ has to test the x-variables in a columnwise manner. A variable ordering for $\mathcal{X}_{BM}(x, y)$ seems to be a good one, if the x-variables are tested in a rowwise, the y-variables in a columnwise manner, and the x- and y-variables in an interleaved variable ordering which means an x-variable, a y-variable and so on.

Using our considerations above we are now able to define an input graph G'_n whose OBDD representation size is polynomial and whose unique maximum flow has exponential OBDD complexity. The third part of G_n is replaced by the edges of the maximum flow considered above between the vertices with property $\mathcal{R}$ and the vertices with property $\mathcal{C}$. It is not difficult to prove that the OBDD size for the characteristic function of G'_n is linear with respect to the variable ordering $x_{11}, y_{11}, x_{12}, y_{21}, \ldots, x_{1n}, y_{n1}, x_{21}, y_{12}, \ldots, x_{2n}, y_{n2}, \ldots, x_{nn}, y_{nn}$ and we are done. Summarizing, we have shown that the maximum flow problem in 0-1-networks needs exponential space on OBDD-represented graphs by generating instances with an exponential gap between the input and the output OBDD size.

Concluding Remarks

Symbolic algorithms on OBDD-represented graphs are implicitly parallel, since vertices or edges can be treated simultaneously. Sawitzki [19] has shown that a problem is in the complexity class NC, which contains the problems that can be solved efficiently in parallel, if it can be solved with a polylogarithmic number of OBDD-operations with respect to the number of the vertices in a given graph. The general maximum flow problem is P-complete, therefore it cannot exist a symbolic algorithm with a polylogarithmic number of OBDD operations, if P $\neq$ NC. It is an open problem whether the restricted maximum flow problem in 0-1 networks is in NC but it is known to be in RNC [13]. Nevertheless, we have seen that symbolic algorithms for this problem need exponential space with respect to the size length of the implicit representation of the input graph.

Note that the value of a maximum flow can be computed in space poly($\log |V|$) by a nondeterministic Turing Machine using the well-known Turing reduction from the value variant to the decision variant of flow maximization. Already Sawitzki [18] has mentioned that the decision problem can be solved efficiently by nondeterministic Turing machines using $\mathcal{X}_E$ as oracle. Each oracle request can be implemented by an OBDD evaluation operation. Therefore, together with the assumption that the size of the implicit representation is at least $\log |V|$ and the fact that NPSPACE = PSPACE the problem can be solved in polynomial space with respect to the size of the implicit representation of the input graph.

Acknowledgment

The author would like to thank the anonymous referees for their helpful comments.

References

1. Balcázar, J.L., Lozano, A.: The complexity of graph problems for succinctly represented graphs. In: Nagl, M. (ed.) WG 1989. LNCS, vol. 411, pp. 277–285. Springer, Heidelberg (1990)
2. Bollig, B.: On the OBDD complexity of the most significant bit of integer multiplication. In: Agrawal, M., Du, D.-Z., Duan, Z., Li, A. (eds.) TAMC 2008. LNCS, vol. 4978, pp. 306–317. Springer, Heidelberg (2008)
3. Bollig, B.: Symbolic OBDD-based reachability analysis needs exponential space. In: Proc. of SOFSEM. LNCS, vol. 5901, pp. 224–234. Springer, Heidelberg (2010)
4. Bryant, R.E.: Graph-based algorithms for Boolean function manipulation. IEEE Trans. on Computers 35, 677–691 (1986)
5. Bryant, R.E.: On the complexity of VLSI implementations and graph representations of Boolean functions with application to integer multiplication. IEEE Trans. on Computers 40, 205–213 (1991)
6. Downey, R.G., Fellows, M.R.: Parameterized Complexity. Monographs in Computer Science. Springer, Heidelberg (1999)
7. Feigenbaum, J., Kannan, S., Vardi, M.V., Viswanathan, M.: Complexity of problems on graphs represented as OBDDs. In: Meinel, C., Morvan, M. (eds.) STACS 1998. LNCS, vol. 1373, pp. 216–226. Springer, Heidelberg (1998)
8. Galperin, H., Wigderson, A.: Succinct representations of graphs. Information and Control 56, 183–198 (1983)
9. Gentilini, R., Piazza, C., Policriti, A.: Computing strongly connected components in a linear number of symbolic steps. In: Proc. of SODA, pp. 573–582. ACM Press, New York (2003)
10. Gentilini, R., Piazza, C., Policriti, A.: Symbolic graphs: linear solutions to connectivity related problems. Algorithmica 50, 120–158 (2008)
11. Gu, T., Xu, Z.: The symbolic algorithms for maximum flow in networks. Computers & Operations Research 34, 799–816 (2007)
12. Hachtel, G.D., Somenzi, F.: A symbolic algorithm for maximum flow in $0-1$ networks. Formal Methods in System Design 10, 207–219 (1997)
13. Karp, R.M., Upfal, E., Wigderson, A.: Constructing a perfect matching is in Random NC. Combinatorica 6, 35–48 (1986)
14. Papadimitriou, C.H., Yannakakis, M.: A note on succinct representations of graphs. Information and Control 71, 181–185 (1986)
15. Sawitzki, D.: Implicit flow maximization by iterative squaring. In: Van Emde Boas, P., Pokorný, J., Bieliková, M., Štuller, J. (eds.) SOFSEM 2004. LNCS, vol. 2932, pp. 301–313. Springer, Heidelberg (2004)
16. Sawitzki, D.: Lower bounds on the OBDD size of graphs of some popular functions. In: Vojtáš, P., Bieliková, M., Charron-Bost, B., Sýkora, O. (eds.) SOFSEM 2005. LNCS, vol. 3381, pp. 298–309. Springer, Heidelberg (2005)
17. Sawitzki, D.: Algorithmik und Komplexität OBDD-repräsentierter Graphen. PhD thesis, University of Dortmund (2006) (in German)
18. Sawitzki, D.: Exponential lower bounds on the space complexity of OBDD-based graph algorithms. In: Correa, J.R., Hevia, A., Kiwi, M. (eds.) LATIN 2006. LNCS, vol. 3887, pp. 781–792. Springer, Heidelberg (2006)
19. Sawitzki, D.: The complexity of problems on implicitly represented inputs. In: Wiedermann, J., Tel, G., Pokorný, J., Bieliková, M., Štuller, J. (eds.) SOFSEM 2006. LNCS, vol. 3831, pp. 471–482. Springer, Heidelberg (2006)
20. Sieling, D., Wegener, I.: NC-algorithms for operations on binary decision diagrams. Parallel Processing Letters 48, 139–144 (1993)
21. Woelfel, P.: Symbolic topological sorting with OBDDs. Journal of Discrete Algorithms 4(1), 51–71 (2006)

Robust Computations with Dynamical Systems

Olivier Bournez[1], Daniel S. Graça[2,3], and Emmanuel Hainry[4,5]

[1] Ecole Polytechnique, LIX, 91128 Palaiseau Cedex, France
Olivier.Bournez@lix.polytechnique.fr
[2] DM/FCT, Universidade do Algarve, C. Gambelas, 8005-139 Faro, Portugal
dgraca@ualg.pt
[3] SQIG/Instituto de Telecomunicações, Lisbon, Portugal
[4] LORIA, BP 239 - 54506 Vandœuvre lès Nancy Cedex, France
Emmanuel.Hainry@loria.fr
[5] Nancy Université, Université Henri Poincaré, Nancy, France

Abstract. In this paper we discuss the computational power of Lipschitz dynamical systems which are robust to infinitesimal perturbations.

Whereas the study in [1] was done only for not-so-natural systems from a classical mathematical point of view (discontinuous differential equation systems, discontinuous piecewise affine maps, or perturbed Turing machines), we prove that the results presented there can be generalized to Lipschitz and computable dynamical systems.

In other words, we prove that the perturbed reachability problem (i.e. the reachability problem for systems which are subjected to infinitesimal perturbations) is co-recursively enumerable for this kind of systems. Using this result we show that if robustness to infinitesimal perturbations is also required, the reachability problem becomes decidable. This result can be interpreted in the following manner: undecidability of verification doesn't hold for Lipschitz, computable and robust systems.

We also show that the perturbed reachability problem is co-r.e. complete even for C^∞-systems.

Keywords: Verification, Model-checking, Computable Analysis, Analog Computations.

1 Introduction

The investigations on the relationships between dynamics and computations attracted the attention of several research communities. One of them is highly motivated by the question of computer aided verification, and in particular by the question of computer aided verification of hybrid systems (see e.g. [2]).

One main motivation of this community is to get some "as automatic as possible" computer systems, that would take as input the description of a continuous or discrete (or hybrid) system, and the description of some property, call it "safety", and that would tell whether the system satisfies it or not.

The point is, by undecidability of the halting problem of Turing machines, there is no hope to get a fully decidable procedure if the provided formalism for

P. Hliněný and A. Kučera (Eds.): MFCS 2010, LNCS 6281, pp. 198–208, 2010.

describing hybrid systems allows the description of Turing machines, and if the formalism for describing the property allows to talk about their halting.

Some classes of models, such as Timed Automata [3], have been shown to provide subclasses of systems for which verification of the reachability property is decidable.

However, unfortunately, very simple classes of linear hybrid automata [4] or piecewise constant derivative systems [5] have been shown of being able to simulate arbitrary Turing machines. As a consequence, verification procedures are semi-decision procedures and not decision procedures. A more general result can be found in [6], where semidecidability is shown for non-linear systems in general, and decidability is proved for systems satisfying some particular condition.

Since the proofs of undecidability or, more generally, of simulation of Turing machines, often involve to encode the configuration of a Turing machine (or of a two counter automata) into some real numbers, and since this require infinite precision, in the hybrid system verification community a folklore conjecture appeared saying that this undecidability is due to non-stability, non-robustness, sensitivity to initial values of the systems, and that it never occurs in "real systems" [1].

For example, Martin Fränzle writes in [7] "Hence, on simple information-theoretic grounds, the undecidability results thus obtained can be said to be artifacts of an overly idealized formalization. However, while this implies that the particular proof pattern sketched above lacks physical interpretation, it does not yield any insight as to whether the state reachability problem for hybrid systems featuring noise is decidable or not. We conjecture that there is a variety of realistic noise models for which the problem is indeed decidable".

There were several attempts to formalize and prove (or to disprove) this conjecture: it has been proved that small perturbations of the trajectory still yields undecidability [8]. Infinitesimal perturbations of the dynamics for a certain model of hybrid systems has shown to rise to decidability [7]. This has been extended to several models by [1]. In [9] it is shown that Turing machines exposed to small stochastic noise can decide the Halting problem, since its computational power when the error converges to 0 is $\approx \Pi_2^0$.

Let us look at the result presented in [1]: they consider several classes of widely used models of dynamical systems: Turing machines, piecewise affine maps, linear hybrid automata, and piecewise constant derivative systems. For each of them a notion of "perturbed" dynamics is introduced and the computational power of the corresponding perturbed systems is studied. Perturbations are defined for each model using a notion of metrics on the state space. For a given model, with reachability relation R, the idea is to perturb the dynamic by a small ε, and then take (as the perturbed dynamics of the system) the limit (intersection) R_ω of the perturbed reachability relations as this ε tends to 0. In that setting, a system is said "robust" if its reachability relation does not change under small perturbations of the dynamics, i.e. R_ω is equal to R [1]. This has a close resemblance with the notion of "structural stability" for dynamical systems: a system

$\mathcal{A}$ is structurally stable if, roughly, ε-perturbed systems converge to $\mathcal{A}$ as $\varepsilon \to 0$, a concept widely studied in the dynamical system theory see e.g. [10], [11].

In [1], the authors show that for Turing machines, piecewise affine maps, linear hybrid automata, and piecewise constant derivative systems, the relation R_ω belongs to the class Π_0^1 (it is co-recursively enumerable), and moreover, any Π_0^1 relation can be reduced to a relation R_ω of a perturbed system: any complement of a recursively enumerable set, can be semi-decided by an infinitesimally perturbed system.

This means that, for any robust system, its reachability problem is decidable. Indeed, as any system, its reachability problem is semi-decidable (recursively enumerable), and since it is robust, the complement of its reachability problem must be recursively enumerable, from which it follows that the reachability problem must be recursive for robust systems.

In other words, this gives a (partial) answer to the above mentioned conjecture: verification is decidable for robust systems, if the notion of robustness is the one considered here. If one prefers, undecidability of verification arises only when non-robust systems are considered.

In this paper we extend the result of [1] for the case of Lipschitz and computable (in the sense of recursive analysis [12]) systems defined on a compact set, considered as a model of computation. We present both continuous-time and discrete-time versions of our results.

Our aim is to reinforce in some sense the previous result: it follows that verification is decidable for robust systems considered in classical mathematics and computer science, that is to say for robust, Lipschitz, and computable dynamics. In other words, undecidability of verification is really a by-product of non-robustness, even if the system does not rely on trivial (piecewise) dynamics as in [1].

In a more provocative way, undecidability of verification for safety properties over a compact domain is indeed an artifact of modelization for very general and natural classes of systems.

2 Formal Setting: Continuous-Time Dynamical Systems

We begin with some definitions.

Definition 1. *A function $f : \mathbb{R}^m \to \mathbb{R}^k$ is said Lipschitz over a set X if there is some $K > 0$ such that for all $\mathbf{x}, \mathbf{y} \in X$ one has*

$$\|f(\mathbf{x}) - f(\mathbf{y})\| \leq K \|\mathbf{x} - \mathbf{y}\| . \tag{1}$$

In particular it is well known that C^1 functions are Lipschitz over a compact set $X \subseteq \mathbb{R}^m$ and that an initial-value problem

$$\begin{cases} x' = f(t, x) \\ x(t_0) = x_0 \end{cases}$$

where f is Lipschitz, have an unique solution (see e.g. [13]).

Definition 2 (Dynamical systems). *Let $X \subset \mathbb{R}^d$, and consider some function $f : X \to X$. Then we can define a (homogeneous inputless) discrete or continuous-time dynamical system associated to (X, f) as follows:*

- *In the discrete-time case, a* trajectory *is a sequence of points $\{\mathbf{x}_0, \mathbf{x}_1, ...\} \in X^{\mathbb{N}}$, satisfying $f(\mathbf{x}_{i+1}) = \mathbf{x}_i$ for all $i \in \mathbb{N}$.*
- *In the continuous-time case, a* trajectory *is a solution of the differential equation $\dot{\mathbf{x}} = f(\mathbf{x})$, $\mathbf{x}(0) = \mathbf{x}_0 \in X$, i.e. a derivable function $\phi : \mathbb{R}_0^+ \to X$, satisfying $\phi(0) = \mathbf{x}_0$, and $\phi'(t) = f(\phi(t))$ for all t.*

Note that dynamical systems are deterministic: there is only one trajectory starting in a given initial point.

In this paper we consider dynamical systems as recognizers of languages: Σ denotes the alphabet $\Sigma = \{0, 1\}$ and Σ^* denotes words over this alphabet. Therefore we need to encode words over Σ as points in X. This is done using the encoding $\nu : \Sigma^* \to [0, 1]$ defined by: if $w = w_1 \dots w_n \in \{0, 1\}^*$, where $w_1, \dots, w_n \in \{0, 1\}$, then $\nu(w) = \sum_{i=1}^{n} \frac{(2w_i+1)}{4^i}$.

This classical encoding is rather arbitrary. Similar encodings would still yield the results proved in this paper.

We want to avoid not-so-interesting ways to get uncomputability:

- First, we restrict ourselves to dynamics over a compact domain. It is known that smooth systems can robustly simulate Turing machines [14], [15] (configurations are coded as integers), if the perturbations are $\leq \epsilon$, for some fixed $\epsilon > 0$. If we do some mapping from an unbounded domain to a bounded domain, these fixed ϵ-perturbations correspond to infinitesimal perturbations in the compact space. Further allowing infinitesimal errors in an unbounded space seems to originate a degree of modelization which is exceedingly artificial to be considered.
  There is no loss in generality in assuming the compact domain to be $[-1, 1]^d$.
- Second, we want to avoid undecidability due to the impossibility of distinguishing two reals in the recursive analysis setting: this is why we assume that when some computation is accepted this can be stated by considering a value that is clearly below some threshold (1/4), and that this quantity is clearly above a bigger threshold (1/2) when the computation is not terminated.

This leads to the following definition:

Definition 3 (Considering a dynamical system as a language recognizer). *Let $\mathcal{H}$ be a discrete/continuous-time dynamical system over space $X = [-1, 1]^d$. Let V_{accept} be the set of $\mathbf{x} \in X$ with $\|\mathbf{x}\| \leq 1/4$ and $V_{compute}$ be the set of $\mathbf{x} \in X$ with $\|\mathbf{x}\| \geq 1/2$. We say that $\mathcal{H}$ computes a language $L \subset \Sigma^*$ (or that L is the language of $\mathcal{H}$), over alphabet $\Sigma = \{0, 1\}$, if the following holds: for all $w \in \Sigma^*$, $w \in L$ iff the trajectory of $\mathcal{H}$ starting from $(\nu(w), 0, \cdots, 0, 1)$ reaches V_{accept}. For robustness reasons, we assume that for any $w \notin L$, the corresponding trajectory always stays in $V_{compute}$.*

3 Formal Setting: Recursive Analysis

Recursive analysis or computable analysis, was introduced by Turing [16], Grzegorczyk [17], Lacombe [18]: see [12] for an up-to-date monograph presentation of recursive analysis using a computability point of view, or [19] for a presentation using a complexity theory point of view.

Following Ker-I Ko [19], let $\nu_{\mathbb{Q}} : \mathbb{N} \to \mathbb{Q}$ be the following representation[1] of dyadic rational numbers by integers: $\nu_{\mathbb{Q}}(\langle p, q, r \rangle) \mapsto \frac{p-q}{2^r}$, where $\langle ., ., . \rangle : \mathbb{N}^3 \to \mathbb{N}$ is an elementarily polynomial time computable bijection.

A sequence of integers $(x_i)_{i \in \mathbb{N}} \in \mathbb{N}^{\mathbb{N}}$ converges quickly toward x (denoted by $(x_i)_{i \in \mathbb{N}} \rightsquigarrow x$) if the following holds for all i: $|\nu_{\mathbb{Q}}(x_i) - x| < 2^{-i}$.

A point $\mathbf{x} = (x_1, \ldots, x_d) \in \mathbb{R}^d$ is said computable if for all j, there is a computable sequence $(x_i)_{i \in \mathbb{N}} \in \mathbb{N}^{\mathbb{N}}$ (i.e. a computable function $a : \mathbb{N} \to \mathbb{N}$ such that $x_i = a(i)$ for all $i \in \mathbb{N}$) satisfying $(x_i)_{i \in \mathbb{N}} \rightsquigarrow x_j$.

A function $f : X \subset \mathbb{R}^d \to \mathbb{R}$, where X is compact, is said computable if there exists some d-oracle Turing machine M such that, for all $\mathbf{x} = (x_1, \ldots, x_d) \in X$, for all sequences $(x_i^j)_{i \in \mathbb{N}} \rightsquigarrow x_j$, M taking as oracles these d sequences computes a sequence $(x'_i)_{i \in \mathbb{N}}$ with $(x'_i)_{i \in \mathbb{N}} \rightsquigarrow f(\mathbf{x})$. A function $f : X \subset \mathbb{R}^d \to \mathbb{R}^d$, where X is compact, is said computable if all its projections are.

4 Formal Setting: Robustness

We now introduce the settings of [1], based on an idea of [20].

Definition 4 (ε-perturbation). *Consider a discrete/continuous-time dynamical system $\mathcal{H} = (X, f)$. Given $\varepsilon > 0$, its ε-perturbation $\mathcal{H}_\varepsilon$ is the discrete/continuous-time system $\mathcal{H}_\varepsilon$ defined over the same space X, where:*

1. *$(\mathbf{x}_0, \mathbf{x}_1, \ldots)$ is a trajectory of $\mathcal{H}_\varepsilon$ (the trajectory may not be unique), in the case where $\mathcal{H}$ is discrete-time, if $\|\mathbf{x}_{i+1} - f(\mathbf{x}_i)\| \leq \varepsilon$ for all $i \in \mathbb{N}$;*
2. *$\phi : \mathbb{R}_0^+ \to X$ is a trajectory of $\mathcal{H}_\varepsilon$ (the trajectory may not be unique), in the case where $\mathcal{H}$ is continuous-time, if $\phi(0) \in X$ and $\|\phi'(t) - f(\phi(t))\| \leq \varepsilon$ for all $t \in \mathbb{R}_0^+$.*

Note that the ε-perturbation $\mathcal{H}_\varepsilon$ of a dynamical system $\mathcal{H}$ is not, in general, a dynamical system since it is not deterministic (several trajectories may start with a given initial point).

Similarly to what is done in Definition 3, we can define a language computed by $\mathcal{H}_\varepsilon$, which we denote as L_ε: $w \in L_\varepsilon$ iff there is some trajectory of $\mathcal{H}_\varepsilon$ which starts from $(\nu(w), 0, \cdots, 0, 1)$ and reaches V_{accept}.

The following result is immediate.

Lemma 1. *For $0 < \varepsilon < \varepsilon'$, $L_\varepsilon \subset L_{\varepsilon'}$.*

Following the idea given in [1], we consider $L_\omega = \cap_{\varepsilon > 0} L_\varepsilon$.

Definition 5 (Robustness). *A dynamical system $\mathcal{H}$ is said robust iff its language L is equal to L_ω.*

---

1 Many other natural representations of rational numbers can be chosen: they still yield the same class of computable functions – see [12,19].

5 Main Results

The following results are extensions from those obtained in [1].

Theorem 1 (L_ω is co-r.e. for Lipschitz and computable Systems I). *Assume that language L is computed by a discrete-time system $\mathcal{H}$ defined over $[-1,1]^d$ which transition function is Lipschitz and computable. Then L_ω is co-recursively enumerable.*

Proof. Let $n \in \mathbb{N}\backslash\{0\}$ and decompose X in d-dimensional hypercubes $V_1, ..., V_s$ of size $\frac{1}{n}$. We build a finite automaton A_n, which states are $V_1, ..., V_s$, that roughly recognizes $L_{\frac{1}{n}}$. To complete the description of this automaton we need to define two things: (i) the set of accepting states and (ii) the transition rule δ_n. The set of accepting states consists of those hypercubes which overlap V_{accept}, i.e. hypercubes that have vertices within distance $\leq 1/4$ of the origin. These hypercubes can easily be identified.

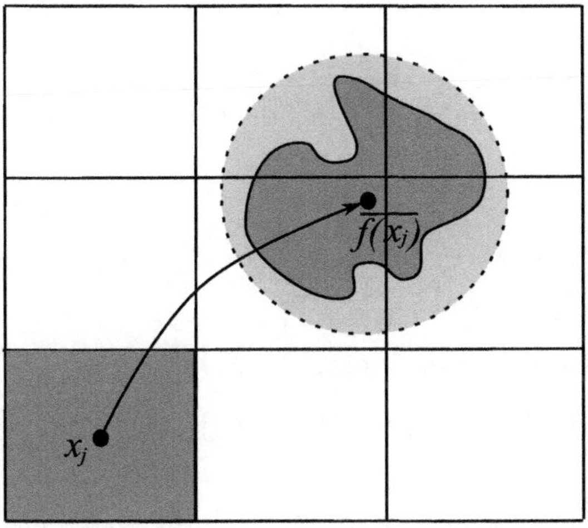

Fig. 1. A figure depicting various elements used in the demonstration of Theorem 1

Now we have to present the transition rule of A_n. The following construction is depicted in Fig. 1. Let V_j be some hypercube. Then pick its central point $\mathbf{x}_j$ (this is an easily computable rational) and compute a rational approximation $\overline{f(\mathbf{x}_j)}$ of $f(\mathbf{x}_j)$ with precision $\frac{1}{n}$. Because f is Lipschitz, there will be some Lipschitz constant $K > 0$ satisfying condition (1) for all $\mathbf{x}, \mathbf{y} \in X$. Then, if $\mathbf{x} \in V_j$ is another point of the same hypercube, we have

$$\|f(\mathbf{x}) - \mathbf{y}\| \leq \frac{1}{n} \quad \Rightarrow \tag{2}$$

$$\left\|\overline{f(\mathbf{x}_j)} - \mathbf{y}\right\| \leq \left\|\overline{f(\mathbf{x}_j)} - f(\mathbf{x}_j)\right\| + \|f(\mathbf{x}_j) - f(\mathbf{x})\| + \|f(\mathbf{x}) - \mathbf{y}\| \leq \frac{K+2}{n}.$$

By other words, if $\mathbf{y}$ is an ε-perturbed image of a point of V_j, then this point will be within distance $\frac{K+2}{n}$ of $\overline{f(\mathbf{x}_j)}$. We use this fact to proceed as follows. After computing $\overline{f(\mathbf{x}_j)}$, determine all the hypercubes which are within distance $\leq (K+2)/n$ of this point (in Fig. 1 this corresponds to all hypercubes covered by the ball of center $\overline{f(\mathbf{x}_j)}$ and radius $(K+2)/n$). This can be done algorithmically, in finite time, since it is only necessary to check which are the hypercubes (which are finitely many) that have vertices within distance $\leq (K+2)/n$ of $\overline{f(\mathbf{x}_j)}$. Let $W_1, ..., W_i$ be these hypercubes. Then we define the transition rule over the hypercubes as follows: $\delta(V_j) = \{W_1, ..., W_i\}$. This defines the automaton A_n.

Now we say that a point $\mathbf{x} \in X$ is accepted by A_n if it lies in an accepted hypercube. Let $L^*_{\frac{1}{n}}$ be the language accepted by A_n. From (2), the dynamics of A_n includes those of the $\frac{1}{n}$-perturbed system $\mathcal{H}_{\frac{1}{n}}$. Hence $L_{\frac{1}{n}} \subseteq L^*_{\frac{1}{n}}$. On the other side, it is not difficult to see that the dynamics of A_n are included in those of $\mathcal{H}_{\frac{K+3}{n}}$.[2] Therefore

$$L_{\frac{1}{n}} \subseteq L^*_{\frac{1}{n}} \subseteq L_{\frac{K+3}{n}} \quad \Rightarrow \quad \cap_{n=1}^{\infty} L^*_{\frac{1}{n}} = \cap_{\varepsilon>0} L_\varepsilon = L_\omega.$$

Let us now show that $L_\omega = \cap_{n=1}^{\infty} L^*_{\frac{1}{n}}$ is co-r.e., as required. Let $w \in \Sigma^*$. Then build a Turing machine which performs the following steps:

```
i=0
Repeat
    i++
    Simulate A_i with input ν_X(w)
Until ν_X(w) is rejected
Reject w
```

This shows that the complement of L_ω is r.e., as required.

We now introduce a (classical) tool to discretize a continuous-time system (see e.g. [21]).

Definition 6 (Stroboscopic map). *Consider the continuous-time system* $\mathcal{H} = (X, f)$. *Then its corresponding* stroboscopic map $g : X \to X$ *is given by* $g(\mathbf{x}_0) = \phi(1)$, *where* $\phi : \mathbb{R}_0^+ \to \mathbb{R}$ *satisfies* $\phi'(t) = f(\phi(t))$, $\phi(0) = \mathbf{x}_0$.

In other words, $g(\mathbf{x}_0)$ gives the point where the trajectory of $\mathcal{H}$, starting on $\mathbf{x}_0$, would reach after one time unit.

Lemma 2. *Let* $\mathcal{H} = (X, f)$ *be a continuous-time system, and let* $\widehat{\mathcal{H}} = (X, g)$ *be its discrete-time counterpart, where* g *is the stroboscopic map of* $\mathcal{H}$. *Then, if* $L, \hat{L}$ *are the languages computed by* $\mathcal{H}$ *and* $\widehat{\mathcal{H}}$, *respectively, under the same input encoding, we have:*

1. $L = \hat{L}$;
2. $L_\omega = \hat{L}_\omega$ *and* $L_\varepsilon = \hat{L}_\varepsilon$, *for all* $\varepsilon > 0$.

[2] Here we suppose that $\|\mathbf{x}\| = \|\mathbf{x}\|_\infty = \max_{1 \leq i \leq n} |x_i|$. However, a similar result holds for other norms since all norms are equivalent in a finite-dimensional space.

Proof. Note that for all $\mathbf{x}_0 \in X$, if $\phi : \mathbb{R}_0^+ \to \mathbb{R}$, $\phi(0) = \mathbf{x}_0$ is a trajectory of $\mathcal{H}$ and $(\mathbf{x}_0, \mathbf{x}_1, ...)$ is a trajectory of $\widehat{\mathcal{H}}$, we must have $\phi(k) = \mathbf{x}_k$ for all $k \in \mathbb{N}$. Therefore, because only the accepting trajectories reach V_{accept} and then stay there, we must have $L = \hat{L}$. Similar arguments can be used to show the identity $L_\varepsilon = \hat{L}_\varepsilon$ that, by its turn, implies $L_\omega = \hat{L}_\omega$.

Theorem 2 (L_ω is co-r.e. for Lipschitz and computable Systems II). *Assume that language L is computed by a continuous-time system $\mathcal{H}$ defined over $[-1,1]^d$ with a Lipschitz and computable transition function. Then L_ω is co-recursively enumerable.*

Proof. Using the same assumptions of the previous lemma, to show that the language L computed by the continuous-time system $\mathcal{H} = (X, f)$, where $X = [-1,1]^d$, is co-r.e., it is enough to show that $\hat{L}$ is a co-r.e. language. If we show that g is Lipschitz and computable, then this can be done as indicated in the proof of Theorem 1. It is well-known (see e.g. [13]) that if f satisfies equation (1) over X, and if ϕ_0, ϕ_1 are solutions of the differential equation $x' = f(x)$ over X, then

$$\|\phi_0(t) - \phi_1(t)\| \leq \|\phi_0(0) - \phi_1(0)\| \, e^{Kt}.$$

By other words, the stroboscopic map g is Lipschitz on X, with Lipschitz constant e^K. Moreover, g is computable by using an algorithmic version of Euler's method on the time interval $[0,1]$ for the differential equation $\dot{\mathbf{x}} = f(\mathbf{x})$.(cf. [19]).

The following lemma is a corollary from the results of [6], [22]: one can semi-compute any trajectory of $\mathcal{H}$. Therefore one can semi-decide if a trajectory finishes in V_{accept}, and thus semi-decide L.

Lemma 3. *Let L be the language computed by a system $\mathcal{H} = (X, f)$, where $X = [-1,1]^d$ and f is Lipschitz and computable. Then L is r.e.*

Since a system is robust iff $L = L_\omega$, the following corollary is immediate.

Corollary 1 (Robust =>Recursive for Lipschitz and Computable Systems). *In the conditions of the lemma above, if $\mathcal{H}$ is robust, then L is recursive.*

We now prove that Turing machines can be simulated robustly with C^∞-systems. Recall that C^∞-systems are also Lipschitz. This is a generalization of a similar result from [1], which holds for piecewise systems (which are not of class C^1).

Theorem 3 (Perturbed reachability is complete in Π_1^0). *Let A be a recursively enumerable language. Then there is a computable and C^∞ dynamical system $\mathcal{H}$ such that $L_\omega = \bar{A}$.*

Proof. For any Turing machine M, Theorems 4.4 and 4.5 from [23] give an explicit way to build a continuous-time dynamical system (X, f) where X is compact and f_M is C^∞ such that $y' = f_M(y)$ simulates M. It is easy to observe that the function f_M built there is computable. Changing coordinates, and reasoning through homothety, we can assume without loss of generality X

to be $[-T,T]^3$ and that the accepting configuration of M correspond to the origin $c = (0,0,0) \in [-T,T]^3$, and that the initial starting point corresponding to some input w is of the form $c = (\nu(w),0,1)$ as in Definition 3. This yields a dynamical system such that a word w is accepted by Turing machine M iff the trajectory starting from $(\nu(w),0,1)$ reaches the origin c. Let C be the cubic neighborhood of c defined by $[-0.1,0.1]^3$, and C' be the cubic neighborhood of c defined by $[-0.2,0.2]^3$. The resulting system is also such that a non-accepted word corresponds to a trajectory that stays forever outside of neighborhood C'.

We construct from this 3-dimensional smooth dynamical system f_M a smooth 4-dimensional dynamical system g_M whose perturbed version semi-recognizes the complement of the language recognized by M.

Actually, robustness will only be shown for one component, the extra component added to the original 3-dimensional system, which is the most critical for the simulation, . The other 3 components can be made robust using the construction presented in [14], [15] which, although originally done for unbounded spaces, can be brought to the compact space $[-T,T]^3$ via a transformation using a function like, e.g. arctan, and can still be used here. However, giving all the details would span the contents of this paper outside allowable size, so we only show robustness for the crucial component.

We will use the notation $\mathbf{x}, \mathbf{y}$ for 3-dimensional vectors, and h for the fourth dimension. This 4-dimensional system is mainly the original smooth dynamical system embedded in the hyperplane $h = 3$ of $\mathbb{R}^4$, however with two changes. First, the neighborhood C of the original dynamical system becomes rejecting in the new system. Second, the zone $h \leq 1$ becomes accepting in the new system.

The idea is that for any word w accepted by M, the original system f_M will eventually come to C, and hence the perturbed dynamical system g_M will eventually go up and reject. For any word w not accepted by M, the system g_M will slowly drift "down" until it reaches the accepting zone $h \leq 1$.

In order to get a smooth dynamics, we start by the following technicalities. This is an easy and classical mathematical exercise to prove that for any computable reals $a < b$, function $h_{a,b}(x) = \exp(-1/(x-a)^2 - 1/(b-x)^2)$ for $a < x < b$ and 0 for $x \geq b$ or $x \leq a$ is $\mathcal{C}^\infty$ and computable. If we denote by γ the constant $\gamma = \int_a^b h_{a,b}(x)dx$, and by $\chi_{a,b}(x)$ the function $1/\gamma \int_a^x h_{a,b}(x)$, one gets a computable and $\mathcal{C}^\infty$ function that values 0 for $x \leq a$, and 1 for $x \geq b$, and a value in interval $[0,1]$ in between. Given $a < b < c$, the function $\chi_{a,b,c}(x) = \chi_{a,b}(x) - \chi_{b,c}(x)$ values 0 for $x \leq a$, 1 in b, and 0 for $x \geq c$, with a value in interval $[0,1]$ in between. In the same spirit, let $\chi_C : \mathbb{R}^3 \to \mathbb{R}$ be some function that values 1 over neighborhood C, and 0 outside neighborhood C'.

Formally the new system g_M is then defined on $[-T,T]^3 \times [-1,5] \subset \mathbb{R}^4$ as follows: $g_M(\mathbf{x}, h) = (\mathbf{y}, h')$ where

$$h' = \chi_{4,5}(h) + \chi_C(\mathbf{x})\chi_{2,3,4}(h) - \chi_{0,1,2}(h) + \chi_{-1,-0.5,0}(h)$$

and

$$\mathbf{y} = f_M(\mathbf{x})(1 - \chi_C(x))\chi_{2,3,4}(h) + C(\mathbf{x})\chi_{0,0.5,1}(h),$$

where $\dot{x} = C(\mathbf{x})$ is a smooth and computable vector field with all the trajectories going to the origin.

In other words,

- if $h \geq 4$, anything that arrives in the layer $h \geq 4$ goes "up" and is rejected.
- if $2 < h < 4$:
  - if $\mathbf{x} \in C$, then it goes "up" and it is ultimately rejected.
  - if $\mathbf{x} \notin C$, then it basically simulates the original system f_M.
- if $1 < h < 2$, then it goes "down" until it reaches $h \leq 1$,
- if $-1 \leq h < 1$, then it goes to the origin.

Consider the trajectory starting from $(\nu(w), 0, 1, 3)$: suppose that word w is accepted by M: the corresponding trajectory of f_M will eventually go to the origin in some finite time t, and then the trajectory of g_M will reach neighborhood C, and hence will go up for ever. Taking ϵ small enough (depending on t) any ϵ-perturbed trajectory of g_M will be close at time t to this trajectory, and hence also in C, and hence going for ever up afterwards.

Suppose that word w is not accepted by M: the corresponding trajectory of f_M will run for ever outside of C'. For any $\epsilon > 0$, we can easily construct a trajectory of g_M that drifts down slowly in the fourth coordinate until reaching $h \leq 1$, and then going to the origin.

Through a change of coordinates, the obtained system fulfills all constraints of Definition 3 and of the statement of the Theorem.

In other words, "perturbed reachability is complete in Π_1^0" as this is termed in [1].

6 Conclusion

In this paper we showed that, on compact sets, the perturbed reachability problem is co-r.e. for Lipschitz and computable systems. We also proved that for Lipschitz and computable systems which are robust to infinitesimal perturbations, the reachability problem is decidable. The results are both for the discrete and continuous-time case. It would be interesting to know what happens at a more refined level, i.e. if from a complexity point of view.

Acknowledgments. D. Graça was partially supported by *Fundação para a Ciência e a Tecnologia* and EU FEDER POCTI/POCI via SQIG - Instituto de Telecomunicações.

References

1. Asarin, E., Bouajjani, A.: Perturbed Turing machines and hybrid systems. In: Proc. 16th Annual IEEE Symposium, Logic in Computer Science, pp. 269–278 (2001)
2. Alur, R., Pappas, G.J. (eds.): HSCC 2004. LNCS, vol. 2993. Springer, Heidelberg (2004)

3. Alur, R., Dill, D.L.: Automata for modeling real-time systems. In: Paterson, M. (ed.) ICALP 1990. LNCS, vol. 443, pp. 322–335. Springer, Heidelberg (1990)
4. Henzinger, T.A., Kopke, P.W., Puri, A., Varaiya, P.: What's decidable about hybrid automata? J. Comput. System Sci. 57(1), 94–124 (1998)
5. Asarin, E., Maler, O., Pnueli, A.: Reachability analysis of dynamical systems having piecewise-constant derivatives. Theoret. Comput. Sci. 138, 35–65 (1995)
6. Collins, P.: Continuity and computability of reachable sets. Theor. Comput. Sci. 341, 162–195 (2005)
7. Fränzle, M.: Analysis of hybrid systems: An ounce of realism can save an infinity of states. In: Flum, J., Rodríguez-Artalejo, M. (eds.) CSL 1999. LNCS, vol. 1683, pp. 126–140. Springer, Heidelberg (1999)
8. Henzinger, T.A., Raskin, J.F.: Robust undecidability of timed and hybrid systems. In: Lynch, N.A., Krogh, B.H. (eds.) HSCC 2000. LNCS, vol. 1790, pp. 145–159. Springer, Heidelberg (2000)
9. Asarin, E., Collins, P.: Noisy Turing machines. In: Caires, L., Italiano, G.F., Monteiro, L., Palamidessi, C., Yung, M. (eds.) ICALP 2005. LNCS, vol. 3580, pp. 1031–1042. Springer, Heidelberg (2005)
10. Guckenheimer, J., Holmes, P.: Nonlinear Oscillations, Dynamical Systems, and Bifurcation of Vector Fields. Springer, Heidelberg (1983)
11. Hirsch, M.W., Smale, S., Devaney, R.: Differential Equations, Dynamical Systems, and an Introduction to Chaos. Academic Press, London (2004)
12. Weihrauch, K.: Computable Analysis: an Introduction. Springer, Heidelberg (2000)
13. Birkhoff, G., Rota, G.C.: Ordinary Differential Equations, 4th edn. John Wiley & Sons, Chichester (1989)
14. Graça, D.S., Campagnolo, M.L., Buescu, J.: Robust simulations of Turing machines with analytic maps and flows. In: Cooper, S.B., Löwe, B., Torenvliet, L. (eds.) CiE 2005. LNCS, vol. 3526, pp. 169–179. Springer, Heidelberg (2005)
15. Graça, D.S., Campagnolo, M.L., Buescu, J.: Computability with polynomial differential equations. Adv. Appl. Math. 40(3), 330–349 (2008)
16. Turing, A.M.: On computable numbers, with an application to the Entscheidungsproblem. Proc. London Math. Soc. (Ser. 2-42), 230–265 (1936)
17. Grzegorczyk, A.: Computable functionals. Fund. Math. 42, 168–202 (1955)
18. Lacombe, D.: Extension de la notion de fonction récursive aux fonctions d'une ou plusieurs variables réelles III. C. R. Acad. Sci. Paris 241, 151–153 (1955)
19. Ko, K.-I.: Computational Complexity of Real Functions. Birkhäuser, Basel (1991)
20. Puri, A.: Dynamical properties of timed automata. In: Ravn, A.P., Rischel, H. (eds.) FTRTFT 1998. LNCS, vol. 1486, pp. 210–227. Springer, Heidelberg (1998)
21. Ott, E.: Chaos in Dynamical Systems, 2nd edn. Cambridge University Press, Cambridge (2002)
22. Collins, P., Graça, D.S.: Effective computability of solutions of differential inclusions — the ten thousand monkeys approach. Journal of Universal Computer Science 15(6), 1162–1185 (2009)
23. Bournez, O., Cosnard, M.: On the computational power of dynamical systems and hybrid systems. Theoret. Comput. Sci. 168(2), 417–459 (1996)

On Factor Universality in Symbolic Spaces

Laurent Boyer and Guillaume Theyssier*

LAMA, (UMR 5127 — CNRS, Université de Savoie), Campus Scientifique,
73376 Le Bourget-du-lac cedex France
laurent.boyer@univ-savoie.fr,
guillaume.theyssier@univ-savoie.fr

Abstract. The study of factoring relations between subshifts or cellular automata is central in symbolic dynamics. Besides, a notion of intrinsic universality for cellular automata based on an operation of rescaling is receiving more and more attention in the literature. In this paper, we propose to study the factoring relation up to rescalings, and ask for the existence of universal objects for that simulation relation.

In classical simulations of a system S by a system T, the simulation takes place on a specific subset of configurations of T depending on S (this is the case for intrinsic universality). Our setting, however, asks for every configurations of T to have a meaningful interpretation in S. Despite this strong requirement, we show that there exists a cellular automaton able to simulate any other in a large class containing arbitrarily complex ones. We also consider the case of subshifts and, using arguments from recursion theory, we give negative results about the existence of universal objects in some classes.

1 Introduction and Definitions

Tilings and cellular automata are two paradigmatic models often considered in the fields of complex systems and natural computing. They are complementary —one is static and non-deterministic and the other is dynamic and deterministic— but they are both formally simple and both related to symbolic spaces. Moreover, many links are now established between the two models (see for instance [10,9]) so it is natural to consider them together.

Both are known to be Turing-powerful since their introduction in the mid-20th century [23,17]. However, analyzing their ability to process information only through translations into the Turing world is very restrictive. Such models of natural computing deserve a natural and intrinsic notion of reduction to compare their objects one to each other. Following this line of thought, several notions of simulations were proposed recently which are intrinsic to each model, and lead to corresponding intrinsic notions of universality [18,22,14,5]: a system is universal if it is able to simulate any other from the same class.

Intrinsic universality for cellular automata is probably the most studied of such notions [19,20,15,2,6]. The underlying relation of simulation uses uniform

* Partially supported by French ANR 'projet blanc' EMC (NT09_555297).

P. Hliněný and A. Kučera (Eds.): MFCS 2010, LNCS 6281, pp. 209–220, 2010.

encodings working at the level of blocks of cells. More precisely, S simulates T if S, when restricted to a suitable subset of 'correct' configurations, is isomorphic to T via such an encoding. Our approach is different and uses redundancy of information instead of restriction to a subset of configurations. In our setting, S simulates T if there is a uniform way of projecting *the whole* phase space of S onto the phase space of T (a precise definition is given below). The question addressed by this paper is the existence of universal objects with respect to that simulation relation, we call them *factor universal* objects.

The first contribution of this paper is the formalism based on the well-known mathematical notion of action: it allows to encompass both subshifts and cellular automata, it gives a new look at the notion of cell grouping which is the root of the simulation relation used in intrinsic universality, and it establishes connections with the work of Hochman [8] where the use of sub-actions is crucial. Our main result is that, although factor-universal objects do not generally exist (theorem 1), it can still be constructed for some large class like the set of cellular automata having a persistent state (theorem 2).

Basic definitions. Given a finite set Q and an integer $d \geq 1$, the *symbolic space* of *dimension d* over *alphabet* Q is the set $Q^{\mathbb{Z}^d}$. It can be seen as an infinite set of cells arranged as a lattice $\mathbb{Z}^d$ and each carrying a value from Q. An element of $Q^{\mathbb{Z}^d}$ is called a *configuration*. $Q^{\mathbb{Z}^d}$ is naturally equipped with the compact Cantor topology [12] which is the product topology of the discrete topology on Q (it can also be defined via a metric).

Another key notion in the context of symbolic spaces is that of *finite patterns* that may occur in infinite configurations. For our purpose, rectangular patterns will be enough. Given $\boldsymbol{z} = (z_1, \cdots, z_d) \in \mathbb{Z}^d$ with $z_i > 0$ for all i, the hyperrectangle $\mathcal{R}_{\boldsymbol{z}}$ is the set of vectors $\boldsymbol{z'} = (z'_1, \cdots, z'_d) \in \mathbb{Z}^d$ such that $0 \leq z'_i < z_i$ for all i. A Q-pattern of shape $\mathcal{R}_{\boldsymbol{z}}$ is a coloring of $\mathcal{R}_{\boldsymbol{z}}$ by Q, that is an element of $Q^{\mathcal{R}_{\boldsymbol{z}}}$. Given a configuration $c \in Q^{\mathbb{Z}^d}$, the pattern of shape $\mathcal{R}_{\boldsymbol{z_s}}$ extracted from c at position $\boldsymbol{z_p} \in \mathbb{Z}^d$, denoted by $\mathcal{P}_{\boldsymbol{z_p}}^{\boldsymbol{z_s}}(c)$, is simply: $z \in \mathcal{R}_{\boldsymbol{z_s}} \mapsto c(z_p + z)$.

The objects we study (subshifts and cellular automata) share the property of being uniform, *i.e.* invariant by translations. Formally, given $z \in \mathbb{Z}^d$ the translation of vector z, denoted σ_z, is the function mapping a configuration $c \in Q^{\mathbb{Z}^d}$ to the configuration $\sigma_z(c)$ such that $\forall z' \in \mathbb{Z}^d$, $\sigma_z(c)(z') = c(z' + z)$.

A *subshift* is a subset of $Q^{\mathbb{Z}^d}$ which is translation invariant and closed for the Cantor topology. Equivalently, a subshift is a set Σ_L of configurations avoiding any occurrence of any finite pattern from a given language of patterns L:

$$\Sigma_L = \left\{c \in Q^{\mathbb{Z}^d} : \forall z, z' \in \mathbb{Z}^d \text{ with } z_i > 0 \text{ for all } i,\ \mathcal{P}_{\boldsymbol{z'}}^{\boldsymbol{z}}(c) \notin L\right\}.$$

A subshift of *finite type* is a subshift of the form Σ_L where L is finite. There are strong connections between subshifts of finite type in dimension 2 and sets of tilings generated by a set of wang tiles. In particular, due to Berger's theorem [1], it is undecidable, given a finite L, to determine whether Σ_L is empty or not.

A *cellular automaton* is a local and uniform map on a symbolic space. Formally, it is given as a 4-tuple by its dimension d, its alphabet Q, its *neighborhood*

$V \subseteq \mathbb{Z}^d$ (finite) and its *local transition map* $f : Q^V \to Q$. To that formal object we associate a *global map* F acting on $Q^{\mathbb{Z}^d}$ as follows:

$$\forall c \in \mathbb{Z}^d, \forall z \in \mathbb{Z}^d,\ F(c)(z) = f\big(z' \in V \mapsto c(z + z')\big).$$

The fundamental theorem of Curtis-Lyndon-Hedlund [7] states that global maps of cellular automata are exactly continuous maps on symbolic spaces which commute with translations.

Actions and rescalings. Let $(\mathbb{M}, +)$ be a monoid (a set equipped with an associative law and a neutral element). An $\mathbb{M}$-*action* on a space X is a function $\Psi : \mathbb{M} \times X \to X$ such that $\Psi(0, x) = x$ (for all $x \in X$ and 0 being the neutral element of $\mathbb{M}$) and

$$\forall x \in X, \forall m, m' \in \mathbb{M},\ \Psi(m + m', x) = \Psi\big(m, \Psi(m', x)\big).$$

We will use the formalism of action to study both subshifts and cellular automata:

- if $\Sigma \subseteq Q^{\mathbb{Z}^d}$ is a subshift, we canonically associate to it the $\mathbb{Z}^d$-action Ψ_Σ on Σ defined by $\Psi_\Sigma(z, x) = \sigma_z(x)$;
- if F is a cellular automaton on the space $Q^{\mathbb{Z}^d}$, we canonically associate to it the $\mathbb{N} \times \mathbb{Z}^d$-action Ψ_F on $Q^{\mathbb{Z}^d}$ defined by $\Psi_F\big((t, z), x\big) = \sigma_z \circ F^t(x)$.

If $\mathbb{M}'$ is a sub-monoid of $\mathbb{M}$, Ψ induces an $\mathbb{M}'$-action by restriction to the domain $\mathbb{M}' \times X$. $\mathbb{M}$ and $\mathbb{M}'$ can be isomorphic or not and both cases might be interesting. For instance, studying a cellular automaton F as a classical dynamical system consists in forgetting the spacial component of Ψ_F and focusing on the pure temporal action of F. This point of view was often adopted in the literature (*e.g.*, topological dynamics of cellular automata [12]) but, interestingly enough, recent work of Sablik [21] tends to re-incorporate the spacial component of actions to better study the dynamics of cellular automata.

In this paper, we will only consider the case where $\mathbb{M}$ and $\mathbb{M}'$ are isomorphic. More precisely, in our context, $\mathbb{M}$ will be of the form $\mathbb{Z}^d$ or $\mathbb{N} \times \mathbb{Z}^d$ and we will consider sub-monoids of the form $\mathbb{M}' = t_0\mathbb{N} \times z_1\mathbb{Z} \times \cdots \times z_d\mathbb{Z}$, with $t_0 > 0$ and $z_i > 0$ for all i. In this case, passing from the $\mathbb{M}$-action to the $\mathbb{M}'$-action can be seen as a neutral change of point of view on the system that we call *rescaling* in the sequel. The intuition is that we change the discrete units of time and space, passing from 1 to t_0 in time and 1 to z_i in direction i. Given a subshift or a cellular automaton, a *scaled action* is simply the restriction of their canonical action to some sub-monoid of the form $\mathbb{M}'$. It is worth noticing that a scaled action associated to a subshift (resp. a cellular automaton) on the alphabet Q is always isomorphic to the canonical action of a subshift (resp. a cellular automaton) on an alphabet of the form Q^k. More concretely, this isomorphism comes from the natural one-to-one map from $Q^{\mathbb{Z}^d}$ to $\big(Q^{\mathcal{R}_{\boldsymbol{z_s}}}\big)^{\mathbb{Z}^d}$, where $\boldsymbol{z_s} = (z_1, \ldots, z_d)$, which maps a configuration c to: $z \mapsto \mathcal{P}^{\boldsymbol{z_s}}_{\boldsymbol{z} \times \boldsymbol{z_s}}(c)$, where the operation $\times$ on $\mathbb{Z}^d$ denotes coordinate-wise multiplication. Our notion of rescaling for cellular automata is similar to the one in [18,22] which is the basic ingredient to define intrinsic universality.

Factors. One of the central notion in symbolic dynamics is that of *factor*. Intuitively, a factor is a uniform continuous projection. This notion has also been used with success in the study of expansive cellular automata [16] and more generally as a classification tools for cellular automata [11,3]. As we study both multi-dimensional subshifts and cellular automata, we give a unified definition using the formalism of actions.

Definition 1. *Let* $\mathbb{M}$ *and* $\mathbb{M}'$ *be isomorphic monoids via* $i : \mathbb{M} \to \mathbb{M}'$. *We say an* $\mathbb{M}'$*-action* ϕ' *on* X' *is a* factor *of a* $\mathbb{M}$*-action* ϕ *on* X *if there is a continuous onto map* $\pi : X \to X'$ *such that:* $\forall x \in X, \forall m \in \mathbb{M},\ \pi\big(\phi(m,x)\big) = \phi'\big(i(m), \pi(x)\big)$.

Two key points are that: (1) any orbit in (ϕ, X) projects onto some orbit of (ϕ', X') via π, and (2) any orbit of ϕ' can be realized as such a projection. In a word, the simulation of (ϕ', X') by (ϕ, X) is *everywhere meaningful* and *complete.*

2 Factor Universality

At this point, we could compare subshifts or cellular automata through the factoring relation between their canonical actions, saying that system S factors onto system T if the canonical action of S factors onto that of T. However, this gives an excessive importance to the alphabet and forbid the existence of universal objects due to entropy considerations (factoring cannot increase entropy). In [8], this limitation is bypassed via dimension changes: a d-dimensional system is compared to k-dimensional systems ($k < d$) via its k-dimensional sub-actions. Our point of view is different. We always work at constant dimension, but we use another kind of sub-actions: *scaled actions* defined above. For a fixed dimension, monoids of scaled actions are all isomorphic and we will consider only canonical component-wise isomorphisms between them. We can now formulate the central definition of the paper.

Definition 2. *Let* S *and* T *be two* d*-dimensional subshifts (resp. CA). We say that* T *is simulated by* S*, denoted* $T \preccurlyeq S$*, if some scaled action of* S *factors onto some scaled action of* T.

As usual when working on symbolic spaces, continuity and uniformity implies locality (Curtis-Lyndon-Hedlund theorem [7]). In our context of rescalings, the locality is no longer expressed at the level of cells, but at the level of groups of cells. More precisely, we say that a map $\phi : Q_1^{\mathbb{Z}^d} \to Q_2^{\mathbb{Z}^d}$ is *local* if there exist: $r \in \mathbb{N}$ (locality radius), two shapes $\mathcal{R}_{\boldsymbol{z_1}}$ and $\mathcal{R}_{\boldsymbol{z_2}}$ (source and destination scales), and a local function $f : Q_1^{\mathcal{R}_{(2r+1)\boldsymbol{z_1}}} \to Q_2^{\mathcal{R}_{\boldsymbol{z_2}}}$ such that

$$\forall c \in Q_1^{\mathbb{Z}^d}, \forall z \in \mathbb{Z}^d,\ \mathcal{P}^{\boldsymbol{z_2}}_{\boldsymbol{z}\times\boldsymbol{z_2}}(\phi(c)) = f\big(\mathcal{P}^{(2r+1)\boldsymbol{z_1}}_{\boldsymbol{z}\times\boldsymbol{z_1}-r\boldsymbol{z_2}}(c)\big).$$

To fix ideas, if $d = z_1 = z_2 = 1$ and $Q_1 = Q_2$, f is just the local map of a cellular automaton of radius r and ϕ is its corresponding global map.

Proposition 1. *Fix a dimension d. Let Σ_1 and Σ_2 be two d-dimensional subshifts and let F_1 and F_2 be two d-dimensional CA of alphabet Q_1 and Q_2 respectively. Then we have:*

- $\Sigma_2 \preccurlyeq \Sigma_1$ *if and only if there is a local map ϕ such that $\phi(\Sigma_1) = \Sigma_2$;*
- $F_2 \preccurlyeq F_1$ *if and only if there is an onto local map ϕ from $Q_1^{\mathbb{Z}^d}$ to $Q_2^{\mathbb{Z}^d}$ and integers $t_1, t_2 \in \mathbb{N}$ such that $\phi \circ F_1^{t_1} = F_2^{t_2} \circ \phi$.*

Besides the work of Hochman [8], notions of simulations similar to $\preccurlyeq$ have already been considered for tilings [14] or for cellular automata [22,4] or more general settings [13]. Each time, one of the main concern is the existence of universal objects: this is precisely the central point of the present paper.

Definition 3. *Let $\mathcal{C}$ be a class of subshifts (resp. cellular automata). A subshift (resp. cellular automaton) U is* $\mathcal{C}$-universal *if $U \in \mathcal{C}$ and $X \preccurlyeq U$ for any $X \in \mathcal{C}$.*

Whatever the fixed dimension, there is no universal subshift for cardinality reasons: there are uncountably many subshifts but for a given subshift U there are at most countably many different subshifts $\preccurlyeq$-simulated by U (by proposition 1). The following theorem uses recursion theoretic arguments to yield other negative results concerning universality (similar arguments where used in [18,8] in different settings).

Theorem 1. *Fix a dimension $d \geq 2$. Then there is no universal subshift of finite type of dimension d and there is no surjective-universal CA of dimension d.*

3 A Large Class with a Universal Object

In this section, we restrict to dimension 1 to make a clear exposition of the main result (theorem 2).

Definition 4. *A CA $\mathcal{A}$ is said to be* persistent *if there is a state $q_0 \in Q_\mathcal{A}$ such that for any configuration $c \in {Q_\mathcal{A}}^\mathbb{Z}$ if $c(i) = q_0$ then $\mathcal{A}(c)\ (i) = q_0$.*
We denote by $\mathbb{P}$ the set of all persistent CA.

Note that for any CA, you may add an extra persistent state and obtain a CA in $\mathbb{P}$ containing the dynamics of the first one.

Theorem 2. *There exists a $\mathbb{P}$-universal cellular automaton.*

In the following, we describe such a $\mathbb{P}$-universal CA $\mathcal{U}$ with radius 1 and alphabet $Q_\mathcal{U}$.

We denote by $\mathbb{P}_0$ the set of CA of $\mathbb{P}$ with radius 1 and alphabet of size 2^p for some $p \in \mathbb{N}$. One may easily describe for any CA $\mathcal{B} \in \mathbb{P}$ a CA $\mathcal{A} \in \mathbb{P}_0$ such that $\mathcal{B} \preccurlyeq \mathcal{A}$. Using transitivity of $\preccurlyeq$, it will be sufficient, in order to prove $\mathbb{P}$-universality of $\mathcal{U}$, to exhibit for any CA $\mathcal{A} \in \mathbb{P}_0$ an onto local map $\phi_\mathcal{A}$ from ${Q_\mathcal{U}}^\mathbb{Z}$ to ${Q_\mathcal{A}}^\mathbb{Z}$ and an integer $\tau_\mathcal{A}$ such that $\mathcal{U}^{\tau_\mathcal{A}} \circ \phi_\mathcal{A} = \phi_\mathcal{A} \circ \mathcal{A}$.

To do so, for each $\mathcal{A}$, we introduce an integer $l_\mathcal{A}$ and a dichotomy on words of ${Q_\mathcal{U}}^{l_\mathcal{A}}$.

- on the one side we have what we call $\mathcal{A}$-*correct macrocells* (or $\mathcal{A}$-macrocells). They encode information about a current state $x \in Q_{\mathcal{A}}$, about the local rule of $\mathcal{A}$, and a machinery used to apply this rule to update the current state. In almost any case, they will be interpreted through $\phi_{\mathcal{A}}$ as x.
- on the other side we call all the other patterns $\mathcal{A}$-*incorrect*, and they will be interpreted as the persistent state of $\mathcal{A}$.

The idea behind the local rule of $\mathcal{U}$ is to make every $\mathcal{A}$-macrocell determine if it is surrounded by other $\mathcal{A}$-macrocells. If this is the case, then interaction is possible, and the current state of the neighbor will be taken into account to compute the new current state, following the rule of $\mathcal{A}$. Otherwise, there is no interaction, the $\mathcal{A}$-macrocell evolves considering every $\mathcal{A}$-incorrect neighborhood as a persistent state neighbor. The difficulty is that although correctness is related to the particular CA being simulated, every configuration must evolve correctly for every possible CA.

The proof of universality uses the combination of two key properties: on one hand, correct patterns remain correct and evolve according to the rule being simulated, even if not surrounded by correct patterns (lemma 4); on the other hand, incorrect patterns are interpreted as the persistent state and never become correct (lemma 5).

To make the construction of $\mathcal{U}$ readable, we describe its state set as a superposition of several layers: the main layer M contains most of the information about the simulation and the macrocells informations; *signals* layers are used to manage the evolution of the main layer; and *clock* layers guarantee synchronizations.

Correct Macrocells Description. In the following we consider a simulated CA $\mathcal{A} \in \mathbb{P}_0$ with radius 1 and state set $Q_{\mathcal{A}}$ of size $n = 2^p$. We use a canonical binary enumeration of the state set, in which the first word $(0^{log(n)})$ represents a persistent state of $\mathcal{A}$, denoted $p_{\mathcal{A}}$. Our $\mathcal{A}$-correct macrocells will be words of length $l_{\mathcal{A}}$ (specified above) whose main layer follow the pattern

$$\#\ C_i\ |Transition\ table|\ |State|\ |memory|\ \#$$

- $\#$ are delimiters, they never appear or disappear during the computation
- C_i is the *control state* used to control the successive steps of computation.
- $|Transition\ table|$ is the binary description of the transition table of $\mathcal{A}$.
- $|State|$ contains the binary value of the current state of the macrocell.
- $|memory|$ is a binary area which will be used to keep the values of the neighbors' current states before computing the new current state of the macrocell.

$|Transition\ table|$, $|State|$ and $|memory|$ are encoded with disjoint binary sub-alphabets. A cell whose state belongs to one of those sub-alphabets will never change sub-alphabet. Moreover, the states of the transition table's cells are never modified.

The current state description is $log(n)$-bits-long. In the transition table, images are ordered canonically, so the length of the description is simply $n^3 log(n)$. The memory should be at least $2log(n)$-bits-long in order to contain current state values of the two neighbors. But in order to simplify some proofs, we chose $l_{\mathcal{A}}$ such that it is at least half the total size of the macrocell, and such that the function $\mathcal{A} \to l_{\mathcal{A}}$ is one-to-one.

Most of the computation will happen on those very constrained patterns. In the next definition, we add an extra constraint on the control state to obtain $\mathcal{A}$-correct macrocells.

By stability of the sub-alphabets, such a correct $\mathcal{A}$-macrocell remains correct, but the state value may change. This is why we take some care when defining the current state value associated to the macrocell.

Definition 5. *A word $u \in {Q_{\mathcal{U}}}^{l_{\mathcal{A}}}$ of length $l_{\mathcal{A}}$ is said to be* a $\mathcal{A}$-correct macrocells, *denoted by $u \in \mathcal{C}_{\mathcal{A}}$, if its main layer follows the structure defined above (correct sub-alphabets for each cell, and correct transition table of $\mathcal{A}$), and if its control state is in C_0.*

For each such $\mathcal{A}$-correct macrocell u, we define its associated state value *$v(u) \in Q_{\mathcal{A}}$, which is the state described by its current state value after $l_{\mathcal{A}}$ steps of computation by $\mathcal{U}$. And this state value $v(u)$ only depends on u.*

To ensure that the sate value only depends on u, the memory area is used as a buffer to prevent modification of the current state value coming from the left, before the computation has been initialized. Details will be given in the following.

By extension we may sometime call $\mathcal{A}$-macrocells words following the general pattern, even with non-C_0 control state, in particular when they are images of a $\mathcal{A}$-correct macrocell.

The Local Rule. We describe the local behavior of $\mathcal{U}$ starting from a correct $\mathcal{A}$-macrocell. The local rule will first determine which neighbors it may interact with (*Check of the neighbor's length and synchronization*, and *Transition table and state encoding check*), and then compute its new current state according to the rule of $\mathcal{A}$ and eventually the value of those neighbors (*New current state computation*).

In order to guarantee the synchronization between $\mathcal{A}$-macrocells, we specify the duration of each step, and even of some sub-steps. It is done by a clock, which use specific layers of the states; their existence is proved by the following lemma:

Lemma 1. *For any $k, h \in \mathbb{N} \setminus \{1\}$, there exist a CA, and two states q_s, and q_f such that the leftmost cell of an area delimited by two # separated by $l-2$ cells turns to state q_f at some time $t > k \cdot l^2 + h \cdot l$ iff this cell was in state q_s exactly $k \cdot l^2 + h \cdot l$ steps before. Moreover, this property is guaranteed independently of what is outside the two #.*

At the beginning of each step, the control state C_i will turn to C_{i+1}, initiate the corresponding clock, and initiate some signals which will manage the evolution. Those signals are distinct states propagating on upper layers of the configuration, and interacting with the main layer and other signals. We say that a signal *belongs* to a macrocell if it was generated in this macrocell's area, between the two #. And, thanks to our evolution rule, a signal always knows if it is in its cell or in the area to the right or left of its cell. It is also useful sometimes to make signals carry one extra bit of information. It is simple to do it using distinct states, since the number of bits is bounded.

Check of the neighbor's length and synchronization. $C_0 \to C_1$ (that is to say that when the control cell's state is C_0 it becomes C_1):

Recall that we are interested to the behavior in the case of an $\mathcal{A}$-correct macrocell. When C_0 becomes C_1, it initializes two *control bits* with value 0, in the main layer of the control cell, and it launches signals. Since the construction is classical, we simply illustrate the desired behaviors by figure 1. Those two pictures illustrate the signal machinery in the case of respectively left and right neighbors of same length and with state C_0 appearing simultaneously (what we call synchronized). Every transition whose image is one of the signal involved in this checking appears on those pictures.

The first signals s_1 and s_4 erase all signals belonging to our macrocell. Together with the length of the memory being bigger than the length between the control state where signals are generated and the end of the current state area, it constitutes the protection of the current state value from eratic signals coming from outside the macrocell, and justifies that in definition 5 $v(u)$ is a function of u.

If the neighbors have same length and are synchronized, this whole step takes 4 times the length of the macrocell, $l_{\mathcal{A}}$. After $4 \cdot l_{\mathcal{A}}$ steps, the control state C_1 becomes C_2, and if it did not receive a positive result from one side, it concludes that the involved neighbor is incorrect. This is managed using a clock signal (with $h = 2$ and $k = 0$ in lemma 1) initialized by C_0 on a specific layer. When q_f is raised on this layer, C_1 becomes C_2. The important point is that we ensure the following property.

Lemma 2. *The control state of an $\mathcal{A}$-correct macrocell becomes C_2 exactly $4 \cdot l_{\mathcal{A}}$ steps after C_0 appeared. At this step each control bit has turned to 1 iff the corresponding neighboring macrocell has same length and synchronisation than the considered macrocell.*

The proof of this lemma is direct for the length but asks to enter into some more (simple but fastidious) details for the synchronization part.

Transition table and state encoding check. $C_2 \to C_3$:

In this step, for each neighboring pattern with same length and synchronization, the macrocell checks whether the transition table and the current state are compatible with its own (same lengths, and same content for the transition table) or not.

When C_2 appears it launches the following test for each neighbor whose corresponding control bit was 1, and initializes two fresh bits to 0. First, a signal is generated and puts a *mark* (that is to say a non-moving signal) on the first cell of the transition table of its macrocell, and another mark on the first cell of the transition table of the neighbor it checks. Then signals are exchanged between

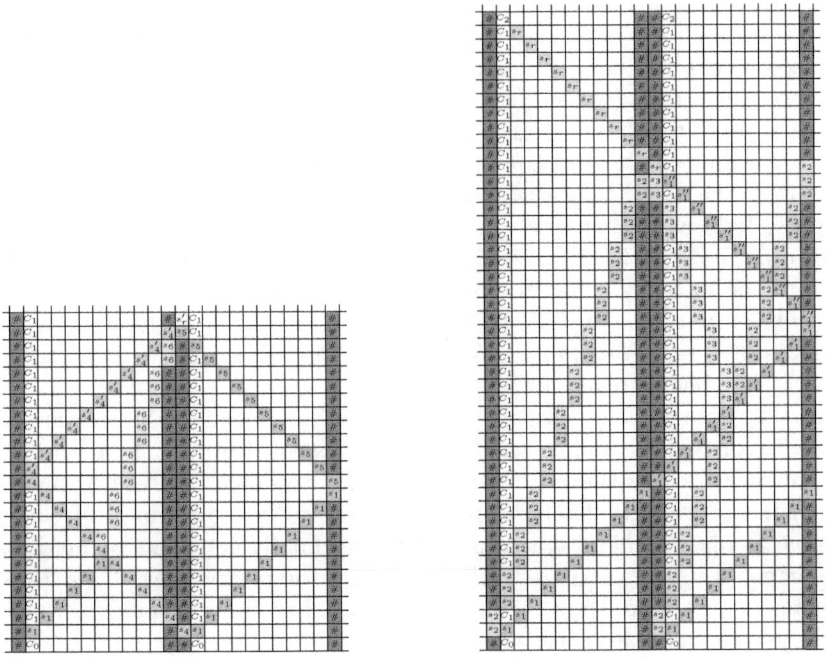

Fig. 1. Successful left (resp. right) neighbor test by the right (resp. left) macrocell (mix of main and signal layers for easier reading)

those two marks that will each time carry the binary state of the cell pointed by one mark to the next unchecked cell of the other macrocell; it compares this cell's binary state to the carried binary state, and push the mark by two cells. If no difference is detected and if both marks reach the end of the transition tables simultaneously, a correctness signal is sent to the control state.

After the transition table has been checked, the same mechanism is used to check that the current state encoding areas have same length. At the end of those tests, the results are sent to the control cell which again keeps the information on two control bits. For each cell of the transition table or the current state, checking takes $2 \cdot l_{\mathcal{A}}$ steps. So checking a whole neighbor takes less than $2 \cdot l_{\mathcal{A}}^2$. Again, a clock is used to make this test last exactly $2 \cdot l_{\mathcal{A}}^2$ steps. Then the control cell is turned to C_4.

Lemma 3. *The control state of our macrocell becomes C_4 exactly $2 \cdot l_{\mathcal{A}}^2$ steps after C_2 appeared. At this step each control bit has turned to* 1 *iff the corresponding neighboring pattern has same length, synchronization, and if the lengths of the transition tables and current states, and the content of the transition table are equal. In this case we say that this pattern is* compatible *with our $\mathcal{A}$-macrocell.*

The proof of this lemma is straightforward. Keep in mind that some signals erased all erratic signals that could interact with our cell at a previous step.

New current state computation. $C_4 \rightarrow C_5$:

After all the tests have been done, the new current state has to be computed. We need to explicit how we consider the neighboring pattern. In the following, what we call *detected state* of one such pattern by our macrocell will be: either the persistent state if the neighbor is non-compatible with our $\mathcal{A}$-macrocell, or its current state if this is a compatible $\mathcal{A}$-macrocell.

At first, the detected states of the left and right neighbors are written to the memory. It is written in the binary memory alphabet. Each detected state is written on $log(n)$ cells. If one neighbor is compatible, we copy its current state value to the memory using marks and signals similarly to the previous step. If it is not compatible, we write $0^{log(n)}$, the length being the same as that of the current state area. We add a clock to specify that copies last exactly $2 \cdot l_{\mathcal{A}}^2$ steps, the neighbor being correct or not.

After $2 \cdot l_{\mathcal{A}}^2$ additional steps, the search for the image in the transition table starts. It consists in reading the binary word formed by the three image states (the current state of the cell followed by the two detected states copied in the memory), and turning it into a unary position in the transition table. It is then possible to place a mark at this position, and finally copy this pointed state to the current state area. We make the reading of the position last $4 \cdot l_{\mathcal{A}}^2$ steps. And copying the new state lasts $2 \cdot l_{\mathcal{A}}^2$. After the whole computation step, which lasts $7 \cdot l_{\mathcal{A}}^2$, the control state turns to C_5.

Finally one step of simulation is completed after exactly $\tau_{\mathcal{A}} = 9 \cdot l_{\mathcal{A}}^2 + 4 \cdot l_{\mathcal{A}}$ steps. After this time the control state turns to C_5.

To become C_0 again, and launch a new step of computation, we add another condition. We ask a clock launched exactly $\tau_{\mathcal{A}}$ steps before to raise a flag. And obviously this clock may only be launched by C_0. It is realized using again signals of the lemma 1 computing on one more layer.

The state set of the universal CA is given by $Q_{\mathcal{U}} = M \times S \times C \cup \{C_f\}$ with

- the main layer : $M = \{C_0, C_5\} \cup \{C_i\}_{i \in \{1,..,4\}} \times \{0,1\}^2 \cup \{0_i, 1_i\}_{i \in \{tt,cs,m\}}$
- the signals layers : $S = \times_{i \in I}\{s_i\} \times \times_{j \in J}(\{s_j\} \times \{0,1\})$
- the clocks layers (see lemma 1), one for each duration needed. $C = (\{0,1\} \times \{s_i\}_{i \in I_c})^4$
- C_f is a single persistent state ensuring that $\mathcal{U} \in \mathbb{P}$

Yet, the transition rule of $\mathcal{U}$ is partially specified, we call correct transitions those defined up to now, in the case of correct macrocells. But the other transitions may not be chosen arbitrarily. We specify the following behaviors:

- C_f is never modified by any transition
- the main layer is never modified by a non-correct transition: they act as the identity on the main layer.
- concerning the signal layer, apart from the collisions corresponding to the behavior described in the previous steps, all signals may cross each other (each kind of signal is evolving on its own layer). However, except for transitions involved in the behavior described above, any signal that crosses a # is destroyed.

Interpretation. We now describe the continuous onto map $\phi_{\mathcal{A}} : Q_{\mathcal{U}}^{\mathbb{Z}} \to Q_{\mathcal{A}}^{\mathbb{Z}}$ associated to $\mathcal{A}$. This map is induced by a local map $f_{\mathcal{A}}$ from patterns of shape $l_{\mathcal{A}}$ to individual states of $\mathcal{A}$. More precisely, using notation from proposition 1, we have $r = 0$, $z_1 = l_{\mathcal{A}}$, $z_2 = 1$, $t_2 = 1$ and $t_1 = \tau_{\mathcal{A}}$.

If $p_{\mathcal{A}}$ is the persistent state of $\mathcal{A}$, the local map $\psi_{\mathcal{A}}$ is defined as follows:

1. $\forall u \notin \mathcal{C}_{\mathcal{A}}, f_{\mathcal{A}}(u) = p_{\mathcal{A}}$
2. $\forall u \in \mathcal{C}_{\mathcal{A}}, f_{\mathcal{A}}(u) = v(u)$, with $v(u)$ the value from definition 5

Proof of Theorem 2. The proof of the theorem relies on the two following lemmas. They are consequences of the construction, the intermediate lemmas and the clock lemma.

Lemma 4. *$\forall\ c \in Q_{\mathcal{U}}^{\mathbb{Z}}$, $t_0 \in \mathbb{N}$, if $\mathcal{U}^{t_0}(c)_{[0,l_{\mathcal{A}}-1]} \in \mathcal{C}_{\mathcal{A}}$, then $v = \mathcal{U}^{t_0+\tau_{\mathcal{A}}}(c)_{[0,l_{\mathcal{A}}-1]} \in \mathcal{C}_{\mathcal{A}}$, and $\psi_{\mathcal{A}}(v) = \delta_{\mathcal{A}}(\psi_{\mathcal{A}}(c_{[-l_{\mathcal{A}},-1]}), \psi_{\mathcal{A}}(c_{[0,l_{\mathcal{A}}-1]}), \psi_{\mathcal{A}}(c_{[l_{\mathcal{A}},2\cdot l_{\mathcal{A}}-1]}))$.*

Lemma 5. *If $\exists\ t \geq \tau_{\mathcal{A}}$, $c \in Q_{\mathcal{U}}^{\mathbb{Z}}$ such that $u = \mathcal{U}^t(c)_{[0,l_{\mathcal{A}}-1]} \in \mathcal{C}_{\mathcal{A}}$ then $v = \mathcal{U}^{t-\tau_{\mathcal{A}}}(c)_{[0,l_{\mathcal{A}}-1]} \in \mathcal{C}_{\mathcal{A}}$.*

We can finally prove our main claim: $\forall \mathcal{A} \in \mathbb{P}_0$, $\mathcal{A} \preccurlyeq \mathcal{U}$. We use the characterization of proposition 1. Let $\mathcal{A} \in \mathbb{P}_0$ the associated length $l_{\mathcal{A}}$ and function $\phi_{\mathcal{A}}$ are defined as explained before. First, $\phi_{\mathcal{A}}$ is local (by definition) and onto, because correct macrocells are enough to encode any state of $\mathcal{A}$ and thus concatenations of correct macrocells allow to encode any configuration of $\mathcal{A}$. Second, we have $\phi_{\mathcal{A}} \circ \mathcal{U}^{\tau_{\mathcal{A}}} = \mathcal{A} \circ \phi_{\mathcal{A}}$. To see this we discuss on the pattern of shape $\mathcal{R}_{l_{\mathcal{A}}}$ at position 0 and the rest follows by translation. If this pattern is not in $\mathcal{C}_{\mathcal{A}}$ its image after $\tau_{\mathcal{A}}$ steps remains out of $\mathcal{C}_{\mathcal{A}}$ (lemma 5). If conversely this central word belongs to $\mathcal{C}_{\mathcal{A}}$, lemma 4 gives the desired property.

4 Perspectives

A natural extension of our work could be to generalize the construction to cellular automata having an equicontinuous point. The idea would be to use blocking words as a replacement for the persistent state. But it seems much harder, if not impossible.

Besides, the main open question left by this paper is the existence of universal CA. We conjecture that they do not exist and more precisely that no CA can simulate all products of shifts. A possible way to obtain this negative result would be to study limit sets: by a compacity argument, one can show that a universal CA must have a universal limit set. The main obstacle is that subshifts that are limit sets of CA are not well characterized.

Finally, we also leave open the existence of universal SFT and universal surjective CA in dimension 1.

References

1. Berger, R.: The undecidability of the domino problem. Mem. Amer. Math Soc. 66 (1966)
2. Boyer, L., Theyssier, G.: On local symmetries and universality in cellular automata. In: STACS, pp. 195–206 (2009)
3. Cervelle, J., Formenti, E., Guillon, P.: Ultimate traces of cellular automata. In: STACS, pp. 155–166 (2010)
4. Delorme, M., Mazoyer, J., Ollinger, N., Theyssier, G.: Bulking ii: Classifications of cellular automata. CoRR, abs/1001.5471 (2010)
5. Doty, D., Lutz, J.H., Patitz, M.J., Summers, S.M., Woods, D.: Intrinsic universality in self-assembly. In: STACS, pp. 275–286 (2010)
6. Durand-Lose, J.O.: Intrinsic universality of a 1-dimensional reversible cellular automaton. In: STACS, pp. 439–450 (1997)
7. Hedlund, G.A.: Endomorphisms and Automorphisms of the Shift Dynamical Systems. Mathematical Systems Theory 3(4), 320–375 (1969)
8. Hochman, M.: A note on universality in multidimensional symbolic dynamics. Discrete Contin. Dyn. Syst. Ser. S 2(2), 301–314 (2009)
9. Hochman, M.: On the dynamics and recursive properties of multidimensional symbolic systems. Inventiones Mathematicae 176(1), 131–167 (2009)
10. Kari, J.: The Nilpotency Problem of One-dimensional Cellular Automata. SIAM Journal on Computing 21, 571–586 (1992)
11. Kůrka, P.: Languages, equicontinuity and attractors in cellular automata. Ergodic Theory and Dynamical Systems 17, 417–433 (1997)
12. Kůrka, P.: Topological and symbolic dynamics. Société Mathématique de France (2003)
13. Kůrka, P.: Zero-dimensional dynamical systems, formal languages, and universality. Theory Comput. Syst. 32(4), 423–433 (1999)
14. Lafitte, G., Weiss, M.: An almost totally universal tile set. In: TAMC, pp. 271–280 (2009)
15. Moreira, A.: Universality and decidability of number-conserving cellular automata. Theor. Comput. Sci. 292(3), 711–721 (2003)
16. Nasu, M.: The dynamics of expansive invertible onesided cellular automata. Trans. Amer. Math. Soc. 354, 4067–4084 (2002)
17. Von Neumann, J.: Theory of Self-Reproducing Automata. University of Illinois Press, Urbana (1966)
18. Ollinger, N.: Automates Cellulaires: structures. PhD thesis, École Normale Supérieure de Lyon (décembre 2002)
19. Ollinger, N.: The quest for small universal cellular automata. In: Widmayer, P., Triguero, F., Morales, R., Hennessy, M., Eidenbenz, S., Conejo, R. (eds.) ICALP 2002. LNCS, vol. 2380, pp. 318–330. Springer, Heidelberg (2002)
20. Ollinger, N.: The intrinsic universality problem of one-dimensional cellular automata. In: Alt, H., Habib, M. (eds.) STACS 2003. LNCS, vol. 2607, pp. 632–641. Springer, Heidelberg (2003)
21. Sablik, M.: Directional dynamics for cellular automata: A sensitivity to initial condition approach. Theor. Comput. Sci. 400(1-3), 1–18 (2008)
22. Theyssier, G.: Automates Cellulaires: un modèle de complexités. PhD thesis, École Normale Supérieure de Lyon (Décembre 2005)
23. Wang, H.: Proving theorems by pattern recognition ii. Bell System Tech. Journal 40(2) (1961)

Toward a Deterministic Polynomial Time Algorithm with Optimal Additive Query Complexity

Nader H. Bshouty and Hanna Mazzawi

Technion - Israel Institute of Technology
{bshouty,hanna}@cs.technion.ac.il

Abstract. In this paper, we study two combinatorial search problems: The coin weighing problem with a spring scale (also known as the vector reconstructing problem using additive queries) and the problem of reconstructing weighted graphs using additive queries. Suppose we are given n identical looking coins. Suppose that m out of the n coins are counterfeit and the rest are authentic. Assume that we are allowed to weigh subsets of coins with a spring scale. It is known that the optimal number of weighing for identifying the counterfeit coins and their weights is at least

$$\Omega\left(\frac{m\log n}{\log m}\right).$$

We give a deterministic polynomial time adaptive algorithm for identifying the counterfeit coins and their weights using

$$O\left(\frac{m\log n}{\log m}+m\log\log m\right)$$

weighings, assuming that the weight of the counterfeit coins are greater than the weight of the authentic coin. This algorithm is optimal when $m \le n^{c/\log\log n}$, where c is any constant. Also our weighing complexity is within $\log\log m$ times the optimal complexity for all m.

To obtain this result, our algorithm makes use of search matrices, the divide and conquer approach and the guess and check approach.

When combining these methods with the technique introduced in [Optimally Reconstructing Weighted Graphs Using Queries. *SODA*, 2010], we get a similar positive result for the problem of reconstructing a hidden weighted graph using additive queries.

1 Introduction

In this paper, we study two well known combinatorial search problems: The coin weighing problem with a spring scale (also called the problem of reconstructing vectors using additive queries) and the problem of reconstructing weighted graphs using additive queries. These two problems have known some progress lately [17,15,27,12,8,7,9,23]. We start with the coin weighing problem.

P. Hliněný and A. Kučera (Eds.): MFCS 2010, LNCS 6281, pp. 221–232, 2010.

1.1 Coin Weighing Problem

The coin weighing problem can be formally introduced as follows: Let $v \in (\mathbb{R}_0^+)^n$ be a hidden vector, where $\mathbb{R}_0^+$ denotes the set of non-negative real numbers. Assume that v contains at most m non-zero entries and that we are allowed to ask queries of the form

$$Q_v(s) = s^T v,$$

where s is a $(0,1)$-vector. The goal is to exactly reconstruct the hidden vector using minimal number of queries.

The coin weighing problem was first studied in the restricted case where the hidden vector is a $(0,1)$-vector and $m = n$. This case was studied in [10] by Cantor, in [29] by Soderberg and Shapiro, in [14] by Erdös and Rényi, in [18,19,20,21] by Lindström and in [11] by Cantor and Mills. For this problem, Erdös and Rényi proved a lower bound of $2n/\log n$ queries. See also [19,24,21,25,22]. Later, Lindström [18] and independently Cantor and Mills [11] showed a non-adaptive polynomial time algorithm for the problem with query complexity that matches the lower bound[1]. Simplifications for Lindström's algorithm were introduced in [21,1,7].

For the case where the hidden vector is a $(0,1)$-vector with no restriction on m, the information theoretic lower bound for reconstructing a hidden vector is

$$\Omega\left(\frac{m\log\frac{n}{m}}{\log m}\right).$$

In [15] Grebinski and Kucherov showed a non-constructive algorithm for the problem that matches this lower bound for all m (that is, they proved the existence of an algorithm with query complexity that matches the information theoretic lower bound without showing an explicit construction).

Several papers in the literature generalize the former results by showing algorithms for reconstructing non-boolean vectors with bounded sum (L_1 norm). These papers study the problem of reconstructing a hidden vector with entries that are non-negative *integers* such that the sum of the entries is bounded by k. In [26] Pippenger gave a non-constructive algorithm in case $k = n$. This algorithm asks $O(n/\log n)$ queries. In [17] Grebinski extended this result by showing an optimal non-constructive algorithm for $n^2 \geq k \geq n$. In [28] Ruszinkó and Vanroose gave the first polynomial time adaptive algorithm for the case of $k = n$. This algorithm asks $O((n\log\log n)/\log n)$ queries. The algorithm runs in polynomial time, however, its query complexity is not optimal.

For many years, the only algorithms reaching the information theoretic lower bound for all the above mentioned problems were non-constructive (except for the case where the hidden vector is a $(0,1)$-vector and $m = n$). It was only until recently that Bshouty [7] gave the first optimal polynomial time adaptive algorithm for reconstructing a hidden $(0,1)$-vector. The algorithm Bshouty gave

[1] Lindström's algorithm works for the more general case of reconstructing a hidden vector from $[d]^n$. In this case, it uses $O(n/\log_d n)$ queries.

works when reconstructing an integer hidden vector with bounded sum k. For $k \le n$, this algorithm asks

$$O\left(\frac{k \log \frac{n}{k}}{\log k}\right)$$

queries. Since in the $(0,1)$ case we have that the sum of entries k equals the number of non-zero entries m, the algorithm matches the information theoretic lower bound when reconstructing a $(0,1)$-vector for all range of m. This is not true for non-boolean vectors since the query complexity is not optimal when $k = \omega(m)$.

Finally, for the general case (where the hidden vector is an n-vector with at most m non-zeros entries that are unbounded positive real numbers), the information theoretic lower bound is

$$\Omega\left(\frac{m \log n}{\log m}\right).$$

In [12] Choi and Han Kim showed an optimal non-constructive algorithm for reconstructing a hidden vector with at most m non-zero entries, where each non-zero entry holds a real number with magnitude bounded by n^{-a} and n^b for any constants a and b. Later, in [9], Bshouty and Mazzawi, extended this result by showing an optimal non-constructive algorithm for reconstructing any hidden vector with at most m non-zero unbounded *real* entries.

Below is a table that summarize all known results.

Table 1. Known results for the coin weighing problem. Here $\mathbb{Z}_0^+$ denotes the set of non-negative integers.

Type	Hidden vector	Restriction	Lower bound	Upper bound	Polynomial time
Non Adaptive	$\{0,1\}^n$	$m = n$	$\Omega\left(\frac{n}{\log n}\right)$	tight [18,11]	tight [18,11]
		–	$\Omega\left(\frac{m \log \frac{n}{m}}{\log m}\right)$	tight [15]	OPEN
	$(\mathbb{Z}_0^+)^n$	–	$\Omega\left(\frac{m \log n}{\log m}\right)$	tight [9]	OPEN
	$(\mathbb{R}_0^+)^n$	$\forall i : v_i \in [n^{-a}, n^b]$	$\Omega\left(\frac{m \log n}{\log m}\right)$	tight [12]	OPEN
		–	$\Omega\left(\frac{m \log n}{\log m}\right)$	tight [9]	OPEN
Adaptive	$\{0,1\}^n$	–	$\Omega\left(\frac{m \log \frac{n}{m}}{\log m}\right)$	tight [15]	tight [7]
	$(\mathbb{Z}_0^+)^n$	$\sum_i v_i = O(m)$	$\Omega\left(\frac{m \log \frac{n}{m}}{\log m}\right)$	tight [17]	tight [7]
		–	$\Omega\left(\frac{m \log n}{\log m}\right)$	tight [9]	Ours
	$(\mathbb{R}_0^+)^n$	–	$\Omega\left(\frac{m \log n}{\log m}\right)$	tight [9]	Ours

For the problem of reconstructing a vector with real number entries, all algorithms that match the information theoretic lower bound are non-constructive and therefore proving an upper bound only. They prove that there exists a set of queries such that the answer to the queries uniquely identify the hidden vector; However, it is unknown how to deterministically construct this set of queries in polynomial time, and moreover, given the answer to such set of queries, it is unknown how to reconstruct the hidden vector in polynomial time. The best polynomial time algorithm for the problem is the trivial recursive search which asks $O(m \log \frac{n}{m})$ queries.

In this paper, we show the following: Let $v \in (\mathbb{R}_0^+)^n$ be a hidden vector that contains at most m non-zero real entries. We give a polynomial time adaptive algorithm for reconstructing v using

$$O\left(\frac{m \log n}{\log m} + m \log \log m\right)$$

additive queries.

Our algorithm is iterative. It uses the divide and conquer approach. The algorithm holds disjoint sets of entries of the hidden vector v. Each such set holds at least one non-zero entry. At the beginning of each iteration, the algorithm knows the sum of the entries in each set. It divides each set into two equal size set (up to ± 1) and its goal is to find the sum of the entries for each new set. The algorithm uses search matrices to obtain this goal. When using the search matrices, the algorithm makes some assumptions on the input of the subproblem. These assumptions are not always true, and therefore, before advancing to the next iteration, the algorithm has an additional stage that checks its outcome and corrects it if needed.

This is the first polynomial time algorithm that beats the trivial $O(m \log \frac{n}{m})$ recursive search. Our algorithm matches the information theoretic lower bound when $m \le n^{c/\log\log n}$, where c is any constant. Also, our query complexity is within $\log \log m$ times the optimal complexity for all m.

1.2 Reconstructing Weighted Graphs Using Additive Queries

The weighted graph reconstructing problem is the following: Let $G = (V, E, w)$ be a weighted hidden graph, where $E \subseteq V \times V$ and $w \in (\mathbb{R}^+)^E$ (here $\mathbb{R}^+$ denotes the set of positive real numbers). Suppose that the set of vertices is known. The goal is to exactly reconstruct the set of edges and their weights using additive queries of the form

$$Q_G(S) = \sum_{e \in (S \times S) \cap E} w(e)$$

where $S \subseteq V$. See applications in [12].

For the unweighted problem (in this case the answer to the query with the set of vertices S simply returns the number of edges in the subgraph induced by S), Grebinski and Kucherov, [15], showed an optimal non-constructive algorithm for reconstructing a d-bounded degree graph. They also showed $O(|V|^2/\log |V|)$ polynomial algorithm for reconstructing an arbitrary graph. Later, Grebinski [17] extended the result by showing a non-constructive optimal algorithm for reconstructing a d-degenerate graph. Finally, Choi and Han Kim [12], extended the later result by showing a non-constructive optimal algorithm for arbitrary graphs.

A recent result by Mazzawi [23] gave the first polynomial time adaptive algorithm that is optimal, that is, matches the information theoretic lower bound for the query complexity.

As for the weighted case where the weights are real numbers, the information theoretic lower bound for reconstructing a hidden weighted graph is

$$O\left(\frac{|E|\log|V|}{\log|E|}\right).$$

Choi and Han Kim [12], showed a non-constructive optimal algorithm for reconstructing hidden graphs with many edges ($|E|$ must be at least $\log^{\alpha}|V|$ for a sufficiently large α) and weights that are bounded by n^{-a} and n^{b}, where a and b are any constants. Later, the condition on the number of edges was removed in [8], and the condition on the weights was removed in [9].

As in the coin weighing problem, also in this problem, when dealing with the case of real number weights, the only optimal algorithms are non-constructive.

When combining our new algorithms for reconstructing hidden vectors presenting in this paper, with the technique introduced in [23], we obtain a new positive result for the problem of reconstructing a weighted (real numbers weights) hidden graph using additive queries. We show the following: Let $G = (V, E, w)$ be any hidden graph, where the weights on the edges are arbitrary positive real numbers. There is a polynomial time algorithm that reconstruct the hidden graph in

$$O\left(\frac{|E|\log|V|}{\log|E|} + |E|\log\log|E|\right)$$

queries. The query complexity matches the information theoretic lower bound when $|E| \le |V|^{c/\log\log|V|}$, where c is any constant. This is the first algorithm that beats the trivial $O\left(|E|\log\frac{|V|^2}{|E|}\right)$ recursive search algorithm.

2 Preliminaries

2.1 Notation

Let $\mathbb{R}$ denote the set of real numbers. We denote by $\mathbb{R}^+$ the set of positive real numbers and by $\mathbb{R}_0^+$ the set $\mathbb{R}^+ \cup \{0\}$. Let r be a positive integer, we denote by $[r]$ the set $\{1, 2, \ldots, r\}$. Let S_i be a set of real numbers where $i \in [t]$, we denote by $\prod_{i=1}^{t} S_i$ the Cartesian product of the sets, that is, the set of t-vectors $\{(w_1, w_2, \ldots, w_t) \mid \forall i : w_i \in S_i\}$.

For a t-vector w, we denote by w_i the ith entry of w. We denote by $|w|$ the t-vector where $|w|_i = |w_i|$ for all $i \in [t]$. For $w \in \{-1, 1\}^t$ we denote by $||w||$ the number of ones in w. For two vectors $a, b \in \{-1, 1\}^t$, we write $a \le b$ if for every $i \in [t]$ we have $a_i \le b_i$. We write $a < b$ if $a \le b$ and $a \ne b$.

2.2 Fourier Representation

In this subsection, we present the well known Fourier representation of functions.

Let $x = (x_1, x_2, \ldots, x_\ell)$ be variables. Define the following basis

$$B = \left\{ \chi_a(x) \triangleq \prod_{i:a_i=1} x_i \,\middle|\, a \in \{-1,1\}^\ell \right\},$$

where $\chi_a : \{-1,1\}^\ell \to \{-1,1\}$. It is known that every function $f : \{-1,1\}^\ell \to \mathbb{R}$ has a unique representation of the form

$$f(x) = \sum_{a\in\{-1,1\}^\ell} \hat{f}(a)\chi_a(x),$$

where for every $a \in \{-1,1\}^\ell$, $\hat{f}(a)$ is a real number and is called the Fourier coefficient of χ_a in f.

3 Search Matrix

In this section we show,

Theorem 1. *Let $y \in (\mathbb{R}^+)^t$ be any vector such that $y_1 \geq y_2 \geq \ldots \geq y_t$. There exists a polynomial time non-adaptive algorithm for reconstructing a hidden vector $v \in \prod_{i=1}^t \{0, y_i\}$ that uses $O(t/\log t)$ queries of the form: $Q(w) = w^T v$ where $w \in \{0,1\}^t$.*

Before proving this theorem, note that this problem is equivalent to finding in polynomial time a (0,1)-matrix M_t of size $O(t/\log t) \times t$ such that, $M_t v \neq M_t u$ for all $u, v \in \prod_{i=1}^t \{0, y_i\}$ where $u \neq v$ and given $M_t v$ one can reconstruct v in polynomial time. Such matrix M_t is called a *search matrix*.

We now show how to construct a search matrix for the problem. The matrix we present was first introduced in [7]. Let $a \in \{-1,1\}^\ell$. Let $j_1, j_2, \ldots, j_{||a||}$ be the indices of the entries in a that are equal to one. For $k \in [||a||]$, define the following function

$$g_{a,k}(x) = \left(2\prod_{i=1}^{k} \frac{x_{j_i}+1}{2} - 1 \right) x_{j_{k+1}} x_{j_{k+2}} \cdots x_{j_{||a||}}.$$

Now, let $f_{a,k}(x) = (g_{a,k}(x)+1)/2$. Define the following family of functions

$$F_\ell = \{ f_{a,k} | a \in \{-1,1\}^\ell \text{ and } k \in [||a||] \}.$$

Finally, define a partial order over F_ℓ in the following way: $f_{a,k_1} \leq f_{b,k_2}$ if and only if $a < b$ or ($a = b$ and $k_1 \geq k_2$).

The following are properties of F_ℓ and its functions [7].

P1. $f_{a,k} \in \{0,1\}^{\{-1,1\}^\ell}$. **P2.** $|F_\ell| = \sum_i i\binom{\ell}{i} = \ell 2^{\ell-1}$.
P3. The Fourier coefficient of χ_a in $f_{a,k}$ is 2^{-k}.
P4. For any $a, b \in \{-1,1\}^\ell$, such that $a \neq b$ and $b \not> a$ the Fourier coefficient of χ_a in $f_{b,k}$ is equal to 0.

The following lemma can be derived immediately from **P3** and **P4**.

Lemma 1. *Let $a \in \{-1,1\}^\ell$ be vector. Let $I \subseteq [||a||]$ be any set. Let $A = \{(a,i)|i \in I\}$. Let $B = \{(b_1,i_1),\ldots,(b_{|B|},i_{|B|})\} \subseteq \{-1,1\}^\ell \times [\ell]$ such that for all j we have $i_j \in [||b_j||]$. If for all j we have $b_j \not\geq a$, then the Fourier coefficient of χ_a in*

$$f(x) = \sum_{(v,u)\in A\cup B} c_{u,v} f_{v,u}(x),$$

where $c_{u,v}$ are real numbers, is equal to $\sum_{(v,u)\in A} \frac{c_{v,u}}{2^u}$.

We now use F_ℓ to construct our search matrix. Define the matrix $M_t \in \{0,1\}^{2^\ell \times \ell 2^{\ell-1}}$ in the following way (we assume w.l.o.g. that $t = \ell 2^{\ell-1}$ for some integer ℓ. This is not always true. In such case, we take the minimal ℓ such that $\ell 2^{\ell-1} > t$ create the matrix and take the first t columns): First, we label the rows of M_t with the elements of $\{-1,1\}^\ell$. Next, we label the columns with element of F_ℓ in a descending order, that is, for every two indices i and j and their corresponding functions $f_{a,k}$ and $f_{b,s}$ (respectively), we have that if $i < j$ then $f_{b,s} \not\geq f_{a,k}$. Let $M_t[x, f_{a,k}] = f_{a,k}(x)$ be the search matrix that algorithm uses. That is, let $z_1 = 1^\ell$, $z_2 = 1^{\ell-1}\cdot -1$, $\ldots$, $z_{2^\ell} = -1^\ell$, where $\cdot$ denotes concatenation, then

$$M_t = \left(\begin{array}{ccc|ccc|c|c} f_{z_1,1}(z_1) & \ldots & f_{z_1,\ell}(z_1) & f_{z_2,1}(z_1) & \ldots & f_{z_2,\ell-1}(z_1) & \ldots & f_{z_{2^\ell-1},1}(z_1) \\ f_{z_1,1}(z_2) & \ldots & f_{z_1,\ell}(z_2) & f_{z_2,1}(z_2) & \ldots & f_{z_2,\ell-1}(z_2) & \ldots & f_{z_{2^\ell-1},1}(z_2) \\ \vdots & & \vdots & \vdots & & \vdots & & \vdots \\ f_{z_1,1}(z_{2^\ell}) & \ldots & f_{z_1,\ell}(z_{2^\ell}) & f_{z_2,1}(z_{2^\ell}) & \ldots & f_{z_2,\ell-1}(z_{2^\ell}) & \ldots & f_{z_{2^\ell-1},1}(z_{2^\ell}) \end{array}\right).$$

We will now give an algorithm showing that given $r = M_t v$ where $v \in \prod_i \{0, y_i\}$ one can exactly reconstruct v. Let $\sigma_t : [\ell 2^{\ell-1}] \to \{(a,k)|a \in \{-1,1\}^\ell$ and $k \in [||a||]\}$ defined as follows: $\sigma_t(j) = (a,k)$, where $f_{a,k}$ is the function corresponding to the jth column of the search matrix M_t.

Algorithm (Proof of Theorem 1). The algorithm **Reconstruct_Vector**, presented in Figure 1, is iterative. We run it for *start* $= 1$, and $r = M_t v$. The goal of the algorithm is to output a vector $h = v$. In each iteration the algorithm determines one entry of the hidden vector v. At iteration i, the entries $1, 2, \ldots, i-1$ are known to the algorithm (stored in $h_1, h_2, \ldots, h_{i-1}$), and the goal is to determine the ith entry. The vector r will be regarded as a function $r : \{-1,1\}^\ell \to \mathbb{R}$. Therefore,

$$r(x) = \sum_{j=1}^{t} v_j f_{\sigma_t(j)}(x).$$

The algorithm first calculates the function

$$f(x) = r(x) - \sum_{j=1}^{i-1} h_j f_{\sigma_t(j)}(x).$$

Let (a, k) be equal to $\sigma_t(i)$. The algorithm calculates the Fourier coefficient of χ_a in f in polynomial time. If $\hat{f}(a) < \frac{y_i}{2^k}$ then, v_i equals 0. Otherwise, we have that $\hat{f}(a) \geq \frac{y_i}{2^k}$ and then, v_i equals y_i.

We now prove the correctness of the algorithm by induction.

Lemma 2. *(Algorithm's Correctness) Assuming correctness of the reconstruction of the first $i-1$ indices, that is, given that $h_1 = v_1, \ldots, h_{i-1} = v_{i-1}$, then, we have that $h_i = v_i$.*

Proof. Since $h_j = v_j$ for all $j \in [i-1]$, then we have

$$f(x) = r(x) - \sum_{j=1}^{i-1} h_j f_{\sigma_t(j)}(x) = r(x) - \sum_{j=1}^{i-1} v_j f_{\sigma_t(j)}(x) = \sum_{j=i}^{t} v_j f_{\sigma_t(j)}(x).$$

Let s be the index such that $\sigma_t(s) = (a, ||a||)$. Since for every $j > s$ and $(b, k') = \sigma_t(j)$, we have that $b \not\geq a$, by Lemma 1, the coefficient of χ_a in f is

$$\hat{f}(a) = \frac{v_i}{2^k} + \frac{v_{i+1}}{2^{k+1}} + \cdots + \frac{v_s}{2^{||a||}}.$$

Now, note that if $v_i = y_i$ then,

$$\hat{f}(a) = \frac{v_i}{2^k} + \frac{v_{i+1}}{2^{k+1}} + \cdots + \frac{v_s}{2^{||a||}} \geq \frac{y_i}{2^k}$$

and therefore $h_i = y_i$. Otherwise, $v_i = 0$, and then

$$\begin{aligned}\hat{f}(a) &= \frac{v_i}{2^k} + \frac{v_{i+1}}{2^{k+1}} + \cdots + \frac{v_s}{2^{||a||}} = \frac{v_{i+1}}{2^{k+1}} + \frac{v_{i+2}}{2^{k+2}} + \cdots + \frac{v_s}{2^{||a||}} \\ &\leq \frac{y_{i+1}}{2^{k+1}} + \frac{y_{i+2}}{2^{k+2}} + \cdots + \frac{y_s}{2^{||a||}} \leq \frac{y_i}{2^{k+1}} + \frac{y_i}{2^{k+2}} + \cdots + \frac{y_i}{2^{||a||}} < \frac{y_i}{2^k},\end{aligned}$$

and therefore, $h_i = 0$. ∎

Corollary 1. *Let $y \in (\mathbb{R}^+)^t$ be any positive real numbers. Then, there exists a polynomial time non-adaptive algorithm for reconstructing a hidden vector $v \in \prod_{i=1}^{t}\{0, y_i\}$ that asks $O\left(\frac{t}{\log t}\right)$ queries.*

Proof. Let $P \in \{0,1\}^{t \times t}$ be a permutation matrix that sorts the vector y. Then, the matrix $M_t P$ is a search matrix for the problem. Given $M_t P v$, we can reconstruct Pv and therefore reconstruct v. ∎

Algorithm **Reconstruct_Vector(**$start$, t, r, h, y**)**

For $i = start$ **to** t **do**
  $f(x) \leftarrow r(x) - \sum_{j=1}^{i-1} h_j f_{\sigma_t(j)}(x)$.
  Let $(a, k) \leftarrow \sigma_t(i)$.
  if $\hat{f}(a) < \frac{y_i}{2^k}$ **then**
    $h_i = 0$.
  Else
    $h_i = y_i$.
  End if.
End for.
Return h.

Fig. 1. Algorithm for reconstructing a vector in $\prod_i \{0, y_i\}$. The algorithm starts reconstructing from entry v_{start}, assuming that $h_1 = v_1, \ldots, h_{start-1} = v_{start-1}$.

4 The Main Algorithm

In this section we prove our main result.

Theorem 2. *There is a polynomial time algorithm for reconstructing a hidden vector $v \in (\mathbb{R}_0^+)^n$ with at most m non-zero entries that uses*

$$O\left(\frac{m \log n}{\log m} + m \log \log m\right)$$

queries.

Proof. The algorithm we present is iterative. It searches for non-zero entries using the divide and conquer approach. At iteration τ, the algorithm holds at most m disjoint sets $S_1, S_2, \ldots, S_q \subseteq [n]$ of indices. Each set S_i contains at least one index j for which $v_j \neq 0$. For a set $S \subseteq [n]$ we denote by $X(S)$ the sum

$$X(S) = \sum_{j \in S} v_j.$$

At the beginning of the iteration τ the algorithm knows $X(S_i)$ for all $i \in [q]$. As before, we may assume that for every $i \in [q-1]$ we have that $X(S_i) \geq X(S_{i+1})$. The algorithm divides each set S_i into two arbitrary equal size (up to ± 1) sets $S_{i,1}$ and $S_{i,2}$. Obviously, $X(S_i) = X(S_{i,1}) + X(S_{i,2})$. Now, the algorithm's goal is to find $X(S_{i,j})$ for all $i \in [q]$ and $j \in [2]$. Instead of asking q queries to achieve this, the algorithm uses the search matrix presented in the previous section in order to reconstruct the hidden vector,

$$u = (X(S_{1,1}), X(S_{1,2}), X(S_{2,1}), X(S_{2,2}), \ldots, X(S_{q,1}), X(S_{q,2})),$$

assuming that $u \in \prod_{i=1}^{2q}\{0, X(S_{\lceil i/2\rceil})\}$. The algorithm simulates queries for u using the fact that

$$Q_u(x) = Q_v(y), \text{ where } y_k = \begin{cases} 1 & \exists i \in [q], \exists j \in [2] : k \in S_{i,j} \\ & \text{and } x_{2(i-1)+j} = 1 \\ 0 & \text{otherwise.} \end{cases}$$

Here x is any $(0,1)$ vector. Denote by h the output of **Reconstruct_Vector** for reconstructing u. The assumption that $u \in \prod_{i=1}^{2q}\{0, X(S_{\lceil i/2\rceil})\}$ is not always true. It may happen that for some index $j \in [q]$ we have $X(S_{j,1}) \neq 0$ and $X(S_{j,2}) \neq 0$, that is, $X(S_{j,1})$ and $X(S_{j,2})$ are not in $\{0, X(S_j)\}$. Such violation can occur at most m times throughout the running of the algorithm. In case where $X(S_{j,1}) \neq 0$ and $X(S_{j,2}) \neq 0$, we say that a *split* have occurred in the set S_j.

The next lemmas show how to find a split. Their proof is similar to the proof of Lemma 2 and will be in the full paper

Lemma 3. *At iteration i of the running of **Reconstruct_Vector** for reconstructing $u = (X(S_{1,1}), X(S_{1,2}), X(S_{2,1}), X(S_{2,2}), \ldots, X(S_{q,1}), X(S_{q,2}))$. If*

$$h_1 = u_1, h_2 = u_2, \ldots, h_{i-1} = u_{i-1},$$

and $u_i \in \{0, X(S_{\lceil i/2\rceil})\}$, then, the algorithm reconstructs u_i correctly. That is, $h_i = u_i$.

Lemma 4. *Let h be the output of **Reconstruct_Vector** for reconstructing $u = (X(S_{1,1}), X(S_{1,2}), X(S_{2,1}), X(S_{2,2}), \ldots, X(S_{q,1}), X(S_{q,2}))$. Let S_j be the set with the minimal index in which a split has occurred. Then, for every $i \le 2(j-1)$ we have that $h_i = u_i$. Moreover, either $h_{2j-1} = 0$ or $h_{2j} = 0$.*

Using the lemmas above, we now show how to find the first split and how to change h as if no split had occurred. Assume, for the sake of simplicity, that $2q = \ell 2^{\ell-1}$ for some integer ℓ (in the previous section we showed how to create a search matrix for the case in which the number of columns is not of the form $\ell 2^{\ell-1}$). For $a \in \{-1,1\}^\ell$, let B_a denote the set of all indices j such that $\sigma_{2q}(j) = (a, k)$ for some k. Also let H_0 denote the set of all indices j for which $h_j = 0$. Given a set $S \subseteq [n]$ we denote by 1_S the (0,1)-vector of size n where the ith entry equals one if and only if $i \in S$. For an integer j, denote $\bar{j} = j-1$ if j is even and $\bar{j} = j+1$, otherwise. Let $z_1 = 1^\ell$, $z_2 = 1^{\ell-1} \cdot -1$, $\ldots$, $z_{2^\ell} = -1^\ell$ (these are the vectors we used to label the rows of the search matrix, see Section 3). The algorithm performs the following steps:

1. Set $z \leftarrow z_1$, $i \leftarrow 1$.
2. While $(Q_u(1_{H_0 \cap B_z}) = 0$ and $i \neq 2^\ell + 1)$ do $i \leftarrow i+1$, $z \leftarrow z_i$.
3. If $i = 2^\ell + 1$, then halt.
4. Using a binary search, find the minimal index $j \in H_0 \cap B_z$ for which $u_j \neq 0$.
5. Ask a query to find u_j (that is, $Q_u(1_{\{j\}})$).
   Update $h_{\bar{j}} = X(S_{\lceil j/2\rceil}) - u_j$ and $h_j = u_j$.

6. Use the algorithm **Reconstruct_Vector** with the new values $h_1, \ldots, h_{\max(j,\bar{j})}$ and reconstruct h starting from entry $\max(j, \bar{j}) + 1$.
7. Update H_0 using the new h and goto 2.

The idea of the algorithm is the following: For each B_z where $z \in \{z_1, \ldots, z_{2^\ell - 1}\}$, the algorithm searches for errors caused by a split (note that $B_{z_1}, B_{z_2}, \ldots, B_{z_{2^\ell-1}}$ are disjoint and that $B_{z_1} \cup B_{z_2} \cup \cdots \cup B_{z_{2^\ell-1}} = [2q]$). Lemma 4 implies that if $Q_u(1_{H_0 \cap B_z}) = 0$ then the algorithm has reconstructed the entries that correspond to the indices B_z correctly (note that the order of searching for errors is important, we must first search for errors in B_{z_1} then in B_{z_2}, etc.). On the other hand, if $Q_u(1_{H_0 \cap B_z}) \neq 0$, then there is an index $j \in B_z \cap H_0$ such that $u_j \neq 0$. Let $j \in B_z \cap H_0$ be the minimal index for which $u_j \neq 0$, Lemma 3 and Lemma 4 imply that a split have occurred in $S_{\lceil j/2 \rceil}$. Line 5 corrects the two entries in h that were effected by this split. In line 6, **Reconstruct_Vector** retraces its steps and cancel errors in following entries that were possibly caused by the erroneous values of h_j and $h_{\bar{j}}$. Now, after reconstructing with the correct h_j and $h_{\bar{j}}$ (again by Lemma 3 and Lemma 4) the next error is guaranteed to occur in the entries that correspond to the next split.

After finding $X(S_{i,j})$ for all $i \in [q]$ and $j \in [2]$. The algorithm throws out all sets $S_{i,j}$ such that $X(S_{i,j}) = 0$ and advances to iteration $\tau + 1$.

As for the complexity analysis, the algorithm runs $\log n$ iterations. At each iteration τ, let β_τ denote the number of splits in this iterations. Let q_τ denote the number of sets at the beginning of the iteration. At each iteration, the algorithm uses search matrices. The complexity of this phase is $O(2q_\tau / \log 2q_\tau)$. Next, the algorithm corrects h. In line 2, the algorithm asks at most $O(2q_\tau / \log 2q_\tau) + \beta_\tau$ queries. Line 4 asks at most $O(\beta_\tau \log\log 2q_\tau)$ queries and line 5 asks β_τ queries. It is easy to see that the algorithm do not need to ask queries in line 6. Therefore, using the fact that $q_\tau \leq m$ for every $\tau \in [\log n]$, the total query complexity is

$$\sum_{\tau=1}^{\log n} O\left(\frac{m}{\log m}\right) + O(\beta_\tau \log\log m).$$

Since $\sum_{\tau=1}^{\log n} \beta_\tau = m$, we get the result. ∎

References

1. Aigner, M.: Combinatorial Search. John Wiley and Sons, Chichester (1988)
2. Alon, N., Asodi, V.: Learning a Hidden Subgraph. SIAM J. Discrete Math 18(4), 697–712 (2005)
3. Alon, N., Beigel, R., Kasif, S., Rudich, S., Sudakov, B.: Learning a Hidden Matching. SIAM J. Comput. 33(2), 487–501 (2004)
4. Angluin, D., Chen, J.: Learning a Hidden Graph Using O(log n) Queries per Edge. In: Conference on Learning Theory, pp. 210–223 (2004)
5. Angluin, D., Chen, J.: Learning a Hidden Hypergraph. Journal of Machine Learning Research 7, 2215–2236 (2006)
6. Bouvel, M., Grebinski, V., Kucherov, G.: Combinatorial Search on Graphs Motivated by Bioinformatics Applications: A Brief Survey. In: Kratsch, D. (ed.) WG 2005. LNCS, vol. 3787, pp. 16–27. Springer, Heidelberg (2005)

7. Bshouty, N.H.: Optimal Algorithms for the Coin Weighing Problem with a Spring Scale. In: Conference on Learning Theory (2009)
8. Bshouty, N.H., Mazzawi, H.: Reconstructing Weighted Graphs with Minimal Query Complexity. In: Gavaldà, R., Lugosi, G., Zeugmann, T., Zilles, S. (eds.) ALT 2009. LNCS (LNAI), vol. 5809, pp. 97–109. Springer, Heidelberg (2009)
9. Bshouty, N.H., Mazzawi, H.: On Parity Check $(0, 1)$-Matrix over $\mathbb{Z}_p$. TR09-067. In: ECCC (2009)
10. Cantor, D.: Determining a set from the cardinalities of its intersections with other sets. Canadian Journal of Mathematics 16, 94–97 (1962)
11. Cantor, D., Mills, W.: Determining a Subset from Certain Combinatorial Properties. Canad. J. Math. 18, 42–48 (1966)
12. Choi, S., Han Kim, J.: Optimal Query Complexity Bounds for Finding Graphs. In: STOC, pp. 749–758 (2008)
13. Du, D., Hwang, F.K.: Combinatorial group testing and its application. Series on applied mathematics, vol. 3. World Science (1993)
14. Erdös, Rényi, A.: On two problems of information theory. Publ. Math. Inst. Hung. Acad. Sci. 8, 241–254 (1963)
15. Grebinski, V., Kucherov, G.: Optimal Reconstruction of Graphs Under the Additive Model. Algorithmica 28(1), 104–124 (2000)
16. Grebiniski, V., Kucherov, G.: Reconstructing a hamiltonian cycle by querying the graph: Application to DNA physical mapping. Discrete Applied Mathematics 88, 147–165 (1998)
17. Grebinski, V.: On the Power of Additive Combinatorial Search Model. In: Hsu, W.-L., Kao, M.-Y. (eds.) COCOON 1998. LNCS, vol. 1449, pp. 194–203. Springer, Heidelberg (1998)
18. Lindström, B.: On a combinatorial problem in number theory. Canad. Math. Bull. 8, 477–490 (1965)
19. Lindström, B.: On a combinatorial detection problem II. Studia Scientiarum Mathematicarum Hungarica 1, 353–361 (1966)
20. Lindström, B.: On Möbius functions and a problem in combinatorial number theory. Canad. Math. Bull. 14(4), 513–516 (1971)
21. Lindström, B.: Determining subsets by unramified experiments. In: Srivastava, J.N. (ed.) A Survey of Statistical Designs and Linear Models, pp. 407–418. North Holland, Amsterdam (1975)
22. Li, M., Vitányi, P.M.B.: Combinatorics and Kolmogorov Complexity. In: Structure in Complexity Theory Conference, pp. 154–163 (1991)
23. Mazzawi, H.: Optimally Reconstructing Weighted Graphs Using Queries. In: Symposium on Discrete Algorithms, pp. 608–615 (2010)
24. Moser, L.: The second moment method in combinatorial analysis. In: Combinatorial Structure and their applications, pp. 283–384. Gordon and Breach, New York (1970)
25. Pippenger, N.: An Informtation Theoretic Method in Combinatorial Theory. J. Comb. Theory, Ser. A 23(1), 99–104 (1977)
26. Pippenger, N.: Bounds on the performance of protocols for a multiple-access broadcast channel. IEEE Transactions on Information Theory 27(2), 145–151 (1981)
27. Reyzin, L., Srivastava, N.: Learning and Verifying Graphs using Queries with a Focus on Edge Counting. In: Hutter, M., Servedio, R.A., Takimoto, E. (eds.) ALT 2007. LNCS (LNAI), vol. 4754, pp. 277–289. Springer, Heidelberg (2007)
28. Ruszinkó, M., Vanroose, P.: How an Erdős-Rényi-type search approach gives an explicit code construction of rate 1 for random access with multiplicity feedback. IEEE Transactions on Information Theory 43(1), 368–372 (1997)
29. Soderberg, S., Shapiro, H.S.: A combinatory detection problem. American Mathematical Monthly 70, 1066–1070 (1963)

Resource Combinatory Algebras

Alberto Carraro[1,2], Thomas Ehrhard[2], and Antonino Salibra[1]

[1] Dipartimento di Informatica, Università Ca' Foscari Venezia, Italia
[2] Laboratoire PPS, Université Paris-Diderot, Paris, France

Abstract. We initiate a purely algebraic study of Ehrhard and Regnier's resource λ-calculus, by introducing three equational classes of algebras: *resource combinatory algebras*, *resource lambda-algebras* and *resource lambda-abstraction algebras*. We establish the relations between them, laying down foundations for a model theory of resource λ-calculus. We also show that the ideal completion of a resource combinatory (resp. lambda-, lambda-abstraction) algebra induces a "classical" combinatory (resp. lambda-, lambda-abstraction) algebra, and that any model of the classical λ-calculus raising from a resource lambda-algebra determines a λ-theory which equates all terms having the same Böhm tree.

1 Introduction

There have been several attempts to reformulate the λ-calculus as a purely algebraic theory. The earliest and best known algebraic models are the *combinatory algebras* of Schönfinkel and Curry [5]. Combinatory algebras, as well as their remarkable subclass of λ-*algebras*, have a purely equational characterization but yield somewhat weak notions of models of the λ-calculus. In fact, the combinatory interpretation of λ-calculus does not satisfy the ξ-rule: under the interpretation, $M = N$ does not necessarily imply $\lambda x.M = \lambda x.N$. Thus, the class of λ-algebras is not sound for λ-theories, and one is forced to consider the non-equational class of λ-models (see [1]). There are many advantages in using algebraic languages rather than languages with binders, particularly in connection with equational reasoning. The former have well-understood model theory, and the models are closed under standard constructions such as cartesian products, subalgebras, quotients and free algebras. The above-mentioned problem with the ξ-rule seems to suggest that the λ-calculus is not quite equivalent to an algebraic theory. The lattice of λ-theories is isomorphic to the congruence lattice of the term algebra of the least λ-theory $\lambda\beta$. This remark is the starting point for studying λ-calculus by universal algebraic methods, through the variety (i.e. equational class of algebras) generated by the term algebra of $\lambda\beta$, which Salibra [19] has shown to be axiomatized by the finite scheme of identities characterizing λ-abstraction algebras. These algebras, introduced by Pigozzi and Salibra [17], are intended as an alternative to combinatory algebras, which keeps the lambda notation and hence all the functional intuitions. In [18] the connections between the variety of λ-abstraction algebras and the other algebraic models of λ-calculus are explained; it is also shown that the free extension of a λ-algebra can be turned

P. Hliněný and A. Kučera (Eds.): MFCS 2010, LNCS 6281, pp. 233–245, 2010.

into a λ-abstraction algebra, thus validating all rules of the λ-calculus, including the ξ-rule. The algebraic approach to λ-calculus has been fruitful in studying the structure of the lattice of λ-theories and in generalizing the Stone representation theorem for Boolean algebras to combinatory and λ-abstraction algebras (see [12, 13, 14]). The Stone theorem has been also applied to provide an algebraic incompleteness theorem that encompasses incompleteness results for all known semantics of λ-calculus.

In the '90s Boudol [2] introduced the *λ-calculus with multiplicities*, an extension of λ-calculus where arguments may come in limited availability and mixed together. After one decade Ehrhard and Regnier [7] introduced the *differential λ-calculus*, a conservative (see [7, Prop. 19]) extension of the λ-calculus with differential constructions, in which the linear application of a term M to an argument roughly corresponds to applying the derivative of M in 0 (which is a linear function) to that argument. The presence of linear application, and linear substitution force the enrichment of the calculus with an operation of sum with a neutral element. In [8, 9] Ehrhard and Regnier introduce a simple subsystem of the differential λ-calculus, that they call *resource λ-calculus*, and establish a correspondence between *differential nets*, a variation of Girard's [10] linear logic proof-nets (without promotion rule), and resource λ-calculus. Very recently, Tranquilli [20] enriched the resource λ-calculus with a promotion operator (bearing strong similarities to Boudol's λ-calculus with multiplicities), establishing a correspondence with differential interaction nets extended with promotion. Tranquilli's resource calculus has been recently studied from the syntactical point of view by Pagani and Tranquilli [16], for confluence results, and by Pagani and Ronchi Della Rocca [15] for results about solvability. Regarding the semantics of these calculi, the first studies were conducted by Boudol et al. [3] for the λ-calculus with multiplicities. In a forthcoming paper Bucciarelli et al. [4] define categorical models for the differential λ-calculus.

In this paper we initiate a purely algebraic study of Ehrhard and Regnier's resource λ-calculus. We follow the lines of the universal-algebraic tradition in the study of λ-calculi, exploring a number of varieties which can be considered as classes of algebraic models of resource λ-calculus. We axiomatize the variety of *resource combinatory algebras* (RCAs) which are to the resource λ-calculus what combinatory algebras are to the classical λ-calculus, in the sense that they contain basic combinators which allow to define an abstraction on polynomials and to obtain a combinatory completeness result. Then establishing a parallel with the work of Curry we isolate the subvariety of *resource lambda-algebras* (RLAs) and prove that the free extension of an RLA validates the so-called *ξ-rule* for the abstraction; this is done by a construction, analogue to that of Krivine [11] for lambda-algebras, which shows that the free extension of an RLA is, up to isomorphism, an object very similar to the graded algebras which appear in module theory. Along the line of the work of Pigozzi and Salibra, we axiomatize the variety of *resource λ-abstraction algebras*. We also establish the relations between these varieties, laying down foundations for a model theory of resource λ-calculus. We then show that the ideal completion of a resource combinatory (resp. lambda-, λ-abstraction) algebra determines a "classical" combinatory (resp. lambda-, λ-abstraction) algebra, and that any model of the classical λ-calculus raising from a resource lambda-algebra induces a λ-theory which equates all terms having the same Böhm tree.

2 Preliminaries

We identify every natural number $n \in \mathbb{N}$ with the set $n = \{0, \dots, n-1\}$. $\mathfrak{S}_n$ denotes the set of all permutations (i.e., bijections) of set $n \in \mathbb{N}$.

Sequences: The overlined letters $\bar{a}, \bar{b}, \bar{c}, \dots$ range over the set A^* of all finite sequences over A. The length of a sequence $\bar{a}$ is denoted by $|\bar{a}|$. If $\bar{a}$ is a sequence then a_i ($i \in \mathbb{N}$) denotes the i-th element of $\bar{a}$. For a sequence $\bar{a}$ of length n and a map $\sigma : k \to n$ ($k, n \in \mathbb{N}$), the composition $\sigma\bar{a}$ is the sequence $(a_{\sigma(0)}, \dots, a_{\sigma(k-1)})$. Given two sequences $\bar{a}$ and $\bar{b}$, their concatenation is denoted by $\bar{a} \cdot \bar{b}$. Sequences of length one and elements of A are identified so that $a \cdot \bar{b}$ is the concatenation of $a \in A$ and $\bar{b} \in A^*$. If $a \in A$, then a^k denotes the sequence $(a, \dots, a)$ of length k. If $\bar{i}$ is a sequence of natural numbers of length k then $\Sigma\bar{i}$ denotes $i_0 + \dots + i_{k-1}$.

Sequences of sequences will be denoted by the double over-line. Thus, $\bar{\bar{a}}$ will be a sequence of sequences, whose elements are the sequences $\bar{a}_0, \dots, \bar{a}_{|\bar{\bar{a}}|-1}$. We denote by $\prod \bar{\bar{a}}$ the sequence $\bar{a}_0 \cdot \bar{a}_1 \cdot \dots \cdot \bar{a}_{|\bar{\bar{a}}|-1}$ that is the juxtaposition of the sequences $\bar{a}_i$.

Partitions of a sequence: Let $\bar{a} \in A^n$ and $\bar{i} \in \mathbb{N}^*$ be sequences. A $\bar{i}$-*partition* of $\bar{a}$ is a sequence $\bar{\bar{b}}$ of sequences such that $|\bar{\bar{b}}| = |\bar{i}| = k+1$, $|\bar{b}_0| = i_0, \dots, |\bar{b}_{k-1}| = i_{k-1}$ and there exists $\sigma \in \mathfrak{S}_n$ such that $\sigma\bar{a} = \prod \bar{\bar{b}}$. We write $\mathcal{Q}_{\bar{a},\bar{i}}$ to denote the set of all $\bar{i}$-partitions of $\bar{a}$ and we agree that $\mathcal{Q}_{\bar{a},\bar{i}} \neq \emptyset$ if, and only if, $\Sigma\bar{i} = |\bar{a}|$. Moreover by $\mathcal{Q}_{\bar{a},k}$ we indicate the set $\bigcup_{\bar{i} \in \mathbb{N}^*, |\bar{i}|=k} \mathcal{Q}_{\bar{a},\bar{i}}$. Let $\bar{x}, \bar{y}$ be sequences of the same length and let $\bar{\bar{a}} \in \mathcal{Q}_{\bar{x},\bar{n}}$. We say that $\bar{\bar{b}} \in \mathcal{Q}_{\bar{y},\bar{n}}$ is the *partition of* $\bar{y}$ *induced by* $\bar{\bar{a}}$ iff $\prod \bar{\bar{a}} = \sigma\bar{x}$ and $\prod \bar{\bar{b}} = \sigma\bar{y}$.

Kronecker's delta: In order to give concise axiomatic presentations, we will use the Kronecker function $\delta_{n,m} : A \to A$ with values in a pointed set A (with distinguished element 0) given by $\delta_{n,m}(a) = a$ if $n = m$ and $\delta_{n,m}(a) = 0$ otherwise. In particular for $\bar{a} \in A^*$ we will adopt the convention that the value of the expression $\delta_{|\bar{a}|,m}(a_0)$ is 0 if $|\bar{a}| = 0$.

Free extensions: Let $\mathcal{V}$ be a variety and $\mathbf{A} \in \mathcal{V}$. The *free extension* $\mathbf{A}[X]$ of $\mathbf{A}$ in $\mathcal{V}$ by X is characterized by: (i) $X \subseteq A[X]$; (ii) for any homomorphism $f : \mathbf{A} \to \mathbf{B} \in \mathcal{V}$ and any function $g : X \to B$ there exists a unique homomorphism $f^* : \mathbf{A}[X] \to \mathbf{B}$ extending both f and g. It is known that $\mathbf{A}[X]$ can be characterized as a free algebra whose universe consists of equivalence classes of terms (on a suitably enriched similarity type).

Direct sums of join-semilattices: Let $(\mathbf{A}_i)_{i \in I}$ be a family of join-semilattices. We say that $\mathbf{B}$ is the *direct sum* of the family $(\mathbf{A}_i)_{i \in I}$, notation $\mathbf{B} = \oplus_{i \in I} \mathbf{A}_i$, if

$\mathbf{B} \leq \prod_{i \in I} \mathbf{A}_i$ is the subalgebra of the sequences $(a_i \in A_i : i \in I)$ such that $\{i : a_i \neq 0\}$ is finite.

3 Bag-Applicative Algebras

Let $\mathbf{R}$ be a semiring with unit. We introduce an *algebraic signature* Γ constituted by a binary operator "+", a nullary operator "0", a family of unary operators r ($r \in \mathbf{R}$), and a family of operators $\cdot_k$ ($k \in \mathbb{N}$) of arity $k+1$, called collectively *applications.*

The prefix notation for application is indeed cumbersome for common use so that each operation $\cdot_n(a, b_0, \ldots, b_{n-1})$ will be replaced by the lighter $a[b_0, \ldots, b_{n-1}]$, so that, for example, $\cdot_0(a) = a[\,]$ and $\cdot_2(a, b, c) = a[b, c]$. Another reason for this choice is that, when we write $a[b_0, \ldots, b_{n-1}]$, we think to the element a applied to the "bag" $[b_0, \ldots, b_{n-1}]$. We will also adopt the usual convention that application associates to the left. For a sequence $\bar{b}$ of length n, we adopt the further notational simplification to write $a\bar{b}$ instead of $a[b_0, \ldots, b_{n-1}]$. By a^k we indicate the sequence $(a, \ldots, a)$ of length k, thus $ba^k = b[a, \ldots, a]$ (a repeated k times), with the convention that $ba^0 = b[\,]$. Note that the above conventions lead to write just ab for $a[b]$: clearly in case b is itself an application cd, we are obliged to write $a[cd]$ in order to avoid any ambiguity.

Definition 1. *A Γ-algebra is called a* bag-applicative algebra *if it satisfies the following axioms, which are universally quantified.*

Commutative Monoid Axioms :
$(x + y) + z = x + (y + z)$; $\quad x + y = y + x$; $\quad 0 + x = x$
Module Axioms $(r, s \in R)$:
$r(x + y) = rx + ry$; $\quad (r + s)x = rx + sx$; $\quad (rs)x = r(sx)$; $\quad 1x = x$; $\quad 0x = 0$
Multiset Axiom : $x[y_0, \ldots, y_{k-1}] = x[y_{\sigma(0)}, \ldots, y_{\sigma(k-1)}]$ $\quad (\sigma \in \mathfrak{S}_k)$
Multilinearity Axioms :
$x[0, y_0, \ldots, y_{k-1}] = 0$; $\quad 0[y_0, \ldots, y_{k-1}] = 0$
$(ax + by)[y_0, \ldots, y_{k-1}] = a(x[y_0, \ldots, y_{k-1}]) + b(y[y_0, \ldots, y_{k-1}])$
$x[\ldots, ay + bz, \ldots] = a(x[\ldots, y, \ldots]) + b(x[\ldots, z, \ldots])$

If a signature Δ extends Γ, we say that a Δ-algebra $\mathbf{A}$ is a bag-applicative Δ-algebra if it is so the Γ-reduct of $\mathbf{A}$.

4 The Linear Resource Lambda Calculus

We will now briefly introduce the linear fragment of resource λ-calculus ($r\lambda$-calculus, for short). Let V be an infinite countable set of *names* which represent the variables of λ-calculus. The set Λ^r of $r\lambda$*-terms* is described by the following grammar: $M, N, L ::= x \mid \lambda x.M \mid M[N_1, \ldots, N_k] \mid rM \mid M + L \mid 0$, with $x \in V$ and r constant for elements in a semiring. As usual, $r\lambda$-terms are to be considered modulo α-conversion. The definition of the linear substitution of a "bag" $[N_1, \ldots, N_k]$ for x in M, notation $M\langle[N_1, \ldots, N_k]/x\rangle$, can be found in [16]. Notice that the linear substitution is a "meta-operation" as the usual substitution $M\{x := N\}$ in classical λ-calculus. Finally we present the system of equations governing the $r\lambda$-calculus:

Axioms of a bag-applicative algebra (instantied by $r\lambda$-terms)
βr-conversion: $(\lambda x.M)[N_1, \ldots, N_k] = M\langle[N_1, \ldots, N_k]/x\rangle\{x := 0\}$
Multilinearity for λ ($r, s \in R$): $\lambda x.0 = 0$; $\lambda x.(rM + sN) = r\lambda x.M + s\lambda x.N$

Fig. 1. Equational axioms for linear $r\lambda$-calculus over the semiring $\mathbf{R}$

Warning: In the remaining part of this paper we take $\mathbf{R}$ to be the semiring $\mathbf{2} = (\{0,1\}, +, \cdot, 0, 1)$ of truth values with $0 < 1$, where $+$ and $\cdot$ are the usual lattice operation. This means that the module axioms can be substituted by the single identity $x + x = x$. In this way, the commutative monoid becomes a join semilattice. Working with coefficients from an arbitrary semiring would increase significantly the complexity of notations in statements and proofs, covering the essence of the results presented in this paper.

Theorem 1. *[7, 21] The $r\lambda$-calculus over the semiring $\mathbf{2}$ is consistent, because it enjoys confluence and strong normalization.*

5 The $r\lambda$-calculus from the Algebraic Point of View

The variable-binding properties of λ-abstraction prevent names in $r\lambda$-calculus from operating as real algebraic variables. The same problem occurs in classic λ-calculus and was solved by Pigozzi and Salibra [17] by introducing the variety of λ-abstraction algebras. We adopt here their solution and transform the names (i.e., elements of V) into constants.

Definition 2. *The signature Γ_λ is an extension of the signature Γ of bag-applicative algebra by a family of nullary operators $x \in V$, one for each element of V, and a family of unary operators λx ($x \in V$), called collectively λ-abstractions.*

The $r\lambda$-terms are just the Γ_λ-terms without occurrences of algebraic variables. The absolutely free Γ_λ-algebra is the algebra $\mathbf{\Lambda^r} = (\Lambda^r, +, 0, \cdot_k, \lambda x, x)_{x \in V, k \in \mathbb{N}}$, where Λ^r is the set of $r\lambda$-terms and the operations are just the syntactical operations of construction of the $r\lambda$-terms.

Definition 3. *A $r\lambda$-theory is any congruence on $\mathbf{\Lambda^r}$ (with respect to all the involved operations) including all the identities of Figure 1.*

The least $r\lambda$-theory, denoted by $\lambda\beta r$, is consistent by Theorem 1. If T is a $r\lambda$-theory, we denote by $\mathbf{\Lambda^r_T} \equiv \mathbf{\Lambda^r}/T$ the quotient of the absolutely free Γ_λ-algebra $\mathbf{\Lambda^r}$ modulo the $r\lambda$-theory T. $\mathbf{\Lambda^r_T}$ is called the *term algebra* of T.

We now abstract the notion of term algebra by introducing the variety of resource λ-abstraction algebras as a pure algebraic theory of $r\lambda$-calculus. The term algebras of $r\lambda$-theories are the first example of resource λ-abstraction algebra. Another example will be given in Section 9.

Definition 4. *A* resource λ-abstraction algebra *(RLAA, for short) is a bag-applicative Γ_λ-algebra satisfying the following identities (for all $a \in A$, $\bar{a}, \bar{b}, \bar{c}, \bar{d} \in A^*$, and $x \neq y \in V$):*

$(r\beta_1)$ $(\lambda x.x)\bar{a} = \delta_{|\bar{a}|,1}(a_0)$
$(r\beta_2)$ $(\lambda x.y)\bar{a} = \delta_{|\bar{a}|,0}(y)$
$(r\beta_3)$ $(\lambda x.\lambda x.a)\bar{b} = \delta_{|\bar{b}|,0}(\lambda x.a)$
$(r\beta_4)$ $(\lambda y.b_i)[\,] = b_i$, *for all* $i < |\bar{b}| \;\Rightarrow\; (\lambda x.\lambda y.a)\bar{b} = \lambda y.(\lambda x.a)\bar{b}$
$(r\beta_5)$ $(\lambda x.a\bar{b})\bar{c} = \Sigma_{\bar{\bar{d}} \in \mathcal{Q}_{\bar{c},k+1}} (\lambda x.a)\bar{d}_0[(\lambda x.b_0)\bar{d}_1, \dots, (\lambda x.b_{k-1})\bar{d}_k]$ $(|\bar{b}| = k)$
$(r\alpha)$ $(\lambda x.a)x^k = a,\; (\lambda y.a)[\,] = a \;\Rightarrow\; \lambda x.a = \lambda y.(\lambda x.a)y^k$
$(r\gamma)$ $(\lambda x.a)x^n + a = a$
$(r\lambda)$ $\lambda x.0 = 0; \quad \lambda x.(a+b) = \lambda x.a + \lambda x.b$

Some of the axioms above are not pure identities, though they can be turned into such as it is done in the case of λ-abstraction algebras [17].

An element a is *finite-dimensional* if there exists a finite subset $X \subseteq V$ such that $(\lambda x.a)[\,] = a$ for all $x \in V - X$ and, for all $x \in X$, there is exactly one $n \neq 0$ such that $(\lambda x.a)x^n = a$, and in such a case $(\lambda x.a)x^k = 0$ for all $k \neq n$; this last statement follows from $(r\beta_5)$,$(r\beta_3)$ and $(r\beta_1)$. Finite-dimensional elements are a generalization of the $r\lambda$-terms. In particular $a \in A$ is *zero-dimensional* if $(\lambda x.a)[\,] = a$ for all $x \in V$. We say that a *name* $x \in V$ *does not occur free in* $a \in A$ if $(\lambda x.a)[\,] = a$. The set of zero-dimensional elements, which generalizes closed $r\lambda$-terms, will be denoted by Zd $\mathbf{A}$. In general a RLAA may have elements where all the names occur free; these elements are a generalization of infinite λ-terms. A RLAA $\mathbf{A}$ is called *locally finite* if it is generated by its finite-dimensional elements (through the join/sum operator). Every RLAA $\mathbf{A}$ contains a canonical locally finite RLAA, which is the subalgebra of $\mathbf{A}$ generated by all its finite-dimensional elements. This algebra will be denoted by Lf$\mathbf{A}$.

Proposition 1. *(i) For any $r\lambda$-theory T the term algebra $\mathbf{\Lambda}^{\mathbf{r}}_{\mathbf{T}}$ is a locally finite* RLAA.
(ii) The minimal subalgebra of a RLAA $\mathbf{A}$ *is isomorphic to $\mathbf{\Lambda}^{\mathbf{r}}_{\mathbf{T}}$ for some T.*

6 Resource Combinatory Algebras

In this section we introduce a class of algebras which are to the $r\lambda$-calculus what combinatory algebras are to the classical λ-calculus. The signature Γ_c of is an extension of the signature Γ of bag-applicative algebras by a nullary operator K and a family of nullary operators $S_{\bar{n}}$ ($\bar{n} \in \mathbb{N}^*$). Recall the definition of the set $\mathcal{Q}_{\bar{z},\bar{n}}$ from the preliminaries.

Definition 5. *A* resource combinatory algebra *(*RCA*, for short) is a bag-applicative Γ_c-algebra satisfying the following identities:*

(K) $K\bar{x}\bar{y} = \delta_{|\bar{y}|,0}(\delta_{|\bar{x}|,1}(x_0))$
($S_{\bar{n}}$) $S_{\bar{n}}\bar{x}\bar{y}\bar{z} = \delta_{|\bar{x}|,1}(\delta_{k,|\bar{y}|}(\delta_{|\bar{y}|,|\bar{n}|-1}(\delta_{|\bar{z}|,\Sigma\bar{n}}(\sum_{\bar{c}\in\mathcal{Q}_{\bar{z},\bar{n}}} x_0\bar{c}_0[y_0\bar{c}_1,\ldots,y_{k-1}\bar{c}_k]))))$

The variety of resource combinatory algebras is denoted by RCA. We secretly think of K and $S_{\bar{n}}$ as the following $r\lambda$-terms:

$$K_\lambda \equiv \lambda xy.x; \quad S_{\bar{n},\lambda} \equiv \lambda xyz.xz^{n_0}[yz^{n_1},\ldots,yz^{n_k}] \quad (|\bar{n}|-1=k) \qquad (1)$$

We define *(resource) monomials* with names in V and constant in A by the following grammar: $t ::= x \mid c_a \mid K \mid S_{\bar{n}} \mid t_0[t_1,\ldots t_n]$ ($\bar{n} \in \mathbb{N}, a \in A$). A *(resource) polynomial* is a finite sum of monomials: $t_1 + \cdots + t_n$. We denote by $P(A)$ the set of all polynomials with names in V and constant in A. For a monomial t we define the *degree* $deg_x(t)$ of $x \in V$ in t as the number of occurrences of the name x in t.

We define an abstraction operation on polynomials, with which the abstraction of $r\lambda$-calculus can be simulated. First of all we need to define the combinator $I \equiv S_{(1)}K[\,]$. It is immediate to see that $I\bar{x} = \delta_{|\bar{x}|,1}(x_0)$.

Definition 6. *Let $t, t_1, \ldots, t_n$ be monomials. We define a new monomial $\lambda^*x.t$ as follows:*

(i) $\lambda^*x.t \equiv Kt$ *if* $deg_x(t) = 0$
(ii) $\lambda^*x.x \equiv I$
(iii) $\lambda^*x.t_0[t_1,\ldots,t_k] \equiv S_{\bar{n}}[\lambda^*x.t_0][\lambda^*x.t_1,\ldots,\lambda^*x.t_k]$ $(\bar{n} = (deg_x(t_0),\ldots,deg_x(t_k)))$ *if* $\exists i\ deg_x(t_i) \neq 0$.

We extend the definition of abstraction to polynomials: $\lambda^*x.\Sigma_{i=1}^n t_i = \Sigma_{i=1}^n \lambda^*x.t_i$.

Let t be a monomial with $deg_x(t) = n$, $\bar{p}$ be a sequence of n polynomials and $\sigma \in \mathfrak{S}_n$ be a permutation. Then the expression $t\{\bar{x} := \sigma\bar{p}\}$ denotes the *simultaneous* substitution of the i-th occurrence x^i of x in t by the polynomial $p_{\sigma(i)}$ ($i = 1,\ldots,n$). As usual we write $\mathbf{A} \models t = u$ to express the fact the equation $t = u$ holds under any valuation in the algebra $\mathbf{A}$.

Lemma 1. *Let* $\mathbf{A}$ *be a* RCA*. For any monomial t, any sequence $\bar{p}$ of polynomials, and any name x we have:* $\mathbf{A} \models (\lambda^*x.t)\bar{p} = \delta_{deg_x(t),|\bar{p}|}(\Sigma_{\sigma\in\mathfrak{S}_{deg_x(t)}} t\{\bar{x} := \sigma\bar{p}\})$.

Let $\mathbf{A}$ be a RLAA. The *combinatory reduct* of $\mathbf{A}$ is defined as the algebra $\mathrm{Cr}\,\mathbf{A} = (A, \cdot_k, K_\lambda^{\mathbf{A}}, S_{\bar{n},\lambda}^{\mathbf{A}})$, where the $r\lambda$-terms K_λ and $S_{\bar{n},\lambda}$ are defined in (1) above. The subalgebra of $\mathrm{Cr}\,\mathbf{A}$ constituted by the zero-dimensional elements of $\mathbf{A}$ will be denoted by $\mathbf{Zd}\,\mathbf{A}$.

Proposition 2. *Let* $\mathbf{A}$ *be a locally finite* RLAA*. Then,* $\mathrm{Cr}\,\mathbf{A}$ *is a* RCA.

The proof of the above proposition is trivial because of the hypothesis of locally finiteness. If we drop this hypothesis, then we cannot always apply α-conversion because elements may exist where all variables occur free.

The $r\lambda$-term t_λ associated with a polynomial t can be easily defined by induction: $K, S_{\bar{n}}$ are respectively translated into K_λ and $S_{\bar{n},\lambda}$ (see (1) above); $(c_a)_\lambda = c_a$; $(t[s_1, \dots, s_n])_\lambda = t_\lambda[s_{1,\lambda}, \dots, s_{n,\lambda}]$; $(\Sigma t_i)_\lambda = \Sigma t_{i,\lambda}$.

The following lemma can be shown by induction over the complexity of the polynomial p. If $\mathbf{A}$ is a RLAA, then $p^{\mathrm{Cr}\,\mathbf{A}}$ denotes the interpretation of p into Cr $\mathbf{A}$.

Lemma 2. *Let* $\mathbf{A}$ *be a* RLAA *and* p *be a polynomial. Then,* $p^{\mathrm{Cr}\,\mathbf{A}} = p_\lambda^{\mathbf{A}}$ *and* $(\lambda^* x.p)^{\mathrm{Cr}\,\mathbf{A}} = \lambda x^{\mathbf{A}}.p_\lambda^{\mathbf{A}}$.

7 Resource λ-algebras

In this section we axiomatize the variety of resource λ-algebras ($r\lambda$-algebras for short), and prove that the free extension of an $r\lambda$-algebra in the variety of $r\lambda$-algebras can be turned into a RLAA, so that it validates all the rules of $r\lambda$-calculus. For the subsequent developments, it turns out very important to isolate a particular family of combinators: for $n \in \mathbb{N}$, the *n-homogenizer* is the combinator $H_n \equiv S_{(0,n)}[KI]$. Using the equation schemata of RCAs we obtain that $H_n \bar{x}\bar{y} = \delta_{|\bar{x}|,1}(\delta_{|\bar{y}|,n}(x_0\bar{y}))$. The elements of the form $H_n a$ are the semantical counterpart of monomials of the form $\lambda^* x.t$, with $deg_x(t) = n$. Via H_n it is in fact possible to give a semantical notion of degree: $a \in A$ is called *homogeneous of degree* n iff $H_n a = a$.

We now define $r\lambda$-algebras. We advice the reader that some identities defining $r\lambda$-algebras are difficult to read, nonetheless they still resemble those for λ-algebras.

Definition 7. *A* RCA $\mathbf{A}$ *is a* $r\lambda$-algebra *if it satisfies the* λ^**-closure of the following identities:*

(R0) $H_n[H_m x] = \delta_{n,m}(H_m x)$
(R1) $K = H_1 K;\quad Kx = H_0[Kx]$
(R2) $S_{\bar{n}} = H_1 S_{\bar{n}};\quad S_{\bar{n}} x = H_{|\bar{n}|-1}[S_{\bar{n}} x];\quad S_{\bar{n}} x\bar{y} = H_{\Sigma\bar{n}}[S_{\bar{n}} x\bar{y}]$
(R3) $S_{\bar{m}}[S_{\bar{n}}[KK]\bar{x}]\bar{y} = \begin{cases} H_{n_1} x_0 \text{ if } |\bar{x}| = 1,\ |\bar{y}| = 0,\ \bar{n} = (0, n_1),\ \bar{m} = (n_1) \\ 0 \qquad \text{otherwise} \end{cases}$

(R4) $S_{\bar{n}}[S_{\bar{m}}[S_{\bar{p}}[KS_{\bar{l}}]\bar{x}]\bar{y}]\bar{z} =$
$\Sigma_{\bar{\bar{s}} \in \mathcal{Q}_{\bar{z},\bar{l}}} S_{(\Sigma(m_0\cdot\bar{o}_0),\dots,\Sigma(m_k\cdot\bar{o}_k))}[S_{m_0\cdot\bar{o}_0} x_0 \bar{s}_0][S_{m_1\cdot\bar{o}_1} y_0 \bar{s}_1, \dots, S_{m_k\cdot\bar{o}_k} y_{k-1}\bar{s}_k]$,
if $|\bar{x}| = 1$, $\bar{p} = (0, p_1)$, $|\bar{y}| + 1 = |\bar{m}| = k$, $\bar{n} = (\Sigma\bar{m}) \cdot \bar{n}'$, $|\bar{z}| = |\bar{n}'| = \Sigma\bar{l}$, $\bar{m} = p_1 \cdot \bar{l}$, *and, for each* $\bar{\bar{s}} \in \mathcal{Q}_{\bar{z},\bar{l}}$, $\bar{\bar{o}} \in \mathcal{Q}_{\bar{n}',\bar{l}}$ *is the partition of* $\bar{n}'$ *induced by* $\bar{\bar{s}}$*; it is equal to* 0*, otherwise*
(R5) $K[x\bar{y}] = S_{0^{k+1}}[Kx][Ky_0, \dots, Ky_{k-1}]$ $(|\bar{y}| = k)$
(R6) $H_k x = S_{0\cdot 1^k}[Kx]I^k$

The variety of $r\lambda$-algebras will be denoted by RLA. The next lemma shows the aforementioned connection between homogenizers and the induced λ-abstraction on polynomials.

Lemma 3. *Let* $\mathbf{A}$ *be a* RLA *and* t *be a monomial. Then* $\mathbf{A} \models H_n[\lambda^* x.t] = \delta_{n,deg_x(t)}(\lambda^* x.t)$.

The following theorem is the main result of the section; its proof, divided into five lemmas, occupies the rest of this section and involves the explicit construction of the free extension of a RLA by one name as a graded algebra (Lemmas 4-7) and two key observations (Lemmas 8,9).

Theorem 2. *The free extension* $\mathbf{A}[V]$ *of a* $r\lambda$*-algebra* $\mathbf{A}$ *by the set* V *of names in the variety* RCA *satisfies the following* ξ*-rule, for all polynomials* $p, q \in P(A)$*:*

$$(\xi) \qquad \mathbf{A}[V] \models p = q \Rightarrow \mathbf{A}[V] \models \lambda^* x.p = \lambda^* x.q.$$

We apply the above theorem to define λ-abstraction operators on $A[V]$. For any $e \in A[V]$, we define $\lambda x.e = \lambda^* x.p$, for some polynomial $p \in e$. Rule ξ validates the above definition of λx. Define the algebra $\mathbf{A}[V]_\lambda = (A[V], +, 0, \cdot_k, \lambda x, x)_{x \in V}$, where $(A[V], +, 0, \cdot_k,)$ is the Γ-reduct of the free extension $\mathbf{A}[V]$, λx is defined as above and the name $x \in V$ is viewed as a nullary operator; $\mathbf{A}[V]_\lambda$ is called the RLAA *freely generated by the* $r\lambda$*-algebra* $\mathbf{A}$.

Corollary 1. $\mathbf{A}[V]_\lambda$ *is a locally finite* RLAA *such that* $K_\lambda^{\mathbf{A}[V]_\lambda} = K^{\mathbf{A}}$ *and* $S_{\bar{n},\lambda}^{\mathbf{A}[V]_\lambda} = S_{\bar{n}}^{\mathbf{A}}$.

Corollary 2. *Let* $\mathbf{A}$ *be a* RCA*. Then,* $\mathbf{A}$ *is a* $r\lambda$*-algebra iff* $\mathbf{A}$ *can be embedded into the combinatory reduct of some* RLAA $\mathbf{B}$.

This corollary is very useful to prove when a RCA is a RLA (see Section 9). The construction of the free extension will turn out to be the construction of a graded algebra as a direct sum of specific join semilattices. We now provide the proof of Theorem 2. The proof is inspired by a construction by Krivine [11].

Lemma 4. *Let* $\mathbf{A}$ *be a* $r\lambda$*-algebra and set* $B_n = \{a \in A : H_n a = a\}$*. Then* $(B_n, +, 0)$ *is a join sub-semilattice of* $(A, +, 0)$ *such that* $B_n \cap B_m = \{0\}$ *if* $n \neq m$.

We now define an algebra $\mathbf{B} = (B, +, 0, \bullet_k, KK, KS_{\bar{n}})_{k \in \mathbb{N}, \bar{n} \in \mathbb{N}^*}$, called the $\mathbb{N}$-*graded algebra generated by* $\mathbf{A}$ in the similarity type of RCA by setting:

1. $(B, +, 0) = \oplus_{n \in \mathbb{N}} (B_n, +, 0)$ is the direct sum of the join semilattices $(B_n, +, 0)$.
2. each application "$\bullet_k$" is the extension by linearity of the following operation: $a_0 \bullet_k [a_1, \ldots, a_k] = S_{\bar{p}} a_0 [a_1, \ldots, a_k]$, with $a_i \in B_{p_i}$.

Lemma 5. *The* $\mathbb{N}$*-graded algebra* $\mathbf{B}$ *is a* RCA *which satisfies the following conditions: (i)* KK *and* $KS_{\bar{n}}$ *are elements of* B_0*; (ii)* $B_{d_0} \bullet_k [B_{d_1}, \ldots, B_{d_{k-1}}] \subseteq B_{\Sigma \bar{d}}$*, for all* $\bar{d} \in \mathbb{N}^{k+1}$.

Lemma 6. *The map* ι*, defined by* $\iota(a) = Ka$*, is an embedding of* $\mathbf{A}$ *into* $\mathbf{B}$.

Proof. By (R5) $\iota(a_0[a_1, \ldots, a_n]) = K[a_0[a_1, \ldots, a_n]] = S_{0^{n+1}}[Ka_0][Ka_1, \ldots, Ka_n] = \iota(a_0) \bullet [\iota(a_1), \ldots, \iota(a_n)]$. The other properties are trivial. □

By Lemma 6 **B** is a RLA. We are now going to show the connection between the $\mathbb{N}$-graded algebra **B** and the free extension $\mathbf{A}[x]$ by one name x.

Lemma 7. *The* $\mathbb{N}$*-graded algebra* **B** *is the free extension of* **A** *by one name in the variety* RLA. *Consequently,* $\mathbf{B} \cong \mathbf{A}[x]$.

Proof. We prove that $\mathbf{B} \cong \mathbf{A}[x]$. Let **C** be a RCA, $c \in C$ and $f : \mathbf{A} \to \mathbf{C}$ be a homomorphism. Define a family of functions $f_k : B_k \to C$ $(k \in \mathbb{N})$ as follows: $f_k(a) = f(a) \cdot^{\mathbf{C}} c^k$ for all $a \in B_k$. Let $f^* : B \to C$ be the unique extension by linearity of the family of functions f_k, that is, $f^*(0) = 0$ and $f^*(\Sigma_{i=1}^m a_i) = \Sigma_{i=1}^m f_{d_i}(a_i)$ for $a_i \in B_{d_i}$.

We now prove that f^* is a homomorphism. It is immediate to check that f^* is a monoid homomorphism, using multi-linearity of application. Since f^* extends f by linearity, it suffices to prove the following:

$$\begin{aligned} f_{\Sigma\bar{e}}(a \bullet \bar{b}) &= f(S_{\bar{e}} a\bar{b})c^{\Sigma\bar{e}} = S^{\mathbf{C}}_{\bar{e}} f(a)[f(b_0), \dots, f(b_{n-1})]c^{\Sigma\bar{e}} \text{ , since } f \text{ is hom,} \\ &= f(a)c^{e_0}[f(b_0)c^{e_1}, \dots, f(b_{n-1})c^{e_n}], \text{by} (S_{\bar{n}}) \text{ and idempotence of sum,} \\ &= f_{e_0}(a)[f_{e_1}(b_0), \dots, f_{e_n}(b_{n-1})]. \end{aligned}$$

We have: $f_0(K^{\mathbf{A}}K^{\mathbf{A}}) = f(K^{\mathbf{A}}K^{\mathbf{A}})[\,] = K^{\mathbf{C}}K^{\mathbf{C}}[\,] = K^{\mathbf{C}}$. A similar argument shows that $f^*(K^{\mathbf{A}}S^{\mathbf{A}}_{\bar{n}}) = S^{\mathbf{C}}_{\bar{n}}$. This shows that f^* is a homomorphism.

We have: $f^*(\iota(a)) = f_0(Ka) = f(Ka)[\,] = K^{\mathbf{C}} f(a)[\,] = f(a)$; and $f^*(I) = f_1(I) = f(I)c = I^{\mathbf{C}}c = c$. This proves $f^* \circ \iota = f$ and $f^*(I) = c$.

Finally suppose $h : \mathbf{B} \to \mathbf{C}$ is another homomorphism satisfying $h \circ \iota = f$ and $h(I) = c$. The uniqueness of f^* is shown as follows:

$$\begin{aligned} h(a) &= h(H_k a), \text{for some } k \in \mathbb{N}, \\ &= h(S_{0 \cdot 1^k}[Ka]I^k), \text{by axiom (R6)} \\ &= h((Ka) \bullet I^k) = h(Ka)(h(I))^k = h(\iota(a))c^k = f_k(a)c^k = f^*(a). \end{aligned}$$ □

We denote by ι^* the unique isomorphism from $\mathbf{A}[x]$ onto **B** extending the embedding $\iota : \mathbf{A} \to \mathbf{B}$ defined in Lemma 6, and such that $\iota^*(x) = I$.

Lemma 8. *For all* $a, b \in A$ *we have* $\mathbf{A}[x] \models ax^n = bx^k$ *iff* $\mathbf{A} \models H_n a = H_k b$.

Proof. $\iota^*(ax^n) = \iota(a) \bullet I^n = S_{0 \cdot 1^n}[Ka]I^n = H_n a$, by (R6). We conclude since ι^* is an isomorphism. Of course, if $n \neq k$, then $H_n a = H_k b = 0$. □

Lemma 9. *For all polynomials* p, q *with at most the name* x *we have that* $\mathbf{A}[x] \models p = q$ *implies* $\mathbf{A}[x] \models \lambda^* x.p = \lambda^* x.q$.

Proof. First we prove the result for monomials t, u. Let $n = deg_x(t)$ and $k = deg_x(u)$. By Lemma 1 $\mathbf{A}[x] \models (\lambda^* x.t)x^n = t = u = (\lambda^* x.u)x^k$. Now by Lemma 8 and by Lemma 3 it follows that $\mathbf{A} \models \lambda^* x.t = H_n[\lambda^* x.t] = H_k[\lambda^* x.s] = \lambda^* x.s$; therefore trivially $\mathbf{A}[x] \models \lambda^* x.t = \lambda^* x.s$. Now for polynomials p, q such that $\mathbf{A}[x] \models p = q$, we have that $\iota^{*-1}(\iota^*(\lambda^* x.p)) = \Sigma_{i \in I} \lambda^* x.t_i$ and $\iota^{*-1}(\iota^*(\lambda^* x.q)) = \Sigma_{i \in I} \lambda^* x.u_i$, where I is finite and for each $i \in I$, t_i and u_i are monomials and $\mathbf{A}[x] \models t_i = u_i$; this allows to conclude, using the previous result. □

The extension of the above lemma to polynomials with an arbitrary number of names is standard, because $\mathbf{A}[x, y] \cong \mathbf{A}[x][y]$ and $\mathbf{A}[x]$ is a RLA.

8 From Resource to Classical Lambda Calculus

After having introduced a number of structures which algebrize the resource λ-calculus, we show how, by some standard constructions, we can recover the algebraic models of classical λ-calculus. This is done, as often happens in mathematics, by the method of *ideal completion*. Let $\mathbf{A}$ be a bag-applicative Γ-algebra. An *ideal* is a downward closed subset X of A closed under join. For a subset $X \subseteq A$, $\downarrow X = \{b : \exists a_1, \dots, a_n \in X.\ b \leq \sum_{i=1}^{n} a_i\}$ (where $a \leq b \iff a + b = b$) is the *ideal generated* by X. We denote by $Ide(\mathbf{A})$ the collection of all ideals of $\mathbf{A}$. Let $\mathbf{A}$ be a RCA. Define an algebra $Ide(\mathbf{A}) = (Ide(A), *, \underline{K}, \underline{S})$ by setting $\underline{K} = \downarrow \{K\}$; $\underline{S} = \downarrow \{S_{\bar{n}} : \bar{n} \in \mathbb{N}^*\}$; $X * Y = \downarrow \{a\bar{b} : a \in X,\ \bar{b} \in Y^*\}$. If $\mathbf{B}$ is a RLAA we define the structure $Ide(\mathbf{B}) = (Ide(B), *, \underline{\lambda x}, \underline{x})_{x \in V}$ by setting $\underline{x} = \downarrow \{x\}$; $\underline{\lambda x}.X = \downarrow \{\lambda x.a : a \in X\}$ and the application $*$ as above.

Theorem 3. *(i) If* $\mathbf{A}$ *is a* RCA*, then* $Ide(\mathbf{A})$ *is a combinatory algebra.*
(ii) Let $\mathbf{A}$ *be a* RLAA *and* Lf$\mathbf{A}$ *be the subalgebra of* $\mathbf{A}$ *generated by its locally finite elements. Then* $Ide($Lf$\mathbf{A})$ *is a* λ*-abstraction algebra.*
(iii) If $\mathbf{A}$ *is a* RLA*, then* $Ide(\mathbf{A})$ *is a* λ*-algebra.*

Note that the λ-abstraction algebra $Ide($Lf$\mathbf{A})$ of point (ii) is not necessarily locally finite: for example the element $\downarrow V$ when V is infinite breaks the property.

According to [6], we now define a translation of ordinary λ-terms into sets of $r\lambda$-terms, which also extends to a translation of Böhm trees. Let $\Lambda^{\perp}$ be the set normal terms in the λ-calculus extended with a constant $\perp$; as customary $\Lambda^{\perp}$ is endowed with a partial order whose bottom element is $\perp$ and where "less or equal" means "possibly more defined". Following [1], we identify the Böhm tree $BT(M)$ of a λ-term M with an ideal (downwards closed and directed subset) of $\Lambda^{\perp}$ quotiented by the equations $\perp N = \perp$ and $\lambda x.\perp = \perp$. This way we can also translate Böhm trees into subsets of Λ^r.

As a matter of terminology, a $r\lambda$-term t is: *simple* if none of its subterms (including t) contains either "+" or "0"; *normal* if none of its subterms (including t) is of the form $(\lambda x.t')\bar{s}$; in *canonical form* if it is a sum of simple terms. By an easy argument involving the multilinearity axioms of the $r\lambda$-calculus and Theorem 1, we can argue that for every term $t \in \Lambda^r$, there exists a unique normal term s in canonical form which is equal to t and we let NF(t) be the (finite) set of all simple terms whose sum is the normal canonical form of t.

Definition 8. *[6] Let* $M \in \Lambda$ *be a* λ*-term, possibly containing* $\perp$*. The set* $\mathcal{T}(M) \subseteq \Lambda^r$ *is defined inductively by the clauses:* $\mathcal{T}(x) = \{x\}$, $\mathcal{T}(\perp) = \emptyset$, $\mathcal{T}(\lambda x.N) = \{\lambda x.t : t \in \mathcal{T}(N)\}$*, and* $\mathcal{T}(PQ) = \{t\bar{s} : t \in \mathcal{T}(P),\ \bar{s} \in \mathcal{T}(Q)^*\}$*. Then* $\mathcal{T}(N)$ *happens to be the support of the Taylor expansion (see [6]) of the* λ*-term* N*. Now* $\mathcal{T}(BT(M)) \subseteq \Lambda^r$ *is defined as* $\cup\{\mathcal{T}(B) : B \in BT(M)\}$*.*

Recall now that λ-terms are ground terms in the similarity type of LAAs: hence for a LAA $\mathbf{B}$ it makes sense to write $M^{\mathbf{B}}$ to indicate the interpretation of M in $\mathbf{B}$. Similarly the notation $t^{\mathbf{A}}$ can be used to denote the interpretation of a resource λ-term t in a RLAA $\mathbf{A}$.

Lemma 10. *Let* **A** *be a locally finite* RLAA *and let* $\mathbf{B} = Ide(\mathbf{A})$ *be the* LAA *built over* **A**. *Then for all* $M \in \Lambda$, $M^{\mathbf{B}} = \downarrow \{u^{\mathbf{A}} : u \in NF(t),\ t \in \mathcal{T}(M)\}$.

Theorem 4. *[6] Let* M *be a* λ*-term and let* u *be a normal simple* $r\lambda$*-term. Then* $u \in \mathcal{T}(BT(M))$ *iff there exists* $s \in \mathcal{T}(M)$ *such that* $u \in NF(s)$.

Lemma 11. *Let* **A** *be a locally finite* RLAA. *Then for all terms* $M, N \in \Lambda$ *we have* $BT(M) = BT(N)$ *implies* $M^{Ide(\mathbf{A})} = N^{Ide(\mathbf{A})}$. *In particular all* λ*-theories induced by the ideal completions of* RLAA*s are sensible.*

Proof. Suppose $BT(M) = BT(N)$. Then obviously $\mathcal{T}(BT(M)) = \mathcal{T}(BT(N))$. We conclude by applying Theorem 4 and Lemma 10 as follows: $M^{\mathbf{B}} = \downarrow \{u^{\mathbf{A}} : u \in \mathrm{NF}(t),\ t \in \mathcal{T}(M)\} = \downarrow \{u^{\mathbf{A}} : u \in \mathrm{NF}(t),\ t \in \mathcal{T}(N)\} = N^{\mathbf{B}}$. □

9 An Example

Multisets: $\mathcal{M}_f(D)$ is the set of all finite multisets with elements in D, where $m \in \mathcal{M}_f(D)$ is a function from D into $\mathbb{N}$ such that $m(a) = 0$ for all a belonging to a cofinite subset of D. The natural number $\sharp m = \Sigma_{a\in D} m(a)$ is the cardinality of m. The union $m \uplus p$ of two finite multisets is defined by $(m \uplus p)(a) = m(a)+p(a)$ for all $a \in D$.

Let D be a set together with an injection $\to: \mathcal{M}_f(D) \times D \to D$. We adopt the convention that the operator "$\to$" associates to the right, i.e., $p \to (q \to \gamma)$ is abbreviated by $p \to q \to \gamma$. We define an algebra $\mathbf{D} = (\mathcal{P}(D), \cup, \emptyset, \cdot_n, K, S_{\bar{k}})_{n\in\mathbb{N},\ \bar{k}\in\mathbb{N}^*}$ in the similarity type of RCA, where $K = \{[\alpha] \to [\,] \to \alpha : \alpha \in D\}$, $S_{\bar{k}} = \{[p_0 \to [\beta_1, \dots, \beta_n] \to \beta_0] \to [p_1 \to \beta_1, \dots, p_n \to \beta_n] \to (\uplus_{i=0}^{n} p_i) \to \beta_0 : \beta_i \in D,\ p_i \in \mathcal{M}_f(D),\ \sharp p_i = k_i, |\bar{k}| = n+1\}$, and application is the extension by linearity of the following map on singleton sets (we write γ for $\{\gamma\}$, etc.): $\gamma[\beta_1, \dots, \beta_n] = \alpha$ if $\gamma = [\beta_1, \dots, \beta_n] \to \alpha$; it is equal to $\emptyset$, otherwise. It is an easy calculation to show that **D** is a RCA. To prove that **D** is indeed a RLA, by Corollary 2 it is sufficient to embed **D** into the combinatory reduct of a suitable RLAA **E** that we define here:

(i) $\mathcal{M}_f(D)^{(V)} = \{\rho : V \to \mathcal{M}_f(D) : \rho(x) = [\,]$ for cofinitely many $x \in V\}$ is the set of *environments*;
(ii) ε, defined by $x \mapsto [\,]$, is the *empty environment*, while, for an environment ρ and a finite multiset m, we define a new environment $\rho\{x := m\}$ as follows: $\rho\{x := m\}(x) = m$ and $\rho\{x := m\}(y) = \rho(y)$ if $y \neq x$.

We now construct the algebra $\mathbf{E} = (\mathcal{P}(\mathcal{M}_f(D)^{(V)} \times D), \cup, \emptyset, \cdot_k, \lambda x^{\mathbf{E}}, x^{\mathbf{E}})_{x\in V, k\in\mathbb{N}}$ by defining application and abstraction as the extension by linearity of the following family of functions defined over the singletons (we write (ρ, α) for $\{(\rho, \alpha)\}$): $\lambda x^{\mathbf{E}}(\rho, \alpha) = (\rho\{x := [\,]\}, \rho(x) \to \alpha)$; $x^{\mathbf{E}} = \{(\varepsilon\{x := [\alpha]\}, \alpha) : \alpha \in D\}$; $(\rho_0, \alpha_0)[(\rho_1, \alpha_1) \dots, (\rho_n, \alpha_n)] = (\uplus_{i=0}^{n} \rho_i, \alpha)$ if $\alpha_0 = [\alpha_1, \dots, \alpha_n] \to \alpha$; it is equal to $\emptyset$, otherwise. Notice that $(\lambda x.(\rho, \alpha))x^n = (\rho, \alpha)$ if, and only if, $\sharp\rho(x) = n$.

Theorem 5. *The algebra* **E** *is a* RLAA *and the map* $h : \mathcal{P}(D) \to \mathrm{Zd}\,E$*, defined by* $h(X) = \{(\varepsilon, \alpha) : \alpha \in X\}$ *is an embedding from* **D** *into* $\mathrm{Cr}\,\mathbf{E}$, *making* **D** *an* RLA.

References

[1] Barendregt, H.P.: The λ -calculus: its syntax and semantics. North Holland, Amsterdam (1984)
[2] Boudol, G.: The lambda-calculus with multiplicities. In: Best, E. (ed.) CONCUR 1993. LNCS, vol. 715, pp. 1–6. Springer, Heidelberg (1993)
[3] Boudol, G., Curien, P.-L., Lavatelli, C.: A semantics for lambda calculi with resources. Math. Structures Comput. Sci. 9(4), 437–482 (1999)
[4] Bucciarelli, A., Ehrhard, T., Manzonetto, G.: Categorical models for simply typed resource calculi. In: MFPS 2010 (2010)
[5] Curry, H.B., Feys, R.: Combinatory logic, vol. I. North-Holland, Amsterdam (1958)
[6] Ehrhard, T., Regnier, L.: Böhm trees, Krivine machine and the Taylor expansion of ordinary λ-terms. In: 2nd Conference on Computability in Europe (2006)
[7] Ehrhard, T., Regnier, L.: The differential lambda-calculus. Theor. Comput. Sci. 309(1), 1–41 (2003)
[8] Ehrhard, T., Regnier, L.: Differential interaction nets. Electron. Notes Theor. Comput. Sci. 123, 35–74 (2005)
[9] Ehrhard, T., Regnier, L.: Uniformity and the Taylor expansion of ordinary lambda-terms. Theor. Comput. Sci. 403(2-3), 347–372 (2008)
[10] Girard, J.-Y.: Linear logic. Theor. Comput. Sci. 50(1), 1–102 (1987)
[11] Krivine, J.-L.: Lambda-Calculus, Types and Models. Ellis Horwood Ltd. (1993)
[12] Lusin, S., Salibra, A.: The lattice of λ-theories. J. Logic Comput. 14, 373–394 (2004)
[13] Manzonetto, G., Salibra, A.: Applying universal algebra to lambda calculus. J. Logic Comput. (to appear)
[14] Manzonetto, G., Salibra, A.: Boolean algebras for lambda calculus. In: LICS 2006 (2006)
[15] Pagani, M., Della Rocca, S.R.: Solvability in resource λ-calculus. In: Ong, L. (ed.) FOSSACS 2010. LNCS, vol. 6014, pp. 358–373. Springer, Heidelberg (2010)
[16] Pagani, M., Tranquilli, P.: Parallel reduction in resource λ-calculus. In: Pagani, M., Tranquilli, P. (eds.) APLAS 2009. LNCS, vol. 5904, pp. 226–242. Springer, Heidelberg (2009)
[17] Pigozzi, D., Salibra, A.: Lambda abstraction algebras: representation theorems. Theor. Comput. Sci. 140, 5–52 (1995)
[18] Pigozzi, D., Salibra, A.: Lambda abstraction algebras: coordinatizing models of lambda calculus. Fundam. Inf. 33(2), 149–200 (1998)
[19] Salibra, A.: On the algebraic models of lambda calculus. Theor. Comput. Sci. 249(1), 197–240 (2000)
[20] Tranquilli, P.: Intuitionistic differential nets and λ-calculus. Theor. Comput. Sci. (to appear)
[21] Vaux, L.: The algebraic λ-calculus. Math. Struct. Comp. Sci. 19, 1029–1059 (2009)

Randomness for Free*

Krishnendu Chatterjee[1], Laurent Doyen[2], Hugo Gimbert[3], and Thomas A. Henzinger[1]

[1] IST Austria (Institute of Science and Technology Austria)
[2] LSV, ENS Cachan & CNRS, France
[3] LaBri & CNRS, Bordeaux, France

Abstract. We consider two-player zero-sum games on graphs. These games can be classified on the basis of the information of the players and on the mode of interaction between them. On the basis of information the classification is as follows: (a) partial-observation (both players have partial view of the game); (b) one-sided complete-observation (one player has complete observation); and (c) complete-observation (both players have complete view of the game). On the basis of mode of interaction we have the following classification: (a) concurrent (players interact simultaneously); and (b) turn-based (players interact in turn). The two sources of randomness in these games are randomness in transition function and randomness in strategies. In general, randomized strategies are more powerful than deterministic strategies, and randomness in transitions gives more general classes of games. We present a complete characterization for the classes of games where randomness is not helpful in: (a) the transition function (probabilistic transition can be simulated by deterministic transition); and (b) strategies (pure strategies are as powerful as randomized strategies). As consequence of our characterization we obtain new undecidability results for these games.

1 Introduction

Games on graphs. Games played on graphs provide the mathematical framework to analyze several important problems in computer science as well as mathematics. In particular, when the vertices and edges of a graph represent the states and transitions of a reactive system, then the synthesis problem (Church's problem) asks for the construction of a winning strategy in a game played on a graph [4,16,15,13]. Game-theoretic formulations have also proved useful for the verification [1], refinement [10], and compatibility checking [7] of reactive systems. Games played on graphs are dynamic games that proceed for an infinite number of rounds. In each round, the players choose moves; the moves, together with the current state, determine the successor state. An outcome of the game, called a *play*, consists of the infinite sequence of states that are visited.

Strategies and objectives. A strategy for a player is a recipe that describes how the player chooses a move to extend a play. Strategies can be classified as follows: *pure* strategies, which always deterministically choose a move to extend the play, vs. *randomized* strategies, which may choose at a state a probability distribution over the available moves. Objectives are generally Borel measurable functions [12]: the objective for

* This research was supported by the European Union project COMBEST and the European Network of Excellence ArtistDesign.

P. Hliněný and A. Kučera (Eds.): MFCS 2010, LNCS 6281, pp. 246–257, 2010.

a player is a Borel set B in the Cantor topology on S^ω (where S is the set of states), and the player satisfies the objective iff the outcome of the game is a member of B. In verification, objectives are usually *ω-regular languages*. The ω-regular languages generalize the classical regular languages to infinite strings; they occur in the low levels of the Borel hierarchy (they lie in $\Sigma_3 \cap \Pi_3$) and they form a robust and expressive language for determining payoffs for commonly used specifications.

Classification of games. Games played on graphs can be classified according to the knowledge of the players about the state of the game, and the way of choosing moves. Accordingly, there are (a) *partial-observation* games, where each player only has a partial or incomplete view about the state and the moves of the other player; (b) *one-sided complete-observation* games, where one player has partial knowledge and the other player has complete knowledge about the state and moves of the other player; and (c) *complete-observation* games, where each player has complete knowledge of the game. According to the way of choosing moves, the games on graphs can be classified into *turn-based* and *concurrent* games. In turn-based games, in any given round only one player can choose among multiple moves; effectively, the set of states can be partitioned into the states where it is player 1's turn to play, and the states where it is player 2's turn. In concurrent games, both players may have multiple moves available at each state, and the players choose their moves simultaneously and independently.

Sources of randomness. There are two sources of randomness in these games. First is the randomness in the transition function: given a current state and moves of the players, the transition function defines a probability distribution over the successor states. The second source of randomness is the randomness in strategies (when the players play randomized strategies). In this work we study when randomness can be obtained for *free*; i.e., we study in which classes of games the probabilistic transition function can be simulated by deterministic transition function, and the classes of games where pure strategies are as powerful as randomized strategies.

Motivation. The motivation to study this problem is as follows: (a) if for a class of games it can be shown that randomness is free for transitions, then all future works related to analysis of computational complexity, strategy complexity, and algorithmic solutions can focus on the simpler class with deterministic transitions (the randomness in transition may be essential for modeling appropriate stochastic reactive systems, but the analysis can focus on the deterministic subclass); (b) if for a class of games it can be shown that randomness is free for strategies, then all future works related to correctness results can focus on the simpler class of deterministic strategies, and the results would follow for the more general class of randomized strategies; and (c) the characterization of randomness for free will allow hardness results obtained for the more general class of games (such as games with randomness in transitions) to be carried over to simpler class of games (such as games with deterministic transitions).

Our contribution. Our contributions are as follows:

1. *Randomness for free in transitions.* We show that randomness in the transition function can be obtained for free for complete-observation concurrent games (and any class that subsumes complete-observation concurrent games) and for one-sided complete-observation turn-based games (and any class that subsumes this class). The reduction is polynomial for complete-observation concurrent games, and

exponential for one-sided complete-observation turn-based games. It is known that for complete-observation turn-based games, a probabilistic transition function cannot be simulated by deterministic transition function (see discussion at end of Section 3 for details), and thus we present a complete characterization when randomness can be obtained for free for the transition function.

2. *Randomness for free in strategies.* We show that randomness in strategies is free for complete-observation turn-based games, and for one-player partial-observation games (POMDPs). For all other classes of games randomized strategies are more powerful than pure strategies. It follows from a result of Martin [12] that for one-player complete observation games with probabilistic transitions (MDPs) pure strategies are as powerful as randomized strategies. We present a generalization of this result to the case of one-player partial-observation games with probabilistic transitions (POMDPs). Our proof is totally different from Martin's proof and based on a new derandomization technique of randomized strategies.
3. *New undecidability results.* As a consequence of our characterization of randomness for free, we obtain new undecidability results. In particular, using our results and results of Baier et al. [2] we show for one-sided complete-observation deterministic games, the problem of almost-sure winning for coBüchi objectives and positive winning for Büchi objectives are undecidable. Thus we obtain the first undecidability result for qualitative analysis (almost-sure and positive winning) of one-sided complete-observation deterministic games with ω-regular objectives.

2 Definitions

In this section we present the definition of concurrent games of partial information and their subclasses, and notions of strategies and objectives. Our model of game is the same as in [9] and equivalent to the model of stochastic games with signals [14,3]. A *probability distribution* on a finite set A is a function $\kappa : A \to [0,1]$ such that $\sum_{a \in A} \kappa(a) = 1$. We denote by $\mathcal{D}(A)$ the set of probability distributions on A.

Games of partial observation. A *concurrent game of partial observation* (or simply a *game*) is a tuple $G = \langle S, A_1, A_2, \delta, \mathcal{O}_1, \mathcal{O}_2 \rangle$ with the following components: (1) *(State space).* S is a finite set of states; (2) *(Actions).* A_i ($i = 1, 2$) is a finite set of actions for Player i; (3) *(Probabilistic transition function).* $\delta : S \times A_1 \times A_2 \to \mathcal{D}(S)$ is a concurrent probabilistic transition function that given a current state s, actions a_1 and a_2 for both players gives the transition probability $\delta(s, a_1, a_2)(s')$ to the next state s'; and (4) *(Observations).* $\mathcal{O}_i \subseteq 2^S$ ($i = 1, 2$) is a finite set of observations for Player i that partition the state space S. These partitions uniquely define functions $\mathsf{obs}_i : S \to \mathcal{O}_i$ ($i = 1, 2$) that map each state to its observation such that $s \in \mathsf{obs}_i(s)$ for all $s \in S$.

Special cases. We consider the following special cases of partial observation concurrent games, obtained either by restrictions in the observations, the mode of selection of moves, the type of transition function, or the number of players:

- *(Observation restriction).* The games with *one-sided complete-observation* are the special case of games where $\mathcal{O}_1 = \{\{s\} \mid s \in S\}$ (i.e., Player 1 has complete observation) or $\mathcal{O}_2 = \{\{s\} \mid s \in S\}$ (Player 2 has complete observation). The *games of complete-observation* are the special case of games where

$\mathcal{O}_1 = \mathcal{O}_2 = \{\{s\} \mid s \in S\}$, i.e., every state is visible to each player and hence both players have complete observation. If a player has complete observation we omit the corresponding observation sets from the description of the game.
- *(Mode of interaction restriction).* A *turn-based state* is a state s such that either (i) $\delta(s, a, b) = \delta(s, a, b')$ for all $a \in A_1$ and all $b, b' \in A_2$ (i.e, the action of Player 1 determines the transition function and hence it can be interpreted as Player 1's turn to play), we refer to s as a Player-1 state, and we use the notation $\delta(s, a, -)$; or (ii) $\delta(s, a, b) = \delta(s, a', b)$ for all $a, a' \in A_1$ and all $b \in A_2$. We refer to s as a Player-2 state, and we use the notation $\delta(s, -, b)$. A state s which is both a Player-1 state and a Player-2 state is called a *probabilistic state* (i.e., the transition function is independent of the actions of the players). We write the $\delta(s, -, -)$ to denote the transition function in s. The *turn-based games* are the special case of games where all states are turn-based.
- *(Transition function restriction).* The *deterministic games* are the special case of games where for all states $s \in S$ and actions $a \in A_1$ and $b \in A_2$, there exists a state $s' \in S$ such that $\delta(s, a, b)(s') = 1$. We refer to such states s as deterministic states. For deterministic games, it is often convenient to assume that $\delta : S \times A_1 \times A_2 \to S$.
- *(Player restriction).* The *1½-player games*, also called *partially observable Markov decision processes* (or POMDP), are the special case of games where A_1 or A_2 is a singleton. Note that 1½-player games are turn-based. Games without player restriction are sometimes called 2½-player games.

The 1½-player games of complete-observation are Markov decision processes (or MDP), and 1½-player deterministic games can be viewed as graphs (and are often called one-player games).

Classes of game graphs. We will use the following abbreviations: we will use Pa for partial observation, Os for one-sided complete-observation, Co for complete-observation, C for concurrent, and T for turn-based. For example, CoC will denote complete-observation concurrent games, and OsT will denote one-sided complete-observation turn-based games. For $\mathcal{C} \in \{\mathsf{Pa}, \mathsf{Os}, \mathsf{Co}\} \times \{\mathsf{C}, \mathsf{T}\}$, we denote by $\mathcal{G}_\mathcal{C}$ the set of all $\mathcal{C}$ games. Note that the following strict inclusion: partial observation (Pa) is more general than one-sided complete-observation (Os) and Os is more general than complete-observation (Co), and concurrent (C) is more general than turn-based (T). We will denote by $\mathcal{G}_D$ the set of all games with deterministic transition function.

Plays. In a game structure, in each turn, Player 1 chooses an action $a \in A_1$, Player 2 chooses an action in $b \in A_2$, and the successor of the current state s is chosen according to the probabilistic transition function $\delta(s, a, b)$. A *play* in G is an infinite sequence of states $\rho = s_0 s_1 \dots$ such that for all $i \geq 0$, there exists $a_i \in A_1$ and $b_i \in A_2$ with $\delta(s_i, a_i, b_i, s_{i+1}) > 0$. The *prefix up to s_n* of the play ρ is denoted by $\rho(n)$, its *length* is $|\rho(n)| = n + 1$ and its *last element* is $\mathsf{Last}(\rho(n)) = s_n$. The set of plays in G is denoted $\mathsf{Plays}(G)$, and the set of corresponding finite prefixes is denoted $\mathsf{Prefs}(G)$. The *observation sequence* of ρ for player i $(i = 1, 2)$ is the unique infinite sequence $\mathsf{obs}_i(\rho) = o_0 o_1 \ldots \in O_i^\omega$ such that $s_j \in o_j$ for all $j \geq 0$.

Strategies. A *pure strategy* in G for Player 1 is a function $\sigma : \mathsf{Prefs}(G) \to A_1$. A *randomized strategy* in G for Player 1 is a function $\sigma : \mathsf{Prefs}(G) \to \mathcal{D}(A_1)$. A (pure or randomized) strategy σ for Player 1 is *observation-based* if for all prefixes $\rho, \rho' \in$

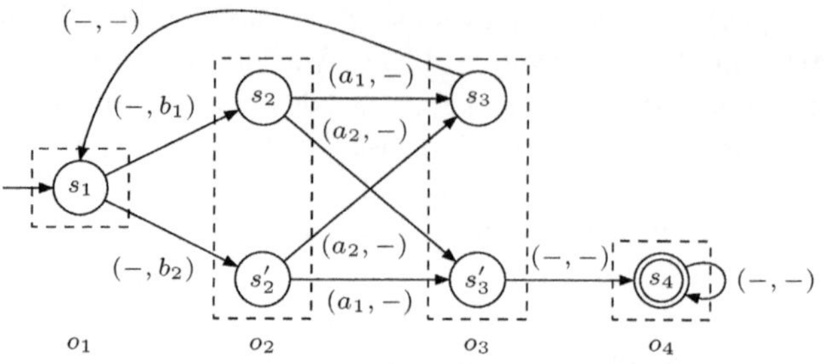

Fig. 1. A game with one-sided complete observation

$\mathsf{Prefs}(G)$, if $\mathsf{obs}_1(\rho) = \mathsf{obs}_1(\rho')$, then $\sigma(\rho) = \sigma(\rho')$. We omit analogous definitions of strategies for Player 2. We denote by Σ_G, Σ_G^O, Σ_G^P, Π_G, Π_G^O and Π_G^P the set of all Player-1 strategies, the set of all observation-based Player-1 strategies, the set of all pure Player-1 strategies, the set of all Player-2 strategies in G, the set of all observation-based Player-2 strategies, and the set of all pure Player-2 strategies, respectively. Note that if Player 1 has complete observation, then $\Sigma_G^O = \Sigma_G$.

Objectives. An *objective* for Player 1 in G is a set $\phi \subseteq S^\omega$ of infinite sequences of states. A play $\rho \in \mathsf{Plays}(G)$ *satisfies* the objective ϕ, denoted $\rho \models \phi$, if $\rho \in \phi$. Objectives are generally Borel measurable: a Borel objective is a Borel set in the Cantor topology on S^ω [11]. We specifically consider ω-regular objectives specified as parity objectives (a canonical form to express all ω-regular objectives [17]). For a play $\rho = s_0 s_1 \ldots$ we denote by $\mathrm{Inf}(\rho)$ the set of states that occur infinitely often in ρ, that is, $\mathrm{Inf}(\rho) = \{s \mid s_j = s \text{ for infinitely many } j\text{'s}\}$. For $d \in \mathbb{N}$, let $p : S \to \{0, 1, \ldots, d\}$ be a *priority function*, which maps each state to a nonnegative integer priority. The *parity* objective $\mathsf{Parity}(p)$ requires that the minimum priority that occurs infinitely often be even. Formally, $\mathsf{Parity}(p) = \{\rho \mid \min\{p(s) \mid s \in \mathrm{Inf}(\rho)\} \text{ is even}\}$. The Büchi and coBüchi objectives are the special cases of parity objectives with two priorities, $p : S \to \{0, 1\}$ and $p : S \to \{1, 2\}$ respectively. We say that an objective ϕ is *visible* for Player i if for all $\rho, \rho' \in S^\omega$, if $\rho \models \phi$ and $\mathsf{obs}_i(\rho) = \mathsf{obs}_i(\rho')$, then $\rho' \models \phi$. For example if the priority function maps observations to priorities (i.e., $p : \mathcal{O}_i \to \{0, 1, \ldots, d\}$), then the parity objective is visible for Player i.

Almost-sure winning, positive winning and value function. An *event* is a measurable set of plays, and given strategies σ and π for the two players, the probabilities of events are uniquely defined [18]. For a Borel objective ϕ, we denote by $\Pr_s^{\sigma,\pi}(\phi)$ the probability that ϕ is satisfied by the play obtained from the starting state s when the strategies σ and π are used. Given a game structure G and a state s, an observation-based strategy σ for Player 1 is *almost-sure winning (almost winning in short)* (resp. *positive winning*) for the objective ϕ from s if for all observation-based randomized strategies π for Player 2, we have $\Pr_s^{\sigma,\pi}(\phi) = 1$ (resp. $\Pr_s^{\sigma,\pi}(\phi) > 0$). The *value function* $\langle\langle 1 \rangle\rangle_{val}^G : S \to \mathbb{R}$ for Player 1 and objective ϕ assigns to every state the maximal probability with which Player 1 can guarantee the satisfaction of ϕ with an observation-based strategy, against all observation-based strategies for Player 2. Formally we have $\langle\langle 1 \rangle\rangle_{val}^G(\phi)(s) = \sup_{\sigma \in \Sigma_G^O} \inf_{\pi \in \Pi_G^O} \Pr_s^{\sigma,\pi}(\phi)$. For $\varepsilon \geq 0$, an observation-based

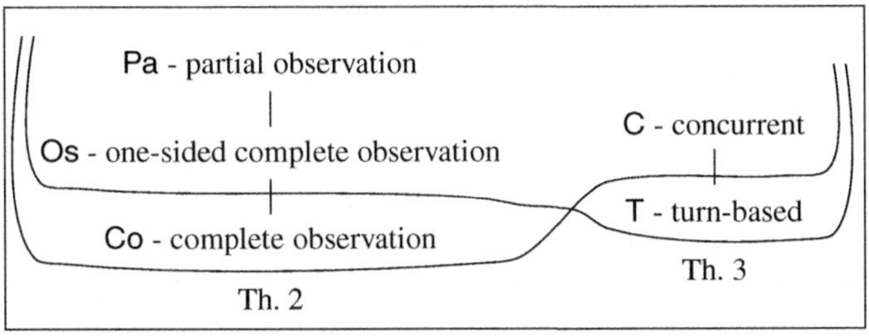

Fig. 2. The various classes of game graphs. The curves materialize the classes for which randomness is for free in transition relation (Theorem 2 and Theorem 3). For $2^1\!/_2$-player games, randomness is not free only in complete-observation turn-based games.

strategy is ε-*optimal* for ϕ from s if we have $\inf_{\pi \in \Pi_G^O} \Pr_s^{\sigma,\pi}(\phi) \geq \langle\langle 1 \rangle\rangle_{val}^G(\phi)(s) - \varepsilon$. An *optimal* strategy is a 0-optimal strategy.

Example 1. Consider the game with one-sided complete observation (Player 2 has complete information) shown in Fig. 1. Consider the Büchi objective defined by the state s_4 (i.e., state s_4 has priority 0 and other states have priority 1). Because Player 1 has partial observation (given by the partition $\mathcal{O}_i = \{\{s_1\}, \{s_2, s_2'\}, \{s_3, s_3'\}, \{s_4\}\}$), she cannot distinguish between s_2 and s_2' and therefore has to play the same actions with same probabilities in s_2 and s_2' (while it would be easy to win by playing a_2 in s_2 and a_1 in s_2', this is not possible). In fact, Player 1 cannot win using a pure observation-based strategy. However, playing a_1 and a_2 uniformly at random in all states is almost-sure winning. Every time the game visits observation o_2, for any strategy of Player 2, the game visits s_3 and s_3' with probability $\frac{1}{2}$, and hence also reaches s_4 with probability $\frac{1}{2}$. It follows that against all Player 2 strategies the play eventually reaches s_4 with probability 1, and then stays there.

3 Randomness for Free in Transition Function

In this section we present a precise characterization of the classes of games where the randomness in transition function can be obtained for *free*: in other words, we present the precise characterization of classes of games with probabilistic transition function that can be reduced to the corresponding class with deterministic transition function. We present our results as three reductions: (a) the first reduction allows us to separate probability from the mode of interaction; (b) the second reduction shows how to simulate probability in transition function with CoC (complete-observation concurrent) deterministic transition; and (c) the final reduction shows how to simulate probability in transition with OsT(one-sided complete-observation turn-based) deterministic transition. All our reductions are *local*: they consist of a gadget construction and replacement locally at every state. Our reductions preserve values, existence of ε-optimal strategies for $\varepsilon \geq 0$, and also existence of almost-sure and positive winning strategies. A visual overview is given in Fig. 2.

3.1 Separation of Probability and Interaction

A concurrent probabilistic game of partial observation G satisfies the *interaction separation* condition if the following restrictions are satisfied: the state space S can be partitioned into (S_A, S_P) such that (1) $\delta : S_A \times A_1 \times A_2 \to S_P$, and (2) $\delta : S_P \times A_1 \times A_2 \to \mathcal{D}(S_A)$ such that for all $s \in S_P$ and all $s' \in S_A$, and for all a_1, a_2, a_1', a_2' we have $\delta(s, a_1, a_2)(s') = \delta(s, a_1', a_2')(s') = \delta(s, -, -)(s')$. In other words, the choice of actions (or the interaction) of the players takes place at states in S_A and actions determine a unique successor state in S_P, and the transition function at S_P is probabilistic and independent of the choice of the players. In this section, we reduce a class of games to the corresponding class satisfying interaction separation.

Reduction to interaction separation. Let $G = \langle S, A_1, A_2, \delta, \mathcal{O}_1, \mathcal{O}_2 \rangle$ be a concurrent game of partial observation with an objective ϕ. We obtain a concurrent game of partial observation $\overline{G} = \langle S_A \cup S_P, A_1, A_2, \overline{\delta}, \overline{\mathcal{O}}_1, \overline{\mathcal{O}}_2 \rangle$ where $S_A = S$, $S_P = S \times A_1 \times A_2$, and:
- *Observation.* For $i \in \{1, 2\}$, if $\mathcal{O}_i = \{\{s\} \mid s \in S\}$, then $\overline{\mathcal{O}}_i = \{\{s'\} \mid s' \in S_A \cup S_P\}$; otherwise $\overline{\mathcal{O}}_i$ contains the observation $o \cup \{(s, a_1, a_2) \mid s \in o\}$ for each $o \in \mathcal{O}_i$.
- *Transition function.* The transition function is as follows:
  1. We have the following three cases: (a) if s is a Player 1 turn-based state, then pick an action a_2^* and for all a_2 let $\overline{\delta}(s, a_1, a_2) = (s, a_1, a_2^*)$; (b) if s is a Player 2 turn-based state, then pick an action a_1^* and for all a_1 let $\overline{\delta}(s, a_1, a_2) = (s, a_1^*, a_2)$; and (c) otherwise, $\overline{\delta}(s, a_1, a_2) = (s, a_1, a_2)$;
  2. for all $(s, a_1, a_2) \in S_P$ we have $\overline{\delta}((s, a_1, a_2), -, -)(s') = \delta(s, a_1, a_2)(s')$.
- *Objective mapping.* Given the objective ϕ in G we obtain the objective $\overline{\phi} = \{\langle s_0 s_0' s_1 s_1' \ldots \rangle \mid \langle s_0 s_1 \ldots \rangle \in \phi\}$ in $\overline{G}$.

It is easy to map observation-based strategies of the game G to observation-based strategies in $\overline{G}$ and vice-versa that preserves satisfaction of ϕ and $\overline{\phi}$ in G and $\overline{G}$, respectively. Let us refer to the above reduction as Reduction: i.e., $\mathsf{Reduction}(G, \phi) = (\overline{G}, \overline{\phi})$. Then we have the following theorem.

Theorem 1. *Let G be a concurrent game of partial observation with an objective ϕ, and let $(\overline{G}, \overline{\phi}) = \mathsf{Reduction}(G, \phi)$. Then the following assertions hold:*
1. *The reduction* Reduction *is* restriction preserving*: if G is one-sided complete-observation, then so is $\overline{G}$; if G is complete-observation, then so is $\overline{G}$; if G is turn-based, then so is $\overline{G}$.*
2. *For all $s \in S$, there is an observation-based almost-sure (resp. positive) winning strategy for ϕ from s in G iff there is an observation-based almost-sure (resp. positive) winning strategy for $\overline{\phi}$ from s in $\overline{G}$.*
3. *The reduction is objective preserving: if ϕ is a parity objective, then so in $\overline{\phi}$; if ϕ is an objective in the k-the level of the Borel hierarchy, then so is $\overline{\phi}$.*
4. *For all $s \in S$ we have $\langle\langle 1 \rangle\rangle_{val}^{G}(\phi)(s) = \langle\langle 1 \rangle\rangle_{val}^{\overline{G}}(\overline{\phi})(s)$. For all $s \in S$ there is an observation-based optimal strategy for ϕ from s in G iff there is an observation-based optimal strategy for $\overline{\phi}$ from s in $\overline{G}$.*

Since the reduction is restriction preserving, we have a reduction that separates the interaction and probabilistic transition maintaining the restriction of observation and mode of interaction.

Uniform-n-ary concurrent probabilistic games. The class of *uniform-n-ary probabilistic games* are the special class of probabilistic games such that every state $s \in S_P$ has n successors and the transition probability to each successor is $\frac{1}{n}$. It follows from the results of [19] that every CoC probabilistic game with rational transition probabilities can be reduced in polynomial time to an equivalent polynomial size uniform-binary (i.e., $n = 2$) CoC probabilistic game for all parity objectives. The reduction is achieved by adding dummy states to simulate the probability, and the reduction extends to all objectives (in the reduced game we need to consider the objective whose projection in the original game gives the original objective).

In the case of partial information, the reduction to uniform-binary probabilistic games of [19] is not valid (see [5] for an example). We reduce a probabilistic game G to a uniform-n-ary probabilistic game with $n = 1/r$ where r is the greatest common divisor of all probabilities in the original game G (a rational r is a divisor of a rational p if $p = q \cdot r$ for some integer q). Note that the number $n = 1/r$ is an integer. We denote by $[n]$ the set $\{0, 1, \ldots, n-1\}$. For a probabilistic state $s \in S_P$, we define the n-tuple $\mathsf{Succ}(s) = \langle s'_0, \ldots, s'_{n-1} \rangle$ in which each state $s' \in S$ occurs $n \cdot \delta(s, -, -)(s')$ times. Then, we can view the transition relation $\delta(s, -, -)$ as a function assigning the same probability $r = 1/n$ to each element of $\mathsf{Succ}(s)$ (and then adding up the probabilities of identical elements). Note that the above reduction is worst-case exponential (because so can be the least common multiple of all probability denominators). This is necessary to have the property that all probabilistic states in the game have the same number of successors. This property is crucial because it determines the number of actions available to Player 1 in the reductions presented in Section 3.2 and 3.3, and the number of available actions should not differ in states that have the same observation.

3.2 Simulating Probability by Complete-Observation Concurrent Determinism

In this section, we show that probabilistic states can be simulated by CoC deterministic gadgets (and hence also by OsC and PaC deterministic gadgets). By Theorem 1, we focus on games that satisfy interaction separation. A probabilistic state with uniform probability over the successors is simulated by a complete-observation concurrent deterministic state where the optimal strategy for both players is to play uniformly over the set of available actions (more details are given in [5]). This gives us Theorem 2.

Theorem 2. *Let $a \in \{\mathsf{Pa}, \mathsf{Os}, \mathsf{Co}\}$ and $b \in \{\mathsf{C}, \mathsf{T}\}$, and let $C = ab$ and $C' = a\mathsf{C}$. Let G be a game in $\mathcal{G}_C$ with probabilistic transition function with rational probabilities and an objective ϕ. A game $\overline{G} \in \mathcal{G}_{C'} \cap \mathcal{G}_D$ (in the class that subsumes $\mathcal{G}_C$ with concurrent interaction) with deterministic transition function can be constructed in (a) polynomial time if $a = \mathsf{Co}$, and (b) in exponential time if $a = \mathsf{Pa}$ or Os, with an objective $\overline{\phi}$ such that the state space of G is a subset of the state space of $\overline{G}$ and we have:*

1. *For all $s \in S$ there is an observation-based almost-sure (resp. positive) winning strategy from s for ϕ in G iff there is an observation-based almost-sure (resp. positive) winning strategy for $\overline{\phi}$ from s in $\overline{G}$.*
2. *For all $s \in S$ we have $\langle\langle 1 \rangle\rangle^{G}_{val}(\phi)(s) = \langle\langle 1 \rangle\rangle^{\overline{G}}_{val}(\overline{\phi})(s)$. For all $s \in S$ there is an observation-based optimal strategy for ϕ from s in G iff there is an observation-based optimal strategy for $\overline{\phi}$ from s in $\overline{G}$.*

3.3 Simulating Probability by One-Sided Complete-Observation Turn-Based Determinism

We show that probabilistic states can be simulated by OsT (one-sided complete-observation turn-based) states, and by Theorem 1 we consider games that satisfy interaction separation. The reduction is as follows: each probabilistic state s is transformed into a Player-2 state with n successor Player-1 states (where n is chosen such that the probabilities in s are integer multiples of $1/n$). Because all successors of s have the same observation, Player 1 has no advantage in playing after Player 2, and because by playing all actions uniformly at random each player can unilaterally decide to simulate the probabilistic state, the value and properties of strategies of the game are preserved. Due to lack of space, the proof of Theorem 3 is given in [5].

Theorem 3. *Let* $a \in \{\mathsf{Pa}, \mathsf{Os}, \mathsf{Co}\}$ *and* $b \in \{\mathsf{C}, \mathsf{T}\}$, *and let* $a' = a$ *if* $a \neq \mathsf{Co}$, *and* $a' = \mathsf{Os}$ *otherwise. Let* $C = ab$ *and* $C' = a'b$. *Let G be a game in* $\mathcal{G}_C$ *with probabilistic transition function with rational transition probabilities and an objective* ϕ. *A game* $G' \in \mathcal{G}_{C'} \cap \mathcal{G}_D$ *(in the class that subsumes one-sided complete-observation turn-based games and the class* $\mathcal{G}_C$*) with deterministic transition function can be constructed in exponential time with an objective* ϕ' *such that the state space of G is a subset of the state space of* G' *and we have:*

1. *For all* $s \in S$ *there is an observation-based almost-sure (resp. positive) winning strategy from* s *for* ϕ *in G iff there is an observation-based almost-sure (resp. positive) winning strategy for* ϕ' *from* s *in* G'.
2. *For all* $s \in S$ *we have* $\langle\langle 1 \rangle\rangle_{val}^{G}(\phi)(s) = \langle\langle 1 \rangle\rangle_{val}^{G'}(\phi')(s)$. *For all* $s \in S$ *there is an observation-based optimal strategy for* ϕ *from* s *in G iff there is an observation-based optimal strategy for* ϕ' *from* s *in* G'.

Role of probabilistic transition in CoT games and POMDPs. We have shown that for CoC games and OsT games, randomness in transition can be obtained for free. We complete the picture by showing that for CoT (complete-observation turn-based) games randomness in transition cannot be obtained for free. It follows from the result of Martin [12] that for all CoT deterministic games and all objectives, the values are either 1 or 0; however, MDPs with reachability objectives can have values in the interval $[0, 1]$ (not value 0 and 1 only). Thus the result follows for CoT games. It also follows that "randomness in transitions" can be replaced by "randomness in strategies" is not true: in CoT deterministic games even with randomized strategies the values are either 1 or 0 [12]; whereas MDPs can have values in the interval $[0, 1]$. For POMDPs, we show in Theorem 5 that pure strategies are sufficient, and it follows that for POMDPs with deterministic transition function the values are 0 or 1, and since MDPs with reachability objectives can have values other than 0 and 1 it follows that randomness in transition cannot be obtained for free for POMDPs. The probabilistic transition also plays an important role in the complexity of solving games in case of CoT games: for example, CoT deterministic games with reachability objectives can be solved in linear time, but for probabilistic transition the problem is in NP $\cap$ coNP and no polynomial time algorithm is known. In contrast, for CoC games we present a polynomial time reduction from probabilistic transition to deterministic transition. Table 1 summarizes our results characterizing the classes of games where randomness in transition can be obtained for free.

Table 1. When randomness is for free in the transition function. In particular, probabilities can be eliminated in all classes of 2-player games except complete-observation turn-based games.

	2½-player			1½-player	
	complete	one-sided	partial	MDP	POMDP
turn-based	not	free	free	not	not
concurrent	free	free	free	(NA)	(NA)

4 Randomness for Free in Strategies

It is known from the results of [8] that in CoC games randomized strategies are more powerful than pure strategies; for example, values achieved by pure strategies are lower than values achieved by randomized strategies and randomized almost-sure winning strategies may exist whereas no pure almost-sure winning strategy exists. Similar results also hold in the case of OsT games (see [6] for an example). By contrast we show that in one-player games, restricting the set of strategies to pure strategies does not decrease the value nor affect the existence of almost-sure and positive winning strategies. We first start with a lemma, then present a result that can be derived from Martin's theorem for Blackwell games [12], and finally present our results precisely in a theorem.

Lemma 1. *Let G be a* POMDP *with initial state s_* and an objective $\phi \subseteq S^\omega$. Then for every randomized observation-based strategy $\sigma \in \Sigma_O$ there exists a* pure *observation-based strategy $\sigma_P \in \Sigma_P \cap \Sigma_O$ such that $\Pr_{s_*}^{\sigma}(\phi) \leq \Pr_{s_*}^{\sigma_P}(\phi)$.*

The main argument in the proof of Lemma 1 relies on showing that the value $\Pr_s^{\sigma}(\phi)$ of any randomized observation-based strategy σ is equal to the average of the values $\Pr_s^{\sigma_i}(\phi)$ of (uncountably many) pure observation-based strategies σ_i. Therefore, one of the pure strategies σ_i has to achieve at least the value of the randomized strategy σ. The theory of integration and Fubini's theorem make this argument precise (see [5] for details).

Theorem 4 ([12]). *Let G be a* CoT *stochastic game with initial state s_* and an objective $\phi \subseteq S^\omega$. Then the following equalities hold: $\inf_{\pi\in\Pi_O}\sup_{\sigma\in\Sigma_O}\Pr_{s_*}^{\sigma,\pi}(\phi) = \sup_{\sigma\in\Sigma_O}\inf_{\pi\in\Pi_O}\Pr_{s_*}^{\sigma,\pi}(\phi) = \sup_{\sigma\in\Sigma_O\cap\Sigma_P}\inf_{\pi\in\Pi_O}\Pr_{s_*}^{\sigma,\pi}(\phi)$.*

We obtain the following result as a consequence of Lemma 1.

Theorem 5. *Let G be a* POMDP *with initial state s_* and an objective $\phi \subseteq S^\omega$. Then the following assertions hold:*

1. $\sup_{\sigma\in\Sigma_O}\Pr_{s_*}^{\sigma}(\phi) = \sup_{\sigma\in\Sigma_O\cap\Sigma_P}\Pr_{s_*}^{\sigma}(\phi)$.
2. *If there is a randomized optimal (resp. almost-sure winning, positive winning) strategy for ϕ from s_*, then there is a pure optimal (resp. almost-sure winning, positive winning) strategy for ϕ from s_*.*

Theorem 4 can be derived as a consequence of Martin's proof of determinacy of Blackwell games [12]: the result states that for CoT stochastic games pure strategies can achieve the same value as randomized strategies, and as a special case the result also

Table 2. When deterministic (ϵ-optimal) strategies are as powerful as randomized strategies. The case $\epsilon = 0$ in complete-observation turn-based games is open.

	2½-player			1½-player	
	complete	one-sided	partial	MDP	POMDP
turn-based	$\epsilon > 0$	not	not	$\epsilon \geq 0$	$\epsilon \geq 0$
concurrent	not	not	not	(NA)	(NA)

holds for MDPs. Theorem 5 shows that the result can be generalized to POMDPs, and a stronger result (item (2) of Theorem 5) can be proved for POMDPs (and MDPs as a special case). It remains open whether result similar to item (2) of Theorem 5 can be proved for CoT stochastic games. The results summarizing when randomness can be obtained for free for strategies is shown in Table 2.

Undecidability result for POMDPs. The results of [2] shows that the emptiness problem for probabilistic coBüchi (resp. Büchi) automata under the almost-sure (resp. positive) semantics [2] is undecidable. As a consequence it follows that for POMDPs the problem of deciding if there is a pure observation-based almost-sure (resp. positive) winning strategy for coBüchi (resp. Büchi) objectives is undecidable, and as a consequence of Theorem 5 we obtain the same undecidability result for randomized strategies. This result closes an open question discussed in [9]. The undecidability result holds even if the coBüchi (resp. Büchi) objectives are visible.

Corollary 1. *Let G be a* POMDP *with initial state s_* and let $\mathcal{T} \subseteq S$ be a subset of states (or subset of observations). Whether there exists a pure or randomized almost-sure winning strategy for Player 1 from s in G for the objective* $\mathsf{coBuchi}(\mathcal{T})$ *is undecidable; and whether there exists a pure or randomized positive winning strategy for Player 1 from s in G for the objective* $\mathsf{Buchi}(\mathcal{T})$ *is undecidable.*

Undecidability result for one-sided complete-observation turn-based games. The undecidability results of Corollary 1 also holds for OsT stochastic games (as they subsume POMDPs as a special case). It follows from Theorem 3 that OsT stochastic games can be reduced to OsT deterministic games. Thus we obtain the first undecidability result for OsT deterministic games (Corollary 2), solving the open question of [6].

Corollary 2. *Let G be an* OsT *deterministic game with initial state s_* and let $\mathcal{T} \subseteq S$ be a subset of states (or subset of observations). Whether there exists a pure or randomized almost-sure winning strategy for Player 1 from s in G for the objective* $\mathsf{coBuchi}(\mathcal{T})$ *is undecidable; and whether there exists a pure or randomized positive winning strategy for Player 1 from s in G for the objective* $\mathsf{Buchi}(\mathcal{T})$ *is undecidable.*

5 Conclusion

In this work we have presented a precise characterization for classes of games where randomization can be obtained for free in transitions and in strategies. As a consequence of our characterization we obtain new undecidability results. The other impact of our characterization is as follows: for the class of games where randomization is free in transition, future algorithmic and complexity analysis can focus on the simpler class of deterministic games; and for the class of games where randomization is free in strategies, future analysis of such games can focus on the simpler class of deterministic strategies. Thus our results will be useful tools for simpler analysis techniques in the study of games.

References

1. Alur, R., Henzinger, T.A., Kupferman, O.: Alternating-time temporal logic. Journal of the ACM 49, 672–713 (2002)
2. Baier, C., Bertrand, N., Größer, M.: On decision problems for probabilistic Büchi automata. In: Amadio, R.M. (ed.) FOSSACS 2008. LNCS, vol. 4962, pp. 287–301. Springer, Heidelberg (2008)
3. Bertrand, N., Genest, B., Gimbert, H.: Qualitative determinacy and decidability of stochastic games with signals. In: Proc. of LICS, pp. 319–328. IEEE Computer Society, Los Alamitos (2009)
4. Büchi, J.R., Landweber, L.H.: Solving sequential conditions by finite-state strategies. Transactions of the AMS 138, 295–311 (1969)
5. Chatterjee, K., Doyen, L., Gimbert, H., Henzinger, T.A.: Randomness for free. CoRR, abs/1006.0673 (2010)
6. Chatterjee, K., Doyen, L., Henzinger, T.A., Raskin, J.-F.: Algorithms for omega-regular games of incomplete information. Logical Methods in Computer Science 3(3:4) (2007)
7. de Alfaro, L., Henzinger, T.A.: Interface theories for component-based design. In: Henzinger, T.A., Kirsch, C.M. (eds.) EMSOFT 2001. LNCS, vol. 2211, pp. 148–165. Springer, Heidelberg (2001)
8. Everett, H.: Recursive games. In: Contributions to the Theory of Games III. Annals of Mathematical Studies, vol. 39, pp. 47–78 (1957)
9. Gripon, V., Serre, O.: Qualitative concurrent stochastic games with imperfect information. In: Albers, S., Marchetti-Spaccamela, A., Matias, Y., Nikoletseas, S., Thomas, W. (eds.) ICALP 2009, Part II. LNCS, vol. 5556, pp. 200–211. Springer, Heidelberg (2009)
10. Henzinger, T.A., Kupferman, O., Rajamani, S.: Fair simulation. Information and Computation 173, 64–81 (2002)
11. Kechris, A.: Classical Descriptive Set Theory. Springer, Heidelberg (1995)
12. Martin, D.A.: The determinacy of Blackwell games. The Journal of Symbolic Logic 63(4), 1565–1581 (1998)
13. McNaughton, R.: Infinite games played on finite graphs. Annals of Pure and Applied Logic 65, 149–184 (1993)
14. Mertens, J.-F., Sorin, S., Zamir, S.: Repeated games. Core Discussion Papers, 9422 (1994)
15. Pnueli, A., Rosner, R.: On the synthesis of a reactive module. In: POPL 1989, pp. 179–190. ACM Press, New York (1989)
16. Ramadge, P.J., Wonham, W.M.: Supervisory control of a class of discrete-event processes. SIAM Journal of Control and Optimization 25(1), 206–230 (1987)
17. Thomas, W.: Languages, automata, and logic. In: Handbook of Formal Languages, Beyond Words, ch. 7, vol. 3, pp. 389–455. Springer, Heidelberg (1997)
18. Vardi, M.Y.: Automatic verification of probabilistic concurrent finite-state systems. In: FOCS, pp. 327–338. IEEE Computer Society Press, Los Alamitos (1985)
19. Zwick, U., Paterson, M.: The complexity of mean payoff games on graphs. Theoretical Computer Science 158, 343–359 (1996)

Qualitative Analysis of Partially-Observable Markov Decision Processes*

Krishnendu Chatterjee[1], Laurent Doyen[2], and Thomas A. Henzinger[1]

[1] IST Austria (Institute of Science and Technology Austria)
[2] LSV, ENS Cachan & CNRS, France

Abstract. We study observation-based strategies for *partially-observable Markov decision processes* (POMDPs) with parity objectives. An observation-based strategy relies on partial information about the history of a play, namely, on the past sequence of observations. We consider qualitative analysis problems: given a POMDP with a parity objective, decide whether there exists an observation-based strategy to achieve the objective with probability 1 (almost-sure winning), or with positive probability (positive winning). Our main results are twofold. First, we present a complete picture of the computational complexity of the qualitative analysis problem for POMDPs with parity objectives and its subclasses: safety, reachability, Büchi, and coBüchi objectives. We establish several upper and lower bounds that were not known in the literature. Second, we give optimal bounds (matching upper and lower bounds) for the memory required by pure and randomized observation-based strategies for each class of objectives.

1 Introduction

Markov decision processes. A *Markov decision process (MDP)* is a model for systems that exhibit both probabilistic and nondeterministic behavior. MDPs have been used to model and solve control problems for stochastic systems: there, nondeterminism represents the freedom of the controller to choose a control action, while the probabilistic component of the behavior describes the system response to control actions. MDPs have also been adopted as models for concurrent probabilistic systems, probabilistic systems operating in open environments [21], and under-specified probabilistic systems [5].

System specifications. The *specification* describes the set of desired behaviors of the system, and is typically an ω-regular set of paths. Parity objectives are a canonical way to define such specifications in MDPs. They include reachability, safety, Büchi and coBüchi objectives as special cases. Thus MDPs with parity objectives provide the theoretical framework to study problems such as the verification and the control of stochastic systems.

Perfect vs. partial observations. Most results about MDPs make the hypothesis of *perfect observation*. In this setting, the controller always knows, while interacting with the system (or MDP), the exact state of the MDP. In practice, this hypothesis is often unrealistic. For example, in the control of multiple processes, each process has only access to

* This research was supported by the European Union project COMBEST and the European Network of Excellence ArtistDesign.

P. Hliněný and A. Kučera (Eds.): MFCS 2010, LNCS 6281, pp. 258–269, 2010.

the public variables of the other processes, but not to their private variables. In control of hybrid systems [13], or automated planning [17], the controller usually has noisy information about the state of the systems due to finite-precision sensors. In such applications, MDPs with *partial observation* (POMDPs) provide a more appropriate model.

Qualitative and quantitative analysis. Given an MDP with parity objective, the *qualitative analysis* asks for the computation of the set of *almost-sure winning* states (resp., *positive winning* states) in which the controller can achieve the parity objective with probability 1 (resp., positive probability); the more general *quantitative analysis* asks for the computation at each state of the maximal probability with which the controller can satisfy the parity objective. The analysis of POMDPs is considerably more complicated than the analysis of MDPs. First, the decision problems for POMDPs usually lie in higher complexity classes than their perfect-observation counterparts: for example, the quantitative analysis of POMDPs with reachability and safety objectives is undecidable [19], whereas for MDPs with perfect observation, this question can be solved in polynomial time [11,10]. Second, in the context of POMDPs, witness winning strategies for the controller need memory even for the simple objectives of safety and reachability. This is again in contrast to the perfect-observation case, where memoryless strategies suffice for all parity objectives. Since the quantitative analysis of POMDPs is undecidable (even for computing approximations of the maximal probabilities [17]), we study the qualitative analysis of POMDPs with parity objective and its subclasses.

Contribution. For the qualitative analysis of POMDPs, the following results are known: (a) the problems of deciding if a state is almost-sure winning for reachability and Büchi objectives can be solved in EXPTIME [1]; (b) the problems for almost-sure winning for coBüchi objectives and positive winning for Büchi objectives are undecidable [1,7]; and (c) the EXPTIME-completeness of almost-sure winning for safety objectives follows from the results on games with partial observation [8,4]. Our new contributions are as follows:

1. First, we show that (a) positive winning for reachability objectives is NLOGSPACE-complete; and (b) almost-sure winning for reachability and Büchi objectives, and positive winning for safety and coBüchi objectives are EXPTIME-hard. We also present a new proof that positive winning for safety and coBüchi objectives can be solved in EXPTIME[1]. It follows that almost-sure winning for reachability and Büchi, and positive winning for safety and coBüchi, are EXPTIME-complete. This completes the picture for the complexity of the qualitative analysis for POMDPs with parity objectives. Moreover our new proofs of EXPTIME upper-bound proofs yield efficient and symbolic algorithms to solve positive winning for safety and coBüchi objectives in POMDPs.
2. Second, we present a complete characterization of the amount of memory required by pure (deterministic) and randomized strategies for the qualitative analysis of POMDPs. For the first time, we present optimal memory bounds (matching upper and lower bounds) for pure and randomized strategies: we show that (a) for positive winning of reachability objectives, randomized memoryless strategies suffice,

[1] A different proof that positive safety can be solved in EXPTIME is given in [14] (see the discussion after Theorem 2 for a comparison).

while for pure strategies linear memory is necessary and sufficient; (b) for almost-sure winning of safety, reachability, and Büchi objectives, and for positive winning of safety and coBüchi objectives, exponential memory is necessary and sufficient for both pure and randomized strategies.

Related work. Though MDPs have been widely studied under the hypothesis of perfect observations, there are a few works that consider POMDPs, e.g., [18,16] for several finite-horizon quantitative objectives. The results of [1] shows the upper bounds for almost-sure winning for reachability and Büchi objectives, and the work of [6] considers a subclass of POMDPs with Büchi objectives and presents a PSPACE upper bound for the subclass. The undecidability of almost-sure winning for coBüchi and positive winning for Büchi objectives is established by [1,7]. We present a solution to the remaining problems related to the qualitative analysis of POMDPs with parity objectives, and complete the picture. Partial information has been studied in the context of two-player games [20,8], a model that is incomparable to MDPs, though some techniques (like the subset construction) can be adapted in the context of POMDPs. More general models of stochastic games with partial information have been studied in [2,14], and lie in higher complexity classes. For example, a result of [2] shows that the decision problem for positive winning of safety objectives is 2EXPTIME-complete in the general model, while for POMDPs, we show that the same problem is EXPTIME-complete.

2 Definitions

A *probability distribution* on a finite set A is a function $\kappa : A \to [0,1]$ such that $\sum_{a \in A} \kappa(a) = 1$. The *support* of κ is the set $\mathsf{Supp}(\kappa) = \{a \in A \mid \kappa(a) > 0\}$. We denote by $\mathcal{D}(A)$ the set of probability distributions on A.

Games and MDPs. A *two-player game structure* or a *Markov decision process (MDP)* (*of partial observation*) is a tuple $G = \langle L, \Sigma, \delta, \mathcal{O} \rangle$, where L is a finite set of states, Σ is a finite set of actions, $\mathcal{O} \subseteq 2^L$ is a set of observations that partition[2] the state space L. We denote by $\mathsf{obs}(\ell)$ the unique observation $o \in \mathcal{O}$ such that $\ell \in o$. In the case of games, $\delta \subseteq L \times \Sigma \times L$ is a set of labeled transitions; in the case of MDPs, $\delta : L \times \Sigma \to \mathcal{D}(L)$ is a probabilistic transition function. For games, we require that for all $\ell \in L$ and all $\sigma \in \Sigma$, there exists $\ell' \in L$ such that $(\ell, \sigma, \ell') \in \delta$. We refer to an MDP of partial observation as a POMDP. We say that G is a game or MDP of *perfect observation* if $\mathcal{O} = \{\{\ell\} \mid \ell \in L\}$. For $\sigma \in \Sigma$ and $s \subseteq L$, define $\mathsf{Post}^G_\sigma(s) = \{\ell' \in L \mid \exists \ell \in s : (\ell, \sigma, \ell') \in \delta\}$ when G is a game, and $\mathsf{Post}^G_\sigma(s) = \{\ell' \in L \mid \exists \ell \in s : \delta(\ell, \sigma)(\ell') > 0\}$ when G is an MDP.

Plays. Games are played in rounds in which Player 1 chooses an action in Σ, and Player 2 resolves nondeterminism by choosing the successor state; in MDPs the successor state is chosen according to the probabilistic transition function. A *play* in G is an infinite sequence $\pi = \ell_0 \sigma_0 \ell_1 \dots \sigma_{n-1} \ell_n \sigma_n \dots$ such that $\ell_{i+1} \in \mathsf{Post}^G_{\sigma_i}(\{\ell_i\})$ for all $i \geq 0$. The infinite sequence $\mathsf{obs}(\pi) = \mathsf{obs}(\ell_0)\sigma_0\mathsf{obs}(\ell_1) \dots \sigma_{n-1}\mathsf{obs}(\ell_n)\sigma_n \dots$ is the *observation* of π.

[2] A slightly more general model with overlapping observations can be reduced in polynomial time to partitioning observations [8].

The set of infinite plays in G is denoted $\mathsf{Plays}(G)$, and the set of finite prefixes $\ell_0\sigma_0\ldots\sigma_{n-1}\ell_n$ of plays is denoted $\mathsf{Prefs}(G)$. A state $\ell \in L$ is *reachable* in G if there exists a prefix $\rho \in \mathsf{Prefs}(G)$ such that $\mathsf{Last}(\rho) = \ell$ where $\mathsf{Last}(\rho)$ is the last state of ρ.

Strategies. A *pure strategy* in G for Player 1 is a function $\alpha : \mathsf{Prefs}(G) \to \Sigma$. A *randomized strategy* in G for Player 1 is a function $\alpha : \mathsf{Prefs}(G) \to \mathcal{D}(\Sigma)$. A (pure or randomized) strategy α for Player 1 is *observation-based* if for all prefixes $\rho, \rho' \in \mathsf{Prefs}(G)$, if $\mathsf{obs}(\rho) = \mathsf{obs}(\rho')$, then $\alpha(\rho) = \alpha(\rho')$. In the sequel, we are interested in the existence of observation-based strategies for Player 1. A *pure strategy* in G for Player 2 is a function $\beta : \mathsf{Prefs}(G) \times \Sigma \to L$ such that for all $\rho \in \mathsf{Prefs}(G)$ and all $\sigma \in \Sigma$, we have $(\mathsf{Last}(\rho), \sigma, \beta(\rho, \sigma)) \in \delta$. A *randomized strategy* in G for Player 2 is a function $\beta : \mathsf{Prefs}(G) \times \Sigma \to \mathcal{D}(L)$ such that for all $\rho \in \mathsf{Prefs}(G)$, all $\sigma \in \Sigma$, and all $\ell \in \mathsf{Supp}(\beta(\rho, \sigma))$, we have $(\mathsf{Last}(\rho), \sigma, \ell) \in \delta$. We denote by $\mathcal{A}_G$, $\mathcal{A}_G^O$, and $\mathcal{B}_G$ the set of all Player-1 strategies, the set of all observation-based Player-1 strategies, and the set of all Player-2 strategies in G, respectively.

Memory requirement of strategies. An equivalent definition of strategies is as follows. Let Mem be a set called *memory*. An observation-based strategy with memory can be described by two functions, a *memory-update* function α_u: $\mathsf{Mem} \times \mathcal{O} \times \Sigma \to \mathsf{Mem}$ that given the current memory, observation and the action updates the memory, and a *next-action* function α_n: $\mathsf{Mem} \times \mathcal{O} \to \mathcal{D}(\Sigma)$ that given the current memory and current observation specifies the probability distribution[3] of the next action, respectively. A strategy is *finite-memory* if the memory Mem is finite and the size of a finite-memory strategy α is the size $|\mathsf{Mem}|$ of its memory. A strategy is *memoryless* if $|\mathsf{Mem}| = 1$. The memoryless strategies do not depend on the history of a play, but only on the current state. Memoryless strategies for player 1 can be viewed as functions α: $\mathcal{O} \to \mathcal{D}(\Sigma)$.

Objectives. An *objective* for G is a set ϕ of infinite sequences of states and actions, that is, $\phi \subseteq (L \times \Sigma)^\omega$. We consider objectives that are Borel measurable, i.e., sets in the Cantor topology on $(L \times \Sigma)^\omega$ [15]. We specifically consider reachability, safety, Büchi, coBüchi, and parity objectives, all of them being Borel measurable. The parity objectives are a canonical form to express all ω-regular objectives [22]. For a play $\pi = \ell_0\sigma_0\ell_1\ldots$, we denote by $\mathsf{Inf}(\pi) = \{\ell \in L \mid \ell = \ell_i \text{ for infinitely many } i\text{'s}\}$ the set of states that appear infinitely often in π.

- *Reachability and safety objectives.* Given a set $\mathcal{T} \subseteq L$ of target states, the *reachability* objective $\mathsf{Reach}(\mathcal{T}) = \{\ \ell_0\sigma_0\ell_1\sigma_1\ldots \in \mathsf{Plays}(G) \mid \exists k \geq 0 : \ell_k \in \mathcal{T}\ \}$ requires that a target state in $\mathcal{T}$ be visited at least once. Dually, the *safety* objective $\mathsf{Safe}(\mathcal{T}) = \{\ \ell_0\sigma_0\ell_1\sigma_1\ldots \in \mathsf{Plays}(G) \mid \forall k \geq 0 : \ell_k \in \mathcal{T}\ \}$ requires that only states in $\mathcal{T}$ be visited; the objective $\mathsf{Until}(\mathcal{T}_1, \mathcal{T}_2) = \{\ell_0\sigma_0\ell_1\sigma_1\ldots \in \mathsf{Plays}(G) \mid \exists k \geq 0 : \ell_k \in \mathcal{T}_2 \wedge \forall j \leq k : \ell_j \in \mathcal{T}_1\}$ requires that only states in $\mathcal{T}_1$ be visited before a state in $\mathcal{T}_2$ is visited.
- *Büchi and coBüchi objectives.* The *Büchi* objective $\mathsf{Büchi}(\mathcal{T}) = \{\pi \mid \mathsf{Inf}(\pi) \cap \mathcal{T} \neq \emptyset\}$ requires that a state in $\mathcal{T}$ be visited infinitely often. Dually, the *coBüchi* objective $\mathsf{coBüchi}(\mathcal{T}) = \{\pi \mid \mathsf{Inf}(\pi) \subseteq \mathcal{T}\}$ requires that only states in $\mathcal{T}$ be visited infinitely often.

[3] For a pure strategy, the next-action function specifies a single action rather than a probability distribution.

– *Parity objectives.* For $d \in \mathbb{N}$, let $p : L \to \{ 0, 1, \ldots, d \}$ be a *priority function* that maps each state to a nonnegative integer priority. The *parity* objective $\mathsf{Parity}(p) = \{ \pi \mid \min\{ p(\ell) \mid \ell \in \mathsf{Inf}(\pi) \} \text{ is even } \}$ requires that the smallest priority that appears infinitely often be even.

Note that the objectives $\mathsf{Büchi}(\mathcal{T})$ and $\mathsf{coBüchi}(\mathcal{T})$ are special cases of parity objectives defined by respective priority functions p_1, p_2 such that $p_1(\ell) = 0$ and $p_2(\ell) = 2$ if $\ell \in \mathcal{T}$, and $p_1(\ell) = p_2(\ell) = 1$ otherwise. An objective ϕ is visible if it depends only on the observations; formally, ϕ is *visible* if, whenever $\pi \in \phi$ and $\mathsf{obs}(\pi) = \mathsf{obs}(\pi')$, then $\pi' \subset \phi$. In this work, all our upper bound results are for the general parity objectives (not necessarily visible), and all the lower bound results for POMDPs are for the special case of visible objectives.

Almost-sure and positive winning. An *event* is a measurable set of plays, and given strategies α and β for the two players (resp., a strategy α for Player 1 in MDPs), the probabilities of events are uniquely defined [23]. For a Borel objective ϕ, we denote by $\mathrm{Pr}_\ell^{\alpha,\beta}(\phi)$ (resp., $\mathrm{Pr}_\ell^{\alpha}(\phi)$ for MDPs) the probability that ϕ is satisfied from the starting state ℓ given the strategies α and β (resp., given the strategy α). Given a game G and a state ℓ, a strategy α for Player 1 is *almost-sure winning* (resp., *positive winning*) for the objective ϕ from ℓ if for all randomized strategies β for Player 2, we have $\mathrm{Pr}_\ell^{\alpha,\beta}(\phi) = 1$ (resp., $\mathrm{Pr}_\ell^{\alpha,\beta}(\phi) > 0$). Given an MDP G and a state ℓ, a strategy α for Player 1 is almost-sure winning (resp. positive winning) for the objective ϕ from ℓ if we have $\mathrm{Pr}_\ell^{\alpha}(\phi) = 1$ (resp., $\mathrm{Pr}_\ell^{\alpha}(\phi) > 0$). We also say that state ℓ is almost-sure winning, or positive winning for ϕ respectively. We are interested in the problems of deciding the existence of an observation-based strategy for Player 1 that is almost-sure winning (resp., positive winning) from a given state ℓ.

3 Upper Bounds for the Qualitative Analysis of POMDPs

In this section, we present upper bounds for the qualitative analysis of POMDPs. We first describe the known results. For qualitative analysis of MDPs, polynomial time upper bounds are known for all parity objectives [11,10]. It follows from the results of [8,1] that the decision problems for almost-sure winning for POMDPs with reachability, safety, and Büchi objectives can be solved in EXPTIME. It also follows from the results of [1] that the decision problem for almost-sure winning with coBüchi objectives and for positive winning with Büchi objectives is undecidable if the strategies are restricted to be pure, and the results of [7] shows that the problem remains undecidable even if randomized strategies are considered. In this section, we complete the results on upper bounds for the qualitative analysis of POMDPs: we present complexity upper bounds for the decision problems of positive winning with reachability, safety and coBüchi objectives. The following result for reachability objectives is simple, and follows from equivalence to the graph reachability problem.

Theorem 1. *Given a* POMDP *G with a reachability objective and a starting state ℓ, the problem of deciding whether there is a positive winning strategy from ℓ in G is NLOGSPACE-complete.*

Positive winning for safety and coBüchi objectives. We now show that the decision problem for positive winning with safety and coBüchi objectives for POMDPs can be solved in EXPTIME. Our result for positive safety and coBüchi objectives is based on the computation of almost-sure winning states for safety objectives, and on the following lemma (proof in [9]).

Lemma 1. *Let $G = \langle L, \Sigma, \delta, \mathcal{O} \rangle$ be a* POMDP *and let $\mathcal{T} \subseteq L$ be the set of target states. If Player 1 has an observation-based strategy in G to satisfy* $\mathsf{Safe}(\mathcal{T})$ *with positive probability from some state ℓ, then there exists a state ℓ' such that (a) Player 1 has an observation-based strategy in G to satisfy* $\mathsf{Until}(\mathcal{T}, \{\ell'\})$ *with positive probability from ℓ, and (b) Player 1 has an observation-based almost-sure winning strategy in G for* $\mathsf{Safe}(\mathcal{T})$ *from ℓ'.*

By Lemma 1, positive winning states can be computed as the set of states from which Player 1 can force with positive probability to reach an almost-sure winning state while visiting only safe states. Almost-sure winning states can be computed using the following subset construction.

Given a POMDP $G = \langle L, \Sigma, \delta, \mathcal{O} \rangle$ and a set $\mathcal{T} \subseteq L$ of states, the *knowledge-based subset construction* for G is the game of perfect observation $G^{\mathsf{K}} = \langle \mathcal{L}, \Sigma, \delta^{\mathsf{K}} \rangle$, where $\mathcal{L} = 2^L \backslash \{\emptyset\}$, and for all $s_1, s_2 \in \mathcal{L}$ (in particular $s_2 \neq \emptyset$) and $\sigma \in \Sigma$, we have $(s_1, \sigma, s_2) \in \delta^{\mathsf{K}}$ iff there exists an observation $o \in \mathcal{O}$ such that either $s_2 = \mathsf{Post}^G_\sigma(s_1) \cap o \cap \mathcal{T}$, or $s_2 = (\mathsf{Post}^G_\sigma(s_1) \cap o) \setminus \mathcal{T}$. We refer to states in G^{K} as *cells*. The following result is established using standard techniques (see e.g., Lemma 3.2 and Lemma 3.3 in [8]).

Lemma 2. *Let $G = \langle L, \Sigma, \delta, \mathcal{O} \rangle$ be a* POMDP *and $\mathcal{T} \subseteq L$ a set of target states. Let G^{K} be the knowledge-based subset construction for G and $F_{\mathcal{T}} = \{s \subseteq \mathcal{T}\}$ be the set of safe cells. Player 1 has an almost-sure winning observation-based strategy in G for* $\mathsf{Safe}(\mathcal{T})$ *from ℓ if and only if Player 1 has an almost-sure winning strategy in G^{K} for* $\mathsf{Safe}(F)$ *from the cell $\{\ell\}$.*

Theorem 2. *Given a* POMDP *G with a safety objective and a starting state ℓ, the problem of deciding whether there exists a positive winning observation-based strategy from ℓ can be solved in EXPTIME.*

Algorithms. The complexity bound of Theorem 2 has been established previously in [14], using an extension of the knowledge-based subset construction which is not necessary (where the state space is $L \times 2^L$). Our proof (of Theorem 2, details in [9]) is simpler and also yield efficient and symbolic algorithms that an be obtained from the antichain algorithm of [8] for almost-sure winning of safety objectives, and simple graph reachability for positive winning of reachability objectives.

The positive winning states for a coBüchi objective are computed as the set of almost-sure winning states for safety that can be reached with positive probability (for details see [9]).

Theorem 3. *Given a* POMDP *G with a coBüchi objective and a starting state ℓ, the problem of deciding whether there exists a positive winning observation-based strategy from ℓ can be solved in EXPTIME.*

4 Lower Bounds for the Qualitative Analysis of POMDPs

In this section we present lower bounds for the qualitative analysis of POMDPs. We first present the lower bounds for MDPs with perfect observation (proofs in [9]).

Theorem 4. *Given an MDP G of perfect observation, the following assertions hold: (a) the positive winning problem for reachability objectives is NLOGSPACE-complete, and the positive winning problem for safety, Büchi, coBüchi and parity objectives is PTIME-complete; and (b) the almost-sure winning problem for reachability, safety, Büchi, coBüchi and parity objectives is PTIME-complete.*

Lower bounds for POMDPs. We have already shown that positive winning with reachability objectives in POMDPs is NLOGSPACE-complete. As in the case of MDPs with perfect observation, for safety objectives and almost-sure winning, a POMDP can be equivalently considered as a game of partial observation where Player 2 makes choices of the successors from the support of the probability distribution of the transition function, and the almost-sure winning set is the same in the POMDP and the game. Since the problem of almost-sure winning in games of partial observation with safety objective is EXPTIME-complete [4], the EXPTIME-completeness result follows. We now show that almost-sure winning with reachability objectives and positive winning with safety objectives is EXPTIME-complete. Before the result we first present a discussion on polynomial-space alternating Turing machines (ATM).
Discussion. Let M be a polynomial-space ATM and let w be an input word. Then, there is an exponential bound on the number of configurations of the machine. Hence if M can accept the word w, then it can do so within some $k_{|w|}$ steps, where $|w|$ is the length of the word w, and $k_{|w|}$ is bounded by an exponential in $|w|$. We construct an equivalent polynomial-space ATM M' that behaves as M but keeps track (in polynomial space) of the number of steps executed by M, and given a word $|w|$, if the number of steps reaches $k_{|w|}$ without accepting, then the word is rejected. The machine M' is equivalent to M and reaches the accepting or rejecting states in a number of steps bounded by an exponential in the length of the input word. The problem of deciding, given a polynomial-space ATM M and a word w, whether M accepts w is EXPTIME-complete.

Reduction from Alternating PSPACE Turing machine. Let M be a polynomial-space ATM such that for every input word w, the accepting or the rejecting state is reached within exponential steps in $|w|$. A polynomial-time reduction R_G of a polynomial-space ATM M and an input word w to a game $G = R_G(M, w)$ of partial observation is given in [8] such that (a) there is a special accepting state in G, and (b) M accepts w iff there is an observation-based strategy for Player 1 in G to reach the accepting state with probability 1. If the above reduction is applied to M, then the game structure satisfies the following additional properties: there is a special rejecting state that is absorbing, and for every observation-based strategy for Player 1, either (a) against all Player 2 strategies the accepting state is reached with probability 1; or (b) there is a pure Player 2 strategy that reaches the rejecting state with positive probability $\eta > 0$ in $2^{|L|}$ steps and the accepting or the rejecting state is reached with probability 1 in $2^{|L|}$ steps. We now present the reduction to POMDPs:

1. *Almost-sure winning for reachability.* Given a polynomial-space ATM M and w an input word, let $G = R_G(M, w)$. We construct a POMDP G' from G as follows: we only modify the transition function in G' by uniformly choosing over the successor choices. Formally, for a state $\ell \in L$ and an action $\sigma \in \Sigma$ the probabilistic transition function δ' in G' is as follows: $\delta'(\ell, \sigma)(\ell') = 0$ if $(\ell, \sigma, \ell') \notin \delta$; and $\delta'(\ell, \sigma)(\ell') = 1/|\{ \ell_1 \mid (\ell, \sigma, \ell_1) \in \delta \}|$ if $(\ell, \sigma, \ell') \in \delta$. Given an observation-based strategy for Player 1 in G, we consider the same strategy in G': (1) if the strategy reaches the accepting state with probability 1 against all Player 2 strategies in G, then the strategy ensures that in G' the accepting state is reached with probability 1; and (2) otherwise there is a pure Player 2 strategy β in G that ensures the rejecting state is reached in $2^{|L|}$ steps with probability $\eta > 0$, and with probability at least $(1/|L|)^{2^{|L|}}$ the choices of the successors of strategy β is chosen in G', and hence the rejecting state is reached with probability at least $(1/|L|)^{2^{|L|}} \cdot \eta > 0$. It follows that in G' there is an observation-based strategy for almost-sure winning the reachability objective with target of the accepting state iff there is such a strategy in G.
2. *Positive winning for safety.* The reduction is same as above. We obtain the POMDP G'' from the POMDP G' above by making the following modification: from the state accepting, the POMDP goes back to the initial state with probability 1. If there is an observation-based strategy α for Player 1 in G' to reach the accepting state, then repeating the strategy α each time the accepting state is visited, it can be ensured that the rejecting state is reached with probability 0. Otherwise, against every observation-based strategy for Player 1, the probability to reach the rejecting state in $k \cdot (2^{|L|}+1)$ steps is at least $1-(1-\eta')^k$, where $\eta' = \eta \cdot (1/|L|)^{2^{|L|}} > 0$ (this is because there is a probability to reach the rejecting state with probability at least η' in $2^{|L|}$ steps, and unless the rejecting state is reached the starting state is again reached within $2^{|L|} + 1$ steps). Hence the probability to reach the rejecting state is 1. It follows that G' is almost-sure winning for the reachability objective with the target of the accepting state iff in G'' there is an observation-based strategy for Player 1 to ensure that the rejecting state is avoided with positive probability. This completes the proof of correctness of the reduction.

A very brief (two line proof) sketch was presented as the proof of Theorem 1 of [12] to show that positive winning in POMDPs with safety objectives is EXPTIME-hard. We were unable to reconstruct the proof: the proof suggested to simulate a nondeterministic Turing machine. The simulation of a polynomial-space nondeterministic Turing machine only shows PSPACE-hardness, and the simulation of a nondeterministic EXPTIME Turing machine would have shown NEXPTIME-hardness, and an EXPTIME upper bound is known for the problem. Our proof presents a different and detailed proof of the result of Theorem 1 of [12]. Hence we have the following theorem, and the results are summarized in Table 1.

Theorem 5. *Given a* POMDP G*, the following assertions hold: (a) the positive winning problem for reachability objectives is NLOGSPACE-complete, the positive winning problem for safety and coBüchi objectives is EXPTIME-complete, and the positive winning problem for Büchi and parity objectives is undecidable; and (b) the almost-sure*

Table 1. Computational complexity of POMDPs with different classes of parity objectives for positive and almost-sure winning. Our contribution of upper and lower bounds are indicated as "up" and "lo" respectively in parenthesis.

	Positive	Almost-sure
Reachability	NLOGSPACE-complete (up+lo)	EXPTIME-complete (lo)
Safety	EXPTIME-complete (up+lo)	EXPTIME-complete [4]
Büchi	Undecidable [1]	EXPTIME-complete (lo)
coBüchi	EXPTIME-complete (up+lo)	Undecidable [1]
Parity	Undecidable [1]	Undecidable [1]

winning problem for reachability, safety and Büchi objectives is EXPTIME-complete, and the almost-sure winning problem for coBüchi and parity objectives is undecidable.

5 Optimal Memory Bounds for Strategies

In this section we present optimal bounds on the memory required by pure and randomized strategies for positive and almost-sure winning for reachability, safety, Büchi and coBüchi objectives.

Bounds for safety objectives. First, we consider positive and almost-sure winning with safety objectives in POMDPs. It follows from the correctness argument of Theorem 2 that pure strategies with exponential memory are sufficient for positive winning with safety objectives in POMDPs, and the exponential upper bound on memory of pure strategies for almost-sure winning with safety objectives in POMDPs follows from the reduction to games. We now present a matching exponential lower bound for randomized strategies.

Lemma 3. *There exists a family $(P_n)_{n \in \mathbb{N}}$ of* POMDPs *of size $O(p(n))$ for a polynomial p with a safety objective such that the following assertions hold: (a) Player 1 has a (pure) almost-sure (and therefore also positive) winning strategy in each of these* POMDPs*; and (b) there exists a polynomial q such that every finite-memory randomized strategy for Player 1 that is positive (or almost-sure) winning in P_n has at least $2^{q(n)}$ states.*

Proof sketch. The set of actions of the POMDP P_n is $\Sigma_n \cup \{\#\}$ where $\Sigma_n = \{1, \ldots, n\}$. The POMDP is composed of an initial state q_0 and n sub-MDPs A_i with state space Q_i, each consisting of a loop over p_i states $q_1^i, \ldots, q_{p_i}^i$ where p_i is the i-th prime number. From each state q_j^i $(1 \le j < p_i)$, every action in Σ_n leads to the next state q_{j+1}^i with probability $\frac{1}{2}$, and to the initial state q_0 with probability $\frac{1}{2}$. The action $\#$ is not allowed. From $q_{p_i}^i$, the action i is not allowed while the other actions in Σ_n lead back the first state q_1^i and to the initial state q_0 both with probability $\frac{1}{2}$. Moreover, the action $\#$ leads back to the initial state (with probability 1). The disallowed actions lead to a bad state. The states of the A_i's are indistinguishable (they have the same observation), while the initial state q_0 is visible. There are two observations, the state $\{q_0\}$ is labelled by observation o_1, and the other states in $Q_1 \cup \cdots \cup Q_n$ (that we call the loops)

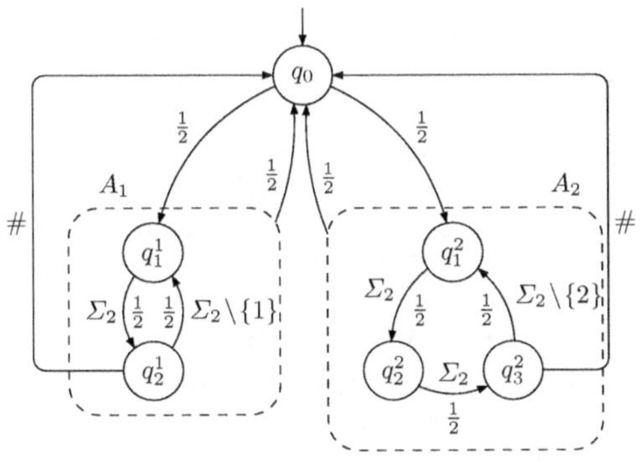

Fig. 1. The POMDP P_2

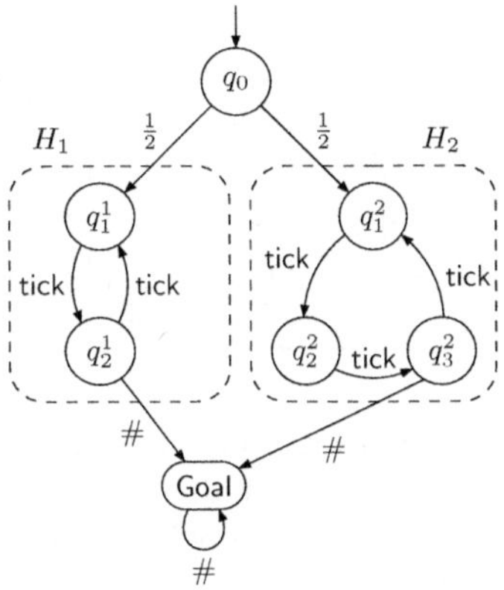

Fig. 2. The POMDP P_2'

by observation o_2. Fig. 1 shows the game P_2: the witness family of POMDPs have similarities with analogous constructions for games [3]. However the construction of [3] shows lower bounds only for pure strategies and in games, whereas we present lower bound for randomized strategies and for POMDPs (the proof and formal definition of the POMDP family $(P_n)_{n\in\mathbb{N}}$ can be found in [9]). Intuitively, exponential memory is required to win in P_n (even positively) because the action $\#$ needs to be played after $p_n^* = \prod_{i=1}^{n} p_i$ steps in the loops, and cannot be played before. Therefore, a winning strategy has to be able to count up to p_n^* which requires exponential memory.

Bounds for reachability objectives. The bounds for positive winning with reachability objectives are as follows: randomized memoryless strategies suffice, and for pure strategies, memory linear in the number of states is both necessary and sufficient (details in [9]). It follows from the results of [1] that for almost-sure winning with reachability objectives in POMDPs pure strategies with exponential memory suffice, and we now prove an exponential lower bound for randomized strategies.

Lemma 4. *There exists a family $(P_n)_{n\in\mathbb{N}}$ of* POMDP*s of size $O(p(n))$ for a polynomial p with a reachability objective such that the following assertions hold: (a) Player 1 has an almost-sure winning strategy in each of these* POMDP*s; and (b) there exists a polynomial q such that every finite-memory randomized strategy for Player 1 that is almost-sure winning in P_n has at least $2^{q(n)}$ states.*

Proof sketch. Fix the action set as $\Sigma = \{\#, \mathsf{tick}\}$. The POMDP P_n' is composed of an initial state q_0 and n sub-MDPs H_i, each consisting of a loop over p_i states $q_1^i, \ldots, q_{p_i}^i$ where p_i is the i-th prime number. From each state in the loops, the action tick can be played and leads to the next state in the loop (with probability 1). The action $\#$ can be played in the last state of each loop and leads to the Goal state. The objective is to reach Goal with probability 1. Actions that are not allowed lead to a sink state from which it is impossible to reach Goal. There is a unique observation that consists of the whole state space. Intuitively, the argument for exponential memory is analogous to the case of Lemma 3. Fig. 2 shows P_2' and see [9] for a proof of Lemma 4.

Table 2. Optimal memory bounds for pure and randomized strategies

	Pure Positive	Randomized Positive	Pure Almost	Randomized Almost
Reachability	Linear	Memoryless	Exponential	Exponential
Safety	Exponential	Exponential	Exponential	Exponential
Büchi	No Bound	No Bound	Exponential	Exponential
coBüchi	Exponential	Exponential	No Bound	No Bound
Parity	No Bound	No Bound	No Bound	No Bound

Bounds for Büchi and coBüchi objectives. An exponential upper bound for memory of pure strategies for almost-sure winning of Büchi objectives follows from the results of [1], and the matching lower bound for randomized strategies follows from our result for reachability objectives. Since positive winning is undecidable for Büchi objectives there is no bound on memory for pure or randomized strategies for positive winning. An exponential upper bound for memory of pure strategies for positive winning of coBüchi objectives follows from the correctness proof of Theorem 3 that iteratively combines the positive winning strategies for safety and reachability to obtain a positive winning strategy for coBüchi objective. The matching lower bound for randomized strategies follows from our result for safety objectives. Since almost-sure winning is undecidable for coBüchi objectives there is no bound on memory for pure or randomized strategies for positive winning. This gives us the following theorem (also summarized in Table 2), which is in contrast to the results for MDPs with perfect observation where pure memoryless strategies suffice for almost-sure and positive winning for all parity objectives.

Theorem 6. *The optimal memory bounds for strategies in* POMDP*s are as follows.*

1. *Reachability objectives: for positive winning randomized memoryless strategies are sufficient, and linear memory is necessary and sufficient for pure strategies; and for almost-sure winning exponential memory is necessary and sufficient for both pure and randomized strategies.*
2. *Safety objectives: for positive winning and almost-sure winning exponential memory is necessary and sufficient for both pure and randomized strategies.*
3. *Büchi objectives: for almost-sure winning exponential memory is necessary and sufficient for both pure and randomized strategies; and there is no bound on memory for pure and randomized strategies for positive winning.*
4. *coBüchi objectives: for positive winning exponential memory is necessary and sufficient for both pure and randomized strategies; and there is no bound on memory for pure and randomized strategies for almost-sure winning.*

References

1. Baier, C., Bertrand, N., Größer, M.: On decision problems for probabilistic Büchi automata. In: Amadio, R.M. (ed.) FOSSACS 2008. LNCS, vol. 4962, pp. 287–301. Springer, Heidelberg (2008)
2. Bertrand, N., Genest, B., Gimbert, H.: Qualitative determinacy and decidability of stochastic games with signals. In: Proc. of LICS, pp. 319–328. IEEE Computer Society, Los Alamitos (2009)

3. Berwanger, D., Chatterjee, K., Doyen, L., Henzinger, T.A., Raje, S.: Strategy construction for parity games with imperfect information. In: van Breugel, F., Chechik, M. (eds.) CONCUR 2008. LNCS, vol. 5201, pp. 325–339. Springer, Heidelberg (2008)
4. Berwanger, D., Doyen, L.: On the power of imperfect information. In: Proc. of FSTTCS. Dagstuhl Seminar Proceedings 08004 (2008)
5. Bianco, A., de Alfaro, L.: Model checking of probabilistic and nondeterministic systems. In: Thiagarajan, P.S. (ed.) FSTTCS 1995. LNCS, vol. 1026, pp. 499–513. Springer, Heidelberg (1995)
6. Chadha, R., Sistla, A.P., Viswanathan, M.: Power of randomization in automata on infinite strings. In: Bravetti, M., Zavattaro, G. (eds.) CONCUR 2009. LNCS, vol. 5710, pp. 229–243. Springer, Heidelberg (2009)
7. Chatterjee, K., Doyen, L., Gimbert, H., Henzinger, T.A.: Randomness for free. In: Proc. of MFCS (2010)
8. Chatterjee, K., Doyen, L., Henzinger, T.A., Raskin, J.-F.: Algorithms for omega-regular games of incomplete information. Logical Methods in Computer Science 3(3:4) (2007)
9. Chatterjee, K., Doyen, L., Henzinger, T.A.: Qualitative analysis of Partially-observable Markov decision processes. CoRR, abs/0909.1645 (2009)
10. Chatterjee, K., Jurdziński, M., Henzinger, T.A.: Quantitative stochastic parity games. In: Proc. of SODA, pp. 114–123 (2004)
11. de Alfaro, L.: Formal Verification of Probabilistic Systems. PhD thesis, Stanford University. Technical Report STAN-CS-TR-98-1601 (1997)
12. de Alfaro, L.: The verification of probabilistic systems under memoryless partial-information policies is hard. In: Proc. of ProbMiV: Probabilistic Methods in Verification (1999)
13. De Wulf, M., Doyen, L., Raskin, J.-F.: A lattice theory for solving games of imperfect information. In: Hespanha, J.P., Tiwari, A. (eds.) HSCC 2006. LNCS, vol. 3927, pp. 153–168. Springer, Heidelberg (2006)
14. Gripon, V., Serre, O.: Qualitative concurrent stochastic games with imperfect information. In: Albers, S., Marchetti-Spaccamela, A., Matias, Y., Nikoletseas, S., Thomas, W. (eds.) ICALP 2009, Part II. LNCS, vol. 5556, pp. 200–211. Springer, Heidelberg (2009)
15. Kechris, A.: Classical Descriptive Set Theory. Springer, Heidelberg (1995)
16. Littman, M.L.: Algorithms for sequential decision making. PhD thesis, Brown University (1996)
17. Madani, O., Hanks, S., Condon, A.: On the undecidability of probabilistic planning and related stochastic optimization problems. Artif. Intell. 147(1-2) (2003)
18. Papadimitriou, C.H., Tsitsiklis, J.N.: The complexity of Markov decision processes. Mathematics of Operations Research 12, 441–450 (1987)
19. Paz, A.: Introduction to probabilistic automata. Academic Press, London (1971)
20. Reif, J.: The complexity of two-player games of incomplete information. Journal of Computer and System Sciences 29, 274–301 (1984)
21. Segala, R.: Modeling and Verification of Randomized Distributed Real-Time Systems. PhD thesis, MIT. Technical Report MIT/LCS/TR-676 (1995)
22. Thomas, W.: Languages, automata, and logic. In: Handbook of Formal Languages, Beyond Words, ch. 7, vol. 3, pp. 389–455. Springer, Heidelberg (1997)
23. Vardi, M.Y.: Automatic verification of probabilistic concurrent finite-state systems. In: Proc. of FOCS, pp. 327–338. IEEE Computer Society Press, Los Alamitos (1985)

All Symmetric Predicates in $NSPACE(n^2)$ Are Stably Computable by the Mediated Population Protocol Model*

Ioannis Chatzigiannakis[1,2], Othon Michail[1,2], Stavros Nikolaou[2], Andreas Pavlogiannis[2], and Paul G. Spirakis[1,2]

[1] Research Academic Computer Technology Institute (RACTI), Patras, Greece
[2] Computer Engineering and Informatics Department (CEID), University of Patras
{ichatz,michailo,spirakis}@cti.gr,
{snikolaou,paulogiann}@ceid.upatras.gr

Abstract. This work focuses on the computational power of the *Mediated Population Protocol* model on complete communication graphs and initially identical edges (*SMPP*). In particular, we investigate the class MPS of all predicates that are stably computable by the SMPP model. It is already known that MPS is in the symmetric subclass of $NSPACE(n^2)$. Here we prove that this inclusion holds with equality, thus, providing the following exact characterization for MPS: *A predicate is in* MPS *iff it is symmetric and is in* $NSPACE(n^2)$.

1 Introduction - Population Protocols

Theoretical models for Wireless Sensor Networks (WSNs) have received great attention recently, mainly because they constitute an abstract but yet formal and precise method for understanding the limitations and capabilities of this widely applicable new technology. The *Population Protocol* model [1] was designed to represent a special category of WSNs which is mainly identified by two distinctive characteristics: each sensor node is an extremely limited computational device and all nodes move according to some mobility pattern over which they have totally no control.

One reason for studying extremely limited computational devices is that in many real WSNs' application scenarios having limited resources is inevitable. For example, power supply limitations may render strong computational devices useless due to short lifetime. In other applications, mote's size is an important constraint that thoroughly determines the computational limitations. The other reason is that the population protocol model constitutes the starting point of a brand new area of research and in order to provide a clear understanding and foundation of the laws and the inherent properties of the studied systems it ought to be minimalistic. In terms of computational characterization each

* This work has been partially supported by the ICT Programme of the European Union under contract number ICT-2008-215270 (FRONTS).

P. Hliněný and A. Kučera (Eds.): MFCS 2010, LNCS 6281, pp. 270–281, 2010.

node is simply a finite-state machine additionally equipped with sensing and communication capabilities and is usually called an *agent*. A *population* is the collection of all agents that constitute the distributed computational system.

The prominent characteristic that diversifies population protocols from classical distributed systems is the total inability of the computational devices to control or predict their underlying mobility pattern. Their movement is usually the result of some unstable environment, like water flow or wind, or the natural mobility of their carriers, and is known as *passive mobility*. The agents interact in pairs and are absolutely incapable of knowing the next pair in the interaction sequence. This inherent nondeterminism of the interaction pattern is modeled by an *adversary* whose job is to select interactions. The adversary is a black-box and the only restriction imposed is that it has to be *fair* so that it does not forever partition the population into non-communicating clusters and guaranteeing that interactions cannot follow some inconvenient periodicity.

As expected, due to the minimalistic nature of the population protocol model, the class of computable predicates was proven [1,2] to be fairly small: it is the class of *semilinear predicates*, or, equivalently, all predicates definable by first-order logical formulas in *Presburger arithmetic* [10], which does not include multiplication of variables, exponentiations, and many other important and natural operations on input variables. Moreover, Delporte-Gallet *et al.* [9] showed that population protocols can tolerate only $\mathcal{O}(1)$ crash failures and not even a single Byzantine agent.

2 Enhancing the Model

The next big step is naturally to strengthen the population protocol model with extra realistic and implementable assumptions, in order to gain more computational power and/or speed-up the time to convergence and/or improve fault-tolerance. Several promising attempts have appeared towards this direction. In each case, the model enhancement is accompanied by a logical question: What is exactly the class of predicates computable by the new model?

An interesting extension was the *Community Protocol* model of Guerraoui and Ruppert [11] in which the agents have read-only industrial unique ids picked from an infinite set of ids. Moreover, each agent can store up to a constant number of other agents' ids. In this model, agents are only allowed to compare ids, that is, no other operation on ids is permitted. The community protocol model was proven to be extremely strong: the corresponding class consists of all symmetric predicates in $NSPACE(n \log n)$. It was additionally shown that if faults cannot alter the unique ids and if some necessary preconditions are satisfied, then community protocols can tolerate $\mathcal{O}(1)$ Byzantine agents.

The *Passively mobile Machines* (*PM*) model [5] made the assumption that each agent is a Turing Machine and defined *PALOMA* protocols as those protocols that use in every agent space that is bounded by a logarithm in the population size. Interestingly, it turned out that the agents are able to assign unique consecutive ids to themselves, get informed of the population size and, by exploiting these, organize themselves into a distributed Nondeterministic TM (*NTM*) that makes full use of the agents' memories. The TM draws its nondeterminism by the nondeterminism inherent in the interaction pattern. The main result of that work was an exact characterization for the class PLM, of all predicates that are stably computable by PALOMA protocols: it is again precisely the class of all symmetric predicates in $NSPACE(n \log n)$.

3 Our Results - Roadmap

This work focuses on the computational power of another extension of the population protocol model that was proposed in [7] (see also [8] and [6]) and is called the *Mediated Population Protocol* (*MPP*) model. The main additional feature in comparison to the population protocol model is that each link (u, v) can be thought of as being itself an agent that only participates in the interaction (u, v). The agents u and v can exploit this joint memory to store pairwise information and to have it available during some future interaction. Another way to think of this system is that agents store pairwise information into some global storage, like, e.g., a base station, called the *mediator*, that provides a small fixed slot to each pair of agents. Interacting agents communicate with the mediator to read and update their collective information.

From [7] we know that the MPP model is strictly stronger than the population protocol model since it can compute a non-semilinear predicate. Moreover, we know that any predicate that is stably computable by the MPP model is also in $NSPACE(n^2)$. In this work, we show that, for complete graphs, this inclusion holds with equality, thus, providing the following exact characterization for the computational power of the MPP model in the fully symmetric case: *A predicate is stably computable by the MPP model iff it is symmetric and is in* $NSPACE(n^2)$. We show in this manner that the MPP model is surprisingly strong.

In section 4, we give a formal definition of the MPP model and introduce a special class of graphs (the *correctly labeled line graphs*) that comes up in our proof later on. Section 5 holds the actual proof. In particular, Subsection 5.1 presents some basic ideas that help us establish a first inclusion. In Subsection 5.2, we show how to extend these ideas in order to prove our actual statement.

4 The Mediated Population Protocol Model

4.1 Formal Definition

A *Mediated Population Protocol* (MPP) is a 7-tuple $(X, Y, Q, S, I, O, \delta)$, where X, Y, Q, and S are all finite sets and X is the *input alphabet*, Y is the *output alphabet*, Q is the set of *agent states*, S is the set of *edge states*, $I : X \to Q$ is the *input function*, $O : Q \to Y$ is the *output function*, and $\delta : Q \times Q \times S \to Q \times Q \times S$ is the *transition function*. If $\delta(a, b, c) = (a', b', c')$, we call $(a, b, c) \to (a', b', c')$ a *transition*, and we define $\delta_1(a, b, c) = a'$, $\delta_2(a, b, c) = b'$ and $\delta_3(a, b, c) = c'$.

An MPP $\mathcal{A}$ runs on the nodes of a *communication graph* $G = (V, E)$, which is directed without self-loops and multiple edges. V is the population, consisting of $n \equiv |V|$ agents. E is the set of permissible interactions between the agents.

In the most general setting, each agent initially senses its environment, as a response to a global start signal, and receives an input symbol from X. Then all agents apply the input function to their input symbols and obtain their initial state. Each edge is initially in one state from S as specified by some *edge initialization function* $\iota : E \to S$, which is not part of the protocol but generally models some preprocessing on the network that has taken place before the protocol's execution.

A *network configuration*, or simply a *configuration*, is a mapping $C : V \cup E \to Q \cup S$ specifying the state of each agent in the population and each edge in the set of permissible interactions. Let C and C' be configurations, and let u, v be distinct agents. We say that C goes to C' via encounter $e = (u, v)$, denoted $C \xrightarrow{e} C'$, if $C'(u) = \delta_1(C(u), C(v), C(e))$, $C'(v) = \delta_2(C(u), C(v), C(e))$, $C'(e) = \delta_3(C(u), C(v), C(e))$, and $C'(z) = C(z)$ for all $z \in (V - \{u, v\}) \cup (E - e)$. We say that C can go to C' in one step, denoted $C \to C'$, if $C \xrightarrow{e} C'$ for some encounter $e \in E$. We write $C \xrightarrow{*} C'$ if there is a sequence of configurations $C = C_0, C_1, \ldots, C_t = C'$, such that $C_i \to C_{i+1}$ for all i, $0 \leq i < t$, in which case we say that C' is *reachable* from C.

An *execution* is a finite or infinite sequence of configurations $C_0, C_1, C_2, \ldots$, where C_0 is an initial configuration and $C_i \to C_{i+1}$, for all $i \geq 0$. We have both finite and infinite kinds of executions since the adversary scheduler may stop after a finite number of steps or continue selecting pairs forever. Moreover, a strong global *fairness condition* is imposed on the adversary to ensure the protocol makes progress. An infinite execution is *fair* if for every pair of configurations C and C' such that $C \to C'$, if C occurs infinitely often in the execution then so does C'. A *computation* is an infinite fair execution. An interaction between two agents is called *effective* if at least one of the initiator's, the responder's, and the edge's states is modified (that is, if C, C' are the configurations before and after the interaction, respectively, then $C' \neq C$).

4.2 Stable Computation

Throughout this work we assume that the communication graph is complete and that all edges are initially in a common state s_0, that is, $\iota(e) = s_0$ for all $e \in E$. Call this for sake of simplicity the SMPP model ('S' standing for "Symmetric"). An SMPP may run on any such communication graph $G = (V, E)$, where $n \geq 2$, and its input (also called an *input assignment*) is any $x = \sigma_1\sigma_2 \ldots \sigma_n \in X^{\geq 2} = \{x \in X^* \mid |x| \geq 2\}$. In particular, by assuming an ordering over V, the input to agent i is the symbol σ_i, for all $i \in \{1, 2, \ldots, n\}$. Let $p : X^{\geq 2} \to \{0, 1\}$ be any predicate over $X^{\geq 2}$. p is called *symmetric* if for every $x \in X^{\geq 2}$ and any x' which is a permutation of x's symbols, it holds that $p(x) = p(x')$ (in words, permuting the input symbols does not affect the predicate's outcome). Any language $L \subseteq X^{\geq 2}$ corresponds to a unique predicate p_L defined as $p_L(x) = 1$ iff $x \in L$. Such a language is symmetric iff p_L is symmetric. Due to this bijection we use the term *symmetric predicate* for both predicates and languages.

Like population protocols, MPPs do not halt. Instead a protocol is required to *stabilize*, in the sense that it reaches a point after which the output of every agent will remain unchanged. A predicate p over $X^{\geq 2}$ is said to be *stably computable* by the SMPP model, if there exists an SMPP $\mathcal{A}$ such that for any input assignment $x \in X^{\geq 2}$, any computation of $\mathcal{A}$ on the complete communication graph of $|x|$ nodes beginning from the initial configuration corresponding to x reaches a configuration after which all agents forever output $p(x)$.

Let MPS (standing for "Mediated Predicates in the fully Symmetric case") be the class of all stably computable predicates by the SMPP model. Note that all predicates in MPS have to be symmetric because the communication graph is complete and all edges are initially in the same state. Let $SSPACE(f(n))$ and $SNSPACE(f(n))$ be $SPACE(f(n))$'s and $NSPACE(f(n))$'s restrictions to symmetric predicates, respectively.

4.3 Correctly Labeled Line Graphs

Let $G = (V, E)$ be a communication graph. A *line (di)graph* $L = (K, A)$ is either an isolated node (that is, $|K| = 1$ and $A = \emptyset$), which is the *trivial* line graph, or a tree such that, if we ignore the directions of the links, two nodes have degree one and all other nodes have degree two. A *line subgraph* of G is a line graph $L \subseteq G$ and is called *spanning* if $K = V$. Let $d(u)$ denote the degree of $u \in K$ w.r.t. to A. Let $C_l(t)$ denote the *label component* of the state of $t \in V \cup E$ under configuration C (in the beginning, we call it *state* for simplicity).

We say that a line subgraph of G is *correctly labeled* under configuration C, if it is trivial and its state is l with no active edges incident to it or if it is non-trivial and all the following conditions are satisfied:

1. Assume that $u, v \in K$ and $d(u) = d(v) = 1$. These are the only nodes in K with degree 1. Then one of u and v is in state k_t (non-leader or right endpoint) and the other is either in state l_t or in state l_h (leader or left endpoint). The unique $e_u \in A$ incident to u, where u is w.l.o.g. in state k_t, is an outgoing edge and the unique $e_v \in A$ incident to v is outgoing if $C_l(v) = l_t$ and incoming if $C_l(v) = l_h$.
2. For all $w \in K - \{u, v\}$ (internal nodes) it holds that $C_l(w) = k$.
3. For all $a \in A$ it holds that $C_l(a) \in \{p, i\}$ and for all $e \in E - A$ such that e is incident to a node in K it holds that $C_l(e) = 0$.
4. Let $v = u_1, u_2, \ldots, u_r = u$ be the path from the leader to the non-leader endpoint (resulting by ignoring the directions of the arcs in A). Let $P_L = \{(u_i, u_{i+1}) \mid 1 \leq i < r\}$ be the corresponding directed path from v to u. Then for all $a \in A \cap P_L$ it holds that $C_l(a) = p$ (proper edges) and for all $a' \in A - P_L$ that $C_l(a') = i$ (inverse edges).

See Figure 1 for some examples of correctly labeled line subgraphs. The meaning of each state will become clear in the proof of Theorem 1 in the following section.

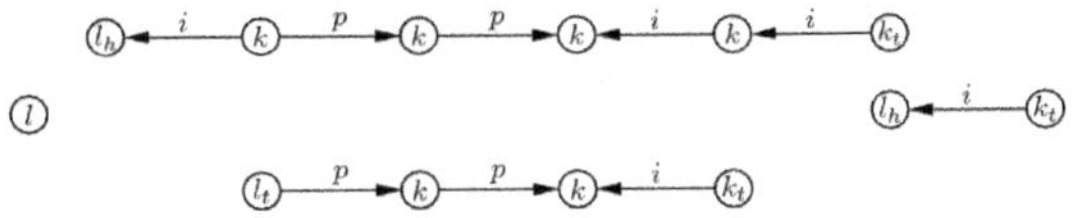

Fig. 1. Some correctly labeled line subgraphs. We assume that all edges not appearing are in state 0 (inactive).

5 The Computational Power of the SMPP Model

In [7], it was shown that MPS is a proper superset of the set of semilinear predicates. Here we are going to establish a much better inclusion. In particular, in Section 5.1 we show that any predicate in $SSPACE(n)$ is also in MPS. In other words, the SMPP model is at least as strong as a linear space TM that computes symmetric predicates. Then in Section 5.2 we extend the ideas used in the proof of this result in order to establish that $SSPACE(n^2)$ is a subset of MPS showing that MPS is a surprisingly wide class. Finally, we improve to $SNSPACE(n^2)$, thus, arriving at an exact characterization for MPS (the inverse inclusion already exists from [7]).

5.1 A First Inclusion: $SSPACE(n) \subseteq MPS$

Theorem 1. *There is an SMPP $\mathcal{A}$ that constructs a correctly labeled spanning line subgraph of any complete communication graph G.*

Proof. We provide a high level description of the protocol $\mathcal{A}$ in order to avoid its many low-level details. All agents are initially in state l, thought of as being *simple leaders*. All edges are in state 0 and we think of them as being *inactive*, that is, not part of the line subgraph to be constructed. An edge in state p is interpreted as *proper* while an edge in state i is interpreted as *inverse* and both are additionally interpreted as *active*, that is, part of the line subgraph to be constructed. An agent in state k is a (*simple*) *non-leader*, an agent in state k_t is a non-leader that is additionally the *tail* of some line subgraph (*tail non-leader*), an agent in state l_t is a leader and a tail of some line subgraph (*tail leader*), and an agent in state l_h is a leader and a *head* of some line subgraph (*head leader*). All these will be clarified in the sequel. A *leader* is a simple, tail, or head leader.

The agents become organized in correctly labeled line subgraphs by the following transitions: $(l, l, 0) \to (k_t, l_h, i)$, $(l_h, l, 0) \to (k, l_h, i)$, $(l, l_h, 0) \to (l_t, k, p)$, $(l_t, l, 0) \to (k, l_h, i)$, and $(l, l_t, 0) \to (l_t, k, p)$.

We now describe how two such line graphs L_1 and L_2 are pieced together. Denote by $l(L) \in V$ and by $k_t(L) \in V$ the leader and tail non-leader endpoints of a correctly labeled line graph L, respectively. When $l(L_1) = u$ interacts as the initiator with $l(L_2) = v$, through an inactive edge, v becomes a non-leader with a special mark, e.g. k', the edge becomes proper with a special mark, and u becomes a leader in a special state l' indicating that this state will travel towards $k_t(L_1)$ while making all proper edges that it meets inverse and all inverse edges proper. In order to know its direction, it marks each edge that it crosses. When it, finally, arrives at the endpoint, it goes to another special state and walks the same path in the inverse direction until it meets v again. This walk can be performed easily, without using the marks, because now all edges have correct labels (states). To diverge from L_1's endpoint it simply follows the proper links as the initiator (moving from their tail to their head) and the inverse links as the responder (moving from their head to their tail) while erasing all marks left from its previous walk. When it reaches v it erases its mark, making its state k, and goes to another special state indicating that it again must walk towards $k_t(L_1)$ for the last time, performing no other operation this time. To do that, it follows the proper links as the responder (from their head to their tail) and the inverse links as the initiator (from their tail to their head). When it, finally, arrives at $k_t(L_1)$ it becomes a normal tail leader and now it is easy to see that L_1 and L_2 have been merged correctly into a common correctly labeled line graph. See Figure 2 for an example. The correctness of this process, called the *merging process*, is captured by Lemma 1.

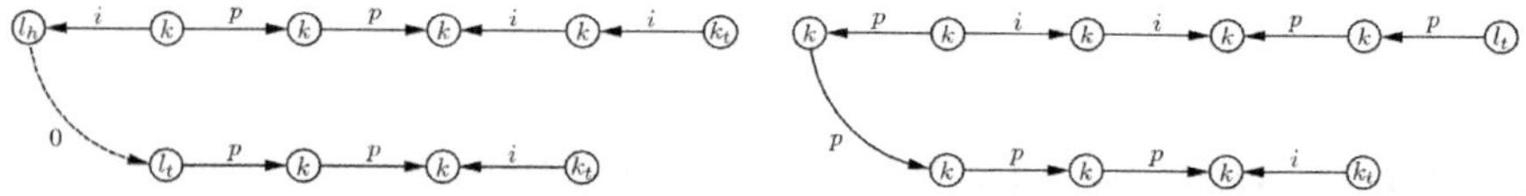

(a) Before merging.

(b) After merging.

Fig. 2. Two line subgraphs just before the execution and after the completion of the merging process

Lemma 1. *When the leader endpoints of two distinct correctly labeled line subgraphs of G, $L_1 = (K_1, A_1)$ and $L_2 = (K_2, A_2)$, interact via $e \in E$, then, in a finite number of steps, L_1 and L_2 are merged into a new correctly labeled line graph $L_3 = (K_1 \cup K_2, A_1 \cup A_2 \cup \{e\})$.*

Initially, G is partitioned into n correctly labeled trivial line graphs. It is easy to see that correctly labeled line graphs never become smaller and, according to Lemma 1, when their leaders interact they are merged into a new line graph. Moreover, given that there are two correctly labeled line subgraphs in the current configuration there is always the possibility (due to fairness) that these line graphs may get merged and there is no other possible effective interaction between them. In simple words, two line graphs can only get merged and there is always the possibility that merging actually takes place. It is easy to see that this process has to end, due to fairness, in a finite number of steps having constructed a correctly labeled spanning line subgraph of G (for simplicity, we call this process the *spanning process*). □

Theorem 2. *Assume that the communication graph $G = (V, E)$ is a correctly labeled line graph of n agents, where each agent takes its input symbol in a second* state component *(the first component is used for the labels of the spanning process*

and is called label component*). Then there is an MPP $\mathcal{A}$ that when running on such a graph simulates a deterministic TM $\mathcal{M}$ of $\mathcal{O}(n)$ (linear) space that computes symmetric predicates.*

Proof. It is already known from [1,3] that the theorem holds for population protocols with no inverse edges. It is easy to see that the correct p and i labels can be exploited by the simulation in order to identify the correct directions. □

It must be clear now, that if the agents could detect termination of the spanning process then they would be able to simulate a deterministic TM of $\mathcal{O}(n)$ space that computes symmetric predicates. But, unfortunately, they are unable to detect termination, because if they could, then termination could also be detected in any non-spanning line subgraph constructed in some intermediate step (it can be proven by symmetry arguments together with the fact that the agents cannot count up to the population size). Fortunately, we can overcome this by applying the *reinitialization* technique of [11,5].

Theorem 3. *$SSPACE(n)$ is a subset of MPS.*

Proof. Take any $p \in SSPACE(n)$. By Theorem 2 we know that there is an MPP $\mathcal{A}$ that stably computes p on a line graph of n nodes. We have to show that there exists an SMPP $\mathcal{B}$ that stably computes p. We construct $\mathcal{B}$ to be the composition of $\mathcal{A}$ and another protocol $\mathcal{I}$ that is responsible for executing the spanning and reinitialization processes. Each agent's state consists of three components: a read-only *input backup*, one used by $\mathcal{I}$, and one used by $\mathcal{A}$. Thus, $\mathcal{A}$ and $\mathcal{I}$ are, in some sense, executed in parallel in different components.

Protocol $\mathcal{I}$ does the following. It always executes the spanning process and when the merging of two line graphs comes to an end it executes the following reinitialization process. The new leader u that resulted from merging becomes marked, e.g. l_t^*. Recall that the new line graph has also correct labels. When u meets its right neighbor, u sets its $\mathcal{A}$ component to its input symbol (by copying it from the input backup), becomes unmarked, and passes the mark to its right neighbor (correct edge labels guarantee that each agent distinguishes its right and left neighbors). When the newly marked agent interacts with its own right neighbor, it does the same, and so on, until the two rightmost agents interact, in which case they are both reinitialized at the same time and the special mark is lost. It is easy to see that this process guarantees that all agents in the line graph become reinitialized and before completion non-reinitialized agents do not have effective interactions with reinitialized ones (the special marked agent acts always as the separator between reinitialized and non-reinitialized agents). Note that if other reinitialization processes are pending from previous reinitialization steps, then the new one may identify and erase them.

From Theorem 1 we know that the spanning process executed by $\mathcal{I}$ results in a correctly labeled spanning line subgraph of G. The spanning process, as already mentioned, terminates when the merging of the last two line subgraphs takes place and merging also correctly terminates in a finite number of steps (Lemma 1). Moreover, from the above discussion we know that, when this happens, the reinitialization process will correctly reinitialize all agents of the spanning line subgraph, thus, all agents in the population. But then, independently of its computation so far, $\mathcal{A}$ will run from the beginning on a correctly labeled line graph of n nodes (this line graph will not be modified again in the future), thus, it will stably compute p. Finally, if we assume that $\mathcal{B}$'s output is $\mathcal{A}$'s output then we conclude that the SMPP $\mathcal{B}$ also stably computes p, thus, $p \in MPS$. □

5.2 An Exact Characterization: $MPS = SNSPACE(n^2)$

We now extend the techniques employed so far to obtain an exact characterization for MPS.

Theorem 4. *Assume that the complete communication graph $G = (V, E)$ contains a correctly labeled spanning line subgraph, where each agent takes its input symbol in a second* state component. *Then there is an MPP $\mathcal{A}$ that when running on such a graph simulates a deterministic TM $\mathcal{M}$ of $\mathcal{O}(n^2)$ space that computes symmetric predicates.*

Proof. For simplicity and w.l.o.g. we assume that $\mathcal{A}$ begins its execution from the leader endpoint, that initially the simulation moves all n input symbols to the leftmost outgoing inactive edges ($n-2$ leaving from the leader and two more leaving from the second agent of the line graph), that the left endpoint is a tail leader, and that the edge states now consist of two components, one used to identify them as active/inactive and the other used by the simulation.

In contrast to Theorem 2 the simulation also makes use of the inactive edges. The agent in control of the simulation is in a special state denoted with a star '$*$'. Since the simulation starts from the left endpoint (tail leader), its state will be l_t^*. When the star-marked leader interacts with its unique right neighbor on the line graph, the neighbor's state is updated to a *r-marked* state (i.e. k^r). The k^r agent then interacts with its own right neighbor which is unmarked and the neighbor updates its state to a special *dot* state (i.e. $\dot{k}$) whereas the other agent (in state k^r) is updated to k. Then the only effective interaction is between the star-marked leader (l_t^*) and the dot non-leader ($\dot{k}$) via the inactive edge joining them. In this way, the inactive edge's state component used for the simulation becomes a part of the TM's tape. In fact $\mathcal{M}$'s tape consists only of the inactive edges and is accessed in a systematic fashion which is described below.

If the simulation has to continue to the right, the interaction $(l_t^*, \dot{k})$ sends the dot agent to state k^r. If it has to proceed left, the dot agent goes to state k^l. An agent in state k^r interacts with its right neighbor sending it to dot state whereas a k^l agent does the same for its left neighbor. In this way, the dot mark is moving left and right between the agents by following the active edges in the appropriate interaction role (initiator or responder) as described in Theorem 1 for the special states traversing through the line graph. The dot mark's (state's) position in the line graph determines which outgoing inactive edge of l_t^* will be used. The sequence in which the dot mark is traversing the graph is the sequence in which l_t^* visits its outgoing inactive edges. Therefore if it has to visit the next inactive edge it moves the dot mark to the right (via a k^r state) or to the left (via a k^l state) if it has to visit the previous one. It should be noted that the dot marked agent plays the role of the TM's head since it points the edge (which would correspond to a tape's cell in $\mathcal{M}$) that is visited. As stated above only the inactive edges hold the contents of the TM's tape. The active ones are used for allowing the special states (symbols) traverse the line graph.

Consider the case where the dot mark reaches the right non-leader endpoint (k_t) and the simulation after the interaction $(l_t^*, \dot{k}_t)$ demands to proceed right. Since l_t^*'s outgoing edges have all been visited by the simulation, the execution must continue on the next agent (right neighbor of leader endpoint l_t) in the line graph. This is achieved by having another special state traversing from right to left (since we are in the right non-leader endpoint) until it finds l_t^*. Then it removes its star mark (state) and assigns it to its right neighbor which now takes control of the simulation visiting its own inactive edges. A similar process takes place when the simulation, controlled by any non-leader agent, reaches the left leader endpoint and needs to proceed to the left cell.

When the control of the simulation reaches a non-leader agent (e.g. from the left leader endpoint side) in order to visit its first edge it places the dot mark to the left leader endpoint and then to the next (on the right) non-leader and so forth. If the dot mark reaches the star-marked agent (in the previous example from the left endpoint side) then it moves the dot to the closer (in the line graph) agent that can "see" via an inactive edge towards the right non-leader endpoint. In this way, each agent visits its outgoing edges in a specific sequence (from leader to non-leader when the simulation moves right and the reverse when it moves left) providing the $\mathcal{O}(n^2)$ space needed for the simulation. □

Theorem 5. *$SSPACE(n^2)$ is a subset of MPS.*

Proof. The main idea is similar to that in the proof of Theorem 3 (based again on the reinitialization technique). We assume that the edge states consist now of two components, one used to identify them as active/inactive and the other used by the simulation (protocol $\mathcal{A}$ from Theorem 4).

This time, the reinitialization process attempts to reinitialize not only all agents of a line graph but also all of their outgoing edges. We begin by describing the reinitialization process in detail. Whenever the merging process of two line graphs comes to an end, resulting in a new line graph L, the leader endpoint of L goes to a special *blocked* state, let it be l^b, blocking L from getting merged with another line graph while the reinitialization process is being executed. Keep in mind that L will only get ready for merging just after the completion of the reinitialization process. By interacting with its unique right neighbor in state k via an active edge it propagates the blocked state towards that neighbor updating its state to k^b and reinitializing the agent. The block state propagates in the same way towards the tail non-leader reinitializing and updating all intermediate non-leaders to k^b from left to right. Once it reaches this endpoint, a new special state k_0 is generated which traverses L in the inverse direction. Once k_0 reaches the leader endpoint, it disappears and the leader updates its state to l^*.

Now reinitialization of the inactive edges begins. When the leader in l^* interacts with its unique right neighbor (via the active edge joining them) it updates its neighbor's state to a special *bar* state (e.g. $\bar{k}$). When the agent with the bar state interacts with its own right neighbor, which is unmarked, the neighbor updates its state to a special *dot* state (e.g. $\dot{k}$). Now the bars cannot propagate and the only effective interaction is between the star leader and the dot non-leader. This interaction reinitializes the state component of the edge used for the simulation and makes the responder non-leader a bar non-leader. Then the new bar non-leader turns its own right neighbor to a dot non-leader, the second outgoing edge of the leader is reinitialized in this manner, and so on, until the edge joining the star leader (left endpoint) with the dot tail non-leader (right endpoint) is reinitialized. What happens then is that the bars are erased one after the other from right to left and finally the star moves one step to the right. So the first non-leader has now the star and it reinitializes its own inactive outgoing edges from left to right in a similar manner. The process repeats the same steps over and over, until the right endpoint of L reinitializes all of its outgoing edges. When this happens, $\mathcal{A}$ will execute its simulation on the correct reinitialized states. The above process is clearly executed correctly when L is spanning (because all outgoing edges have their heads on the line graph). When it isn't, the correctness of the process is captured by the following lemma.

Lemma 2. *Let L and L' be two distinct line subgraphs of G. If L runs a reinitialization process then it always terminates in a finite number of steps.*

Proof. If L' is not running a reinitialization process then there can be no conflict between L and L'. If L' is also running its own reinitialization process, a conflict occurs when a star agent of one graph interacts with a dot agent of the other, but in both cases the reinitialization process is either not affected or cannot be delayed, thus, it always terminates in a finite number of steps. □

We finally ensure that the simulation does not ever alter the agent labels used by the spanning and reinitialization processes. In the proof of Theorem 4 we made $\mathcal{A}$ put marks on the labels in order to get executed correctly. Now we simply think of these marks as being placed in a separate subcomponent of $\mathcal{A}$ that is ignored by the other processes. □

Theorem 6. *$SNSPACE(n^2)$ is a subset of MPS.*

Proof. We modify the deterministic TM of Theorem 5 by adding another component in each agent's state which stores a non-negative integer of value at most equal to the greatest number of non-deterministic choices that the new NTM $\mathcal{N}$ can face at any time. Note that this number is independent of the population size. In every reinitialization each agent obtains this value from its neighbors according to its position (which depends on the distance from the leader endpoint) in the line graph. Nondeterministic choices are mapped to these values and whenever such a choice has to be made, the agent in control of the simulation uses the value of the agent with whom it has the next arbitrary interaction. The inherent nondeterminism of the interaction pattern ensures that choices are made nondeterministically. If the accept state is reached all agents accept whereas if the reject state is reached the TM's computation is reinitialized. Fairness guarantees that all paths in the tree representing $\mathcal{N}$'s nondeterministic computation will eventually (although maybe after a long time) be followed. □

We have now arrived at the following exact characterization for MPS.

Theorem 7. $MPS = SNSPACE(n^2)$.

Proof. Follows from Theorem 6 and Theorem 8 of [7]. □

See the corresponding technical report [4] for a formal constructive proof and some graphical examples.

References

1. Angluin, D., Aspnes, J., Diamadi, Z., Fischer, M.J., Peralta, R.: Computation in networks of passively mobile finite-state sensors. Distributed Computing, 235–253 (March 2006)
2. Angluin, D., Aspnes, J., Eisenstat, D.: Stably computable predicates are semilinear. In: PODC 2006: Proceedings of the 25th annual ACM Symposium on Principles of Distributed Computing, pp. 292–299. ACM Press, New York (2006)
3. Aspnes, J., Ruppert, E.: An introduction to population protocols. Bulletin of the European Association for Theoretical Computer Science 93, 98–117 (2007)
4. Chatzigiannakis, I., Michail, O., Nikolaou, S., Pavlogiannis, A., Spirakis, P.G.: All symmetric predicates in NSPACE(n^2) are stably computable by the mediated population protocol model. Technical Report FRONTS-TR-2010-17, RACTI (2010), `http://fronts.cti.gr/aigaion/?TR=155`
5. Chatzigiannakis, I., Michail, O., Nikolaou, S., Pavlogiannis, A., Spirakis, P.G.: Passively mobile communicating logarithmic space machines. Technical Report FRONTS-TR-2010-16, RACTI (2010), `http://fronts.cti.gr/aigaion/?TR=154`, CoRR. `http://arxiv.org/abs/1004.3395`
6. Chatzigiannakis, I., Michail, O., Spirakis, P.G.: Brief announcement: Decidable graph languages by mediated population protocols. In: Keidar, I. (ed.) DISC 2009. LNCS, vol. 5805, pp. 239–240. Springer, Heidelberg (2009)
7. Chatzigiannakis, I., Michail, O., Spirakis, P.G.: Mediated population protocols. In: Albers, S., Marchetti-Spaccamela, A., Matias, Y., Nikoletseas, S., Thomas, W. (eds.) ICALP 2009. LNCS, vol. 5556, pp. 363–374. Springer, Heidelberg (2009)
8. Chatzigiannakis, I., Michail, O., Spirakis, P.G.: Recent advances in population protocols. In: Královič, R., Niwiński, D. (eds.) MFCS 2009. LNCS, vol. 5734, pp. 56–76. Springer, Heidelberg (2009)
9. Delporte-Gallet, C., Fauconnier, H., Guerraoui, R., Ruppert, E.: When birds die: Making population protocols fault-tolerant. In: Gibbons, P.B., Abdelzaher, T., Aspnes, J., Rao, R. (eds.) DCOSS 2006. LNCS, vol. 4026, pp. 51–66. Springer, Heidelberg (2006)
10. Ginsburg, S., Spanier, E.H.: Semigroups, Presburger formulas, and languages. Pacific Journal of Mathematics 16, 285–296 (1966)
11. Guerraoui, R., Ruppert, E.: Names trump malice: Tiny mobile agents can tolerate byzantine failures. In: Albers, S., Marchetti-Spaccamela, A., Matias, Y., Nikoletseas, S., Thomas, W. (eds.) ICALP 2009, Part II. LNCS, vol. 5556, pp. 484–495. Springer, Heidelberg (2009)

Online Clustering with Variable Sized Clusters*

János Csirik[1], Leah Epstein[2], Csanád Imreh[1,3], and Asaf Levin[4]

[1] Department of Informatics, University of Szeged, 6720 Szeged, Hungary
{csirik,cimreh}@inf.u-szeged.hu
[2] Department of Mathematics, University of Haifa, 31905 Haifa, Israel
lea@math.haifa.ac.il
[3] System Science Innovation Centre PLC, Balatonfüred, H-8230, Hungary
[4] Chaya fellow. Faculty of Industrial Engineering and Management, The Technion, 32000 Haifa, Israel
levinas@ie.technion.ac.il

Abstract. In online clustering problems, the classification of points into sets (called clusters) is done in an online fashion. Points arrive one by one at arbitrary locations, to be assigned to clusters at the time of arrival. A point can be assigned to an existing cluster, or a new cluster can be opened for it. We study a one dimensional variant on a line, where there is no restriction on the length of a cluster, and the cost of a cluster is the sum of a fixed set-up cost and its diameter. The goal is to minimize the sum of costs of the clusters used by the algorithm.

We study several variants, all maintaining the essential property that a point which was assigned to a given cluster must remain assigned to this cluster, and clusters cannot be merged. In the strict variant, the diameter and the exact location of the cluster must be fixed when it is initialized. In the flexible variant, the algorithm can shift the cluster or expand it, as long as it contains all points assigned to it. In an intermediate model, the diameter is fixed in advance while the exact location can be modified. We give tight bounds on the competitive ratio of any online algorithm in each of these variants. In addition, for each one of the models, we consider also the semi-online case, where points are presented sorted by their location.

1 Introduction

In clustering problems, the goal is to partition a set of points into groups, typically called clusters, while optimizing a given objective function. These are fundamental problems, having numerous applications, which in addition to multiple computer related uses, include applications in other fields such as medical diagnosis.

We study online environments, where points are presented one by one to the algorithm. Each point must be assigned to a cluster at the time of arrival,

* This research has been partially supported by the TÁMOP-4.2.2/08/1/2008-0008 program of the Hungarian National Development Agency. Cs. Imreh was supported by the Bolyai Scholarship of the Hungarian Academy of Sciences.

P. Hliněný and A. Kučera (Eds.): MFCS 2010, LNCS 6281, pp. 282–293, 2010.

where a cluster is identified by its name. An assignment of a point to a cluster becomes fixed at this time, and cannot be changed later. Existing clusters cannot be merged or split. We consider a one-dimensional variant, where clusters are intervals on the real line, and the cost of a cluster is composed of a fixed cost of 1 plus the length of the cluster. The cost of the solution is defined as the total cost of all clusters. This clustering problem has a number of applications. One possible application is establishing the positions and types of sensors to be located on the line to service request points, where the sensors are represented by intervals and the diameter of an interval is the distance which can be covered by the corresponding sensor.

We measure the performance of an online algorithm $\mathcal{A}$ by comparing it to an optimal offline algorithm OPT using the standard criterion of the competitive ratio, which is defined as $\sup_\sigma \frac{\mathcal{A}(\sigma)}{\text{OPT}(\sigma)}$, an input σ is a sequence of request points, OPT(σ) is the cost of an optimal offline algorithm for σ and ALG(σ) denotes the cost of an algorithm ALG for this input. The costs are based on the number of clusters and their properties, and depend on the exact definition of the problem. For randomized algorithms, $\mathcal{A}(\sigma)$ is replaced with $E(\mathcal{A}(\sigma))$, and the competitive ratio is $\sup_\sigma \frac{E(\mathcal{A}(\sigma))}{\text{OPT}(\sigma)}$. An algorithm with competitive ratio of at most $\mathcal{R}$ is called $\mathcal{R}$-competitive.

Online partitioning of points into clusters was studied by Charikar et al. [2]. They considered the *online unit covering problem.* In this problem, a set of n points needs to be covered by balls of unit radius, and the goal is to minimize the number of balls used. They gave an upper bound of $O(2^d d \log d)$ and a lower bound of $\Omega(\log d/ \log\log\log d)$ on the competitive ratio of deterministic online algorithms in d dimensions. This problem is strictly online in the sense that points arrive one by one, each point needs to be assigned to a ball upon arrival, and if it is assigned to a new ball, the exact location of this ball is fixed at this time. The tight bounds on the competitive ratio for $d = 1$ and $d = 2$ are respectively 2 and 4.

Chan and Zarrabi-Zadeh [1] introduced the *unit clustering problem.* In this problem the input and goals are very similar to unit covering. This is an online problem as well, but it is more flexible in the sense that the online algorithm is not required to fix the exact position of each ball in advance. The algorithm needs to make sure that a set of points which is assigned to one ball (cluster) can always be covered by that ball, thus the ball can be shifted if necessary. The goal is still to minimize the total number of balls used. Therefore, the algorithm may terminate having clusters that still have more than one option for their location. In an offline scenario, unit covering and unit clustering are the same problem, which is solvable in polynomial time for $d = 1$. However, in the online model, an algorithm for the second problem has more flexibility due to the option of shifting a cluster after a new point arrives, as long as this cluster still covers all the points that are assigned to it. In [1], the authors showed that standard approaches lead to algorithms of competitive ratio 2 (some of which are valid for unit covering). Note that the lower bound of 2 for unit covering in one dimension holds even for randomized algorithms. Two non-trivial randomized algorithms were presented in [1] (a $\frac{15}{8}$-competitive algorithm) and in [19] (a $\frac{11}{6}$-competitive algorithm). Improved deterministic algorithms were later given in [8] (an algorithm of competitive ratio $\frac{7}{4}$) and in [6] (an algorithm of competitive ratio $\frac{5}{3}$). The best currently known lower bound is $\frac{8}{5}$ [8].

In [1,19,8,6], the two-dimensional problem is considered in the L_∞ norm rather than the L_2 norm. Thus "balls", are actually squares or cubes. Most results use the one-dimensional algorithms as building blocks. Note that this problem attains a higher competitive ratio than the one-dimensional case (the best currently known lower bound is $\frac{13}{6}$ [6]). Additional variants of the one-dimensional problem were studied in [7].

Our problem is related to online facility location problems [16,10,9,11,12], where the input is again a stream of points, however in such problems it is sometimes allowed to merge clusters, or to re-assign points, and the cost takes into account the sum of distances of points from the facility (rather than the diameter of a cluster). We are not aware of any work on the online clustering problem, with arbitrary sizes of cluster, where the diameter of a cluster counts towards its cost.

Note that in this paper we consider only the (absolute) competitive ratio and not the asymptotic competitive ratio. This is motivated by the fact that in all the variants that we consider one can repeat the input sequence multiple times in disjoint parts of the real line. These disjoint parts cannot be assigned to the same sets of clusters, and therefore the cost of the solution is the sum of all costs (of the different parts).

We next formally define the clustering problem, CLUSTERING TO MINIMIZE THE SUM OF DIAMETERS WITH A FIXED COST (CSDF) which we consider in this paper. Given an input consisting of n requests which are points on the real line, the goal is to partition the points into groups called *clusters*. The cost of a cluster C is defined as $1 + \max_{i,j \in C} |i - j|$, that is, the sum of a fixed cost which is scaled to 1, and the diameter of the cluster. The goal function is to find a partition of the input into clusters so that the total cost of the clusters is minimized.

The corresponding offline problem was studied on trees [13,14,18,17], where it was shown that the problem is polynomially solvable in this case. The variant of the offline problem where the number of clusters is fixed and the goal is to minimize the sum of diameters or the sum of radii was studied before in [5,3], and a more generalized cost function was studied in [15].

A related offline problem in two dimensions was studied in [4]. Clusters are rectangles which have a fixed area, with their lower edges located on a common base line. However, their aspect ratios can vary. The goal is to cover a given set of points above the base line with a minimal number of clusters.

We consider online scenarios in which the points arrive one by one, and need to be assigned to clusters. In this paper, when we refer to a semi-online algorithm we mean an online algorithm for restricted sets of inputs in which the points arrive one by one when they are sorted according to their coordinates from smallest to largest (i.e., from left to right). We consider three different online scenarios which differ in the abilities of the online algorithm to modify the realization of the clusters as intervals on the real line. We next describe the three models.

1. The strict model. In this model, when a new cluster is opened we need to specify the coordinates of the interval which will be associated with this cluster, and the algorithm is allowed to assign only points belonging to this interval to the cluster. Here the cost of a cluster is defined as 1 plus the length of the interval associated with it. For this model, we show tight bounds of $1 + \sqrt{2} \approx 2.41421$ on the competitive ratio for the online problem, and tight bounds of 2 for the semi-online version. Note that the model used by [2], is the strict model (for a different goal function, and a different set of feasible solutions).

2. The intermediate model. In this model, when a new cluster is opened we need to specify the length of the interval which will be associated with this cluster, but its coordinates might be changed by the algorithm in the future. The algorithm cannot assign a new point to an existing cluster, if this will increase its diameter beyond the length which was specified for this cluster. For this model, we extend the results of the previous model and show that the same bounds are tight for it as well.

3. The flexible model. In this model, when a new cluster is opened we need to specify its label, but its coordinates as well as its diameter might be changed by the algorithm in the future. For this model the cost of a cluster may change as new points are assigned to it. We show that using the flexible model, the best competitive ratio drops to $\phi = \frac{1+\sqrt{5}}{2} \approx 1.618$. The semi-online version of this model is solved optimally using a trivial algorithm which we will discuss as well. The papers [1,19,8,6] study a model which can be both seen as the intermediate model and the flexible model, since the length of a cluster is uniform and cannot vary.

Note that an algorithm for the strict model can be used for the intermediate model, and an algorithm for the intermediate model can be used for the flexible model.

Paper outline. We next consider the flexible model in Section 2. We first show that an optimal solution for this model can be obtained by a semi-online algorithm if the points are sorted from left to right. We design a ϕ-competitive online algorithm and show that no online algorithm of smaller competitive ratio can exist. In Section 3, we consider the strict model, for which we show that there is a semi-online algorithm whose competitive ratio is 2 and that this is best possible, and we present a $1 + \sqrt{2}$-competitive online algorithm and show that this is best possible. In Section 4, we extend the results of Section 3 to the intermediate model. Some proofs are omitted due to space constraints.

2 The Flexible Model

We first show that there is a semi-online algorithm which always returns an optimal solution, i.e. it is 1-competitive. Our algorithm identifies the structure of an optimal solution.

Proposition 1. *In the flexible model there is a semi-online algorithm which always returns an optimal solution.*

Proof. When a new point arrives, we check if its distance from the last opened cluster is at least 1. If so, the algorithm opens a new cluster and assigns the new point to the new cluster. Otherwise the point is assigned to the last opened cluster. To show that this algorithm results in an optimal solution, consider a fixed optimal solution OPT which maximizes the number of clusters (among all optimal solutions). We will show by induction on the number of points in a prefix of the input that the solution returned by the online algorithm is equal to OPT. That is, we need to show that the algorithm opens a new cluster if and only if OPT opens a new cluster. First assume that the algorithm opens a new cluster when point p arrives. Then the distance between p and the points which are smaller than p is at least 1. If OPT does not open a new cluster when p arrives, then partitioning the cluster which contains p in OPT to two parts; one consisting of the points which are smaller than p and the other consists of the remaining points, improves the quality of OPT or increases the number of clusters in the solution without increasing its cost. In both cases we reach a contradiction to the definition of OPT. It remains to consider the case where the algorithm did not open a new cluster for the new point p. In this case, the distance between p and the previous point q is less than 1. Assume by contradiction that OPT opens a new cluster for p, then if we union the cluster which contains p and the cluster which contains q, we decrease the cost of the optimal solution OPT. Clearly, this contradicts the fact that OPT is an optimal solution, and the claim follows. □

We next design a ϕ-competitive algorithm for the online case. The algorithm is summarized as follows.

Algorithm. EXTEND CLOSED CLUSTERS

1. Let p be the new point.
2. If the algorithm has a cluster whose current associated interval contains p, then assign p to that cluster, and do not modify the associated interval of the cluster.
3. Else, let q be the closest point to p which arrives prior to p (break ties arbitrarily).
   (a) If the distance between q and p is at most ϕ, then assigns p to the cluster of q, and extend its associated interval so that it will end at p.
   (b) Otherwise, open a new cluster and assign p to the new cluster. In this case the current interval associated with this new interval is $[p, p]$ (i.e., an interval consisting of a single point p).

Although the algorithm is fairly simple, its analysis is much more involved.

Theorem 1. *Algorithm* EXTEND CLOSED CLUSTERS *is ϕ-competitive.*

Proof. Consider an input sequence and remove from this input all points which caused no action of the algorithm, i.e., points that were already covered by an existing cluster of the algorithm at the time of arrival. A removed point cannot be the closest point to a new point, so its removal has no effect on the actions of the algorithm. Using the same tie breaking policy, the algorithm results in the same output, while the optimal cost cannot increase. In what follows we analyze the modified sequence. Thus, the process of construction of a cluster of the algorithm is as follows. The first request point creates the cluster, which has a length of 0 at this time. Each time that a point joins the cluster, there is a request either on the left hand side or the right hand side of the cluster. The distance of the request to the cluster is $x \leq \phi$, and the cluster is lengthened by x in the direction of the request.

Assume (without loss of generality) that an optimal solution consists of a disjoint set of intervals, and that every endpoint of an interval is a request point. The solution returned by the algorithm satisfies these properties as well.

We next break the input sequence into subsequences. The competitive ratio of the algorithm would be analyzed on each such subsequence separately. To create such a subsequence we first construct the so-called *intersection graph* as follows. This graph is a bipartite graph whose vertex set consists of the set of clusters in the optimal solution (on one side) and the set of clusters resulting by the algorithm (on the other side). There is an edge between two vertices of this graph if the two associated interval intersect (i.e., they serve at least one common request point). In this intersection graph we consider connected components. Each such connected component corresponds to a subsequence in the following way. The subsequence contains every input point assigned to a cluster whose vertex (in the intersection graph) belongs to this connected component. Clearly, each input point belongs to exactly two clusters (one in each of the two solutions), and both these clusters belong to the same connected component. We also note that if the algorithm is applied on such a subsequence, it can apply the same tie breaking policy as before and thus it would result in the same set of clusters as in the connected component of this subsequence in the entire input. Hence, it suffices to prove the competitive ratio on a subsequence, and we assume that the sequence induces a connected intersection graph.

We partition the cost of OPT into the costs of its clusters. The cost of a cluster of OPT is defined as the sum of its opening cost (1 for each cluster) and the diameter of the cluster. Since we would like to prove a competitive ratio of ϕ, we need to show that the cost of the algorithm is at most ϕ times the number of clusters of OPT plus ϕ times their total diameter. We allocate this total cost (which is the cost of OPT times ϕ) to the algorithm and show that this allocation covers the total cost of the algorithm.

Thus, we allocate a budget of ϕ for the opening cost of each cluster of OPT, and a budget of $\phi - 1$ for each unit of length of a cluster of OPT. The total budget will be used for paying the total opening cost of clusters of the algorithm (i.e., the number of such clusters) and for the length of intervals which are covered by the algorithm and not by the optimal solution. This is sufficient because the total cost of the algorithm is at most the sum of the total opening cost of the clusters and the length of the intervals which are covered by the optimal solution or only by the solution returned by the algorithm. That is, the total common length of both algorithms is deducted first, and the remaining cost of $\phi - 1$ times the total length of the clusters of OPT, plus ϕ times the number of clusters of OPT needs to cover only a part of the cost of the algorithm, namely, the number of its clusters, and the total length of intervals where the algorithm has a cluster, while OPT does not.

A cluster of the algorithm C is called *fragmented cluster* if it is not contained in any cluster of OPT. Each fragmented cluster is composed of *parts* where a part is a (non-empty) intersection of the fragmented cluster and an optimal cluster. We say that a cluster of the algorithm which is not a fragmented cluster, is also a part. For each part we identify the first point of the input which arrives in this part. This first point of a part is called its *center*. A center is primary center if it is the first point of the cluster of the algorithm (that is, the algorithm decided to open a new cluster when this point arrived), and otherwise it is a secondary center. We associate an opening cost of 1 for each primary center, and an opening cost for a secondary center which is defined next. Obviously, there is no real cost for the initialization of a secondary center, and the "opening cost" is simply the cost of extending the length of the existing cluster. On the other hand, a new cluster has only a cost of 1, since its initial length is zero. For each secondary center Z, the algorithm has decided to extend an existing cluster to one of the two directions. In this direction the gap between two consecutive intervals of OPT (i.e., the distance between the cluster of OPT which contains Z and the adjacent cluster of OPT, which contains the point from which the cluster of the online algorithm was extended) is denoted by $o(Z)$, and this distance is at most ϕ because it is no larger than the distance between Z and the point from which the cluster was extended towards Z. The opening cost of Z is defined to be $o(Z)$. Note that the total opening cost of the primary and secondary centers is at least the opening cost of clusters of the algorithm (i.e., the number of such clusters) and the length of intervals which are covered by the algorithm and not by the optimal solution. We say that a part is a primary part if its center is a primary center, and otherwise the part is a secondary part.

Consider a cluster c of OPT of length L and assign the total budget of it according to the following rules:

- Assume that there is exactly one part of a cluster of the algorithm which is contained in L. Allocate a budget of ϕ for its opening cost. This is sufficient since if it is a primary part then its opening cost is 1, and otherwise it is at most ϕ. In the following we neglect such clusters of OPT and such centers.
- Otherwise, it assigns $\phi - 1$ of its opening cost budget to a secondary center. The secondary cluster is chosen in a way that each secondary center is allocated $\phi - 1$. Note that this is possible because the number of secondary centers is always at most the number of clusters in OPT (since every part beside the first part of a cluster is created by a cut caused by a gap between two clusters of OPT). The remaining budget of 1, resulting from the opening cost of c, is allocated to the opening cost of the first arriving primary center contained in c (if exists), and otherwise, in which case c must consist of two

parts, each of which is a secondary center, to the first arriving secondary center.
- We allocate $\frac{L}{\phi} = L(\phi - 1)$ to the opening cost of the centers which belong to c.

Claim. If c does not contain a primary part, then the total allocated budget suffices to cover the opening costs of the centers of c.

In the remaining case we assume that c contains $k + 1$ primary centers, where $k \geq 0$. If c contains secondary centers as well, there can be at most two such centers, on each one of the ends of c, and we denote them by x and y. If only one secondary center exists, we denote it by x. We denote the opening costs of x and y by $o(x)$ and $o(y)$, respectively, where we use the convention that if a secondary center does not exists then its opening cost is 0.

Claim. $L \geq k \cdot \phi + o(x) + o(y)$.

The total remaining opening cost of c (after the deduction of the cost of $\phi - 1$, which was assigned to each secondary cluster) is $o(x) + o(y) + k + 1 - 2(\phi - 1)$ if both x and y exist, $o(x) + k + 1 - (\phi - 1)$ if x exists and y does not exist, and otherwise it is $k + 1$. We need to show that we have enough budget to cover it. In all cases the total budget is $L(\phi - 1) + 1$.

Claim. In all three cases, $(o(x) + o(y)) \cdot (\phi - 1) + k + 1 \leq L(\phi - 1) + 1$.

Claim. In all three cases, the remaining opening cost of c is at most $(o(x) + o(y)) \cdot (\phi - 1) + k + 1$.

By the above claims, in all three cases we showed that the total budget is sufficient to pay for the total opening cost of centers in c, and hence the competitive ratio of Algorithm Extend closed clusters is at most ϕ. □

We conclude this section by proving that our Algorithm Extend closed clusters is the best possible online algorithm.

Theorem 2. *There is no deterministic online algorithm for the flexible model whose competitive ratio is strictly smaller than ϕ.*

Proof. Let n be a large integer, and let $\varepsilon = \frac{\phi}{2n}$. Our lower bound construction is carried in two steps. In the first step, we introduce n points in positions $i \cdot \phi$ for $i = 0, 1, \ldots, n - 1$. In the second step, we consider each pair of consecutive points of the first step which were assigned different clusters (if there exists at least one such pair of points). Assume that the algorithm assigns points $i \cdot \phi$ and $(i + 1) \cdot \phi$, for some value of i, to different clusters. Then in the second step we introduce additional $2n - 1$ points between $i \cdot \phi$ and $(i + 1) \cdot \phi$. That is, for $j = 1, 2, \ldots, 2n - 1$ we introduce the point $i \cdot \phi + j \cdot \varepsilon$. We do this procedure for every pair of consecutive points of the first step which the algorithm assigns to distinct clusters. In the full version of the paper we show that this construction proves the claim. □

3 The Strict Model

We first consider the semi-online variant. In the strict model, an optimal algorithm cannot be obtained even if the points arrive sorted from left to right.

Lemma 1. *The competitive ratio of any semi-online algorithm for the strict model is at least* 2.

Proof. The first request point is $p_1 = 0$. Let a_1 be the length of the cluster which is opened by the algorithm. Since all future points are in $[0, +\infty)$, then the first cluster must be $C_1 = [0, a_1]$, and the righteous point covered by any cluster is a_1. We show an iterative construction of the sequence. The sequence consists of N requests for a large integer N. Every new request is defined so that a new cluster is required for it. We use a very small constant $\delta > 0$. We later specify the conditions under which the sequence stops earlier, that is, the sequence may stop after k points for any $k \leq N$.

Assume that the request points $1, 2, \ldots, i-1$, for $i \geq 0$ were defined. Let q_i be the righteous point covered by any cluster. Then the new request is $p_i = q_i + \delta$. Let a_i be the length of the new cluster. Since points arrive sorted from left to right, without loss of generality, the new cluster is $[p_i, p_i + a_i]$, and the righteous point covered by any cluster is $p_i + a_i$. It is not difficult to see that $p_k = \sum_{i=1}^{k-1}(a_i + \delta)$.

Let OPT_k and ALG_k denote the costs of an optimal offline algorithm, and of the online algorithm, respectively, after k requests have arrived. An algorithm which has a cluster of length 0 for each point has a cost of k, so $\text{OPT}_k \leq k$. The algorithm has a cost of $k + \sum_{i=1}^{k} a_i$. If for some $k < N$, $\sum_{i=1}^{k} a_i \geq k$, then the sequence can be stopped, and $\text{ALG}_k \geq 2\text{OPT}_k$. Otherwise, let $\Delta = p_N = \sum_{i=1}^{N-1}(a_i + \delta)$. Among all N request points, the righteous request point is p_N, and the leftmost one is 0, so an algorithm which has a single cluster has a cost of $1 + \sum_{i=1}^{N-1}(a_i + \delta)$, and therefore $\text{OPT}_N \leq 1 + \sum_{i=1}^{N-1}(a_i + \delta) = 1 + \Delta$ while $\text{ALG}_N = N + \sum_{i=1}^{N} a_i = N + a_N + \Delta - (N-1)\delta$.

Recall that the sequence does not stop earlier, after p_j was presented for some $1 \leq j \leq N-1$, and in particular it did not stop after p_{N-1} was presented. Therefore, $\sum_{i=1}^{N-1} a_i < N - 1$, so $\Delta - (N-1)\delta < N - 1$ or $\Delta < (1+\delta)(N-1)$. Therefore, $\text{OPT}_N \leq 1 + \Delta \leq 1 + (1+\delta)(N-1)$. Thus $\text{ALG}_N \geq (\Delta + 1) + (N-1)(1-\delta) \geq \text{OPT}_N + (N-1)(1-\delta)$. Therefore, $\frac{\text{ALG}_N}{\text{OPT}_N} \geq 1 + \frac{(N-1)(1-\delta)}{1+(1+\delta)(N-1)}$. Letting δ be arbitrarily close to 0 and N arbitrarily large, this expression tends to 2. □

We consider the following simple semi-online algorithm. Upon arrival of a new request point p, if it is not already covered by a cluster, open the cluster $[p, p+1]$.

Theorem 3. *The competitive ratio of this semi-online algorithm is* 2, *which is best possible.*

We next consider the online variant. We can show a higher lower bound in this case.

Lemma 2. *The competitive ratio of any online algorithm for the strict model is at least* $1 + \sqrt{2} \approx 2.414$.

Proof. The first request point is $p_1 = 0$. Let $[x, y]$ be the cluster which is opened by the algorithm, and let $a_1 = \min\{y, -x\}$. Without loss of generality, it is possible to assume $a_1 = y$. From this time on, the sequence of requests is increasing (if $a_1 = -x$, then a decreasing sequence should be used instead). From this time, the construction is the same as in Lemma 1. The only difference is that the length of the first cluster of the online algorithm is at least $2 \cdot a_1$ rather than a_1.

We use the same notation and consider only offline algorithms which use a single cluster for all request points. Thus $\text{OPT}_k \leq 1 + p_k = 1 + \sum_{i=1}^{k-1} a_i + (k-1)\delta$, while $\text{ALG}_k \geq k + a_1 + \sum_{i=1}^{k} a_i$. Assume that there exists an algorithm with a competitive ratio of at most $\mathcal{R} = 1 + \sqrt{2} - \varepsilon$, for some small $0 < \varepsilon < 0.01$, the value of δ is chosen so that $\delta \leq \frac{\varepsilon}{4}$.

We can prove by induction that $a_k \leq \frac{\sqrt{2}}{2}$. Thus $\text{OPT}_k \leq 1 + (k-1)\frac{\sqrt{2}}{2}$, while $\text{ALG}_k \geq \text{OPT}_k + (k-1)(1-\delta)$. For a sufficiently large value of k, the ratio $\frac{\text{ALG}_k}{\text{OPT}_k}$ exceeds the value $\mathcal{R}$, since this ratio tends to $1 + \sqrt{2}$ as k grows to infinity and δ tends to zero. □

We design a simple algorithm which turns out to be optimal. Upon arrival of a new request point p, if it is not already covered by a cluster, let ℓ denote the closest point of any cluster located to the left of p (if no such cluster exists, $\ell = -\infty$), and let r denote the closest point of any cluster located to the right of p (if no such cluster exists, $r = \infty$). Let $\alpha = \frac{\sqrt{2}}{2}$. Open the cluster $[\max\{\ell, p - \alpha\}, \min\{p + \alpha, r\}]$. That is, the idea is to open a cluster of length $\sqrt{2}$, with the new point in the middle. However, if this would creat an overlap, the cluster is shorter so that it has a common endpoint with the cluster that it would otherwise overlap.

We observe the following simple properties.

Proposition 2. *No two clusters have an overlap, except for overlap at endpoints. The length of each cluster is at most* $2\alpha = \sqrt{2}$.

Theorem 4. *The competitive ratio of the algorithm is* $1 + \sqrt{2}$, *which is optimal.*

Proof. Due to the lower bound proved in Lemma 2, we only need to prove that the competitive ratio is at most $1 + \sqrt{2}$. Consider a request sequence, and remove all request points which did not result in a new cluster. This does not change the cost of the algorithm, and may only decrease the cost of an optimal offline algorithm. Each cluster has exactly one request point. The distance between two request points is larger than α. To see this, consider two consecutive request points p and q, where $p < q$ (consecutive in the sorted list of request points, but not necessarily in the input sequence). Assume without loss of generality that p arrives before q. If the right endpoint of the cluster of p is $p + \alpha$, then since q needs a new cluster, then $q > p + \alpha$ and we are done. Otherwise, let r be the right endpoint of the cluster of p in the algorithm. There is another cluster $[r, r']$, which was opened for a request point x which arrived before p. We have $p < q < x$, since p and q are consecutive. However, at the time of arrival of q, the interval $[p, x]$ is fully covered by clusters, which contradicts the need of a new cluster for q.

Consider a fixed optimal offline solution OPT, and let C be a cluster of length λ used by this solution. We compute the cost of the algorithm for the service of the points which OPT covers by C. The largest distance between any two such points is λ, and since the distance between two points is more than α, then the number of request points is at most $\frac{\lambda}{\alpha}+1$. Moreover, the clusters of the algorithm cannot overlap, and the leftmost cluster and righteous cluster may extend by at most α out of the interval of C. Thus the total cost of the algorithm is at most $\lambda + 2\alpha + \frac{\lambda}{\alpha} + 1$, while the cost of OPT is $\lambda + 1$. The ratio between the two costs is at most $\max\{2\alpha + 1, 1 + \frac{1}{\alpha}\} = 1 + \sqrt{2}$. □

4 The Intermediate Model

In this section we show that the results of the previous section on the strict model apply also for the intermediate model. First note that an online algorithm for the strict model is also an online algorithm for the intermediate model. Hence, using theorems 3 and 4, we conclude the following.

Proposition 3. *There is an online* $1 + \sqrt{2}$*-competitive algorithm for the intermediate model. There is a semi-online* 2*-competitive algorithm for the intermediate model.*

In the Appendix, we extend the lower bounds as well. While the extension of the lower bound for semi-online algorithms is straightforward, our last result, which extends the lower bound of $1 + \sqrt{2}$ on the competitive ratio of any online algorithm, is more difficult to establish. We basically show that already after a small number of clusters, the algorithm is forced to introduce its new clusters in fixed positions, even if it could postpone the decision regarding the exact positions till later. To show this, we analyze several possible behaviors of the algorithm with respect to the definition and usage of its early clusters.

Theorem 5. *The competitive ratio of any online algorithm for the intermediate model is at least* $1 + \sqrt{2}$*. The competitive ratio of any semi-online algorithm for the intermediate model is at least* 2*.*

Acknowledgment. The authors thank Prof. Arie Tamir of Tel-Aviv University, for pointing out the literature on the offline variant of the problem on trees.

References

1. Chan, T.M., Zarrabi-Zadeh, H.: A randomized algorithm for onine unit clustering. Theory of Computing Systems 45(3), 486–496 (2009)
2. Charikar, M., Chekuri, C., Feder, T., Motwani, R.: Incremental clustering and dynamic information retrieval. SIAM Journal on Computing 33(6), 1417–1440 (2004)
3. Charikar, M., Panigrahy, R.: Clustering to minimize the sum of cluster diameters. Journal of Computer and System Sciences 68(2), 417–441 (2004)
4. Chin, F.Y.L., Ting, H.-F., Zhang, Y.: Variable-size rectangle covering. In: Proc. of the 3rd International Conference on Combinatorial Optimization and Applications (COCOA 2009), pp. 145–154 (2009)
5. Doddi, S., Marathe, M., Ravi, S.S., Taylor, D.S., Widmayer, P.: Approximation algorithms for clustering to minimize the sum of diameters. Nordic Journal of Computing 7(3), 185–203 (2000)
6. Ehmsen, M.R., Larsen, K.S.: Better bounds on online unit clustering. In: Proc. of the 12th Scandinavian Symposium and Workshops on Algorithm Theory, SWAT 2010 (to appear 2010)
7. Epstein, L., Levin, A., van Stee, R.: Online unit clustering: Variations on a theme. Theoretical Computer Science 407(1-3), 85–96 (2008)
8. Epstein, L., van Stee, R.: On the online unit clustering problem. ACM Transactions on Algorithms (to appear)
9. Fotakis, D.: Incremental algorithms for facility location and k-median. Theoretical Computer Science 361(2-3), 275–313 (2006)
10. Fotakis, D.: Memoryless facility location in one pass. In: Durand, B., Thomas, W. (eds.) STACS 2006. LNCS, vol. 3884, pp. 608–620. Springer, Heidelberg (2006)
11. Fotakis, D.: A primal-dual algorithm for online non-uniform facility location. Journal of Discrete Algorithms 5(1), 141–148 (2007)
12. Fotakis, D.: On the competitive ratio for online facility location. Algorithmica 50(1), 1–57 (2008)
13. Hoffman, A.J., Kolen, A., Sakarovitch, M.: Totally balanced and greedy matrices. SIAM Journal on Algebraic and Discrete Methods 6, 721–730 (1985)
14. Kolen, A., Tamir, A.: Covering problems. In: Mirchandani, P.B., Francis, R.L. (eds.) Discrete Location Theory, ch. 6, pp. 263–304. Wiley, Chichester (1990)
15. Levin, A.: A generalized minimum cost k-clustering. ACM Transactions on Algorithms 5(4) (2009)
16. Meyerson, A.: Online facility location. In: Annual Symposium on Foundations of Computer Science (FOCS 2001), pp. 426–431 (2001)
17. Shah, R., Farach-Colton, M.: Undiscretized dynamic programming: faster algorithms for facility location and related problems on trees. In: Proc. of 13th Annual ACM-SIAM Symposium on Discrete Algorithms (SODA 2002), pp. 108–115 (2002)
18. Tamir, A.: Maximum coverage with balls of different radii (2003) (mansucript)
19. Zarrabi-Zadeh, H., Chan, T.M.: An improved algorithm for online unit clustering. Algorithmica 54(4), 490–500 (2009)

Deterministic Rendezvous of Asynchronous Bounded-Memory Agents in Polygonal Terrains

Jurek Czyzowicz[1,*], Adrian Kosowski[2,**], and Andrzej Pelc[1,***]

[1] Département d'informatique, Université du Québec en Outaouais, Gatineau, Québec J8X 3X7, Canada
jurek@uqo.ca, pelc@uqo.ca

[2] Department of Algorithms and System Modeling, Gdańsk University of Technology, 80233 Gdańsk, Poland
adrian@kaims.pl

Abstract. Two mobile agents, modeled as points starting at different locations of an unknown terrain, have to meet. The terrain is a polygon with polygonal holes. We consider two versions of this rendezvous problem: *exact RV*, when the points representing the agents have to coincide at some time, and *ε-RV*, when these points have to get at distance less than ε in the terrain. In any terrain, each agent chooses its trajectory, but the movements of the agent on this trajectory are controlled by an adversary that may, e.g., speed up or slow down the agent. Agents have bounded memory: their computational power is that of finite state machines. Our aim is to compare the feasibility of exact and of ε-RV when agents are anonymous vs. when they are labeled. We show classes of polygonal terrains which distinguish all the studied scenarios from the point of view of feasibility of rendezvous. The features which influence the feasibility of rendezvous include symmetries present in the terrains, boundedness of their diameter, and the number of vertices of polygons in the terrains.

1 Introduction

The problem. Two mobile agents modeled as points starting at different locations of an a priori unknown terrain have to meet. We consider two versions of rendezvous: *exact rendezvous*, when the points representing the agents have to coincide at some time, and *ε-rendezvous*, when we require that these points get at distance less than ε in the terrain. In any terrain, each agent chooses its trajectory, but the movements of the agent on this trajectory are controlled by an adversary that may speed up or slow down the agent.

Agents and their memory. In this paper we consider the question of determining if rendezvous is possible in all terrains when using agents equipped with bounded memory, and if not, in which terrains can rendezvous be achieved. Our study is

* Partially supported by NSERC discovery grant.
** This work was done during this author's visit to the Université du Québec.
*** Partially supported by NSERC discovery grant and by the Research Chair in Distributed Computing at the Université du Québec en Outaouais.

P. Hliněný and A. Kučera (Eds.): MFCS 2010, LNCS 6281, pp. 294–305, 2010.

in some sense motivated by the results from [10] where it is shown that if the agents have different labels and unbounded memory, then in spite of this powerful adversary, rendezvous is possible almost always, provided that the terrain is closed and path-connected. (More precisely, if the agents start at any interior points with rational coordinates in such a terrain, then exact rendezvous is possible, and if they start at any interior points, then ε-rendezvous is possible, for any $\varepsilon > 0$.) However, the size of the memory of the agents used to accomplish rendezvous was large, possibly dependent on e.g. the size of the terrain.

First observe that, if the memory of the agents is bounded, then rendezvous cannot be accomplished in many unbounded terrains, in particular in the empty plane (this is in contrast to the feasibility of rendezvous with unbounded memory, as mentioned above). Indeed, when starting at a point of the plane, the trajectory of an agent with bounded memory occupies a stripe between two parallel lines in the plane: the agent is either confined to a bounded region of the stripe or must progress in one direction of this stripe executing the same moves cyclically. It is easy to see that placing any two such agents in appropriate points in the plane guarantees that their distance will always be larger than any a priori given constant, and thus even ε-rendezvous is impossible. Hence we assume that the terrain in which the agents operate is bounded. It is represented as a polygon, possibly with a finite number of polygonal obstacles. We assume that the boundary of the terrain is included in it. Thus, formally, a terrain is a set $\mathcal{P}_0 \setminus (\mathcal{P}_1 \cup \cdots \cup \mathcal{P}_k)$, where $\mathcal{P}_0$ is a closed polygon and $\mathcal{P}_1, \ldots, \mathcal{P}_k$ are disjoint open polygons, called *holes*, included in $\mathcal{P}_0$. The boundary of the terrain may not have any self-intersections, and the term *polygon* refers to simple polygons, throughout the paper.

In order to capture the assumption that the memory of the agent is bounded, we model an agent as an abstract state machine $\mathcal{A} = (S, \pi, \lambda, s_0)$, where S is a finite set of states among which there is a specified state s_0 called the *initial* state, $\pi : S \times \{0, 1\} \rightarrow S$ is a function describing state transitions, and for any $s \in S$, the value $\lambda(s)$ defines some segment of the plane whose one extremity is the current location of the agent. Initially the agent is at some point u_0 of the terrain in the initial state $s_0 \in S$. The agent constructs consecutive segments of its trajectory depending on its current state. Each segment is prescribed by the value $\lambda(s)$, when the agent is in state s. Upon completing the move in state s, the agent gets a one-bit input: 0, if the move terminates in an interior point of the terrain, and 1, if it terminates at the boundary of the terrain. This input bit b causes the transition from state s to state $s' = \pi(s, b)$. Now the agent executes the move prescribed by $\lambda(s')$. The agent continues moving in this way, possibly infinitely. If it meets the other agent (exactly or within a distance ε, depending on the task), it stops. Otherwise, it executes its moves forever.

Moves of the agents. It remains to describe precisely how the value $\lambda(s)$ determines the current segment of the agent's trajectory. Each agent has a compass. Compasses of both agents need not be coherent, but they have the same chirality, i.e., the same notion of clockwise and counterclockwise. An agent currently located at a point u in a state s chooses some angle $\alpha \in [0, 2\pi)$ and some distance $x \geq 0$. Let v be the point at distance x from u at the angle α from the north of the agent. If the segment $(u, v]$ is included in the interior of the terrain, then $\lambda(s) = [u, v]$ and the agent moves to point v along the segment $[u, v]$. Otherwise, let w be the boundary point of the terrain in $[u, v]$ furthest from u such that the entire segment $[u, w]$ is still contained in the terrain. In this case $\lambda(s) = [u, w]$ and the agent moves to point w along the segment $[u, w]$ and finishes its current move at w, i.e. the agent finishes its current move when it can no longer continue moving within the terrain towards the intended target along the segment. Note that if the segment is immediately directed outside the terrain, then $w = u$, which means that the agent does not move at all, and has no capability of detecting that such a situation has occurred.

The moves described above are called *jumps*. However, they turn out to be too restrictive for agents at boundary points. Indeed, consider an agent in a vertex of a triangle whose angle is so small that it misses all jump directions available to the agent (recall that the set of states is finite). Such an agent would be immobilized at this vertex and hence two agents starting at such vertices of an empty triangle with corresponding angles sufficiently small would be never capable of moving, let alone meeting. Hence we need to add some other possible moves. A natural choice is a *slide* move which enables an agent currently located at a boundary point to slide to the end of the side σ of the polygon on which it is located, in either direction. Thus we add two other possible values of the function λ: $\lambda(s) = \oplus$ causes an agent at a boundary point to slide to the end of the segment σ in the clockwise direction and $\lambda(s) = \ominus$ causes sliding to the end of the segment σ in the counterclockwise direction. If the agent is currently located at a vertex, it slides to the other end of the respective side. When an agent is not at a boundary point, the effect of a slide is null (the agent remains at the same point).

Asynchronous rendezvous. We consider the asynchronous version of the rendezvous problem. The asynchrony of the agents' movements is captured by the assumption that the actual walk of each agent is decided by the adversary. More formally, the *trajectory* in a terrain is a sequence $(\sigma_1, \sigma_2, \dots)$ of segments, where $\sigma_i = [a_i, a_{i+1}]$ is the segment corresponding to step i of the construction, prescribed by the function λ as above. We now describe the walk f of an agent on its trajectory. Let $R = (\sigma_1, \sigma_2, \dots)$ be the trajectory of an agent. Let $(t_1, t_2, \dots)$, where $t_1 = 0$, be an increasing sequence of reals, chosen by the adversary, that represent points in time. Let $f_i : [t_i, t_{i+1}] \rightarrow [a_i, a_{i+1}]$ be any continuous function, chosen by the adversary, such that $f_i(t_i) = a_i$ and $f_i(t_{i+1}) = a_{i+1}$. For any $t \in [t_i, t_{i+1}]$, we set $f(t) = f_i(t)$. The interpretation of the walk f is as follows: at time t the agent is at the point $f(t)$ of its trajectory. This general definition of the walk[1] and the fact that it is constructed by the adversary capture the asynchronous characteristics of the process. Throughout the paper, *rendezvous*

[1] Notice that our definition of the walk allows the adversary not only to speed up or slow down the agent but also to stop it or even move it back and forth, as long as the walk of the agent in each segment of its route is continuous, does not leave it and covers all of it. In fact our impossibility results hold even for an adversary that can only speed up or slow down the agent, without moving it back.

means deterministic asynchronous rendezvous: no random choices are involved and the moves of an agent in a given terrain depend solely on the deterministic automaton that formalizes it and on the adversary.

An exact rendezvous is guaranteed for trajectories R_1 and R_2 if, regardless of the walks f_1 and f_2 chosen by the adversary, the agents using these trajectories meet at some time t, i.e., points $f_1(t)$ and $f_2(t)$ are equal. An ε-rendezvous is guaranteed for trajectories R_1 and R_2 if, regardless of the walks f_1 and f_2 chosen by the adversary, the agents using these trajectories get at distance less than ε *in the terrain*, i.e., for some t, the shortest path included in the terrain between points $f_1(t)$ and $f_2(t)$ has length less than ε. We consider two versions of the rendezvous problem: for labeled and for anonymous agents. The following definition is crucial for our considerations. Let $\mathcal{C}$ be a class of terrains.

- An exact (resp. ε-) rendezvous of labeled agents is feasible for the class $\mathcal{C}$, if there exists an infinite sequence of agents $\mathcal{A}_1, \mathcal{A}_2, \ldots$, such that for any terrain $\mathcal{T}$ of class $\mathcal{C}$ and for any agents $\mathcal{A}_i$ and $\mathcal{A}_j$, where $i \neq j$, starting at any distinct points of $\mathcal{T}$, the trajectories of these agents guarantee an exact (resp. ε-) rendezvous. In this case we say that the sequence of agents $\mathcal{A}_1, \mathcal{A}_2, \ldots$ accomplishes exact (resp. ε-) rendezvous for the class $\mathcal{C}$.
- An exact (resp. ε-) rendezvous of anonymous agents is feasible for the class $\mathcal{C}$, if there exists an agent $\mathcal{A}$, such that for any terrain $\mathcal{T}$ of class $\mathcal{C}$ and for two identical copies of $\mathcal{A}$ starting at any distinct points of $\mathcal{T}$, the trajectories of these agents guarantee an exact (resp. ε-) rendezvous. In this case we say that the agent $\mathcal{A}$ accomplishes exact (resp. ε-) rendezvous for the class $\mathcal{C}$.

Our results. Our aim is to compare feasibility of exact and of ε-RV when agents are anonymous vs. when they are labeled. We first describe our results for exact rendezvous. It turns out that for anonymous agents, rotational symmetries of the terrain preclude the possibility of meeting. Hence, already for the class of equilateral triangles, exact rendezvous is impossible. Moreover, we show that for a known terrain, i.e., a one-element class, a pair of anonymous agents accomplishing exact rendezvous can be constructed, if and only if the terrain does not have non-trivial rotational symmetries.

Labeled agents turn out to be much more powerful, however significant limitations apply in this case as well. (This is in sharp contrast with results from [9] where it is shown, in particular, that with unbounded memory, exact rendezvous is feasible for the class of all polygonal terrains with arbitrary polygonal holes.) First we prove that even labeled agents cannot solve the exact rendezvous problem for the class of regular polygons with arbitrarily many vertices. It turns out that restricting the number of vertices of polygons helps. We show that, for any fixed integer k, labeled agents can solve exact rendezvous in the class of all polygons with at most k vertices, even with an arbitrary number of triangular holes. Surprisingly, the restriction to triangular holes is necessary. Indeed, we prove that exact rendezvous is unfeasible even for all triangles with quadrangular holes. If the terrain is known, rendezvous of labeled agents is always feasible.

Finally, we consider the problem of ε-rendezvous. Allowing approximate meeting changes the situation drastically. For anonymous agents, ε-rendezvous is

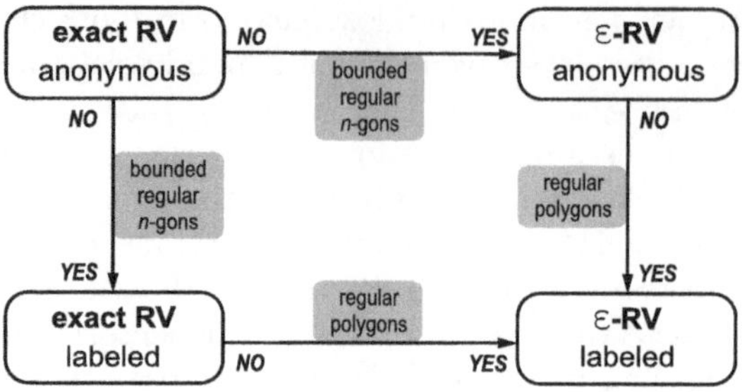

Fig. 1. Feasibility of exact and ε-rendezvous for anonymous and labeled agents

possible for the class of regular polygons of bounded diameter (in contrast to exact rendezvous). However, in the larger class of regular polygons of unbounded diameter, ε-rendezvous of anonymous agents turns out to be impossible. On the other hand, for labeled agents, ε-rendezvous (in contrast to exact rendezvous) is feasible even for the class of regular polygons with unbounded diameter.

A corollary of our results is depicted in Fig. 1. It shows that the four scenarios induced by combining anonymous vs. labeled agents and exact vs. ε-rendezvous have all different rendezvous power: for each comparable pair of these scenarios (in the figure these are pairs joined by an oriented path of arrows) there is a class of terrains distinguishing this pair from the point of view of rendezvous feasibility, i.e., such that for this class rendezvous is possible in the easier scenario but impossible in the harder one. The distinction occurs already for the class of regular polygons without holes, and the features responsible for distinguishing each pair are diameter boundedness and the number of vertices. We have seen above that other characteristics that may influence feasibility of rendezvous are the presence of symmetries and the shape of holes.

Related work. A thorough discussion of the literature on rendezvous can be found in the book [2]. Most of the results on rendezvous can be divided into two groups: those considering the geometric scenario (rendezvous in the line, see, e.g., [5,14,20], or in the plane, see, e.g., [4]), and those discussing rendezvous in graphs, e.g., [1,3]. Some of the authors, e.g., [1,5] consider the probabilistic scenario where inputs and/or rendezvous strategies are random. A generalization of the rendezvous problem is that of gathering [12,15,16], when more than two agents have to meet in one location.

If graphs are unlabeled, deterministic rendezvous requires the breaking of symmetry, which can be accomplished either by allowing marking the nodes or by labeling the agents. Deterministic rendezvous with anonymous agents working in unlabeled graphs but equipped with tokens used to mark nodes was considered, e.g., in [18]. Rendezvous in trees with anonymous agents having bounded memory was studied in [13]. In [11,17,21] deterministic rendezvous in graphs with labeled agents was considered. However, in all the above papers, the synchronous setting was assumed. Asynchronous gathering under geometric scenarios has been studied, e.g., in [7,12,19] in different models than ours: agents could not remember past events, but they were assumed to have at least partial visibility of the scene. The first paper to consider deterministic asynchronous rendezvous in graphs was [8]. The authors assumed that agents have different labels and concentrated on the complexity of rendezvous in simple classes of graphs, such as rings and infinite lines. Further improvements of their results for the infinite line were proposed in [20]. In [10] the feasibility of asynchronous rendezvous for labeled agents was considered both for graphs and in the geometric scenario. It was proved that rendezvous is possible in any connected (even countably infinite) graph and implications mentioned at the beginning of this section were drawn for the feasibility of rendezvous in the geometric scenario. The memory of agents in [10] was assumed to be unbounded. Gathering many robots in a graph, under a different asynchronous model and assuming that the whole graph is seen by each robot, has been studied in [15,16]. The problem of identifying terrains by robots moving around a polygon has also been considered [6].

2 Exact Rendezvous of Anonymous Agents

The following theorem shows that for anonymous agents, rotational symmetries of the terrain preclude exact rendezvous, but if the terrain is known in advance, then in the absence of such symmetries exact rendezvous is possible. This vulnerability to symmetries will be later shown to disappear both for exact rendezvous of labeled agents and for ε-rendezvous of anonymous agents: both these tasks turn out to be possible for some terrains with non-trivial rotational symmetries.

Theorem 1. *For any given terrain $\mathcal{T}$, there exist anonymous agents which solve the rendezvous problem in $\mathcal{T}$, if and only if $\mathcal{T}$ does not have non-trivial rotational symmetries.*

Proof. ($\Leftarrow$) Assume that $\mathcal{T}$ has no non-trivial rotational symmetries. Before defining the agent, we construct a map of the terrain $\mathcal{T}$ oriented with respect to an arbitrarily chosen direction which we will call *absolute North*, and assign unique labels to all the vertices of the external polygon and the holes of $\mathcal{T}$. One arbitrarily chosen vertex v of the external polygon is treated as distinguished. We now describe an agent that performs a finite sequence of actions, split into the following steps:
1. The agent reaches some (unknown) vertex of the terrain.
2. By performing a local exploration, the agent identifies the label u of this vertex and obtains an approximation (with an error value bounded by a small constant) of the angle between the north of its compass and the absolute North.
3. The agent moves from vertex u to vertex v and stops.

Such an agent is sufficient to solve the rendezvous problem, since eventually both agents will meet at vertex v.

In order to reach an arbitrary vertex u in Step 1, the agent performs a jump south from its starting location, of length exceeding the diameter of the terrain (thus hitting a boundary point), and then makes a single clockwise slide.

Next, let δ_2 be a fixed positive real. There exists a terrain $\mathcal{T}_p$ and a positive real $\delta_1 < \delta_2$ which satisfy the following conditions:
• The terrain $\mathcal{T}_p$ is contained within the terrain $\mathcal{T}$.
• The boundary of $\mathcal{T}_p$ is contained within a δ_2-neighborhood of the boundary of $\mathcal{T}$, and no point of the boundary of $\mathcal{T}_p$ is contained within a δ_1-neighborhood of the boundary of $\mathcal{T}$.
• There exists a one-to-one mapping Φ between the vertices of $\mathcal{T}$ and the vertices of $\mathcal{T}_p$, such that the distance between any pair of corresponding vertices does not exceed δ_2.

Fix a terrain $\mathcal{T}_p$ as above. Since terrain $\mathcal{T}$ has no rotational symmetries, there exists a unique (trivial) even isometry of the set of vertices of $\mathcal{T}$. When δ_2 is sufficiently small w.r.t. to the length of the shortest side of the terrain, the mapping Φ must preserve the order of vertices along the external boundaries of the terrains. Moreover, since $\delta_1 > 0$, there must exist an angle $\alpha > 0$ such that a rotation of the terrain $\mathcal{T}_p$ by any angle not greater than α around any of its vertices results in a terrain $\mathcal{T}_p'$ which is still contained within terrain $\mathcal{T}$, and the maximum distance between corresponding vertices of $\mathcal{T}_p$ and $\mathcal{T}_p'$ does not exceed δ_2. Consequently, the distance between corresponding vertices of $\mathcal{T}$ and $\mathcal{T}_p'$ does not exceed $2\delta_2$.

The agent proceeds to identify its current vertex u by detecting the label of the corresponding vertex $u_p = \Phi(u)$ in the terrain $\mathcal{T}_p$. To do this, for each assumed value of u_p, it performs a sequence of jumps forming a traversal of a closed trajectory described on the map of the terrain by a sequence of segments, starting and ending at $\Phi^{-1}(u_p)$, and including all the segments of the boundary of the terrain $\mathcal{T}_p$ (this is possible even for terrains with holes). The lengths of jumps and their angles measured w.r.t. the absolute North are fixed, and the lengths of jumps are sufficiently small. For each assumed value of u_p, the agent traverses the trajectory for all possible relative angles φ in the range $[0, 2\pi)$ between the north of its compass and absolute North, starting from angle 0 with a step of α. Each traversal of the trajectory, associated with the assumed pair (u_p, φ) may have one of the two following outcomes. If the agent completes the traversal of its trajectory without encountering a boundary point of $\mathcal{T}$, then vertex u_p has been uniquely identified. Otherwise, if some jump reaches the boundary, the agent can reverse all of the previously made jumps, attempting to return to vertex v. Notice that since the last jump of the sequence may be shortened by hitting the boundary, the sequence of reverse jumps forms a trajectory in a small neighborhood with respect to the original one. Assuming that all the jumps are of sufficiently small length w.r.t. the lengths of the sides of the terrain, the agent can terminate the sequence of reverse jumps at a point of a uniquely determined side of the polygon incident to u, and then return to u by a slide.

By the properties of the mapping Φ and of terrains $\mathcal{T}_p$ and $\mathcal{T}_p'$, a pair (u_p, φ) for which the agent completes the trajectory without hitting the boundary point will eventually be successfully identified. The agent now proceeds to Step 3, following a trajectory described by jumps along segments of the map: from u to u_p and then within terrain $\mathcal{T}_p$ from u_p to v_p. When performing this trajectory, the angle between absolute North and the agent's north is assumed to be equal to φ. By the properties of terrain $\mathcal{T}_p'$, the agent will reach some point v_p' located within a distance of $2\delta_2$ from v, without touching the boundary. Finally, knowing the angle of the polygon at vertex v, the agent can jump onto a uniquely determined side of the external polygon, incident to v, reach v by a slide, and stop.

($\Rightarrow$) If $\mathcal{T}$ has a non-trivial rotational symmetry by an angle $0 < \gamma < 2\pi$, position the two agents at two points of the boundary of the terrain such that one is the image of the other under rotation of the terrain by γ, and let the relative angle between the north directions of the agents be equal to γ. Moreover, the adversary fixes the angles between the absolute North and the north directions of the agents so that neither of the agents ever passes through the center of symmetry of the terrain. The agents will then follow trajectories which are identical up to rotation by γ around the center of symmetry of the terrain. Rendezvous can be avoided by an adversary which schedules the motion of both agents at the same speed, keeping them at all times at symmetric points of the terrain. □

Theorem 1 implies, in particular, that no anonymous agents can perform exact rendezvous in an equilateral triangle.

3 Exact Rendezvous of Labeled Agents

Although labeled agents will be shown to be much more powerful than anonymous ones for solving exact rendezvous, some significant restrictions apply in their case as well. We start with the following negative result.

Theorem 2. *Exact rendezvous of labeled agents is not feasible for the class of regular polygons without holes, even for polygons of bounded diameter.*

To show our positive result for exact rendezvous we will use the following lemma.

Lemma 1. *Let k be a fixed integer parameter. There exists a sequence $\mathcal{R}_1, \mathcal{R}_2, \ldots$ of labeled agents for the class of polygons with at most k sides, which have the following properties:*
- *If two distinct agents are initially located on the boundary of the polygon, then they meet within a finite number of steps.*
- *Otherwise, if rendezvous is not reached, then both agents stop at their initial locations after a finite number of steps.*

The following positive result shows that what caused the infeasibility proved in Theorem 2 was the unbounded number of vertices of the polygons, rather than their symmetries.

Theorem 3. *For any fixed integer $k \geq 3$ exact rendezvous of labeled agents is feasible for the class of all polygons with at most k vertices, with an arbitrary number of triangular holes.*

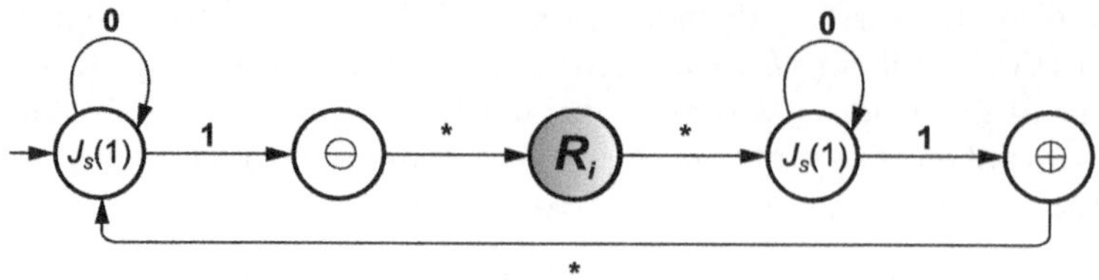

Fig. 2. The transition diagram of the i-th agent for rendezvous in polygons with triangular holes. $J_S(1)$ denotes a jump to the South by distance 1.

Proof. The proof proceeds by defining a sequence $\mathcal{A}_1, \mathcal{A}_2, \ldots$ of agents which will move from an arbitrary initial position in the terrain to the boundary of the external polygon, and then stabilize to a trajectory contained entirely within this boundary. This property is achieved by iteratively applying the following set of rules:

- As long as the agent is located in an interior point, it jumps by distance 1 to the south (with respect to its compass).
- Once the agent has reached a boundary point, it performs the following sequence of moves: a counterclockwise slide, a jump (or sequence of jumps) to the south until reaching a boundary point, a clockwise slide, and another jump (or sequence of jumps) to the south until reaching a boundary point.

Observe that the latter rule allows the agent to leave the boundary of any triangular hole; since there exists a natural partial order on holes w.r.t. the chosen south direction (there exists a point of one hole south of a point of the other), no hole will be visited again once its boundary has been left. Moreover, upon reaching the external boundary of the terrain after a jump, the agent will permanently remain on it, with its trajectory stabilizing to a periodic traversal of one of the sides of the polygon.

The proof is completed by combining the above defined set of rules with the agent $\mathcal{R}_i$ from Lemma 1, which can be treated as a "subroutine" called in each iteration of the agent's cycle. Formally, agent $\mathcal{A}_i$ is presented in Fig. 2. Consider the rendezvous problem with agents $\mathcal{A}_{i_1}$ and A_{i_2}. From some time moment t on, both agents will be permanently located on the boundary of the external polygon. Now, consider the next period of time, starting after t, during which agent $\mathcal{A}_{i_1}$ executes rules from $\mathcal{R}_{i_1}$. Throughout this period of time, agent $\mathcal{A}_{i_2}$ may either be performing a traversal of one side of the polygon, or executing rules from $\mathcal{R}_{i_2}$. By Lemma 1 and in view of the behavior of agents $\mathcal{R}_i$, in both of these cases rendezvous will occur. □

Our next result shows that, rather surprisingly, the restriction to triangular holes in the previous theorem was necessary.

Theorem 4. *Exact rendezvous of labeled agents is not feasible for the class of triangular terrains with quadrangular holes.*

Proof. (Sketch) For any two fixed agents, we construct an example of a terrain for which rendezvous is impossible; the terrain is chosen sufficiently small for all moves of the agents to end on the boundary of some polygon. The prefix and the cycle of state transitions performed by the agents are then independent of the definition of the terrain. We can confine considerations to two agents whose cycles of moves are not composed entirely of jumps. We show that for any such agent placed in the plane, it is possible to define a set of quadrangular holes and a starting point in such a way that the trajectory of the agent stabilizes to a periodic traversal of a set of segments. Consequently, the region of the plane visited by each of the agents is bounded, and rendezvous can always be avoided by positioning the agents in two such disjoint bounded regions and enclosing them in a sufficiently large bounding triangle. The construction of systems of holes satisfying the above property is rather technical; we omit it from this extended abstract. □

We finally show that, as opposed to the case of anonymous agents, if the terrain is known in advance, then labeled agents accomplishing exact rendezvous in it can always be constructed, regardless of the symmetries of the terrain.

Theorem 5. *For any given terrain $\mathcal{T}$, there exists a dedicated sequence $\mathcal{A}_1, \mathcal{A}_2, \ldots$ of labeled agents which accomplishes exact rendezvous in $\mathcal{T}$.*

4 ε-rendezvous

We start our analysis of feasibility of ε-rendezvous by establishing a positive result which should be contrasted with Theorem 1. To prove it we need the following lemma.

Lemma 2. *Let δ be a fixed positive real and consider the class $\mathcal{C}$ of regular polygons of diameter at least 2δ. There exists an agent which reaches, for any polygon $\mathcal{P}$ of the class $\mathcal{C}$, some vertex v of $\mathcal{P}$, called the* home base, *and stops. The choice of v may be affected by the orientation of the compass of the agent, but not by its initial location.*

Theorem 6. *ε-rendezvous of anonymous agents is feasible in the class of regular polygons with a known upper bound on the diameter.*

Proof. We assume that the polygon is of diameter at least ε, otherwise ε-rendezvous occurs immediately. Consider anonymous agents that start by executing the subroutine of locating the home base, as given by Lemma 2 for $\delta = \varepsilon/2$. Below we describe the rest of the behavior of the agents.

Let D be a known upper bound on the diameter of the polygon, let v be the home base of the agent, and let $\mathcal{Q}$ be the set of squares of side $d = \varepsilon^3/(256D^2)$ with vertices on the grid containing v, horizontally and vertically spaced at distance d (the terms horizontal and vertical refer to the east-west and north-south directions, respectively). Denote by $\mathcal{Q}_i$ the subset of $\mathcal{Q}$ consisting of squares whose boundary is contained in the interior of polygon $\mathcal{P}$. Observe that since $\mathcal{P}$ is convex, the set $\mathcal{Q}_i$ induces a connected region. Moreover, the agent can identify the relative coordinates of all vertices of $\mathcal{Q}_i$ w.r.t. v. Indeed, the agent can perform jumps of length d in the north-south and east-west directions, exploring successive segments of the lattice delimiting the squares in a depth-first-search manner. If some jump reaches the boundary, the agent returns to its home base v by applying the procedure from Lemma 2 and continues the exploration of other vertices of the terrain. Since the ratio D/d is bounded, the exploration is completed after a bounded number of steps, and $\mathcal{Q}_i$ is fully identified. From now on, we will identify the set $\mathcal{Q}_i$ with the union of squares in $\mathcal{Q}_i$. Intuitively, $\mathcal{Q}_i$ will serve as a raster approximation (known to the agent) of the polygon $\mathcal{P}$. Let $A(\cdot)$ denote the area of a polygon; we note that the set $\mathcal{P} \setminus \mathcal{Q}_i$ contains no points further than $2d$ from the boundary of $\mathcal{P}$, hence $A(\mathcal{P} \setminus \mathcal{Q}_i) < 8dD$. Moreover, since the diameter of the polygon is at least ε, we have the bound $A(\mathcal{Q}_i) \geq \varepsilon^2/16$.

The agent finally moves to the center of mass of $\mathcal{Q}_i$ and stops. Consider the distance a between this point and the center of mass of $\mathcal{P}$. By linearity of the center of mass, we have $a \leq D\frac{A(\mathcal{P}\setminus\mathcal{Q}_i)}{A(\mathcal{Q}_i)+A(\mathcal{P}\setminus\mathcal{Q}_i)} < D\frac{8dD}{\varepsilon^2/16} \leq \varepsilon/2$. Thus, when both agents stop at the centers of mass of their respective sets $\mathcal{Q}_i$, they will both be at a distance of at most $\varepsilon/2$ from the center of mass of $\mathcal{P}$, thus achieving ε-rendezvous. □

It turns out that boundedness was a crucial assumption in Theorem 6. Indeed, we have the following negative result.

Theorem 7. *ε-rendezvous of anonymous agents is not feasible in the class of regular n-gons, for any fixed $n \geq 3$.*

Our last result shows that, from the point of view of ε-rendezvous feasibility, the class of regular polygons distinguishes anonymous agents from labeled agents.

Theorem 8. *ε-rendezvous of labeled agents is feasible in the class of all regular polygons.*

It is natural to ask in what terrains ε-rendezvous is impossible even for labeled agents, i.e., in the easiest of our four scenarios. An example of such a class of terrains are triangles with quadrangular holes, for which exact rendezvous of labeled agents is impossible by Theorem 4. In fact, the proof of this theorem can be slightly changed to hold for ε-rendezvous as well: it is enough to move the two systems of holes (within which each of the agents is perpetually confined) far enough to preclude the approaching of the agents at distance ε.

5 Conclusion

We have considered four rendezvous scenarios for agents with bounded memory in polygonal terrains: scenarios with anonymous vs. labeled agents and with exact rendezvous vs. ε-rendezvous. While in this paper we focused on differences in rendezvous feasibility between the above scenarios, it would also be interesting to give an exact geometric characterization of terrains permitting rendezvous under each scenario.

References

1. Alpern, S.: The rendezvous search problem. SIAM Journal on Control and Optimization 33, 673–683 (1995)
2. Alpern, S., Gal, S.: The theory of search games and rendezvous. Int. Series in Operations Research and Management Science, vol. 55. Kluwer, Dordrecht (2002)
3. Alpern, J., Baston, V., Essegaier, S.: Rendezvous search on a graph. Journal of Applied Probability 36, 223–231 (1999)
4. Anderson, E., Fekete, S.: Two-dimensional rendezvous search. Operations Research 49, 107–118 (2001)
5. Baston, V., Gal, S.: Rendezvous on the line when the players' initial distance is given by an unknown probability distribution. SIAM Journal on Control and Optimization 36, 1880–1889 (1998)
6. Bilò, D., Disser, Y., Mihalák, M., Suri, S., Vicari, E., Widmayer, P.: Reconstructing Visibility Graphs with Simple Robots. In: Kutten, S., Žerovnik, J. (eds.) SIROCCO 2009. LNCS, vol. 5869, pp. 87–99. Springer, Heidelberg (2010)
7. Cieliebak, M., Flocchini, P., Prencipe, G., Santoro, N.: Solving the Robots Gathering Problem. In: Baeten, J.C.M., Lenstra, J.K., Parrow, J., Woeginger, G.J. (eds.) ICALP 2003. LNCS, vol. 2719, pp. 1181–1196. Springer, Heidelberg (2003)
8. De Marco, G., Gargano, L., Kranakis, E., Krizanc, D., Pelc, A., Vaccaro, U.: Asynchronous deterministic rendezvous in graphs. Theoretical Computer Science 355, 315–326 (2006)
9. Czyzowicz, J., Ilcinkas, D., Labourel, A., Pelc, A.: Asynchronous deterministic rendezvous in bounded terrains. In: Patt-Shamir, B., Ekim, T. (eds.) SIROCCO 2010. LNCS, vol. 6058, pp. 72–85. Springer, Heidelberg (2010)
10. Czyzowicz, J., Labourel, A., Pelc, A.: How to meet asynchronously (almost) everywhere. In: Proc. SODA 2010 (to appear 2010)
11. Dessmark, A., Fraigniaud, P., Kowalski, D., Pelc, A.: Deterministic rendezvous in graphs. Algorithmica 46, 69–96 (2006)
12. Flocchini, P., Prencipe, G., Santoro, N., Widmayer, P.: Gathering of asynchronous oblivious robots with limited visibility. In: Ferreira, A., Reichel, H. (eds.) STACS 2001. LNCS, vol. 2010, pp. 247–258. Springer, Heidelberg (2001)
13. Fraigniaud, P., Pelc, A.: Deterministic rendezvous in trees with little memory. In: Taubenfeld, G. (ed.) DISC 2008. LNCS, vol. 5218, pp. 242–256. Springer, Heidelberg (2008)
14. Gal, S.: Rendezvous search on the line. Operations Research 47, 974–976 (1999)
15. Klasing, R., Kosowski, A., Navarra, A.: Taking advantage of symmetries: gathering of asynchronous oblivious robots on a ring. In: Baker, T.P., Bui, A., Tixeuil, S. (eds.) OPODIS 2008. LNCS, vol. 5401, pp. 446–462. Springer, Heidelberg (2008)
16. Klasing, R., Markou, E., Pelc, A.: Gathering asynchronous oblivious mobile robots in a ring. Theoretical Computer Science 390, 27–39 (2008)
17. Kowalski, D., Malinowski, A.: How to meet in anonymous network. Theoretical Computer Science 399, 141–156 (2008)
18. Kranakis, E., Krizanc, D., Santoro, N., Sawchuk, C.: Mobile agent rendezvous in a ring. In: Proc. ICDCS 2003, pp. 592–599 (2003)
19. Prencipe, G.: Impossibility of gathering by a set of autonomous mobile robots. Theoretical Computer Science 384, 222–231 (2007)
20. Stachowiak, G.: Asynchronous Deterministic Rendezvous on the Line. In: Nielsen, M., Kucera, A., Miltersen, P.B., Palamidessi, C., Tuma, P., Valencia, F.D. (eds.) SOFSEM 2009. LNCS, vol. 5404, pp. 497–508. Springer, Heidelberg (2009)
21. Ta-Shma, A., Zwick, U.: Deterministic rendezvous, treasure hunts and strongly universal exploration sequences. In: Proc. SODA 2007, pp. 599–608 (2007)

Counting Classes and the Fine Structure between NC^1 and L*

Samir Datta[1], Meena Mahajan[2], B.V. Raghavendra Rao[3], Michael Thomas[4], and Heribert Vollmer[4]

[1] Chennai Mathematical Institute, India
sdatta@cmi.ac.in
[2] The Institute of Mathematical Sciences, Chennai, India
meena@imsc.res.in
[3] Universität des Saarlandes, Saarbrücken, Germany
bvrr@cs.uni-sb.de
[4] Leibniz Universität, Hannover, Germany
{thomas,vollmer}@thi.uni-hannover.de

Abstract. The class NC^1 of problems solvable by bounded fan-in circuit families of logarithmic depth is known to be contained in logarithmic space L, but not much about the converse is known. In this paper we examine the structure of classes in between NC^1 and L based on counting functions or, equivalently, based on arithmetic circuits. The classes PNC^1 and $\mathsf{C_=NC}^1$, defined by a test for positivity and a test for zero, respectively, of arithmetic circuit families of logarithmic depth, sit in this complexity interval. We study the landscape of Boolean hierarchies, constant-depth oracle hierarchies, and logarithmic-depth oracle hierarchies over PNC^1 and $\mathsf{C_=NC}^1$. We provide complete problems, obtain the upper bound L for all these hierarchies, and prove partial hierarchy collapses—in particular, the constant-depth oracle hierarchy over PNC^1 collapses to its first level PNC^1, and the constant-depth oracle hierarchy over $\mathsf{C_=NC}^1$ collapses to its second level.

1 Introduction

The class NC^1 occupies a special place in the study of complexity classes inside P, owing to its robustness and multiple characterizations. It is defined as the class of languages accepted by families of circuits of polynomial size and logarithmic depth using bounded fan-in Boolean gates. By uniform NC^1 we mean the subclass where the circuit families have succinct descriptions: given a (unary) size parameter, the circuit for that size from the family can be "easily" computed. Various notions of uniformity give rise to the same class of languages, also coinciding with the class of languages accepted by logarithmic-time alternating machines $\mathsf{ALOGTIME}$. Other characterizations of NC^1 include polynomial-sized formulas, bounded-width branching programs, bounded-width circuits and programs over finite monoids.

* Supported in part by the Indian DST and the German DAAD.

P. Hliněný and A. Kučera (Eds.): MFCS 2010, LNCS 6281, pp. 306–317, 2010.

It is known that all NC^1 languages can be accepted in logarithmic space L, but it is not known whether this containment is strict. All L-complete languages are candidates for membership in L but not in NC^1, and most of these candidates lie in classes defined using the natural counting classes associated with NC^1, namely, $\#\mathsf{NC}^1$ and GapNC^1. The former counts "proving sub-circuits" in an NC^1 circuit (see Section 2 for formal definitions); the latter is its closure under subtraction. It is not yet known whether these functions can be evaluated in NC^1, although the best upper bound is very very close (an $O(\log^*)$ factor in depth). It is known that functions in $\#\mathsf{NC}^1$ and GapNC^1 can be evaluated in function logarithmic space FL; thus languages definable by applying simple predicates to such functions are also in L. The natural choices of predicates are a test for zero and a test for positivity, giving rise to the language classes $\mathsf{C}_=\mathsf{NC}^1$ and PNC^1 sitting between NC^1 and L. (There are also predicates testing for zero modulo a fixed prime; the resulting language classes are already known to coincide with NC^1.) A nice survey of these classes can be found in [1].

It is not clear how much structure is there between NC^1 and L if the classes are distinct. We attempt to explore the structure between NC^1 and L, based on hierarchies of language classes built upon $\mathsf{C}_=\mathsf{NC}^1$ and PNC^1. For a complexity class $\mathcal{C}$, there are three standard ways of defining the hierarchies above $\mathcal{C}$: the Boolean hierarchy $\mathsf{BH}(\mathcal{C})$, the constant-depth hierarchy using oracle gates $\mathsf{AC}^0(\mathcal{C})$, and the NC^1-oracle-gate hierarchy $\mathsf{NC}^1(\mathcal{C})$, with $\mathsf{BH}(\mathcal{C}) \subseteq \mathsf{AC}^0(\mathcal{C}) \subseteq \mathsf{NC}^1(\mathcal{C})$.

Our results: As a first step in our study, we describe the oracle hierarchies in terms of arithmetic circuits augmented with test gates. These are the arithmetic-Boolean circuits defined in [13]; with size and depth restrictions as in NC^1, and with test gates for "$=0$?" or ">0?", we obtain the classes $\mathsf{a}\text{-}\mathsf{NC}^1_=$ and $\mathsf{a}\text{-}\mathsf{NC}^1_>$. We observe that if each path in the circuit has $O(1)$ test gates, then $\mathsf{a}\text{-}\mathsf{NC}^1_=$ and $\mathsf{a}\text{-}\mathsf{NC}^1_>$ coincide with $\mathsf{AC}^0(\mathsf{C}_=\mathsf{NC}^1)$ and $\mathsf{AC}^0(\mathsf{PNC}^1)$ respectively (Proposition 3.4). However, there is a subtlety in similarly characterizing $\mathsf{NC}^1(\mathsf{C}_=\mathsf{NC}^1)$ and $\mathsf{NC}^1(\mathsf{PNC}^1)$. We introduce a syntactic restriction on the arithmetic-Boolean circuits giving rise to a reasonable definition, and show that (1) the classes so defined coincide with $\mathsf{NC}^1(\mathsf{C}_=\mathsf{NC}^1)$ and $\mathsf{NC}^1(\mathsf{PNC}^1)$ (Proposition 3.5), and (2) as expected, are indeed contained in L (Theorem 3.9). On the other hand, without this restriction, the best upper bound we can show for the arithmetic circuit hierarchy is the complexity class TC^1 (Theorem 3.10), which subsumes L and even nondeterministic logspace NL, but is contained in NC^2.

Next, we show that the constant-depth hierarchy over PNC^1 (and hence also the Boolean hierarchy) collapses to PNC^1 (Theorem 4.1). We adapt the proof of [11], where an analogous result for PL is shown. One difficulty in the adaptation is showing the required normal form for GapNC^1 circuits. We use the equivalent characterization of GapNC^1 as arithmetic bounded-width branching programs $\mathsf{GapBWBP}$, and establish the normal form here. Another difficulty is computing an exponential sum; we use the notion of read-once certified circuits and read-once exponential sums, introduced in [10], to carry the proof through.

Finally, we examine the hierarchies over $\mathsf{C}_=\mathsf{NC}^1$. Since $\mathsf{C}_=\mathsf{NC}^1$ is not even known to be closed under complementation, we do not expect a collapse all the way down.

Our first result is a characterization of the Boolean hierarchy over $\mathsf{C_=NC^1}$ as the class of languages described by checking feasibility of small systems of linear equations, where the coefficients themselves are $\mathsf{GapNC^1}$-computable functions of the input word (Theorem 5.5). Our second result is that the constant-depth hierarchy over $\mathsf{C_=NC^1}$ collapses to a class slightly weaker than the second level (Theorem 5.9). Both these results appear as analogues of known results [2] for the corresponding logarithmic-space class $\mathsf{C_=L}$, but require substantially different proofs.

Also, unlike in the case of PL and $\mathsf{C_=L}$, our results do not seem to go through for the $\mathsf{NC^1}$-hierarchies over $\mathsf{PNC^1}$ and $\mathsf{C_=NC^1}$.

2 Background

For any language L, χ_L denotes its characteristic function: $\chi_L(x) := 1$ if $x \in L$, $\chi_L(x) := 0$ if $x \notin L$.

Boolean circuits and language classes: We denote by L the class of languages accepted by deterministic logarithmic-space Turing machines.

We consider Boolean circuits with internal gates labelled $\vee$, $\wedge$, or $\neg$. By $\mathsf{NC^1}$ we denote the class of languages which can be accepted by a family $\{C_n\}_{n\geq 0}$ of Boolean circuits of polynomial size whose depth is bounded by $O(\log n)$, with each gate having constant fan-in. The class $\mathsf{AC^0}$ denotes the set of languages accepted by a Boolean circuit family $\{C_n\}_{n\geq 0}$ of polynomial size and constant depth, with unbounded fan-in. Without loss of generality, we can assume that negation gates appear only at the leaves, and that the $\mathsf{AC^0}$ and $\mathsf{NC^1}$ circuits are actually formulas: every gate has out-degree one. An $\mathsf{NC^0}$ circuit is a Boolean circuit, or formula, of constant size, with each gate having constant fan-in. We denote by $\mathsf{AC^0}_k$ (respectively) $\mathsf{NC^0}_k$ the polynomial size (respectively, constant size) circuit families of depth at most k.

By $\mathsf{TC^0}$ and $\mathsf{TC^1}$ we denote the class of languages decided by circuit families of polynomial size and constant (respectively, logarithmic) depth, where each gate is either a negation gate or an unbounded fan-in majority gate: it outputs 1 if and only if more than half of its inputs are 1. Integer addition and multiplication are known to be in $\mathsf{TC^0}$.

A branching program (BP for short) is a layered acyclic graph G with edges labelled by constants (0 or 1) or literals, and with two special vertices s and t. It accepts an input x if there is an $s \leadsto t$ path where each edge is labelled by a true literal or the constant 1; we call such a path an *accepting path* on input x. BWBP denotes the class of languages that can be accepted by families of polynomial size bounded width branching programs $\{G_n\}_{n\geq 0}$, where the graph G_n considers n variables. It is known that BWBP equals $\mathsf{NC^1}$ ([4]). Restricted to uniform circuits (with appropriate notions of uniformity, see for instance [12]), it is known that $\mathsf{NC^1} = \mathsf{BWBP} \subseteq \mathsf{L}$.

Proposition 2.1 (Known containments).
$\mathsf{AC^0} \subseteq \mathsf{TC^0} \subseteq \mathsf{NC^1} = \mathsf{BWBP} \subseteq \mathsf{L} \subseteq \mathsf{TC^1} \subseteq \mathsf{DSPACE}(\log^2 n) \cap \mathsf{P}$.

Arithmetic circuit classes: For the purposes of this paper, an arithmetic circuit is a circuit where the gates are labelled from the set $\{+, \times, -1, 0, 1, x_1, \ldots, x_n\}$.

The gates $+$ and $\times$ are the addition and multiplication operations over $\mathbb{Z}$. Such a circuit computes a function $f : \{0,1\}^n \to \mathbb{Z}$.

An a-NC^1 circuit family $\{C_n\}_{n\geq 0}$ is a family of bounded fan-in arithmetic circuits where for each n, C_n is of size polynomial in n, depth logarithmic in n, and computes a function $f_n : \{0,1\}^n \to \mathbb{Z}$. The family computes the function $f : \{0,1\}^* \to \mathbb{Z}$ where $f(x) := C_{|x|}(x)$. GapNC^1 is the class of functions computed by a-NC^1 circuit families. The analogous arithmetic class for constant-depth unbounded fan-in circuits is denoted by a-AC^0.

An arithmetic branching program is a BP B where edges are labelled by literals or constants from the set $\{-1,0,1\}$. For an $s \rightsquigarrow t$ path P, let $wt(P(a))$ denote the product of all the edge labels in P under the assignment a. Then the function computed by B is defined as follows:

$$\text{for all } a \in \{0,1\}^n \qquad f(a) := \sum_{P \text{ is an } s\rightsquigarrow t \text{ path in } B} wt(P(a))$$

An a-BWBP family $\{B_n\}_{n\geq 0}$ is a family of arithmetic branching programs of polynomial size and bounded width. $\mathsf{GapBWBP}$ is the class of functions computed by a-BWBP program families.

For a Boolean (no edge labelled -1) BP B and an input assignment a, let $\#[s \rightsquigarrow t](a)$ denote the the number of $s \rightsquigarrow t$ paths in B under the assignment a. $\#\mathsf{BWBP}$ is the class of functions : $\{0,1\}^* \to \mathbb{N}$ computed by BWBP. The class $\mathsf{DiffBWBP}$ is the closure of $\#\mathsf{BWBP}$ under finite subtractions; $\mathsf{DiffBWBP} = \{f - g \mid f, g \in \#\mathsf{BWBP}\}$.

The above three classes coincide:

Proposition 2.2 ([6]). $\mathsf{GapNC}^1 = \mathsf{GapBWBP} = \mathsf{DiffBWBP}$.

We will often use the following equivalent form for GapNC^1 functions: for any GapNC^1 function f, there is a BWBP B with start node s, two target nodes t_1 and t_2, and $f(a) = \#[s \rightsquigarrow t_1](a) - \#[s \rightsquigarrow t_2](a)$. We say that B *gap-represents* the function f.

It is known that NC^1 circuits can be made unambiguous [9]. In terms of arithmetic circuits, this yields:

Proposition 2.3. *Let L be any AC^0 (or NC^1) language. Then there is an a-AC^0 (a-NC^1, respectively) circuit family C that does not use the constant -1 such that for each string w, $C(w) = \chi_L(w)$.*

The classes $\mathsf{C_=NC}^1$ and PNC^1, central to this paper, are defined as follows.

$$\mathsf{C_=NC}^1 := \left\{ L \in \{0,1\}^* \;\middle|\; \begin{array}{l}\text{for some } f \in \mathsf{GapNC}^1, \text{ for all } x \in \{0,1\}^*, \\ x \in L \text{ if and only if } f(x) = 0. \end{array} \right\}$$

$$\mathsf{PNC}^1 := \left\{ L \in \{0,1\}^* \;\middle|\; \begin{array}{l}\text{for some } f \in \mathsf{GapNC}^1, \text{ for all } x \in \{0,1\}^*, \\ x \in L \text{ if and only if } f(x) > 0. \end{array} \right\}$$

Proposition 2.4 **([6])**

1. $\mathsf{NC}^1 \subseteq \mathsf{C_=NC}^1 \subseteq \mathsf{PNC}^1 \subseteq \mathsf{L}$.
2. $\mathsf{C_=NC}^1$ *is closed under union and intersection.*
3. PNC^1 *is closed under union, intersection and complementation.*

Arithmetic-Boolean circuits: Let a *test gate for "=0?"* (respectively *">0?"*) be a unary gate that outputs 1 if its input is equal to 0 (respectively greater than 0) and 0 otherwise. Define an $\mathsf{a\text{-}NC}^1_=$ *circuit* (respectively $\mathsf{a\text{-}NC}^1_>$ *circuit*) to be an arithmetic circuit of logarithmic depth and polynomial size over Boolean input gates, binary +- and ×-gates, constants −1, 0 and 1 as well as test gates for "=0?" (respectively ">0?"). From the definitions, it follows that

Proposition 2.5. *A language L is in $\mathsf{C_=NC}^1$ (or PNC^1) if and only if χ_L can be computed by an $\mathsf{a\text{-}NC}^1_=$ (respectively $\mathsf{a\text{-}NC}^1_>$) circuit family in which each circuit has exactly one test gate appearing as the output gate.*

Miscellaneous We denote by $\mathcal{C}_1 \cdot \mathcal{C}_2$ a circuit which can be split horizontally into two parts, with the top part being a circuit of type $\mathcal{C}_1$, and all its inputs being either circuit inputs (literals or constants) or circuits of type $\mathcal{C}_2$. We denote by $[\mathcal{C}]$ an oracle gate for a language in $\mathcal{C}$. Thus $[\mathcal{C}] \cdot \mathsf{AC}^0$ is the class of all languages accepted by $\mathsf{AC}^0(\mathcal{C})$ oracle circuits such that each circuit has a single oracle gate at the output, and each input bit to the oracle gate is the output of an AC^0 sub-circuit.

3 Hierarchies: Definitions and Upper Bounds

Among the simplest is the Boolean hierarchy, which characterizes the languages expressible as Boolean combinations of any constant number of languages from respectively $\mathsf{C_=NC}^1$ or PNC^1.

Definition 3.1 (The Boolean Hierarchy). *Let $\mathcal{C}$ be a complexity class. The* Boolean hierarchy over *$\mathcal{C}$ is defined as the set of languages L for which there exists an NC^0 circuit C with k inputs and $A_1, \ldots, A_k \in \mathcal{C}$ such that for all $x \in \{0,1\}^*$,*

$$x \in L \Longleftrightarrow C(\chi_{A_1}(x), \chi_{A_2}(x), \ldots, \chi_{A_k}(x)) = 1$$

We denote this class of languages by $\mathsf{NC}^0 \cdot \mathcal{C}$ or $\mathsf{BH}(\mathcal{C})$.

Remark 3.2. A more standard way of defining the Boolean hierarchy is to define the levels $\mathsf{BH}_0(\mathcal{C}) := \mathcal{C}$, $\mathsf{BH}_i(\mathcal{C}) := \{L_1 \triangle L_2 \mid L_1, L_2 \in \mathsf{BH}_{i-1}(\mathcal{C})\}$, and then take the union $\bigcup_{i>0} \mathsf{BH}_i(\mathcal{C})$. If $\mathcal{C}$ is closed under union and intersection, then these definitions coincide with each other and with the definition of $\mathsf{NC}^0 \cdot \mathcal{C}$ above ([8]).

The other way of defining hierarchies is via oracle queries. As shown in [3] (see also [2]), nesting queries above a base machine is equivalent to adding oracle gates in an AC^0 circuit. And it also often turns out to be equivalent to adding oracle gates in an NC^1 circuit. We present the oracle-circuit definitions first introduced by Wilson [14] below.

Let L be any language. An $\mathsf{AC}^0(L)$ *circuit family* is a sequence $\{C_n\}_{n\geq 0}$ of AC^0circuits containing additional oracle gates for L of unbounded fan-in. Similarly, an $\mathsf{NC}^1(L)$ *circuit family* is a sequence $\{C_n\}_{n\geq 0}$ of NC^1 circuits with additional oracle gates for L of unbounded fan-in such that oracle gates of fan-in m account for depth $\lceil \log m \rceil$.

Definition 3.3 (The AC^0 and the NC^1 Hierarchy). *Let $\mathcal{C}$ be a complexity class. Then $\mathsf{AC}^0(\mathcal{C})$ (respectively $\mathsf{NC}^1(\mathcal{C})$) is defined to comprise those problems decidable by an $\mathsf{AC}^0(L)$ (respectively $\mathsf{NC}^1(L)$) circuit family for some $L \in \mathcal{C}$.*

We now characterize the hierarchies using Arithmetic-Boolean Circuits. From Proposition 2.5, we know that $\mathsf{C_=NC}^1$ and PNC^1 have equivalent Arithmetic-Boolean circuits. It is natural to ask whether there are equivalent such circuits for the hierarchies above these classes. For the AC^0 hierarchy, this is easy to see; we show below that $\mathsf{AC}^0(\mathsf{C_=NC}^1)$ and $\mathsf{AC}^0(\mathsf{PNC}^1)$ can be characterized using arithmetic-Boolean circuits. We need the notion of nesting depth: in a circuit C, the nesting depth of gates of a type t is the largest number k such that some path from the output to a leaf of C goes through exactly k gates of type t.

Proposition 3.4. $\mathsf{AC}^0(\mathsf{C_=NC}^1)$ *(respectively* $\mathsf{AC}^0(\mathsf{PNC}^1)$*) equals the class of languages decidable by* $\mathsf{a\text{-}NC}^1_=$ *(respectively* $\mathsf{a\text{-}NC}^1_>$*) circuit families such that the nesting depth of test gates is bounded by a constant and the output gate of each circuit is a test gate.*

It is tempting to believe that dropping the requirement on nesting depth of test gates will characterize $\mathsf{NC}^1(\mathsf{C_=NC}^1)$ and $\mathsf{NC}^1(\mathsf{PNC}^1)$. This, however, is not the case. The conversion from left to right ($\mathsf{NC}^1(\mathsf{C_=NC}^1)$ to $\mathsf{a\text{-}NC}^1_=$) goes through, but for the converse, the requisite depth bound does not follow. We describe a certain condition under which we can obtain an exact characterization.

Let C be an $\mathsf{a\text{-}NC}^1_=$ circuit (respectively $\mathsf{a\text{-}NC}^1_>$ circuit) with n inputs and let $g_1, \dots, g_m$ enumerate all of its test gates. Denote by S_i the maximal connected sub-circuit of C rooted at g_i that consists of $+$, $\times$-gates and the constants -1, 0, 1; these are the "blobs" in the proof of Proposition 3.4. As the depth of C is logarithmic in the number of its inputs, we may without loss of generality assume that $S_1, \dots, S_m$ induce a partition of the non-input gates of C. Thus any path from the output to a leaf in C goes through a chain of these blobs. There can be $O(\log n)$ blobs on any such chain, and the logarithm of the size of a blob can be as large as $\theta(\log n)$, and this causes the problem in replicating the above proof. We "define away" the problem: We say that C has the *small-blob-chains* property if for every path π from the root of C to an input gate or a constant,

$$\sum_{g_i \text{ occurs in } \pi} \log |S_i| \in O(\log n).$$

Now we can characterize exactly the NC^1 hierarchies above $\mathsf{C_=NC}^1$ and PNC^1.

Proposition 3.5. $\mathsf{NC}^1(\mathsf{C_=NC}^1)$ *(respectively* $\mathsf{NC}^1(\mathsf{PNC}^1)$*) equals the class of languages decidable by* $\mathsf{a\text{-}NC}^1_=$ *(respectively* $\mathsf{a\text{-}NC}^1_>$*) circuit families with the small-blob-chains property in which the output gate of each circuit is a test gate.*

It is not hard to see that there exist arithmetic-Boolean circuits violating the small-blob-chains property. Hence, dropping the small-blob-chains property from the circuits in Proposition 3.5 leads to presumably different class of languages. We denote these classes by AH, for arithmetic hierarchy, defined analogously to the classes figuring in Propositions 2.5 and 3.5.

Definition 3.6 (Arithmetic Circuit Hierarchies over $\mathsf{C_=NC^1}$ and $\mathsf{PNC^1}$). *A language L is said to be in $\mathsf{AH(C_=NC^1)}$ (or $\mathsf{AH(PNC^1)}$) if and only if χ_L can be computed by an $\mathsf{a\text{-}NC^1_=}$ (respectively $\mathsf{a\text{-}NC^1_>}$) circuit family such that in each circuit, the output gate is a test gate.*

The following chain of inclusions holds.

Observation 3.7.

$$\begin{array}{ccccccccc} \mathsf{C_=NC^1} & \subseteq & \mathsf{BH(C_=NC^1)} & \subseteq & \mathsf{AC^0(C_=NC^1)} & \subseteq & \mathsf{NC^1(C_=NC^1)} & \subseteq & \mathsf{AH(C_=NC^1)} \\ \rotatebox{90}{$\supseteq$} & & \rotatebox{90}{$\supseteq$} & & \rotatebox{90}{$\supseteq$} & & \rotatebox{90}{$\supseteq$} & & \rotatebox{90}{$\supseteq$} \\ \mathsf{PNC^1} & \subseteq & \mathsf{BH(PNC^1)} & \subseteq & \mathsf{AC^0(PNC^1)} & \subseteq & \mathsf{NC^1(PNC^1)} & \subseteq & \mathsf{AH(PNC^1)}. \end{array}$$

Remark 3.8. We can also augment the $\mathsf{a\text{-}NC^1_=}$ and $\mathsf{a\text{-}NC^1_>}$ circuits in Definition 3.6 by allowing oracle gates, with $\lceil \log(\text{fan-in}(g)) \rceil$ charged to the depth of each such gate g. Since, without loss of generality, we deal with languages over a binary alphabet, the inputs to the oracle gate must be Boolean inputs. But the circuit computes arithmetic values, except at test gates. Thus, we will require that all the inputs to an oracle gate are either Boolean circuit inputs (literals or the constants 0,1, but not -1) or the outputs of test gates. It can be shown that allowing $\mathsf{C_=NC^1}$ oracle gates in $\mathsf{a\text{-}NC^1_=}$ circuits, or $\mathsf{PNC^1}$ oracle gates in $\mathsf{a\text{-}NC^1_>}$ circuits, with this condition, does not add to the power of the circuit families beyond $\mathsf{AH(C_=NC^1)}$ and $\mathsf{AH(PNC^1)}$ respectively.

We now show some upper bounds. We first establish that the $\mathsf{AC^0}$ and the $\mathsf{NC^1}$ hierarchies over $\mathsf{C_=NC^1}$ and $\mathsf{PNC^1}$ are contained in L. By the containments depicted in Observation 3.7, it suffices to show this bound for $\mathsf{NC^1(PNC^1)}$.

Theorem 3.9. $\mathsf{NC^1(PNC^1)} \subseteq \mathsf{L}$.

We give two proofs of this theorem; one works directly with the oracle circuit, and the second works with the $\mathsf{a\text{-}NC^1_=}$ circuit. By Proposition 3.5, $\mathsf{AH(PNC^1)}$ differs from $\mathsf{NC^1(PNC^1)}$ only in the small-blob-chains property. In the absence of this property, the recursive simulation in the second proof of Theorem 3.9 yields only a $O(\log^2 n)$ space bound. Also, since the log-space evaluation of each blob may not be read-once in its inputs, each blob may have to be evaluated several times. So we cannot obtain a polynomial time bound for the recursive procedure. However, using a bottom-up evaluation, we can show that $\mathsf{AH(PNC^1)}$ circuits can be evaluated in $\mathsf{TC^1}$.

Theorem 3.10. $\mathsf{AH(PNC^1)} \subseteq \mathsf{TC^1}$.

4 The PNC^1 Hierarchy Collapses

In this section we show that the constant-depth PNC^1 hierarchy, $\mathsf{AC}^0(\mathsf{PNC}^1)$, collapses to the base level.

Theorem 4.1. $\mathsf{AC}^0(\mathsf{PNC}^1) = \mathsf{PNC}^1$.

Proof. Since PNC^1 is closed under complementation, and since unbounded fan-in $\vee$ and $\wedge$ functions are in NC^1 and hence in PNC^1, we can assume without loss of generality that the $\mathsf{AC}^0(\mathsf{PNC}^1)$ circuit has only oracle gates. Theorem 4.2 below shows how to collapse two adjacent levels of PNC^1 oracle gates into one. Applying this repeatedly gives the desired result. □

The rest of this section is devoted to proving Theorem 4.2:

Theorem 4.2. $[\mathsf{PNC}^1] \cdot [\mathsf{PNC}^1] = \mathsf{PNC}^1$.

We adapt the techniques of [11] to the case of constant width branching programs. Also, as in [11], we use the polynomial technique developed earlier ([5, 7]) to show closure properties of the complexity class PP. A new ingredient we need is read-once certified circuits and exponential sums, from [10].

4.1 Overview of the Collapse Argument

Consider a language L in $[\mathsf{PNC}^1] \cdot [\mathsf{PNC}^1]$. Then there is a language $H \in \mathsf{PNC}^1$ and a circuit family $\{C_n\}$ accepting L where each C_n has depth 2 and has only oracle gates for H. That is, the output gate g is an oracle gate whose inputs are themselves oracle gates or literals or constants. Without loss of generality, we can assume that in fact all inputs to g are outputs of oracle gates. Let g have fan-in t. On input x, its inputs are $\chi_H(Y_1), \chi_H(Y_2), \ldots, \chi_H(Y_t)$, where each Y_i is a projection (re-ordering of bits) of the input x.

Let f be the GapNC^1 function witnessing that $H \in \mathsf{PNC}^1$. Then there is a a-BWBP family computing f. The idea is to consider the a-BWBP B for inputs of length t, say $y_1, \ldots, y_t$, and try to replace each edge labeled y_i by a copy of the a-BWBP on Y_i. However, since Y_i is the input to an oracle gate, we want a 0-1 value for the sign of $f(Y_i)$, not the value of $f(Y_i)$ itself. If the sign function can be computed by a suitable polynomial function, then we can apply this function to each $f(Y_i)$ to get another GapNC^1 function. Unfortunately, the sign function cannot be represented in this fashion. However, it can be approximated by rational functions (ratios of polynomials); this approximation was first used in [5], and later in [7] and [11]. We follow the presentation from [11].

To show that using such approximations is valid, we require that B satisfies a certain condition: All paths should have equal susceptibility to error, so as to not change the overall outcome. In particular, since a y_i edge label corresponds to using the output of an oracle gate, and since different oracle gates can have different errors, we will require that each path has exactly the same multiset of edge labels, independent of the input. This is a strong normal form. Such a normal form was required to collapse $\mathsf{AC}^0(\mathsf{PL})$ to PL, and was shown in [11]. We show a corresponding normal form for a-BWBPs in Lemma 4.3.

Finally, we need to show that there is a GapBWBP function h which has the same sign as the value of the a-BWBP B with the rational approximations in place. In [11], the analogous result is shown by describing an appropriate probabilistic log-space machine. In the GapBWBP setting, things are a bit more complicated since we have only $O(1)$ storage. We get around this by using the notion of exponential sums over read-once certified circuits, introduced in [10]. The GapBWBP family computing the desired h is described in Section 4.2, completing the proof of Theorem 4.2.

4.2 Some Details of the Proof

We introduce a notation here. A node v in a BP B is called a *nondeterministic* node if there is an input assignment for which v has two out-edges labelled 1. We show the following normal form for branching programs computing functions in GapNC^1; this is analogous to Lemma 3.1 in [11] for PL and #L functions.

Lemma 4.3. *Let f be a function in GapNC^1. Then there exists a branching program Q of width $O(1)$ such that*

1. *Q has a single start node s and two terminal nodes t_1 and t_2;*
2. *every path originating from s ends at either t_1 or t_2 and nowhere else;*
3. *any path of Q on any given input x contains exactly q nondeterministic nodes;*
4. *every edge is labelled by a literal y_i or $\neg y_i$;*
5. *on any input y, Q has exactly 2^q paths, originating from s, where $q = q(n) \leq \mathsf{poly}(n)$;*
6. *$f = \#[s \rightsquigarrow t_1] - \#[s \rightsquigarrow t_2]$.*

We use the following characterization of the class PNC^1.

Proposition 4.4. *A language L belongs to PNC^1 if and only if there is a function $f \in \mathsf{GapNC}^1$ such that if $x \in L$ then $f(x) \geq 1$ and if $x \notin L$ then $f(x) \leq -1$.*

We now complete the proof of Theorem 4.2. Let $L \in [\mathsf{PNC}^1]\cdot[\mathsf{PNC}^1]$. As described in Section 4.1, there is a GapNC^1 function f and a circuit family accepting L such that for any word x, $f(x) \neq 0$ and $x \in L \Leftrightarrow f(b_1, \ldots, b_t) > 0$, where $b_i = \chi_H(Y_i)$ and so $b_i = 1$ if $f(Y_i) > 0$; $b_i = 0$ otherwise. Each query string Y_i is obtained from x by a projection and is an oracle query at the bottom layer; b_i is the oracle reply.

Replace each b_i by a variable y_i and apply Lemma 4.3 to get a polynomial size branching program Q, with three special nodes s, t_1, and t_2, computing $f(Y)$ on t-bit inputs via the gap $f = \#[s \rightsquigarrow t_1] - \#[s \rightsquigarrow t_2]$. Note that for every layer k of Q, there is a variable $u_k \in Y$ such that the edges from layer k to layer $k+1$ are labelled from the set $\{u_k, \neg u_k\}$. Note that all the u_k need not be distinct. Henceforth, we denote by y_k and Y_k the variable at layer k of Q and the corresponding query string, respectively. Without loss of generality we can assume that every layer is a nondeterministic layer. Let Q have p layers. Then every pair of bit-strings w, u, each of length p, uniquely represents a path in the BP Q, by considering the ith bit w_i of w as the query answer at the ith layer and ith bit u_i of u as the nondeterministic choice. For w, u, with $|w| = |u| = p$, define the Boolean function $e(x, w, u)$ as follows: $e(x, w, u) = 1$ if and only if the path of Q represented by the strings w and u on input x is an accepting path (that is, it terminates at t_1). Now define the following functions:

$$T(x) := \sum_{u,w\in\{0,1\}^p} e(x,w,u)\tilde{S}(x,w),$$

$$a(x) := \sum_{u,w\in\{0,1\}^p} e(x,w,u)\tilde{\alpha}(x,w), \text{ and}$$

$$h(x) := 4a(x) - 2^{p+1}\beta(x).$$

Here, $\tilde{S}(x, w)$, $\tilde{\alpha}(x, w)$ and $\beta(x)$ are Ogihara's polynomials (see [11]).

As shown in [11], $T(x) = a(x)/\beta(x)$. Using the properties of $\tilde{S}$ we have:

Lemma 4.5. *[11] If $x \in L$ then $T(x) > 2^{p-1}$, and if $x \notin L$ then $T(x) < 2^{p-1}$. Hence, $x \in L$ if and only if $h(x) \geq 0$.*

Now it suffices to prove the following;

Lemma 4.6. $h(x) \in \mathsf{GapBWBP}{=}\mathsf{GapNC}^1$.

5 The Hierarchy above $\mathsf{C_{=}NC}^1$

Since we do not even know if $\mathsf{C_{=}NC}^1$ is closed under complementation, we cannot hope for a direct collapse of the hierarchies above $\mathsf{C_{=}NC}^1$ all the way down to $\mathsf{C_{=}NC}^1$. However, we show here two partial collapses. For the analogous class $\mathsf{C_{=}L}$, it has been shown in [2] that the hierarchy collapses to $\mathsf{L}^{\mathsf{C_{=}L}}$, and that testing feasibility of systems of linear equations FSLE is complete for this class. At the level of NC^1, we show that the analogous situation splits into two counterparts. We define an appropriate non-trivial notion of constant-dimension FSLE and show that it is complete for the Boolean hierarchy over $\mathsf{C_{=}NC}^1$, $\mathsf{BH}(\mathsf{C_{=}NC}^1)$. We then show that the constant-depth hierarchy over $\mathsf{C_{=}NC}^1$, $\mathsf{AC}^0(\mathsf{C_{=}NC}^1)$, collapses to a certain level within the hierarchy that we denote $\mathsf{AC}^0 \cdot \mathsf{C_{=}NC}^1$; this is contained in the second level of the hierarchy.

5.1 The Boolean Hierarchy above $\mathsf{C_{=}NC}^1$

Definition 5.1. *For any $k \in \mathbb{N}$, and any class $\mathcal{C}$ of functions from words to integers, the language class $\mathsf{FSLE}^k[\mathcal{C}]$ is defined as follows: A language L belongs to the class $\mathsf{FSLE}^k[\mathcal{C}]$ if there are functions $A_{ij} \in \mathcal{C}$ for $1 \leq i, j \leq k$ and a vector $b \in \mathbb{Z}^k$ such that for each $w \in \{0,1\}^*$, $w \in L$ if and only if the system $Az = b$ of linear equations in k variables z_j, where $A_{ij} = A_{ij}(w)$, has a feasible solution over the rationals. The class $\mathsf{FSLE}^{bdd}[\mathcal{C}]$ is the union of $\mathsf{FSLE}^k[\mathcal{C}]$ over all k.*

Proposition 5.2. *The following two containments hold:*

$$\mathsf{coC_{=}NC^1} \subseteq \mathsf{FSLE^1[GapNC^1]}, \qquad \mathsf{C_{=}NC^1} \subseteq \mathsf{FSLE^2[GapNC^1]}.$$

We prove something stronger, by showing that $\mathsf{FSLE}^{bdd}[\mathsf{GapNC}^1]$ can express conjunctions and negations.

Lemma 5.3. $\mathsf{NC^0}_d \cdot \mathsf{C_{=}NC^1} \subseteq \mathsf{FSLE}^{3(2^{d+1}-1)}[\mathsf{GapNC^1}]$.

We establish a converse as well, with somewhat different parameters. The proof uses the fact that to check feasibility, the ranks of finitely many sub-matrices need to be computed.

Lemma 5.4. $\mathsf{FSLE}^k[\mathsf{GapNC^1}] \subseteq \mathsf{NC^0}_{3k} \cdot \mathsf{C_{=}NC^1}$.

From Lemmas 5.3 and 5.4, we have shown the following:

Theorem 5.5. $\mathsf{NC^0} \cdot \mathsf{C_{=}NC^1} = \mathsf{FSLE}^{bdd}[\mathsf{GapNC^1}]$.

5.2 The $\mathsf{AC^0}$ Hierarchy above $\mathsf{C_{=}NC^1}$

We now show the collapse of the constant-depth hierarchy over $\mathsf{C_{=}NC^1}$, that is, we prove $\mathsf{AC^0(C_{=}NC^1)} = \mathsf{AC^0}_3 \cdot [\mathsf{C_{=}NC^1}]$. First we set up some notation.

Let $\mathsf{AC^0}_k(\mathcal{C})$ denote the class of languages accepted by $\mathsf{AC^0}$ oracle circuits, where the oracle gates are for a language in $\mathcal{C}$, and where on any root-to-leaf path, the number of oracle gates encountered is at most k. (This is in analogy with $\mathsf{AC^0}_k$ denoting depth-k $\mathsf{AC^0}$ circuits.) Then, $\mathsf{AC^0}_k(\mathcal{C})$ is exactly $\mathsf{AC^0} \cdot [\mathcal{C}] \cdot \mathsf{AC^0} \ldots$ (k times) $\ldots [\mathcal{C}] \cdot \mathsf{AC^0}$. In particular, when $\mathcal{C} = \mathsf{C_{=}NC^1}$, using notation from Proposition 3.4 we can see that $\mathsf{AC^0}_k(\mathsf{C_{=}NC^1})$ equals $\mathsf{a\text{-}NC^1_{=}}$ circuits where the nesting depth of the test gates is at most k.

Proposition 5.6. $[\mathsf{C_{=}NC^1}] \cdot \mathsf{AC^0} = \mathsf{C_{=}NC^1}$ *and* $[\mathsf{coC_{=}NC^1}] \cdot \mathsf{AC^0} = \mathsf{coC_{=}NC^1}$.

The heart of our collapse result is the following lemma, stating that two adjacent levels of $\mathsf{coC_{=}NC^1}$ oracle gates can be combined into one.

Lemma 5.7. $[\mathsf{coC_{=}NC^1}] \cdot [\mathsf{coC_{=}NC^1}] \subseteq \mathsf{AC^0} \cdot [\mathsf{coC_{=}NC^1}]$. *In particular, the* $\mathsf{AC^0}$ *circuitry is of depth 3, with an* OR *of* AND*s and some negations at the leaves.*

The result follows immediately from Lemma 5.8 below.

Lemma 5.8. *Let* $h : \{0,1\}^t \longrightarrow \{0,1\}$, $f_1, f_2, \ldots, f_t : \{0,1\}^n \longrightarrow \{0,1\}$ *be functions in* $\mathsf{GapNC^1}$, *where for all* w, $f_i(w) \geq 0$. *Then for some* $T \in t^{O(1)}$, *there exist* $\mathsf{GapNC^1}$ *functions* $g_1, g_2, \ldots, g_T : \{0,1\}^n \longrightarrow \{0,1\}$ *and an* $\mathsf{AC^0}$ *circuit* H *on* T *inputs such that, for all* $w \in \{0,1\}^n$,

$$h(b_1, b_2, \ldots, b_t) \neq 0 \Longleftrightarrow H(d_1, d_2, \ldots, d_T) = 1$$

where $b_i := \begin{cases} 1 & \text{if } f_i(w) \neq 0, \\ 0 & \text{otherwise,} \end{cases}$ *and* $d_j := \begin{cases} 1 & \text{if } g_j(w) \neq 0, \\ 0 & \text{otherwise.} \end{cases}$

To establish this, we define appropriate symmetric polynomials such that when evaluated at the values $f_i(w)$, a simple predicate involving them reveals the value of h. We then use the fact that the symmetric polynomials are efficiently computable over fields.

Using Proposition 5.6 and Lemma 5.7, we get our collapse result.

Theorem 5.9. *The* AC^0 *hierarchy over* $\mathsf{C_=NC}^1$ *collapses to its first level, requiring a single layer of oracle gates and a depth-3 circuit above it,*

$$\mathsf{AC}^0(\mathsf{C_=NC}^1) = \mathsf{AC}^0 \cdot [\mathsf{C_=NC}^1] = \mathsf{AC}^0 \cdot [\mathsf{coC_=NC}^1] = \mathsf{AC}^0{}_3 \cdot [\mathsf{C_=NC}^1].$$

References

[1] Allender, E.: Arithmetic circuits and counting complexity classes. In: Krajicek, J. (ed.) Complexity of Computations and Proofs. Quaderni di Matematica, vol. 13, pp. 33–72. Seconda Universita di Napoli (2004); An earlier version appeared in the Complexity Theory Column. SIGACT News 28(4), 2–15 (December 1997)

[2] Allender, E., Beals, R., Ogihara, M.: The complexity of matrix rank and feasible systems of linear equations. Computational Complexity 8(2), 99–126 (1999)

[3] Allender, E., Ogihara, M.: Relationships among PL, #L, and the determinant. RAIRO Theoretical Information and Applications 30, 1–21 (1996); Conference version in Proc. 9th IEEE Structure in Complexity Theory Conference, pp. 267–278 (1994)

[4] Barrington, D.A.: Bounded-width polynomial size branching programs recognize exactly those languages in NC^1. Journal of Computer and System Sciences 38, 150–164 (1989)

[5] Beigel, R., Reingold, N., Spielman, D.A.: PP is closed under intersection. Journal of Computer and System Sciences 50(2), 191–202 (1995)

[6] Caussinus, H., McKenzie, P., Thérien, D., Vollmer, H.: Nondeterministic NC^1 computation. Journal of Computer and System Sciences 57, 200–212 (1998); Preliminary version in Proceedings of the 11th IEEE Conference on Computational Complexity, pp. 12–21 (1996)

[7] Fortnow, L., Reingold, N.: PP is closed under truth-table reductions. Inf. Comput. 124(1), 1–6 (1996)

[8] Köbler, J., Schöning, U., Wagner, K.W.: The difference and truth-table hierarchies for NP. Theoretical Informatics and Applications 21(4), 419–435 (1987)

[9] Lange, K.-J.: Unambiguity of circuits. Theor. Comput. Sci. 107(1), 77–94 (1993)

[10] Mahajan, M., Raghavendra Rao, B.V.: Small-space analogues of Valiant's classes. In: Gębala, M. (ed.) FCT 2009. LNCS, vol. 5699, pp. 250–261. Springer, Heidelberg (2009)

[11] Ogihara, M.: The PL hierarchy collapses. SIAM J. Comput. 27(5), 1430–1437 (1998)

[12] Vollmer, H.: Introduction to Circuit Complexity: A Uniform Approach. Springer, New York (1999)

[13] von zur Gathen, J., Seroussi, G.: Boolean circuits versus arithmetic circuits. Information and Computation 91(1), 142–154 (1991)

[14] Wilson, C.B.: Relativized circuit complexity. J. Comput. Syst. Sci. 31(2), 169–181 (1985)

The Average Complexity of Moore's State Minimization Algorithm Is $\mathcal{O}(n \log\log n)$*

Julien David

Institut Gaspard Monge, Université Paris Est
77454 Marne-la-Vallée Cedex 2, France

Abstract. We prove that the average complexity, for the uniform distribution on complete deterministic automata, of Moore's state minimization algorithm is $\mathcal{O}(n \log\log n)$, where n is the number of states in the input automata.

1 Introduction

Due to their efficiency to represent regular languages and perform most of usual computations they involve, finite state automata are used in various fields such as linguistics, bioinformatics, program verification and data compression. A minimal automata is the smallest complete deterministic automata that can be associated to a regular language. Because this automaton is unique, up to isomorphism on the labels of the states, it is a canonical representation of a regular language and permits to test the equality between regular languages and equivalence between automata. Most state minimization algorithms compute the minimal automaton of a regular language taking a deterministic automaton as an input, by identifying the indistinguishable states.

Moore proposed the first minimization algorithm[8], which is based on the calculus of the Myhill-Nerode equivalence, by refinements of partitions of the set of states. There are at most $n-2$ such refinements, each of them requiring a linear running time: in the worst case, the complexity is quadratic. Though, in [1], it is proved that the average complexity of the algorithm is bounded by $\mathcal{O}(n \log n)$. Since this result does not rely on the underlying graph of the automaton, it holds for any probabilistic distribution on this graph. Also, the bound is tight for unary automata.

Hopcroft's state minimization algorithm [6] is the best known algorithm with an $\mathcal{O}(n \log n)$ worst-case complexity. It also uses partition refinements to compute the minimal automaton, but its description is not deterministic, making its analysis complicated. Different proofs of its correctness were given [5,7] and several authors [3,4] proved the tightness of the upper bound of the complexity for different families of automata.

In this paper, we prove that for the uniform distribution on complete deterministic automata, the average complexity of the algorithm due to Moore is

* This work was completed with the support of the ANR project GAMMA number 07 − 2_195422.

P. Hliněný and A. Kučera (Eds.): MFCS 2010, LNCS 6281, pp. 318–329, 2010.

$\mathcal{O}(n \log\log n)$. The article is organized as follows: after recalling the basics of automata minimization (Section 2), we introduce the tools we use for the average analysis (Subsections 2.3 to 2.5). Section 3 is dedicated to the average time complexity analysis of Moore's algorithm. Due to a lack of space, the proof of Lemma 6 is not fully detailed, but an idea of the proof is given. The paper closes with a discussion on Hopcroft's algorithm executions, which are faster than Moore's algorithm ones, for any input automaton and a conjecture on the average complexity of both algorithms, for various distributions on automata.

2 Preliminaries

2.1 Definitions and Notations

A *finite deterministic automaton* $\mathcal{A} = (A, Q, \cdot, q_0, F)$ is a quintuple where Q is a finite set of *states*, A is a finite set of *letters* called *alphabet*, the *transition function* $\cdot$ is a mapping from $Q \times A$ to Q, $q_0 \in Q$ is the *initial state* and $F \subset Q$ is the set of final states. An automaton is *complete* when its transition function is total. The transition function can be extended by morphism to all words of A^*: $p \cdot \varepsilon = p$ for any $p \in Q$ and for any $u, v \in A^*$, $p \cdot (uv) = (p \cdot u) \cdot v$. A word $u \in A^*$ is recognized by an automaton when $p \cdot u \in F$. The set of all words recognized by $\mathcal{A}$ is denoted by $L(\mathcal{A})$. We note A^i the words of length i and $A^{\leq i}$ the word of length less or equal to i. An automaton is *accessible* when for any state $p \in Q$, there exists a word $u \in A^*$ such that $q_0 \cdot u = p$.

A *transition structure* is an automaton where the set of final states is not specified. Given such a transition structure $\mathcal{T} = (A, Q, \cdot, q_0)$ and a subset F of Q, we denote by $(\mathcal{T}, F)$ the automaton $(A, Q, \cdot, q_0, F)$. For a given deterministic transition structure with n states there are exactly 2^n distinct deterministic automata that can be built from this transition structure. Each of them corresponds to a choice of set of final states.

In the following we only consider complete deterministic automata and complete deterministic transition structures, the accessibility is not guaranteed. Consequently these objects will most of the time just be called respectively *automata* or *transition structures*. The set Q of an n-state transition structure will be denoted by $\{1, \ldots, n\}$. The set of automata and the set of transition structures with n states will respectively be denoted $\mathcal{A}_n$ and $\mathcal{T}_n$. Also, since there are kn transitions and since for each transition, there are n possible arrival states, we have $|\mathcal{T}_n| = n^{kn}$ and $|\mathcal{A}_n| = 2^n n^{kn}$ (when $|E|$ is the cardinal of the set E). The term 2^n comes from the choice of the set of final states.

The *military order* on words, noted $<_{mil}$, is defined as follows: $\forall u, v \in A^*$, $u <_{mil} v$ if $|u| < |v|$ or $|u| = |v|$ and u is smaller than v in the lexicographical order. Let *Cond* be a Boolean condition, the Iverson bracket $[\![Cond]\!]$ is equal to 1 if the condition *Cond* is satisfied and 0 otherwise.

For any non-negative integer i, two states $p, q \in Q$ are *i-equivalent*, denoted by $p \sim_i q$, when for all words $u \in A^{\leq i}$, $[\![p \cdot u \in F]\!] = [\![q \cdot u \in F]\!]$. Two states p and q are *equivalent* (noted $p \sim q$) when for all $u \in A^*$, $[\![p \cdot u \in F]\!] = [\![q \cdot u \in F]\!]$. This equivalence relation on Q is called *Myhill-Nerode equivalence* [9]. This relation is said to be *right invariant*, meaning that

$$\text{for all } u \in A^* \text{ and all } p, q \in Q, \quad p \sim q \Rightarrow p \cdot u \sim q \cdot u.$$

Proposition 1. *Let $\mathcal{A} = (A, Q, \cdot, q_0, F)$ be a deterministic automaton with n states. The following properties hold:*

(1) For all $i \in \mathbb{N}$, $\sim_{i+1}$ is a partition refinement of $\sim_i$, that is, for all $p, q \in Q$, if $p \sim_{i+1} q$ then $p \sim_i q$.
(2) For all $i \in \mathbb{N}$ and for all $p, q \in Q$, $p \sim_{i+1} q$ if and only if $p \sim_i q$ and for all $a \in A$, $p \cdot a \sim_i q \cdot a$.
(3) If for some $i \in \mathbb{N}$ $(i+1)$-equivalence is equal to i-equivalence then for every $j \geq i$, j-equivalence is equal to Myhill-Nerode equivalence.

For any integer $n \geq 1$ and any $m \in \mathbb{N}$, we denote by $\mathcal{A}_n^m$ the set of automata with n states for which m is the smallest integer such that the m-equivalence $\sim_m$ is equal to Myhill-Nerode equivalence. It is well known that $m \leq n - 2$.

2.2 Moore's State Minimization Algorithm

In this section we describe Moore's algorithm [8] to compute the minimal automaton of a regular language represented by a deterministic automaton. It builds the partition of the set of states corresponding to Myhill-Nerode equivalence and mainly relies on properties (2) and (3) of Proposition 1: The partition π is initialized according to the 0-equivalence $\sim_0$, then at each iteration the partition corresponding to the $(i+1)$-equivalence $\sim_{i+1}$ is computed from the one corresponding to the i-equivalence $\sim_i$ using property (2). The algorithm halts when no new partition refinement is obtained, and the result is Myhill-Nerode equivalence according to property (3). The minimal automaton can then be computed from the resulting partition since it is the quotient automaton by Myhill-Nerode equivalence.

According to Proposition 1, if an automaton is minimized in more than ℓ partition refinements, then there exists at least a pair of states p, q and a word u of length $\ell + 1$, such that $p \sim_\ell q$ and $p \cdot u \nsim_0 q \cdot u$, that is to say at least two states are separated during the $\ell + 1$-th partition refinement. In the remainder of this section we introduce the **dependency tree** and a modification of the **dependency graph** introduced in [1]. Those tools will allow us to give an upper bound on the number of automata minimized in more than ℓ partition refinements, which is useful for the average complexity analysis.

2.3 The Dependency Tree

In the following, we introduce the dependency tree to model a set of transition structures. To begin with, we explain how a dependency tree $\mathcal{R}(p)$ can be obtained from a fixed transition structure τ and a fixed state p and then how this object will help to estimate the cardinal of a set of transition structures. For a

Algorithm 1. Moore's algorithm

1 **if** $F = \emptyset$ **then**
2   **return** $(A, \{1\}, *, 1, \emptyset)$
3 **if** $F = \{1, \ldots, n\}$ **then**
4   **return** $(A, \{1\}, *, 1, \{1\})$
5 **forall** $p \in \{1, \ldots, n\}$ **do**
6     $\pi'[p] = [\![p \in F]\!]$
7 $\pi =$ undefined
8 **while** $\pi \neq \pi'$ **do**
9     $\pi = \pi'$
10   compute π' from π
11 **return** *the quotient of $\mathcal{A}$ by π*

In this description of Moore's algorithm, $*$ denotes the function such that $1 * a = 1$ for all $a \in A$. Lines 1-4 correspond to the special cases where $F = \emptyset$ or $F = Q$. In the process, π' is the new partition and π the former one. Lines 5-6 is the initialization of π' to the partition of $\sim_0$, π is initially undefined. Lines 8-10 are the main loop of the algorithm where the new partition is computed, using the second algorithm below. The *number of iterations* of Moore's algorithm is the number of times those lines are executed.

The computation of the new partition is done using the following property on associated equivalence relations:

$$p \sim_{i+1} q \Leftrightarrow \begin{cases} p \sim_i q \\ p \cdot a \sim_i q \cdot a \quad \forall a \in A \end{cases}$$

To each state p is associated a signature $s[p]$ such that $p \sim_{i+1} q$ if and only if $s[p] = s[q]$. The states are then sorted according to their signature, in order to compute the new partition. The use of a lexicographic sort provides a complexity of $\Theta(kn)$ for this part of the algorithm.

Algorithm 2. Computing π' from π

forall $p \in \{1, \ldots, n\}$ **do**
    $s[p] = (\pi[p], \pi[p \cdot a_1], \ldots, \pi[p \cdot a_k])$
compute the permutation σ that sorts the states according to $s[]$
$i = 0$
$\pi'[\sigma(1)] = i$
forall $p \in \{2, \ldots, n\}$ **do**
    if $s[p] \neq s[p-1]$ **then** $i = i + 1$
    $\pi'[\sigma(p)] = i$
return π'

Fig. 1. Description of Moore's algorithm

fixed transition structure with n states over a k-letter alphabet and a fixed state p, we define the function `isnode` mapping A^* to $\{0, 1\}$ as follows:

$$isnode(w) = \begin{cases} 0 \text{ if } \exists v \in A^* \text{ such that } p \cdot w = p \cdot v \text{ and } v <_{mil} w, \\ 1 \text{ otherwise.} \end{cases}$$

$\mathcal{R}(p)$ is a tree in which nodes and leaves of depth h are labelled by words of length h. It is built recursively, using a breadth-first traversal of the nodes of the tree starting from the node p. For each node of depth h labelled by w, and each letter a in the alphabet, we add a **node** labelled by wa at depth $h + 1$ if `isnode(wa)` is equal to 1, and a leaf otherwise. Figure 2 gives an example of a dependency tree. Note that this construction resembles the method used in [2] to randomly generate accessible automata, except the authors use a depth-first traversal. It is easy to see that some dependency trees can be obtained from several fixed transition structures and states. In the remainder of the paper, we

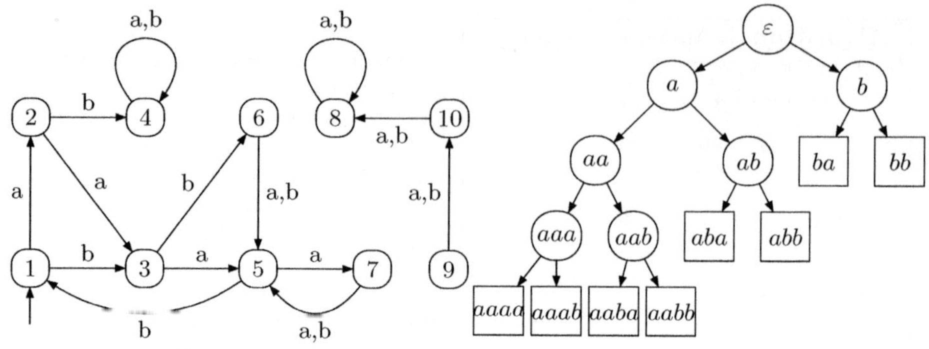

(a) **(b)**

Fig. 2. Let **(a)** be the fixed transition structure and 2 be the fixed state, **(b)** is the associated **dependency tree** $\mathcal{R}(2)$. We have $S_2(2) = \{\varepsilon, a, b, aa, ab\}$, $L_2(2) = \{aa, ab\}$ and $s_2(2) = \{2, 3, 4, 5, 6\}$.

characterize sets of transition structures corresponding to particular dependency trees.

We introduce some notations associated to a dependency tree $\mathcal{R}(p)$: $S_h(p)$ denotes the set of all nodes of depth less or equal to h, $L_h(p)$ denotes the set of all the nodes at given depth h. Since every node in the tree is labelled by a word, we note $w \in S_h(p)$ or $w \in L_h(p)$ if w is a word labelling a node in those sets. We also define the set $s_h(p)$ of all the states that are reached from a state p by following a path labelled by a word of less or equal to h. For all the transition structures associated to a dependency tree $\mathcal{R}(p)$, we have $|s_h(p)| = |S_h(p)|$.

Lemma 1. *For any fixed state p, if a dependency tree $\mathcal{R}(p)$ contains f leaves at a depth less or equal to h, then the number of associated transition structures is bounded above by $|\mathcal{T}_n| \left(\frac{|S_h(p)|}{n}\right)^f$.*

Proof. We recall that $|\mathcal{T}_n|$ is equal to the product of the cardinals of the sets of possible arrival states, for each transition. Let wa be the label of a leaf at depth less than h. For every transition structure associated to the tree $\mathcal{R}(p)$, the transition labelled by a outgoing from the state $p \cdot w$ ends in a state $p \cdot v$, with $v \in S_h(p)$. Therefore, the number of possible arrival states for this transition is bounded above by $|S_h(p)|$ instead of n. This is a rough upper bound but sufficient for the needs of the proof.

2.4 The $\mathcal{T}$-Dependency Graph

We introduce another model for sets of transition structures. For two fixed states p and q, two fixed x-tuples (x is a fixed integer) of non-empty words $\overrightarrow{u} = (u_1, \ldots, u_x)$ and $\overrightarrow{v} = (v_1, \ldots, v_x)$, two fixed sets φ_p and φ_q of pairs of words (w,w') such that $w' <_{mil} w$, we define the set $\mathcal{T}_n(p,q,\varphi_p,\varphi_q,\overrightarrow{u},\overrightarrow{v})$ as follows:

$$\begin{aligned}\mathcal{T}_n(p,q,\varphi_p,\varphi_q,\overrightarrow{u},\overrightarrow{v}) = \{\tau \in \mathcal{T}_n \mid & \forall (w_p,w'_p) \in \varphi_p, p\cdot w_p = p \cdot w'_p,\\ & \forall (w_q,w'_q) \in \varphi_q, q\cdot w_q = q\cdot w'_q,\\ & \forall i \leq x,\ p\cdot u_i = q\cdot v_i\}\end{aligned}$$

We define the x-tuples of words $\overrightarrow{u'} = (u'_1,\ldots,u'_x)$ and $\overrightarrow{v'} = (v'_1,\ldots,v'_x)$ and the x-tuples of letters $\overrightarrow{\alpha'} = (\alpha'_1,\ldots,\alpha'_x)$ and $\overrightarrow{\beta'} = (\beta'_1,\ldots,\beta'_x)$, such that for all $i \leq x$ we have $u_i = u'_i\alpha_i$ and $v_i = v'_i\beta_i$. From $\mathcal{T}_n(p,q,\varphi_p,\varphi_q,\overrightarrow{u},\overrightarrow{v})$, one can define the undirected graph $G_n(p,q,\varphi_p,\varphi_q,\overrightarrow{u},\overrightarrow{v})$, called the *$\mathcal{T}$-dependency graph*, as follows:

- its vertices are pairs (r,a), with $r\in Q$ and $a \in A$, that model transitions.
- There is an edge $((r,a),(t,b))$ in $G_n(p,q,\varphi_p,\varphi_q,\overrightarrow{u},\overrightarrow{v})$ if and only if for all $\tau \in \mathcal{T}_n(p,q,\varphi_p,\varphi_q,\overrightarrow{u},\overrightarrow{v})$, $r\cdot a = t\cdot b$.

The $\mathcal{T}$-dependancy graph $G_n(p,q,\varphi_p,\varphi_q,\overrightarrow{u},\overrightarrow{v})$ satisfies the two following properties:

- For all $i \leq x$, there exists an edge $((p\cdot u'_i,\alpha_i),(q\cdot v'_i,\beta_i))$.
- For all $(w_1,w_2) \in \varphi_p$ (resp. $(w_3,w_4) \in \varphi_q$), we have $w_1 = w'_1a_1$ and $w_2 = w'_2a_2$ with $a_1,a_2 \in A$ and such that there exists an edge $((p\cdot w'_1,a_1),(p\cdot w'_2,a_2))$ (resp. $((q\cdot w'_3,a_3),(q\cdot w'_3,a_3))$).

Lemma 2. *If $G_n(p,q,\varphi_p,\varphi_q,\overrightarrow{u},\overrightarrow{v})$ contains an acyclic subgraph induced by a subset of nodes with j edges, then:*

$$|\mathcal{T}_n(p,q,\varphi_p,\varphi_q,\overrightarrow{u},\overrightarrow{v})| \leq \frac{|\mathcal{T}_n|}{n^j}$$

Proof. Two transitions in the same connected components of $G_n(p,q,\varphi_p,\varphi_q,$-$\overrightarrow{u},\overrightarrow{v})$ share the same arrival state. Hence if x is the number of connected components in the graph, then $|\mathcal{T}_n(p,q,\varphi_p,\varphi_q,\overrightarrow{u},\overrightarrow{v})| \leq n^x$. If $G_n(p,q,\varphi_p,\varphi_q,\overrightarrow{u},\overrightarrow{v})$ contains an acyclic subgraph with exactly j edges, then there is at most $kn-j$ connected components.

2.5 The $\mathcal{F}$-Dependency Graph

In this subsection, we slightly modify the notion of dependency graph introduced in [1]. Let τ be a fixed transition structure with n states and ℓ be an integer such that $1 \leq \ell < n$. Let p,q be two states of τ such that $p \neq q$ and u a word of length ℓ. We define $\mathcal{F}_\tau(p,q,u)$ as the set of sets of final states F for which in the automaton (τ,F) the states p and q are separated by the word u. That is to say :

$$\begin{aligned}\mathcal{F}_\tau(p,q,u) = \{F \subset \{1,\ldots,n\} \mid \text{for all } (\tau,F),\ & p \sim_{|u|-1} q,\\ & [\![p\cdot u \in F]\!] \neq [\![q\cdot u\in F]\!]\}\end{aligned}$$

From the set $\mathcal{F}_\tau(p,q,u)$ one can define the undirected graph $\mathcal{G}_\tau(p,q,u)$, called the *$\mathcal{F}$-dependency graph*, as follows:

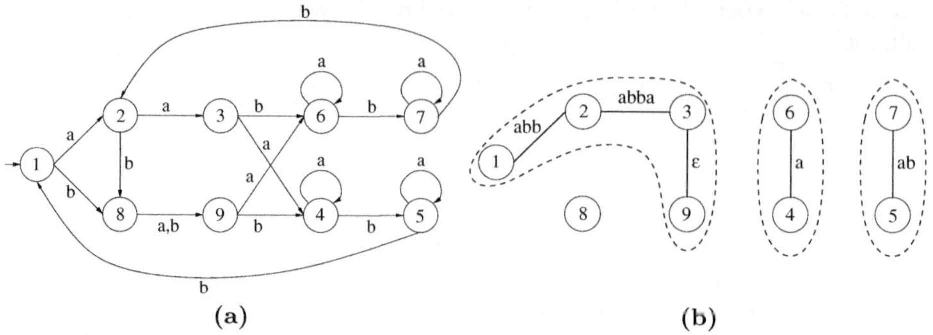

Fig. 3. **(a)** is a fixed transition structure and **(b)** the **$\mathcal{F}$-dependency graph** for $p = 3$, $q = 9$ and $u = abbaa$. Thanks to **(b)**, we know that all states in a same connected component will be either all final or all non-final. Hence, there are at most 2^4 possible sets of final states, instead of 2^9.

- its set of vertices is $\{1, \dots, n\}$, the set of states of $\mathcal{T}$;
- there is an edge (s, t) between two vertices s and t if and only if there exists a word w of length less than ℓ such that $s = p \cdot w$ and $t = q \cdot w$ and for all $F \in \mathcal{F}_{\mathcal{T}}(p, q, u)$, $[\![s \in F]\!] = [\![t \in F]\!]$.

The $\mathcal{F}$-dependency graph contains some information that is a basic ingredient of the proof: it is a convenient representation of necessary conditions for a set of final states to be in $\mathcal{F}_{\mathcal{T}}(p, q, u)$, that is, for Moore's algorithm to require more than $|u|$ iterations because of p, q and u. Figure 3 shows an example of a F-dependency graph.

Lemma 3. *[1] If $\mathcal{G}_{\mathcal{T}}(p, q, u)$ contains an acyclic subgraph with at least i edges, then $|\mathcal{F}_{\mathcal{T}}(p, q, u)| \leq 2^{n-i}$.*

The notions of dependency graphs and dependency tree will be used in subsections 3.2 and 3.3 to obtain upper bounds on the cardinal of sets of automata with given properties and prove that their contribution to the average complexity is negligible.

3 Moore's Algorithm: Average Case Analysis

In [1], it is proved that the average complexity of Moore's state minimization algorithm is $\mathcal{O}(n \log n)$. Since the result is obtained by studying only properties on the sets of final states of automata minimized in a given complexity, it holds for any distribution on the set of transition structures. In this paper, in order to improve the upper bound on the average complexity, we also have to study some properties of transition structures. Since the enumeration of accessible automata with given properties is an open problem, we focus our study on the uniform distribution over the set of complete deterministic automata.

3.1 Main Result and Decomposition of the Proof

The main result of this paper is the following.

Theorem 1. *For any fixed integer $k \geq 2$ and for the uniform distribution over the deterministic and complete automata with n states over a k-letter alphabet, the average complexity of Moore's state minimization algorithm is $\mathcal{O}(n \log \log n)$.*

Recall that the number of partition refinements made by Moore's state minimization algorithm is smaller or equal to $n-2$ and that $\mathcal{A}_n^i$ is the set of automata of $\mathcal{A}_n$ for which i is the smallest integer such that $\sim_i$ is equal to Myhill-Nerode equivalence. The average number of partition refinements is given by:

$$\mathcal{N}_n = \frac{1}{|\mathcal{A}_n|}\left(\sum_{i=0}^{n-2}(i+1) \times |\mathcal{A}_n^i|\right)$$

We define $\lambda = \lceil \log_k \log_2 n^3 + 2 \rceil$, which will be used in the sequel. We gather the sets $\mathcal{A}_n^i$, in order to obtain the following upper bound:

$$\mathcal{N}_n \leq \underbrace{\frac{\lambda+1}{|\mathcal{A}_n|}\sum_{i\leq\lambda}|\mathcal{A}_n^i|}_{\mathbf{S1}} + \underbrace{\frac{(5\log_2 n+1)}{|\mathcal{A}_n|}\sum_{i=\lambda+1}^{5\log_2 n}|\mathcal{A}_n^i|}_{\mathbf{S2}} + \underbrace{\frac{n-1}{|\mathcal{A}_n|}\sum_{i=5\log_2 n+1}^{n-2}|\mathcal{A}_n^i|}_{\mathbf{S3}}$$

S1 is less than λ and equal to $\mathcal{O}(\log \log n)$.
In [1], it is proved that $\sum_{i=5\log_2 n+1}^{n-2}|\mathcal{A}_n^i| \leq \frac{|\mathcal{A}_n|}{n}$. Therefore we know that **S3** is equal to $\mathcal{O}(1)$. Hence, in the following, we prove that:

$$\mathbf{S2} = \frac{(5\log_2 n+1)}{|\mathcal{A}_n|}\sum_{i=\lambda+1}^{5\log_2 n}|\mathcal{A}_n^i| = \mathcal{O}(\log\log n) \tag{1}$$

For any $\ell > 0$, we define the set $\mathcal{A}_n(p,q,\ell)$ as the set of automata with n states, where the states p and q are separated during the ℓ-th partition refinement:

$$\mathcal{A}_n(p,q,\ell) = \{(\tau,F) \in \mathcal{A}_n \mid \tau \subset \mathcal{T}_n,\ F \subseteq \{1,\dots,n\},\ p \sim_{\ell-1} q,\ p \nsim_\ell q\}$$

Remark 1. Note that if in the automaton (τ, F), for all letter $a \in A$, $p \cdot a = q \cdot a$, then either $p \nsim_0 q$ or $p \sim q$. Therefore $(\tau,F) \notin \mathcal{A}_n(p,q,\ell)$ with $\ell > 0$. Consequently, in the remainder of the proof, in all sets of transition structures where p and q are fixed, there exists a letter a such that $p \cdot a \neq q \cdot a$.

The following statement comes from the definition of the sets it involves:

$$\bigcup_{i>\lambda}\mathcal{A}_n^i = \bigcup_{p,q\in\{1,\dots,n\}}\mathcal{A}_n(p,q,\lambda+1)$$

Let τ be a transition structure, and p,q be two distinct states. Recalling that $s_h(p)$ is defined in Section 2.3, if A is a k-letter alphabet, and μ a positive integer, we define two properties associated to transition structures:

(1) $largeTree(\tau, p, \mu)$ is $true$ if and only if $|s_\mu(p)| \geq k^\mu - 1$.
Note that this implies that $|s_{\mu-2}(p)| \geq k^{\mu-2} - 1$.
(2) $noIntersection(\tau, p, q)$ is $true$ if and only if $s_{\lambda-2}(p) \cap s_{\lambda-2}(q) = \emptyset$.

For fixed states p and q, an automaton is in $\mathcal{A}_n(p, q, \lambda + 1)$ if its associated transition structure is in one of the three distinct sets we are about to define:

- **$\mathcal{X}_n$ is the set of all transition structures τ such that:**
  - there exists a state $r \in Q$ such that $largeTree(\tau, r, \lambda)$ is $false$.

  Note that this set does not rely on the values of p and q.
- **$\mathcal{Y}_n(p, q)$ is the set of transition structures τ such that:**
  - for all state $r \in Q$, the property $largeTree(\tau, r, \lambda)$ is $true$,
  - for all words $w \in A^{\leq 2}$, the property $noIntersection(\tau, p \cdot w, q \cdot w)$ is $false$.
- **$\alpha_n(p, q)$ is the set of transition structures τ such that:**
  - for all state $r \in Q$, the property $largeTree(\tau, r, \lambda)$ is $true$,
  - there exists $w \in A^{\leq 2}$ such that $noIntersection(\tau, p \cdot w, q \cdot w)$ is $true$.

3.2 Transition Structures with a Huge $\mathcal{F}$-Dependency Graph

Lemma 4. *For any fixed transition structures $\tau \in \alpha_n(p, q)$, and a fixed word u of length $\lambda+1$, the following property holds: every $\mathcal{F}$-dependency graph $\mathcal{G}_\tau(p, q, u)$ contains an acyclic subgraph with at least $k^{\lambda-2} - 1$ edges.*

Proof. Let $\mathcal{G}'$ be the subgraph $\mathcal{G}_\tau(p, q, u)$ defined as follows: there exists an edge $(p \cdot wv, q \cdot wv)$ in $\mathcal{G}'$, if and only if v labels a node in $S_{\lambda-2}(p \cdot w)$. $\mathcal{G}'$ contains exactly $|s_{\lambda-2}(p \cdot w)|$ edges, since for all $v \in S_{\lambda-2}(p \cdot w)$, the states $p \cdot wv$ are all pairwise distinct. Since $largeTree(\tau, r, \lambda)$ is true for all state $r \in Q$, we have $|s_{\lambda-2}(p \cdot w)| \geq k^{\lambda-2} - 1$. $\mathcal{G}'$ is acyclic: indeed, the property $noIntersection(\tau, p \cdot w, q \cdot w)$ indicates that the set of nodes connected to at least one edge forms a bipartite graph (the nodes of $s_{\lambda-2}(p \cdot w)$ on one side and the nodes of $s_{\lambda-2}(q \cdot w)$ on the other), where all the nodes of $s_{\lambda-2}(p \cdot w)$ are only connected to one edge.

Corollary 1. *For $\lambda = \lceil \log_k \log_2 n^3 + 2 \rceil$, the number of automata in $\mathcal{A}_n(p, q, \lambda + 1)$ whose transition structures are in $\alpha_n(p, q)$ is $\mathcal{O}\left(|\mathcal{T}_n| \frac{2^n \log n}{n^3}\right)$.*

Proof. This follows directly from Lemmas 3 and 4 : for any distinct states p and q, any word u of length $\lambda + 1$, and any fixed transition structure $\tau \in \alpha_n(p, q)$, we have

$$|\mathcal{F}_\tau(p, q, u)| = \mathcal{O}\left(2^{n - \log_2 n^3}\right)$$

For a fixed transition structure $\tau \in \alpha_n(p, q)$, since the number of words in $A^{\lambda+1}$ is $\mathcal{O}(\log n)$, the number of choices of sets of final states such that the automata are in $\mathcal{A}_n(p, q, \lambda + 1)$ is bounded above by:

$$\sum_{u \in A^{\lambda+1}} |\mathcal{F}_\tau(p, q, u)| = \mathcal{O}\left(\frac{2^n \log n}{n^3}\right)$$

3.3 Negligible Sets of Transition Structures

Lemma 5. *The number of transition structure in $\mathcal{X}_n$ is $\mathcal{O}\left(|\mathcal{T}_n|\frac{\log^5 n}{n}\right)$.*

Proof. For a fixed state r and a fixed integer $\mu \in \{1,\ldots,\lambda\}$, we define the sets $\mathcal{X}_n(r,\mu)$ of all transition structures τ for which μ is the smallest integer such that the property $largeTree(\tau,r,\mu)$ is false. We have :

$$\mathcal{X}_n = \bigcup_{r\in\{1,\ldots,n\}} \left(\bigcup_{\mu\in\{1,\ldots,\lambda\}} \mathcal{X}_n(r,\mu) \right)$$

For all transition structures in $\mathcal{X}_n(r,\mu)$, the dependency tree $\mathcal{R}(r)$ contains at least two leaves of depth less or equal to μ. Indeed, if $\mathcal{R}(r)$ contains at most one leaf of depth less than μ, then there exist $k-1$ letters $a \in A$ such that $\mathcal{R}(r\cdot a)$ does not contain any leaf of depth less than μ. Therefore we have $|S_\mu(p)| \geq k^\mu - 1$. We decompose the possible dependency trees $\mathcal{R}(r)$ into two different kinds:

1. All leaves are at level μ. Let k be the size of the alphabet and f the number of leaves, the number of trees of this kind is equal to $\sum_{f=2}^{k^\mu}\binom{k^\mu}{f}$.
2. There exists exactly one leaf of depth h (with $h < \mu$), and at least one of depth μ. The number of trees of this kind is at most $\sum_{h=1}^{\mu-1}\left(k^h \sum_{f=1}^{k^\mu}\binom{k^\mu}{f}\right)$.

Using the upper bound of Lemma 1 on the number of transition structures counted by each tree, we obtain:

$$|\mathcal{X}_n(r,\mu)| \leq \sum_{f=2}^{k^\mu}\left[\binom{k^\mu}{f}|\mathcal{T}_n|\left(\frac{|S_\mu(r)|}{n}\right)^f\right] + \sum_{h=1}^{\mu-1}\sum_{f=1}^{k^\mu}\left[k^h\binom{k^\mu}{f}|\mathcal{T}_n|\left(\frac{|S_\mu(r)|}{n}\right)^{f+1}\right]$$

Since $\mu \leq \lambda$, we have $|S_\mu(r)| < k^{\lambda+1}$ and:

$$|\mathcal{X}_n(r,\mu)| < |\mathcal{T}_n|\left(\sum_{f=2}^{k^\lambda}\left[\binom{k^\lambda}{f}\left(\frac{k^{\lambda+1}}{n}\right)^f\right] + \frac{\lambda k^{2\lambda+1}}{n}\sum_{f=1}^{k^\lambda}\left[\binom{k^\lambda}{f}\left(\frac{k^{\lambda+1}}{n}\right)^f\right]\right)$$

Since we have $\binom{k^\lambda}{f}\left(\frac{k^{\lambda+1}}{n}\right)^f \leq \left(\frac{k^{2\lambda+1}}{n}\right)^f$:

$$|\mathcal{X}_n(r,\mu)| < |\mathcal{T}_n|\left(\frac{k^{4\lambda+2}}{n^2}\sum_{f=0}^{\infty}\left(\frac{k^{2\lambda+1}}{n}\right)^f + \frac{\lambda k^{4\lambda+2}}{n^2}\sum_{f=0}^{\infty}\left(\frac{k^{2\lambda+1}}{n}\right)^f\right)$$

$$|\mathcal{X}_n(r,\mu)| = \mathcal{O}\left(|\mathcal{T}_n|\frac{\lambda k^{4\lambda}}{n^2}\right) = \mathcal{O}\left(|\mathcal{T}_n|\frac{\log^4 n^3 \times \log\log n^3}{n^2}\right)$$

Since this upper bound holds for any $\mu \in \{1,\ldots,\lambda\}$ and any $r \in Q$, we obtain:

$$|\mathcal{X}_n| \leq \left(\sum_{r\in\{1,\ldots,n\}}\sum_{\mu\in\{1,\ldots,\lambda\}}|\mathcal{X}_n(r,\mu)|\right) = \mathcal{O}\left(|\mathcal{T}_n|\frac{\log^4 n^3 \times \log^2\log n^3}{n}\right)$$

Lemma 6. *For any distinct states p and q, the number of transition structures in $\mathcal{Y}_n(p,q)$ is $\mathcal{O}\left(|\mathcal{T}_n|\frac{\log^6 n}{n^3}\right)$.*

Proof. For any transition structure in $\mathcal{Y}_n(p,q)$, for all words $w \in A^{\leq 2}$, there exist two words $u, v \in A^{\leq \lambda - 2}$ such that $p \cdot wu = q \cdot wv$. We partition the set $\mathcal{Y}_n(p,q)$ according to the leaves the dependency trees contain.
Both trees do not contain a leaf of depth less or equal to λ : let E be the set of letters, such that for all $a \in E$, $p \cdot a = q \cdot a$. We define $e = |E|$. According to Remark 1, we have $e < k$. For all $b \in A \setminus E$ and all $c \in A$, $noIntersection(\tau, p \cdot bc, q \cdot bc)$ is $false$. For $\overrightarrow{u}$ and $\overrightarrow{v}$ of size $x = e + k(k-e)$, such that for all $1 \leq j \leq e$, we have $u_j = v_j = a_j$ with $a_j \in E$ and such that for all $e < j' \leq x$, w'_j is a prefix of $u_{j'}$ and $v_{j'}$, where w'_j is a word of the form bc. This subset is included in:

$$\bigcup_{\substack{E \subsetneq A \\ \forall a \in E,\ p \cdot a = q \cdot a}} \quad \bigcup_{\overrightarrow{u}, \overrightarrow{v}} \mathcal{T}_n(p, q, \emptyset, \emptyset, \overrightarrow{u}, \overrightarrow{v})$$

There are $2^k - 1$ possible subsets E. There are less than $k^{2\lambda(x-e)}$ possible choices for $u_i, v_i \in A^{\leq \lambda}$, for $e < i \leq x$. For all $j, l \leq x$, $j \neq l$, since u_j and u_l (resp. v_j and v_l) label nodes in $\mathcal{R}(p)$ (resp. $\mathcal{R}(q)$), setting $u_j = u'_j \alpha_j$ and $u_l = u'_l \alpha_l$, we have $(p \cdot u'_j, \alpha_j) \neq (p \cdot u'_l, \alpha_l)$ and there is no path between $(p \cdot u'_j, \alpha_j)$ and $(p \cdot u'_l, \alpha_l)$ since if would imply that $p \cdot u_j = p \cdot u_l$ and that either u_j or u_l labels a leaf. Therefore, $G_n(p, q, \emptyset, \emptyset, \overrightarrow{u}, \overrightarrow{v})$ contains an acyclic graph with x edges $((p \cdot u'_j, \alpha_j), (q \cdot v'_j, \beta_j))$. Since $x \geq k+1$ (for $e = k-1$), using Lemma 2, we obtain the upper bound stated above.
At least one tree contains a leaf of depth less or equal to λ : due to a lack of space, we will not describe this set. The idea is that, just like in the previous case, we are able to guarantee that a $\mathcal{T}$-dependency graph always contains an acyclic subgraph with $k+1$ edges and that there is $\mathcal{O}(\log^{2(k+1)} n)$ possible graphs.

3.4 Concluding the Proof

Recall that we want to prove Equation 1. We define $\widetilde{\mathcal{X}_n}$, $\widetilde{\alpha_n}(p,q)$ and $\widetilde{\mathcal{Y}_n}(p,q)$ as the sets of automata whose transition structure are respectively in $\mathcal{X}_n$, $\alpha_n(p,q)$ and $\mathcal{Y}_n(p,q)$. We have:

$$\bigcup_{i > \lambda} \mathcal{A}_n^i = \bigcup_{p,q \in \{1,\dots,n\}} \mathcal{A}_n(p,q,\lambda+1) \subseteq \widetilde{\mathcal{X}_n} \cup \left(\bigcup_{p,q \in \{1,\dots,n\}} \widetilde{\alpha_n}(p,q) \cup \widetilde{\mathcal{Y}_n}(p,q) \right)$$

Using Lemmas 4,5 and 6 we obtain:

$$\sum_{i > \lambda} |\mathcal{A}_n^i| \leq |\mathcal{T}_n| \frac{2^n \log^5 n}{n} + n^2 \left(|\mathcal{T}_n| \times \frac{2^n \log n}{n^3} + |\mathcal{T}_n| \frac{2^n \log^6 n}{n^3} \right) \leq |\mathcal{A}_n| \frac{\log^6 n}{n}$$

$$\frac{(5 \log_2 n + 1)}{|\mathcal{A}_n|} \sum_{i=\lambda+1}^{5 \log_2 n} |\mathcal{A}_n^i| = \mathcal{O}\left(\frac{\log^7 n}{n}\right) = \mathcal{O}(\log \log n)$$

Hence $\mathcal{N}_n = \mathcal{O}(\log \log n)$, this concludes the proof of the main theorem.

4 Conclusion

In this paper, we obtained a new upper bound on the average complexity of Moore's state minimization algorithm, for the uniform distribution on complete deterministic automata. Also, it is possible to describe a set of Hopcroft's algorithm executions which, for any deterministic automata, compute the equivalence in less steps than Moore's algorithm (due to a lack of space, we are not giving the description of those executions in this paper). Hence, for the uniform distribution on complete deterministic automata with n states, there exists an execution of Hopcroft's algorithm whose average complexity is $\mathcal{O}(n \log \log n)$. This paper is a first step to prove the conjecture made in the conclusion of [1]: for the uniform distribution on complete deterministic accessible automata, the average complexity of Moore algorithm is $\Theta(n \log \log n)$. To prove this conjecture is not an easy task, since it requires a better knowledge of the average size of the accessible part in a complete deterministic automaton, but also the average number of minimal automata amongst the complete deterministic and accessible.

I would like to thank Phillipe Duchon for the fruitful discussion on upper bound of the cardinal of the set $\mathcal{X}_n$, but also Cyril Nicaud and Frederique Bassino for their advices and comments.

References

1. Bassino, F., David, J., Nicaud, C.: On the average complexity of Moore's state minimization algorithm. In: Albers, S., Marion, J.-Y. (eds.) 26th International Symposium on Theoretical Aspects of Computer Science (STACS 2009), Freiburg, Germany. LIPIcs, vol. 3, pp. 123–134. Schloss Dagstuhl - Leibniz-Zentrum fuer Informatik, Germany (2009)
2. Bassino, F., Nicaud, C.: Enumeration and random generation of accessible automata. Theor. Comput. Sci. 381, 86–104 (2007)
3. Berstel, J., Carton, O.: On the complexity of Hopcroft's state minimization algorithm. In: Domaratzki, M., Okhotin, A., Salomaa, K., Yu, S. (eds.) CIAA 2004. LNCS, vol. 3317, pp. 35–44. Springer, Heidelberg (2005)
4. Castiglione, G., Restivo, A., Sciortino, M.: On extremal cases of Hopcroft's algorithm. In: Maneth, S. (ed.) CIAA 2009. LNCS, vol. 5642, pp. 14–23. Springer, Heidelberg (2009)
5. Gries, D.: Describing an algorithm by Hopcroft. Acta Inf. 2, 97–109 (1973)
6. Hopcroft, J.E.: An n log n algorithm for minimizing states in a finite automaton. Technical report, Stanford University, Stanford, CA, USA (1971)
7. Knuutila, T.: Re-describing an algorithm by Hopcroft. Theor. Comput. Sci. 250(1-2), 333–363 (2001)
8. Moore, E.F.: Gedanken experiments on sequential machines. Automata Studies, pp. 129–153. Princeton U. (1956)
9. Nerode, A.: Linear automaton transformation. In: Proc. American Mathematical Society, pp. 541–544 (1958)

Connected Searching of Weighted Trees

Dariusz Dereniowski*

Department of Algorithms and System Modeling,
Gdańsk University of Technology, Poland
deren@eti.pg.gda.pl

Abstract. In this paper we consider the problem of connected edge searching of weighted trees. Authors claim in [L. Barrière at al., Capture of an intruder by mobile agents, *SPAA*'02 (2002) 200-209] that there exists a polynomial-time algorithm for finding an optimal search strategy. However, due to some flaws in their algorithm, the problem turns out to be open. It is proven in this paper that the considered problem is strongly NP-complete even for node-weighted trees (the weight of each edge is 1). It is also shown that there exists a polynomial-time algorithm for finding an optimal connected search strategy for a given bounded degree tree with arbitrary weights on the edges and on the vertices.

Keywords: connected searching, graph searching, search strategy.

1 Introduction

The background and related work

Given a simple undirected graph G, a fugitive is located on an edge of G. The task is to design a sequence of moves of a team of searchers that results in capturing the fugitive. The fugitive is invisible for the searchers — they can deduce the location of the fugitive only from the history of their moves; the fugitive is fast, i.e. whenever he moves, he can traverse a path of arbitrary length in the graph, as long as the path is free of searchers. Finally, the fugitive has a complete knowledge about the graph and about the strategy of the searchers, which means that he will avoid the capture as long as it is possible. The allowable moves for the searchers are, in general, placing a searcher on a vertex, removing a searcher from a vertex and sliding a searcher along an edge of G. An edge is *clear* if it cannot contain the fugitive, otherwise it is *contaminated*. Capturing the fugitive is then equivalent to clearing all the edges of G. The minimum number of searchers sufficient to clear the graph is the *search number* of G, denoted by $\mathtt{s}(G)$. For a survey on graph searching problems see [7].

A key property of a search strategy is the monotonicity. A search is *monotone* if the strategy ensures that the fugitive cannot reach an edge that has been already cleared. The minimum number of searchers needed to construct a monotone search strategy for G is denoted by $\mathtt{ms}(G)$. A search is *internal* if removing

* Partially supported by the Foundation for Polish Science (FNP) and by MNiSW grant N N206 379337 (2009-2011).

P. Hliněný and A. Kučera (Eds.): MFCS 2010, LNCS 6281, pp. 330–341, 2010.

the searchers from the graph is not allowed, while for the search to be *connected* we require that after each move the subgraph of G that is clear is connected. The smallest number k for which a connected k-search strategy $\mathcal{S}$ exists is the *connected search number* of G, denoted by $\texttt{cs}(G)$. The minimum number of k searchers such that there exists a monotone connected k-search for G is called the *monotone connected search number* of G, and is denoted by $\texttt{mcs}(G)$. If a (monotone) connected search strategy $\mathcal{S}$ uses ($\texttt{mcs}(G)$) $\texttt{cs}(G)$ searchers, then $\mathcal{S}$ is called an *optimal* (monotone) connected search strategy for G.

It has been proven that recontamination does help for connected searching [13], and the difference between $\texttt{cs}(G)$ and $\texttt{mcs}(G)$ can be arbitrarily large for some graphs G [13]. However, if T is a tree then $\texttt{cs}(T) = \texttt{mcs}(T)$ [1]. Several algorithmic results for connected searching of special classes of graphs are known, including chordal graphs [11], hypercubes [6], a pyramid [12], chordal rings and tori [5], or outerplanar graphs [8]. The non-connected searching problem for weighted trees has been proven to be NP-complete [10].

The summary of the results

Authors in [1] provided an efficient algorithm for connected searching of weighted trees. However, due to some flaws in the algorithm, it does not always produce an optimal solution (the tree in Figure 1 in Section 3 may serve as an example), which results in an approximation algorithm. The complexity status of searching weighted trees turns out to be NP-complete, which we prove in this work. This gives a motivation for finding non-trivial subclasses of trees that are computationally tractable. From the NP-completeness proof presented here it follows that if a tree has been partially cleared, i.e. for a given vertex v a subset X of edges incident to v is contaminated, then finding the order of clearing the edges in X in an optimal connected search strategy is in general 'as difficult' as finding the strategy itself. For this reason we focus on an algorithm designed for bounded degree trees. However, unlike in the case of searching unweighted trees (both in the classical and connected models), if several subtrees of the tree to search are contaminated, then, in general, a connected search strategy that clears them sequentially (i.e. clears all the edges of one tree and then proceeds to clearing another tree) cannot be optimal. We use the dynamic programming method (we keep a collection of search strategies for some subtrees) together with a greedy rule to narrow down the search space and derive a polynomial running time for bounded degree trees with arbitrary weight functions.

2 Preliminaries

In the following we assume that all graphs $G = (V(G), E(G), w)$ are connected, i.e. there exists a path between each pair of vertices of G. The sets $V(G)$ and $E(G)$ are, respectively, the vertices and the edges of G, while $w\colon V(G) \cup E(G) \to \mathbb{N}_+$ is a weight function. ($\mathbb{N}_+$ is the set of positive integers.)

Now we define our problems formally. Let $k \geq 0$ be an integer. Initially all the edges of a weighted graph $G = (V(G), E(G), w)$ are *contaminated.* A *connected k-search strategy* $\mathcal{S}$ selects a vertex $v_0 \in V(G)$, called the *homebase*, and places k

searchers on v_0. Each move of $\mathcal{S}$ consists of sliding $j \geq 1$ searchers along an edge $e \in E(G)$. If e is contaminated, then we require $j \geq w(e)$, and e becomes *clear* as a result of the move. A clear edge $uv \in E(G)$ becomes *contaminated* if there exists an edge vy that can contain the fugitive (i.e. vy is contaminated) and less than $w(v)$ searchers occupy v. The set of clear edges has to form a connected subgraph after each move of $\mathcal{S}$. After the last move of $\mathcal{S}$ all the edges of G are clear. In the Connected Searching problem (CS) we ask for an optimal connected k-search strategy for a given graph G. The problem where the homebase is a part of the input is denoted by CSFH (Connected Searching with Fixed Homebase).

Given any strategy $\mathcal{S}$, $\mathtt{s}(\mathcal{S})$ is the number of searchers used by $\mathcal{S}$, $|\mathcal{S}|$ is the number of moves in $\mathcal{S}$ and $\mathcal{S}[i]$ is its ith move, $1 \leq i \leq |\mathcal{S}|$. For each $i = 1, \ldots, |\mathcal{S}|$, $\delta(\mathcal{S}[i])$ is the set of vertices v, occupied by searchers at the end of move i, such that there exists a contaminated edge incident to v. We say that the vertices in $\delta(\mathcal{S}[i])$ are *guarded* in step i. Thus, to avoid recontamination, for each $v \in \delta(\mathcal{S}[i])$ at least $w(v)$ searchers occupy v.

The number of searchers used for guarding at the end of step $\mathcal{S}[i]$ is denoted by $\|\mathcal{S}[i]\|$. Note that $\|\mathcal{S}[i]\| = \sum_{v \in \delta(\mathcal{S}[i])} w(v)$. The searchers which are not used for guarding in a given step $\mathcal{S}[i]$ are called *free* searchers in step i. In particular, if more than $w(v)$ searchers occupy $v \in \delta(\mathcal{S}[i])$, then $w(v)$ of them are guarding v, while the remaining ones are considered to be free. Free searchers can move arbitrarily along the clear edges until the next move $\mathcal{S}[i']$, $i' > i$, which clears an edge uv, where $u \in \delta(\mathcal{S}[i])$. The move $\mathcal{S}[i']$ can be performed only if the required number of j searchers (with j' free searchers among them), which will slide along uv in $\mathcal{S}[i']$, is at u. So, each move among $\mathcal{S}[i+1], \ldots, \mathcal{S}[i'-1]$ which is not necessary for gathering the j searchers for clearing uv in $\mathcal{S}[i']$ can be performed after $\mathcal{S}[i']$. Moreover, each set of j' searchers, which are free at the end of move $\mathcal{S}[i]$, can be used to clear uv in $\mathcal{S}[i']$. For this reason, we do not list the moves of sliding searchers along clear edges. Thus, due to this simplifying assumption, $|\mathcal{S}| = |E(G)|$ for monotone strategies $\mathcal{S}$.

We say that a strategy is *partial* if it clears a subset of the edges of G. Given a search strategy $\mathcal{S}$ for G, the symbol $\mathcal{S}[\preceq i]$, $i \in \{1, \ldots, |\mathcal{S}|\}$, is used to denote the partial search strategy consisting of the moves $\mathcal{S}[1], \ldots, \mathcal{S}[i]$. Clearly, if $\mathcal{S}$ is connected, then $\mathcal{S}[\preceq i]$ is also connected. Given a partial search strategy $\mathcal{S}'$, we extend our notation so that $\delta(\mathcal{S}')$ is the set of guarded vertices after the last move of $\mathcal{S}'$, $\delta(\mathcal{S}') = \delta(\mathcal{S}'[|\mathcal{S}'|])$. Let $C_{\mathrm{E}}(\mathcal{S}')$ denote the set of edges cleared by a partial strategy $\mathcal{S}'$. In particular, if $\mathcal{S}$ clears G, then $\delta(\mathcal{S}) = \emptyset$ and $C_{\mathrm{E}}(\mathcal{S}) = E(G)$.

3 Searching Trees — Basic Properties

We will make several simplifying assumptions on connected search strategies restricted to weighted trees $T = (V(T), E(T), w)$. The symbol $\mathtt{cs}(T, r)$ is used to denote the minimum number of searchers needed to clear T when r is the homebase. Note that

$$\mathtt{cs}(T) = \min\{\mathtt{cs}(T, v) \colon v \in V(T)\}. \tag{1}$$

All considered trees T are rooted at the homebase $r \in V(T)$. E_v is the set of edges between v and its children, $v \in V(T)$, and T_v is the subtree of T rooted at v. For each tree T it holds $\texttt{mcs}(T) = \texttt{cs}(T)$ [1]. Thus, in what follows each connected search strategy is monotone. As mentioned in Section 2, we only list the clearing moves of a search strategy $\mathcal{S}$, which implies $|\mathcal{S}| = |E(T)|$.

Consider a connected search strategy $\mathcal{S}$ for T. Let $\mathcal{S}[i]$ be a move of clearing an edge uv. If v is a leaf and $v \neq r$, then the number of searchers that need to slide along uv to clear it in step $\mathcal{S}[i]$ is $w(uv)$ and is independent of $w(v)$, because there is no need to guard v at the end of $\mathcal{S}[i]$. Similarly, if v is a leaf and $v = r$, then $i = 1$, and $\max\{w(uv), w(u)\}$ searchers suffice to clear uv (r does not have to be guarded at the end of $\mathcal{S}[1]$). So, we may w.l.o.g. assume that

$$w(v) = 1 \text{ for each leaf } v \in V(T). \tag{2}$$

Given a connected search strategy $\mathcal{S}$ for T with homebase r, consider a move $\mathcal{S}[i]$ of clearing an edge uv, where v is a child of u. At the beginning of $\mathcal{S}[i]$ the vertex v is unoccupied and u is guarded by $w(u)$ searchers. To clear uv we need to slide $\max\{w(uv), w(v)\}$ searchers along uv. If $w(uv) < w(v)$, then by (2) v is not a leaf of T, and consequently, v has to be guarded at the end of $\mathcal{S}[i]$, which means that we have to slide $w(v)$ searchers along uv. Thus, for each edge uv, where u is the parent of v we w.l.o.g. obtain

$$w(uv) \geq w(v). \tag{3}$$

Our next simplifying assumption is considering the CS and CSFH problems for node-weighted trees only, and we argue that it does not lead to the loss generality. Consider now a new tree $T' = (V(T'), E(T'), w')$ obtained from T by replacing each edge uv by two edges ux_{uv} and vx_{uv}, where x_{uv} is a new vertex of T' corresponding to the edge uv of T (in other words, we subdivide the edges of T to obtain T'). Let $w'(ux_{uv}) = w'(vx_{uv}) = 1$ and $w(x_{uv}) = w(uv)$ for each $uv \in E(T)$ and let $w'(v) = w(v)$ for each $v \in V(T)$. Clearly, $|E(T')| = 2|E(T)|$.

Lemma 1. *For each T and its corresponding tree T', $\texttt{cs}(T', r) = \texttt{cs}(T, r)$ for each $r \in V(T)$.* □

We skip the proof of the lemma. In the remaining part of this paper we assume that the weight of each edge $e \in E(T)$ is 1.

Definition 1. *Let $\mathcal{S}$ and $\mathcal{S}'$ be partial (not necessarily connected) search strategies for T, where $C_{\mathrm{E}}(\mathcal{S}) \cap C_{\mathrm{E}}(\mathcal{S}') = \emptyset$. $\mathcal{S} \oplus \mathcal{S}'$ is a search strategy such that:*

1. *$(\mathcal{S} \oplus \mathcal{S}')[i] = \mathcal{S}[i]$ for each $i = 1, \ldots, |\mathcal{S}|$,*
2. *$(\mathcal{S} \oplus \mathcal{S}')[|\mathcal{S}| + i]$, $i = 1, \ldots, |\mathcal{S}'|$, clears the edge cleared in the move $\mathcal{S}'[i]$, while the set of guarded vertices at the end of the move $(\mathcal{S} \oplus \mathcal{S}')[|\mathcal{S}| + i]$ is $\delta((\mathcal{S} \oplus \mathcal{S}')[|\mathcal{S}| + i]) = \delta(\mathcal{S}'[i]) \cup (\delta(\mathcal{S}) \setminus X)$, where X is the set of vertices initially guarded in $\mathcal{S}'$.*

In other words, $\mathcal{S} \oplus \mathcal{S}'$ clears all the edges cleared by $\mathcal{S}$ and $\mathcal{S}'$ in the order corresponding to the moves $\mathcal{S}[1], \ldots, \mathcal{S}[|\mathcal{S}|], \mathcal{S}'[1], \ldots, \mathcal{S}'[|\mathcal{S}'|]$. Note that in particular $C_{\mathrm{E}}((\mathcal{S} \oplus \mathcal{S}')[\preceq i]) = C_{\mathrm{E}}(\mathcal{S}[\preceq i])$ for each $i = 1, \ldots, |\mathcal{S}|$, and $C_{\mathrm{E}}((\mathcal{S} \oplus \mathcal{S}')[\preceq (|\mathcal{S}|+i)]) = C_{\mathrm{E}}(\mathcal{S}) \cup C_{\mathrm{E}}(\mathcal{S}'[\preceq i])$ for each $i = 1, \ldots, |\mathcal{S}'|$. Furthermore, for $\mathcal{S} \oplus \mathcal{S}'$ to be a partial connected search with homebase r, $\mathcal{S}$ has to be a partial connected search with homebase r.

Definition 2. *Suppose that we are given a tree T rooted at a homebase r, a vertex $v \in V(T)$, and an integer $k \geq 0$. We say that a partial connected k-search $\mathcal{S}_v$ for T_v, $v \in V(T)$, is* (k, v)-minimal *if $w(\delta(\mathcal{S}_v)) \leq w(v)$ and $w(\delta(\mathcal{S}_v)) \leq w(\delta(\mathcal{S}'_v))$ for each partial connected k-search $\mathcal{S}'_v$ for T_v.*

It follows from the definition that a (k, v)-minimal search strategy is also assumed to be partial and connected. A strategy $\mathcal{S}_v$ is *not minimal* if there exists no k such that $\mathcal{S}$ is (k, v)-minimal. A partial connected search strategy $\mathcal{S}$ for T_r can be *extended* to a (k, r)-minimal search for T_r if there exists a search strategy $\mathcal{S}'$ such that $\mathcal{S} \oplus \mathcal{S}'$ is a (k, r)-minimal search strategy for T_r. The latter in particular implies that $\mathtt{s}(\mathcal{S}) \leq k$. Given a tree T_r and $E' \subseteq E(T_r)$, $T_r - E'$ is the set of maximal rooted subtrees induced by the edges in $E(T_r) \setminus E'$.

Lemma 2. *A partial non-minimal connected search strategy $\mathcal{S}$ for T_r can be extended to a (k, r)-minimal search for T_r if and only if there exist T'_v (rooted at v) in $T - C_{\mathrm{E}}(\mathcal{S})$ and a $(k - w(\delta(\mathcal{S}) \setminus \{v\}), v)$-minimal search $\mathcal{S}_v$ for T'_v, such that $\mathcal{S} \oplus \mathcal{S}_v$ can be extended to a (k, r)-minimal search strategy for T_r.*

Proof. The "only if" part is obvious. To prove the "if" part let $\mathcal{S} \oplus \mathcal{S}_1$ be a (k, r)-minimal search for T_r. For each $v \in \delta(\mathcal{S})$ there exists a contaminated edge in E_v, which implies that there exists a nonempty subtree T'_v in $T_r - C_{\mathrm{E}}(\mathcal{S})$ rooted at v. (If all edges in E_v are contaminated, then $T'_v = T_v$.) First we argue that there exist $v \in \delta(\mathcal{S})$ and a $(k - w(\delta(\mathcal{S}) \setminus \{v\}), v)$-minimal search strategy $\mathcal{S}_v$ for T'_v. For each $v \in \delta(\mathcal{S})$ and for each move $\mathcal{S}_1[i]$ define $B(i, v) = \delta(\mathcal{S}_1[i]) \cap V(T'_v)$. Find the minimum l such that $w(B(l, v)) < w(v)$ for some $v \in \delta(\mathcal{S})$. Such an integer l does exist, because otherwise $w(\delta(\mathcal{S} \oplus \mathcal{S}_1)) \geq \delta(\mathcal{S})$ which contradicts the minimality of $\mathcal{S} \oplus \mathcal{S}_1$. Let $\mathcal{S}'_v$ be $\mathcal{S}_1$ restricted to clearing the edges in $C_{\mathrm{E}}(\mathcal{S}_1[\preceq l]) \cap E(T'_v)$ in the same order as they are cleared by $\mathcal{S}_1$. $\mathcal{S} \oplus \mathcal{S}'_v$ uses at most k searchers (which gives that $\mathtt{s}(\mathcal{S}'_v) \leq k - w(\delta(\mathcal{S}) \setminus \{v\})$), and $w(\delta(\mathcal{S}'_v)) = w(B(l, v)) < w(v)$. So, the set of partial $(k - w(\delta(\mathcal{S}) \setminus \{v\}))$-search strategies $\mathcal{S}'_v$ for T'_v satisfying $w(\delta(\mathcal{S}'_v)) < w(v)$ is nonempty and, by the definition, a strategy $\mathcal{S}_v$ with the minimum $w(\delta(\mathcal{S}_v))$ is $(k - w(\delta(\mathcal{S}) \setminus \{v\}), v)$-minimal.

We will use $\mathcal{S}_1$ to extend $\mathcal{S} \oplus \mathcal{S}_v$ to a (k, r)-minimal search strategy $\mathcal{S} \oplus \mathcal{S}_v \oplus \mathcal{S}_2$ for T_r. To obtain $\mathcal{S}_2$ we simply remove from $\mathcal{S}_1$ all the operations of clearing the edges in $C_{\mathrm{E}}(\mathcal{S}_v)$, preserving the order of clearing the remaining edges in $\mathcal{S}_1$. One can prove that $\mathcal{S} \oplus \mathcal{S}_v \oplus \mathcal{S}_2$ is connected.

It remains to prove that $\mathtt{s}(\mathcal{S} \oplus \mathcal{S}_v \oplus \mathcal{S}_2) \leq k$. By the definition, $\mathtt{s}(\mathcal{S} \oplus \mathcal{S}_v) \leq k$, so let us consider a move $(\mathcal{S} \oplus \mathcal{S}_v \oplus \mathcal{S}_2)[i_2]$ of clearing an edge e, $i_2 > |\mathcal{S} \oplus \mathcal{S}_v|$. Select $i_1 > |\mathcal{S}|$ so that $(\mathcal{S} \oplus \mathcal{S}_1)[i_1]$ is the move of clearing e. It is sufficient to prove that $|(\mathcal{S} \oplus \mathcal{S}_v \oplus \mathcal{S}_2)[i_2]| \leq |(\mathcal{S} \oplus \mathcal{S}_1)[i_1]|$. Let

$$U = \delta((\mathcal{S} \oplus \mathcal{S}_v \oplus \mathcal{S}_2)[i_2]) \setminus \delta((\mathcal{S} \oplus \mathcal{S}_1)[i_1]). \tag{4}$$

In other words, U is the set of vertices guarded at the end of step i_2 of $\mathcal{S}\oplus\mathcal{S}_v\oplus\mathcal{S}_2$ but unguarded at the end of step i_1 of $\mathcal{S}\oplus\mathcal{S}_1$. Clearly, $U\subseteq\delta(\mathcal{S}_v)$. For each $u\in U$ there exists a vertex $x_u\in\delta((\mathcal{S}\oplus\mathcal{S}_1)[i_1])$ on the path connecting v and u in T'_v, because $C_{\mathrm{E}}((\mathcal{S}\oplus\mathcal{S}_1)[\preceq i_1])\cap E(T'_v)\subseteq C_{\mathrm{E}}((\mathcal{S}\oplus\mathcal{S}_v\oplus\mathcal{S}_2)[\preceq i_2])\cap E(T'_v)$. Let X_U be the set of all such vertices x_u, $u\in U$. We argue that

$$w(X_U)\geq w(U). \tag{5}$$

Suppose for a contradiction that (5) does not hold. Find a set X, with minimum $w(X)$, such that each path connecting v and u, $u\in\delta(\mathcal{S}_v)$, contains a vertex in X (possibly u). It holds $w(X)<w(\delta(\mathcal{S}_v))$, because $w((\delta(\mathcal{S}_v)\setminus U)\cup X_U)<w(\delta(\mathcal{S}_v))$ and by the minimality of X, $w(X)\leq w((\delta(\mathcal{S}_v)\setminus U)\cup X_U)$. Define $\mathcal{S}'_v$ which clears the edges that are in $C_{\mathrm{E}}(\mathcal{S}_v)$ but not in $E(T_x)$ for each $x\in X$ in the same order as they are cleared in $\mathcal{S}_v$. Then, $\mathtt{s}(\mathcal{S}'_v)\leq\mathtt{s}(\mathcal{S}_v)$ and $w(\delta(\mathcal{S}'_v))=w(X)<w(\delta(\mathcal{S}_v))$. Thus, $\mathcal{S}_v$ is not $(k-w(\delta(\mathcal{S})\setminus\{v\}),v)$-minimal — a contradiction, which proves (5). Hence, $|(\mathcal{S}\oplus\mathcal{S}_v\oplus\mathcal{S}_2)[i_2]|\leq|(\mathcal{S}\oplus\mathcal{S}_1)[i_1]|\leq\mathtt{cs}(T,r)$. Since i_2 has been chosen arbitrarily, we have proven the thesis. □

As an example consider a tree in Fig. 1(a). Let $\mathcal{S}$ be a partial strategy clearing the edges ru, rv, rt (in this order). $\mathtt{s}(\mathcal{S})=12$ and $\delta(\mathcal{S})=\{u,v,t\}$. Denote by $\mathcal{S}_v$ the $(8,v)$-minimal search strategy for T_v ($\mathcal{S}_v$ completely clears the left branch of T_v first). There exist a $(8,u)$-minimal strategy $\mathcal{S}_u$ for T_u with $\delta(\mathcal{S}_u)=\{x\}$ (this strategy clears the two edges on the path from u to x) and a $(8,t)$-minimal search $\mathcal{S}_t$ for T_t, where $\delta(\mathcal{S}_t)=\{y,z\}$ ($\mathcal{S}_t$ clears the three edges on the paths connecting t and y,z). Fig. 1(b) depicts a partial strategy $\mathcal{S}\oplus\mathcal{S}_t\oplus\mathcal{S}_u\oplus\mathcal{S}_v$, where the dashed arrows represent the moves of the strategy. Their labels $i:c+g$ indicate the number i of the clearing move, while c and g are, respectively, the number of searchers that move along the corresponding edge and guard other vertices in this move. This strategy can be extended to a connected 12-serach strategy for T by clearing the subtrees T_y, T_z and T_x in this order. Finally, note that clearing each of the subtrees T_u, T_v, T_t completely while guarding r results in a

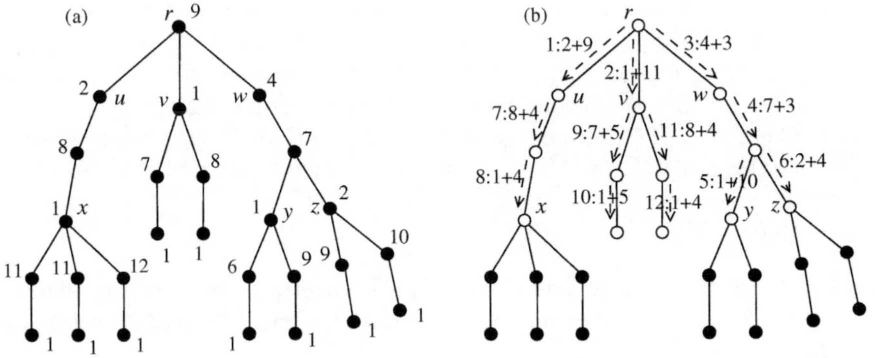

Fig. 1. (a) a node weighted tree T_r; (b) $\mathcal{S}\oplus\mathcal{S}_t\oplus\mathcal{S}_u\oplus\mathcal{S}_v$

search strategy using more than 12 searches, which implies that the algorithm for searching weighted trees presented in [1] is not optimal.

4 An Efficient Algorithm for Bounded-Degree Trees

In this section we provide an algorithm for searching bounded degree weighted trees. In an informal way, our method may be described as follows. We start by placing k searchers at the root r of T_r. Assume that the algorithm calculated a partial search strategy $\mathcal{S}$. If $\delta(\mathcal{S}) = \emptyset$, then $\mathcal{S}$ clears T_r and the computation stops. Otherwise we select a vertex $v \in \delta(\mathcal{S})$ and a partial connected search $\mathcal{S}_v$ for T_v. We continue with $\mathcal{S} \oplus \mathcal{S}_v$. Note that $\mathcal{S} \oplus \mathcal{S}_v$ requires $\mathtt{s}(\mathcal{S})$ searchers to perform $\mathcal{S}$ and then the moves of $\mathcal{S}_v$ follow, where $w(\delta(\mathcal{S}) \setminus \{v\})$ searchers guard the vertices that are not in T_v, that is those in $\delta(\mathcal{S}) \setminus \{v\}$, and, in addition, $\mathtt{s}(\mathcal{S}_v)$ searchers work on T_v. So, if $\mathcal{S}$ can be extended to a connected k-search for T_r and if we find a $(k - w(\delta(\mathcal{S}) \setminus \{v\}), v)$-minimal strategy $\mathcal{S}_v$, then, by Lemma 2, $\mathcal{S} \oplus \mathcal{S}_v$ can be extended to a connected k-search for T_r. The fact that any such vertex v is sufficient reduces the size of the search space for the algorithm. However, it follows immediately from the NP-completeness proof in Section 5 that finding a strategy $\mathcal{S}_v$ is intractable, unless P=NP.

For each $v \in V(T_r)$ a set $\mathcal{C}_v$ is a global variable, a collection of (k, v)-minimal search strategies for T_v, for selected values of k.

We start by describing a procedure, called `MCPS` (*Minimal Connected Partial Strategy*), which for a given integer k, a rooted tree T_r, and an ordering $rv_1, \ldots, rv_d$ of the edges in E_r, finds a (k, r)-minimal search strategy $\mathcal{S}$, which clears the edges in E_r according to the given order, whenever such a strategy exists. Our final algorithm will process T_r in a bottom-up fashion, so when `MCPS` is called, then for each $v \in V(T_r) \setminus \{r\}$ some (k', v)-minimal search strategies for T_v belong to $\mathcal{C}_v$ for some integers k'. Moreover, $w(r)$ searchers already occupy r when `MCPS` starts. The procedure is as follows:

Step 1. For each $i = 1, \ldots, d-1$ repeat the following: (i) if k searchers are sufficient to clear rv_i, then clear rv_i as the next step of $\mathcal{S}$ and find (k', v_i)-minimal search $\mathcal{S}_{v_i} \in \mathcal{C}_{v_i}$ with maximum k' such that $k' \leq k - w(\delta(\mathcal{S}) \setminus \{v_i\})$. If $\mathcal{S}_{v_i}$ exists, then let $\mathcal{S} := \mathcal{S} \oplus \mathcal{S}_{v_i}$, otherwise proceed to $i+1$; (ii) if more than k searchers are needed to clear rv_i, then return 'failure'.

Step 2. Clear rv_d. (If k' searchers are not sufficient to do it, then return 'failure'.) While there exist $v \in \delta(S)$ and $\mathcal{S}_v \in \mathcal{C}_v$ such that $\mathcal{S}_v$ is (k', v)-minimal, $k' \leq k - w(\delta(\mathcal{S}) \setminus \{v\})$, then $\mathcal{S} := \mathcal{S} \oplus \mathcal{S}_v$.

Step 3. Return $\mathcal{S}$.

Lemma 3. *If $\mathcal{S}$ can be extended to a (k, r)-minimal search strategy that clears the edges in E_r according to the order $\pi = (rv_1, \ldots, rv_d)$, then* `MCPS` *returns such a strategy.*

Proof. Assume that there exists a (k, r)-minimal search strategy $\mathcal{S}_{\mathrm{opt}}$ clearing the edges in E_r according to the order π. Let, for brevity, $\mathcal{S}_i$ denote the partial connected search strategy calculated in Steps 1-2 of MCPS, where clearing rv_i is the last move of $\mathcal{S}_i$, $i = 1, \ldots, d$.

We use an induction on $i = 1, \ldots, d$ to prove that $\mathcal{S}_i$ can be extended to (k, r)-minimal search for T_r. The claim follows immediately for $i = 1$, since by assumption, $\mathcal{S}_{\text{opt}}$ starts by clearing rv_1. Assume that rv_i has been cleared by $\mathcal{S}_i$, $i < d$. The procedure MCPS proceeds in Step 1 by finding a $(k - w(\delta(\mathcal{S}_i) \setminus \{v_i\}), v_i)$-minimal search strategy $\mathcal{S}_{v_i}$ for T_{v_i}. If $\mathcal{S}_{v_i}$ exists, then by Lemma 2, $\mathcal{S}_i \oplus \mathcal{S}_{v_i}$ can be extended to a (k, r)-minimal search for T_r. By the definition, there is no $v \in \delta(\mathcal{S}_i \oplus \mathcal{S}_{v_i}) \setminus \{r\}$ for which there exists a $(k - w(\delta(\mathcal{S}_i \oplus \mathcal{S}_{v_i}) \setminus \{v\}), v)$-minimal search strategy for T_v. Thus, the next edge e cleared by $\mathcal{S}_i \oplus \mathcal{S}_{v_i}$ must be in E_r. On the other hand, if $\mathcal{S}_{v_i}$ does not exist, then the next edge e to clear is in E_r. Hence, in both cases $e = rv_{i+1}$ which results in strategy $\mathcal{S}_{i+1}$.

Thus, we obtain that $\mathcal{S}_d$ can be extended to a (k, r)-minimal search for T_r. Then, MCPS finds in Step 2 a sequence of vertices $v_{d+1}, \ldots, v_{d+l}$ and search strategies $\mathcal{S}_{d+1}, \ldots, \mathcal{S}_{d+l}$ such that $\mathcal{S}_{d+i}$ is $(k - w(\delta(\mathcal{S}_d \oplus \cdots \oplus \mathcal{S}_{d+i-1}) \setminus \{v_{d+i}\}), v_{d+i})$-minimal and $v_{d+i} \in \delta(\mathcal{S}_d \oplus \cdots \oplus \mathcal{S}_{d+i-1})$. By Lemma 2, each strategy $\mathcal{S}_d \oplus \cdots \oplus \mathcal{S}_{d+i}$, $i = 0, \ldots, l$, can be extended to a (k, r)-minimal search for T.

Let $\mathcal{S} = \mathcal{S}_d \oplus \cdots \oplus \mathcal{S}_{d+l}$. We obtain that $\mathcal{S}$ is (k, r)-minimal, because otherwise, as proved above, it can be extended to a (k, v)-minimal search for T_r, and consequently, by Lemma 2, there exists $v \in \delta(\mathcal{S})$ and a $(k - w(\delta(\mathcal{S}) \setminus \{v\}), v)$-minimal search $\mathcal{S}_v$ such that $\mathcal{S} \oplus \mathcal{S}_v$ can be extended to a (k, v)-minimal search for T_r, which gives a contradiction with the fact that no such vertex has been found following v_{d+l} by MCPS. □

Now we are ready to give a listing of the algorithm CST (*Connected Searching of a Tree*) for finding an optimal connected search strategy for a tree T_r with homebase r. This algorithm is exponential in the maximum degree of T.

Step 1. For each child v of r call CST(T_r). This step guarantees that for each $v \in V(T) \setminus \{r\}$ the collection $\mathcal{C}_v$ of all minimal search strategies for T_v is calculated (which is required for subsequent calls of MCPS).

Step 2. Fix a permutation $\pi = (rv_1, \ldots, rv_d)$ of the edges in E_r. Set $k := 1$. If Step 3 has been executed for all the $d!$ permutations π, then Exit.

Step 3. Call MCPS(k, T_r, π). If the 'failure' has been returned, then increase k and repeat Step 3. If a search strategy $\mathcal{S}_r$ has been returned and there is no $\mathcal{S} \in \mathcal{C}_r$ such that $w(\delta(\mathcal{S})) \leq w(\delta(\mathcal{S}_r))$ and $\mathtt{s}(\mathcal{S}) \leq \mathtt{s}(\mathcal{S}_r)$ then add $\mathcal{S}_r$ to $\mathcal{C}_r$ and remove from $\mathcal{C}_r$ all search strategies $\mathcal{S} \neq \mathcal{S}_r$ such that $w(\delta(\mathcal{S})) \geq w(\delta(\mathcal{S}))$ and $\mathtt{s}(\mathcal{S}) \geq \mathtt{s}(\mathcal{S}_r)$. If $\delta(\mathcal{S}_r) = \emptyset$, then go to Step 2 to fix the next permutation π. Otherwise increase k and repeat Step 3.

The following theorem can be proved using an induction on the size of the tree to search (we skip the proof due to the space limitations).

Lemma 4. *Let k be an integer. The set $\mathcal{C}_v$ contains a (k, r)-minimal search strategy for T_r whenever such a strategy exists.* □

Lemma 4 in particular implies, that CST finds an optimal solution to the CSFH problem, because an optimal connected search strategy $\mathcal{S}$ is $(\mathtt{cs}(T, r), r)$-minimal and $\delta(\mathcal{S}) = \emptyset$. During an execution of MCPS in Step 3 of CST the algorithm records the minimum $k' > k$ for which the execution of MCPS Step 3 returns a different strategy for the same input permutation π. The strategy corresponding to k clears a proper subset of edges cleared by the strategy corresponding to k'. Therefore, there are at most $|E(T_r)|$ different values of k for which we execute MCPS is Step 3 of CST. We obtain the following.

Lemma 5. *Given a bounded degree tree T, the running time of the algorithm* CST *is $O(n^3 \log n)$, where $n = |V(T)|$.* □

The complexity of solving CS for trees remains the same, because for each vertex v we calculate the search strategies for at most $\deg_T(v)$ subtrees rooted at v when all the n possible homebases are considered.

Theorem 1. *Given a bounded degree weighted tree T, an optimal connected search strategy for T can be computed in $O(n^3 \log n)$ time, $n = |V(T)|$.* □

Our final remark is that CSFH has been designed to clear node-weighted trees. Thus, for clearing a tree T with non-unit edge weights we apply first to T the transformations from Section 3, which results in a tree T' with unit edge weights, $\texttt{cs}(T) = \texttt{cs}(T')$ and an optimal connected search strategy for T can be obtained on the basis of the strategy calculated by CST for T'.

5 Connected Searching of Weighted Trees Is Hard

First we recall a problem of scheduling time-dependent (deteriorating) tasks, where a set of tasks $\mathcal{J} = \{J_1, \ldots, J_n\}$ is given, where each task $J_j \in \mathcal{J}$ has an integer deadline d_j and a non-decreasing function $p_j\colon \{0, \ldots, d_j - 1\} \to \mathbb{N}_+$ defining its the execution time (the execution time of a task J_j depends on s_j, the point of time when the execution of J_j start, i.e. it equals $p_j(s_j)$). The completion time of J_j is $C_j = s_j + p_j(s_j)$. We are interested in the single machine scheduling. A schedule D is *feasible* if the completion time C_j of each task J_j is not greater than its deadline, $C_j \leq d_j$, and the execution intervals of two different tasks do not overlap. The *makespan* of a schedule D is $\text{ms}(D) = \max\{C_j \colon J_j \in \mathcal{J}\}$. Observe that a schedule D can be described by a permutation $\pi_D\colon \{1, \ldots, |\mathcal{J}|\} \to \mathcal{J}$, because the idle times between the execution of two consecutive tasks are not necessary for non-decreasing (in time) execution times. In the Time-Dependent Scheduling (TDS) problem we ask whether there exists a feasible schedule for $\mathcal{J}$. A good survey and a more detailed description of this problem can be found in [4]. For a survey on scheduling problems and terminology see [2].

The integer valued functions p_j imply that in each schedule s_j and C_j are integers, $J_j \in \mathcal{J}$, which also justifies the fact that we consider the values of p_j only at integer points. For each $J_j \in \mathcal{J}$ let f_j be the latest possible integer starting point for J_j, i.e. $f_j = \max\{t \in \mathbb{N} \colon t + p_j(t) \leq d_j\}$. The integer L is selected to be an upper bound for the length of each feasible schedule,

$$L = \max\{d_j \colon J_j \in \mathcal{J}\}. \tag{6}$$

There are several NP-completeness results for very restricted (linear) functions for execution time of a task, see e.g. [3,9]. We may use them to prove that the problem of connected searching of weighted trees is weakly NP-hard, i.e. when the input to the problem is given in binary. However, to obtain a stronger result, i.e. NP-completeness of the CS problem for instances with integer and polynomially bounded weights on the vertices, we use the following theorem (we skip the proof due to the space limitations).

Theorem 2. *Given a set of tasks $\mathcal{J}$ with integer deadlines and integer non-decreasing (in time) execution times, the problem of deciding if there exists a feasible schedule for $\mathcal{J}$ is strongly* NP*-complete.* □

To prove NP-hardness of CS problem we start by reducing TDS to CSFH. Then, we conclude that CS is NP-complete as well.

Given $\mathcal{J}$, we construct a node-weighted tree $T = (V(T), E(T), w)$ rooted at r. For each $J_j \in \mathcal{J}$ define a path P_j with $V(P_j) = \{u_j^i, v_j^i : i = 0, \ldots, f_j\}$ and

$$E(P_j) = \{u_j^i v_j^i : i = 0, \ldots, f_j\} \cup \{v_j^{i+1} u_j^i : i = 0, \ldots, f_j - 1\}.$$

The tree T, in addition to the vertices in $\bigcup_{J_j \in \mathcal{J}} V(P_j)$, contains the vertices r and y_j, z_j, $j = 0, \ldots, |\mathcal{J}|$. The root r is adjacent to y_0 and to the endpoint $u_j^{f_j}$ of each path P_j, $j = 1, \ldots, |\mathcal{J}|$. The other endpoint of P_j, namely the vertex v_j^0, is adjacent to y_j for each $j = 1, \ldots, |\mathcal{J}|$. Finally, for each $0 = 1, \ldots, |\mathcal{J}|$ the vertex y_j is the parent of z_j. The weight function $w \colon V(G) \to \mathbb{N}_+$ is as follows

$$w(r) = 2L, w(y_j) = 3L, w(z_j) = 1 \quad j = 0, \ldots, |\mathcal{J}|, \tag{7}$$

$$w(u_j^i) = 2L - i \text{ and } w(v_j^i) = p_j(i), \quad j = 1, \ldots, |\mathcal{J}|, i = 0, \ldots, f_j. \tag{8}$$

Let $k = 4L$. Note that for each u_j^i and $v_j^{i'}$, $0 \le i, i' \le f_j$, it holds

$$w(u_j^i) > L \ge w(v_j^{i'}), \tag{9}$$

because $f_j < L$ for each $j = 1, \ldots, |\mathcal{J}|$. Moreover,

$$w(u_j^0) > w(u_j^1) > \cdots > w(u_j^{f_j}), \quad j = 1, \ldots, |\mathcal{J}|, \tag{10}$$

$$w(v_j^{f_j}) > w(v_j^{f_j - 1}) > \cdots > w(v_j^0), \quad j = 1, \ldots, |\mathcal{J}|. \tag{11}$$

Given a schedule D for $\mathcal{J}$, we construct a k-search strategy $\mathcal{S}$ for T_r:

Step 1: Initially $4L$ searchers occupy r.
Step 2: For each $i = 1, \ldots, |\mathcal{J}|$ do the following: let $J_j = \pi_D(i)$; clear the path $P_j(D) \subseteq P_j$ consisting of the vertices $u_j^{f_j}, v_j^{f_j}, \ldots, u_j^{s_j}, v_j^{s_j}$. (After this step, by (8), $w(v_j^{s_j}) = p_j(s_j)$ searchers occupy $v_j^{s_j}$ to guard it.)
Step 3: Clear the edges ry_0 and $y_0 z_0$.

Step 4: For each $J_j \in \mathcal{J}$ clear the path $u_j^{s_j-1}, v_j^{s_j-1}, \ldots, u_j^0, v_j^0, y_j, z_j$ (after this step the subtree rooted at $v_j^{f_j}$ is clear).

Lemma 6. *If there exists a valid schedule for $\mathcal{J}$, then there exists a connected $4L$-search strategy for the weighted tree T rooted at r.*

Proof. It is enough to prove that $\mathcal{S}$ defined above is a connected search for T and that $\mathtt{s}(\mathcal{S}) \leq 4L$. It is easy to see that after each step the subtree that is clear is connected. Now we prove that the number of searchers used is at most k. Initially $2L$ searchers guard r. We prove by induction on $i = 1, \ldots, |\mathcal{J}|$ that k searchers suffice to clear the path $P_j(D)$ in Step 2, where $J_j = \pi_D(i)$, and the number of searchers used in $\mathcal{S}$ for guarding when the vertex $v_j^{s_j}$ becomes guarded is

$$x_i = 2L + \sum_{j' : \pi_D^{-1}(J_{j'}) \leq i} p_{j'}(s_{j'}). \tag{12}$$

The cases when $i = 1$ and $i > 1$ are analogous ($x_0 = 2L$), so we prove it for i, assuming that it is true for $i - 1$, $1 \leq i \leq |\mathcal{J}|$.

Let $P_j(D) \subseteq P_j$ be the ith path cleared, i.e. $J_j = \pi_D(i)$. By (9) and (10), $w(u_j^{s_j}) = \max\{w(v) : v \in V(P_j(D))\}$. So, by (12), $w(u_j^{s_j}) + x_{i-1}$ searchers are needed to clear $P_j(D)$. We obtain

$$w(u_j^{s_j}) + x_{i-1} = (2L - s_j) + 2L + \sum_{j' : \pi_D^{-1}(J_{j'}) < i} p_{j'}(s_{j'}) = 4L,$$

because, by the definition of a schedule for time-dependent tasks the execution of a task J_j starts immediately after the execution of the preceding task, which can be stated as

$$s_j = \sum_{j' : \pi_D^{-1}(J_{j'}) < i} p_{j'}(s_{j'}).$$

Thus, $x_i = x_{i-1} + w(v_j^{s_j}) = x_{i-1} + p_j(s_j)$ and (12) follows. This proves that $4L$ searchers are used during search moves defined in Steps 1 and 2 above. When the execution of all search operations constructed in Step 2 is completed, $2L$ searchers are used for guarding r, while for guarding the vertices $v_j^{s_j}$, $j = 1, \ldots, |\mathcal{J}|$ we need $\sum_{j=1,\ldots,|\mathcal{J}|} p_j(s_j) \leq L$ searchers. The last inequality follows from Equation (6) and from the fact that in a valid schedule D each task is completed within interval $[0, L]$. Thus, we can use $3L$ searchers to clear y_0, z_0 and then the remaining subpaths $u_j^{s_j-1}, v_j^{s_j-1}, \ldots, u_j^0, v_j^0, y_j, z_j$. □

Due to the space limitations we omit a technical proof of the following.

Lemma 7. *If there exists a connected $4L$-search strategy $\mathcal{S}$ for the weighted tree T_r, then there exists a valid schedule for $\mathcal{J}$.* □

The CSFH is clearly in NP, and the reduction is polynomial in $n = |\mathcal{J}|$, because L is, by Theorem 2, polynomially bounded in n. Lemmas 6 and 7 give us the theorem.

Theorem 3. *Given a weighted tree T rooted at r and an integer $k \geq 0$, deciding whether* $\texttt{cs}(T, r) \leq k$ *is* NP*-complete.* □

Let $T_r = (V(T), E(T), w)$ and k be an input to the CSFH problem. Define $T_r^2 = (V(T), E(T), 2w)$ to be the tree with the same vertex and edge sets as T_r, while the weight of each vertex v of T_r^2 is two times bigger than the weight of v in T_r. There exists a connected k-search strategy for T_r if and only if there exists a connected $(2k)$-search strategy for T_r^2. Take three copies of T_r^2, and a vertex r' (the weight of r' is 1), and let the roots of the trees T_r^2 be the children of r'. The new tree is denoted by T_r'. We obtain that $\texttt{cs}(T', r) = 2k + 1$. Moreover, if $\mathcal{S}'$ is a connected $(2k+1)$-search strategy for T_r' then regardless of the homebase in $\mathcal{S}'$, the strategy is forced to clear one of the subtrees T_r^2 in T_r' by starting at r and using $2k$ searchers. This leads to the following

Corollary 1. *The problem of connected searching of weighted trees is strongly* NP*-hard.* □

References

1. Barrière, L., Flocchini, P., Fraigniaud, P., Santoro, N.: Capture of an intruder by mobile agents. In: SPAA 2002: Proceedings of the fourteenth annual ACM symposium on parallel algorithms and architectures, pp. 200–209. ACM, New York (2002)
2. Błażewicz, J., Ecker, K., Pesch, E., Schmidt, G., Węglarz, J.: Scheduling computer and manufacturing processes. Springer, New York (1996)
3. Cheng, T.C.E., Ding, Q.: Scheduling start time dependent tasks with deadlines and identical initial processing times on a single machine. Comput. Oper. Res. 30(1), 51–62 (2003)
4. Cheng, T.C.E., Ding, Q., Lin, B.M.T.: A concise survey of scheduling with time-dependent processing times. European Journal of Operational Research 152(1), 1–13 (2004)
5. Flocchini, P., Huang, M.J., Luccio, F.L.: Decontaminating chordal rings and tori using mobile agents. Int. J. Found. Comput. Sci. 18(3), 547–563 (2007)
6. Flocchini, P., Huang, M.J., Luccio, F.L.: Decontamination of hypercubes by mobile agents. Netw. 52(3), 167–178 (2008)
7. Fomin, F.V., Thilikos, D.M.: An annotated bibliography on guaranteed graph searching. Theor. Comput. Sci. 399(3), 236–245 (2008)
8. Fomin, F.V., Thilikos, D.M., Todinca, I.: Connected graph searching in outerplanar graphs. Electronic Notes in Disc. Math. 22, 213–216 (2005)
9. Kubiak, W., van de Velde, S.: Scheduling deteriorating jobs to minimize makespan. Naval Research Logistics 45(5), 511–523 (1998)
10. Mihai, R., Todinca, I.: Pathwidth is NP-hard for weighted trees. In: Deng, X., Hopcroft, J.E., Xue, J. (eds.) FAW 2009. LNCS, vol. 5598, pp. 181–195. Springer, Heidelberg (2009)
11. Nisse, N.: Connected graph searching in chordal graphs. Discrete Applied Math. 157(12), 2603–2610 (2008)
12. Shareghi, P., Imani, N., Sarbazi-Azad, H.: Capturing an intruder in the pyramid. In: Grigoriev, D., Harrison, J., Hirsch, E.A. (eds.) CSR 2006. LNCS, vol. 3967, pp. 580–590. Springer, Heidelberg (2006)
13. Yang, B., Dyer, D., Alspach, B.: Sweeping graphs with large clique number. Discrete Mathematics 309(18), 5770–5780 (2009)

Iterated Regret Minimization in Game Graphs*

Emmanuel Filiot, Tristan Le Gall, and Jean-François Raskin

Université Libre de Bruxelles
efiliot@ulb.ac.be, tlegall@ulb.ac.be, jraskin@ulb.ac.be

Abstract. Iterated regret minimization has been introduced recently by J.Y. Halpern and R. Pass in classical strategic games. For many games of interest, this new solution concept provides solutions that are judged more reasonable than solutions offered by traditional game concepts – such as *Nash equilibrium* –. In this paper, we investigate iterated regret minimization for infinite duration two-player quantitative non-zero sum games played on graphs.

1 Introduction

The analysis of complex interactive systems like embedded systems or distributed systems is a major challenge of computer aided verification. Zero-sum games on graphs provide a good framework to model interactions between a component and an environment as they are strictly competitive. However in the context of modern interactive systems, several components may interact and be controlled independently. Non-zero sum games on graphs are more accurate to model such systems, as the objectives of the components are not necessarily antagonist. Because of the quantitative aspects of the components (like energy consumption), we need some solution concept to be able to synthesis a component that respects some formal specification and is, in some sense, optimal. In the context of non-zero sum games on graphs for verification and synthesis, Nash equilibria or particular classes of Nash equilibria have been studied for qualitative objectives [5,2,3] or quantitative objectives [1]. Recently, J.Y. Halpern and R. Pass defined the notion of *iterated regret minimization* (IRM) [6] in the general framework of (single-shot) *strategic games*, where the players choose in parallel their strategy among a finite number of strategies, and their respective payoffs are given by a finite matrix. They show that for many games of interest, Nash equilibria suggest strategies that are rejected by common sense, while iterated regret is an alternative solution concept that sometimes proposes more intuitive solutions. In this paper we consider games where the matrix is not given explicitly but defined implicitly by a game graph and we study the algorithmic aspects of IRM on such games. While it is easy to compute the iterated regret on a finite matrix, it is not that simple for game

	A_2	B_2
A_1	$(2,1)$	$(3,4)$
B_1	$(1,2)$	$(4,3)$

Fig. 1.

* Work supported by the projects: (*i*) Quasimodo: "Quantitative System Properties in Model-Driven-Design of Embedded", http://www.quasimodo.aau.dk/, (*ii*) Gasics: "Games for Analysis and Synthesis of Interactive Computational Systems", http://www.ulb.ac.be/di/gasics/, and (*iii*) Moves: "Fundamental Issues in Modelling, Verification and Evolution of Software", http://moves.ulb.ac.be, a PAI program funded by the Federal Belgian Government.

P. Hliněný and A. Kučera (Eds.): MFCS 2010, LNCS 6281, pp. 342–354, 2010.

played on trees or graphs, as the number of strategies (and therefore the underlying matrix) is exponential in the size of a tree and even infinite for a graph.

The IRM solution concept assumes that instead of trying to minimize what she has to pay, each player tries to minimize her *regret*. The regret is informally defined as the difference between what a player actually pays and what she could have payed if she knew the strategy chosen by the other player[1]. More formally, if $\mathsf{c}_1(\lambda_1, \lambda_2)$ represents what Player 1 pays[2] when the pair of strategies (λ_1, λ_2) is played, $\mathsf{reg}_1(\lambda_1, \lambda_2) = \mathsf{c}_1(\lambda_1, \lambda_2) - \min_{\lambda_1'} \mathsf{c}_1(\lambda_1', \lambda_2)$. Consider the strategic game defined by the matrix of Figure 1. In this game, Player 1 has two strategies A_1 and B_1 and Player 2 has two strategies A_2 and B_2. The two players choose a strategy at the same time and the pairs of strategies define what the two players have to pay. The regret of playing A_1 for Player 1 when Player 2 plays A_2 is equal to 1 because $\mathsf{c}_1(A_1, A_2)$ is 2 while $\mathsf{c}_1(B_1, A_2)$ is 1. Knowing that Player 2 plays A_2, Player 1 should have played B_1 instead of A_1.

As players have to choose strategies before knowing how the adversary will play, we associate a regret with each strategy as follows. The regret of a strategy λ_1 of Player 1 is : $\mathsf{reg}_1(\lambda_1) = \max_{\lambda_2} \mathsf{reg}_1(\lambda_1, \lambda_2)$. In the example, the regret of strategy A_1 is 1, because when Player 2 plays A_2, Player 1's regret is 1, and when Player 2 plays B_2 her regret is 0. A rational player should minimize her regret, so that Player 1's regret is defined as $\mathsf{reg}_1 = \min_{\lambda_1} \mathsf{reg}_1(\lambda_1)$, summarizing, we get $\mathsf{reg}_1 = \min_{\lambda_1} \max_{\lambda_2} (\mathsf{c}_1(\lambda_1, \lambda_2) - \min_{\lambda_1'} \mathsf{c}_1(\lambda_1', \lambda_2))$. A symmetrical definition can be given for Player 2's regret.

Let us come back to the example. The regret attached to strategy B_1 is 1. So A_1 and B_1 are equivalent for Player 1 w.r.t. regret minimization. On the other hand, for Player 2, the regret of A_2 is 0, and the regret of B_2 is 3. So, if Player 1 assumes that Player 2 tries to minimize her regret, then she must conclude that Player 2 will play A_2. Knowing that, Player 1 recomputes her regret for each action, and in this case, the regret of action A_1 is 1 while the regret of B_1 is 0. So rational players minimizing their regret should end up playing (B_1, A_2) in this game.

Reasoning on rationality is formalized by Halpern and Pass by introducing a *delete operator* that removes strictly dominated strategies. This operator takes sets of strategies (Λ_1, Λ_2) for each player and returns $D(\Lambda_1, \Lambda_2) = (\Lambda_1', \Lambda_2')$ the strategies that minimize regret. Then $D(\Lambda_1', \Lambda_2')$ returns the strategies that minimize regret under the hypothesis that adversaries minimize their regret i.e., choose their strategies in Λ_1' and Λ_2' respectively. We iterate this operator until we reach a fixpoint, which represents the strategies minimizing the *iterated regret*.

In this paper, we consider games where the matrix is not given explicitly but defined implicitly by a game graph. In particular, we use the same definition of iterated regret as Halpern and Pass. More precisely, we consider graphs where vertices are partitioned into vertices that belong to Player 1 and vertices that belong to Player 2. Each edge is annotated by a cost for Player 1 and one for Player 2. Additionally, there are two designated sets of vertices, one that Player 1 tries to reach and the other one that Player 2 tries to reach. The game starts in the initial vertex of the graph and is played for an infinite number of rounds as follows. In each round, the player who owns the vertex on which the pebble is placed moves the pebble to an adjacent vertex using an edge of

[1] We only consider 2-player games, but our work can be easily extended to n-player games.

[2] We could have considered rewards instead of penalties, everything is symmetrical.

the graph, and a new round starts. The infinite plays generate an infinite sequence of vertices and the amount that the players have to pay are computed as follows. Player 1 pays $+\infty$ if the sequence does not reach the target set assigned to her, otherwise she pays the sum of edge costs assigned to her on the prefix up to the first visit to her target set. The amount to pay for Player 2 is defined symmetrically. Strategies in such games are functions from the set of histories of plays (sequences of visited vertices) to the set of edges (choice of moves for the pebble).

Contributions. We first consider target-weighted arenas, where the payoff function is defined for each state of the objectives. We give a PTIME algorithm to compute the regret by reduction to a min-max game (and in linear time for trees). We then consider edge-weighted arenas. Each edge is labeled by a pair of integers – one for each player –, and the payoffs are defined by the sum of the weights along the path until the first visit to an objective. We give a pseudo-PTIME algorithm to compute the regret in an edge-weighted arena, by reduction to a target-weighted arena. We then study the problem of IRM. We provide a *delete operator* that removes strictly dominated strategies. We show how to compute the effect of iterating this operator on tree arenas and on the general class of graphs where the weights are stricly positive. In the first case, we provide a quadratic time algorithm and in the second case, a pseudo-exponential time algorithm.

Related works. Regret mimimization is a popular notion to define a goal for a learning agent [7] (in the *machine learning* theory). One can consider the selection of some strategies as a learning process. IRM is however not a classical concept of machine learning, since this notion is meaningless when there is a single learning agent. Like [6], our work is thus more related to game theory than learning theory.

[9] investigates how to find a strategy that "mimimizes" the regret in extensive games with imperfect information. This kind of games can be considered as games on a probabilistic finite tree arena. We also consider finite tree arenas in Section 5.1. However in that paper, the strategies that achieve the mimimal value for regret are not computed, but a strategy with a regret less than a bound depending on the range of utilities, the number of actions and the number of rounds. IRM is not considered neither.

There are several notions of equilibria for reasoning on 2-player non-zero-sum (strategic) games, for instance *Nash equilibrium, sequential equilibrium, perfect equilibrium* - see [8] for an overview. Those equilibria formalize notions of rational behavior by defining optimality criteria for pairs of strategies. As it has be shown by Halpern and Pass for several examples like the *Centipede game* or the *Traveller's dilemma*, IRM proposes solutions that are more intuitive and natural than Nash equilibria.

2 Weighted Games and Regret

Given a cartesian product $A \times B$ of two sets, we denote by proj_i the i-th projection, $i = 1, 2$. It is naturally extended to sequence of elements of $A \times B$ by $\mathsf{proj}_i(c_1 \dots c_n) = \mathsf{proj}_i(c_1) \dots \mathsf{proj}_i(c_n)$. For all $k \in \mathbb{N}$, we let $[k] = \{0, \dots, k\}$.

Reachability Games. Turn-based two-player games are played on game arenas by two players. A (finite) *game arena* is a tuple $G = (S = S_1 \uplus S_2, s_0, T)$ where S_1, S_2 are finite disjoint sets of player positions (S_1 for Player 1 and S_2 for Player 2), $s_0 \in S$ is

1. $\mathsf{c}_i^G(\pi) = \begin{cases} +\infty & \text{if } \pi \text{ is not winning for Player } i \\ \sum_{j=0}^{\min\{k \mid \pi_k \in \mathsf{C}_i\}-1} \mu_i(\pi_j, \pi_{j+1}) & \text{otherwise} \end{cases}$ *cost of a play* π
2. $\mathsf{c}_i^G(\lambda_1, \lambda_2) = \mathsf{c}_i^G(\mathsf{Out}^G(\lambda_1, \lambda_2))$ *cost of two strategies* λ_1, λ_2.
3. $\mathsf{br}_i^G(\lambda_{-i}) = \min_{\lambda_i \in \Lambda_i(G)} \mathsf{c}_i^G(\lambda_i, \lambda_{\text{-}i})$ *best response of player* i *against* $\lambda_{\text{-}i}$
4. $\mathsf{reg}_i^G(\lambda_i, \lambda_{\text{-}i}) = \mathsf{c}_i^G(\lambda_i, \lambda_{\text{-}i}) - \mathsf{br}_i^G(\lambda_{\text{-}i})$. *regret of two strategies* λ_1, λ_2
5. $\mathsf{reg}_i^G(\lambda_i) = \max_{\lambda_{\text{-}i} \in \Lambda_{\text{-}i}(G)} \mathsf{reg}_i^G(\lambda_i, \lambda_{\text{-}i})$ *regret of a strategy* λ_i
6. $\mathsf{reg}_i^G = \min_{\lambda_i \in \Lambda_i(G)} \mathsf{reg}_i^G(\lambda_i)$. *regret of Player* i

Fig. 2. Cost, best response, and regret

the initial position, and $T \subseteq S \times S$ is the transition relation. A *finite play* on G of length n is a finite word $\pi = \pi_0\pi_1 \dots \pi_n \in S^*$ such that $\pi_0 = s_0$ and for all $i = 0, \dots, n-1$, $(\pi_i, \pi_{i+1}) \in T$. Infinite plays are defined similarly. We denote by $\mathsf{P}_f(G)$ (resp. $\mathsf{P}_\infty(G)$) the set of finite (resp. infinite) plays on G, and we let $\mathsf{P}(G) = \mathsf{P}_f(G) \cup \mathsf{P}_\infty(G)$. For any node $s \in S$, we denote by (G, s) the arena G where the initial position is s.

Let $i \in \{1, 2\}$. We let $-i = 1$ if $i = 2$ and $-i = 2$ if $i = 1$. A *strategy* $\lambda_i : \mathsf{P}_f(G) \to S \cup \{\perp\}$ for Player i is a mapping that maps any finite play π whose last position – denoted last(π) – is in S_i to $\perp$ if there is no outgoing edge from last(π), and to a position s such that $(\text{last}(\pi), s) \in T$ otherwise. The set of strategies of Player i in G is denoted by $\Lambda_i(G)$. Given a strategy $\lambda_{\text{-}i} \in \Lambda_{-i}(G)$, the *outcome* $\mathsf{Out}^G(\lambda_i, \lambda_{\text{-}i})$ is the unique play $\pi = \pi_0 \dots \pi_n \dots$ such that (i) $\pi_0 = s_0$, (ii) if π is finite, then there is no outgoing edge from last(π), and (iii) for all $0 \leq j \leq |\pi|$ and all $\kappa = 1, 2$, if $\pi_j \in S_\kappa$, then $\pi_{j+1} = \lambda_\kappa(\pi_0 \dots \pi_j)$. We also define $\mathsf{Out}^G(\lambda_i) = \{\mathsf{Out}^G(\lambda_i, \lambda_{\text{-}i}) \mid \lambda_{\text{-}i} \in \Lambda_{\text{-}i}(G)\}$. A strategy λ_i is *memoryless* if for any play $h \in \mathsf{P}_f(G)$, $\lambda_1(h)$ only depends on last(h). Thus λ_i can be seen as a function $S_i \mapsto S \cup \{\perp\}$. It is *finite-memory* if $\lambda_1(h)$ only depends on last(h) and on some state of a finite state set (see [4] for formal definitions).

A *reachability winning condition* for Player i is given by a subset of positions $\mathsf{C}_i \subseteq S$ – called the *target set* –. A play $\pi \in \mathsf{P}(G)$ is winning for Player i if some position of π is in C_i. A strategy λ_i for Player i is *winning* if all the plays of $\mathsf{Out}^G(\lambda_i)$ are winning. In this paper, we often consider two target sets $\mathsf{C}_1, \mathsf{C}_2$ for Player 1 and 2 respectively. We write $(S_1, S_2, s_0, T, \mathsf{C}_1, \mathsf{C}_2)$ to denote the game arena G extended with those target sets. Finally, let $\lambda_i \in \Lambda_i(G)$ be a winning strategy for Player i and $\lambda_{\text{-}i} \in \Lambda_{\text{-}i}(G)$. Let $\pi_0\pi_1 \dots \in \mathsf{P}(G)$ be the outcome of $(\lambda_i, \lambda_{\text{-}i})$. The *outcome* of $(\lambda_i, \lambda_{\text{-}i})$ *up to* C_i is defined by $\mathsf{Out}^{G,\mathsf{C}_i}(\lambda_i, \lambda_{\text{-}i}) = \pi_0 \dots \pi_n$ such that $n = \min\{j \mid \pi_j \in \mathsf{C}_i\}$. We extend this notation to sets of plays $\mathsf{Out}^{G,\mathsf{C}_i}(\lambda_i)$ naturally.

Weighted Games. We add weights on edges of arenas and include the target sets. A (finite) *weighted game arena* is a tuple $G = (S = S_1 \uplus S_2, s_0, T, \mu_1, \mu_2, \mathsf{C}_1, \mathsf{C}_2)$ where (S, s_0, T) is a game arena, $\mu_i : T \to \mathbb{N}$ is a weight function for Player i and C_i its target set, $i = 1, 2$. We let M_i^G be the maximal weight of Player i, i.e. $M_i^G = \max_{e \in T} \mu_i(e)$ and $M^G = \max(M_1^G, M_2^G)$.

G is a *target-weighted arena* (TWA for short) if only the edges leading to a target node are weighted by strictly positive integers, and any two edges leading to the same node carry the same weight. Formally, for all $(s, s') \in T$, if $s' \notin \mathsf{C}_i$, then $\mu_i(s, s') = 0$, otherwise for all $(s'', s') \in T$, $\mu_i(s, s') = \mu_i(s'', s')$. Thus for target-weighted arenas, we assume in the sequel that the weight functions map C_i to $\mathbb{N}$.

Plays, cost, best response and regret. Let $\pi = \pi_0\pi_1 \ldots \pi_n$ be a finite play in G. We extend the weight functions to finite plays, so that for all $i = 1, 2$, $\mu_i(\pi) = \sum_{j=0}^{n-1} \mu_i(\pi_j, \pi_{j+1})$. Let $i \in \{1, 2\}$, the *cost* $\mathsf{c}_i^G(\pi)$ of π (for Player i) is $+\infty$ if π is not winning for Player i, and the sum of the weights occuring along the edges defined by π until the first visit to a target position otherwise, see Fig. 2(1). In Fig. 2(2), we extend this notion to the cost of two strategies λ_1, λ_2 of Player 1 and 2 respectively. The *best response* of Player i to λ_{-i}, denoted by $\mathsf{br}_i^G(\lambda_{-i})$, is the least cost Player i can achieve against λ_{-i} (Fig. 2(3)). The *regret* for Player i of playing λ_i against λ_{-i} is the difference between the cost Player i pays and the best response to λ_{-i} (Fig. 2(4)). Note that $\mathsf{reg}_i^G(\lambda_i, \lambda_{-i}) \geq 0$, since $\mathsf{br}_i^G(\lambda_{-i}) \leq \mathsf{c}_i^G(\lambda_i, \lambda_{-i})$. The regret of λ_i for Player i is the maximal regret she gets for all strategies of Player $-i$ (Fig. 2(5)). Finally, the regret of Player i in G is the minimal regret she can achieve (Fig. 2(6)). We let $+\infty - (+\infty) = +\infty$.

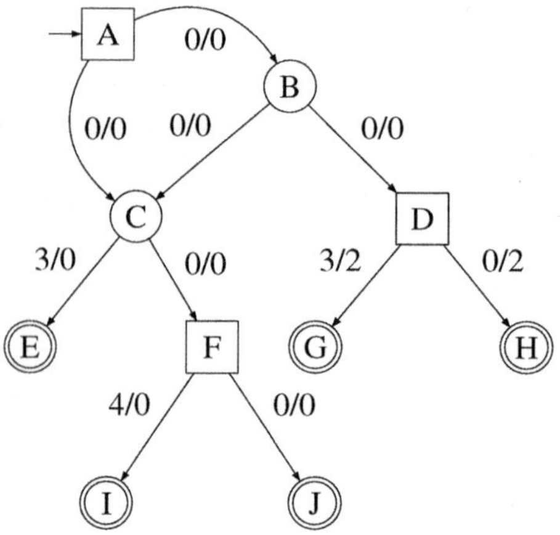

Fig. 3. Weighted Graph Arena

Proposition 1. $\mathsf{reg}_i^G < +\infty$ *iff Player* i *has a winning strategy,* $i = 1, 2$.

Proof. If Player i has no winning strategy, then for all $\lambda_i \in \Lambda_i(G)$, there is $\lambda_{-i} \in \Lambda_{-i}(G)$ s.t. $\mathsf{c}_i^G(\lambda_i, \lambda_{-i}) = +\infty$. Thus $\mathsf{reg}_i^G(\lambda_i, \lambda_{-i}) = +\infty$. Therefore $\mathsf{reg}_i^G = +\infty$. If Player i has a winning strategy λ_i, then for all $\lambda_{-i} \in \Lambda_{-i}(G)$, $\mathsf{c}_i^G(\lambda_i, \lambda_{-i}) < +\infty$ and $\mathsf{br}_i^G(\lambda_{-i}) \leq \mathsf{c}_i^G(\lambda_i, \lambda_{-i}) < +\infty$. Thus $\mathsf{reg}_i^G \leq \mathsf{reg}_i^G(\lambda_i) < +\infty$. □

Example 1. Consider the game arena G of Fig. 3. Player 1's positions are circle nodes and Player 2's positions are square nodes. The target nodes are represented by double circles. The initial node is A. The weights are given by pairs of integers for Player 1 and 2 respectively. In this example, we first compute Player 1's regret.

Let λ_1 be the memoryless strategy defined by $\lambda_1(B) = C$ and $\lambda_1(C) = E$. For all $\lambda_2 \in \Lambda_2(G)$, $\mathsf{Out}_1^G(\lambda_1, \lambda_2)$ is either ACE or $ABCE$, depending on whether Player 2 goes directly to C or passes by B. In both cases, the outcome is winning and $\mathsf{c}_1^G(\lambda_1, \lambda_2) = 3$. What is the regret of playing λ_1 for Player 1? To compute $\mathsf{reg}_G^1(\lambda_1)$, we should consider all possible strategies of Player 2, but a simple observation allows us to restrict this range. Indeed, to maximize the regret of Player 1, Player 2 should cooperate in subtrees where λ_1 prevents to go, i.e. in the subtrees rooted at D and F. Therefore we only have to consider the two following memoryless strategies λ_2 and λ_2': both λ_2 and λ_2' move from F to J and from D to H, but $\lambda_2(A) = B$ while $\lambda_2'(A) = C$.

In both cases, going to F is a best response to λ_2 and λ_2' for Player 1, i.e. $\mathsf{br}_1^G(\lambda_2) = \mathsf{br}_1^G(\lambda_2') = 0$. Therefore we get $\mathsf{reg}_1^G(\lambda_1, \lambda_2) = \mathsf{c}_1^G(\lambda_1, \lambda_2) - \mathsf{br}_1^G(\lambda_2) = 3 - 0 = 3$. Similarly $\mathsf{reg}_1^G(\lambda_1, \lambda_2') = 3$. Therefore $\mathsf{reg}_1^G(\lambda_1) = 3$.

As a matter of fact, the strategy λ_1 minimizes the regret of Player 1. Indeed, if she chooses to go from B to D, then Player 2 moves from A to B and from D to G (so that Player 1 gets a cost 3) and cooperates in the subtree rooted at C by moving from F to J. The regret of Player 1 is therefore 3. If Player 1 moves from B to C and from C to F, then Player 2 moves from A to C and from F to I (so that Player 1 gets a cost 4), and from D to H, the regret of Player 1 being therefore 4. Similarly, one can show that all other strategies of Player 1 have a regret at least 3. Therefore $\mathsf{reg}_1^G = 3$.

Note that the strategy λ_1 does not minimize the regret in the subgame defined by the subtree rooted at C. Indeed, in this subtree, Player 1 has to move from C to F, and the regret of doing this is $4 - 3 = 1$. However the regret of λ_1 in the subtree is 3. This example illustrates a situation where a strategy that minimizes the regret in the whole game does not necessarily minimize the regret in the subgames. Therefore we cannot apply a simple inductive bottom-up algorithm to compute the regret. As we will see in the next section, we first have to propagate some information in the subgames.

3 Regret Minimization on Target-Weighted Graphs

In this section, our aim is to give an algorithm to compute the regret of Player i. This is done by reduction to a min-max game, defined in the sequel. We say that we *solve* the regret minimization problem (RMP for short) if we can compute the minimal regret and a (finite representation of a) strategy that achieves this value. We first introduce the notion of games with minmax objectives. Let $G = (S = S_1 \uplus S_2, s_0, T, \mu_1, \mu_2, \mathsf{C}_1, \mathsf{C}_2)$ be a TWA and $i = 1, 2$. We let $\mathsf{minmax}_i^G = \min_{\lambda_i \in \Lambda_i(G)} \max_{\lambda_{-i} \in \Lambda_{-i}(G)} \mathsf{c}_i^G(\lambda_i, \lambda_{-i})$.

Proposition 2. *Given a* TWA $G = (S, s_0, T, \mu_1, \mu_2, \mathsf{C}_1, \mathsf{C}_2)$, $i \in \{1, 2\}$ *and* $K \in \mathbb{N}$, *one can decide whether* $\mathsf{minmax}_i^G \leq K$ *in time* $O(|S|+|T|)$. *The value* minmax_i^G *and a memoryless strategy that achieves it can be computed in time* $O(\log_2(M_i^G)(|S|+|T|))$.

Since the player have symmetric roles, wlog we can focus on computing the regret of Player 1 only. Therefore we do not consider Player 2's targets and weights. Let $G = (S = S_1 \uplus S_2, s_0, T, \mu_1, \mathsf{C}_1)$ be a TWA (assumed to be fixed from now on). Let $\lambda_1 \in \Lambda_1(G)$ be a winning strategy of Player 1 (if it exists). Player 2 can enforce Player 1 to follow one of the paths of $\mathsf{Out}^{G,\mathsf{C}_1}(\lambda_1)$ by choosing a suitable strategy. When choosing a path $\pi \in \mathsf{Out}^{G,\mathsf{C}_i}(\lambda_1)$, in order to maximize the regret of Player 1, Player 2 cooperates (i.e. she minimizes the cost) if Player 1 would have deviated from π. This leads to the notion of *best alternative* along a path. Informally, the best alternative along π is the minimal value Player 1 could have achieved if she deviated from π, assuming Player 2 cooperates. Since Player 2 can enforce one of the paths of $\mathsf{Out}^{G,\mathsf{C}_1}(\lambda_1)$, to maximize the regret of Player 1, she will choose the path π with the highest difference between $\mathsf{c}_1^G(\pi)$ and the minimal best alternative along π. As an example consider the TWA arena of Fig. 3. In this example, if Player 1 moves from C to E, then along the path ACE, the best alternative is 0. Indeed, the other alternative was to go from C to F and in this case, Player 2 would have cooperated.

We now formally define the notion of best alternative. Let $s \in S_1$. The best value that can be achieved from s by Player 1 when Player 2 cooperates is defined by:

$$\mathsf{best}_1^G(s) = \min_{\lambda_1 \in \Lambda_1(G,s)} \min_{\lambda_2 \in \Lambda_2(G,s)} \mathsf{c}_1^{(G,s)}(\lambda_1, \lambda_2)$$

Let $(s, s') \in T$. The *best alternative* of choosing s' from s for Player 1, denoted by $\mathsf{ba}_1^G(s, s')$, is defined as the minimal value she could have achieved by choosing another successor of s (assuming Player 2 cooperates). Formally:

$$\mathsf{ba}_1^G(s, s') = \begin{cases} +\infty & \text{if } s \in S_2 \\ \min_{(s,s'') \in T, s'' \neq s'} \mathsf{best}_1^G(s'') & \text{if } s \in S_1 \end{cases}$$

with $\min \varnothing = +\infty$. Finally, the best alternative of a path $\pi = s_0 s_1 \dots s_n$ is defined as $+\infty$ if $n = 0$ and as the minimal best alternative of the edges of π otherwise:

$$\mathsf{ba}_G^1(\pi) = \min_{0 \leq j < n} \mathsf{ba}_G^1(s_j, s_{j+1})$$

We first transform the graph G into a graph G' such that all the paths that lead to a node s have the same best alternative. This can be done since the number of best alternatives is bounded by $|\mathsf{C}_1|$. The construction of G' is done inductively by storing the best alternatives in the positions.

Definition 1. *The graph of best alternatives of G is the* TWA $G' = (S' = S'_1 \uplus S'_2, s'_0, T', \mu'_1, \mathsf{C}'_1)$ *where* $S'_i = S_i \times ([M_1^G] \cup \{+\infty\})$, $s'_0 = (s_0, +\infty)$, $\mathsf{C}'_1 = S'_1 \cap (C_1 \times [M_1^G])$, *for all* $(s, b) \in \mathsf{C}'_1$, $\mu'_1(s, b) = \mu_1(s)$, *and for all* $(s, b_1), (s', b'_1) \in S'$, $((s, b_1), (s', b'_1)) \in T'$ *iff* $(s, s') \in T$ *and*

$$b'_1 = \begin{cases} \min(b_1, \mathsf{ba}_1^G(s, s')) & \textit{if } s \in S_1 \\ b_1 & \textit{if } s \in S_2 \end{cases}$$

The best alternative along the paths that lead to the same node is unique. Moreover, as the number of best alternatives is bounded by $|\mathsf{C}_1|$, G' can be constructed in PTime:

Proposition 3. *For all $(s, b) \in S'$ and all finite path π in G' from $(s_0, +\infty)$ to (s, b), $\mathsf{ba}_1^{G'}(\pi) = b$. G' can be constructed in time $O\left(((|\mathsf{C}_1| + \log_2(M_1^G)) \times (|S| + |T|)\right)$.*

Since the best alternative information depends only on the paths, the paths of G and those of G' are in bijection. This bijection can be extended to strategies. In particular, we define two mappings Φ_i from $\Lambda_i(G)$ to $\Lambda_i(G')$, for all $i = 1, 2$. For all path $\pi = s_0 s_1 \dots$ in G (finite or infinite), we denote by $B(\pi)$ the path of G' defined by $(s_0, b_0)(s_1, b_1) \dots$ where $b_0 = +\infty$ and for all $j > 0$, $b_j = \mathsf{ba}_G^1(s_0 \dots s_{j-1})$. The mapping B is bijective, and its inverse corresponds to proj_1.

The mapping Φ_i maps any strategy $\lambda_i \in \Lambda_i(G)$ to a strategy $\Phi_i(\lambda_i) \in \Lambda_i(G')$ such that $\Phi_i(\lambda_i)$ behaves as λ_i on the first projection of the play and adds the best alternative information to the position. Let $h \in S'^*$ such that $\text{last}(h) \in S'_i$. Let $s = \lambda_i(\mathsf{proj}_1(h))$. Then $\Phi_i(\lambda_i)(h) = (s, \mathsf{ba}_G^1(\mathsf{proj}_1(h).s))$. The inverse mapping Φ_i^{-1} just projects the best alternative information. In particular, for all $\lambda'_i \in \Lambda_i(G')$, and all $h \in S^*$ such that $\mathsf{last}(h) \in S_i$, $\Phi_i^{-1}(\lambda_i)(h) = \mathsf{proj}_1(\lambda_i(B(h)))$. Clearly, the mappings Φ_i are bijective.

The best alternative information is crucial. This is a global information that allows us to compute the regret locally. For all $(s,b) \in \mathsf{C}'_1$, we let $\nu_1(s,b) = \mu_1(s) - \min(\mu_1(s), b)$, and extend ν_1 to strategies naturally ($\nu_1(\lambda_1,\lambda_2) = +\infty$ if λ_1 is losing).

Lemma 1. *Let $H = (S', s'_0, T', \nu^1, \mathsf{C}'_1)$ where $S', s'_0, T', \mathsf{C}'_1$ are defined in Definition 1. For all $\lambda_1 \in \Lambda_1(G)$ and all $\lambda'_1 \in \Lambda_1(G')$, the following holds:*

1. $\mathsf{reg}_1^G(\lambda_1) = \mathsf{reg}_1^{G'}(\Phi_1(\lambda_1))$ *2.* $\mathsf{reg}_1^{G'}(\lambda'_1) = \max_{\lambda'_2 \in \Lambda_2(G')} \nu_1(\lambda'_1, \lambda'_2)$ *3.* $\mathsf{reg}_1^G = \mathsf{minmax}_1^H$

Proof. 1 and 2 are in the full version of the paper. For 3, we have

$$\begin{aligned}\mathsf{reg}_1^G &= \min_{\lambda_1 \in \Lambda_1(G)} \mathsf{reg}_1^G(\lambda_1) \text{ (by definition)} = \min_{\lambda_1 \in \Lambda_1(G)} \mathsf{reg}_1^H(\Phi_1(\lambda_1)) \quad \text{(by 1)}\\ &= \min_{\lambda_1 \in \Lambda_1(H)} \mathsf{reg}_1^H(\lambda_1) \text{ (by 1)} \quad = \min_{\lambda_1 \in \Lambda_1(H)} \max_{\lambda_2 \in \Lambda_2(H)} \nu^1(\lambda_1, \lambda_2) \text{ (by 2)} \square\end{aligned}$$

As a consequence of Lemma 1, we can solve the RMP on TWAs. We first compute the graph of best alternatives and solve a minmax game. This gives us a memoryless strategy that achieves the minimal regret in the graph of best alternatives. To compute a strategy in the original graph, we apply the inverse mapping Φ_1^{-1}: this gives a finite-memory strategy whose memory is exactly the best alternative seen along the current finite play. Therefore the needed memory is bounded by the number of best alternatives, which is bounded by $|C_1|$.

Theorem 1. *The* RMP *on a* TWA $G = (S, s_0, T, \mu_1, \mathsf{C}_1)$ *can be solved in time* $O(|\mathsf{C}_1| \cdot log_2(M_1^G) \cdot (|S| + |T|))$, *where* $M_1^G = \max_{e \in T} \mu_i(e)$.

4 Regret Minimization in Edge-Weighted Graphs

In this section, we give a pseudo-polynomial time algorithm to solve the RMP in weighted arenas (with weights on edges). In a first step, we prove that if the regret is finite, the strategies minimizing the regret generates outcomes whose cost is bounded by some value which depends on the graph. This allows us to reduce the problem to the RMP in a TWA, which can then be solved by the algorithm of the previous section.

Let $G = (S = S_1 \uplus S_2, s_0, T, \mu_1, \mathsf{C}_1)$ be a weigthed game arena with objective C_1. As in the previous section, we assume that we want to minimize the regret of Player 1, so we omit the weight function and the target of Player 2.

Definition 2 (Bounded strategies). *Let $B \in \mathbb{N}$ and $\lambda_1 \in \Lambda_1(G)$. The strategy λ_1 is bounded by B if for all $\lambda_2 \in \Lambda_2(G)$, $\mathsf{c}_1^G(\lambda_1, \lambda_2) \leq B$.*

Note that a bounded strategy is necessarily winning, since by definition, the cost of some outcome is infinite iff it is loosing. The following lemma states that the winning strategies that minimize the regret of Player 1 are bounded.

Lemma 2. *For all weighted arena $G = (S, s_0, T, \mu_1, \mathsf{C}_1)$ and for any strategy $\lambda_1 \in \Lambda_1(G)$ winning in G for Player 1 that minimizes her regret, λ_1 is bounded by $2M^G|S|$.*

Proof (Sketch). If there is a winning strategy, there is a memoryless winning strategy γ_1 ([4]). Then $\mathsf{reg}_1^G \le \mathsf{reg}_1^G(\gamma_1)$. For any λ_2, $\mathsf{Out}^{G,\mathsf{C}_1}(\gamma_1, \lambda_2)$ does not contain twice the same position. Thus $\mathsf{c}_G^1(\gamma_1, \lambda_2) \le M^G|S|$. Moreover, $\mathsf{br}_1^G(\lambda_2) \le \mathsf{c}_G^1(\gamma_1, \lambda_2) \le M^G|S|$. Thus $\mathsf{reg}_1^G \le \mathsf{reg}_1^G(\gamma_1, \lambda_2) \le M^G|S|$. Let λ_1 minimizing $\mathsf{reg}_1^G(\lambda_1)$. Then $\mathsf{reg}_1^G(\lambda_1, \lambda_2) \le M^G|S|$, so $\mathsf{c}_1^G(\lambda_1, \lambda_2) \le M^G|S| + \mathsf{br}_1(\lambda_2) \le 2M^G|S|$, for any λ_2.

Let $B = 2M^G|S|$. Thanks to Lemma 2 we can reduce the RMP in a weighted arena into the RMP in a TWA. Indeed, it suffices to enrich every position of the arena with the sum of the weights occuring along the path used to reach this position. A position may be reachable by several paths, therefore it will be duplicated as many times as they are different path utilities. This may be unbounded, but Lemma 2 ensures that it is sufficient to sum the weights up to B only. This may results in a larger graph, but its size is still pseudo-polynomial (polynomial in the maximal weight and the size of the graph).

Definition 3. *Let $G = (S = S_1 \uplus S_2, s_0, T, \mu_1, \mathsf{C}_1)$ be a weigthed game arena. The graph of cost is the* TWA *$G' = (S' = S_1' \uplus S_2', s_0', T', \mu_1', \mathsf{C}_1')$ defined by: (i) $S_i' = S_i \times [B]$ and $s_0' = (s_0, 0)$; (ii) for all $(s, u), (s', u') \in S'$, $((s, u), (s', u')) \in T'$ iff $(s, s') \in T$ and $u' = u + \mu_1(s, s')$; (iii) $\mathsf{C}_1' = (\mathsf{C}_1 \times [B]) \cap S'$ and $\forall (s, u) \in \mathsf{C}_1'$, $\mu_1'(s, u) = u$.*

Lemma 3. $\mathsf{reg}_1^G = \mathsf{reg}_1^{G'}$

Proof (Sketch). We define a mapping Φ that maps the strategies of Player i in G to the strategies of Player i in G', for all $i \in \{1, 2\}$, which satisfies $\Phi(\Lambda_i(G)) = \Lambda_i(G')$ and preserves the regret of Player 1's strategies bounded by B, i.e. $\mathsf{reg}_1^G(\lambda_1) = \mathsf{reg}_1^G(\Phi(\lambda_1))$, for all $\lambda_1 \in \Lambda_1(G)$ bounded by B. □

To solve the RMP for a weighted arena G, we first construct the graph of cost G', and then apply Theorem 1, since G' is a TWA. Correctness is ensured by Lemma 3. This returns a finite-memory strategy of G' that minimizes the regret, whose memory is the best alternative seen so far. To obtain a strategy of G minimizing the regret, one applies the inverse mapping Φ^{-1}. This gives us a finite-memory strategy whose memory is the cost of the current play up to M^G and the best alternative seen so far.

Theorem 2. *The* RMP *on a weighted arena $G = (S = S_1 \uplus S_2, s_0, T, \mu_1, \mathsf{C}_1)$ can be solved in time $O\left((M^G)^2 \cdot log_2(|S| \cdot M^G) \cdot |S| \cdot |\mathsf{C}_1| \cdot (|S| + |T|)\right)$.*

5 Iterated Regret Minimization (IRM)

In this section, we show how to compute the iterated regret for tree arenas and for weighted arenas where weights are strictly positive (by reduction to a tree arena).

Let $G = (S = S_1 \uplus S_2, s_0, T, \mu_1, \mu_2, \mathsf{C}_1, \mathsf{C}_2)$ be a weighted arena. Let $i \in \{1, 2\}$, $P_i \subseteq \Lambda_i(G)$ and $P_{-i} \subseteq \Lambda_{-i}(G)$. The regret of Player i when she plays strategies of P_i and when Player $-i$ plays strategies of P_{-i} is defined by:

$$\mathsf{reg}_i^{G,P_i,P_{-i}} = \min_{\lambda_i \in P_i} \max_{\lambda_{-i} \in P_{-i}} \mathsf{c}_i^G(\lambda_i, \lambda_{-i}) - \mathsf{br}_i^{G,P_i}(\lambda_{-i})$$
$$\mathsf{br}_i^{G,P_i}(\lambda_{-i}) = \min_{\lambda_i^* \in P_i} \mathsf{c}_i^G(\lambda_i^*, \lambda_{-i})$$

For all $\lambda_i \in P_i$ and $\lambda_{-i} \in P_{-i}$, we define $\mathsf{reg}_i^{G,P_i,P_{-i}}(\lambda_i)$ and $\mathsf{reg}_i^{G,P_i,P_{-i}}(\lambda_i, \lambda_{-i})$ accordingly. We now define the strategies of rank j, which are the ones that survived j times the deletion of strictly dominated strategies. The strategies of rank 0 for Player i is $\Lambda_i(G)$. The strategies of rank 1 for both players are those which minimize their regret against strategy of rank 0. More generally, the strategies of rank j for Player i are the strategies of rank $j-1$ which minimize her regret against Player $-i$'s strategies of rank $j-1$. Formally, let $i \in \{1,2\}$, $P_1 \subseteq \Lambda_1(G)$ and all $P_2 \subseteq \Lambda_2(G)$. Then $D_i(P_1, P_2)$ is the set of strategies $\lambda_i \in P_i$ such that $\mathsf{reg}_i^{G,P_i,P_{-i}} = \mathsf{reg}_i^{G,P_i,P_{-i}}(\lambda_i)$. Then the strategies of rank j are obtained via a *delete* operator $D : 2^{\Lambda_1(G)} \times 2^{\Lambda_2(G)} \to 2^{\Lambda_1(G)} \times 2^{\Lambda_2(G)}$ such that $D(P_1, P_2) = (D_1(P_1, P_2), D_2(P_1, P_2))$. We let $D^j = D \circ \cdots \circ D$ (j times).

Definition 4 (j-th regret). *Let $j \geq 0$. The set of strategies of rank j for Player i is $P_i^j = \mathsf{proj}_i(D^j(\Lambda_1(G), \Lambda_2(G)))$. The $j+1$-th regret for Player i is defined by $\mathsf{reg}_i^{G,j+1} = \mathsf{reg}_i^{G,P_i^j,P_{-i}^j}$. In particular, $\mathsf{reg}_i^{G,1} = \mathsf{reg}_i^G$.*

Proposition 4. *Let $i \in \{1,2\}$. For all $j \geq 0$, $P_i^{j+1} \subseteq P_i^j$ and $\mathsf{reg}_i^{G,j+1} \leq \mathsf{reg}_i^{G,j}$. Moreover, there is $\star \geq 1$ such that for all $j \geq \star$, for all $i \in \{1,2\}$, $\mathsf{reg}_i^{G,j} = \mathsf{reg}_i^{G,\star}$.*

Definition 5 (iterated regret). *For all $i = 1, 2$, the iterated regret of Player i is $\mathsf{reg}_i^{G,\star}$.*

Example 2. Consider the example of Fig. 3. We already saw that the strategies that minimize Player 1's regret are $B \mapsto C \mapsto E$ and $B \mapsto D$, in which cases we have $\mathsf{reg}_1^G = \mathsf{reg}_1^{G,1} = 3$. The strategies that minimize Player 2's regret are $\lambda_2 : A \mapsto C, F \mapsto I$ and $\lambda_2' : A \mapsto C, F \mapsto J$, in which cases her regret is 0. If Player 1 knows that Player 2 plays according to λ_2 or λ_2', she can still play $C \mapsto E$ but now her regret is 0, so that $\mathsf{reg}_1^{G,\star} = 0$. Similarly, $\mathsf{reg}_2^{G,\star} = 0$.

5.1 IRM in Tree Arenas

In this section, we let $i \in \{1,2\}$ and $G = (S = S_1 \uplus S_2, s_0, T, \mu_1, \mu_2, \mathsf{C}_1, \mathsf{C}_2)$ be a finite edge-weighted tree arena. We can transform G into a target-weighted tree arena such that $\mathsf{C}_1 = \mathsf{C}_2$ (denoted by C in the sequel) is the set of leaves of the tree, if we allow the functions μ_i to take the value $+\infty$. This transformation results in a new target-weighted tree arena $G' = (S = S_1 \uplus S_2, s_0, T, \mu_1', \mu_2', \mathsf{C})$ with the same set of states and transitions as G and for all leaf $s \in \mathsf{C}$, $\mu_i'(s) = \mathsf{c}_i^{G'}(\pi)$, where π is the root-to-leaf path leading to s. The time complexity of this transformation is $O(|S|)$.

We now assume that $G = (S = S_1 \uplus S_2, s_0, T, \mu_1, \mu_2, \mathsf{C})$ is a target-weighted tree arena where C is the set of leaves. Our goal is to define a delete operator D such that $D(G)$ is a subtree of G such that for all $i = 1, 2$, $\Lambda_i(D(G))$ are the strategies of $\Lambda_i(G)$ that minimize reg_i^G. In other words, any pairs of subsets of strategies for both players in G can be represented by a subtree of G. This is possible since all the strategies in a tree arena are memoryless. A set of strategies $P_i \subseteq \Lambda_i(G)$ is therefore represented by removing from G all the edges (s, s') such that there is no strategy $\lambda_i \in P_i$ such that $\lambda_i(s) = s'$. In our case, one first computes the set of strategies that minimize regret. This is done as in Section 3 by constructing the tree of best alternatives H (but in this case with the best alternative of both players) and by solving a min-max game. From H

we delete all edges that are not compatible with a strategy that minimize the minmax value of some player. We obtain therefore a subtree $D(H)$ of H such that any strategy of H is a strategy of $D(H)$ for Player i iff it minimizes the minmax value in H for Player i. By projecting away the best alternative information in $D(H)$, we obtain a subtree $D(G)$ of G such that any Player i's strategy of G is a strategy of $D(G)$ iff it minimizes Player i's regret in G. We can iterate this process to compute the iterated regret, and we finally obtain a subtree $D^*(G)$ such that any strategy of G minimizes the iterated regret for Player i iff it is a Player i's strategy in $D^*(G)$.

Definition 6. *The tree of best alternatives of* G *is the tree* $H = (S'_1, S'_2, s'_0, T', \mu'_1, \mu'_2, \mathsf{C}')$ *where:* (i) $S' = S'_1 \uplus S'_2$ *with* $S'_i = \{(s, b_1, b_2) \mid s \in S_i, b_\kappa = \mathsf{ba}^G_\kappa(\pi_s), \kappa = 1, 2\}$ *where* π_s *is the path from the root* s_0 *to* s, (ii) $s'_0 = (s_0, +\infty, , +\infty)$, (iii) *for all* $s, s' \in S'$, $(s, s') \in T'$ *iff* $(\mathsf{proj}_1(s), \mathsf{proj}_1(s')) \in T$, (iv) $\mathsf{C}' = \{s \in S' \mid \mathsf{proj}_1(s) \in \mathsf{C}\}$, (v) *for all* $(s, b_1, b_2) \in \mathsf{C}'$, $\mu'_i(s, b_1, b_2) = \mu_i(s) - \min(\mu_i(s), b_i)$.

Note that H is isomorphic to G. There is indeed a one-to-one mapping Φ between the states of G and the states of H: for all $s \in S$, $\Phi(s)$ is the only state $s' \in S'$ of the form $s' = (s, b_1, b_2)$. Moreover, this mapping is naturally extended to strategies. Since all strategies are memoryless, any strategy $\lambda_i \in \Lambda_i(G)$ is a function $S_i \to S$. Thus, for all $s' \in S'_i$, $\Phi(\lambda_i)(s') = \Phi\left(\lambda_i(\Phi^{-1}(s'))\right)$. Without loss of generality and for a technical reason, we assume that any strategy λ_i is only defined for states $s \in S_i$ that are compatible with this strategy, i.e. if s is not reachable under λ_i then the value of λ_i does not need to be defined. The lemmas of Section 3 still hold for the tree H. In particular, for all $i \in \{1, 2\}$, $\Phi(\Lambda_i(G)) = \Lambda_i(H)$ and any strategy $\lambda_i \in \Lambda_i(G)$ minimizes reg_i^G iff $\Phi(\lambda_i)$ minimizes minmax_i^H. Moreover $\mathsf{reg}_i^G = \mathsf{minmax}_i^H$.

As in Section 3, the RMP on a tree arena can be solved by a min-max game. For all $s \in S'$, we define $\mathsf{minmax}_i^H(s) = \mathsf{minmax}_i^{(H,s)}$ and compute these values inductively by a bottom-up algorithm that runs in time $O(|S|)$. This algorithm not only allows us to compute minmax_i^H for all $i \in \{1, 2\}$, but also to compute a subtree $D(H)$ that represents all Player i's strategies that achieve this value. We actually define the operator D in two steps. First, we remove the edges $(s, s') \in T'$, such that $s \in S'_i$ and $\mathsf{minmax}_i^H(s') > \mathsf{minmax}_i^H$ for all $i = 1, 2$. We obtain a new graph H' consisting of several disconnected tree components. In particular, there are some states no longer reachable from the root s'_0. Then we keep the connected component that contains s'_0 and obtain a new tree $D(H)$. Since there is a one-to-one correspondence between the strategies minimizing the regret in G and the strategies minimizing the minmax value in H, we can define $D(G)$ by applying to $D(H)$ the isomorphism Φ^{-1}, in other words by projecting the best alternatives away, and by restoring the functions μ_i.

We obtain a new tree $D(G)$ whose Player i's strategies minimize the regret of Player i, $i = 1, 2$. We can iterate the regret computation on $D(G)$ and get the Player i's strategies that minimize the regret of rank 2 of Player i, $i = 1, 2$. We continue iteration until we get a tree G' such that $D(G') = G'$. We let $D^0(G) = G$ and $D^{j+1}(G) = D(D^j(G))$.

Proposition 5. *Let* $i \in \{1, 2\}$ *and* $j > 0$. *We have* $\mathsf{reg}_i^{G,j} = \mathsf{reg}_i^{D^{j-1}(G)}$.

Theorem 3. *Let* $G = (S = S_1 \uplus S_2, s_0, T, \mu_1, \mu_2, \mathsf{C})$ *be a tree arena. For all* $i = 1, 2$, *the iterated regret of Player* i, $\mathsf{reg}_i^{G,\star}$, *can be computed in time* $O(|S|^2)$.

Proof. There is an integer j such that $\mathsf{reg}_i^{G,\star} = \mathsf{reg}_i^{D^j(G)}$. According to the definition of $D(G)$, $j \leq |S|$ because we remove at least one edge of the tree at each step. Since $D(G)$ can be constructed in time $O(|S|)$, the whole time complexity is $O(|S|^2)$. □

5.2 IRM in Positive Weighted Arenas

A weighted arena G is said to be *positive* if all edges are weighted by strictly positive weights only. In this section, we let $G = (S = S_1 \uplus S_2, s_0, T, \mu_1, \mu_2, \mathsf{C}_1, \mathsf{C}_2)$ be a positive weighted arena. Remind that $P_i^j(G)$ is the set of strategies that minimize $\mathsf{reg}_i^{G,j}$, for all $j \geq 0$ and $i = 1, 2$. As for the regret computation in edge-weighted graphs, we define a notion of boundedness for strategies. Then the iterated regret is computed on the unfolding of the graph, up to some cost bound.

Definition 7 (j-winning and j-bounded strategies). *Let* $i \in \{1, 2\}$ *and* $\lambda_i \in \Lambda_i(G)$. *The strategy* λ_i *is* j*-winning if for all* $\lambda_{-i} \in P_{-i}^j(G)$, $\mathsf{Out}^G(\lambda_i, \lambda_{-i})$ *is winning. It is* j*-bounded by some* $B \geq 0$ *if it is* j*-winning, and for all* $\lambda_{-i} \in P_{-i}^j(G)$ *and all* $\kappa \in \{i, -i\}$, $\mu_\kappa(\mathsf{Out}^{G,\mathsf{C}_i}(\lambda_i, \lambda_{-i})) \leq B$.

Note that j-boundedness differs from boundedness as we require that the utilities of both players are bounded. We let $b^G = 6(M^G)^3|S|$. We get a similar result than the boundedness of strategies that miminize the regret of rank 1, but for any rank:

Lemma 4. *For all* $i = 1, 2$ *and all* $j \geq 0$, *all* j*-winning strategies of Player* i *which minimize the* $(j + 1)$*-th regret are* j*-bounded by* b^G.

Lemma 4 allows us to reduce the problem to the IRM in a weighted tree arena, by unfolding the graph arena G up to some maximal cost value. Lemma 4 suggests to take b^G for this maximal value. However the best responses to a strategy j-bounded by b^G are not necessarily bounded by b^G, but they are necessarily j-bounded by $b^G \cdot M^G$, since the weights are strictly positive. Therefore we let $B^G = b^G \cdot M^G$ and take B^G as the maximal value. Since the j-winning strategies are j-bounded by b^G and the best responses are j-bounded by B^G, we do not lose information by taking the unfolding up to B^G. Finally we apply Theorem 3 on the unfolding. One of the most technical result is to prove the correctness of this reduction.

Theorem 4. *The iterated regret for both players in a positive weighted arena* G *can be computed in pseudo-exponential time (exponential in* $|S|$, $|T|$ *and* M^G*).*

For all $i = 1, 2$, the procedure of Section 5.1 returns a finite-memory strategy λ_i minimizing the iterated regret in G' whose memory is the best alternatives seen so far by both players. From λ_i we can compute a finite-memory strategy in G minimizing the iterated regret of Player i, the needed memory is the best alternatives seen by both players and the current finite play up to B^G. When the cost is greater than B^G, then any move is allowed. Therefore one needs to add one more bit of memory expressing whether the cost is greater than B^G.

Conclusion. The theory of infinite qualitive non-zero sum games over graphs is still in an initial development stage. We adapted a new solution concept from strategic games to game graphs, and gave algorithms to compute the regret and iterated regret. The strategies returned by those algorithms have a finite memory. One open question is to know whether this memory is necessary. In other words, are memoryless strategies sufficient to minimize the (iterated) regret in game graphs? Another question is to determine a lower bound on the complexity of (iterated) regret minimization. Iterated regret minimization over the full class of graphs is still open. In the case of (strictly) positive arenas, the unfolding of the graph arena up to some cost bound can be seen as a finite representation of (possibly infinite) set of strategies of rank j in G. Finding such a representation is not obvious for the full class of weighted arenas, since before reaching its objective, a player can take a 0-cost loop finitely many times without affecting her minimal regret. This suggests to add *fairness* conditions on edges to compute the IR.

References

1. Brihaye, T., Bruyère, V., De Pril, J.: Equilibria in quantitative reachability games. In: International Computer Science Symposium in Russia (2010)
2. Chatterjee, K., Henzinger, T.A., Jurdzinski, M.: Games with secure equilibria. In: LICS, pp. 160–169 (2004)
3. Fisman, D., Kupferman, O., Lustig, Y.: Rational synthesis. In: Esparza, J., Majumdar, R. (eds.) TACAS 2010. LNCS, vol. 6015, pp. 190–204. Springer, Heidelberg (2010)
4. Grädel, E., Thomas, W., Wilke, T. (eds.): Automata, Logics, and Infinite Games. LNCS, vol. 2500. Springer, Heidelberg (2002)
5. Grädel, E., Ummels, M.: Solution concepts and algorithms for infinite multiplayer games. In: New Perspectives on Games and Interaction. Texts In Logic and Games, vol. 4, pp. 151–178 (2008)
6. Halpern, J.Y., Pass, R.: Iterated regret minimization: A more realistic solution concept. In: IJCAI (2009)
7. Hastie, T., Tibshirani, R., Friedman, J.H.: The Elements of Statistical Learning. Springer, Heidelberg (August 2001)
8. Osborne, M.J., Rubinstein, A.: A Course in Game Theory. MIT Press, Cambridge (1994)
9. Zinkevich, M., Johanson, M., Bowling, M., Piccione, C.: Regret minimization in games with incomplete information. In: NIPS (2007)

Properties of Visibly Pushdown Transducers*

Emmanuel Filiot[1], Jean-François Raskin[1], Pierre-Alain Reynier[2],
Frédéric Servais[1], and Jean-Marc Talbot[2]

[1] Université Libre de Bruxelles (U.L.B.)
[2] LIF, Université Aix-Marseille & CNRS

Abstract. Visibly pushdown transducers (VPTs) form a strict subclass of pushdown transducers (PTs) that extends finite state transducers with a stack. Like visibly pushdown automata, the input symbols determine the stack operations. It has been shown that visibly pushdown languages form a robust subclass of context-free languages. Along the same line, we show that word transductions defined by VPTs enjoy strong properties, in contrast to PTs. In particular, functionality is decidable in PTIME, k-valuedness is in NPTIME and equivalence of (non-deterministic) functional VPTs is EXPTIME-C. Those problems are undecidable for PTs. Output words of VPTs are not necessarily well-nested. We identify a general subclass of VPTs that produce well-nested words, which is closed by composition, and for which the type checking problem is decidable.

1 Introduction

Visibly pushdown languages (VPLs) form a robust subclass of context-free languages (CFLs) [2]. This class strictly extends the class of regular languages and still enjoys strong properties: closure under all Boolean operators and decidability of emptiness, universality, inclusion and equivalence. On the contrary, context-free languages are not closed under complement nor under intersection, moreover universality, inclusion and equivalence are all undecidable. Along the same line, we study the class of visibly pushdown transductions, a subclass of pushdown transductions, and we show that while extending regular transductions it also preserves desired properties.

Visibly pushdown automata (VPAs), that characterize VPLs, are obtained as a restriction of pushdown automata. In these automata the input symbols determine the stack operations. The input alphabet is partitioned into call, return and internal symbols: if a call is read, the automaton must push a symbol on the stack; if it reads a return, it must pop a symbol; and while reading an internal symbol, it can not touch, not even read, the stack. *Visibly pushdown transducers* (VPTs) are obtained by adding outputs to VPAs: each time the VPA reads an input symbol it also outputs a word. No restriction is imposed on the output word.

* Work supported by the projects: (i) QUASIMODO (FP7- ICT-STREP-214755), (ii) GASICS (ESF-EUROCORES LogiCCC), (iii) Moves: "Fundamental Issues in Modelling, Verification and Evolution of Software", `http://moves.ulb.ac.be`, a PAI program funded by the Federal Belgian Gouvernment, and (iv) ECSPER (ANR-JC09-472677) and SFINCS (ANR-07-SESU-012), two projects supported by the French National Research Agency.

P. Hliněný and A. Kučera (Eds.): MFCS 2010, LNCS 6281, pp. 355–367, 2010.

A transducer is k-valued if it transforms an input word into at most k (output) words. It is *functional* if it is 1-valued. The functionality problem for finite state (word) transducers has been extensively studied. The first proof of decidability was given in [19], and later in [4]. The first PTIME upper bound has been proved in [10], more generally this algorithm can be used for deciding k-valuedness. Also, this proof can be refined to get an NLOGSPACE upper bound [9]. An efficient procedure for testing functionality has been given in [3]. Those problems are undecidable for PTs.

We claim that VPTs form a rather robust model between finite-state transducers (FSTs) and pushdown transducers (PTs). Our main contribution is to show that interesting problems are decidable for VPTs: inclusion of the domain into a VPL, functionality (in PTIME), equivalence of functional transducers (EXPTIME-C), and most notably k-valuedness (in NPTIME). VPTs are not closed under composition and the type checking is undecidable. We exhibit the class of well-nested VPTs (wnVPTs), a subclass of VPTs, closed under composition and with decidable type checking (EXPTIME-C). As the output words are well-nested, this class is well-suited to model unranked tree transformations. To the best of our knowledge, this is the first class of unranked tree transducers that supports concatenation of tree sequences with decidable k-valuedness.

Visibly pushdown transducers have been first introduced in [18]. In that paper, VPTs allow for ϵ-transitions that can produce outputs and only a single letter can be produced by each transition. Using ϵ-transitions causes many interesting problems to be undecidable, such as functionality and equivalence (even of functional transducers). In contrast to [18], in this paper we consider visibly pushdown transducers where ϵ-transitions are not allowed, but where the transitions can output a word. Moreover, no visibly restriction is imposed on the output word. Therefore in the sequel we call the transducers of [18] ϵ-VPTs, and VPTs will denote the visibly pushdown transducers considered here. VPTs are exactly the so called *nested word to word transducers* of [23]. XML-DPDTs [14], equivalent to left-to-right attribute grammars, correspond to the deterministic VPTs.

Deciding equivalence of deterministic (and therefore functional) VPTs has been shown to be in PTIME [23]. However, functional VPTs can be exponentially more succinct and are strictly more expressive than deterministic VPTs. In particular, non-determinism is often needed to model functional transformations whose current production depends on some input which may be arbitrary far away from the current input. For instance, the transformation that swaps the first and the last input symbols is functional but non-determinism is needed to guess the last input. The proof of [23] is based on a reduction to the morphism equivalence problem on a context-free language, which is known to be decidable in PTIME [17]. In this paper, we show that the same reduction can be applied to prove that functionality is decidable in PTIME for VPTs, and as a consequence, equivalence of functional transducers is decidable. Moreover we extend this result by showing that the k-valuedness problem for VPTs can be reduced to the multiple morphism equivalence problem on a CFL which was proved to be decidable in [11]. Finally, we show this last problem can be decided in NPTIME.

While functionality and equivalence are decidable, the class of visibly pushdown transductions, characterized by VPTs, is not closed under composition and the type checking problem is undecidable. We identify a subclass of VPTs in which a connection is imposed between stack symbols and nesting of outputs. We call the resulting class well-nested visibly pushdown transducers (wnVPTs), as it only accepts well-nested input words and the output condition ensures that output words are also well-nested. As a subclass of VPTs, it enjoys all the good properties stated above, and in addition it is closed under composition and type checking is EXPTIME-C. This class is of particular interest for XML document transformations as well-nested words can naturally model such documents.

Relation to tree transducers. We distinguish *ranked trees* from *unranked trees*, whose nodes may have an arbitrary number of ordered children. There is a strong connection between wnVPTs and unranked tree transducers. Indeed, unranked trees over an arbitrary finite alphabet Σ can be naturally represented by well-nested words over the structured alphabet $\Sigma \times \{c\} \cup \Sigma \times \{r\}$. wnVPTs are therefore well-suited to model tree transformations. To the best of our knowledge, wnVPTs consist in the first (non-deterministic) model of unranked tree transformations (that support concatenation of tree sequences) for which k-valuedness and equivalence of functional transformations are decidable. Finite (ranked) tree transducers [5] on binary encodings of unranked trees do not support concatenation, they are incomparable to wnVPTs, as they allow for copy, which is not the case of wnVPTs, but cannot define all context-free languages as codomain, what wnVPTs can do, as they support concatenation of tree sequences. Functionality is known to be decidable in PTIME for tree transducers [20]. More generally, finite-valuedness of (and equivalence of finite-valued) tree transducers is decidable [21]. However, those results cannot be lifted to unranked trees, as unranked tree transformations have to support concatenation of tree sequences, making usual binary encodings unsuitable. Considering finite tree transducers, their ability to copy subtrees is the main concern when dealing with k-valuedness. However for wnVPTs, it is more their ability to concatenate sequences of trees which makes this problem difficult. While concatenation of tree sequences cannot be modeled by ranked tree transducers on binary encodings, it can be simulated by *macro (ranked) tree transducers* (MTTs) by using parameters [8]. This makes wnVPTs strictly less expressive than MTTs. However, equivalence is decidable for the subclass of linear size increase *deterministic* MTTs, which are equivalent to deterministic MSO tree transductions [7,6]. In particular, k-valuedness is still open for MTTs. There have been several attempts to generalize ranked tree transducers to unranked tree transducers [15,16,14], but again, k-valuedness is still open.

Organization of the paper In Section 2, we define VPTs as an extension of VPAs. In Section 3, we give basic results about properties of VPTs. We prove in Section 4 the PTIME decidability of functionality for VPTs as well as the NPTIME result for deciding k-valuedness of VPTs. Finally, well-nested transducers are introduced in Section 5.

2 Visibly Pushdown Transducers

Let Σ be a finite alphabet partitioned into two disjoint sets Σ_c and Σ_r denoting respectively the *call* and *return* alphabets[1]. We denote by Σ^* the set of (finite) words over Σ

[1] In contrast to [2], we do not consider *internal* symbols i, as they can be simulated by a (unique) call c_i followed by a (unique) return r_i. All our results extend trivially to alphabets with internal symbols. We make this assumption to simplify notations.

and by ϵ the empty word. The length of a word u is denoted by $|u|$. The set of *well-nested* words Σ^*_{wn} is the smallest subset of Σ^* such that $\epsilon \in \Sigma^*_{\mathsf{wn}}$ and for all $c \in \Sigma_c$, all $r \in \Sigma_r$, all $u, v \in \Sigma^*_{\mathsf{wn}}$, $cur \in \Sigma^*_{\mathsf{wn}}$ and $uv \in \Sigma^*_{\mathsf{wn}}$. The *height* of a well-nested word is inductively defined by $h(\epsilon) = 0$, $h(cur) = 1 + h(u)$, and $h(uv) = \max(h(u), h(v))$.

Visibly Pushdown Languages A *visibly pushdown automaton* (VPA) [2] on finite words over Σ is a tuple $A = (Q, I, F, \Gamma, \delta)$ where Q is a finite set of states, $I \subseteq Q$, respectively $F \subseteq Q$, the set of initial states, respectively final states, Γ the (finite) stack alphabet, and $\delta = \delta_c \uplus \delta_r$ where $\delta_c \subseteq Q \times \Sigma_c \times \Gamma \times Q$ are the *call transitions*, $\delta_r \subseteq Q \times \Sigma_r \times \Gamma \times Q$ are the *return transitions*.

On a call transition $(q, a, q', \gamma) \in \delta_c$, γ is pushed onto the stack and the control goes from q to q'. On a return transition $(q, \gamma, a, q') \in \delta_r$, γ is popped from the stack.

Stacks are elements of Γ^*, and we denote by $\perp$ the empty stack. A *run* of a VPA A on a word $w = a_1 \dots a_l$ is a sequence $\{(q_k, \sigma_k)\}_{0 \le k \le l}$, where q_k is the state and $\sigma_k \in \Gamma^*$ is the stack at step k, such that $q_0 \in I$, $\sigma_0 = \perp$, and for each $k < l$, we have either: (i) $(q_k, a_{k+1}, \gamma, q_{k+1}) \in \delta_c$ and $\sigma_{k+1} = \sigma_k \gamma$; or (ii) $(q_k, a_{k+1}, \gamma, q_{k+1}) \in \delta_r$, and $\sigma_k = \sigma_{k+1}\gamma$. A run is *accepting* if $q_l \in F$ and $\sigma_l = \perp$. A word w is *accepted* by A if there exists an accepting run of A over w. $L(A)$, the *language* of A, is the set of words accepted by A. A language L over Σ is a *visibly pushdown language* if there is a VPA A over Σ such that $L(A) = L$.

In contrast to [2] and to ease the notations, we do not allow return transitions on the empty stack and we accept on empty stack only. Therefore the words accepted by a VPA are well-nested (every call symbol has a matching return symbol and conversely).

Visibly Pushdown Transducers As finite-state transducers extend finite-state automata with outputs, visibly pushdown transducers extend VPAs with outputs. To simplify notations, we suppose that the output alphabet is Σ, but our results still hold for an arbitrary output alphabet.

Definition 1 (Visibly pushdown transducers). A *visibly pushdown transducer*[2] (VPT) on finite words over Σ is a tuple $T = (Q, I, F, \Gamma, \delta)$ where Q is a finite set of states, $I \subseteq Q$ is the set of initial states, $F \subseteq Q$ the set of final states, Γ is the stack alphabet, $\delta = \delta_c \uplus \delta_r$ the (finite) transition relation, with $\delta_c \subseteq Q \times \Sigma_c \times \Sigma^* \times \Gamma \times Q$, $\delta_r \subseteq Q \times \Sigma_r \times \Sigma^* \times \Gamma \times Q$.

A *configuration* of a VPT is a pair $(q, \sigma) \in Q \times \Gamma^*$. A *run* of T on a word $u = a_1 \dots a_l \in \Sigma^*$ from a configuration (q, σ) to a configuration (q', σ') is a finite sequence $\rho = \{(q_k, \sigma_k)\}_{0 \le k \le l}$ such that $q_0 = q$, $\sigma_0 = \sigma$, $q_l = q'$, $\sigma_l = \sigma'$ and for each $1 \le k \le l$, there exist $v_k \in \Sigma^*$ and $\gamma_k \in \Gamma$ such that $(q_{k-1}, a_k, v_k, \gamma_k, q_k) \in \delta_c$ and either $a_k \in \Sigma_c$ and $\sigma_k = \sigma_{k-1}\gamma_k$, or $a_k \in \Sigma_r$ and $\sigma_{k-1} = \sigma_k \gamma_k$. The word $v = v_1 \dots v_l$ is called an *output* of ρ. We write $(q, \sigma) \xrightarrow{u/v} (q', \sigma')$ when there exists a run on u from (q, σ) to (q', σ') producing v as output. The transducer T defines a binary word relation $[\![T]\!] = \{(u, v) \mid \exists q \in I, q' \in F,\ (q, \perp) \xrightarrow{u/v} (q', \perp)\}$.

[2] In contrast to [18], there is no producing ϵ-transitions (inserting transitions) but a transition may produce a word and not only a single symbol. Moreover, the images of a word may not be necessarily well-nested.

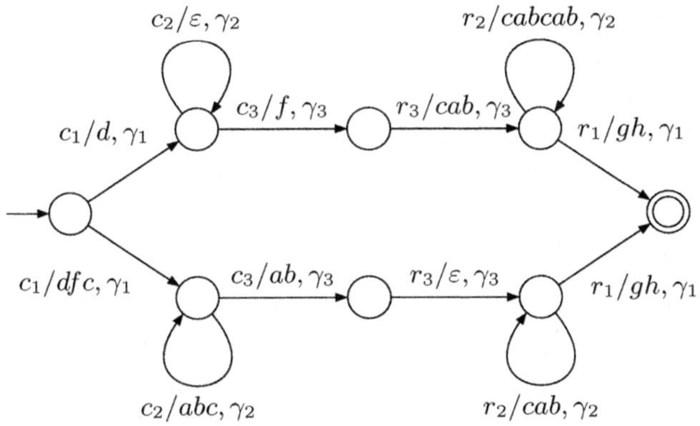

Fig. 1. A functional VPT on $\Sigma_c = \{c_1, c_2, c_3\}$ and $\Sigma_r = \{r_1, r_2, r_3\}$

The *domain* of T (denoted by *Dom*(T)), resp. the *codomain* of T (denoted by *CoDom*(T)), is the domain of $[\![T]\!]$, resp. the codomain of $[\![T]\!]$. Note that the domain of T contains only well-nested words, which is not necessarily the case of the codomain.

Consider the VPT T of Figure 1. Call (resp. return) symbols are denoted by c (resp. r). The domain of T is $\mathit{Dom}(T) = \{c_1(c_2)^n c_3 r_3 (r_2)^n r_1 \mid n \in \mathbb{N}\}$. For each word of *Dom*$(T)$, there are two accepting runs, corresponding respectively to the upper and lower part of T. For instance, when reading c_1, it pushes γ_1 and produces either d (upper part) or *dfc* (lower part). By following the upper part (resp. lower part), it produces words of the form $dfcab(cabcab)^n gh$ (resp. $dfc(abc)^n ab(cab)^n gh$). Therefore T is functional.

3 Properties of VPTs

In this section, we present results about expressiveness of VPTs and decision problems. We let VPLs, resp. CFLs, denote the class of visibly pushdown languages, resp. context-free languages, over Σ.

Proposition 1 (Domain and codomain). *Let T be a* VPT, *let L be a* VPL. *The domain Dom(T) of T is a* VPL *and the language $T(L)$ is a* CFL. *Moreover, for any* CFL *L' over Σ, there exists a* VPT *whose codomain CoDom(T) is L'. All these constructions can effectively be done in polynomial time.*

Proof (Sketch). By projecting the transitions of a VPT T on the input (resp. on the output), we obtain a VPA (resp. a pushdown automaton) which defines *Dom*(T) (resp. *CoDom*(T)). As a consequence, given $L \in$ VPL, by restricting *Dom*(T) to L which can be done by a classical product construction, we obtain $T(L)$ is a CFL. To produce as output a CFL L' on alphabet Σ, it is already known [2] that there exists a VPL L'' on a structured alphabet $\hat{\Sigma}$ and a renaming $\pi : \hat{\Sigma} \to \Sigma$ such that $\pi(L'') = L'$. The VPT implements π. □

As a consequence of Proposition 1 and of the fact that language inclusion for VPAs is EXPTIME-C, we can test whether a VPL is included in the domain of a given VPT and conversely. This is of particular interest for XML transformations as it amounts to decide if any document valid for an input XML schema will be transformed. This is undecidable for ϵ-VPTs and PTs.

Thanks to non-determinism, transductions defined by VPTs are closed under *union*. This is not the case however for *composition* and *inverse*. Non-closure under composition can be simply proved by using Proposition 1 and by producing two VPTs whose composition transforms a VPL into a non CFL language. More formally, let $\Sigma = \{c_1, r_1, c_2, r_2, c_3, r_3\}$ be an alphabet where c_i's are call symbols and r_i's are return symbols. We let $l_i = c_i r_i$ for $i = 1, 2, 3$. First consider the following VPL language: $L_1 = (c_1)^n\ (r_2)^n\ (l_3)^*$. We can easily construct a VPT that transforms L_1 into the language $L_2 = (c_1)^n\ (l_2)^n\ (r_3)^*$. Applying the identity transducer on L_2 produces the non CFL language $L_3 = (c_1)^n\ (l_2)^n\ (r_3)^n$. This identity transducer has a domain which is a VPL and thus it extracts from L_2 the well-nested words which form the non CFL set L_3. Non closure under inverse is a consequence of the fact that, for any VPT T, for any word w, $T(w)$ is a finite set while a word w can be the image of an infinite number of input words. Finally, note that, as in the case of FSTs, VPTs are not closed under *intersection* (easy coding of PCP).

The following problem is known as the *translation membership problem* [13].

Proposition 2 (Translation Membership). *Let T be a* VPT *and $(u, v) \in \Sigma^* \times \Sigma^*$, the problem of deciding whether $(u, v) \in [\![T]\!]$ is in* PTIME.

Proof. We can first restrict T to a transducer $T|_u$ such that $Dom(T|_u) = \{u\}$ and $T|_u(u) = T(u)$. By Proposition 1, membership in $CoDom(T|_u)$ can be tested in PTIME. □

Theorem 1 (Type Checking). *Given a* VPT *T and two* VPAs *A_1, A_2, it is undecidable if $T(L(A_1)) \subseteq L(A_2)$.*

Proof. Given an instance $(u_1, v_1), (u_2, v_2), \ldots, (u_n, v_n)$ of PCP defined on the finite alphabet Σ, we associate with this instance a CFL and a VPL language defined on the alphabet $\Sigma_c = \Sigma$ and $\Sigma_r = \{1 \ldots n\}$. For all j, we let $l_j = |u_j|$. The CFL language is $L_1 = \{v_{i_1} \ldots v_{i_k} \#(i_k)^{l_k} \ldots (i_1)^{l_1} \mid i_1, \ldots, i_k \in \Sigma_r\}$. The VPL language is $L_2 = \{u_{i_1} \ldots u_{i_k} \#(i_k)^{l_k} \ldots (i_1)^{l_1} \mid i_1, \ldots, i_k \in \Sigma_r\}$. Clearly the PCP instance is negative if and only if L_1 is included in $\overline{L_2}$. By Proposition 1, there exists a VPT T whose image is L_1 and by definition, there exists a VPA that accepts $\overline{L_2}$, because VPAs are closed under complementation. □

4 On k-Valuedness of VPTs

Let $k \in \mathbb{N}$. A VPT T is *k-valued* if for all $u \in \Sigma^*$, $|\{v \mid (u, v) \in [\![T]\!]\}| \leq k$. T is *functional* if it is 1-valued. Two VPTs T_1, T_2 are *equivalent* if $[\![T_1]\!] = [\![T_2]\!]$. In this section, we prove that deciding if a VPT is k-valued is decidable in NPTIME (for a fixed k), and in PTIME for $k = 1$. The proof is done via a reduction to the *multiple morphism equivalence problem* on a context-free language, which was proved to be decidable in [11]. This reduction extends the one of [23], which was used to prove the decidability of equivalence of deterministic (and therefore functional) VPTs ($k = 1$). By using a recent result of [24] on the complexity of constructing an existential Presburger formula for the Parikh image of a pushdown automaton, we give an upper bound for the multiple morphism equivalence problem. When there is only one pair of morphisms, this problem is known to be in PTIME [17].

Let Σ_1, Σ_2 be two finite alphabets. A *morphism* is a mapping $\Phi : \Sigma_1^* \to \Sigma_2^*$ such that $\Phi(\epsilon) = \epsilon$ and for all $u, v \in \Sigma_1^*$, $\Phi(uv) = \Phi(u)\Phi(v)$. A morphism can be finitely represented by its restriction on Σ_1, i.e. by the set of pairs $(a, \Phi(a))$ for all $a \in \Sigma_1$. Therefore its size is $|\Sigma| + \Sigma_{a\in\Sigma}|\Phi(a)|$.

Definition 2 (Multiple Morphism Equivalence on Context-Free Languages). *Given ℓ pairs of morphisms $(\Phi_1, \Psi_1), \ldots, (\Phi_\ell, \Psi_\ell)$ from Σ_1^* to Σ_2^* and a context free language L on Σ_1, $(\Phi_1, \Psi_1), \ldots, (\Phi_\ell, \Psi_\ell)$ are equivalent on L if for all $u \in L$, there exists i such that $\Phi_i(u) = \Psi_i(u)$.*

The next result was proved to be decidable in [11] on any class of languages whose Parikh images are effectively semi-linear. In the case of context-free languages, we show that it can be decided in NPTIME. The main part of our proof is to show that, for a fixed k, the emptiness of one-reversal pushdown machine with k counters is in NPTIME.

Theorem 2. *Let ℓ be fixed. Given ℓ pairs of morphisms and a pushdown automaton A, testing whether they are equivalent on $L(A)$ can be done in* NPTIME. *It is in* PTIME *if $\ell = 1$ and if the context-free language is given by a grammar in Chomsky normal form (Plandowski [17]).*

Proof. In order to prove this theorem, we briefly recall the procedure of [11] in the particular case of pushdown machines. It relies on the emptiness problem of reversal-bounded pushdown machines with a fixed number of counters. Let $k, m \in \mathbb{N}$, *an m-reversal pushdown machine with k counters (m-k-RBPM)* on an alphabet Σ is a pushdown automaton on Σ augmented with k counters. Each counter can be incremented or decremented by one and tested for zero, but the number of alternations between a nondecreasing and a non-increasing mode is bounded by m in any computation. The emptiness problem for such machines is decidable [12]. In order to decide the morphism equivalence problem of ℓ pairs of morphisms on a CFL L, the idea is to construct an 1-2ℓ-RBPM that accepts the language $L' = \{w \in L \mid \Phi_i(w) \neq \Psi_i(w) \text{ for all } i\}$. Clearly, $L' = \varnothing$ iff the morphisms are equivalent on L. Let A be the pushdown automaton that accepts L. We construct a pushdown automaton A' augmented with 2ℓ counters $c_{11}, c_{12}, \ldots, c_{\ell 1}, c_{\ell 2}$ that simulates A on the input word and counts the lengths of the outputs by the 2ℓ morphisms. For all $i \in \{1, \ldots, \ell\}$, A' guesses some position p_i where $\Phi_i(w)$ and $\Psi_i(w)$ differ: it increments in parallel (with ϵ-transitions) the counters c_{i1} and c_{i2} and non-deterministically decides to stop incrementing after p_i steps. Then when reading a letter $a \in \Sigma_1$, the two counters c_{i1} and c_{i2} are decremented by $|\Phi_i(a)|$ and $|\Psi_i(a)|$ respectively (by possibly several transitions as the counters can be incremented by at most one at a time). When one of the counter reaches zero A' stores the letter associated with the position (in the control state). At the end of the computation, for all $i \in \{1, \ldots, \ell\}$, one has to check that the two letters associated with the position p_i in $\Phi_i(w)$ and $\Psi_i(w)$ are different. If n is the number of states of A and m is the maximal length of an image of a letter $a \in \Sigma_1$ by the 2ℓ morphisms, then A' has $O(n \cdot m \cdot |\Sigma_2|^{2\ell})$ states, because for all 2ℓ counters one has to store the letters at the positions represented by the counter values. This is polynomial as ℓ is fixed. Note that the resulting machine is 1-reversal bounded (counters are first set to zero and are incremented up to a position in the output word, and then are decremented to zero).

We now show that the emptiness of a one-reversal pushdown machine A with k counters on an alphabet Σ is in NPTIME. Wlog, we assume that each counter starts and ends with zero value, which is the case in the previous reduction. The NPTIME upper bound remains true without this assumption. For this, we recall the construction of [11] for testing emptiness of reversal-bounded machines with counters. The idea is to construct a semi-linear set for the Parikh image of A.[3] The emptiness of A then reduces to the emptiness of its Parikh image. Following [11], one extends the alphabet Σ with $3k$ letters $+_j, -_j, s_j \notin \Sigma$ intended to simulate the increasing, decreasing transitions of the j-th counter, and the transitions that do not change the j-th counter (skip). We denote by Σ_+ this alphabet. We construct a pushdown automaton B on Σ_+ that simulates A. When reading a letter $a \in \Sigma$, B tries to apply a transition of A, and passes into a mode in which it verifies that the next letters correspond to the increasing, decreasing or skip actions on the counters of the transition. Moreover, since A is 1-reversal bounded, B has to ensure that each counter does at most one reversal. The language of B is the set of words of the form $w = a_1 t_1 a_2 t_2 \dots a_n t_n$ where $a_i \in \Sigma$ and each t_i is a word of the form $b_1^i \dots b_k^i$ where $b_j^i \in \{+_j, -_j, s_j\}$, $j \in \{1, \dots, k\}$. Moreover, we require that (i) there exists a run of B on $a_1 \dots a_n$ ending up in a final state such that the counter actions of the transitions corresponds to $t_1 \dots t_n$ (ii) for all $j \in \{1, \dots, k\}$, $b_j^1 \dots b_j^n \in \{+_j, s_j\}^*\{-_j, s_j\}^*$ (one reversal). Let $\psi(w) = a_1 \dots a_n$ and $\psi_j(w) = b_j^1 \dots b_j^n$ for all $j \in \{1, \dots, k\}$. Condition (i) is enforced by a simple simulation of A, and condition (ii) is enforced by adding vectors of $\{+, -\}^k$ to the control states indicating whether the j-th counter is in increasing or decreasing mode. Note that $L(A) \subseteq \psi(L(B))$, but this inclusion may be strict, as we do not require that the counters end up in a zero value. More formally, we have $L(A) = \bigcap_{j=1}^k \{\psi(w) \mid w \in L(B) \text{ and } \psi_j(w) \in s_j^*(+_j.s_j^*)^\ell(-_j.s_j^*)^\ell, \ell \geq 0\}$.

As $L(B)$ is a context-free language, it is known by Parikh's theorem that the Parikh image of $L(B)$ is semi-linear. Therefore there exists an existential Presburger formula ϕ with $|\Sigma| + 3k$ free variables $(x_a)_{a \in \Sigma}$ and $(x_{+_j}, x_{-_j}, x_{s_j})_{j \in \{1,\dots,k\}}$ which defines the Parikh image of $L(B)$. Moreover, this formula can be constructed in time $O(|B|)$ [24]. Finally, the formula $\exists x_{+_1} \exists x_{-_1} \exists x_{s_1} \dots \exists x_{+_k} \exists x_{-_k} \exists x_{s_k} \phi \wedge \bigwedge_{j=1}^k x_{+_j} = x_{-_j}$ defines exactly the Parikh image of $L(A)$. Since B can be constructed in $O(|A| \cdot 2^k)$ (which is polynomial as k is fixed) and the satisfiability of existential Presburger formulas is in NP [24], one gets an NP algorithm to test the emptiness of A. We can conclude the proof by combining this result to the reduction of the multiple morphism equivalence problem described in the first part of the proof. □

[3] The Parikh image of a language $L \subseteq \Sigma^*$ over an ordered alphabet $\Sigma = \{a_1, \dots, a_n\}$ is the set $\{(\#_{a_1}(u), \dots, \#_{a_n}(u)) \mid u \in L\}$ where $\#_{a_i}(u)$ is the number of occurences of a_i in u.

Following ideas introduced in [3] for deciding functionality of FSTs, we define a notion of product for the class of VPTs. The k-power of T simulates k parallel executions on the same input. Note that this construction is possible for VPTs (but not for general PTs) because two runs along the same input have necessarily the same stack behavior. Let $T = (Q, I, F, \Gamma, \delta)$ be a VPT and O_T the set of outputs words occurring in the transitions of T, i.e. $O_T = \{u \mid \exists (p \xrightarrow{a/u} q) \in \delta\}$. As this set is finite, it can be regarded as an alphabet. The k-power of T is a VPT from words over Σ to words over $(O_T)^k$ defined as follows:

Definition 3 (k-Power). *The k-power of T, denoted T^k, is the* VPT *defined from Σ to $(O_T)^k$ by $T^k = (Q^k, I^k, F^k, \Gamma^k, \delta^k)$ where the transition relation $\delta^k = \delta_c^k \uplus \delta_r^k$ is defined for all $\alpha \in \{c, r\}$ and all $a \in \Sigma_\alpha$ by:*

$$(q_1, \dots, q_k) \xrightarrow{a|(u_1,\dots,u_k),(\gamma_1,\dots,\gamma_k)} (q_1', \dots, q_k') \in \delta_\alpha^k \text{ iff } q_i \xrightarrow{a|u_i,\gamma_i} q_i' \in \delta_\alpha \quad \forall 1 \leq i \leq k$$

For all $k \geq 0$, we define the morphisms $\Phi_1, \dots, \Phi_k$ as follows:

$$\begin{aligned} \Phi_i : (O_T)^k &\to \Sigma^* \\ (u_1, \dots, u_k) &\mapsto u_i \end{aligned}$$

Clearly, we obtain the following equivalence:

Proposition 3. *T is k-valued iff $(\Phi_i, \Phi_j)_{1 \leq i \neq j \leq k+1}$ are equivalent on CoDom(T^{k+1}).*

By Proposition 1 the language *CoDom*(T^k) is a context-free language. By Theorem 2, as *CoDom*(T^k) is represented by an automaton of polynomial size if k is fixed, we get:

Theorem 3 (k-valuedness). *Let $k \geq 0$ be fixed. The problem of deciding whether a* VPT *is k-valued is in* NPTIME. *It is in* PTIME *if $k = 1$.*

To get the PTIME bound when $k = 1$, one can construct a context-free grammar G_T in Chomsky normal form whose language is exactly the codomain of T^2.

Given two functional VPTs, they are equivalent iff their union is functional and they have the same domains. The domains being VPLs, testing their equivalence is EXPTIME-C. Therefore as a consequence of Theorem 3, we have:

Theorem 4 (Equivalence). *Testing equivalence of functional* VPTs *is* EXPTIME-C.

We end this section with a result on k-ambiguity of VPTs. A VPT is k-ambiguous if its underlying VPA is k-ambiguous, i.e. for each input word there are at most k accepting runs. The notion of k-ambiguity is stronger than k-valuedness. k-ambiguity can be tested in PTIME for tree automata. The standard construction to obtain a tree automaton (top-down or bottom-up) equivalent to a given VPA preserves the number of accepting runs, however it can yield an exponential blowup [1]. Therefore the PTIME bound cannot be obtained directly from this translation. For a given k, a PTIME construction to test k-ambiguity for VPAs can be obtained with a straightforward generalization of the construction for finite state automata. Basically, one constructs a VPA that simulates $k+1$ runs of the original VPA (this is possible because the stack are synchronized), and records which of these runs are different. It will accept any word that is accepted by the original VPA with $k+1$ different runs. If k is fixed, this construction can be done in PTIME, moreover testing emptiness of VPAs is in PTIME.

Proposition 4 **(k-Ambiguity).** *Let $k \in \mathbb{N}$ be fixed. Given a* VPA *A, resp. a* VPT *T, the problem of deciding whether A, resp. T, is k-ambiguous is in* PTIME.

5 Well-Nested VPTs

We have seen that VPTs are not closed under composition and their type checking problem is undecidable. In this section we introduce a natural subclass of VPTs that is closed for composition and for which the type checking is decidable.

The undecidability of the type checking is a consequence of the fact that the stack of the transducers and the stack of the output VPA are not synchronized, because no (visibly) restriction is imposed on the output words. Similarly, non-closure under composition is a consequence of the fact that the stack of both VPTs are not synchronized. To overcome those problems we introduce a restriction between the stack symbols and the output words.

Definition 4. *A* VPT *$T = (Q, I, F, \Gamma, \delta)$ is* well-nested *(*wnVPT*) if for all $(q_1, c, u, \gamma, q_1') \in \delta_c$ and $(q_2, r, v, \gamma, q_2') \in \delta_r$, we have $uv \in \Sigma^*_{\mathsf{wn}}$.*

This restriction ensures that all output words are well-nested.

Lemma 1. *For all* wnVPTs *T and all words $w \in \Sigma^*_{\mathsf{wn}}$, $T(w) \subseteq \Sigma^*_{\mathsf{wn}}$.*

We now show that this class of transducers is closed under composition and has a decidable type checking problem.

Proposition 5 **(Closure properties).** *The class of* wnVPTs *is effectively closed under union and composition.*

Proof (Sketch).
We first need an additional notion in order to present the construction of the composition of two such transducers. A word is *return-matched* (resp. *call-matched*) if there is no unmatched returns (resp. calls). Let $\mathsf{m}(w)$ be equal to the number of unmatched returns (resp. unmatched calls) if w is call-matched (resp. return-matched).

It is easy to show that a VPT $T = (Q, I, F, \Gamma, \delta)$ is well-nested iff (i) for all $(q, \alpha, w, \gamma, q') \in \delta_c$ (resp. δ_r), the word w is return-matched (resp. call-matched), and (ii) there exists a function $\mathsf{val} : \Gamma \to \mathbb{N}$ (called a *valuation*) such that for all $(q, \alpha, w, \gamma, q') \in \delta$, we have $\mathsf{val}(\gamma) = \mathsf{m}(w)$. This valuation is unique and can be computed in linear time.

Let $T_i = (Q_i, I_i, F_i, \Gamma_i, \delta_i)$, $i \in \{1, 2\}$, be two wnVPTs, and val_i their associated valuation. We define their composition T as the tuple $(Q_1 \times Q_2, I_1 \times I_2, F_1 \times F_2, \Gamma, \delta, \mathsf{val})$. Intuitively, it is a synchronized product in which the synchronization is not letter to letter, but is based on mappings val_i. More precisely, the stack alphabet Γ of T is defined as the finite set $\{(\gamma_1, \sigma_2) \in \Gamma_1 \times \Gamma_2^* \mid \mathsf{val}_1(\gamma_1) = |\sigma_2|\}$. The valuation val is defined by $\mathsf{val}(\gamma_1, \sigma_2) = \mathsf{val}_2(\sigma_2)$ where the extension of val_2 to Γ_2^* is defined as follows: if $\sigma_2 = \gamma_{2,1}\gamma_{2,2} \dots \gamma_{2,n}$ then $\mathsf{val}_2(\sigma_2) = \sum_{i=1}^{n} \mathsf{val}_2(\gamma_{2,i})$. Call transitions are defined, for $c \in \Sigma_c$, by $(q_1, q_2) \xrightarrow{c/w, (\gamma_1, \sigma_2)} (q_1', q_2') \in \delta_c$ if and only if there exists

Table 1. Decision problems and closure properties

	Functionality / k-valuedness	Equivalence of functional	Type checking (against VPL)	$\cup$	$\circ$
FST	NL/P	PSPACE-C	EXP-C	Yes	Yes
dVPT	-	P [23]	Undec	No	No
VPT	**P/NP**	**Exp-c**	**Undec**	**Yes**	**No**
wnVPT	**P/NP**	**Exp-c**	**Exp-c**	**Yes**	**Yes**
ϵ-VPT [18]	Undec	Undec	EXP-C [18]	Yes	No
dPT	-	Dec[22]	Undec	No	No
PT	Undec	Undec	Undec	Yes	No

$v \in \Sigma^*$ such that $q_1 \xrightarrow{c/v,\gamma_1} q'_1 \in \delta^1_c$, and $(q_2, \perp) \xrightarrow{v/w} (q'_2, \sigma_2)$ is a run in the transducer T_2. Note that as T_1 is well-nested, we have $\mathsf{val}_1(\gamma_1) = \mathsf{m}(v)$, and then, as T_2 is a VPT, $\mathsf{val}_1(\gamma_1) = |\sigma_2|$. Return transitions are defined similarly and one can verify that T is a wnVPT and that the construction is correct. □

Theorem 5 (Type Checking). *Given a* wnVPT T, *two* VPAs A_1, A_2, *the problem of deciding if* $T(L(A_1)) \subseteq L(A_2)$ *is* EXPTIME-C. *It is in* PTIME *if* A_2 *is deterministic.*

Proof. For the EXPTIME-HARD part, first note that we can construct a wnVPT T_{id} whose domain is the set of well-nested words on the structured alphabet Σ and whose relation is the identity relation. Given any VPA A_1, A_2, we have that $T_{id}(L(A_1)) \subseteq L(A_2)$ if and only if $L(A_1) \subseteq L(A_2)$. This later problem is EXPTIME-C [2].

To prove it is in EXPTIME, we consider the wnVPT T_2 whose domain is $L(A_2)$ and whose relation is the identity relation. As wnVPTs are closed under composition, we can construct a wnVPT T' such that $T' = T \circ T_2$. Then we can note that $Dom(T') = T^{-1}(L(A_2))$. As $T(L(A_1)) \subseteq L(A_2)$ if and only if $L(A_1) \subseteq T^{-1}(L(A_2))$ and as all those transducers and automata can be constructed in polynomial time, we conclude that we can decide our problem in EXPTIME by checking the former inclusion using the algorithm for language inclusion between VPA. □

6 Conclusion

Table 1 summarizes the known results on several classes of word transducers. The results of this paper are in bold face. PTs denotes the class of pushdown transducers, and deterministic classes are denoted with a preceding d. Undecidability of ambiguity and functionality for PTs is well-known and can for example be proved by reduction of the emptiness problem for the intersection of two CFLs. Undecidability of the equivalence problem for two functional PTs is a direct consequence of the undecidability of the equivalence problem for CFLs. Undecidability of these problems for ϵ-VPTs can be proved in the exact same way since we can embed any CFL into the domain of such a transducer [18]. The undecidability of type checking for dPTs and PTs against VPLs can be proved as in Theorem 1. Finally, note that for all classes where equivalence of functional transducers is decidable, the complexity depends on the complexity of testing equivalence of their domains.

As future works, we would like to investigate several problems. The first problem is the *sequentiality* problem for VPTs [3]. In particular, this problem asks whether a given VPT is equivalent to an *input-deterministic* VPT. Input-determinism is relevant to XML streaming transformations, as very large documents have to be processed on-the-fly without storing the whole document in memory. A second problem is to decide if two k-valued VPTs are equivalent.

Acknowledgments. We are grateful to the referees for their valuable comments and relevant questions, and we warmly thank Oscar H. Ibarra for pointing out some references.

References

1. Alur, R.: Marrying words and trees. In: PODS. LNCS, vol. 5140, pp. 233–242. Springer, Heidelberg (2007)
2. Alur, R., Madhusudan, P.: Visibly pushdown languages. In: STOC, pp. 202–211 (2004)
3. Béal, M.-P., Carton, O., Prieur, C., Sakarovitch, J.: Squaring transducers: an efficient procedure for deciding functionality and sequentiality. TCS 292(1), 45–63 (2003)
4. Blattner, M., Head, T.: Single-valued a-transducers. JCSS 15(3), 310–327 (1977)
5. Comon, H., Dauchet, M., Gilleron, R., Löding, C., Jacquemard, F., Lugiez, D., Tison, S., Tommasi, M.: Tree automata techniques and applications (2007), `http://www.grappa.univ-lille3.fr/tata`
6. Engelfriet, J., Maneth, S.: Macro tree translations of linear size increase are MSO definable. SICOMP 32, 950–1006 (2003)
7. Engelfriet, J., Maneth, S.: The equivalence problem for deterministic MSO tree transducers is decidable. IPL 100(5), 206–212 (2006)
8. Engelfriet, J., Vogler, H.: Macro tree transducers. JCSS 31(1), 71–146 (1985)
9. Filiot, E., Raskin, J.-F., Reynier, P.-A., Servais, F., Talbot, J.-M.: On functionality of visibly pushdown transducers. CoRR, abs/1002.1443 (2010)
10. Gurari, E.M., Ibarra, O.H.: A note on finite-valued and finitely ambiguous transducers. MST 16 (1983)
11. Harju, T., Ibarra, O.H., Karhumaki, J., Salomaa, A.: Some decision problems concerning semilinearity and commutation. JCSS 65 (2002)
12. Ibarra, O.H.: Reversal-bounded multicounter machines and their decision problems. JACM 25(1), 116–133 (1978)
13. Inaba, K., Maneth, S.: The complexity of translation membership for macro tree transducers. CoRR, abs/0910.2315 (2009)
14. Koch, C., Scherzinger, S.: Attribute grammars for scalable query processing on XML streams. VLDB 16(3), 317–342 (2007)
15. Maneth, S., Neven, F.: Structured document transformations based on XSL. In: Connor, R.C.H., Mendelzon, A.O. (eds.) DBPL 1999. LNCS, vol. 1949, pp. 80–98. Springer, Heidelberg (2000)
16. Perst, T., Seidl, H.: Macro forest transducers. IPL 89(3), 141–149 (2004)
17. Plandowski, W.: Testing equivalence of morphisms on context-free languages. In: van Leeuwen, J. (ed.) ESA 1994. LNCS, vol. 855, pp. 460–470. Springer, Heidelberg (1994)
18. Raskin, J.-F., Servais, F.: Visibly pushdown transducers. In: Aceto, L., Damgård, I., Goldberg, L.A., Halldórsson, M.M., Ingólfsdóttir, A., Walukiewicz, I. (eds.) ICALP 2008, Part II. LNCS, vol. 5126, pp. 386–397. Springer, Heidelberg (2008)

19. Schützenberger, M.P.: Sur les relations rationnelles. In: Automata Theory and Formal Languages. LNCS, vol. 33, pp. 209–213. Springer, Heidelberg (1975)
20. Seidl, H.: Single-valuedness of tree transducers is decidable in polynomial time. TCS 106(1), 135–181 (1992)
21. Seidl, H.: Equivalence of finite-valued tree transducers is decidable. MST 27(4), 285–346 (1994)
22. Sénizergues, G.: T(A) = T(B)? In: Wiedermann, J., Van Emde Boas, P., Nielsen, M. (eds.) ICALP 1999. LNCS, vol. 1644, pp. 665–675. Springer, Heidelberg (1999)
23. Staworko, S., Laurence, G., Lemay, A., Niehren, J.: Equivalence of deterministic nested word to word transducers. In: Gębala, M. (ed.) FCT 2009. LNCS, vol. 5699, pp. 310–322. Springer, Heidelberg (2009)
24. Verma, K.N., Seidl, H., Schwentick, T.: On the complexity of equational horn clauses. In: Nieuwenhuis, R. (ed.) CADE 2005. LNCS (LNAI), vol. 3632, pp. 337–352. Springer, Heidelberg (2005)

Second-Order Algebraic Theories
(Extended Abstract)

Marcelo Fiore and Ola Mahmoud

University of Cambridge, Computer Laboratory

Abstract. Fiore and Hur [10] recently introduced a conservative extension of universal algebra and equational logic from first to second order. *Second-order universal algebra* and *second-order equational logic* respectively provide a model theory and a formal deductive system for languages with variable binding and parameterised metavariables. This work completes the foundations of the subject from the viewpoint of categorical algebra. Specifically, the paper introduces the notion of *second-order algebraic theory* and develops its basic theory. Two categorical equivalences are established: at the syntactic level, that of second-order equational presentations and second-order algebraic theories; at the semantic level, that of second-order algebras and second-order functorial models. Our development includes a mathematical definition of syntactic translation between second-order equational presentations. This gives the first formalisation of notions such as encodings and transforms in the context of languages with variable binding.

1 Introduction

Algebra started with the study of a few sample algebraic structures: groups, rings, lattices, *etc.* Based on these, Birkhoff [3] laid out the foundations of a general unifying theory, now known as universal algebra.

Birkhoff's formalisation of the notion of algebra starts with the introduction of equational presentations. These constitute the syntactic foundations of the subject. Algebras are then the semantics or model theory, and play a crucial role in establishing the logical foundations. Indeed, Birkhoff introduced equational logic as a sound and complete formal deductive system for reasoning about algebraic structure.

The investigation of algebraic structure was further enriched by the advent of category theory, with the fundamental work of Lawvere on algebraic theories [18] and of Linton on finitary monads [17]. These approaches give a presentation-independent treatment of the subject. Algebraic theories correspond to the syntactic line of development; monads to the semantic one (see *e.g.* [15]).

We contend that it is only by looking at algebraic structure from all of the above perspectives, and the ways in which they interact, that the subject is properly understood. In the context of computer science, for instance, consider that: (*i*) initial-algebra semantics provides canonical compositional interpretations [14]; (*ii*) free constructions amount to abstract syntax [19], that is amenable to proofs by structural induction and definitions by structural recursion [4];

P. Hliněný and A. Kučera (Eds.): MFCS 2010, LNCS 6281, pp. 368–380, 2010.

(*iii*) equational presentations can be regarded as (bidirectional) rewriting theories, and studied from a computational point of view [16]; (*iv*) algebraic theories come with an associated notion of algebraic translation [18], whose syntactic counterpart provides the right notion of syntactic translation between equational presentations [12, 13]; (*v*) strong monads have an associated metalogic from which equational logics can be synthesised [9, 10].

The realm of universal algebra is restricted to first-order languages. In particular, this leaves out languages with variable binding. Variable-binding constructs are at the core of fundamental calculi and theories in computer science and logic [5, 6], and incorporating them into algebra has been a main foundational research problem. The present work develops such a programme from the viewpoint of algebraic theories.

Our presentation is in two parts. The first part (Sections 2 and 3) sets up the necessary background; the second part (Sections 4 to 6) constitutes the contribution of the paper.

The background material gives an introduction to the work of Fiore and Hur [10] on a conservative extension of universal algebra and its equational logic from first to second order, *i.e.* to languages with variable binding and parameterised metavariables. Our summary recalls: (*i*) the notion of second-order equational presentation, that allows the specification of equational theories by means of schematic identities over signatures of variable-binding operators; (*ii*) the model theory of second-order equational presentations by means of second-order algebras; and (*iii*) the deductive system underlying formal reasoning about second-order algebraic structure.

The crux of our work is the notion of *second-order algebraic theory* (Definition 4.1). At the syntactic level, the correctness of our definition is established by showing a categorical equivalence between second-order equational presentations and second-order algebraic theories (Theorem 5.2). This involves distilling a notion of syntactic translation between second-order equational presentations that corresponds to the canonical notion of morphism between second-order algebraic theories. These syntactic translations provide a mathematical formalisation of notions such as encodings and transforms. On top of the syntactic correspondence, we furthermore establish a semantic one, by which *second-order functorial semantics* is shown to correspond to the model theory of second-order universal algebra (Theorem 6.1 and Corollary 6.1).

2 Second-Order Equational Logic

We briefly present *Second-Order Equational Logic* as introduced by Fiore and Hur [10] together with the syntactic machinery that surrounds it. For succinctness, our exposition is restricted to the unityped setting. The general multi-typed framework can be found in [10].

Signatures. A (unityped second-order) *signature* $\Sigma = (O, |-|)$ is specified by a set of operators O and an arity function $|-| : O \longrightarrow \mathbb{N}^*$, see [1, 2]. For $\mathsf{o} \in O$, we write $\mathsf{o} : (n_1, \ldots, n_k)$ whenever $|\mathsf{o}| = (n_1, \ldots, n_k)$. The intended meaning is that the operator o takes k arguments with the ith argument binding n_i variables.

Example 2.1. The signature of the λ-*calculus* has operators $\mathsf{abs} : (1)$ and $\mathsf{app} : (0, 0)$.

Terms. We consider terms in contexts with two zones, respectively declaring metavariables and variables. Metavariables come with an associated natural number arity. A metavariable M of arity m, denoted $\mathrm{M} : [m]$, is to be parameterised by m terms. We represent contexts as $\mathrm{M}_1 : [m_1], \ldots, \mathrm{M}_k : [m_k] \triangleright x_1, \ldots, x_n$ where the metavariables M_i and the variables x_j are assumed distinct.

Signatures give rise to terms in context. Terms are built up by means of operators from both variables and metavariables, and hence referred to as second-order. The judgement for *terms* in context $(\Theta \triangleright \Gamma \vdash -)$ is defined by the following rules.

(Variables) For $x \in \Gamma$,

$$\frac{}{\Theta \triangleright \Gamma \vdash x}$$

(Metavariables) For $(\mathrm{M} : [m]) \in \Theta$,

$$\frac{\Theta \triangleright \Gamma \vdash t_i \ \ (1 \leq i \leq m)}{\Theta \triangleright \Gamma \vdash \mathrm{M}[t_1, \ldots, t_m]}$$

(Operators) For $\mathsf{o} : (n_1, \ldots, n_k)$,

$$\frac{\Theta \triangleright \Gamma, \vec{x_i} \vdash t_i \ \ (1 \leq i \leq k)}{\Theta \triangleright \Gamma \vdash \mathsf{o}\big((\vec{x_1})\, t_1, \ldots, (\vec{x_k})\, t_k\big)}$$

where $\vec{x_i}$ stands for $x_{i,1}, \ldots, x_{i,n_i}$.

Second-order terms are considered up the α-equivalence relation induced by stipulating that, for every operator o, in the term $\mathsf{o}\big(\ldots, (\vec{x_i})t_i, \ldots\big)$ the $\vec{x_i}$ are bound in t_i.

Example 2.2. Two terms for the λ-calculus signature (Example 2.1) follow:

$$\mathrm{M} : [1], \mathrm{N} : [0] \triangleright \cdot \vdash \mathsf{app}\big(\mathsf{abs}\big((x)\mathrm{M}[x]\big), \mathrm{N}[\,]\big) \ , \quad \mathrm{M} : [1], \mathrm{N} : [0] \triangleright \cdot \vdash \mathrm{M}[\mathrm{N}[\,]] \ .$$

Substitution calculus. The second-order nature of the syntax requires a two-level substitution calculus [1, 8]. Each level respectively accounts for the substitution of variables and metavariables, with the latter operation depending on the former.

The operation of capture-avoiding simultaneous *substitution* of terms for variables maps

$$\Theta \triangleright x_1, \ldots, x_n \vdash t \quad \text{and} \quad \Theta \triangleright \Gamma \vdash t_i \ \ (1 \leq i \leq n)$$

to

$$\Theta \triangleright \Gamma \vdash t[{}^{t_i}\!/\!_{x_i}]_{1 \leq i \leq n}$$

according to the following inductive definition:

- $x_j[{}^{t_i}\!/\!_{x_i}]_{1\leq i\leq n} = t_j$
- $\big(\mathrm{M}[\ldots, s, \ldots]\big)[{}^{t_i}\!/\!_{x_i}]_{1\leq i\leq n} = \mathrm{M}\big[\ldots, s[{}^{t_i}\!/\!_{x_i}]_{1\leq i\leq n}, \ldots\big]$
- $\big(\mathsf{o}(\ldots,(y_1,\ldots,y_k)s,\ldots)\big)[{}^{t_i}\!/\!_{x_i}]_{1\leq i\leq n}$
  $= \mathsf{o}\big(\ldots,(z_1,\ldots,z_k)s[{}^{t_i}\!/\!_{x_i}, {}^{z_j}\!/\!_{y_j}]_{1\leq i\leq n, 1\leq j\leq k},\ldots\big)$
  with $z_j \notin \mathrm{dom}(\Gamma)$ for all $1 \leq j \leq k$

The operation of *metasubstitution* of abstracted terms for metavariables maps

$$\mathrm{M}_1 : [m_1], \ldots, \mathrm{M}_k : [m_k] \rhd \Gamma \vdash t \quad \text{and} \quad \Theta \rhd \Gamma, \vec{x_i} \vdash t_i \quad (1 \leq i \leq k)$$

to

$$\Theta \rhd \Gamma \vdash t\{\mathrm{M}_i := (\vec{x_i})t_i\}_{1\leq i\leq k}$$

according to the following inductive definition:

- $x\{\mathrm{M}_i := (\vec{x_i})t_i\}_{1\leq i\leq k} = x$
- $\big(\mathrm{M}_\ell[s_1,\ldots,s_m]\big)\{\mathrm{M}_i := (\vec{x_i})t_i\}_{1\leq i\leq k} = t_\ell[{}^{s'_j}\!/\!_{x_{i,j}}]_{1\leq j\leq m}$
  where, for $1 \leq j \leq m$, $s'_j = s_j\{\mathrm{M}_i := (\vec{x_i})t_i\}_{1\leq i\leq k}$
- $\big(\mathsf{o}(\ldots,(\vec{x})s,\ldots)\big)\{\mathrm{M}_i := (\vec{x_i})t_i\}_{1\leq i\leq k} = \mathsf{o}\big(\ldots,(\vec{x})s\{\mathrm{M}_i := (\vec{x_i})t_i\}_{1\leq i\leq k},\ldots\big)$

Presentations. An *equational presentation* is specified by a signature together with a set of axioms over it, each of which is a pair of terms in context.

Example 2.3. The equational presentation of the λ-calculus extends the signature of Example 2.1 with the following equations.

$$(\beta) \quad \mathrm{M} : [1], \mathrm{N} : [0] \rhd \cdot \vdash \mathsf{app}\big(\mathsf{abs}(\,(x)\mathrm{M}[x]\,), \mathrm{N}[\,]\big) \equiv \mathrm{M}\big[\mathrm{N}[\,]\big]$$

$$(\eta) \quad \mathrm{F} : [0] \rhd \cdot \vdash \mathsf{abs}\big(\,(x)\mathsf{app}(\mathrm{F}[\,], x)\,\big) \equiv \mathrm{F}[\,]$$

Logic. The rules of *Second-Order Equational Logic* are given in Figure 1. Besides the rules for axioms and equivalence, it consists of just one additional rule stating that the operation of metasubstitution in extended variable contexts is a congruence.

We note the following basic result from [10]: *Second-Order Equational Logic is a conservative extension of (First-Order) Equational Logic.*

3 Second-Order Universal Algebra

The model theory of Fiore and Hur [10] for second-order equational presentations is recalled. This is presented here in concrete elementary terms, but could have also been given in abstract monadic terms. The reader is referred to [10] for the latter perspective.

Semantic universe. We write $\mathbf{F}$ for the free cocartesian category on an object. Explicitly, it has set of objects $\mathbb{N}$ and morphisms $m \longrightarrow n$ given by functions $\|m\| \longrightarrow \|n\|$, where, for $\ell \in \mathbb{N}$, $\|\ell\| = \{1, \ldots, \ell\}$.

We will work within and over the semantic universe $\boldsymbol{\mathcal{S}et}^{\mathbf{F}}$ of sets in variable contexts [11]. We write $\boldsymbol{y}$ for the Yoneda embedding $\mathbf{F}^{\mathrm{op}} \hookrightarrow \boldsymbol{\mathcal{S}et}^{\mathbf{F}}$.

(Axiom)

$$\frac{(\Theta \rhd \Gamma \vdash s \equiv t) \in E}{\Theta \rhd \Gamma \vdash s \equiv t}$$

(Equivalence)

$$\frac{\Theta \rhd \Gamma \vdash t}{\Theta \rhd \Gamma \vdash t \equiv t} \qquad \frac{\Theta \rhd \Gamma \vdash s \equiv t}{\Theta \rhd \Gamma \vdash t \equiv s} \qquad \frac{\Theta \rhd \Gamma \vdash s \equiv t \qquad \Theta \rhd \Gamma \vdash t \equiv u}{\Theta \rhd \Gamma \vdash s \equiv u}$$

(Extended metasubstitution)

$$\frac{\mathrm{M}_1 : [m_1], \ldots, \mathrm{M}_k : [m_k] \rhd \Gamma \vdash s \equiv t \qquad \Theta \rhd \Delta, \vec{x_i} \vdash s_i \equiv t_i \quad (1 \leq i \leq k)}{\Theta \rhd \Gamma, \Delta \vdash s\{\mathrm{M}_i := (\vec{x_i})s_i\}_{1 \leq i \leq k} \equiv t\{\mathrm{M}_i := (\vec{x_i})t_i\}_{1 \leq i \leq k}}$$

Fig. 1. Second-Order Equational Logic

Substitution. We recall the *substitution monoidal structure* in semantic universes [11]. It has tensor unit and tensor product respectively given by $\boldsymbol{y}1$ and $X \bullet Y = \int^{k \in \mathbf{F}} X(k) \times Y^k$.

A monoid $\boldsymbol{y}1 \xrightarrow{\nu} A \xleftarrow{\varsigma} A \bullet A$ for the substitution monoidal structure equips A with substitution structure. In particular, the map $\nu_k = (\boldsymbol{y}k \cong (\boldsymbol{y}1)^k \xrightarrow{\nu^k} A^k)$ induces the embedding

$$\big(A^{\boldsymbol{y}n} \times A^n\big)(k) \longrightarrow A(k+n) \times A^k(k) \times A^n(k) \longrightarrow \big(A \bullet A\big)(k)$$

which together with the multiplication yield a *substitution operation*

$$\varsigma_n : A^{\boldsymbol{y}n} \times A^n \longrightarrow A \ .$$

These substitution operations provide the interpretation of metavariables.

Algebras. Every signature Σ induces a *signature endofunctor* on $\boldsymbol{Set}^{\mathbf{F}}$ given by $\mathcal{F}_\Sigma X = \coprod_{\mathsf{o}:(n_1,\ldots,n_k) \text{ in } \Sigma} \prod_{1 \leq i \leq k} X^{\boldsymbol{y}n_i}$. $\mathcal{F}_\Sigma$-algebras $\mathcal{F}_\Sigma X \longrightarrow X$ provide an interpretation $[\![\mathsf{o}]\!]_X : \prod_{1 \leq i \leq k} X^{\boldsymbol{y}n_i} \longrightarrow X$ for every operator $\mathsf{o} : (n_1, \ldots, n_k)$ in Σ.

We note that there are canonical natural isomorphisms

$$\textstyle\coprod_{i \in I}(X_i \bullet Y) \cong \big(\coprod_{i \in I} X_i\big) \bullet Y$$

$$\textstyle\big(\prod_{1 \leq i \leq n} X_i\big) \bullet Y \cong \prod_{1 \leq i \leq n}(X_i \bullet Y)$$

and, for all points $\eta : \boldsymbol{y}1 \longrightarrow Y$, natural extension maps

$$\eta^{\#_n} : X^{\boldsymbol{y}n} \bullet Y \longrightarrow (X \bullet Y)^{\boldsymbol{y}n} \ .$$

These constructions equip every signature endofunctor with a *pointed strength* $\varpi_{X,\boldsymbol{y}1 \to Y} : \mathcal{F}_\Sigma(X) \bullet Y \longrightarrow \mathcal{F}_\Sigma(X \bullet Y)$. See [8] for details.

Models. The models that we are interested in (referred to as Σ-*monoids* in [11, 8]) are algebras equipped with a compatible substitution structure. For a signature Σ, we let Σ-Mod be the category of Σ-*models* with objects $A \in \boldsymbol{Set}^{\mathbf{F}}$

equipped with an $\mathcal{F}_\Sigma$-algebra structure $\alpha : \mathcal{F}_\Sigma A \longrightarrow A$ and a monoid structure $\boldsymbol{y}1 \xrightarrow{\nu} A \xleftarrow{\varsigma} A \bullet A$ that are compatible in the sense that the diagram

$$\begin{array}{ccccc} \mathcal{F}_\Sigma(A) \bullet A & \xrightarrow{\varpi_{A,\nu}} & \mathcal{F}_\Sigma(A \bullet A) & \xrightarrow{\mathcal{F}_\Sigma \varsigma} & \mathcal{F}_\Sigma(A) \\ {\scriptstyle \alpha \bullet A}\downarrow & & & & \downarrow{\scriptstyle \alpha} \\ A \bullet A & & \xrightarrow[\varsigma]{} & & A \end{array}$$

commutes. Morphisms are maps that are both $\mathcal{F}_\Sigma$-algebra and monoid homomorphims.

Semantics. For $\Theta = (\mathrm{M}_1 : [m_1] \ldots, \mathrm{M}_k : [m_k])$ and $\Gamma = (x_1, \ldots, x_n)$, the interpretation of a term $\Theta \rhd \Gamma \vdash t$ in a model A is a morphism

$$[\![\Theta \rhd \Gamma \vdash t]\!]_A : [\![\Theta \rhd \Gamma]\!]_A \longrightarrow A \ ,$$

where $[\![\Theta \rhd \Gamma]\!]_A = \prod_{1 \leq i \leq k} A^{\boldsymbol{y} m_i} \times \boldsymbol{y} n$, given by structural induction as follows:

- $[\![\Theta \rhd \Gamma \vdash x_j]\!]_A$ is the composite $[\![\Theta \rhd \Gamma]\!]_A \xrightarrow{\pi_2} \boldsymbol{y} n \xrightarrow{\nu_n} A^n \xrightarrow{\pi_j} A$.
- $[\![\Theta \rhd \Gamma \vdash \mathrm{M}_i[t_1, \ldots, t_{m_i}]]\!]_A$ is the composite

  $$[\![\Theta \rhd \Gamma]\!]_A \xrightarrow{\langle \pi_i\, \pi_1, f \rangle} A^{\boldsymbol{y} m_i} \times A^{m_i} \xrightarrow{\varsigma_{m_i}} A$$

  where $f = \big\langle [\![\Theta \rhd \Gamma \vdash t_j]\!] \big\rangle_{1 \leq j \leq m_i}$.
- For $\mathsf{o} : (n_1, \ldots, n_\ell)$,

  $$[\![\Theta \rhd \Gamma \vdash \mathsf{o}\big((\vec{y_1})t_1, \ldots, (\vec{y_\ell})t_\ell\big)]\!]$$

  is the composite $[\![\Theta \rhd \Gamma]\!]_A \xrightarrow{\langle f_j \rangle_{1 \leq j \leq \ell}} \prod_{1 \leq j \leq \ell} A^{\boldsymbol{y} n_j} \xrightarrow{[\![\mathsf{o}]\!]_A} A$ where f_j is the exponential transpose of

$$\textstyle\prod_{1 \leq i \leq k} A^{\boldsymbol{y} m_i} \times \boldsymbol{y} n \times \boldsymbol{y} n_j \cong \prod_{1 \leq i \leq k} A^{\boldsymbol{y} m_i} \times \boldsymbol{y}(n + n_j) \xrightarrow{[\![\Theta \rhd \Gamma, \vec{y_j} \vdash t_j]\!]_A} A \ .$$

Equational models. We say that a model A *satisfies* $\Theta \rhd \Gamma \vdash s \equiv t$, for which we use the notation $A \models (\Theta \rhd \Gamma \vdash s \equiv t)$, iff $[\![\Theta \rhd \Gamma \vdash s]\!]_A = [\![\Theta \rhd \Gamma \vdash t]\!]_A$.

For an equational presentation (Σ, E), we write (Σ, E)-Mod for the full subcategory of Σ-Mod consisting of the Σ-models that satisfy the axioms E.

Soundness and completeness [10].

For an equational presentation (Σ, E), the judgement $\Theta \rhd \Gamma \vdash s \equiv t$ is derivable from E iff $A \models (\Theta \rhd \Gamma \vdash s \equiv t)$ for all (Σ, E)-models A.

4 Second-Order Algebraic Theories

We introduce the notion of unityped second-order algebraic theory and establish it as the categorical counterpart to that of second-order equational presentation. The generalisation to the multi-typed case should be evident.

Remark. Having omitted the monadic view of second-order universal algebra, the important role played by the monadic perspective in our development will not be considered here.

Theory of equality. The theory of equality plays a pivotal role in the definition of algebraic theory. Thus, we proceed first to identify the second-order algebraic theory of equality. This we do both in syntactic and semantic terms. The (first-order) algebraic theory of equality is then considered from this new perspective.

The syntactic viewpoint leads us to define the category **M** with set of objects $\mathbb{N}^*$ and morphisms $(m_1, \ldots, m_k) \longrightarrow (n_1, \ldots, n_\ell)$ given by tuples

$$\langle\, \mathrm{M}_1 : [m_1], \ldots, \mathrm{M}_k : [m_k] \rhd x_1, \ldots, x_{n_i} \vdash t_i \,\rangle_{i \in \|\ell\|}$$

of terms under the empty signature. The identity on $(m_1, \ldots, m_k)$ is given by

$$\langle\, \mathrm{M}_1 : [m_1], \ldots, \mathrm{M}_k : [m_k] \rhd x_1, \ldots, x_{m_i} \vdash \mathrm{M}_i[x_1, \ldots, x_{m_i}] \,\rangle_{i \in \|k\|} \;;$$

whilst the composition of

$$\langle\, \mathrm{M}_1 : [\ell_1], \ldots, \mathrm{M}_i : [\ell_i] \rhd x_1, \ldots, x_{m_p} \vdash s_p \,\rangle_{p \in \|j\|} : (\ell_1, \ldots, \ell_i) \longrightarrow (m_1, \ldots, m_j)$$

and

$$\langle\, \mathrm{M}_1 : [m_1], \ldots, \mathrm{M}_j : [m_j] \rhd x_1, \ldots, x_{n_q} \vdash t_q \,\rangle_{q \in \|k\|} : (m_1, \ldots, m_j) \longrightarrow (n_1, \ldots, n_k)$$

is given by metasubstitution as follows:

$$\langle\, \mathrm{M}_1 : [\ell_1], \ldots, \mathrm{M}_i : [\ell_i] \rhd x_1, \ldots, x_{n_q} \vdash t_q\{\mathrm{M}_p := (x_1, \ldots, x_{m_p})s_p\}_{p \in \|j\|} \,\rangle_{q \in \|k\|} \;.$$

The category **M** is strict cartesian, with terminal object given by the empty sequence and binary products given by concatenation. Furthermore, the object $(0) \in \mathbf{M}$ is exponentiable. Indeed, the exponential object $(0) \Rightarrow (m_1, \ldots, m_k)$ is $(m_1 + 1, \ldots, m_k + 1)$ with evaluation map

$$(m_1 + 1, \ldots, m_k + 1, 0) \longrightarrow (m_1, \ldots, m_k)$$

given by

$$\left\langle \begin{array}{l} \mathrm{M}_1 : [m_1 + 1], \ldots, \mathrm{M}_k : [m_k + 1], \mathrm{M}_{k+1} : [0] \rhd x_1, \ldots, x_{m_i} \\ \quad \vdash \; \mathrm{M}_i\big[x_1, \ldots, x_{m_i}, \mathrm{M}_{k+1}[\,]\big] \end{array} \right\rangle_{i \in \|k\|}$$

In fact, this structure provides a semantic characterisation of **M**.

Lemma 4.1 (Universal property of M). *The category* **M***, together with the object* $(0) \in \mathbf{M}$*, is initial amongst cartesian categories equipped with an exponentiable object (with respect to cartesian functors that preserve the exponentiable object).*

Loosely speaking, then, **M** is the free (strict) cartesian category on an exponentiable object.

Algebraic theories. We extend Lawvere's fundamental notion of (first-order) algebraic theory [18] to second order.

Definition 4.1 (Second-order algebraic theories). *A* second-order algebraic theory *consists of a cartesian category* $\mathbb{T}$ *and a strict cartesian identity-on-objects functor* $\mathbf{M} \longrightarrow \mathbb{T}$ *that preserves the exponentiable object* (0).

The most basic example is the *second-order algebraic theory of equality* given by $\mathbf{M}$ (together with the identity functor).

Every second-order algebraic theory has an underlying (first-order) algebraic theory. To formalise this, recall that the (first-order) algebraic theory of equality $\mathbf{L} = \mathbf{F}^{\mathrm{op}}$ is the free (strict) cartesian category on an object and consider the unique cartesian functor $\mathbf{L} \longrightarrow \mathbf{M}$ mapping the generating object to the exponentiable object. Then, the (first-order) algebraic theory underlying $\mathbf{M} \longrightarrow \mathbb{T}$ is $\mathbf{L} \longrightarrow \mathbb{T}_0$ for $\mathbf{L} \longrightarrow \mathbb{T}_0 \hookrightarrow \mathbb{T}$ the identity-on-objects/full-and-faithful factorisation of $\mathbf{L} \longrightarrow \mathbf{M} \longrightarrow \mathbb{T}$. In particular, $\mathbf{L}$ underlies $\mathbf{M}$.

The theory of a presentation. For a second-order equational presentation $\mathcal{E}$, the *classifying category* $\mathbf{M}(\mathcal{E})$ has set of objects $\mathbb{N}^*$ and morphisms $\vec{m} \longrightarrow \vec{n}$, say with $\vec{m} = (m_1, \ldots, m_k)$ and $\vec{n} = (n_1, \ldots, n_\ell)$, given by tuples

$$\big\langle\, [\, \mathrm{M}_1 : [m_1], \ldots, \mathrm{M}_k : [m_k] \rhd x_1, \ldots, x_{n_i} \vdash t_i \,]_{\mathcal{E}} \,\big\rangle_{i \in \|\ell\|}$$

of equivalence classes of terms under the equivalence relation that identifies two terms iff they are provably equal from $\mathcal{E}$ in *Second-Order Equational Logic*. (Identities and composition are defined on representatives as in $\mathbf{M}$.)

Lemma 4.2. *For a second-order equational presentation $\mathcal{E}$, the category $\mathbf{M}(\mathcal{E})$ together with the canonical functor $\mathbf{M} \longrightarrow \mathbf{M}(\mathcal{E})$ is a second-order algebraic theory.*

We refer to $\mathbf{M} \longrightarrow \mathbf{M}(\mathcal{E})$ as the second-order algebraic theory of $\mathcal{E}$.

The presentation of a theory. The *internal language* $\mathcal{E}(T)$ of a second-order algebraic theory $T : \mathbf{M} \longrightarrow \mathbb{T}$ is the second-order equational presentation defined as follows:

(Operators) For every $f : (m_1, \ldots, m_k) \longrightarrow (n)$ in $\mathbb{T}$, we have an operator o_f of arity $(m_1, \ldots, m_k, \underbrace{0, \ldots, 0}_{n \text{ times}})$.

(Equations) Setting

$$\mathsf{t}_f = \mathsf{o}_f\big((x_1, \ldots, x_{m_1})\mathrm{M}_1[x_1, \ldots, x_{m_1}], \ldots, (x_1, \ldots, x_{m_k})\mathrm{M}_k[x_1, \ldots, x_{m_k}], x_1, \ldots, x_n\big)$$

for every $f : (m_1, \ldots, m_k) \longrightarrow (n)$ in $\mathbb{T}$, we have

- $\mathrm{M}_1 : [m_1], \ldots, \mathrm{M}_k : [m_k] \rhd x_1, \ldots, x_n \vdash s \equiv \mathsf{t}_{T\langle s \rangle}$ for every $\langle s \rangle : (m_1, \ldots, m_k) \longrightarrow (n)$ in $\mathbf{M}$,
- $\mathrm{M}_1 : [m_1], \ldots, \mathrm{M}_k : [m_k] \rhd x_1, \ldots, x_n \vdash \mathsf{t}_h \equiv \mathsf{t}_g\{\mathrm{M}_i := (x_1, \ldots, x_{n_i})\mathsf{t}_{f_i}\}_{1 \leq i \leq \ell}$ for every $h : (m_1, \ldots, m_k) \longrightarrow (n)$, $g : (n_1, \ldots, n_\ell) \longrightarrow (n)$, and $f_i : (m_1, \ldots, m_k) \longrightarrow (n_i)$, $1 \leq i \leq \ell$, such that $h = g \circ \langle f_1, \ldots, f_\ell \rangle$ in $\mathbb{T}$.

Algebraic translations. For second-order algebraic theories $T : \mathbf{M} \longrightarrow \mathbb{T}$ and $T' : \mathbf{M} \longrightarrow \mathbb{T}'$, a second-order *algebraic translation* $T \longrightarrow T'$ is a functor $F : \mathbb{T} \longrightarrow \mathbb{T}'$ such that $T' = FT$. We write $\mathcal{SOAT}$ for the category of second-order algebraic theories and algebraic translations.

Theorem 4.1 (Theory/presentation correspondence). *Every second-order algebraic theory $T : \mathbf{M} \longrightarrow \mathbb{T}$ is isomorphic to the second-order algebraic theory of its associated equational presentation $\mathbf{M} \longrightarrow \mathbf{M}(\mathcal{E}(T))$.*

5 Second-Order Syntactic Translations

We introduce the notion of *syntactic translation* between second-order equational presentations. This we justify by establishing its equivalence with that of algebraic translation between the associated second-order algebraic theories.

Signature translations. A *syntactic translation* $\tau : \Sigma \longrightarrow \Sigma'$ between second-order signatures is given by a mapping from the operators of Σ to the terms of Σ' as follows:

$$\mathsf{o} : (m_1, \ldots, m_k) \quad \longmapsto \quad \mathrm{M}_1 : [m_1], \ldots, \mathrm{M}_k : [m_k] \rhd \cdot \vdash \tau_{\mathsf{o}} \quad .$$

Note that the term associated to an operator has an empty variable context and that the metavariable context is determined by the arity of the operator.

A translation $\tau : \Sigma \longrightarrow \Sigma'$ extends to a mapping from the terms of Σ to the terms of Σ'

$$\Theta \rhd \Gamma \vdash t \quad \longmapsto \quad \Theta \rhd \Gamma \vdash \tau(t)$$

according to the following inductive definition:

- $\tau(x) = x$
- $\tau\big(\mathrm{M}[t_1, \ldots, t_m]\big) = \mathrm{M}\big[\tau(t_1), \ldots, \tau(t_m)\big]$
- $\tau\big(\mathsf{o}\big((\vec{x_1})t_1, \ldots, (\vec{x_k})t_k\big)\big) = \tau_{\mathsf{o}}\{\mathrm{M}_i := (\vec{x_i})\tau(t_i)\}_{1 \le i \le k}$

Lemma 5.1 (Compositionality). *The extension of a syntactic translation between second-order signatures commutes with substitution and metasubstitution.*

Example 5.1 (Continutation Passing Style). A formalisation of the CPS transform for the λ-calculus as a syntactic translation due to Plotkin [20] follows. We provide it in informal notation for ease of readability.

$$\mathsf{app} : (0,0) \quad \longmapsto \quad \mathrm{M} : [0], \mathrm{N} : [0] \rhd \cdot \vdash \lambda k.\, \mathrm{M}[\,]\,\big(\lambda m.\, m\,(\lambda \ell.\, \mathrm{N}[\,]\,\ell)\,k\big)$$

$$\mathsf{abs} : (1) \quad \longmapsto \quad \mathrm{F} : [1] \rhd \cdot \vdash \lambda k.\, k\,\big(\lambda x.\, (\lambda \ell.\, \mathrm{F}[x]\,\ell)\big)$$

Equational translations. A *syntactic translation* between second-order equational presentations $\tau : (\Sigma, E) \longrightarrow (\Sigma,' E')$ is a translation $\tau : \Sigma \longrightarrow \Sigma'$ such that, for every axiom $\Theta \rhd \Gamma \vdash s \equiv t$ in E, the judgement $\Theta \rhd \Gamma \vdash \tau(s) \equiv \tau(t)$ is derivable from E'.

Lemma 5.2. *The extension of a syntactic translation between second-order equational presentations preserves second-order equational derivability.*

We write $\boldsymbol{\mathcal{SOEP}}$ for the category of second-order equational presentations and syntactic translations. (The identity syntactic translation maps an operator $\mathsf{o} : (m_1, \ldots, m_k)$ to the term $\mathsf{o}\big(\ldots, (x_1, \ldots, x_{m_i})\mathrm{M}_i[x_1, \ldots, x_{m_i}], \ldots\big)$; whilst the composition of τ followed by τ' maps o to $\tau'(\tau_{\mathsf{o}})$.)

Theorem 5.1 (Presentation/theory correspondence). *Every second-order equational presentation $\mathcal{E}$ is isomorphic to the second-order equational presentation of its associated algebraic theory $\mathcal{E}(\mathbf{M}(\mathcal{E}))$.*

Syntactic and algebraic translations. A syntactic translation $\tau : \mathcal{E} \longrightarrow \mathcal{E}'$ induces the algebraic translation $\mathbf{M}(\tau) : \mathbf{M}(\mathcal{E}) \longrightarrow \mathbf{M}(\mathcal{E}')$, mapping $\langle\, [t_1]_{\mathcal{E}}, \ldots, [t_\ell]_{\mathcal{E}} \,\rangle$ to $\langle\, [\tau(t_1)]_{\mathcal{E}'}, \ldots, [\tau(t_\ell)]_{\mathcal{E}'} \,\rangle$. This gives a functor $\boldsymbol{\mathcal{SOEP}} \longrightarrow \boldsymbol{\mathcal{SOAT}}$. Conversely, an algebraic translation $F : T \longrightarrow T'$ induces the syntactic translation $\mathcal{E}(F) : \mathcal{E}(T) \longrightarrow \mathcal{E}(T')$, mapping an operator o_f, for $f : (m_1, \ldots, m_k) \longrightarrow (n)$ in $\mathbb{T}$, to the term $\mathsf{t}_{Ff}\big[\mathrm{M}_{k+1}[\,]/x_1, \ldots, \mathrm{M}_{k+n}[\,]/x_n\big]$. This gives a functor $\boldsymbol{\mathcal{SOAT}} \longrightarrow \boldsymbol{\mathcal{SOEP}}$.

Theorem 5.2. *The categories $\boldsymbol{\mathcal{SOAT}}$ and $\boldsymbol{\mathcal{SOEP}}$ are equivalent.*

6 Second-Order Functorial Semantics

We extend Lawvere's functorial semantics for algebraic theories [18] from first to second order.

Functorial models. The category $\boldsymbol{\mathcal{M}od}(T)$ of (set-theoretic) *functorial models* of a second-order algebraic theory $T : \mathbf{M} \longrightarrow \mathbb{T}$ is the category of cartesian functors $\mathbb{T} \longrightarrow \boldsymbol{\mathcal{S}et}$ and natural transformations between them.

Every $\mathcal{E}$-model A, for a second-order equational presentation $\mathcal{E}$, provides a functorial model $\mathbf{M}(\mathcal{E}) \longrightarrow \boldsymbol{\mathcal{S}et}$ as follows:

- on objects, $(m_1, \ldots, m_k)$ is mapped to $\prod_{1 \le i \le k} A(m_i)$;
- on morphisms, $\langle\, [\mathrm{M}_1 : [m_1], \ldots, \mathrm{M}_k : [m_k] \rhd x_1, \ldots, x_{n_i} \vdash t_j]_{\mathcal{E}} \,\rangle_{j \in \|\ell\|}$ is mapped to $\langle\, (f_j)_0 \rangle_{1 \le j \le \ell}$ where $f_j : \prod_{1 \le i \le k} A^{\mathsf{y} m_i} \longrightarrow A^{\mathsf{y} n_j}$ is the exponential transpose of $[\![\mathrm{M}_1 : [m_1], \ldots, \mathrm{M}_k : [m_k] \rhd x_1, \ldots, x_{n_j} \vdash t_j]\!]_A$.

As we proceed to show, every functorial model essentially arises in this manner (see Corollary 6.1).

Clones. We need recall and develop some aspects of the theory of *clones* from universal algebra (see *e.g.* [7]).

Let C be an exponentiable object in a cartesian category $\mathscr{C}$. Recall that the family $\langle C \rangle = \{C^n \Rightarrow C\}_{n \in \mathbb{N}}$ has a canonical clone structure

$$\iota_i^{(n)} : 1 \longrightarrow \langle C \rangle_n \ (1 \le i \le n \in \mathbb{N}) \ , \quad \varsigma_{m,n} : \langle C \rangle_m \times \langle C \rangle_n{}^m \longrightarrow \langle C \rangle_n \ (m, n \in \mathbb{N})$$

known as the *clone of operations* on C. Thus, as it is the case with every clone, the family $\langle C \rangle$ canonically extends to a functor $\mathbf{F} \longrightarrow \mathscr{C} : n \longmapsto \langle C \rangle_n$.

For every $m_1, \ldots, m_k \in \mathbb{N}$ (for $k \in \mathbb{N}$), $n \in \mathbb{N}$, and $f : \prod_{1 \le i \le k} \langle C \rangle_{m_i} \longrightarrow \langle C \rangle_n$ in $\mathscr{C}$ let $\widetilde{f} = \{\widetilde{f}_\ell\}_{\ell \in \mathbb{N}}$ be given by setting

$$\widetilde{f}_\ell = \Big(\prod_{1 \le i \le k} \langle C \rangle_{\ell + m_i} \cong C^\ell \Rightarrow \prod_{1 \le i \le k} \langle C \rangle_{m_i} \xrightarrow{C^\ell \Rightarrow f} C^\ell \Rightarrow \langle C \rangle_n \cong \langle C \rangle_{\ell + n} \Big).$$

The family $\widetilde{f}$ is a natural transformation $\prod_{1 \le i \le k} \langle C \rangle_{(-)+m_i} \longrightarrow \langle C \rangle_{(-)+n}$ and commutes with the clone structure. The latter in the sense that, for

$$w_\ell = \Big(\langle C\rangle_q{}^p \cong \langle C\rangle_q{}^p \times 1 \xrightarrow{\langle C\rangle_\jmath{}^p \times \langle \iota^{(q+\ell)}_{q+i}\rangle_{1\leq i\leq \ell}} \langle C\rangle_{q+\ell}{}^p \times \langle C\rangle_{q+\ell}{}^\ell \cong \langle C\rangle_{q+\ell}{}^{p+\ell} \Big)$$

where $\jmath$ denotes the inclusion $\|q\| \hookrightarrow \|q+\ell\|$, the diagram

$$\begin{array}{ccccc} & & \prod_{1\leq i\leq k} \langle C\rangle_{p+m_i} \times \langle C\rangle_{q+m_i}{}^{p+m_i} & & \\ & \nearrow^{\langle \mathrm{id} \times w_{m_i}\rangle_{1\leq i\leq k}} & & \searrow^{\prod_{1\leq i\leq k} \varsigma_{p+m_i,q+m_i}} & \\ \prod_{1\leq i\leq k} \langle C\rangle_{p+m_i} \times \langle C\rangle_q{}^p & & & & \prod_{1\leq i\leq k} \langle C\rangle_{q+m_i} \\ \downarrow{\scriptstyle f_p \times w_n} & & & & \downarrow{\scriptstyle \widetilde{f}_q} \\ \langle C\rangle_{p+n} \times \langle C\rangle_{q+n}{}^{p+n} & & \xrightarrow[\varsigma_{p+n,q+n}]{} & & \langle C\rangle_{q+n} \end{array}$$

commutes for all $p, q \in \mathbb{N}$.

Let Σ be a second-order signature, and consider a functorial model $S : \mathbf{M}(\Sigma) \longrightarrow \boldsymbol{Set}$. Then, the image under the cartesian functor S of the clone of operations induced by the exponentiable object $(0) \in \mathbf{M}(\Sigma)$ together with the family $\{\widetilde{f_{\mathsf{o}}}\}_{\mathsf{o}:(m_1,\ldots,m_k) \text{ in } \Sigma}$, where $f_{\mathsf{o}} = \langle\, \mathsf{o}(\ldots,(x_1,\ldots,x_{m_i})\mathrm{M}_i[x_1,\ldots,x_{m_i}],\ldots)\,\rangle$, yields a Σ-model $\underline{S} \in \boldsymbol{Set}^{\mathbf{F}}$.

Furthermore, for all $f = \langle\, \mathrm{M}_1 : [m_1], \ldots, \mathrm{M}_k : [m_k] \rhd x_1, \ldots, x_n \vdash t \,\rangle$ in $\mathbf{M}(\Sigma)$ we have that the image of $\widetilde{f}$ under $S : \mathbf{M}(\Sigma) \longrightarrow \boldsymbol{Set}$ amounts to the interpretation of t in $\underline{S}$. Thus, for all second-order equational presentations $\mathcal{E} = (\Sigma, E)$, the Σ-model induced by the restriction of a functorial model $\mathbf{M}(\mathcal{E}) \longrightarrow \boldsymbol{Set}$ to $\mathbf{M}(\Sigma)$ is an $\mathcal{E}$-model.

The above constructions between functorial and algebraic models provide an equivalence.

Theorem 6.1. *For every second-order equational presentation $\mathcal{E}$, the category of algebraic models $\mathcal{E}$-*Mod *and the category of functorial models $\boldsymbol{\mathcal{M}od}(\mathbf{M}(\mathcal{E}))$ are equivalent.*

Corollary 6.1. *For every second-order algebraic theory T, the category of functorial models $\boldsymbol{\mathcal{M}od}(T)$ and the category of algebraic models $\mathfrak{E}(T)$-*Mod *are equivalent.*

Algebraic functors. As in the first-order case, every algebraic translation $F : T \longrightarrow T'$ between second-order algebraic theories contravariantly induces an *algebraic functor* $\boldsymbol{\mathcal{M}od}(T') \longrightarrow \boldsymbol{\mathcal{M}od}(T) : S \longmapsto S\,F$ between the corresponding categories of models. We also have the following fundamental result.

Theorem 6.2. *The algebraic functor $\boldsymbol{\mathcal{M}od}(T') \longrightarrow \boldsymbol{\mathcal{M}od}(T)$ induced by a second-order algebraic translation $T \longrightarrow T'$ has a left adjoint.*

7 Concluding Remarks

We have introduced *second-order algebraic theories* (Section 4): (*i*) showing them to be the presentation-independent categorical syntax of second-order equational presentations (Theorems 4.1, 5.1, and 5.2), and (*ii*) establishing that their functorial semantics amounts to second-order universal algebra (Theorem 6.1 and Corollary 6.1). In the context of (*i*), our development included a notion of second-order syntactic translation (Section 5), which, in the context of (*ii*), contravariantly gives rise to algebraic functors between categories of models (Theorem 6.2).

With this theory in place, one is now in a position to: (*a*) consider constructions on second-order equational presentations in a categorical setting, and indeed the developments for (first-order) algebraic theories on limits, colimits, and tensor product carry over to the second-order setting; (*b*) investigate conservative-extension results for second-order equational presentations in a mathematical framework; and (*c*) study Morita equivalence for second-order algebraic theories.

References

[1] Aczel, P.: A general Church-Rosser theorem. Typescript (1978)
[2] Aczel, P.: Frege structures and the notion of proposition, truth and set. In: The Kleene Symposium, pp. 31–59 (1980)
[3] Birkhoff, G.: On the structure of abstract algebras. P. Camb. Philos. Soc. 31, 433–454 (1935)
[4] Burstall, R.: Proving properties of programs by structural induction. The Computer Journal 12(1), 41–48 (1969)
[5] Church, A.: An unsolvable problem of elementary number theory. Am. J. Math. 58, 354–363 (1936)
[6] Church, A.: A formulation of the simple theory of types. J. Symbolic Logic 5, 56–68 (1940)
[7] Cohn, P.: Universal Algebra. Mathematics and its Applications, vol. 6. Springer, Heidelberg (1981)
[8] Fiore, M.: Second-order and dependently-sorted abstract syntax. In: LICS 2008, pp. 57–68 (2008)
[9] Fiore, M., Hur, C.-K.: Term equational systems and logics. In: MFPS XXIV. LNCS, vol. 218, pp. 171–192. Springer, Heidelberg (2008)
[10] Fiore, M., Hur, C.-K.: Second-order equational logic. In: Dawar, A., Veith, H. (eds.) CSL 2010. LNCS, vol. 6247, pp. 320–335. Springer, Heidelberg (2010)
[11] Fiore, M., Plotkin, G., Turi, D.: Abstract syntax and variable binding. In: LICS 1999, pp. 193–202 (1999)
[12] Fujiwara, T.: On mappings between algebraic systems. Osaka Math. J. 11, 153–172 (1959)
[13] Fujiwara, T.: On mappings between algebraic systems, II. Osaka Math. J. 12, 253–268 (1960)
[14] Goguen, J., Thatcher, J., Wagner, E.: An initial algebra approach to the specification, correctness and implementation of abstract data types. In: Current Trends in Programming Methodology, vol. IV, pp. 80–149. Prentice-Hall, Englewood Cliffs (1978)
[15] Hyland, M., Power, J.: The category theoretic understanding of universal algebra: Lawvere theories and monads. ENTCS 172, 437–458 (2007)

[16] Knuth, D., Bendix, P.: Simple word problems in universal algebras. In: Computational Problems in Abstract Algebra, pp. 263–297 (1970)
[17] Linton, F.: Some aspects of equational theories. In: Proc. Conf. on Categorical Algebra at La Jolla, pp. 84–95 (1966)
[18] Lawvere, F.W.: Functorial Semantics of Algebraic Theories and Some Algebraic Problems in the context of Functorial Semantics of Algebraic Theories. Republished in: Reprints in TAC (5), pp. 1–121 (2004)
[19] McCarthy, J.: Towards a mathematical science of computation. In: IFIP Congress 1962. North-Holland, Amsterdam (1963)
[20] Plotkin, G.: Binding algebras: A step from universal algebra to type theory. Invited talk at RTA 1998 (1998)

Frame Definability for Classes of Trees in the μ-calculus

Gaëlle Fontaine[1] and Thomas Place[2,⋆]

[1] ILLC, Universiteit van Amsterdam
gaelle.fontaine@uva.nl
[2] INRIA and ENS Cachan
place@lsv.ens-cachan.fr

Abstract. We are interested in frame definability of classes of trees, using formulas of the μ-calculus. In this set up, the proposition letters (or in other words, the free variables) in the μ-formulas correspond to second order variables over which universally quantify. Our main result is a semantic characterization of the **MSO** definable classes of trees that are definable by a μ-formula. We also show that it is decidable whether a given **MSO** formula corresponds to a μ-formula, in the sense that they define the same class of trees.

Basic modal logic and μ-calculus can be seen as logical languages for talking about *Kripke models* and *Kripke frames*. On Kripke models every modal formula is equivalent to a first order formula in one free variable and every μ-calculus formula is equivalent to a monadic second order formula in one free first order variable. On Kripke frames, we universally quantify over the free propositional variables occurring in the formulas and each modal formula or μ-formula is equivalent to a sentence of monadic second order logic. For example, the modal formula $p \to \Diamond p$ corresponds locally on Kripke models to the first order formula $\alpha(u, P) = P(u) \to \exists v(uRv \wedge P(v))$ (where P is a unary predicate corresponding to p, R is the binary relation of the model and u is a point of the model). The same modal formula corresponds globally on Kripke frames to the second order sentence $\forall P \forall u \alpha(u, P)$, which happens to be equivalent to the first order sentence $\forall u, uRu$.

The expressive power of modal logic from both perspectives (models and frames) has been extensively studied. For Kripke models, Johan van Benthem characterized modal logic semantically as the bisimulation invariant fragment of first order logic [vB76]. The problem whether a formula of first order logic in one free variable has a modal correspondent on the level of models, is undecidable [vB96].

The expressive power of modal logic on Kripke frames has been studied since the 1970s and this study gave rise to many key results in the modal logic area.

⋆ The research of the first author has been made possible by VICI grand 639.073.501 of the NWO. We would like to thank Balder ten Cate for inspiring us and for helping us continuously during the research and redaction of this paper. We also thank Johan van Benthem, Diego Figueira, Luc Segoufin and Yde Venema for helpful insights.

P. Hliněný and A. Kučera (Eds.): MFCS 2010, LNCS 6281, pp. 381–392, 2010.

When interpreted on frames, modal logic corresponds to a fragment of monadic second order logic, because the definition of validity involves quantifying over all the proposition letters in the formulas. However, most work has concentrated on the first order aspect of modal definability.

A landmark result is the Goldblatt-Thomason Theorem [GT75] which gives a characterization of the first order definable classes of frames that are modally definable, in terms of semantic criteria. It is undecidable whether a given first order sentence corresponds to a modal formula, in the sense that they define the same class of frames.

On the level of Kripke models, the expressive power of the μ-calculus is well understood. In [JW96], David Janin and Igor Walukiewicz showed that the μ-calculus is the bisimulation invariant fragment of **MSO**. It is undecidable whether a class of Kripke models definable in **MSO** is definable by a formula of the μ-calculus. For classes of trees, the problem becomes decidable (see [JW96]).

About the expressive power of the μ-calculus on the level of Kripke frames, nothing is known. This paper contributes to a partial solution of this question by giving a characterization of the **MSO** definable classes of trees that are definable by a μ-formula. Our main result states that an **MSO** definable class of trees is definable in the μ-calculus iff it is closed for subtrees and p-morphic images. We also show that given an **MSO** formula, it is decidable whether there exists a μ-formula which defines the same class of trees as the **MSO** formula.

The proof is in three steps. First, we use the connection between **MSO** and the graded μ-calculus proved by Igor Walukiewicz [Wal02] and establish a correspondence between the **MSO** formulas that are preserved under p-morphic images and a fragment that is between the μ-calculus and the graded μ-calculus (the fragment with a counting $\Box$ operator and a usual $\Diamond$ operator). We call this fragment the $\Box$-graded μ-calculus.

The second step consists in showing that each $\Box$-graded μ-formula φ can be translated into a μ-formula ψ such that locally, the truth of φ (on trees seen as Kripke models) corresponds to the validity of ψ (on trees seen as Kripke frames). So this step is a move from the model perspective to the frame perspective. The last step consists in shifting from the local perspective to the global one (that is, we are interested in validity at all points, not at a given point).

1 Preliminaries

μ-calculus. The set of formulas of the *μ-calculus* (over a set *Prop* of proposition letters and a set *Var* of variables) is given by

$$\varphi ::= \top \mid p \mid \neg p \mid x \mid \varphi \vee \varphi \mid \varphi \wedge \varphi \mid \Diamond\varphi \mid \Box\varphi \mid \mu x.\varphi \mid \nu x.\varphi,$$

where p ranges over the set *Prop*, x ranges over the set *Var*.

A *Kripke frame* over a set *Prop* is a pair (W, R), where W is a set and R a binary relation on W. A *Kripke model* over *Prop* is a triple (W, R, V) where (W, R) is a Kripke frame and $V : Prop \to \mathcal{P}(W)$ a valuation.

Given a formula φ, a Kripke model $\mathcal{M} = (W, R, V)$ and an assignment $\tau : Var \to \mathcal{P}(W)$, we define a subset $[\![\varphi]\!]_{\mathcal{M},\tau}$ that is interpreted as the set of points at which φ is true. We only recall that

$$[\![\mu x.\varphi]\!]_{\mathcal{M},\tau} = \bigcap\{U \subseteq W \mid [\![\varphi]\!]_{\mathcal{M},\tau[x:=U]} \subseteq U\},$$
$$[\![\nu x.\varphi]\!]_{\mathcal{M},\tau} = \bigcup\{U \subseteq W \mid U \subseteq [\![\varphi]\!]_{\mathcal{M},\tau[x:=U]}\},$$

where $\tau[x := U]$ is the assignment τ' such that $\tau'(x) = U$ and $\tau'(y) = \tau(y)$, for all $y \neq x$. The set $[\![\mu x.\varphi]\!]_{\mathcal{M},\tau}$ is the least fixpoint of the map $\varphi_x : \mathcal{P}(W) \to \mathcal{P}(W)$ defined by $\varphi_x(U) := [\![\varphi]\!]_{\mathcal{M},\tau[x:=U]}$, for all $U \subseteq W$.

In case $w \in [\![\varphi]\!]_{\mathcal{M},\tau}$, we write $\mathcal{M}, w \Vdash_\tau \varphi$ and we say that φ is *true* at w. If all the variables in φ are bound, we simply write $\mathcal{M}, w \Vdash \varphi$. A formula φ is *true* in $\mathcal{M}$, notation: $\mathcal{M} \Vdash \varphi$, if for all $w \in W$, we have $\mathcal{M}, w \Vdash \varphi$. Two formulas φ and ψ are *equivalent* if for all models $\mathcal{M}$ and for all $w \in \mathcal{M}$, $\mathcal{M}, w \Vdash \varphi$ iff $\mathcal{M}, w \Vdash \psi$.

If (W, R) is a Kripke frame and w belongs to W, we say that φ is *valid* at w if for all valuations V, φ is true at w in (W, R, V). We use the notation $(W, R), w \Vdash \varphi$. Finally, φ is *valid* in (W, R), notation: $(W, R) \Vdash \varphi$, if φ is valid at w, for all w in W.

Trees. Our characterizations apply only to classes of trees, not classes of arbitrary Kripke frames.

Let (W, R) be a Kripke frame. A point r in W is a *root* if for all w in W, there is a sequence $w_0, \ldots, w_n$ such that $w_0 = r$, $w_n = w$ and (w_i, w_{i+1}) belongs to R, for all $i \in \{0, \ldots, n-1\}$. The frame (W, R) is a *tree* if it has a root, every point distinct from the root has a unique predecessor and the root has no predecessor. If (W, R, V) is a Kripke model over a set *Prop* and (W, R) is a tree, we say that (W, R, V) is a *tree Kripke model* over *Prop* or simply a tree over *Prop* or a tree model. Two formulas φ and ψ over *Prop* are *equivalent on tree models* if for all trees t over *Prop* with root r, $t, r \Vdash \varphi$ iff $t, r \Vdash \psi$.

If the frame (W, R) is a tree, v is *child* of w if $(w, v) \in R$ and we write $Child(w)$ to denote the children of w. A *subtree* of a tree t is a tree consisting of a node in t and all of its descendants in t. If t is a tree and u is a node of t, we let $t|_u$ denote the subtree of t at position u.

A class of trees L over *Prop* is a *regular class of trees* if there exists an **MSO** formula α such that for all trees t, t belongs to L iff α is valid on t. When this happens, we say that α *defines* L.

2 μ-definability on Trees

We are interested in characterizing the regular classes of trees (seen as Kripke frames) that are definable using μ-formulas. The characterization we propose, is very natural and only involves two well-known notions of modal logic: subtree and p-morphism. We recall these notions together with the notion of definability and state our main result (Theorem 1).

μ-definability. A class of trees L is *μ-definable* if there exists a μ-formula φ such that L is exactly the class of trees which make φ valid.

p-morphisms. Let $\mathcal{M} = (W, R, V)$ and $\mathcal{M}' = (W', R', V')$ be two models. A map $f : W \to W'$ is a *p-morphism* between $\mathcal{M}$ and $\mathcal{M}'$ if the two following conditions hold. For all $w, v \in W$ such that wRv, $f(w)R'f(v)$. For all $w \in W$ and $v' \in W'$ such that $f(w)R'v'$, there exists $v \in W$ such that $f(v) = v'$ and wRv.

A class L of Kripke models is *closed under p-morphic images* if for all Kripke models $\mathcal{M} \in L$ and and for all Kripke models $\mathcal{M}'$ such that there is a surjective p-morphism between $\mathcal{M}$ and $\mathcal{M}'$, we have that $\mathcal{M}'$ belongs to L. An **MSO** formula α is *preserved under p-morphic images for tree models* if the class of tree models defined by α is closed under p-morphic images.

Closure for subtrees. A class of trees L is *closed for subtrees* if for all $t \in L$ and all $u \in t$, we have that $t|_u$ belongs to L.

Theorem 1. *A regular class of trees is μ-definable iff it is closed under p-morphic images for tree models and closed for subtrees.*

The rest of this paper is devoted to the proof of Theorem 1. First we characterize the class of regular classes of trees which are preserved under p-morphic images. It corresponds to some fragment of the graded μ-calculus (roughly, the fragment where we allow counting with the $\Box$ operator, but not the $\Diamond$ operator). Next we prove that this fragment defines the regular classes of trees which are of the form $\{t \text{ tree} \mid \text{ for all } V : \mathit{Prop}' \to \mathcal{P}(t), (t, V), r \Vdash \varphi\}$, where r is the root of t and φ is a formula of the μ-calculus. Finally we show how to derive Theorem 1.

3 Graded μ-calculus: Connection with MSO and Disjunctive Normal Form

An important tool for characterizing the class of regular tree languages which are preserved under p-morphic images, is the connection between **MSO** and graded μ-calculus. We also use fact that there is a normal form for the graded μ-formulas (when they are expressed using a ∇-like operator).

Graded μ-calculus. The set μGL of formulas of the *graded μ-calculus* (over a set *Prop* of proposition letters and a set *Var* of variables) is given by

$$\varphi ::= \top \mid p \mid x \mid \varphi \vee \varphi \mid \varphi \wedge \varphi \mid \Diamond^k \varphi \mid \Box^k \varphi \mid \mu x.\varphi \mid \nu x.\varphi,$$

where p ranges over the set *Prop*, x ranges over the set *Var* and k is a natural number. Given a formula φ, a model $\mathcal{M} = (W, R, V)$, an assignment $\tau : \mathit{Var} \to \mathcal{P}(W)$ and $w \in W$, the relation $\mathcal{M}, w \Vdash \varphi$ is defined by induction as in the case of the μ-calculus with the extra conditions:

$\mathcal{M}, w \Vdash_\tau \Diamond^k \varphi$ if there exist successors $w_0, \ldots, w_k$ of w s.t. for all $i \neq j$, $w_i \neq w_j$ and for all $i \in \{0, \ldots, k\}$, $\mathcal{M}, w_i \Vdash \varphi$

$\mathcal{M}, w \Vdash_\tau \Box^k \varphi$ if w has no successors or there exist successors $w_1, \ldots, w_k$ of w s.t. for all $w \notin \{w_1, \ldots, w_k\}$, $\mathcal{M}, w \Vdash \varphi$.

We define a ∇-like operator corresponding to graded μ-calculus (inspired by [Wal02]). The ∇-formulas of the graded μ-calculus correspond exactly to the formulas of the graded μ-calculus.

∇ operator for the graded μ-calculus. Given a multiset Φ of formulas, the *multiplicity* of a formula φ in Φ is the number of occurrences of φ in Φ. The total number of elements in a multiset, including repeated memberships, is the *cardinality* of the multiset. We denote by $card(\Phi)$, the cardinality of Φ. A *literal* over a set *Prop* is a proposition letter in *Prop* or the negation of a proposition letter.

The set $\mu\mathrm{GL}^\nabla$ of ∇-formulas of the graded μ-calculus (over a set *Prop* of proposition letters and a set *Var* of variables) is given by:

$$\varphi ::= x \mid \varphi \vee \varphi \mid \varphi \wedge \varphi \mid \Pi \bullet \nabla^g(\Phi; \Psi) \mid \mu x.\varphi \mid \nu x.\varphi,$$

where x ranges over the set *Var*, Π is a conjunction literals or $\Pi = \top$, Φ is a multiset of formulas and Ψ is a finite set of formulas.

Given a formula φ, a model $\mathcal{M} = (W, R, V)$, an assignment $\tau : Var \to \mathcal{P}(W)$ and $w \in W$, the relation $\mathcal{M}, w \Vdash_\tau \varphi$ is defined by induction as in the case of the μ-calculus with the extra condition: $\mathcal{M}, w \Vdash_\tau \Pi \bullet \nabla^g(\Phi, \Psi)$ iff $\mathcal{M}, w \Vdash_\tau \Pi$ and for some $\{w_\varphi$ successor of $w \mid \varphi \in \Phi\}$, we have

1. the size of the set $\{w_\varphi \mid \varphi \in \Phi\}$ is equal to $card(\Phi)$,
2. $\mathcal{M}, w_\varphi \Vdash_\tau \varphi$,
3. for all successors u of w such that $u \notin \{w_\varphi \mid \varphi \in \Phi\}$, $\mathcal{M}, u \Vdash_\tau \bigvee \Psi$.

A map $m : \mu\mathrm{GL}^\nabla \to \mathcal{P}(R[w])$ is a *∇^g-marking* for (Φ, Ψ) if there exists a set $\{w_\varphi \mid \varphi \in \Phi\}$ of size $card(\Phi)$ such that $w_\varphi \in m(\varphi)$ and for all successors u of w such that $u \notin \{w_\varphi \mid \varphi \in \Phi\}$, there is $\psi \in \Psi$ such that $u \in m(\psi)$.

A formula in $\mu\mathrm{GL}^\nabla$ is in *disjunctive normal form* if its only subformulas of the form $\varphi_0 \wedge \varphi_1$ are such that φ_0 and φ_1 are literals or conjunctions of literals.

The next result is proved by a standard (although a bit tedious) induction on the complexity of the formulas.

Proposition 1. *Each formula in μGL is equivalent to a formula in $\mu\mathrm{GL}^\nabla$. Each formula in $\mu\mathrm{GL}^\nabla$ is equivalent to a formula in μGL.*

In [Wal02], Igor Walukiewicz showed that on trees, **MSO** is equivalent to first order logic extended with the unary fixpoint operator. By adapting[1] the proof of Lemma 44 in [Wal02], we can obtain the following result (see also [JL03]).

[1] This adaptation is mainly based on the following observation (which is immediate from Proposition 1): For all formulas $\varphi \in \mathrm{DBF(n)}$ (as defined in [Wal02]), there is $\psi \in \mu$GL such that for all trees t, for all nodes u in t, ψ is true at u iff the formula obtained from φ by relativizing all the quantifiers to the children of u, holds.

Theorem 2 (from [Wal02]). *For every **MSO** formula α, there is a graded μ-formula φ such that for all trees t with root r, α is valid on t iff φ is true at r. For every graded μ-formula φ, there is an **MSO** formula α such that for all trees t with root r, φ is true at r iff α is valid on t.*

As mentioned earlier, a key result for one of our proofs is the fact that the graded μ-calculus has a normal form. This follows from a result proved by David Janin in [JW95].

Theorem 3. *Each formula of the graded μ-calculus is equivalent to a formula of the graded μ-calculus in disjunctive normal form.*

4 $\Box$-graded μ-Calculus and Preservation under p-morphic Images

We establish a correspondence between the **MSO** formulas that are preserved under p-morphic images and some set of formulas, that is in between the μ-calculus and the graded μ-calculus. We call this set the set of $\Box$-graded formulas, as we are only allowed to count with the $\Box$ operator. For this set of $\Box$-graded formulas, we also introduce a ∇-like operator, that we write ∇'.

$\Box$-graded μ-calculus. The set $\mu\mathrm{GL}^{\Box}$ of fixpoint $\Box$-graded formulas (over a set *Prop* of proposition letters and a set *Var* of variables) is given by

$$\varphi ::= \top \mid p \mid x \mid \varphi \vee \varphi \mid \varphi \wedge \varphi \mid \Diamond\varphi \mid \Box^k\varphi \mid \mu x.\varphi \mid \nu x.\varphi,$$

where p ranges over the set *Prop*, x ranges over the set *Var* and k is a natural number. Given a formula φ, a model $\mathcal{M} = (W, R, V)$, an assignment $\tau : Var \to \mathcal{P}(W)$ and $w \in W$, the relation $\mathcal{M}, w \Vdash_\tau \varphi$ is defined by induction as in the case of the graded μ-calculus, with $\Diamond = \Diamond^0$.

The set $\mu\mathrm{GL}^{\nabla'}$ of ∇'-formulas of the graded μ-calculus are given by:

$$\varphi ::= x \mid \varphi \vee \varphi \mid \varphi \wedge \varphi \mid \Pi \bullet \nabla'(\Phi; \Psi) \mid \mu x.\varphi \mid \nu x.\varphi,$$

where x ranges over the set *Var* of variables, Π is a conjunction of literals or $\Pi = \top$, Φ is a multiset of formulas and Ψ is a finite set of formulas.

Given a formula φ, a Kripke model $\mathcal{M} = (W, R, V)$, an assignment $\tau : Var \to \mathcal{P}(W)$ and a point $w \in W$, the relation $\mathcal{M}, w \Vdash_\tau \varphi$ is defined by induction as in the case of the μ-calculus with the extra condition: $\mathcal{M}, w \Vdash_\tau \Pi \bullet \nabla^g(\Phi, \Psi)$ iff $\mathcal{M}, w \Vdash_\tau \Pi$ and for some $\{w_\varphi$ successor of $w \mid \varphi \in \Phi\}$, we have

1. $\mathcal{M}, w_\varphi \Vdash_\tau \varphi$,
2. for all successors u of w such that $u \notin \{w_\varphi \mid \varphi \in \Phi\}$, $\mathcal{M}, u \Vdash_\tau \bigvee \Psi$.

A map $m : \mu\mathrm{GL}^{\nabla'} \to \mathcal{P}(R[w])$ is a ∇'-*marking* for (Φ, Ψ) if the set there exists a set $\{w_\varphi \mid \varphi \in \Phi\}$ such that $w_\varphi \in m(\varphi)$ and for all successors u of w such that $u \notin \{w_\varphi \mid \varphi \in \Phi\}$, there is $\psi \in \Psi$ such that $u \in m(\psi)$.

The only difference between a ∇'-marking and a ∇^g-marking is that the set of successors associated to a ∇'-marking might not contain points which are pairwise distinct (as in the case for the ∇^g-marking).

A formula $\varphi \in \mathrm{GL}^{\Box} \cup \mu\mathrm{GL}^{\nabla'}$ is *equivalent on tree models* to an **MSO** formula α if for all tree models t, α is valid on t iff φ is true at the root of t.

We also introduce a game semantic for the languages $\mu\mathrm{GL}^{\nabla}$ and $\mu\mathrm{GL}^{\nabla'}$. The fact the the existence of a winning strategy for a player in the game corresponds to the truth of a formula at a given point, is proved using classical methods (as in the case of μ-calculus).

Game semantics. Let φ be a formula in $\mu\mathrm{GL}^{\nabla} \cup \mu\mathrm{GL}^{\nabla'}$ such that each variable in φ is bound. Without loss of generality, we may assume that for all $x \in Var$ which occurs in φ, there is a unique subformula of φ, which is of the form $\eta x.\delta_x$, where $\eta \in \{\mu, \nu\}$. We also fix a model $\mathcal{M} = (W, R, V)$. We define the *evaluation game* $\mathcal{E}(\mathcal{M}, \varphi)$ as a parity game between two players, $\forall$ and $\exists$. The rules of the game are given in the table below.

Position	Player	Possible moves
$(\top, w)$	$\forall$	$\emptyset$
(x, w)	-	$\{(\delta_x, w)\}$
$(\varphi_1 \wedge \varphi_2, w)$	$\forall$	$\{(\varphi_1, w), (\varphi_2, w)\}$
$(\varphi_1 \vee \varphi_2, w)$	$\exists$	$\{(\varphi_1, w), (\varphi_2, w)\}$
$(\eta x.\psi, w)$	-	$\{(\psi, w)\}$
$(\Pi \bullet \nabla^g(\Phi, \Psi), w)$	$\exists$	$\{m : \mu\mathrm{GL}^{\nabla^g} \to \mathcal{P}(R[w]) \mid \mathcal{M}, w \Vdash \Pi$ and m ∇^g-marking for $(\Phi, \Psi)\}$
$(\Pi \bullet \nabla'(\Phi, \Psi), w)$	$\exists$	$\{m : \mu\mathrm{GL}^{\nabla'} \to \mathcal{P}(R[w]) \mid \mathcal{M}, w \Vdash \Pi$ and m ∇'-marking for $(\Phi, \Psi)\}$
$m : \mu\mathrm{GL}^{\nabla} \cup \mu\mathrm{GL}^{\nabla'} \to \mathcal{P}(R[w])$	$\forall$	$\{(\psi, u) \mid u \in m(\psi)\}$

where w belongs to W, x belongs to Var, η belongs to $\{\mu, \nu\}$, φ_1, φ_2 and ψ belongs to $\mu\mathrm{GL}^{\nabla} \cup \mu\mathrm{GL}^{\nabla'}$, Π is a conjunction of literals or $\Pi = \top$, Ψ is a finite subset of $\mu\mathrm{GL}^{\nabla} \cup \mu\mathrm{GL}^{\nabla'}$, Φ is a finite multiset of formulas in $\mu\mathrm{GL}^{\nabla} \cup \mu\mathrm{GL}^{\nabla'}$.

If a match is finite, the player who get stuck, loses. If a match is infinite, we let *Inf* be the set of variables x such that there are infinitely many positions of the form (x, w) in the match. Let x_0 be a variable in *Inf* such that for all variables $x \in \mathit{Inf}$, δ_x is a subformula of δ_{x_0}. If x_0 is bound by a μ-operator, then $\forall$ wins. Otherwise $\exists$ wins.

The notions of *strategy* and *winning strategy* for a player are defined as usual. If h is a strategy for a player P, a match during which P plays according to h is called an *h-conform match.*

Proposition 2. *Let φ be a formula in $\mu\mathrm{GL}^{\nabla} \cup \mu\mathrm{GL}^{\nabla'}$. For all Kripke models $\mathcal{M} = (W, R, V)$ and all $w \in W$, $\mathcal{M}, w \Vdash \varphi$ iff $\exists$ has a winning strategy in the game $\mathcal{E}(\mathcal{M}, \varphi)$ with starting position (w, φ).*

We are now ready to show that modulo equivalence on trees, the **MSO** formulas preserved under p-morphic images are exactly the $\Box$-graded formulas.

Proposition 3. *Let α be an* **MSO** *formula. The following are equivalent:*

- *α is equivalent on tree models to a $\Box$-graded formula,*
- *α is equivalent on tree models to a formula in $\mu\mathrm{GL}^{\nabla'}$,*
- *α is preserved under p-morphic images for tree models .*

Proof. We only give details for the hardest implication which is that an **MSO** formula which is preserved under p-morphic images for tree models is equivalent on tree models to a ∇'-graded fixpoint formula. Let α be an **MSO** formula that is preserved under p-morphic images for tree models. By Theorem 2, there is a graded μ-formula χ such that for all tree models t with root r, α is valid on t iff χ is true at r. By Proposition 3, we may assume the formula χ to be in disjunctive normal form.

Now let δ be the formula χ in which we replace each operator ∇^g by ∇'. We show that under the assumption that α is preserved under p-morphic images for trees, χ and δ are equivalent on tree models. It is easy to check that for all tree models, if χ holds at a node, then δ also holds at the node.

For the other direction, let t be a tree model with root r and suppose that δ is true at r. We have to show that χ is true at r. Since δ is true at r, $\exists$ has a winning strategy h in the evaluation game with starting position (r, δ). We start by fixing some notation. We denote by $\mathbb{N}^*$ the set of finite sequences over $\mathbb{N}$. The empty sequence ϵ belongs to $\mathbb{N}^*$. If φ is a ∇'-formula, we write φ^g for the formula obtained by replacing ∇' by ∇^g in φ. If Φ is set (multiset) of formulas, we write Φ^g for the set (multiset) $\{\varphi^g \mid \varphi \in \Phi\}$. We also let A be the set $\{(u, \boldsymbol{n}) \mid u \text{ is a node of } t \text{ and } \boldsymbol{n} \in \mathbb{N}^*)\}$.

The idea is to define a new tree model t' the domain of which is a subset of A and the child relation of which is such that for all $(u_1, (n_1, \ldots, n_k))$ and $(u_2, \boldsymbol{m})$ in t', $(u_2, \boldsymbol{m})$ is a child of $(u_1, (n_1, \ldots, n_k))$ iff u_2 is a child of u_1 in t and there is a natural number n_{k+1} such that $\boldsymbol{m} = (n_1, \ldots, n_k, n_{k+1})$. The depth of a node $(u_1, (n_1, \ldots, n_k))$ in t' is $k+1$. The tree t' will also be such that its root is (r, ϵ). Finally, we define a positional strategy h' for the evaluation game $\mathcal{E}(t', \chi)$ with starting position $((r, \epsilon), \chi)$ which satisfies the following conditions. For all $(u, \boldsymbol{n}) \in t$, there is exactly one match during which $(u, \boldsymbol{n})$ occurs. Moreover, a position of the form $((u, \boldsymbol{n}), \varphi^g)$ is reached in an h'-conform match iff the position (u, φ) is reached in an h-conform match.

The definitions of t' and h' will be by induction. More precisely, at stage i of the induction, we specify which are the nodes of t' of depth i and we also define $\exists$'s answer (according to h') when a position of the form $((u, \boldsymbol{n}), \varphi^g)$ is reached, where the depth of $(u, \boldsymbol{n})$ in t' is $i - 1$.

For the basic case, the only node of depth 1 in t' is the node (r, ϵ). For the induction step, take $i > 1$ and suppose that we know already which are the nodes in t' of depth at most i and that we also have defined the strategy h' for all positions of the form $((u, \boldsymbol{n}), \varphi)$, where the depth of $(u, \boldsymbol{n})$ in t' is at most $i - 1$. We have to specify which points of the form $(u, (n_1, \ldots, n_i))$ belongs to t' and what is the strategy for the points of depth i. Let $(u, \boldsymbol{n}) = (u, (n_1, \ldots, n_{i-1}))$ be a node in t' of depth i (note that if $i = 1$, then $(n_1, \ldots, n_{i-1})$ is ϵ). Suppose that in a (partially defined) h'-conform match π', a position of the form $((u, \boldsymbol{n}), \varphi^g)$

is reached and if $\boldsymbol{n} = \epsilon$, we can assume that $\varphi = \delta$ and $\varphi^g = \chi$. By induction hypothesis, such a match π' is unique. Moreover, we know (by induction hypothesis if $i > 1$ or trivially if $i = 1$) that the position (u, φ) is reached in an h-conform match π. Now, there are different possibilities depending on the shape of φ. First, suppose that φ is a disjunction $\varphi_1 \vee \varphi_2$. Then, in the h-conform match π, the position following (u, φ) is of the form (u, ψ), where ψ is either φ_1 or φ_2. We define h' such that the position following $((u, \boldsymbol{n}), \varphi^g)$ is $((u, \boldsymbol{n}), \psi^g)$. Finally, suppose that φ^g is of the form $\Pi \bullet \nabla^g(\Phi^g; \Psi^g)$. Then, in the h-conform match π, the position following (u, φ) is a marking $m : \Phi \cup \Psi \to \mathcal{P}(Child(s))$ such that m is a ∇'-marking for (Φ, Ψ). In the h'-conform match π', we first define which are the children of $(u, \boldsymbol{n})$ in t' and then, we give a ∇^g-marking $m^g : Child(u, \boldsymbol{n}) \to \mathcal{P}(\Psi^g \cup \Gamma^g)$ for (Φ^g, Ψ^g).

Since m is a ∇'-marking for (Φ, Ψ), there exists $\{u_\varphi \mid \varphi \in \Phi\}$ such that the two following conditions holds. For all $\varphi \in \Phi$, u_φ is a child of u and u_φ belongs to $m(\varphi)$. For all children v of u such that v does not belong to $\{u_\varphi \mid \varphi \in \Phi\}$, there exists $\psi \in \Psi$ such that v belongs to $m(\psi)$. We let $u_1, \ldots, u_k$ be the children of u such that $\{u_1, \ldots, u_k\} = \{u_\psi \mid \psi \in \Psi\}$ and $u_i \neq u_j$, if $i \neq j$. Fix i in $\{1, \ldots, k\}$. Let Φ_i be the biggest submultiset of Φ such that for all φ in Φ_i, $u_\varphi = u_i$. We let $k(i)+1$ be the size of Φ_i and we fix an arbitrary bijection f_i between Φ_i and the set $\{0, \ldots, k(i)\}$. Now, we add to t' the set of nodes

$$\{(u_i, (n_1, \ldots, n_{i-1}, j)) \mid i \in \{1, \ldots, k\}, 0 \leq j \leq k(i)\} \cup$$
$$\{(v, (n_1, \ldots, n_{i-1}, 0)) \mid v \text{ child of } u, v \notin \{u_1, \ldots, u_k\}\}.$$

These points are the children of $(u, \boldsymbol{n})$ in t'.

We are now going to define a ∇^g-marking $m^g : \Phi^g \cup \Psi^g \to \mathcal{P}(Child(s, \boldsymbol{n}))$ for (Φ^g, Ψ^g). Fix a formula φ^g in Φ^g. We define $m^g(\varphi^g)$ as $\{(u_i, (n_1, \ldots, n_{i-1}, j))\}$, if φ belongs to Φ_i and $f_i(\varphi) = j$. Next fix a formula ψ^g in Ψ^g. We define m^g such that $m^g(\psi^g) = \{(v, (n_1, \ldots, n_{i-1}, 0)) \mid v \notin \{u_1, \ldots, u_k\}, v \in m(\psi)\}$. The proofs that m^g is a ∇^g-marking for (Φ^g, Ψ^g) and that the induction hypothesis remain true are standard. This finishes the definition of t' and h'.

It is easy to check that the strategy h' is winning for $\exists$ in the evaluation game $\mathcal{E}(t', \chi)$ with starting position $((r, \epsilon), \chi)$. Therefore, the formula χ is true at the root of t'. Now the map which sends a node $(u, \boldsymbol{n})$ to u is a surjective p-morphism between t' and t. So t is a p-morphic image of t'. Since α is preserved under p-morphic images for trees, χ is also true at the root of t and this finishes the proof that χ and δ are equivalent on tree models. It follows that χ and α are also equivalent on trees.

5 μ-definability at the Root and μ-definability

μ-definability at the root. A tree language L over *Prop* is *μ-definable at the root* if there are a set *Prop'* of proposition letters and a μ-formula φ over $Prop \cup Prop'$ such that L is equal to $\{t \text{ tree} \mid \text{ for all } V : Prop' \to \mathcal{P}(t), (t, V), r \Vdash \varphi\}$, where r is the root of t.

Proposition 4. *A regular class of trees over Prop is μ-definable at the root iff it is closed under p-morphic images for tree models.*

Proof. The only difficult direction is from right to left. Let L be a regular class of trees over *Prop* that is closed under p-morphic images for trees. By Proposition 3, there is a $\Box$-graded formula φ over *Prop* such that for all trees t over *Prop* with root r, φ is true at r iff t belongs to L. Now we show by induction on the complexity of φ that there exist a set of propositions $Prop'$ and a μ-formula ψ over $Prop \cup Prop'$ such that for all trees t over *Prop* with root r and for all assignments $\tau : Var \to \mathcal{P}(t)$,

$$t, r \Vdash_\tau \varphi \quad \text{iff} \quad \text{for all valuations } V : Prop' \to \mathcal{P}(t),\ (t, V), r \Vdash_\tau \psi. \tag{1}$$

When this happens, we will say that ψ is a *μ-translation* of φ. Moreover, we prove that for all trees t over *Prop* and for all assignments $\tau : Var \to \mathcal{P}(t)$, there is a valuation $V_\psi(\tau) : Prop' \to \mathcal{P}(t)$ such that for all nodes u in t,

$$(t, V_\psi(\tau)), u \Vdash_\tau \psi \quad \text{iff} \quad \text{for all valuations } V : Prop' \to \mathcal{P}(t),\ (t, V), u \Vdash_\tau \psi. \tag{2}$$

When a valuation $V_\psi(\tau) : Prop' \to \mathcal{P}(t)$ satisfies condition (2), we say that $V_\psi(\tau)$ is a *distinctive valuation* for t, τ and ψ.

We only treat the most two difficult cases where φ is a formula of the form $\Box^k \varphi_1$ or of the form $\eta x.\varphi_1$, where η belongs to $\{\mu, \nu\}$.

Suppose that φ is of the form $\Box^k \varphi_1$. By induction hypothesis, there is a set $Prop'_1$ and there is a μ-formula ψ_1 over $Prop \cup Prop'_1$ such that ψ_1 is a μ-translation of φ_1. We let $p_0, \ldots, p_k$ be fresh proposition letters and we define $Prop'$ as $Prop'_1 \cup \{p_0, \ldots, p_k\}$. We let ψ be the formula

$$\bigvee\{\Box(\neg p_i \vee \psi_1) \mid 0 \leq i \leq k\} \vee \bigvee\{\Diamond(p_i \wedge p_{i'}) \mid 0 \leq i, i' \leq k, i \neq i'\}.$$

Fix a tree model t and an assignment $\tau : Var \to \mathcal{P}(t)$. We define $V_\psi(\tau)$. Let V_0 be the assignment $V_{\psi_1}(\tau)$ and let u be a point in t. We define $U = \{u_j \mid j \in J\}$ as the biggest set of successors of u such that for all $j \in J$, we have $(t, V_0), u_j \nVdash \psi_1$. Suppose first that the size of U is less or equal to k. So U is a set of the form $\{u_1, \ldots, u_n\}$, where $n \leq k$. Then we can fix a valuation V_u which satisfies the following. For all $i \in \{0, \ldots, k\}$, $V_u(p_i)$ is a set of the form $\{u_i\}$, where $u_i \in U$. Moreover, for all $v \in U$, there is $i \in \{0, \ldots, k\}$ such that $V_u(p_i) = \{v\}$.

Next suppose that U is infinite or U is a finite set of size strictly greater than k. Then we can fix an arbitrary valuation V_u such that the following hold. For all $i \in \{0, \ldots, k\}$, $V_u(p_i)$ is a set of the form $\{u_i\}$, where $u_i \in U$. Moreover, for all $i, i' \in \{0, \ldots, k\}$, if $i \neq i'$, then $u_i \neq u_{i'}$. That is, for all $i, i' \in \{0, \ldots, k\}$, if $i \neq i'$, then $V_u(p_i) \cap V_u(p_j) = \emptyset$.

We are now ready to define $V_\psi(\tau)$. For all propositions $p' \in Prop'$, we have

$$V_\psi(\tau)(p') = \begin{cases} \bigcup\{V_u(p_i) \mid u \in t\} & \text{if } p' = p_i \text{ for some } i \in \{0, \ldots, k\}, \\ V_0(p') & \text{otherwise.} \end{cases}$$

The proofs ψ is a μ-translation of φ and that $V_\psi(\tau)$ is a distinctive valuation for t, are left as an exercise to the reader.

Next, suppose that φ is a formula of the form $\eta x.\varphi_1$, where η belongs to $\{\mu, \nu\}$. We will assume that $\eta = \mu$, but the proof can be easily adapted to the case where $\eta = \nu$. By induction hypothesis, there is a set $Prop'_1$ and there is a μ-formula ψ_1 over $Prop \cup Prop'_1$ such that ψ_1 is a μ-translation of φ_1. Moreover, for all trees t and all assignments $\tau : Var \to \mathcal{P}(t)$, there is a distinctive valuation $V_{\psi_1}(\tau)$ for t, τ and ψ_1.

Now we define ψ as $\mu x.\psi_1$ and given a tree model t and an assignment $\tau : Var \to \mathcal{P}(t)$, we define $V_\psi(\tau)$ as the valuation $V_{\psi_1}(\tau[x \mapsto U_0])$, where $U_0 := [\![\mu x.\varphi_1]\!]_{t,\tau}$ $(= [\![\varphi]\!]_{t,\tau})$. We have to show that $V_\psi(\tau)$ is a distinctive valuation for t, τ and ψ and that ψ is a μ-translation of φ. It is sufficient to prove that for all trees t, all assignments $\tau : Var \to \mathcal{P}(t)$ and all $u \in t$, we have

$$t, u \Vdash_\tau \varphi \quad \Rightarrow \quad \text{for all valuations } V : Prop' \to \mathcal{P}(t),\ (t, V), u \Vdash_\tau \psi, \qquad (3)$$

$$(t, V_\psi(\tau)), u \Vdash_\tau \psi \quad \Rightarrow \quad t, u \Vdash_\tau \varphi. \qquad (4)$$

First, we show that condition (3) holds. So suppose that $t, u \Vdash_\tau \mu x.\varphi_1$ and let $V : Prop' \to \mathcal{P}(t)$ be a valuation. We have to verify that $(t, V), u \Vdash_\tau \mu x.\psi_1$. That is, for all subsets U of t such that $[\![\psi_1]\!]_{(t,V),\tau[x\mapsto U]} \subseteq U$, we have $u \in U$.

So fix a subset U of t such that $[\![\psi_1]\!]_{(t,V),\tau[x\mapsto U]} \subseteq U$. Since ψ_1 is a μ-translation of φ_1, we know that $[\![\varphi_1]\!]_{t,\tau[x\mapsto U]}$ is a subset of $[\![\psi_1]\!]_{(t,V),\tau[x\mapsto U]}$. It follows that $[\![\varphi_1]\!]_{t,\tau[x\mapsto U]}$ is a subset of U. That is, U is a pre-fixpoint of $(\varphi_1)_x$ in t under the assignment τ. Since $t, u \Vdash_\tau \mu x.\varphi_1$, this implies that u belongs to U and this finishes the proof of implication (3).

Next we show that implication (4) is true. Assume that $(t, V_\psi(\tau)), u \Vdash_\tau \mu x.\psi_1$. We have to prove that u belongs to $[\![\varphi]\!]_{t,\tau}$. That is, u belongs to U_0. Since $(t, V_\psi(\tau)), u \Vdash_\tau \mu x.\psi_1$, u belongs to all the pre-fixpoints of the map $(\psi_1)_x$ in the tree $(t, V_\psi(\tau))$ under the assignment τ. So it is sufficient to show that the set U_0 is a pre-fixpoint of the map $(\psi_1)_x$ in the tree $(t, V_\psi(\tau))$ under the assignment τ. That is,

$$[\![\psi_1]\!]_{(t,V_\psi(\tau)),\tau_0} \subseteq U_0, \qquad (5)$$

where τ_0 is the assignment $\tau[x \mapsto U_0]$. By definition of $V_\psi(\tau)$, we have that $[\![\psi_1]\!]_{(t,V_\psi(\tau)),\tau_0}$ is equal to $[\![\psi]\!]_{(t,V_{\psi_1}(\tau_0)),\tau_0}$. Recall that for all assignments $\tau' : Var \to \mathcal{P}(t)$, we have that $[\![\psi_1]\!]_{(t,V_{\psi_1}(\tau')),\tau'}$ is equal to $[\![\varphi_1]\!]_{t,\tau'}$, as $V_{\psi_1}(\tau')$ is a distinctive valuation for t, τ' and ψ_1. In particular, $[\![\psi_1]\!]_{(t,V_{\psi_1}(\tau_0)),\tau_0}$ is equal to $[\![\varphi_1]\!]_{t,\tau_0}$. Since $U_0 = [\![\mu x.\varphi_1]\!]_{t,\tau}$, we also have that $[\![\varphi_1]\!]_{t,\tau_0}$ is equal to $[\![\mu x.\varphi_1]\!]_{t,\tau}$. That is, $[\![\varphi_1]\!]_{t,\tau_0}$ is equal to U_0. Putting everything together, we obtain that $[\![\psi_1]\!]_{(t,V_\psi(\tau)),\tau_0}$ is equal to U_0. Condition (5) immediately follows.

We can now prove Theorem 1.

Proof. We concentrate on the hardest direction, which is from left to right. Let L be a regular tree language, which is closed under p-morphic images for tree models and closed for subtrees. It follows from Proposition 4 that there is a μ-formula φ such that for all trees t, t belongs to L iff the formula φ is valid at the root of t. Now we prove that for all trees t, t belongs to L iff the formula φ is valid at all the nodes of t. It will immediately follow that L is μ-definable.

For the direction from right to left, let t be a tree such that φ is valid at all the nodes of t. In particular, φ is valid at the root. Therefore, t belongs to L.

For the other direction, let t be a tree in L. We have to show that for all nodes u in t, φ is valid at u. Fix a node u in t. Since t belongs to L and since L is closed under subtrees, the tree $t|_u$ belongs to L. That is, the formula φ is valid at the root of $t|_u$. It follows that the formula φ is valid at u in t.

Corollary 1. *It is decidable whether a regular class of trees is μ-definable.*

In order to derive this corollary from Theorem 1, it is sufficient to show that both closure for subtrees and closure under p-morphic images are decidable properties of regular classes of trees. Given a regular class of trees L, it is possible to find an **MSO** formula α such that L is closed for subtrees iff α is valid on trees. Decidability of closure for subtrees follows then from the decidability of **MSO** on trees. Decidability of closure under p-morphic images follows from a careful inspection of the proof of Proposition 3 together with the decidability of **MSO** on trees.

6 Discussion

A natural further question is to investigate the μ-definability for classes of frames, not only classes of trees. Unlike on trees, graded μ-calculus does not have the same expressive power as **MSO** on models: it corresponds to the fragment of **MSO** invariant under counting bisimilation (see [JL03] and [Wal02]). Moreover, the proof of Proposition 3 does not work for classes of frames, as it relies on the fact that given a disjunctive formula, a strategy for $\exists$ in the game associated to the formula and a tree t, there is at most one match conform to the strategy during which a given node occurs.

References

[GT75] Goldblatt, R., Thomason, S.: Axiomatic classes in propositional modal logic. In: Algebra and Logic. Lecture Notes in Mathematics, vol. 450, pp. 163–173. Springer, Heidelberg (1975)

[JL03] Janin, D., Lenzi, G.: On the relationship between moandic and weak monadic second order logic on arbitrary trees, with application to the mu-calculus. Fudamenta Informaticae 61, 247–265 (2003)

[JW95] Janin, D., Walukiewicz, I.: Automata for the mu-calculus and related results. In: Hájek, P., Wiedermann, J. (eds.) MFCS 1995. LNCS, vol. 969. Springer, Heidelberg (1995)

[JW96] Janin, D., Walukiewicz, I.: On the expressive completeness of the propositional modal mu-calculus and related results. In: Sassone, V., Montanari, U. (eds.) CONCUR 1996. LNCS, vol. 1119. Springer, Heidelberg (1996)

[vB76] van Benthem, J.: Modal correspondance theory. PhD thesis, Mathematisch Instituut en Instituut voor Grondslagenonderzoek, University of Amsterdam (1976)

[vB96] van Benthem, J.: Exploring logical dynamics. CSLI publications, Stanford (1996)

[Wal02] Walukiewicz, I.: Monadic second order logic on tree-like structures. Theoretical Computer Science 275, 311–346 (2002)

Evaluating Non-square Sparse Bilinear Forms on Multiple Vector Pairs in the I/O-Model

Gero Greiner and Riko Jacob

Technische Universität München

Abstract. We consider evaluating one bilinear form defined by a sparse $N_y \times N_x$ matrix $\mathbf{A}$ having h entries on w pairs of vectors The model of computation is the semiring I/O-model with main memory size M and block size B. For a range of low densities (small h), we determine the I/O-complexity of this task for all meaningful choices of N_x, N_y, w, M and B, as long as $M \geq B^2$ (tall cache assumption). To this end, we present asymptotically optimal algorithms and matching lower bounds. Moreover, we show that multiplying the matrix $\mathbf{A}$ with w vectors has the same worst-case I/O-complexity.

1 Introduction

We consider the problem of computing w scalars $z^{(i)} = {\mathbf{y}^{(i)}}^T \mathbf{A}\mathbf{x}^{(i)}$, $0 \leq i \leq w$, where $\mathbf{A}$ is a sparse matrix with h non-zero entries, and all $\mathbf{x}^{(i)}$ and $\mathbf{y}^{(i)}$ are (dense) vectors. This is highly related to the matrix vector products $\mathbf{A}\mathbf{x}^{(i)}$, $0 \leq i \leq w$, and we show that both tasks actually have asymptotically the same complexity in our model. While, from a traditional point of view, bilinear form and matrix vector product are easily obtained with a number of multiplications equal to the number of matrix entries, the sparseness sometimes induces irregular access patterns that lead to situations where memory access becomes the bottleneck of computation. Empirical studies show that for the naive algorithm, CPU-usage can be as low as 10% [6,9].

One way of dealing with this problem is the construction of algorithms where, instead of CPU-cycles, the movement of data between layers of the memory hierarchy is optimized. In this paper, we use a slight modification of the I/O-model [1], the semiring I/O-model with the parameters M and B, denoting the memory size, and the size of a block, see Section 2 for details. In this model, Bender, Brodal, Fagerberg, Jacob and Vicari [2] determined the I/O-complexity of computing the sparse matrix vector product for square matrices. These results are generalised here to the case of non-square matrices. Furthermore, we extend the results to the evaluation of multiple products, i.e., the matrix vector products of multiple vectors with the same matrix. Considering the evaluation of matrix vector products on multiple vectors is a step towards closing the gap between sparse matrix vector multiplication and sparse matrix dense matrix multiplication since the set of w vectors $\mathbf{x}^{(i)}$ constitutes a dense $N_x \times w$ matrix $\mathbf{X}$.

P. Hliněný and A. Kučera (Eds.): MFCS 2010, LNCS 6281, pp. 393–404, 2010.

Related work. Evaluating the matrix vector product $\mathbf{Ax}$ for an $N \times N$ matrix $\mathbf{A}$ with kN entries has been investigated in [2]. They show that the I/O-complexity of this task for matrices in the so called column major layout is $\Theta\left(\min\left\{\frac{kN}{B}\,\overline{\log}_{\frac{M}{B}}\frac{N}{\max\{M,k\}}, kN\right\}\right)^1$, and $\Theta\left(\min\left\{\frac{kN}{B}\,\overline{\log}_{\frac{M}{B}}\frac{N}{Mk}, kN\right\}\right)$ for a layout chosen by the program.

The multiplication of two dense matrices has been examined by Hong and Kung in [5], showing a bound of $\Theta\left(\frac{N^3}{B\sqrt{M}}\right)$ on the number of I/Os. Very recently, in [4] the multiplication of a sparse matrix with a dense matrix was considered.give the formulas? For sparse $N \times N$ matrices with kN entries, it is shown that for certain ranges of k, the performance can be increased by finding denser than average submatrices.

The evaluation of bilinear forms in the I/O-model has been considered as an optimisation problem in [7]. There it has been shown that an optimal program for the evaluation of matrix vector products is $\mathcal{NP}$-hard to find, even for $B = 1$. In [8], the I/O-complexity of evaluating the bilinear form for a (non-square) $N_y \times N_x$ matrix with h entries that form a diagonal band, i.e., entries are only placed near the diagonal, is determined to be $\Theta\left(\frac{h}{BM} + \frac{N_x+N_y}{B}\right)$.

Our results. In this work, we consider the case where the number of entries for each submatrix in $\mathbf{A}$ is proportional to the number of rows and columns of the submatrix. This is possible if the average number of entries per column in $\mathbf{A}$ is some $k \leq \frac{N_y}{M^{1-\epsilon}N_x^{\epsilon}}$ with constant $\epsilon > 0$. For such k, a modification of the proof of [4] shows that the I/O-complexity of $\mathbf{AX}$ for any $w \geq B$ is $\Theta\left(\frac{wh}{B}\right)$. As a lower bound, this extends directly to the case of multiplying a sparse matrix with w vectors, even if the program is allowed to choose the layout of the vectors.

The case of $w \leq B$ is examined here in detail, forming a bridge between the results of [2] and [4]. For matrices of the described density ($h/N_x \leq \frac{N_y}{M^{1-\epsilon}N_x^{\epsilon}}$), we cover all dimensions of $\mathbf{A}$, and products with an arbitrary number of vectors. However, for all other choices of h, we present upper bounds in form of algorithms. For certain cases where B/w is small the algorithms are indeed optimal for all ranges of h. Furthermore, we show that evaluating the bilinear form has the same I/O-complexity as multiplying vectors with the same matrix.

Theorem 1. *Given a matrix* $\mathbf{A}$ *with fixed layout and fixed parameters* M *and* B*. Evaluating* w *bilinear forms with* $\mathbf{A}$ *has the same semiring I/O-complexity as evaluating the matrix vector product of* $\mathbf{A}$ *with* w *vectors if at least* $\ell = \Omega\left(\frac{hw}{B}\right)$ *I/Os are required.*

This result is explained in Section 3 and allows us to extend the results from [2] to matrices in row major layout, i.e., where the entries are given in external memory in a consecutive ordering by their row index, and their column index to break ties. Moreover, our main results hold for bilinear forms and matrix vector products, both, in column major and row major layout. Note that for

---

[1] $\overline{\log}_b(x) := \max\{\log_b(x), 1\}$.

Theorem 2, the dimensions N_x and N_y have to be swapped if $\mathbf{A}$ is given in row major layout.

For the complexity of the task, we are going to prove the following theorems depending on the layout of the matrix $\mathbf{A}$. Similar to [4], our bounds are only asymptotically tight if we assume a tall cache, i.e., $M \geq B^2$. However, this is only necessary to transpose $\mathbf{X}$ for some of the presented algorithms.

Our results for the case that the matrix $\mathbf{A}$ is stored in column major layout, are given in the following theorem.

Theorem 2. *Given an $N_y \times N_x$ matrix $\mathbf{A}$ with h entries in column major layout and parameters B, M. Assume $M \geq 4B$, $h/N_x \leq N_y^{\epsilon}$, and $\log N_x \leq N_y^{\epsilon}$ for some $0 < \epsilon < 1$. Then, evaluating w bilinear products with $\mathbf{A}$ has (worst-case) complexity in the semiring I/O-model*

$$\Theta\left(\min\left\{h\,,\ \frac{h\log N_y}{\log N_x}\,,\ \frac{h}{B}\log_{\frac{M}{B}}\min\left\{\frac{N_y}{M},\frac{N_xN_y}{h}\right\}+\frac{hw}{B}\log_{\frac{M}{B}}\frac{N_xN_y}{hM}\right\}\right)$$

unless this term is asymptotically smaller than $\frac{hw}{B}$.

The terms are obtained by a modification of the proof in [2] for a single matrix vector multiplication, also keeping track of the different matrix dimensions. Additionally, the lower bounds of Theorem 3 apply which yields the second term of the sum.

On the algorithmic side, for very asymmetric matrices $\mathbf{A}$, where the number of columns is much higher than the number of rows, building a table of tuples with multiple dimensions of all $\mathbf{y}^{(i)}$ vectors can be superior to the direct algorithm. The direct algorithm simply scans over $\mathbf{A}$ and loads for each a_{ij} the corresponding vector elements $x_j^{(0)},\ldots,x_j^{(w)}$ and $y_i^{(0)},\ldots,y_i^{(w)}$ to create products.

In [2], an algorithm is presented based on sorting the matrix entries to build row sums. This sorting approach can be used to initially change the layout of $\mathbf{A}$ to the best-case layout. Then, the sorting algorithm for best-case layout can be applied for one vector pair after another.

For the best-case layout, i.e., if the algorithm is allowed to choose the layout of the matrix, the following theorem holds.

Theorem 3. *Given an $N_y \times N_x$ matrix $\mathbf{A}$ with h entries in best-case layout and parameters B, M. Assume $M \geq 4B$, for $N_y \leq N_x$, and $h/N_x \leq \sqrt[6]{N_y}$. Then, evaluating w bilinear products with $\mathbf{A}$ has (worst-case) complexity in the semiring I/O-model*

$$\Theta\left(\min\left\{\frac{hw}{B}\log_{\frac{M}{B}}\frac{N_xN_y}{hM}\,,\ h\,,\ h\log\left(\frac{wN_xN_y\log N_x}{BhM}\right)/\log N_x\right\}\right)$$

unless this term is asymptotically smaller than $\frac{hw}{B}$.

All our algorithms require at least hw/B I/Os. In contrast, we do not know a corresponding lower bound that would hold for all choices of the parameters. However, if the dimensions are polynomially bounded in each other (which is expressed using a condition on the density in the lemma), the results of [4] can be extended to obtain a lower bound of $\Omega\left(hw/B\right)$ by the following lemma.

Lemma 1. *Let* $\mathbf{A}$ *be a sparse* $N_y \times N_x$ *matrix with* h *non-zero entries for* $N_x \geq N_y$*. For an average number of entries per column* $h/N_x \leq N_y/(M^{1-\epsilon}N_x^{\epsilon})$ *with constant* $\epsilon > 0$*, evaluating* w *bilinear products over a semiring requires* $\Omega\left(\frac{hw}{B}\right)$ *I/Os.*

The proof of this lemma can be found in [3]. Hence, for a $N_y \times N_x$ matrix $\mathbf{A}$ with h non-zero entries, for $N_x \geq N_y$ but $N_x^{2\epsilon} \leq N_y$, and average number of entries per column $h/N_x \leq \sqrt{N_y}/M$ a lower bound of $\Omega\,(hw/B)$ is given.

1.1 Outline

The outline of the paper is as follows: In Section 2, the model of computation is introduced along with the terminology used in this paper. The main results are then proven by providing (optimal) algorithms for upper bounding the complexity in Section 4 and up to constant factors matching lower bounds in Section 5. Due to size limitations, not all proofs are presented in full detail here, but they can be found the full version of the paper [3].

2 Model of Computation

We use a combination of the I/O-model described in [1] and the model used in [5], the so called *semiring I/O-model* introduced in [2]. It models two layers of the memory hierarchy, namely a fast memory of limited capacity M called *internal memory*. Calculations, namely addition, multiplication, copying, and deletion, can only be performed on the elements residing in internal memory, whereas the program inputs and any (intermediate) results are stored on an *external memory* (aka disk) of unbounded size which is organised in blocks (aka tracks) of size B. The I/O cost of a program is the number of transfers of a block between internal and external memory. Programs are assumed to work for an arbitrary semiring defining addition and multiplication, i.e., subtraction and division may not be used. Hence, all intermediate results have one of the following forms: $a_{jk}x_k^{(i)}$, $y_j^{(i)}a_{jk}$, $x_k^{(i)}y_j^{(i)}$ and $y_j^{(i)}a_{jk}x_k^{(i)}$, for $1 \leq i \leq w$, $j \in [N_x]$, $k \in [N_y]$, are referred to as *elementary products*, using the notation $[N] = \{1, \ldots, N\}$. Sums of the form $\sum_{k\in S'} a_{jk}x_k^{(i)}$, $\sum_{j\in S} y_j^{(i)}a_{jk}$, and $\sum_{j\in S}\sum_{k\in S'} y_j^{(i)}a_{jk}x_k^{(i)}$, with $1 \leq i \leq w$, $S \subseteq [N_x]$ and $S' \subseteq [N_y]$, are called *partial sums*. Altogether, the term *canonical partial result* refers to any of these forms. The detailed definition and the argument leading to this classification can be found in [3]. For the lower bounds we use the non-uniform notion that an *algorithm* is a family of programs where the program can be chosen according to the parameters underlying the problem. By $\ell(\mathcal{A})$, we denote the maximum number of I/Os induced by the algorithm $\mathcal{A}$ for all choices of the parameters, unless otherwise noted. In particular, for matrix multiplications, the parameters are the dimensions, the sparseness, the *conformation* of the matrix $\mathbf{A}$, i.e., the position of the non-zero entries in $\mathbf{A}$, the memory size M and the block size B.

Since we can assume that every program requires at least one I/O, when writing complexity using $\mathcal{O}$, Θ, or Ω at least 1 is meant.

3 Transformations

In this section, we discuss how a program that evaluates bilinear forms can be transformed into one that computes matrix vector multiplications.

Lemma 2. *In a normalised semiring I/O program evaluating a bilinear form on multiple vector pairs, no elementary product is created twice.*

For the proof of Theorem 1, we present transformations in both directions in the following lemmas. Note that the layout of the matrix $\mathbf{A}$ is not of concern, it just has to be the same for both tasks. We state the easy transformation without a proof:

Lemma 3. *If the matrix vector products $\mathbf{A}\mathbf{x}^{(i)}$ for $1 \leq i \leq w$ can be computed for an arbitrary semiring with ℓ I/Os, then the bilinear forms ${\mathbf{y}^{(i)}}^T \mathbf{A}\mathbf{x}^{(i)}$ can be evaluated with at most 3ℓ I/Os.*

Lemma 4. *If the bilinear forms ${\mathbf{y}^{(i)}}^T \mathbf{A}\mathbf{x}^{(i)}$, $1 \leq i \leq w$ can be evaluated in the semiring I/O-model with internal memory size M and block size B using ℓ I/Os, then the w products $\mathbf{c}^{(i)} = \mathbf{A}\mathbf{x}^{(i)}$ can be computed using $2\ell + 4wh/B$ I/Os with internal memory size $M + B$ and block size B.*

Proof. Here, we will only give a short description of the transformation of a program. A more detailed analysis can be found in [3].

Let P be a program to evaluate the w bilinear forms using ℓ I/Os. By Lemma 2, there is a program $\hat{P}$ for the same task which computes only canonical partial results with at most ℓ I/Os. We can then use $\hat{P}$ to construct a program for the matrix vector products. This construction is based on the following idea. During a simulation of $\hat{P}$, canonical partial results are extracted, and temporarily stored on disk. In a second phase, $\hat{P}$ is simulated time-reversed, as will be described later on, and the movement of $y_j^{(i)}$ variables in $\hat{P}$ can be used to lead the previously extracted results to the corresponding position in $\mathbf{y}^{(i)}$. In the end, the memory cells, where $\mathbf{y}^{(i)}$ is expected for P, constitute $\mathbf{c}^{(i)}$.

Construction. For the first phase, we create a program P_F for the semiring I/O-model with internal memory size $M+B$. We use the first M cells in memory for a simulation of $\hat{P}$, and reserve the last B cells $(m_{M+1}, \ldots, m_{M+B}) =: \mathcal{B}$ for further output operations. As soon as $\mathcal{B}$ is entirely full, i.e., no element in $\mathcal{B}$ is 0, the block is moved to disk.

During the simulation of $\hat{P}$, the following additional operations are performed. If a computation σ in $\hat{P}$ performs a multiplication of an element $m_l = y_j^{(i)}$ or $m_l = y_j^{(i)} x_k^{(i)}$ with an element $m_{l'}$ then $m_{l'}$ is copied into an empty position of $\mathcal{B}$ immediately before σ is performed. Furthermore, whenever in $\hat{P}$ a computation σ involves an element $m_l = x_k^{(i)}$, the result can only be of the form $x_k^{(i)} a_{jk}$, $x_k^{(i)} y_j^{(i)}$, $x_k^{(i)} (y_j^{(i)} a_{jk})$, or $x_k^{(i)} (\sum_{j \in S} y_j^{(i)} a_{jk})$. For the latter two cases, $x_k^{(i)}$ is copied into an empty cell of $\mathcal{B}$ before performing σ. We call these newly created copy operation *snapshot* and σ its associated operation.

The program P_F is executed with input $\mathbf{A}$ and $\mathbf{x}^{(i)}$ as given, but $\mathbf{y}^{(i)} = (1,\ldots,1)$ for all $1 \le i \le w$. For each elementary product created in $\hat{P}$, there are at most two elements copied to $\mathcal{B}$ (the corresponding $x_k^{(i)}$ and a_{jk}). Recall that in $\hat{P}$ elementary products are only created once. Since there are at most wh elementary products necessary, P_F performs no more than $\ell + \lceil \frac{2wh}{B} \rceil$ I/Os.

For the second phase, we have to time-reverse P_F. In this phase, we consider only the elements that consist of a polynomial containing a $y_j^{(i)}$, all other elements are ignored. When time is inverted, naturally an input becomes an output and vice versa. Internal computation operations are mapped in the following way. To this end, each copy operation of $\hat{P}$ that sets $m_k := m_l$ becomes a sum operation $m_l := m_l + m_k$ in the time-reversed program P_B. Each sum operation $m_i := m_j + m_k$ in P_F becomes a copy operation $m_j := m_k := m_i$ in P_B. Delete operations of P_F are simply ignored in P_B, i.e., nothing is created. The additional snapshot operations introduced in P_F are only made when an $y_j^{(i)}$ is involved in a computation operation σ. Considering the different cases of associated operations, the elements that were extracted in P_F are now copied, multiplied, or added into one of the cells that are accessed by σ.

Correctness. Since every canonical partial result that includes some $y_j^{(i)}$ has an input of the element $y_j^{(i)}$ as its predecessor in P_F, in the time-reversed P_B, all created partial results can be transferred to the initial position of $y_j^{(i)}$. Furthermore, since all hw elementary products have to be created for the bilinear product, and an $y_j^{(i)}$ is a predecessor for each, the created vectors $\mathbf{c}^{(i)}$, $1 \le i \le w$ are complete, i.e., the summation does not lack summands.

4 Algorithms

Note that all the presented algorithms can be used for both, evaluating matrix vector products and bilinear forms, by Theorem 1.

4.1 Direct Algorithm

The computation of $w \le B$ bilinear forms is possible with $\mathcal{O}(h)$ I/Os by considering the non-zero entries of $\mathbf{A}$ in an arbitrary order. For every entry a_{jk} the elementary products $x_k^{(0)} a_{jk} y_j^{(0)}, \ldots, x_k^{(w)} a_{jk} y_j^{(w)}$ are added to the respective current partial sums $z^{(0)}, \ldots, z^{(w)}$. For this to occur only a constant number of I/Os, the values $x_k^{(0)}, \ldots, x_k^{(w)}$ need to be stored in one block (or at least consecutively on disk), similarly to $y_j^{(0)}, \ldots, y_j^{(w)}$. This can be achieved by transposing the matrices $\mathbf{X} = \left[\mathbf{x}^{(0)} \ldots \mathbf{x}^{(w)}\right]$ and $\mathbf{Y} = \left[\mathbf{y}^{(0)} \ldots \mathbf{y}^{(w)}\right]$, which takes $\mathcal{O}((N_x + N_y)/B) = \mathcal{O}(h)$ I/Os [1], given the tall cache assumption $M \ge B^2$.

4.2 Sorting Based Algorithm

In [2] a sorting based approach for evaluating the product $\mathbf{Ax}$ for square matrices is presented. These algorithms can be extended straightforwardly to the matrix vector product of a non-square matrix $\mathbf{A}$ with one vector $\mathbf{x}$. For column major layout the vector $\mathbf{c} := \mathbf{Ax}$ can be created with $\mathcal{O}\left(\frac{h}{B}\,\overline{\log}_{M/B} \min\left\{\frac{N_x N_y}{h}, \frac{N_y}{M}\right\}\right)$ I/Os.

For the best-case layout the algorithm uses $\mathcal{O}\left(\frac{h}{B}\,\overline{\log}_{M/B} \frac{N_x N_y}{Mh}\right)$ I/Os. With slight modifications, this algorithm can also be described for a broader class of layouts. For this class of best-case layouts, the matrix $\mathbf{A}$ is given as a split-up of its columns into meta-columns where each meta-column is written in row major layout. Columns of a meta-column have to be continuous, each column is assigned to only one meta-column, and each meta-columns consist of an arbitrary number of columns, but at most $M - B$. Additionally, the number of meta-columns is at most $\lceil N_x/B \rceil + 2\lceil h/N_y \rceil$. Since each meta-column consists of no more than $M - B$ continuous columns, for each meta-column, the corresponding elements of $\mathbf{x}$ can all be loaded into internal memory, and the meta-column is scanned to create elementary products which are then written back to $\mathbf{A}$. Afterwards, if $N_x/B > h/N_y$, meta-columns are merged together using Merge sort until there are at most h/N_y runs. Since in this case there are no more than $3\lceil N_x/B \rceil$ meta-columns, $\mathcal{O}\left(\frac{h}{B}\,\overline{\log}_{M/B} \frac{N_x N_y}{Bh}\right) = \mathcal{O}\left(\frac{h}{B}\,\overline{\log}_{M/B} \frac{N_x N_y}{Mh}\right)$ I/Os are sufficient. Otherwise, if $h/N_y \geq N_x/B$, there are at most $3\lceil h/N_y \rceil$ meta-columns. Since meta-columns and runs are in row major layout, with one scan of each meta-column / run, elements from the same row can be summed together, and meta-columns / runs become a single column. All created columns can then be summed into the first column with $\mathcal{O}\left(\frac{h}{N_y} \cdot \frac{N_y}{B}\right)$ I/Os. Hence, in all cases, the matrix vector product for a matrix $\mathbf{A}$ given in a layout meeting the conditions described can be determined with $\mathcal{O}\left(\frac{h}{B}\,\overline{\log}_{M/B} \frac{N_x N_y}{Mh}\right)$ I/Os.

Multiple Vectors. For the evaluation of w matrix vector products, the algorithms can simply be run for each single vector, which increases the running time by a factor w. However, for column major layout, it can be faster to transform the layout of $\mathbf{A}$ into one belonging to the class of generalized best-case layouts described above, and then use that algorithm for each single vector.

The transformation of the layout has two cases, depending on the parameters. The first case handles situations with $N_x \leq h/(M - B)$, where the average column consists of more than $M - B$ already sorted entries. Then the N_x columns are bottom up merged using the M/B-way Merge sort, each time reducing the number of meta-columns by a factor of M/B. This is continued as long as the resulting meta columns have width $\leq M - B$, and the number of meta columns is greater than h/N_y. Hence, the running time of this merging is $\mathcal{O}\left(\frac{h}{B}\log_{M/B} \min\left\{M, \frac{N_x N_y}{h}\right\}\right)$ I/Os, and there are at most $\max\{\lceil N_x/(M - B)\rceil, \lceil h/N_y \rceil\}$ meta-columns that can contain less than $N_y/2$ entries, i.e., they form a generalised best case layout.

The second case assumes $h/(M - B) \leq N_x$, and mimics the creation of initial runs of length $M - B$. The possibility of columns having vastly different number of entries makes this slightly more involved. First, the columns of $\mathbf{A}$ are split into $2h/N_y$ continuous groups such that each group contains at most N_y entries. Then, each group that spans at most $M - B$ columns is transformed into row major layout using the classical M/B-way Merge sort in $\mathcal{O}\left(\frac{h}{B}\log_{M/B}\frac{N_y}{M}\right)$ I/Os Groups that span more than $M - B$ columns are divided into subgroups that span at most $M - B$. This can be achieved by greedily assigning the blocks of a group to subgroups such that each subgroup spans no more than $M-B$ columns. Splitting blocks that belong to two columns is possible with $\mathcal{O}(N_x/M)$ I/Os. The subgroups are then transformed into row major layout using Merge sort. Since there are at most N_y/B blocks per group, this can be done with another $\mathcal{O}\left(\frac{h}{B}\log_{M/B}\frac{N_y}{M}\right)$ I/Os.

Because one block spans at most B columns, each of the subgroups, except at most one per group, spans at least $M - 2B$ columns. Since we assume $M \geq 4B$ in this paper, for each group, there can be at most one meta-column with span less than $2B$. Hence, in the created layout, we have at most $N_x/(2B) + 2h/N_y$ meta-columns, each spanning at most $M - B$ columns, and the algorithm for the class of generalised best-case layouts can be applied.

4.3 Table Based Algorithm

For very asymmetric cases of $\mathbf{A}$ where $N_x \gg N_y$, the construction of tuples of rows of $\mathbf{Y}$, such that arbitrary dimensions of each vector can be loaded within one I/O. We present the following algorithms in the setting of evaluating of bilinear products.

Column Major Layout. The bilinear forms $\mathbf{y}^{(i)}\mathbf{A}\mathbf{x}^{(i)}$ for w vector pairs can be evaluated with $\mathcal{O}\left(\max\left\{\frac{wh}{B}, h\frac{\log N_y}{\log N_x}\right\}\right)$ I/Os as follows. The algorithm starts by creating a table of all c-tuples of rows of $\mathbf{Y}$ in lexicographical order of the row indices. To this end, define $c := \min\left\{\left\lfloor\frac{B}{w}\right\rfloor - 1, \left\lfloor\frac{\log N_x}{2\log N_y}\right\rfloor\right\}$ such that a c-tuple of rows of $\mathbf{Y}$ does not exceed one block. Since we assume that $\mathbf{Y}$ is in column major layout, it first has to be transposed which is possible with $\mathcal{O}(wN_y/B)$ I/Os (cf. Section 4.1). Further, since we assume a tall cache, i.e. $M \geq B^2$, internal memory can hold c blocks at a time and one for the output of a created tuple.

A table of all c-tuples has size $wcN_y^c \leq wc\sqrt{N_x} \leq wN_x$, where the last inequality relies upon $c \leq \frac{1}{2}\log N_x\sqrt{N_x}$, which is true for all N_x. The table can easily be created in $\mathcal{O}(wN_x/B)$ I/Os, a term dominated by the I/Os needed to read $\mathbf{X}$.

After creating a table of all c-tuples of rows of $\mathbf{Y}$, the algorithm simultaneously scans the entries of $\mathbf{A}$ and the corresponding elements of $\mathbf{X}$. Since we have $M \geq 4B$, we use one block for the scanning of $\mathbf{A}$, one for elements of $\mathbf{X}$, one for a c-tuple of $\mathbf{Y}$, and the last block to sum elementary results together for each of the $w \leq B$ vectors. Throughout the scanning of $\mathbf{A}$, for each $c \leq B$ entries $a_{i_1,j_1}, \ldots, a_{i_c,j_c}$, the c-tuple containing the corresponding rows $i_1, \ldots i_c$ of $\mathbf{Y}$ is loaded, and elementary products for each pair of vectors are created. These can be summed immediately into the block reserved for the results. Hence, $\mathcal{O}(h/c)$ I/Os are sufficient to evaluate the w bilinear forms.

Best-case Layout. If $w < B$ and $N_x \geq N_y^2$, the table-based approach for column major layout can be improved using a different layout of $\mathbf{A}$. In the following, we assume $c \leq B/w$, otherwise the algorithm for column major layout is applied. Similarly, the matrices $\mathbf{X}$ and $\mathbf{Y}$ are transposed in the beginning.

Again, we create a table of c-tuples of rows of $\mathbf{Y}$, but for different c. This time, the matrix $\mathbf{A}$ is read in tiles such that each tile contains on average $(M-B)/w$ entries, and the layout of $\mathbf{A}$ reflects these tiles. Using this, a tile can be loaded and all elementary products can be created while still one free block is available in internal memory. For a tile of height $\frac{wc^2}{B}\frac{2N_xN_y}{hM}$ rows and width $\frac{MB}{2wc}$ columns, we get a performance of

$$\mathcal{O}\left(h\frac{\log\left(\frac{wN_xN_y}{hBM}\log N_x\right)}{\log N_x}\right).$$

The details of the calculations can be found in the full version of the paper [3].

5 Lower Bounds

For the lower bounds, we only consider matrix vector products. By Theorem 1 this also implies lower bounds for bilinear products.

5.1 Column Major Layout

The following lower bound is only for single matrix vector products. However, together with the lower bound for the best-case layout, multiple evaluations are covered too.

Lemma 5. *Computing over an arbitrary semiring the bilinear product with an $N_x \times N_y$ matrix A with h entries, stored in column major layout has (worst-case) I/O-complexity for B, M with $M \geq 4B$*

$$\Omega\left(\min\left\{\frac{h}{B}\log_{\frac{M}{B}}\frac{N_x}{M}\quad,\quad\frac{h}{B}\log_{\frac{M}{B}}\frac{N_xN_y}{h}\quad,\quad h\quad,\quad h\max\left\{\frac{1}{B},\frac{\log N_x}{\log N_y}\right\}\right\}\right).$$

To proof this lemma, the dimensions of $\mathbf{A}$ have to be simply replaced in the proof in [2]. The calculations can be found in [3].

5.2 Best-Case Layout

As described in Section 2, for the best-case layout it is up to the program to choose the layout of $\mathbf{A}$. The proof of Lemma 5 is based on the task of computing row sums. To obtain a lower bound for the best-case layout, we have to use a different approach because producing row sums is trivial when using a row major layout. Therefore, we consider the sequence of configurations of a program and follow the movement of input variables of $\mathbf{X}$ and partial results of $\mathbf{Y}$. Furthermore, we allow accessing $\mathbf{A}$ for free. This can only weaken the lower bound.

We count the number of different matrix conformations that can be handled by programs for matrix vector multiplication with ℓ I/Os. For a given program, the conformation of a matrix can be identified by considering multiplication operations including input variables, and their results: When there is an input variable $x_j^{(i)}$ loaded, and it is used to form an elementary product that is a predecessor of $c_k^{(i)}$, this describes the existence of a non-zero entry a_{ik} in $\mathbf{A}$. Hence, by tracking all copies of input variables $x_j^{(i)}$ and all elements that are predecessors of a unique result $c_k^{(i)}$ (this can be elementary products or partial sums), and by choosing the positions in a program where multiplications involving such elements are performed, the conformation of a matrix is uniquely determined. To do this, it suffices to consider the tracking of elements only for one of the w matrix vectors multiplication. All these information will be called *trace* in the following.

In order to describe the trace, we normalise programs which changes the number of I/Os only by constant factors. The following normalisation is a variation of [5, Theorem 3.1].

Lemma 6. *Assume there is an I/O program $\mathcal{A}$ performing ℓ I/Os for parameters M and B. Then there is an I/O program $\mathcal{B}$ computing the same function performing at most $3\ell + M/B$ I/Os for parameters $2M$ and B, that works in rounds: Each round consists of $2M/B$ input operations, an arbitrary number of computation operations followed by $2M/B$ output operations such that after each round internal memory is empty.*

In the following, we consider programs in rounds according to the above lemma. To determine the traces of input variables and result predecessors in a round-based program, we consider the transfer of blocks between rounds, i.e. a block that is output by one round and input by another.

The movements of input variables can be described as follows. For a vector $\mathbf{x}^{(k)}$, we consider the subset $\mathcal{T}_{\mathcal{V},i} \subseteq [N_x]$ of indices of elements $x_i^{(k)}$ in a block i and trace the copying and deletion of variables in each round. For the trace of a predecessor of a unique result $c_j^{(k)}$, we abstract from the element itself, and consider only the index of the result j. Hence, we have the subset $\mathcal{T}_{\mathcal{R},i} \subseteq [N_y]$ of indices of unique result predecessors transferred by block i.

As written before, it suffices to consider the traces for one pair of input and result vector only. Every block of an I/O can be separated into values belonging to the w different tasks implied by the different pairs of vectors. Hence, for the l-th I/O, we have the number $u_l^{(k)}$ of elements belonging to vector pair k. By averaging we have $\sum_{0 \le l \le \ell} u_l^{(k)} \le B\ell/w$ for some k, and we determine the traces for this pair of vectors in the following.

Describing the traces. Because we will describe the traces of programs by blocks transferred between rounds, we view the input variables as output of rounds with no cost. Further, we are only interested in lower bounds below h such that we can assume $\ell \le h$. Let R be the total number of rounds. For each block that is an output of a round, and input of another round there are $R^2 \le h^2$

possibilities to choose the origin and destination of the block. Because there are $\ell/2$ blocks transferred, h^ℓ is an upper bound on the total number of possible macroscopic structures of how blocks travel between rounds.

Further, the values of $u_l^{(k)}$ can be chosen which yields at most B^ℓ possibilities. Every traced element that is transferred by a block can terminate at the destination, i.e., it is not copied further. Hence, there are $2^{\ell B/w}$ choices of terminating elements. Each of the s_i non-terminal incoming elements of round i can appear up to M/B times in the outgoing blocks, namely once per outgoing block. Hence, there are $\binom{s_i M/B}{t_i}$ possibilities to choose the t_i outgoing elements of round i, for some $t_i \geq s_i$. Since we have $\sum_{1 \leq i \leq R} t_i \leq B\ell/w$, the total number of possibilities for this is bounded by $\binom{M\ell/w}{B\ell/w}$.

Finally, we have to specify the subset of possible multiplications that are actually performed. To this end, let W_i be the number of partial results output by round i. Together with the number of vector variables U_i loaded in round i, there are $\sum_{i \leq \ell \frac{B}{M}} U_i W_i$ possible multiplications with matrix entries during the program. Additionally, we have the conditions $U_i \leq M$, $W_i \leq M$, and $\sum_{i \leq \ell \frac{B}{M}} (U_i + W_i) \leq \frac{\ell B}{w}$. The term $\sum_{i \leq \ell \frac{B}{M}} U_i W_i$ is hence maximised for $U_i = W_i = M$, for some indices $i \in \mathcal{I}$, $\mathcal{I} \subseteq [\ell \frac{B}{M}]$ with $|\mathcal{I}| = \frac{\ell B}{2Mw}$, and the size of the set of possible multiplications is at most $\frac{\ell B}{2Mw} \cdot M^2 = \frac{\ell M B}{2w}$. From this, we select a subset of size h, yielding no more than $\binom{\ell M B/(2w)}{h}$ possibilities.

Calculations. With the above discussion, we get

$$\binom{N_x N_y}{h} \leq h^\ell \cdot B^\ell \cdot 2^{\ell B/w} \cdot \binom{eM}{B} \cdot \binom{\frac{\ell M B}{2w}}{h}.$$

W.l.o.g. we assume $N_x \geq N_y$. Define $k = h/N_x$, i.e., the average number of entries per column. Rearranging terms yields $\ell \geq h \frac{\log \frac{N_y}{k} - \log \frac{e \ell M B}{wh}}{\log h + \log B + \frac{B}{w}(1 + \log \frac{eM}{B})}$. In the following, we can assume $h \geq M \geq B$, and thus $\frac{\ell}{h} \geq \frac{\log\left(\frac{N_y}{k} \cdot \frac{wh}{e\ell M B}\right)}{2 \log h + \frac{B}{w}(\log \frac{6M}{B})}$. Otherwise, if $h \leq M$, the task is trivial and a scanning bound of $\Omega\left(\frac{h}{B}\right)$ for reading $\mathbf{A}$ suffices. Applying Lemma B.2 in [2] ($x = \ell/h$, $t = 2\log h + \frac{B}{w}(\log \frac{6M}{B})$, $s = \frac{wN_y}{ekMB}$), and estimating $t \geq \log N_x + 3\frac{B}{w}$, we get $\frac{2\ell}{h} \geq \frac{\log\left(\frac{wN_y}{ekMB} \cdot \left(\frac{3B}{w} + \log N_x\right)\right)}{2 \log h + \frac{B}{w}(\log \frac{6M}{B})}$. Now, it remains to distinguish according to the leading term in the denominator.

Case 1 ($2 \log h \leq \frac{B}{w}(\log \frac{6M}{B})$): $\frac{\ell}{h} \geq \frac{\log \frac{N_y}{kM}}{4\frac{B}{w}(\log \frac{6M}{B})}$ Using $M > 4B$ yields $\ell \geq \frac{hw}{B} \frac{\log \frac{N_y}{kM}}{4 \cdot \frac{3}{2} \log \frac{M}{B}} = \frac{hw}{6B} \log_{\frac{M}{B}} \frac{N_y}{kM}$ which matches the sorting based bound.

Case 2 ($2 \log h > \frac{B}{w}(\log \frac{6M}{B})$): $\frac{2\ell}{h} \geq \frac{\log\left(\frac{wN_y}{BekM} \log N_x\right)}{4 \log h}$ matches the table based bound. Recall that the table based upper bound requires $N_x \geq N_y^2$.

However, for $N_x \leq N_y^2$, a linear lower bound is obtained as follows. Assuming $k \leq \sqrt[6]{N_y}$, $B \geq 16w$, and $N_x \geq 2^{30}$ implies $\frac{N_y}{BkeM} \geq \sqrt[8]{N_x}$ (for details see [3]), such that $\frac{2\ell}{h} \geq \frac{\log \frac{wN_y}{BekM}}{2 \log h} \geq \frac{\log \sqrt[8]{N_y}}{4(1+1/2) \log N_y} \geq \frac{1}{48}$.

Otherwise, if $N_y \leq N_x < 2^{30}$, also h is bounded from above, and thus, the task is trivially possible in $\mathcal{O}(1)$. Together with the presented algorithms and the lower bound from Lemma 1, this discussion yields Theorem 3. Furthermore, together with the lower bounds for column major layout, the proof of Theorem 2 is completed.

Acknowledgement

Thanks to Dan Roche and Clement Pernet for asking the question about multiplying many vector pairs. Many thanks to an anonymous referee for suggestions on an earlier draft of this article.

References

1. Aggarwal, A., Vitter, J.S.: The input/output complexity of sorting and related problems. Communications of the ACM 31(9), 1116–1127 (1988)
2. Bender, M.A., Brodal, G.S., Fagerberg, R., Jacob, R., Vicari, E.: Optimal sparse matrix dense vector multiplication in the I/O-model. In: Proceedings of SPAA 2007, pp. 61–70. ACM, New York (2007)
3. Greiner, G., Jacob, R.: Evaluating non-square sparse bilinear forms on multiple vector pairs in the I/O-model. Technical report, Technische Universität München (June 2010)
4. Greiner, G., Jacob, R.: The I/O complexity of sparse matrix dense matrix multiplication. In: López-Ortiz, A. (ed.) LATIN 2010. LNCS, vol. 6034, pp. 143–156. Springer, Heidelberg (2010)
5. Hong, J.-W., Kung, H.T.: I/O complexity: The red-blue pebble game. In: Proceedings of STOC 1981, pp. 326–333. ACM, New York (1981)
6. Jacob, R., Schnupp, M.: Experimental performance of I/O-optimal sparse matrix dense vector multiplication algorithms within main memory. Technical report, Technische Universität München (June 2010)
7. Lieber, T.: Combinatorial approaches to optimizing sparse matrix dense vector multiplication in the I/O-model. Master's thesis, Informatik Technische Universität München (2009)
8. Roos, F.F., Jacob, R., Grossmann, J., Fischer, B., Buhmann, J.M., Gruissem, W., Baginsky, S., Widmayer, P.: Pepsplice: cache-efficient search algorithms for comprehensive identification of tandem mass spectra. Bioinformatics 23(22), 3016–3023 (2007)
9. Vuduc, R.W.: Automatic Performance Tuning of Sparse Matrix Kernels. PhD thesis, University of California, Berkeley (Fall 2003)

Finding and Counting Vertex-Colored Subtrees

Sylvain Guillemot[1] and Florian Sikora[2]

[1] Lehrstuhl für Bioinformatik, Friedrich-Schiller Universität Jena, Ernst-Abbe Platz 2, 00743 Jena, Germany
sylvain.guillemot@uni-jena.de
[2] Université Paris-Est, LIGM - UMR CNRS 8049, France
sikora@univ-mlv.fr

Abstract. The problems studied in this article originate from the Graph Motif problem introduced by Lacroix et al. [17] in the context of biological networks. The problem is to decide if a vertex-colored graph has a connected subgraph whose colors equal a given multiset of colors M. Using an algebraic framework recently introduced by Koutis et al. [15,16], we obtain new FPT algorithms for Graph Motif and variants, with improved running times. We also obtain results on the counting versions of this problem, showing that the counting problem is FPT if M is a set, but becomes #W[1]-hard if M is a multiset with two colors.

1 Introduction

An emerging field in the modern biology is the study of the biological networks, which represent the interactions between biological elements [1]. A network is modeled by a vertex-colored graph, where nodes represent the biological compounds, edges represent their interactions, and colors represent functionalities of the graph nodes. Networks are often analyzed by studying their *network motifs*, which are defined as small recurring subnetworks. Motifs generally correspond to a set of elements realizing a same function, and which may have been evolutionarily preserved. Therefore, the discovery and the querying of motifs is a crucial problem [20], since it can help to decompose the network into functional modules, to identify conserved elements, and to transfer biological knowledge across species.

The initial definition of network motifs involves conservation of the topology and of the node labels; hence, looking for topological motifs is roughly equivalent to subgraph isomorphism, and thus is a computationally difficult problem. However, in some situations, the topology is not known or is irrelevant, which leads to searching for *functional* motifs instead of *topological* ones. In this setting, we still ask for the conservation of the node labels, but we replace topology conservation by the weaker requirement that the subnetwork should form a connected subgraph of the target graph. This approach was advocated by [17] and led to the definition of the Graph Motif problem [10]: given a vertex-colored graph $G = (V, E)$ and a multiset of colors M, find a set $V' \subseteq V$ such that the induced

P. Hliněný and A. Kučera (Eds.): MFCS 2010, LNCS 6281, pp. 405–416, 2010.

subgraph $G[V']$ is connected, and the multiset of colors of the vertices of V' is equal to M. In the literature, a distinction is made between the *colorful* case (when M is a set), and the *multiset* case (when M is an arbitrary multiset). Although this problem has been introduced for biological motivations, [3] points out that it may also be used in social or technical networks.

Not surprisingly, GRAPH MOTIF is NP-hard, even if G is a bipartite graph with maximum degree 4 and M is built over two colors only [10]. The problem is still NP-hard if G is a tree, but in this case it can be solved in $\mathcal{O}(n^{2c+2})$ time, where c is the number of distinct colors in M, while being W[1]-hard for the parameter c [10]. The difficulty of this problem is counterbalanced by its fixed-parameter tractability when the parameter is k, the size of the solution [17,10,3]. The currently fastest FPT algorithms for the problem run in $\mathcal{O}^*(2^k)$ time for the colorful case, $\mathcal{O}^*(4.32^k)$ time for the multiset case, and use exponential space [1].

Our contribution is twofold. First, we consider in Section 3 the *decision* versions of the GRAPH MOTIF problem, as well as some variants: we obtain improved FPT algorithms for these problems, by using the algebraic framework of *multilinear detection* for arithmetic circuits [15,16], presented in the next section. Second, we investigate in Section 4 the *counting* versions of the GRAPH MOTIF problem: instead of deciding if a motif appears in the graph, we now want to count the occurrences of this motif. This allows to assess if a motif is over- or under- represented in the network, by comparing the actual count of the motif to its expected count under a null hypothesis [19]. We show that the counting problem is FPT in the colorful case, but becomes #W[1]-hard for the multiset case with two colors. We refer the reader to [12,11] for definitions related to parameterized counting classes.

2 Definitions

This section contains definitions related to arithmetic circuits, and to the MULTILINEAR DETECTION (MLD) problem. It concludes by stating Theorem 1, which will be used throughout the paper.

2.1 Arithmetic Circuits

In the following, a capital letter X will denote a set of variables, and a lowercase letter x will denote a single variable. If X is a set of variables and $\mathbb{A}$ is a commutative ring, we denote by $\mathbb{A}[X]$ the ring of multivariate polynomials with coefficients in $\mathbb{A}$ and involving variables of X. Given a monomial $m = x_1...x_k$ in $\mathbb{A}[X]$, where the x_is are variables, its *degree* is k, and m is *multilinear* iff its variables are distinct.

An *arithmetic circuit* over X is a pair $\mathcal{C} = (C, r)$, where C is a labeled directed acyclic graph (dag) such that (i) the children of each node are totally ordered,

[1] We use the notations $\mathcal{O}^*$ and $\tilde{\mathcal{O}}$ to suppress polynomial and polylogarithmic factors, respectively.

(ii) the nodes are labeled either by $op \in \{+, \times\}$ or by an element of X, (iii) no internal node is labeled by an element of X, and where r is a distinguished node of C called the *root*. We denote by V_C the set of nodes of C, and for a given node u we denote by $N_C(u)$ the set of children (*i.e.* out-neighbors) of u in C. We recall that a node u is called a *leaf* of C iff $N_C(u) = \emptyset$, an *internal node* otherwise. We denote by $T(\mathcal{C})$ the *size of* $\mathcal{C}$ (defined as the number of arcs), and we denote by $S(\mathcal{C})$ the number of nodes of $\mathcal{C}$ of indegree ≥ 2.

Given a commutative ring $\mathbb{A}$, *evaluating* $\mathcal{C}$ *over* $\mathbb{A}$ *under a mapping* $\phi : X \to \mathbb{A}$ consists in computing, for each node u of C, a value $val(u) \in \mathbb{A}$ as follows: 1. for a leaf u labeled by $x \in X$, we let $val(u) = \phi(x)$, 2. for an internal node u labeled by $+$ (resp. $\times$), we compute $val(u)$ as the sum (resp. product) of the values of its children. The result of the evaluation is then $val(r)$. The *symbolic evaluation* of $\mathcal{C}$ is the polynomial $P_\mathcal{C} \in \mathbb{Z}[X]$ obtained by evaluating $\mathcal{C}$ over $\mathbb{Z}[X]$ under the identity mapping $\phi : X \to \mathbb{Z}[X]$.

We stress that the above definition of arithmetic circuits does not allow constants, a restriction which is necessary for the algorithms. However, we can safely allow the two constants $0_\mathbb{A}$ and $1_\mathbb{A}$, the zero and the unit of $\mathbb{A}$ (which is assumed to be a unital ring). For simplicity, these two constants will be represented by an empty sum and an empty product, respectively.

2.2 Multilinear Detection

Informally, the Multilinear Detection problem asks, for a given arithmetic circuit $\mathcal{C}$ and an integer k, if the polynomial $P_\mathcal{C}$ has a multilinear monomial of degree k. However, this definition does not give a certificate checkable in polynomial-time, so for technical reasons we define the problem differently.

A *monomial-subtree* of $\mathcal{C}$ is a pair $T = (\mathcal{C}', \phi)$, where $\mathcal{C}' = (C', r')$ is an arithmetic circuit over X whose underlying dag C' is a directed tree, and where $\phi : V_{\mathcal{C}'} \to V_\mathcal{C}$ is such that (i) $\phi(r') = r$, (ii) if $u \in V_{\mathcal{C}'}$ is labeled by $x \in X$, then so is $\phi(u)$, (iii) if $u \in V_{\mathcal{C}'}$ is labeled by $+$ then so is $\phi(u)$, and $N_{\mathcal{C}'}(u)$ consists of a single element $v \in N_\mathcal{C}(\phi(u))$, (iv) if $u \in V_{\mathcal{C}'}$ is labeled by $\times$, then so is $\phi(u)$, and ϕ maps bijectively $N_{\mathcal{C}'}(u)$ into $N_\mathcal{C}(\phi(u))$ by preserving the ordering on siblings. The *variables* of T are the leaves of C' labeled by variables in X. We say that T is *distinctly-labeled* iff its variables are distinct.

Intuitively, a monomial-subtree tells us how to construct a multilinear from the circuit: Condition (i) tells us to start at the root, Condition (iii) tells us that when reaching a $+$ node we are only allowed to pick one child, and Condition (iv) tells us that when reaching a $\times$ node we have to pick all children. The (distinctly-labeled) monomial-subtrees of $\mathcal{C}$ with k variables will then correspond to the (multilinear) monomials of $P_\mathcal{C}$ having degree k. Therefore, we formulate the Multilinear Detection problem as follows:

Name: Multilinear Detection (MLD)
Input: An arithmetic circuit $\mathcal{C}$ over a set of variables X, an integer k
Solution: A distinctly-labeled monomial-subtree of $\mathcal{C}$ with k variables.

Solving MLD amounts to decide if $P_{\mathcal{C}}$ has a multilinear monomial of degree k (observe that there are no possible cancellations), and solving #MLD amounts to compute the sum of the coefficients of multilinear monomials of $P_{\mathcal{C}}$ having degree k. The restriction of MLD when $|X| = k$ is called EXACT MULTILINEAR DETECTION (XMLD). In this article, we will rely on the following far-reaching result from [21,16] to obtain new algorithms for GRAPH MOTIF:

Theorem 1 ([21,16]). MLD *can be solved by a randomized algorithm which uses* $\tilde{O}(2^k T(\mathcal{C}))$ *time and* $\tilde{O}(S(\mathcal{C}))$ *space.*

3 Finding Vertex-Colored Subtrees

In this section, we consider several variants of the GRAPH MOTIF problem, and we obtain improved FPT algorithms for these problems by reduction to MLD. Notably, we obtain $\mathcal{O}^*(2^k)$ time algorithms for problems involving *colorful* motifs, and $\mathcal{O}^*(4^k)$ time algorithms for *multiset* motifs.

3.1 The Colorful Case

In the colorful formulation of the problem, the graph is vertex-colored, and we seek a subtree with k vertices having distinct colors. This leads to the following formal definition:

Name: COLORFUL GRAPH MOTIF (CGM)
Input: A graph $G = (V, E)$, $k \in \mathbb{N}$, a set C, a function $\chi : V \to C$
Solution: A subtree $T = (V_T, E_T)$ of G s.t. (i) $|V_T| = k$ and (ii) for each $u, v \in V_T$ distinct, $\chi(u) \neq \chi(v)$.

The restriction of COLORFUL GRAPH MOTIF when $|C| = k$ is called EXACT COLORFUL GRAPH MOTIF (XCGM). Note that this restriction requires that the vertices of T are bijectively labeled by the colors of C. In [7], the XCGM problem was shown to be solvable in $\mathcal{O}^*(2^k)$ time and space, while it is not difficult to see that the general CGM problem can be solved in $\mathcal{O}^*((2e)^k)$ time and $\mathcal{O}^*(2^k)$ space by color-coding. By using a reduction to MULTILINEAR DETECTION, we improve upon these complexities. In the following, we let n and m denote the number of vertices and the number of edges of G, respectively.

Proposition 1. CGM *is solvable by a randomized algorithm in* $\tilde{O}(2^k k^2 m)$ *time and* $\tilde{O}(kn)$ *space.*

Proof (Sketch). Let I be an instance of CGM. We construct the following circuit $\mathcal{C}_I$: its set of variables is $\{x_c : c \in C\}$, and we introduce intermediary nodes $P_{i,u}$ for $1 \leq i \leq k, u \in V$, as well as a root node P. Informally, the multilinear monomials of $P_{i,u}$ will correspond to colorful subtrees of G having i vertices, including u. The definitions are as follows:

$$P_{i,u} = \sum_{i'=1}^{i-1} \sum_{v \in N_G(u)} P_{i',u} P_{i-i',v} \text{ if } i > 1, \qquad P_{1,u} = x_{\chi(u)}$$

and $P = \sum_{u \in V} P_{k,u}$. The resulting instance of MLD is $I' = (\mathcal{C}_I, k)$. By applying Theorem 1, and by observing that $T(\mathcal{C}_I) = \mathcal{O}(k^2 m)$ and $S(\mathcal{C}_I) = \mathcal{O}(kn)$, we solve I' in $\tilde{\mathcal{O}}(2^k k^2 m)$ time and $\tilde{\mathcal{O}}(kn)$ space. The correctness of the construction follows by showing by induction on $1 \leq i \leq k$ that: $x_{c_1} ... x_{c_d}$ is a multilinear monomial of $P_{i,u}$ iff (i) $d = i$ and (ii) there exists $T = (V_T, E_T)$ colorful subtree of G such that $u \in V_T$ and $\chi(V_T) = \{c_1, ..., c_d\}$. □

3.2 The Multiset Case

We consider the multiset formulation of the problem: we now allow some colors to be repeated but impose a maximum number of occurrences for each color. This problem can be seen as a generalization of the original Graph Motif problem.

Given a multiset M over a set A, and given an element $x \in A$, we denote by $n_M(x)$ the number of occurrences of x in M. Given two multisets M, M', we denote their inclusion by $M \subseteq M'$. We denote by $|M|$ the size of M, where elements are counted with their multiplicities. Given two sets A, B, a function $f : A \to B$ and a multiset X over A, we let $f(X)$ denote the multiset containing the elements $f(x)$ for $x \in X$, counted with multiplicities; precisely, given $y \in B$ we have $n_{f(X)}(y) = \sum_{x \in A: f(x)=y} n_X(x)$.

We now define the following two variants of Colorful Graph Motif, which allow for multiset motifs:

Name: Multiset Graph Motif (MGM)
Input: A graph $G = (V, E)$, an integer k, a set C, a function $\chi : V \to C$, a multiset M over C.
Solution: A subtree $T = (V_T, E_T)$ of G s.t. (i) $|V_T| = k$ and (ii) $\chi(V_T) \subseteq M$.

Name: Multiset Graph Motif With Gaps (MGMG)
Input: A graph $G = (V, E)$, integers k, r, a set C, a function $\chi : V \to C$, a multiset M over C.
Solution: A subtree $T = (V_T, E_T)$ of G s.t. (i) $|V_T| \leq r$ and (ii) there exists $S \subseteq V_T$ of size k such that $\chi(S) \subseteq M$.

The restriction of Multiset Graph Motif when $|M| = k$ is called Exact Multiset Graph Motif (XMGM). Note that in this case we require that T contains every occurrence of M, *i.e.* $\chi(V_T) = M$. In this way, the XMGM problem coincides with the Graph Motif problem defined in [10,3], while the MGM problem is the parameterized version of the Max Motif problem considered in [9]. The notion of gaps is introduced in [17], and encompasses the notion of insertions and deletions of [7].

Previous algorithms for these problems relied on color-coding [2]; these algorithms usually have an exponential space complexity, and a high time complexity. For the Graph Motif problem, [10] gives a randomized algorithm with an implicit $\mathcal{O}(87^k km)$ running time, while [3] describes a first randomized algorithm running in $\mathcal{O}(8.16^k m)$, and shows a second algorithm with $\mathcal{O}(4.32^k k^2 m)$ running time, using two different speed-up techniques ([4] and [13]). For the MAX MOTIF problem, [9] presents a randomized algorithm with an implicit $\mathcal{O}((32e^2)^k km)$ running time. Here again, we can apply Theorem 1 to improve the time and space complexities:

Proposition 2. *1.* MGM *is solvable by a randomized algorithm in* $\tilde{\mathcal{O}}(4^k k^2 m)$ *time and* $\tilde{\mathcal{O}}(kn)$ *space.*
2. MGMG *is solvable by a randomized algorithm in* $\tilde{\mathcal{O}}(4^k r^2 m)$ *time and* $\tilde{\mathcal{O}}(rn)$ *space.*

Proof. Point 1. We modify the circuit of Proposition 1 as follows. For each color $c \in C$ with $n_M(c) = m$, we introduce variables $y_{c,1}, ..., y_{c,m}$, and we introduce a node $Q_c = y_{c,1} + ... + y_{c,m}$. For each vertex $u \in V$, we introduce a variable x_u, and we define:

$$P_{i,u} = \sum_{i'=1}^{i-1} \sum_{v \in N_G(u)} P_{i',u} P_{i-i',v} \text{ if } i > 1, \qquad P_{1,u} = x_u Q_{\chi(u)}$$

and $P = \sum_{u \in V} P_{k,u}$. Note that we changed only the base case in the recurrence of Proposition 1. The intuition is that the variables x_u will ensure that we choose different vertices to construct the tree, and that the variables $y_{c,i}$ will ensure that a given color cannot occur more than required. The resulting instance of MLD is $I' = (\mathcal{C}_I, 2k)$, and since $T(\mathcal{C}_I) = \mathcal{O}(k^2 m)$ and $S(\mathcal{C}_I) = \mathcal{O}(kn)$, we solve it in the claimed bounds by Theorem 1. A similar induction as in Proposition 1 shows that: for every $1 \leq i \leq k$, a multilinear monomial of $P_{i,u}$ has the form $x_{v_1} y_{c_1,j_1} ... x_{v_i} y_{c_i,j_i}$, and it is present iff there is a subtree (V_T, E_T) of G such that $u \in V_T$, $V_T = \{v_1, ..., v_i\}$ and $\chi(V_T) = \{\{c_1, ..., c_i\}\} \subseteq M$.

Point 2. We modify the construction of Point 1 by now setting $P_{1,u} = 1 + x_u Q_{\chi(u)}$ for each $u \in V$, and $P = \sum_{u \in V} \sum_{i=1}^{r} P_{i,u}$. Informally, adding the constant 1 to each $P_{1,u}$ permits to ignore some vertices of the subtree, allowing to only select a set S of k vertices such that $\chi(S) \subseteq M$. The correctness of the construction is shown by a similar induction as above. The catch here is that when considering two trees T_1, T_2 obtained from $P_{i',u}, P_{i-i',v}$, their selected vertices will be distinct, but they may have "ignored" vertices in common; we can then find a subset of $E(T_1) \cup E(T_2) \cup \{uv\}$ which forms a tree containing all selected vertices from T_1, T_2. □

3.3 Edge-Weighted Versions

We consider an edge-weighted variant of the problem, where the subtree is now required to have a given total weight, in addition to respecting the color constraints. This variant has been studied in [6] under the name EDGE-WEIGHTED GRAPH MOTIF. In our case, we define two problems, depending on whether we consider colorful or multiset motifs.

Name: WEIGHTED COLORFUL GRAPH MOTIF (WCGM)
Input: A complete graph $G = (V, E)$, a function $\chi : V \rightarrow C$, a weight function $w : E \rightarrow \mathbb{N}$, integers k, r

Solution: A subtree $T = (V_T, E_T)$ of G such that (i) $|V_T| = k$, (ii) χ is injective on V_T, (iii) $\sum_{e \in E_T} w(e) \leq r$.

Name: Weighted Multiset Graph Motif (WMGM)
Input: A complete graph $G = (V, E)$, a function $\chi : V \to C$, a weight function $w : E \to \mathbb{N}$, integers k, r, a multiset M
Solution: A subtree $T = (V_T, E_T)$ of G such that (i) $|V_T| = k$, (ii) $\chi(V_T) \subseteq M$, (iii) $\sum_{e \in E_T} w(e) \leq r$.

We observe that the WMGM problem contains as a special case the Min-CC problem introduced in [8], which seeks a subgraph respecting the multiset motif, and having at most r connected components. Indeed, we can easily reduce Min-CC to WMGM: given the graph G, we construct a complete graph G' with the same vertex set, and we assign a weight 0 to edges of G, and a weight 1 to non-edges of G.

Proposition 3. *1. WCGM is solvable by a randomized algorithm in $\tilde{O}(2^k k^2 r^2 m)$ time and $\tilde{O}(krn)$ space.*
2. WMGM is solvable by a randomized algorithm in $\tilde{O}(4^k k^2 r^2 m)$ time and $\tilde{O}(krn)$ space.

Proof. We only prove 1, since 2 relies on the same modification as in Proposition 2. The construction of the arithmetic circuit is similar to the construction in Proposition 1. The set of variables is $\{x_c : c \in C\}$, and we introduce nodes $P_{i,j,u}$, for $1 \leq i \leq k$ and $0 \leq j \leq r$, whose multilinear monomials will correspond to colorful subtrees having i vertices including u, and with total weight $\leq j$. The definitions are as follows:

$$P_{1,j,u} = x_{\chi(u)}$$
$$P_{i,j,u} = \sum_{i'=1}^{i-1} \sum_{v \in V} \sum_{j'=0}^{j-w(uv)} P_{i',j',u} P_{i-i',j-j'-w(uv),v} \text{ if } i > 1$$

and $P = \sum_{u \in V} P_{k,r,u}$. The resulting instance of MLD is $I' = (\mathcal{C}_I, k)$, and since $T(\mathcal{C}_I) = \mathcal{O}(k^2 r^2 m)$ and $S(\mathcal{C}_I) = \mathcal{O}(krn)$, we solve it in the claimed bounds by Theorem 1. The correctness of the construction follows by showing that: given $1 \leq i \leq k, 0 \leq j \leq r$, $u \in V$, $x_{c_1} ... x_{c_d}$ is a multilinear monomial of $P_{i,j,u}$ iff (i) $d = i$ and (ii) there exists $T = (V_T, E_T)$ colorful subtree of G with $u \in V_T, \chi(V_T) = \{c_1, ..., c_d\}$ and $\sum_{e \in E_T} w(e) \leq j$. □

4 Counting Vertex-Colored Subtrees

In this section, we consider the counting versions of the problems XCGM and XMGM introduced in Section 3. For the former, we show that its counting version #XCGM is FPT; for the latter, we prove that its counting version #XMGM is #W[1]-hard.

4.1 FPT Algorithms for the Colorful Case

We show that #XCGM is fixed-parameter tractable (Proposition 5). We rely on a general result for #XMLD (Proposition 4), which uses inclusion-exclusion as in [14].

Say that a circuit $\mathcal{C}$ is *k-bounded* iff $P_{\mathcal{C}}$ has only monomials of degree $\leq k$. Observe that given a circuit $\mathcal{C}$, we can efficiently transform it in a k-bounded circuit $\mathcal{C}'$ such that (i) $\mathcal{C}$ and $\mathcal{C}'$ have the same monomials of degree k, (ii) $|\mathcal{C}'| \leq (k+1)^2|\mathcal{C}|$; the details of the construction are omitted [2]. The following result shows that we can efficiently count solutions for k-bounded circuits with k variables (and thus for general circuits, with an extra $\mathcal{O}(k^2)$ factor in the complexity).

Proposition 4. #XMLD *for k-bounded circuits is solvable in* $\mathcal{O}(2^k T(\mathcal{C}))$ *time and* $\mathcal{O}(S(\mathcal{C}))$ *space.*

Proof. Let $\mathcal{C}$ be the input circuit on a set X of k variables. For a monomial m let $Var(m)$ denote its set of variables. Given $S \subseteq X$, let N_S, resp. N'_S, be the number of monomials m of $P_{\mathcal{C}}$ such that $Var(m) = S$, resp. $Var(m) \subseteq S$. Observe that for every $S \subseteq X$, we have $N'_S = \sum_{T \subseteq S} N_T$. Therefore, by Möbius inversion it holds that for every $S \subseteq X$, $N_S = \sum_{T \subseteq S} (-1)^{|S \setminus T|} N'_T$.

Since $\mathcal{C}$ is k-bounded, N_X is the number of multilinear monomials of $P_{\mathcal{C}}$ having degree k. Now, each value N'_S can be computed by evaluating $\mathcal{C}$ under the mapping $\phi : X \rightarrow \mathbb{Z}$ defined by $\phi(v) = 1$ if $v \in S$, $\phi(v) = 0$ if $v \notin S$. By the Möbius inversion formula, we can thus compute the desired value N_X in $\mathcal{O}(2^k T(\mathcal{C}))$ time and $\mathcal{O}(S(\mathcal{C}))$ space. □

It is worth mentioning that Proposition 4 generalizes several counting algorithms based on inclusion-exclusion, such as the well-known algorithm for #Hamiltonian Path of [14], as well as results of [18]. Indeed, the problems considered in these articles can be reduced to counting multilinear monomials of degree n for circuits with n variables (where n is usually the number of vertices of the graph), which leads to algorithms running in $\mathcal{O}^*(2^n)$ time and polynomial space.

Let us now turn to applying Proposition 4 to the #XCGM problem. Recall that we defined in Proposition 1 a circuit $\mathcal{C}_I$ for the general CGM problem; we will have to modify it slightly for the purpose of counting solutions.

Proposition 5. #XCGM *is solvable in* $\mathcal{O}(2^k k^3 m)$ *time and* $\mathcal{O}(k^2 n)$ *space.*

Proof. Let I be an instance of XCGM. A *rooted solution* for I is a pair (u, T) where T is a solution of XCGM on I and u is a vertex of T (which should be seen as the root of the tree). The solutions of XCGM on I are also called *unrooted solutions*. Let $N_r(I)$ and $N_u(I)$ be the number of rooted, resp. unrooted,

[2] The idea is to assume w.l.o.g. that $\mathcal{C}$ has outdegree 2. Then, we create $k+1$ copies $u_0, ..., u_k$ of each node u of $\mathcal{C}$, such that the monomials of u_i correspond to the degree i-monomials of u. If r is the root node of $\mathcal{C}$, then r_k becomes the root node of $\mathcal{C}'$.

solutions for I. We will show how to compute $N_r(I)$ in the claimed time and space bounds; since $N_u(I) = \frac{N_r(I)}{k}$, the result will follow.

To compute N_r, observe first that we cannot apply Proposition 4 to the circuit $\mathcal{C}_I$ of Proposition 1. Indeed, the circuit $\mathcal{C}_I$ counts the ordered subtrees, and not the unordered ones. Therefore, we need to modify the circuit in the following way: at each vertex v of V_T, we examine its children by increasing color. This leads us to define the following circuit $\mathcal{C}'_I$: suppose w.l.o.g. that $C = \{1, ..., k\}$, introduce nodes $P_{i,j,u}$ for each $1 \leq i \leq k, 1 \leq j \leq k+1, u \in V$, variables x_i for each $1 \leq i \leq k$, and define:

$$P_{1,j,u} = x_{\chi(u)}, \quad P_{i,j,u} = 0 \text{ if } i \geq 2, j = k+1$$

$$P_{i,j,u} = P_{i,j+1,u} + \sum_{i'=1}^{i-1} \sum_{v \in N_G(u): \chi(v)=j} P_{i',j+1,u} P_{i-i',1,v} \text{ if } i \geq 2, 1 \leq j \leq k$$

Let us also introduce a root node $P = \sum_{u \in V} P_{k,1,u}$. Given $1 \leq i,j \leq k$ and $u \in V$, let $\mathcal{S}_{i,j,u}$ denote the set of pairs (u, T) where (i) T is a properly colored subtree of I containing u and having i vertices, (ii) the neighbors of u in T have colors $\geq j$. It can be shown by induction on i that: there is a bijection between $\mathcal{S}_{i,j,u}$ and the multilinear monomials of $P_{i,j,u}$. Therefore, the number of multilinear monomials of P is equal to N_r; since $T(\mathcal{C}'_I) = \mathcal{O}(k^3 m), S(\mathcal{C}'_I) = \mathcal{O}(k^2 n)$ and since $\mathcal{C}'_I$ is k-bounded, it follows by Proposition 4 that N_r can be computed in $\mathcal{O}(2^k k^3 m)$ time and $\mathcal{O}(k^2 n)$ space. □

4.2 Hardness of the Multiset Case

In this subsection, we show that #XMGM is #W[1]-hard. For convenience, we first restate the problem in terms of *vertex-distinct embedded subtrees.*

Let $G = (V, E)$ and $H = (V', E')$ be two multigraphs. An *homomorphism* of G into H is a pair $\phi = (\phi_V, \phi_E)$ where $\phi_V : V \to V'$ and $\phi_E : E \to E'$, such that if $e \in E$ has endpoints x, y then $\phi_E(e)$ has endpoints $\phi_V(x), \phi_V(y)$. An *embedded subtree* of G is denoted by $\mathcal{T} = (T, \phi_V, \phi_E)$ where $T = (V_T, E_T)$ is a tree, and (ϕ_V, ϕ_E) is an homomorphism from T into G. We say that $\mathcal{T}$ is a *vertex-distinct* embedded subtree of G (a "vdst" of G) if ϕ_V is injective. We say $\mathcal{T}$ is an *edge-distinct* embedded subtree of G (an "edst" of G) iff ϕ_E is injective. We restate XMGM as follows:

Name: Exact Multiset Graph Motif (XMGM)
Input: A graph $G = (V, E)$, an integer k, a set C, a function $\chi : V \to C$, a multiset M over C s.t. $|M| = k$.
Solution: A vdst (T, ϕ_V, ϕ_E) of G s.t. $\chi \circ \phi_V(V_T) = M$.

We first show the hardness of two intermediate problems (Lemma 1). Before defining these problems, we need the following notions. Consider a multigraph $G = (V, E)$. Consider a partition $\mathcal{P}$ of V into $V_1, ..., V_k$, and a tuple $t \in [r]^k$. A $(\mathcal{P}, t)$-*mapping* from a set A is an injection $\psi : A \to V \times [r]$ such that for every $x \in A$, if $\psi(x) = (v, i)$ with $v \in V_j$, then $1 \leq i \leq t_j$. From ψ, we define its *reduction* as the function $\psi^r : A \to V$ defined by $\psi^r(x) = v$ whenever $\psi(x) = (v, i)$. We also define a tuple $T(\psi) = (n_1, ..., n_k) \in [r]^k$ such that for each $i \in [k]$, $n_i = \max_{v \in V_i} |\{x \in A : \psi^r(x) = v\}|$.

Given two tuples $t, t' \in [r]^k$, denote $t \leq t'$ iff $t_i \leq t'_i$ for each $i \in [k]$. Note that for a $(\mathcal{P}, t)$-mapping ψ, we always have $T(\psi) \leq t$ since ψ is injective. We say that a *$(\mathcal{P}, t)$-labeled edst for G* is a tuple (T, ψ_V, ψ_E) where (i) $T = (V_T, E_T)$ is a tree, (ii) ψ_V is a $(\mathcal{P}, t)$-mapping from V_T, (iii) (T, ψ_V^r, ψ_E) is an edst of G. Our intermediate problems are defined as follows:

Name: MULTICOLORED EMBEDDED SUBTREE-1 (MEST − 1)
Input: Integers k, r, a k-partite multigraph G with partition $\mathcal{P}$, a tuple $t \in [r]^k$
Solution: A $(\mathcal{P}, t)$-labeled edst (T, ψ_V, ψ_E) for G s.t. $|V_T| = r$ and $T(\psi_V) = t$.

The MEST − 2 problem is defined similarly, except that we do not require that $T(\psi_V) = t$ (and thus we only have $T(\psi_V) \leq t$). While we will only need #MEST − 2 in our reduction for #XMGM, we first show the hardness of #MEST − 1, then reduce it to #MEST − 2.

Lemma 1. *#MEST − 1 and #MEST − 2 are #W[1]-hard for parameter (k, r).*

The proof is omitted due to space constraints.

Proposition 6. *#XMGM is #W[1]-hard for parameter k.*

Proof. We reduce from #MEST − 2, and conclude using Lemma 1. Let $I = (k, r, G, t)$ be an instance of #MEST − 2, where $G = (V, E)$ is a multigraph, and let $\mathcal{S}_I$ be its set of solutions. From G, we construct a graph H as follows: (i) we subdivide each edge $e \in E$, creating a new vertex $a[e]$, (ii) we substitute each vertex $v \in V_i$ by an independent set formed by t_i vertices $b[v, 1], ..., b[v, t_i]$. We let A be the set of vertices $a[e]$ and B the set of vertices $b[v, i]$, we therefore have a bipartite graph $H = (A \cup B, F)$. We let $I' = (H, 2r - 1, C, \chi, M)$, where $C = \{1, 2\}$, χ maps A to 1 and B to 2, and M consists of $r - 1$ occurrences of 1 and r occurrences of 2.

Then I' is our resulting instance of #XMGM, and we let $\mathcal{S}_{I'}$ be its set of solutions. Notice that by definition of χ and M, $\mathcal{S}_{I'}$ is the set of vdst (T, ϕ_V, ϕ_E) of H containing $r - 1$ vertices mapped to A and r vertices mapped to B. We now show that we have a parsimonious reduction, by describing a bijection $\Phi : \mathcal{S}_I \to \mathcal{S}_{I'}$. Consider $\mathcal{T} = (T, \psi_V, \psi_E)$ in $\mathcal{S}_I$; we define $\Phi(\mathcal{T}) = (T', \phi_V, \phi_E)$ as follows:

- For each edge $e = uv \in E(T)$, we have $f_e := \psi_E(e) \in E(G)$: we then subdivide e, creating a new vertex x_e. Let T' be the resulting tree;
- For each vertex x_e, we define $\phi_V(x_e) = a[f_e]$. For each other vertex u of T', we have $u \in V(T)$, let $(v, i) = \psi_V(u)$; we then set $\phi_V(u) = b[v, i]$ (this is possible since if $v \in V_j$ then $1 \leq i \leq t_j$, by definition of ψ_V).

From ϕ_V, we then define ϕ_E in a natural way. Then $\mathcal{T}' = \Phi(\mathcal{T})$ is indeed in $\mathcal{S}_{I'}$: (i) T' is a vertex distinct subtree of H (by definition of ϕ_V and since $\mathcal{T}$ was edge-distinct, the values $\phi_V(x_e)$ are distinct; by injectivity of ψ_V, the other values $\phi_V(u)$ are distinct); (ii) it has $r-1$ vertices mapped to A and r vertices mapped to B. To prove that Φ is a bijection, we describe the inverse correspondence $\Psi : \mathcal{S}_{I'} \to \mathcal{S}_I$. Consider $\mathcal{T}' = (T', \phi_V, \phi_E)$ in $\mathcal{S}_{I'}$; we define $\Psi(\mathcal{T}') = (T, \psi_V, \psi_E)$ as follows. Let A', B' be the vertices of T' mapped to A, B respectively. Let i be the number of nodes of A' which are leaves: since the nodes of A' have degree 1 or 2 in T' depending on whether they are leaves or internal nodes, we then have $|E(T')| \leq i + 2(r-1-i) = 2r - i - 2$; since $|E(T')| = 2r-2$, we must have $i = 0$. It follows that all leaves of T' belong to B'; from T', by contracting each vertex of A' in T' we obtain a tree T with r vertices. We then define ψ_V, ψ_E as follows: (i) given $u \in B'$, if $\phi_V(u) = b[v,j]$, then $\psi_V(u) = (v,j)$; (ii) given $e = uv \in E(T)$, there corresponds two edges $ux, vx \in E(T')$ with $x \in A'$, and we thus have $\phi_V(x) = a[f]$, from which we define $\psi_E(e) = f$. It is easily seen that the resulting $\mathcal{T} = \Psi(\mathcal{T}')$ is in $\mathcal{S}_I$, and that the operations Φ and Ψ are inverse of each other. □

5 Conclusion

In this paper, we have obtained improved FPT algorithms for several variants of the Graph Motif problem. Reducing to the Multilinear Detection problem resulted in faster running times and a polynomial space complexity. We have also considered the counting versions of these problems, for the first time in the literature. Our results demonstrate that the algebraic framework of [16] has potential applications to computational biology, though a practical evaluation of the algorithms remains to be done. In particular, how do they compare to implementations based on color-coding or ILPs [7,5]?

We conclude with some open questions. A first question concerns our results of Section 3.2 for multiset motifs: is it possible to further reduce the $\mathcal{O}^*(4^k)$ running times? Another question relates to the edge-weighted problems considered in Section 3.3: our algorithms are only pseudopolynomial in the maximum weight r, can this dependence in r be improved? Finally, is approximate counting possible for the #XMGM problem? We believe that some of these questions may be solved through an extension of the algebraic framework of Koutis and Williams.

References

1. Alm, E., Arkin, A.P.: Biological networks. Curr. Opin. Struct. Biol. 13(2), 193–202 (2003)
2. Alon, N., Yuster, R., Zwick, U.: Color-coding. J. of ACM 42(4), 844–856 (1995)
3. Betzler, N., Fellows, M.R., Komusiewicz, C., Niedermeier, R.: Parameterized Algorithms and Hardness Results for Some Graph Motif Problems. In: Ferragina, P., Landau, G.M. (eds.) CPM 2008. LNCS, vol. 5029, pp. 31–43. Springer, Heidelberg (2008)

4. Björklund, A., Husfeldt, T., Kaski, P., Koivisto, M.: Fourier meets möbius: fast subset convolution. In: STOC, pp. 67–74 (2007)
5. Blin, G., Sikora, F., Vialette, S.: GraMoFoNe: a Cytoscape plugin for querying motifs without topology in Protein-Protein Interactions networks. In: BICoB 2010, pp. 38–43 (2010)
6. Böcker, S., Rasche, F., Steijger, T.: Annotating Fragmentation Patterns. In: Salzberg, S.L., Warnow, T. (eds.) WABI 2009. LNCS (LNBI), vol. 5724, pp. 13–24. Springer, Heidelberg (2009)
7. Bruckner, S., Hüffner, F., Karp, R.M., Shamir, R., Sharan, R.: Topology-Free Querying of Protein Interaction Networks. In: Batzoglou, S. (ed.) RECOMB 2009. LNCS, vol. 5541, pp. 74–89. Springer, Heidelberg (2009)
8. Dondi, R., Fertin, G., Vialette, S.: Weak pattern matching in colored graphs: Minimizing the number of connected components. In: ICTCS, pp. 27–38 (2007)
9. Dondi, R., Fertin, G., Vialette, S.: Maximum Motif Problem in Vertex-Colored Graphs. In: Kucherov, G., Ukkonen, E. (eds.) CPM 2009. LNCS, vol. 5577, pp. 221–235. Springer, Heidelberg (2009)
10. Fellows, M.R., Fertin, G., Hermelin, D., Vialette, S.: Sharp Tractability Borderlines for Finding Connected Motifs in Vertex-Colored Graphs. In: Arge, L., Cachin, C., Jurdziński, T., Tarlecki, A. (eds.) ICALP 2007. LNCS, vol. 4596, pp. 340–351. Springer, Heidelberg (2007)
11. Flum, J., Grohe, M.: The Parameterized Complexity of Counting Problems. SIAM Journal on Computing 33(4), 892–922 (2004)
12. Flum, J., Grohe, M.: Parameterized Complexity Theory. Springer, Heidelberg (2006)
13. Hüffner, F., Wernicke, S., Zichner, T.: Algorithm Engineering For Color-Coding To Facilitate Signaling Pathway Detection. In: APBC 2007, pp. 277–286 (2007)
14. Karp, R.M.: Dynamic-programming meets the principle of inclusion and exclusion. Oper. Res. Lett. 1, 49–51 (1982)
15. Koutis, I.: Faster Algebraic Algorithms for Path and Packing Problems. In: Aceto, L., Damgård, I., Goldberg, L.A., Halldórsson, M.M., Ingólfsdóttir, A., Walukiewicz, I. (eds.) ICALP 2008, Part I. LNCS, vol. 5125, pp. 575–586. Springer, Heidelberg (2008)
16. Koutis, I., Williams, R.: Limits and Applications of Group Algebras for Parameterized Problems. In: Albers, S., et al. (eds.) ICALP 2009, Part I. LNCS, vol. 5555, pp. 653–664. Springer, Heidelberg (2009)
17. Lacroix, V., Fernandes, C.G., Sagot, M.-F.: Motif Search in Graphs: Application to Metabolic Networks. Trans. Comput. Biol. Bioinform. 3(4), 360–368 (2006)
18. Nederlof, J.: Fast Polynomial-Space Algorithms Using Möbius Inversion: Improving on Steiner Tree and Related Problems. In: Albers, S., Marchetti-Spaccamela, A., Matias, Y., Nikoletseas, S., Thomas, W. (eds.) ICALP 2009. LNCS, vol. 5555, pp. 713–725. Springer, Heidelberg (2009)
19. Schbath, S., Lacroix, V., Sagot, M.-F.: Assessing the exceptionality of coloured motifs in networks. In: EURASIP JBSB, pp. 1–9 (2009)
20. Sharan, R., Ideker, T.: Modeling cellular machinery through biological network comparison. Nature Biotechnology 24, 427–433 (2006)
21. Williams, R.: Finding paths of length k in $O^*(2^k)$ time. IPL 109(6), 315–318 (2009)

Limiting Negations in Bounded Treewidth and Upward Planar Circuits

Jing He, Hongyu Liang, and Jayalal M.N. Sarma

Institute for Theoretical Computer Science,
Tsinghua University, Beijing, China
{hejing2929,hongyuliang86,jayalal.sarma}@gmail.com

Abstract. The decrease of a Boolean function $f : \{0,1\}^n \to \{0,1\}$, denoted by $d(f)$ is the maximum number of inverse indices in any increasing chain of inputs $x_1, \ldots, x_\ell \in \{0,1\}^n$, where i is an inverse index if $f(x_i) > f(x_{i+1})$. It follows from a theorem of Markov (JACM 1958) that the minimum number of negation gates in a circuit necessary and sufficient to compute any Boolean function f is $\lceil \log(d(f)+1) \rceil$. A recent result due to Morizumi (ICALP 2009) proves that $d(f)$ negations are necessary and sufficient when the computation is done by formulas. We explore the situation in between formulas (directed trees) and general circuits (DAGs), and related models. We obtain the following results:

1. We argue that for any Boolean function f, there is a circuit computing f, that uses $\lceil \log(d(f)+1) \rceil$ negations and has treewidth at most $\lceil \log(d(f)+1) \rceil + 1$. For $1 \leq k \leq \lceil \log(d(f)+1) \rceil$, we prove that $d(f) \cdot 8k/2^k$ negations are sufficient to compute any Boolean function f by circuits of treewidth at most k. Moreover, if there is a circuit family of size $s = s(n)$ and treewidth $k = k(n)$ computing $\{f_n\}$, then there exists a circuit family of size $s \cdot n^{O(1)} \cdot 2^{O(\min\{k, \log n\})}$ and treewidth at most $2k$ which computes $\{f_n\}$ and contains $O(\max\{nk/2^{2k}, \log n\})$ negation gates.
2. We obtain tight bounds on the number of negation gates required to compute specific functions such as PARITY$_n$, $\overline{\text{PARITY}_n}$ and INVERTER$_n$ by one-input-face upward planar circuits. We extend these lower bounds to a larger class of functions (which also includes natural functions like ADD and SUBTRACT) and we show a direct sum theorem for this class with respect to the number of negations.
3. We demonstrate the limitations of the one-input-face constraint in the upward planar circuits by showing the explicit function which can be computed by a monotone upward planar circuit, but cannot be computed by any montone one-input-face upward planar circuit.
4. We prove that for every Boolean function f, there exists a multilective upward planar circuit which uses at most $\lceil \frac{d(f)+1}{2} \rceil$ negation gates for computing f.

1 Introduction

Proving super-polynomial size lower bounds for circuits computing explicit functions in NP is a central problem in circuit complexity theory. Theory of monotonicity in Boolean circuits became fruitful in this context and it culminated in

P. Hliněný and A. Kučera (Eds.): MFCS 2010, LNCS 6281, pp. 417–428, 2010.

the exponential size lower bound due to Razborov [13] for any monotone circuit computing the clique function. Tardos [17] demonstrated that there are explicit functions in P which requires exponential size for monotone circuits computing them. This area received a lot of attention and several important resource lower bounds were proved against monotone Boolean circuits [6,12]. Relaxing the monotonicity constraint, Amano and Maruoka [1] proved exponential size lower bounds for an explicit function when the circuit is allowed to use only $\frac{1}{6}\log\log n$ negations.

How far can one improve the above lower bound result in terms of the number of negations allowed in the circuit? This highlights the importance of exploring the power of negation gates in Boolean circuits. A very fundamental question in this direction is about the number of negations that is required to compute any Boolean function, called the inversion complexity of the function. Historically much earlier (in 1958), Markov [8] came up with a surprisingly tight bound for the inversion complexity of any Boolean function. Let $f : \{0,1\}^n \rightarrow \{0,1\}$, and let $(x_1, x_2, \ldots, x_n)$ and $(y_1, y_2, \ldots, y_n)$ be two Boolean vectors in $\{0,1\}^n$. Define $x \leq y$ if $x_i \leq y_i$ for all i. The decrease of function f with respect to an increasing chain of Boolean vectors $v_1, \ldots, v_m \in \{0,1\}^n$, is defined as the number of i such that $f(v_i) > f(v_{i+1})$. The decrease of the function, denoted by $d(f)$ is the maximum decrease over all increasing chains of Boolean vectors. Thus, the decrease of the function f can at most be n. Markov [8] showed a tight characterization of the inversion complexity of a function f as $\lceil \log(d(f)+1) \rceil$. However, the circuits that Markov constructed are of exponential size although they use only $O(\log n)$ negations. Complementing this, Fischer [4] showed that for every poly-sized circuit, there is an equivalent poly-sized circuit which uses at most $\lceil \log(n+1) \rceil$ negations.

Many years after Markov's and Fischer's results, Santha and Wilson [14] (see also [16]) showed a contrasting picture in the constant depth world: there are functions requiring super-logarithmic number of negation gates in any poly-sized constant-depth circuit computing them. Recently, Morizumi [11] studied the case of formulas and proved tight lower and upper bounds for the inversion complexity of Boolean functions. More precisely, he proved that the inversion complexity in formulas computing the function f is exactly $d(f)$. He also proved an analogue of Fischer's result in this context: if there is a polynomial size formula computing a function f, then there is a polynomial size formula for f which uses at most $\lceil \frac{n}{2} \rceil$ number of negations.

In this paper, we study circuit classes between formulas (directed trees) and general circuits (directed acyclic graphs). Many parameters interpolate between the two. Two important ones we consider here are upward planarity and the treewidth of the underlying undirected graph of the circuit. We defer the formal definition of these parameters to Section 2.

Roughly speaking, treewidth measures how formula-like the circuit is (the treewidth of a formula is 1). We consider circuits of treewidth k, and parameterize the number of negations in the circuit in terms of k. The power of such circuits has been considered earlier in the context of classical [5] and quantum computation [9]. As noted in [5], leveled circuits (or graphs) of width k has treewidth at most $2k-1$, and they showed that width bounded and treewidth bounded circuit classes roughly interleave in terms of computational power. As noted earlier, the treewidth bounded classes also generalizes formulas. It was shown in [5] that the circuits of treewidth k and size s can also be simulated by formulas of size roughly s^{k^2}. However, the number of negations used in the construction is large even if the original circuit is monotone.

We explore the power of such circuits in the context of inversion complexity. To begin with, we argue that treewidth beyond $O(\log n)$ does not help in general(Theorem 5). We prove the following parameterized upper bound:

Theorem 1. *Let f be a Boolean function, $1 \leq k \leq \lceil \log(d(f)+1) \rceil$. There exists a circuit of treewidth at most k computing f using at most $d(f) \cdot 8k/2^k$ negations.*

However, as in the case of Markov's theorem, the size of the circuit in the above theorem could be exponentially large. We also obtain an analogue of Fisher's result in our context; i.e. the inverse complexity under size constraints.

Theorem 2. *Let $\{f_n\}$ be a family of Boolean functions. If there is a circuit family of size $s = s(n)$ and treewidth $k = k(n)$ computing $\{f_n\}$, then there exists a circuit family of size $s \cdot n^{O(1)} \cdot 2^{O(\min\{k, \log n\})}$ and treewidth at most $2k$ which computes $\{f_n\}$ and contains $O(\max\{nk/2^{2k}, \log n\})$ negation gates.*

Upward planar circuits are circuits whose underlying graph is upward planar (see Section 2). These circuit classes have been considered in the literature in many contexts [10,3,2,7]. In this model, we require that each input label appears at most once in the circuit and that all input vertices are in one face and all edges in the circuit go upwards in the plane. These are better known in the literature as *one-input-face upward planar circuits* (and are also considered by McColl [10] and Beynon and Buckle [3] in the context of monotone circuits.) For this model, we explore the inversion complexity of specific functions and prove tight upper and lower bounds for the PARITY function and the INVERTER function. More specifically we prove the following:

Theorem 3. *Let $n \geq 2$, $f \in \{\text{PARITY}_n, \overline{\text{PARITY}_n}\}$. The inversion complexity of the function f with respect to one-input-face upward planar circuits is $n-1$. The inversion complexity of INVERTER_n with respect to such circuits is n.*

We generalize this argument further to a non-trivial collection of Boolean functions (Theorem 7). For functions in this class, we show a direct sum theorem (Theorem 8) by proving a tight lower bound on the number of negations required to compute t functions simultaneously using a one-input-face upward planar circuit. We also exhibit the limitation of the one-input-face constraint by showing an explicit function which can be computed by a monotone upward planar circuit, but cannot be computed by any monotone one-input-face upward planar circuit (Theorem 9).

Although formulas are planar, the above model does not include them since formulas allow input labels to be duplicated. A planar circuit model where the inputs can be duplicated is called *multilective planar circuit* (which was introduced in [15]). Formulas are clearly multilective planar circuits. For this more powerful class of circuits, we are able to improve the upper bound on the inversion complexity (denoted as $I_{M-UP}(f)$) slightly.

Theorem 4. *For every Boolean function f, $I_{M-UP}(f) \leq \lceil \frac{d(f)+1}{2} \rceil$.*

2 Preliminaries

We introduce some basic definitions in this section. Let $\mathbb{B}_{n,m}$ denote the set of Boolean functions $f : \{0,1\}^n \to \{0,1\}^m$. $\mathbb{B}_n$ stands for $\mathbb{B}_{n,1}$. For a function $f = f(x_1, \ldots, x_n)$, we say f *essentially depends on* x_i if $f|_{x_i=0} \neq f|_{x_i=1}$.

A circuit is an acyclic directed graph, in which all vertices of fan-in 0 (*input gates*) are associated with some variable $x \in \{x_1, \ldots, x_n\}$ or a constant $c \in \{0,1\}$, and all other nodes are either $\wedge, \vee$ or $\neg$. The *size* of a circuit is the number of gates contained in it, and the *depth* of a circuit is the length of the longest directed path from any input vertex to any output vertex. We refer to a standard text book [19] for more definitions.

A circuit is called *semilective* if for all $x \in \{x_1, \ldots, x_n\}$, at most 1 input vertices in the circuit is associated with x. It is called *multilective* otherwise. A *formula* is a multilective circuit all vertices of which have fan-out at most 1.

For a Boolean function f, the *inversion complexity* of f, denoted by $I(f)$, is the minimum number of negation gates contained in any circuit computing f. If restricting the circuits to be formulas (then f should be a single-output function), we get the definition of *inversion complexity of f in formulas*, denoted by $I_F(f)$. The inversion complexity of a family of Boolean functions can be similarly defined, as a function of n. In this notation, Markov [8] proved that $I(f) = \lceil \log(d(f)+1) \rceil$ for every Boolean function f and Morizumi [11] proved that $I_F(f) = d(f)$.

A *tree decomposition* of a graph $G = (V, E)$ is given by a tuple $(T, (X_d)_{d \in V[T]})$, where T is a tree, each X_d is a subset of V called a *bag*, satisfying 1) $\bigcup_{d \in V[T]} X_d = V$, 2) For each edge $(u, v) \in E$, there exists a tree node d with $\{u, v\} \subseteq X_d$, and 3) For each vertex $u \in V$, the set of tree nodes $\{d : u \in X_d\}$ forms a connected subtree of T. Equivalently, for any three vertices $t_1, t_2, t_3 \in V[T]$ such that t_2 lies in the path from t_1 to t_3, it holds that $X_{t_1} \cap X_{t_3} \subseteq X_{t_2}$. The *width* of the tree decomposition is defined as $\max_d |X_d| - 1$. The *treewidth* $tw(G)$ of a graph G is the minimum width of a tree decomposition of G.

A *planar circuit* is a circuit in which each input label appears exactly once and the underlying undirected graph can be embedded on the plane without edge crossings. It is further called an *upward planar circuit* if it has some planar embedding in which all edges go upwards (monotonically increasing in the vertical direction), and is called *one-input-face upward planar* if besides upward planarity all input nodes are placed at the lowest level.

For every function f, let $I_{OUP}(f)$ denote the minimum number of negation gates required for computing f by any one-input-face upward planar circuit.

3 Bounded Treewidth Circuits

In this section we consider circuits with bounded treewidth, which naturally generalizes formulas. We allow circuits to be multilective, i.e., they may contain duplicated input variables. By a Boolean function we will mean a single-output Boolean function.

3.1 Inversion Complexity in Bounded Treewidth Circuits

Recall that Markov's theorem states that for every Boolean function f, the minimum number of negation gates contained in a circuit computing f is precisely $\lceil \log(d(f)+1) \rceil$; that is, $I(f) = \lceil \log(d(f)+1) \rceil$. We will first show that, if we only care the number of negation gates, then treewidth beyond $I(f)$ is useless.

Theorem 5. *For every Boolean function f, there is a circuit computing f that contains $\lceil \log(d(f)+1) \rceil$ negations and has treewidth at most $\lceil \log(d(f)+1) \rceil + 1$.*

We need the following definition of *connectors*. For any two Boolean functions f_0 and f_1 with the same set of input variables, a *connector of f_0 and f_1* is a function $\mu(y, y', x)$ satisfying that $\mu(i, 1-i, x) = f_i(x)$ for both $i = 0, 1$, where x is the input vector of f_0 and f_1. Markov showed that there always exists a connector of f_0 and f_1 containing at most $\max\{I(f_0), I(f_1)\}$ negation gates, which is then used in the construction of negation-limited circuits. We first demonstrate the existence of a special connector for which we can show a bound on the treewidth.

Lemma 1. *Every pair of Boolean functions $f_0(x)$ and $f_1(x)$ has a connector $\mu(y, y', x)$ which can be computed by a circuit containing* $\max\{I(f_0), I(f_1)\}$ *negation gates and having treewidth at most* $1 + \max\{I(f_0), I(f_1)\}$.

The outline of the proof of the above lemma is similar to that of Markov's connector, and proceeds by induction on m. The main observation is that at the induction step we can optimize on the treewidth while the negation gates are being combined. The proof of Theorem 5 also uses some observations about the treewidth while implementing Markov's construction. We skip the details.

Let $I_k(f)$ denote the minimum number of negation gates contained in any circuit computing f with treewidth at most k. In this notation, $I_1(f) = I_F(f) = d(f)$ and $I_{\lceil \log(d(f)+1) \rceil + 1}(f) = I(f) = \lceil \log(d(f)+1) \rceil$. Now we prove Theorem 1, which gives an upper bound on $I_k(f)$ for any $1 \le k \le \lceil \log(d(f)+1) \rceil$.

Proof of Theorem 1: We will prove, by induction on $d(f)$, that $I_k(f) \le d(f) \cdot 8k/2^k - 1$ for all $1 \le k \le \lceil \log(d(f)+1) \rceil$. The statement is obvious for $d(f) = 1$. Suppose $d(f) \ge 2$ and the theorem holds for all f' such that $d(f') < d(f)$. Let $S \subseteq \{0,1\}^n$ be the set of all input vectors x such that every chain Y starting with x satisfies that $d_Y(f) \le d(f)/2$. Then, for every chain Y ending at a vector $x \notin S$, $d_Y(f) \le d(f)/2$ (otherwise we can find a chain with decrease $\ge d(f)+1$ by concatenating Y and the chain witnessing $x \notin S$). We also have $(x \in S$ and $x \le y) \Rightarrow y \in S$.

Define two functions $f_0(x)$ and $f_1(x)$ as follows:

$$f_0(x) = \begin{cases} 1 & \text{if } x \in S, \\ f(x) & \text{if } x \notin S, \end{cases} \qquad \text{and} \qquad f_1(x) = \begin{cases} f(x) & \text{if } x \in S, \\ 0 & \text{if } x \notin S. \end{cases}$$

It is easy to see that $\max\{d(f_0), d(f_1)\} \le d(f)/2$ and $f(x) = (h_S(x) \wedge f_1(x)) \vee (\overline{h_S(x)} \wedge f_0(x))$, where $h_S(x)$ is the characteristic function of the set S, which is monotone and hence can be computed by a monotone formula. If there exists C_i computing f_i for both $i = 0, 1$ such that C_i has treewidth at most k and contains at most t negation gates, we can obtain a circuit with treewidth at most k which computes f and contains at most $2t + 1$ negation gates. Having $t = d(f) \cdot 4k/2^k - 1$ will suffice.

If $k \le \lceil \log(d(f_i) + 1) \rceil$ for both $i = 0, 1$, we are done by induction hypothesis. If $k > \lceil \log(d(f_i) + 1) \rceil$ for some $i \in \{0, 1\}$, we can get directly from Theorem 5 that there exists a circuit computing f_i with treewidth at most k which contains exactly $I(f_i) = \lceil \log(d(f_i) + 1) \rceil < k$ negation gates. Since $k \le \lceil \log(d(f) + 1) \rceil$, we have $2^k \le 2(d(f) + 1) \le 4d(f)$, from which it follows that $I(f_i) \le k - 1 \le d(f) \cdot 4k/2^k - 1$. Therefore, for both $i = 0, 1$, there exists a circuit C_i computing f_i which has treewidth $\le k$ and contains $\le d(f) \cdot 4k/2^k - 1$ negation gates. This finishes the induction proof. □

3.2 Inversion Complexity under Polynomial Size Constraints

In this subsection we show Theorem 2 stated in the introduction. Let $f \in \mathbb{B}_n$. Borrowing the notation from [11], we say f' is a *pseudo* i^{th} *slice* of f iff $f'(x) = f(x)$ for all $x = (x_1, \ldots, x_n)$ such that $\sum_{j=1}^{n} x_j = i$. Let C be a circuit computing f of treewidth k and size s. For $i = 0, 1, \ldots, n$, we construct a circuit $C^{(i)}$, which computes a monotone pseudo i^{th} slice of f, by pushing all negations in C down to the input nodes by the De Morgan's law and then replacing $\overline{x_i}$ with $Th_i^{n-1}(x_{-i})$. Here we use x_{-i} to denote $(x_1, \ldots, x_{i-1}, x_{i+1}, \ldots, x_n)$, and Th_m^n is the threshold function which equals 1 iff at least m out of n input variables are 1. It can be seen that pushing the negations down (with possible duplications of gates) can cause at most a doubling of the treewidth of the circuit (we skip the details of this construction). Using $n^{O(1)}$-sized formulas [18] for threshold functions in the above construction gives the following:

Lemma 2. *For a Boolean function $f \in \mathbb{B}_n$ which can be computed by a circuit of size s and treewidth k, and an integer $i \in \{0, 1, \ldots, n\}$, there exists a monotone circuit of size $s \cdot n^{O(1)}$ and treewidth $2k$ which computes a pseudo i^{th} slice of f.*

The basic idea is to make use of these pseudo slices to reconstruct f and hence save on the number of negation gates in the process. For this, we divide them into groups containing some consecutive slices of f. We will use groups of 2^{2k} concecutive slices each (which is a generalization of Morizumi's argument [11] which puts two neighboring slices into one group). Among each group, we find a circuit of limited treewidth and limited number of negation gates which plays the role of a "selector", in the sense that it selects which slice to use according to the number of 1's in the inputs. Finally, a formula serving as a "universal selector" is applied to choose the correct group, which will not increase the treewidth.

Before presenting the proof, we need to introduce some notations. Let $X = \{x_1, x_2, \ldots, x_n\}$ be a set of variables, and $\mathcal{H} = \{h_1, h_2 \ldots, h_m\}$ be a set of Boolean functions each taking $x_1, \ldots, x_n$ as inputs. A circuit *augmented with* $\mathcal{H}$ is similarly defined as a normal circuit, except that every fan-in 0 vertex is now assigned with some function $h_i \in \mathcal{H}$. Functions in $\mathcal{H}$ are called *help functions*, and circuits augmented with $\mathcal{H}$ are also called $\mathcal{H}$*-circuits*. The normal definition of circuits can be seen as a special case in which $\mathcal{H} = \{x_1, x_2, \ldots, x_n\}$; that is, the help functions are exactly the variables themselves.

In the following we fix $\mathcal{H} = \{Th_i^n : i = 0, 1, \ldots, n\} \cup \{f^{(i)} : i = 0, 1, \ldots, n\}$, where $f^{(i)}$ is a pseudo i^{th} slice function of f. For every pair of integers a, b such that $0 \le a \le b \le n$, a Boolean function g is called a (a, b)*-selector of* f iff g is a pseudo i^{th} slice function of f for every i such that $a \le i \le b$. It follows that $f^{(i)}$ is a (i, i)-selector of f, and f is a $(0, n)$-selector of itself.

For a $\mathcal{H}$-circuit C, we say C is a (a, b)*-selector circuit of* f if it computes a (a, b)-selector of f. We call it a *good* (a, b)*-selector circuit of* f if in addition it satisfies the following "replacement rule": For every integer r such that $-a \le r \le n - b$, if we replace every input vertex Th_i^n of C by Th_{i+r}^n, and replace every input vertex $f^{(i)}$ of C by $f^{(i+r)}$, for every $0 \le i \le n$, then the resulting circuit is a $(a + r, b + r)$-selector circuit of f. Now we show the following existence lemma.

Lemma 3. *For every $a, b \in \{0, 1, \ldots, n\}$ such that $b - a + 1 = 2^k$ for some integer $k \ge 1$, there is a good (a, b)-selector circuit of f which has size at most 5^k, treewidth at most k and contains at most k negation gates.*

Proof. By induction on k. When $k = 1$, the circuit be $(Th_b^n \wedge f^{(b)}) \vee (\overline{Th_b^n} \wedge f^{(a)})$ is a good (a, b)-selector circuit of f with size 4, treewidth 1 and contains 1 negation gate. Now suppose $k \ge 2$ and the theorem holds for all smaller k. Let C_{k-1} be a good $(a, a + 2^{k-1} - 1)$-selector circuit of f which has size at most 5^{k-1}, treewidth at most $k - 1$ and contains at most $k - 1$ negation gates. Let v be an arbitrary input vertex of C_{k-1}. If v is Th_i^n for some i, we replace it with $\left(Th_{a+2^{k-1}}^n \wedge Th_{i+2^{k-1}}^n\right) \vee \left(\overline{Th_{a+2^{k-1}}^n} \wedge Th_i^n\right)$. If v is $f^{(i)}$ for some i, we replace it with $\left(Th_{a+2^{k-1}}^n \wedge f^{(i+2^{k-1})}\right) \vee \left(\overline{Th_{a+2^{k-1}}^n} \wedge f^{(i)}\right)$. After we replacing every input node v, we combine all the negations connected to nodes assigned with $Th_{a+2^{k-1}}^n$ together to form a new node. Call the new circuit C_k. Since only one copy of $\overline{Th_{a+2^{k-1}}^n}$ is used, C_k contains exactly 1 more negation gates than C_{k-1}. The treewidth of C_k is at most k, as we first replace each input vertex with a tree (which will preserve the treewidth of C_{k-1}) and then combine some vertices together to form a new one (which will increase the treewidth by at most 1). To bound the size of C_k, we note that each input node of C_{k-1} is replaced with a circuit of size 4, and the number of input nodes does not exceed the number of gates in C_{k-1}. Therefore, the size of C_k is at most $5^{k-1} + 4 \cdot 5^{k-1} = 5^k$.

It remains to show that C_k is a good (a, b)-selector circuit of f. Due to our construction, when $\sum_{j=1}^{n} x_j < a + 2^{k-1}$ (that is, $Th_{a+2^{k-1}}^n(x) = 0$), C_k is equivalent to C_{k-1}, and hence C_k is a $(a, a+2^{k-1}-1)$-selector circuit of f. When $\sum_{j=1}^{n} x_j \geq a + 2^{k-1}$, C_k becomes the circuit obtained from C_{k-1} by replacing Th_i^n with $Th_{i+2^{k-1}}^n$ and replacing $f^{(i)}$ with $f^{(i+2^{k-1})}$, which is of course a $(a+2^{k-1}, a+2^k-1)$-selector circuit of f because C_{k-1} is good by the induction hypothesis. Hence, C_k is a (a,b)-selector circuit of f (remember that $b = a + 2^k - 1$). To see why C_k is good, we shift all the parameters (except n) by r for every $-a \leq r \leq b$, and can similarly prove that the resulting circuit is a $(a+r, b+r)$-selector circuit of f. This completes the induction step. □

Proof of Theorem 2: Let $f \in \mathbb{B}_n$ and C be a circuit computing it with size s and treewidth k. Let t be the minimum integer such that $n \leq 2^t - 1$, and let $n' = 2^t - 1$. Let $k' = \min\{k, t/2\}$. It is clear that $n' \leq 2n - 1$ and $\log n \leq t \leq \lceil \log n \rceil + 1$. We add $n' - n$ dummy input variables to f and regard it as a function in $\mathbb{B}_{n'}$. For $l = 0, 1, \ldots, 2^{t-2k'} - 1$, let C'_l be a $(2^{2k'}l, 2^{2k'}(l+1)-1)$-selector circuit of f constructed by Lemma 3, which has size at most $5^{2k'}$, treewidth at most $2k'$ and contains at most $2k'$ negations. Let C' be a formula-like circuit of the form $\bigvee_{l=0}^{2^{t-2k'}} \left(Th_{2^{2k'}l}^{n'} \wedge \overline{Th_{2^{2k'}(l+1)}^{n'}} \wedge C'_l\right)$ (different terms will use different copies of input vertices). It is easy to see that C computes exactly f and contains at most $(2k'+1)2^{t-2k'} = (2k'+1)(n'+1)/2^{2k'} \leq 2n(2k'+1)/2^{2k'}$ negation gates. Since $k' = \min\{k, t/2\}$ and $\log n \leq t \leq \lceil \log n \rceil + 1$, this is at most $\max\{2n(2k+1)/2^{2k}, 2n(t+1)/2^t\} = \max\{O(nk/2^{2k}), O(\log n)\} = O(\max\{nk/2^{2k}, \log n\})$. Furthermore, C' has size at most $2^{t-2k'}(4+5^{2k'}) = n \cdot 2^{O(k')} = n \cdot 2^{O(\min\{k,t/2\})} \leq n \cdot 2^{O(\min\{k, \log n\})}$, and has treewidth at most $2k'$. Note that C' is not a "true" circuit, but one augmented with help functions $\{Th_i^n : i = 0, 1, \ldots, n\} \cup \{f^{(i)} : i = 0, 1, \ldots, n\}$. We replace every input node of C' with a circuit computing the corresponding help function. Since Th_i^n is computable by a poly-sized formula, and by Lemma 2 $f^{(i)}$ is computable by a monotone circuit of treewidth $2k$ and size $s \cdot n^{O(1)}$, the resulting circuit has size $s \cdot n^{O(1)} \cdot 2^{O(\min\{k, \log n\})}$ and treewidth at most $\max\{2k, 2k'\} = 2k$. This finishes the proof of Theorem 2. □

4 Inversion Complexity in Planar Circuits

4.1 Lower Bounds for One-Input-Face Upward Planar Circuits

In this section we will focus on the inversion complexity in one-input-face upward planar circuits. We will prove $\Omega(n)$ lower bounds of $I_{OUP}(f)$ for many functions f, including some tight results. Since $I(f) = O(\log n)$ for all $f \in \mathbb{B}_{n,m}$, an exponential gap between the number of negation gates used in general circuits and one-input-face upward planar circuits is obtained for a number of natural functions, including PARITY, INVERTER, ADD and SUBTRACT.

Theorem 6. *For* $f \in \{\textsc{Parity}_n, \overline{\textsc{Parity}_n}\}$ *where* $n \geq 2$, $I_{OUP}(f) = n - 1$.

Proof. To prove that $n-1$ is an upper bound, just notice that $\overline{\textsc{Parity}_n} = \textsc{Equiv}(\textsc{Parity}_{n-1}(x_1, \ldots, x_{n-1}), x_n)$ ($\textsc{Equiv}(x,y)$ computes $x \equiv y$), and that both the XOR gate and the EQUIV gate can be simulated by planar circuits each containing one negation gate (by standard constructions).

Next we prove that $I_{OUP}(f) \geq n-1$ for $f \in \{\text{PARITY}_n, \overline{\text{PARITY}_n}\}$ where $n \geq 2$. The statement will be proved by induction on n. It is obvious when $n = 2$, since both functions are non-monotone. Now suppose $n \geq 3$ and the statement holds for all $n' < n$. Let $f \in \{\text{PARITY}_n, \overline{\text{PARITY}_n}\}$. For the sake of contradiction, let C be a one-input-face upward planar circuit computing f which contains at most $n-2$ negation gates. Without loss of generality, let the order of the input variables (from left to right in the input face) be $x_1, \ldots, x_n$.

Suppose G is the predecessor of the first negation gate in C (under some topological order of the underlying graph of C). Let g denote the function computed at the gate G. Let $S = \{x_i \mid g \text{ essentially depends on } x_i, 1 \leq i \leq n\}$. Let $l = \min\{i \mid x_i \in S\}$ and $r = \max\{i \mid x_i \in S\}$. Denote by PI the set of all prime implicants of g. Since g is monotone, every prime implicant of it only contains positive literals. First we argue that $S = \{x_i \mid l \leq i \leq r\}$. For this, assume the contrary that $x_i \notin S$ for some $l < i < r$. We set all variables in S to 1. Since g is monotone, this will fix G to be 1. Hence we can find one path from x_l to G on which each gate is fixed to be 1, and another path from x_r to G with the same property. The output gate of C (denoted by O) must lie out of the area defined by these two paths and the input face due to its non-monotonicity. Since C is upward planar, we know that every path connecting x_i and O must intersect the two 1-paths previously found. Thus the variable x_i is "disconnected" from O, and gives a contradiction since the function being computed is PARITY. Similarly we can show that any prime implicant is an "interval".

Lemma 4. *Let $p = x_{i_1}x_{i_2}\ldots x_{i_k} \in PI$ be any prime implicant of g, where $l \leq i_1 < \ldots < i_k \leq r$. Then $i_m = i_{m-1}+1$ for all $2 \leq m \leq k$.*

Using the induction hypothesis (we skip the details) we can show that each prime implicant of g has size at least 2. In addition, we can prove the following: Let p_l and p_r be the prime implicants of g containing x_l and x_r, respectively. Then p_l and p_r do not intersect with each other; that is, they contain no common variables. Thus, PI contains at least 2 different prime implicants.

Now we are ready to prove Theorem 6. Let C_G be the induced sub-circuit of C with output gate G. More precisely, C_G contains all vertices of C (variables and gates) from which G is reachable, and all edges spanning them. So C_G computes the function g. Let $p_l = x_1x_2\ldots x_j$ and $p_r = x_kx_{k+1}\ldots x_r$ be the prime implicants of g containing x_l and x_r, respectively. By the arguments in the previous paragraph, we have $l < j < k < r$. Imagine the scenario where all variables except x_k are set to 0. Since f still essentially depends on x_k after this restriction, there exists a "switching path" $\mathcal{P}$ from x_k to the output gate of C such that flipping the value of x_k will cause all gates on $\mathcal{P}$ to change their values (given that other variables are set to 0). Also note that the output gate is not contained within the area of C_G. Due to the upward planarity of C, $\mathcal{P}$ must intersect the boundary of C_G. Let $\mathcal{P}_{in}$ denote the inside part of $\mathcal{P}$ respect to C_G. As C_G is monotone, any gate on $\mathcal{P}_{in}$ switches from 0 to 1 if x_k switches from 0 to 1, given that other variables are 0. Therefore, setting x_k to 1 will fix

all gates on $\mathcal{P}_{in}$ to evaluate to 1, regardless of the assignment to other variables. $\mathcal{P}_{in}$. The following two cases finishes the induction proof.
Case 1: $\mathcal{P}_{in}$ intersects the right boundary of C_G. We set $x_l = x_{l+1} = \ldots = x_{k-1} = 0$ and $x_k = x_{k+1} = \ldots = x_{r-1} = 1$. Under this restriction G will compute exactly x_r. But x_r cannot affect G now, since any path from x_r to G must intersect $\mathcal{P}_{in}$ This leads to a contradiction.
Case 2: $\mathcal{P}_{in}$ intersects the left boundary of C_G. We set $x_1 = x_2 = \ldots = x_{j-1} = 1$, $x_{j+1} = \ldots = x_{k-1} = 0$, $x_k = 1$ and $x_{k+1} = \ldots = x_r = 0$. Under this restriction G will compute exactly x_j. But any path from x_j to G must intersect $\mathcal{P}_{in}$, which again gives a contradiction. □

We next introduce the classes of functions for which we can apply a similar argument and prove linear lower bounds on $I_{OUP}(f)$. Although the definition seems restrictive, it can be shown that several natural functions fall into these classes. Define the class $\mathbb{W}^n_{k,n_0}$ for $k, n_0, n, m \in \mathbb{N}, k \geq 1$ of functions $f \in \mathbb{B}_{n,m}$ as the ones with the two properties: (1) If $n > n_0$, then for any variable $x \in X$ and any "uniform" restriction σ which maps all variables in $X \setminus \{x\}$ to the same constant $b \in \{0,1\}$, there exists a non-monotone output $y \in Y$ such that $y|_\sigma$ essentially depends on x. (2) If $n > n_0$, then for any variable $x_0 \in X$ and any constant $c_0 \in \{0,1\}$, there exists a set of k' variables $\{x_1, \ldots, x_{k'}\} \subseteq X \setminus \{x_0\}$ and a sequence of Boolean constants $c_1, \ldots, c_{k'}$, where $0 \leq k' \leq \min\{k-1, n-1\}$, such that $f|_\gamma \in \mathbb{W}^{n-k'-1}_{k,n_0}$, where γ denotes the restriction which maps x_i to c_i for all $0 \leq i \leq k'$. Now we state the following theorem about the functions in this class, the proof of which is essentially a generalisation of the proof of Theorem 6. We skip the details due to space constraints.

Theorem 7. *$I_{OUP}(f) \geq \lceil \frac{n-n_0}{k} \rceil$ if $f \in \mathbb{W}^n_{k,n_0}$.*

It can be verified that the class $\mathbb{W}$ contains some natural functions such as {INVERTER, ADD, SUBTRACT, OROFPARITY} (OROFPARITY is the OR of t PARITY's defined on pairwise-disjoint variables). Combined with the trivial upper-bound for INVERTER, this gives:

Corollary 1. – *$I_{OUP}(\text{INVERTER}_n) = n$ for all $n \geq 1$.*
– *$I_{OUP}(\text{OROFPARITY}_{n_1,\ldots,n_t}) \geq \frac{\sum_{i=1}^t n_i}{2}$ where all n_i's are even.*
– *For $f \in \{\text{ADD}_{2n}, \text{SUBTRACT}_{2n}\}$ where $n \geq 1$, $I_{OUP}(f) \geq n$.*

A Direct Sum Theorem: We consider the direct sum of Boolean functions, i.e., a collection of different Boolean functions on pairwise disjoint variable sets. If f is the direct sum of $f_1, \ldots, f_t$, then each of the functions is called an *element* of f. What is the relationship between the inversion complexity of f and that of its elements? Trivially $I(f) \leq \sum_{i=1}^t I(f_i)$, but the result is far from tight; for example, when each f_i is $\text{PARITY}_{\sqrt{n}}$ and $t = \sqrt{n}$, the LHS is $O(\log n)$ but the RHS is $\sum_{i=1}^t I(f_i) = n^{\Theta(1)}$. This indicates that computing the direct sum of functions (with large decreases) can benefit from interconnections between its seemingly independent elements. We will show that this is not always the case if we adopt planar circuits as our computation model.

Theorem 8. *Let* $f_i \in \mathbb{W}^{n_i}_{k_i,m_i}, \forall i \in [t]$. *If* f *is the direct sum of* f_i*'s, then* $I_{OUP}(f) \geq \sum_{i=1}^{t} \lceil \frac{n_i - m_i}{k_i} \rceil$.

The proof of this theorem is again based on induction, and Theorem 7 forms the base case of it. The induction step further generalizes the proof of Theorem 6. The only difference is that, when dealing with a special input variable x_i, we need to identify which function it belongs to. We skip the details of the proof in this short version. It is straightforward from Theorems 6,8 and the bounds for the functions that $I(f) = \sum_{i=1}^{t} I(f_i)$ if each f_i is PARITY, $\overline{\text{PARITY}}$ or INVERTER, and f is the direct sum of all f_i's. This differs from general circuits.

Limitation of the One-Input-Face Constraint: We show that restricting all input vertices to be on the same face (or equivalently on the exterior face, or at the lowest level in the plane) may increase the number of negation gates used for computing some functions. We prove a stronger result, by showing a monotone (multi-output) function which has a monotone upward planar circuit computing it, but cannot be computed by any monotone one-input-face cylindrical circuits. Here a circuit is called *cylindrical* if it can be embedded on a cylinder surface without edge crossings and every edge goes upwards. It is easy to see that cylindricality generalizes upward planarity. Let $\text{MINMAX}_n(x_1, x_2, \ldots, x_n) = (\bigwedge_{i=1}^{n} x_i, \bigvee_{i=1}^{n} x_i)$. Thus $\text{MINMAX}_n \in \mathbb{B}_{n,2}$ and computes the minimum and maximum values among all input variables. It is easy to construct a monotone upward planar circuit for it. Using an argument similar to that of the proof of Theorem 6, in the context of cylindricality, we can prove the following theorem:

Theorem 9. *Let* $n \geq 3$. *Then* MINMAX_n *can be computed by a monotone upward planar circuit, but cannot be computed by any monotone one-input-face cylindrical circuit.*

4.2 Multilective Upward Planar Circuits

In this subsection, we will outline the proof of Theorem 4. That is, for every single-output Boolean function f, $I_{M-UP}(f) \leq \lceil \frac{d(f)+1}{2} \rceil$. The proof follows a similar line to that of Markov's. Given a circuit C, we say a vertex (an input variable or a gate) in G is *out* if it lies in the outer-face of C. We say C is *out-negationed* if either C is monotone, or there is a negation gate in C which is out, and the sub-circuit below it is monotone.

The following statement essentially captures the argument. For $i = 0, 1$, let $f_i(x)$ be a Boolean function which can be computed by a out-negationed multilective upward planar circuit containing t_i negation gates. In addition suppose $t_1 = 1$. Then f_0 and f_1 have a connector $\mu(y, y', x)$ which can be computed by a multilective upward planar circuit containing at most $\max\{t_0, t_1\}$ negation gates and exactly one input node assigned with the variable y'. Moreover, this y'-node is out. The proof of Theorem 4 now follows by a careful induction combined with some crucial observations in Markov's original proof.

Acknowledgements. This work was supported in part by the National Natural Science Foundation of China Grant 60553001, and the National Basic Research Program of China Grant 2007CB807900,2007CB807901.

References

1. Amano, K., Maruoka, A.: A superpolynomial lower bound for a circuit computing the clique function with at most (1/6)log log n negation gates. SIAM Journal of Computing 35(1), 201–216 (2005)
2. Barrington, D.A.M., Lu, C.-J., Miltersen, P.B., Skyum, S.: On Monotone Planar Circuits. In: Proceedings of the 14th Annual IEEE Conference on Computational Complexity (CCC), pp. 24–31 (1999)
3. Beynon, M., Buckle, J.: On the planar monotone computation of boolean functions. Theor. Comput. Sci. 53(2-3), 267–279 (1987)
4. Fischer, M.: The complexity of negation-limited networks (a brief survey). LNCS, vol. 33, pp. 71–82. Springer, Heidelberg (1974)
5. Jansen, M., Sarma, M.N., Jayalal: Balancing Bounded Treewidth Circuits. In: Proeedings of CSR 2010 (to appear 2010)
6. Karchmer, M., Wigderson, A.: Monotone circuits for connectivity require super-logarithmic depth. In: Proc. of STOC 1988, pp. 539–550 (1988)
7. Limaye, N., Mahajan, M., Sarma, M.N.J.: Upper bounds for monotone planar circuit value and variants. Computational Complexity 18(3), 377–412 (2009)
8. Markov, A.A.: On the inversion complexity of a system of functions. J. ACM 5(4), 331–334 (1958)
9. Markov, I.L., Shi, Y.: Simulating quantum computation by contracting tensor networks. SIAM J. Comput. 38(3), 963–981 (2008)
10. McColl, W.F.: On the planar monotone computation of threshold functions. In: Mehlhorn, K. (ed.) STACS 1985. LNCS, vol. 182, pp. 219–230. Springer, Heidelberg (1984)
11. Morizumi, H.: Limiting negations in formulas. In: Albers, S., et al. (eds.) ICALP 2009, Part I. LNCS, vol. 5555, pp. 701–712. Springer, Heidelberg (2009)
12. Raz, R., Wigderson, A.: Monotone circuits for matching require linear depth. In: STOC 1990: Proceedings of the twenty-second annual ACM symposium on Theory of computing, pp. 287–292 (1990)
13. Razborov, A.A.: Lower bounds on the monotone complexity of some boolean functions. Soviet Math. Dokl. 281, 798–801 (1985)
14. Santha, M., Wilson, C.: Limiting negations in constant depth circuits. SIAM J. Comput. 22(2), 294–302 (1993)
15. Savage, J.E.: The performance of multilective vlsi algorithms. Journal of Computer and System Science 29(2), 243–273 (1984)
16. Sung, S.C., Tanaka, K.: Limiting negations in bounded-depth circuits: An extension of markovs theorem. In: Ibaraki, T., Katoh, N., Ono, H. (eds.) ISAAC 2003. LNCS, vol. 2906, pp. 108–116. Springer, Heidelberg (2003)
17. Tardos, E.: The Gap Between Monotone and Non-monotone Circuit Complexity is Exponential. Combinatorica 7, 393–394 (1987)
18. Valiant, L.G.: Short monotone formulae for the majority function. J. Algorithms 5(3), 363–366 (1984)
19. Vollmer, H.: Introduction to Circuit Complexity: A Uniform Approach. Springer, New York (1999)

On the Topological Complexity of MSO+U and Related Automata Models

Szczepan Hummel, Michał Skrzypczak, and Szymon Toruńczyk*

University of Warsaw

Abstract. We show that Monadic Second Order Logic on ω-words extended with the unbounding quantifier (MSO+U) can define non-Borel sets. We conclude that there is no model of nondeterministic automata with a Borel acceptance condition which captures all of MSO+U. We also give an exact topological complexity of the classes of languages recognized by nondeterministic ωB-, ωS- and ωBS-automata studied by Bojańczyk and Colcombet in [BC06]. Furthermore, we show that corresponding alternating automata have higher topological complexity than nondeterministic ones — they inhabit all finite levels of the Borel hierarchy.

Introduction

Motivation and background. The notion of an ω-regular language is well established in the theory of automata. This class of languages carries over to ω-words many of the good properties of regular languages of finite words. It can be described using automata, namely by nondeterministic Büchi automata, or the equivalent deterministic Muller automata, and also alternating automata. In terms of logic, they are equivalent to both Monadic Second Order Logic (MSO) and Weak Monadic Second Order Logic (Weak MSO) – the fragment of MSO where quantifiers may only bind finite sets. Such connections between logic and automata are extremely important in the field of verification and specification.

Recently, in [Boj10, Boj09, BT09] it has been suggested that there are other robust classes of languages of ω-words, extending the canonical notion. It has been advocated that natural examples of languages that *might* be seen as regular (for instance, because of a finite Myhill-Nerode index) are languages such as

$$L_B = \{a^{n_1}ba^{n_2}b\ldots \mid \limsup n_i < \infty\},$$
$$L_S = \{a^{n_1}ba^{n_2}b\ldots \mid \liminf n_i = \infty\}.$$

which are not ω-regular in the usual sense.

These papers describe several such classes of languages, but none of them is known to have all the robust properties of ω-regular languages. On one hand, automata models often allow deciding emptiness, on the other hand, a class

* All authors have been supported by the Polish government grant no. N206 008 32/0810.

P. Hliněný and A. Kučera (Eds.): MFCS 2010, LNCS 6281, pp. 429–440, 2010.

described in terms of logic usually has good closure properties. Ideally, one would like to have a class of languages which can be described both in terms of automata and in terms of logic, as in the case of ω-regular languages.

The connection between automata and logic is better understood when restricted to deterministic automata and weak logics. Deterministic max-automata, introduced in [Boj09], have an alternative description in terms of logic, namely Weak MSO extended with the unbounding quantifier U, which is defined so that the formula $\mathsf{U}X.\varphi(X)$ is equivalent to writing:

"*$\varphi(X)$ is satisfied by arbitrarily large finite sets X of positions*"

Thus U is suited to capture (the complement of) L_B.

As shown in [BT09], the correspondence between deterministic automata and weak logics extends to various other classes of languages, for instance deterministic min-max-automata are equivalent to a logic called Weak MSO $+\mathsf{U}+\mathsf{R}$, and embrace both L_B and L_S.

Problems arise when we look for classes closed under set-theoretic projection, which corresponds to full existential quantification in logic or to nondeterminism on the automata side. In [BC06], ωBS-automata were defined as automata equipped with counters which can be incremented or reset, but not read. The acceptance condition may require a counter to be bounded (the B-condition) or convergent to ∞ (the S-condition). Although nondeterministic ωBS-automata are not closed under complementation, there is a partial fix to this problem. The main technical result of [BC06] shows that the complement of a language defined by an ωB-automaton is accepted by an ωS-automaton and vice versa, where the two are subclasses of ωBS-automata using only the B-condition or the S-condition, respectively. However, boolean combinations of ωB-automata are not closed under existential quantification.In consequence, it seems unlikely to find any sensible logic corresponding to either ωBS-automata or boolean combinations of ωB-automata. To try to overcome these issues, one might consider alternating ωBS-automata. So far, it was not known whether in the ωBS-setting there is any advantage of alternation over nondeterminism.

From the logic side, in order to capture nondeterminism, it seems natural to consider the logic MSO+U, which extends the class of languages recognized by ωBS-automata. However, we face the essential question, whether this logic is decidable. In this paper we analyze large classes of automata to seek for a model capable of capturing MSO+U.

Topological complexity. Our approach is to investigate from the topological viewpoint the classes of languages mentioned above and also explore other large classes of automata. Such an analysis can guide in constructing a suitable model of automata for MSO+U, or show that such a model cannot exist. For instance, we prove that a large class of models of nondeterministic automata cannot capture MSO+U. We also discover that alternating ωBS-automata are strictly more expressive than boolean combinations of nondeterministic ωBS-automata.

Let us illustrate these techniques here with some elementary examples. The language L_S corresponds to a property of a sequence of numbers $n_1, n_2, \ldots$ – namely, being convergent to ∞ – which is equivalent to:

$$\bigwedge_k \bigvee_m \bigwedge_{i \geq m} (n_i > k).$$

Being able to define L_S with a formula with three alternations of logical conectives directly translates into its topological complexity — we say that L_S is (at most) *in the third level of the Borel hierarchy.* Similarly, it is easy to see that L_B is in the second level of the Borel hierarchy.

A run of a *deterministic* automaton is a *continuous* function which maps an input word into a sequence of states. The recognized language is the inverse image of the set of accepting runs under this mapping. A basic property of continuous mappings proves that the language is topologically not more complex than this set of accepting runs.

This immediately yields several results: ω-regular languages occupy at most the first two levels of the Borel hierarchy, since such is the complexity of the Muller acceptance condition; max-automata (equivalently, Weak MSO+U) also fall into the first two levels of the hierarchy, as their acceptance condition is L_B.

As a sample impossibility result (stated in [Boj09]), observe that deterministic max-automata do not recognize the language L_S since it is in the third level of the Borel hierarchy and provably not lower.

In Section 2 we exhibit an example of a language M definable in MSO+U which is analytic-complete, i.e. lays beyond the infinite Borel hierarchy. This instantly proves that there can be no deterministic model of automata with a Borel acceptance condition which captures all of MSO+U.

The above method does not give upper bounds for the complexity of languages defined by *nondeterministic* automata. Such bounds require some nontrivial combinatorial results, usually determinization (as for ω-regular languages). We, in turn, use two difficult results to give upper bounds concerning nondeterministic automata.

In Section 2, Corollary 1, we use a strong topological result of Souslin to conclude that *no model of nondeterministic automata* with a Borel acceptance condition can capture all of MSO+U.

In Section 3, we use the combinatorial complementation result of [BC06] which allows us to compute the topological complexity of languages defined by nondeterministic ωB-, ωS- and ωBS-automata. In particular, ωBS-automata, which form the largest among all the subclasses of MSO+U that we know to have decidable emptiness, reach only the fourth level of the Borel hierarchy.

However, there is still some hope in alternating automata. In Section 4 we define the alternating variant of ωBS-automata and we prove that they inhabit at least all finite levels of the Borel hierarchy. This implies that alternating ωBS-automata are strictly more powerful than (boolean combinations of) nondeterministic ωBS-automata. However, we do not know whether they can recognize analytic languages, such as M. This leaves open the question whether this model captures MSO+U.

1 Basic Notions

Logic. We assume familiarity with the *Monadic Second Order Logic* (MSO). Fix an alphabet A. We denote positions of ω-words using symbols $x, y, \ldots$ and sets of positions with symbols $X, Y, \ldots$. For $a \in A$, the unary predicate P_a holds in all positions of the word where an a stands. It is well known that languages that can be described by this logic are exactly ω-regular languages.

MSO+U allows building formulas using MSO constructs and an additional quantifier U, called the *unbounding quantifier*, defined as follows. The formula $\mathsf{U}X.\varphi(X)$ holds in a word w if $\varphi(X)$ is satisfied for arbitrarily large finite sets X of positions. Formally, $\mathsf{U}X.\varphi(X)$ is equivalent to:

$$\bigwedge_{n\in\mathbb{N}} \exists X.\ (\varphi(X)\ \wedge\ n < |X| < \infty)$$

The canonical examples of languages that can be described are the languages L_B and L_S defined in the introduction.

Topology. For a fixed alphabet A, we treat A^ω as a topological space. A basic open set is determined by a prefix $s \in A^*$ and is of the form $s \cdot A^\omega$. Other open sets are obtained by taking unions of basic open sets. If A is finite, this topological space is homeomorphic (i.e. topologically isomorphic) to the Cantor space.

The Borel hierarchy. The Borel hierarchy is defined inductively. We assist the definition with the following diagram[1].

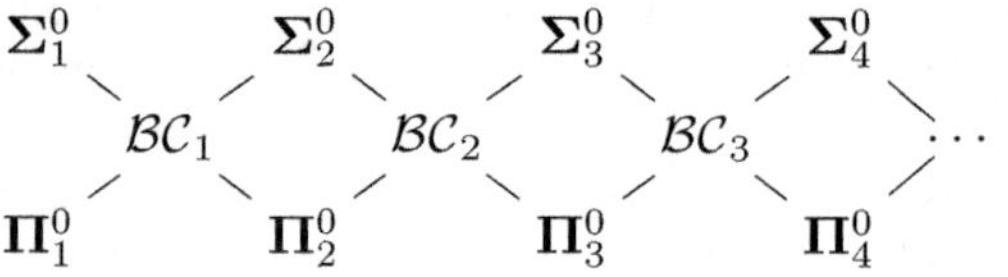

$\mathbf{\Sigma}^0_1$ denotes the class of open sets and $\mathbf{\Pi}^0_1$ denotes the class of closed sets, i.e. complements of open sets. Having defined $\mathbf{\Sigma}^0_n$ and $\mathbf{\Pi}^0_n$, we define $\mathcal{BC}_n$ as (finite) boolean combinations of $\mathbf{\Sigma}^0_n$-sets and $\mathbf{\Pi}^0_n$-sets. In the next step, we define $\mathbf{\Sigma}^0_{n+1}$ as unions of countable families of $\mathcal{BC}_n$-sets and $\mathbf{\Pi}^0_{n+1}$ as intersections of countable families of $\mathcal{BC}_n$-sets. Note that for each n, $\mathbf{\Pi}^0_n$ consists of complements of $\mathbf{\Sigma}^0_n$-sets, and vice versa.

This way we define all *finite* levels of the Borel hierarchy, which is all we will need in this paper. Note that for each n, both $\mathbf{\Sigma}^0_n$ and $\mathbf{\Pi}^0_n$ are strictly contained in both $\mathbf{\Sigma}^0_{n+1}$ and $\mathbf{\Pi}^0_{n+1}$. In fact, in order to obtain a class which is closed under both complements and countable unions, one should continue the construction using transfinite induction up to level ω_1, where we arrive at the class of *Borel sets*. For these and other facts concerning the Borel hierarchy see e.g. [Sri98, Chapter 3.6].

[1] This diagram is more commonly presented with the larger class $\mathbf{\Delta}^0_{n+1} = \mathbf{\Sigma}^0_{n+1} \cap \mathbf{\Pi}^0_{n+1}$ in place of $\mathcal{BC}_n$. However, we will not use the classes $\mathbf{\Delta}^0_n$.

Analytic sets. The direct image of a Borel set under a continuous mapping may no longer be a Borel set. We call such sets *analytic*, and the class of all analytic sets is denoted $\mathbf{\Sigma}^1_1$. Complements of analytic sets are called *coanalytic* and form the class $\mathbf{\Pi}^1_1$. An important result in the theory, the theorem of Souslin (see e.g. [Kec95, Chapter 14.C]) states that if both a set and its complement are analytic then they are in fact Borel. It is worth mentioning that the Borel hierarchy and analytic sets are part of a bigger hierarchy of classes, called the projective hierarchy.

Topological complexity. A *topological complexity class* $\mathbf{C}$, for the needs of this paper is any of the classes $\mathbf{\Sigma}^0_n$, $\mathbf{\Pi}^0_n$ where n is a finite number (although the full Borel hierarchy has levels above ω), and the classes $\mathbf{\Sigma}^1_1$ and $\mathbf{\Pi}^1_1$. Analogously to complexity theory, we have the notions of *reductions* and *completeness*. Let A, B be two alphabets and let $K \subseteq A^\omega$ and $L \subseteq B^\omega$. We say that a continuous mapping $f\colon A^\omega \to B^\omega$ is a *reduction* of K to L if $K = f^{-1}(L)$. It is a simple property of continuous mappings that if L belongs to a topological complexity class $\mathbf{C}$ then so does K. The language L is called $\mathbf{C}$-*hard* iff any set $K \in \mathbf{C}$ can be reduced to L. We say that L is $\mathbf{C}$-*complete* if additionally $L \in \mathbf{C}$.

2 Non-borel Sets in MSO+U

In this section we show that the set B of trees on $\mathbb{N}$ (i.e. prefix closed subsets of $\mathbb{N}^*$) with an infinite branch is MSO+U-definable modulo some encoding. The set B is well known to be $\mathbf{\Sigma}^1_1$-complete (see [Kec95, Theorem 27.1]).

To simplify notation, we consider B as a subset of $\mathcal{T}$ — the set of infinite trees. Since $B \subseteq \mathcal{T}$ and $\mathcal{T} \subseteq 2^{\mathbb{N}^*}$ is in $\mathbf{\Pi}^0_2$, this restriction doesn't affect the topological complexity of B.

Let $\prec$ be some fixed order of type ω on $\mathbb{N}^*$. We continuously embed $\mathcal{T}$ into A^ω, where $A = \{a, b, c\}$. For a given vertex $v = (n_1, n_2, \ldots, n_m) \in \mathbb{N}^*$, let $K(v) = a^{n_1}ba^{n_2}b \ldots ba^{n_m} \in A^*$. For a given tree $T \in \mathcal{T}$, let $K(T) = K(v_1)cK(v_2)c \ldots \in A^\omega$, where v_i is the i'th vertex of T in the order $\prec$. It is easy to see that $K\colon \mathcal{T} \to A^\omega$ defined above is a homeomorphism onto its image.

Proposition 1. *There exists an MSO+U formula φ such that*

$$T \in B \quad \Leftrightarrow \quad K(T) \models \varphi.$$

Proof. Below, pre(G) denotes the set of prefixes of elements of G.

Lemma 1. *Let $T \in \mathcal{T}$ be a tree. The following conditions are equivalent:*

1. *T has an infinite branch,*
2. *T has an infinite set of vertices G such that for any subset S of* pre(G)*, if S has bounded height then S is finite.*

The proof of the lemma is an easy application of König's Lemma.

To prove Proposition 1 it suffices to show that the second condition can be verified by a formula of MSO+U on $K(T)$. The construction of such a formula is described below.

For fixed $w \in A^\omega$, $S \subseteq \omega$, let $\psi(S)$ express the following properties:

- for each *block* of the form $c(a^*b)^*a^*c$ in w, S contains some initial segment of its positions,
- there is a bound r such that within every block the number of b's contained in S is bounded by r. Let r_S denote the minimal bound.

Let $\gamma(S)$ express that all a-blocks inside S are jointly bounded in length.

Let φ be an MSO+U formula expressing that there exists an infinite set $G \subseteq \omega$ containing only whole blocks of the form $c(a^*b)^*a^*c$, such that

$$\forall S \subseteq G.\ \psi(S) \Rightarrow \gamma(S) \tag{1}$$

The formula φ verifies the second condition of the lemma. □

Let us denote $M = L(\varphi) = \{w \in A^\omega : w \models \varphi\}$. Therefore we have shown that K is a reduction of B to M.

Proposition 2. *$M \subseteq A^\omega$ is a $\mathbf{\Sigma}_1^1$-set.*

Proof. Let Z be a set of pairs $(w, G) \subseteq A^\omega \times 2^\omega$ with G being a witness for φ and (w, G) satisfying the formula (1).

Note that for each $r_S \in \mathbb{N}$ and $G \subseteq \omega$, there is a maximal set $S \subseteq G$, satisfying $\psi(S)$ with given r_S bound. Such S depends continuously on (w, G). If we take $S \subseteq S' \subseteq G$, then $\gamma(S') \Rightarrow \gamma(S)$. Therefore, to check the validity of (1) it is enough to consider only countably many maximal S's (one per each $r_S \in \mathbb{N}$). The formula $\gamma(S)$ defines a Borel set. So Z is a countable intersection of Borel sets. Since $M = \pi_1(Z)$, M is $\mathbf{\Sigma}_1^1$. □

Therefore, we have proved:

Theorem 1. *$M \subseteq A^\omega$ is a $\mathbf{\Sigma}_1^1$-complete set (in particular, non-Borel) and is definable in MSO+U.*

Corollary 1. *There is no model of nondeterministic automata with a Borel acceptance condition, capturing all of MSO+U.*

Proof. Assume that for each MSO+U formula ψ, there exists a nondeterministic automaton $\mathcal{A}$ with a Borel acceptance condition $F \subseteq Q^\omega$, such that $L(\mathcal{A}) = L(\psi)$. In this proof the set Q can be even infinite (but countable), taking into account values of counters or any other additional state information. Then, $L(\psi)$ is a $\mathbf{\Sigma}_1^1$-set — it is the projection of the Borel set of pairs (w, ρ) such that ρ is a run of $\mathcal{A}$ on w.

Therefore, both $L(\varphi) = M$ and $L(\neg\varphi) = A^\omega \setminus M$ are $\mathbf{\Sigma}_1^1$-sets. By the theorem of Souslin, M must be Borel, which is a contradiction. □

3 ωBS-Automata

We will now define ωBS-automata as described in [BC06, Boj10]. They define a strict subclass of MSO+U, but it is the greatest subclass of MSO+U of which we know to have decidable emptiness.

An *ωBS-automaton* $\mathcal{A}$, as other nondeterministic finite automata, has a finite input alphabet A, a finite set Q of states and an initial state q_I. Apart from that it is equipped with a finite set Γ of counters. The counters can only be updated and cannot be read during the run. They will be used by the acceptance condition. A transition of the automaton is a transformation of states, as in standard NFA's, and additionally a finite sequence of counter updates. A counter update can be either an increment or a reset of a counter $c \in \Gamma$.

The value of a counter c is initially set to 0 and is incremented or reset according to the transitions in a run. For $c \in \Gamma$ we define a sequence $val_\rho(c)$, where $val_\rho(c)_i$ is the value of counter c right before its i-th reset in the run ρ. Note that if a counter c is reset only finitely many times then the sequence $val_\rho(c)$ is finite.

The acceptance condition of ωBS-automaton is a boolean combination of constraints that can be of one of the forms:

$$\limsup_i val_\rho(c)_i < \infty \qquad\qquad \liminf_i val_\rho(c)_i = \infty$$

The first constraint is called the *B-condition* (bounded), the second — the *S-condition* (strongly unbounded). In order that lim inf and lim sup make sense, the constraints implicitly require the corresponding sequences to be infinite.

It is a simple observation that the negation of a B-condition can be simulated using an S-condition and nondeterminism, and vice versa. Thanks to this fact we can consider automata with acceptance conditions that are *positive* boolean combinations of S- and B-conditions, without loss of expressive power.

We will use the notation $B(c)$ for the B-condition and $S(c)$ for the S-condition imposed on a counter c.

If the acceptance condition of an automaton is a positive boolean combination of B-conditions, the automaton is called an *ωB-automaton.* We similarly define *ωS-automata.*

Languages recognized by ωBS-automata (ωB-automata, ωS-automata) are called *ωBS-regular* (resp. *ωB-regular, ωS-regular*). An important result of [BC06] is that the complement of an ωB-regular language is an ωS-regular language and vice versa. Both classes are extensions of the class of ω-regular languages since the Büchi condition can be simulated by either a B-condition or an S-condition.

Example 1. The language L_S defined in the introduction can be recognized by an ωS-automaton. The automaton has one state and uses one counter that is increased when reading a letter a and is reset after each b. The acceptance condition is simply an S-condition on the only counter.

3.1 Complexity of ωB- and ωS-Regular Languages

Theorem 2. *Each ωB-regular language is in* $\mathbf{\Sigma}^0_3$.

Proof. Fix an ωB-automaton $\mathcal{A}$ recognizing a language L, and let us first assume that its accepting condition is a conjunction of B-conditions, i.e. is of the form:

$$\bigwedge_{c \in \Gamma_B} B(c)$$

Each of the considered counters is bounded iff there is a common bound k for all of them. Therefore L can be defined as:

$$L = \{w : \exists\rho. \bigwedge_{c\in\Gamma_B} val_\rho(c) \text{ is bounded but infinite}\}$$
$$= \bigcup_k \underbrace{\{w : \exists\rho. \bigwedge_{c\in\Gamma_B} val_\rho(c) \text{ is bounded by } k \text{ and infinite}\}}_{L_k},$$

where the quantification is over the set of all runs of $\mathcal{A}$ on w.

It is easy to see that for a fixed k, L_k can be recognized by a nondeterministic Büchi automaton. We simply store the counter values in the state and do not allow them to be incremented above k. The acceptance condition requires each of the counters $c \in \Gamma_B$ to be reset infinitely often. Hence L_k is ω-regular. Since all ω-regular languages are in $\mathcal{BC}_2$, $L \in \mathbf{\Sigma}^0_3$ as a countable union of $\mathcal{BC}_2$-sets.

In the general form, the acceptance condition of an ωB-automaton is a positive boolean combination of B-conditions. We can write such a condition in disjunctive normal form (DNF). The language accepted by this automaton is a union of languages corresponding to each disjunct. Hence it is in $\mathbf{\Sigma}^0_3$. □

Thanks to the complementation result of [BC06], we have:

Corollary 2. *Each ωS-regular language is in $\mathbf{\Pi}^0_3$.*

The complexity bounds given by Theorem 2 and Corollary 2 are tight.

Theorem 3. *There is a $\mathbf{\Sigma}^0_3$-complete set that is ωB-regular and a $\mathbf{\Pi}^0_3$-complete set that is ωS-regular.*

Proof. Because ωB-regular languages are complements of ωS-regular languages, it suffices to show only one of the claims.

We recall that the language L_S is in $\mathbf{\Pi}^0_3$ and ωS-regular. $\mathbf{\Pi}^0_3$-completeness of L_S follows from [Kec95, Exercise 23.2] via an obvious reduction. □

3.2 Complexity of ωBS-Regular Languages

Now we switch to languages recognized by automata that can use both S- and B-conditions. We prove the following.

Theorem 4. *Each ωBS-regular language is in $\mathbf{\Sigma}^0_4$.*

Proof. The proof, on the one hand, will use the result of Theorem 2 and, on the other hand, will repeat a similar reasoning.

Let us fix an ωBS-regular language L and an automaton $\mathcal{A}$ recognizing it. First consider an acceptance condition of the form:

$$\bigwedge_{c\in\Gamma_B} B(c) \quad\wedge\quad \bigwedge_{c\in\Gamma_S} S(c)$$

The language L can then be defined by:

$$L = \bigcup_k \underbrace{\left\{ w : \exists \rho. \begin{array}{l} \bigwedge_{c \in \Gamma_B} \ val_\rho(c) \text{ is bounded by } k \text{ and infinite} \\ \bigwedge_{c \in \Gamma_S} \ val_\rho(c) \text{ converges to } \infty \end{array} \right\}}_{L_k}$$

Note that each L_k language is ωS-regular, hence (by Theorem 2), it is in $\mathbf{\Pi}^0_3$. So L, as a countable union of such languages, is in $\mathbf{\Sigma}^0_4$.

A general acceptance condition can be written in disjunctive normal form (DNF). Again, the language accepted by such an automaton is a union of languages corresponding to each disjunct, so it is in $\mathbf{\Sigma}^0_4$. □

Now we show that the bound is tight. For that we consider the language, that was used in [BC06, Corollary 2.8] to show that the class of ωBS-regular languages is not closed under complements. Let

$$G = \left\{ a^{n_1} b a^{n_2} b \ldots : \begin{array}{l} \text{the sequence } n_1, n_2, \ldots \text{ can be partitioned into} \\ \text{a (possibly empty) bounded subsequence and} \\ \text{a (possibly empty) subsequence tending to } \infty \end{array} \right\}$$

The following fact is presented as an example in [TL93, page 595].

Lemma 2. *The language G is $\mathbf{\Sigma}^0_4$-complete.*

Now it suffices to note that the language G is ωBS-regular. It is proven in [BC06] (by showing an appropriate ωBS-regular expression), but it is straightforward to construct an automaton recognizing it.

4 Alternating ωBS-Automata

On the way towards finding a model of automata for the logic MSO+U we consider alternating ωBS-automata.

Alternating ωBS-automata are defined similarly as nondeterministic ωBS-automata. The difference is that the state space Q is split into $Q_\forall$ (universal states) and $Q_\exists$ (existential states).

We use standard game semantics for such automata. For a given alternating automaton $\mathcal{A}$ and word $w \in A^\omega$ we define a two-player game. A play in this game starts in the initial state of the automaton and in the first position of the word and proceeds by applying transitions of the automaton on the word w consistent with current state and a letter in current position in the word. Player $\forall$ chooses transitions when the automaton is in a state from $Q_\forall$, Player $\exists$ — from $Q_\exists$. Finally the play produces an infinite sequence of transitions consistent with consecutive letters of the word. The word w is accepted by the automaton iff Player $\exists$ has a winning strategy in the game with the winning condition of exactly the same form as an acceptance condition of nondeterministic ωBS-automata, i.e. a boolean combination of B- and S-conditions.

4.1 Languages Complete for the Classes $\mathbf{\Pi}^0_{2n}$

We will now present examples of languages of infinite words complete for the Borel classes $\mathbf{\Pi}^0_{2n}$, which are recognized by alternating ωBS-automata.

To make proofs easier, we will work with the spaces of sequences of vectors of numbers $\mathcal{N}_n = (\mathbb{N}^n)^\omega$. An easy embedding will transfer the results into the space of infinite words. For $n = 0$, the above definition gives a space of sequences of empty tuples, i.e. $\mathcal{N}_0 = \{\omega\}$.

Let us fix an alphabet $A = \{a, b, c\}$. We use encoding of a sequence of vectors into the space A^ω, where each vector $(z_n, z_{n-1}, \ldots, z_1)$ is mapped to the word $a^{z_n}ba^{z_{n-1}}b\ldots a^{z_1}c$. We will call the embedding defined this way $W_n \colon \mathcal{N}_n \to A^\omega$.

We will use the following notations to easily operate on sequences of vectors.

- For $\eta \in \mathcal{N}_n$ and $m \in \mathbb{N}$, let $\eta\restriction_m$ be a subsequence of η consisting of those vectors that have value m at the first coordinate.
- Let $\pi_n \colon \mathcal{N}_n \to \mathcal{N}_{n-1}$ be the projection which cuts off the first coordinate from each vector in a given sequence.

Definition 1. *Let $L_n \subseteq \mathcal{N}_n$ for $n > 0$ be the set of all $\eta \in \mathcal{N}_n$ such that*

$$\exists^\infty_{m_n} \exists^\infty_{m_{n-1}} \cdots \exists^\infty_{m_1} \exists^\infty_{x\in\omega} \eta(x) = (m_n, m_{n-1}, \ldots, m_1),$$

where $\exists^\infty$ stands for "exists infinitely many". Additionally, let $L_0 = \{\omega\} = \mathcal{N}_0$.

The following lemma describes the languages L_n in an inductive fashion.

Lemma 3. *For $n > 0$, $\eta \in L_n$ iff there exist infinitely many $m \in \mathbb{N}$ such that $\eta\restriction_m$ is an infinite sequence and $\pi_n(\eta\restriction_m) \in L_{n-1}$.*

The topological complexity of the languages $W_n(L_n)$ is presented as an example in [TL93, pages 595–596], here we only recall it.

Theorem 5. *For every $n>0$, the language $W_n(L_n)$ is $\mathbf{\Pi}^0_{2n+2}$-complete.*

Logic. Now we present MSO+U formulas describing the languages L_n. We do not formally prove that the formulas define exactly the desired sets, but they will serve as a guideline for us in the construction of alternating automata recognizing the languages.

First define a formula over $\mathcal{N}_n$, expressing boundedness of the first coordinates of vectors marked by X:

$$\mathrm{Bnd}_n(X) \equiv \exists_{k\in\mathbb{N}} \forall_{x\in X}\, \eta(x)_1 \leq k.$$

Now we build the formula for the language L_n inductively:

$$\varphi_n \equiv \forall X.\, \mathrm{Bnd}_n(X) \implies \exists Y.\, \mathrm{Bnd}_n(Y) \wedge (X \cap Y = \emptyset) \wedge (\varphi_{n-1}|_Y), \qquad (2)$$

where $\varphi_{n-1}|_Y$ is φ_{n-1} with all quantifiers restricted to Y and operating on $\mathcal{N}_n$ by ignoring the first coordinates of vectors, and φ_0 simply states that a sequence is infinite.

The formula (2) deals with sequences of vectors, but it is easy to rewrite it in such a way that it works on ω-words over A and defines $W_n(L_n)$. It is possible because properties like "being a maximal block of consecutive a's that correspond to the k-th coordinate of one of the vectors in a sequence" are expressible in MSO. Expressing Bnd_n in this context requires U.

4.2 Automata Construction

Theorem 6. *For each $n \in \mathbb{N}$ there is an alternating ωBS-automaton recognizing a $\mathbf{\Pi}^0_{2n+2}$-hard language.*

Proof. For a fixed n, it is possible to construct an ωBS-automaton recognizing exactly the language $W_n(L_n)$. However, to avoid some technical inconveniences, we construct an automaton $\mathcal{A}_n$ for which we only require that it accepts a word $W_n(\eta)$ iff $\eta \in L_n$.

The automaton will mimic the formula φ_n. The problem that we face is that alternation in automata and quantifier alternation in logic have different semantics. In logic, using the second order quantifier refers to choosing a set all at once, while in automata, players make decisions step by step (position by position). We will be able to overcome this problem using properties of the B-condition.

Automaton. The automaton $\mathcal{A}_n$ will be defined in the following way.

While reading the code of a sequence of vectors, before reading each vector Player $\forall$ decides if he selects the first component of the vector. If $\forall$ has not chosen the component, $\exists$ can choose it. If the component was chosen by $\forall$, counter a_n counts its length and then resets. If the component was chosen by $\exists$, counter e_n counts its length and then resets.

If the first component was chosen by $\exists$ then the procedure is repeated for the second component and for the counters a_{n-1} and e_{n-1}. We continue with the next components until Player $\exists$ does not select a component or all components of the vector are selected by $\exists$.

The whole process is repeated for all the vectors in a word.

Player $\forall$ can additionally reset any of a_i counters at any time (except the moment when it is actually incremented). This is to allow Player $\forall$ to select a finite (even empty) set.

The acceptance condition (winning condition for $\exists$ in the game) requires that among the counters $a_n, e_n, a_{n-1}, e_{n-1}, \ldots, a_1, e_1$, the left-most which is unbounded (or reset finitely many times) is an a-counter, or all counters are reset infinitely many times and are bounded.

Soundness. For a given word $w = W_n(\eta)$ such that $\eta \in L_n$, we have to prove that the existential player has a winning strategy in $\mathcal{A}_n$ on w. We proceed by induction. As stated above, $\eta \in L_n$ if and only if there exist infinitely many $m \in \mathbb{N}$ such that

$$\eta\restriction_m \text{ is infinite} \quad \text{and} \quad \pi_n(\eta\restriction_m) \in L_{n-1} \tag{3}$$

Player $\exists$ uses the following strategy. Let k be the greatest value of the first component among vectors selected by $\forall$ so far. Let m_k be the least m greater than k, for which condition (3) holds. Player $\exists$ selects a vector if its first component is equal to m_k.

Note that we may assume that k is increased only finitely many times during the run (otherwise Player $\forall$ loses). Hence, there exists a value m_{k0} that occurs at the first component of almost all vectors selected by Player $\exists$. By the assumption, $\pi_n(\eta\restriction_{m_{k0}}) \in L_{n-1}$. Since the set L_n is prefix-independent (i.e. for all $\nu \in (\mathbb{N}^n)^*$,

$\eta \in L_n$ iff $\nu\eta \in L_n$), also η restricted to the vectors selected by $\exists$, with the first component erased, belongs to L_{n-1}. It follows by inductive assumption that $\exists$ has a strategy on further components of vectors of this restricted sequence.

Induction basis: The automaton recognizing the set L_0 simply accepts all infinite sequences.

Correctness. Now let us take $w = W_n(\eta)$ such that $\eta \notin L_n$. We prove that the universal player has a winning strategy in $\mathcal{A}_n$ on w. We, again, proceed by induction. Note that if $\eta \notin L_n$ there exists m_0 such that for all $m \geq m_0$

$$\eta\restriction_m \text{ is finite} \quad \text{or} \quad \pi_n\left(\eta\restriction_m\right) \notin L_{n-1} \tag{4}$$

Player $\forall$ marks all the vectors whose first coordinate is less than m_0. If there are only finitely many such vectors, $\forall$ uses additional resets. During the game, Player $\forall$ remembers the largest first coordinate M of a vector selected by Player $\exists$.

For every $i \in \{m_0, \ldots, M\}$ we have $\eta\restriction_i$ is finite or $\pi_n\left(\eta\restriction_i\right) \notin L_{n-1}$, so

$$\overline{\eta_M} := \eta\restriction_{\{m_0, m_0+1, \ldots, M\}} \quad \text{is finite} \quad \text{or} \quad \eta_M := \pi_n\left(\overline{\eta_M}\right) \notin L_{n-1}.$$

If $\overline{\eta_M}$ is finite $\exists$ will lose the game (if he doesn't increase M). Otherwise, at every moment, Player $\forall$ assumes that M will not increase and uses the winning strategy from the inductive assumption for η_M at the next coordinates.

The value M can increase only finitely many times during the game (otherwise $\exists$ loses). Using prefix independence, we obtain that $\forall$ wins the game.

The inductive basis is trivial here, since there is no $\eta \in \mathcal{N}_0 \setminus L_0$. □

5 Conclusion

Our results seem to indicate that deciding MSO+U (if possible at all) might require considering the emptiness problem of some rather complicated model of automata, such as alternating ωBS-automata.

Acknowledgements. We would like to thank the anonymous referees for their constructive comments.

References

[BC06] Bojańczyk, M., Colcombet, T.: Bounds in ω-regularity. In: LICS, pp. 285–296 (2006)

[Boj09] Bojańczyk, M.: Weak MSO with the unbounding quantifier. In: STACS, pp. 159–170 (2009)

[Boj10] Bojańczyk, M.: Beyond ω-regular languages. In: STACS, pp. 11–16 (2010)

[BT09] Bojańczyk, M., Toruńczyk, S.: Deterministic automata and extensions of Weak MSO. In: FSTTCS, pp. 73–84 (2009)

[Kec95] Kechris, A.S.: Classical Descriptive Set Theory. Graduate Texts in Mathematics, vol. 156. Springer, Heidelberg (1995)

[Sri98] Srivastava, S.M.: A Course on Borel Sets. Graduate Texts in Mathematics, vol. 180. Springer, Heidelberg (1998)

[TL93] Thomas, W., Lescow, H.: Logical specifications of infinite computations. In: REX School/Symposium, pp. 583–621 (1993)

Least and Greatest Solutions of Equations over Sets of Integers

Artur Jeż[1,⋆] and Alexander Okhotin[2,3,⋆⋆]

[1] Institute of Computer Science, University of Wrocław, Poland
aje@cs.uni.wroc.pl
[2] Department of Mathematics, University of Turku, Finland
alexander.okhotin@utu.fi
[3] Academy of Finland

Abstract. Systems of equations with sets of integers as unknowns are considered, with the operations of union, intersection and addition of sets, $S + T = \{m + n \mid m \in S,\ n \in T\}$. These equations were recently studied by the authors ("On equations over sets of integers", *STACS 2010*), and it was shown that their unique solutions represent exactly the hyperarithmetical sets. In this paper it is demonstrated that greatest solutions of such equations represent exactly the Σ_1^1 sets in the analytical hierarchy, and these sets can already be represented by systems in the *resolved form* $X_i = \varphi_i(X_1, \ldots, X_n)$. Least solutions of such resolved systems represent exactly the recursively enumerable sets.

1 Introduction

Consider equations $\varphi(X_1, \ldots, X_n) = \psi(X_1, \ldots, X_n)$, in which the unknowns X_i are sets of integers, and the expressions φ, ψ may contain addition $S + T = \{m + n \mid m \in S,\ n \in T\}$, Boolean operations and ultimately periodic constants. At a first glance, they might appear as a simple arithmetical object. However, already their simple special case, *expressions* and *circuits* over sets of integers, have a non-trivial computational complexity, studied by McKenzie and Wagner [10] in the case of nonnegative integers and by Travers [18] in the case of all integers.

If only nonnegative integers are allowed in the equations, they become isomorphic to *language equations* [8] over a one-letter alphabet. Language equations over multiple-letter alphabets are known to be computationally complete [15,14]: their unique solutions represent exactly the recursive sets, while their least and greatest solutions represent exactly the recursively enumerable sets and their complements, respectively. This result has been subsequently re-created by the authors [4,5] for the one-letter case, that is, for equations over sets of natural

⋆ Supported by MNiSW grants N206 259035 2008–2010 and N206 492638 2010-2012 and by personal grant FNP START of Polish Foundation for Science.
⋆⋆ Supported by the Academy of Finland under grant 134860.

P. Hliněný and A. Kučera (Eds.): MFCS 2010, LNCS 6281, pp. 441–452, 2010.

numbers. As recently shown by Lehtinen and Okhotin [9], this computational universality extends to systems of such a simple form as $\{X + X + C = X + X + D, X + E = F\}$, with a unique unknown X.

The first study of equations over sets of integers, both positive and negative, was recently conducted by the authors [6]. The main result was that a set is representable by a unique solution of such a system if and only if it is *hyper-arithmetical.* Hyper-arithmetical sets are defined as the intersection $\Sigma_1^1 \cap \Pi_1^1$ of the two bottom classes of the analytical hierarchy, and are accordingly a proper superset of the sets representable in first-order Peano arithmetic. The results on unique solutions of such systems are recalled and commented in Section 2. Concerning least and greatest solutions of these equations, one can easily see that they must belong to Π_1^1 and to Σ_1^1, respectively, though no lower bounds are yet known.

This paper begins the study of least and greatest solutions of equations over sets of integers with systems of the following form:

$$\begin{cases} X_1 = \varphi_1(X_1, \ldots, X_n) \\ \quad\vdots \\ X_n = \varphi_n(X_1, \ldots, X_n) \end{cases} \tag{*}$$

This is the same general form as in the most well-known kind of language equations used to define context-free grammars [1]. It is known that such a system has a least solution corresponding to the context-free derivation; it is a folklore knowledge that greatest solutions are context-free as well. Least and greatest solutions are obtained by the fixpoint iteration, in which a solution is always reached after ω iterations. These results extend to a natural generalization of the context-free grammars, the *conjunctive grammars* [11,12].

In this paper, the unknowns in a system (*) are sets of integers, and the operations are union, intersection and addition. Tarski's fixpoint theorem [17] guarantees the existence of a least and a greatest solution, and, as explained in Section 3, an iterative version of Tarski's theorem asserts that a fixpoint is always reachable in ω_1 iterations, that is, iterating over countable ordinals. In Section 4 it is shown that in the case of greatest solutions, all ω_1 iterations are actually used, and that every set in Σ_1^1 can be represented by a greatest solution of such a system. On the other hand, Section 5 demonstrates that least solutions can be always reached in only ω iterations, and the family of sets represented by these solutions is exactly the family of recursively enumerable sets.

2 Equations over Sets of Integers

Consider systems of equations of the *resolved form* $X_i = \varphi_i(X_1, \ldots, X_n)$ with $i \in \{1, \ldots, n\}$, where the unknowns X_i are sets of integers, and the expressions φ_i may use the operations of union, intersection and addition of sets, as well as ultimately periodic constants[1]. When such a system has a unique solution, it

[1] A set of integers $S \subseteq \mathbb{Z}$ is *ultimately periodic* if there exist such numbers $n_0 \geqslant 0$ and $p \geqslant 1$, that $n \in S$ if and only if $n + p \in S$ for all n with $|n| \geqslant n_0$.

can be regarded as a definition of the sets in that solution. When a system of this form has multiple solutions, it is known from Tarski's fixpoint theorem [17] that among them there is the *least* and the *greatest solution* with respect to the partial order of componentwise inclusion.

If the unknowns are sets of natural numbers, such equations were first studied by Jeż [2], who established their nontriviality by representing the set $\{4^n \mid n \geqslant 0\}$:

Example 1 (Jeż [2]). The system of equations

$$\begin{cases} X_1 = \big[(X_1 + X_3) \cap (X_2 + X_2)\big] \cup \{1\} \\ X_2 = \big[(X_1 + X_1) \cap (X_2 + X_6)\big] \cup \{2\} \\ X_3 = \big[(X_1 + X_2) \cap (X_6 + X_6)\big] \cup \{3\} \\ X_6 = (X_1 + X_2) \cap (X_3 + X_3) \end{cases}$$

over sets of natural numbers has a least solution with $X_1 = \{4^n \mid n \geqslant 0\}$, $X_2 = \{2 \cdot 4^n \mid n \geqslant 0\}$, $X_3 = \{3 \cdot 4^n \mid n \geqslant 0\}$ and $X_6 = \{6 \cdot 4^n \mid n \geqslant 0\}$.

To understand this construction, it is useful to consider positional notation of numbers. Let $\Gamma_k = \{0, 1, \ldots, k-1\}$ be digits in base-k notation. For every $w \in \Gamma_k^*$, let $(w)_k$ be the number defined by this string of digits. For a language $L \subseteq \Gamma_k^*$ of positional notations, define $(L)_k = \{(w)_k \mid w \in L\}$. Now the solution of the above system can be conveniently represented in base-4 notation as $\big((10^*)_4, (20^*)_4, (30^*)_4, (120^*)_4\big)$. Substituting these four sets into the first equation, one obtains

$$\big((10^*)_4 + (30^*)_4\big) \cap \big((20^*)_4 + (20^*)_4\big) = \\ = \big((10^+)_4 \cup (10^*30^*)_4 \cup (30^*10^*)_4\big) \cap \big((10^+)_4 \cup (20^*20^*)_4\big) = (10^+)_4,$$

that is, both sums contain some "garbage", yet the garbage in the sums is disjoint, and is accordingly "filtered out" by the intersection. Finally, the union with $\{1\}$ yields the set $\{4^n \mid n \geqslant 0\}$, turning the first equation into an equality. The rest of the equations are verified similarly [2].

The idea of this example was generalised by the authors [3] by representing every set of numbers with their positional notation recognised by a certain kind of cellular automata. These are one-way real-time cellular automata, known under a proper name of *trellis automata* [13].

Proposition 1 (Jeż, Okhotin [3, Thm. 3]). *For every $k \geqslant 2$ and for every trellis automaton M over $\Gamma_k = \{0, \ldots, k-1\}$, such that $L(M) \cap 0\Gamma_k^* = \varnothing$, there exists and can be effectively constructed a resolved system of equations over sets of natural numbers using the operations of union, intersection and addition and singleton constants, such that its least solution contains a component $(L(M))_k$.*

Trellis automata are notable, in particular, for recognising the language of *computation histories of a Turing machine*, which is generally defined in the form $\mathrm{VALC}(T) = \{C_T(w)\natural w \mid w \in L(T)\}$, where $C_T(w)$ is a sequence of consecutive configurations in the accepting computation of T on w, encoded in a suitable way. This follows from the fact that trellis automata can recognise any finite intersections of linear context-free languages, and VALC(T) is representable as such an intersection. Assume that VALC(T) is defined over an alphabet of k-ary digits Γ_k. Then, any computation represents a number $(C_T(w)\natural w)_k$, and Proposition 1 asserts that the set of such numbers is a solution of some system of equations [3]. A more complicated construction on top of $(\mathrm{VALC}(T))_k$ allows *extracting* $(L)_k$ *out of* VALC(T), leading to a representation of every recursive (r.e., co-r.e.) set by unique (least, greatest, respectively) solution of a system $\varphi_i(X_1, \ldots, X_n) = \psi_i(X_1, \ldots, X_n)$ over sets of natural numbers [4].

When constructing equations over sets of integers, applying Proposition 1 to VALC(T) remains a useful technique. As in the authors' previous work on systems of equations over sets of integers [6], VALC(T) shall be defined over the alphabet of digits in base-7 notation, with each computation encoded by a string $C_T(w) \in \{3,6\}^+$, and with

$$\mathrm{VALC}(T) = \{C_T(w)\mathtt{1}w \mid w \in T\}.$$

The exact details of the encoding are not important, as trellis automata are flexible enough to recognise such a variant of VALC(T). Then the corresponding set of numbers

$$\{(C_T(w)\mathtt{1}w)_7 \mid (w)_7 \in L(T)\}$$

is representable by the unique solution of a resolved system of equations over sets of natural numbers with union, intersection and addition [3, Thm. 3]. If every occurrence of every variable X is replaced with $X \cap (\mathbb{N}+1)$, the system will have the same unique solution if interpreted over sets of integers.

Using equations over sets of integers, the set $(L(T))_7$ can be obtained out of $(\mathrm{VALC}(T))_7$ generally by subtracting the computation history from each number in VALC(T) as follows: $(C_T(w)\mathtt{1}w)_7 - (C_T(w)\mathtt{1}\mathtt{0}^{|w|})_7 = (w)_7$. This has to be done by adding a set of negative numbers to VALC(T), and filtering out numbers of the form $(C_T(w)\mathtt{1}w)_7 - (x)_7$ with $x \neq (C_T(w)\mathtt{1}\mathtt{0}^{|w|})_7$. Since $C_T(w)$ is a string of digits 3 and 6, this subtraction can be regarded as the removal of the prefix $\{3,6\}^+$, or as an existential quantification over such prefixes:

Lemma 1 (Representing the existential quantifier [6, Lemma E]). *The value of the expression*

$$\left[(X \cap (\{3,6\}^+\mathtt{1}\Gamma_7^*)_7) + (-(\{3,6\}^+\mathtt{0}^*)_7)\right] \cap (\mathtt{1}\Gamma_7^*)_7$$

on any $S \subseteq (\{3,6\}^+\mathtt{1}\Gamma_7^*)_7$ *is* $E(S) = \{(\mathtt{1}w)_7 \mid \exists x \in \{3,6\}^* \ (x\mathtt{1}w)_7 \in S\}$.

Then $E(\mathrm{VALC}(T)) = \{(\mathtt{1}w)_7 \mid w \in L(T)\}$, and it is left to remove the leading digit 1, which is performed by the expression in the next lemma:

Lemma 2 (Removing leading digit 1 [6]). *The value of the expression*

$$\bigcup_{i \in \Gamma_7 \setminus \{0\}} \bigcup_{t \in \{0,1\}} \left[(X \cap (\mathtt{1}i\Gamma_7^t(\Gamma_7^2)^*)_7) + (-\mathtt{1}\mathtt{0}^*)_7\right] \cap (i\Gamma_7^t(\Gamma_7^2)^*)_7$$

on any $S \subseteq (\mathtt{1}(\Gamma_7^+ \setminus \mathtt{0}\Gamma_7^*))_7$ *is* $\{(w)_7 \mid (\mathtt{1}w)_7 \in S\}$.

The two above lemmata yield a representation of r.e. sets:

Theorem 1. *Every r.e. set $S \subseteq \mathbb{Z}$ is the unique solution of a resolved system of equations over sets of integers using union, intersection and addition, as well as singleton constants and the constants $\mathbb{N}$, $-\mathbb{N}$.*

Proof (sketch). Assume first that $S \subseteq \mathbb{N}$ and let T be a Turing machine accepting S. Then, as long as the constant VALC(T), and the constants in Lemmata 1 and 2 are given, the expression $Remove_1(E(\text{VALC}(T)))$ yields the set S.

The constant $\text{VALC}(T) \subseteq \mathbb{N}$, as well as the constant sets of natural numbers in the Lemmata, are representable by equations over sets of natural numbers by Proposition 1. This construction is replicated for equations over sets of integers, by applying an intersection with a constant $\mathbb{N}$. The constant sets of negative integers in Lemmata 1 and 2 are represented as if the sets of the opposite numbers, negating all constants in the system.

This construction can be applied to any r.e. set of negative integers by representing the set of opposite numbers as above, and then by replacing every constant C by $-C$. Finally, any r.e. set of integers $S \subseteq \mathbb{Z}$ is represented as a union of its positive and negative subsets. □

The natural counterpart of the "existential quantifier" $E(X)$ is the function $A(X)$, defined as $A(S) = \{(1w)_7 \mid \forall x \in \{3,6\}^* \ (x1w)_7 \in S\}$. Equations of the general form $\varphi_i(X_1, \ldots, X_n) = \psi(X_1, \ldots, X_n)$ representing $A(X)$ were constructed by the authors [6]. Then, applying $A(X)$ and $E(X)$ to a recursive set finitely many times allowed constructing every set from the arithmetical hierarchy, and doing this iteratively led to the representation of every hyper-arithmetical set as a unique solution of such a system [6]. Intuitively, that system implemented an equation $X = A(E(X)) \cup C$, for a recursive constant $C \subseteq ((1\{3,6\}^+)^*10\Gamma_7^*)_7$, in which the digit blocks $\{3,6\}^+$ correspond to the quantified variables, 1 is a separator, while 10 marks the end of the quantifier prefix. Processing the latter requires an extra equation:

Lemma 3 (Removing leading digits 10 [6]). *The value of the expression*

$$Remove_{10}(Z) = (Z \cap \{(10)_7\} - \{(10)_7\}) \cup \bigcup_{i \in \Gamma_7 \setminus \{0\}} \bigcup_{t \in \{0,1,2\}} (Z \cap (10i\Gamma_7^t(\Gamma_7^3)^*)_7) - (10^*)_7 \cap (i\Gamma_7^t(\Gamma_7^3)^*)_7$$

on any $S \subseteq (10(\Gamma_7^* \setminus 0\Gamma_7^*))_7$ is $Remove_{10}(S) = \{(w)_7 \mid (10w)_7 \in S\}$.

Proposition 2 (Jeż, Okhotin [6, Thm. 2]). *For every hyper-arithmetical set $S \subseteq \mathbb{Z}$ there is a system of equations over subsets of $\mathbb{Z}$ using union, addition, singleton constants and the constants $\mathbb{N}$ and $-\mathbb{N}$, with a unique solution $(S, \ldots)$.*

This representation result has a matching upper bound: whenever such a system has a unique solution, it is a hyper-arithmetical set [6]. The proof of this upper bound can actually be split into two statements: first, least solutions are demonstrated to be in the class Π_1^1, and second, greatest solutions always belong to

Σ_1^1. As unique solutions are both least and greatest at the same time, they are in the class $\Pi_1^1 \cap \Sigma_1^1 = \Delta_1^1$, that is, are hyper-arithmetical. These bounds are based upon the following translation of equations into an arithmetical formula:

Proposition 3 (Jeż, Okhotin [6]). *For every system of equations in variables $X_1, \ldots, X_n$ using operations expressible in first-order arithmetic there exists an arithmetical formula $Eq(X_1, \ldots, X_n)$, where $X_1, \ldots, X_n$ are free second-order variables, such that $Eq(S_1, \ldots, S_n)$ is true if and only if $X_i = S_i$ is a solution of the system.*

Constructing this formula is only a matter of reformulation. As an example, an equation $X_i = X_j + X_k$ is represented by

$$(\forall n)\big[n \in X_i \leftrightarrow (\exists \ell)(\exists m)\ n = \ell + m \ \wedge\ \ell \in X_j \ \wedge\ m \in X_k\big].$$

Applying existential quantification to the set variables produces a Σ_1^1-formula $\varphi(x) = (\exists X_1)\ldots(\exists X_n)Eq(X_1, \ldots, X_n) \wedge (x \in X_1)$ representing the greatest solution, while universal quantification leads to a Π_1^1-formula $\varphi'(x) = (\forall X_1)\ldots(\forall X_n)Eq(X_1, \ldots, X_n) \rightarrow (x \in X_1)$ for the least solution:

Proposition 4. *For every system of equations in variables $X_1, \ldots, X_n$ using operations expressible in first-order arithmetic that has a least (greatest) solution $X_i = S_i$, the sets S_i are in the class Π_1^1 (in Σ_1^1, respectively).*

3 Resolved Systems and Their Properties

A system of equations is called *explicit* or *resolved* if it is of the form

$$X_i = \varphi_i(X_1, \ldots, X_n) \quad (1 \leqslant i \leqslant n). \tag{1}$$

When the unknowns are formal languages, such equations are used to define the context-free grammars and their generalization, the conjunctive grammars [11].

It is convenient to regard (1) as a single equation $X = \varphi(X)$, where X is an unknown n-tuple of sets, while $\varphi = (\varphi_1, \ldots, \varphi_n)$ is an operator on the set of such n-tuples. A solution of such equation is known as a *fixpoint* of the operator φ. As long as φ is *monotone* under some partial ordering, that is, if

$$A \preccurlyeq A' \Longrightarrow \varphi(A) \preccurlyeq \varphi(A'),$$

a least and a greatest fixpoint exists by Tarski's [17] theorem.

In case of vectors of sets of integers, the partial ordering is defined by $(S_1, \ldots, S_n) \sqsubseteq (T_1, \ldots, T_n)$ if $S_i \subseteq T_i$ for each i. The operations of union, intersection and addition are all monotone with respect to this ordering.

Another general property of operators is continuity. A sequence of sets $\{A_n\}_{n \geqslant 0}$ is *convergent* if for every element $x \in \bigcup_n A_n$ the set $\{n | x \in A_n\}$ is either finite or co-finite; in such a case $\lim_{n\to\infty} A_n = \{x \mid x \text{ is in infinitely many } A_n\text{'s}\}$. Now φ is *continuous*, if for every convergent sequence $\{A_n\}_{n=1}^{\infty}$,

$$\lim_{n\to\infty} \varphi(A_n) = \varphi(\lim_{n\to\infty} A_n).$$

A composition of monotone (continuous) operators is monotone (continuous). Provided that a system (1) has monotone and continuous right-hand sides, its least solution is reached by ω iterations of φ, beginning with a vector of empty sets: $\bigsqcup_{k=1}^{\infty} \varphi^k(\varnothing, \ldots, \varnothing)$. If the iteration begins with the top element $(\overline{\varnothing}, \ldots, \overline{\varnothing})$, then the greatest solution is similarly reached after ω steps of a similar iteration, with intersection instead of union. This is the case with language equations using concatenation, union and intersection [1,12], or similar equations over sets of natural numbers [3].

However, when equations over sets of *integers* are considered (that is, if negative numbers are allowed), the addition of such sets is no longer continuous: consider $\varphi(X) = X + X$ and a sequence $X_n = \{-n, n\}$. Then $\lim_{n\to\infty} X_n = \varnothing$ and $\varphi(\lim_{n\to\infty} X_n) = \varnothing$. On the other hand, $0 \in X_n + X_n$ for each n, and accordingly $0 \in \lim_{n\to\infty} \varphi(X_n)$. This makes the above ω-step fixpoint iteration inapplicable to such systems, as the vector obtained after ω steps need not be a solution.

When all is known about a system (1) is that its right-hand sides are monotone, Tarski's [17] fixpoint theorem asserts that it has a least and a greatest solution. This result can be shown using a transfinite induction as follows. Denote by ω_1 the first uncountable ordinal. For each ordinal $\alpha \leqslant \omega_1$, define the vector of sets after α iterations of φ:

$$S^{(0)} = (\varnothing, \ldots, \varnothing) \tag{2a}$$

$$S^{(\alpha+1)} = \varphi(S^{(\alpha)}) \tag{2b}$$

$$S^{(\alpha)} = \bigsqcup_{\gamma<\alpha} S^{(\gamma)} \quad \text{when } \alpha \text{ is a limit ordinal} \tag{2c}$$

Lemma 4. *$S^{(\omega_1)}$ is the least fixpoint of the system* (1).

The proof proceeds along the following steps. First it is shown that $S^{(\alpha)}$ is a weakly increasing sequence, that is, $S^{(\alpha)} \sqsubseteq S^{(\gamma)}$ for all ordinals $\alpha < \gamma$. Then it is proved that there are countably many ordinals α with $S^{(\alpha)} \sqsubset S^{(\alpha+1)}$, and accordingly the sequence converges to a fixpoint in fewer than ω_1 steps. This fixpoint is then proved to be the least. All arguments are by a transfinite induction on the ordinals.

Similarly, for the greatest solutions define

$$T^{(0)} = (\mathbb{Z}, \ldots, \mathbb{Z}) \tag{3a}$$

$$T^{(\alpha+1)} = \varphi(T^{(\alpha)}) \tag{3b}$$

$$T^{(\alpha)} = \bigsqcap_{\gamma<\alpha} T^{(\gamma)} \quad \text{when } \alpha \text{ is a limit ordinal} \tag{3c}$$

Lemma 5. *$T^{(\omega_1)}$ is the greatest fixpoint of the system* (1).

4 Greatest Solutions

The greatest solution of any system of equations over sets of integers is in Σ_1^1 in the analytical hierarchy. It shall now be proved that, conversely, every set $S \subseteq \mathbb{N}$ in Σ_1^1 is representable. The construction combines the definition of a certain Σ_1^1-complete set $\mathcal{T} \subseteq \mathbb{N}$ with a reduction function from S to $\mathcal{T}$. Representing any Σ_1^1-subset of $\mathbb{Z}$ is achieved by a simple additional step.

The announced Σ_1^1-hard set contains the yes-instances of the following problem (its complement is Π_1^1-complete [16, THM. 16-XX]): "Given a Turing machine M working on natural numbers, determine whether there exists an infinite sequence of strings $\{x_i\}_{i=1}^{\infty}$ with $x_i \in \{3,6\}^+$, such that, for all $k \geqslant 0$, the number $(1x_k1x_{k-1}1\dots1x_11)_7$ is in $L(M)$". Base-7 notations of these numbers encode finite sequences of natural numbers, and are formatted for processing by Lemmata 1 and 2. Now for any Σ_1^1 set S there exists a total recursive reduction function f_S, such that

$$n \in S \iff \exists\{x_i\}_{i=1}^{\infty} \forall k \geqslant 0\ (1x_k1x_{k-1}1\dots1x_11)_7 \in L(M_{f_S(n)}),$$

where $M_0, M_1, \dots, M_i, \dots$ is any effective enumeration of Turing machines.

Fix S and its reduction f_S witnessing $S \leqslant_{rec} \mathcal{T}$. Define the set

$$C = \big\{(1x_k1x_{k-1}1\dots1x_110s)_7 \mid s \in \Gamma_7^* \setminus 0\Gamma_7^*,\\ \forall k' \leqslant k\ (1x_{k'}1x_{k'-1}1\dots1x_11)_7 \in L(M_{f_S((s)_7)})\big\},$$

which is r.e.: given a number $(1x_k1x_{k-1}1\dots1x_110s)_7$, a Turing machine calculates its base-7 notation, extracts $(s)_7$, constructs $M_{f_S((s)_7)}$ and simulates it on each input $(1)_7, \dots, (1x_k1x_{k-1}1\dots1x_11)_7$. If they are all accepted, this number belongs to C. By Theorem 1, C can be represented as a unique (and, in particular, the greatest) solution of a resolved system of equations.

For any fixed number $(s)_7 \in \mathbb{N}$, the set C induces a set of finite sequences

$$\big\{(n_1, \dots, n_{k-1}, n_k) \mid (1x_k1x_{k-1}1\dots1x_110s)_7 \in C, \text{ where each } x_i \text{ represents}\\ \text{the binary notation of } n_i, \text{ using } 3 \text{ for zero and } 6 \text{ for one}\big\}.$$

This set of sequences is closed under taking prefixes, and thus may be regarded as a *tree*. Each sequence is a *node* of the tree. A node $(n_1, n_2, \dots, n_{k-1}, n_k)$ is a *child* of the node $(n_1, n_2, \dots, n_{k-1})$, which is its *parent*. The empty sequence is the unique node without a parent, that is, the *root* of the tree; a node is a *leaf* if it has no children. A tree has an *infinite path* if there exists such a sequence $(n_1, n_2, \dots, n_k, \dots)$ that all of its finite prefixes belong to the tree. This tree terminology shall be adopted for a fixed $(s)_7$ when referring to C: for example, $(1x_k1x_{k-1}1\dots1x_110s)_7 \in C$ is the parent of $(1x_{k+1}1x_k1\dots1x_110s)_7 \in C$, etc.

In this terminology, an element $(1x_k1x_{k-1}1\dots1x_110s)_7 \in C$ is said to *have an infinite path* if the tree corresponding to s has an infinite path beginning with the node corresponding to this element; or, equivalently, if

$$\exists\{x_{k+i}\}_{i=0}^{\infty}\ \forall \ell \geqslant 0\ (1x_{k+\ell}1x_{k+\ell-1}1\dots1x_110s)_7 \in C$$

In particular, a number $(s)_7$ is in S if and only if the element $(\mathtt{10}s)_7$ has an infinite path. The goal is to construct an equation with the greatest solution comprised exactly of numbers with an infinite path. Since the greatest solution is a limit of a descending chain of sets, see Lemma 5 and (3), the equation shall iteratively shorten finite paths, so that the numbers without an infinite path are eventually eliminated.

For every node with finitely many descendants there is a well-defined *height* of its subtree. This concept is generalised to trees with infinite paths and infinite degrees of nodes as follows. The *rank* of an element of C, see Rogers [16, §16], is an ordinal defined by

$$r(x) = \begin{cases} 1, & \text{if } x \text{ is a leaf,} \\ \sup\{r(y)+1 \mid y \text{ is a child of } x\}, & \text{otherwise.} \end{cases} \tag{4a}$$

For some elements of C the recursion does not terminate, and the definition is extended by

$$r(x) = \omega_1, \qquad \text{when } r(x) \text{ is not defined by (4a).} \tag{4b}$$

Lemma 6. *The rank of an element* $(\mathtt{1}x_k\mathtt{1}x_{k-1}\mathtt{1}\dots\mathtt{1}x_1\mathtt{10}s)_7 \in C$ *is not defined by* (4a) *if and only if it has an infinite path.*

As argued by Rogers [16, Thm. 16-XVIII(a)], all ordinals assigned by (4a) are countable. By definition, $\omega_1 > \alpha$ for every countable ordinal α, that is, for every rank defined in (4a). Now it can be said that the elements without an infinite path are those with a countable rank. There exists a natural approach of removing these elements by an iterative removal of the leaves. While it is easily seen that this works for elements with a finite rank, it is not so obvious, what happens for elements ranked with an infinite ordinal. Nevertheless, it turns out that this approach works in the general case of countable ordinals.

Consider an equation

$$X = C \cap E(Remove_1(X)),$$

Denote its right-hand side by $\varphi(X) = C \cap E(Remove_1(X))$, and consider the sequence $T^{(\alpha)}$ corresponding to this equation, see (3). Note, that $T^{(0)} = \mathbb{Z}$, $T^{(1)} = C$ and $T^{(\alpha)} \subseteq C$ for every ordinal α. Every step of this sequence contains the fathers of all elements occurring at the previous step:

Lemma 7. *For every countable ordinal* α, $x \in T^{(\alpha+1)}$ *if and only if* $x = (\mathtt{1}x_k\mathtt{1}x_{k-1}\mathtt{1}\dots\mathtt{1}x_1\mathtt{10}s)_7$ *and there is* x_{k+1} *with* $(\mathtt{1}x_{k+1}\mathtt{1}x_k\mathtt{1}\dots\mathtt{1}x_1\mathtt{10}s)_7 \in T^{(\alpha)}$.

Intuitively, the rank of an element specifies how many times this transformation can be applied until the element disappears. This is formalised as follows:

Lemma 8. *For every countable ordinal* α, $(\mathtt{1}x_k\mathtt{1}x_{k-1}\mathtt{1}\dots\mathtt{1}x_1\mathtt{10}s)_7 \in T^{(\alpha)}$ *if and only if* $r((\mathtt{1}x_k\mathtt{1}x_{k-1}\mathtt{1}\dots\mathtt{1}x_1\mathtt{10}s)_7) \geqslant \alpha$.

The proof is by an iterative application of Lemma 7 in a transfinite induction on α.

After ω_1 iterations, all elements with a countable rank are eliminated, and the greatest fixed point $T^{(\omega_1)}$ consist exactly of the elements with an infinite path, as they are invariant under φ.

Lemma 9. $(1x_k \ldots 1x_1 10s)_7 \in T^{(\omega_1)}$ *if and only if* $r((1x_k \ldots 1x_1 10s)_7) = \omega_1$.

Taking Lemma 6 into account, $(1x_k \ldots 1x_1 10s)_7 \in T^{(\omega_1)}$ if and only if there exists an infinite sequence $x_{k+1}, \ldots x_{k+\ell}, \ldots$, such that for each $\ell \geqslant 0$, $(1x_{k+\ell} 1x_{k+\ell-1} \ldots 1x_1 10s)_7 \in C$. It remains to extract the set S out of $T^{(\omega_1)}$. This is done using the expression $Remove_{10}(F) = \{(w)_7 \mid (10w)_7 \in F\}$ defined in Lemma 3. Consider a new variable Y with an new equation, which forms the following system:

$$\begin{cases} X = C \cap E(Remove_1(X)) \\ Y = Remove_{10}(X) \end{cases} \tag{5}$$

Main Lemma. *The system* (5) *has a greatest solution with* $Y = S$.

The system constructed in this section uses a recursively enumerable constant set $C \subseteq \mathbb{N}$, as well as several constants required by Lemmata 1, 2 and 3. The former constant is representable by Theorem 1, while the rest of the constants are expressed as in the proof of that theorem. The method in the proof of Theorem 1 is also used to represent a set of integers from its positive and negative part. This yields the following result:

Theorem 2. *Every* Σ^1_1*-set* $S \subseteq \mathbb{Z}$ *is a unique solution of a resolved system of equations over sets of integers using union, intersection and addition, as well as singleton constants and the constants* $\mathbb{N}$, $-\mathbb{N}$.

The construction in the this section essentially used the infinite constants $\mathbb{N}$ and $-\mathbb{N}$. It turns out that at least one infinite constant is needed, as otherwise only trivial greatest solutions can be obtained.

Lemma 10. *For every solution of a resolved system of equations over* $\mathbb{Z}$ *using union, intersection, addition and finite constants, there is a greater solution with each component either finite or equal to* $\mathbb{Z}$.

5 Least Solutions of Resolved Systems

As mentioned in Section 3, whenever a monotone operator is also continuous, reaching its least fixed point does not require a transfinite number of iterations: $S^{(\omega)}$ is always the least solution. In fact, this holds for a weaker property than continuity.

An operator φ is said to be $\cup$-*continuous* if $\varphi(\bigsqcup_{i \in \mathbb{N}} B_i) = \bigsqcup_{i \in \mathbb{N}} \varphi(B_i)$ holds for every *increasing* sequence B_i. A composition of $\cup$-continuous operators is $\cup$-continuous as well. It turns out that while addition of sets of integers is not continuous, it possesses this weaker property.

Table 1. Expressive power of solutions

	least	unique	greatest
unresolved over $2^{\mathbb{N}}$, with $\{+,\cup\}$	Σ_1^0 (r.e.) [4]	Δ_1^0 (rec.) [4]	Π_1^0 (co-r.e.) [4]
resolved over $2^{\mathbb{Z}}$, with $\{+,\cup,\cap\}$	Σ_1^0	Σ_1^0	Σ_1^1
unresolved over $2^{\mathbb{Z}}$, with $\{+,\cup\}$	?	Δ_1^1 (HA) [6]	Σ_1^1

Lemma 11. *A function over sets of integers defined as a composition of union, intersection, addition and any constants is* $\cup$*-continuous.*

Then it is known that the least fixpoint of any such function is reached in ω iterations. This leads to the following theorem:

Theorem 3. *The least solution of every resolved system of equations* $X_i = \varphi_i(X_1, \ldots, X_n)$ *over sets of integers using union, intersection, addition and r.e. constants is an r.e. set.*

For singleton constants, an algorithm constructs $S^{(\alpha)}$ for all $\alpha < \omega$, until the input number is found. The case of r.e. constants is reduced to the former case by encoding the constants as in Theorem 1.

Conversely, by Theorem 1, every r.e. set is represented by such a unique solution of a system with singleton constants and constants $\mathbb{N}$ and $-\mathbb{N}$, and hence by a least solution of such a system. Furthermore, the sets $\mathbb{N}$ and $-\mathbb{N}$ can be expressed as least solutions of the following equations:

$$X = (X + 1) \cup \{0\} \qquad X' = (X' + \{-1\}) \cup \{0\}.$$

Altogether, the following characterization is obtained:

Corollary 1. *Least solutions of resolved systems of equations* $X_i = \varphi_i(X_1, \ldots, X_n)$ *over sets of integers using union, intersection, addition and constants* $\{1\}$ *and* $\{-1\}$ *represent exactly the r.e. sets. If all r.e. constants are allowed, only r.e. sets can be represented.*

6 Conclusion

The new results on the expressive power of least and greatest solutions of equations over sets of integers are summarised and compared to related results in Table 1. The same results extend to a slightly different model: equations over sets of natural numbers with union, intersection, addition and subtraction: $A \dot{-} B = \{a - b \mid a \in A,\, b \in B,\, a \geqslant b\}$ their least solutions represent exactly the r.e. sets, while their greatest solutions represent all sets in Σ_1^1. These equations are isomorphic to language equations over a unary alphabet, with the operations of union, intersection, concatenation and quotient. Furthermore, the same results could be extended to language equations over multiple-letter alphabets, by a technically much simpler construction than presented in this paper.

Of the decision problems for these equations, solution existence is trivial (as there is always a least and a greatest solution), while the complexity of testing whether a system has a unique solution is left as an open problem.

References

1. Ginsburg, S., Rice, H.G.: Two families of languages related to ALGOL. Journal of the ACM 9, 350–371 (1962)
2. Jeż, A.: Conjunctive grammars can generate non-regular unary languages. International Journal of Foundations of Computer Science 19(3), 597–615 (2008)
3. Jeż, A., Okhotin, A.: Conjunctive grammars over a unary alphabet: undecidability and unbounded growth. Theory of Computing Systems 46(1), 27–58 (2010), http://dx.doi.org/10.1007/s00224-008-9139-5
4. Jeż, A., Okhotin, A.: On the computational completeness of equations over sets of natural numbers. In: Aceto, L., Damgård, I., Goldberg, L.A., Halldórsson, M.M., Ingólfsdóttir, A., Walukiewicz, I. (eds.) ICALP 2008, Part II. LNCS, vol. 5126, pp. 63–74. Springer, Heidelberg (2008)
5. Jeż, A., Okhotin, A.: Equations over sets of natural numbers with addition only. In: STACS 2009, Freiburg, Germany, February 26-28, pp. 577–588 (2009)
6. Jeż, A., Okhotin, A.: On equations over sets of integers. In: STACS 2010, Nancy, France, March 4-6, pp. 477–488 (2010)
7. Kleene, S.C.: Introduction to metamathematics. North-Holland, Amsterdam (1952)
8. Kunc, M.: What do we know about language equations? In: Harju, T., Karhumäki, J., Lepistö, A. (eds.) DLT 2007. LNCS, vol. 4588, pp. 23–27. Springer, Heidelberg (2007)
9. Lehtinen, T., Okhotin, A.: On language equations $XXK = XXL$ and $XM = N$ over a unary alphabet. In: Yu, S. (ed.) DLT 2010. LNCS, vol. 6224, pp. 291–302. Springer, Heidelberg (2010)
10. McKenzie, P., Wagner, K.: The complexity of membership problems for circuits over sets of natural numbers. Computational Complexity 16(3), 211–244 (2007)
11. Okhotin, A.: Conjunctive grammars. Journal of Automata, Languages and Combinatorics 6(4), 519–535 (2001)
12. Okhotin, A.: Conjunctive grammars and systems of language equations. Programming and Computer Software 28(5), 243–249 (2002)
13. Okhotin, A.: On the equivalence of linear conjunctive grammars to trellis automata. Informatique Théorique et Applications 38(1), 69–88 (2004)
14. Okhotin, A.: Strict language inequalities and their decision problems. In: Jedrzejowicz, J., Szepietowski, A. (eds.) MFCS 2005. LNCS, vol. 3618, pp. 708–719. Springer, Heidelberg (2005)
15. Okhotin, A.: Decision problems for language equations. Journal of Computer and System Sciences 76(3-4), 251–266 (2010)
16. Rogers Jr., H.: Theory of Recursive Functions and Effective Computability. McGraw-Hill, New York (1967)
17. Tarski, A.: A lattice theoretical fixpoint theorem and its applications. Pacific Journal of Mathematics 5, 285–310 (1955)
18. Travers, S.D.: The complexity of membership problems for circuits over sets of integers. Theoretical Computer Science 369(1-3), 211–229 (2006)

Improved Simulation of Nondeterministic Turing Machines

Subrahmanyam Kalyanasundaram[1], Richard J. Lipton[1], Kenneth W. Regan[2], and Farbod Shokrieh[3]

[1] College of Computing, Georgia Tech
{subruk,rjl}@cc.gatech.edu
[2] Department of Computer Science and Engg., University at Buffalo
regan@buffalo.edu
[3] School of Electrical and Computer Engg., Georgia Tech
farbod@ece.gatech.edu

Abstract. The standard simulation of a nondeterministic Turing machine (NTM) by a deterministic one essentially searches a large bounded-degree graph whose size is exponential in the running time of the NTM. The graph is the natural one defined by the configurations of the NTM. All methods in the literature have required time linear in the size S of this graph. This paper presents a new simulation method that runs in time $\tilde{O}(\sqrt{S})$. The search savings exploit the one-dimensional nature of Turing machine tapes. In addition, we remove most of the time-dependence on nondeterministic choice of states and tape head movements.

1 Introduction

How fast can we deterministically simulate a nondeterministic Turing machine (NTM)? This is one of the fundamental problems in theoretical computer science. Of course, the famous P $\neq$ NP conjecture, as most believe, would answer that we cannot hope to simulate nondeterministic Turing machines very fast. However, the best known result to date is the famous theorem of Paul, Pippenger, Szemerédi, and Trotter [11] that NTIME($O(n)$) is not contained in DTIME($o((n\log^* n)^{1/4})$). This is a beautiful result, but it is a long way from the current belief that the deterministic simulation of a nondeterministic Turing machine *should* in general take exponential time.

We look at NTM simulations from the opposite end: rather than seeking better lower bounds, we ask how far can one improve the upper bound? We suspect even the following could be true:

For any $\varepsilon > 0$,

$$\mathrm{NTIME}(t(n)) \subseteq \mathrm{DTIME}(2^{\varepsilon t(n)}).$$

To our knowledge, this does not contradict any of the current strongly held beliefs. This interesting question has been raised before, see e.g., [3].

Our main theorem is:

P. Hliněný and A. Kučera (Eds.): MFCS 2010, LNCS 6281, pp. 453–464, 2010.

Theorem 1. *Any k-tape NTM N with tape alphabet size a that runs in time $t(n)$, can be simulated by a deterministic Turing machine in time*

$$a^{kt(n)/2} \cdot H_N^{\sqrt{t(n)}\log t(n)},$$

up to polynomial factors, and where H_N is a constant that depends only on a.

Our bound has two key improvements. First, all nondeterminism arising from the choice of the next state or tape head movements is subsumed into the factor $H_N^{\sqrt{t(n)}\log t(n)}$ with much smaller time dependence, compared to the main exponential term. Second, while N may write any of $S = a^{kt(n)}$ strings nondeterministically on its k tapes, our simulator needs to search only $\sqrt{S}$ of that space. Thus, we search the NTM graph in the *square-root* of its size.

There is no general deterministic procedure that can search a graph of size S in $\sqrt{S}$ time, even if the graph has a simple description. Hence to prove our theorem we must use the special structure of the graph: we must use that the graph arises from an NTM. We use several simple properties of the operation of Turing tapes and the behavior of guessing to reduce the search time by the square root.

We believe that while the actual theorem is interesting, the techniques that are used to prove the theorem may be of use in other problems. We speculate that our methods may be extended to lower the exponent further.

In section 5, we consider NTMs with limited nondeterminism, and prove:

Theorem 2. *Suppose $t(n) = nr(n)$, where $r(n)$ is constructible in unary in $O(n)$ time, and fix an input alphabet of size b. Then for any NTM N that runs in time $t(n)$ with $o(n)$ nondeterminism and computes a function f, there exist circuits C_n of size $O(t(n)\log r(n))$ that compute f correctly on $b^{n-o(n)}$ inputs.*

2 Model and Problem Statement

We use a standard model of a nondeterministic multitape Turing machine, in which nondeterminism may arise through characters written, head motions on the tapes, and/or the choice of next state. Heads may stay stationary as well as move one cell left or right in any step. We stipulate that an NTM N runs in time $t(n)$ if all branches of computations on inputs of length n halt within $t(n)$ steps. Since our results involve bounds $t(n)$ that are fully time and space constructible, this is equivalent to definitions that apply the time bound only to accepting computations. Throughout this paper, we use q for the number of states, k for the number of tapes, and a for the alphabet size of N. Our results hold for all sufficiently large input lengths n.

Our question is, in terms of a, k, q, what is the most efficient simulation of N by a deterministic Turing machine (DTM)? We identify three basic strategies:

1. *Tracing the computation tree:* Since we do not limit N to be binary-branching, individual nodes of the tree may have degree as high as $v = a^k 3^k q$,

where the "3" allows each head on each tape to move left, right, or stationary. This is reflected in classic theorem statements of the form

Theorem 3. *Any NTM N with time complexity $t(n)$ can be simulated by a DTM M in time $c(N)^{t(n)}$, where $c(N)$ is a constant depending on N.*

According to proofs such as that in [10], $c(N)$ depends on q as well as k and a. There is thus a factor q^t in the running time of M. It will be our goal to eliminate such a factor.

2. *Enumerating a witness predicate:* That is, one finds a predicate $R(x, y)$ that is deterministically efficient to decide, such that for all x, N accepts x iff for some y of a given length, $R(x, y)$ holds. Then one tries all possible y. This may be specific to the language $L(N)$ rather than simulate N in the direct sense of strategy 1. However, when $R(x, y)$ is the predicate "y codes an accepting path in the computation tree" it is the same as strategy 1.
3. *Searching the configuration graph:* A *configuration* of a Turing machine is an encoding of the current state, the non-blank contents of the tapes, and current position of the tape heads. Configurations form a directed graph where there are directed edges from a configuration to a valid successor configuration, with sources being the initial configurations I_x on given inputs x and sinks being accepting configurations I_a (perhaps including non-accepting halting configurations too). When N uses at most space s on any one tape, the number S of nodes in the graph (below I_x) is at most

$$S = qa^{ks}s^k.$$

Notice that $s \leq t$ holds trivially, where t is the running time of N. Using a look up table for simulating the transition function of the machine N, the dominant term in the running time is

$$O(Sv \cdot \log(Sv) \cdot \log S) = q^2(3at)^k a^{kt} \mathsf{poly}(\log q, k, t, a).$$

Note that the dependence on q is at most q^2, not q^t.

The classic tradeoff between strategy 1 and strategy 3 concerns the space requirement. Tracing the tree requires storing only the current path from the root and some local information, though it may waste time by re-computing nodes that are reached by multiple paths when the computation is treated as a graph. Breadth-first search of the graph avoids redundant expansion at the expense of storing the whole list of visited nodes. In this paper we find that by judicious mixing-in of strategy 2, there is also mileage to be gained on the running time alone. The following preliminary result illustrates the basic idea:

Proposition 1. *Any NTM N with time complexity $t(n)$ can be simulated by a DTM M in time $c(N)^{t(n)}$, where the constant $c(N)$ depends on the alphabet size a and the number of tapes k of N, but is* ***independent*** *of q.*

Proof. We define a *weak trace* as comprising the move labels on an accepting path in the computation tree, but *omitting* the next-state information. There are only $(a^k 3^k)^t$ such potential witnesses to enumerate. We call a path "compatible with the weak trace y" if it adds states $q_0, \ldots, q_t$ to the parts specified by y to make a legal computation. Below, we show that each of these weak traces can be verified in time $q^2 a^{2k} 3^k \mathsf{poly}(\log q, a, k, t)$.

For each step j in the computation, define Q_j to be the set of states N can be in at step j on some full path that is compatible with y. Initially $Q_0 = \{ q_0 \}$, the start state of N. Given Q_{j-1}, to compute Q_j we take each state r in Q_{j-1} and look up all possible next states r' in a pre-computed lookup table based on the transition relation of N. After computing each Q_j, M needs to sort and remove the duplicate states in Q_j, else the size could explode by the end of the simulation. The simulation finally accepts if and only if Q_t contains the accepting state q_a, which we may suppose persists for each step once it is entered.

Our deterministic machine M has $k+3$ tapes, k to re-create the tapes of N guided by the weak trace, one to code the transition function of N serially as a lookup table, plus two for bookkeeping. The lookup table rows are indexed by the current state, the k symbols currently read, the k symbols that would be written, and movements left, right or stay for each tape head. The entries give all possible next-states for N in such a transition. There are $q(3a^2)^k$ rows, and each row can have at most q states. The cost of a serial lookup is upper-bounded[1] by $q(3a^2)^k \cdot [k \log(3a^2) + \log q + q \log q]$.

After the lookups, we need to sort and remove duplicates from a set (of states) which could be potentially q^2 in size. This takes $q^2 \log q$ comparisons, where each comparison costs $\log q$, yielding a running time of $q^2 \log^2 q$. Multiplying the whole expression by t, we get that the running time per weak trace is

$$[q(3a^2)^k \cdot [k \log(3a^2) + \log q + q \log q] + q^2 \log^2 q] \cdot t,$$

which can be upper bounded by

$$h(a, q, k, t) = q^2 a^{2k} 3^k \mathsf{poly}(\log q, a, k, t).$$

The overall running time is $(3^k a^k)^t$ multiplied by the function h. The factor h is majorized by $(1+\delta)^t$ for any $\delta > 0$ as t becomes sufficiently large. The whole time is thus bounded by $(3^k a^k + \delta')^t$, where $\delta' = 3^k a^k \delta$. Note that δ' is independent of q and can likewise be made arbitrarily small when a and k are fixed. Hence the deterministic simulation time is asymptotically bounded by $c(N)^{t(n)}$ where $c(N)$ is independent of q. □

Our further improvements come from (a) slimming witnesses y further, (b) more-sophisticated precomputation, and (c) trading off strategies 1 and 3 according to the space actually used by N on the one-dimensional Turing tapes.

[1] One can remove the $q \log q$ inside the brackets by organizing the rows in canonical order of the subsets of states they produce, and having M count special aliased dividers up to 2^q in binary as it scans serially, to determine which subset goes with a given row. A final $q \log q$ outside the brackets can be for wriings out the indexed subset as a list of states. However, this extra efficiency does not matter to our results.

3 Block-Trace Simulation

We begin the push for faster simulations by breaking computations by NTMs N into "blocks" of d steps, where d will be specified later.

Definition 1. *A* segment of size d *for a k-tape NTM N with alphabet of size a is a sequence of 4-tuples*

$$\tau = [(r_1, f_1, \ell_1, u_1), \ldots, (r_k, f_k, \ell_k, u_k)]$$

where for each tape j, $1 \leq j \leq k$:

- $r_j \in \{0, \ldots, d\}$ *stands for the maximum number of cells to the right of its starting position the tape head will ever be over the next d steps,*
- $f_j \in \{0, \ldots, d - r_j\}$ *is the number of cells left of the position of r_j that the tape head ends up after the d-th step, and*
- $\ell_j \in \{1, \ldots, d\}$ *is the number of distinct cells that shall be changed over the next d steps on tape j. For a given r_j and f_j we have the bound $\ell_j \leq d + 1 - \min\{r_j, f_j\}$.*
- u_j *is a string of length ℓ_j, which is interpreted as the final contents of those cells.*

Technically ℓ_j can always be set to the stated bound, but we keep it separate for clarity.

Definition 2. *A* block trace of block-size d, *for an NTM N, is a sequence of segments of size d.*

Definition 3. *An accepting full path is* compatible *with a block trace if the latter has $\lceil t/d \rceil$ blocks where t is the total number of steps in the path, and in every block each 4-tuple (r_j, f_j, ℓ_j, u_j) correctly describes the head locations after the corresponding d steps of the full path,* and *every character in u_j is the correct final content of its cell after the d steps.*

Our witness predicate now asserts the existence of a block trace y with which some accepting computation is compatible. Clearly every accepting computation gives rise to such a y, so the predicate is correct. The running time of the resulting simulation is a consequence of the following lemmas. Notice that the above definition includes all the possible head movements of N over the next d steps.

Lemma 1. *The number B of valid segments is at most $(32a^d)^k$. Hence the number of potential block trace witnesses is at most $B^{t/d} = a^{kt}32^{kt/d}$.*

Proof. We first bound the number of 4-tuples per each tape. We note that for ℓ cells affected for a particular segment, there are a^ℓ possible strings u. We sum over all the possible values of ℓ – ranging from d to 1. Direct calculation gives us that for $\ell = d$, there are at most 6 possible sets of (r, f), for $\ell = d - 1$ at most

14, etc. The bound on number of possible sets for $\ell = d+1-i$ is i^2+5i. A total number of distinct 4-tuples is upper bounded by

$$\sum_{\ell=d}^{1}[(d+1-\ell)^2 + 5(d+1-\ell)]a^\ell = a^d \cdot \sum_{i=1}^{d}(i^2+5i)/a^{i-1} \leq 32a^d$$

where the last inequality follows by the worst case value $a = 2$. Since we have k tapes, we obtain $B \leq (32a^d)^k$. (In fact, we can get $B \leq (C_a a^d)^k$ where $C_a \longrightarrow 6$ as $a \longrightarrow \infty$, but we do not need this tighter counting.) □

Lemma 2. *Whether there is an accepting computation that is compatible with a given block trace witness can be decided by a deterministic Turing machine in time* $q^2 a^{3kd}\mathsf{poly}(\log q, k, t, a, d)$.

Proof. We generalize the ideas in Proposition 1. We are given a block trace witness, i.e., t/d segments of size d each. The idea is to maintain the set Q_i of states that N on input x can possibly be in, this time after the i-th segment of d steps in some computation path. We precompute a lookup table T_d whose values are sets of states, and whose rows are indexed by the following information:

- An initial state p entering the segment of d steps.
- Strings w_j of length at most $2d-1$ indicating the true contents in the cells surrounding the head on tape j. The cases where a segment of cells on the right or left are blank (through never having been visited before) are handled by adjoining integers b_j indicating such cells.
- The string u_j and integers r_j, f_j for each tape j, representing a segment in a block trace.

The lookup table is the d-length segment equivalent of the lookup table in Proposition 1. There are $qa^{(3d-1)k}d^2$ rows of the table, the length of each index in binary being thus asymptotic to $\log_2 q + (3d-1)k\log_2 a + 2\log_2 d$. The cost of each lookup is thus upper bounded by $qa^{3kd}d^2(\log q + 3kd\log a + 2\log d) + q\log q$. By including the time for sorting the states, and multiplying by the running time of t/d segments, we get

$$[qa^{3kd}d^2(\log q + 3kd\log a + 2\log d) + q\log q + q^2\log^2 q] \cdot t/d.$$

which is upper bounded by

$$q^2 a^{3kd}\mathsf{poly}(\log q, k, t, a, d).$$

□

Theorem 4. *A nondeterministic k-tape TM with q states and alphabet size a can be simulated by a multi-tape deterministic TM in time*

$$a^{kt}C_N^{\sqrt{t}} \cdot q^2\mathsf{poly}(\log q, k, t, a),$$

where C_N is a constant that depends only on a and k.

Proof. This follows from Lemmas 1 and 2. The simulator machine tries out all the possible block witnesses, with a running time

$$q^2 a^{kt+3kd} 32^{kt/d} \mathsf{poly}(\log q, k, t, a, d).$$

The two factors in the above expression that depend on d in a big way are a^{3kd} and $32^{kt/d}$. We can choose d to be such that these the product of these two factors are minimized. Direct calculation gives us that this happens when $d = \sqrt{5t/(3\log_2 a)}$. Setting $C_N = 2^{2k\sqrt{15\log_2 a}}$, we get a running time of

$$a^{kt} C_N^{\sqrt{t}} \cdot q^2 \mathsf{poly}(\log q, k, t, a).$$ □

4 Main Theorem

We have seen two simulations of an NTM where the dominant term in the running time is a^{kt}. One is strategy 3, searching the configuration graph, discussed in Section 2, with a running time of $q^2(3at)^k a^{kt} \mathsf{poly}(\log q, k, t, a)$. The other is the block trace method, with a running time of $a^{kt} C_N^{\sqrt{t}} \cdot q^2 \mathsf{poly}(\log q, k, t, a)$. Even though the time bounds seem similar, the approaches are quite different – a difference that we shall exploit in this section.

Our goal in this section is to reduce the exponent of the simulation time by half. In the graph search method, the dominating part in the running time is caused by the number of configurations. There are at most $qa^{kt}t^k$ of them. If the NTM used only a tape space of $kt/2$ over all the k tapes, then the dominating part in counting the number of configurations would have reduced. We have only a maximum possible $a^{kt/2}$ combinations of tape contents. This would lead to a simulation which requires $q^2(3at)^k a^{kt/2} \mathsf{poly}(\log q, k, t, a)$ time.

But of course, not all NTM simulations use less than $kt/2$ tape space. Here we will use the block trace method to exploit an interesting property of the Turing machines. We make the following observation: the last time we visit a location in the NTM tape, we need not write any character there. This is because the tape head is not going to return to that position. If the NTM visits at least $kt/2$ locations on all tapes together, then there are at least $kt/2$ locations visited for a last time. Now, when we consider block traces, we do not need to have a symbol to write down, if we are visiting a tape location for a last time. We could potentially save on a factor of $a^{kt/2}$ on the running time. This brings down the main factor in the running time in Theorem 4 to $a^{kt/2}$ as well.

For the final theorem, we need one more definition.

Definition 4. *A* directional segment of size d *for a k-tape NTM N with alphabet size a is a segment of size d, omitting the strings u_j, that is*

$$\tau = [(r_1, f_1, \ell_1), \ldots, (r_k, f_k, \ell_k)]$$

where r_j, f_j, ℓ_j are defined as in Definition 1.

A directional trace *of block size d, is a sequence of directional segments of size d.*

Lemma 3. *The number of segments of block size d is upper bounded by d^3. The number of potential directional trace witnesses is at most $(d^3)^{t/d}$.*

Proof. The calculations are similar to those in the proof of Lemma 1. The difference here is that we do not need to count the number of possible strings u for each tape. This bounds the number of directional segments to $\sum_{i=1}^{d}(i^2+5i) = \frac{1}{3}d(d+1)(d+8) \leq d^3$, for d large enough. The bound on directional traces follows. □

We are now ready to prove the main theorem.

Theorem 1. *(**Restated.**) A nondeterministic k-tape TM N with q states and alphabet size a can be simulated by a multi-tape deterministic TM in time*

$$a^{kt(n)/2} H_N^{\sqrt{t(n)}\log t(n)} \cdot q^2 \mathsf{poly}(\log q, k, t(n), a),$$

where $t(n)$ is the running time of N and H_N is a constant that depends only on a.

Proof. We assume that we know an upper bound $t = t(n)$ as a function of the input length n. (If not, one can run the simulations for $t = 1, 2, 3, \cdots$, and this will introduce a multiplicative factor $t(t-1)/2$, which is polynomial in t anyway.)

The simulation consists of three parts. First, preprocessing the directional traces. Second, running the block trace simulation for those traces which have tape usage $\geq kt/2$. And third, running the graph search simulation restricting the tape usage to $kt/2$.

1. In the preprocessing stage, the simulator lists down all the possible directional traces. There are $d^{3t/d}$ such traces by Lemma 3. For $d = \sqrt{5t/(3\log_2 a)}$, as optimized in Theorem 4, we get that the number of traces is $(\sqrt{t})^{O(\sqrt{t})}$ or $H_N^{\sqrt{t}\log t}$, where H_N is a constant that depends on only a. Using the directional trace, the simulator calculates the total tape usage of N. In particular, the simulator decides if the total tape usage is $\leq kt/2$ or $\geq kt/2$. The simulator also calculates the time of the last visit to each of the tape locations. This data is stored in a lookup table, which is stored in another tape of the simulator. All of the above operations can be performed in time $\mathsf{poly}(k,t)$ per directional trace.
2. If the total tape usage is $\geq kt/2$ for a given directional trace, the block trace simulation is performed. All the block traces which match the (r, f, ℓ) parts of the directional trace are generated—with a twist. For those time instances for which the tape head is visiting the location for the last time, the block trace is generated with a ␣ character in the corresponding location. The preprocessed data from the directional traces would be used to determine if the location is being visited for the last time or not.
There are at least $kt/2$ locations visited for the last time, so the number of block traces that correspond to a given directional trace is $\leq a^{kt/2}$. So the total number of relevant block traces here is upper bounded by $H_N^{\sqrt{t}\log t} a^{kt/2}$. The running time in the Lemma 2 holds essentially by the following observation. The lookup table could be expanded (slightly) to accommodate

one more symbol in the alphabet, the '␣' symbol. The set of states that are possible in the lookup table after a doing block trace move with a ␣ are the union of the set possible states after a move with the block trace with one of the original a characters in place of the ␣.
The running time contribution of this stage is $a^{kt/2} H_N^{\sqrt{t}\log t} \cdot q^2 \mathsf{poly}(\log q, k, t, a)$.
3. For the cases when the total tape usage is $\leq kt/2$, the directional trace is discarded. For all such cases combined, one call to the graph search simulation is enough. The simulator needs to keep track of the configurations, and reject a branch as soon as the tape usage exceeds $kt/2$. This gives a running time of $a^{kt/2} q^2 (3at)^k \mathsf{poly}(\log q, k, t, a)$.

The theorem follows by observing that if the NTM has an accepting computation path, at least one of the two simulations, the block trace, or the graph search method would yield an accepting path. The running time is

$$T(n) = a^{kt/2} H_N^{\sqrt{t}\log t} \cdot q^2 \mathsf{poly}(\log q, k, t, a). \qquad \square$$

We remark that a similar bound applies in a uniform simulation, meaning a single DTM M that takes an NTM N and its input x as arguments. Reducing from the k tapes of N to the fixed tapes of M via [14] incurs a factor of $\log T(n)$ penalty, but it gets absorbed into the $\mathsf{poly}(\log q, k, t, a)$ term. The program size of N is bounded by $3q^2a^2 \log q$, and even if the largest value $n' = n + 3q^2a^2 \log q$ is used for the length of the input $\langle N, x\rangle$ to M, expressing the bound $T(n)$ in terms of n' does not change its nature much.

5 Sub-linear Nondeterminism and Small Circuits

Now we consider NTMs N that have $o(n)$ nondeterministic steps in any computation path on inputs of length n, where the inputs are over an alphabet Σ of size b. For each n, it follows that some nondeterministic choice string α_n is used for a set of at least $b^{n-o(n)}$ strings. When N is a language acceptor, the computation on α_n also gives the correct answer for all rejected strings, so we add them when defining S to be the set of inputs on which N-with-α_n works correctly. When N computes a partial multi-valued function f, S includes all strings not in the domain of f, and for all other $x \in S$, N with α_n outputs a legal value of $f(x)$. We can hard-wire α_n into deterministic circuits C_n that work correctly on S. The main theorem of [14] gives C_n size $O(t(n) \log t(n))$. We show that for $t(n)$ near linear time we can improve the size of C_n considerably.

Theorem 2. ***(Restated.)*** *Suppose $t(n) = nr(n)$, where $r(n)$ is constructible in unary in $O(n)$ time. Then for any NTM N that runs in time $t(n)$ with $o(n)$ nondeterminism and computes a function f, there exist circuits C_n of size $O(t(n) \log r(n))$ that compute f correctly on $b^{n-o(n)}$ inputs.*

The size improves on [14] when $r(n) = n^{o(1)}$. When $r(n) = (\log n)^{O(1)}$, meaning $t(n)$ is *quasi-linear* time, this reduces the size of C_n to $t(n) \log\log t(n)$. When

$t(n) = O(n)$, this says we can reduce the overhead to any constructible slow-growing unbounded function, in a sense getting the circuit size as close to linear as desired. Of course the circuits C_n work only on a sizable fraction of the inputs—on other $x \in dom(f)$ they may incorrectly fail to output a legal value.

The proof employs Wolfgang Paul's notion of a *block respecting* Turing Machine, from his paper with Hopcroft and Valiant [4] separating time and space. The result of [4] were later extended to multi-dimensional and tree-structured tapes in [13] and [12]. The notion of block-respecting Turing machines has been used a number of times to prove other results, e.g. in [8]. We refer the reader to [15] for a discussion on the results of [4].

Proof. Given n, take $B = r(n)^2$. Let the Turing machine N computing f have k tapes, and regard those tapes as segmented into "blocks" of length B. By the block-respecting construction in [4], we can modify N into N' computing f in time $t'(n) = O(t(n))$ such that on all inputs of length n, all tape heads of $N'(x)$ cross a block boundary only at time-steps that are multiples of B.

For all length-n strings x, and nondeterministic choice strings α_n, we define the "block-respecting graph" $G_{x,\alpha}$ to have vertices $V_{\ell,i}$ standing for the ith block on tape ℓ, and W_j for $0 \leq j < t'(n)/B$—note also $i < t'(n)/B$ since N' runs in $t'(n)$ space. We use the notation $i(j,\ell)$ to denote the block that N is on the ℓth tape, during the time block from $(j-1)B$ to jB. For all time steps jB, if the heads before that step were in blocks $V_{\ell,i(j-1,\ell)}$ and are in blocks $V_{\ell,i(j,\ell)}$ afterward, then $G_{x,\alpha}$ has edges from all $V_{\ell,i(j-1,\ell)}$ to W_j and from W_j to the nodes $V_{\ell,i(j,\ell)}$. Because there are at most 3 choices of next-block per tape at any j, there are at most $R(n) = (3^k)^{t'(n)/B}$ different block-respecting graphs. By the choice of B, $R(n) = b^{O(n/r(n))}$. There are also $A(n) = |A|^{o(n)}$-many possible α_n. Hence, by the pigeonhole principle, there is some block-respecting graph G_n that equals G_{x,α_n} for at least $b^n/R(n)A(n) = b^{n-O(n/r(n))-o(n)} = b^{n-o(n)}$-many x's.

Now from G_n we define the circuits g_n as a cascade of $t'(n)/B$-many segments S_j. Each S_j represents a time-B computation whose input x_j is the current contents of the r-many blocks $V_{\ell,i(j,\ell)}$, with output written to those blocks. By the result of [14], S_j needs circuit size only $O(B \log B)$. So the entire circuit has size $O\left(\frac{t'(n)}{B}\right) B \log B = O(t(n) \log r(n))$.

To finish the proof, we note that there are also junctures between segments that represent any cases head on tape crossing a block boundary at time jB. If in fact the head does not cross the boundary, then the juncture generates a null value '$*$', which then propagates through all remaining segments to produce a rejecting output. The sizes for the junctures are negligible, so the above bound on the size of the circuits holds. □

6 Conclusions

We have shown techniques by which we can search the computation tree of an NTM in time square root of the size of the graph. It would be interesting to see if these techniques can be used to push the running time even lower. Also, it would be interesting to see lower bounds for the problem, i.e., to understand the limitations of determinism as compared to nondeterminism.

6.1 Some Related Work

The only separation of nondeterministic from deterministic time known is DTIME(n) $\neq$ NTIME(n) proved in [11], which is also specific to the multi-tape Turing machine model. It is also known that nondeterministic two-tape machines are more powerful than deterministic one-tape machines [6], and nondeterministic multi-tape machines are more powerful than deterministic multi-tape machines with additional space bound [7]. Limited nondeterminism was analyzed in [3], which showed that achieving it for certain problems implies a general subexponential simulation of nondeterministic computation by deterministic computation. In [18] an unconditional simulation of time-$t(n)$ probabilistic multi-tape Turing machines Turing machines operating in deterministic time $o(2^t)$ is given.

For certain NP-complete problems, improvements over exhaustive search that involve the constant in the exponent were obtained in [17], [16], and [1], while [9] and [5] also found NP-complete problems for which exhaustive search is not the quickest solution. Williams [19] showed that having such improvements in all cases would collapse other complexity classes. Drawing on [18], Williams [19] showed that the exponent in the simulation of NTM by DTM can be reduced by a multiplicative factor smaller than 1. The NTMs there are allowed only the string-writing form of nondeterminism, but may run for more steps; since the factor is not close to 1/2, the result in [19] is incomparable with ours.

Finally there remains the question asked at the beginning: Is

$$\mathrm{NTIME}(t(n)) \subseteq \mathrm{DTIME}(2^{\varepsilon t(n)})$$

for all $\varepsilon > 0$? We have not found any "dire" collapses of complexity classes that would follow from a 'yes' answer, but it would show that nondeterminism is weaker than we think. David Doty [2] showed that there is an oracle relative to which the answer is no. Our techniques do not resolve this question as yet, but may provide new leads.

Acknowledgments. We thank David Doty, Bart de Keijzer, Ryan Williams, and the anonymous referees for suggestions and helpful comments.

References

1. Beigel, R., Eppstein, D.: 3-coloring in time $O(1.3289^n)$. J. Algorithms 54(2), 168–204 (2005)
2. Doty, D.: An oracle a such that $\mathrm{NTIME}^A(t(n)) \not\subseteq \mathrm{DTIME}^A(2^{\varepsilon t(n)})$, via Kolmogorov complexity. Private Communication (2009)

3. Feige, U., Kilian, J.: On limited versus polynomial nondeterminism. Chicago J. Theoret. Comput. Sci., Article 1, 20 p. (1997) (electronic)
4. Hopcroft, J., Paul, W.J., Valiant, L.: On time versus space. J. Assoc. Comput. Mach. 24(2), 332–337 (1977)
5. Itai, A., Rodeh, M.: Finding a minimum circuit in a graph. SIAM J. Comput. 7(4), 413–423 (1978)
6. Kannan, R.: Towards separating nondeterministic time from deterministic time. In: 22nd Annual Symposium on Foundations of Computer Science, SFCS 1981, pp. 235–243 (October 1981)
7. Kannan, R.: Alternation and the power of nondeterminism. In: STOC 1983: Proceedings of the fifteenth annual ACM symposium on Theory of computing, pp. 344–346. ACM, New York (1983)
8. Lipton, R.J., Viglas, A.: Non-uniform depth of polynomial time and space simulations. In: Lingas, A., Nilsson, B.J. (eds.) FCT 2003. LNCS, vol. 2751, pp. 311–320. Springer, Heidelberg (2003)
9. Nešetřil, J., Poljak, S.: On the complexity of the subgraph problem. Comment. Math. Univ. Carolin. 26(2), 415–419 (1985)
10. Papadimitriou, C.H.: Computational complexity. Addison-Wesley Publishing Company, Reading (1994)
11. Paul, W.J., Pippenger, N., Szemeredi, E., Trotter, W.T.: On determinism versus non-determinism and related problems. In: 24th Annual Symposium on Foundations of Computer Science, pp. 429–438 (November 1983)
12. Paul, W.J., Reischuk, R.: On time versus space. II. J. Comput. System Sci. 22(3), 312–327 (1981), Special issued dedicated to Michael Machtey
13. Pippenger, N.: Probabilistic simulations (preliminary version). In: STOC 1982: Proceedings of the fourteenth annual ACM symposium on Theory of computing, pp. 17–26. ACM, New York (1982)
14. Pippenger, N., Fischer, M.J.: Relations among complexity measures. J. Assoc. Comput. Mach. 26(2), 361–381 (1979)
15. Santhanam, R.: Relationships among time and space complexity classes (2001), `http://citeseerx.ist.psu.edu/viewdoc/summary?doi=10.1.1.24.5170`
16. Schroeppel, R., Shamir, A.: A $T \cdot S^2 = O(2^n)$ time/space tradeoff for certain NP-complete problems. In: 20th Annual Symposium on Foundations of Computer Science, San Juan, Puerto Rico, pp. 328–336. IEEE, New York (1979)
17. Tarjan, R.E., Trojanowski, A.E.: Finding a maximum independent set. SIAM J. Comput. 6(3), 537–546 (1977)
18. van Melkebeek, D., Santhanam, R.: Holographic proofs and derandomization. SIAM J. Comput. 35(1), 59–90 (2005) (electronic)
19. Williams, R.: Improving exhaustive search implies superpolynomial lower bounds. In: STOC 2010: Proceedings of the fortysecond annual ACM symposium on Theory of computing (to appear, 2010)

The Prize-Collecting Edge Dominating Set Problem in Trees

Naoyuki Kamiyama*

Department of Information and System Engineering, Chuo University
kamiyama@ise.chuo-u.ac.jp

Abstract. In this paper, we consider the prize-collecting edge dominating set problem, which is a generalization of the edge dominating set problem. In the prize-collecting edge dominating set problem, we are not forced to dominate all edges, but we need to pay penalties for edges which are not dominated. It is known that this problem is $\mathcal{NP}$-hard, and Parekh presented a $\frac{8}{3}$-approximation algorithm. To the best of our knowledge, no polynomial-time solvable case is known for this problem. In this paper, we show that the prize-collecting edge dominating set problem in trees can be solved in polynomial time.

1 Introduction

Throughout this paper, we denote by $\mathbb{Z}_+$ and $\mathbb{R}_+$ the sets of nonnegative integers and nonnegative real numbers, respectively. Given a function or a vector f on a ground set U, we use the notation $f(X) = \sum_{e \in X} f(e)$ for each $X \subseteq U$.

Let $G = (V, E)$ be an undirected graph with a vertex set V and an edge set E. In this paper, we regard an edge as a set of exactly two vertices. For each $X \subseteq V$, we denote by $\delta(X)$ be the set of $e \in E$ such that $e \cap X \neq \emptyset$. For each $v \in V$, we use the notation $\delta(v)$ instead of $\delta(\{v\})$. We say that $F \subseteq E$ *dominates* $e \in E$ if $e \in \delta(f)$ for some $f \in F$. We call $F \subseteq E$ an *edge dominating set of* G if F dominates all the edges of E. The *edge dominating set problem* asks for finding an edge dominating set of G with minimum cardinality. This problem is one of fundamental covering problems such as the vertex cover problem.

Yannakakis and Gavril [1] proved that the edge dominating set problem is $\mathcal{NP}$-hard in a graph which is planar or bipartite of maximum degree 3. Gotthilf, Lewenstein and Rainshmidt [2] presented a $(2 - c\frac{\log n}{n})$-approximation algorithm which is based on the local search technique, where c is an arbitrary constant and n is the number of vertices. As a special case, several classes of graph in which this problem can be solved in polynomial time are known, e.g., trees [3].

In the *weighted edge dominating set problem*, we are given a weight function $w\colon E \to \mathbb{R}_+$, and this problem asks for finding an edge dominating set of G with minimum weight, where the weight of $F \subseteq E$ is defined by $w(F)$. Fujito and Nagamochi [4] and Parekh [5] independently presented a 2-approximation

* Supported by Grants-in-Aid from the Ministry of Education, Culture, Sports, Science and Technology of Japan.

P. Hliněný and A. Kučera (Eds.): MFCS 2010, LNCS 6281, pp. 465–476, 2010.

algorithm for this problem. As a special case, Berger and Parekh [6] gave a polynomial-time algorithm for this problem in trees.

In this paper, we consider the *prize-collecting edge dominating set problem*, which is a generalization of the weighted edge dominating set problem. Recall that in the edge dominating set problem we have to dominate *all* edges. However, in the prize-collecting edge dominating set problem we are not forced to dominate all edges, but we need to pay *penalties* for edges which are not dominated. More formally, the prize-collecting edge dominating set problem is defined as follows. We are given a graph $G = (V, E)$, a weight function $w\colon E \to \mathbb{R}_+$ and a penalty function $\pi\colon E \to \mathbb{R}_+$. The *cost* of $F \subset E$ is defined by $w(F) + \pi(F')$, where F' denotes the set of the edges of E which are not dominated by F. The prize-collecting edge dominating set problem asks for finding a subset of E with minimum cost. Prize-collecting type variants of combinatorial optimization problems have been extensively studied (for example, see [7,8]).

For the prize-collecting edge dominating set problem, Parekh [9] gave a $\frac{8}{3}$-approximation algorithm. To the best of our knowledge, no polynomial-time solvable case is known for this problem. In this paper, we show that the prize-collecting edge dominating set problem in trees can be solved in polynomial time. Our algorithm is based on the algorithm of Berger and Parekh [6] for the weighted edge dominating set problem in trees, but we should emphasize that the extension is non-trivial.

The rest of this paper is organized as follows. Section 2 gives our algorithm for the prize-collecting edge dominating set problem in trees. In Section 3, we show the correctness of our algorithm. In Section 4, we consider the time complexity of our algorithm. In Section 5, we consider the total dual integrality of a polyhedron related to our problem and some generalization.

Notations. Let $G = (V, E)$ be a tree, and we specify an arbitrary vertex of G as a root. For each $v \in V$, the *depth* of v is the number of the edges contained in the (unique) path from the root to v (denoted by $d(v)$), and the *parent* of v is the (unique) vertex $u \in V$ such that $d(u) = d(v) - 1$ and $\{u, v\} \in E$. For each $v \in V$, let pv and e_v be the parent of v and the edge $\{v, pv\}$, respectively. We say that $v \in V$ is a *child* of pv. For each $v \in V$, we denote by gv the parent of pv.

2 Algorithm

In this section, we present an algorithm for the prize-collecting edge dominating set problem in trees. More precisely, we give an algorithm for a more general problem, called the *prize-collecting b-edge dominating set problem*. In this problem, we are given a graph $G = (V, E)$, a weight function $w\colon E \to \mathbb{R}_+$, a penalty function $\pi\colon E \to \mathbb{R}_+$ and a demand function $b\colon E \to \{0, 1\}$. The cost of $F \subseteq E$ is defined by $w(F) + \pi(\overline{F})$, where $\overline{F}$ is the set of $e \in E$ such that $b(e) = 1$ and e is not dominated by F. If $b(e) = 1$ for all $e \in E$, the prize-collecting b-edge dominating set problem is equivalent to the prize-collecting edge dominating set problem.

In the sequel, let G be a tree with a root. For each $v \in V$, let $\delta_1(v)$ be the set of $e \in \delta(v)$ such that $b(e) = 1$, let $\delta^c(v)$ be the set of $e \in E$ between v and its children, and let $\delta_1^c(v)$ (resp., $\delta_0^c(v)$) be the set of $e \in \delta^c(v)$ such that $b(e) = 1$ (resp., $b(e) = 0$). We denote by M the set of $v \in V$ such that $d(v) = \max_{u \in V} d(u)$, and let M_1 be the set of $v \in M$ such that $b(e_v) = 1$.

Our algorithm is recursively defined according to the number of vertices $v \in V$ such that $d(v) > 1$. For the base case, we consider the case where there exists no $v \in V$ such that $d(v) > 1$, i.e., G is a star. Let E_1 be the set of $e \in E$ such that $b(e) = 1$. If $\min_{e \in E} w(e) \leq \pi(E_1)$, the algorithm outputs a minimizer of $\min_{e \in E} w(e)$. Otherwise, the algorithm outputs $\emptyset$.

From here we consider the case where there exists $v \in V$ such that $d(v) > 1$. We divide this case into the following subcases.

Case A: $M_1 \neq \emptyset$.
Case A1: There exists $v \in M_1$ such that $b(e_{pv}) = 0$.
Case A2: There exists $v \in M_1$ such that $b(e_{pv}) = 1$ and

$$\min_{e \in \delta(pv)} w(e) \leq \sum_{e \in \delta_1^c(pv)} \pi(e). \tag{1}$$

Case A3: For all $v \in M_1$ such that $b(e_{pv}) = 1$, (1) does not hold.
Case B: $M_1 = \emptyset$.

In the rest of this section, we give the detail of our algorithm for each case. Given a subgraph $G' = (V', E')$ of G, a weight function $w' \colon E' \to \mathbb{R}_+$ and a demand function $b' \colon E' \to \{0, 1\}$, an *instance* $I' = (G', w', b')$ is the prize-collecting b'-edge dominating set problem in G' with w' and the restriction of π on E'. Namely, the original problem is an instance $I = (G, w, b)$. Let AI be the output of our algorithm for the instance I. (We adopt this notation for any instance I', i.e., we denote by AI' the output of our algorithm to I'.)

2.1 Case A1

In this subsection, we give an algorithm for Case A1. Let $v \in M_1$ be a vertex such that $b(e_{pv}) = 0$. We define α by

$$\alpha = \min\Big(\min_{e \in \delta(pv)} w(e), \sum_{e \in \delta_1^c(pv)} \pi(e)\Big).$$

Let $G' = (V', E')$ be the graph obtained from G by removing the children of pv and the edges of $\delta^c(pv)$. We define a weight function w' on E' by

$$w'(e) = \begin{cases} w(e) - \alpha, & \text{if } e = e_{pv}, \\ w(e), & \text{otherwise.} \end{cases}$$

We define a demand function b' on E' by $b'(e) = b(e)$ for all edges of E'. Let I' be an instance (G', w', b'). If $e_{pv} \in AI'$, our algorithm outputs AI', i.e., $AI = AI'$. If $e_{pv} \notin AI'$ and

$$\min_{e \in \delta(pv)} w(e) \leq \sum_{e \in \delta_1^c(pv)} \pi(e), \tag{2}$$

the algorithm outputs $AI' \cup \{e^*\}$, where e^* is a minimizer of the left-hand side of (2). If $e_{pv} \notin AI'$ and (2) does not hold, our algorithm outputs AI'.

2.2 Case A2

In this subsection, we give an algorithm for Case A2. Let $v \in M_1$ be a vertex such that $b(e_{pv}) = 1$ and (1) holds. Let $G' = (V', E')$ be the graph obtained from G by removing the children of pv and the edges of $\delta^c(pv)$. We define a weight function w' on E' by

$$w'(e) = \begin{cases} w(e) - \min_{e \in \delta(pv)} w(e), & \text{if } e = e_{pv}, \\ w(e), & \text{otherwise.} \end{cases}$$

We define a demand function b' on E' by $b'(e_{pv}) = 0$ and $b'(e) = b(e)$ for the other edges of E'. Let I' be an instance (G', w', b'). If $e_{pv} \in AI'$, our algorithm outputs AI'. Otherwise, the algorithm outputs $AI' \cup \{e^*\}$, where e^* is a minimizer of the left-hand side of (1).

2.3 Case A3

In this subsection, we give an algorithm for Case A3. Let $v \in M_1$ be a vertex such that $b(e_{pv}) = 1$ and (1) does not hold. We define β and γ by

$$\beta = \min\Big\{ \min_{e \in \delta(pv)} w(e), \sum_{e \in \delta_1(pv)} \pi(e) \Big\} - \sum_{e \in \delta_1^c(pv)} \pi(e),$$

$$\gamma = \min\Big(\min_{e \in \delta(gv) \setminus \{e_{pv}\}} w(e), \beta \Big).$$

Let $G' = (V', E')$ be the graph obtained from G by removing the children of pv and the edges of $\delta^c(pv)$. We define a weight function w' on E' by

$$w'(e) = \begin{cases} w(e) - \Big(\gamma + \sum_{e \in \delta_1^c(pv)} \pi(e)\Big), & \text{if } e = e_{pv}, \\ w(e) - \gamma, & \text{if } e \in \delta(gv) \setminus \{e_{pv}\}, \\ w(e), & \text{otherwise.} \end{cases}$$

We define a demand function b' on E' by $b'(e_{pv}) = 0$ and $b'(e) = b(e)$ for the other edges of E'. Let I' be an instance (G', w', b'). If AI' contains an edge of $\delta(gv)$, our algorithm outputs AI'. From here, we consider the case where AI' does not contain an edge of $\delta(gv)$. If

$$\min_{e \in \delta(gv) \setminus \{e_{pv}\}} w(e) \leq \beta, \tag{3}$$

our algorithm outputs $AI' \cup \{e_1^*\}$, where e_1^* is a minimizer of the left-hand side of (3). If (3) does not hold and

$$\min_{e \in \delta(pv)} w(e) \leq \sum_{e \in \delta_1(pv)} \pi(e), \tag{4}$$

our algorithm outputs $AI' \cup \{e_2^*\}$, where e_2^* is a minimizer of the left-hand side of (4). If (3) and (4) do not hold, our algorithm outputs AI'.

2.4 Case B

In this subsection, we give an algorithm for Case B. Let v be a vertex of M. Without loss of generality we can assume that $b(e_{pv}) = 1$. Otherwise, we can remove edges of $\delta^c(pv)$. We define η by

$$\eta = \min\Big(\min_{e \in \delta(e_{pv})} w(e), \pi(e_{pv})\Big)$$

Let $G' = (V', E')$ be the graph obtained from G by removing the children of pv and the edges of $\delta^c(pv)$. We define a weight function w' on E' by

$$w'(e) = \begin{cases} w(e) - \eta, & \text{if } e \in \delta(gv), \\ w(e), & \text{otherwise.} \end{cases}$$

We define a demand function b' on E' by $b'(e_{pv}) = 0$ and $b'(e) = b(e)$ for the other edges of E'. Let I' be an instance (G', w', b'). If AI' contain an edge of $\delta(gv)$, our algorithm outputs AI'. If AI' does not contain an edge of $\delta(gv)$ and

$$\min_{e \in \delta(e_{pv})} w(e) \leq \pi(e_{pv}), \tag{5}$$

our algorithm outputs $AI' \cup \{e^*\}$, where e^* is a minimizer of the left-hand side of (5). If AI' does not contains an edge of $\delta(e_{pv})$ and (5) does not hold, our algorithm outputs AI'.

3 Correctness

In this section, we prove the correctness of the algorithm presented in Section 2. The integer programming formulation IP(I) of an instance $I = (G, w, b)$ can be described as follows.

$$\text{IP}(I) \left| \begin{array}{ll} \min & \langle w, x\rangle + \langle \pi, y\rangle \\ \text{s.t.} & x(\delta(e)) + y(e) \geq b(e) \quad (\forall e \in E) \\ & x, y \in \mathbb{Z}_+^E, \end{array} \right.$$

where let $\langle f, g\rangle$ be the inner product of f and g for vectors f and g on the same ground set. The dual problem DUAL(I) of the linear programming relaxation of IP(I) can be described as follows.

$$\text{DUAL}(I) \left| \begin{array}{ll} \max & \langle b, z\rangle \\ \text{s.t.} & z(e) \leq \pi(e) \quad (\forall e \in E) \\ & z(\delta(e)) \leq w(e) \quad (\forall e \in E) \\ & z \in \mathbb{R}_+^E. \end{array} \right.$$

By the weak duality theorem [10], a feasible solution z to DUAL(I) such that $w(AI) + \pi(\overline{AI}) \leq \langle b, z\rangle$, AI is optimal to an instance I. Hence, in order to show the correctness, it suffices to prove the following lemma.

Lemma 1. *Given an instance* $I = (G, w, b)$*, there exists a feasible solution* z *to* $\text{DUAL}(I)$ *such that* $w(AI) + \pi(\overline{AI}) \leq \langle b, z \rangle$.

In the sequel, we show Lemma 1 for each case by induction on the number of $v \in V$ such that $d(v) > 1$. For an instance I, we denote by $c(I)$ the cost of a solution which our algorithm outputs, i.e., $c(I) = w(AI) + \pi(\overline{AI})$.

3.1 The Star Case

In this subsection, we consider the case where G is a star. We first consider the case of $\min_{e \in E} w(e) \leq \pi(E_1)$. We arbitrarily name the edges of E_1 as $e_1, \ldots, e_m$, where $m = |E_1|$. Since $\min_{e \in E} w(e) \leq \pi(E_1)$, there exists the index $l \in \{1, \ldots, m\}$ such that

$$\sum_{i=1}^{l-1} \pi(e_i) < \min_{e \in E} w(e) \leq \sum_{i=1}^{l} \pi(e_i).$$

We define $z \in \mathbb{R}_+^E$ as follows. Set $z(e_i) = \pi(e_i)$ for all $i \in \{1, \ldots, l-1\}$, and

$$z(e_l) = \min_{e \in E} w(e) - \sum_{i=1}^{l-1} \pi(e_i).$$

Set $z(e) = 0$ for the other edges e. By the definition, $z(e) \leq \pi(e)$ and $z(\delta(e)) = \min_{f \in E} w(f) \leq w(e)$ for all $e \in E$. Hence, z is feasible for $\text{DUAL}(I)$. Furthermore, $c(I) = \min_{e \in E} w(e)$ and $\langle b, z \rangle = \min_{e \in E} w(e)$ clearly holds.

Next we consider the case of $\min_{e \in E} w(e) > \pi(E_1)$. In this case, set $z(e) = \pi(e)$ for all $e \in E_1$ and $z(e) = 0$ for the other edges e. By the definition, $z(e) \leq \pi(e)$ for all $e \in E$. Since $\min_{e \in E} w(e) > \pi(E_1)$, $z(\delta(e)) = \pi(E_1) < w(e)$ for all $e \in E$. Hence, z is feasible to $\text{DUAL}(I)$. Furthermore, $c(I) = \pi(E_1)$ and $\langle b, z \rangle = \pi(E_1)$ clearly holds. This completes the proof.

3.2 Case A1

By the induction hypothesis, there exists a feasible solution z' to $\text{DUAL}(I')$ such that $c(I') \leq \langle b', z' \rangle$. We define $z \in \mathbb{R}_+^E$ as follows. For each $e \in \delta_1^c(pv)$, define $z(e)$ so that $z(e) \leq \pi(e)$ and

$$\sum_{f \in \delta_1^c(pv)} z(f) = \alpha.$$

By the definition α, we can do this in the same manner for the star case. For each $e \in \delta_0^c(pv)$, set $z(e) = 0$. For the other edges e of E', define $z(e) = z'(e)$.

Now we consider the feasibility of z. By the definition, $z(e) \leq \pi(e)$ for all $e \in \delta^c(pv)$. Hence, by the induction hypothesis, $z(e) \leq \pi(e)$ for all $e \in E$. Next we consider the second condition. Here it should note that we can assume that

$z'(e_{pv}) = 0$ since $b'(e_{pv}) = 0$ holds. Since $z(e_{pv}) = 0$ holds by $z(e_{pv}) = z'(e_{pv})$, $z(\delta(e)) = \alpha \leq w(e)$ for each $e \in \delta^c(pv)$. By the definition of w',

$$z(\delta(e_{pv})) = z'(\delta(gv)) + \alpha \leq w(e_{pv}).$$

For the other edges, the condition is satisfied by the induction hypothesis, which implies the feasibility of z.

Finally, we show that $c(I) \leq \langle b, z\rangle$. In every case, $c(I) - c(I') \leq \alpha$ and $\langle b, z\rangle - \langle b', z'\rangle = \alpha$. This completes the proof.

3.3 Case A2

By the induction hypothesis, there exists a feasible solution z' to $\textsc{Dual}(I')$ such that $c(I') \leq \langle b', z'\rangle$. We define $z \in \mathbb{R}^E_+$ as follows. For each $e \in \delta^c_1(pv)$, define $z(e)$ so that $z(e) \leq \pi(e)$ and

$$\sum_{f \in \delta^c_1(pv)} z(f) = \min_{f \in \delta(pv)} w(f).$$

Since (1) holds, we can do this in the same manner for the star case. For each $e \in \delta^c_0(pv)$, set $z(e) = 0$. For the other edges e of E', define $z(e) = z'(e)$.

Now we consider the feasibility of z. By the definition, $z(e) \leq \pi(e)$ for all $e \in \delta^c(pv)$. Hence, by the induction hypothesis, $z(e) \leq \pi(e)$ for all $e \in E$. Next we consider the second condition. We can assume that $z'(e_{pv}) = 0$ since $b'(e_{pv}) = 0$ holds. Since $z(e_{pv}) = 0$ holds by $z(e_{pv}) = z'(e_{pv})$,

$$z(\delta(e)) = \sum_{f \in \delta^c_1(pv)} z(f) = \min_{f \in \delta(pv)} w(f) \leq w(e)$$

for each $e \in \delta^c(pv)$. By the definition of w',

$$z(\delta(e_{pv})) = z'(\delta(gv)) + \min_{e \in \delta(pv)} w(e) \leq w(e_{pv}).$$

For the other edges, the condition is satisfied by the induction hypothesis, which implies the feasibility of z.

Finally, we show that $c(I) \leq \langle b, z\rangle$. In both cases,

$$c(I) - c(I') \leq \min_{e \in \delta(pv)} w(e), \quad \langle b, z\rangle - \langle b', z'\rangle = \min_{e \in \delta(pv)} w(e).$$

This completes the proof.

3.4 Case A3

By the induction hypothesis, there exists a feasible solution z' to $\textsc{Dual}(I')$ such that $c(I') \leq \langle b', z'\rangle$. We define $z \in \mathbb{R}^E_+$ by

$$z(e) = \begin{cases} \pi(e), & \text{if } e \in \delta^c_1(v), \\ 0, & \text{if } e \in \delta^c_0(v), \\ \gamma, & \text{if } e = e_{pv}, \\ z'(e), & \text{otherwise.} \end{cases}$$

First we consider the feasibility of z. By the definition, $z(e) \le \pi(e)$ for all $e \in \delta^c(pv)$. By the definition of γ, $z(e_{pv}) = \gamma \le \pi(e_{pv})$ holds. Hence, by the induction hypothesis, $z(e) \le \pi(e)$ for all $e \in E$. Next we consider the second condition. For each $e \in \delta^c(pv)$.

$$z(\delta(e)) = \gamma + \sum_{e \in \delta_1^c(pv)} \pi(e) \le \min_{f \in \delta(pv)} w(f) \le w(e). \tag{6}$$

Furthermore, by the definition of w',

$$z(\delta(e_{pv})) = z'(\delta(gv)) + \gamma + \sum_{e \in \delta_1^c(pv)} \pi(e) \le w(e_{pv}). \tag{7}$$

For each $e \in \delta(gv) \setminus \{e_{pv}\}$,

$$z(\delta(e)) = z'(\delta(e)) + \gamma \le w(e) \tag{8}$$

Notice that in (6)-(8) the first equation follows from that we can assume $z'(e_{pv}) = 0$ since $b'(e_{pv}) = 0$. For the other edges, the condition is satisfied by the induction hypothesis, which implies the feasibility of z.

Next we show that $c(I) \le \langle b, z \rangle$. For this, we first show that we can assume that AI' contains at most one edge of $\delta(gv)$. Suppose that AI' contains more than one edge of $\delta(gv)$. In this case, AI' contains at least one edge e of $\delta^c(gv)$. When e is incident to a leaf, $c(I')$ does not increase by removing e from AI'. If e is not incident to a leaf, $c(I')$ does not increase removing e from AI' since (2) does not hold. Hence, by the optimality of AI', we can assume that AI' contains at most one edge of $\delta(gv)$. By this fact, in every case $c(I) - c(I')$ is at most

$$\gamma + \sum_{e \in \delta_1^c(pv)} \pi(e), \tag{9}$$

by $z'(e_{pv}) = 0$ and $\langle b, z \rangle - \langle b', z' \rangle$ is equal to (9). This completes the proof.

3.5 Case B

By the induction hypothesis, there exists a feasible solution z' to DUAL(I') such that $c(I') \le \langle b', z' \rangle$. We define $z \in \mathbb{R}_+^E$ as follows.

$$z(e) = \begin{cases} 0, & \text{if } e \in \delta^c(v), \\ \eta, & \text{if } e = e_{pv}, \\ z'(e), & \text{otherwise.} \end{cases}$$

First we consider the feasibility of z. By the definition of η, $z(e_{pv}) = \eta \le \pi(e_{pv})$ holds. Hence, by the induction hypothesis, $z(e) \le \pi(e)$ for all $e \in E$. Next we consider the second condition. For each $e \in \delta^c(pv)$, $z(\delta(e)) = \eta \le w(e)$. By the definition of w',

$$z(\delta(e)) = z'(\delta(e)) + \eta \le w(e)$$

for each $e \in \delta(gv) \setminus \{e_{pv}\}$. The first equation follows from that we can assume $z'(e_{pv}) = 0$ since $b'(e_{pv}) = 0$ holds. Furthermore, by the definition of w',

$$z(\delta(e_{pv})) = z'(\delta(gv)) + \eta \leq w(e_{pv}).$$

For the other edges, the condition is satisfied by the induction hypothesis, which implies the feasibility of z.

Next we show that $c(I) \leq \langle b, z\rangle$. For this, we first show that we can assume that AI' contains at most one edge of $\delta(gv)$. Suppose that AI' contains more than one edge of $\delta(gv)$. In this case, AI' contains at least one edge e of $\delta^c(gv)$. Since $b(e_v) = 0$ for all $v \in M$, $c(I')$ does not increase by removing e from AI'. Hence, by the optimality of AI', we can assume that AI' contains at most one edge of $\delta(gv)$. By this fact, in the both cases $c(I) - c(I') \leq \eta$ and $\langle b, z\rangle - \langle b', z'\rangle = \eta$. This completes the proof.

4 Time Complexity

In this section, we consider the time complexity of our algorithm. Our algorithm, called **Algorithm PEDS**, can be described as follows.

Algorithm PEDS

1. Compute $d(v)$ for all $v \in V$, and set $l = \max_{v \in V} d(v)$.
2. If $l > 1$, remove all the vertices $v \in V$ such that $d(v) = l$, and change the weight function and the demand function in the manner describe in Section 2, whiling keeping the followings in mind.
   (a) We give priority to vertices for which the conditions of Case A1, Case A2, Case A3 and Case B in this order.
   (b) When we remove a vertex for which the conditions of Case A3, some vertices may become to satisfy the conditions of Case A2, i.e., (1). In this case, we remove these vertices before other vertices for which the conditions of Case A3.
   (c) After removing all the vertices $v \in V$ such that $d(v) = l$, update $l = l - 1$.
3. Notice that in this step the input graph becomes a star. Hence, compute an optimal solution to the star, and construction an optimal solution to the original problem.

Theorem 1. *Given a tree $G = (V, E)$, a weight function $w\colon E \to \mathbb{R}_+$, a penalty function $\pi\colon E \to \mathbb{R}_+$ and a demand function $b\colon E \to \{0, 1\}$,* **Algorithm PEDS** *can solve the prize-collecting b-edge dominating set problem in $O(|V|^2)$ time.*

Proof. Since the correctness of our algorithm is proved in the previous section, we consider the time complexity. Clearly, we can complete Step 1 in $O(|V|)$ time. Also Step 3 can be done in $O(|V|)$ time by storing the order in which vertices are removed. Hence, the time required to complete Step 2 is dominating factor of our algorithm. Although the problem is that we have to check some vertex become to satisfy the conditions of Case A2, we can do this in constant time by storing a difference of the both sides of (1). Hence, we can do Step 4 in $O(|V_l|^2)$ time for each l, where V_l is the set of $v \in V$ such that $d(v) = l$. This completes the proof. □

Here we give a bad example for which our algorithm requires $\Omega(|V|^2)$ time. A vertex set V consists of a root r and vertices $X = \{x_1, \dots, x_k\}$ and $Y = \{y_1, \dots, y_k\}$. An edge set E consists of $\{r, x_i\}$ and $\{x_i, y_i\}$ for all $i \in \{1, \dots, k\}$. For all $e \in E$, we define $b(e) = 1$, and $w(e) = +\infty$ and $\pi(e) = 1$. Then, whenever we remove a vertex of Y, we have to update the weights of all the edges of $\delta(r)$. Hence, our algorithm requires $\Omega(|V|^2)$ time for this instance.

5 Total Dual Integrality and Generalization

In this section, we consider the total dual integrality of a polytope related to our problem and some generalization.

5.1 Total Dual Integrality of a Related Polyhedron

In our algorithm, we have an *integral* dual optimal solution if a weight and a penalty of each edge are integral. Hence, we can obtain the following polyhedral result. Let A be a $p \times q$-matrix in which every entry is rational, and let b be a p-dimensional rational vector. Then, a system $Ax \geq b$ for $x \in \mathbb{R}^q_+$ is called *total dual integral* when for each $l \in \mathbb{Z}^q$ the dual problem of minimizing $\langle l, x\rangle$ over $Ax \geq b$ has an integer optimal solution if the dual problem has a feasible solution and the optimal objective value of the dual problem is finite. It is known [11] that if a system $Ax \geq b$ for $x \in \mathbb{R}^p_+$ is total dual integral and $b \in \mathbb{Z}^p$, every vertex of the polyhedron which is determined by a system $Ax \geq b$ has integer coordinates.

Theorem 2. *Given a tree $G = (V, E)$ and a demand function $b\colon E \to \{0, 1\}$, a system $\{x(\delta(e)) + y(e) \geq b(e) \mid e \in E\}$ for $x, y \in \mathbb{R}^E_+$ is total dual integral.*

Proof. For a weight function $w\colon E \to \mathbb{Z}_+$ and a penalty function $\pi\colon E \to \mathbb{Z}_+$, it follows from the proof of the correctness of our algorithm that DUAL(I) has an integral optimal solution, where I is an instance (G, w, b). For a weight function w and a penalty function π such that there exists $e \in E$ such that $w(e) < 0$ or $\pi(e) < 0$, DUAL(I) has no solution. This completes the proof. □

5.2 Generalization

Here we consider a generalization of the prize collecting b-edge dominating set problem. It is natural to generalize an image of a demand function b from $\{0, 1\}$ to $\mathbb{Z}_+$. More precisely, the *generalized prize-collecting b-edge dominating set problem* is defined as follows. We are given a graph $G = (V, E)$, a weight function $w\colon E \to \mathbb{R}_+$, a penalty function $\pi\colon E \to \mathbb{R}_+$ and a demand function $b\colon E \to \mathbb{Z}_+$. The cost of a vector $x \in \mathbb{Z}^E_+$ is defined by $\langle w, x\rangle + \langle \pi, \overline{x}\rangle$, where $\overline{x} \in \mathbb{R}^E_+$ is defined $\max\{0, b(e) - x(\delta(e))\}$ for each $e \in E$. Notice that the integer programming formulation of this problem is also IP(I). Although it is open whether the generalized prize-collecting b-edge dominating set problem in trees can be solved in polynomial time, by using the theory of *total unimodular* we can show that this problem in a path G can be solved in polynomial time. A matrix A is called totally unimodular when if every square submatrix has determinant $0, \pm 1$. The following theorem is known (see also [10, Corollary 19.2a]).

Theorem 3 (Hoffman and Kruskal [12]). *Let A be a totally unimodular $p \times q$-matrix. Then, for each $b \in \mathbb{Z}^p$, every extreme point of the polyhedron determined by a system $Ax \geq b$ for $x \in \mathbb{R}_+$ has integer coordinates.*

We define the *edge-edge adjacency matrix* A_G of G as follows. Letting $|E| = m$ and $E = \{e_1, \ldots, e_m\}$, A_G is an $m \times m$-matrix whose entry corresponding to a i-th row and a j-th column is defined by 1 if $e_i \cap e_j \neq \emptyset$, and 0 otherwise. By Theorem 3, if $[A_G, \Delta]$ is totally unimodular, the generalized prize-collecting b-edge dominating set problem can be solved in polynomial time by solving the linear programming relaxation of IP(I), where Δ is an identity matrix and $[A_G, \Delta]$ is a matrix obtained by combining A_G and Δ. If G is a path, we can prove this by using the following theorems.

Theorem 4 (Schrijver [10, Example 7 in p.279]). *If every entry of a matrix A is 0 or 1 and each row of A has its 1's consecutively, A is totally unimodular.*

Theorem 5 (Ghoulia-Houri [13]). *A matrix A is totally unimodular if and only if each collection R of rows of A can be partitioned into classes R_1 and R_2 such that the sum of the rows in R_1 minus the sum of the rows in R_2 is a vector with entries $0, \pm 1$ only.*

If G is a path, it follows from Theorem 4 that A_G is totally unimodular, and then each collection R of rows of A_G can be partitioned into classes R_1 and R_2 satisfying the condition in Theorem 5. It is clear that for the classes R_1 and R_2, Δ satisfies the condition in Theorem 5. Hence, it follows from Theorem 5 that $[A_G, \Delta]$ is totally unimodular, which implies the polynomial-time solvability of the generalized prize-collecting b-edge dominating set problem in paths.

Theorem 6. *The generalized prize-collecting b-edge dominating set problem in paths can be solved in polynomial time.*

References

1. Yannakakis, M., Gavril, F.: Edge dominating sets in graphs. SIAM Journal on Applied Mathematics 38(3), 364–372 (1980)
2. Gotthilf, Z., Lewenstein, M., Rainshmidt, E.: A $(2 - c\frac{\log n}{n})$ approximation algorithm for the minimum maximal matching problem. In: Bampis, E., Skutella, M. (eds.) WAOA 2008. LNCS, vol. 5426, pp. 267–278. Springer, Heidelberg (2009)
3. Mitchell, S., Hedetniemi, S.: Edge domination in trees. In: Proceedings of the Eighth Southern Conference on Combinatorics, Graph Theory, and Computing, pp. 489–509 (1977)

4. Fujito, T., Nagamochi, H.: A 2-approximation algorithm for the minimum weight edge dominating set problem. Discrete Applied Mathematics 118(3), 199–207 (2002)
5. Parekh, O.: Edge dominating and hypomatchable sets. In: Proceedings of the Thirteenth Annual ACM-SIAM Symposium on Discrete Algorithms (SODA 2002), pp. 287–291 (2002)
6. Berger, A., Parekh, O.: Linear time algorithms for generalized edge dominating set problems. Algorithmica 50(2), 244–254 (2008)
7. Archer, A., Bateni, M., Hajiaghayi, M.T., Karloff, H.J.: Improved approximation algorithms for prize-collecting steiner tree and TSP. In: Proceedings of the Fiftieth Annual IEEE Symposium on Foundations of Computer Science (FOCS 2009), pp. 427–436 (2009)
8. Hochbaum, D.S.: Solving integer programs over monotone inequalities in three variables: A framework for half integrality and good approximations. European Journal of Operational Research 140(2), 291–321 (2002)
9. Parekh, O.: Approximation algorithms for partially covering with edges. Theoretical Computer Science 400(1-3), 159–168 (2008)
10. Schrijver, A.: Theory of Linear and Integer Programming. J. Wiley & Sons, Chichester (1986)
11. Edmonds, J., Giles, R.: A min–max relation for submodular functions on graphs. Annals of Discrete Mathematics 1, 185–204 (1977)
12. Hoffman, A.J., Kruskal, J.B.: Integral boundary points of convex polyhedra. In: Kuhn, H.W., Tucker, A.W. (eds.) Linear Inequalities and Related Systems, pp. 223–246. Princeton University Press, Princeton (1956)
13. Ghouila-Houri, A.: Caractérisation des matrices totalement unimodulaires. Comptes Redus Hebdomadaires des Séances de l'Académie des Sciences (Paris) 254, 1192–1194 (1962)

The Multivariate Resultant Is NP-hard in Any Characteristic

Bruno Grenet, Pascal Koiran, and Natacha Portier*

LIP**, École Normale Supérieure de Lyon, Université de Lyon
and Department of Computer Science, University of Toronto
{Bruno.Grenet,Pascal.Koiran,Natacha.Portier}@ens-lyon.fr

Abstract. The multivariate resultant is a fundamental tool of computational algebraic geometry. It can in particular be used to decide whether a system of n homogeneous equations in n variables is satisfiable (the resultant is a polynomial in the system's coefficients which vanishes if and only if the system is satisfiable). In this paper we present several NP-hardness results for testing whether a multivariate resultant vanishes, or equivalently for deciding whether a square system of homogeneous equations is satisfiable. Our main result is that testing the resultant for zero is NP-hard under deterministic reductions in any characteristic, for systems of low-degree polynomials with coefficients in the ground field (rather than in an extension). We also observe that in characteristic zero, this problem is in the Arthur-Merlin class AM if the generalized Riemann hypothesis holds true. In positive characteristic, the best upper bound remains PSPACE.

1 Introduction

Given two univariate polynomials, their Sylvester matrix is a matrix built on the coefficients of the polynomials which is singular iff the polynomials have a common root. The determinant of the Sylvester matrix is known as the resultant of the polynomials. This determinant is easy to compute since the size of the Sylvester matrix is the sum of the degrees of the polynomials. The study of the possible generalizations to multivariate systems comes within the scope of the theory of elimination [33, 26, 11, 27, 32, 12]. This theory proves that the only case where a unique polynomial can testify to the existence of a common root to the system is the case of n homogeneous polynomials in n variables: the resultant of a square system of homogeneous polynomials $f_1, \ldots, f_n \in \mathbb{K}[x_1, \ldots, x_n]$ is a polynomial in the indeterminate coefficients of $f_1, \ldots, f_n$ which vanishes iff $f_1, \ldots, f_n$ have a nonzero common root in the algebraic closure of $\mathbb{K}$. The resultant of such a system is known as the *multivariate resultant* in the literature. This captures

* A part of this work was done during visits to the Fields Institute and the University of Toronto. It was partially funded by the Fields Institute and the European Community (7th PCRD Contract: PIOF-GA-2009-236197).

** UMR 5668 École Normale Supérieure de Lyon – CNRS – UCBL – INRIA.

P. Hliněný and A. Kučera (Eds.): MFCS 2010, LNCS 6281, pp. 477–488, 2010.

the case of two univariate polynomials *modulo* their homogenization. Furthermore, in many cases a system of more than n homogeneous polynomials in n variables can be reduced to a system of n homogeneous polynomials, so that the square case is an important one. This result is sometimes known as Bertini's theorem (as explained toward the end of this section, we will use an effective version of this result in one of our NP-hardness proofs). In this paper, we focus on the multivariate resultant which we simply refer to as the resultant.

The resultant has been extensively used to solve polynomial systems [25, 29, 7, 9] and for the elimination of quantifiers in algebraically or real-closed fields [30, 17]. More recently, the multivariate resultant has been of interest in pure and applied domains. For instance, the problem of robot motion planning is closely related to the multivariate resultant [3, 4, 8], and more generally the multivariate resultant is used in real algebraic geometry [5, 22]. Finally, in the domain of symbolic computation progress has been made for finding explicit formulations for the resultant [21, 6, 10, 2, 9, 18], see also [20].

Definition 1. *Let $\mathbb{K}$ be a field and $f_1, \ldots, f_n$ be n homogeneous polynomials in $\mathbb{K}[x_1, \ldots, x_n]$, $f_i(\bar{x}) = \sum_{|\alpha|=d_i} \gamma_{i,\alpha} x^\alpha$. The* multivariate resultant *R of $f_1, \ldots, f_n$ is an irreducible polynomial in $\mathbb{K}[\overline{\gamma_{i,\alpha}}]$ such that*

$$R(\overline{\gamma_{i,\alpha}}) = 0 \iff \exists \bar{x} \in \bar{\mathbb{K}}, f_1(\bar{x}) = \cdots = f_n(\bar{x}) = 0. \quad (1)$$

The multivariate resultant is unique up to a constant factor.

The problem we are interested in is testing the resultant for zero. This is the same as deciding whether a square system of homogeneous polynomials (that is n polynomials in n variables) has a non-trivial root. This is closely related to the decision problem problem for the existential theory of an algebraically closed field. This problem is sometimes called the *Hilbert Nullstellensatz* problem:

Definition 2. *Let $\mathbb{K}$ be a field and $\bar{\mathbb{K}}$ be an algebraic closure of $\mathbb{K}$. The* Hilbert Nullstellensatz *problem over $\mathbb{K}$, $\mathtt{HN}(\mathbb{K})$, is the following: Given a system f of s polynomials in $\mathbb{K}[x_0, \ldots, x_n]$, does there exist a root of f in $\bar{\mathbb{K}}^{n+1}$?*

Let us now assume that the s components of f are homogeneous polynomials. Then the homogeneous *Hilbert Nullstellensatz problem over $\mathbb{K}$, $\mathtt{H_2N}(\mathbb{K})$, is to decide whether a* non trivial *(that is, nonzero) root exists in $\bar{\mathbb{K}}$.*

If f is supposed to contain as many homogeneous polynomials as variables, the problem is called the square *homogeneous Hilbert Nullstellensatz over $\mathbb{K}$, $\mathtt{H_2N}^{\square}(\mathbb{K})$.*

In the case of the field $\mathbb{Q}$, it is more natural to have coefficients in $\mathbb{Z}$. We shall use the notations $\mathtt{HN}$, $\mathtt{H_2N}$ and $\mathtt{H_2N}^{\square}$ for this case where the system is made of integer polynomials. In the sequel, for any prime number p, the finite field with p elements is denoted by $\mathbb{F}_p$. The notation is extended to characteristic zero, and $\mathbb{F}_0 = \mathbb{Q}$.

In the case of polynomials with coefficients in $\mathbb{Z}$, Canny [4] gave in 1987 a PSPACE algorithm to compute the resultant. To the authors' knowledge, this is the best known upper bound. In this paper we show that testing the resultant for zero is NP-hard in any characteristic. In other words, $\mathtt{H_2N}(\mathbb{K})$ is NP-hard for any field $\mathbb{K}$.

Main Results and Proof Techniques

In Section 2 we observe that for polynomials with integer coefficients, testing the resultant for zero is a problem in the Arthur-Merlin (AM) class. This result assumes the generalized Riemann hypothesis, and follows from a simple reduction to the Hilbert Nullstellensatz. For this problem, membership in AM assuming GRH was established in [23]. The remainder of the paper is devoted to hardness results.

In characteristic zero, it seems to be a "folklore" result that testing the resultant for zero is NP-hard. We give a proof of this fact in the full version of this paper [16] since we have not been able to find one in the literature. In fact, we give two proofs of two results of incomparable strength. The first proof is based on a reduction from the `Partition` problem [14, problem SP12]. The second proof is based on a result of Plaisted [28] and shows that the problem remains NP-hard for systems of only two homogeneous polynomials. For the latter result to be true, we need to use a sparse encoding for our two polynomials (their degree can therefore be exponential in the input size).

The first proof does not carry over to positive characteristic since the NP-hardness of `Partition` relies in an essential way on the fact that the data are integers (in fact, in any finite field the analogue problem can be solved in polynomial time by dynamic programming).

Plaisted's result can be adapted to positive characteristic [34, 19] but this requires randomization. By contrast, our ultimate goal is NP-hardness for deterministic reductions and low degree polynomials. We therefore need to use different techniques. Our starting point is a fairly standard encoding of $\mathtt{3-SAT}$ by systems of polynomial equations. Using this encoding we show at the beginning of Section 3 that deciding the existence of a nontrivial solution to a system of homogeneous equations is NP-hard in any characteristic. The resulting system has in general more equations than variables. In order to obtain a square system two basic strategies can be explored:

(i) Decrease the number of equations.
(ii) Increase the number of variables.

In [16], we give a randomized NP-hardness result based on the first strategy. The idea is to replace the initial system by a random linear combinations of the system's equations (the fact this does not change the solution set is sometimes called a "Bertini's theorem").

In Section 3 we use the second strategy to obtain two NP-hardness results for deterministic reductions. The main difficulty is to make sure that the introduction of new variables does not create spurious solutions (we do not want to turn an unsatisfiable system into a satisfiable system). Our solution to this problem can be viewed as a derandomization result. Indeed, it can be shown that the coefficients of the monomials where the new variables occur could be chosen at random. It would be interesting to find out whether the proof based on the first strategy can also be derandomized.

2 Complexity of the Resultant in Characteristic 0

In this section we show that testing the resultant for zero is reducible to $\mathtt{HN}(\mathbb{K})$. In the case $\mathbb{K} = \mathbb{Z}$, this allows us to conclude (under the Generalized Riemann Hypothesis) that our problem is in the polynomial hierarchy, and more precisely in the Arthur-Merlin class. In fact, we show that this applies more generally to the satisfiability problem for homogeneous systems (recall that testing the resultant for zero corresponds to the square case).

Proposition 1. *For any field* $\mathbb{K}$*, the problem* $\mathtt{H_2N}(\mathbb{K})$ *is polynomial-time many-one reducible to* $\mathtt{HN}(\mathbb{K})$*.*

Proof. Consider an instance $\mathcal{S}$ of $\mathtt{H_2N}(\mathbb{K})$, that is s homogeneous polynomials $f_1, \ldots, f_s \in \mathbb{K}[x_1, \ldots, x_n]$. The polynomials $f_1, \ldots, f_s$ can be viewed as elements of $\mathbb{K}[x_1, \ldots, x_n, y_1, \ldots, y_n]$ where $y_1, \ldots, y_n$ are new variables which do not appear in the f_i. Let $\mathcal{T}$ be the system containing all the f_i and the new (non-homogeneous) polynomial $\sum_{i=1}^{n} x_i y_i - 1$. This is an instance of the problem $\mathtt{HN}(\mathbb{K})$. It remains to prove that $\mathcal{S}$ and $\mathcal{T}$ are equivalent.

Given a root $(a_1, \ldots, a_n, b_1, \ldots, b_n)$ of $\mathcal{T}$, the new polynomial ensures that there is at least one nonzero a_i. So $(a_1, \ldots, a_n)$ is a non trivial root of $\mathcal{S}$. Conversely, suppose that $\mathcal{S}$ has a non trivial root $(a_1, \ldots, a_n)$, and let i be such that $a_i \neq 0$. Then the tuple $(a_1, \ldots, a_n, 0, \ldots, 0, a_i^{-1}, 0, \ldots, 0)$ where a_i^{-1} corresponds to the variable y_i is a root of $\mathcal{T}$.

Thus $\mathtt{H_2N}(\mathbb{K})$ is polynomial-time many-one reducible to $\mathtt{HN}(\mathbb{K})$.

Koiran [23] proved that $\mathtt{HN} \in \mathsf{AM}$ under the Generalized Riemann Hypothesis. We denote here by AM the Arthur-Merlin class, defined by *interactive proofs with public coins* (see [1]). Thereby,

Corollary 1. *Under the Generalized Riemann Hypothesis,* $\mathtt{H_2N}$ *is in the class* AM.

In positive characteristic, the best upper bound on the complexity of the Hilbert Nullstellensatz known to this day remains PSPACE (in particular it is not known whether the problem lies in the polynomial hierarchy, even assuming some plausible number-theoretic conjecture such as the generalized Riemann hypothesis).

We now give our first NP-hardness result, for the satisfiability of square systems of homogeneous polynomial equations. As explained in the introduction, this seems to be a "folklore" result. We give the (short) proof in the full version [16] since finding an explicit statement (and proof) of this result in the literature appears to be difficult. The second part of the theorem shows that the problems remains NP-hard even for systems with small integer coefficients (i.e., coefficients bounded by 2). This is achieved by a standard trick: we introduce new variables in order to "simulate" large integers coefficients. It is interesting to note, however, that a similar trick for reducing degrees does not seem to apply to the resultant problem (more on this in the full version [16]).

Theorem 1. *The problem* $\mathtt{H_2N}^{\square}$ *of deciding whether a square system of homogeneous polynomials with coefficients in* $\mathbb{Z}$ *has a non trivial root is* NP*-hard.*

The problem remains NP*-hard even if no polynomial has degree greater that* 2 *and even if the coefficients are bounded by* 2.

The reduction is done from `Partition` which is known to be NP-hard [14, problem SP12]. For a full proof of this theorem, please refer to [16].

A related result is Plaisted's [28] on the NP-hardness of deciding whether the gcd of two sparse univariate polynomials has degree greater than one. By homogenization of the polynomials, this is the same problem as in Theorem 1 for only two bivariate polynomials. Note that the polynomials are sparse and can be of very high degree since exponents are written in binary (this polynomial representation is sometimes called "supersparse" [19]). If both polynomials were dense, the resultant could be computed in polynomial time since it is equal to the determinant of their Sylvester matrix. Plaisted's theorem stated in the same language as Theorem 1 is the following:

Theorem 2. *Given two sparse homogeneous polynomials in* $\mathbb{Z}[x, y]$*, it is* NP*-hard to decide whether they share a common root in* $\mathbb{C}^2$.

3 The Resultant is NP-hard in Arbitrary Characteristic

In this section we give two increasingly stronger NP-hardness results for testing the resultant. We prove these two NP-hardness results for deterministic reductions: the first one applies to systems with coefficients in an extension of the ground field, and the second (stronger) result to systems with coefficients in the ground field only. Both results are based on the strategy of increasing the number of variables to make the system square. Note that the other strategy consisting in decreasing the number of polynomials can also be used. We use this strategy to give another hardness result, but for randomized reduction, in [16]. The starting point for these three NP-hardness results is the following lemma.

Lemma 1 ([24]). *Given a field* $\mathbb{K}$ *of any characteristic, it is* NP*-hard to decide whether a system of* s *homogeneous polynomials in* $\mathbb{K}[x_0, \ldots, x_n]$ *has a non trivial root. That is,* $\mathtt{H_2N}(\mathbb{K})$ *is* NP*-hard.*

In [24], $\mathtt{H_2N}(\mathbb{K})$ was proven NP-hard by reduction from `Boolsys`. An input of `Boolsys` is a system of boolean equations in the variables $X_1, \ldots, X_n$ where each equation is of the form $X_i = \text{True}$, $X_i = \neg X_j$, or $X_i = X_j \vee X_k$. The question is the existence of a valid assignment for the system, that is an assignment of the variables such that each equation is satisfied. This problem is easily shown NP-hard by reduction from $3 - \mathtt{SAT}$. We now give a proof of this lemma since the specific form of the systems that we construct in the reduction will be useful in the sequel. This proof is a slight variation on the proof from [24].

Proof. Let $\mathbb{K}$ be a field of any characteristic p, p being either zero or a prime number. At first, p is supposed to be different from 2. The proof has to be slightly changed in the case $p = 2$ and this case is explained at the end of the proof.

Let $\mathcal{B}$ be an instance of `Boolsys`. Let us define a system of homogeneous polynomials from this instance with the property that $\mathcal{B}$ is satisfiable iff the polynomial system has a non trivial common root. The variables in the system are $x_0, \ldots, x_n$ where x_i, $1 \leq i \leq n$, corresponds to the boolean variable X_i in `Boolsys`, and x_0 is a new variable. The system contains four kinds of polynomials:

- $x_0^2 - x_i^2$, for each $i > 0$;
- $x_0 \cdot (x_i + x_0)$, for each equation $X_i =$ True in `Boolsys`;
- $x_0 \cdot (x_i + x_j)$, for each equation $X_i = \neg X_j$;
- $(x_i + x_0)^2 - (x_j + x_0) \cdot (x_k + x_0)$, for each equation $X_i = X_j \vee X_k$.

Let us denote by f the polynomial system obtained from $\mathcal{B}$. The first kind of polynomials ensures that if $(a_0, \ldots, a_n)$ is a non trivial root of f, then $a_0^2 = a_1^2 = \cdots = a_n^2$. Now if f has a non trivial root $(a_0, \ldots, a_n)$, then one can readily check that the assignment $X_i =$ True if $a_i = -a_0$ and $X_i =$ false if $a_i = a_0$ satisfies $\mathcal{B}$. Conversely, if there is a valid assignment $X_1, \ldots, X_n$ for $\mathcal{B}$, any $(n+1)$-tuple $(a_0, \ldots, a_n)$ where $a_0 \neq 0$ and $a_i = -a_0$ if $X_i =$ True and $a_i = a_0$ if $X_i =$ false is a non trivial root of f.

This proof works for any field of characteristic different from 2. The problem in characteristic 2 is the implementation of `Boolsys` in terms of a system of polynomials. Indeed, for the other characteristics, the truth is represented by $-a_0$ and the falseness by a_0. In characteristic 2, those values are equal. Yet, one can just change the polynomials and define in the case of characteristic 2 the following system:

- $x_0 x_i - x_i^2$, for each $i > 0$;
- $x_0(x_i + x_0)$, for each equation $X_i =$ True in `Boolsys`;
- $x_0(x_i + x_j + x_0)$, for each equation $X_i = \neg X_j$;
- $x_i^2 + x_j x_k + x_0 \cdot (x_j + x_k)$, for each equation $X_i = X_j \vee X_k$.

Now, given any nonzero value a_0 for x_0, the truth of a variable X_i is represented by $x_i = a_0$ whence the falseness is represented by $x_i = 0$. A root of the system is in particular a root of the polynomials defined by the first item. Therefore each x_i has to be set either to a_0 or to 0. The system has a non trivial root iff the instance of `Boolsys` is satisfiable. □

We now show that testing the resultant for zero is NP-hard *for deterministic reductions.*

Theorem 3. *Let p be either zero or a prime number. The following problem is* NP-*hard under deterministic reductions:*

- *INPUT: a square system of homogeneous equations with coefficients in a finite extension of* $\mathbb{F}_p$.
- *QUESTION: is the system satisfiable in the algebraic closure of* $\mathbb{F}_p$*?*

In the case $p = 0$, *the results also holds for systems with coefficients in* $\mathbb{Z}$.

Proof. The proof of Lemma 1 gives a method to implement an instance of Boolsys with a system f of s homogeneous polynomials in $n+1$ variables with coefficients in $\mathbb{F}_p$. It remains to explain how to construct a square system g that has a non trivial root iff f does. Let us denote by $f_1, \ldots, f_s$ the components of f, with for each $i = 1, \ldots, n$, $f_i = x_0^2 - x_i^2$ if $p \neq 2$ and $f_i = x_0 x_i - x_i^2$ if $p = 2$. A new system g of s polynomials in s variables is built. The s variables are $x_0, \ldots, x_n$ and $y_1, \ldots, y_{s-n-1}$, that is $(s-n-1)$ new variables are added. The system g is the following:

$$g(\bar{x}, \bar{y}) = \begin{pmatrix} f_1(\bar{x}) & & \\ \vdots & & \\ f_n(\bar{x}) & & \\ f_{n+1}(\bar{x}) & & +\lambda y_1^2 \\ f_{n+2}(\bar{x}) & -y_1^2 & +\lambda y_2^2 \\ & \vdots & \\ f_{n+i}(\bar{x}) & -y_{i-1}^2 & +\lambda y_i^2 \\ & \vdots & \\ f_{s-1}(\bar{x}) & -y_{s-n-2}^2 & +\lambda y_{s-n-1}^2 \\ f_s(\bar{x}) & -y_{s-n-1}^2 & \end{pmatrix} \quad (2)$$

The parameter λ is to be defined later. Clearly, if f has a non trivial root $\bar{a}$, then $(\bar{a}, \bar{0})$ is a non trivial root of g. Let us now prove that the converse also holds true for some λ: if g has a non trivial root, then so does f. Note that a suitable λ has to be found in polynomial time.

Let $(a_0, \ldots, a_n, b_1, \ldots, b_{s-n-1})$ be any non trivial root of g. Since $\bar{a}$ must be a common root of $f_1, \ldots, f_n$, we have $a_0^2 = \cdots = a_n^2$ if $p \neq 2$, and $a_i \in \{0, a_0\}$ for every i if $p = 2$. Now, either $a_0 = 0$ and $f_i(\bar{a}) = 0$ for every i, or a_0 can be supposed to equal 1. Therefore, if $p \neq 2$ either $\bar{a} = \bar{0}$ or $a_i = \pm 1$ for every i, and if $p = 2$ either $\bar{a} = \bar{0}$ or $a_i \in \{0, 1\}$ for every i. Let us define $\epsilon_i = f_{n+i}(\bar{a}) \in \mathbb{F}_p$. As $(\bar{a}, \bar{b})$ is a root of g, the b_i^2 satisfy the linear system

$$\begin{cases} \epsilon_1 & & + \lambda Y_1 & = 0, \\ \epsilon_2 & - Y_1 & + \lambda Y_2 & = 0, \\ & & \vdots & \\ \epsilon_{s-n-1} & - Y_{s-n-2} & + \lambda Y_{s-n-1} & = 0, \\ \epsilon_{s-n} & - Y_{s-n-1} & & = 0. \end{cases} \quad (3)$$

This system can be homogenized by replacing each ϵ_i by $\epsilon_i Y_0$ where Y_0 is a fresh variable. This gives a square homogeneous linear system. The determinant of the matrix of this system is equal to $(-1)^{s-n-1}\left(\epsilon_1 + \epsilon_2 \lambda + \cdots + \epsilon_{s-n} \lambda^{s-n-1}\right)$.

Let us consider this determinant as a polynomial in λ. This polynomial vanishes identically iff all the ϵ_i are zero. In that case, the only solutions satisfy $Y_i = 0$ for $i > 0$, that is $(\bar{a}, \bar{0})$ is a root of g and therefore $\bar{a}$ is a root of f. If some ϵ_i are nonzero, this is a nonzero polynomial of degree $(s-n-1)$. If λ can be chosen such that it is not a root of this polynomial (for any possible nonzero value of $\bar{\epsilon}$), then the only solution to the linear system is the trivial one. This means that the only non trivial root of g is $(\bar{a}, \bar{0})$ where $\bar{a}$ is a root of f.

If the polynomials have coefficients in $\mathbb{Z}$, $\lambda = 3$ (or any other integer $\lambda > 2$) satisfies the condition. Indeed, one can check that $\epsilon_i = f_{n+i}(\bar{a}) \in \{-4, 0, 2, 4\}$ when $a_0 = 1$. The determinant is zero iff $\epsilon'_1 + \epsilon'_2\lambda + \cdots + \epsilon'_{s-n}\lambda^{s-n-1} = 0$ where $\epsilon'_i = \epsilon_i/2 \in \{-2, 0, 1, 2\}$. For each i, let $\epsilon_i^+ = \max\{\epsilon'_i, 0\}$ and $\epsilon_i^- = \max\{-\epsilon'_i, 0\}$. Then $\epsilon'_i = \epsilon_i^+ - \epsilon_i^-$, and $0 \leq \epsilon_i^+, \epsilon_i^- \leq 2$. Now the determinant is zero iff $\sum_i \epsilon_i^+ 3^i = \sum_i \epsilon_i^- 3^i$. By the unicity of base-3 representation, this means that for all i, $\epsilon_i^+ = \epsilon_i^-$, and so $\epsilon'_i = 0$.

For a field of positive characteristic, this argument cannot be applied. The idea is to find a λ that is not a root of any polynomial of degree $(s - n - 1)$. Nothing else can be supposed on the polynomial because if $\mathbb{F}_p = \mathbb{F}_3$ for example, any polynomial of $\mathbb{F}_p[\lambda]$ can appear. This also shows that λ cannot be found in the ground field. Suppose an extension of degree $(s - n)$ is given as $\mathbb{F}_p[X]/(P)$ where P is an irreducible degree-$(s-n)$ polynomial with coefficients in $\mathbb{F}_p$. Then a root of P in $\mathbb{F}_p[X]/(P)$ cannot be a root of a degree-$(s-n-1)$ polynomial with coefficients in $\mathbb{F}_p$. Thus, if one can find such a P, taking for λ the indeterminate X is sufficient. For any fixed characteristic p, Shoup gives a deterministic polynomial-time algorithm [31] that given an integer N outputs a degree-N irreducible polynomial P in $\mathbb{F}_p[X]$. Thus, the system g is now a square system of polynomials in $(\mathbb{F}_p[X]/(P))[\bar{x}, \bar{y}]$ and this system has a non trivial root iff f has a non trivial root. And Shoup's algorithm allows us to build g in polynomial time from f.

For any field $\mathbb{F}_p$, it has been shown that from an instance $\mathcal{B}$ of `Boolsys` a square system g of polynomials with coefficients in an extension of $\mathbb{F}_p$ (in $\mathbb{Z}$ for integer polynomials) can be built in deterministic polynomial time such that g has a non trivial root iff $\mathcal{B}$ is satisfiable. This shows that the problem is NP-hard. □

The previous result is somewhat unsatisfactory as it requires, in the case of positive characteristic, to work with coefficients in an extension field rather than in the ground field. A way to get rid of this limitation is now shown. Yet, a property of the previous result is lost. Instead of having constant-degree (even degree-2) polynomials, our next result uses linear-degree polynomials. It is not clear whether the same result can be obtained for degree-2 polynomials (for instance, as explained in the full version [16] reducing the degree by introducing new variables can create unwanted solutions at infinity).

The basic idea behind Theorem 4 is quite simple (we put the irreducible polynomial used to build the extension field into the system), but some care is required in order to obtain an equivalent homogeneous system.

Recall from the introduction that $\mathtt{H_2N}^{\square}(\mathbb{F}_p)$ is the following problem:

- INPUT: a square system of homogeneous equations with coefficients in $\mathbb{F}_p$.
- QUESTION: is the system satisfiable in the algebraic closure of $\mathbb{F}_p$?

Theorem 4. *For any prime p, $\mathtt{H_2N}^{\square}(\mathbb{F}_p)$ is* NP*-hard under deterministic reductions.*

Proof. The idea for this result is to turn coefficient λ in the previous proof into a variable and to add the polynomial P as a component of the system. Of course, considering λ as a variable implies that the polynomials are not homogeneous anymore. Thus, it remains to explain how to keep the system homogeneous.

First, the polynomial P needs to be homogenized. This is done through the variable x_0 in the canonical way. As $P(\lambda)$ is irreducible, it is in particular not divisible by λ. Hence, the homogenized polynomial $P(\lambda, x_0)$ contains a monomial $\alpha\lambda^d$ and another one βx_0^d where d is the degree of P. Hence x_0 is zero iff λ is.

The other polynomials have the form $f_{n+i}(\bar{x}) - y_{i-1}^2 + \lambda y_i^2$. It is impossible to homogenize those polynomials by multiplying f_{n+i} and y_{i-1}^2 by x_0 (or any other variable) because then the variable y_{i-1} never appears alone in a monomial, and a s-tuple with all variables set to 0 but y_{i-1} would be a non trivial solution. Moreover, in the previous proof, the fact that the y_i all appear with degree 2 is used to consider the system as a linear system in the y_i^2. Thus replacing the monomial λy_i^2 by λy_i does not work either. Instead, we construct the slightly more complicated homogeneous system:

$$g_h(\bar{x}, \bar{y}, \lambda) = \begin{pmatrix} f_1(\bar{x}) & & \\ \vdots & & \\ f_n(\bar{x}) & & \\ x_0^{s-n-1} f_{n+1}(\bar{x}) & & +\lambda y_1^{s-n} \\ x_0^{s-n-2} f_{n+2}(\bar{x}) & -y_1^{s-n} & +\lambda y_2^{s-n-1} \\ & \vdots & \\ x_0^{s-n-i} f_{n+i}(\bar{x}) & -y_{i-1}^{s-n-i+2} & +\lambda y_i^{s-n-i+1} \\ & \vdots & \\ x_0 f_{s-1}(\bar{x}) & -y_{s-n-2}^3 & +\lambda y_{s-n-1}^2 \\ f_s(\bar{x}) & -y_{s-n-1}^2 & \\ P(\lambda, x_0) & & \end{pmatrix} \qquad (4)$$

Contrary to the previous proof, the y_i do not appear all at the same power. Yet, all the occurrences of each y_i have the same degree, and we shall prove that this is sufficient.

Let us prove that if f does not have any non trivial root, then neither does g_h. Some of the observations made for g in the previous proof remain valid. Hence, it is sufficient to prove that a non trivial $(s+1)$-tuple $(\bar{a}, \bar{b}, \ell)$ cannot be solution of g_h whenever $a_0 = 1$, $\bar{b} \neq \bar{0}$ and $a_0^2 = \cdots = a_n^2$ if $p \neq 2$ or $a_i \in \{0, a_0\}$ if $p = 2$. By a previous remark on the polynomial P, ℓ can also be supposed to be nonzero.

So, similarly as in the previous proof, let us define $\epsilon_i = a_0^{s-n-i} f_{n+i}(\bar{a}) \in \mathbb{F}_p$. In the system g_h, the variable y_i only appears at the power $(s-n-i+1)$. Therefore, given a value of $\bar{a}$ and ℓ, the tuple $(\bar{a}, \bar{b}, \ell)$ is a root of g_h iff the $b_i^{s-n-i+1}$ satisfy the linear system

$$\begin{cases} \epsilon_1 & & + & \ell Y_1 & = 0 \\ \epsilon_2 & - & Y_1 & + & \ell Y_2 & = 0 \\ & & & \vdots & & \\ \epsilon_{s-n-1} & - & Y_{s-n-2} & + & \ell Y_{s-n-1} & = 0 \\ \epsilon_{s-n} & - & Y_{s-n-1} & & & = 0 \end{cases} \tag{5}$$

This is the same system as in the previous proof. Now if $(\ell, 1)$ is supposed to be a root of P, as P is an irreducible polynomial of degree $(s-n)$, ℓ cannot be a root of a univariate polynomial of degree less than $(s-n)$ with coefficient in $\mathbb{F}_p$. But the determinant of the linear system is such a polynomial, and thus cannot be zero. This determinant is then 0 iff all the $\epsilon_i = 0$. The same arguments as in the previous proof can be used to conclude that $(\bar{a}, \bar{b}, \ell)$ can be a root of g_h iff $\bar{a}$ is a root of f.

Thus, from an instance $\mathcal{B}$ of Boolsys, a square homogeneous system g_h of polynomials with coefficients in the ground field $\mathbb{F}_p$ is built in deterministic polynomial time. This system has a non trivial root iff $\mathcal{B}$ is satisfiable. The result is proved. □

4 Final Remarks

In characteristic zero, the upper and lower bounds on $\mathtt{H_2N}^{\square}$ are in a sense close to each other. Indeed, $\mathsf{NP} \subseteq \mathsf{AM} \subseteq \Pi_2\mathsf{P}$, that is, AM lies between the first and the second level of the polynomial hierarchy. Furthermore, "under plausible complexity conjectures, AM = NP" [1, p157]. Improving the NP lower bound may be challenging as the proof of Proposition 1 shows that this would imply the same lower bound for *Hilbert's Nullstellensatz.*

In positive characteristic, the situation is quite different. Indeed, the best known upper bound for *Hilbert's Nullstellensatz* as well as for the resultant is PSPACE. As in characteristic zero, the known upper and lower bounds are therefore the same for both problems. But as the gap between the NP lower bound and the PSPACE upper bound is rather big, these problems might be of widely different complexity (more precisely, testing the resultant for zero could in principle be much easier than deciding whether a general polynomial system is satisfiable). Canny's algorithm for computing the resultant [4] involves the computation of the determinants of exponential-size matrices, known as Macaulay matrices, in polynomial space. Those matrices admit a succinct representation (in the sense of [13]). One can prove that computing the determinant of a general succinctly represented matrix is FPSPACE-complete (and even testing for zero is PSPACE-complete) [15]. It follows that the FPSPACE upper bound could be improved only by exploiting the specific structure of the Macaulay matrices in an essential way, or by finding an altogether different (non Macaulay-based) approach to this problem. As pointed out in Section 2, in characteristic zero a different approach is indeed possible for *testing whether the resultant vanishes* (rather than for computing it). This problem is wide open in positive characteristic.

Finally, an interesting open question is whether the randomized reduction mentioned in Section 3 that we give in [16] can be derandomized.

Acknowledgments. We thank Bernard Mourrain and Maurice Rojas for sharing their insights on the complexity of the resultant in characteristic 0.

References

1. Arora, S., Barak, B.: Computational Complexity: A Modern Approach, 1st edn. Cambridge University Press, Cambridge (2009)
2. Busé, L., D'Andrea, C.: On the irreducibility of multivariate subresultants. CR Math. 338(4), 287–290 (2004)
3. Canny, J.F.: A new algebraic method for robot motion planning and real geometry. In: Proc. FOCS 1987, pp. 39–48 (1987)
4. Canny, J.F.: The complexity of robot motion planning. In: ACM Doctoral Dissertation Award, vol. 1987. MIT Press, Cambridge (1988)
5. Canny, J.F.: Some algebraic and geometric computations in PSPACE. In: Proc. STOC 1988, pp. 460–469 (1988)
6. Canny, J.F.: Generalized characteristic polynomials. In: Gianni, P. (ed.) ISSAC 1988. LNCS, vol. 358. Springer, Heidelberg (1989)
7. Canny, J.F., Kaltofen, E., Yagati, L.: Solving systems of nonlinear polynomial equations faster. In: Proc. SIGSAM 1989, pp. 121–128 (1989)
8. Canny, J.F., Reif, J.H.: New lower bound techniques for robot motion planning problems. In: Proc. FOCS 1987, pp. 49–60 (1987)
9. Cattani, E., Dickenstein, A.: Introduction to residues and resultants. In: Dickenstein, A., Emiris, I. (eds.) Solving Polynomial Equations, pp. 1–61. Springer, Heidelberg (2005)
10. D'Andrea, C., Dickenstein, A.: Explicit formulas for the multivariate resultant. J. Pure Appl. Algebra 164(1-2), 59–86 (2001)
11. Dixon, A.: The eliminant of three quantics in two independent variables. Proc. Lond. Math. Soc. 6, 468–478 (1908)
12. Emiris, I., Mourrain, B.: Matrices in elimination theory. J. Symb. Comput. 28(1-2), 3–43 (1999)
13. Galperin, H., Wigderson, A.: Succinct representations of graphs. Inform. Control 56(3), 183–198 (1984)
14. Garey, M.R., Johnson, D.S.: Computers and Intractability: A Guide to the Theory of NP-Completeness. Series of Books in the Mathematical Sciences. W.H. Freeman, New York (1979)
15. Grenet, B.: Difficulté du résultant et des grands déterminants. Technical report, RRLIP2009-32, LIP (2009), http://prunel.ccsd.cnrs.fr/ensl-00431714/
16. Grenet, B., Koiran, P., Portier, N.: The multivariate resultant is NP-hard in any characteristic. Technical report, RRLIP2009-34, LIP (2009), http://prunel.ccsd.cnrs.fr/ensl-00440842/
17. Ierardi, D.: Quantifier elimination in the theory of an algebraically-closed field. In: Proc. STOC 1989, pp. 138–147 (1989)
18. Jeronimo, G., Sabia, J.: Computing multihomogeneous resultants using straight-line programs. J. Symb. Comput. 42(1-2), 218–235 (2007)
19. Kaltofen, E., Koiran, P.: On the complexity of factoring bivariate supersparse (lacunary) polynomials. In: Proc. ISSAC 2005, pp. 208–215. ACM, New York (2005)
20. Kaltofen, E., Koiran, P.: Expressing a fraction of two determinants as a determinant. In: Proc. ISSAC 2008, pp. 141–146 (2008)

21. Kapur, D., Saxena, T.: Comparison of various multivariate resultant formulations. In: Proc. ISSAC 1995, pp. 187–194 (1995)
22. Kapur, D., Saxena, T., Yang, L.: Algebraic and geometric reasoning using Dixon resultants. In: Proc. ISSAC 1994, pp. 99–107 (1994)
23. Koiran, P.: Hilbert's Nullstellensatz is in the polynomial hierarchy. J. Complexity 12(4), 273–286 (1996)
24. Koiran, P.: The complexity of local dimensions for constructible sets. J. Complexity 16(1), 311–323 (2000)
25. Lazard, D.: Résolution des systémes d'èquations algébriques. Theor. Comput. Sci. 15(1), 77–110 (1981)
26. Macaulay, F.S.: Some formulae in elimination. Proc. Lond. Math. Soc. 1(1), 3 (1902)
27. Macaulay, F.S.: The Algebraic Theory of Modular Systems. Cambridge Mathematical Library, Cambridge University Press, Cambridge (1994)
28. Plaisted, D.A.: New NP-hard and NP-complete polynomial and integer divisibility problems. Theor. Comput. Sci. 31(1-2), 125–138 (1984)
29. Renegar, J.: On the worst-case arithmetic complexity of approximating zeros of systems of polynomials. SIAM J. Comput. 18, 350 (1989)
30. Seidenberg, A.: A new decision method for elementary algebra. Ann. Math. 60(2), 365–374 (1954)
31. Shoup, V.: New algorithms for finding irreducible polynomials over finite fields. Math. Comput. 54(189), 435–447 (1990)
32. Sturmfels, B.: Sparse elimination theory. In: Eisenbud, D., Robbiano, L. (eds.) Proc. Comput. Algebr. Geom. Commut. Algebra (1991)
33. van der Waerden, B.L.: Modern Algebra, 3rd edn. F. Ungar Publishing Co, New York (1950)
34. von zur Gathen, J., Karpinski, M., Shparlinski, I.: Counting curves and their projections. Comput. Complex 6(1), 64–99 (1996)

Parameterized Complexity and Kernelizability of Max Ones and Exact Ones Problems*

Stefan Kratsch[1], Dániel Marx[2], and Magnus Wahlström[1]

[1] Max-Planck-Institut für Informatik, 66123 Saarbrücken, Germany
{skratsch,wahl}@mpi-inf.mpg.de
[2] Tel Aviv University, Israel
dmarx@cs.bme.hu

Abstract. For a finite set Γ of Boolean relations, Max Ones SAT(Γ) and Exact Ones SAT(Γ) are generalized satisfiability problems where every constraint relation is from Γ, and the task is to find a satisfying assignment with at least/exactly k variables set to 1, respectively. We study the parameterized complexity of these problems, including the question whether they admit polynomial kernels. For Max Ones SAT(Γ), we give a classification into 5 different complexity levels: polynomial-time solvable, admits a polynomial kernel, fixed-parameter tractable, solvable in polynomial time for fixed k, and NP-hard already for $k = 1$. For Exact Ones SAT(Γ), we refine the classification obtained earlier by having a closer look at the fixed-parameter tractable cases and classifying the sets Γ for which Exact Ones SAT(Γ) admits a polynomial kernel.

1 Introduction

The constraint satisfaction problem (CSP) provides a framework in which it is possible to express, in a natural way, many combinatorial problems encountered in artificial intelligence and computer science. A CSP instance is represented by a set of variables, a domain of values for each variable, and a set of constraints on the values that certain collections of variables can simultaneously take. The basic aim is then to find an assignment of values to the variables that satisfies the constraints. Boolean CSP (when all variables have domain $\{0, 1\}$) generalizes satisfiability problems such as 2SAT and 3SAT by allowing that constraints are given by arbitrary relations, not necessarily by clauses.

As Boolean CSP problems are NP-hard in general, there have been intensive efforts at finding efficiently solvable special cases of the general problem. One well-studied type of special cases is obtained by restricting the allowed constraint relations to a fixed set Γ; we denote by SAT(Γ) the resulting problem. We expect that if the relations in Γ are simple, then SAT(Γ) is easy to solve. For example, if Γ contains only binary relations, then SAT(Γ) is polynomial-time solvable, as it can be expressed by 2SAT. On the other hand, if Γ contains all the ternary relations, then SAT(Γ) is more general than 3SAT, and hence it is NP-hard.

* The second author is supported by ERC Advanced Grant DMMCA and the Hungarian National Research Fund (OTKA grant 67651).

P. Hliněný and A. Kučera (Eds.): MFCS 2010, LNCS 6281, pp. 489–500, 2010.

A celebrated classical result of T.J. Schaefer [18] from 1978 characterizes the complexity of SAT(Γ) for *every* finite set Γ: it shows that if Γ has certain simple combinatorial properties, then SAT(Γ) is polynomial-time solvable, and if Γ does not have these properties, then SAT(Γ) is NP-hard. This result is surprising for two reasons. First, Ladner's Theorem [14] states that if P $\neq$ NP, then there are problems in NP that are neither in P nor NP-complete. Therefore, it is surprising that every SAT(Γ) problem is either in P or NP-complete, and no intermediate complexity appears for this family of problems. Second, it is surprising that the borderline between the P and NP-complete cases of SAT(Γ) can be conveniently characterized by simple combinatorial properties.

Schaefer's result has been generalized in various directions. Bulatov [3] generalized it from Boolean CSP to CSP over a 3-element domain and it is a major open question if it can be generalized to arbitrary finite domains (see [4,10]). Creignou et al. [6] classified the polynomial-time solvable cases of the problem Exact Ones SAT(Γ), where the task is to find a satisfying assignment such that exactly k variables have value 1, for some integer k given in the input. Natural optimization variants of SAT(Γ) were considered in [5,7,12] with the goal of classifying the approximability of the different problems. In Max SAT(Γ) we have to find an assignment maximizing the number of satisfied constraints, while in Min UnSAT(Γ) we have to find an assignment minimizing the number of unsatified constraints. Min Ones SAT(Γ) and Max Ones SAT(Γ) ask for a satisfying assignment minimizing and maximizing, respectively, the number of variables having value 1.

Parameterized complexity. Recently, there have been investigations of the hardness of CSP from the viewpoint of parameterized complexity [15,13]. This paradigm investigates hardness in finer detail than classical complexity, which focuses mostly on polynomial-time algorithms. A *parameterization* of a problem is assigning an integer k to each input instance. Consider, for example, two standard NP-complete problems Vertex Cover and Clique. Both have the natural parameter k: the size of the required vertex cover/clique. Both problems can be solved in time $n^{O(k)}$ on n-vertex graphs by complete enumeration. Notice that the degree of the polynomial grows with k, so the algorithm becomes useless for large graphs, even if k is as small as 10. However, Vertex Cover can be solved in time $O(2^k \cdot n^2)$ [11,9]. In other words, for every fixed cover size there is a polynomial-time (in this case, quadratic in the number of vertices) algorithm solving the problem where the degree of the polynomial is independent of the parameter. Problems with this property are called *fixed-parameter tractable.* The notion of W[1]-hardness in parameterized complexity is analogous to NP-completeness in classical complexity. Problems that are shown to be W[1]-hard, such as Clique [11,9], are very unlikely to be fixed-parameter tractable.

Kernelization. One of the most basic techniques for showing that a problem is fixed-parameter tractable is to show that the computationally hard "core" of the problem can be extracted in polynomial time. Formally, *kernelization* is a polynomial-time transformation that, given an instance I of problem P with parameter k, creates an equivalent instance I' of problem P with parameter $k' \leq f(k)$ such that the size of I' is at most $g(k)$ for some functions f, g

(usually, $k' \leq k$ is achievable). For example, a classical result of Nemhauser and Trotter [16] shows that every instance I of Vertex Cover with parameter k can be transformed into an instance I' with parameter $k' \leq k$ such that I' has at most $g(k) = 2k$ vertices. Observe that the existence of a kernelization algorithm for P immediately implies that P is FPT, assuming that P is decidable: performing the kernelization and then doing a brute force solution on I' clearly takes only $n^{O(1)} + f(k)$ time for some function f. From the practical point of view, *polynomial kernels,* i.e., kernelization algorithms where $g(k)$ is a polynomial, are of particular interest. If a problem has this property, then this means that there is an efficient preprocessing algorithm for the problem with a provable bound on the way it shrinks the instance. Such a preprocessing can be an invaluable opening step in any practical solution for the problem. Very recently, however, it has been shown that under standard complexity assumptions, not every FPT problem has a polynomial kernel: e.g., the k-Path problem can be solved in (randomized) time $2^k \cdot n^{O(1)}$ [19], but has no polynomial kernel unless NP $\subseteq$ co-NP/poly [1]. The negative toolkit developed in [1] has been successfully applied to a number of other problems [2,8].

Results. The parameterized complexity of Exact Ones SAT(Γ) was studied in [15], where it was shown that a property called weak separability characterizes the complexity of the problem: Exact Ones SAT(Γ) is FPT if Γ is weakly separable, and W[1]-complete otherwise. The problem Min Ones SAT(Γ) is FPT for every Γ by a simple branching algorithm, but it is not obvious to see for which Γ there is a polynomial kernel. This question has been resolved in [13] by showing that (unless NP $\subseteq$ co-NP/poly) Min Ones SAT(Γ) has a polynomial kernel if and only if Min Ones SAT(Γ) is in P or Γ has a property called mergebility.

We continue this line of research by considering the so far unexplored problem Max Ones SAT(Γ) and revisit Exact Ones SAT(Γ). We will characterize (under

Table 1. Examples of sets of relations Γ and the properties for Min Ones SAT(Γ), Exact Ones SAT(Γ), and Max Ones SAT(Γ). Problems marked PK have polynomial kernels; problems marked FPT are FPT but admit no polynomial kernelization unless NP $\subseteq$ co-NP/poly.

Γ	Min Ones	Exact Ones	Max Ones
width-2 affine	P	P	P
$\{\mathrm{ODD}_3\}$	PK	PK	P
$\{\mathrm{EVEN}_3\}$	P	FPT	PK
$\{\mathrm{EVEN}_3, (x)\}$	FPT	FPT	PK
$\{\mathrm{ODD}_4\}$, general affine	FPT	FPT	PK
$\{(x \vee y), (x \neq y)\}$	PK	PK	PK
$\{((x \rightarrow y) \wedge (y \neq z))\}$	PK	FPT	FPT
$\{(x \vee y), (x \neq y), (x \rightarrow y)\}$	PK	W[1]-complete	FPT
bijunctive	PK	W[1]-complete	W[1]-hard, XP
$\{R_{1\text{-IN-}3}\}$	PK	PK	not in XP
$\{\sum_i x_i = p \pmod q\}$	FPT	FPT	not in XP
general	FPT	W[1]	not in XP

standard complexity assumptions) parameterized Max Ones SAT(Γ) problems for finite constraint languages Γ as the following 5 types: solvable in polynomial time; NP-hard, but having polynomial kernelization; being FPT but admitting no polynomial kernelization; being W[1]-hard and in XP; and not being in XP. The characterization uses results of Nordh and Zanuttini [17] on frozen co-clones. For Exact Ones SAT(Γ), we refine the classification of [15] by precisely characterizing those weakly separable sets Γ for which Exact Ones SAT(Γ) is not only FPT, but admits a polynomial kernel. Table 1 shows some examples.

The kernelization lower bounds for both problems use reductions from a maximization problem Multiple Compatible Patterns, which is FPT but admits no polynomial kernelization unless NP $\subseteq$ co-NP/poly. This problem may be useful for other hardness reductions as well.

2 Preliminaries and Notation

Boolean CSP. A *formula* ϕ is a pair (V, C) consisting of a set V of *variables* and a set C of *constraints*. Each constraint $c_i \in C$ is a pair $\langle \overline{s}_i, R_i \rangle$, where $\overline{s}_i = (x_{i,1}, \ldots, x_{i,r_i})$ is an r_i-tuple of variables (the *constraint scope*) and $R_i \subseteq \{0,1\}^{r_i}$ is an r_i-ary Boolean relation (the *constraint relation*). A function $f : V \to \{0,1\}$ is a *satisfying assignment* of ϕ if $(f(x_{i,1}), \ldots, f(x_{i,r_i}))$ is in R_i for every $c_i \in C$. Let Γ be a set of Boolean relations. A formula is a *Γ-formula* if every constraint relation R_i is in Γ. In this paper, Γ is always a finite set containing only non-empty relations. For a fixed finite Γ, every Γ-formula $\phi = (V, C)$ can be represented with length polynomial in $|V|$ and $|C|$: each constraint relation can be represented by constant number of bits (depending only on Γ). The *weight* $w(f)$ of an assignment f is the number of variables x with $f(x) = 1$.

We also use some definitions from [17]. Let $\phi = (V, C)$ be a formula and $x \in V$ a variable. Then x is said to be *frozen* in ϕ if x takes the same value in every satisfying assignment of ϕ. Further, let Γ be a set of relations, and R an n-ary relation. Then Γ *freezingly implements* R if there is a formula ϕ over $\Gamma \cup \{=\}$ such that $R(x_1, \ldots, x_n) \equiv \exists X \phi$, where ϕ uses variables $X \cup \{x_1, \ldots, x_n\}$ only, and all variables in X are frozen in ϕ. If only relations of Γ are used, then we have a *frozen implementation without equality*. This will be our standard notion of implementation in the paper, and as such is shortened to simply "implements".

We recall some standard definitions concerning Boolean constraints (cf. [5]):

- R is *0-valid* if $(0, \ldots, 0) \in R$.
- R is *1-valid* if $(1, \ldots, 1) \in R$.
- R is *Horn* or *weakly negative* if it can be expressed as a conjunction of clauses such that each clause contains at most one positive literal. It is known that R is Horn if and only if it is *AND-closed:* if $(a_1, \ldots, a_r) \in R$ and $(b_1, \ldots, b_r) \in R$, then $((a_1 \wedge b_1), \ldots, (a_r \wedge b_r)) \in R$.
- R is *anti-Horn* or *weakly positive* if it can be expressed as the conjunction of clauses such that each clause contains at most one negated literal. It is known that R is anti-Horn if and only if it is *OR-closed:* if $(a_1, \ldots, a_r) \in R$ and $(b_1, \ldots, b_r) \in R$, then $((a_1 \vee b_1), \ldots, (a_r \vee b_r)) \in R$.

- R is *bijunctive* if it can be expressed as the conjunction of constraint such that each constraint is the disjunction of two literals.
- R is *affine* if it can be expressed as a conjunction of constraints of the form $x_1+x_2+\cdots+x_t = b$, where $b \in \{0,1\}$ and addition is modulo 2. The number of tuples in an affine relation is always an integer power of 2. We denote by EVEN_r the r-ary relation $x_1 + x_2 + \cdots + x_r = 0$ and by ODD_r the r-ary relation $x_1 + x_2 + \cdots + x_r = 1$.
- R is *width-2 affine* if it can be expressed as a conjunction of constraints of the form $x = y$, $x \neq y$, (x), and $(\neg x)$.
- R is *monotone* if $a \in R$ and $b \geq a$ implies $b \in R$, where $\geq$ is applied component-wise. Such a relation is implementable by positive clauses, by adding a clause over the false positions of every maximal false tuple.
- The relation $R_{p\text{-IN-}q}$ (for $1 \leq p \leq q$) has arity q and $R_{p\text{-IN-}q}(x_1, \ldots, x_q)$ is true if and only if exactly p of the variables $x_1, \ldots, x_q$ have value 1.

The above is extended to properties of sets of relations, by saying that a set of relations Γ is 0-valid (1-valid, Horn, ...) if this holds for every $R \in \Gamma$.

Theorem 1 (Schaefer [18]). *Let Γ be a set of Boolean relations. Then $SAT(\Gamma)$ is in* P *if Γ has one of the following properties: 0-valid, 1-valid, Horn, anti-Horn, bijunctive, or affine. Otherwise, $SAT(\Gamma)$ is* NP*-complete.*

Max Ones SAT(Γ) and Exact Ones SAT(Γ). For a fixed set of relations Γ, Max Ones SAT(Γ) is the following problem:

> **Input:** A formula ϕ over Γ; an integer k.
> **Parameter:** k.
> **Task:** Decide whether there is a satisfying assignment for ϕ of weight at least k.

For example, Max Ones SAT($\neg x \vee \neg y$) is equivalent to Independent Set, and is thus W[1]-complete. Further examples can be found in Table 1. Similarly, Exact Ones SAT(Γ), for a fixed set of relations Γ, is the following problem.

> **Input:** A formula ϕ over Γ; an integer k.
> **Parameter:** k.
> **Task:** Decide whether there is a satisfying assignment for ϕ of weight exactly k.

Parameterized complexity and kernelization. A parameterized problem $\mathcal{Q}$ is a subset of $\Sigma^* \times \mathbb{N}$; the second component is called the *parameter*. The problem $\mathcal{Q}$ is *fixed-parameter tractable* (FPT) if there is an algorithm that decides $(I,k) \in \mathcal{Q}$ in time $f(k) \cdot n^{O(1)}$, where f is some computable function. A *kernelization* is a polynomial-time mapping $K : (I,k) \mapsto (I',k')$ such that (I,k) and (I',k') are equivalent, $k' \leq f(k)$, and $|I'| \leq g(k)$, for some functions f and g. Usually, f can be taken as the identity function, i.e., $k' \leq k$; this will be the case throughout this paper. If $|I'|$ is bounded by a polynomial in k, then K is a *polynomial kernelization*. It is well-known that every decidable parameterized problem is fixed-parameter tractable if and only if it has a (not necessarily polynomial) kernelization [11]. A *polynomial time and parameter reduction* from $\mathcal{Q}$ to $\mathcal{Q}'$ is a polynomial-time mapping $\Phi : (I, k) \mapsto (I', k')$ such that $(I, k) \in \mathcal{Q}$ if and only if $(I', k') \in \mathcal{Q}$ and such that k' is polynomially bounded in k; we denote the existence of such a reduction by $\mathcal{Q} \leq_{Ptp} \mathcal{Q}'$. These reductions were introduced by Bodlaender et al. [2], who also showed that they preserve polynomial kernelizability.

The MCP problem. Our kernelization lower bounds will use the problem Multiple Compatible Patterns (MCP), defined as follows:

> **Input:** A set of patterns from $\{0, 1, \bigstar\}^r$, where $\bigstar$ (the wildcard character) matches 0 or 1; an integer k.
> **Parameter:** $r + k$.
> **Task:** Decide whether there is a string in $\{0, 1\}^r$ that matches at least k patterns.

A kernelization lower bound for MCP follows from the methods of [1]. Briefly, we get NP-completeness by a reduction from Clique, and compositionality by adding $\log t$ bits to compose t instances.

Lemma 2. Multiple Compatible Patterns *(MCP) is* FPT, NP*-complete, and admits no polynomial kernelization unless* NP $\subseteq$ co-NP/poly.

3 Max Ones Characterization

This section contains the our characterization of the parameterized complexity properties of Max Ones SAT(Γ) problems.

As a very first distinction, observe that if SAT(Γ) is NP-complete, then Max Ones SAT(Γ) is NP-complete even for a parameter $k = 0$. Thus, we know by Schaefer (Theorem 1) that Γ has to fall in one of the classes 0-valid, 1-valid, affine, Horn, anti-Horn, or bijunctive for the problem to be in XP. Of these, the classes of 1-valid relations and anti-Horn relations are polynomial-time solvable, leaving four classes to examine. The cases of affine, Horn, and 0-valid relations can be characterized without too much difficulty, and will be treated summarily, as we will focus on the more interesting cases that occur when Γ is bijunctive.

We begin with the polynomial cases, as proven by Khanna et al. [12].

Theorem 3 ([12]). Max Ones SAT(Γ) *is in* P *if* Γ *is 1-valid, weakly positive (i.e. anti-Horn), or width-2 affine, and APX-hard in all other cases.*

The following lemma covers the properties of every set of relations Γ except the bijunctive cases; full proofs will be found in the full version.

Lemma 4. *Let Γ be a set of relations; the following hold.*

1. *If Γ is affine, then* Max Ones SAT(Γ) *has a kernel with $O(k)$ variables.*
2. *If Γ is Horn, but not anti-Horn and not 1-valid, then* Max Ones SAT(Γ) *is W[1]-hard.*

3. *If Γ is 0-valid, but neither anti-Horn, 1-valid, affine, nor Horn, then* Max Ones SAT(Γ) *is NP-hard for $k = 1$.*

Proof (sketches). 1. We can check in polynomial time which variables have to be set to false in every solution, and remove these. For the rest, we can by a greedy procedure find a solution which sets at least half the remaining variables to be true. Thus, we either find a solution with weight at least k, or leave a kernel with at most $2k$ variables.

2. If Γ is Horn, and the listed cases do not apply, then Γ admits a reduction from Independent Set by implementing $(\neg x \vee \neg y)$; either directly, or, e.g., via relations $(x \wedge y \rightarrow z)$ and $(\neg z)$.

3. Let Γ be 0-valid such that no listed case applies. It can be shown that Γ implements $R(x, y, z) = \{(0, 0, 0), (1, 1, 0), (1, 0, 1)\}$; we will show that Max Ones SAT(R) is NP-hard for $k = 1$. By a trick of splitting variables, we can adjust a given formula to add a universal variable z_1 such that $z_1 = 1$ in any solution where at least one variable is true. Relations $R(z_1, x, y)$ then become $(x \neq y)$ in any such solution, effectively constructing a reduction from SAT($R, (x \neq y)$). By Theorem 1, this problem is NP-complete, and the claim follows. □

3.1 Bijunctive Cases

In this subsection we treat the cases of Max Ones SAT(Γ) where Γ is bijunctive but not Horn, anti-Horn, or width-2 affine (or 0-valid, or 1-valid, but this follows implicitly). This corresponds to the sets Γ which, using existentially quantified variables, can implement all 2SAT clauses; see [17]. See also Table 1 for a summary of the maximal cases.

For the result, we will need the results of Nordh and Zanuttini [17]. Recall the definition of a frozen implementation (with equality). The *frozen partial co-clone* $\langle \Gamma \rangle_{fr}$ generated by Γ is the set of all relations that can be freezingly implemented by Γ. We will use the characterization of [17] of the frozen partial co-clones that our Γ can generate. The free use of equality constraints is somewhat more general than what we wish to allow, but we will find that it causes no problems.

We need the following special cases.

1. $\Gamma_2^{p\neq} = \{(x \vee y), (x \neq y)\}$
2. $R_3^n = (\neg x \vee \neg y) \wedge (x \neq z)$; $\Gamma_3^n = \{R_3^n\}$
3. $\Gamma_2^{p\neq i} = \{(x \vee y), (x \neq y), (x \rightarrow y)\}$

Finally, we need a technical lemma to show that we can assume that we have access to the constants. We refer to the full version for a proof.

Lemma 5. *If Γ is neither 0-valid, 1-valid, nor affine, but SAT(Γ) is not NP-hard, then the constants can be implemented.*

Let us now proceed with settling the remaining cases of Max Ones SAT(Γ).

Lemma 6. *Assume that Γ is bijunctive but not Horn or anti-Horn. Then the following hold.*

1. *If $\Gamma \subseteq \langle \Gamma_2^{p\neq} \rangle_{fr}$, where $\Gamma_2^{p\neq} = \{(x \vee y), (x \neq y)\}$, then* Max Ones SAT($\Gamma$) *has a polynomial kernel (of $O(k^2)$ variables). Otherwise,* Max Ones SAT(Γ) *admits no polynomial kernelization, unless* NP $\subseteq$ co-NP/poly.
2. *If $\Gamma \subseteq \langle \Gamma_2^{p\neq i} \rangle_{fr}$, where $\Gamma_2^{p\neq i} = \{(x \vee y), (x \neq y), (x \to y)\}$, then* Max Ones SAT($\Gamma$) *is FPT (with running time $O^*(2^k)$). Otherwise,* Max Ones SAT(Γ) *is W[1]-hard.*

Proof. Let (ϕ, k) be a Max Ones SAT(Γ) instance. Assume throughout that the instance is feasible (as otherwise, the problem is trivial). We split the proof into proofs of feasibility (1a, 2a), and lower bound proofs (1b, 2b).

1a. By [17], every relation in Γ, and thus all of ϕ, has a frozen implementation over $\Gamma_2^{p\neq} \cup \{=\}$. We will refer to this implementation when inferring a kernel, but the kernelization will apply for the original Γ as well. Let a set of at least two variables which are connected by disequality or equality, with at least one disequality, be referred to as a *class* of variables. If there are at least k variable classes, then any solution will contain at least k true variables, and can be found in polynomial time. If any class contains at least $2k$ variables, then either the variables of this class have fixed values, in which case we make the corresponding assignments, or we can find a solution with at least k true variables. Finally, if any variable does not occur in a variable class, it can safely be set to 1. These observations leave a kernel with $O(k)$ variable classes and $O(k^2)$ variables in total. Finally, as the only changes we made to the formula were assignments, we can apply the kernelization using only relations in Γ by replacing all assigned variables by the constant variables z_1 or z_0.

1b. By [17], there is an implementation of R_3^n over $\Gamma \cup \{=\}$. As the equality constraint will not be useful in such an implementation, there is also an implementation directly over Γ, showing Max Ones SAT(Γ_3^n) $\leq_{Ptp}$ Max Ones SAT(Γ); we will in turn show that MCP $\leq_{Ptp}$ Max Ones SAT(Γ_3^n) (the problem MCP was defined in Section 2).

Observe that, renaming variables, R_3^n can be written as $(x \neq y) \wedge (z \to x)$. Let (I, k) be an instance of MCP, with string length r. Create variables $(x_i \neq y_i)$ for $1 \leq i \leq r$, coding the entries of the string; these variables contribute weight exactly r to any solution. Now for every pattern i, create a variable z_i, and for every position j of pattern i containing 0, add a constraint $(x_j \neq y_j) \wedge (z_i \to x_j)$. For positions containing 1, create the same constraint with an implication instead to y_j. Any solution with $r + k$ true variables corresponds one-to-one to a string in $\{0, 1\}^r$ and k patterns matching it. Thus (by [2]), Max Ones SAT(Γ) admits no polynomial kernelization unless NP $\subseteq$ co-NP/poly.

2a. As before, there is an implementation of ϕ over $\Gamma_2^{p\neq i} \cup \{=\}$. Again consider the variable classes; if they number at least k, then find a solution in polynomial time. Otherwise, we check all $O(2^k)$ assignments to variables of the variable classes. For each such assignment, propagate assignments to the remaining variables. Any formula that remains after this is 1-valid.

2b. By [17], there is an implementation of $(\neg x \vee \neg y)$ over $\Gamma \cup \{=\}$, and again the equality constraint would not be useful. Thus there is an FPT reduction from Independent Set to Max Ones SAT(Γ).

Our results for Max Ones SAT(Γ) summarize into the following picture.

Theorem 7. *Let Γ be a finite set of Boolean relations. Then* Max Ones SAT(Γ) *falls into one of the following cases.*

1. *If Γ is 1-valid, anti-Horn, or width-2 affine, then* Max Ones SAT(Γ) *is in* P*; in the remaining cases, it is* NP*-complete.*
2. *If Γ is affine, or if $\Gamma \subseteq \langle (x \vee y), (x \neq y) \rangle_{fr}$, then* Max Ones SAT($\Gamma$) *has a polynomial kernel.*
3. *If $\Gamma \subseteq \langle (x \vee y), (x \neq y), (x \rightarrow y) \rangle_{fr}$, then* Max Ones SAT($\Gamma$) *is in FPT, with a running time of $O^*(2^k)$, but if the previous case does not apply, then there is no polynomial kernelization unless* NP $\subseteq$ co-NP/poly.
4. *If none of these cases applies, then* Max Ones SAT(Γ) *is W[1]-hard; if Γ is Horn or bijunctive, then* Max Ones SAT(Γ) *is in XP.*
5. *Otherwise* Max Ones SAT(Γ) *is* NP*-complete for $k = 1$.*

4 Exact Ones CSP

In this section we classify Exact Ones SAT(Γ) into admitting or not admitting a polynomial kernelization depending on the set of allowed relations Γ. We start from the characterization of its fixed-parameter tractability [15] as well as the characterization of when Min Ones SAT(Γ) admits a polynomial kernelization [13]. To this end we recall the invariants called weak separability and mergeability used for the respective characterization. We also introduce a joined, stronger version of the two partial polymorphisms defining weak separability; this will be used to characterize kernelizability of Exact Ones SAT(Γ).

Definition 8. *A t-ary* partial polymorphism *is a partially defined function $f : \{0,1\}^t \rightarrow \{0,1\}$. For an r-ary relation R, we say that R is* invariant under f *if for any t tuples $\alpha_1, \dots, \alpha_t \in R$, such that $f(\alpha_1(i), \dots, \alpha_t(i))$ is defined for every $i \in [r]$, we have $(f(\alpha_1(1), \dots, \alpha_t(1)), \dots, f(\alpha_1(r), \dots, \alpha_t(r))) \in R$.*

We present partial polymorphisms in a matrix form, where the columns represent the tuples for which f is defined, and the value below the horizontal line is the corresponding value of f.

Definition 9 ([15,13]). *Let FPT(1), FPT(2), and FPT($1 \bowtie 2$) denote the following partial polymorphisms:*

FPT(1)			
0	0	0	1
0	0	1	1
0	1	0	1
0	1	1	1

FPT(2)			
0	1	0	1
0	1	1	1
0	0	0	1
0	0	1	1

FPT($1 \bowtie 2$)				
0	1	0	0	1
0	1	0	1	1
0	0	1	0	1
0	0	1	1	1

A boolean relation R is weakly separable *if it is invariant under* FPT(1) and FPT(2). *It is* semi-separable *if it is invariant under* FPT($1 \bowtie 2$). *Finally, a relation is* mergeable *if it is invariant under the following partial polymorphism:*

Mergeable
0 1 0 1 1 0 1
0 1 0 0 0 0 1
0 0 1 1 0 1 1
0 0 1 0 0 0 1
0 1 0 1 0 0 1

Theorem 10 ([15]). Exact Ones SAT(Γ) *is fixed-parameter tractable if every relation* $R \in \Gamma$ *is weakly separable. In the remaining cases it is* W[1]-complete.

Since any kernelization for a problem also implies fixed parameter tractability, we will only need to further classify the fixed-parameter tractable cases.

By a simple observation, Min Ones SAT(Γ) reduces to Exact Ones SAT(Γ) by a polynomial time and parameter reduction. This allows us to transfer lower bounds from the min ones to the exact ones setting.

Lemma 11. Min Ones SAT(Γ) *reduces to* Exact Ones SAT(Γ) *by a polynomial time and parameter reduction.*

Thus, using the kernelization dichotomy for Min Ones SAT(Γ) [13], we may exclude further cases.

Theorem 12 ([13]). *Unless* NP $\subseteq$ co-NP/poly, Min Ones SAT(Γ) *admits a polynomial kernel if and only if every relation in* Γ *is mergeable or* Min Ones SAT(Γ) *is in* P.

Corollary 13. *If* Γ *is not mergeable and* Min Ones SAT(Γ) *is* NP*-hard then* Exact Ones SAT(Γ) *does not admit a polynomial kernel unless the polynomial hierarchy collapses.*

According to Khanna et al. [12] Min Ones SAT(Γ) is in P when Γ is 0-valid, weakly negative, or width-2 affine; in all other cases it is NP-hard (APX-hard).

Theorem 14. *Let* Γ *be a finite set of weakly separable relations.*

1. *If* Γ *is width-2 affine then* Exact Ones SAT(Γ) *is in* P*; this includes the cases where* Γ *is Horn, or both 0-valid and mergeable. In the remaining cases, the problem is* NP*-complete.*
2. *If* Γ *is anti-Horn, or both mergeable and semi-separable, then* Exact Ones SAT(Γ) *admits a polynomial kernelization.*
3. *In all other cases* Exact Ones SAT(Γ) *does not admit a polynomial kernelization unless* NP $\subseteq$ co-NP/poly.

We only give an outline of the proof; the full proof will be given in the full version.

Proof (outline). We first consider the cases when Min Ones SAT(Γ) is in P, i.e., when Γ is zero-valid, Horn, or width-2 affine [12]. In all other cases, due to Corollary 13, we may then use that Γ is mergeable (since otherwise Exact Ones SAT(Γ) does not admit a polynomial kernel).

If Γ is width-2 affine, then by Creignou et al. [6], Exact Ones SAT(Γ) is in P; otherwise it is NP-complete. *If Γ is Horn* we show that it can be implemented by $\{=, (x), (\neg x)\}$ and Exact Ones SAT(Γ) is in P. The same is true *if Γ is zero-valid and mergeable.*

If Γ is zero-valid but not mergeable then (unless Γ is Horn) we are able to reduce Exact Ones SAT(Γ') to Exact Ones SAT(Γ) where $\Gamma' = \Gamma \cup \{(x), (\neg x)\}$ by a polynomial time and parameter reduction. Since Γ' is neither zero-valid, Horn, nor width-2 affine we conclude that Min Ones SAT(Γ') is NP-hard. This implies that Exact Ones SAT(Γ') does not admit a polynomial kernelization by Corollary 13, which extends also to Exact Ones SAT(Γ) through our reduction.

For all further choices of Γ (i.e., neither zero-valid, Horn, nor width-2 affine) we have that Min Ones SAT(Γ) and Exact Ones SAT(Γ) are NP-hard. Therefore, by Corollary 13, we assume that Γ is mergeable.

If Γ is anti-Horn (and weakly separable) we show that it can be implemented by equality, negative assignments, and positive clauses. This also means that Γ is semi-separable and mergeable. Now one of two cases applies. If Γ is monotone, then Exact Ones SAT(Γ) reduces to d-Hitting Set and we are done. Otherwise, Exact Ones SAT($\Gamma \cup \{(x), (\neg x)\}$) reduces to Exact Ones SAT(Γ), implying that we have constants available. We will later show that for any semi-separable and mergeable Γ that contains (x) and $(\neg x)$ Exact Ones SAT(Γ) admits a polynomial kernel.

Otherwise, in particular, if Γ is not Horn or anti-Horn, we show that Exact Ones SAT($\Gamma \cup \{\neq, (x), (\neg x)\}$) reduces to Exact Ones SAT(Γ) by a polynomial time and parameter reduction; i.e., as above we may assume to have disequality and constants available in Γ. Then if Γ is not semi-separable, we show that Exact Ones SAT(Γ) does not admit a polynomial kernel by a polynomial time and parameter reduction from the MCP problem: The central fact is that we must have a witness against semi-separability (i.e., invariant under FPT($1 \bowtie 2$)), but all relations in Γ are weakly separable (i.e., invariant under FPT(1) and FPT(2)). Using disequality this witness permits us to implement $(x \to y) \wedge (y \neq z)$; we then use the reduction from MCP as in Lemma 6.

To conclude our proof it now suffices to give a polynomial kernelization for the case that Γ is mergeable, semi-separable, and contains positive and negative assignments. To this end we use a sunflower lemma for tuples to repeatedly find and simplify sunflowers while there are too many non-zero-valid constraints. The crucial part is that semi-separability allows us to essentially split constraints that form a sunflower into a core constraint and independent petal constraints: The core assignment and the petal assignment are independent for all feasible assignments to the core variables. Mergeability of Γ restricts zero-valid constraints to be implementable by equality and assignments, which can be handled in a straightforward way. □

Corollary 15. *Let Γ be a finite set of relations. Then* Exact Ones SAT(Γ) *is FPT if and only if Γ is weakly separable, unless FPT = W[1]; and admits a polynomial kernel if and only if Γ is semi-separable and mergeable, unless* NP $\subseteq$ co-NP/poly.

References

1. Bodlaender, H.L., Downey, R.G., Fellows, M.R., Hermelin, D.: On problems without polynomial kernels (extended abstract). In: Aceto, L., Damgård, I., Goldberg, L.A., Halldórsson, M.M., Ingólfsdóttir, A., Walukiewicz, I. (eds.) ICALP 2008, Part I. LNCS, vol. 5125, pp. 563–574. Springer, Heidelberg (2008)
2. Bodlaender, H.L., Thomassé, S., Yeo, A.: Kernel bounds for disjoint cycles and disjoint paths. In: Fiat, A., Sanders, P. (eds.) ESA 2009. LNCS, vol. 5757, pp. 635–646. Springer, Heidelberg (2009)
3. Bulatov, A.A.: A dichotomy theorem for constraints on a three-element set. In: Proc. 43th Symp. Foundations of Computer Science, pp. 649–658. IEEE, Los Alamitos (2002)
4. Bulatov, A.A.: Tractable conservative constraint satisfaction problems. In: LICS, p. 321. IEEE, Los Alamitos (2003)
5. Creignou, N., Khanna, S., Sudan, M.: Complexity Classifications of Boolean Constraint Satisfaction Problems. In: SIAM Monographs on Discrete Mathematics and Applications, vol. 7 (2001)
6. Creignou, N., Schnoor, H., Schnoor, I.: Non-uniform boolean constraint satisfaction problems with cardinality constraint. In: Kaminski, M., Martini, S. (eds.) CSL 2008. LNCS, vol. 5213, pp. 109–123. Springer, Heidelberg (2008)
7. Crescenzi, P., Rossi, G.: On the Hamming distance of constraint satisfaction problems. Theoretical Computer Science 288(1), 85–100 (2002)
8. Dom, M., Lokshtanov, D., Saurabh, S.: Incompressibility through colors and ids. In: Albers, S., Marchetti-Spaccamela, A., Matias, Y., Nikoletseas, S., Thomas, W. (eds.) ICALP 2009. LNCS, vol. 5555, pp. 378–389. Springer, Heidelberg (2009)
9. Downey, R.G., Fellows, M.R.: Parameterized Complexity. In: Monographs in Computer Science. Springer, New York (1999)
10. Feder, T., Vardi, M.Y.: The computational structure of monotone monadic SNP and constraint satisfaction: a study through Datalog and group theory. SIAM J. Comput. 28(1), 57–104 (1999)
11. Flum, J., Grohe, M.: Parameterized Complexity Theory. Springer, Heidelberg (2006)
12. Khanna, S., Sudan, M., Trevisan, L., Williamson, D.P.: The approximability of constraint satisfaction problems. SIAM J. Comput. 30(6), 1863–1920 (2000)
13. Kratsch, S., Wahlström, M.: Preprocessing of min ones problems: A dichotomy. In: ICALP (2010) (to appear)
14. Ladner, R.E.: On the structure of polynomial time reducibility. J. Assoc. Comput. Mach. 22, 155–171 (1975)
15. Marx, D.: Parameterized complexity of constraint satisfaction problems. Computational Complexity 14(2), 153–183 (2005)
16. Nemhauser, G.L., Trotter Jr., L.E.: Vertex packings: structural properties and algorithms. Math. Programming 8, 232–248 (1975)
17. Nordh, G., Zanuttini, B.: Frozen boolean partial co-clones. In: ISMVL, pp. 120–125 (2009)
18. Schaefer, T.J.: The complexity of satisfiability problems. In: STOC, pp. 216–226. ACM, New York (1978)
19. Williams, R.: Finding paths of length k in $O^*(2^k)$ time. Inf. Process. Lett. 109(6), 315–318 (2009)

Meta-Envy-Free Cake-Cutting Protocols

Yoshifumi Manabe[1] and Tatsuaki Okamoto[2]

[1] NTT Communication Science Laboratories, 3-1 Morinosato-Wakamitya, Atsugi, Kanagawa 243-0198 Japan
manabe.yoshifumi@lab.ntt.co.jp
[2] NTT Information Sharing Platform Laboratories, 3-9-11, Midori-cho, Musashino-shi, 180-8585 Japan
okamoto.tatsuaki@lab.ntt.co.jp

Abstract. This paper discusses cake-cutting protocols when the cake is a heterogeneous good that is represented by an interval in the real line. We propose a new desirable property, the meta-envy-freeness of cake-cutting, which has not been formally considered before. Though envy-freeness was considered to be one of the most important desirable properties, envy-freeness does not prevent envy about role assignment in the protocols. We define meta-envy-freeness that formalizes this kind of envy. We show that current envy-free cake-cutting protocols do not satisfy meta-envy-freeness. Formerly proposed properties such as strong envy-free, exact, and equitable do not directly consider this type of envy and these properties are very difficult to realize. This paper then shows meta-envy-free cake-cutting protocols for two and three party cases.

1 Introduction

Cake-cutting is an old problem in game theory. It can be employed for such purposes as dividing territory on a conquered island or assigning jobs to members of a group. This paper discusses the cake-cutting problem when the cake is a heterogeneous good that is represented by an interval $[0, 1]$ in the real line. The most famous cake-cutting protocol is 'divide-and-choose' for two players. Player 1 (Divider) cuts the cake into two equal size pieces. Player 2 (Chooser) takes the piece that she prefers. Divider takes the remaining piece. This protocol is proved to be envy-free. Envy-freeness is defined as: after the assignment is finished, no player wants to exchange his/her part for that of the other player. Divider must cut the cake into two equal size pieces (using Divider's utility function), otherwise Chooser might take the larger piece and Divider will obtain less than half. Since Divider cuts the cake into equal size pieces, she never envies Chooser whichever piece Chooser selects. Chooser never envies Divider because she chooses first.

Although it appears that the 'divide-and-choose' protocol is perfect, actually it is not, because it is not a complete protocol. When Alice and Bob execute this protocol, they must first decide who will be Divider and Chooser. Chooser is the better choice as mentioned in several papers [3][9]. If the utility functions of Alice and Bob are the same, Divider and Chooser obtain exactly half of the cake by using their utility function. Next we consider a case where the utility functions of

P. Hliněný and A. Kučera (Eds.): MFCS 2010, LNCS 6281, pp. 501–512, 2010.

Alice and Bob differ. Let us assume that Bob is Divider. Let us also assume that by using Bob's utility function, $[0, 1/4]$ and $[1/4, 1]$ is an exact division, because the cake is chocolate coated near 0 and Bob likes chocolate. Alice does not have such a preference, thus by choosing $[1/4, 1]$, Alice's utility is $3/4$. If Alice is Divider, she cuts to $[0, 1/2]$ and $[1/2, 1]$. Then Bob chooses $[0, 1/2]$ and obtains more than half by his utility. Therefore, Chooser is never worse than Divider, and Chooser is properly better than Divider if their utility functions differ. If both Alice and Bob know this fact, they both want to be Chooser. Therefore, they must employ a method such as coin-flipping to decide who will be Divider. If Alice is assigned the role of Divider, she definitely *envies* Bob who is Chooser.

Some readers might think that coin-flipping will result in a fair decision between Alice and Bob, and so it is not a problem. If this supposition is accepted, the following protocol must also be accepted: 'Flip a coin and the winner takes the whole cake and the loser gets nothing.' This is an unfair (envy) assignment using fair coin-flipping. Game-theory researchers have discussed cake-cutting protocols where the unfairness (envy) is minimized. If there is the possibility of unfair assignment, we need to consider a better way that eliminates it. Now that we know 'divide-and-choose' is unfair, we must consider eliminating this kind of envy. Although this type of envy is known, it has not been formally defined. This paper defines this type of envy for the first time as meta-envy and proposes new protocols that eliminate it for two-party case and three-party case.

Previous studies defined stronger properties for the obtained portion such as strong envy-free, super envy-free, exact, and equitable [6][13]. These properties are hard to realize and do not directly consider this type of envy. We can obtain a three-party meta-envy-free protocol by modifying a three player envy-free protocol.

Note that we do not eliminate every coin-flip. For the above example of 'divide-and-choose', if Alice and Bob's utility functions are exactly the same, their cutting points are the same. Thus, both Alice and Bob think that the values of the two pieces are the same. To complete the protocol, we must assign each party either piece. Coin-flipping is necessary for such a case, but can only be allowed if its result causes no envy.

2 Preliminaries

Throughout this paper, the cake is a heterogeneous good that is represented by an interval $[0, 1]$ in the real line. Each party P_i has a utility function μ_i that has the following three properties. (1) For any non-empty $X \subseteq [0, 1]$, $\mu_i(X) > 0$. (2) For any X_1 and X_2 such that $X_1 \cap X_2 = \emptyset$, $\mu_i(X_1 \cup X_2) = \mu_i(X_1) + \mu_i(X_2)$. (3) $\mu_i([0, 1]) = 1$. The tuple of $P_i (i = 1, \dots, n)$'s utility function is denoted by $(\mu_1, \dots, \mu_n)$. Utility functions might differ among parties. No party has knowledge of the other parties' utility functions.

In this paper, 'party' indicates a person such as Alice, Bob, etc. and is denoted by P. 'Player' is a role in a protocol and is denoted by p. We sometimes state that 'party X is assigned to player y' if a person X executes the role of player y in the protocol.

An n-player cake-cutting protocol f assigns several portions of $[0,1]$ to the players such that every portion of $[0,1]$ is assigned to one player. We denote $f_i(\mu_1,\ldots,\mu_n)$ as the set of portions assigned to player p_i by f, when party $P_i(i=1,\ldots,n)$ is assigned to player $p_i(i=1,\ldots,n)$ in f. When f is a randomized algorithm, let us denote $f_i(\mu_1,\ldots,\mu_n;r)$ as the assignment to p_i when the sequence of random values used in f is r.

All parties are risk averse, namely they avoid gambling. They try to maximize the worst case utility they can obtain.

A desirable property for cake-cutting protocols is strategy-proofness [6]. A protocol is strategy-proof if there is no incentive for any player to lie about his utility function. A protocol defines what to do for each player p_i according to its utility function μ_i. Since μ_i is unknown to any other player, p_i can execute some action that differs from the protocol's definition (by pretending that p_i's utility function is $\mu'_i(\neq \mu_i)$). If p_i obtains more utility by lying about his utility function, the protocol is not strategy-proof. If a protocol is not strategy-proof, each player has to consider what to do and the result might differ from the intended result. If a protocol is strategy-proof, the best policy for each player is simply observing the rule of the protocol. Thus strategy-proofness is very important. As for 'divide-and-choose', the protocol requires Divider to cut the cake in half by using Divider's true utility function. Divider can cut the cake other than in half. However, if Divider does so, Chooser might take the larger portion and Divider might obtain less than half. Thus a risk averse party honestly executes the protocol, and 'divide-and-choose' is strategy-proof.

3 Meta-Envy-Freeness

This section provides the definition of meta-envy-freeness. We offer two definitions and show that they are equivalent.

Definition 1. *A cake-cutting protocol f is meta-envy-free if for any $(\mu_1,\ldots,\mu_n)$, i,j, and r,*

$$\mu_i(f_i(\mu_1,\ldots,\mu_i,\ldots,\mu_j,\ldots,\mu_n;r)) \geq \mu_i(f_j(\mu_1,\ldots,\mu_j,\ldots,\mu_i,\ldots,\mu_n;r)) \quad (1)$$

This definition considers the following two executions of f. (1) party P_i (whose utility function is μ_i) plays the role of player p_i and party P_j (whose utility function is μ_j) plays the role of player p_j in f. (2) party P_i plays the role of player p_j and party P_j plays the role of player p_i in f, that is, P_i and P_j swap role assignments. If the swap does not increase the utility of the obtained portions, P_i will not want to swap the role assignment, thus the protocol is envy-free as regards the role assignment.

Next we show a stronger definition.

Definition 2. *A cake-cutting protocol f is meta-envy-free if for any $(\mu_1,\ldots,\mu_n)$, permutation $\pi:\{1,\ldots,n\}\rightarrow\{1\ldots,n\}$, i, and r,*

$$\mu_i(f_i(\mu_1,\ldots,\mu_n;r)) = \mu_i(f_{\pi^{-1}(i)}(\mu_{\pi(1)},\ldots,\mu_{\pi(n)};r)) \quad (2)$$

This definition allows any permutation of the role assignment, which includes the case where P_i's role is unchanged. In addition, the utility must be unchanged for any permutation.

Theorem 1. *Definition 1 and Definition 2 are equivalent.*

Proof. If the condition of Definition 2 is satisfied, the condition of Definition 1 is obviously satisfied. Thus we prove the opposite direction.

Suppose that f satisfies the condition of Definition 1 and for some $(\mu_1,\ldots,\mu_i,\ldots,\mu_j,\ldots,\mu_n)$, i,j, and r,

$$\mu_i(f_i(\mu_1,\ldots,\mu_i,\ldots,\mu_j,\ldots,\mu_n;r)) > \mu_i(f_j(\mu_1,\ldots,\mu_j,\ldots,\mu_i,\ldots,\mu_n;r)) \quad (3)$$

is satisfied. Then consider another execution of f with $(\mu_1,\ldots,\mu_j,\ldots,,\mu_i,\ldots,\mu_n)$, that is, P_i's utility function is μ_j and P_j's utility function is μ_i. Since the condition of Definition 1 is satisfied, swapping the roles of P_i and P_j does not increase P_j's utility, that is,

$$\mu_i(f_j(\mu_1,\ldots,\mu_j,\ldots,\mu_i,\ldots,\mu_n;r)) \geq \mu_i(f_i(\mu_1,\ldots,\mu_i,\ldots,\mu_j,\ldots,\mu_n;r)) \quad (4)$$

This contradicts Eq. (3). Thus, for any $(\mu_1,\ldots,\mu_i,\ldots,\mu_j,\ldots,\mu_n)$, i,j, and r,

$$\mu_i(f_i(\mu_1,\ldots,\mu_i,\ldots,\mu_j,\ldots,\mu_n;r)) = \mu_i(f_j(\mu_1,\ldots,\mu_j,\ldots,\mu_i,\ldots,\mu_n;r)) \quad (5)$$

is satisfied.

Next we consider a general permutation of the role assignment. Any permutation π can be realized by a sequence in which two elements are swapped. As shown above, P_i's utility is unchanged when the swap involves P_i, thus we discuss P_i's utility when there is a swap between the other parties. Consider two utilities $\mu_i(f_i(\ldots,\mu_i,\ldots,\mu_j,\ldots,\mu_k,\ldots;r))$ and $\mu_i(f_i(\ldots,\mu_i,\ldots,\mu_k,\ldots,\mu_j,\ldots;r))$.

The roles of P_j and P_k can be swapped by the sequence of (S1) swapping P_i and P_j, (S2) swapping P_i (current role is p_j) and P_k, and (S3) swapping P_i (current role is p_k) and P_j (current role is p_i).

For each swap, Eq. (5) must be satisfied. From these equalities, we obtain

$$\mu_i(f_i(\ldots,\mu_i,\ldots,\mu_j,\ldots,\mu_k,\ldots;r)) = \mu_i(f_j(\ldots,\mu_j,\ldots,\mu_i,\ldots,\mu_k,\ldots;r))$$
$$\mu_i(f_j(\ldots,\mu_j,\ldots,\mu_i,\ldots,\mu_k,\ldots;r)) = \mu_i(f_k(\ldots,\mu_j,\ldots,\mu_k,\ldots,\mu_i,\ldots;r))$$
$$\mu_i(f_k(\ldots,\mu_j,\ldots,\mu_k,\ldots,\mu_i,\ldots;r)) = \mu_i(f_i(\ldots,\mu_i,\ldots,\mu_k,\ldots,\mu_j,\ldots;r)).$$

From these equalities, we obtain

$$\mu_i(f_i(\ldots,\mu_i,\ldots,\mu_j,\ldots,\mu_k,\ldots;r)) = \mu_i(f_i(\ldots,\mu_i,\ldots,\mu_k,\ldots,\mu_j,\ldots;r)).$$

Since this equality holds for any single swap, the equality holds for any permutation π. □

Several desirable properties have been defined as shown below [6][13], but these definitions do not take role assignment into consideration.

Simple fair For any i, $\mu_i(f_i(\mu_1, \ldots, \mu_n)) \geq 1/n$.
Strong fair For any i, $\mu_i(f_i(\mu_1, \ldots, \mu_n)) > 1/n$.
Envy-free For any $i, j(i \neq j)$, $\mu_i(f_i(\mu_1, \ldots, \mu_n)) \geq \mu_i(f_j(\mu_1, \ldots, \mu_n))$.
Strong envy-free For any $i, j(i \neq j)$, $\mu_i(f_i(\mu_1, \ldots, \mu_n)) > \mu_i(f_j(\mu_1, \ldots, \mu_n))$.
Super envy-free For any $i, j(i \neq j)$, $\mu_i(f_j(\mu_1, \ldots, \mu_n)) < 1/n$.
Exact For any i, j, $\mu_i(f_j(\mu_1, \ldots, \mu_n)) = 1/n$.
Equitable For any i, j, $\mu_i(f_i(\mu_1, \ldots, \mu_n)) = \mu_j(f_j(\mu_1, \ldots, \mu_n))$.

Simple fair division can be achieved for any number of parties by using the moving-knife protocol [8]. Strong fair division cannot be achieved if every party has an identical utility function μ. Woodall [14] proposed an algorithm for achieving strong fair division provided that there is a portion $X \subset [0,1]$ such that $\mu_1(X) \neq \mu_2(X)$, when $n = 2$. The algorithm for obtaining such a portion X is an open problem. Envy-free division can be achieved for any number of parties [5], however the protocol is very complicated.

As regards strong envy-free cake-cutting, the lower bound of the number of cuts is shown [10]. Super envy-free division can be achieved if utility functions $\mu_1, \ldots, \mu_n$ are linearly independent, however the algorithm for obtaining an actual assignment is not shown[2]. An exact division algorithm has been reported for two players using a moving knife method [1]. Though existence of exact division was proved [11], no algorithm has been shown for $n \geq 3$. An equitable division algorithm between two parties has been described [9]. The case where $n \geq 3$ is an open problem.

As shown above, stronger properties than envy-free such as strong-envy-free, super-envy-free, exact, and equitable are very hard to realize.

A definition, similar to ours, called 'anonymous,' is provided in [12]. A cake-cutting protocol is anonymous if for any $(\mu_1, \ldots, \mu_i, \ldots, \mu_j, \ldots, \mu_n)$, i, and j,

$$f_i(\mu_1, \ldots, \mu_i, \ldots, \mu_j, \ldots, \mu_n) = f_j(\mu_1, \ldots, \mu_j, \ldots, \mu_i, \ldots, \mu_n)$$

holds. This is a severe definition that requires the assigned portion to be identical for any role swapping. In meta-envy-freeness the assigned portions need not be identical but their utilities must be identical for any role swapping. In addition, randomization is not explicitly considered in the definition of anonymity.

Equitability does not imply meta-envy-freeness. There can be an (artificial) protocol that is equitable but not meta-envy-free. Party P_1's utility μ_1 satisfies $\mu_1([0,1/4]) = 0.3$, $\mu_1([1/4,1/2]) = 0.3$, $\mu_1([1/2,3/4]) = 0.2$, and $\mu_1([3/4,1]) = 0.2$. Party P_2's utility μ_2 satisfies $\mu_2([0,1/4]) = 0.2$, $\mu_2([1/4,1/2]) = 0.2$, $\mu_2([1/2,3/4]) = 0.3$, and $\mu_2([3/4,1]) = 0.3$. A protocol f initially assigns $[0,1/4]$ to the first player and $[3/4,1]$ to the second player. The result of $f(\mu_1,\mu_2)$ is $f_1(\mu_1,\mu_2) = [0,1/2]$ and $f_2(\mu_1,\mu_2) = [1/2,1]$ and the utilities are 0.6 for both parties. On the other hand, $f(\mu_2,\mu_1)$ might result in $f_1(\mu_2,\mu_1) = ([0,1/4], [1/2,3/4])$ and $f_2(\mu_2,\mu_1) = ([3/4,1],[1/4,1/2])$, thus the utilities are 0.5 for both parties. Therefore this (artificial) protocol is equitable, but not meta-envy-free, since P_1 prefers the first player. On the other hand, the meta-envy-free protocols shown in the next section are not equitable. Note that meta-envy-freeness does not imply envy-freeness. As shown in the introduction, the following holds.

1: **begin**
2: p_1 cuts into three pieces (so that p_1 considers their sizes are the same)
3: Let X_1, X_2, X_3 be the pieces where X_1 is the largest and X_3 is the smallest for p_2.
4: **if** X_1 is larger than X_2 for p_2 **then**
5: p_2 cuts L from X_1 so that $X_1' = X_1 - L$ is the same as X_2 for p_2.
6: p_3 selects the largest (for p_3) among X_1', X_2, and X_3.
7: **if** X_1' remains **then**
8: **begin**
9: p_2 must select X_1'.
10: Let (p_a, p_b) be (p_3, p_2).
11: **end**
12: **else**
13: **begin**
14: p_2 selects X_2 (the largest for p_2).
15: Let (p_a, p_b) be (p_2, p_3).
16: **end**
17: p_1 obtains the remaining piece.
18: **if** L is not empty **then**
19: p_a cuts L into three pieces (such that p_a considers their sizes are the same) and p_b, p_1, and p_a selects one piece in this order.
20: **end.**

Fig. 1. Three-player envy-free protocol

Theorem 2. *The 'divide-and-choose' protocol is not meta-envy-free.*

Next, we consider the envy-free cake-cutting protocol for three players, found independently by Selfridge and Conway (introduced in [6]), and shown in Fig. 1.

Note that without loss of envy-freeness, we assume that when a player cuts L from $X_1 = [x_1, x_2]$, L must be cut as $[x_1, x_3]$ for some x_3.

Theorem 3. *The protocol in Fig. 1 is not meta-envy-free.*

Proof. Let there be three parties P_x, P_y, and P_z whose utility functions are μ_x, μ_y, and μ_z, respectively.

We show that party P_x prefers the role of player p_3 to that of p_2 in this protocol. Let us consider the following two executions:
(Ex1) $(p_1, p_2, p_3) = (P_z, P_y, P_x)$ and (Ex2) $(p_1, p_2, p_3) = (P_z, P_x, P_y)$.

The result of the initial cut by P_z at line 2 is the same in (Ex1) and (Ex2). Let the three pieces be Z_1, Z_2, and Z_3. Without loss of generality, Z's are ordered from the largest to the smallest for P_y. All possible cases are categorized as follows.

(Case 1) P_y does not cut L in (Ex1).
(Case 1-1) P_x cuts L' from some piece Z in (Ex2).
(Case 1-2) P_x does not cut L in (Ex2).
(Case 2) P_y cuts L from Z_1 in (Ex1).

(Case 2-1) P_x also cuts L' from Z_1 in (Ex2).

(Case 2-1-1) L' is larger[1] than L. **(Case 2-1-2)** L' is smaller than L. **(Case 2-1-3)** $L' = L$.

(Case 2-2) P_x cuts L' from another piece Z in (Ex2).

(Case 2-3) P_x does not cut L' in (Ex2).

(Case 1-1) Let the largest piece for P_x be Z_1'. P_x selects Z_1' at line 6 in (Ex1) and obtains utility $\mu_x(Z_1')$. In contrast, at lines 7-16 of (Ex2), P_x obtains a piece whose utility equals $\mu_x(Z_1' - L')$, because there are two pieces with utility $\mu_x(Z_1' - L')$ after cutting L'. At line 19 of (Ex2), P_x obtains a cut of L' whose utility is smaller than $\mu_x(L')$. Thus, the total utility of P_x is smaller than $\mu_x(Z_1')$. Therefore, (Ex1) is better for P_x.

(Case 1-2) There are at least two largest pieces for P_x among Z_1, Z_2, and Z_3. P_x selects the largest piece at line 6 in (Ex1). In contrast, after P_y has selected Z_1 at line 6 in (Ex2), P_x can select one of the largest pieces at lines 7-16. Thus P_x obtains the same utility in (Ex1) and (Ex2).

(Case 2-1-1) At line 6 in (Ex1), the largest piece for P_x is $Z_1 - L$, since L' is larger than L. At line 19, P_x obtains at least $\mu_x(L)/3$. Thus, P_x obtains at least $\mu_x(Z_1) - 2\mu_x(L)/3$ in total. In contrast, P_y selects Z_2, which is larger than $Z_1 - L'$, at line 6 in (Ex2). Thus P_x selects $Z_1 - L'$ at line 9. In addition, P_x obtains at least $\mu_x(L')/3$. P_x obtains at least $\mu_x(Z_1) - 2\mu_x(L')/3$ in total. Thus, (Ex1) is better for risk averse party P_x.

(Case 2-1-2) At line 6 in (Ex1), P_x does not select $Z_1 - L$, since it is not greater than the second largest piece, whose utility is $\mu_x(Z_1 - L')$, for P_x. P_x chooses the piece and obtains $\mu_x(Z_1 - L')$. In addition, at line 19, P_x obtains $\mu_x(L)/3$ because P_x cuts L. P_x obtains $\mu_x(Z_1) - \mu_x(L') + \mu_x(L)/3$ in total. In contrast, at line 6 in (Ex2), P_y selects $Z_1 - L'$, which is the largest for P_y. Thus P_x selects Z_2 or Z_3 whose utility is $\mu_x(Z_1 - L')$. P_x then obtains $\mu_x(L')/3$ at line 19 because P_x cuts L'. P_x obtains $\mu_x(Z_1) - 2\mu_x(L')/3$ in total, which is smaller than that in (Ex1), since L' is smaller than L.

(Case 2-1-3) In both (Ex1) and (Ex2), P_x obtains a piece whose utility is $\mu_x(Z_1 - L)$. The only difference is who cuts L. As shown in the proof of 'divide-and-choose', being Chooser is the better than being Divider at line 19. In (Ex1), P_x can select $Z_1 - L$ and become Chooser. In (Ex2), if P_y selects $Z_1 - L$, P_x must become Divider. Thus (Ex1) is better than (Ex2).

(Case 2-2) In (Ex1), P_x selects the largest piece, which is not $Z_1 - L$, at line 6 and obtains $\mu_x(Z)$. At line 19, P_x obtains at least $\mu_x(L)/3$. In (Ex2), P_y selects Z_1 not $Z - L'$ at line 6. Thus P_x obtains $\mu_x(Z) - \mu_x(L')$ at line 9. At line 19, P_x obtains less than $\mu_x(L')$. P_x obtains less than $\mu_x(Z)$ in total, which is worse than in (Ex1).

(Case 2-3) There are at least two largest pieces among Z_1, Z_2, and Z_3 for P_x. Let $\mu_x(Z)$ be the utility of the largest piece. In (Ex1), P_x can obtain $\mu_x(Z)$ at line 6. In addition, P_x obtains $\mu_x(L)/3$ at line 19. In contrast, in (Ex2), P_x obtains $\mu_x(Z)$. Thus (Ex1) is better than (Ex2) for P_x. □

---

[1] To compare the sizes of L and L', they must be cut in a canonical way. Thus the additional rule for cutting L is necessary.

begin
$P_i (i = 1, 2)$ simultaneously declare c_i that satisfies $\mu_i([0, c_i]) = 1/2$.
if $c_1 = c_2$ **then**
    Cut at c_1, coin-flip and decide which party obtains $[0, c_1]$ or $[c_1, 1]$.
else
    Cut as $[0, (c_1 + c_2)/2]$, $[(c_1 + c_2)/2, 1]$. P_i obtains the piece which contains c_i.
end.

Fig. 2. Two-party meta-envy-free protocol

4 Meta-Envy-Free Protocols for Two and Three Parties

This section shows meta-envy-free cake-cutting protocols for two and three parties. Note that the word 'party' is used in the descriptions in this section because every player's role is identical. When there are two parties, the protocol proposed in [4], shown in Fig. 2, is meta-envy-free.

The simultaneous declaration of values by multiple parties can be realized in several ways, (1) Trusted third party (TTP): P_i sends c_i to the TTP. After the TTP receives all the values, he broadcasts them to all parties. (2) Commitment scheme [7]: P_i first sends $com_i(c_i)$, which is a commitment of c_i. The other parties cannot obtain the value c_i from $com_i(c_i)$. After P_i has obtained the other parties' committed values, P_i opens its commitment (that is, sends c_i and a proof that $com_i(c_i)$ is really made by c_i). P_i cannot provide a false proof that $com_i(c_i)$ is made by $c'_i(\neq c_i)$.

Theorem 4. *The protocol in Figure 2 is meta-envy-free, envy-free, and strategy-proof.*

Proof. The cut point depends only on the parties' declared values. The result is independent of the role assignment or the order of declaration. Thus the protocol is meta-envy-free. The protocol is envy-free because both parties obtain at least half evaluated by their utility functions. The protocol is strategy-proof since if P_1 declares false cut point c'_1, P_2's true cut point c_2 might satisfy $c_2 = c'_1$ and P_1 might obtain less than half by coin-flipping. Thus, risk adverse parties obey the rule and declare their true cut points. □

There is another method for assigning portions when the declared values differ. Without loss of generality, assume that $c_1 < c_2$. Assign $[0, c_1]$ to P_1, $[c_2, 1]$ to P_2, and execute the same protocol again for the remaining piece $[c_1, c_2]$. Although this method might need an infinite number of declaration rounds and each party might obtain multiple fragments of the cake, the assignment guarantees $\mu_1(f_1(\mu_1, \mu_2)) = \mu_2(f_2(\mu_1, \mu_2))$.

Avoiding multiple declaration is possible if P_i simultaneously declares the utility density function u_i. Utility density function u_i satisfies $u_i(z) > 0$ for $[0, 1]$ and $\int_0^1 u_i(z)dz = 1$.

When the remaining piece is $[l^{(j)}, r^{(j)}]$ at round j ($l^{(1)} = 0$ and $r^{(1)} = 1$), The cut point declaration at round j is the point $c_i^{(j)}$ that satisfies

$$\int_{l^{(j)}}^{c_i^{(j)}} u_i(z)dz = \int_{c_i^{(j)}}^{r^{(j)}} u_i(z)dz. \tag{6}$$

If $c_1^{(j)} \neq c_2^{(j)}$, let $l^{(j+1)} = min(c_1^{(j)}, c_2^{(j)})$, $r^{(j+1)} = max(c_1^{(j)}, c_2^{(j)})$, and execute next round.

A protocol that uses a utility density function is also proposed in [3]. Here the cake is cut into two pieces. However, the protocol has the disadvantage that it is not strategic-proof, that is, a party can obtain more utility by declaring a false utility density function.

Next we show a protocol for a three-party case in Fig. 3. The protocol is outlined as follows. First, each party P_i simultaneously declares cut point l_i such that $[0, l_i]$ is 1/3 for P_i. Cases are switched according to how many of l_1, l_2, and l_3 are the same. If at least two of them are the same, the parties with the same value simultaneously declare cut point r_i such that $[r_i, 1]$ is 1/3 for P_i. Envy-free assignment can be easily obtained using the declared values when at least two of l_1, l_2, and l_3 are the same. The remaining case is when l_1, l_2, and l_3 are all different (without loss of generality, assume that $l_1 < l_2 < l_3$). Here, we execute the three-player envy-free protocol in Fig. 1 with the role assignment (p_1, p_2, p_3)=(P_3, P_2, P_1), that is, P_3 plays the role of p_1 in the protocol, and so on, with the restriction that P_3 must use l_3 as a cut. Note that this role assignment is executed by the declared value l_i, thus the protocol is meta-envy-free.

Although (p_1, p_2, p_3)=(P_3, P_2, P_1) is not a unique acceptable role assignment, there are unacceptable role assignments. Let us consider the following role assignment: (p_1, p_2, p_3)=(P_2, P_1, P_3), namely, the cake is cut at l_2, r_2 and P_1 cuts L from the largest piece. Suppose that $[0, l_2]$ is the largest for P_1. P_1 cuts L from $[0, l_2]$. In this case, $[0, l_2]$ is less than 1/3 for P_3 because $l_3 > l_2$. After P_1 cuts L from $[0, l_2]$, P_3 will never select $[0, l_2] - L$ as the largest piece for P_3. P_1 knows this fact from $l_3 > l_2$, thus P_1 will not cut L honestly from $[0, l_2]$. In this case, P_3 will select some piece other than $[0, l_2]$. P_1 then selects $[0, l_2]$ and obtains more utility than when honestly cutting L. Thus, the protocol is not strategy-proof.

Theorem 5. *The protocol in Fig. 3 is meta-envy-free, envy-free, and strategy-proof.*

Proof. The protocol is meta-envy-free because the role is decided solely by the declared values. Next let us consider envy-freeness. All possible cases are categorized as follows. **(Case 1)** $l_1 = l_2 = l_3$ and $r_1 = r_2 = r_3$. **(Case 2)** $l_1 = l_2 = l_3$, $r_1 = r_2$, and $r_3 > r_1$. **(Case 3)** $l_1 = l_2 = l_3$, $r_1 = r_2$, and $r_1 > r_3$. **(Case 4)** $l_1 = l_2 = l_3$ and $r_1 < r_2 < r_3$. **(Case 5)** $l_1 = l_2 (\neq l_3)$ and $r_1 = r_2$. **(Case 6)** $l_1 = l_2 (\neq l_3)$ and $r_1 < r_2$. **(Case 7)** $l_1 < l_2 < l_3$.

(Case 1) Since the utilities of $[0.l_1]$, $[l_1, r_1]$, and $[r_1, 1]$ are 1/3 for all parties, no assignment causes envy.

(Case 2) The utilities of $[0, l_1]$, $[l_1, r_1]$, and $[r_1, 1]$ are the same for P_1 and P_2. $[r_1, 1]$ is the largest for P_3 since $r_3 > r_1$ and $l_3 = l_1$. Thus assigning $[r_1, 1]$ does

Each party P_i simultaneously declares l_i such that $[0, l_i]$ is 1/3 for P_i.
if $l_1 = l_2 = l_3$ **then**
  begin
  Each party P_i simultaneously declares r_i such that $[r_i, 1]$ is 1/3 for P_i.
  if $r_1 = r_2 = r_3$ **then**
    Cut at l_1 and r_1. Coin-flip and assign $[0, l_1], [l_1, r_1], [r_1, 1]$ to the parties.
  else
    if two of r_1, r_2, r_3 are the same **then**
      begin /* Without loss of generality, let $r_1 = r_2$. */
      Cut at l_1 and r_1.
      if $r_3 > r_1$ **then** Assign $[r_1, 1]$ to P_3.
      else /* $r_3 < r_1$ */
        Assign $[l_1, r_1]$ to P_3.
      Coin-flip and assign the remaining two pieces to P_1 and P_2.
      end
    else /* Without loss of generality, let $r_1 < r_2 < r_3$. */
      Cut at l_1 and r_2. Assign $[0, l_1]$ to P_2, $[l_1, r_2]$ to P_1, and $[r_2, 1]$ to P_3.
  end /* end of case $l_1 = l_2 = l_3$. */
else
  if two among l_1, l_2, and l_3 are the same **then**
    begin /* Without loss of generality, let $l_1 = l_2$. */
    P_1 and P_2 simultaneously declare r_i such that $[r_i, 1]$ is 1/3 for P_i.
    if $r_1 = r_2$ **then**
      begin
      Cut at l_1 and r_1. P_3 selects one piece among $[0, l_1], [l_1, r_1]$, and $[r_1, 1]$.
      Coin-flip and assign the remaining two pieces to P_1 and P_2.
      end
    else /* $r_1 \neq r_2$. */
      begin /* Without loss of generality, let $r_1 < r_2$. */
      Cut at l_1, r_1, r_2. $L \leftarrow [r_1, r_2]$. P_3 selects one among $[0, l_1]$, $[l_1, r_1]$, $[r_2, 1]$.
      if P_3 selects $[0, l_1]$ **then**
        begin
        Assign $[l_1, r_1]$ and $[r_2, 1]$ to P_1 and P_2, respectively.
        P_3 cuts L into three pieces. P_1, P_2, P_3 selects one in this order.
        end
      else
        if P_3 selects $[l_1, r_1]$ **then**
          begin
          Assign $[0, l_1]$ and $[r_2, 1]$ to P_1 and P_2, respectively.
          P_3 cuts L into three pieces. P_2, P_1, P_3 selects one in this order.
          end
        else /* P_3 selects $[r_2, 1]$. */
          begin
          Assign $[l_1, r_1]$ and $[0, l_1]$ to P_1 and P_2, respectively.
          P_3 cuts L into three pieces. P_1, P_2, P_3 selects one in this order.
          end
      end
    end
  else /* l_1, l_2 and l_3 are different. Without loss of generality, let $l_1 < l_2 < l_3$. */
    Execute Fig. 1 with $(p_1, p_2, p_3) = (P_3, P_2, P_1)$ and l_3 is used as a cut.

Fig. 3. Three party meta-envy-free protocol

not cause any party envy. Assigning the remaining pieces to P_1 and P_2 can be arbitrary.

(Case 3) The utilities of $[0, l_1]$, $[l_1, r_1]$, and $[r_1, 1]$ are the same for P_1 and P_2. $[l_1, r_1]$ is the largest for P_3 since $r_3 < r_1$ and $l_3 = l_1$. Thus assigning $[l_1, r_1]$ does not cause any party envy. Assigning the remaining pieces to P_1 and P_2 can be arbitrary.

(Case 4) Among $[0, l_1]$, $[l_1, r_2]$, and $[r_2, 1]$, $[l_1, r_2]$ is the largest for P_1 since $r_1 < r_2$. $[r_2, 1]$ is the largest for P_3 since $r_2 < r_3$ and $l_1 = l_3$. P_2 feels the three pieces are the same size, thus assigning $[0, l_1]$ to P_2 does not cause envy.

(Case 5) The utilities of $[0, l_1]$, $[l_1, r_1]$, and $[r_1, 1]$ are the same for P_1 and P_2. Thus, P_3's selection from these pieces does not cause envy.

(Case 6) The utilities of $[0, l_1]$, $[l_1, r_1]$, and $[r_1, 1]$ are the same for P_1. The utilities of $[0, l_1]$, $[l_1, r_2]$, and $[r_2, 1]$ are the same for P_2. Cutting the cake into four pieces, $[0, l_1]$, $[l_1, r_1]$, $[r_2, 1]$, and $L = [r_1, r_2]$ is exactly the same situation as during three-player envy-free cutting (Case 6-1) P_1 executes the initial cut ($[0, l_1]$, $[l_1, r_1]$, and $[r_1, 1]$) and P_2 cuts L from the largest piece $[r_1, 1]$ so that its size becomes that of the second largest piece $[0, l_1]$ and (Case 6-2) P_2 executes the initial cut ($[0, l_1]$, $[l_1, r_2]$, and $[r_2, 1]$) and P_1 cuts L from the largest piece $[l_1, r_2]$ so that its size becomes that of the second largest piece $[0, l_1]$.

When P_3 selects $[0, l_1]$ from the three pieces, we can regard this as (Case 6-2) being executed. With the three-player envy-free protocol, next P_1 must select $[l_1, r_1]$ and P_2 selects the remaining piece $[r_2, 1]$. P_3 cuts L into three pieces. P_1, P_2, and P_3 each select one piece in this order. Because of the envy-freeness of the three-player protocol, the result is envy-free.

When P_3 selects $[l_1, r_1]$ from the three pieces, we can regard this as (Case 6-1) being executed. With the three-player envy-free protocol, next P_2 must select $[r_2, 1]$ and P_1 selects the remaining piece $[0, l_1]$. P_3 cuts L into three pieces. P_2, P_1, and P_3 each select one piece in this order. Because of the envy-freeness of the three-player protocol, the result is envy-free.

Lastly, when P_3 selects $[r_2, 1]$ from the three pieces, we can regard this as (Case 6-2) being executed. With the three-player envy-free protocol, next P_1 must select $[l_1, r_1]$ and P_2 selects the remaining piece $[0, l_1]$. P_3 cuts L into three pieces. P_1, P_2, and P_3 each select one piece in this order. Because of the envy-freeness of the three-player protocol, the result is envy-free.

(Case 7) Since the players execute the three-player envy-free protocol, the result is envy-free.

Lastly, let us discuss strategy-proofness. When P_i declares a cut point l_i (or r_i) simultaneously with some other process P_j, declaring a false value l'_i (or r'_i) might result in a worse utility, since P_j's true value l_j (or r_j) might satisfy $l_j = l'_i$ (or $r_j = r'_i$) and P_i might obtain a smaller piece by coin-flipping.

When P_3 selects one piece at line 30, a false selection results in a worse utility for P_3. Note that this selection does not affect who will be the divider of L.

Next, consider the execution of the three-player envy-free protocol with extra information $l_1 < l_2 < l_3$. When P_3 cuts as $[0, l_3], [l_3, r_3]$, and $[r_3, 1]$, a false cut r'_3 might result in P_3 obtaining less than $1/3$. When P_2 cuts L from the largest piece, information of l_1 does not help P_2 to obtain greater utility with false cut L' even if P_2 cuts L from $[0, l_3]$. The reason is as follows. For any true cut L, either of the two cases can happen according to P_1's utility (that is unknown to P_2): (1) $[l_3, r_3]$ or $[r_3, 1]$ is the largest for P_1 or (2) $[0, l_3] - L$ is the largest for P_1. Thus, if P_2 cuts L' that is smaller than L, P_1 might select $[0, l_3] - L'$ and P_2's utility might become worse. If P_2 cuts L'' that is larger than L, P_1 might select $[l_3, r_3]$ and P_2's utility might become worse. With respect to cutting L into three pieces, the strategy-proofness is exactly the same as that of the original three-player envy-free protocol. Therefore, the protocol is strategy-proof. □

Acknowledgment. We thank Dr. Hiro Ito for his valuable comments.

References

1. Austin, A.K.: Sharing a Cake. Mathematical Gazette 66(437), 212–215 (1982)
2. Barbanel, J.B.: Super Envy-Free Cake Division and Independence of Measures. J. of Mathematical Analysis and Applications 197(1), 54–60 (1996)
3. Brams, S.J., Jones, M.A., Klamler, C.: Better Ways to Cut a Cake. Notices of the AMS 53(11), 1314–1321 (2006)
4. Brams, S.J., Jones, M.A., Klamler, C.: Divide-and-Conquer: A Proportional, Minimal-Envy Cake-Cutting Procedure. In: Proc. of Dagstuhl Seminar (2007)
5. Brams, S.J., Taylor, A.D.: An Envy-Free Cake Division Protocol. American Mathematical Monthly 102(1), 9–18 (1995)
6. Brams, S.J., Taylor, A.D.: Fair Division: From Cake-Cutting to Dispute Resolution. Cambridge University Press, Cambridge (1996)
7. Brassard, G., Chaum, D., Crépeau, C.: Minimum Disclosure Proofs of Knowledge. Journal of Computer and System Sciences 37(2), 156–189 (1988)
8. Dubins, L.E., Spanier, E.H.: How to Cut a Cake Fairly. American Mathematical Monthly 85(1), 1–17 (1961)
9. Jones, M.A.: Equitable, Envy-free, and Efficient Cake Cutting for Two People and its Application to Divisible Goods. Mathematics Magazine 75(4), 275–283 (2002)
10. Magdon-Ismail, M., Busch, C., Krishnamoorthy, M.S.: Cake Cutting is Not a Piece of Cake. In: Alt, H., Habib, M. (eds.) STACS 2003. LNCS, vol. 2607, pp. 596–607. Springer, Heidelberg (2003)
11. Neyman, J.: Un theoreme d'existence. C. R. Acad. Sci. Paris 222, 843–845 (1946)
12. Nicolò, A., Yu, Y.: Strategic Divide and Choose. Games and Economic Behavior 64(1), 268–289 (2008)
13. Robertson, J., Webb, W.: Cake-Cutting Algorithms: Be Fair If You Can. A.K. Peters, Wellesley (1998)
14. Woodall, D.R.: A Note on the Cake-Division Problem. J. of Combbinatorial Theory, A 42(2), 300–301 (1986)

Two Variables and Two Successors

Amaldev Manuel

Institute of Mathematical Sciences
Taramani, Chennai, India
amal@imsc.res.in

Abstract. We look at the finite satisfiability problem of the two variable fragment of first order logic with the successors of two linear orders. While the logic with both the successors and their transitive closures remains undecidable, we prove that the logic with only the successors is decidable.

1 Introduction

The two variable fragment of the first order logic is known to have an elementarily decidable satisfiability problem [1,2] and reasonably good expressive power (it contains several propositional modal logics). Recently, in the context of verification and semi-structured data, several extensions of FO^2 have been studied [3,4,5]. Strongly inspired by these results, we consider the decidability of FO^2 over finite structures with two linear orders.

A detailed context of the results in this paper is presented later. In the following we give a short account of the results on finite satisfiability problem (henceforth referred as FinSat) of several FO^2 extensions. In [6] it is shown that FinSat of FO^2 over words is Nexptime-complete. [7] shows that FinSat is undecidable for FO^2 with (at least) eight orders. In [8] it is shown that FO^2 with two transitive relations (without equality) is undecidable. In [9] it is shown that FO^2 is undecidable with three equivalence relations, but is decidable when the number of equivalence relations is two. Later in [10] it is shown that in the case of two equivalence relations, FinSat is decidable in 3-Exptime. In the same paper the undecidability is sharpened to one equivalence relation and one transitive relation. In [3] it is shown that FO^2 over words with an equivalence relation is decidable. The same authors in [4] showed that FO^2 over trees with an equivalence relation is decidable when only the successor relation in the tree is used.

Recently, considerations from semi-structured data and infinite state verification motivated the notion of datawords : linear orders labelled by a pair of elements, one from a finite set and one from an infinite domain. Formally, a dataword $w = (a_1, d_1) \ldots (a_n, d_n)$, $a_i \in \Sigma$, $d_i \in D$ where Σ is a finite alphabet and D is an infinite set. When operations on D is limited to only equality checking, it is easy to see that w can also be seen as a word with an equivalence relation on its positions. In a landmark paper, the authors of [3] showed that $FO^2(\Sigma, \prec, \prec^+, \sim)$ is decidable. The idea behind the proof, which goes back to

P. Hliněný and A. Kučera (Eds.): MFCS 2010, LNCS 6281, pp. 513–524, 2010.

Büchi, is to define a suitable automaton mechanism (called Data Automata) over datawords and to reduce the formulas of the logic to equivalent automata. Thus, the satisfiability problem of the logic is reduced to a reachability problem of the automaton, which is then shown decidable. In [3] it is also shown that a lot of natural extensions of $\mathrm{FO}^2(\Sigma, \prec, \prec^+, \sim)$ are undecidable, in particular $\mathrm{FO}^2(\Sigma, \prec, \prec^+, \sim, <')$ where $<'$ is a linear order on the data values. When all the data values are distinct, this logic deals with structures with two linear orders.

Almost all the proofs in this paper extensively follow the technique developed in [3]. However, there is a significant departure in the technical details, which we believe will make the results interesting. One crucial aspect where the proof differs is in the use of marking. In the case of Data automata, as shown in [11] the marking can be dispensed with. In the case of text automata it is not clear how to do this. Also, the translation from logic to automata depends on the fact that the marked string projection to the second order can be computed on the fly from the marked string projection to the first order.

Results. Our main result is the following : The finite satisfiability problem of FO^2 over two linear orders where the vocabulary contains only the successor relations of the orders is decidable. Our proof is automata theoretic. We define the notion of a text automaton and study its properties. We show that the emptiness checking of the automaton is decidable in NP. Finally, we show that a formula of the logic can be converted to an equivalent text automaton in 2-EXPTIME.

Related work. Recently, Thomas Schwentick and Thomas Zeume proved that finite satisfiability of two-variable logic over structures with a linear order and a total preorder $\mathrm{FO}^2(\Sigma, <, \lesssim)$ is EXPSPACE-Complete [12]. Since it is expressible in FO^2 that a total preorder is a linear order, this result implies that FINSAT of $\mathrm{FO}^2(\Sigma, \prec_1^+, \prec_2^+)$ is in EXPSPACE. Their results deal with the order relations and not successor relations, thus renting them incomparable to our results, but certainly complementing the results in this paper. In [13] the author showed that existential MSO over two successors is strictly weaker than over linear orders. (We give an alternative proof of this fact in this paper).

2 Preliminaries

Let T be a finite set. We say a binary relation R on T is a (strict) partial order if it is (1) irreflexive (2) transitive and hence asymmetric. We say a partial order R is a (strict) linear order if it is (3) total. We call the binary relation $S \subseteq R$, defined as $\{(a, b) \mid aRb, \neg\exists c \in T, aRc \wedge cRb\}$ the successor or covering relation of R.

We denote by $\mathbb{N}$ the set of natural numbers and we represent the successor relation on $\mathbb{N}$ by $\prec^{\mathbb{N}}$. For an $n \in \mathbb{N}$, we denote by $[n]$ the set $\{1, \ldots n\}$. We represent by $\prec_n^{\mathbb{N}}$ the successor relation on $\mathbb{N}$ restricted to the set $[n]$, that is $\prec_n^{\mathbb{N}} = \prec^{\mathbb{N}} \cap ([n] \times [n])$.

Let Σ be an alphabet, a non-empty finite set. A word w over Σ is any finite sequence of characters from Σ. For a word w, we denote the length of w as $|w|$. Given a word $w = a_1 a_2 \ldots a_n$ over Σ, we can represent the word as a first order structure $\mathfrak{W} = ([n], \lambda_w, \prec_n^{\mathbb{N}})$, where $\lambda_w : [n] \to \Sigma$ is the labeling function, defined as $\lambda_w(i) = a_i$.

A text $\mathfrak{T}$ over Σ is a first order structure $\mathfrak{T} = (T, \lambda, \prec_1, \prec_2)$ where T is a finite set, $\lambda : T \to \Sigma$ is a labeling function, $\prec_1, \prec_2$ are successor relations of two linear orders over T. We denote the the linear order corresponding to $\prec_1$ (alternatively $\prec_2$) by the symbol $\prec_1^+$ (alternatively $\prec_2^+$). Restricting the structure $\mathfrak{T}$ to either of the orders yields a word, we call the word $(T, \lambda, \prec_1)$ the projection of $\mathfrak{T}$ to the order $\prec_1$.

Given any text $\mathfrak{T} = (T, \lambda, \prec_1, \prec_2)$ where $|T| = n$ we can rewrite $\mathfrak{T}$ uniquely as $([n], \lambda', \prec_n^{\mathbb{N}}, \prec_2')$ such that $\lambda' = \kappa^{-1} \circ \lambda$ and $\prec_2' = \{(\kappa(x), \kappa(y)) \mid x \prec_2 y\}$ where κ is the unique isomorphism from $(T, \prec_1)$ to $([n], \prec_n^{\mathbb{N}})$. Similarly, it can be also rewritten uniquely as $([n], \lambda'', \prec_1', \prec_n^{\mathbb{N}})$.

By $\mathrm{FO}^n(\tau)$ we mean the n-variable first order logic with the vocabulary τ (including equality) over texts. The predicate $\prec_1$ (alt. $\prec_2$) is interpreted as the successor relation in the first (alt. second) linear order. The predicate $\prec_1^+$ (alt. $\prec_2^+$) is interpreted as the first (alt. second) linear order. The predicate $\prec_1^i$ (alt. $\prec_2^i$) is interpreted as the 'i-th successor' relation of the first (alt. second) linear order.

3 Automata on Texts

Given a text of the form $\mathfrak{T} = ([n], \lambda, \prec_1 = \prec_n^{\mathbb{N}}, \prec_2)$, let $([n], \lambda, \prec_n^{\mathbb{N}}) = a_1 a_2 \ldots a_n$ be the projection of $\mathfrak{T}$ to the order $\prec_1$. We define the marked string projection of $\mathfrak{T}$ to $\prec_1$, abbreviated as $msp_{\prec_1}(\mathfrak{T})$, as the word $(a_1, b_1)(a_2, b_2) \ldots (a_n, b_n)$ where $b_i \in \{-1, 0, 1\}$, such that

$$b_i = \begin{cases} -1 \text{ if } 1 \leq i < n \text{ and } i+1 \prec_2 i \\ 1 \text{ if } 1 \leq i < n \text{ and } i \prec_2 i+1 \\ 0 \qquad\qquad \text{otherwise} \end{cases}$$

Given any text $\mathfrak{T}$ we can define its $msp_{\prec_1}(\mathfrak{T})$ by converting it into the above form.

Similarly, we can define the marked string projection of $\mathfrak{T}$ to $\prec_2$ denoted as $msp_{\prec_2}(\mathfrak{T})$. For this, we first convert it into the form $\mathfrak{T}' = ([n], \lambda, \prec_1, \prec_2 = \prec_n^{\mathbb{N}})$ where $|T| = n$. let $([n], \lambda, \prec_n^{\mathbb{N}}) = a_1 a_2 \ldots a_n$ be the projection of $\mathfrak{T}'$ to the order $\prec_2$. $msp_{\prec_2}(\mathfrak{T}) = msp_{\prec_2}(\mathfrak{T}')$ is defined as the word $(a_1, b_1)(a_2, b_2) \ldots (a_n, b_n)$ where $b_i \in \{-1, 0, 1\}$, such that

$$b_i = \begin{cases} -1 \text{ if } 1 \leq i < n \text{ and } i+1 \prec_1 i \\ 1 \text{ if } 1 \leq i < n \text{ and } i \prec_1 i+1 \\ 0 \qquad\qquad \text{otherwise} \end{cases}$$

In the following we define the notion of a text automaton. Fix an alphabet Σ. A text automaton $A = (B, C)$ is a composite automaton consisting of two word automata B and C. The automaton B is a non-deterministic letter-to-letter word transducer with the input alphabet $\Sigma \times \{-1,0,1\}$ and an output alphabet Σ' (included in the definition of B). The automaton C is a non-deterministic finite state recognizer accepting words over the alphabet Σ'. Given a text $\mathfrak{T} = (T, \lambda, \prec_1, \prec_2)$ the automaton works as follows. The transducer B runs over the $msp_{\prec_1}(\mathfrak{T})$ yielding a string $w = (T, \lambda', \prec_1)$ in Σ'^*, where $\lambda' : T \to \Sigma'$. The automaton C runs over the string $w' = (T, \lambda', \prec_2)$, notice that w is permuted to the order $\prec_2$. Finally, the automaton A accepts $\mathfrak{T}$ if both B and C have a successful run, that is they both finish in one of their final states respectively.

Formally, let $A = (B, C)$, where B is a word transducer given by the tuple $B = (Q_b, (\Sigma \times \{-1,0,1\}), \Sigma', O_b, \Delta_b, I_b, F_b)$, where Q_b is the finite set of states, $(\Sigma \times \{-1,0,1\})$ is the input alphabet, Σ' is the output alphabet, $I_b \subseteq Q_b$ is the set of initial states, $F_b \subseteq Q_b$ is the set of final states, $\Delta_b \subseteq Q_b \times (\Sigma \times \{-1,0,1\}) \times Q_b$ is the set of transitions and $O_b : Q_b \times (\Sigma \times \{-1,0,1\}) \to \Sigma'$ is the output function. Given a marked string $w = (a_1, b_1)(a_2, b_2) \ldots (a_n, b_n)$ we define a run ρ_B of B as a sequence $q_0 q_1 \ldots q_n$ such that $q_0 \in I_b$ and for every $i \in [n]$ there is a transition $(p, (a, b), q)$ in Δ_b such that $q_{i-1} = p$, $q_i = q$, $a_i = a$ and $b_i = b$. The run ρ_b is accepting if $q_n \in F_b$. Given an accepting run ρ_b of B on w, it uniquely defines an output string $w' = a'_1 a'_2 \ldots a'_n \in \Sigma'^*$ where $a'_i = O_b(q_{i-1}, a_i)$. The automaton C is given by the tuple $C = (Q_c, \Sigma', \Delta_c, I_c, F_c)$ where Q_c is the finite set of states, Σ' is the alphabet, $I_c \subseteq Q_c$ is the set of initial states, $F_c \subseteq Q_c$ is the set of final states, $\Delta_c \subseteq Q_c \times \Sigma' \times Q_c$ is the set of transitions. Given a word $w' = a'_1 a'_2 \ldots a'_n \in \Sigma'^*$ a run ρ_c of C is a sequence $q_0 q_1 \ldots q_n$ such that $q_0 \in I_c$ and for every $i \in [n]$ there is a transition (p, a', q) in Δ_c such that $q_{i-1} = p$, $q_i = q$ and $a'_i = a'$. The run ρ_c is accepting if $q_n \in F_c$. Now, we define the run ρ of A on the text $\mathfrak{T} = (T, \lambda, \prec_1, \prec_2)$ as a pair (ρ_b, ρ_c) such that (i) ρ_b is an accepting run of B on $msp_{\prec_1}(\mathfrak{T})$ yielding a word $(T, \lambda', \prec_1)$ and (ii) ρ_c is an accepting run of C on the word $(T, \lambda', \prec_2)$.

We look at some example languages. Let L_1 be the language of texts where both the orders coincide, that is $L_1 = \{\mathfrak{T} \mid \mathfrak{T} \models \forall xy\ x \prec_1^+ y \Leftrightarrow x \prec_2^+ y\}$. This is easily done by checking the markings. What is more interesting is that we can accept texts whose string projections are non-regular. Consider the formula, $\varphi = \varphi_{\prec_1} \wedge \varphi_{\prec_2}$, where $\varphi_{\prec_1}$ says that the word projection of the text to the order $\prec_1$ belongs to the language $a^*b^*c^*$, whereas $\varphi_{\prec_2}$ says that the projection to $\prec_2$ belongs to the language $(abc)^*$. Hence, the projection of the text to $\prec_1$ has to be the language $\{a^n b^n c^n \mid n \in \mathbb{N}\}$. A text automaton can accept $L(\varphi)$ in the following way, the transducer projects the marked string to Σ and checks if it belongs to $a^*b^*c^*$. The automaton C checks if its input is in $(abc)^*$.

Lemma 1. *Given a regular language $L \subseteq \Sigma^*$, there is a text automaton accepting all texts whose projections to $\prec_1$ is in L. Similarly, there is a text automaton accepting all texts whose projections to $\prec_2$ is in L.*

Proof. In the first case the transducer B checks if the projection of the text to $\prec_1$ (ignoring the markings) is in L and C accepts Σ'^*. For the second case, the transducer B simply copies the string (again ignoring the markings) and C accepts if its input is in L.

Lemma 2. *Languages recognized by text automata are closed under union, intersection and renaming.*

Proof. Closure under union and intersection is obtained from usual product construction (using a composed output alphabet). Closure under renaming is achieved using the non-determinism of the transducer.

Consider the language L_M, the collection of texts such that (1) the projection to $\prec_1$ is in $\$_1a^+\#_1\$_2b^+\#_2$. (2) projection to $\prec_2$ is in $\$_1\$_2(ab)^+\#_1\#_2$. (3) There exists two positions x_0, x_1 having the same label from $\{a, b\}$ such that $x_0 \prec_1^+ x_1$ and $x_1 \prec_2^+ x_0$. The language, L_M is accepted by a text automaton. Conditions 1 and 2 can be checked easily by B and C. For condition 3, the transducer B non-deterministically chooses two positions having the same label (either a or b), $x_0 \prec_1^+ x_1$ and outputs 0 at x_0 and 1 at x_1 and \$ at every other position. The automaton C verifies that its input is of the form $\$^*1\$^*0\$^*$.

But $\overline{L_M}$, the complement of the above language is not accepted by any text automaton. For the sake of contradiction, assume that there is a text automaton $A = (B, C)$ accepting $\overline{L_M}$. Let the number of states in B be n. Consider the text $\mathfrak{T}_1 = \left([2k+4], \lambda, \prec_{2k+4}^{\mathbb{N}}, \prec_2\right)$ such that $\left([2k+4], \lambda, \prec_{2k+4}^{\mathbb{N}}\right)$ is $\$_1a^k\#_1\$_2b^k\#_2$ and $\prec_2$ is the order $\{(1, k+3), (k+3, 2), (2, k+4), (k+4, 3) \ldots (k+2, 2k+4)\}$ where $k > n$. Note that in the $\mathrm{msp}_{\prec_1}(\mathfrak{T})$ all the markings are zero. Since $\mathfrak{T}$ is in $\overline{L_M}$, there is an accepting run of B such that there exists two positions $i < j$ with label a and $q_{i-1} = q_{j-1}$ in the run. We define the order $\prec_2'$ as

$$\begin{aligned}\{(l, k+2+l) \mid 1 \le l \le k+2, l \ne i, l \ne j\}\\ \cup \{(k+2+l, l+1) \mid 1 \le l < k+2, l+1 \ne i, l+1 \ne j\}\\ \cup \{(k+1+i, j), (j, k+2+i), (k+1+j, i), (i, k+2+j)\}\end{aligned}$$

In the relation $\prec_2'$ only the positions i and j are switched from $\prec_2$. Let $\mathfrak{T}_1' = \left([2k+4], \lambda, \prec_{2k+4}^{\mathbb{N}}, \prec_2'\right)$. It is the case that $\mathrm{msp}_{\prec_1}(\mathfrak{T}) = \mathrm{msp}_{\prec_1}(\mathfrak{T}')$ and B has an accepting run on $\mathrm{msp}_{\prec_1}(\mathfrak{T}')$ outputting the same string as in the case of $\mathfrak{T}$, which then permuted to $\prec_2$ and $\prec_2'$ gives the same string. Hence C also has an accepting run. But, $\mathfrak{T}'$ does not belong to $\overline{L_M}$, leading to a contradiction. This shows that,

Lemma 3. *The class of languages accepted by text automata are not closed under complementation.*

Using a similar argument, we can show that the class of text automata where the transducer B is deterministic is strictly weaker.

We can prove the following proposition analogous to the $\mathrm{EMSO}^2(\Sigma, \prec)$ characterisation (Büchi–Elgot–Trakhtenbrot) of finite state automata.

Proposition 1. *For every text automaton A, there is an* $\mathrm{EMSO}^2(\Sigma, \prec_1, \prec_2)$ *formula φ_A such that $L(A) = L(\varphi_A)$.*

4 Decidability of the Text Automaton

In the definition of the automaton A, the transducer has access to $\mathrm{msp}_{\prec_1}(\mathfrak{T})$, whereas C can only access the output of B permuted to $\prec_2$. In the following, we show that it is possible for B to output a string which then permuted to $\prec_2$, yields $\mathrm{msp}_{\prec_2}(\mathfrak{T})$. Let $\mathfrak{T} = (T, \lambda, \prec_1, \prec_2)$ be a text and $\mathrm{msp}_{\prec_1}(\mathfrak{T}) = (a_1, b_1)(a_2, b_2)\dots(a_n, b_n)$. Let b'_i be the marking of the position i in $\mathrm{msp}_{\prec_2}(\mathfrak{T})$. It is easy to verify that b'_i is a function of b_i and b_{i-1} as evidenced by the following table.

b_{i-1}	b_i	b'_i	b_{i-1}	b_i	b'_i	b_{i-1}	b_i	b'_i
$-$	0	0	0	1	1	-1	0	-1
$-$	-1	0	1	0	0	-1	-1	-1
$-$	1	1	1	1	1	-1	1	$\perp$
0	0	0	0	-1	0	1	-1	$\perp$

Note that the configurations $b_{i-1} = -1$, $b_i = 1$ and $b_{i-1} = 1$, $b_i = -1$ does not constitute a valid marking. The above table immediately gives a strategy for outputting $\mathrm{msp}_{\prec_2}(\mathfrak{T})$. The automaton B always remembers the previous position's marking in its states, and computes b'_i. Once the output of B is permuted to $\prec_2$, the string becomes $\mathrm{msp}_{\prec_2}(\mathfrak{T})$.

Using the above fact, we can show that the emptiness problem for a text automaton is decidable in NPTIME. The idea is the following: Given a text automaton $A = (B, C)$, $L(A)$ is non-empty if there is a marked word w accepted by B such that a permutation of output of B on w, 'consistent' with the marking of w, is accepted by C.

Fix an alphabet Σ. Given a marked word $w \in (\Sigma \times \{0, 1, -1\})^*$ we denote the projection of w to Σ by $w \downarrow \Sigma$. Let $w = ([|w|], \lambda, \prec^{\mathbb{N}})$ be a marked word. A permutation $\pi : [|w|] \to [|w|]$ is a bijection and $\pi(w)$ is defined as the word $\left([|w|], \pi^{-1} \circ \lambda, \prec^{\mathbb{N}}_{|w|}\right)$. Note that π defines an order $\prec_\pi = \pi^{-1}(1)\dots\pi^{-1}(|w|)$ on the positions of w. We say that the permutation π is consistent with the marking if w is the marked string projection of the text $\mathfrak{T} = \left([|w|], \lambda, \prec^{\mathbb{N}}_{|w|}, \prec_\pi\right)$ to the order $\prec^{\mathbb{N}}_{|w|}$. Given a word $w' \in \Sigma^*$, by $\mathrm{mperm}(w')$, we denote the set of all the marked words w such that there is a permutation π consistent with the marking such that $\pi(w \downarrow \Sigma) = w'$. Given $L \subseteq \Sigma^*$, we define $\mathrm{mperm}(L) = \bigcup_{w' \in L} \mathrm{mperm}(w')$. Next we show that if L is regular then $\mathrm{mperm}(L)$ is accepted by a Presburger automaton.

A Presburger automaton A is a tuple (B, φ) where B is a finite state automaton with states $q_0, \dots q_n$ and φ is a Presburger formula with free variables $|q_0|, \dots, |q_n|$. A word w is accepted by the automaton A if there is an accepting run ρ of B on w such that $\varphi(|q_0|, \dots, |q_n|)$ is true, where $|q_i|$ is the number of times q_i appears in ρ.

Lemma 4. *If L is regular then* $\mathrm{mperm}(L)$ *is accepted by a Presburger automaton.*

Proof. The idea is to adapt the technique from [14]. Let $A = (Q, \Sigma, \Delta, q_0, F)$ be a non-deterministic finite state automaton accepting L. Given a marked word $w = (a_0, b_0)\dots(a_n, b_n) \in (\Sigma \times \{0, -1, 1\})^*$, the Presburger automaton P_A checks non-deterministically if there is a run of A on some consistent permutation of w. To achieve this, the automaton P_A assigns a transition $\delta = (p, a_i, q) \in \Delta$

to each position i of the marked word. The information whether the transition should or should not be 'joined' with the transitions of the neighbouring positions is also remembered. All this information is stored in the state occuring immediately after position i. If the marking is 1 or -1 the automaton ensures that the successive transitions associated with the positions are consistent with the marking. We can define a flow f where each transition δ of A is labelled by the number of times it is associated with a position. Finally, we can write a Presburger formula which checks that,

1. f is locally consistent.
2. the subgraph induced by the states with a non-zero flow is connected.
3. f is consistent with the marking.

Properties 1 and 2 is encoded as in the proof of theorem 1 in [14]. Property 3 is easily encoded as a summation. The resulting automata P_A is poly-sized in terms of the size of A.

Theorem 1. *Emptiness checking of a text automaton is in* NP.

Proof. Given the text automaton $A = (B, C)$, we construct a Presburger automaton P_C which accepts mperm$(L(C))$. We take the intersection of the transducer B and P_C such a way that the output of B is supplied as the input of P_C. Finally we check the emptiness of the resulting automaton which is in NP.

Next we show that the emptiness problem of Presburger automata which is NP-hard reduces to the emptiness problem for text automata, yielding a lower bound.

Theorem 2. *Emptiness checking of a Presburger automaton is polynomial time reducible to the emptiness checking of a text automaton.*

Proof. We use an alternative definition of Presburger automaton from [5], which is a type of automata with counters. A Presburger automaton P_A is a finite state automaton with a set of counters $\{1 \dots k\}$ holding integer (possibly negative) values. Formally it is a tuple $(Q, \Sigma, \Delta, q_0, F)$ where Q is the finite set of states, q_0 is the initial state and F is the set of final states. The transition relation Δ is a subset of $Q \times \Sigma \times \{I(j), D(j) \mid 1 \leq j \leq k\} \times Q$. A configuration of the automaton is of the form $(p, \bar{u})$ where $p \in Q$ and $\bar{u} : [k] \to \mathbb{Z}$. The automaton starts in the initial state with all the counters being zero. On a state p with counter values $\bar{u}$, the automaton can make a transition $(p, a, I(j), q)$ (alt. $(p, a, D(j), q)$) on the letter a resulting in the state q with counter values being the same, except for the counter j which is incremented (alt. decremented) by one. Finally at the end of the word, the automaton accepts if it reaches an accepting state with all the counters being zero.

For emptiness checking, without loss of generality we can ignore the labels on the transitions. We construct a text automaton whose alphabet is $Q \cup \{I_1, D_1, \dots, I_k, D_k\}$. We can represent the transition $(p, I(j), q)$ as the word pI_jq over this alphabet. A run ρ of P_A is represented by a text where (1) the string projection to the first order represents a sequence of transitions $\delta_1 \ldots \delta_n$, $\delta_i \in \Delta$ which is consistent — the run starts in the initial state, ends in a final state and two successive transitions have a common state (2) the string projection to the second order belongs to the language $Q^* \cdot (I_1 D_1 + \ldots + I_k D_k)^*$. This language can be recognized using a text automaton $A = (B, C)$ where B verifies (1) and C verifies (2). It is easy to see that $L(P_A)$ is non-empty if and only if $L(A)$ is non-empty.

5 Reduction from Logic to Automata

In this section we show that given an $\mathrm{FO}^2(\Sigma, \prec_1, \prec_2)$ formula we can transform it into an equivalent text automaton. Below, we shorthand the formula $\neg x \prec_1 y$ as $x \not\prec_1 y$, similarly the others too. First of all, given a formula $\varphi \in \mathrm{FO}^2(\Sigma, \prec_1, \prec_2)$ we transform it into an equivalent formula in *Scott normal form* which is of the form,

$$\exists R_1 \ldots R_n \left(\forall x \forall y \ \chi \wedge \bigwedge_i \forall x \exists y \ \psi_i \right)$$

where the predicates R_i are unary, and χ and ψ_i are quantifier-free formulas in $\mathrm{FO}^2(\Sigma, \prec_1, \prec_2)$. The resulting formula is linear in terms of the size of the original formula. Earlier we showed that text automata are closed under renaming and intersection. Therefore it suffices to show that we can construct a text automaton for each of the formulas $\forall x \exists y \ \chi$ and $\forall x \exists y \ \psi_i$. The following two lemmas show precisely that.

In the following a *type* is a one variable quantifier-free formula containing only unary predicates.

Lemma 5. *Given an* $\mathrm{FO}^2(\Sigma, \prec_1, \prec_2)$ *formula of the form* $\varphi = \forall x \forall y \ \chi$ *where* χ *is quantifier free, an equivalent text automaton of doubly exponential size can be constructed.*

Proof. Firstly, we write φ in CNF causing an exponential blowup in the size of the formula, followed by distributing the universal quantification over the conjunctions and rewriting the formula as $\bigwedge_i \forall x \forall y \ \chi_i$ where each χ_i is of the form,

$$\chi_i = \alpha(x) \vee \beta(y) \vee \epsilon(x, y) \vee \delta_1(x, y) \vee \delta_2(x, y).$$

Above $\alpha(x)$ and $\beta(y)$ are unary types. The formulas in the group $\epsilon(x, y)$ are $x = y$ and $x \neq y$. The formula δ_1 is a disjunction of literals from the set O_1 and δ_2 is a disjunction of literals from the set O_2, where $O_1 = \{x \prec_1 y, x \not\prec_1 y, y \prec_1 x, y \not\prec_1 x\}$ and $O_2 = \{x \prec_2 y, x \not\prec_2 y, y \prec_2 x, y \not\prec_2 x\}$. It is enough to construct a text automaton for each χ_i since the automata are closed under intersection. The alphabet Σ of the automata is going to be bit vectors which represent the evaluation of the unary predicates including R_i used in the formula at a given position. Hence, the size of the alphabet is exponential in the length of the formula. The automaton we construct in each case has constant number of states, but may have exponentially many transitions. Finally the intersection of these automata is of size doubly exponential.

We note that whenever χ_i describes a regular property, we can construct an equivalent text automaton by converting the finite state automaton equivalent to χ_i. If one of $\delta_1(x,y)$ and $\delta_2(x,y)$ is absent, the formula χ_i describes a regular property over one linear order. Therefore we restrict our attention to those χ_i where both δ_1 and δ_2 are present. Suppose $\epsilon(x,y) \equiv x = y \vee x \neq y$. In this case, the formula is tautology hence we construct a text automaton which accepts all texts.

Suppose $\epsilon(x,y) \equiv x \neq y$. In this case we can rewrite χ_i as,

$$\chi_i \equiv [\alpha'(x) \wedge \beta'(y) \wedge x = y] \rightarrow [\delta_1(x,y) \vee \delta_2(x,y)] .$$

Substituting x for y, we can see that the consequent reduces to either $\top$ or $\bot$. In both cases, the formula describes a regular property.

When $\epsilon(x,y) \equiv x = y$ and $\delta_1(x,y)$ or $\delta_2(x,y)$ contains a negative literal, we can rewrite χ_i in one of the following two forms,

$$\chi_i \equiv [\alpha'(x) \wedge \beta'(y) \wedge x \neq y \wedge \delta'_1(x,y)] \rightarrow \delta_2(x,y),$$
$$\chi_i \equiv [\alpha'(x) \wedge \beta'(y) \wedge x \neq y \wedge \delta'_2(x,y)] \rightarrow \delta_1(x,y),$$

where $\alpha', \beta', \delta'_1, \delta'_2$ are the negations of $\alpha, \beta, \delta_1, \delta_2$ respectively. We choose the appropriate form such that $\delta'_1(x,y)$ or $\delta'_2(x,y)$ contains a positive literal. In this case the automaton can verify the formula χ_i by looking at the marked string projection to $\prec_1$ or $\prec_2$ depending upon the chosen form.

The only remaining case is when $\epsilon(x,y) \equiv x = y$, and neither δ_1 nor δ_2 contains a negative literal, that is when δ_1 and δ_2 are disjunctions of positive literals. We rewrite χ_i in the following form,

$$\chi_i = [\alpha'(x) \wedge \beta'(y) \wedge x \neq y] \rightarrow [\delta_1(x,y) \vee \delta_2(x,y)] .$$

The formula says the following. Whenever α' holds at x and β' holds at y and x, y are distinct then either they are neighbours in $\prec_1$ as dictated by δ_1 or neighbours in $\prec_2$ as dictated by δ_2. If there is no α' in the word there can be any number of β'. Similarly there can be any number of α' if there is no β' occurring in the word. The automaton B can guess both these cases and verify them easily. When there is at least one α' and β' present in the word the number of α' and β' are bounded. Therefore in this case the formula χ_i can be checked by a text automaton by labelling the α' and β'.

This completes the proof.

Lemma 6. *For each* $\mathrm{FO}^2(\Sigma, \prec_1, \prec_2)$ *formula of the form* $\forall x \exists y\ \psi$ *where* ψ *is quantifier free, an equivalent text automaton of doubly exponential size can be constructed.*

Proof. First of all, ψ can be written (using the truth table for ψ) as an exponential size conjunction of disjunctions of the form $\forall x \exists y \bigwedge_i \bigvee_j [\alpha_i(x) \rightarrow \theta_{ij}(x,y)]$, where α_i enumerates through all possible maximal types, that is $\bigvee_i(\alpha_i(x))$ is a tautology and $\neg(\alpha_i(x) \wedge \alpha_j(x))$ for all $i \neq j$. The formula θ_{ij} is either $\perp$ or of the form,

$$\beta(y) \wedge \epsilon(x, y) \wedge \delta_1(x, y) \wedge \delta_2(x, y)$$

where, β is a type, ϵ is one of $x = y$, $x \neq y$, δ_1 is in O_1 and δ_2 is in O_2.

We notice that the premise occurring in distinct conjuncts are distinct (and mutually exclusive). Hence it is possible to distribute the $\forall x \exists y$ over the conjunction. The resulting formula is of the form, $\wedge_i \forall x \exists y \ \vee_j (\alpha_i(x) \supset \theta_{ij}(x, y))$. We eliminate the disjunction by adding to every disjunct a new unary predicate $\Lambda_{ij}(x)$ which denotes that at the position x, the j-th disjunct is witnessing α_i. We can rewrite every conjunct in the above formula as,

$$\exists \Lambda_{i1} \Lambda_{i2} \ldots (\forall x \vee_j \Lambda_{ij}(x)) \wedge \bigwedge_j \forall x \exists y \ [(\alpha_i(x) \wedge \Lambda_{ij}(x)) \rightarrow \theta_{ij}(x, y)]$$

A text automaton can guess the predicates Λ_{ij}. So it is enough to construct a text automaton for each formula of the form $\forall x \exists y \ [\alpha(x) \supset \theta_{ij}(x, y)]$. If the consequent is false, the language is regular. So we concentrate on the cases where the consequent is satisfiable.

$$\forall x \exists y \ [\alpha(x) \rightarrow (\beta(y) \wedge \epsilon(x, y) \wedge \delta_1(x, y) \wedge \delta_2(x, y))]$$

We do a case analysis. If $\epsilon(x, y) \equiv x = y$, the language is regular. Hence now onwards we fix ϵ to be $x \neq y$.

As in the previous proof, we have two cases, when δ_1 or δ_2 contains a positive literal and when they do not. If δ_1 or δ_2 contains a positive literal, we can easily verify the formula by looking at the marked string projection to the appropriate order.

The only remaining case is when neither δ_1 nor δ_2 contains a positive literal. Consider the case when $\delta_1 \equiv x \not\prec_1 y$ and $\delta_2 \equiv x \not\prec_2 y$. The formula says that if there is an α at x there should be a witness y with β holding there, such that y is not a successor of x in both the orders. Notice that if there are at least four β occurring in the word we will be able to find a witness for any α. The automaton guesses whether the word contains at least four β and verifies it, in which case the formula is taken care of. If the automaton guesses that the word contains fewer than four β, it labels each β distinctly and verifies that for every αthere is atleast one β witnessing it. Since the number of β is bounded, this can be done easily.

In the cases where $\delta_1 \wedge \delta_2$ is one of $y \not\prec_1 x \wedge x \not\prec_2 y$, $x \not\prec_1 y \wedge y \not\prec_2 x$, $y \not\prec_1 x \wedge y \not\prec_2 x$, the sufficient number of β is three. When $\delta_1 \wedge \delta_2$ is $x \not\prec_1 y \wedge y \not\prec_1 x \wedge x \not\prec_2 y \wedge y \not\prec_2 x$, the sufficient number of β is five. In all other cases the sufficient number of β is four. In all the above cases, the construction is similar.

This completes the proof.

Now, we can state the main result.

Theorem 3. FinSat *of* $\mathrm{FO}^2(\Sigma, \prec_1, \prec_2)$ *is in 2-*NExpTime.

In [15], it is shown that FinSat of $\mathrm{FO}^2(\Sigma)$ and FinSat of $\mathrm{FO}^2(\prec)$ with one unary predicate are NExpTime-hard. This lower bound applies to FinSat of $\mathrm{FO}^2(\Sigma, \prec_1, \prec_2)$ as well.

Using a reduction to PCP similar to the one in [3] we can show the following.

Theorem 4. *The satisfiability problems for the following logics are undecidable.*

(a) $\mathrm{FO}^2\left(\Sigma, \prec_1, \prec_1^+, \prec_2, \prec_2^+\right)$
(b) $\mathrm{FO}^3\left(\Sigma, \prec_1, \prec_2\right)$
(c) $\mathrm{FO}^2\left(\Sigma, \prec_1, \prec_1^2, \prec_1^3, \prec_2, \prec_2^2\right)$

The above theorem has to be compared with Proposition 29 from [3].

6 Discussion and Conclusion

In this paper we introduced a class of automata working over texts. Using this we proved that $\mathrm{FO}^2(\Sigma, \prec_1, \prec_2)$ is decidable in 2-NExpTime. The present lower bound is the obvious one, NExpTime, leaving a gap of one exponent.

In the beginning we showed that $\mathrm{FO}^2(\Sigma, \prec_1, \prec_2, \prec_1^+, \prec_2^+)$ is undecidable. From [12] we know that $\mathrm{FO}^2(\Sigma, \prec_1^+, \prec_2^+)$ is decidable in ExpSpace. This leaves open the decidability question of the following fragments (we omit the symmetric cases) $\mathrm{FO}^2(\Sigma, \prec_1, \prec_2^+)$, $\mathrm{FO}^2(\Sigma, \prec_1, \prec_2, \prec_2^+)$. However, it is interesting to note that these fragments contain languages which are not accepted by any text automaton. Consider the following languages,

$$L_1 = \{\mathfrak{T} \mid \mathfrak{T} \models \forall x \forall y \ \left(a(x) \wedge a(y) \wedge x \prec_1^+ y \supset x \prec_2^+ y\right)\}$$

$$L_2 = \{\mathfrak{T} \mid \mathfrak{T} \models \forall x \forall y \ \left(a(x) \wedge a(y) \wedge x \prec_1 y \supset x \prec_2^+ y\right)\}$$

The language L_1 is definable in $\mathrm{FO}^2(\Sigma, \prec_1^+, \prec_2^+)$ and L_2 in $\mathrm{FO}^2(\Sigma, \prec_1, \prec_2^+)$ [also in $\mathrm{FO}^2(\Sigma, \prec_1, \prec_2, \prec_2^+)$]. To show that these languages are not accepted by any text automaton, we refer back to lemma 3. We considered a text $\mathfrak{T}_1$ which belongs to $\overline{L_M}$, note that $\mathfrak{T}_1$ also belongs to both L_1 and L_2. Using a similar argument we can construct the text $\mathfrak{T}_1'$ which does not belong to L_1 (also L_2), but is accepted by any automata accepting L_1 (alternatively L_2). This essentially shows that our automaton is unable to check any transitive relations. This also implies the theorem which is proved in [13].

We want to draw the attention to why the decidability proof does not generalize to $\mathrm{FO}^2\left(\Sigma, \prec_1, \prec_2, \prec_3\right)$. The reason is that we relied on $\mathrm{msp}_{\prec_1}$ to compute $\mathrm{msp}_{\prec_2}$. This step is not possible in the case of three successor relations. This can be however overcome by providing the component automata (each running on $\prec_1, \prec_2$ and $\prec_3$) with their own marked string projection, in which case it is not clear how to prove the decidability of the text automaton.

Acknowledgements. We wish to emphasize that a suggestion by Luc Segoufin — to use Presburger Arithmetic to show the decidability — has helped to greatly improve the complexity results in this paper. The previous version of this paper used Multicounter Automata for showing the decidability, yielding no elementary complexity bounds. We thank him for the invaluable discussions and suggestions. We also thank Thomas Schwentick and Thomas Zeume for sharing their results with us and in particular Thomas Schwentick for very informative discussions and pointing out an error in the previous version.

References

1. Mortimer, M.: On languages with two variables. Zeitschr. f. Logik und Grundlagen d. Math. 21, 135–140 (1975)
2. Grädel, E., Kolaitis, P.G., Vardi, M.Y.: On the decision problem for two-variable first-order logic. Bulletin of Symbolic Logic 3(1), 53–69 (1997)
3. Bojanczyk, M., Muscholl, A., Schwentick, T., Segoufin, L., David, C.: Two-variable logic on words with data. In: LICS, pp. 7–16 (2006)
4. Bojanczyk, M., Muscholl, A., Schwentick, T., Segoufin, L.: Two-variable logic on data trees and XML reasoning. J. ACM 56(3) (2009)
5. Björklund, H., Bojanczyk, M.: Shuffle expressions and words with nested data. In: Kučera, L., Kučera, A. (eds.) MFCS 2007. LNCS, vol. 4708, pp. 750–761. Springer, Heidelberg (2007)
6. Etessami, K., Vardi, M.Y., Wilke, T.: First-order logic with two variables and unary temporal logic. In: LICS, pp. 228–235 (1997)
7. Otto, M.: Two variable first-order logic over ordered domains. J. Symb. Log. 66(2), 685–702 (2001)
8. Kieronski, E.: Results on the guarded fragment with equivalence or transitive relations. In: Ong, L. (ed.) CSL 2005. LNCS, vol. 3634, pp. 309–324. Springer, Heidelberg (2005)
9. Kieronski, E., Otto, M.: Small substructures and decidability issues for first-order logic with two variables. In: LICS, pp. 448–457. IEEE Computer Society, Los Alamitos (2005)
10. Kieronski, E., Tendera, L.: On finite satisfiability of two-variable first-order logic with equivalence relations. In: LICS, pp. 123–132. IEEE Computer Society, Los Alamitos (2009)
11. Björklund, H., Schwentick, T.: On notions of regularity for data languages. Theor. Comput. Sci. 411(4-5), 702–715 (2010)
12. Schwentick, T., Zeume, T.: Two-variable logic with two order relations. In: Dawar, A., Veith, H. (eds.) CSL 2010 (to appear 2010)
13. Mathissen, C.: Existential mso over two successors is strictly weaker than over linear orders. Theor. Comput. Sci. 410(38-40), 3982–3987 (2009)
14. Seidl, H., Schwentick, T., Muscholl, A., Habermehl, P.: Counting in trees for free. In: Díaz, J., Karhumäki, J., Lepistö, A., Sannella, D. (eds.) ICALP 2004. LNCS, vol. 3142, pp. 1136–1149. Springer, Heidelberg (2004)
15. Etessami, K., Vardi, M.Y., Wilke, T.: First-order logic with two variables and unary temporal logic. Inf. Comput. 179(2), 279–295 (2002)

Harnessing ML^F with the Power of System F

Giulio Manzonetto[1,*] and Paolo Tranquilli[2,**]

[1] Laboratoire LIPN, CNRS UMR 7030
Université Paris Nord, France
giulio.manzonetto@lipn.univ-paris13.fr
[2] LIP, CNRS UMR 5668, INRIA,
ENS de Lyon, Université Claude Bernard Lyon 1, France
paolo.tranquilli@ens-lyon.fr

Abstract. We provide a strong normalization result for ML^F, a type system generalizing ML with first-class polymorphism as in system F. The proof is achieved by translating ML^F into a calculus of coercions, and showing that this calculus is just a decorated version of system F.

Keywords: ML^F, strong normalization, coercions, polymorphic types.

1 Introduction

One of the most efficient techniques for assuring that a program "behaves well" is *static type-checking*: types are assigned to every subexpression of a program, so that consistency of such an assignment (checked at compile time) implies well-behavedness. Such assignment may be *explicit*, i.e. requiring the programmer to annotate the types at key points in the program (e.g. variables), as in C or Java. Otherwise we can free the programmer of the hassle and leave to an automatic type reconstructor, part of the compiler, the boring task of scattering the code with types. One of the most prominent examples of this approach is ML [1] and its dialects, a functional programming language, as such based on λ-calculus.

In this context *type polymorphism* allows greater flexibility, making it possible to reuse code that works with elements of different types. For example an identity function will have type $\alpha \to \alpha$ for any α, so one can give it the type $\forall(\alpha)(\alpha \to \alpha)$. However full polymorphism (like in system F [2]) leads to undecidable type systems: no automatic reconstructor would be available. For this reason ML has the so called second-class polymorphism (i.e. available only for named variables), more restricted but allowing a type inference procedure. Unfortunately, the programmer is also *forced* to use second-class polymorphism only. One could wish for a more flexible approach, where one would write just enough type annotations to let the compiler's type reconstructor do the job, while still being able to employ first-class polymorphism, if desired.

ML^F [3] answers this call by providing a partial type annotation mechanism with an automatic type reconstructor. This extension allows to write system F

* Supported by COLLODI (2009-28HD) and CALMOC (612.000.936) projects.
** Supported by ANR project COMPLICE (ANR-08-BLANC-0211-01).

P. Hliněný and A. Kučera (Eds.): MFCS 2010, LNCS 6281, pp. 525–536, 2010.

programs, which is not possible in general in ML, while remaining conservative: ML programs still type-check without needing any annotation. An important feature are principal type schemata, lacking in system F, which are obtained by employing a downward bounded quantification $\forall(\alpha \geq \sigma)\tau$, called a *flexible* quantifier. Such a type intuitively denotes that τ may be instantiated to any $\tau\{\sigma'/\alpha\}$, *provided that* σ' *is an instantiation of* σ. Usual quantification is recovered by allowing $\bot$ (morally equivalent to $\forall\alpha.\alpha$) as bound. $\mathsf{ML^F}$ also uses a *rigid* quantifier $\forall(\alpha = \sigma)\tau$, fundamental for type inference but not for the semantics[1].

One of the properties of well-behavedness that a type system can assure is *strong normalization* (SN), that is the termination of all typable programs whatever execution strategy is used. For example, system F is strongly normalizing. As already pointed out, system F is contained in $\mathsf{ML^F}$; it is not yet known, but it is conjectured [3], that the inclusion is strict. This makes the question of SN of $\mathsf{ML^F}$ a non-trivial one, to which we answer positively in this paper. The result is proved via a suitable simulation in system F, with additional decorations dealing with the complex type instantiations possible in $\mathsf{ML^F}$.

Our starting point is $\mathsf{xML^F}$ [4], the Church version of $\mathsf{ML^F}$, briefly presented in section 2. In $\mathsf{xML^F}$ type inference and the rigid quantifier $\forall(\alpha = \sigma)\tau$ are abandoned, with the aim of providing an internal language to which a compiler might map the surface language briefly presented above (which in fact is denoted more precisely by $\mathsf{eML^F}$[2]). Compared to Church-style system F, the type reduction $\rightarrow_\iota$ of $\mathsf{xML^F}$ is more complex, and may *a priori* cause unexpected glitches: it could cause non-termination, or block the reduction of a β-redex. To prove that none of this happens, we use as target language of our translation a decoration of system F, the *coercion calculus* $\mathsf{F_c}$, which has its own interest. Indeed, $\mathsf{xML^F}$ has syntactic entities (the *instantiations* ϕ) testifying an instance relation between types, and it is natural to regard them as coercions. The delicate point is that some of these instantiations (the "abstractions" $!\alpha$) behave in fact as variables, abstracted when introducing a bounded quantifier: in a way, $\forall(\alpha \geq \sigma)\tau$ expects a coercion from σ to α, whatever the choice for α may be.

A question naturally arising is: what does it mean to be a coercion in this context, where such operations of coercion abstraction and substitution are available? Our answer, which works for $\mathsf{xML^F}$, is in the form of a type system ($\mathsf{F_c}$, Figure 2). In section 3 we will show the good properties enjoyed by $\mathsf{F_c}$: it is a decoration of system F, so it is SN; moreover it has a *coercion erasure* which ideally recovers the actual semantics of a term, and establishes a *weak bisimulation* with system F, where coercion reductions $\rightarrow_{\mathsf{c}}$ take the role of silent actions, while β-reduction $\rightarrow_\beta$ remains the observable one.

The generality of coercion calculus allows then to lift these results to $\mathsf{xML^F}$ via the above mentioned translation (section 4). Its main idea is the same as for the one shown for $\mathsf{eML^F}$ in [5], where however no dynamic property was studied.

---

[1] Indeed $\forall(\alpha = \sigma)\tau$ can be regarded as being $\tau\{\sigma/\alpha\}$.

[2] There is also a completely annotation-free version, $\mathsf{iML^F}$, clearly at the cost of loosing type inference. For details on the different versions of $\mathsf{ML^F}$, the reader may be referred to `http://gallium.inria.fr/~remy/mlf/`

Here we produce a proof of SN for all versions of $\mathsf{ML^F}$. Moreover the bisimulation result establishes that $\mathsf{xML^F}$ can indeed be used as an internal language for $\mathsf{eML^F}$, as the additional structure cannot block reductions of the intended program.

As a final note before entering the details of the work, one may wonder whether the candidates of reducibility deliver the same result, and indeed it was the first approach we tried. A naïve interpretation[3] fails in general. A more elaborate interpretation works for the recast version [6], and in any case only for the β-reduction, leaving outside the type reduction (the ι one). Contrary to system F the latter is non trivial, so its presence is another reason for abandoning the reducibility approach.

Notations. Given reductions $\rightarrow_1$ and $\rightarrow_2$, we write $\rightarrow_1\rightarrow_2$ (resp. $\rightarrow_{12}$) for their concatenation (resp. their union). Moreover $\leftarrow$, $\xrightarrow{+}$, $\xrightarrow{=}$ and $\xrightarrow{*}$ denote the transpose, the transitive, the reflexive and the transitive-reflexive closures of $\rightarrow$ respectively. In confluence diagrams, solid arrows denote reductions one starts with, while dashed arrows are the entailed ones.

2 A Short Presentation of $\mathsf{xML^F}$

Currently, $\mathsf{ML^F}$ comes in a Curry-style version $\mathsf{iML^F}$, where no type information is needed, and a type-inference version $\mathsf{eML^F}$ requiring partial type information, though a large amount of type information is still inferred. A truly Church-style version of $\mathsf{ML^F}$, called $\mathsf{xML^F}$, has been recently introduced in [4] and will be our main object of study in this paper. However, we will draw conclusions for $\mathsf{iML^F}$ and $\mathsf{eML^F}$ too. We warn the reader that we we will only present the definition we need, while we refer to [4] for an in-depth discussion on $\mathsf{xML^F}$.

All the syntactic definitions of $\mathsf{xML^F}$ can be found in Figure 1. Types include: usual variable and arrow types; a type $\bot$ corresponding to system F's type $\forall\alpha.\alpha$; the *flexible quantification* $\forall(\alpha \geq \sigma)\tau$ generalizing $\forall\alpha.\tau$ of system F. Intuitively, $\forall(\alpha \geq \sigma)\tau$ restricts the variable α to range just over instances of σ. The variable α is bound in τ but not in σ. The instantiation ϕ maps a type σ to a type τ which is an instance of σ. Thus ϕ can be seen as a 'witness' of the instance relation holding between σ and τ. In $\forall(\alpha \geq)\phi$, α is bounded in ϕ.

Environments $\varGamma$ are finite maps assigning types (resp. bounds) to term (resp. type) variables. We write: $\mathrm{dom}(\varGamma)$ for the set of all term and type variables that are bound by $\varGamma$; $\mathrm{ftv}(\tau)$ for the set of type variables appearing free in τ. Environments of shape $\varGamma, \alpha \geq \tau, \varGamma'$ or $\varGamma, x : \tau, \varGamma'$ are *well-formed* if $\mathrm{ftv}(\tau) \subseteq \mathrm{dom}(\varGamma)$. All environments in this paper are supposed to be well-formed. Reduction rules are divided into $\rightarrow_\beta$ (regular β-reductions) and $\rightarrow_\iota$, reducing instantiations. We recall (from [4, Sec. 2.1]) that both $\rightarrow_\beta$ and $\rightarrow_\iota$ enjoy subject reduction. One of the ι-steps uses the definition of type instantiation $\tau\phi$, giving the unique type such that $\varGamma \vdash \phi : \tau \leq \tau\phi$, if ϕ type-checks.

The original calculus contains a `let` construct which is however added mainly to accommodate $\mathsf{eML^F}$'s type reconstructor. In the whole paper we suppose that

[3] Namely interpreting flexible quantification by $[\![\forall\alpha \geq \sigma.\tau]\!]_\varSigma = \bigcap_{S \supseteq [\![\sigma]\!]_\varSigma} [\![\tau]\!]_{\varSigma[\alpha \mapsto S]}$.

Syntactic definitions

$$
\begin{array}{lll}
\alpha, \beta, \ldots & \text{(type variables)} & x, y, z, \ldots \quad \text{(variables)} \\
\sigma, \tau ::= \alpha \mid \sigma \to \tau \mid \bot \mid \forall(\alpha \geq \sigma)\tau & \text{(types)} & a, b, c ::= x \mid \lambda(x : \tau)a \mid ab \\
\phi, \psi ::= \tau \mid \phi; \psi \mid \mathbf{1} \mid \& \mid \parr & & \quad \mid \Lambda(\alpha \geq \tau)a \mid a\phi \quad \text{(terms)} \\
\quad \mid\ !\alpha \mid \forall(\geq \phi) \mid \forall(\alpha \geq)\phi & \text{(instantiations)} & \quad \mid \mathtt{let}\, x = a\, \mathtt{in}\, b \\
\Gamma ::= \emptyset \mid \Gamma, \alpha \geq \tau \mid \Gamma, x : \tau & \text{(environments)} & A, B ::= a \mid \phi \quad \text{(expressions)}
\end{array}
$$

Instantiation rules

$$
\frac{}{\Gamma \vdash \tau : \bot \leq \tau}\,\textsc{IBot} \qquad
\frac{\Gamma, \alpha \geq \tau \vdash \phi : \tau_1 \leq \tau_2}{\Gamma \vdash \forall(\alpha \geq)\phi : \forall(\alpha \geq \tau)\tau_1 \leq \forall(\alpha \geq \tau)\tau_2}\,\textsc{IUnder}
$$

$$
\frac{\alpha \geq \tau \in \Gamma}{\Gamma \vdash\ !\alpha : \tau \leq \alpha}\,\textsc{IAbs} \qquad
\frac{\Gamma \vdash \phi : \tau_1 \leq \tau_2}{\Gamma \vdash \forall(\geq \phi) : \forall(\alpha \geq \tau_1)\tau \leq \forall(\alpha \geq \tau_2)\tau}\,\textsc{IInside}
$$

$$
\frac{\alpha \notin \mathrm{ftv}(\tau)}{\Gamma \vdash \parr : \tau \leq \forall(\alpha \geq \bot)\tau}\,\textsc{IIntro} \qquad
\frac{}{\Gamma \vdash \& : \forall(\alpha \geq \sigma)\tau \leq \sigma\{\tau/\alpha\}}\,\textsc{IElim}
$$

$$
\frac{\Gamma \vdash \phi : \tau_1 \leq \tau_2 \qquad \Gamma \vdash \psi : \tau_2 \leq \tau_3}{\Gamma \vdash \phi; \psi : \tau_1 \leq \tau_3}\,\textsc{IComp} \qquad
\frac{}{\Gamma \vdash \mathbf{1} : \tau \leq \tau}\,\textsc{IId}
$$

Typing rules

$$
\frac{\Gamma(x) = \tau}{\Gamma \vdash x : \tau}\,\textsc{Var} \qquad
\frac{\Gamma \vdash a : \tau \qquad \Gamma, x : \tau \vdash b : \sigma}{\Gamma \vdash \mathtt{let}\, x = a\, \mathtt{in}\, b : \sigma}\,\textsc{Let}
$$

$$
\frac{\Gamma, x : \tau \vdash a : \sigma}{\Gamma \vdash \lambda(x : \tau)a : \tau \to \sigma}\,\textsc{Abs} \qquad
\frac{\Gamma \vdash a : \sigma \to \tau \qquad \Gamma \vdash b : \sigma}{\Gamma \vdash ab : \tau}\,\textsc{App}
$$

$$
\frac{\Gamma, \alpha \geq \sigma \vdash a : \tau \qquad \alpha \notin \mathrm{ftv}(\Gamma)}{\Gamma \vdash \Lambda(\alpha \geq \sigma)a : \forall(\alpha \geq \sigma)\tau}\,\textsc{TAbs} \qquad
\frac{\Gamma \vdash a : \tau \qquad \Gamma \vdash \phi : \tau \leq \sigma}{\Gamma \vdash a\phi : \sigma}\,\textsc{TApp}
$$

Type instantiation

$$
\begin{array}{llll}
\tau(!\alpha) := \alpha, & \bot\tau := \tau, & \tau\mathbf{1} := \tau, & \tau(\phi; \psi) := (\tau\phi)\psi, \\
\tau\parr := \forall(\alpha \geq \bot)\tau, & \alpha \notin \mathrm{ftv}(\tau), & & (\forall(\alpha \geq \sigma)\tau)\& := \tau\{\sigma/\alpha\}, \\
\multicolumn{2}{l}{(\forall(\alpha \geq \sigma)\tau)(\forall(\geq \phi)) := \forall(\alpha \geq \sigma\phi)\tau,} & \multicolumn{2}{l}{(\forall(\alpha \geq \sigma)\tau)(\forall(\alpha \geq)\phi) := \forall(\alpha \geq \sigma)(\tau\phi).}
\end{array}
$$

Reduction rules

$$
\begin{array}{ll}
(\lambda(x : \tau)a)b \to_\beta a\{x/b\} & a\parr \to_\iota \Lambda(\alpha \geq \bot)a, \quad \alpha \notin \mathrm{ftv}(\tau) \\
\mathtt{let}\, x = b\, \mathtt{in}\, a \to_\beta a\{x/b\} & (\Lambda(\alpha \geq \tau)a)\& \to_\iota a\{\mathbf{1}/!\alpha\}\{\tau/\alpha\} \\
a\mathbf{1} \to_\iota a & (\Lambda(\alpha \geq \tau)a)(\forall(\alpha \geq)\phi) \to_\iota \Lambda(\alpha \geq \tau)(a\phi) \\
a(\phi; \psi) \to_\iota (a\phi)\psi & (\Lambda(\alpha \geq \tau)a)(\forall(\geq \phi)) \to_\iota \Lambda(\alpha \geq \tau\phi)a\{\phi; !\alpha/!\alpha\}
\end{array}
$$

Fig. 1. Syntactic definitions, typing and reduction rules of $\mathsf{xML^F}$

in all $\mathsf{xML^F}$ terms every $\mathtt{let}\, x = a\, \mathtt{in}\, b$ has been replaced by $(\lambda(x : \sigma)b)a$, with σ the correct type of a.

The *type erasure* $\lceil a \rceil$ of an $\mathsf{xML^F}$ (or $\mathsf{eML^F}$) term a is straightforwardly defined by erasing all type and instantiation annotations, mapping a to an ordinary λ-term. From [4, Lemma 7, Theorem 6 and §4.2] we know the following.

Theorem 1. *For every* $\mathsf{iML^F}$ *or* $\mathsf{eML^F}$ *term* a*, there is an* $\mathsf{xML^F}$ *term* $[\![a]\!]$ *such that* $\lceil [\![a]\!] \rceil = \lceil a \rceil$[4].

[4] We only need to apply type erasure on the right for partially annotated $\mathsf{eML^F}$ terms.

3 The Coercion Calculus $\mathsf{F_c}$

In this section we will introduce the *coercion calculus* $\mathsf{F_c}$, which is (as shown in subsection 3.2) a decoration of system F accompanied by a type system. Before introducing the details, we point out that the version of $\mathsf{F_c}$ presented here is tailored down to suit $\mathsf{xML^F}$. As such, there are natural choices that have been intentionally left out or restrained. If $\mathsf{F_c}$ is to serve as a good meta-theory of coercions, more liberal choices and constructs are needed, as discussed at page 536. The syntax, the type system and the reduction rules of $\mathsf{F_c}$[5] are presented in Figure 2. In this calculus the notion of 'coercion' is captured by suitable types.

Syntactic definitions

$\alpha, \beta, \ldots$	(type variables)	$\Gamma ::= \emptyset \mid x : \tau, \Gamma$ $\mid\ x : \sigma \multimap \alpha, \Gamma$	(regular env.)
$\sigma, \tau ::= \alpha \mid \sigma \to \tau$ $\mid\ \kappa \to \tau \mid \forall \alpha.\tau$	(types)	$L ::= \emptyset \mid z : \tau$	(linear env.)
$\kappa ::= \sigma \multimap \tau$	(coercion types)	$\Gamma; L$	(environments)
$\zeta ::= \tau \mid \kappa$	(type expr.)	$\Gamma; \vdash_{\mathsf{t}} a : \sigma$	(term judgements)
$x, y, z, \ldots$	(variables)	$\Gamma; \vdash_{\mathsf{c}} a : \sigma \multimap \tau$	(coercion judgements)
$a, b ::= x \mid \lambda x.a \mid \underline{\lambda} x.a \mid \underline{\not\lambda} x.a$ $\mid\ ab \mid a \rhd b \mid a \lhd b$	(terms)	$\Gamma; z : \tau \vdash_{\ell} a : \sigma$	(linear judgements)
$u, v ::= \lambda x.a \mid \underline{\not\lambda} x.u \mid x \rhd u$	(c-values)	$\vdash_{\mathsf{xy}}$, $\mathsf{x}, \mathsf{y} \in \{\mathsf{t}, \mathsf{c}, \ell\}$ stands for $\vdash_{\mathsf{x}}$ or $\vdash_{\mathsf{y}}$.	

Typing rules

$$\frac{\Gamma(y) = \zeta}{\Gamma; \vdash_{\mathsf{t}\ell} y : \zeta}\,\textsc{Ax} \qquad \frac{\Gamma, x : \tau; \vdash_{\mathsf{t}} a : \sigma}{\Gamma; \vdash_{\mathsf{t}} \lambda x.a : \tau \to \sigma}\,\textsc{Abs} \qquad \frac{\Gamma; \vdash_{\mathsf{t}} a : \sigma \to \tau \quad \Gamma; \vdash_{\mathsf{t}} b : \sigma}{\Gamma; \vdash_{\mathsf{t}} ab : \tau}\,\textsc{App}$$

$$\frac{}{\Gamma; z : \tau \vdash_{\ell} z : \tau}\,\textsc{LAx} \qquad \frac{\Gamma; z : \tau \vdash_{\ell} a : \sigma}{\Gamma; \vdash_{\mathsf{c}} \underline{\lambda} z.a : \tau \multimap \sigma}\,\textsc{LAbs} \qquad \frac{\Gamma, x : \kappa; L \vdash_{\mathsf{t}\ell} a : \sigma}{\Gamma; L \vdash_{\mathsf{t}\ell} \underline{\not\lambda} x.a : \kappa \to \sigma}\,\textsc{CAbs}$$

$$\frac{\Gamma; \vdash_{\mathsf{c}} a : \sigma_1 \multimap \sigma_2 \quad \Gamma; L \vdash_{\mathsf{t}\ell} b : \sigma_1}{\Gamma; L \vdash_{\mathsf{t}\ell} a \rhd b : \sigma_2}\,\textsc{LApp} \qquad \frac{\Gamma; L \vdash_{\mathsf{t}\ell} a : \kappa \to \sigma \quad \Gamma \vdash_{\mathsf{c}} b : \kappa}{\Gamma; L \vdash_{\mathsf{t}\ell} a \lhd b : \sigma}\,\textsc{CApp}$$

$$\frac{\Gamma; L \vdash_{\mathsf{t}\ell} a : \sigma \quad \alpha \notin \mathrm{ftv}(\Gamma; L)}{\Gamma; L \vdash_{\mathsf{t}\ell} a : \forall \alpha.\sigma}\,\textsc{Gen} \qquad \frac{\Gamma; L \vdash_{\mathsf{t}\ell} a : \forall \alpha.\sigma}{\Gamma; L \vdash_{\mathsf{t}\ell} a : \sigma\{\tau'/\alpha\}}\,\textsc{Inst}$$

Reduction rules

$$(\lambda x.a)b \to_{\beta} a\{b/x\}, \qquad (\underline{\not\lambda} x.a) \lhd b \to_{\mathsf{c}} a\{b/x\}, \qquad (\underline{\lambda} x.a) \rhd b \to_{\mathsf{c}} a\{b/x\},$$

$$(\underline{\not\lambda} x.u) \lhd b \to_{\mathsf{cv}} u\{b/x\}, \qquad (\underline{\lambda} x.a) \rhd u \to_{\mathsf{cv}} a\{u/x\}, \qquad \text{if } u \text{ is a c-value.}$$

Fig. 2. Syntactic definitions, typing and reduction rules of coercion calculus

Definition 2 (Coercion). *An* $\mathsf{F_c}$ *term* a *is a* coercion *if* $\Gamma; \vdash_{\mathsf{c}} a : \sigma \multimap \tau$.

The use of linear implication for the type of coercions is not casual. Indeed the type system can be seen as a fragment of DILL, the dual intuitionistic linear logic [7]. This captures an aspect of coercions: they consume their argument

[5] We present the coercion calculus in Curry-style, whereas arguably its usefulness outside of this work would rather be in Church-style (which is easy to define).

without erasing it (as they must preserve it) nor duplicate it (as there is no true computation, just a type recasting). *Environments* are of shape $\Gamma; L$, where Γ is a map from variables to type expressions, and L is the *linear* part of the environment, containing (contrary to DILL) *at most* one assignment. Notice the restriction to $\sigma \multimap \alpha$ for coercion variables, which might at first seem overtly restrictive. However, Theorem 19 relies on this restriction, though the preceding results do not. Alternative, more permissive restrictions preserving the bisimulation result are left for future work.

Typing judgments come in three sorts. However, the subscripts we use to distinguish them ($\vdash_{\mathfrak{t}}$, $\vdash_{\mathfrak{c}}$ and $\vdash_{\ell}$) are only for easy recognition, as the sort of the judgment can be recovered from the shape of the environment and the type.

A note on DILL *and λ-calculus.* The language presented in [7] is the term calculus of the logical system, and as such has a constructor for every logical rule. Notably, that work provides no intuitionistic arrow, as the translation $A \to B \cong {!A} \multimap B$ is preferred. Employing DILL as a type system for ordinary λ-terms leads to a system (which we might call F_ℓ) using types rather than terms to strictly differentiate between linear and regular constructs. This system is known as *folklore*[6] but, as far as we know, it has never been studied in the literature. The absence of a thorough presentation of F_ℓ prevents us from deriving properties such as subject reduction (Theorem 7) more or less directly from a more general framework. We leave to further work the rather straightforward presentation of such a system together with a more general version of F_c, along the lines hinted at page 536.

Syntax. F_c terms are extensions of usual λ-terms with two abstractions $\underline{\lambda}$, $\underline{\measuredangle}$ and two applications $\triangleright$, $\triangleleft$. The *linear abstraction* $\underline{\lambda}$ (whose application is $\triangleright$) is used by coercions to ask for the regular term to coerce, so they cannot erase or duplicate it. The *coercion abstraction* $\underline{\measuredangle}$ (whose application is $\triangleleft$) can be used in regular or coercion terms to ask for a coercion, so it is not subject to particular restrictions. The applications $\triangleright$, $\triangleleft$ locate coercions within the terms without carrying the typing around: the triangle's side indicates where the coercion is.

Reductions. Reduction steps are divided into $\to_\beta$ (the actual computation) and $\to_\mathsf{c}$ (the coercion reduction). The reduction $\to_\mathsf{c}$ has a conditional subreduction $\to_{\mathsf{cv}}$ that fires c-redexes only when c-values are at the right of the $\triangleright$ or left of the $\triangleleft$. Intuitively, this reduction is what is strictly necessary to "unearth" a λ-abstraction. Its main role here is that it is general enough to have bisimulation (Theorem 19) and small enough to correspond to xML^F's ι-steps (Theorem 26). As usual, rules are closed by context.

3.1 Some Basic Properties of F_c

We start presenting some basic properties of the coercion calculus. The first statements restrain the shape and the behaviour of coercions.

[6] As an example we might cite [8], where a fragment of F_ℓ is used to characterize poly-time functions.

Remark 3. A coercion a is necessarily either a variable or a coercion abstraction, as AX and LABS are the only rules having a coercion type in the conclusion.

Lemma 4. *If $\Gamma; L \vdash_{\mathsf{c}\ell} a : \zeta$ then no subterm of a is of the form $\lambda x.b$ or bc. In particular a is β-normal.*

Lemma 5. *Let a be an $\mathsf{F_c}$ term. If $\Gamma; \vdash_\mathsf{c} a : \sigma \multimap \tau$, then a is* cv*-normal.*

Proof. Immediate by Theorem 4: there cannot be any subterm $\lambda x.a'$ of a, so in particular a does not contain any c-value. □

Following are basic properties of type systems. Note that though there are two substitution results (points (ii), (iii) below) to accommodate the two types of environment, no weakening property is available to add the linear assignment.

Lemma 6 (Weakening and substitution). *We have the following:*

(i) $\Gamma; L \vdash_{\mathsf{tc}\ell} a : \zeta$ *and* $x \notin \mathrm{dom}(\Gamma; L)$ *entail* $\Gamma, x : \zeta'; L \vdash_{\mathsf{tc}\ell} a : \zeta$;
(ii) $\Gamma; \vdash_{\mathsf{tc}} a : \zeta'$ *and* $\Gamma, x : \zeta'; L \vdash_{\mathsf{tc}\ell} b : \zeta$ *entail* $\Gamma; L \vdash_{\mathsf{tc}\ell} b\{a/x\} : \zeta$;
(iii) $\Gamma; L \vdash_{\mathsf{t}\ell} a : \sigma$ *and* $\Gamma; x : \sigma \vdash_\ell b : \zeta$ *entail* $\Gamma; L \vdash_{\mathsf{t}\ell} b\{a/x\} : \zeta$.

Proposition 7 (Subject reduction). *If $a \rightarrow_{\beta\mathsf{c}} b$ and $\Gamma; L \vdash_{\mathsf{t}\ell\mathsf{c}} a : \zeta$, then $\Gamma; L \vdash_{\mathsf{t}\ell\mathsf{c}} b : \zeta$.*

Proposition 8 (Confluence). *All of $\rightarrow_\beta$, $\rightarrow_\mathsf{c}$, $\rightarrow_\mathsf{cv}$ and $\rightarrow_{\beta\mathsf{c}}$ are confluent.*

Proof. The proof by Tait-Martin Löf's technique of parallel reductions does not pose particular issues. □

3.2 Coercion Calculus as a Decoration of System F

The following definition presents the coercion calculus as a simple decoration of usual Curry-style system F. The latter can be recovered by just collapsing the extraneous constructs $\multimap$, $\underline{\lambda}$, $\measuredangle$, $\triangleleft$ and $\triangleright$ to their regular counterpart. Notably this will lead to a strong normalization result.

Definition 9. *The* decoration erasure *is defined by:*

$$|\alpha| := \alpha, \quad |\zeta \to \tau| := |\zeta| \to |\tau|, \quad |\sigma \multimap \tau| := |\sigma| \to |\tau|,$$
$$|x| := x, \quad |\lambda x.a| = |\underline{\lambda} x.a| = |\measuredangle x.a| := \lambda x.|a|, \quad |a \triangleleft b| = |a \triangleright b| = |ab| := |a||b|,$$
$$|\Gamma|(y) := |\Gamma(y)| \ \ \textit{for } y \in \mathrm{dom}(\Gamma), \qquad |\Gamma; z : \tau| := |\Gamma|, z : |\tau|.$$

The next lemma ensures that the decoration erasure preserves typability (with system F's typability denoted by $\vdash_\mathsf{F}$).

Lemma 10. *Let a be an $\mathsf{F_c}$ term. If $\Gamma; L \vdash_{\mathsf{t}\ell} a : \zeta$ then $|\Gamma; L| \vdash_\mathsf{F} |a| : |\zeta|$.*

Proof. It suffices to see that through $|\,.\,|$ all the new rules collapse to their regular counterpart: LAX becomes AX, CABS, LABS become ABS, and CAPP, LAPP become APP. In the latter cases the weakening lemma for $\vdash_\mathsf{F}$ may have to be applied to add the missing $z : |\tau|$ to one of the two branches. □

Lemma 11. *Given an* $\mathsf{F_c}$ *term* a *we have* $|a|\{|b|/x\} = |a\{b/x\}|$. *Moreover, if* $a \to_{\beta\mathsf{c}} b$ *then* $|a| \to |b|$. *If* a *is typable, then* $|a| \to c$ *implies* $c = |b|$ *with* $a \to_{\beta\mathsf{c}} b$.

Proof. The first two claims are immediate. The converse needs the typability hypothesis: take $|a| = (\lambda x.b_1')b_2'$, then there are b_i with $|b_i| = b_i'$ and a is one of nine combinations $((\lambda x.b_1)b_2, (\underline{\lambda} x.b_1)b_2, (\lambda x.b_1) \lhd b_2$, etc.$)$. However as a is typable only the three matching combinations are possible, giving rise to the three possible redexes in the coercion calculus. □

Corollary 12 (Termination). *The coercion calculus is strongly normalizing.*

Proof. Immediate by Lemmas 10 and 11, using the strong normalization of system F [2, Sec. 14.3]. □

3.3 Preservation of the Semantics

We will now turn to establishing why coercions $a : \tau \multimap \sigma$ can be truly called such. First, we need a way to extract the semantics of a term, i.e., a way to strip it of the structure one may have added to it in order to manage coercions.

Definition 13. *The* coercion erasure *is defined by*

$$\lfloor x \rfloor := x, \quad \lfloor \lambda x.a \rfloor := \lambda x.\lfloor a \rfloor, \quad \lfloor ab \rfloor := \lfloor a \rfloor \lfloor b \rfloor,$$
$$\lfloor \underline{\lambda} x.a \rfloor = \lfloor \underline{\measuredangle} x.a \rfloor := \lfloor a \rfloor, \quad \lfloor a \lhd b \rfloor := \lfloor a \rfloor, \quad \lfloor a \rhd b \rfloor := \lfloor b \rfloor.$$

Lemma 14.

(i) If $\Gamma, x : \kappa; L \vdash_{\mathsf{t}\ell} a : \sigma$ *then* $x \notin \mathrm{fv}(\lfloor a \rfloor)$*;*
(ii) if $\Gamma; z : \tau \vdash_{\ell} a : \sigma$ *then* $\lfloor a \rfloor = z$.

Notice that property (i) above entails that $\lfloor . \rfloor$ is well-defined with respect to α-equivalence on regular, typed terms: given a term $\underline{\measuredangle} x.a$ issued from a coercion abstraction, $\lfloor \underline{\measuredangle} x.a \rfloor = \lfloor a \rfloor$ is independent from x. This is not the case for coercions, as for example $\lfloor \underline{\lambda} x.x \rfloor = x$.

As for property (ii), it greatly restricts the form of a coercion: if $a : \sigma \multimap \tau$ then it is either a variable or an abstraction $\underline{\lambda} x.a'$ (as already written in Theorem 3), with $\lfloor a' \rfloor = x$. Apart when they are variables, coercions are essentially identities.

The problem whether the erasure maps $\mathsf{F_c}$ to a larger set of terms than system F is an open one, probably related to the open question whether $\mathsf{ML^F}$ types more terms than System F.

A note on unrestricted coercion variables. If we dropped the condition on coercion variables, namely that they are typed $\sigma \multimap \alpha$ in the context, we would get way too many terms: indeed the coercion erasure would cover the whole of the untyped λ-calculus. It would suffice to use two coercion variables $y_{o\to o} : o \multimap (o \to o)$ and $y_o : (o \to o) \multimap o$ modelling the recursive type $o \to o \simeq o$. For example, we would have $a_\delta := y_o \rhd (\lambda x.(y_{o\to o} \rhd x)x) : o$ and $a_\Delta := (y_{o\to o} \rhd a_\delta)a_\delta : o$, though $\lfloor a_\Delta \rfloor = (\lambda x.xx)(\lambda x.xx)$ is the renown divergent and untypable term.

Lemma 15. $\lfloor a\{b/x\}\rfloor = \lfloor a\rfloor\{\lfloor b\rfloor/x\}$.

Lemma 16. *If $\Gamma; x:\tau \vdash_\ell a:\sigma$ and $b \to_\beta c$, then $a\{b/x\} \to_\beta a\{c/x\}$.*

Proof (sketch). Essentially the proof is by linearity of x in a. Formally it is carried out by an easy induction on the derivation. □

The following will state some basic dynamic properties of coercion reductions. Intuitively we will prove that β-steps are actual steps of the semantics (point (ii)) and that c-steps preserves it in a strong sense: they are collapsed to the equality (point (iii)) and they preserve β-steps (point (i)).

Proposition 17. *Suppose that a is an $\mathsf{F_c}$ term. Then:*

(i) if $b_1 \leftarrow_{\mathsf{c}} a \to_\beta b_2$ then there is c with $b_1 \to_\beta c \overset{*}{\leftarrow}_{\mathsf{c}} b_2$;
(ii) if $a \to_\beta b$ then $\lfloor a\rfloor \to \lfloor b\rfloor$;
(iii) if $a \to_{\mathsf{c}} b$ then $\lfloor a\rfloor = \lfloor b\rfloor$.

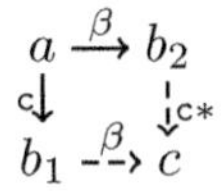

In order to truly see coercions as additional information that is not strictly needed for reduction, one may ask that some converse of property (ii) should also hold. Here the condition on coercion variables ($x:\sigma \multimap \alpha$) starts to play a role[7]. Indeed in general this is not the case: take $a = \measuredangle y.(y \triangleright I)I$ with $I = \lambda x.x$, that would be typable with $; \vdash a : (\sigma_{\mathrm{id}} \multimap \sigma_{\mathrm{id}}) \to \sigma_{\mathrm{id}}$ (where $\sigma_{\mathrm{id}} := \forall\alpha.(\alpha \to \alpha)$). Its coercion erasure is typable but it has a redex that is blocked by a coercion variable.

With the condition on coercion variables in place we are ready to prove a complete correspondence between the β-reductions of the coerced terms and the ones of their coercion erasure. In fact Theorem 19 states that $a \mapsto \lfloor a\rfloor$ is a weak bisimulation for $\to_\beta$, taking $\to_{\mathsf{cv}}$ as the silent actions on the side of coercion calculus. The proof uses the following lemma.

Lemma 18. *Every typable cv-normal term a such that $\lfloor a\rfloor = \lambda x.b$ is a c-value. In particular if a has an arrow type then $a = \lambda x.c$ with $\lfloor c\rfloor = b$.*

Theorem 19 (Bisimulation of $\lfloor\,.\,\rfloor$). *If $\Gamma;\vdash_{\mathsf{t}} a:\sigma$, then $\lfloor a\rfloor \to_\beta b$ iff $a \overset{*}{\to}_{\mathsf{cv}}\to_\beta c$ with $\lfloor c\rfloor = b$.*

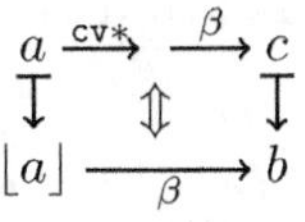

Proof. The if part is given by Theorem 17. For the only if part we can suppose that $a = a_1 a_2$ with $\lfloor a_1\rfloor = \lambda x.d$, so that $(\lambda x.d)\lfloor a_2\rfloor$ is the redex fired in $\lfloor a\rfloor$, i.e. $b = d\{\lfloor a_2\rfloor/x\}$. We can reduce to such a case reasoning by structural induction on a, discarding all the parts of the context where the reduction does not occur.

As a_1 is applied to a_2 there is a derivation giving $\Gamma';\vdash_{\mathsf{t}} a_1 : \tau \to \tau'$ for some Γ', τ, τ'. We can then cv-normalize a_1 to a_1' (Theorem 12), which by subject reduction has the same type. Moreover by Theorem 17(iii) $\lfloor a_1'\rfloor = \lfloor a_1\rfloor = \lambda x.d$, and we conclude by Theorem 18 that $a_1' = \lambda x.e$ with $\lfloor e\rfloor = d$, and we finally get $a_1 a_2 \overset{*}{\to}_{\mathsf{cv}} (\lambda x.e)a_2 \to_\beta e\{a_2/x\}$. Now by Theorem 15 $\lfloor e\{a_2/x\}\rfloor = \lfloor e\rfloor\{\lfloor a_2\rfloor/x\} = d\{\lfloor a_2\rfloor/x\} = b$ and we are done. □

Notice that the above result entails bisimulation with $\to_{\mathsf{c}}$ as a more general silent action: Theorem 17 gives the if part, while $\to_{\mathsf{cv}} \subseteq \to_{\mathsf{c}}$ gives the only if one.

[7] All the results shown so far are valid also without such a condition.

Types and contexts

$$\alpha^\bullet := \alpha, \qquad (\sigma \to \tau)^\bullet := \sigma^\bullet \to \tau^\bullet, \qquad (x : \tau)^\bullet := x : \tau^\bullet,$$
$$\bot^\bullet := \forall\alpha.\alpha, \quad (\forall(\alpha \geq \sigma)\tau)^\bullet := \forall\alpha.(\sigma^\bullet \multimap \alpha) \to \tau^\bullet, \quad (\alpha \geq \tau)^\bullet := i_\alpha : \tau^\bullet \multimap \alpha.$$

Instantiations

$$\tau^\circ := \underline{\lambda} x.x, \qquad (\parr)^\circ := \underline{\lambda} x.\underline{\Lambda} i_\alpha.x, \qquad (\phi;\psi)^\circ := \underline{\lambda} z.\psi^\circ \rhd (\phi^\circ \rhd z),$$
$$(!\alpha)^\circ := i_\alpha, \qquad (\&)^\circ := \underline{\lambda} x.x \lhd \underline{\lambda} z.z, \qquad (\mathbf{1})^\circ := \underline{\lambda} z.z,$$
$$(\forall(\geq \phi))^\circ := \underline{\lambda} x.\underline{\Lambda} i_\alpha.x \lhd (\underline{\lambda} z.i_\alpha \rhd (\phi^\circ \rhd z)),$$
$$(\forall(\alpha \geq)\phi)^\circ := \underline{\lambda} x.\underline{\Lambda} i_\alpha.\phi^\circ \rhd (x \lhd i_\alpha).$$

Terms

$$x^\circ := x, \qquad (\lambda(x:\tau)a)^\circ := \lambda x.a^\circ, \qquad (ab)^\circ := a^\circ b^\circ,$$
$$(\Lambda(\alpha \geq \tau)a)^\circ := \underline{\Lambda} i_\alpha.a^\circ, \qquad (a\phi)^\circ := \phi^\circ \rhd a^\circ.$$

Fig. 3. Translation of types, instantiations and terms into the coercion calculus. For every type variable α we suppose fixed a fresh term variable i_α.

4 The Translation

A translation from $\mathsf{xML^F}$ terms and instantiations into the coercion calculus is given in Figure 3. The idea is that instantiations can be seen as coercions; thus a term starting with a type abstraction $\Lambda(\alpha \geq \tau)$ becomes a term waiting for a coercion of type $\tau^\bullet \multimap \alpha$, and a term $a\phi$ becomes a° coerced by ϕ°. The rest of this section is devoted to showing how this translation and the properties of the coercion calculus lead to the main result of this work, SN of both $\mathsf{xML^F}$ and $\mathsf{eML^F}$. First one needs to show that the translation maps to typed terms. Then with a substitution lemma we will be close to our result.

Lemma 20. *Let a be an $\mathsf{xML^F}$ term and ϕ be an instantiation:*

(i) if $\Gamma \vdash \phi : \sigma \leq \tau$ then $\Gamma^\bullet; \vdash_{\mathsf{c}} \phi^\circ : \sigma^\bullet \multimap \tau^\bullet$.
(ii) if $\Gamma \vdash a : \sigma$ then $\Gamma^\bullet; \vdash_{\mathsf{t}} a^\circ : \sigma^\bullet$.

Lemma 21. *Let A be an $\mathsf{xML^F}$ term or an instantiation. Then we have:*

(i) $(A\{b/x\})^\circ = A^\circ\{b^\circ/x\}$,
(ii) $(A\{\mathbf{1}/!\alpha\}\{\tau/\alpha\})^\circ = A^\circ\{\underline{\lambda} z.z/i_\alpha\}$,
(iii) $(A\{\phi;!\alpha/!\alpha\})^\circ = A^\circ\{(\underline{\lambda} z.i_\alpha \rhd (\phi^\circ \rhd z))/i_\alpha\}$.

Theorem 22 (Coercion calculus simulates $\mathsf{xML^F}$). *If $a \to_\beta b$ (resp. $a \to_\iota b$) in $\mathsf{xML^F}$, then $a^\circ \to_\beta b^\circ$ (resp. $a^\circ \xrightarrow{+}_{\mathsf{c}} b^\circ$) in coercion calculus.*

Proof. (Sketch) As the translation is contextual, it suffices to analyze each reduction rule, perform the reductions and apply Theorem 21 where needed. □

Corollary 23 (Termination). $\mathsf{xML^F}$ *is strongly normalizing.*

The above already shows SN of $\mathsf{xML^F}$, however in order to prove that $\mathsf{eML^F}$ is also normalizing we need to make sure that ι-redexes cannot block β ones: in other words, a bisimulation result. We first need some technical lemmas, proved by structural induction. We recall that $\lceil a \rceil$ is the type erasure of a (page 528).

Lemma 24. *The type erasure of an* $\mathsf{xML^F}$ *term* a *coincides with the coercion erasure of its translation, i.e.* $\lceil a \rceil = \lfloor a^\circ \rfloor$.

Lemma 25.

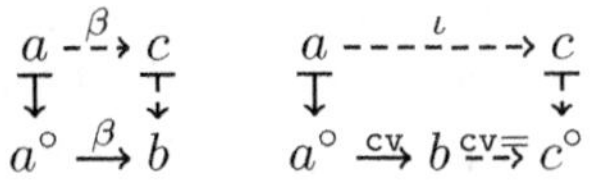

(i) If $a^\circ \to_\beta b$ *then* $a \to_\beta c$ *with* $c^\circ = b$;
(ii) if $a^\circ \to_{\mathsf{cv}} b$ *then* $a \to_\iota c$ *with* $b \overset{=}{\to}_{\mathsf{cv}} c^\circ$.

Notice that the above is not true in general for $\to_{\mathsf{c}}$ in place of $\to_{\mathsf{cv}}$: for example $x\&$ is normal in $\mathsf{xML^F}$, but $(x\&)^\circ = (\lambda y.y) \rhd x \to_{\mathsf{c}} x$.

The following lemma allows us lift to $\mathsf{xML^F}$ the reduction in coercion calculus that bisimulates β-steps (see Theorem 19).

Lemma 26 (Lifting). *Given a typed* $\mathsf{xML^F}$ *term* a, *we have that if* $a^\circ \overset{*}{\to}_{\mathsf{cv}}\to_\beta b$ *then* $a \overset{*}{\to}_\iota\to_\beta c$ *with* $b \overset{*}{\to}_{\mathsf{c}} c^\circ$.

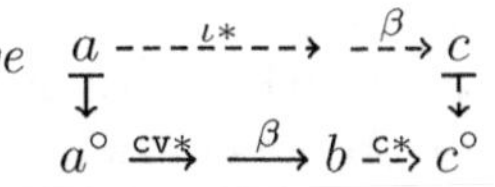

Proof. As $\to_{\mathsf{cv}}$ is strongly normalizing (Theorem 12), we can reason by well-founded induction on a° with respect to $\to_{\mathsf{cv}}$.

First let us suppose that $a^\circ \to_\beta b$: we then apply Theorem 25(i) and get the result directly. Suppose then that $a^\circ \overset{+}{\to}_{\mathsf{cv}}\to_\beta b$. We have the following diagram:

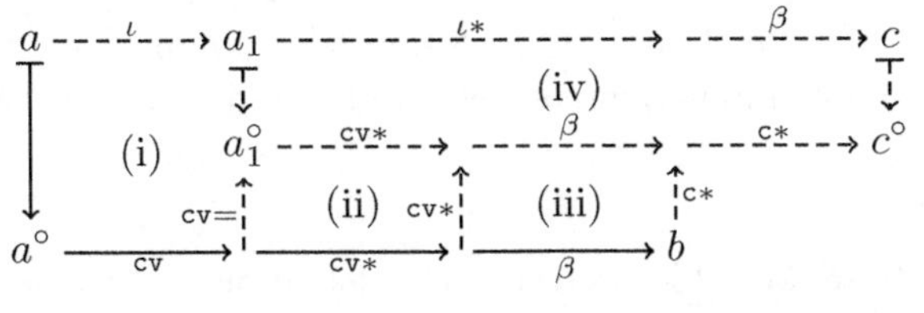

where (i) comes from Theorem 25(ii), (ii) is by confluence (Theorem 8), (iii) is by Theorem 17(i) and (iv) is by inductive hypothesis, as $a^\circ \overset{+}{\to}_{\mathsf{cv}} a_1^\circ$. □

Theorem 27 (Bisimulation of $\lceil . \rceil$). *Given a typed* $\mathsf{xML^F}$ *term* a, *we have that* $\lceil a \rceil \to_\beta b$ *iff* $a \overset{*}{\to}_\iota\to_\beta c$ *with* $\lceil c \rceil = b$.

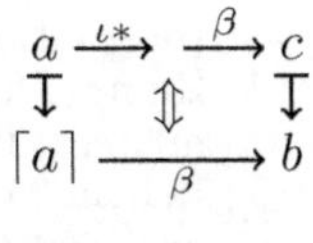

Proof. For the if part, by Theorem 22 we have $a^\circ \overset{*}{\to}_{\mathsf{c}}\to_\beta c^\circ$, which by Theorem 24 and Theorem 17 implies $\lceil a \rceil = \lfloor a^\circ \rfloor \to_\beta \lfloor c^\circ \rfloor = \lceil c \rceil$. For the only if part, as $\lfloor a^\circ \rfloor = \lceil a \rceil \to_\beta b$, by Theorem 19 $a^\circ \overset{*}{\to}_{\mathsf{cv}}\to_\beta b'$ with $\lfloor b' \rfloor = b$. Now by Theorem 26 we have that $b' \overset{*}{\to}_{\mathsf{c}} c^\circ$ with $a \overset{*}{\to}_\iota\to_\beta c$. To conclude, we see that $\lceil c \rceil = \lfloor c^\circ \rfloor = \lfloor b' \rfloor = b$, where we used Theorem 24 and Theorem 17(iii). □

The above proof may be completely carried out within $\mathsf{xML^F}$, by applying a suitably modified version of Theorem 18. However, we preferred this formulation since it provides a better understanding of what happens on the side of the coercion calculus.

Corollary 28. *Terms typed in* $\mathsf{iML^F}$ *and* $\mathsf{eML^F}$ *are strongly normalizing.*

Proof. Immediate by the above result and Theorem 1. □

Further work. We were able to prove new results for $\mathsf{ML^F}$ (namely SN and bisimulation of $\mathsf{xML^F}$ with its type erasure) by employing a more general calculus of coercions. It becomes natural then to ask whether its type system may be a framework to study coercions in general. A first natural target are the coercions arising from Leijen's translation of $\mathsf{ML^F}$ [5], which is more optimized than ours, in the sense that it does not add additional and unneeded structure to system F types. We plan then to study the coercions arising in F_{η} [9] or when using subtyping [10]. As explained at the beginning of section 3, $\mathsf{F_c}$ was purposely tailored down to suit $\mathsf{xML^F}$, stripping it of natural features.

A first, easy extension would consist in more liberal types and typing rules, allowing coercion polymorphism, coercion abstraction of coercions or even coercions between coercions (i.e. allowing types $\forall\alpha.\kappa$, $\kappa_1 \to \kappa_2$ and $\kappa_1 \multimap \kappa_2$). To progress further however, one would need a way to build coercions of arrow types, which are unneeded in $\mathsf{xML^F}$. Namely, given coercions $c_1 : \sigma_2 \multimap \sigma_1$ and $c_2 : \tau_1 \multimap \tau_2$, there should be a coercion $c_1 \Rightarrow c_2 : (\sigma_1 \to \tau_1) \multimap (\sigma_2 \to \tau_2)$, allowing a reduction $(c_1 \Rightarrow c_2) \triangleright \lambda x.a \to_{\mathsf{c}} \lambda x.c_2 \triangleright a\,\{c_1 \triangleright x/x\}$. This could be achieved either by introducing it as a primitive, by translation or by special typing rules. Indeed, if some sort of η-expansion would be available while building a coercion, one could write $c_1 \Rightarrow c_2 := \underline{\lambda} f.\lambda x.(c_2 \triangleright (f(c_1 \triangleright x)))$. However how to do this without loosing bisimulation is under investigation.

Acknowledgements. We thank Didier Rémy for stimulating discussions and remarks, and the anonymous referees for careful reading and detailed comments.

References

1. Milner, R., Tofte, M., Macqueen, D.: The definition of Standard ML. MIT Press, Cambridge (1997)
2. Girard, J.Y., Lafont, Y., Taylor, P.: Proofs and Types. Cambridge tracts in theoretical computer science, vol. 7. Cambridge University Press, Cambridge (1989)
3. Le Botlan, D., Rémy, D.: MLF: Raising ML to the power of System F. In: Proc. of International Conference on Functional Programming (ICFP 2003), pp. 27–38 (2003)
4. Rémy, D., Yakobowski, B.: A Church-style intermediate language for MLF. Submitted (July 2009)
5. Leijen, D.: A type directed translation of MLF to System F. In: Proc. of International Conference on Functional Programming (ICFP 2007). ACM Press, New York (2007)
6. Botlan, D.L., Rémy, D.: Recasting MLF. Inf. Comput. 207(6), 726–785 (2009)
7. Barber, A., Plotkin, G.: Dual intuitionistic linear logic. Technical report LFCS-96-347, University of Edinburgh (1997)
8. Baillot, P., Terui, K.: Light types for polynomial time computation in lambda calculus. Inf. Comput. 207(1), 41–62 (2009)
9. Mitchell, J.C.: Coercion and type inference. In: Proc. of 11th Symposium on Principles of Programming Languages (POPL 1984), pp. 175–185. ACM, New York (1984)
10. Crary, K.: Typed compilation of inclusive subtyping. In: Proc. of International Conference on Functional Programming (ICFP 2000), pp. 68–81 (2000)

Describing Average- and Longtime-Behavior by Weighted MSO Logics

Manfred Droste and Ingmar Meinecke

Institut für Informatik, Universität Leipzig,
D-04109 Leipzig, Germany
{droste,meinecke}@informatik.uni-leipzig.de

Abstract. Weighted automata model quantitative aspects of systems like memory or power consumption. Recently, Chatterjee, Doyen, and Henzinger introduced a new kind of weighted automata which compute objectives like the average cost or the longtime peak power consumption. In these automata, operations like average, limit superior, limit inferior, limit average, or discounting are used to assign values to finite or infinite words. In general, these weighted automata are not semiring weighted anymore. Here, we establish a connection between such new kinds of weighted automata and weighted logics. We show that suitable weighted MSO logics and these new weighted automata are expressively equivalent, both for finite and infinite words. The constructions employed are effective, leading to decidability results for the weighted logic formulas considered.

1 Introduction

In the last years, there has been increasing interest in quantitative features of the specification and analysis of systems. Such quantitative aspects are the consumption of a certain resource or the output of a benefit. Important considerations concern the overall peak of power consumption, the long-time maximal output, or the average consumption of some resource. Weighted automata modeling the average and long-time behavior of systems were introduced recently by Chatterjee, Doyen, and Henzinger [5,6,7,8]. These weighted automata do not fit into the framework of other weighted automata like semiring weighted automata, cf. [1,11,22,23,29], or lattice automata [24]. Here, we present weighted logics which are expressively equivalent to such new weighted automata.

The connection between MSO logic and finite automata as established by Büchi and Elgot [4,17] has proven to be most fruitful. A weighted version of such a result was shown for a new semiring weighted MSO logic a few years ago [9]. The concept of weighted MSO logics worked not only fine for words but also for other structures like trees, traces, pictures, nested words or timed words, see e.g. [14,19,26,27,28]. Such weighted logics also encompass several versions of probabilistic CTL, cf. [3].

The weighted MSO logic used here combines the one from [9,10] (usual MSO logic enriched by the elements of the weight structure) and an idea from [3]. Concerning the semantics, disjunction and existential quantification were interpreted by the sum, whereas the semantics of conjunction and universal quantification were defined by the product of the semiring. If we use the max-plus-semiring for example, the semantical interpretation of $\forall x\varphi$ is the sum of all weights (rewards or time) defined by φ for all

P. Hliněný and A. Kučera (Eds.): MFCS 2010, LNCS 6281, pp. 537–548, 2010.

different positions x. But there are other objectives of high interest: the average reward over all positions, the long-run average reward for infinite sequences, the long-run maximal reward, or a discounted sum of the rewards gained at the single positions. Therefore, in the weighted MSO logic which we consider here, the semantics of universal quantification will be defined by such kinds of valuation functions. This corresponds precisely to the valuation functions of Chatterjee, Doyen, and Henzinger used to determine the weight of a run in a weighted automaton [5]. The second novelty concerns conjunction. For a semiring weighted logic, the semantics of both universal quantification and conjunction was computed by means of the semiring product. Here, we may use completely different operations for conjunction and universal quantification. If the last one is interpreted by average, we may use average, sum, or infimum to interpret conjunction.

The weighted automata which we use are a combination of classical finite automata and of those used in [5]. They are more general than those from [5] because they are equipped with an acceptance condition (accepting states for finite words, resp. a Muller condition for infinite words) and the valuation function can have values in more general structures than $\mathbb{R}$ which we call valuation monoids. The weight of a run is computed by a valuation function and non-determinism is resolved by the monoid operation.

Our main results are as follows. Both for finite and infinite words, we consider three fragments of weighted MSO logic. In all fragments, the use of universal quantification is restricted as it had to be done already for semiring weighted MSO logic [9,10]. The fragments differ in the use of conjunction. All the restrictions for the use of conjunction are purely syntactic ones. Our main results show that weighted automata and these three fragments of weighted MSO logic are expressively equivalent, under suitable assumptions on the underlying valuation monoid, see Theorems 4.4 and 6.2. The properties of the valuation monoid used for these results are nearly the same for finite and infinite words. Our main theorems comprise and generalize results obtained before for conditionally complete commutative semirings [10,13] and for a discounted setting [12] and give new characterizations for a variety of new weighted settings such as longtime peak or average cost. Moreover, for special weight structures using limit superior or limit inferior [5,6,7,8] we show that the whole weighted MSO logic is expressively equivalent to weighted automata, see Theorem 6.4.

We would like to point out that some of our Büchi-like results do not need distributivity of multiplication over addition or commutativity or even associativity of multiplication. Before, these properties were assumed to be essential for weighted automata [1,11,22,23,29]. Such results were also obtained in [15], but the weight structures considered there have strong (locally) finiteness conditions. With an interpretation by average for universal quantification and for conjunction, no such finiteness holds anymore. Thus, our results show that the theory of weighted automata can be extended to a new kind of automata. Logics with such non-standard properties were also considered in other areas like in multi-valued logics [21,25] or in quantum logics where the failure of distributivity of conjunction over disjunction is crucial [2].

All our automata constructions are effective. Thus, decision problems for weighted logic translate into decision problems of weighted automata. Such decision problems were solved in [5] and we can carry over some of the algorithms to our setting whereas for others the details still have to be explored.

2 Valuation Monoids and Weighted Automata on Finite Words

Let Σ be an alphabet and Σ^+ the set of non-empty finite words.

Definition 2.1. *A* valuation monoid $(D, +, \mathrm{Val}, \mathbb{0})$ *consists of a commutative monoid* $(D, +, \mathbb{0})$ *and a* valuation function $\mathrm{Val} : D^+ \to D$ *with* $\mathrm{Val}(d) = d$ *and* $\mathrm{Val}(d_1, \ldots, d_n) = \mathbb{0}$ *whenever* $d_i = \mathbb{0}$ *for some* i.

$(D, +, \mathrm{Val}, \diamond, \mathbb{0}, \mathbb{1})$ *is a* product valuation monoid, *or a* pv-monoid *for short, if* $(D, +, \mathrm{Val}, \mathbb{0})$ *is a valuation monoid,* $\mathbb{1} \in D$, *and* $\diamond : D^2 \to D$ *with* $\mathrm{Val}(\mathbb{1})_{1 \leq i \leq n} = \mathbb{1}$ *for all* $n \geq 1$ *and* $\mathbb{0} \diamond d = d \diamond \mathbb{0} = \mathbb{0}$, $\mathbb{1} \diamond d = d \diamond \mathbb{1} = d$ *for all* $d \in D$.

Whereas valuation monoids suffice to define the behavior of weighted automata, pv-monoids will be used to define a semantics for weighted MSO logic. Especially, the semantics of conjunction will be defined by the operation $\diamond$. Note that the operation $\diamond$ has to be neither associative nor commutative. However, if $\diamond$ is restricted to $\{\mathbb{0}, \mathbb{1}\}$ then it is both associative and commutative and is modeling conjunction in the two-valued boolean algebra.

Definition 2.2. *A* weighted automaton $\mathcal{A} = (Q, I, T, F, \gamma)$ *over the alphabet* Σ *and a valuation monoid* $(D, +, \mathrm{Val}, \mathbb{0})$ *consists of a finite state set* Q, *a set* $I \subseteq Q$ *of initial states, a set* $F \subseteq Q$ *of final states, a set* $T \subseteq Q \times \Sigma \times Q$ *of transitions, and a weight function* $\gamma : T \to D$.

In other words, a weighted automaton is a usual finite automaton equipped with a weight function for the transitions. *Runs* $r = (t_i)_{0 \leq i \leq n}$ are defined as usual as finite sequences of matching transitions, say $t_i = (q_i, a_i, q_{i+1})$. Then we call the word $w = \ell(r) = a_0 a_1 \ldots a_n$ the *label* of the run r and r a run on w. Furthermore, $\gamma(r) = \big(\gamma(t_i)\big)_{0 \leq i \leq n}$ is the sequence of the transition weights of r, and $\mathrm{Val}(\gamma(r))$ is the *weight* of r. A run is *successful* if it starts in I and ends in F. Let $\mathrm{succ}(\mathcal{A})$ be the set of successful runs of $\mathcal{A}$. The *behavior* of $\mathcal{A}$ is the function $\|\mathcal{A}\| : \Sigma^+ \to D$ given by $\|\mathcal{A}\|(w) = \sum\big(\mathrm{Val}(\gamma(r)) \mid r \in \mathrm{succ}(\mathcal{A})$ and $\ell(r) = w\big)$; if there is no successful run on w, then $\|\mathcal{A}\|(w) = \mathbb{0}$. Any function $S : \Sigma^+ \to D$ is called a *series* (or a *quantitative language* as in [5]) over Σ^+. If $S = \|\mathcal{A}\|$ for a weighted automaton $\mathcal{A}$, then S is called *recognizable*.

Example 2.3. $(\mathbb{R} \cup \{-\infty\}, \sup, \mathrm{avg}, -\infty)$ with $\mathrm{avg}(d_1, \ldots, d_n) = \frac{1}{n}\sum_{i=1}^{n} d_i$ is a valuation monoid. As an example automaton, let $\mathcal{A} = (\{q\}, \{q\}, T, \{q\}, \gamma)$ over the alphabet $\Sigma = \{!, ?\}$ (! stands for a send event, ? for a receive) with $T = \{(q, !, q), (q, ?, q)\}$, $\gamma(q, !, q) = 1$, and $\gamma(q, ?, q) = -1$. Then $\mathcal{A}$ computes the average difference of sends and receives, e.g., $\|\mathcal{A}\|(!?!?) = 0$, $\|\mathcal{A}\|(!!!) = 1$, and $\|\mathcal{A}\|(!?!) = \frac{1}{3}$.

A pv-monoid $(D, +, \mathrm{Val}, \diamond, \mathbb{0}, \mathbb{1})$ is *left-+-distributive* if $d \diamond (d_1 + d_2) = (d \diamond d_1) + (d \diamond d_2)$ for any $d, d_1, d_2 \in D$; right-+-distributivity is defined analogously. If D is both left- and right-+-distributive, then it is *+-distributive*. If $\diamond$ is associative, then D is called *associative*. Moreover, D is *left-multiplicative* if for all $n \geq 1$ and $d, d_i \in D$

$$d \diamond \mathrm{Val}(d_1, d_2, \ldots, d_n) = \mathrm{Val}(d \diamond d_1, d_2, \ldots, d_n) \tag{1}$$

We call D *left*-Val-*distributive* if for all $n \geq 1$ and $d, d_i \in D$

$$d \diamond \mathrm{Val}(d_1, d_2, \ldots, d_n) = \mathrm{Val}(d \diamond d_1, d \diamond d_2, \ldots, d \diamond d_n) \tag{2}$$

If $d \diamond d' = d' \diamond d$, we say that d and d' commute. Let $C, C' \subseteq D$. If $d \diamond d' = d' \diamond d$ for all $d \in C$ and $d' \in C'$, then C and C' commute. D is *conditionally commutative* if for any $n \geq 1$ and any two sequences $(d_1, \ldots, d_n)$ and $(d'_1, \ldots, d'_n)$ from D with $d_i \diamond d'_j = d'_j \diamond d_i$ for all $1 \leq j < i \leq n$, we have

$$\mathrm{Val}(d_1, \ldots, d_n) \diamond \mathrm{Val}(d'_1, \ldots, d'_n) = \mathrm{Val}(d_1 \diamond d'_1, \ldots, d_n \diamond d'_n) \tag{3}$$

A pv-monoid D is *left distributive* if it is left-+-distributive and, moreover, left-multiplicative or left-Val-distributive. Whenever D is +-distributive and associative, then $(D, +, \diamond, \mathbb{0}, \mathbb{1})$ is a semiring and we call $(D, +, \mathrm{Val}, \diamond, \mathbb{0}, \mathbb{1})$ a *valuation semiring*. A valuation semiring which is also left-multiplicative or left-Val-distributive, and conditionally commutative is a *cc-valuation semiring*.

We call $d \in D$ *regular* if for every alphabet Σ there is a weighted automaton $\mathcal{A}_d$ such that $\|\mathcal{A}_d\|(w) = d$ for every $w \in \Sigma^+$. D is *regular* if every $d \in D$ is regular.

As is easy to see, any left-distributive pv-monoid is regular.

Example 2.4. $(\mathbb{R} \cup \{-\infty\}, \sup, \mathrm{avg}, +, -\infty, 0)$ is a pv-monoid which is associative, +-distributive, left-Val-distributive, and conditionally commutative, thus, a cc-valuation semiring. But we can put also $d \diamond d' = \frac{1}{2}(d + d')$. In this case we have to add a new $\mathbb{1}$ to the structure which is neutral with respect to $\diamond$. Moreover, we put $\sup(d, \mathbb{1}) = \mathbb{1}$ for all d and $\mathrm{avg}(d_1, \ldots, d_n)$ is the arithmetic mean of all $d_i \neq \mathbb{1}$ (where we put $\mathrm{avg}(\emptyset) = \mathbb{1}$). In this way, we get a regular pv-monoid which, however, is neither associative nor +-distributive nor Val-distributive but conditionally commutative.

Example 2.5. Let $\lambda \in \mathbb{R}$ with $\lambda > 0$. Then $(\mathbb{R} \cup \{-\infty\}, \sup, \mathrm{disc}_\lambda, -\infty)$ with the discounted sum $\mathrm{disc}_\lambda(d_0, \ldots, d_n) = \sum_{i=0}^{n} \lambda^i d_i$ is a valuation monoid. We put $\mathbb{1} = 0$ and $\diamond = +$ to obtain a left-multiplicative cc-valuation semiring.

Remark 2.6. Classical weighted automata work over *semirings* $\mathbb{K} = (K, +, \cdot, \mathbb{0}, \mathbb{1})$. In such automata, weights are multiplied along a run and summed up over all possible runs. This setting fits into our framework: We define $\mathrm{Val}(d_1, \ldots, d_n) = d_1 \cdot \ldots \cdot d_n$. Then $(K, +, \mathrm{Val}, \cdot, \mathbb{0}, \mathbb{1})$ is a left-multiplicative cc-valuation semiring. Examples for such semirings are the natural numbers with addition and multiplication, or the reals together with sup as the sum operation and inf, sup, or addition as multiplication.

Let $(D, +, \mathrm{Val}, \diamond, \mathbb{0}, \mathbb{1})$ be a pv-monoid and $S, S' : \Sigma^+ \to D$. Then $S + S'$ and $S \diamond S'$ are defined pointwise as $(S + S')(w) = S(w) + S'(w)$ and $(S \diamond S')(w) = S(w) \diamond S'(w)$ for all $w \in \Sigma^+$.

Now $S : \Sigma^+ \to D$ is a *recognizable step function* if there are finitely many pairwise disjoint recognizable languages $L_1, \ldots, L_k \subseteq \Sigma^+$ (recognizable by a finite automaton) and values $d_1, \ldots, d_k \in D$ such that $S(w) = d_i$ iff $w \in L_i$ $(1 \leq i \leq k)$; we write $S = \sum_{i=1}^{k} d_i \mathbb{1}_{L_i}$. We call S a *boolean step function* if S is a recognizable step function with $S(\Sigma^+) \subseteq \{\mathbb{0}, \mathbb{1}\}$. We write $\mathbb{1}_L$ for $\mathbb{1}\,\mathbb{1}_L$.

Lemma 2.7. *Let $(D, +, \mathrm{Val}, \mathbb{0})$ be a regular valuation monoid. Every recognizable step function S over D is a recognizable series. For a pv-monoid D, the class of recognizable step functions is closed under + and $\diamond$.*

Table 1. The semantics of weighted MSO-formulas

$$[\![d]\!]_{\mathcal{V}}(w,\sigma) = d$$

$$[\![P_a(x)]\!]_{\mathcal{V}}(w,\sigma) = \begin{cases} \mathbb{1} & \text{if } w_{\sigma(x)} = a, \\ \mathbb{0} & \text{otherwise} \end{cases}$$

$$[\![x \leq y]\!]_{\mathcal{V}}(w,\sigma) = \begin{cases} \mathbb{1} & \text{if } \sigma(x) \leq \sigma(y), \\ \mathbb{0} & \text{otherwise} \end{cases}$$

$$[\![x \in X]\!]_{\mathcal{V}}(w,\sigma) = \begin{cases} \mathbb{1} & \text{if } \sigma(x) \in \sigma(X), \\ \mathbb{0} & \text{otherwise} \end{cases}$$

$$[\![\neg\beta]\!]_{\mathcal{V}}(w,\sigma) = \begin{cases} \mathbb{1} & \text{if } [\![\beta]\!]_{\mathcal{V}}(w,\sigma) = \mathbb{0}, \\ \mathbb{0} & \text{otherwise} \end{cases}$$

$$[\![\varphi \vee \psi]\!]_{\mathcal{V}}(w,\sigma) = [\![\varphi]\!]_{\mathcal{V}}(w,\sigma) + [\![\psi]\!]_{\mathcal{V}}(w,\sigma)$$

$$[\![\varphi \wedge \psi]\!]_{\mathcal{V}}(w,\sigma) = [\![\varphi]\!]_{\mathcal{V}}(w,\sigma) \diamond [\![\psi]\!]_{\mathcal{V}}(w,\sigma)$$

$$[\![\exists x\, \varphi]\!]_{\mathcal{V}}(w,\sigma) = \sum_{i \in \mathrm{dom}(w)} [\![\varphi]\!]_{\mathcal{V}}(w,\sigma[x/i])$$

$$[\![\forall x\, \varphi]\!]_{\mathcal{V}}(w,\sigma) = \mathrm{Val}\big([\![\varphi]\!]_{\mathcal{V}}(w,\sigma[x/i])\big)_{i \in \mathrm{dom}(w)}$$

$$[\![\exists X\, \varphi]\!]_{\mathcal{V}}(w,\sigma) = \sum_{I \subseteq \mathrm{dom}(w)} [\![\varphi]\!]_{\mathcal{V}}(w,\sigma[X/I])$$

$$[\![\forall X\, \beta]\!]_{\mathcal{V}}(w,\sigma) = \mathrm{Val}\big([\![\beta]\!]_{\mathcal{V}}(w,\sigma[X/I])\big)_{I \subseteq \mathrm{dom}(w)}$$

3 Weighted MSO Logic

We provide a countable set $\mathcal{V}$ of first order and second order variables. Lower-case letters like x, y denote first order variables whereas capital letters like X, Y etc. denote second order variables. The syntax of *weighted MSO logics* over a pv-monoid $(D, +, \mathrm{Val}, \diamond, \mathbb{0}, \mathbb{1})$ is a combination of the one in [9] with an idea from [3]:

$$\beta ::= P_a(x) \mid x \leq y \mid x \in X \mid \neg\beta \mid \beta \wedge \beta \mid \forall x\, \beta \mid \forall X\, \beta$$
$$\varphi ::= d \mid \beta \mid \varphi \vee \varphi \mid \varphi \wedge \varphi \mid \exists x\, \varphi \mid \forall x\, \varphi \mid \exists X\, \varphi$$

where $d \in D$, $a \in \Sigma$, $x, y, X \in \mathcal{V}$. The formulas β are called *boolean* formulas and the formulas φ *weighted MSO*-formulas, for short *wMSO*. Note that negation and universal second order quantification are allowed in boolean formulas only.

To define the semantics of formulas, we follow the usual approach for MSO [31] adapted to weighted settings as in [9,10]. The set $\mathrm{free}(\varphi)$ of free variables in φ is defined as usual. Let $w \in \Sigma^+$, $\mathrm{dom}(w) = \{0, \ldots, |w| - 1\}$. Let $\mathcal{V}$ be a finite set of variables with $\mathrm{free}(\varphi) \subseteq \mathcal{V}$. A $(\mathcal{V}, w)$-*assignment* is a mapping $\sigma : \mathcal{V} \to \mathrm{dom}(w) \cup 2^{\mathrm{dom}(w)}$ where every first order variable is mapped to an element of $\mathrm{dom}(w)$ and every second order variable to a subset of $\mathrm{dom}(w)$. The update $\sigma[x/i]$ for $i \in \mathbb{N}$ is defined as $\sigma[x/i](x) = i$ and $\sigma[x/i] \restriction_{\mathcal{V}\setminus\{x\}} = \sigma \restriction_{\mathcal{V}\setminus\{x\}}$. The update $\sigma[X/I]$ for $I \subseteq \mathbb{N}$ is defined similarly. We encode $(\mathcal{V}, w)$-assignments as usual within an extended alphabet $\Sigma_{\mathcal{V}} = \Sigma \times \{0,1\}^{\mathcal{V}}$. Hereby, we refer to a word over the alphabet $\Sigma_{\mathcal{V}}$ by (w, σ). A word (w, σ) over $\Sigma_{\mathcal{V}}$ represents an assignment if and only if for every first order variable the respective row in the extended word contains exactly one 1; then we call (w, σ) *valid*.

For $(w, \sigma) \in \Sigma_{\mathcal{V}}^+$ we define the *semantics* $[\![\varphi]\!]_{\mathcal{V}}(w,\sigma)$ *of* φ *under the* $(\mathcal{V}, w)$-*assignment* σ: if (w, σ) is not valid, then $[\![\varphi]\!]_{\mathcal{V}}(w,\sigma) = \mathbb{0}$; for all valid (w, σ) the semantics is defined inductively as shown in Table 1 where for $\forall X \varphi$ the subsets $I \subseteq \mathrm{dom}(w)$ are enumerated in some fixed order, e.g. lexicographically. By induction, we can show easily that $[\![\beta]\!]_{\mathcal{V}} \in \{\mathbb{0}, \mathbb{1}\}$ for any boolean formula β. Also, the value $[\![\varphi]\!]_{\mathcal{V}}(w,\sigma)$ is determined by $\sigma \restriction_{\mathrm{free}(\varphi)}$. We put $[\![\varphi]\!] = [\![\varphi]\!]_{\mathrm{free}(\varphi)}$.

Example 3.1. Let $\Sigma = \{!, ?\}$ and $D = (\mathbb{R} \cup \{-\infty\}, \sup, \mathrm{avg}, +, -\infty, 0)$ be the pv-monoid from Example 2.4. Consider the wMSO-formula

$$\varphi = \forall x \Big((P_!(x) \to 1) \wedge (P_?(x) \to -1) \Big)$$

where $P_!(x) \to 1$ is an abbreviation for $(P_!(x) \wedge 1) \vee (\neg P_!(x) \wedge 0)$ and, similarly, for $P_?(x) \to -1$. Hence, $[\![P_!(x) \to 1]\!](w, \sigma)$ equals 1 if $w_{\sigma(x)} = !$ and 0 otherwise. Thus, $[\![(P_!(x) \to 1) \wedge (P_?(x) \to -1)]\!](w, \sigma)$ is 1 if $w_{\sigma(x)} = !$ and equals -1 if $w_{\sigma(x)} = ?$. Now universal quantification is interpreted by average, i.e., $[\![\varphi]\!](w)$ equals the average difference of send events ! and receive events ? in w. Hence, $[\![\varphi]\!] = \|\mathcal{A}\|$ for the weighted automaton from Example 2.3.

Next, we will define suitable fragments of weighted MSO because already for semiring weighted automata, the full weighted MSO logics is expressively stronger than weighted automata [9]. Since a boolean formula β does not contain any $d \in D$, we can regard it also as a classical boolean MSO formula defining the language $L(\beta)$. Now one can show easily: $[\![\beta]\!] = \mathbb{1}_{L(\beta)}$, i.e., its semantics is a boolean step function. Additionally, for every classical boolean MSO-formula α there is a boolean weighted MSO formula β with $[\![\beta]\!] = \mathbb{1}_{L(\alpha)}$. Now the class of *almost boolean formulas* of weighted MSO is the smallest class containing all constants $d \in D$ and all boolean formulas and which is closed under disjunction and conjunction. Almost boolean formulas define series which take only finitely many values over recognizable languages (therefore the name 'almost boolean'):

Proposition 3.2. *Let φ be an almost boolean formula. Then $[\![\varphi]\!]$ is a recognizable step function. Conversely, if $S : \Sigma^+ \to D$ is a recognizable step function, then $S = [\![\varphi]\!]$ for some almost boolean formula φ.*

Now we define the crucial fragments. Let $\mathrm{const}(\varphi)$ be the set of elements from D occurring in φ. A weighted MSO formula φ is $\forall$*-restricted*, if whenever it contains a sub-formula $\forall x \psi$, then ψ is almost boolean. Furthermore, φ is

1. *strongly* $\wedge$*-restricted*, if for every sub-formula $\psi_1 \wedge \psi_2$ of φ either both ψ_1 and ψ_2 are almost boolean, or ψ_1 or ψ_2 is boolean,
2. $\wedge$*-restricted*, if for every sub-formula $\psi_1 \wedge \psi_2$ of φ either the sub-formula ψ_1 is almost boolean or ψ_2 is boolean,
3. *commutatively* $\wedge$*-restricted*, if for all sub-formulas $\psi_1 \wedge \psi_2$ of φ either the sets $\mathrm{const}(\psi_1)$ and $\mathrm{const}(\psi_2)$ commute, or ψ_1 is almost boolean.

If φ is strongly $\wedge$-restricted, then it is $\wedge$-restricted. If φ is $\wedge$-restricted, it is commutatively $\wedge$-restricted. The fragment of $\forall$- and commutatively $\wedge$-restricted formulas is similar to the fragment of *syntactically restricted* formulas as defined in [10].

4 Characterization of Recognizable Series

In this section, we wish to characterize the collection of recognizable series by suitable fragments of wMSO. For this, first we build for weighted MSO formulas φ weighted automata $\mathcal{A}_\varphi$ recognizing $[\![\varphi]\!]$. For cc-valuation semirings we can choose the weights of $\mathcal{A}_\varphi$ to belong to the sub-semiring generated by $\mathrm{const}(\varphi)$ with $+$ and $\diamond$.

Proposition 4.1. *Let D be a pv-monoid and φ, ψ be two wMSO-formulas such that $[\![\varphi]\!]$ and $[\![\psi]\!]$ are recognizable. Then $[\![\varphi \vee \psi]\!]$, $[\![\exists x \varphi]\!]$, and $[\![\exists X \varphi]\!]$ are recognizable.*

Proof idea. For disjunction, we build the disjoint union of the weighted automata recognizing $[\![\varphi]\!]$ and $[\![\psi]\!]$, respectively. Concerning existential quantification, we can use the closure of recognizable series under projection of alphabets. We show this closure by a new automaton construction because we have no distributivity law between sum and Val. Therefore, we have to code the projected alphabet into the states of the constructed automaton to preserve the right number of runs for each word. □

Proposition 4.2. *Let D be a pv-monoid and φ, ψ be two wMSO formulas.*
(a) *If $[\![\varphi]\!]$ is recognizable and ψ is boolean, then $[\![\varphi \wedge \psi]\!]$ and $[\![\psi \wedge \varphi]\!]$ are recognizable.*
(b) *Let D be left-distributive. If φ is almost boolean and $[\![\psi]\!]$ is recognizable, then $[\![\varphi \wedge \psi]\!]$ is recognizable.*
(c) *Let D be a cc-valuation semiring and φ, ψ be $\forall$-restricted and commutatively $\wedge$-restricted. If $[\![\varphi]\!]$ and $[\![\psi]\!]$ are recognizable, and $\mathrm{const}(\varphi)$ and $\mathrm{const}(\psi)$ commute, then $[\![\varphi \wedge \psi]\!]$ is recognizable.*

Proof sketch. **(a)** Since ψ is boolean, $[\![\psi]\!]$ can be recognized by a deterministic weighted automaton which has only weights $\mathbb{0}$ and $\mathbb{1}$. Then the product automaton of the two weighted automata recognizing $[\![\psi]\!]$ and $[\![\varphi]\!]$ recognizes $[\![\varphi \wedge \psi]\!] = [\![\psi \wedge \varphi]\!]$.
(b) Let $d \in D$ and L be a regular language. Since D is left-distributive, $(d\mathbb{1}_L) \diamond [\![\psi]\!]$ can be recognized by a product automaton of a deterministic automaton recognizing L and the weighted automaton recognizing $[\![\psi]\!]$. Hence, $[\![\varphi \wedge \psi]\!]$ is recognizable.
(c) Since D is a cc-valuation semiring and $\mathrm{const}(\varphi)$ commutes with $\mathrm{const}(\psi)$, using the above observation we can construct inductively weighted automata $\mathcal{A}_\varphi$ and $\mathcal{A}_\psi$ recognizing $[\![\varphi]\!]$ and $[\![\psi]\!]$, respectively, such that the weights occurring in $\mathcal{A}_\varphi$ commute with those occurring in $\mathcal{A}_\psi$. Now we build the product automaton $\mathcal{A}$ of $\mathcal{A}_\varphi$ and $\mathcal{A}_\psi$. Since D is conditionally commutative, $\mathcal{A}$ recognizes $[\![\varphi \wedge \psi]\!]$. □

Proposition 4.3. *Let D be a pv-monoid and let φ be an almost boolean formula. Then $[\![\forall x\varphi]\!]$ is recognizable.*

Proof sketch. We follow ideas of [9,10]. Since φ is almost boolean, $[\![\varphi]\!] = \sum_{i=1}^{n} d_i \mathbb{1}_{L_i}$ is a recognizable step function. We can encode the information for which update of x a word is in L_i into a recognizable language $\tilde{L}$ over an extended alphabet $\tilde{\Sigma}$. Then we enrich an unweighted deterministic automaton $\tilde{\mathcal{A}}$ recognizing $\tilde{L}$ in a suitable way with the weights d_i and obtain a weighted automaton $\mathcal{A}$. Using the valuation function we can show that $[\![\forall x\varphi]\!]$ is a projection of $||\mathcal{A}||$ which is again recognizable. □

The following is our main result for finite words.

Theorem 4.4. *Let D be a pv-monoid and $S : \Sigma^+ \to D$ a series.*
(a) *Let D be regular. Then S is recognizable if and only if $S = [\![\varphi]\!]$ for a $\forall$-restricted and strongly $\wedge$-restricted wMSO-sentence φ.*
(b) *Let D be left-distributive. Then S is recognizable if and only if $S = [\![\varphi]\!]$ for a $\forall$-restricted and $\wedge$-restricted wMSO-sentence φ.*
(c) *Let D be a cc-valuation semiring. Then S is recognizable if and only if $S = [\![\varphi]\!]$ for a $\forall$-restricted and commutatively $\wedge$-restricted wMSO-sentence φ.*

Proof idea. For a regular pv-monoid D, the semantics of atomic formulas are recognizable by Proposition 3.2 and Lemma 2.7. Now Propositions 3.2 – 4.3 imply the recognizability of $[\![\varphi]\!]$ for formulas φ from the respective fragments. For the other direction, the formula of weighted MSO logic which defines the behavior of some weighted automaton $\mathcal{A}$ is built as in [10]. It is $\forall$-restricted because the requirements for universal quantification were already satisfied in [10], and the requirements for strongly $\wedge$-restriction can be guaranteed since the weights have to be used in the formula only in a very special way. □

5 Omega-Valuation Monoids and Weighted Automata on Infinite Words

Let Σ^ω be the set of infinite words over Σ. For a set D, let $(D_{fin})^\omega = \bigcup_{C \subseteq_{fin} D} C^\omega$.

A monoid $(D, +, \mathbb{0})$ is *complete*, cf. [16], if it has infinitary sum operations $\sum_I : D^I \to D$ for any index set I such that

- $\sum_{i\in\emptyset} d_i = \mathbb{0}$, $\sum_{i\in\{k\}} d_i = d_k$, $\sum_{i\in\{j,k\}} d_i = d_j + d_k$ for $j \neq k$, and
- $\sum_{j\in J}\left(\sum_{i\in I_j} d_i\right) = \sum_{i\in I} d_i$ if $\bigcup_{j\in J} I_j = I$ and $I_j \cap I_k = \emptyset$ for $j \neq k$.

Note that every complete monoid is commutative.

Definition 5.1. *An* ω-valuation monoid $(D, +, \mathrm{Val}^\omega, \mathbb{0})$ *is a complete monoid* $(D, +, \mathbb{0})$ *equipped additionally with an* ω-valuation function $\mathrm{Val}^\omega : (D_{fin})^\omega \to D$ *such that* $\mathrm{Val}^\omega(d_i)_{i\in\mathbb{N}} = \mathbb{0}$ *whenever* $d_i = \mathbb{0}$ *for some* $i \in \mathbb{N}$.

$(D, +, \mathrm{Val}^\omega, \diamond, \mathbb{0}, \mathbb{1})$ *is a* product ω-valuation monoid , *for short* ω-pv-monoid, *if* $(D, +, \mathrm{Val}^\omega, \mathbb{0})$ *is an* ω*-valuation monoid,* $\mathbb{1} \in D$*, and* $\diamond : D^2 \to D$ *with* $\mathrm{Val}^\omega(\mathbb{1}^\omega) = \mathbb{1}$, $\mathbb{0} \diamond d = d \diamond \mathbb{0} = \mathbb{0}$, *and* $\mathbb{1} \diamond d = d \diamond \mathbb{1} = d$ *for all* $d \in D$.

For ω-pv-monoids, we define associativity, left-+-distributivity, left-multiplicativity (1), left-Val^ω-distributivity (2), and conditional commutativity (3) as for pv-monoids by replacing $+$ by $\sum$, Val by Val^ω, and finite sequences by infinite sequences. Especially, we have *left-distributive* ω*-pv-monoids,* ω*-valuation semirings*, and *cc-*ω*-valuation semirings*.

For the next examples, considered in [5], we put $\bar{\mathbb{R}} = \mathbb{R} \cup \{-\infty, \infty\}$.

Example 5.2. The structure $(\bar{\mathbb{R}}, \sup, \limsup, -\infty)$ is an ω-valuation monoid if we put $\limsup(\ldots, -\infty, \ldots) = -\infty$. For the product $\diamond$ we have several possibilities: The ω-pv-monoids $(\bar{\mathbb{R}}, \sup, \limsup, \inf, -\infty, \infty)$ and $(\bar{\mathbb{R}}, \sup, \limsup, +, -\infty, 0)$ are left-Val^ω-distributive ω-valuation semirings, but they are not conditionally commutative.

Example 5.3. $(\bar{\mathbb{R}}, \sup, \lim\mathrm{avg}, -\infty)$ is an ω-valuation monoid where

$$\lim\mathrm{avg}(d_n)_{n\in\mathbb{N}} = \liminf_{n\to\infty} \frac{1}{n}\sum_{i=0}^{n-1} d_i = \lim_{n\to\infty} \inf\left(\frac{1}{k}\sum_{i=0}^{k-1} d_i \mid k \geq n\right).$$

For $\diamond$ and $\mathbb{1}$ there are several possibilities:
$(\bar{\mathbb{R}}, \sup, \lim\mathrm{avg}, \inf, -\infty, \infty)$ is an ω-valuation semiring but neither left-distributive nor conditionally commutative. $(\bar{\mathbb{R}}, \sup, \lim\mathrm{avg}, +, -\infty, 0)$ is a left-distributive ω-valuation-semiring. But $(\bar{\mathbb{R}}, \sup, \lim\mathrm{avg}, \mathrm{avg}, -\infty, \infty)$ with $\mathrm{avg}(d, \infty) = d$ for every $d \in \bar{\mathbb{R}}$ and otherwise $\mathrm{avg}(d, d') = \frac{d+d'}{2}$ yields an ω-pv-monoid which is not even left-+-distributive.

In the literature, *discounting* has found much interest recently (cf. [5,12,20] and the references cited there).

Example 5.4. Let $0 < \lambda < 1$. The structure $(\bar{\mathbb{R}}, \sup, \mathrm{disc}_\lambda, -\infty)$ with $\mathrm{disc}_\lambda(d_n)_{n\in\mathbb{N}} = \lim_{n\to\infty}\sum_{i=0}^{n}\lambda^i d_i$ is an ω-valuation monoid. Now $(\bar{\mathbb{R}}, \sup, \mathrm{disc}_\lambda, \inf, -\infty, \infty)$ is an ω-valuation semiring but neither left-multiplicative nor left-Val^ω-multiplicative. But $(\bar{\mathbb{R}}, \sup, \mathrm{disc}_\lambda, +, -\infty, 0)$ is a left-multiplicative cc-ω-valuation semiring.

Remark 5.5. Various authors have also considered *totally complete semirings* $(\mathbb{K}, +, \cdot, \mathbb{0}, \mathbb{1})$ which have both infinitary sum operations $\sum$ and countably infinite products $\prod$ satisfying several natural axioms, cf. [18] for an overview. This fits into our setting: $(\mathbb{K}, +, \prod, \cdot, \mathbb{0}, \mathbb{1})$ is a left-multiplicative ω-valuation semiring. But in a totally complete semiring there is, moreover, an infinitary associativity law for $\prod$ and also an infinitary distributivity law of $\prod$ over $\sum$ which we do not need here. A totally complete semiring which is conditionally commutative is called a *conditionally complete commutative* semiring in [10]. Hence, these structures are particular instances of cc-ω-valuation semirings.

Now we turn to the automaton model for infinite words.

Definition 5.6. *A weighted Muller automaton (WMA) $\mathcal{A} = (Q, I, T, \mathcal{F}, \gamma)$ over the alphabet Σ and an ω-valuation monoid $(D, +, \mathrm{Val}^\omega, \mathbb{0})$ consists of a finite state set Q, a set $I \subseteq Q$ of initial states, a set $T \subseteq Q \times \Sigma \times Q$ of transitions, a set $\mathcal{F} \subseteq 2^Q$ of accepting sets, and a weight function $\gamma : T \to D$.*

A *run* r is an infinite sequence of matching transitions $r = (t_i)_{i\in\mathbb{N}}$ with $t_i = (q_i, a_i, q_{i+1})$. The label of r is $\ell(r) = w = a_0 a_1 a_2 \ldots$ and r is then a run on w. We put $\gamma(r) = \big(\gamma(t_i)\big)_{i\in\mathbb{N}}$, i.e., $\gamma(r)$ is the infinite sequence of the weights occurring along r and $\mathrm{Val}^\omega(\gamma(r))$ its weight. A run is *successful* if it starts in an initial state and $\{q \in Q \mid q = q_i \text{ for infinitely many } i \in \mathbb{N}\} \in \mathcal{F}$. The *behavior* of $\mathcal{A}$ is the function $\|\mathcal{A}\| : \Sigma^\omega \to D$ defined by $\|\mathcal{A}\|(w) = \sum\big(\mathrm{Val}^\omega(\gamma(r)) \mid r \in \mathrm{succ}(\mathcal{A}) \text{ and } \ell(r) = w\big)$; if there is no successful run for w, then $\|\mathcal{A}\|(w) = \mathbb{0}$. Any function $f : \Sigma^\omega \to D$ is called an *ω-series*. Every ω-series $S : \Sigma^\omega \to D$ which is the behavior of some WMA over D is called *ω-recognizable*.

Remark 5.7. The weighted automata on infinite words as defined in [5] can be understood as total weighted Muller automata with $\mathcal{F} = 2^Q$ and $\gamma(T) \subseteq \mathbb{R}$.

We call an ω-valuation monoid $(D, +, \mathrm{Val}^\omega, \mathbb{0})$ *regular* if for every $d \in D$ there is a WMA $\mathcal{A}_d$ such that $\|\mathcal{A}_d\|(w) = d$ for all $w \in \Sigma^\omega$. All ω-valuation monoids given in the examples above and all left-distributive ω-pv-monoids are regular.

6 A Characterization of ω-Recognizable Series

We define weighted MSO logics as we have done for finite words, cf. Section 3. Also the semantics is defined in the same manner, this time as ω-series over an extended alphabet. The inductive definition is given in Table 1 where we have to replace Val by Val^ω. For $w \in \Sigma^\omega$, we let $\mathrm{dom}(w) = \{0, 1, 2, \ldots\} = \mathbb{N}$. To define the semantics of

$\forall X\varphi$, we have to extend Val^ω to index sets of size continuum such that $\mathbb{0}$ is still annihilating and $\mathrm{Val}^\omega(\mathbb{1})_{i\in I} = \mathbb{1}$. However, we will use these products only for sequences over $\{\mathbb{0}, \mathbb{1}\}$, i.e., we need uncountable products only for boolean values.

Example 6.1. Consider $\Sigma = \{!, ?\}$, $(\bar{\mathbb{R}}, \sup, \lim\mathrm{avg}, +, -\infty, 0)$ from Example 5.3, and the wMSO-formula $\varphi = \forall x\big((P_! \to 1) \wedge (P_? \to -1)\big)$ which is defined as in Example 3.1. Since universal quantification is interpreted as limit average, $[\![\varphi]\!](w)$ equals the long-run average difference between send events ! and receive events ? in $w \in \Sigma^\omega$. For example, $[\![\varphi]\!](!!!!!!?!?!? \ldots) = 0$ and $[\![\varphi]\!](!!?!!?!!? \ldots) = \frac{1}{3}$.

Here, $[\![\varphi]\!] = \|\mathcal{A}\|$ for the weighted Muller automaton $\mathcal{A} = (\{q\}, \{q\}, T, \{\{q\}\}, \gamma)$ with $T = \{(q, !, q), (q, ?, q)\}$, $\gamma(q, !, q) = 1$, and $\gamma(q, ?, q) = -1$.

The different fragments of weighted MSO are defined as before. If φ is an almost boolean formula, then $[\![\varphi]\!] = \sum_{i=1}^{k} d_i \mathbb{1}_{L_i}$ is an *ω-recognizable step function* where every L_i is an ω-recognizable language. If φ is a boolean formula, then $[\![\varphi]\!]$ is an *ω-boolean step function*, i.e., an ω-recognizable step function with $S(\Sigma^\omega) \subseteq \{\mathbb{0}, \mathbb{1}\}$. Using these concepts we obtain our second main result:

Theorem 6.2. *Let D be an ω-pv-monoid and $S : \Sigma^\omega \to D$ an ω-series.*
(a) *Let D be regular. Then S is ω-recognizable if and only if $S = [\![\varphi]\!]$ for a $\forall$-restricted and strongly $\wedge$-restricted wMSO-sentence φ.*
(b) *Let D be left-distributive. Then S is ω-recognizable if and only if $S = [\![\varphi]\!]$ for a $\forall$-restricted and $\wedge$-restricted wMSO-sentence φ.*
(c) *Let D be a cc-ω-valuation semiring. Then S is ω-recognizable if and only if $S = [\![\varphi]\!]$ for a $\forall$-restricted and commutatively $\wedge$-restricted wMSO-sentence φ.*

Proof idea. The proof follows the same lines as the one of Theorem 4.4. When constructing inductively the WMA recognizing $[\![\varphi]\!]$ for the different fragments, now we have to keep track of the Muller acceptance condition which can be done for disjoint unions, product automata, or the automaton recognizing a projection. For the closure under universal first-order quantification, it is important that we use Muller automata because we need a deterministic device in the proof to recognize the corresponding ω-language $\tilde{L}$. Conversely, we can find for every WMA $\mathcal{A}$ a $\forall$- and strongly $\wedge$-restricted sentence φ with $[\![\varphi]\!] = \|\mathcal{A}\|$. This time, we have to define the Muller acceptance condition by a boolean formula, cf. [10,13]. □

Remark 6.3. By Theorem 6.2, the different wMSO-fragments all define the class of ω-recognizable series over D, regardless of the choice of $\diamond$ for interpretation of conjunction, provided the assumptions on D are satisfied.

For special ω-valuation monoids considered in [5,6,7,8], even every weighted MSO formula can be translated into an equivalent weighted Muller automaton because both formalisms define exactly the class of ω-recognizable step functions.

Theorem 6.4. *Let D be one of the following ω-pv monoids:*

$$(\bar{R}, \sup, \lim\sup, \inf, -\infty, \infty), \quad (\bar{R}, \sup, \lim\sup, +, -\infty, 0),$$
$$(\bar{R}, \sup, \lim\inf, \inf, -\infty, \infty), \quad (\bar{R}, \sup, \lim\inf, +, -\infty, 0),$$

and let $S : \Sigma^\omega \to D$. Then S is ω-recognizable if and only if $S = [\![\varphi]\!]$ for some wMSO-sentence φ.

Conclusion

We have extended the use of weighted MSO logic to a new class of quantitative settings comprising average- and long-run objectives. We succeeded in a translation of different fragments to weighted automata. We would like to note that we can enlarge the used fragments of weighted MSO logic under additional idempotency conditions which is, due to space restrictions, not elaborated here in detail.

The translation of formulas into automata opens the way to algorithms deciding questions like satisfiability or equivalence. The proofs of Theorems 4.4, 6.2, and 6.4 are constructive. Thus, satisfiability or equivalence of formulas of weighted logics reduce to the emptiness or equivalence problem for weighted automata. In [5], decision algorithms are given for these problems. But the model of weighted automata used in [5] does not have an acceptance condition like final states or a Muller condition as we use here. For infinite words and $\limsup$, e.g., emptiness can be decided by an adaptation of the algorithm listed in [5]. In our setting, we can synchronously check the Muller condition. We conjecture that we can also decide emptiness for $\lim$ avg-WMA. This works if the $\lim$ avg-automaton does not have an additional Muller condition [5], due to a result about the existence of positional strategies for Markov decision procecces with a limit average reward. But further work has to be done in elaborating these algorithms for the setting considered here.

Another open question is the expressive power of the first order fragment for these new settings, cf. [9]. The development of weighted temporal logic and model checking procedures would be interesting, cf. [20,24].

A subject of on-going work by the authors is the characterization of this new kind of weighted automata by regular expressions [30].

References

1. Berstel, J., Reutenauer, C.: Rational Series and Their Languages. Springer, Heidelberg (1988)
2. Birkhoff, G., von Neumann, J.: The logic of quantum mechanics. Annals of Math. 37, 823–843 (1936)
3. Bollig, B., Gastin, P.: Weighted versus probabilistic logics. In: Diekert, V., Nowotka, D. (eds.) DLT 2009. LNCS, vol. 5583, pp. 18–38. Springer, Heidelberg (2009)
4. Büchi, J.: Weak second-order arithmetic and finite automata. Z. Math. Logik Grundlagen Math. 6, 66–92 (1960)
5. Chatterjee, K., Doyen, L., Henzinger, T.A.: Quantitative languages. In: Kaminski, M., Martini, S. (eds.) CSL 2008. LNCS, vol. 5213, pp. 385–400. Springer, Heidelberg (2008)
6. Chatterjee, K., Doyen, L., Henzinger, T.A.: Alternating weighted automata. In: Gębala, M. (ed.) FCT 2009. LNCS, vol. 5699, pp. 3–13. Springer, Heidelberg (2009)
7. Chatterjee, K., Doyen, L., Henzinger, T.A.: Expressiveness and closure properties for quantitative languages. In: 24th LICS 2009, pp. 199–208. IEEE Comp. Soc. Press, Los Alamitos (2009)
8. Chatterjee, K., Doyen, L., Henzinger, T.A.: Probabilistic weighted automata. In: Bravetti, M., Zavattaro, G. (eds.) Concurrency Theory. LNCS, vol. 5710, pp. 244–258. Springer, Heidelberg (2009)
9. Droste, M., Gastin, P.: Weighted automata and weighted logics. Theoretical Computer Science 380, 69–86 (2007)

10. Droste, M., Gastin, P.: Weighted automata and weighted logics. In: Droste, M., et al. (eds.) [11], ch. 5
11. Droste, M., Kuich, W., Vogler, H. (eds.): Handbook of Weighted Automata. EATCS Monographs in Theoretical Computer Science. Springer, Heidelberg (2009)
12. Droste, M., Rahonis, G.: Weighted automata and weighted logics with discounting. Theoretical Computer Science 410, 3481–3494 (2009)
13. Droste, M., Rahonis, G.: Weighted automata and weighted logics on infinite words. Izvestiya VUZ. Matematika 54, 26–45 (2010)
14. Droste, M., Vogler, H.: Weighted tree automata and weighted logics. Theoretical Computer Science 366, 228–247 (2006)
15. Droste, M., Vogler, H.: Kleene and Büchi theorems for weighted automata and multi-valued logics over arbitrary bounded lattices. In: DLT 2010. LNCS. Springer, Heidelberg (2010)
16. Eilenberg, S.: Automata, Languages, and Machines, vol. A. Academic Press, London (1974)
17. Elgot, C.: Decision problems of finite automata design and related arithmetics. Trans. Amer. Math. Soc. 98, 21–52 (1961)
18. Ésik, Z., Kuich, W.: Finite automata. In: Droste, et al. (eds.) [11], ch. 3
19. Fichtner, I., Kuske, D., Meinecke, I.: Traces, series-parallel posets, and pictures: A weighted study. In: Droste, et al. (eds.) [11], ch. 10
20. Fischer, D., Grädel, E., Kaiser, Ł.: Model checking games for the quantitative μ-calculus. In: STACS 2008, pp. 301–312 (2008)
21. Kreinovich, V.: Towards more realistic (e.g., non-associative) "and"- and "or"-operations in fuzzy logic. Soft Comput. 8(4), 274–280 (2004)
22. Kuich, W.: Semirings and formal power series: their relevance to formal languages and automata. In: Handbook of Formal Languages. Word, Language, Grammar, vol. 1, pp. 609–677. Springer, Heidelberg (1997)
23. Kuich, W., Salomaa, A.: Semirings, Automata, Languages. In: EATCS Monographs in Theoretical Computer Science, vol. 5. Springer, Heidelberg (1986)
24. Kupferman, O., Lustig, Y.: Lattice automata. In: Cook, B., Podelski, A. (eds.) VMCAI 2007. LNCS, vol. 4349, pp. 199–213. Springer, Heidelberg (2007)
25. Mallya, A.: Deductive multi-valued model checking. In: Gabbrielli, M., Gupta, G. (eds.) ICLP 2005. LNCS, vol. 3668, pp. 297–310. Springer, Heidelberg (2005)
26. Mathissen, C.: Weighted logics for nested words and algebraic formal power series. In: Aceto, L., Damgård, I., Goldberg, L.A., Halldórsson, M.M., Ingólfsdóttir, A., Walukiewicz, I. (eds.) ICALP 2008, Part II. LNCS, vol. 5126, pp. 221–232. Springer, Heidelberg (2008)
27. Quaas, K.: Weighted timed MSO logics. In: Diekert, V., Nowotka, D. (eds.) DLT 2009. LNCS, vol. 5583, pp. 419–430. Springer, Heidelberg (2009)
28. Rahonis, G.: Weighted Muller tree automata and weighted logics. Journal of Automata, Languages and Combinatorics 12(4), 455–483 (2007)
29. Salomaa, A., Soittola, M.: Automata-Theoretic Aspects of Formal Power Series. In: Texts and Monographs in Computer Science. Springer, Heidelberg (1978)
30. Schützenberger, M.: On the definition of a family of automata. Information and Control 4, 245–270 (1961)
31. Thomas, W.: Languages, automata, and logic. In: Rozenberg, G., Salomaa, A. (eds.) Handbook of Formal Languages, vol. A, pp. 389–455. Springer, New York (1997)

Solving MINONES-2-SAT as Fast as VERTEX COVER

Neeldhara Misra, N.S. Narayanaswamy*, Venkatesh Raman, and Bal Sri Shankar

Institute of Mathematical Sciences, Chennai, India
{neeldhara,swamy,vraman,balsri}@imsc.res.in

Abstract. The problem of finding a satisfying assignment for a 2-SAT formula that minimizes the number of variables that are set to 1 (MIN ONES 2–SAT) is NP-complete. It generalizes the well-studied problem of finding the smallest vertex cover of a graph, which can be modeled using a 2-SAT formula with no negative literals. The natural parameterized version of the problem asks for a satisfying assignment of weight at most k.

In this paper, we present a polynomial-time reduction from MIN ONES 2–SAT to VERTEX COVER without increasing the parameter and ensuring that the number of vertices in the reduced instance is *equal* to the number of variables of the input formula. Consequently, we conclude that this problem also has a simple 2-approximation algorithm and a $2k$ variables kernel subsuming these results known earlier. Further, the problem admits algorithms for the parameterized and optimization versions whose runtimes will always match the runtimes of the best-known algorithms for the corresponding versions of vertex cover.

1 Introduction and Motivation

Satisfiability is a fundamental problem that encodes several computational problems. Variations of the problem appear as canonical complete problems for several complexity classes. While it is well known that the satisfiability of a formula in CNF form is a canonical NP–complete problem, testing whether a CNF formula has a satisfying assignment with weight[1] at least k is a canonical complete problem for the parameterized complexity class $W[2]$ [DF99]. If the number of variables in each clause is bounded, it is a canonical $W[1]$-complete problem [DF99]. These results imply that it is unlikely that these problems are *fixed parameter tractable* (FPT). In other words, it is unlikely that they have an algorithm with running time $O(f(k)n^{O(1)})$ on input formulas of size n.

On the other hand, if the question is whether a d-CNF formula (for fixed d) has a satisfying assignment with weight *at most* k, then this generalizes the well-studied d-hitting set problem and independently, turns out to be fixed parameter tractable with the weight as a parameter ([Nie06, MR99], cf. Section 2). When we restrict our attention to 2-CNF formulas (MIN ONES 2–SAT) this

* This work was done when the author was on sabbatical from IIT Madras.
[1] The *weight* of an assignment is the number of variables assigned 1 by the assignment.

P. Hliněný and A. Kučera (Eds.): MFCS 2010, LNCS 6281, pp. 549–555, 2010.

problem generalizes the well-studied VERTEX COVER problem. For, given a graph $G = (V, E)$, a satisfying assignment of weight at most k on the formula $\wedge(u \vee v)$, where the $\wedge$ runs over all edges (u, v) in E, where u and v are variables corresponding to vertices u and v of G, corresponds to a vertex cover of size at most k in G. However, notice that we do not require negated literals to encode VERTEX COVER using 2-CNF formulas, and thus it appears that MIN ONES 2–SAT is a more general version of the vertex cover problem.

Gusfield and Pitt[GP92] considered this MIN ONES 2–SAT problem and gave a 2-approximation algorithm. The algorithm that follows a greedy approach, gives a solution whose weight is at most twice that of the optimum (assuming that the formula is satisfiable). As satisfiability of 2-CNF-SAT is well known to be polynomial time solvable, we can assume without loss of generality that the given 2-SAT formula is satisfiable. Hochbaum et al [HMNT93] showed that the classical Nemhauser-Trotter theorem for vertex cover [NT75] holds for MIN ONES 2–SAT as well. This implies a 2-approximation algorithm for the optimization version, and a $2k$-variable kernel for the parameterized version.

There is a reduction from MIN ONES 2–SAT to VERTEX COVER, pointed out by Seffi Naor (see [Hoc97]). This reduction takes an instance F of MIN ONES 2–SAT on n variables and returns a graph $G(F)$ that has one vertex for every literal participating in F (i.e. with up to $2n$ vertices), and an edge between a pair of literals whenever they appear together in a clause of F, and an edge $(x, \bar{x})$ for every variable x. Further, the reduction introduces, for every variable x, the edges (u, v) — where $u \in N(x)$ and $v \in N(\bar{x})$, and $N(l)$ is the set of all literals l' such that (l, l') is a clause of F. It can be shown that if there is a satisfying assignment of weight at most k for F, then there is a vertex cover of size n in $G(F)$. Conversely, a vertex cover of size n in $G(F)$ translates to a satisfying assignment for F (i.e. that the vertex cover never needs to choose both x and $\bar{x}$ of a variable — however, the weight of such an assignment can be as large as n.

Observe that this reduction produces a graph with the number of vertices equal to twice the number of variables and, in the parameterized setting, does not transform k into a function of k alone. Since the reduction loses track of the weight of the solution, it does not enable us to employ VERTEX COVER to solve an instance of MIN ONES 2–SAT.

In this paper, we demonstrate a simple extension of this reduction that preserves both k and n, and allows us to carry over everything we know about VERTEX COVER to the more general setting of MIN ONES 2–SAT. Thus, we have that the apparently more general problem of minones can be handled as easily as vertex cover, in both the optimization and parameterized settings. In particular, the problem now has a $2k$-variable kernel, a 2-approximation algorithm, and FPT and exact algorithms that will run as fast as the best algorithms for the corresponding versions of the vertex cover problem (the current best being $O^*(1.27^k)$ [CKX06][2] and $O(1.2132^n)$ [KLR09]). In particular, our reduction subsumes the earlier results (2-approximation algorithms, and Nemhauser-Trotter theorem) on this problem.

[2] We use the notation $O^*()$ to "hide" functions that are polynomial in the variables.

2 Preliminaries

A parameterized problem is denoted by a pair $(Q, k) \subseteq \Sigma^* \times \mathbb{N}$. The first component Q is a classical language, and the number k is called the parameter. Such a problem is *fixed–parameter tractable* (FPT) if there exists an algorithm that decides it in time $O(f(k)n^{O(1)})$ on instances of size n. A *kernelization algorithm* takes an instance (x, k) of the parameterized problem as input, and in time polynomial in $|x|$ and k, produces an equivalent instance (x', k') such that $|x'|$ is a function purely of k. The output x' is called the kernel of the problem and its size is $|x'|$. We refer the reader to [DF99, Nie06] for more details on the notion of fixed-parameter tractability.

Let P be an arbitrary set, whose elements we shall refer to as *variables.* A *literal* is either a variable or its negation. An assignment for P is a function $t : P \rightarrow \{0, 1\}$. Sometimes, we also refer to an assignment setting (mapping) a variable to 'true' or 'false' when we mean to say 1 or 0 respectively.

A formula is in *conjunctive normal form* (CNF) if it is a conjunction of clauses, where a clause is a disjunction of literals. A c-SAT formula has at most c literals in any clause. The *weight* of an assignment is the number of variables that are set to one by that assignment. We refer to the problem of finding a smallest weight satisfying assignment for c-SAT formulae as MIN ONES c–SAT.

Simple FPT algorithm for weight at most k assignments. The natural parameterized version of MIN ONES c–SAT is FPT for any fixed c, when parameterized by the weight: pick a clause that contains only positive literals (as long as one exists) and branch by setting each of the variables to 1. This results in a c-way branch of depth at most k. Notice that at the leaves, every clause has at least one negated literal and the assignment that sets all the remaining variables to 0 satisfies all such clauses. This results in an $O(c^k m)$ algorithm where m is the number of clauses in the formula.

3 Reduction of MIN ONES 2–SAT to VERTEX COVER

In this section, we present a reduction from MIN ONES 2–SAT to VERTEX COVER. Throughout, we use F to denote an instance of MIN ONES 2–SAT, and $C(F)$ denotes the set of clauses in F. Also, let $D(F)$ denote the implication graph of F, which has one vertex for every literal of F, and the directed arcs $(\bar{l_1}, l_2)$ and $(l_1, \bar{l_2})$ for every clause $(l_1, l_2) \in C(F)$. Also, let $A(D(F))$ denote the set of arcs in $D(F)$.

The implication graph of a 2-CNF formula is very well-studied — for example, see Section 1.10 in [BJG08]. We begin by recalling Lemma 1.10.2 from [BJG08] (the proof is reiterated here for completeness).

Lemma 1 ([BJG08]). *If $D(F)$ contains a path from l_1 to l_2, then, for every satisfying truth assignment t, $t(l_1) = 1$ implies that $t(l_2) = 1$.*

Proof. Observe that F contains a clause of the form $\bar{x} \vee y$ when $D(F)$ contains the arc (x, y). Further, every clause takes the value 1 under any satisfying truth assignment. Thus, by the fact that t is a satisfying truth assignment and by the definition of $D(F)$, we have that for every arc $(x, y) \in A(D(F))$, $t(x) = 1$ implies $t(y) = 1$. Now the claim follows easily by induction on the length of the shortest (l_1, l_2)-path in $D(F)$.

We now describe a formula that is easier to work with, because we ensure that such paths are also witnessed by edges. Let F^* be the smallest formula which contains all the clauses of F, and the clause(s) $(l_1 \vee l_2)$, for each pair of literals l_1 and l_2 such that there is a directed path from $\bar{l_1}$ to l_2 in $D(F)$. We refer to F^* as the *closure* of F. One way to compute the closure of F is to compute the transitive closure of the implication graph of F (in polynomial time, see [CLRS01]). The formula corresponding to the graph thus obtained (when treated also as an implication graph) is the closure of F. We work with the closed formula F^* in the discussion that follows.

Theorem 1. *Given a 2-CNF formula F, let F^* be the* closure *of F, and $(F^*)_+$ denote the set of all clauses of F^* where both literals occur positively. Let G be the graph that has one vertex for every variable in $(F^*)_+$, and $(u, v) \in E(G)$ if and only if $(u \vee v) \in C((F^*)_+)$. Then F has a satisfying assignment of weight at most k if and only if G has a vertex cover of size at most k.*

Proof. Suppose that F has a satisfying assignment of weight at most k. Then it implies that the same satisfying assignment is a satisfying assignment of F^* as well. For, if $c = (l_1 \vee l_2)$ is in $C(F^*) \setminus C(F)$, then there is a directed path from $\bar{l_1}$ to l_2, by construction. Hence if the satisfying assignment of F sets l_1 to false, then $\bar{l_1}$ is set to true and hence by Lemma 1, l_2 is set to 1 by the assignment, thus satisfying c. Hence $(F^*)_+$, a sub formula of F^* has a satisfying assignment of weight at most k, which means that the graph G has a vertex cover of size at most k.

Conversely let G have a vertex cover of size k. Let t be the truth assignment corresponding to a minimal vertex cover, say K, of size at most k in G, i.e. let $t(x) = 1$ if and only if $x \in K$, and $t(x) = 0$ otherwise. Clearly, t is a satisfying assignment of $(F^*)_+$ and is of weight at most k. We now show that t is indeed a satisfying assignment of F^*. The proof is by contradiction. Let us assume that F^* is not satisfied by t. This implies there is a clause $C \in F^*$ that is not satisfied by t. Clearly, $C \notin (F^*)_+$. There are two possibilities for C: either $C = (x \vee \bar{y})$, or $C = (\bar{x} \vee \bar{y})$, where x and y are variables. In either case, we arrive at a contradiction to the assumption that t is a satisfying assignment of $(F^*)_+$.

1. $C = (x \vee \bar{y})$: Since C is falsified by t, it follows that $t(x) = 0$ (or equivalently, $t(\bar{x}) = 1$) and $t(y) = 1$. Since t is obtained from a minimal vertex cover K (that containts y as $t(y) = 1$), there is a clause $(y \vee z) \in (F^*)_+$ such that $t(z) = 0$.

   Notice that $D(F^*)$ has arcs from $\bar{x}$ to $\bar{y}$ and from $\bar{y}$ to z, and therefore a path from $\bar{x}$ to z. By Lemma 1, therefore, $t(\bar{x}) = 1$ must imply $t(z) = 1$, a contradiction.

2. $C = (\bar{x} \vee \bar{y})$: Since C is falsified by t, it follows that $t(x) = 1$ and $t(y) = 1$. Since t is obtained from a minimal vertex cover K, there are clauses $(y \vee z_1), (x \vee z_2) \in (F^*)_+$ such that $t(z_1) = 0, t(z_2) = 0$ (Note that z_1 could be equal to z_2).

   As before, we observe that $D(F^*)$ has a path from x to z_1 (through $\bar{y}$), and again by an application of Lemma 1, we observe that $t(x) = 1$ implies that $t(z_1) = 1$, a contradiction.

Consequently, our assumption that t is not a satisfying assignment of F^* is wrong and hence F^* has a satisfying assignment of weight at most k. Since F is a sub-formula of F, it follows that so does F.

Corollary 1. *Given a 2-CNF formula F on n variables and a positive integer k, it can be checked if F admits a satisfying assignment of weight at most k in time $O^*(1.27^k)$ [CKX06]. A satisfying assignment of minimum weight may be obtained in time $O(1.2132^n)$ [KLR09].*

Observe that the reduction stated in Theorem 1 is valid for the weighted version of the problem (where each variable has a non-negative real weight, and the weight of an assignment is the sum of the weights of the variables that it sets to one), and that the proof remains the same is easily verified.

Corollary 2. *Given a 2-CNF formula F on n variables with a weight function $w : V(F) \rightarrow \mathrm{R}^+$ such that $w(v) \geq 1$, for all $v \in V(F)$ and a positive integer k, it can be checked if F admits a satisfying assignment of weight at most k in time $O^*(1.37^k)$ [NR03]. A satisfying assignment of minimum weight may be obtained in time $O(1.32^n)$ [DJ02].*

The problem of solving a $0-1$ integer program which has at most two variables per constraint with an assignment of weight at most k is known to be equivalent to MIN ONES 2–SAT. This is due to a reduction that does not increase the number of variables or the weight of the solution (Section 4, [HMNT93]). The reduction in [HMNT93] is from a more general integer program, one that assumes a bounded range (not necessarily $0-1$) for each variable. However, in the general case, the number of variables created in the reduced instance is a function of the ranges. For $0-1$ integer programs, the number of variables remains the same as that of the original. Thus, we also have that a binary integer program may be solved as fast as weighted vertex cover[3].

Corollary 3. *Consider a binary integer program where the objective function is to be minimized, and every constraint has at most two variables. Given such a program and a positive integer k, it can be checked if the optimum feasible assignment is at most k in time $O^*(1.37^k)$ [NR03], and the optimum assignment may be obtained in time $O(1.32^n)$ [DJ02].*

[3] In case the coefficients for all variables in the objective function are one, then the problem may in fact be solved as fast as the unweighted vertex cover.

4 Concluding Remarks

We show MIN ONES 2–SAT to be equivalent to VERTEX COVER in both the parameterized and optimization settings, by demonstrating a polynomial-time reduction from MIN ONES 2–SAT to VERTEX COVER that preserves the optimum value and keeps the number of vertices of the graph to the number of variables in the formula. This allows us to employ the best known algorithms for VERTEX COVER to MIN ONES 2–SAT incurring only an additional polynomial cost.

The complexity of MIN ONES c–SAT for $c > 2$ is an interesting line of research. In this case, the problem is a natural generalization of c-hitting set. While c-hitting set has a $k^{O(c)}$ kernel [AK07], a polynomial sized kernel is unlikely for MIN ONES c–SAT even for $c = 3$, as a special case of MIN ONES 3–SAT (not-1-in-3 SAT) is unlikely to have a polynomial sized kernel [KW09]. See [KWar] for a classification of the types of bounded variable constraints for which polynomial sized kernel is possible. Improving the obvious $O(c^k m)$ time bound (mentioned in Section 2) for the parameterized question is a natural open problem.

References

[AK07] Abu-Khzam, F.N.: Kernelization algorithms for d-hitting set problems. In: Dehne, F.K.H.A., Sack, J.-R., Zeh, N. (eds.) WADS 2007. LNCS, vol. 4619, pp. 434–445. Springer, Heidelberg (2007)

[BJG08] Bang-Jensen, J., Gutin, G.Z.: Digraphs: Theory, algorithms and applications. Springer Publishing Company, Heidelberg (2008) (incorporated)

[CKX06] Chen, J., Kanj, I.A., Xia, G.: Improved parameterized upper bounds for vertex cover. In: Královič, R., Urzyczyn, P. (eds.) MFCS 2006. LNCS, vol. 4162, pp. 238–249. Springer, Heidelberg (2006)

[CLRS01] Cormen, T.H., Leiserson, C.E., Rivest, R.L., Stein, C.: Introduction to algorithms. MIT Press, Cambridge (2001)

[DF99] Downey, R.G., Fellows, M.R.: Parameterized complexity. Springer, Heidelberg (November 1999)

[DJ02] Dahllöf, V., Jonsson, P.: An algorithm for counting maximum weighted independent sets and its applications. In: SODA, pp. 292–298 (2002)

[GP92] Gusfield, D., Pitt, L.: A bounded approximation for the minimum cost 2-sat problem. Algorithmica 8, 103–117 (1992)

[HMNT93] Hochbaum, D., Meggido, N., Naor, J., Tamir, A.: Tight bounds and 2-approximation algorithms for integer programs with two variables per inequality. Mathematical Programming 62, 69–83 (1993)

[Hoc97] Hochbaum, D.S. (ed.): Approximation algorithms for NP-hard problems. PWS Publishing Co., Boston (1997)

[KLR09] Kneis, J., Langer, A., Rossmanith, P.: A fine-grained analysis of a simple independent set algorithm. In: Kannan, R., Kumar, N. (eds.) IARCS Annual Conference on Foundations of Software Technology and Theoretical Computer Science (FSTTCS 2009), Leibniz International Proceedings in Informatics (LIPIcs), Dagstuhl, Germany, vol. 4, pp. 287–298. Schloss Dagstuhl–Leibniz-Zentrum fuer Informatik (2009)

[KW09] Kratsch, S., Wahlström, M.: Two edge modification problems without polynomial kernels. In: Chen, J., Fomin, F.V. (eds.) IWPEC 2009. LNCS, vol. 5917, pp. 264–275. Springer, Heidelberg (2009)

[KWar] Kratsch, S., Wahlström, M.: Preprocessing of min ones problems: A dichotomy. In: 37th International Colloquium on Automata, Languages and Programming, ICALP (to appear 2010)

[MR99] Mahajan, M., Raman, V.: Parameterizing above guaranteed values: Maxsat and maxcut. J. Algorithms 31(2), 335–354 (1999)

[Nie06] Niedermeier, R.: Invitation to fixed parameter algorithms (oxford lecture series in mathematics and its applications). Oxford University Press, USA (March 2006)

[NR03] Niedermeier, R., Rossmanith, P.: On efficient fixed-parameter algorithms for weighted vertex cover. J. Algorithms 47(2), 63–77 (2003)

[NT75] Nemhauser, G.L., Trotter, L.E.: Vertex packings: Structural properties and algorithms. Mathematical Programming 8, 232–248 (1975)

Unambiguous Finite Automata over a Unary Alphabet

Alexander Okhotin*

Department of Mathematics, University of Turku, Turku FI–20014, Finland
Academy of Finland
alexander.okhotin@utu.fi

Abstract. Nondeterministic finite automata (NFA) with at most one accepting computation on every input string are known as unambiguous finite automata (UFA). This paper considers UFAs over a unary alphabet, and determines the exact number of states in DFAs needed to represent unary languages recognized by n-state UFAs: the growth rate of this function is $e^{\Theta(\sqrt[3]{n \ln^2 n})}$. The conversion of an n-state unary NFA to a UFA requires UFAs with $g(n)+O(n^2) = e^{\sqrt{n \ln n}(1+o(1))}$ states, where $g(n)$ is Landau's function. In addition, it is shown that the complement of n-state unary UFAs requires up to at least $n^{2-o(1)}$ states in an NFA, while the Kleene star requires up to exactly $(n-1)^2+1$ states.

1 Introduction

This paper is concerned with a noteworthy family of automata located between deterministic finite automata (DFA) and nondeterministic finite automata (NFA): the *unambiguous finite automata* (UFA), that is, NFAs that have at most one accepting computation for every string. Apparently, this family was first studied by Schmidt [19], whose unpublished thesis contains an interesting method of proving lower bounds for UFAs based upon the rank of certain matrices, and a $2^{\Omega(\sqrt{n})}$ lower bound on the tradeoff between UFAs and DFAs. These methods were further elaborated by Leung [10,11] and by Hromkovič et al. [6], who studied degrees of nondeterminism in finite automata. In particular, Leung [11] established a precise $2^n - 1$ UFA–DFA tradeoff. Computational complexity of testing properties of UFAs was studied by Stearns and Hunt [21] and recently by Björklund and Martens [2].

This paper considers UFAs in the special case of a *unary alphabet* $\Sigma = \{a\}$. The main properties of DFAs and NFAs over a unary alphabet are quite different from the case of a general alphabet. Lyubich [12] and Chrobak [3] have shown that in the unary case the DFA–NFA tradeoff is $g(n) + O(n^2)$, where $g(n) = e^{(1+o(1))\sqrt{n \ln n}}$ is the maximum order of an element in the group of permutations on n objects, known as *Landau's function* [9]. State complexity of basic operations on unary DFAs was first studied by Yu, Zhuang and K. Salomaa [22], and elaborated by Pighizzini and Shallit [17]. A similar study for unary

* Supported by the Academy of Finland under grant 134860.

P. Hliněný and A. Kučera (Eds.): MFCS 2010, LNCS 6281, pp. 556–567, 2010.

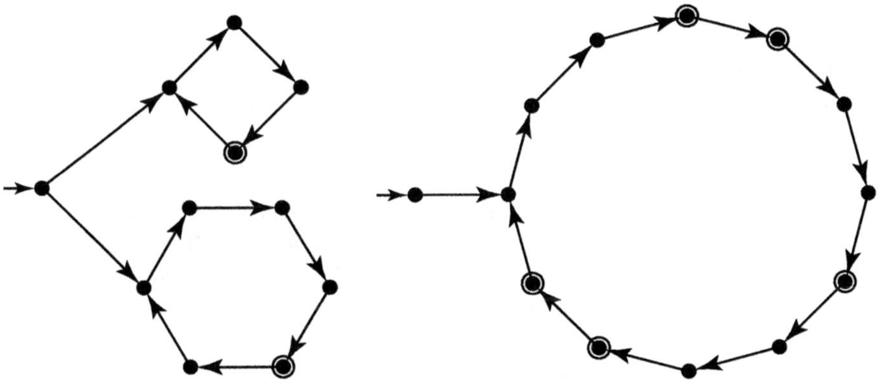

Fig. 1. An 11-state unary UFA and the 13-state minimal equivalent DFA

NFAs was carried out by Holzer and Kutrib [5], and the hardest languages for complementation were further studied by Mera and Pighizzini [14]. Succinctness of two-way automata over a unary alphabet has received particular attention in the works of Chrobak [3], Mereghetti and Pighizzini [15] and Geffert et al. [4].

The first natural question about unary UFAs is whether they are nontrivial: that is, any more succinct than unary DFAs. The smallest example of a nontrivial UFA is presented in Figure 1, left; it is unambiguous, because only strings of even length are accepted in the first cycle, and only strings of odd length are accepted in the second cycle. This UFA has $1 + 4 + 6 = 11$ states, while the smallest equivalent DFA shown on the right requires $1 + \text{lcm}(4, 6) = 13$ states. This example motivates the study of unary UFAs, which is undertaken in the present paper.

It should be noted that the existing methods of proving lower bounds on the size of UFAs, based upon the matrix methods of Schmidt [19], are quite hard to apply in the case of a unary alphabet. For unary inputs, Schmidt's matrix belongs to a class of *circulant matrices*, and the problem of determining the rank of a circulant matrix of 0s and 1s, studied by Ingleton [7], is surprisingly hard in the general case. Unless the matrix for a particular language happens to be of some special form, finding its rank is difficult.

New methods of analysis are thus required, and they shall be derived from the earlier work on unary NFAs. Perhaps the most important basic result on unary NFAs is the *Chrobak normal form*, in which there is one tail of states, ending with transitions into one or more disjoint cycles. It was proved by Chrobak [3] that every n-state NFA can be transformed to this normal form, with the cycles of combined length at most n and with the tail of length $O(n^2)$. For the case for a UFA, Chrobak's transformation was refined by Jiang, McDowell and Ravikumar [8, Thm. 2.2] to a transformation to the same normal form, but *without increasing the number of states.*

This paper begins with formulating an additional condition in the Chrobak normal form that is specific to UFAs. This condition is then used to determine the precise tradeoff between UFAs and DFAs, which is expressed in terms of a more complicated variant of Landau's function, denoted $\widetilde{g}$. In particular, the UFA–DFA tradeoff is asymptotically equivalent to $\widetilde{g}$, and the growth rate of the latter is determined as $2^{\Theta(\sqrt[3]{n \ln^2 n})}$. A close lower bound on the tradeoff between NFAs and UFAs is established using the matrix methods of Schmidt [19], and the tradeoff is found to be of the order of the original Landau's function, that is, $e^{(1+o(1))\sqrt{n \ln n}}$. Finally, the complexity of operations on UFAs is approached, and an $n^{2-o(1)}$ lower bound for complementation is established, which shows for the first time that the complement of a UFA sometimes requires additional states. The complexity of Kleene star is determined precisely as $(n-1)^2+1$.

2 Chrobak Normal Form for Unambiguous Automata

A *nondeterministic finite automaton* (NFA) is a quintuple $A = (\Sigma, Q, Q_0, \delta, F)$, where Σ is an input alphabet, Q is a finite nonempty set of states; $Q_0 \subseteq Q$ is the set of initial states; $\delta : Q \times \Sigma \to 2^Q$ is the transition function; $F \subseteq Q$ is the set of accepting states. The automaton A is said to accept a string $w = a_1 \ldots a_n$ if there exists a sequence of states $r_0, \ldots, r_n \in Q$, in which $r_0 \in Q_0$, $r_i \in \delta(r_{i-1}, a_i)$ for all i, and $r_n \in F$. The language recognized by an NFA, denoted by $L(A)$, is the set of all strings it accepts. The transition function shall be extended to $\delta : Q \times \Sigma^* \to 2^Q$ by $\delta(q, \varepsilon) = \{q\}$ and $\delta(q, aw) = \bigcup_{q' \in \delta(q,a)} \delta(q', w)$.

In some literature, NFAs are defined with a unique initial state, that is, with $Q_0 = \{q_0\}$. Every NFA can be converted to an NFA with a unique initial state by adding a new initial state.

A *deterministic finite automaton* (DFA) is an NFA with a unique outgoing transition from each state by each symbol ($|\delta(q, a)| = 1$ for all q, a) and with a unique initial state ($|Q_0| = 1$). An NFA A is a *partial DFA* if $|Q_0| = 1$ and $|\delta(q, a)| \leqslant 1$ for all q and a.

An NFA is *unambiguous* if for every $w \in L(A)$ the corresponding sequence of states $r_0, \ldots, r_{|w|}$ in the definition of acceptance is unique. An unambiguous NFA is called an *unambiguous finite automaton* (UFA).

The first lower bound argument for UFAs was given by Schmidt [19, Thm. 3.9] in his proof of a $2^{\Omega(\sqrt{n})}$ lower bound on the NFA–UFA tradeoff. The following general statement of Schmidt's lower bound method is due to Leung [11]:

Schmidt's Theorem [19,11]. *Let $L \subseteq \Sigma^*$ be a regular language and let $\{(u_1, v_1), \ldots, (u_n, v_n)\}$ with $n \geqslant 1$ and $u_i, v_i \in \Sigma^*$ be a finite set of pairs of strings. Consider the integer matrix $M \in \mathbb{Z}^{n \times n}$ defined by $M_{i,j} = 1$ if $u_i v_j \in L$, and $M_{i,j} = 0$ otherwise. Then every UFA recognizing L has at least* $\operatorname{rank} M$ *states.*

The study of NFAs over a unary alphabet $\Sigma = \{a\}$ is founded upon the following normal form:

Definition 1 (Chrobak [3]). *An NFA over $\{a\}$ is said to be in Chrobak normal form if its set of states is $\{q_0, \ldots, q_{\ell-1}\} \cup \bigcup_{i=1}^{k} R_i$, with $\ell \geqslant 0$, $k \geqslant 0$,*

$R_i = \{r_{i,0}, \ldots, r_{i,p_i-1}\}$ and $1 \leqslant p_1 < p_2 < \ldots < p_k$, the unique initial state is q_0 if $\ell \geqslant 1$ or there is a set of initial states $\{r_{1,0}, \ldots, r_{k,0}\}$ if $\ell = 0$, and the transitions are:

$$\begin{aligned} \delta(q_i, a) &= \{q_{i+1}\} \quad (0 \leqslant i \leqslant \ell - 2) \\ \delta(q_{\ell-1}, a) &= \{r_{1,0}, r_{2,0}, \ldots, r_{k,0}\} \quad (\textit{if } \ell \geqslant 1) \\ \delta(r_{i,j}, a) &= \{r_{i,j+1 \bmod p_i}\} \quad (1 \leqslant i \leqslant k,\ 0 \leqslant j \leqslant p_i - 1), \end{aligned}$$

The set of accepting states may be arbitrary.

It is known from Chrobak [3] that every NFA with n states can be transformed to an equivalent NFA in this normal form, with $\ell = O(n^2)$ and $\sum_{i=1}^{k} p_n \leqslant n$. The growth in the number of states is thus at most quadratic.

Turning to unary UFAs, in this case the transformation to the Chrobak normal form can be always done without increasing the number of states:

Proposition 1 (Jiang, McDowell, Ravikumar [8, Thm. 2.2]). *For every UFA recognizing an infinite language over $\{a\}$ there exists (and can be effectively constructed) a UFA in Chrobak normal form with the same number of states recognizing the same language. Furthermore, if the original UFA has a unique initial state, then so does the resulting UFA.*

Once a UFA is converted to the Chrobak normal form, the following key restriction of unambiguous automata is exposed:

Main Condition. *An NFA $(\{a\}, Q, q_0, \delta, F)$ in Chrobak normal form recognizing an infinite language over $\{a\}$ is unambiguous if and only if for every two accepting states $r_{i,f}, r_{j,f'} \in F$ with $i \neq j$, the offsets f and f' are different modulo $\gcd(p_i, p_j)$.*

The proof is based upon the Chinese Remainder Theorem. The Main Condition, in particular, implies that the lengths of the cycles cannot be primes (unless there is a unique cycle), and that $\gcd(p_i, p_j) \geqslant 2$ for any two distinct cycles. For example, the UFA in Figure 1 in the introduction has $\gcd(4, 6) = 2$, and accepting states are separated by the parity of their offsets.

3 UFA–DFA Tradeoff

An upper bound on the number of states in a DFA needed to represent unary languages recognized by n-state unary NFAs has been established by Lyubich [12]. It is asymptotically equivalent to the maximum order of a permutation on n elements:

$$g(n) = \max\{\, \operatorname{lcm}(p_1, \ldots, p_k) \mid k \geqslant 1,\ p_1 + \ldots + p_k \leqslant n \,\}.$$

This function is known as *Landau's function*, as its $e^{\sqrt{n \ln n}(1+o(1))}$ asymptotics was determined by Landau [9], see also Miller [16] for a more accessible argument.

Twenty years after Lyubich, an asymptotically matching lower bound on the unary NFA to DFA tradeoff was obtained by Chrobak [3], who also gave a new, combinatorial proof of Lyubich's upper bound. These results can be stated as follows:

Proposition 2 (Lyubich [12], Chrobak [3]). *For every n-state unary NFA there exists a DFA with at most $g(n)+n^2$ states recognizing the same language. Conversely, for every n there is a language recognized by an n-state NFA, such that every equivalent DFA requires $g(n)$ states.*

The essense of this result is a natural correspondence between unary NFAs and Landau's function. The numbers $p_1, \ldots, p_k$ in the definition of $g(n)$ correspond to lengths of cycles of an NFA in Chrobak normal form, the sum $p_1 + \ldots + p_k$ represents the number of states in an NFA, and an equivalent DFA has to have $\operatorname{lcm}(p_1, \ldots, p_k)$ states.

This analysis of NFAs can be extended to UFAs, if the constraints on their Chrobak normal form given in the Main Condition are embedded into the definition of Landau's function. This leads to the following variant of this function:

$$\begin{aligned}\widetilde{g}(n) = \max \{ \operatorname{lcm}(p_1, \ldots, p_k) \mid k \geqslant 1,\ & p_1 + \ldots + p_k \leqslant n, \\ & \exists f_1, \ldots, f_k \text{ with } f_i \in \{0, \ldots, p_i - 1\} : \\ & \forall i, j\ (i \neq j)\ f_i \not\equiv f_j \ (\mathrm{mod}\ \gcd(p_i, p_j))\}\end{aligned}$$

For n up to 9 the value of $\widetilde{g}(n)$ is n. The next value is $\widetilde{g}(10) = 12$, given by $k = 2$, $p_1 = 4$, $p_2 = 6$, $f_1 = 0$ and $f_2 = 1$ with $0 \not\equiv 1 \ (\mathrm{mod}\ \gcd(4,6))$. This function can be asymptotically estimated as $e^{\Theta(\sqrt[3]{n \ln^2 n})}$, and this estimation will be the subject of the next section. Now the task is to express the tradeoff between UFAs and DFAs using this function, which can be done as follows:

Theorem 1. *For every $n \geqslant 1$, the following number of states is sufficient and in the worst case necessary for a DFA to recognize a unary language recognized by an n-state UFA with multiple initial states:*

$$f_{UFA\text{-}DFA}(n) = \begin{cases} n+1, & \text{if } n \leqslant 9 \\ \max\limits_{0 \leqslant \ell < n} \widetilde{g}(n-\ell) + \ell, & \text{if } n \geqslant 10 \end{cases}$$

For UFAs with a unique initial state, the tradeoff function takes the following form:

$$f_{UFA_1\text{-}DFA}(n) = \begin{cases} n+1, & \text{if } n \leqslant 10 \\ \max\limits_{1 \leqslant \ell < n} \widetilde{g}(n-\ell) + \ell, & \text{if } n \geqslant 11 \end{cases}$$

For $n \leqslant 9$, UFAs are not yet any more powerful than partial DFAs, and thus can be simulated by DFAs with $n+1$ states, with the lower bound witnessed by a finite language. Once there are sufficiently many states to reach nontrivial values of $\widetilde{g}$, the following witness languages can be represented:

Lemma 1. *Let $k \geqslant 2$, $\ell \geqslant 0$, $p_1, \ldots, p_k \geqslant 2$ and $f_1, \ldots, f_k \geqslant 0$ with $0 \leqslant f_i < p_i$ be any numbers, such that (a) $f_i \neq f_j \pmod{\gcd(p_i, p_j)}$ for all $1 \leqslant i < j \leqslant k$, (b) $\operatorname{lcm}(p_1, \ldots, p_{i-1}, p_{i+1}, \ldots, p_k)$ is not divisible by p_i for any $1 \leqslant i \leqslant k$, and (c) $f_i = p_i - 1$ for some i. Then the language*

$$L = a^\ell \cdot \bigcup_{i=1}^{k} a^{f_i}(a^{p_i})^*$$

has a UFA with $\ell+p_1+\ldots+p_k$ states, while every DFA for this language requires $\ell + \operatorname{lcm}(p_1, \ldots, p_m)$ states.

The matching upper bound is implied by the following lemma:

Lemma 2. *For every n-state UFA in Chrobak normal form with a tail of length $\ell \geqslant 0$ there exists a DFA with at most $\ell + \widetilde{g}(n - \ell)$ states recognizing the same language.*

Theorem 1 is as a consequence of the above lemmata. The exact values of the tradeoff can now be computed: for instance, $f_{\text{UFA-DFA}}(50) = \widetilde{g}(50) = 560$ with a witness language $a^{13}(a^{14})^* \cup a^{12}(a^{16})^* \cup a^{14}(a^{20})^*$, and $f_{\text{UFA}_1\text{-DFA}}(50) = \widetilde{g}(46) + 4 = 424$, witnessed by $a^{15}(a^{12})^* \cup a^{16}(a^{14})^* \cup a^{17}(a^{20})^*$.

4 Estimation of $\widetilde{g}$

The function $\widetilde{g}$ characterizes the expressive power of unary UFAs, and estimating the growth rate of this function, especially in comparison with g, is essential to understand the power of ambiguity in finite automata over a unary alphabet. So what is the asymptotic behaviour of the function $\widetilde{g}$? The first step towards determining its growth rate is estimating the maximum number of cycles k for a given sum of cycle lengths.

Lemma 3. *Let $k \geqslant 1$ and let $\pi_1, \ldots, \pi_k \geqslant 2$ be any integers, for which (a) there exist $f_1, \ldots, f_k \in \mathbb{N}$ with $f_i \neq f_j \pmod{\gcd(\pi_i, \pi_j)}$ for all $i \neq j$, and (b) $\operatorname{lcm}(\pi_1, \ldots, \pi_{i-1}, \pi_{i+1}, \ldots, \pi_k)$ is not divisible by p_i for any $1 \leqslant i \leqslant k$. Then $\pi_1 + \ldots + \pi_k > \frac{4}{9}k^3 \ln k - \frac{8}{27}k^3\sqrt{\ln k}$.*

As in Lemma 1, the condition of each cycle contributing something to the least common multiple is essential: if it is lifted, then taking k cycles each of length k gives $\sum \pi_i = k^2$, and the statement does not hold.

For each i, let $r_i = \frac{\operatorname{lcm}(\pi_1, \ldots, \pi_k)}{\operatorname{lcm}(\pi_1, \ldots, \pi_{i-1}, \pi_{i+1}, \ldots, \pi_k)}$ and let $\pi_i = r_i s_i$. Then the numbers $r_1, \ldots, r_k$ are pairwise relatively prime, each of them is at least 2 by the condition (b), and hence $\gcd(\pi_i, \pi_j) = \gcd(s_i, s_j)$ for $i \neq j$. In this notation, the statement of the lemma can be equivalently reformulated as follows:

$$\min_{\substack{r_1, \ldots, r_k \geqslant 2 \\ \text{relatively prime}}} \quad \min_{\substack{s_1, \ldots, s_k \in \mathbb{N} \\ \exists f_1, \ldots, f_k \in \mathbb{N} \\ f_i \neq f_j \pmod{\gcd(s_i, s_j)}}} \quad \sum_{i=1}^{k} r_i s_i > \frac{4}{9}k^3 \ln k - \frac{8}{27}k^3\sqrt{\ln k}.$$

The proof proceeds by simplifying the expression in the left-hand side, decreasing its value, but in the end still obtaining a value greater than $\frac{4}{9}k^3 \ln k - \frac{8}{27}k^3\sqrt{\ln k}$. The first simplification step is replacing the condition on $s_1, \ldots, s_k$ involving the numbers $f_1, \ldots, f_k$ with the following simpler consequence of this condition:

Claim 3.1 *For any* $s_1, \ldots, s_k \geqslant 1$, *if there exist* $f_1, \ldots, f_k \in \mathbb{N}$ *with* $f_i \neq f_j$ *modulo* $\gcd(s_i, s_j)$, *then* $\frac{1}{s_1} + \ldots + \frac{1}{s_k} \leqslant 1$.

Aiming to estimate the smallest values of the sum $\sum r_i s_i$, it is convenient to allow the values of s_i to be any positive real numbers. This will slightly reduce the value of the minimum, but will make it analytically calculable as follows:

Claim 3.2 *Let* $a_1, \ldots, a_m > 0$ *be any positive real numbers. Then*

$$\min_{\substack{x_1,\ldots,x_k \in \mathbb{R}_+ \\ \frac{1}{x_1}+\ldots+\frac{1}{x_k}=1}} \sum_{i=1}^{k} a_i x_i = \left(\sqrt{a_1} + \ldots + \sqrt{a_k}\right)^2$$

and the minimum is reached at the point $x_i = \frac{\sqrt{a_1}+\ldots+\sqrt{a_k}}{\sqrt{a_i}}$.

Therefore, a lower bound on the sum $\sum_{i=1}^{k} r_i s_i$ is $(\sqrt{r_1} + \ldots + \sqrt{r_k})^2$, and the next task is to estimate the least value of this sum for all applicable r_i, that is, for every choice of pairwise relatively prime $r_1, \ldots, r_k \geqslant 2$. In fact, the minimum is achieved by taking the first k primes.

Claim 3.3 *Let* $2 \leqslant r_1 < \ldots < r_k$ *be any pairwise relatively prime natural numbers. Then* $p_i \leqslant r_i$, *where* p_i *is the* i*th prime.*

Therefore, the sum is decreased (or unaltered) by replacing each r_i with the ith prime:

$$(\sqrt{r_1} + \ldots + \sqrt{r_k})^2 \geqslant (\sqrt{p_1} + \ldots + \sqrt{p_k})^2.$$

In order to estimate the sum $\sum_{i=1}^{k} \sqrt{p_i}$, consider the following known fact:

Proposition 3. $p_n > n \ln n$ *for all* $n \geqslant 1$.

It remains to calculate the resulting sum:

Claim 3.4 $\sum_{n=1}^{k} \sqrt{n \ln n} > \frac{2}{3}k\sqrt{k \ln k} - \frac{2}{9}k\sqrt{k}$ *for all* $k \geqslant 1$.

The sum is approximated by the integral $\int_1^k \sqrt{x \ln x}\, dx$, which can be estimated as stated in the claim. With all these auxiliary results established, Lemma 3 is proved by the following chain of inequalities.

$$\min_{\substack{r_1,\ldots,r_k \geqslant 2 \\ \text{relatively prime}}} \min_{\substack{s_1,\ldots,s_k \in \mathbb{N} \\ \exists f_1,\ldots,f_k \in \mathbb{N} \\ f_i \neq f_j \pmod{\gcd(s_i,s_j)}}} \sum_{i=1}^{k} r_i s_i \geqslant \min_{\substack{r_1,\ldots,r_k \geqslant 2 \\ \text{relatively prime}}} \min_{\substack{s_1,\ldots,s_k \in \mathbb{N} \\ \frac{1}{s_1}+\ldots+\frac{1}{s_k} \leqslant 1}} \sum_{i=1}^{k} r_i s_i \geqslant$$

$$\geqslant \min_{\substack{r_1,\ldots,r_k \geqslant 2 \\ \text{relatively prime}}} \min_{\substack{x_1,\ldots,x_k \in \mathbb{R}_+ \\ \frac{1}{x_1}+\ldots+\frac{1}{x_k} \leqslant 1}} \sum_{i=1}^{k} r_i x_i = \min_{\substack{r_1,\ldots,r_k \geqslant 2 \\ \text{relatively prime}}} \left(\sqrt{r_1} + \ldots + \sqrt{r_k}\right)^2 =$$

$$= \left(\sqrt{p_1} + \ldots + \sqrt{p_k}\right)^2 > \left(\sum_{i=1}^{k} \sqrt{i \ln i}\right)^2 > \left(\tfrac{2}{3}k\sqrt{k \ln k} - \tfrac{2}{9}k\sqrt{k}\right)^2 >$$

$$> \tfrac{4}{9}k^3 \ln k - \tfrac{8}{27}k^3\sqrt{\ln k}.$$

This estimation can be reformulated as a lower bound on k as a function of n.

Lemma 4. *Under the assumptions of Lemma 3,* $k < \frac{3}{\sqrt[3]{4}} \sqrt[3]{\frac{n}{\ln n - 2\sqrt{\ln n}}}$, *where* $n = \pi_1 + \ldots + \pi_k \geqslant 55$.

The following upper bound of $\widetilde{g}(n)$ can be inferred from this bound on k.

Theorem 2 (Upper bound). $\widetilde{g}(n) < e^{\sqrt[3]{2n \ln^2 n}(1+o(1))}$.

The proof of the theorem relies *only* on the upper bound on k, and otherwise ignores the additional constraints in the definition of $\widetilde{g}$ as compared to g. The first step is to replace the least common multiple of $\pi_1, \ldots, \pi_k$ with their product, and then allow the cycle lengths to be real numbers. Then the maximum of the product is reached for all factors being identical:

Proposition 4. $\max\limits_{x_1+\ldots+x_k \leqslant x} x_1 \ldots x_k = (\frac{x}{k})^k$ *for every* $k \in \mathbb{N}$ *and* $x \in \mathbb{R}_+$.

Another fact about elementary functions is that $(\frac{n}{k})^k$ reaches its maximum at $k = \frac{n}{e}$, and since the values of k allowed by Lemma 4 are much smaller, one should choose k as large as possible to obtain the greatest value of $(\frac{n}{k})^k$.

Proposition 5. *The function* $f(y) = (\frac{n}{y})^y$ *increases on* $0 < y \leqslant \frac{n}{e}$, *has a maximum at* $y = \frac{n}{e}$ *and decreases on* $\frac{n}{e} \leqslant y$.

The rest of the proof of Theorem 2 is a straightforward calculation.

The second task is to establish a lower bound on $\widetilde{g}$, and the following asymptotically close estimation shall be obtained:

Theorem 3 (Lower bound). $\widetilde{g}(n) > e^{\sqrt[3]{\frac{2}{9}}\sqrt[3]{n \ln^2 n}(1+o(1))}$.

Given a number n, the proof begins with finding the largest k, such that the sum $s_k = \sum_{i=1}^{k} kp_i$, where p_i is the ith prime, is at most n. The numbers $\pi_i = kp_i$ satisfy the definition of $\widetilde{g}$ with $f_i = i - 1$ for each i, and therefore the value of $\widetilde{g}$ on s_k must be at least $\operatorname{lcm}(kp_1, \ldots, kp_k) = k \prod_{i=1}^{k} p_i$. Consider the following known facts about primes:

Proposition 6 ([1]). $\sum_{i=1}^{k} p_i = (1+o(1))\frac{1}{2}k^2 \ln k$.

Proposition 7. $\prod_{i=1}^{k} p_i = e^{(1+o(1))k \ln k}$.

Using these facts to estimate the numbers s_k and $k\prod_{i=1}^{k} p_i$ and carrying out some straightforward calculations proves Theorem 3.

According to Theorems 2–3, the values of the function $\widetilde{g}$ are confined within the following bounds:

$$e^{\sqrt[3]{\frac{2}{9}}\sqrt[3]{n \ln^2 n}(1+o(1))} < \widetilde{g}(n) < e^{\sqrt[3]{2}\sqrt[3]{n \ln^2 n}(1+o(1))}.$$

Corollary 1. $\widetilde{g}(n) = e^{\Theta\left(\sqrt[3]{n(\ln n)^2}\right)}$.

Returning to the UFA–DFA tradeoff, note that the tradeoff function satisfies $\widetilde{g}(n) \leqslant f_{\text{UFA–DFA}} \leqslant \widetilde{g}(n)+n$, while in the case of UFAs with a unique initial state, $\widetilde{g}(n-1) \leqslant f_{\text{UFA}_1\text{–DFA}} \leqslant \widetilde{g}(n-1)+n$. Therefore, both functions asymptotically behave as $\widetilde{g}$:

Corollary 2. *$f_{UFA-DFA} = e^{\Theta\left(\sqrt[3]{n(\ln n)^2}\right)}$ and $f_{UFA_1-DFA} = e^{\Theta\left(\sqrt[3]{n(\ln n)^2}\right)}$.*

5 NFA–UFA Tradeoff

An NFA can be transformed to an equivalent UFA simply by converting it to a DFA. It turns out that for some NFAs no better transformation is possible:

Lemma 5. *For all $k \geqslant 1$ and $p_1, \ldots, p_k \geqslant 2$, the language*

$$L = \{\varepsilon\} \cup a \bigcup_{i=1}^{k} \{\varepsilon, a, a^2, \ldots, a^{p_i-2}\}(a^{p_i})^* = \overline{(a^{\operatorname{lcm}(p_1,\ldots,p_k)})^*} \cup \{\varepsilon\}$$

has an NFA with $1+\sum_{i=1}^{k} p_i$ states, while the smallest UFA for L needs at least $1+\operatorname{lcm}(p_1,\ldots,p_k)$ states.

This smallest UFA is actually a DFA. The lower bound on the size of any UFA for L is proved using the method of Schmidt [19]. Consider the strings $u_i = v_i = a^{i-1}$ for $1 \leqslant i \leqslant \operatorname{lcm}(p_1,\ldots,p_k)+1 = n+1$. The corresponding $(n+1)\times(n+1)$ matrix M is defined by $M_{i,j} = 0$ for $i+j = n+2$ and for $i = j = n+1$, and $M_{i,j} = 1$ for the rest of the entries. It is easy to check that it is a full-rank matrix, and the lower bound follows by Schmidt's Theorem.

Thus $g(n-1)+1$ is a lower bound on the NFA to UFA transformation. An asymptotically matching upper bound of $g(n-1)+O(n^2)$ is given by Chrobak's [3, Thm. 4.4] construction, which begins by converting an n-state NFA to the Chrobak normal form with a tail of length $O(n^2)$ and with at most $n-1$ states in the cycles, and then proceeds by determinizing the cycles, making at most $g(n-1)$ states.

Theorem 4. *For every $n \geqslant 1$, the number of states in a UFA sufficient and, in the worst case, necessary to represent languages recognized by n-state NFAs is $g(n-1)+O(n^2) = e^{\sqrt{n \ln n}(1+o(1))}$.*

6 Complementing Unary UFAs

With respect to DFAs, complementation has state complexity n, since in order to represent the complement of a language recognized by a DFA, it is sufficient to complement its set of accepting states. For unary NFAs, Holzer and Kutrib [5] have shown that complementation may require a blowup of up to $g(n)$ states: that is, complementing some unary NFAs basically requires determinizing them.

The situation with UFAs is quite nontrivial. On one hand, for a substantial class of UFAs, the complement can be constructed by changing the set of accepting states, like in the case of DFAs. On the other hand, it will be proved that complementing some UFAs requires additional states.

The following subclass of UFAs allows efficient complementation:

Lemma 6. *Let $A = (\Sigma, Q, q_0, \delta, F)$ be a unary UFA in Chrobak normal form recognizing an infinite language, and assume that there exists a number p that divides the length of every cycle, such that for every two accepting states $r_{i,f}, r_{j,f'} \in F$ with $i \neq j$, it holds that $f \not\equiv f' \pmod{p}$. Then there exists a set F', such that $A' = (\Sigma, Q, q_0, \delta, F')$ is a UFA recognizing $\overline{L(A)}$.*

The new set of accepting states is defined as follows: $F' = \{\, q_i \mid q_i \notin F \,\} \cup \{\, r_{i,f} \mid (f \bmod p) \in S_i,\ r_{i,f} \notin F \,\}$.

In particular, this lemma is applicable to all UFAs with $k = 2$ cycles, such as the one in Figure 1. But for $k \geqslant 3$ the lengths of the cycles need not have a common divisor, which leads to examples of UFAs not covered by the above lemma. Sometimes the lengths of the cycles may have a common divisor, yet the separation of offsets required by the Main Condition would not be possible. The following example illustrates the latter case.

Example 1. Let $k = 3$ and consider cycle lengths $p_1 = 8$, $p_2 = 10$ and $p_3 = 12$, where $\gcd(8, 10) = 2$, $\gcd(8, 12) = 4$ and $\gcd(10, 12) = 2$. Then the numbers $f_1 = 7$, $f_2 = 8$ and $f_3 = 9$ satisfy the Main Condition, as $7 \not\equiv 8 \pmod{2}$, $7 \not\equiv 9 \pmod{4}$ and $8 \not\equiv 9 \pmod{2}$. This leads to a UFA with $1 + 8 + 10 + 12 = 31$ states recognizing the language $a^8(a^8)^* \cup a^9(a^{10})^* \cup a^{10}(a^{12})^*$. However, $\gcd(8, 10, 12) = 2$ and $7 \equiv 9 \pmod{2}$, and thus Lemma 6 is not applicable to this UFA, and would not be applicable for any choice of offsets f_1, f_2, f_3.

The idea of this example can be generalized to the following lower bound:

Lemma 7. *Let $k \geqslant 1$ and let $p_1, \ldots, p_{2k+1}$ be any pairwise distinct primes. Then the language $L = \bigcup_{i=1}^{2k+1} L_i$, where*

$$L_i = \{\, a^{1+n} \mid n \not\equiv 0 \pmod{p_i},\ n \equiv 0 \pmod{p_{i+1} \ldots p_{i+k}} \,\}$$

(with addition modulo $2k+1$ in subscripts), has a UFA with $1 + \bigcup_{i=1}^{2k+1} p_i \ldots p_{i+k}$ states, while every NFA for $\overline{L}$ contains at least $p_1 \ldots p_{2k+1}$ states.

The key element of the proof is the establishing that every infinite periodic subset of $\overline{L}$ containing any string a^{1+n} with $n \equiv 0 \pmod{p_1 \ldots p_{2k+1}}$ has period divisible by $p_1 \dots p_{2k+1}$. Now the witness language from Lemma 7 can be used to obtain the following modest lower bound on the complexity of complementing UFAs:

Lemma 8. *Let $k \geqslant 1$. Then the number of states in an NFA necessary to represent complements of n-state UFAs over a unary alphabet is at least $\frac{1}{2^{2k+1}(2k+1)^2} \cdot n^{2-\frac{1}{k+1}}$ for all $n \geqslant (2k+1)(4(2k+1)\ln 4(2k+1))^{k+1}$.*

For a given n, the goal is to find $2k+1$ distinct primes $p_1, \dots, p_{2k+1}$, which should be as large as possible, as long as $1 + \sum_{i=1}^{2k+1} p_i p_{i+1} \dots p_{i+k} \leqslant n$, so that the language in Lemma 7 has an n-state UFA. These primes are chosen by a well-known theorem of Ramanujan [18], which asserts that for every m large enough there are at least $2k+1$ primes between $\frac{m}{2}$ and m. To be more precise, m should be greater or equal to the $(2k+1)$-th *Ramanujan prime* r_{2k+1}, and the proof of Lemma 8 relies upon the asymptotic estimation of r_i due to Sondow [20].

Finally, letting k increase with n, the following lower bound can be obtained:

Theorem 5. *The state complexity of complementation for UFAs over a unary alphabet is at least $n^{2-o(1)}$ and at most $f_{UFA\text{-}DFA}(n)$.*

7 State Complexity of Intersection and Star

Consider the operation of *intersection*, which has state complexity mn both for DFAs [13,22] and for NFAs [5], and both over unary and larger alphabets. It maintains the same complexity for UFAs: the upper bound is by the standard *direct product construction*, which always produces a UFA for UFA arguments, and a matching lower bound for select values of m, n is already known from Holzer and Kutrib [5]: for all relatively prime $m, n \geqslant 2$, the language $(a^{mn})^* = (a^m)^* \cap (a^n)^*$ requires an NFA with at least mn states.

Theorem 6. *The state complexity of intersection for UFAs over a unary alphabet is at most mn. This bound is reachable for all relatively prime m, n.*

Turning to the *Kleene star*, its state complexity for unary DFAs is $(n-1)^2+1$, obtained by Yu, Zhuang and Salomaa [22, Thm. 5.3]. An identical result holds for UFAs, in spite of the differences between the two models. The lower bound argument uses a language with a co-finite star, and for such languages UFAs are no more succinct than DFAs:

Lemma 9. *Let $L \subseteq a^*$ be a co-finite language, let a^m be the longest string not in L. Then the smallest NFA in Chrobak normal form for L contains $m+2$ states and coincides with the smallest DFA for L.*

Theorem 7. *For every $n \geqslant 1$, star of an n-state UFA is representable by a UFA with $(n-1)^2+1$ states, and this number of states is in the worst case necessary.*

Establishing the complexity of union and concatenation and improving the bounds on the complementation are left as main open problems.

References

1. Bach, E., Shallit, J.: Algorithmic Number Theory. MIT Press, Cambridge (1996)
2. Björklund, H., Martens, W.: The tractability frontier for NFA minimization. In: Aceto, L., Damgård, I., Goldberg, L.A., Halldórsson, M.M., Ingólfsdóttir, A., Walukiewicz, I. (eds.) ICALP 2008, Part II. LNCS, vol. 5126, pp. 27–38. Springer, Heidelberg (2008)
3. Chrobak, M.: Finite automata and unary languages. Theoretical Computer Science 47, 149–158 (1986); Errata: vol. 302(1-3), pp. 497–498 (2003)
4. Geffert, V., Mereghetti, C., Pighizzini, G.: Complementing two-way finite automata. Information and Computation 205(8), 1173–1187 (2007)
5. Holzer, M., Kutrib, M.: Nondeterministic descriptional complexity of regular languages. Intl. J. of Foundations of Computer Science 14, 1087–1102 (2003)
6. Hromkovič, J., Seibert, S., Karhumäki, J., Klauck, H., Schnitger, G.: Communication complexity method for measuring nondeterminism in finite automata. Information and Computation 172(2), 202–217 (2002)
7. Ingleton, A.W.: The rank of circulant matrices. Journal of the London Mathematical Society 31, 445–460 (1956)
8. Jiang, T., McDowell, E., Ravikumar, B.: The structure and complexity of minimal NFA's over a unary alphabet. International Journal of Foundations of Computer Science 2(2), 163–182 (1991)
9. Landau, E.: Über die Maximalordnung der Permutationen gegebenen Grades (On the maximal order of permutations of a given degree). Archiv der Mathematik und Physik 3(5), 92–103 (1903)
10. Leung, H.: Separating exponentially ambiguous finite automata from polynomially ambiguous finite automata. SIAM Journal on Computing 27(4), 1073–1082 (1998)
11. Leung, H.: Descriptional complexity of NFA of different ambiguity. International Journal of Foundations of Computer Science 16(5), 975–984 (2005)
12. Lyubich, Y.: Bounds for the optimal determinization of nondeterministic autonomic automata. Sibirskii Matematicheskii Zhurnal 2, 337–355 (1964) (in Russian)
13. Maslov, A.N.: Estimates of the number of states of finite automata. Soviet Mathematics Doklady 11, 1373–1375 (1970)
14. Mera, F., Pighizzini, G.: Complementing unary nondeterministic automata. Theoretical Computer Science 330(2), 349–360 (2005)
15. Mereghetti, C., Pighizzini, G.: Optimal simulations between unary automata. SIAM Journal on Computing 30(6), 1976–1992 (2001)
16. Miller, W.: The maximum order of an element of a finite symmetric group. American Mathematical Monthly 94(6), 497–506 (1987)
17. Pighizzini, G., Shallit, J.: Unary language operations, state complexity and Jacobsthal's function. Intl. J. of Foundations of Computer Sci. 13(1), 145–159 (2002)
18. Ramanujan, S.: A proof of Bertrand's postulate. Journal of the Indian Mathematical Society 11, 181–182 (1919)
19. Schmidt, E.M.: Succinctness of Description of Context-Free, Regular and Unambiguous Languages, Ph. D. thesis, Cornell University (1978)
20. Sondow, J.: Ramanujan primes and Bertrand's postulate. American Mathematical Monthly 116, 630–635 (2009)
21. Stearns, R.E., Hunt III, H.B.: On the equivalence and containment problems for unambiguous regular expressions, regular grammars and finite automata. SIAM Journal on Computing 14, 598–611 (1985)
22. Yu, S., Zhuang, Q., Salomaa, K.: The state complexity of some basic operations on regular languages. Theoretical Computer Science 125, 315–328 (1994)

The Complexity of Finding Reset Words in Finite Automata

Jörg Olschewski[1,2,⋆] and Michael Ummels[2,3,⋆⋆]

[1] Lehrstuhl Informatik 7, RWTH Aachen University, Germany
olschewski@automata.rwth-aachen.de
[2] LSV, CNRS & ENS Cachan, France
[3] Mathematische Grundlagen der Informatik,
RWTH Aachen University, Germany
ummels@logic.rwth-aachen.de

Abstract. We study several problems related to finding reset words in deterministic finite automata. In particular, we establish that the problem of deciding whether a shortest reset word has length k is complete for the complexity class DP. This result answers a question posed by Volkov. For the search problems of finding a shortest reset word and the length of a shortest reset word, we establish membership in the complexity classes $\mathrm{FP}^{\mathrm{NP}}$ and $\mathrm{FP}^{\mathrm{NP}[\log]}$, respectively. Moreover, we show that both these problems are hard for $\mathrm{FP}^{\mathrm{NP}[\log]}$. Finally, we observe that computing a reset word of a given length is FNP-complete.

1 Introduction

A *synchronising automaton* is a deterministic finite automaton that can be reset to a single state by reading a suitable word. More precisely, we require needs to exist a word w such that, no matter at which state of the automaton we start, w takes the automaton to the same state q; we call any such word w a *reset word* or a *synchronising word.* Although it is easy to decide whether a given automaton is synchronising and to compute a reset word, finding a *shortest* reset word seems to be a hard problem.

The motivation to study reset words does not only come from automata theory: There are applications in the fields of many-valued logics, biocomputing, set theory, and many more [12]. A purely mathematical viewpoint can be obtained by identifying letters with their associated transition functions, which act on a finite set. The task is then to find a composition of these functions such that the resulting function is constant.

The theory of synchronising automata has been established in the 1960s and is still actively developed. The famous *Černý Conjecture* was formulated in 1971 [3]. The conjecture claims that every synchronising automaton with n states has a reset word of length $(n-1)^2$. As of now, the conjecture has neither been

⋆ Supported by the ESF project GASICS.
⋆⋆ Supported by the French project DOTS (ANR-06-SETI-003).

P. Hliněný and A. Kučera (Eds.): MFCS 2010, LNCS 6281, pp. 568–579, 2010.

proved nor disproved; the best known upper bound on the length of a reset word is $(n^3 - n)/6$, as shown by Pin [8].

While Eppstein [4] showed that the problem of deciding whether there exists a reset word of a given length k is NP-complete, the complexity of deciding whether a *shortest* reset word has length k is not known to be in NP. In his survey paper [12], Volkov asked for the precise complexity of this problem. In this paper, we show that deciding whether a shortest reset word has length k is complete for the class DP, the closure of NP ∪ coNP under finite intersections. In particular, since every DP-complete problem is both NP-hard and coNP-hard, it is unlikely that the problem of deciding the length of a shortest reset word lies in NP ∪ coNP.[1]

The class DP is contained in the class P^{NP}, i.e. every problem in DP can be solved by a deterministic polynomial-time Turing machine that has access to an oracle for an NP-complete problem. In fact, two oracle queries suffice for this purpose. If one restricts the number of oracle queries to be logarithmic in the size of the input, one arrives at the class $P^{NP[\log]}$, which is believed to be a proper superclass of DP. We show that the problem of computing the length of a shortest reset word (as opposed to deciding whether it is equal to a given integer) is, in fact, complete for $FP^{NP[\log]}$, the functional analogue of $P^{NP[\log]}$. Hence, this problem seems to be even harder than deciding the length of a shortest reset word. Our result complements a recent result by Berlinkov [1], who showed that, unless P = NP, there is no polynomial-time algorithm that *approximates* the length of a shortest reset word within a constant factor.

For the more general problem of computing a shortest reset word (not only its length), we prove membership in FP^{NP}, the functional analogue of P^{NP}. While our lower bound of $FP^{NP[\log]}$ on computing the length of a shortest reset word carries over to this problem, we leave it as an open problem whether computing a shortest reset word is also FP^{NP}-hard.

Apart from studying problems related to computing a *shortest* reset word, we also consider the problem of computing a reset word of a given length (represented in unary). We observe that this problem is complete for the class FNP of search problems for which a solution can be verified in polynomial time. In other words: the problem is as hard as computing a satisfying assignment for a given Boolean formula.

2 Preliminaries

Let $\mathcal{A} = \langle Q, \Sigma, \delta \rangle$ be a deterministic finite automaton (DFA) with finite state set Q, finite alphabet Σ and transition function $\delta \colon Q \times \Sigma \to Q$. The transitive closure of δ can be defined inductively by $\delta^*(q, \varepsilon) = q$ and $\delta^*(q, wa) = \delta(\delta^*(q, w), a)$ for each $q \in Q$, $w \in \Sigma^*$ and $a \in \Sigma$. We call any word $w \in \Sigma^*$ such that $|\{\delta^*(q, w) \mid q \in Q\}| = 1$ a *reset word* for $\mathcal{A}$, and we say that $\mathcal{A}$ is *synchronising*

[1] We have been informed that Gawrychowski [5] has shown DP-completeness of SHORTEST-RESET-WORD earlier, but his proof has never been published. While his reduction uses a five-letter alphabet, we prove hardness even over a binary alphabet.

if such a word exists. Note that, if w is a reset word for $\mathcal{A}$, then so is xwy for all $x, y \in \Sigma^*$.

We assume that the reader is familiar with basic concepts of complexity theory, in particular with the classes P, NP and coNP. We will introduce the other classes that play a role in this paper on the fly. For details, see [7,10].

3 Decision Problems

The most fundamental decision problem concerning reset words is to decide whether a given deterministic finite automaton is synchronising. Černý [2] noted that it suffices to check for each pair (q, q') of states whether there exists a word $w \in \Sigma^*$ with $\delta^*(q, w) = \delta^*(q', w)$. The latter property can obviously be decided in polynomial time. The best known algorithm for *computing* a reset word is due to Eppstein [4]: his algorithm runs in time $O(|Q|^3 + |Q|^2 \cdot |\Sigma|)$. Computing a *shortest* reset word, however, cannot be done in polynomial time unless the following decision problems are in P.

SHORT-RESET-WORD: Given a DFA $\mathcal{A}$ and a positive integer k, decide whether there exists a reset word for $\mathcal{A}$ of length k.

SHORTEST-RESET-WORD: Given a DFA $\mathcal{A}$ and a positive integer k, decide whether the minimum length of a reset word for $\mathcal{A}$ equals k.

If the parameter k is given in unary, it is obvious that SHORT-RESET-WORD is in NP. However, even if k is given in binary, this problem is in NP: since every synchronising automaton has a reset word of length $p(|Q|)$ (where p is a low-degree polynomial, e.g. $p(n) = (n^3 - n)/6$), to establish whether there exists a reset word of length k, it suffices to guess a reset word of length $\min\{p(|Q|), k\}$. Eppstein [4] gave a matching lower bound by proving that SHORT-RESET-WORD is also NP-hard.

Regarding SHORTEST-RESET-WORD, Samotij [9] showed that the problem is NP-hard. We prove that SHORTEST-RESET-WORD is complete for DP, the class of all languages of the form $L = L_1 \setminus L_2$ with $L_1, L_2 \in$ NP. Since DP is a superclass of both NP and coNP, our result implies hardness for both of these classes. In fact, we show that SHORTEST-RESET-WORD is DP-hard even over a binary alphabet.

Theorem 1. SHORTEST-RESET-WORD *is* DP-*complete.*

Proof. It is easy to see that SHORTEST-RESET-WORD belongs to DP: indeed, we can write SHORTEST-RESET-WORD as the difference of SHORT-RESET-WORD and SHORT-RESET-WORD$^-$, where

$$\text{SHORT-RESET-WORD}^- = \{(\mathcal{A}, k+1) \mid (\mathcal{A}, k) \in \text{SHORT-RESET-WORD}\},$$

a problem which is obviously in NP (even if k is given in binary).

It remains to prove that SHORTEST-RESET-WORD is DP-hard. We reduce from the canonical DP-complete problem SAT-UNSAT: given two Boolean formulae φ and ψ (in CNF), decide whether φ is satisfiable and ψ is unsatisfiable. More precisely, we show how to construct (in polynomial time) from a pair (φ, ψ) of Boolean formulae in CNF over propositional variables $X_1, \ldots, X_k$ a synchronising automaton $\mathcal{A}$ over the alphabet $\Sigma = \{0, 1\}$ with the following properties:

1. If φ and ψ are satisfiable, then there exists a reset word of length $k + 2$.
2. If φ is satisfiable and ψ is unsatisfiable, then a shortest reset word has length $k + 3$.
3. If φ is unsatisfiable, then every reset word has length at least $k + 4$.

From 1.–3. we get that φ is satisfiable and ψ is unsatisfiable if and only if a shortest reset word has length $k + 3$.

Given formulae $\varphi = C_1 \wedge \ldots \wedge C_n$ and $\psi = D_1 \wedge \ldots \wedge D_n$ where, without loss of generality, φ and ψ have the same number n of clauses, and no propositional variable occurs in both φ and ψ, the automaton $\mathcal{A}$ consists of the states s, t_1, t_2, $p_{i,j}$ and $q_{i,j}$, $i \in \{1, \ldots, n\}$, $j \in \{\bot, \top, 1, \ldots, k\}$; the transitions are depicted in Fig. 1: an edge from p to q labelled with $\Sigma' \subseteq \Sigma$ has the meaning that $\delta(p, a) = q$ for each $a \in \Sigma'$. The sets $\Sigma_i^j \subseteq \Sigma$ are defined by $0 \in \Sigma_i^j \Leftrightarrow \neg X_j \in C_i$ and $1 \in \Sigma_i^j \Leftrightarrow X_j \in C_i$, and the sets $\Gamma_i^j \subseteq \Sigma$ are defined by $0 \in \Gamma_i^j \Leftrightarrow \neg X_j \in D_i$ and $1 \in \Gamma_i^j \Leftrightarrow X_j \in D_i$. Hence, e.g. $0 \in \Sigma_i^j$ if we can satisfy the ith clause of φ by setting variable X_j to false.

Clearly, $\mathcal{A}$ can be constructed in polynomial time from φ and ψ. To establish our reduction, it remains to verify 1.–3.

To prove 1., assume that φ and ψ are both satisfiable. Since φ and ψ share no variable, there exists an assignment $\alpha\colon \{X_1, \ldots, X_k\} \to \{\text{true}, \text{false}\}$ that satisfies both φ and ψ. We claim that the word $01w$, where $w = w_1 \ldots w_k \in \{0, 1\}^k$ is defined by $w_j = 1 \Leftrightarrow \alpha(X_j) = \text{true}$, resets $\mathcal{A}$ to s. Clearly, $\delta^*(q, w) = s$ for all states q that are not of the form $q = p_{i,\bot}$, $q = p_{i,\top}$, $q = q_{i,\bot}$ or $q = q_{i,\top}$. Since $\delta^*(p_{i,\bot}, 01) = \delta^*(p_{i,\top}, 01) = p_{i,1}$ and $\delta^*(q_{i,\bot}, 01) = \delta^*(q_{i,\top}, 01) = q_{i,1}$ for each $i = 1, \ldots, n$, it suffices to show that $\delta^*(p_{i,1}, w) = \delta^*(q_{i,1}, w) = s$ for all i. To prove that $\delta^*(p_{i,1}, w) = s$, consider the least j such that either $X_j \in C_i$ and $\alpha(X_j) = \text{true}$ or $\neg X_j \in C_i$ and $\alpha(X_j) = \text{false}$ (such j exists since α satisfies φ). We have $\delta^*(p_{i,1}, w_1 \ldots w_{j-1}) = p_{i,j}$ and $\delta(p_{i,j}, w_j) = s$ and therefore also $\delta^*(p_{i,1}, w) = s$. The argument for $\delta^*(q_{i,1}, w) = s$ is analogous.

Towards proving 2., assume that φ is satisfiable but ψ is not. Consider an assignment $\alpha\colon \{X_1, \ldots, X_k\} \to \{\text{true}, \text{false}\}$ that satisfies φ. It follows with the same reasoning as above that the word $01w1$, where $w \in \{0, 1\}^k$ is defined by $w_j = 1 \Leftrightarrow \alpha(X_j) = \text{true}$, resets $\mathcal{A}$ to s.

To show that a *shortest* reset word has length $k + 3$, it remains to show that there exists no reset word of length $k + 2$. Towards a contradiction, assume that $w = w_1 \ldots w_{k+2}$ is such a word. Note that w resets $\mathcal{A}$ to s and that there exists $l \geq 2$ such that $\delta^*(q_{i,\bot}, w_1 \cdots w_l) = q_{i,1}$ and $\delta^*(q_{i,1}, w_{l+1} \cdots w_{k+2}) = s$ for all $i = 1, \ldots, n$. Define $\alpha\colon \{X_1, \ldots, X_k\} \to \{\text{true}, \text{false}\}$ by setting $\alpha(X_j) = \text{true} \Leftrightarrow w_{l+j} = 1$. Since $l \geq 2$ but $\delta^*(q_{i,1}, w_{l+1} \cdots w_{k+2}) = s$, for each i there

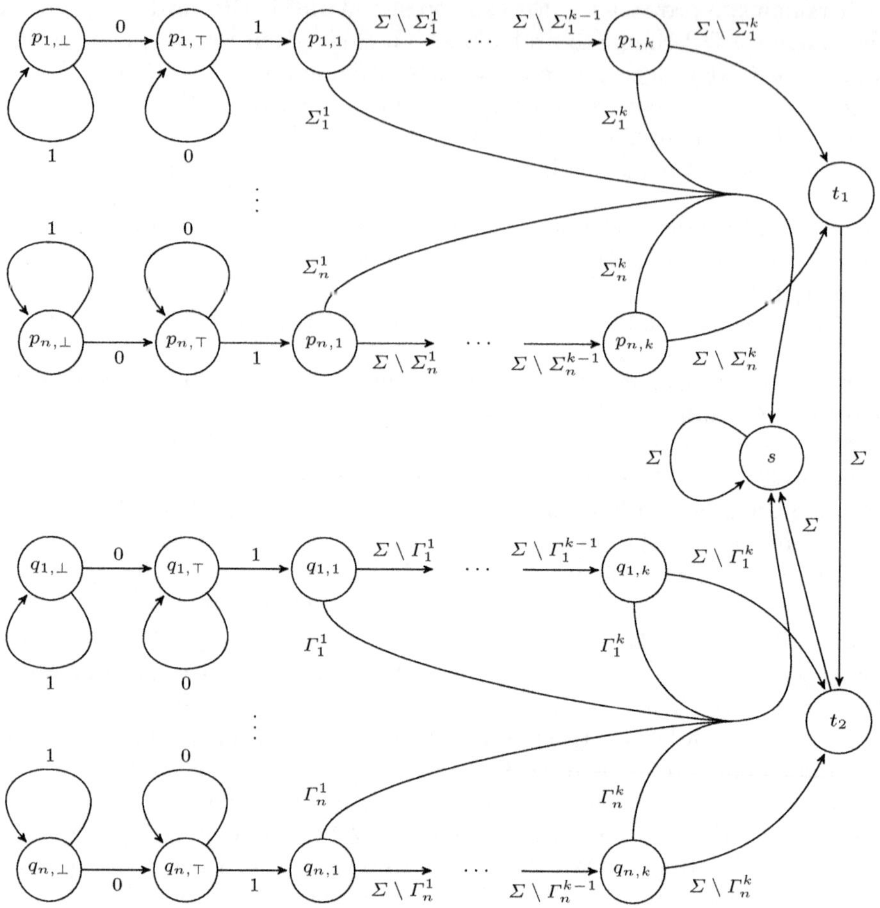

Fig. 1. Reducing SAT-UNSAT to SHORTEST-RESET-WORD

must exist $j \in \{1, \dots, k\}$ such that $\delta(q_{i,j}, w_{l+j}) = s$. But then either $X_j \in D_i$ and $\alpha(X_j) = \text{true}$ or $\neg X_j \in D_i$ and $\alpha(X_j) = \text{false}$. Hence, α is a satisfying assignment for ψ, contradicting our assumption that ψ is unsatisfiable.

Finally, assume that φ is unsatisfiable. With the same reasoning as in the previous case, it follows that there is no reset word of length $k + 3$. □

The above reduction shows DP-hardness for an alphabet size of $|\Sigma| = 2$. For the special case of only one input letter, note that each reset word is of the form 1^n for some n. Asking whether there exists a reset word of length k thus collapses to the question whether 1^k is a reset word for $\mathcal{A}$. This property can be decided with logarithmic space. Hence, both problems, SHORT-RESET-WORD and SHORTEST-RESET-WORD, are in LOGSPACE for $|\Sigma| = 1$.

4 Search Problems

In this section, we leave the realm of decision problems and enter the (rougher) territory of search problems, where the task is not only to decide whether a reset word of some length exists, but to compute a suitable word (or its length). More precisely, we deal with the following search problems:

- Given a DFA $\mathcal{A}$ and a positive integer k in unary, compute a reset word for $\mathcal{A}$ of length k.
- Given a DFA $\mathcal{A}$, compute the length of a shortest reset word for $\mathcal{A}$.
- Given a DFA $\mathcal{A}$, compute a shortest reset word for $\mathcal{A}$.

Let us start with the first problem of computing a reset word of a given length. It turns out that this problem is complete for the class FNP of search problems where the underlying binary relation is both polynomially balanced and decidable in polynomial time.

Proposition 2. *The problem of computing a reset word of a given length is* FNP*-complete.*

Proof. Membership in FNP follows from the fact that the binary relation

$$\{((\mathcal{A}, 1^k), w) \mid w \text{ is a reset word for } \mathcal{A} \text{ of length } k\}$$

is polynomially balanced and polynomial-time decidable.

To prove hardness, we reduce from FSAT, the problem of computing a satisfying assignment for a given Boolean formula in conjunctive normal form. To this end, we describe two polynomial-time computable functions f and g, where f computes from a CNF formula φ a synchronising automaton $\mathcal{A} = f(\varphi)$ over the alphabet $\{0,1\}$ and a unary number $k \in \mathbb{N}$, and g computes from φ and $w \in \Sigma^*$ an assignment for φ, such that, if w is a reset word for $\mathcal{A}$ of length k, then the generated assignment satisfies φ.

Eppstein [4] showed how to compute in polynomial time, given a CNF formula $\varphi = C_1 \wedge \ldots \wedge C_n$ over the variables $X_1, \ldots, X_k$, an automaton $\mathcal{A}_\varphi$ over the alphabet $\{0,1\}$ with the following two properties:

1. A word $w = w_1 \cdots w_k$ is a reset word for $\mathcal{A}$ if and only if the assignment α, defined by $\alpha(X_j) = \text{true} \Leftrightarrow w_j = 1$, satisfies φ.
2. An assignment $\alpha\colon \{X_1, \ldots, X_k\} \to \{\text{true}, \text{false}\}$ satisfies φ if and only if the word $w \in \{0,1\}^k$, defined by $w_j = 1 \Leftrightarrow \alpha(X_j) = \text{true}$, is a reset word for $\mathcal{A}$.

(Note that the reduction we use to prove Theorem 1 has similar properties and could also be used.)

Hence, we can choose f to be the function that maps φ to $(\mathcal{A}_\varphi, 1^k)$ and g to be the function that maps (φ, w) to the corresponding assignment α. (If $|w| \neq k$, then α can be chosen arbitrarily.) □

Remark 3. Note that the mapping $f : \{0,1\}^k \to \{\text{true}, \text{false}\}^{\{X_1,\ldots,X_k\}}$, defined by $f(w)(X_j) = \text{true} \Leftrightarrow w_j = 1$, is a bijection. Eppstein's reduction shows that

Algorithm 1. Computing the length of a shortest reset word

```
if A is not synchronising then reject
low := −1
high := (n^3 − n)/6
while high − low > 1 do
  k := ⌈(low + high)/2⌉
  if A has a reset word of length k then
    high := k
  else
    low := k
end while
return high
```

one can compute from a Boolean formula φ over the variables $\{X_1, \ldots, X_k\}$ an automaton $\mathcal{A}$ such that f remains a bijection when one restricts the domain to reset words for $\mathcal{A}$ and the range to assignments that satisfy φ. Therefore, his reduction can be viewed as a parsimonious reduction from #SAT, the problem of counting all satisfying assignments of a given Boolean formula, to the problem of counting all reset words of a given length (represented in unary). Since the first problem is complete for #P [11], the second problem is #P-hard. On the other hand, it is easy to see that the second problem is in #P. Hence, this problem is #P-complete.

Next, we consider the problem of computing the *length* of a shortest reset word for a given automaton: we establish that this problem is complete for the class $\mathrm{FP}^{\mathrm{NP}[\log]}$ of all problems that are solvable by a polynomial-time algorithm with access to an oracle for a problem in NP where the number of queries is restricted to $\mathrm{O}(\log n)$.

Theorem 4. *The problem of computing the length of a shortest reset word is* $\mathrm{FP}^{\mathrm{NP}[\log]}$*-complete.*

Proof. To prove membership in $\mathrm{FP}^{\mathrm{NP}[\log]}$, consider Algorithm 1 which is a binary-search algorithm for determining the length of a shortest reset word for an automaton $\mathcal{A}$ with n states. The algorithm is executed in polynomial time: the while loop is repeated $\mathrm{O}(\log n)$ times and asks $\mathrm{O}(\log n)$ queries to the oracle, which is used for determining whether $\mathcal{A}$ has a reset word of a given length.

Krentel [6] showed that MAX-SAT-SIZE, the problem of computing the maximum number of simultaneously satisfiable clauses of a CNF formula, is complete for $\mathrm{FP}^{\mathrm{NP}[\log]}$. Therefore, to establish $\mathrm{FP}^{\mathrm{NP}[\log]}$-hardness, it suffices to give a reduction from MAX-SAT-SIZE to our problem. Such a reduction consists of two polynomial-time computable functions f and g with the following properties: f computes from a CNF formula φ a (synchronising) automaton $\mathcal{A} = f(\varphi)$, and g computes from φ and $l \in \mathbb{N}$ a new number $g(\varphi, l) \in \mathbb{N}$ such that, if l is the length of a shortest reset word for $\mathcal{A}$, then the maximum number of simultaneously satisfiable clauses in φ equals $g(\varphi, k)$.

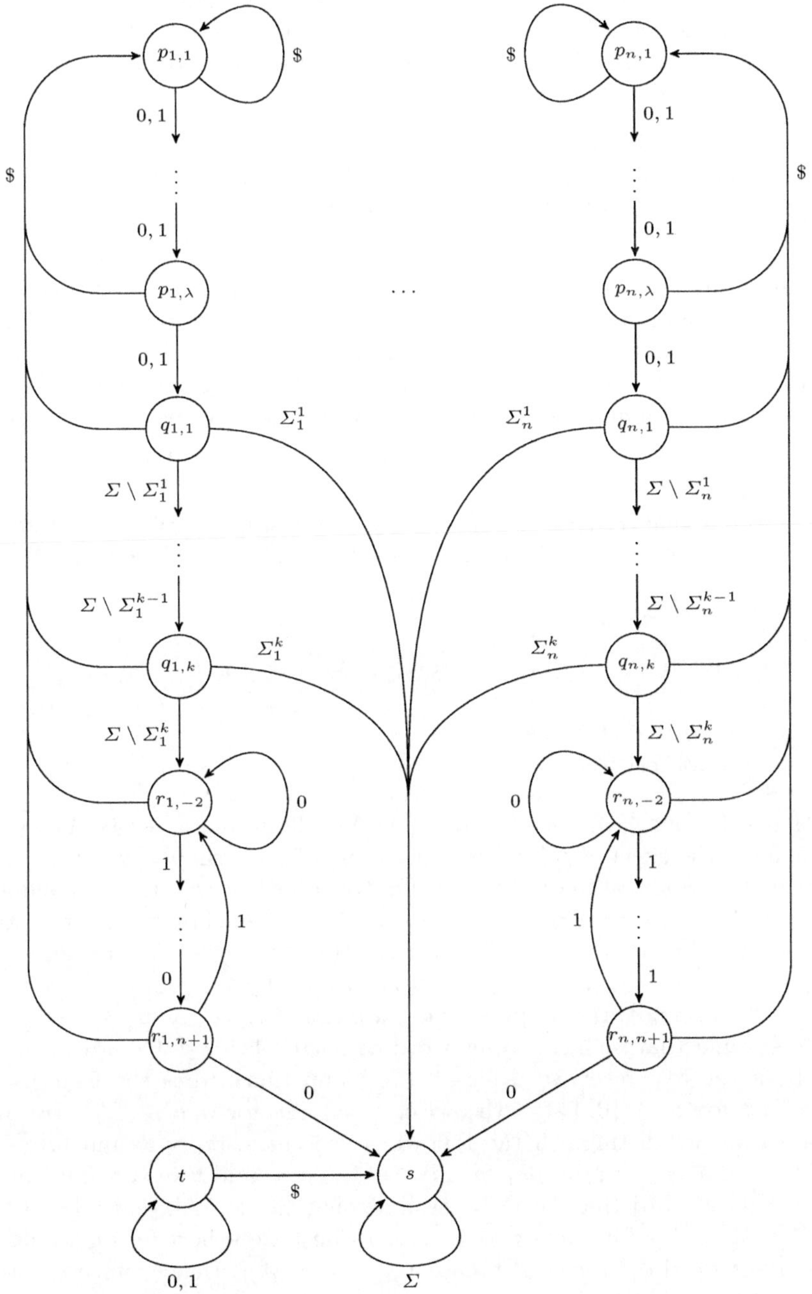

Fig. 2. Reducing MAX-SAT-SIZE to computing the length of a shortest reset word

Given a formula $\varphi = C_1 \wedge \ldots \wedge C_n$ over propositional variables $X_1, \ldots, X_k$, the resulting automaton $\mathcal{A}$ is depicted in Fig. 2: The input alphabet is $\Sigma := \{0, 1, \$\}$, and the sets $\Sigma_i^j \subseteq \Sigma$ are defined as in the proof of Theorem 1; we set $\lambda := k + n(n+4)$. The behaviour of the transition function on vertices of the form $r_{i,j}$ is defined as follows:

- $\delta(r_{i,j}, \$) = p_{i,1}$ for all $j \in \{-2, \ldots, n+1\}$;
- $\delta(r_{i,j}, 1) = r_{i,j+1}$, $\delta(r_{i,j}, 0) = r_{i,-2}$ for all $j \in \{-2, -1, i\}$;
- $\delta(r_{i,j}, 1) = r_{i,-2}$, $\delta(r_{i,j}, 0) = r_{i,j+1}$ for all $j \in \{0, \ldots, i-1, i+1, \ldots, n\}$;
- $\delta(r_{i,n+1}, 1) = r_{i,-2}$, $\delta(r_{i,n+1}, 0) = s$.

It is not difficult to see that $\mathcal{A}$ can be constructed in polynomial time from φ. Moreover, we claim that, for each $m \in \{0, 1, \ldots, n\}$, there exists an assignment that satisfies at least $n - m$ clauses of φ if and only if $\mathcal{A}$ has a reset word of length $1+\lambda+k+m(n+4)$. Hence, if l is the length of a shortest reset word for $\mathcal{A}$, then the maximal number of simultaneously satisfiable clauses of φ is given by $n - \left\lceil \frac{\max\{0, l-1-\lambda-k\}}{n+4} \right\rceil$. Clearly, this number can be computed in polynomial time from φ and l.

($\Rightarrow$) Assume that $\alpha\colon \{X_1, \ldots, X_k\} \to \{\text{true}, \text{false}\}$ is an assignment that satisfies all clauses except (possibly) the clauses $C_{i_1}, \ldots, C_{i_m}$, and consider the word

$$w := \$1^\lambda x_1 \ldots x_k z_{i_1} \ldots z_{i_m},$$

where $z_i = 110^i 10^{n-i+1} \in \{0,1\}^{n+4}$ for $i \in \{1, \ldots, n\}$ and

$$x_j := \begin{cases} 1 & \text{if } \alpha(X_j) = \text{true}, \\ 0 & \text{otherwise.} \end{cases}$$

Note that w has length $1+\lambda+k+m(n+4)$. We claim that w resets $\mathcal{A}$ to s. Since reading \$ has the effect of going from each state of the form $p_{i,j}$, $q_{i,j}$ or $r_{i,j}$ to $p_{i,1}$ and from t to s, and reading 1^λ has the effect of going from $p_{i,1}$ to $q_{i,1}$, it suffices to show that $\delta^*(q_{i,1}, x_1 \ldots x_k z_{i_1} \ldots z_{i_m}) = s$. If C_i is satisfied by α, then this follows from the fact that there exists j such that $\delta(q_{i,j}, x_j) = s$. Otherwise, we have $\delta^*(q_{i,1}, x_1 \ldots x_k) = r_{i,-2}$, $\delta^*(r_{i,-2}, z_j) = r_{i,-2}$ for all $j \neq i$, but $\delta^*(r_{i,-2}, z_i) = s$. Since $i \in \{i_1, \ldots, i_m\}$, this implies that $\delta^*(q_{i,1}, x_1 \ldots x_k z_{i_1} \ldots z_{i_m}) = s$.

($\Leftarrow$) Assume that $\mathcal{A}$ has a reset word of length $1 + \lambda + k + m(n+4)$, and let w be a shortest reset word for $\mathcal{A}$. We claim that w has the form $w = \$u$ or $w = u\$$ for $u \in \{0,1\}^*$. Otherwise, $w = u\$v$ for $u, v \in \Sigma^+$. Towards a contradiction, we distinguish the following two cases: $|u| \leq \lambda$ and $|u| > \lambda$. If $|u| \leq \lambda$, then $\delta^*(p_{i,1}, u\$) = p_{i,1}$ for all $i = 1, \ldots, n$, and the word $\$v$ would be a shorter reset word than w. Now assume that $|u| > \lambda$. It must be the case that $\delta^*(p_{i,1}, u) \neq s$ for some $i \in \{1, \ldots, n\}$ because otherwise $\$u$ would be a shorter reset word than w. But then $\delta^*(p_{i,1}, u\$) = p_{i,1}$. Hence, since w resets $\mathcal{A}$ to s and the shortest path from $p_{i,1}$ to s has length greater than λ, $|v| > \lambda$ and $|w| > 1 + 2\lambda \geq 1 + \lambda + k + n(n+4) \geq 1 + \lambda + k + m(n+4)$, a contradiction.

Now, if φ is satisfiable, we are done. Otherwise, let us fix $u \in \{0,1\}^*$ such that $w = \$u$ or $w = u\$$. Since φ is not satisfiable, $|u| \geq \lambda + k$. Let $u = yx_1 \ldots x_k z$

where $y, z \in \{0,1\}^*$, $|y| = \lambda$, and $x_j \in \{0,1\}$ for all $j = 1, \ldots, k$. Now consider the assignment α defined by

$$\alpha(X_j) = \begin{cases} \text{true} & \text{if } x_j = 1, \\ \text{false} & \text{otherwise.} \end{cases}$$

Moreover, let

$$I := \{i \in \{1, \ldots, n\} \mid C_i \text{ is not satisfied by } \alpha\}.$$

We claim that $|I| \leq m$ (so α satisfies at least $n - m$ clauses of φ). To see this, first note that $\delta^*(p_{i,1}, yx_1 \ldots x_k) = r_{i,-2}$ for all $i \in I$. Hence, we must have that $\delta^*(r_{i,-2}, z) = s$ for all such i. By the construction of $\mathcal{A}$, this is only possible if z contains the word 110^i10^{n-i+1} as an infix for each $i \in I$. Since these infixes cannot overlap, $|z| \geq |I| \cdot (n+4)$. On the other hand, since $|u| \leq \lambda + k + m(n+4)$, we must have $|z| \leq m(n+4)$. Hence, $|I| \leq m$. □

The construction we have presented to prove Theorem 4 uses a three-letter alphabet. With a little more effort, we can actually reduce the alphabet to an alphabet with two letters 0 and 1: For each state $q \notin \{s,t\}$ of $\mathcal{A}$, there are three states $(q,0)$, $(q,1)$ and $(q,2)$ in the new automaton $\mathcal{A}'$. Additionally, $\mathcal{A}'$ contains the states $(t,0)$, $(t,1)$ and s. The new transition function δ' is defined as follows:

$$\begin{aligned} \delta'((q,0),0) &= (q,1), & \delta'((q,0),1) &= (q,2), \\ \delta'((q,1),0) &= (q,1), & \delta'((q,1),1) &= (\delta(q,\$),2), \\ \delta'((q,2),0) &= (\delta(q,0),0), & \delta'((q,2),1) &= (\delta(q,1),0) \end{aligned}$$

for all $q \notin \{s,t\}$, and

$$\begin{aligned} \delta'((t,0),0) &= s, & \delta'((t,0),1) &= (t,1), \\ \delta'((t,1),0) &= (t,0), & \delta'((t,1),1) &= (t,1), \\ \delta'(s,0) &= s, & \delta'(s,1) &= s. \end{aligned}$$

Intuitively, taking a transition in $\mathcal{A}$ corresponds to taking two transitions in $\mathcal{A}'$. It is not difficult to see that a shortest reset word for $\mathcal{A}'$ has length $2l$ if a shortest reset word for $\mathcal{A}$ has length l.

For the potentially harder problem of computing a shortest reset word (not only its length), we can only prove membership in FP^{NP}, the class of all search problems that are solvable in polynomial time using an oracle for a problem in NP (without any restriction on the number of queries). Of course, hardness for $\text{FP}^{\text{NP[log]}}$ carries over from our previous result. We have not been able to close the gap between the two bounds. To the best of our knowledge, the same situation occurs e.g. for MAX-SAT, where the aim is to find an assignment of a given Boolean formula that satisfies as many clauses as possible.

Theorem 5. *The problem of computing a shortest reset word is in* FP^{NP} *and hard for* $\text{FP}^{\text{NP[log]}}$.

Algorithm 2. Computing a shortest reset word

if $\mathcal{A}$ is not synchronising **then reject**
Compute the length l of a shortest reset word for $\mathcal{A}$
$w := \varepsilon$
while $|w| < l$ **do**
  for each $a \in \Sigma$ **do**
    if $\mathcal{A}$ has a reset word of length l with prefix wa **then**
      $w := wa$; **break for**
    end if
  end for
end while
return w

Proof. To prove membership in $\mathrm{FP}^{\mathrm{NP}}$, consider Algorithm 2 for computing a shortest reset word for an automaton $\mathcal{A}$ over any finite alphabet Σ. The algorithm obviously computes a reset word of length l, which is the length of a shortest reset word. To see that the algorithm runs in polynomial time if it has access to an NP oracle, note that deciding whether $\mathcal{A}$ has a reset word of a given length with a given prefix is in NP (since a nondeterministic polynomial-time algorithm can guess such a word). Moreover, as we have shown above, computing the *length* of a shortest reset word can be done by a polynomial-time algorithm with access to an NP oracle.

Hardness for $\mathrm{FP}^{\mathrm{NP}[\log]}$ follows from Theorem 4 since the problem of computing the length of a shortest reset word is trivially reducible to the problem of computing a shortest reset word: an instance of the former problem is also an instance of the latter problem, and a solution of the latter problem can be turned into a solution of the former problem by computing its length. □

5 Conclusion

We have investigated several decision problems and search problems about finding reset words in finite automata. The results we have obtained shed more light on the difficulty of computing such words. In particular, deciding whether for a given automaton a shortest reset word has length k is DP-complete, and computing the length of a shortest reset word is $\mathrm{FP}^{\mathrm{NP}[\log]}$-complete, i.e. as hard as calculating the maximum number of simultaneously satisfiable clauses of a Boolean formula. A summary of all our results is depicted in Fig. 3. (See [7,10] for the relationships between the referred complexity classes.)

Acknowledgements. We thank an anonymous reviewer for pointing out [5]. Moreover, we are grateful to Christof Löding and Wolfgang Thomas for helpful comments on an early draft of this paper.

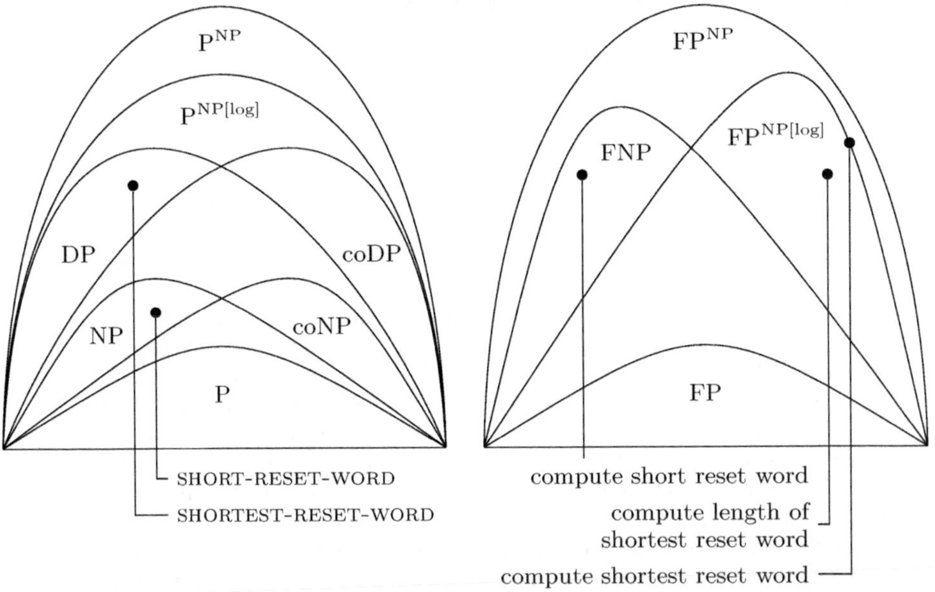

Fig. 3. Summary of results

References

1. Berlinkov, M.V.: Approximating the length of synchronizing words. In: Ablayev, F., Mayr, E.W. (eds.) Computer Science – Theory and Applications. LNCS, vol. 6072, pp. 37–47. Springer, Heidelberg (2010)
2. Černý, J.: Poznámka k. homogénnym experimentom s konecnými automatmi. Matematicko-fyzikalny Časopis Slovensk. Akad. Vied 14(3), 208–216 (1964)
3. Černý, J., Pirická, A., Rosenauerová, B.: On directable automata. Kybernetica 7(4), 289–298 (1971)
4. Eppstein, D.: Reset sequences for monotonic automata. SIAM Journal on Computing 19(3), 500–510 (1990)
5. Gawrychowski, P.: Complexity of shortest synchronizing word. Unpublished manuscript (April 2008)
6. Krentel, M.W.: The complexity of optimization problems. Journal of Computer and System Sciences 36, 490–509 (1988)
7. Papadimitriou, C.H.: Computational complexity. Addison-Wesley, Reading (1994)
8. Pin, J.-É.: On two combinatorial problems arising from automata theory. Annals of Discrete Mathematics 17, 535–548 (1983)
9. Samotij, W.: A note on the complexity of the problem of finding shortest synchronizing words. In: Proc. AutoMathA 2007, University of Palermo, CD (2007)
10. Selman, A.L.: A taxonomy of complexity classes of functions. Journal of Computer and System Sciences 48(2), 357–381 (1994)
11. Valiant, L.G.: The complexity of computing the permanent. Theoretical Computer Science 8, 189–201 (1979)
12. Volkov, M.V.: Synchronizing automata and the Černý conjecture. In: Martín-Vide, C., Otto, F., Fernau, H. (eds.) LATA 2008. LNCS, vol. 5196, pp. 11–27. Springer, Heidelberg (2008)

Does Treewidth Help in Modal Satisfiability?
(Extended Abstract)

M. Praveen

The Institute of Mathematical Sciences, Chennai, India

Abstract. Many tractable algorithms for solving the Constraint Satisfaction Problem (CSP) have been developed using the notion of the treewidth of some graph derived from the input CSP instance. In particular, the *incidence graph* of the CSP instance is one such graph. We introduce the notion of an incidence graph for modal logic formulae in a certain normal form. We investigate the parameterized complexity of modal satisfiability with the modal depth of the formula and the treewidth of the incidence graph as parameters. For various combinations of Euclidean, reflexive, symmetric and transitive models, we show either that modal satisfiability is FPT, or that it is W[1]-hard. In particular, modal satisfiability in general models is FPT, while it is W[1]-hard in transitive models. As might be expected, modal satisfiability in transitive and Euclidean models is FPT.

1 Introduction

Treewidth as a parameter has been very successful in obtaining Fixed Parameter Tractable (FPT) algorithms for many classically intractable problems. One such class of problems is constraint satisfaction and closely related problems like satisfiability in propositional logic and the homomorphism problem [8, 30]. There have been recent extensions to quantified constraint satisfaction [6, 27]. In such problems, treewidth is used as a measure of modularity inherent in the given problem instance and algorithms make use of the modularity to increase their efficiency. Understanding the extent to which treewidth can be stretched in such problems is an active area of research [24, 15]. This work explores the applicability of such techniques to modal satisfiability.

Apart from having many applications (reasoning about knowledge [10], programming [28] and hardware verification [29] etc.), modal logics have nice computational properties [32, 14]. Many tools have been built for checking satisfiability of modal formulae [21, 26], despite being intractable in the classical sense (PSPACE-complete or NP-complete in most cases). Complexity of modal logic decision problems is well studied [23, 17, 16]. Another motivation for this work is to strengthen the complexity classification of modal logics through the refined analysis offered by parameterized complexity.

Our results: It is known that any modal logic formula can be effectively converted into a Conjunctive Normal Form (CNF) [9, 20]. Given a modal logic formula in

P. Hliněný and A. Kučera (Eds.): MFCS 2010, LNCS 6281, pp. 580–591, 2010.

CNF, we associate a graph with it. Restricted to propositional CNF formulae (which are modal formulae with modal depth 0), this graph is precisely the *incidence graph* associated with propositional CNF formulae (see [30] for details). We prove that

1. with the treewidth of the graph and the modal depth of the formula as parameters, satisfiability in general models is FPT,
2. with treewidth and modal depth as parameters, satisfiability in transitive models is W[1]-hard and
3. with treewidth as the parameter, satisfiability in models that are Euclidean[1] and any combination of reflexive, symmetric and transitive is FPT.

Since modal formulae of modal depth 0 contain all propositional formulae, bounding modal depth alone will not give FPT results (unless PTIME=NP). The main idea behind our FPT results is to express satisfiability of a modal formula in Monadic Second Order (MSO) logic over the formula's associated graph and then apply Courcelle's theorem [7]. Modal formulae with low treewidth are quite powerful, capable of encoding complex problems (see the conclusion for relevant pointers). On the other hand, modal formulae with low treewidth contain propositional CNF formulae of low treewidth, which arise naturally in many practical applications. See [12, Section 1.4] and references therein for some context on this.

Related work: In [16], Halpern considers the effect of bounding different parameters (such as the number of propositional variables, modal depth etc., but not treewidth) on complexity. In [25], Nguyen shows that satisfiability of many modal logics reduce to PTIME under the restriction of Horn fragment and bounded modal depth. In [1], Achilleos et. al. consider parameterized complexity of modal satisfiability in general models with the number of propositional variables and other structural aspects (but not treewidth) as parameters. In [2], Adler et. al. associate treewidth with First Order (FO) formulae and use it to obtain a FPT algorithm for model checking.

The Complexity of satisfiability of modal logics follow a pattern. In [18], Halpern et. al. prove that with the addition of Euclidean property, complexity of (infinitely) many modal logics drop from PSPACE-hard to NP-complete. [19] is another work in this direction. Similar pattern is observed in graded modal logics [22]. With treewidth and modal depth as parameters, our results indicate similar behaviour in the world of parameterized complexity — satisfiability in transitive models is W[1]-hard, while satisfiability in Euclidean and transitive models is FPT, even with treewidth as the only parameter. However, more work is needed in this direction. First, the results in [18, 19] hold for infinitely many cases while we consider only a few fixed cases. Second, satisfiability in general models is PSPACE-complete and drops to NP-complete with the addition of Euclidean property. In our setting, satisfiability in general models is already FPT (but see conclusion for a discussion about why satisfiability in general models is not FPT unless PTIME=NP, when treewidth is the only parameter).

[1] A binary relation $\mapsto$ is Euclidean if $\forall x, y, z$, $x \mapsto y$ and $x \mapsto z$ implies $y \mapsto z$.

2 Preliminaries

Let $\mathbb{N}$ denote the set of natural numbers. For $k \in \mathbb{N}$, we denote the set $\{1, \ldots, k\}$ by $[k]$. We use standard notation about parameterized complexity like FPT algorithms, FPT reductions and W[1]-hardness from [13]. We will also use notation and definitions of relational structures and their tree decompositions from [13]: a *relational vocabulary* τ is a set of relation symbols. Each relation symbol R has an arity $arity(R) \geq 1$. A τ-structure $\mathcal{S}$ consists of a set D called the *domain* and an interpretation $R^{\mathcal{S}} \subseteq D^{arity(R)}$ of each relation symbol $R \in \tau$. A *graph* is an $\{E\}$-structure, where E is a binary edge relation. A *tree* is a graph without cycles. A *path decomposition* is a tree decomposition [13, Definition 11.23] whose underlying tree is a path. The pathwidth of a structure is the minimum of the widths of all path decompositions. It is known that computing optimal tree and path decompositions of a relational structure is FPT when parameterized by treewidth; cf. [13, Corollary 11.28] and [5].

Courcelle's theorem ([13, Theorem 11.37]) states that given a relational structure and a MSO sentence, checking whether the MSO sentence is true in the structure is FPT when parameterized by the treewidth of the structure and the length of the sentence.

We use standard notation for modal logic from [3]: well formed modal logic formulae are defined by the grammar $\phi ::= q \in \Phi \mid \bot \mid \neg\phi \mid \phi \vee \psi \mid \Diamond\phi \mid \Box\phi$, where Φ is a set of propositional variables. A *Kripke model* for the basic modal language is a triple $\mathcal{M} = (W, \mapsto, V\iota)$, where W is a set of worlds, $\mapsto$ is a binary *accessibility* relation on W and $V\iota : W \times \Phi \to \{\top, \bot\}$ is a valuation function. For $w, v \in W$, if $w \mapsto v$, v is said to be a *successor* of w. The pair $(W, \mapsto)$ is called the *frame* $\mathcal{A}$ underlying $\mathcal{M}$. If $\mapsto$ is reflexive, then $\mathcal{A}$ and $\mathcal{M}$ are said to be a reflexive frame and a reflexive model respectively. Similar nomenclature is followed for other properties of $\mapsto$. The relation $\mapsto$ is Euclidean if for all w_1, w_2, w_3, $w_1 \mapsto w_2$ and $w_1 \mapsto w_3$ implies $w_2 \mapsto w_3$. We denote the fact that a modal formula ϕ is satisfied at a world w in a model $\mathcal{M}$ by $\mathcal{M}, w \models \phi$. For $q \in \Phi$, $\mathcal{M}, w \models q$ iff $V\iota(w, q) = \top$. Negation $\neg$ and disjunction $\vee$ are treated in the standard way. For any formula ϕ, $\mathcal{M}, w \models \Diamond\phi$ ($\mathcal{M}, w \models \Box\phi$) iff some (all) successor(s) v of w satisfy $\mathcal{M}, v \models \phi$. A modal formula ϕ is *satisfiable* if there is a model $\mathcal{M}$ and a world w in $\mathcal{M}$ such that $\mathcal{M}, w \models \phi$. Satisfiability in general, reflexive and transitive models are all PSPACE-complete [23], while in equivalence models, it is NP-complete [23].

The modal depth $\mathrm{md}(\phi)$ of a modal formula ϕ is inductively defined as follows. $\mathrm{md}(q) = \mathrm{md}(\bot) = 0$. $\mathrm{md}(\neg\phi) = \mathrm{md}(\phi)$. $\mathrm{md}(\phi \vee \psi) = \max\{\mathrm{md}(\phi), \mathrm{md}(\psi)\}$. $\mathrm{md}(\Diamond\phi) = \mathrm{md}(\Box\phi) = \mathrm{md}(\phi) + 1$. We will use the Conjunctive Normal Form (CNF) for modal logic defined in [20]:

$$\begin{aligned} \textit{literal} &::= q \mid \neg q \mid \Box \textit{clause} \mid \Diamond \textit{CNF} \\ \textit{clause} &::= \textit{literal} \mid \textit{clause} \vee \textit{clause} \mid \bot \\ \textit{CNF} &::= \textit{clause} \mid \textit{CNF} \wedge \textit{CNF} \end{aligned}$$

where q ranges over Φ. Any arbitrary modal formula ϕ can be effectively transformed into CNF preserving satisfiability [9]. A *CNF* is a conjunction of clauses and a *clause* is a disjunction of literals. A *literal* is either a propositional variable, a negated propositional variable or a formula of the form $\Box clause$ or $\Diamond CNF$. If one of the many literals in a clause is $\bot$, then $\bot$ can be ignored without affecting satisfiability. A literal of the form $\Diamond\bot$ can similarly be ignored. However, a clause that has $\bot$ as the only literal cannot be ignored since $\Box\bot$ is satisfied by a world in some Kripke model iff that world has no successors. Henceforth, we will assume that $\bot$ occurs only inside sub-formulae of the form $\Box\bot$.

Suppose ϕ is a modal formula in CNF. If ϕ is of the form $clause_1 \wedge clause_2 \wedge \cdots \wedge clause_{\mathrm{m}}$, then $clause_1, clause_2, \ldots, clause_{\mathrm{m}}$ and all literals appearing in these clauses are said to be at *level* $\mathrm{md}(\phi)$. If $\Box clause_1$ is a *literal* at some level i, then $clause_1$ and all literals occurring in $clause_1$ are said to be at level $i-1$. If $\Diamond CNF$ is a literal at some level i and *CNF* is of the form $clause_1 \wedge clause_2 \wedge \cdots \wedge clause_{\mathrm{m}'}$, then $clause_1, clause_2, \cdots, clause_{\mathrm{m}'}$ and all literals appearing in these clauses are said to be at level $i-1$. Note that a single propositional variable can occur in the form of a *literal* at different levels. The concept of level is similar to the concept of distance defined in [26]. The process of checking satisfiability we describe in section 3 can be considered a variant of the level-based bottom-up algorithm given in [26], which is also implicitly used in [1, Theorem 5]. It requires more work and combination of other ideas to prove that this process can be formalized in MSO logic.

Proofs of lemmata marked with (*) are skipped due to lack of space. A full version of this paper with the same title is available at arXiv, which contains all the proofs.

3 Modal Satisfiability in General Models

In this section, we will associate a relational structure with a modal CNF formula. We show that checking satisfiability of a modal CNF formula is FPT, parameterized by modal depth and the treewidth of the associated relational structure. We begin with an example modal CNF formula.

Consider the modal CNF formula $\{\neg q \vee \Box[r \vee \neg s]\} \wedge \{q \vee \Diamond\bot\} \wedge \{r \vee \Diamond[\neg s]\} \wedge \{\neg r \vee \Diamond[(t \vee \neg s) \wedge (r)]\}$. Its modal depth is 1 and has 4 clauses at level 1. Figure 1 shows a graphical representation of this formula, which is very similar to the formula's syntax tree. The 4 clauses at level 1 are represented by e_1, e_2, e_3 and e_4. e_1 represents the clause $\{\neg q \vee \Box[r \vee \neg s]\}$. Since $\neg q$ occurs as a literal in this clause, there is a dotted arrow from e_1 to q. $\Box[r \vee \neg s]$ (represented by e_9) also occurs as a literal in clause e_1 and hence there is an arrow from e_1 to e_9. e_4 represents the fourth clause at level 1, which contains $\Diamond[(t \vee \neg s) \wedge (r)]$ as a literal. This $\Diamond CNF$ formula is represented by e_{10}. The two clauses $(t \vee \neg s)$ and (r) are represented by e_7 and e_8 respectively and are connected to e_{10} by arrows. The propositional variable r occurs as literal at 2 levels, indicated as Lv_0 and Lv_1.

Now we will formalize the above example. The intuition behind the following definition is to represent all clauses and literals of a modal CNF formula by the

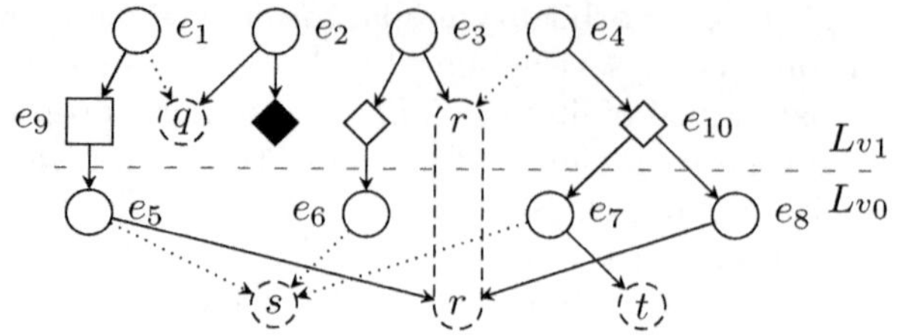

Fig. 1. Relational structure associated with the modal formula $\{\neg q \vee \Box [r \vee \neg s]\} \wedge \{q \vee \Diamond \bot\} \wedge \{r \vee \Diamond [\neg s]\} \wedge \{\neg r \vee \Diamond [(t \vee \neg s) \wedge (r)]\}$

domain elements of a relational structure. Binary relations are used to indicate which literals occur in which clause (and which clauses occur in which literal). Unary relations are used to indicate which elements represent literals and which elements represent clauses. This will enable us to reason about clauses, literals and their dependencies using MSO formulae over the relational structure.

Definition 3.1. *Given a modal CNF formula ϕ, we associate with it a relational structure $\mathcal{S}(\phi)$. It will have one domain element for every clause in ϕ. It will have one domain element for every literal of the form $\Box$clause or $\Diamond$CNF in ϕ. It will also have one domain element for every propositional variable used in ϕ. There are no domain elements representing the propositional constant $\bot$. They will be handled as special cases.*

The relational structure will have two binary relations O_c (occurs) and $\overline{O_c}$ (occurs negatively). $\overline{O_c}(e_1, e_2)$ iff e_1 represents a clause and e_2 represents a propositional variable occurring negated as a literal in the clause represented by e_1. If e_1 represents a clause, then $O_c(e_1, e_2)$ iff e_2 represents a literal (occurring in the clause represented by e_1) of the form $\Box$clause, $\Diamond$CNF or a non-negated propositional variable. If e_1 represents a literal of the form $\Box$clause, then $O_c(e_1, e_2)$ iff e_2 represents the corresponding clause. If e_1 represents a literal of the form $\Diamond$CNF, then $O_c(e_1, e_2)$ iff e_2 represents a clause in the corresponding CNF. Finally, the following unary relations are present:

Cl : *contains all domain elements representing clauses*
Lt : *all domain elements representing literals*
U : *all literals of the form $\Box\bot$*
$B_\Box$: *all literals of the form $\Box$clause*
$D_\Diamond$: *all literals of the form $\Diamond$CNF*
$(Lv_i)_{0 \leq i \leq md(\phi)}$: *all clauses and literals at level i*

For clauses and literals of the form $\Box$*clause* or $\Diamond$*CNF*, there is one domain element for every occurrence of the clause or literal. For example, if the literal $\Diamond(q_1 \wedge q_2)$ occurs in two different positions of a big formula ϕ, the two occurrences will be represented by two different domain elements in $\mathcal{S}(\phi)$. In contrast, different occurrences of a literal that is just a propositional variable will be represented by the same domain element. In the rest of the paper, whenever we refer to the treewidth of a modal CNF formula ϕ, we mean the treewidth of $\mathcal{S}(\phi)$.

If e_1 represents a *clause*, $Oc(e_1, e_2)$ means that the clause represented by e_1 can be satisfied by satisfying the literal represented by e_2. $\overline{Oc}(e_1, e_2)$ means that the clause represented by e_1 can be satisfied by setting the propositional variable represented by e_2 to false.

If $C_{\ell 0} \subseteq Cl \cap Lv_0$ is a subset of domain elements representing clauses at level 0, let $CNF(C_{\ell 0})$ be the modal CNF formula that is the conjunction of clauses represented by domain elements in $C_{\ell 0}$. We will now see how to check satisfiability of $CNF(\{e_7, e_8\})$ in Fig. 1 and describe the generalization of this process given in (1) below. We use c_ℓ and lt for first order variables intended to represent clauses and literals respectively. First of all, there must be a subset $Tr_0 \subseteq \{r, s, t\} = Lt \cap Lv_0$ that will be set to $\top$, as written in the beginning of (1). Then, we must check that this assignment satisfies each clause c_ℓ in $C_{\ell 0}$, written as $\forall c_\ell \in C_{\ell 0}$ in (1). To check that the clause represented by e_7 is satisfied, either a positively occurring literal like t must be set to $\top$ and hence in Tr_0 (written as "$\exists lt \in Tr_0 : Oc(c_\ell, lt)$" in (1)) or a negatively occurring literal like s must be set to $\bot$ and hence not in Tr_0 ("$\exists lt \in (Lt \cap Lv_0) \setminus Tr_0 : \overline{Oc}(c_\ell, lt)$" in (1)). A similar argument applies to e_8 as well.

$$\begin{aligned}\xi[0](C_{\ell 0}) \triangleq\ & \exists Tr_0 \subseteq (Lt \cap Lv_0) : \forall c_\ell \in C_{\ell 0} : \\ & \big[(\exists lt \in Tr_0 : Oc(c_\ell, lt)) \vee (\exists lt \in (Lt \cap Lv_0) \setminus Tr_0 : \overline{Oc}(c_\ell, lt))\big]\end{aligned} \tag{1}$$

$$\begin{aligned}\xi[i](C_{\ell i}) \triangleq\ & \exists Tr_i \subseteq (Lt \cap Lv_i) : \forall c_\ell \in C_{\ell i} : \\ & \big[(\exists lt \in Tr_i : Oc(c_\ell, lt)) \vee (\exists lt \in (Lt \cap Lv_i) \setminus Tr_i : \overline{Oc}(c_\ell, lt))\big] \\ & \wedge[Cm_{i-1} = \{c_\ell' \in (Cl \cap Lv_{i-1}) \mid \exists lt' \in Tr_i \cap B_\Box, Oc(lt', c_\ell')\} \Rightarrow \\ & \quad \forall lt \in Tr_i \cap D_\Diamond : Dm_{i-1} = \{c_\ell \in (Cl \cap Lv_{i-1}) \mid Oc(lt, c_\ell)\} \Rightarrow \\ & \quad\quad \xi[i-1](Dm_{i-1} \cup Cm_{i-1})]\end{aligned} \tag{2}$$

Checking satisfiability at higher levels is slightly more complicated. Suppose $C_{\ell i} \subseteq Cl \cap Lv_i$ is a subset of clauses at level i. We will take $C_{\ell 1} = \{e_1, e_3, e_4\}$ from Fig. 1 as an example. If some world w in some Kripke model $\mathcal{M}$ satisfies $CNF(C_{\ell 1})$, there must be some subset Tr_1 of literals at level 1 satisfied at w ("$\exists Tr_i \subseteq (Lt \cap Lv_i)$" in (2)). As before, we check that for every clause represented in $C_{\ell 1}$ ("$\forall c_\ell \in C_{\ell i}$" in (2)), there is either a positively occurring literal in Tr_1 ("$\exists lt \in Tr_i : Oc(c_\ell, lt)$" in (2)) or a negatively occurring literal not in Tr_1 ("$\exists lt \in (Lt \cap Lv_i) \setminus Tr_i : \overline{Oc}(c_\ell, lt)$" in (2)). Next, we must check that the literals we have chosen to be satisfied at w (by putting them into Tr_1) can actually be satisfied. Suppose Tr_1 was $\{e_9, q, r, e_{10}\}$. Since e_9 represents a literal of the form $\Box$*clause* (with the *clause* represented by domain element e_5), we are committed to satisfy the clause represented by e_5 in any world succeeding w. Let $Cm_0 = \{e_5\}$ be the set of clauses occurring at level 0 that we have committed to as a result of choosing corresponding $\Box$*clause* literals to be in Tr_1 ("$Cm_{i-1} = \{c_\ell' \in (Cl \cap Lv_{i-1}) \mid \exists lt' \in Tr_i \cap B_\Box, Oc(lt', c_\ell')\}$" in (2)). Now, since we have also chosen e_{10} to be in Tr_1 and e_{10} represents a $\Diamond$*CNF* formula, there is a demand to create a world w' that succeeds w and satisfies the corresponding *CNF* formula. We have to check that every such demand in Tr_1 can be satisfied ("$\forall lt \in Tr_i \cap D_\Diamond$" in (2)) by creating successor worlds. In case of the demand created by e_{10}, $\{e_7, e_8\} = Dm_0$ is the set of clauses in the demanded *CNF* formula ("$Dm_{i-1} = \{c_\ell \in (C_l \cap Lv_{i-1}) \mid Oc(l_t, c_\ell)\}$" in (2)). Our aim now is to create a successor world w' in which all clauses represented in Dm_0 are satisfied. However, w' is a successor world and we have already committed to satisfying all clauses represented in Cm_0 in all successor worlds. Hence, we actually check if the clauses represented in $Cm_0 \cup Dm_0$ are satisfiable by inductively invoking $\xi[0](Dm_0 \cup Cm_0)$ ("$\xi[i-1](Dm_{i-1} \cup Cm_{i-1})$" in (2)).

For the sake of clarity, we have skipped handling literals of the form $\Box\bot$ in the above discussion. They will be handled in the formal arguments that follow.

Lemma 3.2 (*). *The property $\xi[i](C_{\ell_i})$ can be written in a MSO logic formula of size linear in i. If ϕ is any modal formula in CNF and C_{ℓ_i} is any subset of domain elements representing clauses at level i, then CNF(C_{ℓ_i}) is satisfiable iff $\xi[i](C_{\ell_i})$ is true in $\mathcal{S}(\phi)$.*

Theorem 3.3. *Given a modal CNF formula ϕ, there is a* FPT *algorithm that checks if ϕ is satisfiable in general models, with treewidth of $\mathcal{S}(\phi)$ and modal depth of ϕ as parameters.*

Proof. Given ϕ, $\mathcal{S}(\phi)$ can be constructed in polynomial time. To check that all clauses of ϕ at level md(ϕ) are satisfiable in some world w of some Kripke model $\mathcal{M}$, we check whether the formula $\exists C_{\ell_{\mathrm{md}(\phi)}} \forall c_\ell (C_{\ell_{\mathrm{md}(\phi)}}(c_\ell) \Leftrightarrow (C_l(c_\ell) \wedge Lv_{\mathrm{md}(\phi)}(c_\ell))) \wedge \xi[\mathrm{md}(\phi)](C_{\ell_{\mathrm{md}(\phi)}})$ is true in $\mathcal{S}(\phi)$. By Lemma 3.2, this is possible iff ϕ is satisfiable and length of the above formula is linear in md(ϕ). An application of Courcelle's theorem will give us the FPT algorithm. □

4 Models with Euclidean Property

In this section, we will investigate the parameterized complexity of satisfiability in Euclidean models. The main observation leading to the FPT algorithm is the fact that if a modal formula is satisfied in a Euclidean model, then it is satisfied in a rather simple model. As proved in [22], if a modal formula is satisfied at some world w_0 in some Euclidean model $\mathcal{M}$, then it is satisfied in a model whose underlying frame is of the form $(W \cup \{w_0\}, \mapsto)$ where $W \times W \subseteq \mapsto$. Therefore, almost all worlds are successors of almost all other worlds. If one world satisfies a formula $\Box clause_1$, then almost all worlds satisfy the formula $clause_1$ (and hence satisfy $\Box clause_1$ as well). If one world satisfies a formula $\Diamond CNF_1$, then almost all worlds satisfy $\Diamond CNF_1$ as well. Thus, most of the worlds are very similar to each other and we can reason about them using small MSO formulae. This holds even if we add more properties like reflexivity, transitivity etc. The technical details needed for the following result can be found in the full version.

Theorem 4.1. *Let ϕ be a modal CNF formula. With treewidth of $\mathcal{S}(\phi)$ as parameter, there is a* FPT *algorithm for checking whether ϕ is satisfiable in a Kripke model that satisfies Euclidean property and any combination of reflexivity, symmetry and transitivity.*

5 Transitive Models

In transitive models, formulae with small modal depth can check properties of all worlds reachable from a given world. To formalize this into a W[1]-hardness proof, we introduce the parameterized Partitioned Weighted Satisfiability (p-Pw-Sat) problem. An instance of p-Pw-Sat problem is a triple $(\mathcal{F}, part : \Phi \to [k], tg : [k] \to \mathbb{N})$, where $\mathcal{F}$ is a propositional CNF formula, *part* partitions the set of propositional variables into k parts and we need to check if there is a satisfying assignment that sets exactly $tg(p)$ variables to $\top$ in each part p. Parameters are k and pathwidth of the primal graph of $\mathcal{F}$ (one vertex for each propositional variable, an edge between two variables iff they occur together in a clause). The following lemma can be proved by a Fpt reduction from the Number List Coloring Problem [11].

Lemma 5.1 (*). *The p-*Pw-Sat *problem is* W[1]*-hard when parameterized by the number of parts k and the pathwidth of the primal graph.*

Theorem 5.2. *With treewidth and modal depth as parameters, modal satisfiability in transitive models is* W[1]*-hard.*

The rest of this section is devoted to a proof of the above theorem, which is by a Fpt reduction from p-Pw-Sat to satisfiability of modal CNF formulae in transitive models. Given an instance $(\mathcal{F}, part : \Phi \to [k], tg : [k] \to \mathbb{N})$ of p-Pw-Sat problem with the pathwidth of the primal graph of $\mathcal{F}$ being pw, we construct a modal CNF formula $\phi_{\mathcal{F}}$ of modal depth 2 in Fpt time such that the pathwidth (and hence the treewidth) of $\mathcal{S}(\phi_{\mathcal{F}})$ is bounded by a function of pw and k and p-Pw-Sat is a Yes instance iff $\phi_{\mathcal{F}}$ is satisfiable in a transitive model. Suppose the propositional variables used in $\mathcal{F}$ are $q_1, q_2, \ldots, q_n$. The idea is that if $\phi_{\mathcal{F}}$ is satisfied at some world w_0 in some transitive model $\mathcal{M}$, then $\mathcal{M}, w_0 \models \mathcal{F}$. To check that the required targets of the number of variables set to true in each partition are met, $\phi_{\mathcal{F}}$ will force the existence of worlds $w_1, w_2, \ldots, w_n$ arranged as $w_0 \mapsto w_1 \mapsto w_2 \mapsto \cdots \mapsto w_n$. In the formula $\phi_{\mathcal{F}}$, we will maintain a counter for each partition of the propositional variables. At each world w_i, if q_i is true, we will force the counter corresponding to $part(q_i)$ to increment. At the world w_n, the counters will have the number of variables set to $\top$ in each partition. We will then verify in the formula $\phi_{\mathcal{F}}$ that these counts meet the given target. Such counting tricks have come under standard usage in complexity theoretic arguments of modal logic. The challenge here is to implement the counting in a modal formula of small pathwidth.

In a p-Pw-Sat instance containing n propositional variables and k partitions, we will denote the number of variables in partition p by $n[p]$. We first construct an optimal path decomposition of the primal graph of $\mathcal{F}$ in Fpt time. We will name the variables occurring in the first bag as $q_1, \ldots, q_i$. We will name the variables newly introduced in the second bag as $q_{i+1}, \ldots, q_{i'}$ and so on. In the rest of the construction, we will use this same ordering $q_1, \ldots, q_n$ of the propositional variables. This will be important to maintain the pathwidth of the resulting modal formula low. The modal CNF formula $\phi_{\mathcal{F}}$ will use all the propositional variables $q_1, \ldots, q_n$ used by $\mathcal{F}$ and also use the following additional variables:

- $t\uparrow_1, \ldots, t\uparrow_k, f\uparrow_1, \ldots, f\uparrow_k$: partition indicators.
- For each partition p, $tr_p^0, \ldots, tr_p^{n[p]}, fl_p^0, \ldots, fl_p^{n[p]}$: counters to count the number of variables set to $\top$ and $\bot$ in partition p.
- $d_0, \ldots, d_{n+1}$: depth indicators.

The modal CNF formula $\phi_{\mathcal{F}}$ is the conjunction of the formulae described below. For clarity, we have used the shorthand notation $\Rightarrow$ but they can be easily converted to CNF. Also for notational convenience, we will use $part(i)$ instead of $part(q_i)$. $\Phi(p)$ is the set of variables among $\{q_1, \ldots, q_n\}$ in partition p. The formula *determined* ensures that all successors of w_0 preserve the assignment of $q_1, \ldots, q_n$. The formula *depth* ensures that for all i, $d_i \wedge \neg d_{i+1}$ holds in the world w_i.

In w_{i-1}, if q_i is set to $\top$, we want to indicate that in w_i, the counter for partition $part(i)$ should be incremented. We will indicate this in the formula *setCounter* by setting the variable $t\uparrow_{part(i)}$ to $\top$. Similar indication is done for the counter keeping track of variables set to $\bot$ in partition p.

$$determined \triangleq \bigwedge_{i=1}^{n} q_i \Rightarrow \Box q_i \wedge \bigwedge_{i=1}^{n} \neg q_i \Rightarrow \Box \neg q_i$$

$$depth \triangleq \Diamond(d_1 \wedge \neg d_2) \wedge \bigwedge_{i=1}^{n-1} \Box\left[(d_i \wedge \neg d_{i+1}) \Rightarrow \Diamond(d_{i+1} \wedge \neg d_{i+2})\right]$$

$$setCounter \triangleq (q_1 \Rightarrow t\uparrow_{part(1)}) \wedge (\neg q_1 \Rightarrow f\uparrow_{part(1)})$$
$$\wedge \bigwedge_{i=2}^{n} \Box\left\{[d_{i-1} \wedge \neg d_i] \Rightarrow [(q_i \Rightarrow t\uparrow_{part(i)}) \wedge (\neg q_i \Rightarrow f\uparrow_{part(i)})]\right\}$$

$$incCounter \triangleq (t\uparrow_{part(1)} \Rightarrow \Box tr^1_{part(1)}) \wedge (f\uparrow_{part(1)} \Rightarrow \Box fl^1_{part(1)})$$
$$\wedge \bigwedge_{p=1}^{k} \bigwedge_{j=0}^{n[p]-1} \Box[t\uparrow_p \Rightarrow (tr_p^j \Rightarrow \Box tr_p^{j+1})] \wedge \Box[f\uparrow_p \Rightarrow (fl_p^j \Rightarrow \Box fl_p^{j+1})]$$

$$targetMet \triangleq \bigwedge_{p=1}^{k} \Box[d_n \Rightarrow (tr_p^{tg(p)} \wedge \neg tr_p^{tg(p)+1})]$$
$$\wedge \bigwedge_{p=1}^{k} \Box[d_n \Rightarrow (fl_p^{n[p]-tg(p)} \wedge \neg fl_p^{n[p]-tg(p)+1})]$$

Variables $tr_p^0, \ldots, tr_p^{n[p]}$ implement the counter keeping track of variables set to $\top$ in partition p. If j variables in $\Phi(p) \cap \{q_1, \ldots, q_i\}$ are set to $\top$, then we want tr_p^j to be set to $\top$ in w_i. To maintain this, in w_{i-1}, if it is indicated that a counter is to be incremented (by setting $t\uparrow_p$ to $\top$), we will force all successors of w_{i-1} to increment the tr_p counter in the formula *incCounter*. Finally, we check that at w_n, all the targets are met in the formula *targetMet*.

The modal CNF formula $\phi_{\mathcal{F}}$ we need is the conjunction of $\mathcal{F}$, the formulae defined above and the miscellaneous formulae below (which ensure that counters are initiated properly and are monotonically non-decreasing).

$$determined' \triangleq \bigwedge_{p=1}^{k} tr_p^0 \Rightarrow \Box tr_p^0 \wedge \bigwedge_{p=1}^{k} fl_p^0 \Rightarrow \Box fl_p^0$$

$$countInit \triangleq d_0 \wedge \neg d_1 \wedge \bigwedge_{p=1}^{k} (\neg tr_p^1 \wedge \neg fl_p^1 \wedge tr_p^0 \wedge fl_p^0)$$

$$depth' \triangleq \bigwedge_{p=1}^{k} \bigwedge_{j=0}^{n[p]} [\Box(tr_p^j \Rightarrow \Box tr_p^j) \wedge \Box(fl_p^j \Rightarrow \Box fl_p^j)]$$

$$countMonotone \triangleq \bigwedge_{i=1}^{n} \Box(d_i \Rightarrow d_{i-1}) \wedge \bigwedge_{p=1}^{k} \bigwedge_{j=2}^{n[p]} [\Box(tr_p^j \Rightarrow tr_p^{j-1}) \wedge \Box(fl_p^j \Rightarrow fl_p^{j-1})]$$

Lemma 5.3 (*). *If a p-*PW-SAT *instance is a* YES *instance, then the modal formula constructed above is satisfied in a transitive Kripke model.*

Lemma 5.4 (*). *Suppose the modal CNF formula $\phi_{\mathcal{F}}$ constructed above is satisfied at some world w_0 of some transitive Kripke model $\mathcal{M}$. Then $\mathcal{M}$ contains distinct worlds $w_1, \ldots, w_n$ such that for each i between 1 and n, w_i is a successor of w_{i-1}. Moreover, $\{d_0, \ldots, d_i\}$ are set to $\top$ and $\{d_{i+1}, \ldots, d_{n+1}\}$ are set to $\bot$ in w_i. For any partition p, if j variables in $\Phi(p) \cap \{q_1, \ldots, q_i\}$ are set to $\top$ in w_0, then $\{tr_p^0, \ldots, tr_p^j\}$ are all set to $\top$ in w_i. If j' variables in $\Phi(p) \cap \{q_1, \ldots, q_i\}$ are set to $\bot$ in w_0, then $\{fl_p^0, \ldots, fl_p^{j'}\}$ are all set to $\top$ in w_i.*

Theorem 5.5 (*). *If $\phi_{\mathcal{F}}$ constructed above is satisfied in a transitive model, then the p-*PW-SAT *instance is a* YES *instance.*

Given an instance of p-PW-SAT problem, the formula $\phi_{\mathcal{F}}$ described above can be constructed in FPT time. To complete the proof of Theorem 5.2, we will prove that the pathwidth of $\phi_{\mathcal{F}}$ is bounded by some function of k and pw. $\phi_{\mathcal{F}}$ has been carefully constructed to keep pathwidth low.

Lemma 5.6 (*). *Pathwidth of $\mathcal{S}(\phi_{\mathcal{F}})$ is at most $4pw + 2k + 5$.*

In the absence of transitivity, the above reduction would require a formula of modal depth that depends on n (and hence it would no longer be a FPT reduction). The above hardness proof will however go through for any class of transitive frames that has paths of unbounded length of the form $w_1 \mapsto w_2 \mapsto \cdots \mapsto w_n$ without any reverse paths[2]. See [31] for some context on such classes of transitive frames of unbounded depth.

6 Conclusions and Future Work

By expressing satisfiability of modal formulae as a MSO property, we obtained a FPT algorithm for modal satisfiability in general models with treewidth and

[2] The author acknowledges an anonymous referee for pointing this out.

modal depth as parameters. Due to the dependence of the constructed MSO sentence on modal depth, the FPT algorithm obtained in section 3 has a running time with a tower of 2's whose height is $\mathcal{O}(\text{md}(\phi))$. Unless, PTIME=NP, such dependence on modal depth cannot be avoided due to the following observation. In [1, Lemma 1], it is shown how to encode an arbitrary propositional CNF formula into an equivalent modal formula (the propositional formula is satisfiable iff the modal formula is satisfiable in a general model). This modal formula has some very low modal depth h such that any function growing slower than a tower of 2's of height $h-5$ is a polynomial in the size of the propositional formula. The treewidth of this modal formula can be verified to be a constant. This also proves that unless PTIME=NP, modal satisfiability in general models is not FPT when treewidth is the only parameter.

We can work out a composition algorithm [4], and hence conclude that with treewidth and modal depth as parameters, there is no polynomial kernel for modal satisfiability in general models.

One direction for future research is towards meta classification as done in [19], instead of the case by case analysis of this work. We can also consider variations in treewidth, such as having different domain elements representing same propositional variable at different levels in $\mathcal{S}(\phi)$. Other variations are modal circuits instead of modal formulae and generalizations of primal/dual graphs instead of incidence graphs.

Acknowledgements. The author wishes to thank Kamal Lodaya, Geevarghese Philip and Saket Saurabh for helpful discussions, pointers to related work and feedback on the draft. The author also thanks anonymous referees for catching some subtle errors and suggesting extensions.

References

[1] Achilleos, A., Lampis, M., Mitsou, V.: Parameterized modal satisfiability. In: ICALP, arXiv:0912.4941v1 (to appear 2010)
[2] Adler, I., Weyer, M.: Tree-width for first order formulae. In: Grädel, E., Kahle, R. (eds.) CSL 2009. LNCS, vol. 5771, pp. 71–85. Springer, Heidelberg (2009)
[3] Blackburn, P., de Rijke, M., Venema, Y.: Modal Logic. CUP, Cambridge (2001)
[4] Bodlaender, H.L., Downey, R.G., Fellows, M.R., Hermelin, D.: On problems without polynomial kernels. J. Comput. Syst. Sci. 75(8), 423–434 (2009)
[5] Bodlaender, H.L., Kloks, T.: Efficient and constructive algorithms for the pathwidth and treewidth of graphs. J. Alg. 21(2), 358–402 (1996)
[6] Chen, H.: Quantified constraint satisfaction and bounded treewidth. In: de Mántaras, R.L., Saitta, L. (eds.) ECAI, pp. 161–165. IOS Press, Amsterdam (2004)
[7] Courcelle, B.: The monadic second-order logic of graphs III: tree-decompositions, minors and complexity issues. ITA 26, 257–286 (1992)
[8] Dalmau, V., Kolaitis, P.G., Vardi, M.Y.: Constraint satisfaction, bounded treewidth, and finite-variable logics. In: Van Hentenryck, P. (ed.) CP 2002. LNCS, vol. 2470, pp. 310–326. Springer, Heidelberg (2002)
[9] Enjalbert, P., del Cerro, L.F.: Modal resolution in clausal form. Theor. Comp. Sc. 65(1), 1–33 (1989)

[10] Fagin, R., Halpern, J.Y., Moses, Y., Vardi, M.Y.: Reasoning About Knowledge. MIT Press, Cambridge (1995)
[11] Fellows, M., Fomin, F.V., Lokshtanov, D., Rosamond, F., Saurabh, S., Szeider, S., Thomassen, C.: On the complexity of some colorful problems parameterized by treewidth. In: Dress, A.W.M., Xu, Y., Zhu, B. (eds.) COCOA. LNCS, vol. 4616, pp. 366–377. Springer, Heidelberg (2007)
[12] Fischer, E., Makowsky, J.A., Ravve, E.V.: Counting truth assignments of formulas of bounded tree-width or clique-width. Disc. App. Math. 156(4), 511–529 (2008)
[13] Flum, J., Grohe, M.: Parameterized Complexity Theory. Springer, Heidelberg (2006)
[14] Grädel, E.: Why are modal logics so robustly decidable? In: Current Trends in Theor. Comp. Sc., pp. 393–408 (2001)
[15] Grohe, M.: The structure of tractable constraint satisfaction problems. In: Královič, R., Urzyczyn, P. (eds.) MFCS 2006. LNCS, vol. 4162, pp. 58–72. Springer, Heidelberg (2006)
[16] Halpern, J.Y.: The effect of bounding the number of primitive propositions and the depth of nesting on the complexity of modal logic. Artif. Intell. 75(2), 361–372 (1995)
[17] Halpern, J.Y., Moses, Y.O.: A guide to completeness and complexity for modal logics of knowledge and belief. Artif. Intell. 54(3), 319–379 (1992)
[18] Halpern, J.Y., Rêgo, L.C.: Characterizing the np-pspace gap in the satisfiability problem for modal logic. In: IJCAI, pp. 2306–2311 (2007)
[19] Hemaspaandra, E., Schnoor, H.: On the complexity of elementary modal logics. In: STACS, pp. 349–360 (2008)
[20] Herzig, A., Mengin, J.: Uniform interpolation by resolution in modal logic. In: Hölldobler, S., Lutz, C., Wansing, H. (eds.) JELIA 2008. LNCS (LNAI), vol. 5293, pp. 219–231. Springer, Heidelberg (2008)
[21] Hustadt, U., Schmidt, R.A.: An empirical analysis of modal theorem provers. J. Applied Non-Classical Logics 9(4) (1999)
[22] Kazakov, Y., Pratt-Hartmann, I.: A note on the complexity of the satisfiability problem for graded modal logics. In: LICS 2009, pp. 407–416. IEEE Computer Society, Los Alamitos (2009)
[23] Ladner, R.E.: The computational complexity of provability in systems of modal propositional logic. SIAM J. Comput. 6(3), 467–480 (1977)
[24] Marx, D.: Can you beat treewidth? In: FOCS, pp. 169–179 (2007)
[25] Nguyen, L.A.: On the complexity of fragments of modal logics. Advances in Modal logic, vol. 5, pp. 249–268 (2005)
[26] Pan, G., Vardi, M.Y.: Optimizing a BDD-based modal solver. In: Baader, F. (ed.) CADE 2003. LNCS (LNAI), vol. 2741, pp. 75–89. Springer, Heidelberg (2003)
[27] Pan, G., Vardi, M.Y.: Fixed-parameter hierarchies inside pspace. In: LICS 2006, pp. 27–36. IEEE Computer Society, Los Alamitos (2006)
[28] Pratt, V.R.: Application of modal logic to programming. Studia Logica 39(2-3), 257–274 (1980)
[29] Reif, J.H., Sistla, A.P.: A multiprocess network logic with temporal and spatial modalities. In: Díaz, J. (ed.) ICALP 1983. LNCS, vol. 154, pp. 629–639. Springer, Heidelberg (1983)
[30] Samer, M., Szeider, S.: Constraint satisfaction with bounded treewidth revisited. J. Comput. Syst. Sci. 76(2), 103–114 (2010)
[31] ten Cate, B.: A note on the expressibility problem for modal logics and star-free regular expressions. Inf. Process. Lett. 109(10), 509–513 (2009)
[32] Vardi, M.Y.: Why is modal logic so robustly decidable? In: Descriptive Complexity and Finite Models, pp. 149–184. AMS, Providence (1996)

Asynchronous Omega-Regular Games with Partial Information

Bernd Puchala*

Mathematische Grundlagen der Informatik, RWTH Aachen University
puchala@logic.rwth-aachen.de

Abstract. We address the strategy problem for ω-regular two-player games with partial information, played on finite game graphs. We consider two different kinds of observability on a general model, a standard synchronous and an asynchronous one. In the asynchronous setting, moves which have no visible effect for a player are hidden completely from that player. We generalize the usual powerset construction for eliminating partial information to arbitrary, not necessarily observation based, winning conditions, both in the synchronous and in the asynchronous case, and we show that this generalized construction effectively preserves ω-regular winning conditions. From this we infer decidability of the strategy problem for arbitrary ω-regular winning conditions, in both cases. We also show that our ω-regular framework is sufficient for reasoning about synchronous and asynchronous knowledge by proving that any formula of the epistemic temporal specification formalism ETL can be effectively translated into an S1S-formula defining the same specification.

1 Introduction

In a two-player graph game the players, called 0 and 1, move a token along the edges of a labeled nonterminating graph by choosing appropriate edge labels, also called actions. This results in an infinite sequence of positions and actions called play, and each such play is either won by player 0 or by player 1 and lost by the other player. We consider games played on finite graphs where the set of plays won by player 0 is ω-regular, i.e., recognizable by a nondeterministic Büchi automaton. These specifications play a key role in modern computer science. They generalize parity objectives and capture fundamental properties of nonterminating reactive systems, cf. [12]. Such a system can be modeled as a two-player game where changes of the system state correspond to changes of the game position. Situations where the change of the system can be controlled correspond to positions of player 0, uncontrollable situations correspond to positions of player 1. A *winning strategy* for player 0 then yields a controller that forces the system into satisfying an ω-regular specification. The problem to determine, for a given game G and a position v, whether player 0 has a winning strategy for G from v is called the strategy problem. An important special case

* This work was partially supported by the ESF-project Logic for Interaction (LINT).

P. Hliněný and A. Kučera (Eds.): MFCS 2010, LNCS 6281, pp. 592–603, 2010.

of ω-regular games are parity games. Parity games with full information have received much attention during the past years, cf. [7]. The key property of parity games is memoryless determinacy which proves that the strategy problem is in NP $\cap$ co-NP. In general, ω-regular games with full information are determined with finite memory and winning strategies can be synthesized effectively.

However, assuming that both players have full information about the history of events in an ω-regular game is not always realistic. For example, if the information about the system state is acquired by imprecise sensors or the system encapsulates private states which cannot be read from outside, then a controller for this system must rely on the information about the state and the change of the system to which it has access. I.e., in the game model, player 0 has uncertainties about the positions and actions in the game, so we have to add partial information to games in order to model this kind of problems. Solving the strategy problem for such games is much harder than solving ω-regular games with full information since we have to keep track of the knowledge of player 0 during the course of events. Such a knowledge tracking is inherently unavoidable and leads to an exponential lower bound for the time complexity of the strategy problem for reachability games with partial information [11] and a super-polynomial lower bound for the memory needed to implement winning strategies in reachability games [1,10]. To keep track of the knowledge of player 0 during a play we compute, for any finite history, the set of positions that player 0 considers possible in this situation. This method, which is called *powerset construction*, has originally been suggested by John H. Reif in [11] to solve the strategy problem for reachability games with partial information. We note that this is not the only possibility for tracking knowledge. Tree-automata techniques have been used to solve synthesis problems with partial information from linear [15] and branching [8] temporal specifications in input-output frameworks. A tree-automaton processes the tree-representation of a strategy, so the information on which the strategy can rely is always given implicitly by the history.

We consider two different kinds of observability on a general model, a synchronous and an asynchronous one. In the standard synchronous case it is assumed that both players know how many moves have been performed. Intuitively, the players share a common clock. If a player moves from position u to v and the opponent cannot distinguish u from v, then this move has no visible effect for the opponent. However, the common clock tells him that a move has been performed. Now, if the system under consideration is not adapted to synchronization, that means, it cannot be reasonably timed by a global signal, then this assumption is not adequate. The asynchronous case, where we do not require the existence of such a common clock but hide a move completely from a player if it has no visible effect for him, is suited for such settings. Several other cases of asynchronous synthesis have already been studied. For example in [9], asynchronism is imposed by requiring a program to be robust against all possible schedules and observability is given by the points at which the system is scheduled to read. In [15], an action-based model with blind components is considered, where all that a player knows is the past history of his own actions. Asynchronism is given by the possibility of the components to remain idle, at any point of a run for an arbitrary number of steps. In our setting, which comprehends the model of [15], uncertainties about positions and actions can be arbitrary and asynchronism is defined via observability. Moreover, the behavior of both players is a priori constrained by an arbitrary finite graph. The same notion of asynchronism has also been used in [2] where stutter closed winning conditions (cf. Section 2) have been considered. The main novelty there is a powerset construction for timed games and an efficient symbolic implementation.

Here, we develop a unified solution for arbitrary winning conditions on a general model and we apply this to the powerful specification formalism of ω-regular languages. We also show that the expressiveness of ω-regular languages captures all ETL-specifications. In Section 3 we adapt the powerset construction to our model and we generalize it to arbitrary, not necessarily observation based, winning conditions. We prove that ω-regular winning conditions are effectively preserved by this generalized construction. From this we infer that the synchronous strategy problem for ω-regular specifications is decidable on our model and finite memory strategies can be synthesized which has, in the synchronous case, also been obtained for various related models in [15,8,14] using tree automata. It has to be strongly emphasized that ω-regular games with partial information are *not determined*, even for reachability conditions. So this result does not imply that from each position one of the players actually has a winning strategy. It merely says that *if* one of the players has a winning strategy, then he has a finite memory winning strategy which can be constructed effectively. In Section 3.1 we modify the construction in order to obtain the same results for the asynchronous case. In Section 4 we consider the epistemic temporal specification formalism ETL which is a valuable tool in the presence of partial information, since it allows to refer explicitly to the knowledge of the players when specifying the intended behavior of a system. We investigate a version of ETL which contains knowledge operators for both synchronous and asynchronous knowledge and we give some examples for the usefulness of ETL. We furthermore discuss an underlying assumption on the evaluation of knowledge operators which, in particular, has interesting consequences on the synthesis of distributed systems from ETL-specifications, cf. [14]. Our main result is that any ETL-formula can be effectively translated into an S1S-formula, so our ω-regular framework is sufficient for reasoning about synchronous and asynchronous knowledge. In particular, this adapts the decidability result of [13] to the asynchronous case.

2 Preliminaries

Games and Strategies. A *deterministic turn based two-player win-loss game* has the form $G = (V, V_0, (f_a)_{a\in A}, W_0)$, where V is the set of positions, A is the set of actions and for $a \in A$, $f_a : \text{dom}(f_a) \subseteq V \to V$ is a function. Furthermore, $V_0 \subseteq V$ are the positions of player 0 and $W_0 \subseteq V(AV)^\omega$ is the winning condition of player 0. Let $V_1 = V \setminus V_0$ and $W_1 = V(AV)^\omega \setminus W_0$. For each position $v \in V$ the set $\text{act}(v) = \{a \in A \mid v \in \text{dom}(f_a)\}$ of actions available at v has to be nonempty. For $i = 0, 1$ let $A_i = \bigcup\{\text{act}(v) \mid v \in V_i\}$. The directed labeled graph $(V, (E_a)_{a\in A})$

with $E_a = \{(u,v) \in V \times V \mid u \in \text{dom}(f_a) \text{ and } f_a(u) = v\}$ is called the *game graph* of G. We consider games played on *finite* graphs, i.e., V and A are finite.

A *play* in G from v_0 is an infinite sequence $\pi = v_0 a_1 v_1 \ldots \in V(AV)^\omega$ such that $v_i \in \text{dom}(f_{a_{i+1}})$ and $f_{a_{i+1}}(v_i) = v_{i+1}$ for each $i < \omega$. A prefix $\rho \preceq \pi$ of a play is called a *history* and $\mathcal{H}_{\text{fin}}$ denotes the set of all finite histories $\pi \in V(AV)^*$ of plays in G. If $\pi \in V(AV)^\alpha$, we denote $\alpha \leq \omega$ by $l(\pi)$ and $\pi(\leq j)$ denotes the unique finite prefix of π with $\pi(\leq j) \in V(AV)^j$. Moreover, if $l(\pi) < \omega$, by $\text{last}(\pi)$ we denote the last position in π. A *strategy* for player i for G is a function $g : \{\pi \in \mathcal{H}_{\text{fin}} \mid \text{last}(\pi) \in V_i\} \to A$ such that $g(\pi) \in \text{act}(\text{last}(\pi))$ for all $\pi \in \text{dom}(g)$. A history $\pi = v_0 a_1 v_1 \ldots$ is *compatible* with g if for all $j < l(\pi)$ such that $v_j \in V_i$ we have $a_{j+1} = g(\pi(\leq j))$. A strategy g for player i is *winning* from v_0 if each play π in G from v_0 that is compatible with g is won by player i, that means, $\pi \in W_i$. A *memory structure* for G has the form $M = (S, \delta_0, \delta)$ where S is a set of states, $\delta_0 : V \to S$ is the initializing function and $\delta : S \times (A \times V) \to S$ is the update function. A *strategy for player i with memory M* is a function $g : S \times V_i \to A$ such that $g(s,v) \in \text{act}(v)$ for all $(s,v) \in S \times V_i$. A history $\pi = v_0 a_1 v_1 \ldots$ is called *compatible* with g if for all $j < l(\pi)$ such that $v_j \in V_i$ we have $a_j = g(\delta^*(\pi(\leq j)), v_j)$, where the function $\delta^* : \mathcal{H}_{\text{fin}} \to S$ is inductively defined by $\delta^*(v) = \delta_0(v)$ for $v \in V$ and $\delta^*(\pi a_j v_j) = \delta(\delta^*(\pi), (a_j, v_j))$ for $\pi \in \mathcal{H}_{\text{fin}}$, $a_j \in \text{act}(\text{last}(\pi))$ and $f_{a_j}(\text{last}(\pi)) = v_j$.

Partial Information. The *knowledge* of player i is given by an equivalence relation $\sim_i$ on $\mathcal{H}_{\text{fin}}$. Equivalent histories are indistinguishable for player i, i.e., if $\pi \sim_i \rho$ then after π has been played and after ρ has been played, player i has exactly the same information, so he cannot distinguish one situation from the other. We refer to [4] for an extensive treatment of knowledge in multi-agent systems. A strategy $g : \{\pi \in \mathcal{H}_{\text{fin}} \mid \text{last}(\pi) \in V_i\} \to A$ for player i for G is called a *partial information strategy* with respect to $\sim_i$ ($\sim_i$-strategy, for short) if $g(\pi) = g(\pi')$ for all $\pi, \pi' \in \mathcal{H}_{\text{fin}}$ with $\pi \sim_i \pi'$. Notice that a $\sim_i$-strategy g for player i is winning from all positions in a set $U \subseteq V$ if and only if it is winning from a simulated initial position v_0 which belongs to player $1-i$ and from which he can secretly choose any position $v \in U$. Moreover, any $\sim_i$-strategy g which is only defined on histories from some initial position v_0 can be extended to a $\sim_i$-strategy g' with $\text{dom}(g) = \{\pi \in \mathcal{H}_{\text{fin}} \mid \text{last}(\pi) \in V_i\}$ by giving g' appropriate value on histories from some initial position $v_0' \neq v_0$. So in our *antagonistic* two-player setting, it suffices to consider strategies which are winning from single initial positions v_0 and only defined on histories from v_0. If we are given a game G, a position v_0 in G and some equivalence relation $\sim_i$ on $\mathcal{H}_{\text{fin}}$, the question whether player i has a *winning $\sim_i$-strategy* for G from v_0 is independent of the partial information of player $1-i$. Therefore, when solving the strategy problem we consider games with partial information only for player 0. However, the logic ETL explicitly refers to the knowledge of player 1 via the knowledge operator K_1 so we introduce the general model with partial information for both players.

We consider games played on finite graphs where the players have uncertainties about the positions and actions in the game, modeled by equivalence relations. The relation $\sim_i$ is then obtained by extending these equivalence relations to an equivalence relation on $\mathcal{H}_{\text{fin}}$. We consider two different types of extensions, resulting in the synchronous and the asynchronous case respectively. A *two-player game with partial information* has the form $\mathcal{G} = (G, (\sim_i^V)_{i=0,1}, (\sim_i^A)_{i=0,1})$ where $G = (V, V_0, (f_a)_{a\in A}, W_0)$ is a two-player game and for $i = 0, 1$, $\sim_i^V \subseteq V \times V$ and $\sim_i^A \subseteq A \times A$ are equivalence relations such that the following conditions hold.

(1) If $u, v \in V$ with $u \sim_i^V v$ then $u, v \in V_i$ or $u, v \notin V_i$.

(2) If $a, b \in A_i$ with $a \neq b$ then $a \not\sim_i^A b$.

(3) If $u, v \in V_i$ with $u \sim_i^V v$, then $\text{act}(u) = \text{act}(v)$.

Condition (1) says that player i always knows when it is his turn and condition (2) says that player i can distinguish all the actions that are available to him at some position of the game. Condition (3) ensures that player i always knows which actions are available to him when it is his turn.

Now we consider two ways to extend the equivalence relations on positions and actions to equivalence relations on finite histories. For positions $u, v \in V$ we say that $u \to v$ is a *private move of player* i, if $u \in V_i$ and $u \sim_{1-i}^V v$. For $\pi = v_0 a_1 v_1 \dots a_n v_n$, $\rho = w_0 b_1 w_1 \dots b_m w_m \in V(AV)^*$ let

– $\pi \sim_i^* \rho \;:\Longleftrightarrow\; n = m$ and $v_j \sim_i^V w_j$ and $a_j \sim_i^A b_j$ for all j

– $\pi \overleftarrow{\sim}_i^* \rho \;:\Longleftrightarrow\; \overleftarrow{\pi} \sim_i^* \overleftarrow{\rho}$

where $\overleftarrow{\pi}$ is obtained from π by contracting each maximal sequence $v_r a_{r+1} v_{r+1} \dots a_s v_s$ of private moves of player $1 - i$ in π to v_r and analogously for $\overleftarrow{\rho}$.

Games where indistinguishability of histories is given by $\sim_i^*$ model synchronous systems. Intuitively, the players share a common clock which tells them how many moves have been performed, even if some of those moves did not have any effect which they could observe. On the other hand, $\overleftarrow{\sim}_i^*$ defines an asynchronous case where we hide such private moves of the opponent completely from a player. If there is no common clock, this is intrinsic to the system.

Remark. Consider the interaction between components of a system where the behavior of each component is prescribed by a controller which has to rely on the information available to this component. In such settings it might seem more appropriate to ask for a $\sim_0^*$-strategy for player 0 which is winning against all $\sim_1^*$-strategies of player 1 rather than a winning $\sim_0^*$-strategy for player 0. However, it is easy to see that in our perfect recall setting, this is equivalent. Therefore, we use the reduced form $\mathcal{G} = (G, \sim^V, \sim^A)$ of our model with partial information only for player 0. The relations $\sim_0^*$ and $\overleftarrow{\sim}_0^*$ are accordingly denoted $\sim^*$ and $\overleftarrow{\sim}^*$.

The winning region $\text{Win}_0^{\mathcal{G}}$ of player 0 in $\mathcal{G}$ is the set of all positions $v \in V$ such that player 0 has a winning $\sim^*$-strategy for $\mathcal{G}$ from v. The winning region $\text{Win}_0^{\mathcal{G},h}$ of player 0 in $\mathcal{G}$ if private moves are hidden is the set of all positions $v \in V$ such that player 0 has a winning $\overleftarrow{\sim}^*$-strategy for $\mathcal{G}$ from v. Notice that we are really considering two different notions of strategies. Of course, $\pi \sim^* \pi'$ implies $\pi \overleftarrow{\sim}^* \pi'$ for all finite histories π, π'. Therefore, a $\overleftarrow{\sim}^*$-strategy is a $\sim^*$-strategy. In particular, $\text{Win}_0^{\mathcal{G},h} \subseteq \text{Win}_0^{\mathcal{G}}$. However, as one can easily see the converse inclusion does not hold in general, even for reachability games.

A $\sim^*$-*memory structure* for $\mathcal{G}$ is a memory structure $M = (S, \delta_0, \delta)$ for G such that $\delta(s, (a, v)) = \delta(s, (b, w))$ for all $s \in S$ and all $(a, v), (b, w) \in A \times V$ with

$v \sim^V w$ and $a \sim^A b$. M is called a $\leftrightharpoons^*$*-memory structure* for $\mathcal{G}$ if additionally, for all $\pi, \pi a v \in \mathcal{H}_{\text{fin}}$ such that $\text{last}(\pi) \to v$ is a private move of player 1, we have $\delta^*(\pi a v) = \delta^*(\pi)$. A *strategy for player* 0 *for* $\mathcal{G}$ *with memory* M is a strategy $g : S \times V_0 \to A$ for player 0 with memory M such that for all $s \in S$ and all $u, v \in V$ with $u \sim^V v$ we have $g(s,u) = g(s,v)$.

Notation. For a position $v \in V$ let $[v] := \{w \in V \mid w \sim^V v\}$ be the equivalence class of v and for $a \in A$ let $[a] := \{b \in A \mid b \sim^A a\}$ be the equivalence class of a. Moreover, for a set $U \subseteq V$ of positions and a set $B \subseteq A$ of actions we define $\text{Post}_B(U) = \{v \in V \mid \exists\, u \in U,\ \exists\, b \in B\ :\ b \in \text{act}(u) \wedge f_b(u) = v\}$.

Winning Conditions. The winning condition W_0 is called observation based if there is a coloring $\text{col} : V \to C$ for a set $C \subseteq \mathbb{N}$ of colors with $\text{col}(u) = \text{col}(v)$ for all $u \sim^V v$ and a set $W \subseteq C^\omega$ such that $\pi = v_0 a_1 v_1 \ldots \in W_0$ if and only if $\text{col}(v_0)\,\text{col}(v_1) \ldots \in W$. W_0 is called stutter closed if, additionally, W is invariant under extending and contracting finite sequences of identical colors, i.e., stutter closed winning conditions are asynchronously observation based. In this case, the private moves of player 1 do not really matter. He just makes those moves to reach some position, but what happens during the moves does not affect the winner of the play. So we can let player 0 observe the private moves of player 1 and in return give player 1 the possibility to skip these moves: For any $v \in V_1$ we insert new edges with appropriate labels to all positions, which are reachable from v by a sequence of private moves, followed by one non-private move. W_0 remains unchanged. This yields a polynomial time reduction of the asynchronous strategy problem for stutter closed specifications to the synchronous strategy problem for observation based specifications. However, assuming that the specification is completely based on the observations of player 0 is often not realistic. So we do not make such assumptions but allow arbitrary winning conditions.

Nondeterministic Games. The powerset construction which we present in Section 3 yields a nondeterministic game in general. A nondeterministic two-player game has the form $G = (V, V_0, (E_a)_{a \in A}, W_0)$ with edge relations $E_a \subseteq V \times V$ for $a \in A$ and all the other components are as before. Plays, strategies, memory strategies and winning strategies are defined as before. Nondeterministic games are not determined in general, even for reachability conditions, and hence not equivalent to deterministic games. However, for each nondeterministic game G and each player $i \in \{0, 1\}$, we can construct a deterministic game G^i such that winning strategies for player i are preserved. We simply resolve the nondeterminism by giving player $1 - i$ control of nondeterministic choices. Technically, for any $v \in V$ and any $a \in \text{act}(v)$ we add a unique a-successor of v to the game graph which belongs to player $1 - i$ and from which he can choose any a-successor of v in the original game graph. The winner of a play in G^i is the winner of the corresponding play in G where we delete all positions and actions which do not belong to G. It is easy to see that this construction preserves ω-regular winning conditions and all special cases like parity conditions.

3 Powerset Construction

First, we consider the standard synchronous strategy problem. We present a powerset construction which turns a game with partial information into a game with full information such that the existence of winning strategies for player 0 is preserved. The idea of this construction has originally been suggested by John H. Reif in [11] where he considered reachability objectives on a somewhat restricted game model with finitely branching game graphs. The construction can also easily be applied to observation based parity conditions on finitely branching graphs and for this case, improved methods for solving the resulting games with full information have been developed, cf. [3,1]. We apply the construction to our model and we generalize the definition of the winning condition of the resulting game, so that it works for arbitrary, not necessarily observation based, winning conditions. Notice that, although we consider only finite game graphs here, this generalized construction can be applied to arbitrary graphs.

The construction yields a game in which both players always know the recent position, so if winning strategies for player 0 shall be preserved, any position of this game must precisely capture the uncertainties about the recent position that player 0 actually has after some finite history $\pi \in \mathcal{H}_{\text{fin}}$. So the positions of the new game are of the form $\overline{v}(\pi) = \{\text{last}(\pi') \mid \pi' \sim^* \pi\}$ for $\pi \in \mathcal{H}_{\text{fin}}$. Each such set is a subset of some equivalence class $[v]$ of positions, so as the set of positions of the new game we simply take the set of all such subsets. Moreover, in the new game graph, there is an edge with label a from $\overline{v}(\pi)$ to $\overline{v}(\rho)$ if and only if $\rho = \pi a v$ for some $v \in V$. The following proposition provides an update mechanism which computes, for finite histories π and $\pi a v$, the set $\overline{v}(\pi a v)$ from the set $\overline{v}(\pi)$, the action a and the position v, without knowing the history π. In particular, the set of a-successors of $\overline{v}(\pi)$ in the new game graph depends only on the set $\overline{v}(\pi)$ and the action a, but not on the history π.

Proposition 1. $\overline{v}(\pi a w) = \text{Post}_{[a]}(\overline{v}(\pi)) \cap [w]$.

Let $\mathcal{G} = (G, \sim^V, \sim^A)$, $G = (V, V_0, (f_a)_{a\in A}, W_0)$ be a game with partial information We define the corresponding game $\overline{G} = (\overline{V}, \overline{V}_0, (\overline{E}_a)_{a\in A}, \overline{W}_0)$ with full information as follows. First, $\overline{V} = \{\overline{v} \in 2^V \mid \exists v \in V : \overline{v} \subseteq [v]\}$ and $\overline{V}_0 = \overline{V} \cap 2^{V_0}$. Moreover, for $a \in A$ we have $(\overline{v}, \overline{w}) \in \overline{E}_a$ if and only if there is some $w \in \text{Post}_a(\overline{v})$ such that $\overline{w} = \text{Post}_{[a]}(\overline{v}) \cap [w]$. Finally,

$$\begin{aligned}&\overline{\pi} = \overline{v}_0 a_1 \overline{v}_1 a_2 \overline{v}_2 \ldots \in \overline{W}_0 \quad :\Longleftrightarrow \\ &\text{for each play } \pi = v_0 a'_1 v_1 a'_2 v_2 \ldots \text{ in } \mathcal{G}\text{, the following holds:} \\ &\text{if} \quad v_i \in \overline{v}_i \text{ and } a'_{i+1} \sim^A a_{i+1} \text{ for all } i < \omega \\ &\text{then} \quad \pi \in W_0.\end{aligned}$$

Proposition 2. *Player 0 has a winning $\sim^*$-strategy f for $\mathcal{G}$ from v_0 if and only if he has a winning strategy $\overline{f}$ for $\overline{G}$ from $\overline{v}_0 = \{v_0\}$.*

Omega-Regular Winning Conditions. Now we prove that ω-regular winning conditions are effectively preserved by this powerset construction, i.e., if the winning condition of the given game $\mathcal{G}$ with partial information is recognized by a Büchi-automaton, then the winning condition of the corresponding game with full information is again recognized by a Büchi-automaton which can be constructed effectively. From this we infer decidability of the synchronous strategy problem for arbitrary ω-regular winning conditions as formulated in Theorem 4.

For the proof we use the fact that Büchi automata can be complemented effectively. That means, from a given Büchi automaton $\mathcal{B}$, we can effectively construct a Büchi automaton $\mathcal{B}^c$ such that $L(\mathcal{B}^c)$ is the complement of $L(\mathcal{B})$, see for example [5]. Therefore, it suffices to construct, from a Büchi automaton $\mathcal{A} = (\mathrm{VA}, Q, q_0, \Delta, F)$ recognizing W_1, a Büchi automaton $\overline{\mathcal{A}}$ recognizing $\overline{W}_1$. As usual Q is the finite set of states, $q_0 \in Q$ is the initial state, $\Delta \subseteq Q \times \Sigma \times Q$ is the transition relation and $F \subseteq Q$ is the set of Büchi-states. By definition of $\overline{W}_0$, for each play $\overline{\pi} = \overline{v}_0 a_1 \overline{v}_1 a_2 \overline{v}_2$ in $\overline{G}$ from $\overline{v}_0 = \{v_0\}$ we have $\overline{\pi} \in \overline{W}_1$ if and only if there is some play $\pi = v_0 a'_1 v_1 a'_2 v_2 \ldots$ in G from v_0 with $a'_i \sim^A a_i$ and $v_i \in \overline{v}_i$ for all $i < \omega$ such that $\pi \in W_1$. The automaton $\overline{\mathcal{A}}$ that recognizes $\overline{W}_1$, guesses such a play π which is a potential witness for $\overline{\pi} \in \overline{W}_1$ and at the same time simulates $\mathcal{A}$ on π. Formally, we define $\overline{\mathcal{A}} = (\overline{\mathrm{V}}\mathrm{A}, \overline{Q}, (q_0, v_0), \overline{\Delta}, \overline{F})$ as follows. First $\overline{Q} = Q \times V$ and $\overline{F} = F \times V$. Moreover,

$$((p,v), \overline{v}a, (q,w)) \in \overline{\Delta} \quad :\Longleftrightarrow \quad v \in \overline{v} \text{ and there is some action } b \sim^A a \text{ such that } b \in \mathrm{act}(v),\ f_b(v) = w \text{ and } (p, vb, q) \in \Delta.$$

Proposition 3. *For each play $\overline{\pi} = \overline{v}_0 a_1 \overline{v}_1 a_2 \overline{v}_2 \ldots$ in $\overline{G}$ from $\overline{v}_0 = \{v_0\}$ we have $\overline{\pi} \in \overline{W}_1$ if and only if $\overline{\pi} \in L(\overline{\mathcal{A}})$.*

Theorem 4. *Given an ω-regular game $\mathcal{G}$ with partial information and some position v_0, we can decide whether $v_0 \in \mathrm{Win}_0^{\mathcal{G}}$. If $v_0 \in \mathrm{Win}_0^{\mathcal{G}}$ we can effectively construct a finite memory winning $\sim^*$-strategy for player 0 for $\mathcal{G}$ from v_0.*

3.1 The Asynchronous Case

To solve the asynchronous strategy we use a modified version of the powerset construction. The idea of the construction is the same as in the synchronous case, i.e., any position of the new game captures the uncertainties about the recent position that player 0 has after some finite history $\pi \in \mathcal{H}_{\mathrm{fin}}$. These uncertainties are now given by the equivalence relation $\leftrightarrows^*$, so the positions of the new game are of the form $\tilde{v}(\pi) = \{\mathrm{last}(\pi') \mid \pi' \leftrightarrows^* \pi\}$ for $\pi \in \mathcal{H}_{\mathrm{fin}}$. Again, each such set is a subset of some equivalence class $[v]$, so as the set of positions of the new game we take the set of all such subsets. And again, in the new game graph, there is an edge with label a from $\tilde{v}(\pi)$ to $\tilde{v}(\rho)$ if and only if $\rho = \pi a v$ for some $v \in V$. The corresponding update mechanism is, however, more involved than in the synchronous case. For a set $U \subseteq V$ of positions, let $\mathrm{Reach}_p(U) \supseteq U$ be the set of all positions which are reachable from some position in U by a (possibly empty) sequence of private moves of player 1. Moreover, for a set $U \subseteq V$, an action a and a position $w \in \mathrm{Post}_a(U)$, we define $U[aw]$ to be U, if $U \subseteq V_1$ and $w \sim^V v$ for some $v \in U$ and we define $U[aw]$ to be $\mathrm{Reach}_p(\mathrm{Post}_{[a]}(U) \cap [w])$, otherwise.

Proposition 5. $\tilde{v}(\pi a w) = \tilde{v}(\pi)[aw]$.

Let $\mathcal{G} = (G, \sim^V, \sim^A)$, $G = (V, V_0, (f_a)_{a \in A}, W_0)$ be a game with partial information. We define the corresponding game $\tilde{G} = (\tilde{V}, \tilde{V}_0, (\tilde{E}_a)_{a \in A}, \tilde{W}_0)$ with full information as follows. First, $\tilde{V} = \{\tilde{v} \in 2^V \mid \exists v \in V : \tilde{v} \subseteq [v]\}$ and $\tilde{V}_0 = \tilde{V} \cap 2^{V_0}$. Moreover, for $a \in A$ we have $(\tilde{v}, \tilde{w}) \in \overline{E}_a$ if and only if there is some $w \in \mathrm{Post}_a(\tilde{v})$ such that $\tilde{w} = \tilde{v}[aw]$. Finally,

$\tilde{\pi} = \tilde{v}_0 a_1 \tilde{v}_1 a_2 \tilde{v}_2 \ldots \in \tilde{W}_0 \quad :\Longleftrightarrow$
for each play $\pi = v_0 a'_1 v_1 a'_2 v_2 \ldots$ in $\mathcal{G}$, the following holds:
if there are numbers $0 = k_0 < k_1 < \ldots$ with $v_{k_i}, \ldots, v_{k_{i+1}-1} \in \tilde{v}_i$
and $a'_{k_{i+1}} \sim^A a_{i+1}$ for all i and $k_{i+1} - k_i = 1$, if $\tilde{v}_i \in \tilde{V}_0$
then $\pi \in W_0$.

Proposition 6. *Player 0 has a winning $\overleftarrow{\sim}^*$-strategy f for $\mathcal{G}$ from v_0 if and only if he has a wining strategy $\tilde{f}$ for $\tilde{G}$ from $\tilde{v}_0 = \tilde{v}(v_0)$.*

Omega-Regular Winning Conditions. Now we prove that ω-regular winning conditions are effectively preserved by the powerset construction in the asynchronous case. Again, we use complementation. Let $\mathcal{A} = (\mathrm{VA}, Q, q_0, \Delta, F)$ be a Büchi automaton with $L(\mathcal{A}) = W_1$. The idea for the automaton $\tilde{\mathcal{A}}$ recognizing $\tilde{W}_1$ is as in the synchronous case, that means, for a given play $\tilde{\pi}$ in $\tilde{G}$ it guesses a play π in $\mathcal{G}$ which is a potential witness for $\tilde{\pi} \in W_1$ and at the same time checks whether $\mathcal{A}$ accepts this play. Such potential witnesses are more complicated objects than in the synchronous case, so the construction of $\tilde{A}$ is more involved. Moreover, it is not obvious that the construction is effective, which we shall prove separately. We define $\tilde{\mathcal{A}} = (\tilde{\mathrm{V}}\mathrm{A}, \tilde{Q}, (q_0, v_0, 0), \tilde{\Delta}, \tilde{F})$ as follows. First, $\tilde{Q} = Q \times V \times \{0,1\}$ and $\tilde{F} = \{(q, v, i) \mid q \in F\} \cup \{(q, v, 1) \mid q \in Q\}$. Moreover,

$((p, v, i), \tilde{v}a, (q, w, j)) \in \tilde{\Delta} \quad :\Longleftrightarrow$
there is a finite history $v_1 a_2 v_2 \ldots a_n v_n$ in G such that:

1. $n = 1$, if $\tilde{v} \in \tilde{V}_0$
2. $v_1 = v$ and $v_l \in \tilde{v}$ for all $1 \leq l \leq n$
3. there is some $b \in act(v_n)$ with $b \sim^A a$ and $f_b(v_n) = w$
4. there are $q_1, \ldots, q_{n-1} \in Q$ such that
   4.1 $(p, v_1 a_2, q_1), \ldots, (q_{n-2}, v_{n-1} a_n, q_{n-1}), (q_{n-1}, v_n b, q) \in \Delta$
   4.2 $q_l \in F$ for some $0 < l < n$ if $j = 1$.

Proposition 7. *For each play $\tilde{\pi} = \tilde{v}_0 a_1 \tilde{v}_1 a_2 \tilde{v}_2 \ldots$ in $\tilde{G}$ from $\tilde{v}_0 = \tilde{v}(v_0)$ we have $\tilde{\pi} \in \tilde{W}_1$ if and only if $\tilde{\pi} \in L(\tilde{A})$.*

Proposition 8. *The following problem is decidable.*
Given a transition $t = ((p, v, i), \tilde{v}a, (q, w, j))$, is $t \in \tilde{\Delta}$?

Proof. (Sketch) If $\tilde{v} \in \tilde{V}_0$, $t \in \tilde{\Delta}$ is clearly decidable, so let $\tilde{v} \in \tilde{V}_1$. We construct an NFA $\mathcal{B}_{\mathrm{fin}}$ over VA accepting precisely those words which are finite histories in G such that conditions 2. and 3. hold. Let $\mathcal{C}_{\mathrm{fin}}$ be the product automaton of $\mathcal{B}_{\mathrm{fin}}$ and the NFA $\mathcal{A}_{\mathrm{fin}}$ obtained from $\mathcal{A}$, by defining all states to be accepting and augmenting $\mathcal{A}$ by a state to remember whether some state in F has been seen if $j = 1$. Then $\mathcal{C}_{\mathrm{fin}}$ accepts precisely those words which are finite histories in G such that conditions 2. - 4. hold. So $t \in \tilde{\Delta}$ if, and only if, $L(\mathcal{C}_{\mathrm{fin}}) \neq \emptyset$. □

Theorem 9. *Given an ω-regular game $\mathcal{G}$ with partial information and some position v_0, we can decide whether $v_0 \in \mathrm{Win}_0^{\mathcal{G},h}$. If $v_0 \in \mathrm{Win}_0^{\mathcal{G},h}$ we can* *effectively construct a finite memory winning* $\overleftarrow{\sim}^*$*-strategy for player* 0 *for* $\mathcal{G}$ *from* v_0.

4 Epistemic Temporal Logic

In this section we consider games $\mathcal{G} = (G, (\sim_i^V)_{i=0,1}, (\sim_i^A)_{i=0,1})$ with partial information for both players. Basically, epistemic temporal logic ETL is obtained from LTL (linear temporal logic) by adding the epistemic operators K_i where $K_i\varphi$ for some formula φ means that agent i in the system knows φ. Many incarnations of this concept have been investigated, for an overview and discussion we refer to [6]. Of course the semantics of the knowledge operators has a different character than the semantics of LTL operators, since we cannot evaluate these formulas by considering a single play but we have to take into account all plays where the history up to the current point is indistinguishable for player i from the actual one. Formally, the syntax of ETL is defined by the grammar

$$\varphi ::= P_{va} \mid \varphi \wedge \varphi \mid \neg\varphi \mid X\varphi \mid \varphi U \varphi \mid K_i\varphi \mid \overleftarrow{K}_i\varphi$$

where $i \in \{0,1\}$ and for $va \in \mathrm{VA}$, P_{va} is an atomic proposition. Notice that since we consider two kinds of observability, we also have two knowledge operators for each agent i, K_i for the synchronous and $\overleftarrow{K}_i$ for the asynchronous case. Incorporating both operators into the same logic enables us to express properties of systems where, e.g., player 0 has a synchronous view and player 1 has an asynchronous view. We evaluate formulas at points during plays in G, so for a formula φ, a play $\pi = v_0a_1v_1 \ldots \in (\mathrm{VA})^\omega$ and some $t < \omega$ we have to define the truth of $\pi, t \models \varphi$. First, $\pi, t \models P_{va}$ if $v_ta_{t+1} = va$. The semantics of the boolean connectives is as usual. Furthermore, $\pi, t \models X\varphi$ if $\pi, t+1 \models \varphi$ and $\pi, t \models \varphi_1 U \varphi_2$ if there is some $s \geq t$ such that $\pi, s \models \varphi_2$ and $\pi, u \models \varphi_1$ for all $t \leq u < s$. Finally $\pi, t \models K_i\varphi$ if and only if for all plays ρ and all $s < \omega$ with $\pi(\leq t) \sim_i^* \rho(\leq s)$ we have $\pi, s \models \varphi$ and $\pi, t \models \overleftarrow{K}_i\varphi$ if and only if for all plays ρ and all $s < \omega$ with $\pi(\leq t) \overleftarrow{\sim}_i^* \rho(\leq s)$ we have $\rho, s \models \varphi$. Notice that $\pi(\leq t) \sim_i^* \rho(\leq s)$ implies $s = t$ which is not necessarily the case for $\overleftarrow{\sim}_i^*$.

Now a formula $\varphi \in \mathrm{ETL}$ defines a set $W_0 = \{\pi \in (\mathrm{VA})^\omega \mid \pi, 0 \models \varphi\}$. In [13], it is shown that the synchronous strategy problem for ETL-specifications with only the knowledge operator K_0 is decidable. In this section we prove that W_0 is ω-regular for any ETL-formula φ. This result is of general interest since it shows that ω-regular languages capture the ability to reason about knowledge. Moreover, our techniques can be used to synthesize finite memory winning strategies from ETL specifications, both in the synchronous and in the asynchronous case. Notice that we should clearly assume player 0 to know his own strategy, so if player 0 plays according to some strategy f then in particular he always considers only those histories possible, which are consistent with f. Thus, when given an ETL-formula φ, we are actually interested in a strategy f for player 0, such that $\pi, 0 \models \varphi$ for each play which is consistent with f, where the evaluation of the knowledge operator K_0 is *relative to histories which are consistent with* f. However, in our two-player setting this makes no difference since if $\pi \sim_0 \rho$ for $\sim_0 \in \{\sim_0^*, \overleftarrow{\sim}_0^*\}$, then π is consistent with f if, and only if, ρ is consistent with f. This is due to the fact that player 0 can distinguish any two of his own actions.

On the other hand, for the knowledge operator K_1 to make sense, we have to assume that player 1 does not know the strategy of player 0, since otherwise player 1 has full information about the history, so $K_1\varphi$ is equivalent to φ. Whether this is appropriate depends highly on the given application. If player 0 models a controller of an environment which is not actually antagonistic but merely unpredictable, then given any strategy f for player 0, the joint system is constrained by f, so the knowledge of both players should be relative to f. On the contrary, if player 0 models a network server which might interact with a user, then the protocol of the server may be off-limits to the user, so player 1 does not know f. One requirement for the protocol might be that the user is never able to learn the value of some internal variables of the server. This can be expressed using K_1. Moreover, using K_1 we can talk about higher-level knowledge, i.e., knowledge about knowledge about knowledge ... Consider, e.g., a parity game $\mathcal{G}$ with coloring col : $V \to \{1, \ldots, r\}$. For $U \subseteq V$ we abbreviate $\bigvee_{v \in U, a \in A} P_{va}$ by U and we use the usual abbreviations G and F defining *globally* and *finally*. Notice that all formulas in the example can be constructed effectively from a given game. The LTL-formula parity $:= \bigvee_{c \text{ even}} GF \operatorname{col}^{-1}(c) \wedge FG \bigwedge_{c'<c} \neg \operatorname{col}^{-1}(c')$ defines the parity objective for player 0. Moreover, the ETL-formula $K_0^{\text{col}} := \bigvee_{c \in C} K_0 \operatorname{col}^{-1}(c)$ says that player 0 knows the color of the recent position. So the formula $\varphi = \text{parity} \wedge G(\neg K_1 K_0^{\text{col}} \wedge \neg K_1 \neg K_0^{\text{col}})$ additionally requires that player 1 never knows whether player 0 knows the recent color.

Theorem 10. *For any epistemic temporal formula $\varphi \in$ ETL we can effectively construct an* S1S$(\mathcal{T}_{VA})$*-sentence $\hat{\varphi}$ such that $L(\hat{\varphi}) = \{\pi \in (VA)^\omega \mid \pi, 0 \models \varphi\}$.*

Proof. (Sketch) By induction over φ we construct an S1S$(\mathcal{T}_{\text{VA}})$-formula $\hat{\varphi}(x)$ such that for all plays π and all $t < \omega$ we have $\pi, t \models \varphi$ if and only if $\pi \models \hat{\varphi}(t)$. The interesting case is $\varphi = Z_i\psi$ for $i \in \{0,1\}$ and $Z \in \{K, \overleftarrow{K}\}$. We only consider $Z_i = K_0$, the other cases are analog. Let $\mathcal{A}$ be a Büchi automaton over VA such that $L(\mathcal{A}) = L(\neg\psi)$. We construct a Büchi automaton $\mathcal{B}$ over VA $\times \{0,1\}$ such that for all plays $\pi = v_0 a_1 v_1 a_2 v_2 \ldots \in (\text{VA})^\omega$ and all $t < \omega$ we have $\pi, t \models \neg K_0\psi$ if and only if $\pi^\wedge\alpha_t \in L(\mathcal{B})$. Where $\pi^\wedge\alpha_t = (v_0 a_1, \alpha_t(0)) \ldots$ and for $t < \omega$, $\alpha_t(t) = 1$ and $\alpha_t(s) = 0$, if $s \neq t$. The idea is as follows. We have $\pi, t \models \neg K_0\psi$ if and only if there is some play ρ and some $s < \omega$ such that $\rho(\leq s) \sim_0^* \pi(\leq t)$ and $\rho, s \models \neg\psi$. So $\mathcal{B}$ guesses such a ρ and checks that $\rho, s \models \neg\psi$ by simulating $\mathcal{A}$ on ρ from position s. To ensure that $\rho(\leq s) \sim_0^* \pi(\leq t)$, while reading π, $\mathcal{B}$ keeps track of the set of positions that player 0 considers possible, using the update mechanism from Proposition 1. Now let $\vartheta \in$ S1S$(\mathcal{T}_{\text{VA}\times\{0,1\}})$ with $L(\vartheta) = L(\mathcal{B})$. By refined projection we obtain from ϑ a formula $\hat{\varphi}(x)$ such that $\pi \models \hat{\varphi}(t)$ if and only if $\pi^\wedge\alpha_t \not\models \vartheta$ i.e., $\pi_t^\alpha \notin L(\mathcal{B})$ which is equivalent to $\pi, t \models K_0\psi = \varphi$ □

Consequences on Multiplayer Games. In [14] it is shown, that it is undecidable whether two players 0 and 1 can cooperate against a player 2 to ensure satisfaction of an ETL specification, even if player 1 is omniscient and player 0 is blind and the strategy f for player 0 is fixed in advance. However, the proof makes heavy use of the underlying assumption that player 0 *knows* the strategy g which player 1 uses, i.e., evaluation of K_0 is relative to histories which are consistent with *both* f and g. But g is yet to be synthesized and histories which are consistent with g and those which are not may very well be indistinguishable for player 0, which makes the problem undecidable. If we assume that player 0 is completely ignorant about the strategy of player 1, then evaluation of K_0 is relative only to plays which are consistent with f. So, by Theorem 10, the given ETL-specification is ω-regular. This generalizes to the case of n cooperating players and ETL-formulas with knowledge operators K_i for any $i \in \{1, \ldots, n\}$, where each player i is completely ignorant about the strategy of any player $j \neq i$. Hence, in this case ETL-specifications can be reduced to parity conditions which are LTL-definable. For LTL-specifications, however, [14] proves that the cooperation problem is decidable for *synchronous hierarchical* games. So our result shows that modeling the cooperation problem for ETL-specifications under the assumption that cooperating players are ignorant about each others strategies, turns the problem decidable for synchronous hierarchical games.

References

1. Berwanger, D., et al.: Strategy Construction for Parity Games with Imperfect Information. In: van Breugel, F., Chechik, M. (eds.) CONCUR 2008. LNCS, vol. 5201, pp. 325–339. Springer, Heidelberg (2008)
2. Cassez, F., et al.: Timed Control with Observation Based and Stuttering Invariant Strategies. In: Namjoshi, K.S., Yoneda, T., Higashino, T., Okamura, Y. (eds.) ATVA 2007. LNCS, vol. 4762, pp. 192–206. Springer, Heidelberg (2007)
3. Chatterjee, K., et al.: Algorithms for Omega-Regular Games with Imperfect Information. Logical Methods in Computer Science 3(3) (2007)
4. Fagin, R., et al.: Reasoning About Knowledge. The MIT Press, Cambridge (2003)
5. Grädel, E., Thomas, W., Wilke, T. (eds.): Automata, Logics, and Infinite Games. LNCS, vol. 2500. Springer, Heidelberg (2002)
6. Halpern, J.Y., van der Meyden, R., Vardi, M.Y.: Complete Axiomatizations for Reasoning about Knowledge and Time. SIAM J. on Comp. 33, 674–703 (2004)
7. Jurdziński, M.: Games for Verification: Algorithmic Issues. PhD thesis, University of Aarhus (2000)
8. Kupferman, O., Vardi, M.Y.: Church's Problem Revisited. The Bulletin of Symbolic Logic 5(2) (1999)
9. Pnueli, A., Rosner, R.: On the Synthesis of an Asynchronous Reactive Module. In: Ronchi Della Rocca, S., Ausiello, G., Dezani-Ciancaglini, M. (eds.) ICALP 1989. LNCS, vol. 372, pp. 652–671. Springer, Heidelberg (1989)
10. Puchala, B.: Infinite Two Player Games with Partial Information: Logic and Algorithms. Diploma Thesis, RWTH Aachen (2008)
11. Reif, J.H.: The Complexity of Two-player Games of Incomplete Information. Journal of Computer and System Sciences 29, 274–301 (1984)
12. Thomas, W.: On the Synthesis of Strategies in Infinite Games. In: Mayr, E.W., Puech, C. (eds.) STACS 1995. LNCS, vol. 900, pp. 1–13. Springer, Heidelberg (1995)
13. van der Meyden, R., Vardi, M.Y.: Synthesis from Knowledge-Based Specifications (Extended Abstract). In: Sangiorgi, D., de Simone, R. (eds.) CONCUR 1998. LNCS, vol. 1466, pp. 34–49. Springer, Heidelberg (1998)
14. van der Meyden, R., Wilke, T.: Synthesis of Distributed Systems from Knowledge-based Specifications. In: Abadi, M., de Alfaro, L. (eds.) CONCUR 2005. LNCS, vol. 3653, pp. 562–576. Springer, Heidelberg (2005)
15. Wong-Toi, H., Dill, D.L.: Synthesizing Processes and Schedulers from Temporal Specifications. In: Larsen, K.G., Skou, A. (eds.) CAV 1991. LNCS, vol. 575, pp. 272–281. Springer, Heidelberg (1992)

Parity Games with Partial Information Played on Graphs of Bounded Complexity

Bernd Puchala* and Roman Rabinovich**

Mathematische Grundlagen der Informatik, RWTH Aachen University
{puchala,rabinovich}@logic.rwth-aachen.de

Abstract. We address the strategy problem for parity games with partial information and observable colors, played on finite graphs of bounded graph complexity. We consider several measures for the complexity of graphs and analyze in which cases, bounding the measure decreases the complexity of the strategy problem on the corresponding classes of graphs. We prove or disprove that the usual powerset construction for eliminating partial information preserves boundedness of the graph complexity. For the case where the partial information is unbounded we prove that the construction does not preserve boundedness of any measure we consider. We also prove that the strategy problem is EXPTIME-hard on graphs with directed path-width at most 2 and PSPACE-complete on acyclic graphs. For games with bounded partial information we obtain that the powerset construction, while neither preserving boundedness of entanglement nor of (undirected) tree-width, does preserve boundedness of directed path-width. Therefore, parity games with bounded partial information, played on graphs with bounded directed path-width can be solved in polynomial time.

1 Introduction

Parity games are played by two players which move a token along the edges of a labeled graph by choosing appropriate edge labels, also called actions. The vertices of the graph, also called positions, have priorities and the winner of an infinite play of the game is determined by the parity of the least priority which occurs infinitely often. Parity games play a key role in modern approaches to verification and synthesis of state-based systems. They are the model-checking games for the modal μ-calculus, a powerful specification formalism for verification problems. Moreover, parity objectives are a canonical form to express ω-regular objectives and therefore capture fundamental properties of non-terminating reactive systems, cf. [17]. Such a system can be modeled as a two-player game where changes of the system state correspond to changes of the game position. Situations where the change of the system can be controlled correspond to positions of player 0, uncontrollable situations correspond to positions of player 1. A *winning strategy*

* Supported by the LINT project, ESF, www.esf.org
** Supported by the LINT and the GASICS projects, ESF, www.esf.org

P. Hliněný and A. Kučera (Eds.): MFCS 2010, LNCS 6281, pp. 604–615, 2010.

for player 0 then yields a controller that forces the system into satisfying an ω-regular specification.

The problem to determine, for a given parity game G and a position v, whether player 0 has a winning strategy for G from v, is called the strategy problem. The algorithmic theory of parity games with full information has received much attention during the past years, cf. [10]. The most important property of parity games with full information is the memoryless determinacy which proves that the strategy problem for parity games is in NP $\cap$ co-NP.

However, assuming that both players have full information about the history of events in a parity game is not always realistic. For example, if the information about the system state is acquired by imprecise sensors or the system encapsulates private states which cannot be read from outside, then a controller for this system must rely on the information about the state and the change of the system to which it has access. I.e., in the game model, player 0 has uncertainties about the positions and actions in the game, so we have to add partial information to parity games in order to model this kind of problems. The uncertainties are represented by equivalence relations on the positions and actions in the game graph meaning that equivalent positions respectively actions are indistinguishable for player 0. Solving the strategy problem for such games is much harder than solving parity games with full information, since we have to keep track of the knowledge of player 0 during a play of the game. For this we compute, for any finite history, the set of positions that player 0 considers possible in this situation. This procedure is often referred to as *powerset construction.*

Such a knowledge tracking is inherently unavoidable and leads to an exponential lower bound for the time complexity of the strategy problem for reachability games with partial information [15] and a super-polynomial lower bound for the memory needed to implement winning strategies in reachability games [3,14]. Therefore, it is expedient to look for classes of games with partial information, where the strategy problem has a lower complexity. A simple while effective approach is to bound the partial information in the game, i.e., the size of the equivalence classes of positions which model the uncertainties of player 0 about the current position. This is appropriate in situation where, e.g., the imprecision of the sensors or the amount of private information of the system does not grow if the system grows. Then, the game which results from the powerset construction has polynomial size, so partial information parity games with a *bounded* number of observable priorities can be solved in polynomial time. Hereby *observable* means, that the priorities are constant over equivalence classes. However, if the number of priorities is not bounded, we cannot prove this approach to be efficient, since the question whether full information parity games with arbitrarily many priorities can be solved in polynomial time is still open.

To obtain a class of parity games with partial information that can be solved in polynomial time, one has to bound certain other parameters. A natural approach is to bound the complexity of the game graphs with respect to appropriate measures. Such graph complexity measures have proven enormous usefulness in algorithmic graph theory. Several problems which are intractable in general can be solved efficiently on classes of graphs where such measures are bounded. The key note here is that bounded complexity with respect to appropriate measures allows to decompose the graph into small parts which are only sparsely related within the graph in a certain sense. One can then solve the problem on these small parts which requires, for each part, only a fixed amount of time, and combine the partial solutions in an efficient way. This has proven to be applicable to a large number of graph theoretic decision problems, e.g., all MSO-definable graph properties [5]. More recently, it has also been applied to the strategy problem for (full information) parity games. It has been shown that parity games played on graphs with bounded tree-width or bounded (monotone) DAG width or bounded entanglement can be solved in polynomial time [2,4,12]. The natural question is whether such results can also be obtained for games with partial information.

Since the direction of the edges is inherently important when solving games and when performing the powerset construction, we primarily consider measures for directed graphs. However, we prove a negative result about (undirected) tree-width, which is the most important measure for undirected graphs, as a prototype witness for the high potential of the powerset construction to create graph complexity when the direction of edges is neglected. From the large variety of measures for directed graphs we focus on DAG-width, directed path-width and entanglement. Two other important measures are directed tree-width [9] and Kelly-width [8]. For those measures, however, our techniques cannot be applied directly, due to somewhat inconvenient conditions in the definitions.

In Section 3 we prove that in the case where the partial information is unbounded, there are classes of graphs G with complexity at most 2 such that the complexity of the corresponding powerset graphs is exponential in the *size* of G for any measure we consider. We also prove that the strategy problem for reachability games with partial information is EXPTIME-hard on graphs with entanglement at most 2 and directed path-width at most 2 and that the problem is PSPACE-complete on acyclic graphs. Notice that reachability games form a subclass of parity games. Roughly speaking, these results show that bounding the graph complexity does not decrease the complexity of the strategy problem, as long as the partial information is unbounded.

In Section 4 we consider parity games with bounded partial information. In this case, the graphs which result from the powerset construction have polynomial size, so if the construction additionally preserves boundedness of appropriate graph complexity measure, then the corresponding strategy problem is in PTIME. For the case of tree-width and entanglement, however, we disprove this preservation of boundedness. Finally, we prove that the construction does preserve boundedness of directed path-width and of *non-monotone* DAG-width. So, parity games with bounded partial information, played on graphs of bounded directed path-width can be solved in polynomial time. Moreover, if DAG-width has bounded monotonicity cost, which is an open question, the same result holds for the case of bounded DAG-width. Detailed proofs of all results appear in the full version of the paper.

2 Preliminaries

Games and Strategies. A *parity game* has the form $G = (V, V_0, (f_a)_{a \in A}, \mathrm{col})$, where V is the set of positions, A is the set of actions and for each action $a \in A$, $f_a : \mathrm{dom}(f_a) \subseteq V \to V$ is a function. We write $v \xrightarrow{a} w$ if $f_a(v) = w$. Furthermore, $V_0 \subseteq V$ are the positions of player 0 and $\mathrm{col} : V \to C$ is a function into a finite set $C \subseteq \mathbb{N}$ of colors (also called priorities). We define $V_1 := V \setminus V_0$ and $A_i := \bigcup\{\mathrm{act}(v) \mid v \in V_i\}$ for $i = 0, 1$. The directed graph (V, E) with $E = \bigcup\{E_a \mid a \in A\}$ where $E_a = \{(u, v) \in V \times V \mid u \in \mathrm{dom}(f_a)$ and $f_a(u) = v\}$ for each $a \in A$ is called the *game graph* of G. Here we consider only *finite* games, i.e., games where V and A are finite.

For a finite sequence $\pi \in V(AV)^*$, by $\mathrm{last}(\pi)$ we denote the last position in π. For $v \in V$, a *play* in G from v is a *maximal* finite or infinite sequence $\pi = v_0 a_0 v_1 \ldots \in v(AV)^* \cup v(AV)^\omega$ such that $v_i \in \mathrm{dom}(f_{a_i})$ and $f_{a_i}(v_i) = v_{i+1}$ for each i. A finite play π is won by player 0 if $\mathrm{last}(\pi) \in V_1$. An infinite play π is won by player 0 if $\min\{c \in C \mid \mathrm{col}(v_i) = c$ for infinitely many $i < \omega\}$ is even. A play π is won by player 1 if and only if it is not won by player 0. A *reachability game* is a parity game with $C = \{1\}$, i.e., player 0 wins only finite plays which end in positions $v \in V_1$. Now let $\mathcal{H}_{\mathrm{fin}}$ be the set of all finite histories $\pi \in V(AV)^*$ of plays in G from v. A *strategy* for player i for G is a function $g : \{\pi \in \mathcal{H}_{\mathrm{fin}} \mid \mathrm{last}(\pi) \in V_i\} \to A$ such that $g(\pi) \in \mathrm{act}(\mathrm{last}(\pi))$ for all $\pi \in \mathrm{dom}(g)$. A history $\pi = v_0 a_0 v_1 a_1 v_2 \ldots$ is called *compatible* with g if for all j such that $v_j \in V_i$ we have $a_j = g(v_0 a_0 \ldots a_{j-1} v_j)$. We call a strategy g for player i a *winning strategy* from v_0 if each play π in G from v_0 that is compatible with g is won by player i.

Partial Information. The *knowledge* of player i after some history $\pi \in \mathcal{H}_{\mathrm{fin}}$ is given by an equivalence relation $\sim_i \subseteq \mathcal{H}_{\mathrm{fin}} \times \mathcal{H}_{\mathrm{fin}}$ where $\pi \sim_i \pi'$ if π and π' are indistinguishable for player i by means of his given information. So, after π has been played, to the best of player i's knowledge, it is possible that instead, π' has been played. A strategy $g : \{\pi \in \mathcal{H}_{\mathrm{fin}} \mid \mathrm{last}(\pi) \in V_i\} \to A$ for player i for G is called a *partial information strategy* with respect to $\sim_i$ ($\sim_i$-strategy, for short) if $g(\pi) = g(\pi')$ for all $\pi, \pi' \in \mathcal{H}_{\mathrm{fin}}$ with $\pi \sim_i \pi'$. Notice that a $\sim_i$-strategy g for player i is winning from all positions in a set $U \subseteq V$ if and only if it is winning from a simulated initial position v_0 which belongs to player $1-i$ and from which he can secretly choose any position $v \in U$. Moreover, any $\sim_i$-strategy g which is only defined on histories from some initial position v_0 can be extended to a $\sim_i$-strategy g' with $\mathrm{dom}(g) = \{\pi \in \mathcal{H}_{\mathrm{fin}} \mid \mathrm{last}(\pi) \in V_i\}$ by giving g' appropriate value on histories from some initial position $v_0' \neq v_0$. So in our *antagonistic* two-player setting, it suffices to consider strategies which are winning from single initial positions v_0 and only defined on histories from v_0.

Now, if we are given a game G, a position v in G and some equivalence relation $\sim_i$ on $\mathcal{H}_{\mathrm{fin}}$, then the question whether player i has a *winning $\sim_i$-strategy* for G from v is independent of the partial information of player $1-i$. Therefore, in this work we investigate games with partial information only for player 0.

We consider games played on finite graphs where player 0 has uncertainties about the positions and actions in the game, modeled by equivalence relations. The relation $\sim_0$ is then obtained by extending these equivalence relations to an equivalence relation on $\mathcal{H}_{\text{fin}}$. In particular, $\sim_0$ is finitely represented which is necessary when considering decision problems for games with partial information. A *parity game with partial information* has the form $\mathcal{G} = (G, \sim^V, \sim^A)$, where $G = (V, V_0, (f_a)_{a\in A}, \text{col})$ is a parity game and $\sim^V \subseteq V \times V$ and $\sim^A \subseteq A \times A$ are equivalence relations such that the following conditions hold:

(1) If $u, v \in V$ with $u \sim^V v$ then $u, v \in V_0$ or $u, v \notin V_0$,
(2) If $a, b \in A_0$ with $a \neq b$ then $a \not\sim^A b$,
(3) If $u, v \in V_0$ with $u \sim^V v$, then $\text{act}(u) = \text{act}(v)$.
(4) If $u, v \in V$ with $u \sim^V v$ then $\text{col}(u) = \text{col}(v)$.

Condition (1) says that player 0 always knows when it is his turn and condition (2) says that player 0 can distinguish all the actions that are available to him at some position of the game. Condition (3) ensures that player 0 always knows which actions are available to him when it is his turn. Finally, condition (4) says that the colors of the game are observable for player 0.

We say that a game $\mathcal{G}$ has *bounded partial information*, if there is some $k \in \mathbb{N}$, such that for any position $v \in V$ the equivalence class $[v] := \{w \in V \mid v \sim^V w\}$ of v has size at most k. Notice that the equivalence classes $[a] := \{b \in A \mid a \sim^A b\}$ of actions $a \in A$ may, however, be arbitrarily large.

The equivalence relation on finite histories is defined as follows. For $\pi = v_0a_0 \dots a_{n-1}v_n, \rho = w_0b_0 \dots b_{m-1}w_m \in V(AV)^*$, let

$$\pi \sim^* \rho :\Longleftrightarrow n = m \text{ and } v_j \sim^V w_j \text{ and } a_j \sim^A b_j \text{ for all } j.$$

The winning region $\text{Win}_0^{\mathcal{G}}$ of player 0 in $\mathcal{G}$ is the set of all positions $v \in V$ such that player 0 has a winning $\sim^*$-strategy for $\mathcal{G}$ from v.

Remark. Consider the interaction between components of a system where the behavior of each component is prescribed by a controller which has to rely on the information available to this component. In such settings it might seem more appropriate to ask for a $\sim_0^*$-strategy for player 0 which is winning against all $\sim_1^*$-strategies of player 1 rather than a winning $\sim_0^*$-strategy for player 0. However, it is easy to see that in our perfect recall setting, this is equivalent.

Powerset Construction. The usual method to solve games with partial information is a powerset construction originally suggested by John H. Reif in [15]. The construction turns a game with partial information into a *nondeterministic* game with full information such that the existence of winning strategies for player 0 is preserved.

A *nondeterministic parity game* has the form $G = (V, V_0, (E_a)_{a\in A}, \text{col})$ where V, V_0, A, and col are as in a deterministic game and for $a \in A$, E_a is a binary relation on V. Plays, strategies and winning strategies are defined as before. Nondeterministic games are not determined in general and hence not equivalent to deterministic games. However, for each nondeterministic game G and each player $i \in \{0, 1\}$, we can construct a deterministic game G^i such that the existence of winning strategies for player i is preserved. We simply resolve the nondeterminism by giving player $1-i$ control of nondeterministic choices. Technically, for any $v \in V$ and any $a \in \mathrm{act}(v)$ we add a unique a-successor of v to the game graph which belongs to player $1-i$ and from which he can choose any a-successor of v in the original game graph. The coloring of such a new position is the coloring of its unique predecessor. This construction does not increase the complexity of the game graph with respect to any measure we consider here.

Now for a parity game $\mathcal{G} = (G, \sim^V, \sim^A)$, $G = (V, V_0, (f_a)_{a\in A}, \mathrm{col})$ with partial information, we construct the corresponding game $\overline{G} = (\overline{V}, \overline{V}_0, (\overline{E}_a)_{a\in A}, \overline{\mathrm{col}})$ with full information as follows. First, for $S \subseteq V$ and $B \subseteq A$ we define the set $\mathrm{Post}_B(S) := \{v \in V \mid \exists\, s \in S, \exists\, b \in B \;:\; b \in \mathrm{act}(s) \wedge f_b(s) = v\}$. The components of $\overline{G}$ are defined as follows.

- $\overline{V} = \{\overline{v} \in 2^V \mid \exists v \in V : \overline{v} \subseteq [v]\}$ and $\overline{V}_0 = \overline{V} \cap 2^{V_0}$
- $\forall\, a \in A$: $(\overline{v}, \overline{w}) \in \overline{E}_a :\Longleftrightarrow \exists\, w \in \mathrm{Post}_a(\overline{v})$: $\overline{w} = \mathrm{Post}_{[a]}(\overline{v}) \cap [w]$
- $\mathrm{col}(\overline{v}) = \mathrm{col}(v)$ for some $v \in V$.

It can be shown that this construction in fact preserves winning strategies for player 0, that means, for any $v_0 \in V$, player 0 has a winning $\sim^*$-strategy for $\mathcal{G}$ from v_0 if and only if he has a winning strategy for $\overline{G}$ from $\{v_0\}$. So when asking for a winning $\sim^*$-strategy for player 0 from a given position v_0, we are only interested in the part of the graph $\overline{G}$ which is reachable from $\{v_0\}$. We denote this subgraph of $\overline{G}$ by $\overline{G}_{v_0}$. The key-property for the correctness of the construction is given in the following lemma which is proved straightforwardly.

Lemma 1. *For each finite history $\overline{\pi} = \overline{v}_0 a_1 \overline{v}_1 \ldots a_n \overline{v}_n$ in $\overline{G}$ and all $v_n \in \overline{v}_n$, there is a finite history $\pi = v_0 a'_1 v_1 \ldots a'_n v_n$ in $\mathcal{G}$ such that $v_i \in \overline{v}_i$ for all $i \in \{0, \ldots, n\}$ and $a'_i \sim^A a_i$ for all $i \in \{1, \ldots, n\}$.*

Graph Complexity. We consider only directed graphs without multi-edges, but possibly with self-loops, i.e., a graph is a pair $G = (V, E)$ where $E \subseteq V \times V$. An undirected graph is a graph with a symmetric edge relation.

All measures we consider can be characterized in terms of cops and robber games, where several cops try to catch a robber on a graph. Technically, these games are reachability games. We do not give formal descriptions of the games but merely describe them in an informal way. In a *graph searching game* there are two players, a cop player and a robber player. Basically, the robber player moves a robber token along cop free paths of the graph. The cop player has a number k of cops at his disposal and he can place and move them on and between vertices. At the very moment a cop is moving he does not block any vertex. The goal of the cop player is to place a cop on the robber, the robber player's goal is to elude capture. That means, infinite plays are won by the robber and finite plays, which end in a position where the robber has no legal moves available, are won by the cops. The number k of cops is a parameter of the game, that means, for any natural number k we have a k-cops and robber game.

Tree-width, see [16], denoted tw, is a measure defined for undirected graphs and the tree-width of a directed graph is the tree-width of his symmetric closure.

In the tree-width game, the cops can be placed and moved in the graph arbitrarily. When a new set U of at most k vertices is chosen to be occupied by the cops, the robber may move along any cop-free path, i.e. no vertex on this path which has been occupied by a cop is also occupied by a cop according to U. The robber may move at unlimited speed, i.e., he may move along a whole path in one step.

DAG-width, introduced in [2,13] and denoted dw, is a generalization of tree-width to directed graphs. The DAG-width game has exactly the same rules as the tree-width game with the only difference that now, the robber has to respect the direction of the edges. The **directed path-width**, denoted dpw, is defined by the same game where now, the robber is *invisible*. So, a strategy for the cop player has to yield the same decision in any two situations which differ only in the current position of the robber.

Finally, in the **entanglement**-game [4] the cop player may, in each round, do nothing or place one of the k cops on the current position of the robber. No matter what the cops do, the robber must go from his recent vertex to a new vertex, which is not occupied by a cop, along an edge of the graph.

For $X \in \{\mathrm{tw}, \mathrm{dw}, \mathrm{dpw}\}$, $X(G)$ is the least natural number k such that $k+1$ cops (k cops, if $X = \mathrm{dw}$) have a (robber-)monotone winning strategy for the X-game on G. A strategy f for the cops is called (robber-)monotone, if in any play compatible with f, the robber can never reach any vertex that has previously been unavailable for him. Such a monotone winning strategy for k cops yields a decomposition of G into (possibly complex) parts of size at most k which are only sparsely related among each other. Such decompositions often allow for efficient dynamic solutions of hard graph problems. Notice that clearly, $\mathrm{dw}(G) \leq \mathrm{dpw}(G) + 1$ for any graph G.

The entanglement of a graph G, denoted $\mathrm{ent}(G)$, is the least natural number k such that k cops have an *arbitrary* winning strategy for the entanglement game on G. For entanglement, only for $k = 2$, a decomposition in the above sense is known [6]. Nevertheless, parity games can be solved efficiently on graph classes of bounded entanglement.

In the following, let $\mathcal{M} = \{\mathrm{tw}, \mathrm{dw}, \mathrm{dpw}, \mathrm{ent}\}$. We say that a measure $X \in \mathcal{M}$ has *monotonicity cost* at most f for a function $f : \mathbb{N} \to \mathbb{N}$ if, for any graph G such that k cops have a winning strategy for the X-game on G, $k + f(k)$ cops have a monotone winning strategy for the X-game on G. We say that X has *bounded monotonicity cost* if there is a function $f : \mathbb{N} \to \mathbb{N}$ such that X has monotonicity cost at most f. Tree-width has monotonicity cost 0 [16] and the same holds for directed path-width, [1,7]. On the contrary, DAG-width does not have monotonicity cost 0 [11]. Whether DAG-width has bounded monotonicity cost, is an important open problem in structure theory of directed graphs.

3 Unbounded Partial Information

First, when the partial information is unbounded, it is easy to prove that boundedness of graph complexity measures is not preserved by the powerset construction. We show that they grow even exponentially.

Proposition 2. *There are games $\mathcal{G}_n$, $n \in \mathbb{N}$ with partial information and with $X(\mathcal{G}_n) \leq 2$ for all $n \in \mathbb{N}$ and for any $X \in \mathcal{M}$, such that the powerset graphs $\overline{G}_n$ have exponential measure X in the size of $\mathcal{G}_n$ for any $X \in \mathcal{M}$.*

Proof. (Sketch) Consider a disjoint union of n cycles of length 2 where all $2n$ positions are equivalent. Additionally, we have an initial position v_0 with an edge to exactly one position from each cycle. Applying the powerset construction from v_0, we obtain a set which contains exactly one element from each cycle. Such sets represent binary numbers with n bits and for each bit we have an action which causes exactly this bit to flip. So, using the Gray-code, we can generate all binary numbers with n bits by successively flipping each bit. If we do this independently for the first $n/2$ bits and for the last $n/2$ bits, it is easy to see that the resulting positions are connected in such a way, that they form an undirected grid $\overline{G}_n$ of size $2^{n/2} \times 2^{n/2}$, for which $X(\overline{G}_n) \geq 2^{n/2}$ for any $X \in \mathcal{M}$. □

Towards our analysis of the complexity of the strategy problem for games with partial information on graphs of bounded complexity, we first note that on trees, solving games with partial information is not harder than solving games with full information. Performing the powerset construction on a tree, we again obtain a tree, where the set of positions on each level partitions the set of positions on the corresponding level of the original tree. This new tree can be computed in polynomial time and has at most as many vertices as the original tree. In the following we prove that, as soon as we consider at least DAGs, the strategy problem for reachability games with partial information becomes intractable as long as we do not bound any other parameters.

Theorem 3. *The following problem is* EXPTIME*-hard. Given a partial information reachability game $\mathcal{G} = (G, \sim^V, \sim^A)$ with $\mathrm{ent}(G) \leq 2$ and $\mathrm{dpw}(G) \leq 2$ and a position $v_0 \in V(G)$, is $v_0 \in \mathrm{Win}_0^{\mathcal{G}}$?*

Proof. (Sketch) Let $M = (Q, \Gamma, \Sigma, \delta, q_0, q_+, q_-)$ be a 1-tape TM with space bound n^k. For simplicity, we consider the case where M is deterministic and prove the claim for $\mathrm{ent}(G) \leq 1$ and $\mathrm{dpw}(G) \leq 1$. The proof can be carried out similarly for alternating machines and $\mathrm{ent}(G) \leq 2$ and $\mathrm{dpw}(G) \leq 2$. For $\Delta = \Sigma \uplus (Q \times \Sigma) \uplus \{\#\}$ we can describe each configuration of M by a word $C = \#w_0 \ldots w_{i-1}(qw_i)w_{i+1} \ldots w_t\# \in \Delta^{n^k+2}$. Moreover, for a configuration C of M and some $2 \leq i \leq n^k+1$ the symbol number i of C' where $C' = \mathrm{Next}(C)$ only depends on the symbols number $i-1$, i and $i+1$ of C. Now let $u = u_1 \ldots u_n \in \Gamma^*$ be an input word for M. The idea for the game corresponding to u is that player 0 constructs the unique run of M on u by selecting symbols from Δ while player 1 checks the correctness of the construction. For this he may, *exactly once*, *secretly* memorize the last three symbols that player 0 has selected and verify the corresponding next symbol locally, according to the above remark.

We implement this as follows. The game graph G consists of $|\Delta|^3 \cdot n^k + 1$ augmented DAGs t_0 and $t_i^{\overline{\delta}}$, $i = 1, \ldots, n^k$, $\overline{\delta} \in \Delta^3$, each of which has a unique top node, edges from any non-bottom level only to the level below and $2 \cdot n^k + 1$ levels in total. For any fixed $i \in \{1, \ldots, n^k\}$, the union of all $t_i^{\overline{\delta}}$ for $\overline{\delta} \in \Delta^3$ is denoted by t_i. We also have a unique root v_0 for the whole graph, from which there is an edge to the top of t_0 and, for any $1 \leq i \leq n^k$, an edge to exactly one top node of t_i. From any leaf node of t_i, $0 \leq i \leq n^k$, there is a back-edge to some top node of t_i. Finally, the i-th level of $t_i^{\overline{\delta}}$ for $i \geq 1$, has no outgoing edges. Obviously $\mathrm{ent}(G) \leq 1$ and $\mathrm{dpw}(G) \leq 1$. In any of the t_i, the players alternate in making moves. From an odd level of t_i, player 0 moves to the next level of t_i by choosing a symbol $\delta \in \Delta$ or, from the bottom level, to the top of t_i. During a traverse of t_0, player 0 constructs one configuration. From an even level $2j$, $j \in \{0, \ldots, n^k\}$ of t_0, player 1 can go to the same position at level $2j+1$ of t_0 or of $t_j^{\overline{\delta}}$, where $\overline{\delta}$ is the sequence of the last three symbols chosen by player 0, which is stored in the current position. In the latter case, player 1 memorizes the last three symbols chosen by player 0: after player 1 has left t_0, the information $\overline{\delta}$ does not change anymore. (If player 1 goes from v_0 to a top node of some t_i with $i > 0$ then he memorizes the symbols $i-1$, i and $i+1$ of the initial configuration.) From a position in some t_j with $j \geq 1$, player 1 can only move to the same position on the next level of t_j. At level i of $t_i^{\overline{\delta}}$, the correctness of the construction that player 0 provides is checked according to the memorized information i and $\overline{\delta}$, yielding a winner of the game. If player 0 provides a faulty construction, player 1 can detect this by memorizing the symbols of player 0 at the appropriate step. To ensure that the construction player 0 provides is not affected by this trick, we define the equivalence relations $\sim^V$ and $\sim^A$ in such a way that the only information player 0 has, is the player whose turn it is, the current level and the last character that he has chosen. Hence, player 0 does not observe whether player 1 leaves t_0 or not. The goal of player 0 is to reach a position where player 1 has tried to detect an error but has failed, or a position where he can safely write down (q_+, a) for some $a \in \Sigma$. This defines the partial information reachability game $\mathcal{G}_u$ with $u \in L(M)$ if, and only if, $v_0 \in \mathrm{Win}_0^{\mathcal{G}}$. □

Remark. It easy to see, that the tree-width of the game graphs constructed in the proof of Theorem 3 is bounded by some $k \in \mathbb{N}$ which is independent of the input u. Therefore, the strategy problem for reachability games with partial information on graphs of tree-width at most k is EXPTIME-hard.

The cases of entanglement and directed path-width at most 1 are still open. However, the strategy problem for sequence-forcing games with partial information, where player 0 tries to enforce a certain sequence of fixed length of positions, is EXPTIME-hard on graphs of entanglement and directed path-width at most 1. Sequence-forcing games can be polynomially reduced to reachability games so we obtain, roughly speaking, the following result.

Theorem 4. *Adding partial information to games played on graphs of entanglement at most* 1 *and directed path-width at most* 1 *can cause an unavoidable exponential blow-up of the time complexity of the corresponding strategy problem.*

Finally, if the we consider acyclic game graphs, the strategy problem for partial information reachability games is PSPACE-complete.

Theorem 5. *The strategy problem for reachability games with partial information on acyclic graphs is* PSPACE*-complete.*

Proof. (Sketch) Carrying out the powerset construction on an acyclic graph G we again obtain an acyclic graph $\overline{G}$ where, by Lemma 1, the paths in $\overline{G}$ are not longer than the paths in G. So, we can solve the reachability game on $\overline{G}$ by an (PSPACE =) APTIME algorithm. Conversely, if the machine from the proof of Theorem 3 is a PTIME ATM, we do not insert edges back to the top of t_i but we go to roots of a new copy of t_i. Since M is polynomially time bounded, the resulting graph has polynomial size and is, by construction, acyclic. □

4 Bounded Partial Information

We turn to the case where the size of the equivalence classes of positions is bounded. The first observation is that bounded tree-width may become unbounded when applying the powerset construction. Afterwards we shall see, that the same results holds for entanglement.

Proposition 6. *There are games $\mathcal{G}_n$, $n \in \mathbb{N}$ with bounded partial information and $X(\mathcal{G}_n) \leq 2$ for all $n \in \mathbb{N}$ and any $X \in \mathcal{M}$ such that the corresponding powerset graphs $\overline{G}_n$ have unbounded tree-width.*

Proof. (Sketch) The game graph G_n is a disjoint union of n (undirected) paths P_i of length n together with a vertex v_0 and (undirected) edges from v_0 to any other vertex. For every even positive (resp. odd) j and every odd (resp. even positive) i, the j-th vertices of P_i and P_{i+1} are not distinguishable. The powerset graph $\overline{G}_n$ is again a union of isomorphic copies of the paths P_i, augmented by the following gadgets inserted between each two *distinguishable* j-th vertices v_j^i and v_j^{i+1} of the paths P_i and P_{i+1}. A gadget is one vertex with directed edges to v_j^i and v_j^{i+1} (and some other edges). While $\mathrm{tw}(G_n) = 2$, $\mathrm{tw}(\overline{G}_n) \geq n/2$. □

Proposition 7. *There are games $\mathcal{G}_n$, $n \in \mathbb{N}$ with bounded partial information and $X(G) \leq 2$ for all $n \in \mathbb{N}$ and any $X \in \mathcal{M} \setminus \{\mathrm{dpw}\}$ such that the corresponding powerset graphs $\overline{G}_n$ have unbounded entanglement.*

Proof. (Sketch) The graph G_n consists of two disjoint copies $\mathcal{T}_1$ and $\mathcal{T}_2$ of the full undirected binary tree. From a vertex in $\mathcal{T}_1$, a directed edge leads to the corresponding vertex in $\mathcal{T}_2$ and there are no edges from $\mathcal{T}_2$ to $\mathcal{T}_1$. Undirected trees have entanglement two, so $\mathrm{ent}(G) = 2$. The edges from $\mathcal{T}_1$ to $\mathcal{T}_2$ are implemented by gadgets which create, when the powerset construction is performed, a back edge while also preserving the original edge. So the graph $\overline{G}_n$ again consists of two disjoint copies of the full undirected binary tree but corresponding vertices are now connected in both directions. Adapting a proof from [4] for similar graphs we obtain that $\mathrm{ent}(\overline{G}_n) \geq n/2 - 2$. □

Now we prove that in contrast to tree-width and entanglement, *non-monotone* DAG-width is preserved by the powerset construction. In the following, let $\mathcal{G} = (G, \sim^V, \sim^A)$ be a game with bounded partial information, i.e., there is some $r \in \mathbb{N}$ such that $|[u]| \leq r$ for all $u \in V(G)$. Let $\overline{G}$ be the powerset graph of $\mathcal{G}$.

Proposition 8. *If k cops win the DAG-width game on G, then $k \cdot r \cdot 2^{r-1}$ cops win the DAG-width game on $\overline{G}$.*

Proof. (Sketch) We translate strategies for k cops from G to $\overline{G}$ and robber's strategies in the opposite direction. Consider positions in games on both graphs. When the robber makes a move on $\overline{G}$ to a vertex $\{v_1, \ldots, v_l\}$ we consider l plays in the game on G where he moves to v_1, $v_2, \ldots$, v_l. For each of these moves, the strategy for the cops for the game on G supplies an answer, moving the cops from U to U'. All these moves are translated into a move in which the cops occupy precisely the vertices of $\overline{G}$ that include a vertex from some U'. These moves of the cop player on $\overline{G}$ can be realized with $k \cdot r \cdot 2^{r-1}$ cops and guarantee that moves of the robber can always be translated back to the game on G. The key argument here is that by Lemma 1, for any path $\overline{u}^0 \xrightarrow{\overline{E}} \overline{u}^1 \xrightarrow{\overline{E}} \ldots \xrightarrow{\overline{E}} \overline{u}^t$ in $\overline{G}$ and for any $u^t \in \overline{u}^t$, there is a path $u^0 \xrightarrow{E} u^1 \xrightarrow{E} \ldots \xrightarrow{E} u^t$ in G such that $u^i \in \overline{u}^i$ for any $i \in \{0, \ldots, t\}$. So if a play continues infinitely on $\overline{G}$ then at least one corresponding play on G continues infinitely. Hence, if we start from a winning strategy for k cops for the game on G, no strategy for the robber can be winning against $k \cdot r \cdot 2^{r-1}$ cops on $\overline{G}$. By determinacy, the result follows. □

Unfortunately, this strategy translation does not necessarily preserve monotonicity, so boundedness of DAG-width does not follow. On the other hand, boundedness of directed path-width can be proved by translating strategies non-monotonously, as it has monotonicity cost 0. However, since directed path-width is defined via partial information strategies, the translation has to be refined.

Proposition 9. *If* $\mathrm{dpw}(G) \le k$, *then* $\mathrm{dpw}(\overline{G}) \le (k+1) \cdot 2^{r-1}$.

Corollary 10. *Parity games with bounded partial information can be solved in polynomial time on graphs of bounded directed path-width.*

An interesting special case where an equiped translation of strategies does preserve monotonicity is given by games with strongly connected equivalence classes of positions. Intuitively this means that for any characteristic of the current state which player 0 is unsure about, it is possible for player 1 to change the value of this characteristic into any other possible value privately, i.e., without changing any characteristics visible for player 0 in between. This is appropriate for situations where, e.g., the uncertainties of player 0 concern some private states of player 1 which are independent of the states visible for player 0.

Proposition 11. *If* $\mathrm{dw}(G) \le k$ *and each equivalence class of positions is strongly connected, then* $\mathrm{dw}(\overline{G}) \le k \cdot r^2 \cdot 2^{r-1}$.

Finally, we remark that our direct translation of the robber's moves back to the game on G cannot be immediately applied to the games which define Kelly-width and directed tree-width. In the Kelly-width game, the robber can only move if a cop is about to occupy his vertex. It can happen that the cops occupy a vertex $\{v_1, \ldots, v_l\}$ in $\overline{G}$ but not all vertices $v_1, \ldots, v_l$ in G. In the directed tree-width game, the robber is not permitted to leave the strongly connected component in which he currently is, which again obstructs a direct translation of the robber's moves from $\overline{G}$ back to G.

References

1. Barát, J.: Directed path-width and monotonicity in digraph searching. Graphs and Combinatorics 22, 161–172 (2006)
2. Berwanger, D., Dawar, A., Hunter, P., Kreutzer, S.: DAG-Width and Parity Games. In: Durand, B., Thomas, W. (eds.) STACS 2006. LNCS, vol. 3884, pp. 524–536. Springer, Heidelberg (2006)
3. Berwanger, D., et al.: Strategy Construction for Parity Games with Imperfect Information. In: van Breugel, F., Chechik, M. (eds.) CONCUR 2008. LNCS, vol. 5201, pp. 325–339. Springer, Heidelberg (2008)
4. Berwanger, D., Grädel, E.: Entanglement – a measure for the complexity of directed graphs with applications to logic and games. In: Baader, F., Voronkov, A. (eds.) LPAR 2004. LNCS (LNAI), vol. 3452, pp. 209–223. Springer, Heidelberg (2005)
5. Courcelle, B.: Graph rewriting: An algebraic and logic approach. In: Handbook of Theoretical Computer Science, pp. 193–242. Elsevier, Amsterdam (1990)
6. Grädel, E., Kaiser, Ł., Rabinovich, R.: Directed Graphs of Entanglement Two. In: Gębala, M. (ed.) FCT 2009. LNCS, vol. 5699, pp. 169–181. Springer, Heidelberg (2009)
7. Hunter, P.: Losing the +1 or directed path-width games are monotone (2006), `http://www.comlab.ox.ac.uk/people/paul.hunter/papers/losing.pdf`
8. Hunter, P., Kreutzer, S.: Digraph measures: Kelly decompositions, games, and orderings. In: SODA, pp. 637–644 (2007)
9. Johnson, T., Robertson, N., Seymour, P.D., Thomas, R.: Directed tree-width. J. Comb. Theory, Ser. B 82(1), 138–154 (2001)
10. Jurdziński, M.: Games for Verification: Algorithmic Issues. PhD thesis, University of Aarhus (2000)
11. Kreutzer, S., Ordyniak, S.: Digraph Decompositions and Monotonicity in Digraph Searching. In: Broersma, H., Erlebach, T., Friedetzky, T., Paulusma, D. (eds.) WG 2008. LNCS, vol. 5344, pp. 336–347. Springer, Heidelberg (2008)
12. Obdržálek, J.: Algorithmic Analysis of Parity Games. PhD thesis, School of Informatics, University of Edinburgh (2006)
13. Obdržálek, J.: Dag-width: connectivity measure for directed graphs. In: SODA, pp. 814–821 (2006)
14. Puchala, B.: Infinite Two Player Games with Partial Information: Logic and Algorithms. Diploma Thesis, RWTH Aachen (2008)
15. Reif, J.H.: The Complexity of Two-player Games of Incomplete Information. JCSS 29, 274–301 (1984)
16. Seymour, P.D., Thomas, R.: Graph searching and a min-max theorem for tree-width. J. Comb. Theory Ser. B 58(1), 22–33 (1993)
17. Thomas, W.: On the Synthesis of Strategies in Infinite Games. In: Mayr, E.W., Puech, C. (eds.) STACS 1995. LNCS, vol. 900, pp. 1–13. Springer, Heidelberg (1995)

Revisiting Ackermann-Hardness for Lossy Counter Machines and Reset Petri Nets*

Philippe Schnoebelen

LSV, ENS Cachan, CNRS
61, av. Pdt. Wilson, F-94230 Cachan, France
www.lsv.ens-cachan.fr/ phs

Abstract. We prove that coverability and termination are not primitive-recursive for lossy counter machines and for Reset Petri nets.

1 Introduction

Lossy counter machines [16,19] and Reset Petri nets [8] are two computational models that can be seen as weakened versions of Minsky counter machines. This weakness explains why some problems (e.g., termination) are decidable for these two models, while they are undecidable for the Turing-powerful Minsky machines.

While these positive results have been used in the literature, there also exists a negative side that has had much more impact. Indeed, we showed in [18] that decidable verification problems for lossy channel systems are Ackermann-hard and hence cannot be answered in primitive-recursive time or space. We also claimed that the construction used for lossy channels could be adapted for lossy counters and Reset Petri nets.

Hardness Theorem (in the Introduction of [18]). *Reachability, termination and coverability for lossy counter machines are Ackermann-hard.*
Termination and coverability for Reset Petri nets are Ackermann-hard.

These hardness results turned out to be relevant in several other areas. Using lossy counter machines, hardness results relying on the first half of the Hardness Theorem have been derived for a variety of logics and automata dealing with data words or data trees [6,7,14,12,20]. Ackermann-hardness has also been shown by reductions from Reset and Transfer nets, relying on the second half of the Hardness Theorem. Examples can be found in, e.g., [1,13]. We refer to [3,4] and the references therein for hardness inherited from lossy channel systems.

Our contribution. In this paper we prove the Hardness Theorem with a simplified construction. Compared to [18], we introduce three main simplifications:
1. We use counter machines and not channel systems, which is more direct since the crux of the construction is the computation of numerical functions.
2. We use a tail-recursive presentation of the F_m functions from the Fast-Growing Hierarchy. Thus we do not build our counter machines in nested stages like in [18]. As a

* Work supported by the Agence Nationale de la Recherche, grant ANR-06-SETIN-001.

P. Hliněný and A. Kučera (Eds.): MFCS 2010, LNCS 6281, pp. 616–628, 2010.

consequence, the correctness of the numerical computations is obvious and we obtain a clearer view of how many counters are really used.
3. We do not define, nor compute, inverses of the F_m functions as done in [18]. Instead, the tail-recursive definition is a simple rewrite loop that can easily be run backwards.

In addition, we strove for extra simplicity. E.g., we use counter machines extended with simple primitives that make computing Ackermann's function less cumbersome.

There are several reasons for providing a new proof of an old result. First, the results are important and influential as demonstrated by the number and the variety of applications we listed above: they definitely deserve being revisited, polished, advertised, etc. Second, our original proof has already been adapted to yet other computational models (e.g., in [15]) and a simplified proof will probably be easier to adapt to further models. Finally, we note that the main contents of [18] is now obsolete since Ackermann-hardness is not optimal for lossy channel systems [3]. However, for lossy counter machines and Reset nets, the Hardness Theorem is optimal (see [17,11]) and will not become obsolete.

Outline of the paper. Section 2 defines counter machines, both reliable and lossy. Section 3 builds counter machines that compute Ackermann's function. Section 4 puts Minsky machines *on a budget*, a gadget that is essential in Section 5 where the main reduction is given and the hardness of reachability and coverability for lossy counter machines is proved. We then show how to deal with reset nets in Section 6 and how to prove hardness of termination in Section 7. Some proofs have been omitted for lack of space: they can be found in the full version of this paper.

2 Counter Machines, Reliable and Lossy

Counter machines are a model of computation where a finite-state control acts upon a finite number of *counters*, i.e., storage locations that hold a natural number. The computation steps are usually restricted to simple tests and updates. For Minsky machines, the tests are zero-tests and the updates are increments and decrements.

For our purposes, it will be convenient to use a slightly extended model that allows more concise constructions, and that will let us handle Reset nets smoothly.

2.1 Extended Counter Machines and Minsky Machines

Formally, an *extended counter machine with n counters*, often just called a "counter machine" (a CM), is a tuple $M = (Loc, C, \Delta)$ where $Loc = \{\ell_1, \ldots, \ell_p\}$ is a finite set of *locations*, $C = \{\mathtt{c}_1, \ldots, \mathtt{c}_n\}$ is a finite set of *counters*, and $\Delta \subseteq Loc \times OP(C) \times Loc$ is a finite set of transition rules. The transition rules are depicted as directed edges (see Fig. 1, 2, and 3 below) between control locations labeled with an instruction $op \in OP(C)$ that is either a *guard* (a condition on the current contents of the counters for the rule to be firable), or an *update* (a method that modifies the contents of the counters), or both.

For CM's, the instruction set $OP(C)$ is given by the following abstract grammar:

$$
\begin{array}{llll}
OP(C) \ni op ::= \texttt{c=0?} & \text{/* zero test */} & \mid \texttt{c:=0} & \text{/* reset */} \\
\quad \mid \texttt{c>0? c--} & \text{/* decrement */} & \mid \texttt{c=c}'\texttt{?} & \text{/* equality test */} \\
\quad \mid \texttt{c++} & \text{/* increment */} & \mid \texttt{c:=c}' & \text{/* copy */}
\end{array}
$$

where $\texttt{c}, \texttt{c}'$ are any two counters in C. (We also allow a no_op instruction that does not test or modify the counters.)

A *Minsky machine* is a CM that only uses instructions among zero tests, decrements and increments (the first three types). Petri nets and Vector Addition Systems with States (VASS's) can be seen as counter machines that only use decrements and increments (i.e., Minsky machines without zero-tests).

2.2 Operational Semantics

The operational semantics of a CM $M = (Loc, C, \Delta)$ is given under the form of transitions between its configurations. Formally, a *configuration* (written $\sigma, \theta, \ldots$) of M is a tuple $(\ell, \boldsymbol{a})$ with $\ell \in Loc$ representing the "current" control location, and $\boldsymbol{a} \in \mathbb{N}^C$, a C-indexed vector of natural numbers representing the current contents of the counters. If C is some $\{\texttt{c}_1, \ldots, \texttt{c}_n\}$, we often write $(\ell, \boldsymbol{a})$ under the form $(\ell, a_1, \ldots, a_n)$. Also, we sometimes use labels in vectors of values to make them more readable, writing, e.g., $\boldsymbol{a} = (0, \ldots, 0, \texttt{c}_k : 1, 0, \ldots, 0)$.

Regarding the behavior induced by the rules from Δ, there is a *transition* (also called a *step*) $\sigma \xrightarrow{\delta}_{\text{std}} \sigma'$ if, and only if, σ is some $(\ell, a_1, \ldots, a_n)$, σ' is some $(\ell', a'_1, \ldots, a'_n)$, $\Delta \ni \delta = (\ell, op, \ell')$ and either:
– op is $\texttt{c}_k\texttt{=0?}$ (zero test): $a_k = 0$, and $a'_i = a_i$ for all $i = 1, \ldots, n$, or
– op is $\texttt{c}_k\texttt{>0? c}_k\texttt{--}$ (decrement): $a'_k = a_k - 1$, and $a'_i = a_i$ for all $i \neq k$, or
– op is $\texttt{c}_k\texttt{++}$ (increment): $a'_k = a_k + 1$, and $a'_i = a_i$ for all $i \neq k$, or
– op is $\texttt{c}_k\texttt{:=0}$ (reset): $a'_k = 0$, and $a'_i = a_i$ for all $i \neq k$, or
– op is $\texttt{c}_k\texttt{=c}_p\texttt{?}$ (equality test): $a_k = a_p$, and $a'_i = a_i$ for all $i = 1, \ldots, n$, or
– op is $\texttt{c}_k\texttt{:=c}_p$ (copy): $a'_k = a_p$, and $a'_i = a_i$ for all $i \neq k$.
(The steps carry a "std" subscript to emphasize that we are considering the usual, standard, operational semantics of counter machines, where the behavior is *reliable*.)

As usual, we write $\sigma \xrightarrow{\Delta}_{\text{std}} \sigma'$, or just $\sigma \rightarrow_{\text{std}} \sigma'$, when $\sigma \xrightarrow{\delta}_{\text{std}} \sigma'$ for some $\delta \in \Delta$. Chains $\sigma_0 \rightarrow_{\text{std}} \sigma_1 \rightarrow_{\text{std}} \cdots \rightarrow_{\text{std}} \sigma_m$ of consecutive steps, also called *runs*, are denoted $\sigma_0 \xrightarrow{*}_{\text{std}} \sigma_m$, and also $\sigma_0 \xrightarrow{+}_{\text{std}} \sigma_m$ when $m > 0$. Steps may also be written more precisely under the form $M \vdash \sigma \rightarrow_{\text{std}} \sigma'$ when several counter machines are at hand and we want to be explicit about where the steps take place.

For a vector $\boldsymbol{a} = (a_1, \ldots, a_n)$, or a configuration $\sigma = (\ell, \boldsymbol{a})$, we let $|\boldsymbol{a}| = |\sigma| = \sum_{i=1}^{n} a_i$ denote its *size*. For $N \in \mathbb{N}$, we say that a run $\sigma_0 \rightarrow \sigma_1 \rightarrow \cdots \rightarrow \sigma_m$ is N-bounded if $|\sigma_i| \leq N$ for all $i = 0, \ldots, n$.

2.3 Lossy Counter Machines

Lossy counter machines are counter machines where the contents of the counters can decrease non-deterministically (the machine can "leak", or "lose data").

Technically, it is more convenient to see lossy machines as counter machines with a different operational semantics (and not as a special class of machines): thus it is possible to use simultaneously the two semantics and relate them.

Formally, this is defined via the introduction of a partial ordering between the configurations of M:

$$(\ell,a_1,...,a_n) \leq (\ell',a'_1,...,a'_n) \stackrel{\text{def}}{\Leftrightarrow} \ell = \ell' \wedge a_1 \leq a'_1 \wedge \cdots \wedge a_n \leq a'_n.$$

$\sigma \leq \sigma'$ can be read as "σ is σ' after some losses (possibly none)".

Now "lossy" steps, denoted $M \vdash \sigma \xrightarrow{\delta}_{\text{lossy}} \sigma'$, are given by the following definition:

$$\sigma \xrightarrow{\delta}_{\text{lossy}} \sigma' \stackrel{\text{def}}{\Leftrightarrow} \exists\theta,\theta' : (\sigma \geq \theta \wedge \theta \xrightarrow{\delta}_{\text{std}} \theta' \wedge \theta' \geq \sigma'). \qquad (*)$$

Note that reliable steps are a special case of lossy steps:

$$M \vdash \sigma \rightarrow_{\text{std}} \sigma' \text{ implies } M \vdash \sigma \rightarrow_{\text{lossy}} \sigma'. \qquad (\dagger)$$

2.4 Behavioral Problems on Counter Machines

We consider the following decision problems:

Reachability: given a CM M and two configurations σ_{ini} and σ_{goal}, is there a run $M \vdash \sigma_{\text{ini}} \rightarrow^* \sigma_{\text{goal}}$?

Coverability: given a CM M and two configurations σ_{ini} and σ_{goal}, is there a run $M \vdash \sigma_{\text{ini}} \rightarrow^* \sigma$ for some configuration $\sigma \geq \sigma_{\text{goal}}$ that covers σ_{goal}?

(Non-)Termination: given a CM M and a configuration σ_{ini}, is there an infinite run $M \vdash \sigma_{\text{ini}} \rightarrow \sigma_1 \rightarrow \cdots \rightarrow \sigma_n \rightarrow \cdots$?

These problems are parameterized by the class of counter machines we consider and, more importantly, by the operational semantics that is assumed. Recall that reachability and termination are decidable for lossy counter machines, i.e., counter machines assuming lossy steps [16,19]. Observe that, for lossy machines, reachability and coverability coincide (except for runs of length 0). For the standard semantics, the same problems are undecidable for Minsky machines but become decidable for VASS's and, except for reachability, for Reset nets (see Section 6).

3 The Fast-Growing Hierarchy

The *Fast-Growing Hierarchy* [10] turns the class of all primitive-recursive functions into a strict cumulative hierarchy built from a sequence $(F_k)_{k=0,1,2,...}$ of number-theoretic functions. The functions $F_k : \mathbb{N} \rightarrow \mathbb{N}$ are defined by induction over $k \in \mathbb{N}$:

$$F_0(n) \stackrel{\text{def}}{=} n+1, \qquad F_{k+1}(n) \stackrel{\text{def}}{=} F_k^{n+1}(n) = \overbrace{F_k(F_k(\ldots F_k}^{n+1 \text{ times}}(n)\ldots)). \qquad \text{(D)}$$

This induces $F_1(n) = 2n-1$ and $F_2(n) = (n+1)2^{n+1}-1$, hence F_2 is not polynomial. Writing down an expression for $F_3(n)$ needs a tower of n exponents and F_3 is non-elementary. Note that, for all k and n, $F_k(n+1) > F_k(n)$ and that F_{k+1} dominates F_k.

Each F_k is primitive-recursive. A classic result is that every primitive-recursive function $f : \mathbb{N} \to \mathbb{N}$ is eventually dominated by some F_k.

It is possible to define a variant of Ackermann's function by a diagonalisation process: $Ack(n) \stackrel{\text{def}}{=} F_n(n)$. The Ack function is recursive but it eventually dominates any F_k, so it is not primitive-recursive.

A tail-recursive definition. The functions $(F_k)_{k\in\mathbb{N}}$ can be defined in a convenient tail-recursive way via the introduction of a generalized, so-called "*vectorial*", function F with two arguments. Formally, for a vector $\boldsymbol{a} = (a_m,\ldots,a_0) \in \mathbb{N}^{m+1}$, we define

$$F(\boldsymbol{a};n) = F(a_m,\ldots,a_0;n) \stackrel{\text{def}}{=} F_m^{a_m}(\ldots F_1^{a_1}(F_0^{a_0}(n))). \tag{V}$$

Hence $Ack(m)$ is $F(1,\mathbf{0};m)$, i.e., $F(1,0,\ldots,0;n)$ with m zeroes, and (D) can be reformulated in vectorial form:

$$F(\mathbf{0};n) = F(0,\ldots,0;n) = n, \tag{D$_0$}$$

$$F(a_m,\ldots,a_0+1;n) = F(a_m,\ldots,a_0;n+1), \tag{D$_1$}$$

$$F(a_m,\ldots,a_k+1,\overbrace{0,\ldots,0}^{k>0\text{ zeroes}};n) = F(a_m,\ldots,a_k,n+1,\overbrace{0,\ldots,0}^{k-1\text{ zeroes}};n). \tag{D$_2$}$$

Fact 3.1 (Monotonicity). *If $\boldsymbol{a} \le \boldsymbol{a}'$ and $n \le n'$ then $F(\boldsymbol{a};n) \le F(\boldsymbol{a}';n')$.*

Reading (D$_{0-2}$) as left-to-right rewrite rules turns them into a functional program for evaluating F: Write $\langle \boldsymbol{a};n\rangle \xrightarrow{D} \langle \boldsymbol{a}';n'\rangle$ when (D$_1$) or (D$_2$) transforms the term $F(\boldsymbol{a};n)$ into $F(\boldsymbol{a}';n')$. Clearly, $\langle \boldsymbol{a};n\rangle \xrightarrow{D} \langle \boldsymbol{a}';n'\rangle$ implies $F(\boldsymbol{a};n) = F(\boldsymbol{a}';n')$.

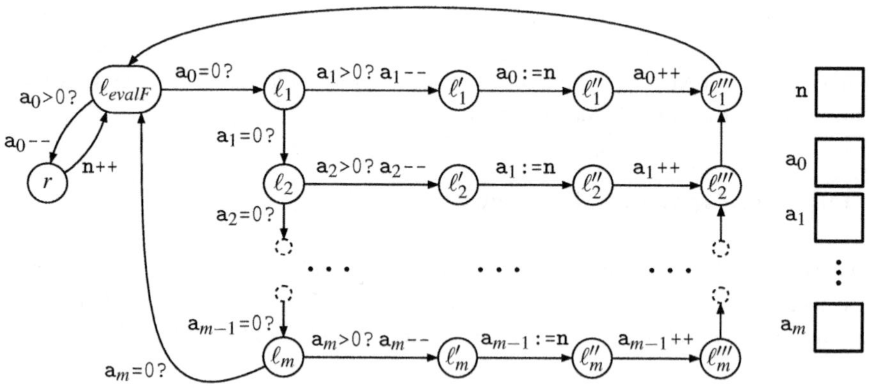

Fig. 1. $M_{evalF}(m)$, a counter machine evaluating F vectorially on $\mathbb{N}^{m+1}$

Now, $\xrightarrow{D}$ terminates since $\langle \boldsymbol{a};n\rangle \xrightarrow{D} \langle \boldsymbol{a}';n'\rangle$ implies $\boldsymbol{a} >_{\text{lexico}} \boldsymbol{a}'$ (recall that the lexicographical ordering is a linear extension of $\leq$, hence a well-ordering of $\mathbb{N}^{m+1}$). Furthermore, if $\boldsymbol{a} \neq \mathbf{0}$, one of the rules among (D_1) and (D_2) can be applied to $\langle \boldsymbol{a};n\rangle$. Hence for all $\boldsymbol{a}$ and n there exists some n' such that $\langle \boldsymbol{a};n\rangle \xrightarrow{D}{}^{*} \langle \mathbf{0};n'\rangle$, and then $n' = F(\boldsymbol{a};n)$ since $F(\boldsymbol{a};n)$ and $F(\mathbf{0};n')$, i.e., n', must coincide. (The reverse relation $\xrightarrow{D}{}^{-1}$ terminates too since, in a step $\langle \boldsymbol{a}';n'\rangle \xrightarrow{D}{}^{-1} \langle \boldsymbol{a};n\rangle$, either n' is decreased, or it stays constant and the number of zeroes in $\boldsymbol{a}'$ is increased.)

A counter machine evaluating F vectorially. Being tail-recursive, the vectorial F can be evaluated via a simple while-loop that implements the $\xrightarrow{D}$ rewriting. Fix a level $m \in \mathbb{N}$: we need $m+2$ counters, one for the n argument, and $m+1$ for the $\boldsymbol{a} \in \mathbb{N}^{m+1}$ argument.

We define a counter machine $M_{evalF}(m) = (Loc_{evalF}, C, \Delta_{evalF})$, or M_{evalF} for short, with $C = \{\mathtt{a}_0, \mathtt{a}_1, ..., \mathtt{a}_m, \mathtt{n}\}$. Its rules are defined pictorially in Fig. 1: they implement $\xrightarrow{D}$ as a loop around a central location ℓ_{evalF}, as captured by the following lemma.

Lemma 3.2 (Behavior of M_{evalF}). *For all $\boldsymbol{a}, \boldsymbol{a}' \in \mathbb{N}^{m+1}$ and $n, n' \in \mathbb{N}$:*
a. If $\langle \boldsymbol{a};n\rangle \xrightarrow{D} \langle \boldsymbol{a}';n'\rangle$ then $M_{evalF} \vdash (\ell_{evalF}, \boldsymbol{a}, n) \rightarrow^{*}_{std} (\ell_{evalF}, \boldsymbol{a}', n')$.
b. If $M_{evalF} \vdash (\ell_{evalF}, \boldsymbol{a}, n) \rightarrow^{*}_{std} (\ell_{evalF}, \boldsymbol{a}', n')$ then $F(\boldsymbol{a};n) = F(\boldsymbol{a}';n')$.
c. If $M_{evalF} \vdash (\ell_{evalF}, \boldsymbol{a}, n) \rightarrow^{*}_{lossy} (\ell_{evalF}, \boldsymbol{a}', n')$ then $F(\boldsymbol{a};n) \geq F(\boldsymbol{a}';n')$.

A counter machine inverting F. The rules (D_0–D_2) can also be used from right to left. Used this way, they *invert F*. This is implemented by another counter machine, $M_{backF}(m) = (Loc_{backF}, C, \Delta_{backF})$, or M_{backF} for short, defined pictorially in Fig. 2.

M_{backF} implements $\xrightarrow{D}{}^{-1}$ as a loop around a central location ℓ_{backF}, as captured by Lemma 3.3. Note that M_{backF} may deadlock if it makes the wrong guess as whether $\mathtt{a}_i$ contains $n+1$, but this is not a problem with the construction.

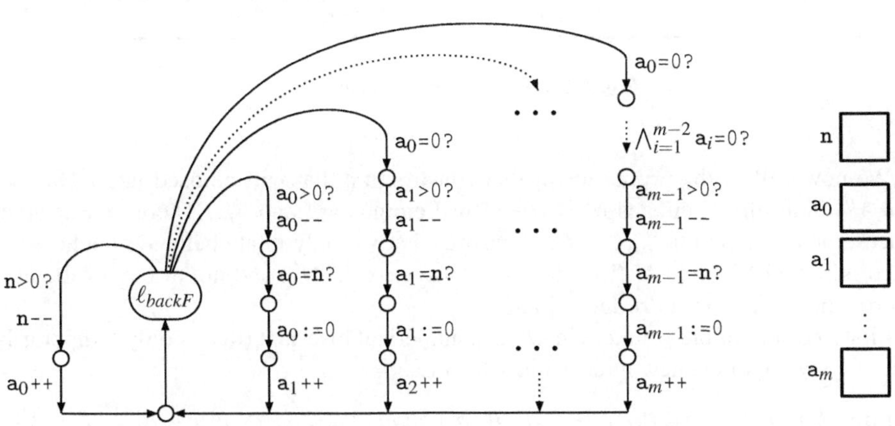

Fig. 2. $M_{backF}(m)$, a counter machine inverting F vectorially on $\mathbb{N}^{m+1}$

Lemma 3.3 (Behavior of M_{backF}). *For all $\boldsymbol{a},\boldsymbol{a}' \in \mathbb{N}^{m+1}$ and $n,n' \in \mathbb{N}$:*
a. If $\langle \boldsymbol{a};n\rangle \xrightarrow{D} \langle \boldsymbol{a}';n'\rangle$ then $M_{backF} \vdash (\ell_{backF},\boldsymbol{a}',n') \rightarrow^*_{std} (\ell_{backF},\boldsymbol{a},n)$.
b. If $M_{backF} \vdash (\ell_{backF},\boldsymbol{a},n) \rightarrow^*_{std} (\ell_{backF},\boldsymbol{a}',n')$ then $F(\boldsymbol{a};n) = F(\boldsymbol{a}';n')$.
c. If $M_{backF} \vdash (\ell_{backF},\boldsymbol{a},n) \rightarrow^*_{lossy} (\ell_{backF},\boldsymbol{a}',n')$ then $F(\boldsymbol{a};n) \geq F(\boldsymbol{a}';n')$.

4 Minsky Machines on a Budget

With a Minsky machine $M = (Loc, C, \Delta)$ we associate a Minsky machine $M^{\text{on\_budget}} = (Loc_b, C_b, \Delta_b)$, called M^b for short. (Note that we are only considering Minsky machines here, and not the extended counter machines from earlier sections.)

$M^{\text{on\_budget}}$ is obtained by adding to M an extra "budget" counter B and by adapting the rules of Δ so that any increment (resp. decrement) in the original counters is balanced by a corresponding decrement (resp. increment) on the new counter B, so that the sum of the counters remains constant. This is a classic idea in Petri nets. The construction is described on a schematic example (Fig. 3) that is clearer than a formal definition. Observe that extra intermediary locations (in gray) are used, and that a rule in M that increments some c_i will be forbidden in M^b when the budget is exhausted.

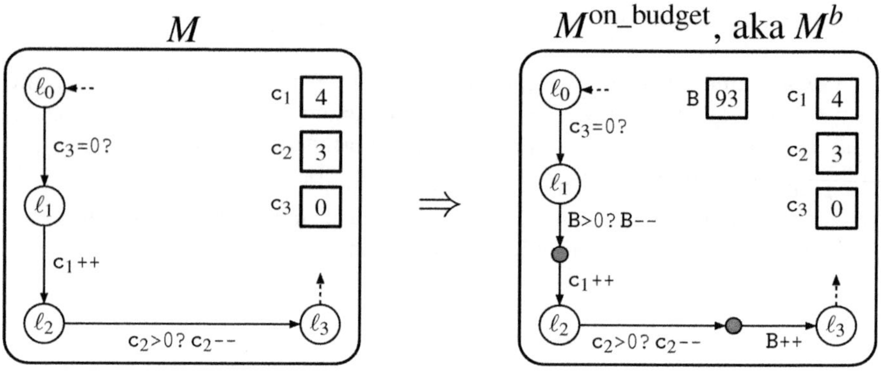

Fig. 3. From M to M^b (schematically)

We now collect the properties of this construction that will be used later. The fact that M^b faithfully simulates M is stated in Lemmas 4.2 and 4.3. There and at other places, the restriction to "$\ell, \ell' \in Loc$" ensures that we only relate behavior anchored at the original locations in M (locations that also exist in M^b) and not at one of the new intermediary locations introduced in M^b.

First, the sum of the counters in M^b is a numerical invariant (that is only temporarily disrupted while in the new intermediary locations).

Lemma 4.1. *If $M^b \vdash (\ell, B, \boldsymbol{a}) \rightarrow^*_{std} (\ell', B', \boldsymbol{a}')$ and $\ell, \ell' \in Loc$, then $B + |\boldsymbol{a}| = B' + |\boldsymbol{a}'|$.*

Observe that M^b can only do what M would do:

Lemma 4.2. *If $M^b \vdash (\ell, B, \boldsymbol{a}) \to^*_{std} (\ell', B', \boldsymbol{a}')$ and $\ell, \ell' \in Loc$ then $M \vdash (\ell, \boldsymbol{a}) \to^*_{std} (\ell', \boldsymbol{a}')$.*

Reciprocally, everything done by M can be mirrored by M^b *provided that a large enough budget is allowed*. More precisely:

Lemma 4.3. *If $M \vdash (\ell, \boldsymbol{a}) \to^*_{std} (\ell', \boldsymbol{a}')$ is an N-bounded run of M, then M^b has an N-bounded run $M^b \vdash (\ell, B, \boldsymbol{a}) \to^*_{std} (\ell', B', \boldsymbol{a}')$ for $B \stackrel{def}{=} N - |\boldsymbol{a}|$ and $B' \stackrel{def}{=} N - |\boldsymbol{a}'|$.*

Now, the point of the construction is that M^b can distinguish between lossy and non-lossy runs in ways that M cannot. More precisely:

Lemma 4.4. *Let $M^b \vdash (\ell, B, \boldsymbol{a}) \to^*_{lossy} (\ell', B', \boldsymbol{a}')$ with $\ell, \ell' \in Loc$. Then $M^b \vdash (\ell, B, \boldsymbol{a}) \to^*_{std} (\ell', B', \boldsymbol{a}')$ if, and only if, $B + |\boldsymbol{a}| = B' + |\boldsymbol{a}'|$.*

Proof (Idea). The "($\Leftarrow$)" direction is an immediate consequence of (†).
For the "($\Rightarrow$)" direction, we consider the hypothesized run $M^b \vdash (\ell, B, \boldsymbol{a}) = \sigma_0 \to_{\text{lossy}} \sigma_1 \to_{\text{lossy}} \cdots \to_{\text{lossy}} \sigma_n = (\ell', B', \boldsymbol{a}')$. Coming back to definition ($*$), these lossy steps require, for $i = 1, \ldots, n$, some reliable steps $\theta_{i-1} \to_{\text{std}} \theta'_i$ with $\sigma_{i-1} \geq \theta_{i-1}$ and $\theta'_i \geq \sigma_i$, and hence $|\theta'_i| \geq |\theta_i|$ for $i < n$. Combining with $|\theta_{i-1}| = |\theta'_i|$ (by Lemma 4.1), and $|\sigma_0| = |\sigma_n|$ (from the assumption that $B + |\boldsymbol{a}| = B' + |\boldsymbol{a}'|$), proves that all these configurations have same size. Hence $\theta'_i = \sigma_i = \theta_i$ and the lossy steps are also reliable steps. □

Corollary 4.5. *Assume $M^b \vdash (\ell, B, \boldsymbol{0}) \to^*_{lossy} (\ell', B', \boldsymbol{a})$ with $\ell, \ell' \in Loc$. Then:*
1. $B \geq B' + |\boldsymbol{a}|$, and
2. $M \vdash (\ell, \boldsymbol{0}) \to^*_{std} (\ell', \boldsymbol{a})$ if, and only if, $B = B' + |\boldsymbol{a}|$. Furthermore, this reliable run of M is B-bounded.

5 Ackermann-Hardness for Lossy Counter Machines

We now collect the ingredients that have been developed in the previous sections.

Let M be a Minsky machine with two fixed "initial" and "final" locations ℓ_{ini} and ℓ_{fin}. With M and a level $m \in \mathbb{N}$ we associate a counter machine $M(m)$ obtained by stringing together $M_{evalF}(m)$, $M^{\text{on\_budget}}$, and $M_{backF}(m)$ and fusing the extra budget counter $\mathtt{B}$ from $M^{\text{on\_budget}}$ with the accumulator $\mathtt{n}$ of $M_{evalF}(m)$ and $M_{backF}(m)$ (these two share their counters). The construction is depicted in Fig. 4.

Theorem 5.1. *The following are equivalent:*
1. $M(m)$ has a lossy run $(\ell_{evalF}, \mathtt{a}_m : 1, \boldsymbol{0}, \mathtt{n} : m, \boldsymbol{0}) \to^*_{lossy} \theta$ for some $\theta \geq (\ell_{backF}, 1, \boldsymbol{0}, m, \boldsymbol{0})$.
2. M^{on_budget} has a lossy run $(\ell_{\text{ini}}, \mathtt{B} : Ack(m), \boldsymbol{0}) \to^*_{lossy} (\ell_{\text{fin}}, Ack(m), \boldsymbol{0})$.
3. M^{on_budget} has a reliable run $(\ell_{\text{ini}}, Ack(m), \boldsymbol{0}) \to^*_{std} (\ell_{\text{fin}}, Ack(m), \boldsymbol{0})$.
4. $M(m)$ has a reliable run $(\ell_{evalF}, 1, \boldsymbol{0}, m, \boldsymbol{0}) \to^*_{std} (\ell_{backF}, 1, \boldsymbol{0}, m, \boldsymbol{0})$.
5. M has a reliable run $(\ell_{\text{ini}}, \boldsymbol{0}) \to^*_{std} (\ell_{\text{fin}}, \boldsymbol{0})$ that is $Ack(m)$-bounded.

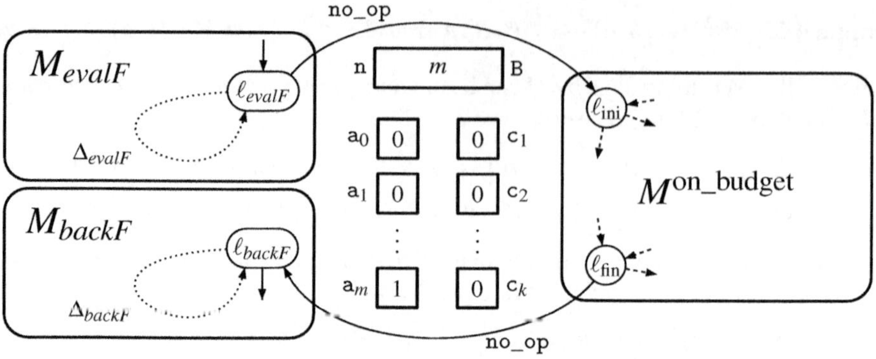

Fig. 4. Constructing $M(m)$ from M^b, M_{evalF} and M_{backF}

Proof (Sketch).
— For "*1* ⇒ *2*", and because coverability implies reachability by (∗), we may assume w.l.o.g. that $M(m)$ has a run $(\ell_{evalF}, 1, \mathbf{0}, m, \mathbf{0}) \to^*_{\text{lossy}} (\ell_{backF}, 1, \mathbf{0}, m, \mathbf{0})$. This run must go through $M^{\text{on\_budget}}$ and be in three parts of the following form:

$$(\ell_{evalF}, 1, \mathbf{0}, m, \mathbf{0}) \xrightarrow{\Delta_{evalF}}{}^*_{\text{lossy}} (\ell_{evalF}, \boldsymbol{a}, \mathtt{n}: x, \mathbf{0}) \qquad \text{(starts in } M_{evalF})$$

$$\to_{\text{lossy}} (\ell_{\text{ini}}, \ldots, B, \mathbf{0}) \xrightarrow{\Delta_b}{}^*_{\text{lossy}} (\ell_{\text{fin}}, \ldots, B', \boldsymbol{c}) \qquad \text{(goes through } M^{\text{on\_budget}})$$

$$\to_{\text{lossy}} (\ell_{backF}, \boldsymbol{a}', x', \ldots) \xrightarrow{\Delta_{backF}}{}^*_{\text{lossy}} (\ell_{backF}, 1, \mathbf{0}, m, \mathbf{0}). \qquad \text{(ends in } M_{backF})$$

The first part yields $F(1,\mathbf{0};m) \geq F(\boldsymbol{a};x)$ (by Lemma 3.2.*c*), the third part $F(\boldsymbol{a}';x') \geq F(1,\mathbf{0};m)$ (by Lemma 3.3.*c*), and the middle part $B \geq B' + |\boldsymbol{c}|$ (by Coro. 4.5.*1*). Lossiness further implies $x \geq B$, $B' \geq x'$ and $\boldsymbol{a} \geq \boldsymbol{a}'$. Now, the only way to reconcile $F(\boldsymbol{a};x) \leq F(1,\mathbf{0};m) = Ack(m) \leq F(\boldsymbol{a}';x')$, $\boldsymbol{a}' \leq \boldsymbol{a}$, $x' \leq x$, and the monotonicity of F (Fact 3.1) is by concluding $x = B = B' = x' = Ack(m)$ and $\boldsymbol{c} = \mathbf{0}$. Then the middle part of the run witnesses $M^{\text{on\_budget}} \vdash (\ell_{\text{ini}}, Ack(m), \mathbf{0}) \to^*_{\text{lossy}} (\ell_{\text{fin}}, Ack(m), \mathbf{0})$.
— "*2* ⇒ *5*" is Coro. 4.5.2.
— "*5* ⇒ *3*" is given by Lemma 4.3.
— "*3* ⇒ *4*" is obtained by stringing together reliable runs of the components, relying on Lemmas 3.2.*a* and 3.3.*a* for the reliable runs of M_{evalF} and M_{backF}.
— Finally "*3* ⇒ *2*" and "*4* ⇒ *1*" are immediate from (†). □

With Theorem 5.1, we have a proof of the Hardness Theorem for reachability and coverability in lossy counter machines: Recall that, for a Minsky machine M, the existence of a run between two given configurations is undecidable, and the existence of a run bounded by $Ack(m)$ is decidable but not primitive-recursive when m is part of the input. Therefore, Theorem 5.1, and in particular the equivalence between its points 1 and 5, states that our construction reduces a nonprimitive-recursive problem to the reachability problem for lossy counter machines.

6 Handling Reset Petri Nets

Reset nets [2,5] are Petri nets extended with special reset arcs that empty a place when a transition is fired. They can equally be seen as special counter machines, called "reset machines", where actions are restricted to decrements, increments, and resets. This is the view we adopt in this paper. *Note that zero-tests are not allowed in reset machines.*

It is known that termination and coverability are decidable for reset machines while other properties like reachability of a given configuration, finiteness of the reachability set, or recurrent reachability, are undecidable [8,9].

Our purpose is to prove the Ackermann-hardness of termination and coverability for reset machines. We start with coverability and refer to section 7 for termination.

6.1 $R(M)$: Replacing Zero-Tests with Resets

For a counter machine M, we let $R(M)$ be the counter machine obtained by replacing every zero-test instruction `c=0?` with a corresponding reset `c:=0`. Note that $R(M)$ is a reset machine when M is a Minsky machine.

Clearly, the behavior of M and $R(M)$ are related in the following way:

Lemma 6.1.
1. $M \vdash \sigma \rightarrow_{std} \sigma'$ *implies* $R(M) \vdash \sigma \rightarrow_{std} \sigma'$.
2. $R(M) \vdash \sigma \rightarrow_{std} \sigma'$ *implies* $M \vdash \sigma \rightarrow_{lossy} \sigma'$.

In other words, the reliable behavior of $R(M)$ contains the reliable behavior of M and is contained in the lossy behavior of M.

We now consider the counter machine $M(m)$ defined in Section 5 and build $R(M(m))$.

Theorem 6.2. *The following are equivalent:*
1. $R(M(m))$ *has a reliable run* $(\ell_{evalF}, \mathtt{a}_m : 1, \mathbf{0}, \mathtt{n} : m, \mathbf{0}) \rightarrow^*_{std} (\ell_{backF}, 1, \mathbf{0}, m, \mathbf{0})$.
2. $R(M(m))$ *has a reliable run* $(\ell_{evalF}, 1, \mathbf{0}, m, \mathbf{0}) \rightarrow^*_{std} \theta$ *for some* $\theta \geq (\ell_{backF}, 1, \mathbf{0}, m, \mathbf{0})$.
3. M *has a reliable run* $(\ell_{\text{ini}}, \mathbf{0}) \rightarrow^*_{std} (\ell_{\text{fin}}, \mathbf{0})$ *that is* $Ack(m)$*-bounded.*

Proof. *1* $\Rightarrow$ *3*: The reliable run in $R(M(m))$ gives a lossy run in $M(m)$ (Lemma 6.1.*2*), and we conclude using "*1*$\Rightarrow$*5*" in Theorem 5.1.
3 $\Rightarrow$ *2*: We obtain a reliable run in $M(m)$ ("*5*$\Rightarrow$*4*" in Theorem 5.1) which gives a reliable run in $R(M(m))$ (Lemma 6.1.*1*) which in particular witnesses coverability.
2 $\Rightarrow$ *1*: The covering run in $R(M(m))$ gives a lossy covering run in $M(m)$ (Lemma 6.1.*2*), hence also a lossy run in $M(m)$ that reaches exactly $(\ell_{backF}, 1, \mathbf{0}, m, \mathbf{0})$ (e.g., by losing whatever is required at the last step). From there we obtain a reliable run in $M(m)$ ("*1*$\Rightarrow$*4*" in Theorem 5.1) and then a reliable run in $R(M(m))$ (Lemma 6.1.*1*). □

We have thus reduced an Ackermann-hard problem (point *3* above) to a coverability question (point *2* above).

This almost proves the Hardness Theorem for coverability in Reset nets, except for one small ingredient: $R(M(m))$ is not a reset machine properly because $M(m)$ is an extended counter machine, not a Minsky machine. I.e., we proved hardness for "extended" reset machines. Before tackling this issue, we want to point out that something as easy as the proof of Theorem 6.2 will prove Ackermann-hardness of reset machines by reusing the hardness of lossy counter machines.

In order to conclude the proof of the Hardness Theorem for Reset nets, we only need to provide versions of M_{evalF} and M_{backF} in the form of Minsky machines (M and M^b already are Minsky machines) and plug these in Figure 4 and Theorem 5.1. This is an easy and unsurprising exercise that we only tackle in the full version of this paper.

7 Hardness for Termination

We can prove hardness for termination by a minor adaptation of the proof for coverability. This adaptation, sketched in Fig. 5, is similar to the one used in [18]. It applies to both lossy counter machines and reset machines.

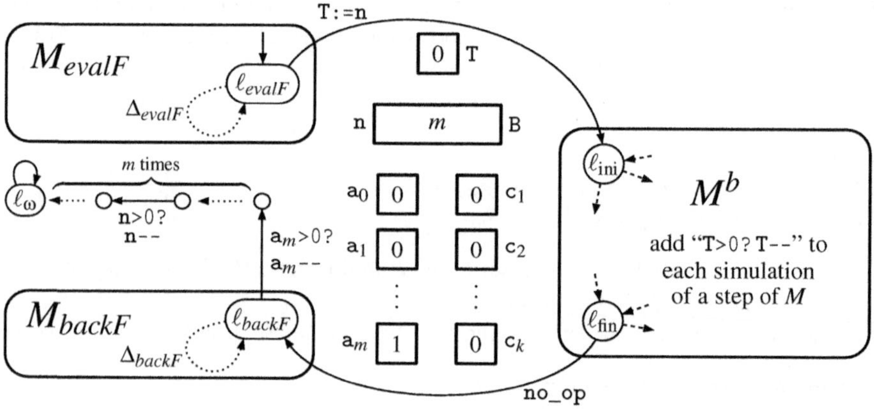

Fig. 5. Hardness for termination: A new version of $M(m)$

Basically, M^b now uses two copies of the initial budget. One copy in B works as before: its purpose is to ensure that *losses will be detected by a budget imbalance* as in Lemma 4.4. The other copy, in a new counter T, is a time limit that is initialized with n and is decremented with every simulated step of M: its purpose is to ensure that the new M^b always terminates. Since M_{evalF} and M_{backF} cannot run forever (because $\xrightarrow{D}$ and $\xrightarrow{D}^{-1}$ terminate, see Section 3), we now have a new $M(m)$ that always terminate when started in ℓ_{evalF} and that satisfies the following variant of Theorems 5.1 and 6.2:

Theorem 7.1. *The following are equivalent:*
1. $M(m)$ *has a lossy run* $(\ell_{evalF}, \mathtt{a}_m : 1, \mathbf{0}, \mathtt{n} : m, \mathbf{0}) \rightarrow^*_{lossy} \theta \geq (\ell_{backF}, 1, \mathbf{0}, m, \mathbf{0})$.
2. $R(M(m))$ *has a reliable run* $(\ell_{evalF}, 1, \mathbf{0}, \mathtt{n} : m, \mathbf{0}) \rightarrow^*_{lossy} \theta \geq (\ell_{backF}, 1, \mathbf{0}, m, \mathbf{0})$.
3. M *has a reliable run* $(\ell_{\text{ini}}, \mathbf{0}) \rightarrow^*_{std} (\ell_{\text{fin}}, \mathbf{0})$ *of length at most* $Ack(m)$.

Finally, we add a series of $m+1$ transitions that leave from ℓ_{backF}, and check that $\sigma_{\text{goal}} \stackrel{\text{def}}{=} (\ell_{backF}, 1, \mathbf{0}, m, \mathbf{0})$ is covered, i.e., that $\mathtt{a}_m$ contains at least 1 and n at least m.

If this succeeds, one reaches a new location ℓ_ω, *the only place where infinite looping is allowed unconditionally*. This yields a machine $M(m)$ that has an infinite lossy run if, and only if, it can reach a configuration that covers σ_{goal}, i.e., if, and only if, M has a reliable run of length at most $Ack(m)$, which is an Ackermann-hard problem.

8 Concluding Remarks

We proved Ackermann-hardness for lossy counter machines and, with very minor adaptations to the proof, for Reset Petri nets. These results are important in the field of algorithmic verification. Indeed, they have been abundantly cited in recent years even though they were only claimed in the introduction of [18]. The proof we present has several simplifications over the one that was given in [18] for channel systems instead of counter machines. We hope that these improvements will facilitate the wider dissemination of these results.

Acknowledgements. We thank Pierre Chambart and Sylvain Schmitz who greatly helped by proof-reading this paper at various stages.

References

1. Amadio, R., Meyssonnier, C.: On decidability of the control reachability problem in the asynchronous π-calculus. Nordic Journal of Computing 9(2), 70–101 (2002)
2. Araki, T., Kasami, T.: Some decision problems related to the reachability problem for Petri nets. Theoretical Computer Science 3(1), 85–104 (1977)
3. Chambart, P., Schnoebelen, P.: The ordinal recursive complexity of lossy channel systems. In: Proc. LICS 2008, pp. 205–216. IEEE Comp. Soc. Press, Los Alamitos (2008)
4. Chambart, P., Schnoebelen, P.: Pumping and counting on the Regular Post Embedding Problem. In: Proc. ICALP 2010. LNCS. Springer, Heidelberg (2010)
5. Ciardo, G.: Petri nets with marking-dependent arc cardinality: Properties and analysis. In: Valette, R. (ed.) ICATPN 1994. LNCS, vol. 815, pp. 179–198. Springer, Heidelberg (1994)
6. Demri, S.: Linear-time temporal logics with Presburger constraints: An overview. J. Applied Non-Classical Logics 16(3-4), 311–347 (2006)
7. Demri, S., Lazić, R.: LTL with the freeze quantifier and register automata. In: Proc. LICS 2006, pp. 17–26. IEEE Computer Society Press, Los Alamitos (2006)
8. Dufourd, C., Finkel, A., Schnoebelen, P.: Reset nets between decidability and undecidability. In: Larsen, K.G., Skyum, S., Winskel, G. (eds.) ICALP 1998. LNCS, vol. 1443, pp. 103–115. Springer, Heidelberg (1998)
9. Dufourd, C., Jančar, P., Schnoebelen, P.: Boundedness of Reset P/T nets. In: Wiedermann, J., Van Emde Boas, P., Nielsen, M. (eds.) ICALP 1999. LNCS, vol. 1644, pp. 301–310. Springer, Heidelberg (1999)
10. Fairtlough, M.V., Wainer, S.S.: Hierarchies of provably recursive functions. In: Handbook of Proof Theory. Studies in Logic, vol. 137, pp. 149–207. North-Holland, Amsterdam (1998)
11. Figueira, D., Figueira, S., Schmitz, S., Schnoebelen, P.: Ackermann and primitive-recursive upper bounds with Dickson's lemma. In: Preparation (2010)
12. Figueira, D., Segoufin, L.: Future-looking logics on data words and trees. In: Královič, R., Niwiński, D. (eds.) MFCS 2009. LNCS, vol. 5734, pp. 331–343. Springer, Heidelberg (2009)

13. Finkel, A., Raskin, J.-F., Samuelides, M., Van Begin, L.: Monotonic extensions of Petri nets: Forward and backward search revisited. In: Proc. INFINITY 2002. Electronic Notes in Theoretical Computer Science, vol. 68(6), pp. 121–144 (2003)
14. Jurdziński, M., Lazić, R.: Alternation-free modal μ-calculus for data trees. In: Proc. LICS 2007, pp. 131–140. IEEE Comp. Soc. Press, Los Alamitos (2007)
15. Jurdziński, T.: Leftist grammars are nonprimitive recursive. In: Aceto, L., Damgård, I., Goldberg, L.A., Halldórsson, M.M., Ingólfsdóttir, A., Walukiewicz, I. (eds.) ICALP 2008, Part II. LNCS, vol. 5126, pp. 51–62. Springer, Heidelberg (2008)
16. Mayr, R.: Undecidable problems in unreliable computations. Theoretical Computer Science 297(1-3), 337–354 (2003)
17. McAloon, K.: Petri nets and large finite sets. Theoretical Computer Science 32(1-2), 173–183 (1984)
18. Schnoebelen, P.: Verifying lossy channel systems has nonprimitive recursive complexity. Information Processing Letters 83(5), 251–261 (2002)
19. Schnoebelen, P.: Lossy counter machines decidability cheat sheet. In: RP 2010. LNCS, vol. 6227. Springer, Heidelberg (2010)
20. Tan, T.: On pebble automata for data languages with decidable emptiness problem. In: Královič, R., Niwiński, D. (eds.) MFCS 2009. LNCS, vol. 5734, pp. 712–723. Springer, Heidelberg (2009)

Enumeration of the Monomials of a Polynomial and Related Complexity Classes

Yann Strozecki

Université Paris 7 - Denis Diderot
strozecki@logique.jussieu.fr

Abstract. We study the problem of generating monomials of a polynomial in the context of enumeration complexity. We present two new algorithms for restricted classes of polynomials, which have a good delay between two generated monomials and the same global running time as the classical ones. Moreover they are simple to describe, use small evaluation points and one of them is parallelizable. We introduce **TotalPP**, **IncPP** and **DelayPP**, which are probabilistic counterparts of the most common classes for enumeration problems, hoping that randomization will be a tool as strong for enumeration as it is for decision. Our interpolation algorithms prove that a few interesting problems are in these classes like the enumeration of the spanning hypertrees of a 3-uniform hypergraph. Finally we give a method to interpolate degree 2 polynomials with an acceptable (incremental) delay. We also prove that finding a specified monomial in a degree 2 polynomial is hard unless RP = NP. It suggests that there is no algorithm with a delay as good (polynomial) as the one we achieve for multilinear polynomials.

1 Introduction

Enumeration, the task of generating all solutions of a given problem, is an interesting generalization of decision and counting. Since a problem typically has an exponential number of solutions, the way we study enumeration complexity is quite different from decision. In particular, the delay between two solutions and the time taken by an algorithm relative to the number of solutions seem to be the most considered complexity measures. In this paper, we revisit the famous problem of polynomial interpolation, i.e. finding the monomials of a polynomial from its values, with these measures in mind.

It has long been known that a finite number of evaluation points is enough to interpolate a polynomial and efficient procedures (both deterministic and probabilistic) have been studied by several authors [1,2,3]. The complexity depends mostly on the number of monomials of the polynomial and on an a priori bound on this number which may be exponential in the number of variables. The deterministic methods rely on prime numbers as evaluation points, with the drawback that they are very large. The probabilistic methods crucially use the Schwarz-Zippel lemma, which is also a tool in this article, and efficient solving of particular linear systems.

P. Hliněný and A. Kučera (Eds.): MFCS 2010, LNCS 6281, pp. 629–640, 2010.

As a consequence of a result about random efficient identity testing [4], Klivans and Spielman give an interpolation algorithm, which happens to have an incremental delay. In this vein, the present paper studies the problem of generating the monomials of a polynomial with the best possible delay. In particular, we consider natural classes of polynomials such as multilinear polynomials, for which we prove that interpolation can be done efficiently. Similar restrictions have been studied in works about identity testing for a quantum model [5] or for depth 3 circuits which thus define almost linear polynomials [6]. Moreover, a lot of interesting polynomials are multilinear like the Determinant, the Pfaffian, the Permanent, the elementary symmetric polynomials or anything defined by a syntactically multilinear arithmetic circuit.

In Sec. 4 we present an algorithm which works for polynomials such that no two of their monomials use the same set of variables. It is structured as in [4] but is simpler and has better delay, though polynomially related. In Sec. 5 we propose a second algorithm for multilinear polynomials. It has a polynomial delay in the number of variables, which makes it exponentially better than the previous one and is also easily parallelizable. In addition, both algorithms enjoy a global complexity as good as the algorithms of the literature, are deterministic for monotone polynomials, and use only small evaluation points making them suitable to work over finite fields.

We describe in Sec. 6 three complexity classes for enumeration, namely **TotalP**, **IncP**, **DelayP** which are now commonly used [7,8,9] to formalize what is an efficiently enumerable problem. We introduce probabilistic variants of these classes, which happen to characterize the enumeration complexity of the different interpolation algorithms. Their use on polynomials computable in polynomial time enable us to prove that well-known problems belong to these classes. Those problems already have better enumeration algorithms except the last, enumeration of the spanning hypertrees of a 3-uniform hypergraph, for which our method gives the first efficient enumeration algorithm.

In the last section, we explain how to combine the two presented algorithms to interpolate degree 2 polynomials with incremental delay. We then encode a restricted version of the Hamiltonian path problem in a polynomial given by the Matrix-Tree theorem (see [10]), to prove that the problem of finding a monomial in a degree 2 polynomial is hard. Thus, a polynomial delay interpolation algorithm for degree 2 polynomials similar to the one for degree 1 would solve the latter problem and it implies RP = NP. Finally, we compare our two algorithms with several classical ones and show that they are good with regard to parameters like number of calls to the black box or size of the evaluation points.

2 Enumeration Problems

In this section, we recall basic definitions about enumeration problems and complexity measures and we introduce the central problem of this article.

The computation model is a RAM machine as defined in [9] which has, in addition to the classical definition, an instruction $Write(A)$ which outputs the content of the register A. The result of a computation of a RAM machine is the sequence of integers which were in A when the instructions $Write(A)$ were executed. For simplicity we consider that these integers encode words, and that the input of the machine is also a word represented by suitable integers in the input registers. Let M be such a machine and x a word, we write $M(x)$ the result of the computation of M on x. The order in which the outputs are given does not matter, therefore $M(x)$ will denote the set of outputs as well as the sequence. We choose a RAM machine instead of a Turing machine since it may be useful to deal with an exponential amount of memory in polynomial time, see for instance the enumeration of the maximal independent sets of a graph [7].

Definition 1 (Enumeration Problem). *Let A be a polynomially balanced binary predicate, i.e. $A(x,y) \Rightarrow |y| \leq Q(|x|)$, for a polynomial Q. We write $A(x)$ for the set of y such that $A(x,y)$. We say that a RAM machine M solves the enumeration problem associated to A,* Enum·A *for short, if $M(x) = A(x)$ and there is no repetition of solutions in the computation.*

Let $T(x,i)$ be the time taken by a machine M to return i outputs from the instance x. Like for decision problems, we are interested by the total time taken by M. We are also interested by the delay between two solutions, that is to say $T(x,i+1) - T(x,i)$. M has an *incremental delay* when it is polynomial in $|x|$ and i, and M has a *polynomial delay* when it is polynomial in $|x|$ only.

A probabilistic RAM machine has a special instruction *rand* which writes in a specific register the integer 0 or 1 with equal probability. All outcomes of the instruction *rand* during a run of a RAM machine are independent.

Definition 2 (Probabilistic enumeration). *We say that the probabilistic RAM machine M solves* Enum·A *with probability p if $P[A(x) = M(x)] > p$ and there is no repetition of solutions in the computation.*

We adapt the model to the case of a computation with an oracle, by a special instruction which calls the oracle on a word contained in a specific register and then writes the answer in another register in unit time.

In this article, we interpret the problem of interpolating a polynomial given by a black box as an enumeration problem. It means that we generate the monomials of a polynomial given by the number of its variables and an oracle which allows to evaluate it on any point in unit time. The general problem is denoted by Enum·*Poly* but we solve it only for restricted classes of polynomials.

3 Finding One Monomial

In this section we introduce all the basic tools we need to build interpolation algorithms. One consider polynomials with n variables and rational coefficients. A sequence of n positive integers $\boldsymbol{e} = (e_1, \ldots, e_n)$ characterizes the monomial $\boldsymbol{X}^{\boldsymbol{e}} = X_1^{e_1} X_2^{e_2} \ldots X_n^{e_n}$. We call t the number of monomials of a polynomial P written $P(\boldsymbol{X}) = \sum_{1 \leq j \leq t} \lambda_j \boldsymbol{X}^{\boldsymbol{e}_j}$.

The degree of a monomial is the maximum of the degrees of its variables and the total degree is the sum of the degrees of its variables. Let d (respectively D) denote the degree (respectively the total degree) of the polynomial we consider, that is to say the maximum of its monomial's degree (respectively total degree). In Sec. 5 we assume that the polynomial is multilinear, i.e. $d = 1$ and D is thus bounded by n.

We assume that the maximum of the bitsize of the coefficients appearing in a polynomial is $O(n)$ to simplify the statement of some results, in the examples of Sec. 6 it is even $O(1)$. When analyzing the delay of an algorithm solving Enum·*Poly* we are interested in both the number of calls to the black box and the time spent between two generated monomials. We are also interested in the size of the integers used in the calls to the oracle, since in real cases the complexity of the evaluation depends on it.

The *support* of a monomial is the set of indices of variables which appears in the monomial. Let L be a set of indices of variables, for instance a support, then f_L is the homomorphism of $\mathbb{Q}[X_1, \ldots, X_n]$ defined by $\begin{cases} X_i \rightarrow X_i & \text{if } i \in L, \\ X_i \rightarrow 0 & \text{otherwise.} \end{cases}$

From now on, we denote $f_L(P)$ by P_L. It is the polynomial obtained by substituting 0 to every variable of index not in L, that is to say the sum of monomials of P which have their support in L. We call $\boldsymbol{X}^L$ the multilinear term of support L, which is the product of all X_i with i in L.

Lemma 1. *Let P be a polynomial without constant term and whose monomials have distinct supports and L a minimal set (for inclusion) of variables such that P_L is not identically zero. Then P_L is a single monomial of support L.*

The first problem we want to solve is to decide if a polynomial given by a black box is the zero polynomial, a problem called *Polynomial Identity Testing*. We are especially interested in the corresponding search problem, i.e. giving explicitly one term and its coefficient. Indeed, we show in Sec. 4 how to turn any algorithm solving this problem into an incremental interpolation algorithm.

It is easy to see [2] that a polynomial with t monomials has to be evaluated in t points to be sure that it is zero. If we do not have any a priori bound on t, then we must evaluate the polynomial on at least $(d+1)^n$ n-tuples of integers to determine it. As we are not satisfied with this exponential complexity, we introduce probabilistic algorithms, which nonetheless have a good and manageable bound on the error.

Lemma 2 (Schwarz-Zippel [11]). *Let P be a non zero polynomial with n variables of total degree D, if $x_1, \ldots, x_n$ are randomly chosen in a set of integers S of size $\frac{D}{\epsilon}$ then the probability that $P(x_1, \ldots, x_n) = 0$ is bounded by ϵ.*

A classical probabilistic algorithm to decide if a polynomial P is identically zero can be derived from this lemma. It picks $x_1, \ldots, x_n$ randomly in $[\frac{D}{\epsilon}]$[1] and calls the oracle to compute $P(x_1, \ldots, x_n)$. If the result is zero, the algorithm decides

[1] We write $[x]$ for the set of integers between 1 and $\lceil x \rceil$.

that the polynomial is zero otherwise it decides that it is non zero. Remark that the algorithm never gives a false answer when the polynomial is zero. The probability of error when the polynomial is non zero is bounded by ϵ thanks to Lemma 2: *Polynomial Identity Testing* is thus in the class RP.

This procedure makes exactly one call to the black box on points of size $\log(\frac{D}{\epsilon})$. The error rate may then be made exponentially smaller by increasing the size of the points. There is an other way to achieve the same reduction of error. Repeat the previous algorithm k times for $\epsilon = \frac{1}{2}$, that is to say the points are randomly chosen in $[2D]$. If all runs return zero, then the algorithm decides that the polynomial is zero else it decides it is non zero. The probability of error of this algorithm is bounded by 2^{-k}, thus to achieve an error bound of ϵ, we have to set $k = \log(\frac{1}{\epsilon})$. We denote by *not_zero*(P, ϵ) the latter procedure, which is given as inputs P, a black box polynomial, and ϵ, a bound on the probability of error. It uses slightly more random bits but it only involves numbers less than $2D$.

Up to Sec. 5, all polynomials have monomials with distinct supports and no constant term. This class of polynomials contains the multilinear polynomials but is much larger. Moreover being without constant term is not restrictive since we can always replace a polynomial by the same polynomial minus its constant term that we compute beforehand by a single oracle call to $P(0, \ldots, 0)$.

We now give an algorithm which finds, in randomized polynomial time, a monomial of a polynomial P thanks to the previous lemmas. In this algorithm, L is a set of indices of variables and i an integer used to denote the index of the current variable.

Algorithm 1. Find_monomial

```
Data: A polynomial P with n variables and the error bound ε
Result: A monomial of P
L ⟵ {1, ..., n}
if not_zero(P, ε/(n+1)) then
    for i = 1 to n do
        if not_zero(P_{L\{i}}, ε/(n+1)) then
            L ⟵ L \ {i}
    return The monomial of support L
else
    return "Zero"
```

Once a set L is found such that P_L is a monomial $\lambda \boldsymbol{X}^{\boldsymbol{e}}$, we must compute λ and $\boldsymbol{e}$. The evaluation of P_L on $(1, \ldots, 1)$ returns λ. For each $i \in L$, the evaluation of P_L on $X_i = 2$ and for $j \neq i$, $X_j = 1$ returns $\lambda 2^{e_i}$. From these n calls to the black box, we compute $\boldsymbol{e}$ in linear time and thus output $\lambda \boldsymbol{X}^{\boldsymbol{e}}$.

We analyze this algorithm, assuming first that the procedure *not_zero* never makes a mistake. We also assume that P is not zero, which means that the algorithm has not answered "Zero". Therefore, at the end of the algorithm, P_L is not zero because an element is removed from L only if this condition is respected. Since removing any element from L would make P_L zero by construction, the set L is minimal for the property of P_L being non zero. Then by Lemma 1 we know that P_L is a monomial of P, which allows us to output it as previously explained.

Errors only appear in the procedure *not_zero* with probability $\frac{\epsilon}{n+1}$. Since we use this procedure $n+1$ times we can bound the total probability of error by ϵ. The total complexity of this algorithm is $O(n\log(\frac{n}{\epsilon}))$ since each of the n calls to the procedure *not_zero* makes $O(\log(\frac{n}{\epsilon}))$ calls to the oracle in time $O(1)$. We summarize the properties of this algorithm in the next proposition.

Proposition 1. *Given a polynomial P as a black box, whose monomials have distinct supports, Algorithm 1 finds, with probability $1-\epsilon$, a monomial of P by making $O(n\log(\frac{n}{\epsilon}))$ calls to the black box on entries of size* $\log(2D)$.

4 An Incremental Algorithm for Polynomials with Distinct Supports

We build an algorithm which enumerates the monomials of a polynomial incrementally by using the procedure *find_monomial* defined in Proposition 1. Recall that incrementally means that the delay between two consecutive monomials is bounded by a polynomial in the number of already generated monomials.

We use a procedure *subtract*(P, Q), which acts as a black box for the polynomial $P-Q$ where P is a black box and Q is an explicit set of monomials. Let D be the total degree of Q, C a bound on the size of its coefficients and i the number of its monomials. One evaluates the polynomial *subtract*(P, Q) on points of size m as follows:

1. compute the value of each monomial of Q in time $O(D\max(C,m))$
2. add the values of the i monomials in time $O(iD\max(C,m))$
3. call the black box to compute P on the same points and return this value minus the one we have computed for Q

Algorithm 2. Incremental computation of the monomials of P

Data: A polynomial P with n variables and the error bound ϵ
Result: The set of monomials of P
$Q \longleftarrow 0$
while not_zero(subtract(P,Q), $\frac{\epsilon}{2n+1}$) **do**
  $M \longleftarrow$ find_monomial(subtract(P,Q), $\frac{\epsilon}{2n+1}$)
  Write(M)
  $Q \longleftarrow Q+M$

Theorem 1. *Let P be a polynomial whose monomials have distinct supports with n variables, t monomials and a total degree D. Algorithm 2 computes the set of monomials of P with probability $1-\epsilon$. The delay between the ith and $i+1^{th}$ outputted monomials is bounded by $O(iDn^2(n+\log(\frac{1}{\epsilon})))$ in time and $O(n(n+\log(\frac{1}{\epsilon})))$ calls to the oracle. The algorithm performs $O(tn(n+\log(\frac{1}{\epsilon})))$ calls to the oracle on points of size* $\log(2D)$.

5 A Polynomial Delay Algorithm for Multilinear Polynomials

In this section we introduce an algorithm which enumerates the monomials of a multilinear polynomial with a polynomial delay. This algorithm has the interesting property of being easily parallelizable, which is obviously not the case of the incremental one.

Let P be a multilinear polynomial with n variables and a total degree D. Let L_1 and L_2 be two disjoint sets of indices of variables and l the cardinal of L_2. We can write $P_{L_1 \cup L_2} = \boldsymbol{X}^{L_2} P_1(\boldsymbol{X}) + P_2(\boldsymbol{X})$, where $\boldsymbol{X}^{L_2}$ does not divide $P_2(\boldsymbol{X})$. We want to decide if there is a monomial of P, whose support contains L_2 and is contained in $L_1 \cup L_2$. It is equivalent to decide wether or not $P_1(\boldsymbol{X})$ is the zero polynomial. To this purpose, we define a univariate polynomial $H(Y)$ from $P_{L_1 \cup L_2}$:

1. substitute a randomly chosen value x_i in $[2D]$ to X_i for all $i \in L_1$
2. substitute the variable Y to each X_i with $i \in L_2$

The polynomial $H(Y)$ can be written $Y^l P_1(\boldsymbol{x}) + P_2(\boldsymbol{x}, Y)$. If P_1 is a non zero polynomial then $P_1(\boldsymbol{x})$ is a non zero constant with probability at least $\frac{1}{2}$ because of Lemma 2. Moreover $P_2(\boldsymbol{x}, Y)$ is a polynomial of degree strictly less than l. Hence, to decide if the polynomial P_1 is not zero, we have to decide if $H(Y)$ is of degree l.

To this aim we do the interpolation of $H(Y)$: for this we need to make l oracle calls on values from 1 to l. The time needed to do the interpolation thanks to these values, with s a bound on the size of $H(i)$ for $1 \leq i \leq l$, is $O(l^2 \log(s))$. We improve the probability of error of the described procedure from $\frac{1}{2}$ to ϵ by repeating it $\log(\frac{1}{\epsilon})$ times and name it *not_zero_improved*(L_1, L_2, P, ϵ).

We now describe a binary tree which contains informations about the monomials of P. The set of node of this tree is the pairs of list (L_1, L_2) such that there exists a monomial of support L in P with $L_2 \subseteq L \subseteq L_1 \cup L_2$. Consider a node labeled by (L_1, L_2), we note i the smallest element of L_1, it has for left child $(L_1 \setminus \{i\}, L_2)$ and for right child $(L_1 \setminus \{i\}, L_2 \cup \{i\})$ if they exist. The root of this tree is $([n], \emptyset)$ and the leaves are of the form $(\emptyset, L_2)$. There is a bijection between the leaves of this tree and the monomials of P: a leaf $(\emptyset, L_2)$ represents the monomial of support L_2.

To enumerate the monomials of P, Algorithm 3 does a depth first search in this tree using *not_zero_improved* and when it visits a leaf, it outputs the corresponding monomial thanks to the following procedure *coefficient*(P, L). We have L of cardinality l, the support of a term and we want to find its coefficient. Consider $H(Y)$ built from $L_1 = \emptyset$ and $L_2 = L$, the coefficient of Y^l in $H(Y)$ is the coefficient of the monomial of support L. We interpolate $H(Y)$ with l calls to the oracle and return this coefficient.

Theorem 2. *Let P be a multilinear polynomial with n variables, t monomials and a total degree D. Algorithm 3 computes the set of monomials of P with probability $1 - \epsilon$. The delay between the ith and $i + 1^{th}$ outputted monomials is*

Algorithm 3. A depth first search of the monomials of P, recursively written

Data: A multilinear polynomial P with n variables and the error bound ϵ
Result: All monomials of P

```
Monomial(L_1, L_2, i) =
if i = n + 1 then
  | Write(coefficient(P, L_2))
else
  | if not_zero_improved(L_1 \ {i}, L_2, P, ε/(2^n n)) then
  |   Monomial(L_1 \ {i}, L_2, i + 1)
  | if not_zero_improved(L_1 \ {i}, L_2 ∪ {i}, P, ε/(2^n n)) then
  |   Monomial(L_1 \ {i}, L_2 ∪ {i}, i + 1)
in Monomial([n], ∅, 1)
```

bounded in time by $O(D^2n^2\log(n)(n+\log(\frac{1}{\epsilon})))$ *and by* $O(nD(n+\log(\frac{1}{\epsilon})))$ *oracle calls. The whole algorithm performs* $O(tnD(n+\log(\frac{1}{\epsilon})))$ *calls to the oracle on points of size* $\log(2D)$.

There is a possible trade-off in the way *not_zero_improved* and *coefficient* are implemented: if one knows a bound on the size of the coefficients of the polynomial and uses exponentially bigger evaluations points then one only needs one oracle call. The number of calls in the algorithm is then less than tn which is close to the optimal $2t$.

Remark that when a polynomial is monotone (coefficients all positive or all negative) and is evaluated on positive points, the result is zero if and only if it is the zero polynomial. Algorithms 2 and 3 may then be modified to work deterministically for monotone polynomials with an even better complexity.

Moreover both algorithms can be extended to work over finite fields. Since they only use evaluation points less than $2D$, polynomials over any field of size more than $2D$ can be interpolated with very few modifications, which is better than most classical algorithms.

6 Complexity Classes for Enumeration

In this part the results about interpolation in the black box formalism are transposed into more classical complexity results. We are interested in enumeration problems defined by predicates $A(x, y)$ such that, for each x, there is a polynomial P_x whose monomials are in bijection with $A(x)$. If P_x is efficiently computable, an interpolation algorithm gives an effective way of enumerating its monomials and thus to solve ENUM·A.

Example 1. We associate to each graph G the determinant of its adjacency matrix. The monomials of this multilinear polynomial are in bijection with the cycle covers of G. Hence the problem of enumerating the monomials of $\det(M)$ is equivalent to enumerating the cycle covers of G.

The specialization of different interpolation algorithms to efficiently computable polynomials naturally correspond to three "classical" complexity classes for enumeration and their probabilistic counterparts. We present several problems related to polynomials, like in Example 1, to illustrate how easily the interpolation methods described in this article produce enumeration algorithms for combinatorial problems. Although the first two examples already had efficient enumeration algorithms, the last did not, which shows that this technique can bring new results in enumeration complexity.

In all the following definitions, we assume that a predicate which defines an enumeration problem is decidable in polynomial time, that is to say the corresponding decision problem is in P.

Definition 3. *A problem* ENUM·A *is decidable in polynomial total time* **TotalP** *(resp. probabilistic polynomial total time* **TotalPP***) if there is a polynomial* $Q(x, y)$ *and a machine* M *which solves* ENUM·A *(resp. with probability greater than* $\frac{2}{3}$*) and satisfies for all* x, $T(x, |M(x)|) < Q(|x|, |M(x)|)$.

The class **TotalPP** is very similar to the class **BPP** for decision problems. By repeating a polynomial number of times an algorithm working in total polynomial time and returning the set of solutions generated in the majority of runs, one decreases exponentially the probability of error. The choice of $\frac{2}{3}$ is hence arbitrary, everything greater than $\frac{1}{2}$ would do. This property holds for the aftermentioned probabilistic classes, and unlike **TotalPP**, the predicate which defines the enumeration problem has to be decidable in polynomial time.

Early termination versions of Zippel's algorithm [3] solve ENUM·*Poly* in a time polynomial in the number of monomials. If we now use this algorithm on the Determinant which is computable in polynomial time, we enumerate its monomials in probabilistic polynomial total time. Thanks to Example 1, the enumeration of the cycle covers of a graph is in **TotalPP**.

Definition 4. *A problem* ENUM·A *is decidable in incremental polynomial time* **IncP** *(resp. probabilistic incremental polynomial time* **IncPP***) if there is a polynomial* $Q(x, y)$ *and a machine* M *which solves* ENUM·A *(resp. with probability greater than* $\frac{2}{3}$*) and satisfies for all* x, $T(x, i + 1) - T(x, i) \leq Q(|x|, i)$.

Since Zippel's algorithm finds all monomials in its last step, even when the polynomial is multilinear, it seems hard to turn it into an incremental algorithm. On the other hand Algorithm 2 does the interpolation with incremental delay.

Example 2. To each graph we associate a polynomial PerfMatch, whose monomials represent the perfect matchings of this graph. For graphs with a "Pfaffian" orientation, such as the planar graphs, this polynomial is related to a Pfaffian and is then efficiently computable. Moreover all the coefficients of this graph are positive, therefore we can use Algorithm 2 to interpolate it deterministically in incremental delay. We have then proved that the enumeration of perfect matching in planar graphs is in **IncP**.

Definition 5. *A problem* ENUM·A *is decidable in polynomial delay* **DelayP**, *(resp. probabilistic polynomial delay* **DelayPP***) if there is a polynomial* $Q(x,y)$ *and a machine* M *which solves* ENUM·A *(resp. with probability greater than* $\frac{2}{3}$*) and satisfies for all* x, $T(x,i+1) - T(x,i) \leq Q(|x|)$.

Example 3 (Spanning Hypertrees). The notion of a spanning tree of a graph has several interesting generalizations to hypergraphs. Nevertheless, deciding if there is a spanning hypertree is polynomially computable only for the notion of Berge acyclicity and 3-uniform hypergraphs [12], thanks to an adaptation of the Lovász matching algorithm in linear polymatroids [13].

A polynomial Z is defined for each 3-uniform hypergraph [14] with coefficients -1 or 1, whose monomials are in bijection with the spanning hypertrees of the hypergraph. A new Matrix-Tree theorem [14] shows that Z is the Pfaffian of a matrix, whose coefficients are linear polynomials depending on the hypergraph. Thus Z is efficiently computable by first evaluating a few linear polynomials and then a Pfaffian. This has been used to give a simple RP algorithm [15] to decide the existence of a spanning hypertree in a 3-uniform hypergraph.

If we use Algorithm 3 we can enumerate the monomials of Z with probabilistic polynomial delay. The delay is good in practice, since the total degree of the monomials is small and the size of the coefficients is 1. As a conclusion, the problem of enumerating the spanning hypertrees of a 3-uniform hypergraph is in **DelayPP**.

7 Degree 2 Polynomials

7.1 An Incremental Algorithm for Degree 2 Polynomials

We now give an incremental algorithm for the case of polynomials of degree $d = 2$. We describe, in the following, a procedure which returns a monomial of a polynomial P, since Algorithm 2 turns it into an incremental algorithm for ENUM·*Poly*.

First, let us remark that Algorithm 1 may be used on a polynomial P of arbitrary degree to find a minimal support L. Since it is minimal, all monomials of P_L have L as support and $P_L(\boldsymbol{X}) = \boldsymbol{X}^L Q(\boldsymbol{X})$ with Q a multilinear polynomial. Therefore if we find a monomial of $Q(\boldsymbol{X})$ and multiply it by $\boldsymbol{X}^L$, we have a monomial of P.

One may simulate an oracle call to $Q(\boldsymbol{X})$ by a call to P_L and a division by the value of $\boldsymbol{X}^L$ as long as no X_i is chosen to be 0. The procedure *not_zero_improved*$(L', L \setminus L', Q, \epsilon)$ calls the black box only on strictly positive values, since $L = L' \cup (L \setminus L')$, and decides if Q has a monomial whose support contains L'. One then grows L' until it is maximal for the last property. Since $Q(\boldsymbol{X})$ is multilinear there is only one term with support L' and one finds its coefficient by the procedure *coefficient(Q,L')*. Therefore, we have described an algorithm which can be used to implement the procedure *find_monomial* for degree 2 polynomials in Algorithm 2.

7.2 Limit to the Polynomial Delay Approach

We study the problem of deciding if a term has coefficient zero in a polynomial. When the polynomial is multilinear, the procedure *not_zero_improved* solves the problem in polynomial time, but for degree 2 polynomials we prove it is impossible unless RP = NP. Therefore there is no generalization of Algorithm 3 to higher degree polynomials, although a polynomial delay algorithm may exist.

Proposition 2. *Assume there is an algorithm which, given a polynomial of degree 2 and a term, can decide in probabilistic polynomial time if the term has a non zero coefficient in the polynomial then* RP = NP.

Proof. Let G be a directed graphs on n vertices, the Laplace matrix $L(G)$ is defined by $L(G)_{i,j} = -X_{i,j}$ when $(i,j) \in E(G)$, $L(G)_{i,i} = \sum_{(i,j) \in E(G)} X_{i,j}$ and 0 otherwise. The Matrix-Tree theorem is the following equality where $\mathcal{T}_s$ is the set of spanning trees of G whose all edges are oriented away from the vertex s and $L(G)_{s,t}$ is the minor of $L(G)$ where row s and column t have been deleted:

$$\det(L(G)_{s,t})(-1)^{s+t} = \sum_{T \in \mathcal{T}_s} \prod_{(i,j) \in T} X_{i,j}$$

We substitute to $X_{i,j}$ the product of variables $Y_i Z_j$ in the polynomial $\det(L(G)_{s,t})$ which makes it a polynomial in $2n$ variables still computable in polynomial time. Every monomial represents a spanning tree whose maximum outdegree is the degree of the polynomial. We assume that every vertex of G has indegree and outdegree less or equal to 2 therefore $\det(L(G)_{s,t})$ is of degree 2.

Remark now that a spanning tree, all of whose vertices have outdegree and indegree less or equal to 1 is an Hamiltonian path. Therefore G has an Hamiltonian path beginning by s and finishing by a vertex v if and only if $\det(L(G)_{s,t})$ contains the monomial $Y_s Z_v \prod_{i \neq s,v} Y_i Z_j$.

There is only a polynomial number of pairs (s, v), thus if we assume there is a probabilistic polynomial time algorithm to test if a term has a non zero coefficient in a degree 2 polynomial, we can use it to decide if G – of outdegree and indegree at most 2 – has an Hamiltonian path. Since this problem is NP-complete [16], we have RP = NP.

8 Conclusion

Let us compare our method to three classical interpolation algorithms, which unlike our method can interpolate polynomials of any degree. Once restricted to multilinear polynomials, Algorithm 3 is really efficient compared to the algorithm of Klivans and Spielman (KS), which is the only known method with a bound on the delay. In the next table, T is a bound on t, the number of monomials, that Ben-Or Tiwari and Zippel algorithms need to do the interpolation. In the row labeled Enumeration is written the kind of enumeration algorithm the interpolation method gives when the polynomial is polynomially computable.

	Ben-Or Tiwari [1]	Zippel [2]	KS [4]	Algorithm 3
Algorithm type	Deterministic	Probabilistic	Probabilistic	Probabilistic
Number of calls	$2T$	tnD	$t(nD)^6$	$tnD(n + \log(\frac{1}{\epsilon}))$
Total time	Quadratic in T	Quadratic in t	Quadratic in t	Linear in t
Enumeration	Exponential	**TotalPP**	**IncPP**	**DelayPP**
Size of points	$T\log(n)$	$\log(\frac{nT^2}{\epsilon})$	$\log(\frac{nD}{\epsilon})$	$\log(D)$

Acknowledgements. Thanks to Hervé Fournier "l'astucieux", Guillaume Malod, Sylvain Perifel and Arnaud Durand for their helpful comments about this article.

References

1. Ben-Or, M.: A deterministic algorithm for sparse multivariate polynomial interpolation. In: Proceedings of the Twentieth Annual ACM Symposium on Theory of Computing, pp. 301–309. ACM, New York (1988)
2. Zippel, R.: Interpolating polynomials from their values. Journal of Symbolic Computation 9(3), 375–403 (1990)
3. Kaltofen, E., Lee, W.: Early termination in sparse interpolation algorithms. Journal of Symbolic Computation 36(3-4), 365–400 (2003)
4. Klivans, A., Spielman, D.: Randomness efficient identity testing of multivariate polynomials. In: Proceedings of the Thirty-Third Annual ACM Symposium on Theory of Computing, pp. 216–223. ACM, New York (2001)
5. Arvind, V., Mukhopadhyay, P., Albers, S., Marion, J.: Quantum query complexity of multilinear identity testing. In: Symposium on Theoretical Aspects of Computer Science, vol. 3, pp. 87–98 (2009)
6. Karnin, Z., Shpilka, A.: Black box polynomial identity testing of generalized depth-3 arithmetic circuits with bounded top fan-in. In: Conference on Computational Complexity, pp. 280–291 (2008)
7. Johnson, D.S., Papadimitriou, C.H., Yannakakis, M.: On generating all maximal independent sets. Inf. Process. Lett. 27(3), 119–123 (1988)
8. Kavvadias, D.J., Sideri, M., Stavropoulos, E.C.: Generating all maximal models of a boolean expression. Inf. Process. Lett. 74(3-4), 157–162 (2000)
9. Bagan, G.: Algorithmes et Complexité des Problèmes d'Énumération pour l'Évaluation de Requêtes Logiques. PhD thesis, Université de Caen (2009)
10. Aigner, M.: A course in enumeration. Springer, Heidelberg (2007)
11. Schwartz, J.: Fast probabilistic algorithms for verification of polynomial identities. Journal of the ACM 27(4), 717 (1980)
12. Duris, D.: Acyclicité des hypergraphes et liens avec la logique sur les structures relationnelles finies. PhD thesis, Université Paris Diderot - Paris 7 (2009)
13. Lovász, L.: Matroid matching and some applications. J. Combin. Theory Ser. B 28(2), 208–236 (1980)
14. Masbaum, G., Vaintrob, A.: A new matrix-tree theorem. International Mathematics Research Notices 2002(27), 1397 (2002)
15. Caracciolo, S., Masbaum, G., Sokal, A., Sportiello, A.: A randomized polynomial-time algorithm for the Spanning Hypertree Problem on 3-uniform hypergraphs. Arxiv preprint arXiv:0812.3593 (2008)
16. Plesník, J.: The np-completeness of the hamiltonian cycle problem in planar digraphs with degree bound two. Inf. Process. Lett. 8(4), 199–201 (1979)

Faster Approximation Schemes and Parameterized Algorithms on H-Minor-Free and Odd-Minor-Free Graphs

Siamak Tazari

Humboldt Universität zu Berlin
tazari@informatik.hu-berlin.de

Abstract. We improve the running time of the general algorithmic technique known as Baker's approach (1994) on H-minor-free graphs from $\mathcal{O}(n^{f(|H|)})$ to $\mathcal{O}(f(|H|)n^{O(1)})$. The numerous applications include, e.g. a 2-approximation for coloring and PTASes for various problems such as dominating set and max-cut, where we obtain similar improvements.

On classes of odd-minor-free graphs, which have gained significant attention in recent time, we obtain a similar acceleration for a variant of the structural decomposition theorem proved by Demaine et al. (2010). We use these algorithms to derive faster 2-approximations; furthermore, we present the first PTASes and subexponential FPT-algorithms for independent set and vertex cover on these graph classes using a novel dynamic programming technique.

We also introduce a technique to derive (nearly) subexponential parameterized algorithms on H-minor-free graphs. Our technique applies, in particular, to problems such as Steiner tree, (directed) subgraph with a property, (directed) longest path, and (connected/independent) dominating set, on some or all proper minor-closed graph classes. We obtain as a corollary that all problems with a minor-monotone subexponential kernel and amenable to our technique can be solved in subexponential FPT-time on H-minor free graphs. This results in a general methodology for subexponential parameterized algorithms outside the framework of bidimensionality.

Keywords: Subexponential Algorithms; Graph Minors; Odd Minors.

1 Introduction

One of the seminal results in algorithmic graph theory is Baker's approach [1] for designing polynomial-time approximation schemes (PTAS) for a wide range of problems on planar graphs. Ever since its discovery, it has been applied and generalized in various ways, see e.g. [2,3,4,5]. The essence of the idea is the following: for any given t, one can partition a planar graph into t parts, so that removing any one of the parts results in a graph of bounded treewidth. Now, to obtain a PTAS, we observe that if t is appropriately chosen, there must exist a part that contains at most an ϵ-fraction of an optimal solution; this can often

P. Hliněný and A. Kučera (Eds.): MFCS 2010, LNCS 6281, pp. 641–652, 2010.

be combined with the solution in the remainder of the graph to obtain a $(1+\epsilon)$-approximation.

H-minor-free graphs, i.e. proper graph classes that are closed under building minors, have gained significant attention in the past two decades, especially due to Robertson and Seymour's graph minor theory, one of the deepest and most far-reaching theories in discrete mathematics in the past few decades. These classes include, e.g. planar graphs, bounded-genus graphs, linklessly embeddable graphs and apex graphs. Using the deep Robertson-Seymour (RS-) decomposition theorem [6], Grohe [3] generalized Baker's technique to H-minor-free graphs and Demaine et al. [5] showed the partitioning theorem mentioned above for all these graph classes. However, both their methods result in algorithms with running time $\mathcal{O}(n^{f(|H|)})$, for some computable function f; since H is assumed to be fixed, this is considered polynomial.

Improving Baker's Decomposition. We provide the first algorithm for Baker's decomposition running in time $\mathcal{O}(g(|H|)n^{\mathcal{O}(1)})$, for some computable function g. This is a significant acceleration of the previous results, especially considering the fact that the constants in graph minor theory, such as the functions f, g above, are usually huge. This immediately implies similar improvements on all the consequences of this algorithm, especially *all* the generic approximation algorithms and schemes in [3,5] and Baker's original problems [1]. In particular, we obtain the first 2-approximation for Coloring H-minor-free graphs in the given time bound and the first PTAS for Independent Set, Max-Cut, Min Color Sum, Max P-Matching, and Dominating Set on these graph classes while avoiding $|H|$ in the exponent of n in their running time. Our main idea is derived from Dawar et al.'s approach [7] of finding a certain tree decomposition of H-minor-free graphs that is more tractable than the RS-decomposition.

Parameterized Complexity. In the language of parameterized complexity, our result above shows that partitioning H-minor-free graphs in the described way is *fixed-parameter tractable* (FPT) when parameterized by $|H|$. In this framework, for a given problem of size n and parameter k, we are interested in algorithms with a running time of $\mathcal{O}(f(k)n^{\mathcal{O}(1)})$, where f is some computable function depending solely on k; we then say that the problem at hand is in FPT. The theory also provides negative evidence that some problems are most likely not FPT; we refer to the books of Downey and Fellows [8] and Flum and Grohe [9] for more background on parameterized complexity.

Once a problem is shown to be FPT, the challenge is to provide algorithms that have the smallest dependence on the parameter k, i.e. make the function f in the running time as small as possible. It is especially desirable to obtain *subexponential* functions and thus provide particularly fast algorithms. Whereas this is often not possible in general graphs, a plethora of results exist that show the existence of such algorithms on restricted graph classes, such as H-minor-free graphs. Perhaps the most general technique to obtain subexponential parameterized algorithms on these graph classes is the theory of *bidimensionality* [10] that captures almost all known results of this type on H-minor-free graphs.

Still, this theory does not apply to a number of prominent problems, such as Steiner Tree, Connected Dominating Set, and Directed k-Path.

One way to show fixed-parameter tractability of a problem is to show the existence of a *kernel* for the problem, which is a polynomial-time algorithm that transforms any instance into an equivalent instance whose size is bounded by a function g solely dependent on k called the *size* of the kernel. Kernelization can be seen as polynomial-time pre-processing with a quality guarantee and has gained much theoretical importance in the recent years – besides its natural practical importance. For an introduction to kernels we refer to the survey by Guo and Niedermeier [11].

Guess and Conquer. In this work, we provide a new framework, that we call *guess and conquer* to obtain (nearly) subexponential parameterized algorithms on H-minor-free graphs for an abundant number of parameterized problems. We obtain an algorithm with a running time $\mathcal{O}(2^{\mathcal{O}(\sqrt{k \log n})} n^{\mathcal{O}(1)}) = \inf_{0<\epsilon\leq 1} \mathcal{O}((1+\epsilon)^k + n^{\mathcal{O}(1/\epsilon)})$ which we call *nearly* subexponential. Note that if $k = \mathcal{O}(\log n)$, our running time is fully polynomial in the input and if $k = \omega(\log n)$, it is subexponential FPT in k. Hence, except for a "small range" of possible parameter values, we have a subexponential FPT algorithm. In fact, we show that the problems we consider admit a minor-monotone subexponential kernel on H-minor-free graphs if and only if they admit a subexponential FPT algorithm on these graph classes. Note that in general graphs, even a linear kernel results only in an exponential FPT-algorithm.

Our technique applies in particular to Connected Dominating Set and Steiner Tree (at least) in bounded-genus graphs and Directed k-Path in all H-minor-free graphs, none of which are known to admit subexponential FPT-algorithms in H-minor-free graphs; for the latter two, such algorithms are not even known for planar graphs.

At the time of preparation of this paper, we became aware that Dorn et al. [12] recently and independently obtained similar nearly subexponential algorithms for some problems, albeit only on apex-minor-free graphs – whereas our techniques apply to general H-minor-free graphs. Also, they discuss it as a solution method for a particular problem and not in the general setting in which we introduce the technique. Furthermore, our technique for domination and covering problems is completely new.

Odd-Minor-Free Graphs. The class of odd-minor-free graphs has attained extensive attention in the graph theory literature [13,14] and recently, in theoretical computer science [15,16,17]. They are strictly more general than H-minor-free graphs as they include, for example, all bipartite graphs and may contain a quadratic number of edges. In addition to their role in graph minor theory and structural graph theory, they bear important connections to the Max-Cut problem [13] and Hadwiger's conjecture [15]. We refer to [16] for a more thorough introduction to odd-minor-free graphs and their significance.

Demaine et al. [16] prove a decomposition theorem for odd-H-minor-free graphs similar to the RS-decomposition of H-minor-free graphs and present an $\mathcal{O}(n^{f(|H|)})$ algorithm to compute such a decomposition. From this, they derive a

Baker-style decomposition of odd-minor-free graphs into two graphs of bounded treewidth. We identify an intermediate decomposition implicit in [16] that is computable in FPT-time and proves to be very useful algorithmically: on one hand, we deduce the Baker-style decomposition into two parts and a number of 2-approximation algorithms (most notably for COLORING) in FPT-time as a corollary; on the other hand, we can answer a question that is posed by Demaine et al. in [16], affirmatively: namely, whether the PTASes and subexponential FPT-algorithms for VERTEX COVER and INDEPENDENT SET can be generalized from H-minor-free graphs to odd-minor-free graphs. These are the first PTASes and subexponential FPT-algorithms developed on odd-minor-free graphs.

2 Preliminaries

We denote graphs by letters G, H, and refer to their vertex/edge sets by $V(G)$ and $E(G)$, respectively. Unless otherwise mentioned, our graphs have n vertices and m edges. For a subset $U \subseteq V(G)$, we write $G[U]$ to denote the subgraph of G induced by U. We denote the standard parameterization of a problem Π by k-Π, i.e. the problem Π parameterized by the solution size k, which is usually the number of vertices or edges in the solution; this applies in particular to k-STEINER TREE.

The Classes SUBEPTand SUBEPT$^+$. Recall that the classes EPT and SUBEPT are defined to be the bounded parameterized complexity classes $2^{\mathcal{O}(k)}$-FPT, and $2^{o^{\text{eff}}(k)}$-FPT. A problem is *subexponential fixed-parameter tractable* if it is in SUBEPT. Observe that if a problem is in SUBEPT then there exists an algorithm for the problem, so that for any fixed $\alpha > 0$ the algorithm runs in time $\mathcal{O}(2^{\alpha k} n^{\mathcal{O}(1)})$. A parameterized problem k-Π is said to be in SUBEPT$^+$ if it can be solved by an algorithm $\mathcal{A}$ such that for any fixed $\alpha > 0$, the running time of $\mathcal{A}$ is bounded by $\mathcal{O}(2^{\alpha k} n^{\mathcal{O}(1/\alpha)})$. In this case, $\mathcal{A}$ is called a *nearly subexponential time* algorithm. Observe that we require a *single* (uniform) algorithm to have this property for the considered problem. Clearly, SUBEPT $\subseteq$ SUBEPT$^+$ $\subseteq$ EPT. Note that the *non-uniform exponential time hypothesis* (ETH) implies that SUBEPT$^+$ $\neq$ EPT.

Minors and Odd Minors. A *model* of H in G is a map that assigns to every vertex v of H, a connected subtree T_v of G such that the images of the vertices of H are all disjoint in G and there is an edge between them if there is an edge between the corresponding vertices in H. A graph H is a minor of G if and only if G contains a model of H. Recall that H-minor-free graphs have bounded average degree (depending on $|H|$), i.e. they fulfill $m = \mathcal{O}_H(n)$ [18]. We use the notation $\mathcal{O}_H$ to denote that the constants hidden in the big-$\mathcal{O}$ depend on $|H|$; this is necessary since in graph minor theory, the dependence is often not known.

A graph H is an *odd-minor* of a graph G if H is a minor of G, and additionally the vertices of the trees in the model of H in G can be 2-colored in such a way that (i) the edges of each tree T_v are bichromatic; and (ii) every edge e_G in G that connects two trees T_u and T_v and corresponds to an edge $e_H = uv$ of H is monochromatic. A graph is *odd-H-minor-free* if it excludes H as an odd minor. For example, bipartite graphs are odd-K_3-minor-free.

Tree Decompositions. We denote a tree decomposition of a graph G by a pair $(T, \mathcal{B})$, where T is a tree and $\mathcal{B} = \{B_u | u \in V(T)\}$ is the family of the bags of the tree decomposition. For a vertex $v \in V(G)$, we let T_v be the connected subtree of T whose bags contain v. The *adhesion* of a tree decomposition is defined as $\max\{|B_u \cap B_t| \mid \{u,t\} \in E_T\}$. We denote the treewidth of a graph G by $\mathrm{tw}(G)$.

3 Partitioning H-Minor-Free Graphs

In [5], Demaine et al. show how to decompose H-minor free graphs into parts, so that upon removal of any part, the problem at hand becomes tractable. In this section, we show how this decomposition can be achieved in FPT-time with $|H|$ as parameter; furthermore, we introduce a refinement of this method. Due to space restrictions, we only state our results. The proofs heavily draw from the work of Dawar et al. [7], Grohe [3,19], and the original proof of Demaine et al. [5]. We refer the interested reader to the full version of our work for further details.

Theorem 1. *For every graph H there is a constant c_H such that for any integer $p \geq 1$ and for every H-minor-free graph G, the vertices (edges) of G can be partitioned into p sets such that any $p-1$ of the sets induce a graph of treewidth at most $c_H p$. Furthermore,*

(i) for every H, there exists an algorithm that computes such a partition in time $\mathcal{O}_H(n^5)$; and
(ii) there exists an explicit uniform algorithm that computes such a partition in time $\mathcal{O}_H(n^{\mathcal{O}(1)})$.

The first algorithm mentioned above running in time $\mathcal{O}_H(n^5)$ uses the graph minor theorem [20] and is hence *non-uniform*, meaning that for every H, an algorithm is guaranteed to exist but is not known explicitly. The second algorithm is slower but is uniform and is based on the decomposition theorem given by Dawar et al. [7].

In some of our applications, we need a more specific version of Theorem 1; we would like to obtain a partition of the *edges* while still being able to bound the number of parts in which each *vertex* might appear. To this end, we shall bound the number of distinct edge-labels incident to each vertex in an edge-partition of the graph. A closer look at Demaine et al.'s [5] proof of Theorem 1 reveals that this number is indeed bounded by $\mathcal{O}_H(1)$; so, we are able to prove the following version as well:

Theorem 2. *For any fixed graph H there are constants c_H and d_H such that for any integer $p \geq 1$ and every H-minor-free graph G, the edges of G can be partitioned into p parts such that any $p-1$ of the parts induce a graph of treewidth at most $c_H p$ and every vertex appears in at most d_H of the parts. Furthermore, such a partition can be found in explicit uniform* FPT*-time, i.e. $\mathcal{O}_H(n^{\mathcal{O}(1)})$.*

3.1 Approximation Algorithms and PTAS

We improve *all* the generic approximation and PTAS results given by Demaine et al. in [5] (specifically, Theorems 3.3–3.7) and also by Grohe in [3] by removing the dependence on $|H|$ from the exponent of n in the presented algorithms. We refrain from re-stating all the generic results and highlight only some important concrete corollaries below.

Corollary 3. *There exists a 2-approximation algorithm for* Coloring *an H-minor-free graph in time $\mathcal{O}_H(n^{\mathcal{O}(1)})$.*

Corollary 4. *There exists a PTAS for* Independent Set, Vertex Cover, Max-Cut, Dominating Set, Min Color Sum, *and* Max P-Matching *in H-minor-free graphs running in time $\mathcal{O}_{H,\epsilon}(n^{\mathcal{O}(1)})$.*

For all the problems mentioned above, our method results in the first algorithm with this running time. The class of problems to which these techniques apply is very large and includes all the problems originally considered by Baker [1] and also most minor-bidimensional problems, whereas for the latter, other known techniques also result in such PTASes [21,22].

4 (Nearly) Subexponential FPT-Algorithms

In this section, we introduce the technique of *guess and conquer* that for a wide range of problems shows their membership in SUBEPT$^+$.

4.1 The Technique of Guess and Conquer

We state our main technique for a broad class of parameterized problems. Given a *graph property* π, which is simply a set of directed or undirected graphs, we consider the following generic problem:

> k-Subgraph with Property π: Given a graph G, does G contain a subgraph with at most k vertices that has property π, i.e. is isomorphic to some graph in π?

The problem is abbreviated as k-SP (π). If we insist on finding *induced* subgraphs with property π, we use the notation k-ISP (π) and if we want k to be the number of edges in an edge-induced subgraph then the problem is denoted by k-EISP (π). We allow that some vertices in the graphs in π have *fixed labels*, in which case, the task becomes to find a subgraph of a (partially) labeled graph G isomorphic to a graph in π, so that the labels match. Another variant is that we are additionally given a set $R \subseteq V(G)$ of *roots* (or terminals) in G and we are seeking a subgraph with property π that contains all the roots. We use the letters L and R to account for the labeled and rooted version of the problem, respectively, and the letter D to emphasize that we are dealing with directed graphs. Finally, we might be given a vertex- or edge-weighted graph and our goal is to find among all subgraphs of G with property π and at most k vertices (or edges), the one of *minimum* or *maximum* weight. We denote this whole class of problems by ({Min,Max})k-{d,r,l,e,i}sp(π).

For example, k-Steiner Tree can be seen as a Min k-rsp(π) problem, where π is the set of all trees and R is the set of terminals that are to be connected in G. Directed st-k-Path is an instance of k-dleisp (π) where π contains only a directed path of length k, in which the first vertex is labeled s and the last vertex is labeled t. Other interesting choices for π include being chordal, bipartite, edge-less (Independent Set), of maximum degree $r \geq 1$, a clique, planar, or containing only/avoiding cycles of specified length [5]. We obtain the following general result:

Theorem 5. *Let π be a graph property such that on graphs of treewidth t one can find a (maximum/minimum weight/rooted/labeled/induced) subgraph with property π in time $\mathcal{O}(2^{\mathcal{O}(t)}n^{\mathcal{O}(1)})$. For any (directed/partially labeled) H-minor-free graph G, there exists an algorithm solving* ({Min,Max})k-{d,r,l,e,i}sp(π) *and that for any $\alpha \geq 1$ and fixed $\delta > 0$ runs in time $\mathcal{O}(2^{\mathcal{O}_H(\sqrt{k \log n})}n^{\mathcal{O}(1)})$ $= \mathcal{O}(2^{\mathcal{O}_H(k/\alpha)} + n^{\mathcal{O}(\alpha)}) = \inf_{0<\epsilon\leq 1} \mathcal{O}((1+\epsilon)^k + n^{\mathcal{O}_H(1/\epsilon)}) = o(n^{\mathcal{O}(1)+\delta\sqrt{k}})$. In particular, the considered problem belongs to* SUBEPT$^+$.

Proof. Let p be some fixed integer; apply Theorem 1 to G to obtain a partition $V_1, \ldots, V_p$ of the vertex set of the graph, so that the graph induced by any $p-1$ of the sets has treewidth at most $c_H p$; such partition can be found in time $\mathcal{O}_H(n^{\mathcal{O}(1)})$. Now, consider an optimal subgraph $S^\star$ fulfilling the requirements of the problem; since $S^\star$ is assumed to have at most k vertices, there exists an $i^\star \in \{1, \ldots, p\}$, so that $V_{i^\star}$ contains at most $\lfloor k/p \rfloor$ vertices of $S^\star$. Since we do not know the value of $i^\star$, we simply *guess* it; there are at most p possibilities to do so and we try all of them. Hence, for each $i \in \{1, \ldots, p\}$, we repeat the following:

For a fixed i, we have to determine which vertices of V_i belong to $S^\star$; once more, since we do not know these vertices, we simply *guess* them; there are at most $n^{\lfloor k/p \rfloor}$ possible subsets to try because we assumed that V_i contains at most $\lfloor k/p \rfloor$ vertices of $S^\star$. For each such subset $T \subseteq V_i$, we consider the subgraph $G' = (G - V_i) \cup T$. The treewidth of this subgraph is at most $c_H p + \lfloor k/p \rfloor$, and hence, we can find an optimal solution in G' in time $\mathcal{O}(2^{\mathcal{O}(c_H p + k/p)}n^{\mathcal{O}(1)})$ and we are done. This expression is minimized for $p = \left\lfloor \sqrt{k \log n / c_H} \right\rfloor$. See the full paper for further details. □

Using the result of Dorn et al. [23] that the following problems are in EPT on (some) H-minor-free graphs when parameterized by treewidth, we immediately obtain:

Corollary 6. *For any graph H, the problem* Directed k-Path *is in* SUBEPT$^+$ *when restricted to H-minor-free graphs; the same is true for* k-Steiner Tree *at least on bounded-genus graphs.*[1]

[1] In [23] it is claimed that Steiner Tree is in EPT on H-minor-free graphs when parameterized by treewidth; however, I know by private communication that at this time, a proof actually exists only up to bounded-genus graphs. The same is true for Connected Dominating Set.

The two problems mentioned above are prominent problems that were not known to admit FPT-algorithms with running time better than $\mathcal{O}(2^k n^{\mathcal{O}(1)})$ before, even on planar graphs. Besides improving on the best known FPT-algorithms for these problems, our result shows that it is very likely that they indeed admit subexponential FPT-algorithms.

4.2 Guess and Conquer for Domination, Covering, and More

We introduced our technique for the class of k-{D,R,L,E,I}SP (π) problems, where we are looking for a subgraph with a certain property. Whereas many problems can be formulated as an instance of this generic problem class, some others like k-Vertex Cover, k-Dominating Set, or k-Leaf Tree and variants can not. We capture another class of problems below by using Theorem 2.

Theorem 7. *Let* Π *be a problem that takes as input a graph* G *and outputs a set* $S \subseteq V$ *of vertices, and let* k-Π *be its parameterization by* $|S|$*. Suppose that*

(i) on graphs of treewidth t, Π *can be solved in time* $\mathcal{O}(2^{\mathcal{O}(t)} n^{\mathcal{O}(1)})$*; and*
(ii) if for an edge $e \in E(G)$ *it is known that some solution of* S *excludes both endpoints of* e *then* Π *can be reduced to finding a solution in* $G - e$*; that is, there exists a* $k' \leq k$ *such that given a solution for* $(G - e, k')$*, one can compute a solution for* (G, k) *in polynomial time.*

Then for any graph G *excluding a fixed minor* H*, there exists an algorithm* $\mathcal{A}$ *solving* k-Π *on instance* (G, k) *such that for any* $\alpha \geq 1$ *and fixed* $\delta > 0$*, algorithm* $\mathcal{A}$ *runs in time* $\mathcal{O}(2^{\mathcal{O}_H(\sqrt{k \log n})} n^{\mathcal{O}(1)}) = \mathcal{O}(2^{\mathcal{O}_H(k/\alpha)} + n^{\mathcal{O}(\alpha)}) = \inf_{0<\epsilon\leq 1} \mathcal{O}((1+\epsilon)^k + n^{\mathcal{O}_H(1/\epsilon)}) = o(n^{\mathcal{O}(1)+\delta\sqrt{k}})$*. In particular,* k-Π *belongs to* SUBEPT$^+$.

The k-Vertex Cover problem satisfies property (ii) above because if for an edge e, we know that both endpoints do not belong to the solution, then we can reject, since e is not covered. For k-Dominating Set, such an edge is simply irrelevant, even for the connected version. That Connected k-Dominating Set fulfills property (i) was shown by Dorn et al. [23] (see footnote on page 647). Hence, we have

Corollary 8. (Connected, Independent) k-Dominating Set *and* (Connected, Independent) k-Vertex Cover *(at least) in bounded-genus graphs belong to the class* SUBEPT$^+$.

Still, Theorems 5 and 7 do not capture all problems to which the basic idea of our technique applies; for example, a modification of the proof of Theorem 7 shows that the technique also works for the undirected k-Leaf Tree or k-Bounded Degree Deletion(d) problem.

4.3 Further Analysis and Relation to Kernels

The analysis in the proof of Theorem 5 reveals that if $k = \mathcal{O}(\log n)$, then our algorithm runs in polynomial time; on the other hand, if $k = \omega(\log n)$, i.e. if

k is known to be at least $\Omega(\iota(n) \log n)$ for any computable, non-decreasing and unbounded function $\iota : \mathbb{N} \to \mathbb{N}$, then we have a SUBEPT algorithm with time complexity $2^{\mathcal{O}_H(k/\sqrt{\iota(k)})}$. But the condition $k = \omega(\log n)$ is nothing else but asking for a *subexponential kernel*; hence, we obtain

Corollary 9. *Let k-Π be a parameterized problem on H-minor-free graphs that can be solved in time $\mathcal{O}(2^{\mathcal{O}_H(\sqrt{k \log n})} n^{\mathcal{O}(1)})$ and admits a minor-monotone subexponential kernel. Then k-Π belongs to* SUBEPT. *In particular, if k-Π admits a minor-monotone polynomial kernel then it can be solved in* SUBEPT*-time $\mathcal{O}(2^{\mathcal{O}_H(\sqrt{k \log k})} n^{\mathcal{O}(1)})$.*

Any parameterized problem that can be solved in time $\mathcal{O}(f(k)n^{\mathcal{O}(1)})$ admits a kernel of size $f(k)$ [24]. It follows that all problems in SUBEPT also have a subexponential kernel. Our corollary above shows the reverse direction of this observation for the problems that admit our technique on H-minor-free graphs; for these problems, we obtain that a subexponential FPT algorithm exists if and only if a minor-monotone subexponential kernel can be constructed.

5 Algorithms on Odd-Minor-Free Graphs

In [16], Demaine et al. prove a structural decomposition theorem for odd-minor-free graphs that is very similar to the RS-decomposition theorem for H-minor-free graphs [6]. They also present an algorithm running in time $n^{\mathcal{O}_H(1)}$ to compute such a decomposition. However, upon inspecting their proof, we obtain a simpler intermediate result that turns out to be more useful for algorithmic purposes when combined with known results on H-minor-free graphs; in particular, it can be used to obtain FPT-versions of various algorithms when combined with our results from Section 3. The precise phrasing of the decomposition needs a lot of preparation which we are forced to omit due to space restrictions; but basically, it is as follows: one can obtain a tree decomposition with constant adhesion of a given odd-H-minor-free graph such that each bag contains either (i) a bipartite graph with at most a constant number of additional vertices called apices; or (ii) an H-minor-free graph appearing in the leaf of the tree decomposition. Furthermore, each bag intersects the bipartite graph of its parent bag in at most one vertex. See the full version of this work for more details. Analogous to Theorem 1, we obtain

Theorem 10. *For any fixed graph H there is a constant c_H such that for every odd-H-minor-free graph G, the vertices of G can be partitioned into two parts such that each of the parts induces a graph of treewidth at most c_H. Furthermore, such a partition can be found in explicit uniform* FPT*-time, i.e. $\mathcal{O}_H(n^{\mathcal{O}(1)})$.*

This is the best possible analog to the Baker-style decomposition of Theorem 1 for odd-minor-free graphs since these graph classes include all bipartite graphs; and complete bipartite graphs can not be partitioned into more than two parts of bounded treewidth. A direct corollary is the following:

Corollary 11. *There exists a 2-approximation algorithm for* Coloring *an odd-H-minor-free graph in time $\mathcal{O}_H(n^{\mathcal{O}(1)})$.*

Also, 2-approximations with the same FPT-running time for various other problems, such as many of the ones mentioned in Section 3.1, can be obtained. See [5] and [16] for more details.

5.1 PTASes on Odd-Minor-Free Graphs

Grohe [3] showed that various problems admit a PTAS on H-minor-free graphs. Most of these PTASes *can not* be generalized to odd-minor-free graphs as they would imply corresponding PTASes on bipartite or even general graphs for APX-hard problems. However, Demaine et al. ask in [16] whether the PTASes for Vertex Cover and Independent Set can be generalized to odd-minor-free graphs; this seems plausible since these two problems can be solved in polynomial time on bipartite graphs. Indeed, we are able to answer this question affirmatively. To this end, we define the *take-or-leave* version of these problems as follows: every vertex of the graph is associated with two numbers w^+ and w^-; if a vertex is chosen to be in the solution, i.e. in the vertex cover or independent set, it contributes a value of w^+ to the objective function; if it is not included in the solution, it contributes w^- to the objective function (the usual unweighted variants are then special cases of the take-or-leave version where $w^+ = 1$ and $w^- = 0$ for every vertex).

Lemma 12. *The take-or-leave versions of* Vertex Cover *and* Independent Set *can be solved in polynomial time on bipartite graphs.*

Theorem 13. *There exists a PTAS for* Vertex Cover *and* Independent Set *in odd-H-minor-free graphs running in time $\mathcal{O}_{H,\epsilon}(n^{\mathcal{O}(1)})$.*

The proof is based on performing dynamic programming on the tree decomposition of odd-minor-free graphs mentioned above and crucially uses the fact that the bags of the tree decomposition intersect the bipartite graphs of their parent bags in at most one vertex; once we have solutions for all the children of a bag, this enables us to define a take-or-leave version of the problem in the bag and obtain solutions in polynomial time. Once more, we refer to the full paper for details. Also note that Theorem 13 holds also for the vertex-weighted versions of these problems; the proof is analogous.

5.2 Subexponential FPT for Odd-Minor-Free Graphs

Another question that is asked by Demaine et al. [16] is whether k-Vertex Cover and k-Independent Set admit SUBEPT-algorithms on odd-minor-free graphs. As in the case of the PTASes, these are basically the only problems for which this seems possible as such algorithms for most other prominent problems would contradict hardness results in parameterized complexity. Indeed, we can obtain subexponential parameterized algorithms for these problems in a similar way as the PTASes above. First, let us state the following known result.[2]

[2] I would like to thank Fedor Fomin for a helpful discussion on this matter.

Lemma 14 (partly taken from [10,25]). *There exists an algorithm that, given an H-minor-free graph G and an integer k, runs in time $\mathcal{O}(2^{\mathcal{O}_H(\sqrt{k})}n^{\mathcal{O}(1)})$ and*

(i) decides if G contains a vertex cover of size at most k and in this case, returns a minimum vertex cover of G; and
(ii) decides if G contains an independent set of size at least k and if this is not *the case, returns an independent set of maximum size in G.*

Theorem 15. *There exists an algorithm that, given an odd-H-minor-free graph G and an integer k, runs in time $\mathcal{O}(2^{\mathcal{O}_H(\sqrt{k})}n^{\mathcal{O}(1)})$ and*

(i) decides if G contains a vertex cover of size at most k and in this case, returns a minimum vertex cover of G; and
(ii) decides if G contains an independent set of size at least k and if this is not *the case, returns an independent set of maximum size in G.*

Acknowledgment. I would like to thank Holger Dell, Martin Grohe, and Matthias Mnich for helpful discussions and comments on this work.

References

1. Baker, B.S.: Approximation algorithms for NP-complete problems on planar graphs. J. ACM 41(1), 153–180 (1994)
2. Eppstein, D.: Diameter and treewidth in minor-closed graph families. Algorithmica 27(3), 275–291 (2000)
3. Grohe, M.: Local tree-width, excluded minors, and approximation algorithms. Combinatorica 23(4), 613–632 (2003)
4. Klein, P.N.: A linear-time approximation scheme for TSP in undirected planar graphs with edge-weights. SIAM J. Comput. 37(6), 1926–1952 (2008)
5. Demaine, E.D., Hajiaghayi, M., Kawarabayashi, K.: Algorithmic graph minor theory: Decomposition, approximation, and coloring. In: FOCS 2005: Proceedings of the 46th Annual IEEE Symposium on Foundations of Computer Science, pp. 637–646. IEEE Computer Society, Los Alamitos (2005)
6. Robertson, N., Seymour, P.: Graph minors. XVI. Excluding a non-planar graph. J. Comb. Theory Ser. B 89(1), 43–76 (2003)
7. Dawar, A., Grohe, M., Kreutzer, S.: Locally excluding a minor. In: LICS 2007: Proceedings of the 22nd Annual IEEE Symposium on Logic in Computer Science, pp. 270–279. IEEE Computer Society, Los Alamitos (2007)
8. Downey, R.G., Fellows, M.R.: Parameterized Complexity. Springer, Heidelberg (1999)
9. Flum, J., Grohe, M.: Parameterized Complexity Theory. Springer, Heidelberg (2006)
10. Demaine, E.D., Fomin, F.V., Hajiaghayi, M., Thilikos, D.M.: Subexponential parameterized algorithms on bounded-genus graphs and H-minor-free graphs. J. ACM 52(6), 866–893 (2005)
11. Guo, J., Niedermeier, R.: Invitation to data reduction and problem kernelization. SIGACT News 38(1), 31–45 (2007)

12. Dorn, F., Fomin, F.V., Lokshtanov, D., Raman, V., Saurabh, S.: Beyond bidimensionality: Parameterized subexponential algorithms on directed graphs. In: STACS 2010: Proceedings of the 27th Symposium on Theoretical Aspects of Computer Science, pp. 251–262 (2010)
13. Guenin, B.: A characterization of weakly bipartite graphs. J. Comb. Theory Ser. B 83(1), 112–168 (2001)
14. Geelan, J., Gerards, B., Goddyn, L., Reed, B., Seymour, P., Vetta, A.: The odd case of Hadwiger's conjecture (Submitted, 2004)
15. Kawarabayashi, K., Mohar, B.: Approximating chromatic number and list-chromatic number of minor-closed and odd-minor-closed classes of graphs. In: STOC 2006: Proceedings of the 38th Annual ACM Symposium on Theory of Computing, pp. 401–406. ACM Press, New York (2008)
16. Demaine, E.D., Hajiaghayi, M., Kawarabayashi, K.: Decomposition, approximation, and coloring of odd-minor-free graphs. In: SODA 2010: Proceedings of the 21st Annual ACM-SIAM Symposium on Discrete Algorithms, pp. 329–344. SIAM, Philadelphia (2010)
17. Kawarabayashi, K., Li, Z., Reed, B.: Recognizing a totally odd K_4-subdivision, parity 2-disjoint rooted paths and parity cycle through specified elements. In: SODA 2010: Proceedings of the 21st Annual ACM-SIAM Symposium on Discrete Algorithms, pp. 318–328. SIAM, Philadelphia (2010)
18. Mader, W.: Homomorphieeigenschaften und mittlere Kantendichte von Graphen. Math. Ann. 174, 265–268 (1967)
19. Grohe, M.: Logic, graphs, and algorithms. In: Logic and Automata – History and Perspectives. Amsterdam University Press, Amsterdam (2007)
20. Robertson, N., Seymour, P.D.: Graph minors. XX. Wagner's conjecture. J. Comb. Theory Ser. B 92(2), 325–357 (2004)
21. Demaine, E.D., Hajiaghayi, M.: Bidimensionality: new connections between FPT algorithms and PTASs. In: SODA 2005: Proceedings of the 16th Annual ACM-SIAM Symposium on Discrete Algorithms, pp. 590–601 (2005)
22. Feige, U., Hajiaghayi, M., Lee, J.R.: Improved approximation algorithms for minimum weight vertex separators. SIAM J. Comput. 38(2), 629–657 (2008)
23. Dorn, F., Fomin, F.V., Thilikos, D.M.: Catalan structures and dynamic programming in H-minor-free graphs. In: SODA 2008: Proceedings of the 19th Annual ACM-SIAM Symposium on Discrete Algorithms, pp. 631–640. SIAM, Philadelphia (2008)
24. Niedermeier, R.: Invitation to fixed-parameter algorithms. Habilitation thesis, Universität Tübingen, Germany (2002)
25. Demaine, E.D., Hajiaghayi, M.: Graphs excluding a fixed minor have grids as large as treewidth, with combinatorial and algorithmic applications through bidimensionality. In: SODA 2005: Proceedings of the 16th Annual ACM-SIAM Symposium on Discrete Algorithms, pp. 682–689. SIAM, Philadelphia (2005)

Semi-linear Parikh Images of Regular Expressions via Reduction

Bahareh Badban[1] and Mohammad Torabi Dashti[2]

[1] Universität Konstanz
[2] ETH Zürich

Abstract. A reduction system for regular expressions is presented. For a regular expression t, the reduction system is proved to terminate in a state where the most-reduced expression readily yields a semi-linear representation for the Parikh image of the language of t.

1 Introduction

Let $A = \{a_1, \cdots, a_n\}$ be a finite set, and fix an order on the elements of A, say $a_1 < \cdots < a_n$. The *commutative* image, or *Parikh* image, of $w \in A^*$, denoted $\Psi(w)$, is a vector in $\mathbb{N}^n$ in which the ith element encodes how many times a_i appears in w. Hence, Ψ is a morphism from the monoid $(A^*, \cdot, \epsilon)$ to the monoid $(\mathbb{N}^n, +, \mathbf{0})$. The commutative image of any regular subset of A^* (in general, any context-free language) is a *semi-linear* set [6], i.e. it can be written as the union of finitely many *linear* sets. A linear set is of the form $\{v \mid v = v_0 + \lambda_1 v_1 + \cdots + \lambda_k v_k,\ \text{with}\ k, \lambda_i \in \mathbb{N}, v_i \in \mathbb{N}^n\}$. In this paper, we study the following problem:

> Given a regular expression t, compute a semi-linear representation for the Parikh image of the regular language that t denotes.

Contributions. A terminating reduction system is proposed which reduces a regular expression t to a simpler regular expression t', with certain constraints, such that t and t' have the same Parikh image. The reduction system is proved to terminate in a state where the resulting most-reduced expression is of star-height at most one. A semi-linear representation of the Parikh image of the initial expression t is then computed from the most-reduced expression by resolving the accumulated constraints.

As an example, consider the regular expression $t = (a^* \cdot b)^*$, with alphabet $A = \{a, b\}$, and $a < b$. In order to compute a semi-linear representation for $\Psi(t)$, we introduce the notion of *constrained regular expressions*. In the example above, we reduce t to $(a^*)^* \cdot b^*$, which is in turn reduced to $a^* \cdot b^*$, while a constraint is added to the expression asserting that "if there is no b, then there is no a". Intuitively, adding constraints to regular expressions allows us to encode the dependency between nested Kleene stars. The reduction system then propagates these constraints downwards while descending and pushing the Kleene stars

P. Hliněný and A. Kučera (Eds.): MFCS 2010, LNCS 6281, pp. 653–664, 2010.

down the parse tree of the regular expression. This procedure eventually leads to a constrained regular expression of star-height at most one. In a second phase, the accumulated constraints are resolved into linear conditions using case distinction. In the example above, either there is no b, and therefore no a, or there is at least one b and an arbitrary number of as. The resulting constraints are linear, hence their union semi-linear.

Related work. Algorithms for computing semi-linear Parikh images of regular grammars (hence, regular expressions) can be extracted from Parikh's original proof [6]. However, Parikh's proof, as well as some of the later variants [4,2], do not specifically aim at *constructing* semi-linear images. For instance, in [6], after a careful rearrangement of derivation trees for context-free grammars one reads: "We claim there is only a finite number of trees satisfying [a certain condition], since in any such tree the length of any chain cannot be greater than the square of the number of symbols [of the grammar]". Intuitively, this proof relies on finiteness of a certain set, and in effect, derives an upper bound on the number of possible elements in another set.

Roughly speaking, semi-linear Parikh images of regular expressions can be constructed via two different strategies: *bottom-up* and *top-down*.

– The bottom-up strategy is perhaps the first method that comes to mind. Start with the symbols of the alphabet, and proceed inductively: Assume that semi-linear images of t_1 and t_2 are given, and use the following rules for computing semi-linear images of more complex expressions.
  - $\Psi(t_1 \cdot t_2) = \Psi(t_1) \oplus \Psi(t_2)$, where $A \oplus B = \{a + b \mid a \in A, b \in B\}$ for two sets A and B.
  - $\Psi(t_1 \cup t_2) = \Psi(t_1) \cup \Psi(t_2)$
  - $\Psi(t_1^*) = S_1^* \oplus \cdots \oplus S_r^*$, if $\Psi(t_1) = S_1 \cup \cdots \cup S_r$, with S_i being linear sets, and $S_i^* = \{\mathbf{0}\} \cup \{v \mid v = v_0 + \lambda_0 v_0 + \lambda_1 v_1 + \cdots + \lambda_k v_k,\ \text{with } k, \lambda_i \in \mathbb{N}, v_i \in \mathbb{N}^n\}$ if $S_i = \{v \mid v = v_0 + \lambda_1 v_1 + \cdots + \lambda_k v_k,\ \text{with } k, \lambda_i \in \mathbb{N}, v_i \in \mathbb{N}^n\}$.

  The inductive method gives a terminating algorithm for computing semi-linear Parikh images, e.g. see [3] for a correctness proof for these rules.
– The main idea of the top-down strategy is to obtain from regular expression t a regular expression t' which has star-height of at most one, while $L(t) = L(t')$ modulo commutativity. Deriving semi-linear Parikh images of the languages of regular expressions of star-height at most one is immediate. A prominent example of the top-down approach is the equational axioms of Piling [7], which can be used to convert any regular expression t to an equivalent regular expression t', modulo commutativity, such that t' is of star-height at most one. See [1] for an excellent overview. These equational axioms however do not immediately lend themselves to a terminating algorithm for computing semi-linear Parikh images. This is because the axioms contain the equation $x \cdot y = y \cdot x$, among others. This equation can cause a cyclic behavior, and hence non-termination, in automated procedures that use it for reducing the star-height of regular expressions.

Our reduction system also follows the top-down strategy. What distinguishes the proposed reduction system from the aforementioned equational axioms is termination – the reduction system is proved to always terminate. The difference between the reduction system and the (bottom-up) inductive method however lies not in termination, but in the *width* of the produced images.

Consider a semi-linear set S represented as $S = S_1 \cup \cdots \cup S_\ell$, with $S_{i \in 1..\ell}$ being linear sets. The width of S, denoted $\omega(S)$, is the number of linear components in S, that is ℓ in this case. Obviously a semi-linear set may have different representations with different widths. The width of semi-linear representations for Parikh images of regular expressions is pertinent to certain decision procedures for regular languages, e.g. see [5,10]. In these procedures, various properties of regular languages are decided by solving a linear Diophantine equation for each linear component of the semi-linear Parikh image of the language. In this context, images with lower width are preferred, because then fewer Diophantine equations need to be solved for answering the decision problem at hand.

The width of the semi-linear representation that the aforementioned inductive method produces for regular expression t grows fast as the widths of the images of the subterms of t grow. This is intuitively due to the "multiplicative" nature of the inductive method. Suppose semi-linear representations for Parikh images of regular expressions t_1 and t_2 are given, resp. with widths w_1 and w_2. The width of the representations that the inductive method produces can then be calculated as $\omega(\Psi(t_1 \cdot t_2)) = w_1 w_2$, $\omega(\Psi(t_1 \cup t_2)) = w_1 + w_2$ and $\omega(\Psi(t_1^*)) = 2^{w_1}$.

Our proposed reduction system produces semi-linear Parikh images which are of (exponentially) lower width, compared to the inductive method. As a simple example, for the Parikh image of the regular expression $a^* \cdot (b \cup c^*)^* \cdot d$, with alphabet $A = \{a, b, c, d\}$, the reduction system produces a semi-linear representation of width 2, while the inductive method produces an image of width 16. Formally, for *unit form* regular expressions t_1 and t_2, whose images have widths w_1 and w_2 respectively, the width of the representations that the reduction system produces is given by $\omega(\Psi(t_1 \cdot t_2)) = w_1 w_2$, $\omega(\Psi(t_1 \cup t_2)) = w_1 + w_2$ and $\omega(\Psi(t_1^*)) = 2w_1$. The definition of unit form expressions, as well as precise bounds for the widths of Parikh images that the reduction system produces are given in the following sections.

Note that, w.r.t. the width of produced Parikh images, the equational axioms of [7,1] cannot be directly compared to the inductive method and the reduction system. This is because, in [7,1], depending on the axioms which are chosen for simplification, and their order, the resulting images may have different widths.

In a related work, Verma, Seidl and Schwentick give a linear-time algorithm for generating semi-linear Parikh images of context-free grammars, represented as existential Presburger formulae [9]. Converting existential Presburger formulae into the set representation that we use is of exponential-time complexity.

Structure of the paper. Section 2 presents the preliminaries. Section 3 describes our reduction system, and contains the proofs of its termination and correctness. Extracting semi-linear representations of Parikh images from the most-reduced expressions is explained in section 4.

2 Preliminaries

An *alphabet* is a finite set $A = \{a_1, \cdots, a_n\}$. Throughout the paper we assume alphabets are totally (and, lexicographically) ordered. The set of all finite sequences of elements of A is denoted by A^*, while ϵ stands for the empty sequence. A *language* over A is a subset of A^*. For two sequences $w_1, w_2 \in A^*$, let $w_1 \cdot w_2$ denote the concatenation of w_1 and w_2. Clearly ϵ is the neutral element of concatenation. For two sets $L_1, L_2 \subseteq A^*$, define $L_1 \cdot L_2 = \{w_1 \cdot w_2 \mid w_1 \in L_1, w_2 \in L_2\}$. For $L \subseteq A^*$, we define $L^0 = \{\epsilon\}$ and $L^n = L \cdot L^{n-1}$, with $n \in \mathbb{N}, n > 0$.

A *regular expression* t, for short *regexp*, over A is defined as usual, $t ::= \emptyset \mid \epsilon \mid a \mid t \cdot t \mid t \cup t \mid t^*$, with $a \in A$. Here $\cdot$, $\cup$ and * denote, respectively, concatenation, union, and Kleene star operators. Any regular expression t defines a *regular language*, $L(t)$, in the standard way: $L(\emptyset) = \emptyset$, $L(\epsilon) = \{\epsilon\}$, $L(a) = \{a\}$ for all $a \in A$, $L(t_1 \cdot t_2) = L(t_1) \cdot L(t_2)$, $L(t_1 \cup t_2) = L(t_1) \cup L(t_2)$, and $L(t^*) = \bigcup_{i \geq 0} L^i$. We write $t \approx t'$ for regexps t and t' iff $L(t_1) = L(t_2)$. When confusion is unlikely we may use a regexp t and its corresponding language $L(t)$ interchangeably.

Any regular expression has a unique parse tree, up to isomorphism. Nodes of a parse tree are either elements of the alphabet (besides ϵ and $\emptyset$), or the symbols $\cdot$, $\cup$ or *. Nodes labeled with elements of the alphabet (and ϵ and $\emptyset$) are terminal, nodes labeled with either of $\cdot$ or $\cup$ have two descendants, while nodes labeled with * have only one descendant. The parse trees of regular expressions correspond to the ground term algebra induced by the signature *Sig*: the elements of the alphabet are nullary function symbols in *Sig*, $\cdot$ and $\cup$ are binary function symbols, and * is a unary function symbol.

A regexp is said to be of *unit form*, for short $\mathcal{U}$-form, iff in its parse tree each path from the root to a node labeled with $\cup$ contains at least one node labeled with *. The following lemma is immediate as $\cdot$ is distributive over $\cup$ modulo $\approx$.

Lemma 1. *Any regexp t can be rewritten into finitely many regexps $t_1, \cdots, t_n$, such that $t \approx t_1 \cup t_2 \cup \cdots \cup t_n$, with t_i being of $\mathcal{U}$-form.*

For $L \subseteq A^*$, let $\Psi(L) = \{\Psi(w) \mid w \in L\}$. When computing semi-linear Parikh images of languages of regexps, lemma 1 allows us to confine to $\mathcal{U}$-form expressions. This is because $\Psi(L_1 \cup L_2) = \Psi(L_1) \cup \Psi(L_2)$. In the following, thus, we focus on $\mathcal{U}$-form regexps.

Below, we define a simple extension of regexps which is useful in proving our results. Fix a countable set of *name tags* $\mathsf{Names} = \{n_1, n_2, \ldots\}$. A *tagged* regexp t is a regexp where some of the nodes labeled with * in the parse tree of t are assigned with elements of Names. Note that any regexp is a tagged regexp, where *none* of its parse tree nodes are tagged. As a convention, in the text we subscript a * operator of a tagged regexp with its name. For instance, suppose the node labeled with * in the parse tree of $(a \cdot (b \cup c))^*$ is assigned with n_1. We denote this by $(a \cdot (b \cup c))^{*_{n_1}}$. An *evaluation* function is a total function $e : \mathsf{Names} \to \mathbb{N}$. For a tagged regexp t, we write t_e for the regexp that corresponds to t when all *s assigned with n are replaced with $e(n)$, for all $n \in \mathsf{Names}$. For example, with $e(n_1) = 2$ and $t = (a \cdot (b \cup c))^{*_{n_1}}$, t_e corresponds to regexp $(a \cdot (b \cup c))^2$. Recall that $r^k = r \cdot r^{k-1}$ for $k \geq 1$, and $r^0 = \epsilon$, for regexp r.

A *constrained* regexp is a tuple $(t;\phi)$, where t is a tagged regexp, and ϕ is a finite set of *constraints*. Each constraint is of either of these forms: $n_i+n_j=n_k$, or $n_i=0 \implies n_j=0$, with $n_i,n_j,n_k \in \mathsf{Names}$.

Given an evaluation function $e:\mathsf{Names}\to\mathbb{N}$, and a set of constraints $\phi(n_1,\cdots,n_\ell)$, we say e satisfies ϕ, denoted $e\models\phi$, iff all the constraints in ϕ evaluate to true when each $n\in\mathsf{Names}$ in ϕ is substituted with $e(n)$. That is, $\phi(e(n_1),\cdots,e(n_\ell))$ evaluates to true. The language of a constrained regexp $(t;\phi)$, denoted $L(t;\phi)$, is the set of all $w\in A^*$ where there exists an evaluation function e such that $w\in t_e$ and $e\models\phi$. We observe that $(t;\phi)$ can in general correspond to a non-regular language.

The example below shows that simply assigning names to * nodes of a regexp can affect the language of the corresponding constrained regexp, even if those names are not bound by any constraint.

Example 1. Consider the constrained regexps $(t;\emptyset)$, with $t=(a\cdot b^{*_{n_1}})^{*_{n_2}}$. Here the constraint set is empty. Note that $w=a\cdot b\cdot a\cdot b\cdot b$ does *not* belong to the language of $(t;\emptyset)$, while obviously w does belong to the language of $(a\cdot b^*)^*$. •

3 The Reduction System

Fix a signature S, and a countable set of variables $\mathcal{V}$. The term algebra induced by S with variables $\mathcal{V}$ is denoted by $\mathcal{T}_{S(\mathcal{V})}$. We write $var(t)$ for the set of variables appearing in term t. A *ground* term is an element of $\mathcal{T}_{S(\emptyset)}$.

A reduction system R is a finite set of ordered pairs, written as $l\to r$, where $l,r\in\mathcal{T}_{S(\mathcal{V})}$, and $var(r)\subseteq var(l)$. Let $l\to r\in R$ and $t\in\mathcal{T}_{S(\emptyset)}$ be a ground term. If for some substitution σ, $\sigma(l)$ is a subterm of t, then we can *reduce* t by replacing one occurrence of $\sigma(l)$ with $\sigma(r)$ in t. If the resulting term is t', we write $t\to t'$ and call it a *reduction step*. A term t is *irreducible* if $\neg\exists t'.\ t\to t'$; otherwise t is *reducible*. A *reduction sequence* is a sequence $t_1,t_2,\ldots$, such that $t_i\to t_{i+1}$, for $i\geq 1$. A reduction system is *terminating* iff it admits no infinite reduction sequences. See [8] for more on reduction systems.

Below, we introduce $\mathcal{R}$, our set of reduction rules for constrained regexps, that is the signature of $\mathcal{R}$ is *Sig* introduced in the previous section. As already mentioned, the ground term algebra $\mathcal{T}_{Sig(\emptyset)}$ corresponds to the set of all parse trees of regular expressions. Therefore, the following reduction rules are described in terms of parse trees. This greatly improves the presentation. The definitions and proofs can however be formalized in the corresponding term algebra as well.

The following reduction rules primarily work on parse trees of regexps of $\mathcal{U}$-form. If a reduction rule is *applicable* on an expression t, i.e. $t\to t'$, then the rule transforms the constrained regexp $(t;\phi)$ into $(t';\phi')$. In the following, how ϕ' is constructed using ϕ is also specified.

Definition 1 (Reduction rules $\mathcal{R}$). *The reduction rules depicted in figure 1 will be applied on constrained regexps $(t;\phi)$ where t is of $\mathcal{U}$-form. Here, in rules R_2 and R_3 fresh name tags are retrieved from* Names *as n_1 and n_2, and n' respectively.*

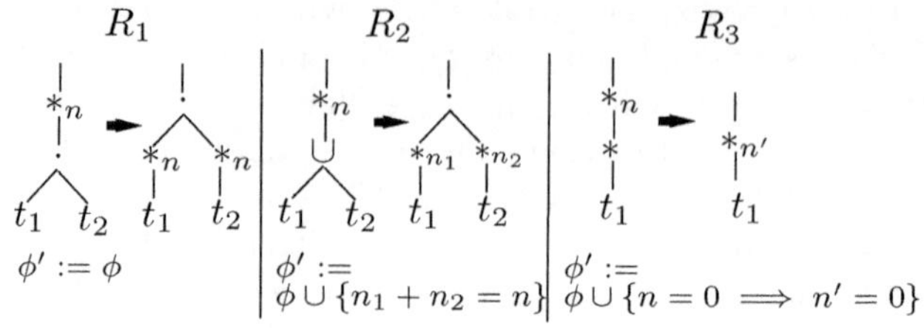

Fig. 1. Reduction rules; name tags of $*$ operators are written beside them

Intuitively, the rules of the reduction system $\mathcal{R}$ eliminate * functions that are applied on other functions (that is $\cdot, \cup$, and other *). Informally, rule (1), that is R_1, states $(t_1 \cdot t_2)^*$ can be rewritten to $t_1^* \cdot t_2^*$. The newly created stars (on top of t_1 and t_2) are assigned with the same name as the initial star. No constraint is added to ϕ in this case. Rule (2) states that $(t_1 \cup t_2)^*$ can be rewritten to $t_1^* \cdot t_2^*$. The newly created stars (on top of t_1 and t_2) are assigned with new names. A constraint is added to the constraint set ϕ, which requires the sum of the new names to be equal to the name of the initial star. Finally, rule (3) states that from $(t_1^*)^*$ the higher star can be removed. A constraint is however added to ϕ which requires if the name of the higher star equals zero, the name of the lower one should also be equal to zero.

Remark that the reduction rules of $\mathcal{R}$ can be consequently applied, because if t is of $\mathcal{U}$-form, and $t \rightarrow t'$ according to $\mathcal{R}$, then t' is of $\mathcal{U}$-form.

Applying $\mathcal{R}$ on $\mathcal{U}$-form regular expressions. Given a regexp t, we first *seed* t; this results in a constrained regexp. Then $\mathcal{R}$ is applied on the resulting constrained regexp, following the *top-most* reduction strategy. These notions are described below.

Seeding A node in the parse tree of t is called a *-node iff the node is labeled with symbol *. Given a $\mathcal{U}$-form regexp t, all the nodes in the parse tree of t which are labelled with * and have no ancestor *-node are assigned with fresh name tags. By abuse of notation, we write t also for the resulting tagged regexp. Then, the constrained regexp $(t; \emptyset)$ is constructed. This process is called seeding t. Note that the language of $(t, \emptyset)$ right after seeding is the same as the language of regexp t.

Top-most reduction strategy We assume a top-most reduction strategy, i.e. in the parse tree of a regexp the top-most reducible subterm of t is reduced first. The top-most reduction strategy corresponds to the *left-most* reduction strategy in the term algebra $\mathcal{T}_{Sig(\emptyset)}$, cf. [8]. The correctness of the reduction system, proved in theorem 3, assumes this reduction strategy.

We proceed with an example.

Example 2. Fix the alphabet $A = \{a, b, c\}$. Below we demonstrate how $\mathcal{R}$ is applied on $t = ((b \cdot a^*) \cup c)^*$. Note that t is indeed of $\mathcal{U}$-form. Seeding t yields the pair $((b \cdot a^*) \cup c)^{*_n}; \emptyset$. Applying the top-most reduction strategy results in the following reduction sequence.

$$\begin{array}{lll} & ((b \cdot a^*) \cup c)^{*_n} & ; \emptyset \\ \rightarrow_{R_2} & (b \cdot a^*)^{*_{n_1}} \cdot c^{*_{n_2}} & ; \{n = n_1 + n_2\} \\ \rightarrow_{R_1} & b^{*_{n_1}} \cdot (a^*)^{*_{n_1}} \cdot c^{*_{n_2}} & ; \{n = n_1 + n_2\} \\ \rightarrow_{R_3} & b^{*_{n_1}} \cdot a^{*_{n_3}} \cdot c^{*_{n_2}} & ; \{n = n_1 + n_2,\ n_1 = 0 \implies n_3 = 0\} \end{array}$$

•

3.1 Termination

The reduction system $\mathcal{R}$ is terminating when applied to seeded $\mathcal{U}$-form regexps, intuitively because by applying each of the rules of $\mathcal{R}$, the * operators are pushed down the parse tree, see figure 1. Ultimately there will be * operators whose operands are constants (elements of A, or the symbols ϵ and $\emptyset$). These cannot be further reduced according to $\mathcal{R}$; the reduction process thus halts.

Theorem 1. *The reduction system $\mathcal{R}$ is terminating for $\mathcal{U}$-form regexps.*

For proving this theorem we need a few definition and lemmas. First, we define a mapping $\aleph$ from the set of regular expressions, over alphabet A, to $\mathbb{N}^3$. Intuitively, $\aleph$ counts the number of non-reduced subterms of a regexp, w.r.t. $\mathcal{R}$.

Definition 2. *Function $\aleph$ from the set of regexps over alphabet A to $\mathbb{N}^3$ is defined recursively by:*

$$\begin{array}{ll} \aleph(\emptyset) = \aleph(\epsilon) = \aleph(a) = \aleph(\epsilon^*) = \aleph(a^*) & = (0,0,0) \\ \aleph(t \cup t') = \aleph(t \cdot t') & = \aleph(t) + \aleph(t') \\ \aleph((t \cdot t')^*) & = (1,0,0) + \aleph(t^*) + \aleph(t'^*) \\ \aleph((t \cup t')^*) & = (0,1,0) + \aleph(t^*) + \aleph(t'^*) \\ \aleph((t^*)^*) & = (0,0,1) + \aleph(t^*) \end{array}$$

where $a \in A$. For any expression t, norm *of t is defined as $||t|| = \sum_{i=1}^{3} \aleph_i(t)$, where $\aleph_i(t)$ is the ith element of the $\aleph(t)$-triplet.*

The reduction system $\mathcal{R}$ strictly reduces the norms of regexps. This is proved in the following lemma. Remark that $\aleph$ is agnostic to the tagging applied to regexps. Indeed, tagging plays no role in proving termination of $\mathcal{R}$.

Lemma 2. *If $t \rightarrow t'$ by the reduction system $\mathcal{R}$, then $||t'|| < ||t||$.*

Proof. The proof is by case analysis. Imagine $t = (r^*)^*$ for some regexp r. Observe that $\aleph_3(t) \geq 1$, and clearly R_3 of $\mathcal{R}$ is applicable on t. Using R_3, t is reduced to r^*, that is $t' = r^*$. Here $||t|| > ||t'||$, because $||t|| = \Sigma_{i=1}^{3} \aleph_i((r^*)^*) = 1 + \aleph_3(r^*) + \Sigma_{i=1}^{2} \aleph_i(r^*) = 1 + \Sigma_{i=1}^{3} \aleph_i(r^*) = 1 + ||r^*|| = 1 + ||t'||$. Similarly, $||(t_1 \cdot t_2)^*|| = 1 + ||t_1^* \cdot t_2^*||$ and $||(t_1 \cup t_2)^*|| = 1 + ||t_1^* \cdot t_2^*||$. The argument immediately carries over to the case a proper subterm of t is reduced. □

Now we are ready to present the proof of theorem 1.

Proof (Theorem 1). According to lemma 2, all reduction rules of $\mathcal{R}$ strictly reduce the norms of regexps. Hence, for any $\mathcal{U}$-form regexp t, after at most $||t||$ steps the resulting term cannot be any further reduced. Thus, $\mathcal{R}$ terminates. □

Time complexity of the reduction system (the number of steps performed before the reduction process halts) is quadratic in the *length* of the input regexp. The length of regexp t, denoted $len(t)$, is measured by counting the number of nodes in t's parse tree. The proof of the theorem is straightforward.

Theorem 2. *Time complexity of the reduction system $\mathcal{R}$ is of $\mathcal{O}(len(t)^2)$, where t is the input $\mathcal{U}$-form regexp.*

3.2 Correctness: Preserving Parikh Images

The reduction system $\mathcal{R}$ preserves the Parikh images of regexps.

Theorem 3. *Let $(t;\phi)$ be a constrained regexp, and $(t;\phi) \to (t';\phi')$ by applying one of the rules of $\mathcal{R}$. Then, $\Psi((t;\phi)) = \Psi((t';\phi'))$.*

Proof. Below, we assume that t coincides with the subterm that is reduced. The case a proper subterm of t is reduced follows immediately by structural induction. The proof goes by case analysis.

- Rule (1) is applicable, i.e. $t = (t_1 \cdot t_2)^{*n}$ and $t' = t_1^{*n} \cdot t_2^{*n}$, and $\phi' = \phi$. Suppose e is an evaluation function, and $e(n) = k$ for some k. Note that in general $\Psi(t_1^k \cdot t_2^k) = \Psi((t_1 \cdot t_2)^k)$. Since $\phi' = \phi$, $e \models \phi$ iff $e \models \phi'$. Therefore, $\Psi((t;\phi)) = \Psi((t';\phi'))$.
- Rule (2) is applicable, i.e. $t = (t_1 \cup t_2)^{*n}$ and $t' = t_1^{*n_1} \cdot t_2^{*n_2}$, and $\phi' = \phi \cup \{n_1 + n_2 = n\}$. Suppose e is an evaluation function, and $e(n) = k$ for some k. In general $\Psi((t_1 \cup t_2)^k) = \bigcup_{(i,j)\in\mathbb{N}\times\mathbb{N},\ i+j=k} \Psi(t_1^i \cdot t_2^j)$. Let e' be an evaluation function equal to e, except that $e'(n_1) = i$ and $e'(n_2) = j$, for some $i, j \in \mathbb{N}$, such that $i + j = k$. Now, if $e \models \phi$, then $e' \models \phi'$. Therefore, $\Psi((t;\phi)) \subseteq \Psi((t';\phi'))$.
  For the other direction, suppose e is any evaluation function such that $e \models \phi'$, that is $e(n_1) = i$, $e(n_2) = j$, $e(n) = k$, for some $i, j, k \in \mathbb{N}$, such that $i+j = k$. Note $e' \models \phi$, simply because n_1 and n_2 are chosen freshly. It is obvious that the language of t'_e is a subset (or equal) to the language of t_e. Therefore, $\Psi((t';\phi')) \subseteq \Psi((t;\phi))$.
- Rule (3) is applicable, that is $t = (t_1^*)^{*n}$ and $t' = t_1^{*n'}$, and $\phi' = \phi \cup \{n = 0 \implies n' = 0\}$. Suppose e is an evaluation function, and $e(n) = k$ for some k. We consider two cases: (1) $k \neq 0$, and (2) $k = 0$. Case (1): Take an arbitrary $w \in t_e$. Note that in general $(r^*)^k = r^*$, for any regexp r and $k \neq 0$. Therefore, $k \neq 0$ implies $w \in t_1^i$, for some $i \in \mathbb{N}$. Define e' as an evaluation function that is equal to e, except that $e'(n') = i$. Clearly $w \in t'_{e'}$. Since $k \neq 0$, $e \models \phi$ entails $e' \models \phi'$. Case (2): If $k = 0$, then the language of r^k is the set $\{\epsilon\}$ for any regexp r. Define e' as an evaluation function that is equal to e, except that $e'(n') = 0$. Note that the only constraint that is in

ϕ' and not in ϕ (i.e. $n = 0 \implies n' = 0$), is satisfied by e'. Thus, from $e \models \phi$ we derive $e' \models \phi'$. Therefore, $\Psi((t;\phi)) \subseteq \Psi((t';\phi'))$.
For the other direction, suppose e' is any evaluation function such that $e' \models \phi'$, that is $e'(n) = k$, $e'(n') = k'$, for some $k, k' \in \mathbb{N}$, such that $(k \neq 0) \vee (k' = 0)$. It is obvious that $t'_{e'} = t_{e'}$ when $k = 0$ (and consequently $k' = 0$). Now, let $k \neq 0$. Remark that not only n' is freshly chosen from Names and hence not present in t, but also the *-node tagged with t' is not tagged in t at all. This is due to our top-most reduction strategy; see figure 1. This is a crucial fact here, since simply tagging nodes, even when those tags are not bound in any constraint, affects the language of constrained regexps, cf. example 1. Now, from $(r^*)^k = r^*$, with $k \neq 0$, it follows that $\Psi((t';\phi')) \subseteq \Psi((t;\phi))$.

This completes our proof. □

4 Extracting Parikh Images from Most-Reduced Regexps

Suppose $(t_i;\emptyset)$ is a seeded $\mathcal{U}$-form regexp, and $(t_i;\emptyset)$ is reduced using $\mathcal{R}$ to $(t;\phi)$, where t is irreducible. According to theorem 1, such a t is always reached in a finite number of reduction steps, and due to theorem 3, $\Psi(t_i;\emptyset) = \Psi(t;\phi)$. This in particular implies $\Psi(t_i) = \Psi(t;\phi)$, due to the seeding procedure. Our goal here is to extract a semi-linear representations of $\Psi(t_i)$ from $(t;\phi)$. We consider two cases: (i) ϕ contains no implication constraints (i.e. no constraints of the form $n_i = 0 \implies n_j = 0$), and (ii) ϕ contains at least one implication constraint. Computing a semi-linear Parikh image of t_i in case (i) is straightforward, as described below. For case (ii), we give a procedure, called *remove_imp*, which removes the implication constraints from ϕ, and returns a finite set of implication-free constraint sets. It is then proved (theorem 4) that $\Psi(t_i) = \cup_{\varphi \in remove\_imp(\phi)} \Psi(t;\varphi)$. Intuitively, case (ii) is reduced to case (i).

Case (i). Constraint set with no implication constraint. In case ϕ contains no implication constraints, computing semi-linear representations of the Parikh image of $(t;\phi)$ is straightforward. This is because constraints of the form $n_i + n_j = n_k$ are inherently linear constraints. We demonstrate this via a number of examples.

Example 3. Let $t_i = (a \cdot (b \cup c))^{*n}$, with alphabet $A = \{a, b, c\}$. The following steps show how $\mathcal{R}$ reduces $(t_i;\emptyset)$.

$$\begin{array}{ll} (a \cdot (b \cup c))^{*n} & ;\emptyset \\ \rightarrow_{R_1} a^{*n} \cdot (b \cup c)^{*n} & ;\emptyset \\ \rightarrow_{R_2} a^{*n} \cdot b^{*n_1} \cdot c^{*n_2} & ;\{n_1 + n_2 = n\} \end{array}$$

From the final irreducible constrained regexp we get $\Psi(t_i) = \{v \in \mathbb{N}^3 \mid v = (\lambda_b + \lambda_c)\mathbf{e}_1 + \lambda_b\mathbf{e}_2 + \lambda_c\mathbf{e}_3,\ \lambda_b, \lambda_c \in \mathbb{N}\}$. This is indeed a linear representation. •

Example 4. Let $t_i = ((a \cup b) \cdot (c \cup d) \cdot e)^{*n}$, with alphabet $A = \{a, b, c, d, e\}$. The following steps show how $\mathcal{R}$ reduces $(t_i; \emptyset)$.

$$\begin{array}{ll} ((a \cup b) \cdot (c \cup d) \cdot e)^{*n} & ; \emptyset \\ \rightarrow_{R_1} (a \cup b)^{*n} \cdot ((c \cup d) \cdot e)^{*n} & ; \emptyset \\ \rightarrow_{R_1} (a \cup b)^{*n} \cdot (c \cup d)^{*n} \cdot e^{*n} & ; \emptyset \\ \rightarrow_{R_2} a^{*n_1} \cdot b^{*n_2} \cdot (c \cup d)^{*n} \cdot e^{*n} & ; \{n_1 + n_2 = n\} \\ \rightarrow_{R_2} a^{*n_1} \cdot b^{*n_2} \cdot c^{*n_3} \cdot d^{*n_4} \cdot e^{*n} & ; \{n_1 + n_2 = n; n_3 + n_4 = n\} \end{array}$$

Notice that equations of the form $n_1 + n_2 = n_3 + n_4$, with $n_i \in \mathbb{N}$, are satisfiable iff there exist $\lambda_1, \lambda_2, \lambda_3, \lambda_4 \in \mathbb{N}$, such that $n_1 = \lambda_1 + \lambda_2$, $n_2 = \lambda_3 + \lambda_4$, $n_3 = \lambda_1 + \lambda_3$ and $n_4 = \lambda_2 + \lambda_4$. In general, any finite number of equations of the form $A_1 + B_1 = \cdots = A_\ell + B_\ell$, with $A_{i \in 1..\ell}, B_{i \in 1..\ell} \in \mathbb{N}$, are simultaneously satisfiable iff there exist 2^ℓ natural numbers $\lambda_1, \cdots, \lambda_{2^\ell}$, whose combinations constitute A_i and B_i. Therefore, from the final irreducible constrained regexp we get $\Psi(t_i) = \{v \in \mathbb{N}^5 \mid v = (\lambda_1 + \lambda_2)\mathbf{e}_1 + (\lambda_3 + \lambda_4)\mathbf{e}_2 + (\lambda_1 + \lambda_3)\mathbf{e}_3 + (\lambda_2 + \lambda_4)\mathbf{e}_4 + (\lambda_1 + \lambda_2 + \lambda_3 + \lambda_4)\mathbf{e}_5,\ \lambda_{i \in 1..4} \in \mathbb{N}\}$. This is indeed a linear representation. •

Case (ii). Constraint set with at least one implication constraint. Implication constraints, i.e. constraints of the form $n_i = 0 \implies n_j = 0$, are intuitively "non-linear". In the following, we give a case splitting procedure to remove such constraints from ϕ, without affecting the set of evaluation functions that satisfy ϕ. This implies that the procedure does not change the language of $(t; \phi)$.

Let ϕ be a set of constraints of the form defined in section 2 over Names. We give an algorithm to remove implication constraints from ϕ. For this purpose, we extend the definition of constraint sets (given in section 2) to include constraints of the form $n = 0$, and $n = n' + 1$ as eligible members. The related definitions (satisfiability, etc.) are extended in the obvious way. Furthermore, in the following we assume that all implication constraints in ϕ which have the same antecedent are lumped together. For example, two constraints $(n = 0 \implies n_1 = 0)$ and $(n = 0 \implies n_2 = 0)$ are lumped into the constraint $n = 0 \implies n_1 = 0 \wedge n_2 = 0$. This merely syntactical convention decreases the number of times the procedure *remove_imp* recurs on ϕ.

Theorem 4. *Fix a constrained regexp $(t; \phi)$. Given the set of constraints ϕ, algorithm 1 returns a finite set of constraint sets $\Phi = \{\phi_1, \cdots, \phi_n\}$ such that $L(t; \phi) = \cup_{\phi_i \in \Phi} L(t; \phi_i)$.*

Proof. Remark that Φ is finite because the number of implication constraints in ϕ is strictly decreasing in each recursion of algorithm 1. The claim of the theorem is immediate by noting that for any evaluation function e, $e \models \phi \iff \bigvee_{\phi_i \in \Phi} e \models \phi_i$. This is because $\forall n \in \mathbb{N}.\ (n = 0 \vee \exists n_\nu \in \mathbb{N}.\ n = n_\nu + 1)$. □

Example 5. Let $\phi = \{n_3 = 0 \implies n_4 = 0; n_2 + n_3 = n; n = 0 \implies n_1 = 0\}$. Suppose the first implication chosen by algorithm 1 is $n_3 = 0 \implies n_4 = 0$. This results in two constraint sets $\phi_1 = \{n_3 = 0; n_4 = 0; n_2 + n_3 = n; n = 0 \implies n_1 = 0\}$ and $\phi_2 = \{n_3 = n' + 1; n_2 + n_3 = n; n = 0 \implies n_1 = 0\}$. Recursively,

Algorithm 1. Removes implications from constraint sets

procedure *remove_imp*(ϕ)
  if there exists a constraint $(n = 0 \implies n_1 = 0 \wedge \cdots \wedge n_\ell = 0)$ in ϕ **then**
    let $\phi := \phi \setminus \{n = 0 \implies n_1 = 0 \wedge \cdots \wedge n_\ell = 0\}$
    return
      remove_imp($\{n = 0, n_1 = 0, \cdots, n_\ell = 0\} \cup \phi$)
      $\cup$
      remove_imp($\{n = n_\nu + 1\} \cup \phi$) where $n_\nu \in$ Names is freshly chosen
  end if
  return $\{\phi\}$

on ϕ_1 the algorithm returns $\phi_3 = \{n_3 = 0; n_4 = 0; n_2 + n_3 = n; n = 0; n_1 = 0\}$ and $\phi_4 = \{n_3 = 0; n_4 = 0; n_2 + n_3 = n; n = n'' + 1\}$. Similarly, on ϕ_2 the algorithm returns $\phi_5 = \{n_3 = n' + 1; n_2 + n_3 = n; n = 0; n_1 = 0\}$ and $\phi_6 = \{n_3 = n' + 1; n_2 + n_3 = n; n = n'' + 1\}$. Ultimately, *remove_imp*(ϕ) $= \{\phi_3, \phi_4, \phi_5, \phi_6\}$. •

Algorithm 1 effectively transforms constraint set ϕ into a finite set of constraint sets $\Phi = \{\phi_1 \cdots, \phi_\ell\}$, such that ϕ_i contains no implication constraints. That is, case (ii) is reduced to case (i), except for constraints of the form $n = n_\nu + 1$ that algorithm 1 may introduce. To cast constraints of the form $n = n_\nu + 1$ into case (i) we need another round of case distinction. This is best explained via an example.

Example 6. Let $t_i = ((a \cup b) \cdot c^* \cdot d)^*$, with $A = \{a, b, c, d\}$. The most-reduced constrained regexp that $\mathcal{R}$ produces from t_i is then $(t; \phi)$ with $t = a^{*n_1} \cdot b^{*n_2} \cdot c^{*n_3} \cdot d^{*n}$ and $\phi = \{n = 0 \implies n_3 = 0, n_1 + n_2 = n\}$. Then, *remove_imp*($\phi$) $= \{\phi_1, \phi_2\}$ where $\phi_1 = \{n = 0, n_3 = 0, n_1 + n_2 = n\}$ and $\phi_2 = \{n = n_\nu + 1, n_1 + n_2 = n\}$. The constraints in ϕ_1 are all linear, hence fall into case (i). In ϕ_2 however the constraint $n_1 + n_2 = n_\nu + 1$ results in $n_1 = \lambda_1 + \lambda_2$, $n_2 = \lambda_3 + \lambda_4$, $n_\nu = \lambda_1 + \lambda_3$ and $1 = \lambda_2 + \lambda_4$ (cf. example 4). The last constraint needs a case distinction; namely, either $\lambda_2 = 1$ and $\lambda_4 = 0$, or $\lambda_2 = 0$ and $\lambda_4 = 1$. With this case distinction, ϕ_2 also falls into case (i). •

Width of produced images. We now turn to the width of produced semi-linear representations of Parikh images of regexps.

Theorem 5. *Let t be a $\mathcal{U}$-form regexp. Then, the width of $\Psi(t)$, produced by the reduction process, is (at worst case) given by*

$$\omega(\Psi(t)) = 2^{\aleph_2(t) + \aleph_3(t)}$$

Proof. Take seeded constrained regexp $(t; \emptyset)$ and suppose it is reduced using $\mathcal{R}$ to $(r; \phi)$, where r is irreducible. The key observation is that the width of $\Psi(t)$ is equal to $|$*remove_imp*(ϕ)$|$ multiplied by the number of case distinctions performed for linearizing $n = n_\nu + 1$ constraints.

Note that $|$*remove_imp*(ϕ)$| = 2^{\aleph_3(t)}$, by definition 2. Now, all the sets in *remove_imp*(ϕ) (might) require further case distinctions for constraints of the form $n = n_\nu + 1$ (cf. example 6). The number of these case distinctions is $2^{\aleph_2(t)}$, according to definition 2. The number of linear components in $\Psi(t)$ is therefore bounded by $2^{\aleph_2(t)+\aleph_3(t)}$. □

To relate theorem 5 to the measures given in the introduction (section 1), let us assume that the widths of images of $\mathcal{U}$-form regexps t_1 and t_2 are, respectively, w_1 and w_2. The reduction process generates an image of width $w_1 + w_2$ for $t_1 \cup t_2$ (cf. lemma 1). For $t_1 \cdot t_2$, we note that $\aleph_i(t_1 \cdot t_2) = \aleph_i(t_1) + \aleph_i(t_2)$, for $i \in \{2, 3\}$. Then, due to theorem 5, $\omega(\Psi(t_1 \cdot t_2)) = w_1 w_2$. For t_1^*, we remark that reducing t_1^* using $\mathcal{R}$ results in a constraint set which has at most one implication constraint more than the constraint set generated for t_1. This is due to lumping the implication constraints (see the discussion right before theorem 4). Then, due to theorem 5, $\omega(\Psi(t_1^*)) = 2^{1+\aleph_2(t_1)+\aleph_3(t_1)}$, that is $\omega(\Psi(t_1^*)) = 2w_1$.

Acknowledgement. We are grateful to B. Conchinha Montalto, W. Fokkink and L. Schrijver for fruitful discussions. Mohammad Torabi Dashti has been supported by AVANTSSAR, FP7-ICT-2007-1 Project no. 216471.

References

1. Aceto, L., Ésik, Z., Ingólfsdóttir, A.: A fully equational proof of Parikh's theorem. ITA 36(2), 129–153 (2002)
2. Blattner, M., Latteux, M.: Parikh-bounded languages. In: Even, S., Kariv, O. (eds.) ICALP 1981. LNCS, vol. 115, pp. 316–323. Springer, Heidelberg (1981)
3. Delgado, M.: Commutative images of rational languages and the Abelian kernel of a monoid. ITA 35(5), 419–435 (2001)
4. Goldstine, J.: A simplified proof of Parikh's theorem. Discrete Math. 19, 235–239 (1977)
5. Ibarra, O., Kim, C.: A useful device for showing the solvability of some decision problems. In: STOC 1976, pp. 135–140. ACM, New York (1976)
6. Parikh, R.: On context-free languages. J. ACM 13(4), 570–581 (1966)
7. Pilling, D.: Commutative regular equations and Parikh's theorem. J. London Math. Soc. 6, 633–666 (1973)
8. Terese: Term Rewriting Systems. Cambridge Tracts in Theoretical Computer Science, vol. 55. Cambridge University Press, Cambridge (2003)
9. Verma, K.N., Seidl, H., Schwentick, T.: On the complexity of equational Horn clauses. In: Nieuwenhuis, R. (ed.) CADE 2005. LNCS (LNAI), vol. 3632, pp. 337–352. Springer, Heidelberg (2005)
10. Xie, G., Li, C., Deng, Z.: Linear reachability problems and minimal solutions to linear Diophantine equation systems. TCS 328, 203–219 (2004)

Breaking the Rectangle Bound Barrier against Formula Size Lower Bounds

Kenya Ueno

The Young Researcher Development Center and
Graduate School of Informatics,
Kyoto University
kenya@kuis.kyoto-u.ac.jp

Abstract. Karchmer, Kushilevitz and Nisan formulated the formula size problem as an integer programming problem called the rectangle bound and introduced a technique called the LP bound, which gives a formula size lower bound by showing a feasible solution of the dual problem of its LP-relaxation. As extensions of the LP bound, we introduce novel general techniques proving formula size lower bounds, named a quasi-additive bound and the Sherali-Adams bound. While the Sherali-Adams bound is potentially strong enough to give a lower bound matching to the rectangle bound, we prove that the quasi-additive bound can surpass the rectangle bound.

1 Introduction

Proving formula size lower bounds is a fundamental problem in complexity theory as a weaker version of the circuit size lower bound problem and $\mathbf{P} \neq \mathbf{NP}$. A super-polynomial formula size lower bound for a function in **NP** implies $\mathbf{NC}^1 \neq \mathbf{NP}$ [19]. As generalizations of the classical result of Khrapchenko [8] who proved an n^2 formula size lower bound for the parity function, there are a lot of techniques studied to improve formula size lower bounds. With all the efforts, improvements are far and few between. Karchmer, Kushilevitz and Nisan [6] formulated the formula size problem as an integer programming problem called the rectangle bound and introduced a technique called the LP bound, which gives a lower bound by showing a feasible solution of the dual problem of its LP-relaxation. At the same time, they also showed that it cannot prove a lower bound larger than $4n^2$ for non-monotone formula size in general. Lee [14] proved that the LP bound [6] subsumes the quantum adversary bound of Laplante, Lee and Szegedy [10], which in turn subsumes most of known techniques such as Khrapchenko [8], its extension by Koutsoupias [9] and a key lemma used in the proof of Håstad [3] showing the current best formula size lower bound $n^{3-o(1)}$. Ueno [20] devised a stronger version of the LP bound by a cutting plane approach utilizing the theory of stable set polytope. However, it is difficult to determine the complete facet structure of the polytope associated with the formula size problem. Thus, improvements of the approach are limited.

P. Hliněný and A. Kučera (Eds.): MFCS 2010, LNCS 6281, pp. 665–676, 2010.

Lift-and-project methods systematically incorporate tighter and tighter constraints into any LP formulation. There are several lift-and-project methods such as Sherali and Adams [16,17], Balas, Ceria and Cornuejols [2] and Lovász and Schrijver [15] and Lasserre [11]. Laurent [12] gave a comparison among these techniques. Among the several techniques, the technique of Sherali and Adams [16,17] has some advantages as the strongest one for LP formulations with relatively simpler descriptions. These techniques have attracted much attention from several contexts. In Section 5, we devise further another stronger version of the LP bound using the lift and project method of Sherali and Adams [16,17] and its application to the set partition polytope by Sherali and Lee [18]. It yields a convex hull of integral solutions and completely closes the integrality gap which causes the limit of the original technique. It is potentially strong enough to prove a lower bound matching to the rectangle bound.

More recently, Hrubeš, Jukna, Kulikov and Pudlák [5] discussed a notion of subadditive rectangle measures on combinatorial rectangles as a conceptual extension of a well-known notion of formal complexity measures. Inspired by this notion and the Sherali-Adams bound, we introduce yet another stronger version of the LP bound, which we name a quasi-additive bound, in Section 3. It directly gives a lower bound for formula size and the protocol partition number of the Karchmer-Wigderson game. In Section 4, we show that the quasi-additive bound can surpass the rectangle bound. So, the quasi-additive bound is not upper-bounded by the rectangle bound in general. This is quite surprising because the quasi-additive bound can be seen as a simple extension of the LP bound of Karchmer, Kushilevitz and Nisan [6], which is originally defined as a relaxation of the rectangle bound. In fact, we can prove that the quasi-additive bound is potentially strong enough to prove the matching formula size lower bound for any Boolean function and the matching protocol partition number for any relation. Another interesting property of the quasi-additive bound is that we can derive a formula size lower bound for any Boolean function from a solution of the quasi-additive bound for the universal relation.

Since the Sherali-Adams and quasi-additive bounds are pure extensions of the LP bound, they can prove formula size lower bounds provable by any techniques subsumed by the LP bound. To extend a solution space for them, we introduce two useful techniques, named a cross argument and a triplet argument in Section 4 and Section 6, respectively. The cross argument is useful to break the rectangle bound barrier and applicable only for the quasi-additive bound. On the other hand, the triplet argument can cover a main part of techniques discussed in [20] and is applicable for both the Sherali-Adams and quasi-additive bounds. Breaking the rectangle bound barrier against formula size lower bounds implies a strong potential of the quasi-additive bound because the rectangle bound is not so far from formula size in general (See Theorem 2 and [13]) and almost all Boolean functions require formula size of at least $\Omega(2^n/\log n)$ (See, e.g., [21]). We hope that our generic techniques will be useful to surpass the best formula size lower bound $n^{3-o(1)}$ of Håstad [3].

2 Preliminaries

We assume that the readers are familiar with the basics of Boolean functions and linear programming. A Boolean function f is called monotone if $x \leq y$ implies $f(x) \leq f(y)$ for all $x, y \in \{0,1\}^n$.

Definition 1 (Formula Size). *A formula is a binary tree with each leaf labeled by a literal and each internal node labeled by either of the binary connectives* $\wedge$ *and* $\vee$. *A literal is either a variable or its negation. The size of a formula is its number of literals. We define formula size* $L(f)$ *of a Boolean function* f *as the size of the smallest formula computing* f. *We also define* $L_m(f)$ *as the monotone formula size of a monotone Boolean function* f *where a monotone formula is a formula without negations.*

Karchmer and Wigderson [7] characterize formula size of any Boolean function in terms of a communication game. In the game, given a Boolean function f, Alice gets an input x such that $f(x) = 1$ and Bob gets an input y such that $f(y) = 0$. The goal of the game is to find an index i such that $x_i \neq y_i$. Here, x_i and y_i denote the i-th bits of x and y, respectively. The number of leaves in a best communication protocol for the Karchmer-Wigderson game is equal to the formula size of f. From the Karchmer-Wigderson game, we consider the following matrix called the communication matrix.

Definition 2 (Communication Matrix). *Given a Boolean function* f, *its communication matrix is defined as a matrix whose rows and columns are indexed by* $X = f^{-1}(1)$ *and* $Y = f^{-1}(0)$, *respectively. Each cell of the matrix contains indices* i *such that* $x_i \neq y_i$. *A combinatorial rectangle is a direct product* $X' \times Y'$ *where* $X' \subseteq X$ *and* $Y' \subseteq Y$. *A combinatorial rectangle* $X' \times Y'$ *is called monochromatic if every cell* $(x, y) \in X' \times Y'$ *contains the same index* i. *To describe it simply, we define a relation* $R_f \subseteq X \times Y \times \{1, 2, \cdots, n\}$ *as* $R_f = \{(x, y, i) \mid x \in X,\ y \in Y,\ x_i \neq y_i\}$. *We can also define the monotone version of the communication matrix and the relation associated with a monotone Boolean function* f *as* $R_f^m = \{(x, y, i) \mid x \in X,\ y \in Y,\ x_i = 1,\ y_i = 0\}$.

To prove a lower bound, we sometimes restrict rows and columns of the communication matrix as $R' = \{(x, y, i) \mid (x, y, i) \in R,\ x \in X',\ y \in Y'\}$ for some $X' \subseteq X$ and $Y' \subseteq Y$. The number of leaves in a best communication protocol for the Karchmer-Wigderson game is equivalent to the following bound.

Definition 3 (Protocol Partition Number). *For any combinatorial rectangle* $X' \times Y'$, *we call its partition a pair of* $X'_1 \times Y'$ *and* $X'_2 \times Y'$ *where* $X' = X'_1 \cup X'_2$ *and* $X'_1 \cap X'_2 = \emptyset$, *or a pair of* $X' \times Y'_1$ *and* $X' \times Y'_2$ *where* $Y' = Y'_1 \cup Y'_2$ *and* $Y'_1 \cap Y'_2 = \emptyset$. *The minimum number of disjoint monochromatic rectangles which recursively partition the communication matrix associated with a relation* R *is defined as* $C^P(R)$, *called the protocol partition number.*

Then, the theorem of Karchmer and Wigderson [7] can be stated as follows.

Theorem 1 ([7]). $\forall f$, $C^P(R_f) = L(f)$ *and* $C^P(R_f^m) = L_m(f)$.

The minimum number of disjoint monochromatic rectangles which exactly cover all cells in the communication matrix gives a lower bound for the protocol partition number because a protocol partition itself is one of exact covers by disjoint monochromatic rectangles. We call it the rectangle bound defined as follows.

Definition 4 (Rectangle Bound). *The minimum size of an exact cover by disjoint monochromatic rectangles for the communication matrix associated with a relation R is defined as $C^D(R)$, called the rectangle bound.*

On the relation between the protocol partition number and the rectangle bound, we know the following result. For the proof, we recommend [13].

Theorem 2 ([6]). $\forall R,\ C^D(R) \le C^P(R) \le 2^{O(\log^2 C^D(R))}$.

Karchmer, Kushilevitz and Nisan [6] formulate the rectangle bound as an integer programming problem and give its LP relaxation.

Definition 5 (LP Bound). *We define* $\mathbf{LP}(R)$ *as the optimal value of the following linear programming formulation associated with a relation R. Let C be the set of all defined cells, M be the set of all monochromatic rectangles and Z_r be a variable associated with each monochromatic rectangle $r \in M$. Then, the LP-relaxation can be written as* $\min \sum_{r \in M} Z_r$ *such that* $\sum_{r \ni c} Z_r = 1$ *for each cell $c \in C$ and $Z_r \ge 0$ for each $r \in M$. The dual problem can be written as* $\max \sum_{c \in C} W_c$ *such that* $\sum_{c \in r} W_c \le 1$ *for each $r \in M$. Here, W_c is a variable indexed by a cell $c \in C$.*

From the duality theorem, showing a feasible solution of the dual problem gives a formula size lower bound.

Theorem 3 ([6]). $\forall f,\ \mathbf{LP}(R_f) \le L(f)$ *and* $\mathbf{LP}(R_f^m) \le L_m(f)$.

They define the universal relation to show the limitation of their technique.

Definition 6 (Universal Relation). *The universal relation U_n represents a matrix whose rows and columns are indexed by $X = Y = \{0,1\}^n$ and each cell (x, y) is indexed by $\{i \mid x_i \ne y_i\}$. It is defined as $U_n = \{(x, y, i) \mid x \in \{0,1\}^n, y \in \{0,1\}^n, x_i \ne y_i\}$. $C^P(U_n)$ and $C^D(U_n)$ are defined in the same way. Note that any cell (x, y) where $x = y$ is undefined and are not counted for any partition. That is, monochromatic rectangles partition all defined cells without covering undefined cells.*

It subsumes any relation R_f as a submatrix. Its protocol partition number and rectangle bound also subsume those of any Boolean function as $C^P(R_f) \le C^P(U_n)$ and $C^D(R_f) \le C^D(U_n)$ for any Boolean function f. They show a limitation of their technique by showing the following theorem.

Theorem 4 ([6]). $\forall f,\ \mathbf{LP}(R_f) \le \mathbf{LP}(U_n) \le 4n^2$.

Thus, the LP bound and all the subsumed techniques [10] cannot prove a formula size lower bound larger than $4n^2$. Limits inherent in previously known proof techniques which get stuck around $\Omega(n^2)$ heavily rely on the above theorem.

3 A Quasi-additive Bound for Formula Size Lower Bounds

In this section, we devise a stronger version of the LP bound, which is derived from a concept of subadditive rectangle measures by Hrubeš, Jukna, Kulikov and Pudlák [5] and inspired by the Sherali-Adams bound discussed in Section 5. We write Γ as the set of combinatorial rectangles and $\Re$ as the set of real numbers. Hrubeš, Jukna, Kulikov and Pudlák [5] introduce a notion of rectangle measures. We call $\mu : \Gamma \mapsto \Re$ a subadditive rectangle measure if it satisfies the following two properties.

1. Normalization: $\mu(m) \leq 1$ for each monochromatic rectangle $m \in M$.
2. Subadditivity: $\mu(r) \leq \mu(r_1) + \mu(r_2)$ for each combinatorial rectangle $r \in \Gamma$ and its arbitrary partition into r_1 and r_2.

They show that $\mu(r)$ gives a lower bound for the protocol partition number $C^P(R)$ by a simple inductive argument and any relation R where r is the whole rectangle associated with the relation R. We can simply extend it for the universal relation as follows.

Lemma 1 ([5]). *If μ is a subadditive rectangle measure for a relation R, then $\mu(X \times Y) \leq C^P(R)$, even in the case of the universal relation in which $\mu(r) = 0$ for every combinatorial rectangle r containing only an undefined cell, where $X \times Y$ is the whole matrix associated with the relation R.*

A remarkable fact is that, if we strengthen the condition "Subadditivity" as

3. Additivity: $\mu(r) = \mu(r_1) + \mu(r_2)$ for each combinatorial rectangle $r \in \Gamma$ and its arbitrary partition into r_1 and r_2,

then it is equivalent to the dual problem of the original LP formulation of Karchmer, Kushilevitz and Nisan [6]. Then, we consider the following LP formulation.

Definition 7 (Quasi-Additive Bound). *Let C be the set of all cells, M be the set of all monochromatic rectangles and Γ be the set of all combinatorial rectangles associated with a relation R. We define* $\mathbf{QA}(R)$ *as the optimal value of the following linear program formulation.*

$$
\begin{aligned}
\max \ & \sum_{c \in C} V_c \\
s.t. \ & \sum_{c \in r} V_c + \sum_{c \notin r} V_{c,r} \leq 1, \quad (\textit{for each } r \in M) \\
& \sum_{c \notin r_1} V_{c,r_1} + \sum_{c \notin r_2} V_{c,r_2} \geq \sum_{c \notin r} V_{c,r}. \\
& (\textit{for each } r \in \Gamma \textit{ and its arbitrary partition into } r_1 \textit{ and } r_2)
\end{aligned}
$$

When we consider the universal relation, we fix $V_c = 0$ for every undefined cell and $V_{c,r} = 0$ for every combinatorial rectangle r which only contains an undefined cell.

We refer to it as the quasi-additive bound. It is stronger than the LP bound and gives a lower bound for the protocol partition number.

Lemma 2. $\forall R,\ \mathbf{LP}(R) \leq \mathbf{QA}(R) \leq C^P(R)$.

Proof. If we set $V_{c,r} = 0$ for each c and r of the quasi-additive bound, it is equivalent to the original LP bound. So, we have $\mathbf{LP}(R) \leq \mathbf{QA}(R)$. To see $\mathbf{QA}(R) \leq C^P(R)$, we regard $\mu(r) = \sum_{c\in r} V_c + \sum_{c\notin r} V_{c,r}$ as a rectangle measure. Then, we have $\mu(X \times Y) = \sum_{c\in C} V_c$, which is equal to the objective value of the quasi-additive bound, because $C = X\times Y$ is the whole rectangle associated with R. From the additivity of V_c, the conditions "Normalization" and "Subadditivity" is equivalent to the first and second constraints of the quasi-additive bound, respectively. Thus, if assignments of V_c and $V_{c,r}$ satisfy all of the first and second constraints of the quasi-additive bound, $\mu(r)$ is a subadditive rectangle measure. Consequently, we have $\mathbf{QA}(R) \leq C^P(R)$. □

We can derive a formula size lower bound for any Boolean function from a solution for $\mathbf{QA}(U_n)$ by calculating $\sum_{c\in r} V_c + \sum_{c\notin r} V_{c,r}$ where r is $f^{-1}(1) \times f^{-1}(0)$. We can eliminate the redundancy of the quasi-additive bound by summarizing variables as $\overline{V}_r = \sum_{c\notin r} V_{c,r}$ for each combinatorial rectangle r and adding a constraint $\overline{V}_{X\times Y} = 0$. However, as we will show, this redundancy is useful to construct a solution for the quasi-additive bound. We can also prove that it is potentially strong enough to give the matching formula size lower bounds.

Theorem 5. $\forall R,\ \mathbf{QA}(R) = C^P(R)$.

Proof. From the information of the protocol partition number $P(r)$ for each combinatorial rectangle r, we can construct a feasible solution whose objective value is equal to $P(X \times Y)$ $(= C^P(R))$. More precisely, we assign V_c so as to satisfy $\sum_{c\in C} V_c = P(X \times Y)$. Then, we assign $V_{c,r}$ so as to satisfy $\sum_{c\notin r} V_{c,r} = P(r) - \sum_{c\in r} V_c$. These assignments satisfy all the constraints of the quasi-additive bound and give the matching lower bound. □

Corollary 1. $\forall f,\ \mathbf{QA}(R_f) = L(f)$ *and* $\mathbf{QA}(R_f^m) = L_m(f)$.

4 A Cross Argument for the Quasi-additive Bound

In this section, we give an example of a relation for which the quasi-additive bound can surpass the rectangle bound. For this purpose, we devise a novel technique named a cross argument to give a solution of the quasi-additive bound.

Theorem 6. $\exists R,\ \mathbf{QA}(R) > C^D(R)$.

Proof. We take 2 disjoint subsets of $\{0,1\}^8$ as

$$X = \{10011000, 00101001, 00010110, 01100100\},$$

$$Y = \{01011101, 10110101, 01111010, 10101110\}.$$

	01011101	10110101	01111010	10101110
10011000	1	5	1	4
00101001	3	5	8	8
00010110	7	7	6	4
01100100	3	2	6	2

Fig. 1. The Monotone Communication Matrix of the Relation R

	01011101	10110101	01111010	10101110
10011000	a	b	c	d
00101001	e	f	g	h
00010110	h	g	f	e
01100100	d	c	b	a

Fig. 2. 8 pairs of 2 cells from 16 cells

Then, we consider the monotone relation R of X and Y as Figure 1. For the relation R, it is easy to see that $C^P(R) \leq 10$, $C^D(R) \leq 8$ by a cover with 8 maximal monochromatic rectangles and $\mathbf{LP}(R) \geq 8$ by assigning a weight $\frac{1}{2}$ for each cell.

Now, we prove $\mathbf{QA}(R) \geq 10$. We assign $V_c = \frac{5}{8}$ for each cell. So, the total weight is 10. To give an assignment rule of $V_{c,r}$, we consider 2 sorts of 8 pairs from 16 cells. One is composed of 8 pairs each of which has 2 cells with the same index in Figure 1. The other is composed of 8 pairs each of which has 2 cells with the same alphabet in Figure 2. For any 2 cells c_1 and c_2 and any combinatorial rectangle r such that $c_1 \notin r_1$ and $c_2 \in r_2$, we define $\Delta_{c_1,r}(c_2)$ as follows. Let c_2' be the other cell which has the same index with c_2.

- If c_1 and c_2 have the same alphabet and r contains c_2', we define $\Delta_{c_1,r}(c_2) = -\frac{1}{8}$.
- If c_1 and c_2 have the same alphabet and r does not contain c_2', we define $\Delta_{c_1,r}(c_2) = \frac{3}{8}$.
- If c_1 and c_2 have different alphabets, we define $\Delta_{c_1,r}(c_2) = 0$.

Then, we assign $V_{c_1,r} = \sum_{c_2 \in r} \Delta_{c_1,r}(c_2)$ for any $c_1 \notin r$. To verify the first constraints of the quasi-additive bound, it is sufficient to consider the 2 cases as monochromatic rectangles with either 1 or 2 cells. In both cases, we have $\sum_{c\in r} V_c + \sum_{c \notin r} V_{c,r} = 1$.

From now on, we consider the second constraint $\sum_{c\notin r_1} V_{c,r_1} + \sum_{c\notin r_2} V_{c,r_2} \geq \sum_{c \notin r} V_{c,r}$ where r_1 and r_2 be an arbitrary partition of a combinatorial rectangle r. We can assume that there is a pair of 2 cells c_1 and c_2 with the same alphabet in r. Otherwise, the constraint cannot be violated because $\Delta_{c_1,r}(c_2) = \Delta_{c_1,r_1}(c_2) = \Delta_{c_1,r_2}(c_2) = 0$ for any $c_1, c_2 \in r$. We can also assume that the pair is partitioned into the 2 combinatorial rectangles as $c_1 \in r_1$ and $c_2 \in r_2$ in the following argument. We consider the case in which any pair of 2 cells c_1 and c_2 in r with the same alphabet are also in r_1. In this case, the assignments of Δ concerned with any pair of c_1 and c_2 in r do not decrease by the partition because $\frac{3}{8} \geq -\frac{1}{8}$. Moreover, no (negative) assignments of Δ do not appear by the partition. Hence, the constraint cannot be violated. The same thing is true for the case of r_2 instead of r_1.

Then, we define the diagonal pair p^d of a pair p in Figure 2 as the pair such that the 4 cells in p^d and p compose a 4×4 combinatorial rectangle. As an example, we consider the case when $c_1 = [1, a] \in r_1$ and $c_2 = [2, a] \in r_2$ without loss of generality. Here, we identify each cell by its index and alphabet in the figures. In this case, r_1 and r_2 must also partition the diagonal pair $c_3 = [3, d]$ and $c_4 = [4, d]$. Then, we take $c'_1 = [1, c]$, $c'_2 = [2, c]$, $c'_3 = [3, e]$ and $c'_4 = [4, e]$ be the 4 cells which have the same index with c_1, c_2, c_3 and c_4, respectively. If at least one of 4 pairs (c_1, c'_1) , (c_2, c'_2), (c_3, c'_3) and (c_4, c'_4) has been already divided at the time of the partition into r_1 and r_2, then the changes of assignments of Δ concerned with only c_1, c_2, c_3 and c_4 are represented as either $\frac{3}{8} - \frac{1}{8} - \frac{1}{8} - \frac{1}{8} \geq 0$, $\frac{3}{8} + \frac{3}{8} - \frac{1}{8} - \frac{1}{8} \geq 0$, $\frac{3}{8} + \frac{3}{8} + \frac{3}{8} - \frac{1}{8} \geq 0$ or $\frac{3}{8} + \frac{3}{8} + \frac{3}{8} + \frac{3}{8} \geq 0$. Any of these does not cause a violation of the constraint. The same thing is true for any other pairs with the same alphabet. So, we can assume either $c_1, c'_1, c_3, c'_3 \in r_1$ and $c_2, c'_2, c_4, c'_4 \in r_2$, or $c_1, c'_1, c_4, c'_4 \in r_1$ and $c_2, c'_2, c_3, c'_3 \in r_2$. In both cases, we have $\sum_{c \notin r_1} V_{c,r_1} = 0$ and $\sum_{c \notin r_2} V_{c,r_2} = 0$ where r_1 and r_2 are 2×4 combinatorial rectangles. Since r is the whole rectangle, we also have $\sum_{c \notin r} V_{c,r} = 0$. Consequently, all the constraints of the quasi-additive bound are satisfied. □

We also know a smaller relation R' of a 3×3 matrix having the gap between the rectangle bound and the protocol partition number. In this case, the gap is also smaller as $C^P(R') = 6$ and $C^D(R') = 5$.

The communication matrix discussed in the above proof is monotone and restricted. We can give an example of a non-monotone and whole communication matrix with the gap between the rectangle bound and the protocol partition number. Laplante, Lee and Szegedy [10] defined a 4-bit Boolean function f_A called Ambainis' function following a similar construction of Ambainis [1]. (See [10] and [4] for more detailed treatment of Ambainis' function.) It outputs 1 when $x_1 \leq x_2 \leq x_3 \leq x_4$ or $x_1 \geq x_2 \geq x_3 \geq x_4$. In Figure 3, we write the whole communication matrix R_{f_A} of Ambainis' function. We can prove $\mathbf{LP}(R_{f_A}) \geq 8$ by assigning a weight $\frac{1}{2}$ for each cell whose number of indices is 1 and a weight 0 otherwise. In Figure 4, we show an upper bound of its rectangle bound $C^D(R_{f_A}) \leq 8$. (Distinct numbers and numbers with apostrophe represent distinct monochromatic rectangles.) Thus, any rectangle bound based techniques cannot improve the LP bound. On the other hand, we know a smallest formula of size 10 for Ambainis' function as

$$((x_1 \vee \neg x_2) \wedge (x_2 \vee \neg x_3) \wedge \neg x_4) \vee ((\neg x_1 \vee x_2) \wedge (\neg x_2 \vee x_3) \wedge x_4).$$

Thus, $L(f_A) = C^P(R_{f_A}) \leq 10$. To the best of our knowledge, the quasi-additive bound is the first generic technique which can prove the matching lower bound $L(f_A) \geq \mathbf{QA}(R_{f_A}) \geq 10$.

	0100	0010	1010	0110	1001	0101	1101	1011
0000	2	3	1,3	2,3	1,4	2,4	1,2,4	1,3,4
1000	1,2	1,3	3	1,2,3	4	1,2,4	2,4	3,4
1100	1	1,2,3	2,3	1,3	2,4	1,4	4	2,3,4
1110	1,3	1,2	2	1	2,3,4	1,3,4	3,4	2,4
0001	2,4	3,4	1,3,4	2,3,4	1	2	1,2	1,3
0011	2,3,4	4	1,4	2,4	1,3	2,3	1,2,3	1
0111	3,4	2,4	1,2,4	4	1,2,3	3	1,3	1,2
1111	1,3,4	1,2,4	2,4	1,4	2,3	1,3	3	2

Fig. 3. The Communication Matrix of Ambainis' function

	0100	0010	1010	0110	1001	0101	1101	1011
0000	2	3	3	2	1'	2	2	1'
1000	1	3	3	1	4'	4'	4'	4'
1100	1	3	3	1	4'	4'	4'	4'
1110	1	2'	2'	1	2'	3'	3'	2'
0001	2	3	3	2	1'	2	2	1'
0011	4	4	4	4	1'	3'	3'	1'
0111	4	4	4	4	1'	3'	3'	1'
1111	1	2'	2'	1	2'	3'	3'	2'

Fig. 4. An Upper Bound 8 for the Rectangle Bound of Ambainis' function

5 Applying Sherali-Adams' Method to the LP Bound

In this section, we strengthen the technique of Karchmer, Kushilevitz and Nisan [6] by a lift and project technique of Sherali and Adams [16,17]. While it is upper-bounded by the rectangle bound, it is worthwhile to introduce because it has a similar structure with the quasi-additive bound. So, a solution for the Sherali-Adams bound may be useful to give a solution for the quasi-additive bound.

We write $P \sqsubseteq M$ when all monochromatic rectangles in $P \subseteq M$ are disjoint. We also write $r \perp P$ when a monochromatic rectangle r does not intersect any monochromatic rectangle in P. Let C be the set of all defined cells, M be the set of all monochromatic rectangles and Z_P be a variable associated with a set of disjoint monochromatic rectangles P in M. Now, we apply the lift and project technique to the LP-relaxation of the rectangle bound. The primal problem is written as follows.

$$\begin{aligned} &\min \sum_{r \in M} Z_{\{r\}} \\ &\text{s.t.} \sum_{r: c \in r \perp P} Z_{P \cup \{r\}} = Z_P, \quad \text{(for each } c \in C \text{ and each } P \sqsubseteq M \text{ s.t. } c \notin \bigcup_{r \in P} r) \\ &\qquad Z_P \geq 0, \qquad\qquad \text{(for each } P \sqsubseteq M) \\ &\qquad Z_\emptyset = 1. \end{aligned}$$

This is a simple modification of Sherali and Lee [18], which discuss the set partition polytope in general. The dual problem is written as follows.

$$\begin{aligned}
\max & \sum_{c \in C} W_{c,\emptyset} \\
\text{s.t.} & \sum_{c \in r} W_{c,\emptyset} - \sum_{c \notin r} W_{c,\{r\}} \leq 1, \qquad \text{(for each } r \in M) \\
& \sum_{r \in P} \sum_{c \in r} W_{c,P \setminus \{r\}} - \sum_{r \in P} \sum_{c \notin r} W_{c,P} \leq 0. \text{ (for each } P \sqsubseteq M \text{ s.t. } |P| > 1)
\end{aligned}$$

Note that $W_{c,P}$ is defined only when $P \sqsubseteq R$ and $c \notin \bigcup_{r \in P} r$.

From the theory of Sherali and Adams [16,17] and Sherali and Lee [18], giving the optimal solution for this dual problem shows the rectangle bound of the corresponding relation.

Definition 8 (Sherali-Adams Bound). *We define* $\mathbf{SA}_h(R)$ *as the optimal value of the h-th level of the Sherali-Adams relaxation associated with R, which corresponds to the restriction of* $W_{c,P} = 0$ *where* $|P| > h$.

If we restrict the Sherali-Adams bound to the first level as $\mathbf{SA}_1(R)$, we have the following simplification of the dual problem by replacing $W_{c,\emptyset}$ by V_c and $-W_{c,\{r\}}$ by $V_{c,r}$.

$$\begin{aligned}
\max & \sum_{c \in C} V_c \\
\text{s.t.} & \sum_{c \in r} V_c + \sum_{c \notin r} V_{c,r} \leq 1, \qquad \text{(for each } r \in M) \\
& \sum_{c \in r_1} V_{c,r_2} + \sum_{c \in r_2} V_{c,r_1} \geq 0. \text{ (for each } \{r_1, r_2\} \sqsubseteq M)
\end{aligned}$$

Giving a feasible solution of the above formulation shows a lower bound for the rectangle bound and hence formula size. Since $C^D(R)$ is upper-bounded by the number of cells, we have the following theorem.

Theorem 7 ([16,17,18]).

$$\forall R,\ \mathbf{LP}(R) = \mathbf{SA}_0(R) \leq \mathbf{SA}_1(R) \leq \cdots \leq \mathbf{SA}_{|C|}(R) = C^D(R).$$

6 A Triplet Argument for the Sherali-Adams and Quasi-additive Bounds

In this section, we introduce a general technique named a triplet argument for giving a solution for the Sherali-Adams bound and the quasi-additive bound. The technique is applicable for any relation. For the explanation, we look at the 3-bit majority function. A majority function $\mathbf{MAJ}_{2l+1}$ with $2l+1$ input bits outputs 1 if the number of 1's in the input bits is greater than or equal to $l+1$ and 0 otherwise. Note that we know $\mathbf{LP}(R_{\mathbf{MAJ}_3}) \leq 4.5$.

Proposition 1. $\mathbf{SA}_1(R_{\mathbf{MAJ}_3}) \geq 5$ *and* $\mathbf{QA}(R_{\mathbf{MAJ}_3}) \geq 5$.

Proof. We consider a communication matrix of the 3-bit majority function whose rows and columns are restricted to minterms and maxterms, respectively. We consider a triplet (c_1, c_2, c_3) composed of 3 cells. Here, in the case of the 3-bit majority function, we consider a triplet of 3 cells each of which has 3 indices.

We firstly assume $V_{c,r} = 0$ for any c and r and change assignments of $V_{c,r}$ (for each triplet sequentially in general case) without violating the second constraints of the Sherali-Adams bound and the quasi-additive bound in the following way. If a combinatorial rectangle r contains 2 cells of the triplet, e.g., c_2 and c_3, we increment $V_{c_1,r}$ by -2δ for the remaining 1 cell. If a combinatorial rectangle r contains 1 cell of the triplet, e.g., c_1, we increment $V_{c_2,r}$ and $V_{c_3,r}$ by δ for the remaining 2 cells. In the case of 3-bit majority function, we set $\delta = \frac{1}{6}$ for the triplet. Then, it is easy to verify the changes of assignments do not violate a constraint $\sum_{c \notin r_1} V_{c,r_1} + \sum_{c \notin r_2} V_{c,r_2} \geq 0$ for any 2 disjoint combinatorial rectangles r_1 and r_2. We also have $\sum_{c \notin r}(V_{c,r} - V_{c,r_1} - V_{c,r_2}) \leq 0$ for any combinatorial rectangle r and its arbitrary partition into r_1 and r_2. Thus, all the second constraints cannot be violated.

We give a weight $-\frac{1}{3}$ for each cell in the triplet and a weight 1 for each cell from the other 6 cells with 1 index. Then, we can also verify the assignments satisfies all the first constraints of the quasi-additive bound and the Sherali-Adams bound. As a consequence, we have the lower bound of 5. □

Combining the triplet argument explained in the above proof and the idea of [20], we can prove the same lower bound $L(\mathbf{MAJ}_{2l+1}) \geq \frac{(l+1)^2}{1-\epsilon(l)}$ where $\epsilon(l) = \frac{l^2(l+1)}{6\cdot\binom{2l+1}{l}}$ as that of [20] for the majority function by both of the first level of the Sherali-Adams bound and the quasi-additive bound.

7 Conclusions

In this paper, we introduced the novel general techniques proving formula size lower bounds, the Sherali-Adams bound and the quasi-additive bound, as extensions of the LP bound of Karchmer, Kushilevitz and Nisan [6]. In particular, we proved that the quasi-additive bound can surpass the rectangle bound by the cross argument. We also showed that the quasi-additive and Sherali-Adams bounds can cover a main part of techniques discussed in the previous paper [20] by the triplet argument.

Acknowledgment

The author thanks Troy Lee for joyful discussions leading to the triplet argument and Kazuyuki Amano for sharing an efficient program computing the protocol partition number. This research is supported by the Kyoto University Hakubi Project and the Japan Society for the Promotion of Science.

References

1. Ambainis, A.: Polynomial degree vs. quantum query complexity. Journal of Computer and System Sciences 72(2), 220–238 (2006)
2. Balas, E., Ceria, S., Cornuéjols, G.: A lift-and-project cutting plane algorithm for mixed 0–1 programs. Mathematical Programming 58, 295–324 (1993)
3. Håstad, J.: The shrinkage exponent of De Morgan formulas is 2. SIAM Journal on Computing 27(1), 48–64 (1998)
4. Høyer, P., Lee, T., Špalek, R.: Negative weights make adversaries stronger. In: Proceedings of the 39th Annual ACM Symposium on Theory of Computing (STOC 2007), pp. 526–535 (2007)
5. Hrubeš, P., Jukna, S., Kulikov, A., Pudlák, P.: On convex complexity measures. Theoretical Computer Science 411, 1842–1854 (2010)
6. Karchmer, M., Kushilevitz, E., Nisan, N.: Fractional covers and communication complexity. SIAM Journal on Discrete Mathematics 8(1), 76–92 (1995)
7. Karchmer, M., Wigderson, A.: Monotone circuits for connectivity require super-logarithmic depth. SIAM Journal on Discrete Mathematics 3(2), 255–265 (1990)
8. Khrapchenko, V.M.: Complexity of the realization of a linear function in the case of π-circuits. Mathematical Notes 9, 21–23 (1971)
9. Koutsoupias, E.: Improvements on Khrapchenko's theorem. Theoretical Computer Science 116(2), 399–403 (1993)
10. Laplante, S., Lee, T., Szegedy, M.: The quantum adversary method and classical formula size lower bounds. Computational Complexity 15(2), 163–196 (2006)
11. Lasserre, J.B.: Global optimization with polynomials and the problem of moments. SIAM Journal on Optimization 11(3), 796–817 (2001)
12. Laurent, M.: A comparison of the Sherali-Adams, Lovász-Schrijver and Lasserre relaxations for $0-1$ programming. Mathematics of Operations Research 28(3), 470–496 (2003)
13. Lee, T.: Kolmogorov Complexity and Formula Size Lower Bounds. PhD thesis, University of Amsterdam (January 2006)
14. Lee, T.: A new rank technique for formula size lower bounds. In: Thomas, W., Weil, P. (eds.) STACS 2007. LNCS, vol. 4393, pp. 145–156. Springer, Heidelberg (2007)
15. Lovász, L., Schrijver, A.: Cones of matrices and set-functions and 0-1 optimization. SIAM Journal on Optimization 1(2), 166–190 (1991)
16. Sherali, H.D., Adams, W.P.: A hierarchy of relaxation between the continuous and convex hull representations for zero-one programming problems. SIAM Journal on Discrete Mathematics 3, 411–430 (1990)
17. Sherali, H.D., Adams, W.P.: A hierarchy of relaxations and convex hull characterizations for mixed-integer zero-one programming problems. Discrete Applied Mathematics 52(1), 83–106 (1994)
18. Sherali, H.D., Lee, Y.: Tighter representations for set partitioning problems. Discrete Applied Mathematics 68(1-2), 153–167 (1996)
19. Spira, P.: On time-hardware complexity tradeoffs for boolean functions. In: Proceedings of the 4th Hawaii Symposium on System Sciences, pp. 525–527 (1971)
20. Ueno, K.: A stronger LP bound for formula size lower bounds via clique constraints. In: Proceedings of the 26th Annual Symposium on Theoretical Aspects of Computer Science (STACS 2009). Leibniz International Proceedings in Informatics (LIPIcs), vol. 3, pp. 685–696. Schloss Dagstuhl–Leibniz-Zentrum fuer Informatik (2009)
21. Wegner, I.: The Complexity of Boolean Functions. Wiley-Teubner, Chichester (1987)

Mesh Deformation of Dynamic Smooth Manifolds with Surface Correspondences

Ho-Lun Cheng and Ke Yan

National University of Singapore
hcheng@comp.nus.edu, yanke2@comp.nus.edu

Abstract. Maintaining a moving mesh of a deforming surface is widely studied in various disciplines. However, difficulties arise with requirements of topology changes, homeomorphism between mesh and surface, and guarantees of triangle quality. We propose a mesh deformation algorithm to satisfy the above requirements. We employ the skin surface by Edelsbrunner that approximates objects in fields like computer graphics, molecular modeling and engineering. We complete the general deformation framework by introducing a new mesh point movement and scheduling function to satisfy the requirements.

1 Introduction

Given a differentiable manifold $\mathcal{F}(t)$ in $\mathbb{R}^3$ that changes its shape, curvature and topology with a parameter $t \in [0..1]$ as time, this work develops an algorithm to approximate $\mathcal{F}(t)$ with a deforming triangulation $\mathcal{M}(t)$ while maintaining a homeomorphism to $\mathcal{F}(t)$. This algorithm guarantees triangle quality and surface coordinate correspondence between $\mathcal{M}(t)$ and $\mathcal{M}(t + \Delta t)$. We work on *general deformation* [6,10] (See Figure 1) and is considered as a generalization of and "completion" of the previous work that is limited to only the growth model [5,13].

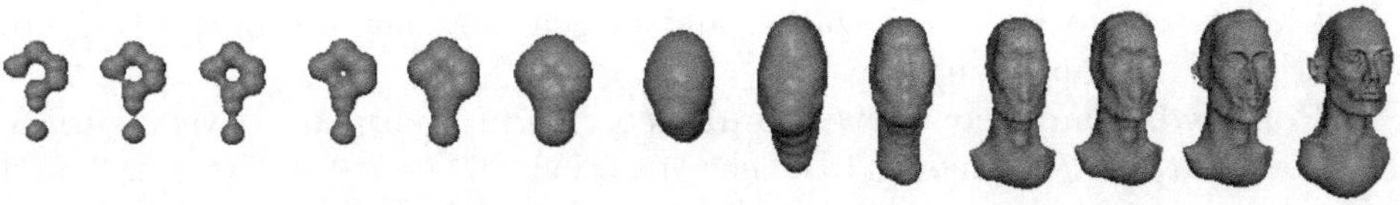

Fig. 1. A general skin deformation into a mannequin skin model (12,684 spheres) from a question mark skin model (7 spheres)

1.1 Motivation

Skin surface deformation, introduced by Edelsbrunner [12], is useful in various fields such as computer graphics, molecular modeling and physical simulations because of its five advantages. In computer graphics and animation, one object

P. Hliněný and A. Kučera (Eds.): MFCS 2010, LNCS 6281, pp. 677–688, 2010.

can deform to another by their skin representations [9]. We can even interpolate more than two objects to perform shape synthesis by their skin models [4]. Apart from the "virtual world", realistic physical simulations for engineering purposes employ deforming surfaces as boundaries of spatial domains that vary with time. An example is molecular surface diffusion that moves each point on a surface according to the normal velocity proportional to Laplacian of mean curvature of the surface in protein folding processes [2,11]. Another example is modeling the deforming boundaries in simulations such as the isosurface of a certain threshold of temperature in a heat dynamic system, or the isosurface of the solidification of liquid formed at the "freezing point". For such deformation applications, the skin surface deformation has five advantages over other surface representations and deformations:

1. **Automatic topology change handling.** When a surface is deforming, its topology changes when its components are split or merged, or when the surface creates or destroys voids and tunnels. Controlling these changes are not trivial with explicit surfaces [1]. Skin surface deformation addresses this issue, and pre-schedules all topology change handling in the process of deformation. For example, in Figure 1, a creation of tunnel and a destruction of void are automatically handled in the process of deformation.
2. **Intuitive object representation and editing.** On one hand, explicit surfaces are not suitable for topology changes. On the other hand, implicit surfaces face difficulties in local manipulation, namely, a local change in a small part of the surface may cause the whole shape to change unexpectedly. However, small local changes can be performed on skin surfaces independently. Moreover, skin surfaces are able to approximate objects within an Hausdorff distance that is better than implicit surfaces [9].
3. **Quality triangulation.** The triangulation of the surface requires good triangle quality to help numerical analysis in physical simulations and maintain homeomorphism to the surface while deforming. Our previous works [5,8] show that we can triangulate the skin surface with good quality triangles. However, CGAL [17] (version 3.5) gives triangles with very small angles, which leads to bad visualization and inaccurate computation (See Figure 3(b) for a comparison).
4. **Point-wise surface correspondence.** During morphing, every surface point $p(t) \in \mathcal{F}(t)$ moves to a new position $p(t + \Delta t) \in \mathcal{F}(t + \Delta t)$ and the pair $p(t)$ and $p(t + \Delta t)$ establishes correspondence for relating different portions between the two surfaces $\mathcal{F}(t)$ and $\mathcal{F}(t + \Delta t)$. This relationship provides point-wise surface correspondence between the same surface in different times during deformation, which is a crucial feature for computation and visualization purposes in fields like medical imaging [16,18], animation [14] (e.g. texture and bump mappings) and physical simulations [3] (e.g. finite element analysis).
5. **Efficiency Improvement.** Shape deformation requires visually continuous frames of meshes during morphing. Current static skin meshing algorithms [6,8,10,15,17] build each intermediate frame from scratch. For example, in

Figure 1, each construction of the intermediate frame requires more than five minutes with any existing static skin mesh approach. We improve the skin deformation algorithm so that each intermediate skin mesh is obtained from the previous time frame. This improvement makes real time visualization of skin deformation possible.

In this paper, we present a new approach of skin surface deformation which generates intermediate skin meshes efficiently with point-wise correspondence. During the deformation, we maintain the triangle quality of the skin mesh by local modifications to the triangulations. All topology changes are scheduled and therefore are handled automatically.

1.2 Problem Definition and Related Work

General skin surface deformation is desired. However, there is still no applications that can handle the general skin deformation efficiently. All current algorithms are not suitable for creating deforming meshes for real time visualization purposes. Several algorithms have been proposed to create a good quality mesh of a certain instance of a skin surface [6,8,10,15]. They are good algorithms but not suitable for creating deformation mesh sequences. First, creating each frame in a sequence individually is not fast enough for real time applications. Second, there is no surface correspondence between each pair of consecutive frames of meshes. The better idea is to modify an instance of the mesh in the deformation sequence from the previous mesh instead of creating each frame individually.

The closest algorithm that adopts this approach is the Dynamic Skin Triangulation (DST) [5,13], but it is only limited to the growth model. DST maintains the triangle quality and topological correctness during a "deformation". However, this deformation is limited to the growth model and the purpose is only to construct the mesh of a certain skin surface by growing. This growth model cannot handle the general deformation that has a totally different structure, such as new additional types of mixed cells, new types of topology changes and new ways of surface point movement, etc. Although we can adopt the overall idea of the algorithm in DST to our problem, new designs and reconsiderations are needed to adapt to the new *intermediate complex* that is more sophisticated than the underlying structure of the growth model.

1.3 Approach and Contribution

The overall approach is similar to DST [5,13]. However, more sophisticated intermediate complex, surface point movements and scheduling are introduced. We provide solutions to these new issues and complete the general skin surface deformation problem.

Approach. We maintain the triangle quality of the mesh at any time t to a certain quality that guarantees homeomorphism between the mesh and the skin surface. This is described in Section 2.4. Assuming at time t, a triangle τ is in good quality. We schedule a time $t + \Delta t$ in the future into a priority queue such that τ may fail the quality check but is not beyond repair. Checks and refinements (if necessary) are performed at time $t + \Delta t$, and all the involved triangles are rescheduled again after refinement. At the same time, topological change operations are also scheduled into the priority queue to maintain the mesh.

New Issues. Firstly, the nature of the mixed cells in the general deformation is more complex than in the growth model. The entire space is partitioned into a finite number of convex polytopes called *mixed cells* and the intersection of each of them and the skin surface is a portion of a quadratic surface. In the growth model, there are only four types of mixed cells and they are fixed in the space. However, there are seven types of mixed cells in the general deformation and they are moving and deforming. Thus, computing the trajectory of each surface point within a mixed cell and the *escaping time* when it transfers from one mixed cell to another are the new issues that have to be reconsidered.

Secondly, movement of these new mixed cells make trajectories of surface points complicated. In the growth model, if we follow the surface normals, each surface point within a mixed cell moves in a straight line or a quadratic curve and this makes the prediction of the triangle distortion possible. The mixed cells in the general deformation deform and the surface points do not move in such simple manners. Following surface normals is not an easy task for general deformation and we propose a new way of surface point movement. We claim that this new movement improves the efficiency of DST.

Thirdly, the topology changes are different. In the growth model, there is at most one topology change within one mixed cell but there are at most two in the general deformation.

Contribution. Our algorithm solves the general skin surface deformation problem based on the DST with new remedies. New types of mixed cell and their transformations are addressed. New surface point movement is proposed to deal with more complicated surface movement. New topology changes are handled by scheduling. In overall, our algorithm improves the efficiency in generating intermediate skin meshes.

2 Background

In this section, we introduce the necessary background for the skin [12], the intermediate Voronoi complexes during deformation [4], the quality criteria of the mesh [5], and the mesh deformation algorithm for growth model [13]. This section serves the purpose of stating notations for later sections. For more details, readers may refer to the specific references.

2.1 Delaunay and Voronoi Complexes

A *weighted point* in $\mathbb{R}^d$ can be written as $b_i = (z_i, w_i) \in \mathbb{R}^d \times \mathbb{R}$, where $z_i \in \mathbb{R}^d$ is the position and $w_i \in \mathbb{R}$ is the weight. Given a finite set $B = \{b_1, b_2, ..., b_n\}$,

$z(B) \subset \mathbb{R}^d$ is denoted as the set of positions of the weighted points in B. The weighted distance of a point $x \in \mathbb{R}^d$ from a weighted point b_i is defined as $\pi_{b_i}(x) = \|x - z_i\|^2 - w_i$. The *Voronoi region* ν_i for each weighted point $b_i \in B$ is defined as,

$$\nu_i = \{x \in \mathbb{R}^d \mid \pi_{b_i}(x) \leq \pi_{b_j}(x), b_j \in B\}.$$

For a set of weighted points $X \subseteq B$, the *Voronoi cell* of X is defined as $\nu_X = \bigcap_{b_i \in X} \nu_i$. The collection of all the non-empty Voronoi cells is called the *Voronoi complex* of B, denoted as, V_B. For each $\nu_X \in V_B$, its corresponding *Delaunay cell*, δ_X, is the convex hull of the set of centers of X, namely, $\text{conv}\,(z(X))$. The collection of all the Delaunay cells is called the *Delaunay complex* of B, denoted as D_B. The Delaunay complex is simplicial under the following general position assumption in $\mathbb{R}^d$, $\forall \nu_X \in V_{B_0}, \text{card}(X) = \dim(\delta_X) + 1$.

2.2 Skin

A skin surface is specified by a set of weighted points $B = \{b_i \in \mathbb{R}^d \times \mathbb{R} \mid i = 1, 2, ..., n\}$. First, three operations on weighted points are defined in the sphere algebra. For $b_i, b_j \in B$ and $\gamma \in \mathbb{R}$, the *addition*, *scalar multiplication* and *square root* of weighted points are defined as,

$$\begin{aligned} b_i + b_j &= (z_i + z_j, w_i + w_j + 2\langle z_i, z_j \rangle), \\ \gamma b_i &= (\gamma z_i, \gamma w_i + (\gamma^2 - \gamma)||z_i||^2), \\ \sqrt{b_i} &= (z_i, w_i/2), \end{aligned}$$

where $\langle z_i, z_j \rangle$ is the dot product of z_i and z_j. The *convex hull* of B is defined as

$$\text{conv}\,(B) = \left\{\sum \lambda_i b_i \mid \sum \lambda_i = 1 \text{ and } \lambda_i \geq 0, i = 1, ..., n\right\}.$$

the skin surface is the boundary of the union of all shrunken balls in $\text{conv}\,(B)$ [5]. Formally,

$$\text{skin}\,(B) = \partial\left(\bigcup\left\{\sqrt{\hat{b}} \mid \hat{b} \in \text{conv}\,(B)\right\}\right).$$

Skin Decomposition. The skin of can be decomposed by *mixed cells.* A mixed cell μ_X is the Minkowski sum of a Delaunay cell and its corresponding Voronoi cell, formally $\mu_X = (\delta_X + \nu_X)/2$. With $\text{card}(X) = 1, 2, 3, 4$, the four types of mixed cells are convex polyhedrons, prisms, triangular prisms and tetrahedra respectively. The center and size of a mixed cell are defined as

$$\begin{aligned} z_X &= \text{aff}\,(\delta_X) \cap \text{aff}\,(\nu_X), \text{and} \\ w_X &= w_i - \|z_X z_i\|^2. \end{aligned}$$

where $b_i = (z_i, w_i)$ is any weighted point in X. Indeed, w_X is the negative square radius of the ball orthogonal to X.

Within each mixed cell μ_X, skin$(B) \cap \mu_X$ is a quadratic surface. In $\mathbb{R}^3$, skin patches are pieces of spheres and hyperboloids of revolution, which can be expressed in standard forms

$$x_1^2 + x_2^2 + x_3^2 = R^2, \quad (1)$$
$$x_1^2 + x_2^2 - x_3^2 = \pm R^2, \quad (2)$$

after the translation of z_X to the origin and the oriented axis to x_3-axis if they are hyperboloid patches. In Equation (2), the plus sign on the right hand side gives the one-sheeted hyperboloid and the minus sign gives the two-sheeted hyperboloid.

2.3 General Skin Deformation and Intermediate Complexes

The deformation starts with an initial shape, ends with a goal shape and generates frames of *intermediate shapes* in between. We first discuss the definition and the generation of intermediate shapes and then state the theorem of invariance of the intermediate Voronoi complexes [4].

Let B_0, B_1 be the weighted point sets of initial shape skin(B_0) and goal shape skin(B_1) respectively. Denote $\mathbb{B} = \{B_0, B_1\}$. The intermediate weighted points set at time $t \in [0,1]$ is defined as,

$$B(t) = \{b_{ij}(t) = (1-t)b_i + tb_j \mid b_i \in B_0, b_j \in B_1\}, \text{for } t \in [0,1].$$

The intermediate shape is defined as skin$(B(t))$ and it deforms smoothly as the changing of t. Here, card$(\mathbb{B}) = 2$ but $\mathbb{B}$ can be extended to $\mathbb{B}'$ such that card$(\mathbb{B}') > 2$ and the intermediate shapes become the skin of convex combinations of $\mathbb{B}'$. The intermediate shape can be further extended to be the skin of the affine combination of $\mathbb{B}'$. However, all the extensions are based on the case of card$(\mathbb{B}) = 2$ and we only consider this case in this paper.

The *intermediate Voronoi complexes* V is defined as the Voronoi complex of $B(t)$ and is required for the generation of skin$(B(t))$. Chen and Cheng [4] prove that all intermediate surfaces share the same intermediate Voronoi complex V which is the superimposing of all Voronoi complexes of $\mathbb{B}$. Let V_{B_0}, V_{B_1} be the Voronoi complexes of B_0 and B_1 respectively, we have

$$V = \{\nu_{XY} = \nu_X \cap \nu_Y \mid \nu_X \in V_{B_0}, \nu_Y \in V_{B_1}\}.$$

Each combination of various cardinalities of X and Y produces a type of mixed cells.

2.4 Dynamic Skin Triangulation

The triangulation of the skin surface guarantees homeomorphism and good quality of triangles if the following Conditions [U] and [L] are satisfied when the mesh is deforming. At any point x on the skin surface, denote $\kappa(x)$ as the maximum curvature at x and the *local length scale* at x as $\varrho(x) = 1/\kappa(x)$. For an edge ab, let $R_{ab} = \|a - b\|/2$ be the half length, and for a triangle abc, let R_{abc} be the circumcircle radius of the triangle. The mesh is homeomorphic to the surface if it satisfies the Lower bound Condition [L] and the Upper bound Condition [U] below. Let ϱ_{ab} be the maximum of $\varrho(a)$ and $\varrho(b)$, and ϱ_{abc} be the minimum of $\varrho(a)$, $\varrho(b)$ and $\varrho(c)$. Then the two conditions are,

[L] $R_{ab}/\varrho_{ab} > C/Q$ for every edge ab, and
[U] $R_{abc}/\varrho_{abc} < CQ$ for every triangle abc,

where C and Q are positive constants chosen empirically. The triangle quality is guaranteed by the minimum angle of the triangles, namely $\sin^{-1}\frac{1}{Q^2}$. For $C = 0.08$ and $Q = 1.65$, the minimal angle of the mesh is proven to be larger than $21.54°$ [5].

During the deformation, mesh refinement and maintenance operations are scheduled with a priority queue from time to time. The two basic operations are edge contraction and point insertion. These two operations fix an edge or a triangle that violates the two Conditions to maintain quality and homeomorphism.

3 Surface Deformation

In this section, we describe the trajectory of each point on the skin surface during the deformation within and across mixed cells. First, we present the deformation of each mixed cell. Second, we describe the trajectory of a point within such a deforming mixed cell in a local coordinate system. Finally, we compute the *escaping time* when a point moves across one mixed cell to another. This happens when the point hits the boundary of its deforming mixed cell.

3.1 Mixed Cell Deformation

The skin patch within a mixed cell $\mu_X(t)$ is determined by the position, the size and the shape of $\mu_X(t)$. The position and size are determined by the center $z_X(t)$ and the size $w_X(t)$ and the later defines $R = \sqrt{w_X(t)/2}$ in Equations (1) and (2).

We first compute $z_X(t)$ that is the intersection of the affine hulls of $\nu_X(t)$ and $\delta_X(t)$. Because $\nu_X(t)$ is unchanged and all the centers of $X(t)$ moves linearly according to t, $\delta_X(t)$ moves linearly and, thus, the intersection $z_X(t)$ moves linearly along $\nu_X(t)$ also. We first compute the two position $z_X(0)$ and $z_X(1)$, and we have

$$z_X(t) = (1-t)\cdot z_X(0) + t\cdot z_X(1), \tag{3}$$
$$w_X(t) = w_i(t) - \|z_i(t)z_X(t)\|^2, \tag{4}$$

where $b_i(t) = (z_i(t), w_i(t)) \in X(t)$.

For the shape of $\mu_X(t)$, it is a convex polytope formed by the intersection of finitely many halfspaces. Each halfspace $H_{XY}(t)$ is bounded by a plane $h_{XY}(t)$ that separates the two mixed cells $\mu_X(t)$ and $\mu_Y(t)$ with $Y(t) \subset B(t)$ and $\delta_Y(t)$

is a face or coface of $\delta_X(t)$. Namely, the symmetric difference $X(t) \ominus Y(t)$ is a singleton set. Formally,

$$H_{XY}(t) = \{x \in \mathbb{R}^3 \mid \langle x, n_{XY}(t)\rangle \leq \langle m_{XY}(t), n_{XY}(t)\rangle\},$$

where the normal of $h_{XY}(t)$ and a point on $h_{XY}(t)$ are

$$n_{XY}(t) = z_Y(t) - z_X(t),$$
$$m_{XY}(t) = \frac{z_X(t) + z_Y(t)}{2}.$$

3.2 Computing the Trajectory for a Sample Point

We employ a new way of moving the surface vertices on the mesh that is more efficient and "intuitive" than the previous work that moves vertices following their normal directions [5,13]. Firstly, the previous method creates a lot of "stretching" or "compressing" near the tips of two-sheet hyperboloids and the waists of one-sheet hyperboloids. The concentration of triangle modification (creation and deletion) at these areas leads to unreasonable or undesirable texture distortion in animation or numerical instability in computations. Secondly, *all* surface triangles in Type 1 and 2 mixed cells require frequent checks on the triangles and edges to ensure their qualities to ensure the homeomorphism between the triangulation and the skin surface. Lastly but the most importantly, the previous work has a simpler movement of surface vertices because the deformation is limited to the growth model and mixed cells do not move nor deform. This makes the scheduling of checks and updates is easier than our general deformation in this paper. Therefore, we propose a new type of vertex movement in order to improve and make it possible for general deformation.

We decide to move vertices only towards and away from their corresponding mixed cell centers to improve efficiency. According to the local coordinate system of a mixed cell, its surface patch within at time t_1 is a scaling with a factor of $\sqrt{w_X(t_1)/w_X(t_0)}$ of the patch at time t_0. Therefore, surface vertices are only shrunken towards their mixed cell centers with that factor. Given a surface vertex $p(t_0)$ in a mixed cell $\mu_X(t_0)$ at time t_0, $p(t)$ moves with a scaling factor within its mixed cell and a translation of its mixed cell center. Using the notation $p'(t) = p(t) - z_X(t)$ as the local coordinates of $p(t)$ in its mixed cell, the location of $p(t)$ is

$$p(t) = z_X(t) + \sqrt{\frac{w_X(t)}{w_X(t_0)}}p'(t_0). \tag{5}$$

Note that every point on the surface is moving along a straight line within a mixed cell locally and the mixed cell is translating in a linear manner. However, it does not cause the point to move on a straight line after the composition of these two movements above. Thus, we still need to handle and compute the time when a point reaches the boundary of its mixed cell and enters another one. We call this moment the *escaping time* and it happens when a surface vertex $p(t)$ in $\mu_X(t)$ transfers to another nearby mixed cell $\mu_Y(t)$ when $p(t)$ is on the boundary of $H_{XY}(t)$.

There are two superior benefits in this new movement scheme over the previous work in DST where surface points move along their normal direction. First, it is faster and simpler to move a point in a straight line manner within a mixed cell. In previous work, every point is moving along a hyperbola and it requires the solving of a quartic equation to compute a new location of the point. Secondly, an element (a triangle or an edge) in a single mixed cell does not need checks and updates on its quality because they are only under scaling according to Conditions [U] and [L]. Thus, only the elements that are across more than one mixed cell requires scheduling and it greatly reduces the priority queue size. By experimental results, the triangles that span a few mixed cells only occupy less than about 8% of the total triangle population.

4 Handling Topology Changes

The topology changes in the general deformation model is different from the growth model in DST because there may be at most two changes in each mixed cell. In the former case, there is at most one topology change for each patch in a mixed cell during the growth but it is different in the case of general deformation while each mixed cell size $w_X(t)$ is a quadratic function of t. Moreover, we schedule a topology changing operation only if the mixed cell center is in the mixed cell. Note that both the center and the cell are moving and deforming, which is different from that in the growth model. This indicates that there is a possibility that the center is not in the mixed cell when $w_X(t) = 0$.

In order to preserve the homeomorphism between the mesh and the skin, metamorphosis are scheduled around the time of topology changes. There are two types of topology changes, namely,

- creating or destroying a sphere,
- changing from a one-sheet to two-sheet hyperboloid or vice versa.

These topology changes in a mixed cell occur if and only if the size of a mixed cell reaches a zero value and that is the time to schedule metamorphosis, namely, $w_X(t') = 0$. However, we use a special sampling within a *hot sphere* [5] at the time slightly before and after the time t'. The triangulation within the hot sphere is specially restructured and controlled from the time $t' - \varepsilon_X$ to $t' + \varepsilon_X$ in $\mu_X(t)$ for a small value of ε_X. For the details of how the mesh changes by metamorphosis, please refer to the dynamic skin triangulation of the growth model [5].

5 Experimental Results

Experimental results show that our skin deformation algorithm is much faster than existing static skin mesh algorithms in terms of generating each frame of intermediate skin mesh model with triangle quality guaranteed. In Table 1, we show the time taken in three different sets of skin models deformations: a mannequin skin model to a question mark skin model; a question mark skin model to a woman skin model; a woman skin model to a fish skin model. All these experiments are performed with Intel Duo Core 2.33GHz and 4GB RAM. In order to show smooth deformation processes, we generate one thousand frames (intermediate skin model meshes) for each example. Therefore, if a short video clip requires 24 frames per second, we are producing a deformation process of approximately one minute. We compare our general skin deformation algorithm (GSDA) with a modern computational geometry library CGAL [17] (version 3.5) and the static skin section of dynamic skin deformation algorithm (DST) developed by Cheng [5,13].

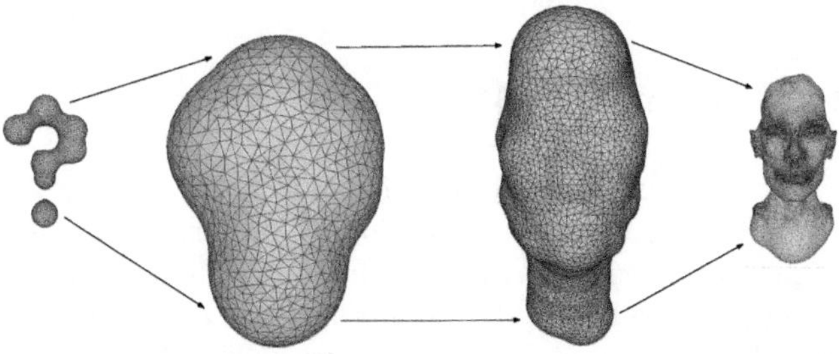

Fig. 2. A zoom-in snapshot of the general deformation between a mannequin skin model and a question mark skin model, triangles are shown in wireframe

In the first experiment (Figure 1), a mannequin skin mesh model (built by 12, 684 spheres) deforms to a question mark (7 spheres). In Figure 2 we show the details of our triangulation quality. In Table 1, we show that our GSDA is hundreds times faster than both static skin mesh algorithms (CGAL-3.5 and DST). These static skin algorithms are slow because they generate each intermediate skin mesh model from scratch. On the other hand, they can be faster if the skin models are built by fewer spheres. In the examples of Question_Mark.skn (7 spheres) deforming to Woman.skn (8 Spheres) and Woman.skn deforming to Fish.skn(8 spheres), our algorithm is only about two times faster than CGAL-3.5. However, our program produces meshes with better triangle quality as compared to CGAL-3.5 (See Figure 3(b)).

The faster performance of GSDA is also due to the new way of linear vertex movement. Time is saved when we only test the triangles that cross more than one mixed cell. Figure 3(a) shows the minority of triangles that are being tested and refined during the time interval of two frames. Note that the total number of triangles being refined is *constant* for each deformation no matter how many frames are generated in the whole deformation process.

Table 1. Average time taken by different algorithms for different examples. Each deformation process runs with 1,000 frames.

Experiments	**GSDA**	**CGAL-3.5**	**DST**
Mannequin.skn ↔ Question_Mark.skn	**1.748 sec**	**323.8 sec**	**695.0 sec**
Question_Mark.skn ↔ Woman.skn	**0.708 sec**	**1.047 sec**	**2.025 sec**
Woman.skn ↔ Fish.skn	**0.615 sec**	**1.025 sec**	**1.875 sec**

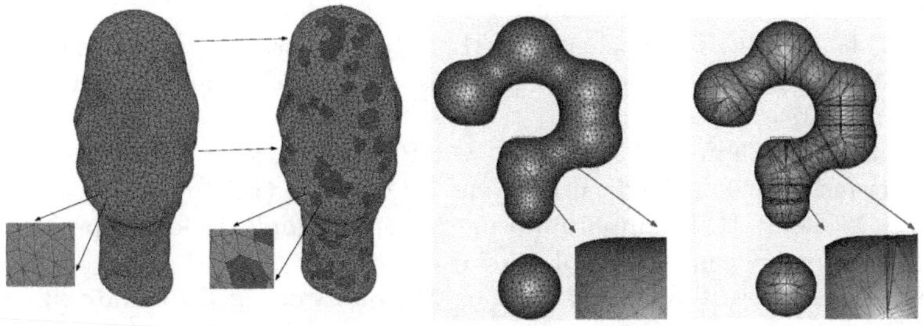

(a) (b)

Fig. 3. (a) we show frame 344 and 345 during the deformation. The zoom-in snap shots show triangle and vertex correspondences. The red regions in frame 345 are triangles being refined from frame 344. (b) we show A comparison of the triangle quality produced by our algorithm and CGAL-3.5. Our algorithm produces better quality triangles.

6 Conclusion

With the ability to approximate any object [9] and the speed boost for the deformation [4], we give an algorithm that allows automatic free form deformation between two objects or even mixing several objects into one. This enhancement does not only make shape synthesis possible in computer animation, engineering and biogeometry applications, but also facilitate shapes manipulations such as shape space searching, simplification or compression. In terms of performance, our algorithm performs the best when the time difference between two frames of deformation are small. The surface correspondence, guaranteed triangle quality and homeomorphism enable robust computation for engineering simulations. As a by-product, this algorithm can be used to replace the dynamic skin triangulation to construct a skin mesh. Namely, we can construct a skin mesh by deforming from empty into the desired skin mesh.

Future research includes controlling the deformation locally, for example, reducing the topology changes during the deformation so that they can be further controlled. The other interesting area is to investigate new types of possible vertex movement scheme such as following orthogonal and parallel directions with the rotational axis of the hyperboloids to enhance performance and quality. Another exciting challenge is to extend our algorithm to construct a deforming *volume tetrahedral mesh* for physical simulation purposes. The tetrahedral refinement in a deforming body is more difficult than the surface mesh. A previous work shows that a good surface mesh of the boundary of an object helps the construction of the volumetric mesh [7].

References

1. Bajaj, C., Blinn, J., Bloomenthal, J., Cani-Gascuel, M.-P., Rockwood, A., Wyvill, B., Wyvill, G.: Introduction to implicit surfaces. Morgan-Kaufmann, San Francisco (1997)
2. Bänsch, E., Morin, P., Nochetto, R.H.: A parametric finite element method for fourth order geometric evolution equations. Journal of Computational Physics 222, 441–467 (2007)
3. Bertalmio, M., Sapiro, G., Randall, G.: Region tracking on level-sets methods. IEEE Transactions on Medical Imaging 18, 448–451 (1999)
4. Chen, C., Cheng, H.-L.: Superimposing Voronoi complexes for shape deformation. Int. J. Comput. Geometry Appl. (2006)
5. Cheng, H.-L., Dey, T.K., Edelsbrunner, H., Sullivan, J.: Dynamic skin triangulation. Discrete Comput. Geom. (2001)
6. Cheng, H.-L., Edelsbrunner, H., Fu, P.: Shape space from deformation. Comput. Geom. Theory Appl., 191–204 (2001)
7. Cheng, H.-L., Shi, X.: Quality tetrahedral mesh. generation for macromolecules. In: Asano, T. (ed.) ISAAC 2006. LNCS, vol. 4288, pp. 203–212. Springer, Heidelberg (2006)
8. Cheng, H.-L., Shi, X.W.: Quality mesh generation for molecular skin surfaces using restricted union of balls. IEEE Visualization (2005)
9. Cheng, H.-L., Tan, T.: Approximating polyhedral objects with deformable smooth surfaces. Computational Geometry, Thoery and Applications 18, 104–117 (2008)
10. Cheng, S.-W., Edelsbrunner, H., Fu, P., Lam, P.: Design and analysis of planar shape deformation. Comput. Geom. Theory Appl., 205–218 (2001)
11. Creighton, T.E.: Proteins structures and molecular principles. Freeman, New York (1984)
12. Edelsbrunner, H.: Deformable smooth surface design. Discrete Comput. Geom., 87–115 (1999)
13. Edelsbrunner, H., Ungor, A.: Relaxed scheduling in dynamic skin triangulation. In: Japanese Conf. Comput. Geom. (2002)
14. Karan, S.: Skinning characters using surface-oriented free-form deformations. In: Graphics Interface 2000, pp. 35–42 (2000)
15. Kruithof, N.G.H., Vegter, G.: Meshing skin surfaces with certified topology. Computational Geometry 36, 166–182 (2007)
16. Starck, J., Hilton, A.: Correspondence labelling for wide-timeframe free-form surface matching, pp. 1–8 (2007)
17. CGAL (version 3.5.1), `http://www.cgal.org`
18. Wang, Y., Peterson, B., Staib, L.: Shape-based 3d surface correspondence using geodesics and local geometry. In: CVPR, pp. 2644–2651 (2000)

Counting Dependent and Independent Strings

Marius Zimand*

Department of Computer and Information Sciences, Towson University, Baltimore, MD, USA

Abstract. We derive quantitative results regarding sets of n-bit strings that have different dependency or independency properties. Let $C(x)$ be the Kolmogorov complexity of the string x. A string y has α dependency with a string x if $C(y) - C(y \mid x) \geq \alpha$. A set of strings $\{x_1, \ldots, x_t\}$ is pairwise α-independent if for all $i \neq j$, $C(x_i) - C(x_i \mid x_j) \leq \alpha$. A tuple of strings $(x_1, \ldots, x_t)$ is mutually α-independent if $C(x_{\pi(1)} \ldots x_{\pi(t)}) \geq C(x_1) + \ldots + C(x_t) - \alpha$, for every permutation π of $[t]$. We show that:

- For every n-bit string x with complexity $C(x) \geq \alpha + 7 \log n$, the set of n-bit strings that have α dependency with x has size at least $(1/\text{poly}(n))2^{n-\alpha}$. In case α is computable from n and $C(x) \geq \alpha + 12 \log n$, the size of same set is at least $(1/C)2^{n-\alpha} - \text{poly}(n)2^{\alpha}$, for some positive constant C.
- There exists a set of n-bit strings A of size $\text{poly}(n)2^{\alpha}$ such that any n-bit string has α-dependency with some string in A.
- If the set of n-bit strings $\{x_1, \ldots, x_t\}$ is pairwise α-independent, then $t \leq \text{poly}(n)2^{\alpha}$. This bound is tight within a $\text{poly}(n)$ factor, because, for every n, there exists a set of n-bit strings $\{x_1, \ldots, x_t\}$ that is pairwise α-dependent with $t = (1/\text{poly}(n)) \cdot 2^{\alpha}$ (for all $\alpha \geq 5 \log n$).
- If the tuple of n-bit strings $(x_1, \ldots, x_t)$ is mutually α-independent, then $t \leq \text{poly}(n)2^{\alpha}$ (for all $\alpha \geq 7 \log n + 6$).

1 Introduction

A fact common to many mathematical settings is that in a sufficiently large set some relationship emerges among its elements. Generically, these are called Ramsey-type results. We list just a few examples: any $n+1$ vectors in an n-dimensional vector space must be dependent; for every k and sufficiently large n, any subset of $[n]$ of constant density must have k elements in arithmetic progression; any set of 5 points in the plane must contain 4 points that form a convex polygon. All these results show that in a sufficiently large set, some attribute of one element is determined by the other elements.

We present in this paper a manifestation of this phenomenon in the very general framework of algorithmic information theory. We show that in a sufficiently large set some form of algorithmical dependency among its elements must exist. Informally speaking, $\text{poly}(n) \cdot 2^{\alpha}$ binary strings of length n must share at least

* `http://triton.towson.edu/~mzimand`

P. Hliněný and A. Kučera (Eds.): MFCS 2010, LNCS 6281, pp. 689–700, 2010.

α bits of information. For one interpretation of "share", we also show that this bound is tight within a poly(n) factor.

Central to our investigation are the notions of information in a string and the derived notion of dependency between strings. The information in a string x is captured by its Kolmogorov complexity $C(x)$. A string y has α-dependency with string x if $C(y) - C(y \mid x) \geq \alpha$. The expression $C(y) - C(y \mid x)$, denoted usually more concisely as $I(x : y)$, represents *the quantity of information in x about y* and is a key concept in information theory. It is known that $I(x : y) = I(y : x) \pm O(\log n)$ (Symmetry of Information Theorem [20]), where n is the length of the longer between the strings x and y, and therefore $I(x : y)$ is also called the mutual information of x and y. For any n-bit string x and positive integer α, we are interested in estimating the size of the set $A_{x,\alpha}$ of n-bit strings y such that $C(y) - C(y \mid x) \geq \alpha$. One can see by a standard counting argument that $|A_{x,\alpha}| \leq 2^{n-\alpha+c}$ for some constant c. Regarding a lower bound for $|A_{x,\alpha}|$, it is easy to see that if $C(x) \preceq \alpha$, then $A_{x,\alpha}$ is empty (intuitively, in order for x to have α bits of information about y, it needs to have α bits of information to start with, regardless of y). The lower bound that we establish holds for any string having Kolmogorov complexity $\succeq \alpha$.[1] For such strings x, we show that $|A_{x,\alpha}| \geq (1/\text{poly}(n))2^{n-\alpha}$. A related set is $B_{x,\alpha}$ consisting of the n-bit strings y with the property $C(y \mid n) - C(y \mid x) \geq \alpha$. This is the set of n-bit strings about which x has α bits of information besides the length. Note that $B_{x,\alpha} \subseteq A_{x,\alpha}$. The same observations regarding an upper bound for $|B_{x,\alpha}|$ and the emptiness of $B_{x,\alpha}$ in case $C(x) \preceq \alpha$ remain valid. For x with $C(x) \succeq \alpha$ and α computable from n, we show the lower bound $|B_{x,\alpha}| \geq (1/C) \cdot 2^{n-\alpha} - \text{poly}(n) \cdot 2^{\alpha}$, for some positive constant C.

We turn to the Ramsey-type results announced above. A set of n-bit strings $\{x_1, \ldots, x_t\}$ is pairwise α-independent if for all $i \neq j$, $C(x_i) - C(x_i \mid x_j) \leq \alpha$. Intuitively, this means that any two strings in the set have in common at most α bits of information. For the notion of mutual independence we propose the following definition (but other variants are conceivable). The tuple of n-bit strings $(x_1, \ldots, x_t) \in (\{0,1\}^n)^t$ is mutually α-independent if $C(x_{\pi(1)} \ldots x_{\pi(t)}) \geq C(x_1) + \ldots + C(x_t) - \alpha$, for every permutation π of $[t]$. Intuitively this means that $x_1, \ldots, x_t$ share at most α bits of information. We show that if $\{x_1, \ldots, x_t\}$ is pairwise α-independent or if $(x_1, \ldots, x_t)$ is mutually α-independent then $t \leq \text{poly}(n)2^{\alpha}$. The bound in the pairwise independent case is tight within a polynomial factor.

We also show that there exists a set B of size poly$(n)2^{\alpha}$ that "α-covers" the entire set of n-bit strings, in the sense that for each n-bit string y there exists a string x in B that has α bits of information about y (i.e., y is in $A_{x,\alpha}$).

The main technical novelty of this paper is the technique used to lower bound the size of $B_{x,\alpha} = \{y \in \{0,1\}^n \mid C(y \mid n) - C(y \mid x) \geq \alpha\}$, which should be contrasted with a known and simple approach. This "normal" and simple approach is best illustrated when x is random. In this case, the prefix $x(1 : \alpha)$ of

[1] We use notation poly(n) for $n^{O(1)}$ and $\approx$, $\preceq$ and $\succeq$ to denote that the respective equality or inequality holds with an error of at most $O(\log n)$.

x of length α is also random and, therefore, if we take z to be an $(n-\alpha)$ long string that is random conditioned by $x(1:\alpha)$, then $C(zx(1:\alpha)) = n - O(\log n)$, $C(zx(1:\alpha) \mid x(1:\alpha)) = n - \alpha - O(\log n)$, and thus, $zx(1:\alpha) \in B_{x,\alpha+O(\log n)}$. There are approximately $2^{n-\alpha}$ strings z as above, and this leads to a lower bound of $2^{n-\alpha}$ for $|B_{x,\alpha+O(\log n)}|$, which implies a lower bound of $(1/\text{poly}(n))2^{n-\alpha}$ for $|B_{x,\alpha}|$. This method is so basic and natural that it looks hard to beat. However, using properties of Kolmogorov complexity extractors, we derive a better lower bound for $|B_{x,\alpha}|$ that does not have the slack of $1/\text{poly}(n)$, in case α is computable from n (even if α is not computable from n, the new method gives a tighter estimation than the above "normal" method). A Kolmogorov complexity extractor is a function that starting with several strings that have Kolmogorov complexity relatively small compared to their lengths, computes a string that has Kolmogorov complexity almost close to its length. A related notion, namely multi-source randomness extractors, has been studied extensively in computational complexity (see[3,1,2,12,11]). Hitchcock, Pavan and Vinodchandran [8] have shown that Kolmogorov complexity extractors are equivalent to a type of functions that are close to being multisource randomness extractors. Fortnow, Hitchcock, Pavan, Vinodchandran and Wang [7] have constructed a polynomial-time Kolmogorov complexity extractor based on the multi-source randomness constractor of Barak, Impagliazzo and Wigderson [1]. The author has constructed Kolmogorov complexity extractors for other settings, such as extracting from infinite binary sequences [18,16] or from binary strings that have a bounded degree of dependence [16,19,17]. The latter type of Kolmogorov complexity extractors is relevant for this paper. Here we modify slightly an extractor E from [17], which, on inputs two n-bit strings x and y that have Kolmogorov complexity at least s and dependency at most α, constructs an m-bit string z with $m \approx s$ and Kolmogorov complexity equal to $m - \alpha - O(1)$ even conditioned by any one of the input strings. Let us call a pair of strings x and y with the above properties as *good-for-extraction*. We fix $x \in \{0,1\}^n$ with $C(x) \geq s$. Let z be the most popular image of the function E restricted to $\{x\} \times \{0,1\}^n$. Because it is distinguishable from all other strings, given x, z can be described with only $O(1)$ bits (we only need a description of the function E and of the input length). Choosing m just slightly larger than α we arrange that $C(z \mid x) < m - \alpha - O(1)$. This implies that all the preimages of z under E restricted as above are *bad-for-extraction*. Since the size of $E^{-1}(z) \cap (\{x\} \times \{0,1\}^n)$ is at least 2^{n-m}, we see that at least 2^{n-m} pairs (x,y) are bad-for-extraction. A pair of strings (x,y) is bad-for-extraction if either y has Kolmogorov complexity below s (and it is easy to find an upper bound on the number of such strings), or if $y \in B_{x,\alpha}$. This allows us to find the lower bound for the size of $B_{x,\alpha}$.

2 Preliminaries

We work over the binary alphabet $\{0,1\}$; $\mathbb{N}$ is the set of natural numbers. A string x is an element of $\{0,1\}^*$; $|x|$ denotes its length; $\{0,1\}^n$ denotes the set of strings of length n; $|A|$ denotes the cardinality of a finite set A; for $n \in \mathbb{N}$, $[n]$ denotes the set $\{1,2,\ldots,n\}$. We recall the basics of (plain) Kolmogorov complexity (for an extensive coverage, the reader should consult one of the monographs by Calude [4], Li and Vitányi [10], or Downey and Hirschfeldt [6]; for a good and concise introduction, see Shen's lecture notes [13]). Let M be a standard Turing machine. For any string x, define the *(plain) Kolmogorov complexity* of x with respect to M, as $C_M(x) = \min\{|p| \mid M(p) = x\}$. There is a universal Turing machine U such that for every machine M there is a constant c such that for all x, $C_U(x) \leq C_M(x) + c$. We fix such a universal machine U and dropping the subscript, we let $C(x)$ denote the Kolmogorov complexity of x with respect to U. We also use the concept of conditional Kolmogorov complexity. Here the underlying machine is a Turing machine that in addition to the read/work tape which in the initial state contains the input p, has a second tape containing initially a string y, which is called the conditioning information. Given such a machine M, we define the Kolmogorov complexity of x conditioned by y with respect to M as $C_M(x \mid y) = \min\{|p| \mid M(p, y) = x\}$. Similarly to the above, there exist universal machines of this type and they satisfy the relation similar to the above one, but for conditional complexity. We fix such a universal machine U, and dropping the subscript U, we let $C(x \mid y)$ denote the Kolmogorov complexity of x conditioned by y with respect to U.

There exists a constant c_U such that for all strings x, $C(x) \leq |x| + c_U$. Strings $x_1, x_2, \ldots, x_k$ can be encoded in a self-delimiting way (i.e., an encoding from which each string can be retrieved) using $|x_1| + |x_2| + \ldots + |x_k| + 2\log|x_2| + \ldots + 2\log|x_k| + O(k)$ bits. For example, x_1 and x_2 can be encoded as $\overline{(bin(|x_2|)}01x_1x_2$, where $bin(n)$ is the binary encoding of the natural number n and, for a string $u = u_1 \ldots u_m$, $\overline{u}$ is the string $u_1u_1 \ldots u_mu_m$ (i.e., the string u with its bits doubled).

Given a string x and its Kolmogorov complexity $C(x)$, one can effectively enumerate all descriptions y of x of length $C(x)$, i.e., the set $\{y \in \{0,1\}^{C(x)} \mid U(y) = x\}$. We denote x^* the first string in this enumeration. Note that $C(x) - O(1) \leq C(x^*) \leq |x^*| + O(1) = C(x) + O(1)$.

The Symmetry of Information Theorem [20] states that for any two strings x and y,

(a) $C(xy) \leq C(y) + C(x \mid y) + 2\log C(y) + O(1)$.

(b) $C(xy) \geq C(x) + C(y \mid x) - 2\log C(xy) - 4\log\log C(xy) - O(1)$.

(c) If $|x| = |y| = n$, $C(y) - C(y \mid x) \geq C(x) - C(x \mid y) - 5\log n$.

(The proof of this variant follows the standard technique and is available in the full version of this paper.)

As discussed in the Introduction, our main focus is on sets of strings having certain dependency or independency properties. For convenience, we restate here the main definitions.

Definition 1. *The string y has α-dependency (where $\alpha \in \mathbb{N}$) with the string x if $C(y) - C(y \mid x) \geq \alpha$ or if x coincides with y.*

We have included the case "x coincides with y" to make a string dependent with itself even in case it has low Kolmogorov complexity.

Definition 2. *The strings $x_1, \ldots, x_t$ are pairwise α-independent if for all $i \neq j$, $C(x_i) - C(x_i \mid x_j) \leq \alpha$.*

Definition 3. *The tuple of strings $(x_1, \ldots, x_t)$ is mutually α-independent (where $\alpha \in \mathbb{N}$) if $C(x_{\pi(1)} x_{\pi(2)} \ldots x_{\pi(t)}) \geq C(x_1) + C(x_2) + \ldots + C(x_t) - \alpha$, for every permutation π of $[t]$.*

3 Strings Dependent with a Given String

Given a string $x \in \{0,1\}^n$, and $\alpha \in \mathbb{N}$, how many strings have dependency with x at least α? That is we are interested in estimating the size of the set

$$A_{x,\alpha} = \{y \in \{0,1\}^n \mid C(y) - C(y \mid x) \geq \alpha\}.$$

This is the set of strings about which, roughly speaking, x has at least α bits of information. A related set is

$$B_{x,\alpha} = \{y \in \{0,1\}^n \mid C(y \mid n) - C(y \mid x) \geq \alpha\},$$

consisting of the n-bit strings about which x provides α bits of information besides the length n. Clearly, $B_{x,\alpha} \subseteq A_{x,\alpha}$, and thus an upper bound for $|A_{x,\alpha}|$ also holds for $|B_{x,\alpha}|$, and a lower bound for $|B_{x,\alpha}|$ also holds for $|A_{x,\alpha}|$.

We show that for some polynomial p and for some constant C, for all x and α except some special values,

$$(1/p(n)) \cdot 2^{n-\alpha} \leq |A_{x,\alpha}| \leq C 2^{n-\alpha},$$

and, in case $\alpha(n)$ is computable from n,

$$(1/C) \cdot 2^{n-\alpha} - p(n) 2^{\alpha} \leq |B_{x,\alpha}| \leq C 2^{n-\alpha},$$

The upper bounds for the sizes of $A_{x,\alpha}$ and $B_{x,\alpha}$ can be readily derived. Observe that the set $A_{x,\alpha}$ is included in $\{y \in \{0,1\}^n \mid C(y \mid x) < n - \alpha + c\}$ for some constant c, and therefore

$$|A_{x,\alpha}| \leq C \cdot 2^{n-\alpha},$$

for $C = 2^c$.

We move to finding a lower bound for the size of $A_{x,\alpha}$. A first observation is that for $A_{x,\alpha}$ to be non-empty, it is needed that $C(x) \succeq \alpha$. Indeed, it is immediate to observe that for any strings x and y of length n,

$$C(y) \leq C(x) + C(y \mid x) + 2 \log C(x) + O(1) \leq C(x) + C(y \mid x) + 2 \log n + O(1),$$

and thus, if $C(y) - C(y \mid x) \geq \alpha$, then $C(x) \geq \alpha - 2 \log n - O(1)$. Intuitively, if the information in x is close to α, not too many strings can be α-dependent with it.

We provide a lower bound for $|A_{x,\alpha}|$, for every string x with $C(x) \geq \alpha + 7 \log n$. The proof uses the basic "normal" approach presented in the Introduction. To simplify the discussion, suppose $C(x) = \alpha$. Then if we take a string z of length $n - \alpha$ that is random conditioned by x^*, it holds that $C(x^*z) \approx n$ and $C(x^*z \mid x^*) \approx n - \alpha$. Thus, $C(x^*z) - C(x^*z \mid x^*) \succeq \alpha$. Note that there are approximately $2^{n-\alpha}$ such strings x^*z. Since x^* can be obtained from x and $C(x)$, we can replace x^* by x in the conditioning at a small price. We obtain approximately $2^{n-\alpha}$ strings in $A_{x,\alpha}$.

Theorem 1. *For every natural number n, for every natural number α and for every $x \in \{0,1\}^n$ such that $C(x) \geq \alpha + 7\log n$,*

$$|A_{x,\alpha}| \geq \frac{1}{2n^7} 2^{n-\alpha},$$

provided n is large enough.

Proof. Let $k = C(x)$ and let $\beta = \alpha + 7\log n$. Let x^* be the smallest description of x as described in the Preliminaries. Let x^*_β be the prefix of x^* of length β. Since x^* is described by x^*_β and by its suffix of length $k - \beta$, $C(x^*) \leq C(x^*_\beta) + (k - \beta) + 2\log C(x^*_\beta) + O(1)$ and, thus

$$\begin{aligned} C(x^*_\beta) &\geq C(x^*) - (k-\beta) - 2\log C(x^*_\beta) - O(1) \\ &\geq (k - O(1)) - (k-\beta) - 2\log C(x^*_\beta) - O(1) \\ &\geq \beta - 2\log\beta - O(1). \end{aligned}$$

The set $B = \{z \in \{0,1\}^{n-\beta} \mid C(z \mid x^*_\beta) \geq n - \beta - 1\}$ has size at least $(1/2) \cdot 2^{n-\beta}$ (using a standard counting argument). Consider a string $y \in \{0,1\}^n$ of the from $y = x^*_\beta z$ with $z \in B$. There are at least $(1/2) \cdot 2^{n-\beta}$ such strings.

By symmetry of information,

$$\begin{aligned} C(y) = C(x^*_\beta z) &\geq C(x^*_\beta) + C(z \mid x^*_\beta) - (2\log n + 4\log\log n + O(1)) \\ &\geq (\beta - 2\log\beta) + (n - \beta - 1) - (2\log n + 4\log\log n + O(1)) \\ &\geq n - (4\log n + 4\log\log n + O(1)) \geq n - 5\log n. \end{aligned}$$

On the other hand, $C(y \mid x^*_\beta) = C(x^*_\beta z \mid x^*_\beta) \leq C(z) + O(1) \leq (n - \beta) + O(1)$. Note that

$$C(y \mid x) \leq C(y \mid x^*_\beta) + 2\log n + 4\log\log n + O(1),$$

because one can effectively construct x^*_β from x, k and β. Therefore,

$$C(y \mid x) \leq (n - \beta) + 2\log n + 4\log\log n + O(1),$$

and thus

$$C(y) - C(y \mid x) \geq \beta - (6\log n + 8\log\log n + O(1)) \geq \beta - 7\log n.$$

So, $y \in A_{x,\beta-7\log n} = A_{x,\alpha}$. Since this holds for all the strings y mentioned above, it follows that $|A_{x,\alpha}| \geq (1/2)2^{n-\beta} = (1/(2n^7)) \cdot 2^{n-\alpha}$. ∎

The lower bound for $|B_{x,\alpha}|$ is obtained using a technique based on Kolmogorov complexity extractors, as explained in the Introduction. We use the following theorem which can be obtained by a simple modification of a result from [17].

Theorem 2. *For any computable functions $s(n), m(n)$ and $\alpha(n)$ with $n \geq s(n) \geq \alpha(n) + 7\log n$ and $m(n) \leq s(n) - 7\log n$, there exists a computable ensemble of functions $E : \{0,1\}^n \times \{0,1\}^n \to \{0,1\}^{m(n)}$ such that for all x and y in $\{0,1\}^n$*

- *if $C(x) \geq s(n), C(y \mid n) \geq s(n)$ and $C(y \mid n) - C(y \mid x) \leq \alpha(n)$*
- *then $C(E(x,y) \mid x) \geq m(n) - \alpha(n) - O(1)$.*

Theorem 3. *Let $\alpha(n)$ be a computable function. For every sufficiently large natural number n, for every $x \in \{0,1\}^n$ such that $C(x) \geq \alpha(n) + 8\log n$,*

$$|B_{x,\alpha(n)}| \geq \frac{1}{C} \cdot 2^{n-\alpha(n)} - n^8 2^{\alpha(n)},$$

for some positive constant C.

Proof. Let $m = \alpha(n) + c$ and $s = \alpha(n) + 8\log n$, where c is a constant that will be specified later. Consider $E : \{0,1\}^n \times \{0,1\}^n \to \{0,1\}^m$ the Kolmogorov extractor given by Theorem 2 for these parameters. Let $z \in \{0,1\}^m$ be the string that has the largest number of E preimages in the set $\{x\} \times \{0,1\}^n$. Note that, for some constant c_1, $C(z \mid x) \leq c_1$, because, given x, z can be constructed from a table of E, which at its turn can be constructed from n which is given because it is the length of x. On the other hand, if $y \in \{0,1\}^n$ is a string with $C(y \mid n) \geq s$ and $C(y \mid n) - C(y \mid x) \leq \alpha(n)$, then Theorem 2 guarantees that, for some constant c_2, $C(E(x,y) \mid x) \geq m - \alpha(n) - c_2 = c - c_2 > c_1$, for an appropriate c. Therefore all the strings y such that $E(x,y) = z$ are bad for extraction, i.e., they belong to the union of $\{y \in \{0,1\}^n \mid C(y \mid n) < s\}$ and $\{y \in \{0,1\}^n \mid C(y \mid n) \geq s$ and $C(y \mid n) - C(y \mid x) \geq \alpha\}$. Since there are at least 2^{n-m} such strings y and the first set above has less than 2^s elements, it follows that

$$|\{y \in \{0,1\}^n \mid C(y \mid n) - C(y \mid x) \geq \alpha(n)\}| \geq 2^{n-m} - 2^s = \frac{1}{2^c} \cdot 2^{n-\alpha(n)} - n^8 2^{\alpha(n)}.$$

This concludes the proof. ∎

The proof of Theorem 1 actually shows more: The lower bound applies even to a subset of $A_{x,\alpha}$ containing only strings with high Kolmogorov complexity. More precisely, if we denote $A_{x,\alpha,s} = \{y \in \{0,1\}^n \mid C(y) \geq s$ and $C(y) - C(y \mid x) \geq \alpha\}$, then $|A_{x,\alpha,n-5\log n}| \geq \frac{1}{2n^7} 2^{n-\alpha}$. Note that there is an interesting "zone" for the parameter s that is not covered by this result. Specifically, it would be interesting to lower bound the size of $A_{x,\alpha,n}$. This question remains open. Nevertheless, the technique from Theorem 3 can be used to tackle the variant in which access to the set $R = \{u \in \{0,1\}^n \mid C(u) \geq |u|\}$ is granted for free. Thus, let $A^R_{x,\alpha,n} = \{y \in \{0,1\}^n \mid C^R(y) \geq n$ and $C^R(y) - C^R(y \mid x) \geq \alpha\}$.

Proposition 1. *For the same setting of parameters as in Theorem 3, $|A^R_{x,\alpha,n}| \geq \frac{1}{C} \cdot 2^{n-\alpha(n)}$, for some positive constant C.*

Proof. Omitted from this extended abstract. ∎

4 Pairwise Independent Strings

We show that if the n-bit strings $x_1, \ldots, x_t$ are pairwise α-independent, then $t \leq \text{poly}(n)2^\alpha$. This upper bound is relatively tight, since there are sets with $(1/\text{poly}(n)) \cdot 2^\alpha$ n-bit strings that are pairwise α-independent.

Theorem 4. *For every sufficiently large n and for every natural number α, the following holds. If $x_1, \ldots, x_t$ are n-bit strings that are α-independent, then $t < 2n^3 \cdot 2^\alpha$.*

Proof. There are less than $2^{\alpha+3\log n}$ strings with Kolmogorov complexity less than $\alpha + 3\log n$. We discard such strings from $x_1, \ldots, x_t$ and assume that $x_1, \ldots, x_{t'}$ are the strings that are left. Since $t < 2^{\alpha+3\log n} + t'$, we need to show that $t' \leq n^3 2^\alpha$.

For $1 \leq i \leq t'$, let $k_i = C(x_i)$ and let x_i^* be the shortest description of x_i as described in the Preliminaries. Let $\beta = \alpha + 3\log n$ (we assume that $\alpha \leq n - 3\log n$, as otherwise the statement is trivial). We show that the prefixes of length β of the strings $x_1, \ldots, x_{t'}$ are all distinct, from which we conclude that $t' \leq 2^\beta = n^3 \cdot 2^\alpha$.

Suppose that there are two strings in the set that have equal prefixes of length β. W.l.o.g. we can assume that they are x_1 and x_2. Then

$$C(x_1^* \mid x_2^*) \leq (k_1 - \beta) + \log\beta + 2\log\log\beta + O(1),$$

because, given x_2^*, x_1^* can be constructed from β and the suffix of length $k_1 - \beta$ of x_1^*. Note that

$$C(x_1^* \mid x_2) \leq C(x_1^* \mid x_2^*) + \log k_2 + 2\log\log k_2 + O(1),$$

because x_2^* can be constructed from x_2 and k_2. Also note that $C(x_1 \mid x_2) \leq C(x_1^* \mid x_2) + O(1)$. Thus,

$$C(x_1 \mid x_2) \leq C(x_1^* \mid x_2^*) + \log k_2 + 2\log\log k_2 + O(1).$$

Therefore,

$$\begin{aligned} C(x_1) - C(x_1 \mid x_2) &\geq k_1 - (C(x_1^* \mid x_2^*) + \log k_2 + 2\log\log k_2 + O(1)) \\ &\geq k_1 - (k_1 - \beta) - \log\beta - 2\log\log\beta - \log k_2 - 2\log\log k_2 \\ &\qquad - O(1) \\ &\geq \beta - 3\log n = \alpha, \end{aligned}$$

which is a contradiction. ∎

The next result shows that the upper bound in Theorem 4 is relatively tight. It relies on the well-known Turán's Theorem in Graph Theory [14], in the form due to Caro (unpublished) and Wei [15] (see [9, page 248]): Let G be a graph with n vertices and let d_i be the degree of the i-th vertex. Then G contains an independent set of size at least $\sum \frac{1}{d_i+1}$.

Theorem 5. *For every natural number n and for every natural number α satisfying $5\log n \le \alpha \le n$, there exists a constant C and $t = \frac{1}{Cn^5} \cdot 2^\alpha$ n-bit strings $x_1, \ldots, x_t$ that are pairwise α-independent.*

Proof. Let $\beta = \alpha - 5\log n$. Consider the graph $G = (V, E)$, where $V = \{0,1\}^n$ and $(u, v) \in E$ iff $C(u) - C(u \mid v) \ge \beta$ and $C(v) - C(v \mid u) \ge \beta$. Note that for every $u \in \{0,1\}^n$, the degree of u is bounded by $|A_{u,\beta}| \le 2^{n-\beta+c}$, for some constant c. Therefore, by Turán's theorem, the graph G contains an independent set I of size at least $2^n \cdot \frac{1}{2^{n-\beta+c}+1} \ge 2^{\beta-c-1} = \frac{1}{Cn^5} \cdot 2^\alpha$. For any two elements u, v in I, we have either $C(u) - C(u \mid v) < \beta$ or $C(v) - C(v \mid u) < \beta$. In the second case, by symmetry of information, $C(u) - C(u \mid v) < \beta + 5\log n = \alpha$. It follows that the strings in I are pairwise α-independent. ∎

5 Mutually Independent Strings

In this section we show that the size of a mutually α-independent tuple of n-bit strings is bounded by $\text{poly}(n)2^\alpha$.

For $u \in \{0,1\}^n$, we define $D_\alpha(u) = \{x \in \{0,1\}^n \mid u \in A_{x,\alpha}\} = \{x \in \{0,1\}^n \mid C(u) - C(u \mid x) \ge \alpha\}$ and $d_\alpha(u) = |D_\alpha(u)|$.

Lemma 1. *For every natural number n sufficiently large, for every natural number α, and for every $u \in \{0,1\}^n$, with $C(u) \ge \alpha + 12\log n$,*

$$\frac{1}{2n^{12}} 2^{n-\alpha} \le d_\alpha(u) \le n^5 \cdot 2^{n-\alpha}.$$

Proof. For every $x \in A_{u,\alpha+5\log n}$, $C(x) - C(x \mid u) \ge \alpha + 5\log n$ which by symmetry of information implies $C(u) - C(u \mid x) \ge \alpha + 5\log n - 5\log n = \alpha$, and therefore, $u \in A_{x,\alpha}$. Thus

$$d_\alpha(u) \ge |A_{u,\alpha+5\log n}| \ge \frac{1}{2n^7} 2^{n-\alpha-5\log n} = \frac{1}{2n^{12}} 2^{n-\alpha}.$$

For every $u \in \{0,1\}^n$,

$$\begin{aligned} x \in D_{u,\alpha} &\Rightarrow u \in A_{x,\alpha} \\ &\Rightarrow C(u) - C(u \mid x) \ge \alpha \\ &\Rightarrow C(x) - C(x \mid u) \ge \alpha - 5\log n \\ &\Rightarrow C(x \mid u) \le n - \alpha + 5\log n. \end{aligned}$$

Thus, $d_\alpha(u) \le |\{x \in \{0,1\}^n \mid C(x \mid u) \le n - \alpha + 5\log n\}| \le n^5 \cdot 2^{n-\alpha}$. ∎

Since for any string x and natural number α, $|A_{x,\alpha}| \le 2^{n-\alpha-c}$, for some constant c, it follows that we need at least $T = 2^{\alpha-c}$ strings $x_1, \ldots, x_T$ to "α-cover" the set of n-bit strings, in the sense that for each n-bit string y, there exists x_i, $i \in [T]$ such that y is α-dependent with x_i. The next theorem shows that $\text{poly}(n)2^\alpha$ strings are enough to α-cover the set of n-bit strings.

Theorem 6. *For every natural number n sufficiently large, for every natural number α, there exists a set $B \subseteq \{0,1\}^n$ of size* $\text{poly}(n)2^\alpha$ *such that each string in $\{0,1\}^n$ is α-dependent with some string in B, i.e., $\{0,1\}^n = \bigcup_{x \in B} A_{x,\alpha}$. More precisely the size of B is bounded by $(2n^{13} + n^{12}) \cdot 2^\alpha$.*

Proof. (a) We choose $T = 2n^{13}2^\alpha$ strings $x_1, \ldots, x_T$, uniformly at random in $\{0,1\}^n$. The probability that a fix u with $C(u) \geq \alpha + 12\log n$ does not belong to any of the sets $A_{x_i,\alpha}$, for $i \in [T]$, is at most $(1 - \frac{1}{2n^{12}2^\alpha})^T < e^{-n}$ (by Lemma 1). By the union bound, the probability that there exists $u \in \{0,1\}^n$ with $C(u) \geq \alpha + 12\log n$, that does not belong to any of the sets $A_{x_i,\alpha}$, for $i \in [T]$, is bounded by $2^n \cdot e^{-n} < 1$. Therefore there are strings $x_1, \ldots, x_T$ in $\{0,1\}^n$ such that $\bigcup A_{x_i,\alpha}$ contains all the strings $u \in \{0,1\}^n$ having $C(u) \geq \alpha + 12\log n$. By adding to $x_1, \ldots, x_T$, the strings that have Kolmogorov complexity $< \alpha + 12\log n$, we obtain the set B that α-covers the entire $\{0,1\}^n$. ∎

To estimate the size of a mutually α-independent tuple of strings, we need the following lemma (whose proof is available in the full version of the paper).

Lemma 2. *Let $\alpha, \beta \in \mathbb{N}$ and let the tuple of n-bit strings $(x_1, x_2, \ldots, x_k)$ satisfy $C(x_1 \ldots x_k) \geq C(x_1) + \ldots + C(x_k) - \beta$. Then there exists a constant d such that*

$$|A_{x_1,\alpha} \cap \ldots \cap A_{x_k,\alpha}| \leq dn^{7k+5}k^3 2^{n-k\alpha+\beta}.$$

Finally, we prove the upper bound for the size of a mutually α-independent tuple of n-bit strings.

Theorem 7. *For every sufficiently large natural number n the following holds. Let α be an integer such that $\alpha > 7\log n + 6$. Let $(x_1, \ldots, x_t)$ be a mutually α-independent tuple of n-bit strings. Then $t \leq \text{poly}(n)2^\alpha$.*

Proof. By Theorem 6, there exists a set B of size at most $\text{poly}(n)2^{\alpha+5\log n}$ such that every n-bit string x is in $A_{y,\alpha+5\log n}$, for some $y \in B$. We view $\{x_1, \ldots, x_t\}$ as a multiset. Let y be the string in B that achieves the largest size of multiset $A_{y,\alpha+5\log n} \cap \{x_1, \ldots, x_t\}$ (we take every common element with the multiplicity in $\{x_1, \ldots, x_t\}$). Let k be the size of the above intersection. Clearly, $k \geq t/|B|$. We will show that $k = \text{poly}(n)$, and, therefore, $t \leq k \cdot |B| = \text{poly}(n) \cdot 2^\alpha$.

Without loss of generality suppose $A_{y,\alpha+5\log n} \cap \{x_1, \ldots, x_t\} = \{x_1, \ldots, x_k\}$ (as multisets). Since, for every $i \in [k]$, $C(x_i) - C(x_i \mid y) \geq \alpha + 5\log n$, by symmetry of information, it follows that $C(y) - C(y \mid x_i) \geq \alpha$. Thus $y \in A_{x_1,\alpha} \cap \ldots \cap A_{x_k,\alpha}$. In particular, $A_{x_1,\alpha} \cap \ldots \cap A_{x_k,\alpha}$ is not empty. We want to use Lemma 2 but before we need to estimate the difference between $C(x_1 \ldots x_k)$ and $C(x_1) + \ldots + C(x_k)$.

Claim. $C(x_1 \ldots x_k) \geq C(x_1) + \ldots + C(x_k) - \beta$, where $\beta = \alpha + 4\log(nt/2)$.

The proof of this claim is available in the full version of the paper. ∎

Now, by Lemma 2,

$$\begin{aligned}|A_{x_1,\alpha} \cap \ldots \cap A_{x_k,\alpha}| &\le dn^{7k+5}k^3 2^{n-k\alpha+\beta} \\ &= dn^{7k+5}k^3 2^{n-(k-1)\alpha+4\log t+4\log(n/2)} \\ &\le dn^{7k+5}k^3 2^{5n-(k-1)\alpha+4\log(n/2)},\end{aligned}$$

where in the last line we used the fact that $t \le 2^n$.

It can be checked that if $\alpha > 7\log n + 6$ and $k \ge n$, then the above upper bound is less than 1, which is a contradiction. It follows that $k < n$. ∎

6 Final Remarks

This paper provides tight bounds (within a polynomial factor) for the size of $A_{x,\alpha}$ (the set of n-bit strings that have α-dependency with x) and for the size of sets of n-bit strings that are pairwise α-independent.

The size of a mutually α-independent tuple of n-bit strings is at most $\text{poly}(n)2^\alpha$. We do not know how tight this bound is and leave this issue as an interesting open problem.

We have recently learned about the paper [5], which obtains similar results regarding the size of sets of pairwise and k-independence strings, for a notion of independence that is suitable for strings with large Kolmogorov complexity.

References

1. Barak, B., Impagliazzo, R., Wigderson, A.: Extracting randomness using few independent sources. In: Proceedings of the 36th ACM Symposium on Theory of Computing, pp. 384–393 (2004)
2. Barak, B., Kindler, G., Shaltiel, R., Sudakov, B., Wigderson, A.: Simulating independence: new constructions of condensers, ramsey graphs, dispersers, and extractors. In: Proceedings of the 37th ACM Symposium on Theory of Computing, pp. 1–10 (2005)
3. Bourgain, J.: More on the sum-product phenomenon in prime fields and its applications. International Journal of Number Theory 1, 1–32 (2005)
4. Calude, C.: Information and Randomness: An Algorithmic Perspective., 2nd edn. Springer, Heidelberg (2002); 1st edn. (1994)
5. Chang, C., Lyuu, Y., Ti, Y., Shen, A.: Sets of k-independent sets. International Journal of Foundations of Computer Science (2009) (to appear)
6. Downey, R., Hirschfeldt, D.: Algorithmic randomness and complexity. Springer, Heidelberg (2010)
7. Fortnow, L., Hitchcock, J., Pavan, A., Vinodchandran, N., Wang, F.: Extracting Kolmogorov complexity with applications to dimension zero-one laws. In: Bugliesi, M., Preneel, B., Sassone, V., Wegener, I. (eds.) ICALP 2006. LNCS, vol. 4051, pp. 335–345. Springer, Heidelberg (2006)
8. Hitchcock, J., Pavan, A., Vinodchandran, N.: Kolmogorov complexity in randomness extraction. In: Electronic Colloquium on Computational Complexity (ECCC) (09-071) (2009)
9. Jukna, S.: Extremal Combinatorics. Springer, Heidelberg (2001)

10. Li, M., Vitanyi, P.: An introduction to Kolmogorov complexity and its applications, 3rd edn. Springer, Heidelberg (2008); 1st edn. (1993)
11. Rao, A.: Extractors for a constant number of polynomially small min-entropy independent sources. In: Proceedings of the 38th ACM Symposium on Theory of Computing, pp. 497–506 (2006)
12. Raz, R.: Extractors with weak random seeds. In: Gabow, H.N., Fagin, R. (eds.) STOC, pp. 11–20. ACM, New York (2005)
13. Shen, A.: Algorithmic information theory and Kolmogorov complexity. Tech. Rep. 2000-034, Uppsala Universitet (December 2000)
14. Turán, P.: On an extremal problem in graph theory. Math. Fiz. Lapok 48, 436–452 (1941) (in Hungarian)
15. Wei, V.: A lower bound on the stability number of a simple graph. Tech. Rep. 81-11217-9, Bell Laboratories (1981)
16. Zimand, M.: Extracting the Kolmogorov complexity of strings and sequences from sources with limited independence. In: Proceedings 26th STACS, Freiburg, Germany, February 26-29 (2009)
17. Zimand, M.: Impossibility of independence amplification in Kolmogorov complexity theory. In: MFCS (2010)
18. Zimand, M.: Two sources are better than one for increasing the Kolmogorov complexity of infinite sequences. In: Hirsch, E.A., Razborov, A.A., Semenov, A., Slissenko, A. (eds.) CSR 2008. LNCS, vol. 5010, pp. 326–338. Springer, Heidelberg (2008)
19. Zimand, M.: On generating independent random strings. In: Ambos-Spies, K., Löwe, B., Merkle, W. (eds.) CiE. LNCS, vol. 5635, pp. 499–508. Springer, Heidelberg (2009)
20. Zvonkin, A., Levin, L.: The complexity of finite objects and the development of the concepts of information and randomness by means of the theory of algorithms. Russian Mathematical Surveys 25(6), 83–124 (1970)

Impossibility of Independence Amplification in Kolmogorov Complexity Theory

Marius Zimand*

Department of Computer and Information Sciences, Towson University, Baltimore, MD, USA

Abstract. The paper studies randomness extraction from sources with bounded independence and the issue of independence amplification of sources, using the framework of Kolmogorov complexity. The dependency of strings x and y is $\text{dep}(x, y) = \max\{C(x) - C(x \mid y), C(y) - C(y \mid x)\}$, where $C(\cdot)$ denotes the Kolmogorov complexity. It is shown that there exists a computable Kolmogorov extractor f such that, for any two n-bit strings with complexity $s(n)$ and dependency $\alpha(n)$, it outputs a string of length $s(n)$ with complexity $s(n) - \alpha(n)$ conditioned by any one of the input strings. It is proven that the above are the optimal parameters a Kolmogorov extractor can achieve. It is shown that independence amplification cannot be effectively realized. Specifically, if (after excluding a trivial case) there exist computable functions f_1 and f_2 such that $\text{dep}(f_1(x, y), f_2(x, y)) \leq \beta(n)$ for all n-bit strings x and y with $\text{dep}(x, y) \leq \alpha(n)$, then $\beta(n) \geq \alpha(n) - O(\log n)$.

Keywords: Kolmogorov complexity, random strings, independent strings, randomness extraction.

1 Introduction

Randomness extraction is an algorithmical process that improves the quality of a source of randomness. A source of randomness can be modeled as a finite probability distribution, or a finite binary string, or an infinite binary sequence and the randomness quality is measured, respectively, by min-entropy, Kolmogorov complexity, and constructive Hausdorff dimension. All the three settings have been studied (the first one quite extensively).

It is desirable to have an extractor that can handle very general classes of sources. Ideally, we would like to have an extractor that obtains random bits from a single defective source under the single assumption that there exists a certain amount of randomness in the source. Unfortunately, this is not possible. In the case of finite distributions, impossibility results for extraction from a single source have been established by Santha and Vazirani [19] and Chor and Goldreich [6]. In the case of finite binary strings and Kolmogorov complexity randomness, Vereshchagin and Vyugin [22] show that there exists strings x with

* http://triton.towson.edu/~mzimand

P. Hliněný and A. Kučera (Eds.): MFCS 2010, LNCS 6281, pp. 701–712, 2010.

relatively high Kolmogorov complexity so that any string shorter than x by a certain amount and which has small Kolmogorov complexity conditioned by x (in particular any such shorter string effectively constructed from x) has small Kolmogorov complexity unconditionally. The issue of extraction from one infinite sequence has been first raised by Reimann and Terwijn [18], and after a series of partial results [18,14,3], Miller [13] has given a strong negative answer, by constructing a sequence x with $\dim(x) = 1/2$ such that, for any Turing reduction f, $\dim(f(x)) \leq 1/2$ (or $f(x)$ does not exist; $\dim(x)$ is the constructive Hausdorff dimension of the sequence x).

Therefore, for extraction from a general class of sources, one has to consider the case of $t \geq 2$ sources, and in this situation, positive results are possible. Computable extractors from $t = 2$ distributions with min-entropy $k = O(\log n)$ are constructed in [6,8]. The construction of polynomial-time multisource extractors is a difficult problem. Currently, for $t = 2$, the best results are by Bourgain [4] who achieves $k = (1/2 - \alpha)n$ for a small constant α, and Raz [17] who achieves $k = \text{polylog} n$ for one distribution and $k = (1/2 + \alpha)n$ for the other one. Polynomial-time extractors for 3 or more distributions with lower values of k for all distributions are constructed in [1,2,17,16,15]. Dodis et al. [7] construct a polynomial-time 2-source extractor for $k > n/2$, where the extracted bits are random conditioned by one of the sources. Kolmogorov extractors for $t \geq 2$ sources also exist. Fortnow et al. [10] actually observe that any randomness extractor for distributions is a Kolmogorov extractor and Hitchcock et al. [11] show that a weaker converse holds, in the sense that any Kolmogorov extractor is a randomness condenser with very good parameters ("almost extractor"). For $t = 2$, the works [23,25] construct computable Kolmogorov extractors with better properties than those achievable by converting the randomness extractors from [6] and [8]. The case of infinite sequences is studied in [24], which shows that it is possible to effectively increase the constructive dimension if the input consists of two sources.

All the positive results cited above require that the sources are independent. At a first glance, without independence, even the distinction between one and two (or more) sources is not clear. However, independence can be quantified and then we can consider two sources having bounded independence. It then becomes important to determine to what extent randomness extraction is possible from sources with a limited degree of independence and whether the degree of independence can be amplified.

We address these questions for the case of finite strings and Kolmogorov complexity-based randomness. The level of dependency of two strings is based on the notion of mutual information. The information that string x has about string y is $I(x : y) = C(y) - C(y \mid x)$, where $C(y)$ is the Kolmogorov complexity of y and $C(y \mid x)$ is the Kolmogorov complexity of y conditioned by x. By the symmetry of information theorem, $I(x : y) \approx I(y : x) \approx C(x) + C(y) - C(xy)$.[1] We define the dependency of strings x and y as $\text{dep}(x, y) = \max\{I(x : y), I(y : x)\}$. Let

---

[1] We use $\approx$, $\preceq$ and $\succeq$ for equalities and inequalities that hold within an additive error bounded by $O(\log n)$.

$S_{k,\alpha}$ be the set of all pairs of strings (x, y) such that $C(x) \geq k$, $C(y) \geq k$ and $\mathrm{dep}(x, y) \leq \alpha$. A Kolmogorov extractor for the class of sources $S_{k,\alpha}$ is a function $f : \{0,1\}^n \times \{0,1\}^n \to \{0,1\}^m$ such that for all $(x, y) \in S_{k,\alpha}$, $C(f(x, y))$ is "close" to m. In other words, if we define the randomness deficiency of a string z as $|z| - C(z)$, we would like that the randomness deficiency of $f(x, y)$ is small. Our first result shows that the randomness deficiency of $f(x, y)$ cannot be smaller than essentially the dependency of x and y.

Result 1 (informal statement; see full statement in Theorem 2). There exists no computable function f with the property that, for all $(x, y) \in S_{k,\alpha}$, the randomness deficiency of $f(x, y)$ is less than $\alpha - \log n - O(\log \alpha)$. This holds true even for high values of k such as $k \succeq n - \alpha$. The only condition is that $m \geq \alpha$ (m is the length of the ouput of f).

We observe that the similar result holds for the case of finite distributions. Let $S_{k,\alpha}$ be the set of all random variables over $\{0,1\}^n$ that have min-entropy at least k and dependency at most α. (The min-entropy of X is $H_\infty(X) = \min_{a \in \{0,1\}^n, X(a)>0} \log(1/\mathrm{Prob}[X = a])$ and the dependency of X and Y is $H_\infty(X) + H_\infty(Y) - H_\infty(X, Y)$.) Then, for every α and $m \geq \alpha$ and for every function $f : \{0,1\}^n \times \{0,1\}^n \to \{0,1\}^m$ (even non-computable), there exists $(X, Y) \in S_{k,\alpha}$ with dependency at most α and min-entropy of $f(X, Y)$ at most $m - \alpha$.

Our next result (and the main technical contribution of this paper) is a positive one. Keeping in mind *Result 1*, the best one can hope for is a Kolmogorov extractor that from any strings x and y having dependency at most α obtains a string z whose randomness deficiency is $\approx \alpha$. We show that this is possible in a strong sense.

Result 2 (informal statement; see full statement in Theorem 4). For every $k > \alpha$, there exists a computable function $f : \{0,1\}^n \times \{0,1\}^n \to \{0,1\}^m$, where $m \approx k$, and such that for every $(x, y) \in S_{k,\alpha}$, $C(f(x, y) \mid x) = m - \alpha - O(1)$ and $C(f(x, y) \mid y) = m - \alpha - O(1)$.

Thus, optimal Kolmogorov extraction from sources with bounded independence can be achieved effectively and in a strong form. Namely, the randomness deficiency of the extracted string z is minimal (i.e., within an additive constant of α) even conditioned by any one of the input strings and furthermore the length of z is maximal. In [23] a similar but weaker theorem has been established. The difference is that in [23] the length of the output is only $\approx k/2$ and k has to be at least 2α. The proof method of *Result 2* extends the one used in [23] in a non-trivial way (the novel technical ideas are described in Section 4.1). We note that the Kolmogorov extractor that can be obtained from the randomness extractor from [8] using the technique in [10] would have weaker parameters (more precisely, the output length would be $m \approx k - 2\alpha$).

The dependency of two strings x and y is another measure of the non-randomness in (x, y) considered as a joint source. Similarly to Kolmogorov extractors that reduce randomness deficiency, it would be desirable to have an algorithm that reduces dependency (equivalently, amplifies independence). The main result of the paper shows that effective independence amplification is essentially impossible. We say that two functions $f_1, f_2 : \{0,1\}^n \times \{0,1\}^n \to \{0,1\}^{l(n)}$ amplify independence from level $\alpha(n)$ to level $\beta(n)$ (for $\beta(n) < \alpha(n)$) if $\text{dep}(f_1(x,y), f_2(x,y)) \le \beta(n)$ whenever $\text{dep}(x,y) \le \alpha(n)$. Note that this is trivial to achieve if $f_1(x,y)$ or $f_2(x,y)$ have Kolmogorov complexity at most $\beta(n)$. Therefore, we also request that $f_1(x,y)$ and $f_2(x,y)$ have Kolmogorov complexity at least $\beta(n) + c \log n$, for some constant c. However, as a consequence of *Result 1* and *Result 2*, this is impossible for any reasonable choice of parameters.

Result 3 (informal statement; see full statement in Theorem 5). Let f_1 and f_2 be computable functions such that for all $(x,y) \in S_{k,\alpha}$, $\text{dep}(f_1(x,y), f_2(x,y)) \le \beta(n)$ (and $C(f_1(x,y)) \succeq \beta(n), C(f_2(x,y)) \succeq \beta(n)$). Then $\beta(n) \succeq \alpha(n)$. This holds true for any $\alpha(n) \preceq n/2$ and any $k \preceq n - \alpha(n)$.

Discussion of some technical aspects. As it is typically the case in probabilistic analysis, handling sources with bounded independence is difficult. In this discussion, an (n,k) source is a random variable over $\{0,1\}^n$ with min-entropy k. Chor and Goldreich [6] show that a random function starting from any two independent sources of type (n,k) extracts $\approx k/3$ bits that are close to random. Dodis and Oliveira [8] using a more refined probabilistic analysis (based on a martingale construction) show the existence of an extractor that from two independent sources X and Y of type (n,k_1) and respectively (n,k_2) obtains $\approx k_1$ bits that are close to random even conditioned by Y. Both constructions use in an essential way the independence of the two input distributions. The independence property allows one to reduce the analysis to the simpler case in which the two input distributions are so called *flat distributions.* A flat distribution with min-entropy k assigns equal probability mass to a subset of size 2^k of $\{0,1\}^n$ and probability zero to the elements outside this set. Extractors that extract from flat distributions admit a nice combinatorial description. Namely, an extractor $E : \{0,1\}^n \times \{0,1\}^n \to \{0,1\}^m$ for two flat distributions X, Y with min-entropy k corresponds to an N-by-N table (where $N = 2^n$) whose cells are colored with M colors (where $M = 2^m$) that satisfy the following balancing property: For any set of colors $A \subseteq [M]$ and for any K-by-K subrectangle of the table (where $K = 2^k$), the number of A-colored cells is close to $|A|/M$. Such tables can be obtained with the probabilistic method.

If the two input distributions are not independent, then the reduction to flat distributions is not known to be possible and the above approach fails. This is why almost all of the currently known randomness extractors (whether running in polynomial time, or merely computable) assume that the weak sources are perfectly independent (one exception is the paper [21]).

In this light, it is surprising that Kolmogorov extractors for input strings that are not fully independent (actually with arbitrarily large level of dependency) can be obtained via balanced tables, as we do in this paper. This approach succeds because the Kolmogorov complexity-based analysis views the level of independence of sources as just another parameter and there is no need for any additional machinery to handle sources that are not fully independent. We believe (based on some partial results) that Kolmogorov complexity is a useful tool not only for analyzing Kolmogorov extractors but also for circumventing some of the technical difficulties in the investigation of multi-source extractors for sources with bounded independence.

Note. Because of the space constraints, this extended abstract has no proofs. They are available from the full version of this work.

2 Preliminaries

We work over the binary alphabet $\{0,1\}$; $\mathbb{N}$ is the set of natural numbers. A string x is an element of $\{0,1\}^*$; $|x|$ denotes its length; $\{0,1\}^n$ denotes the set of strings of length n; $|A|$ denotes the cardinality of a finite set A; for $n \in \mathbb{N}$, $[n]$ denotes the set $\{1,2,\ldots,n\}$. We recall the basics of (plain) Kolmogorov complexity (for an extensive coverage, the reader should consult one of the monographs by Calude [5], Li and Vitányi [12], or Downey and Hirschfeldt [9]; for a good and concise introduction, see Shen's lecture notes [20]). Let M be a standard Turing machine. For any string x, define the *(plain) Kolmogorov complexity* of x with respect to M, as $C_M(x) = \min\{|p| \mid M(p) = x\}$. There is a universal Turing machine U such that for every machine M there is a constant c such that for all x, $C_U(x) \leq C_M(x) + c$. We fix such a universal machine U and dropping the subscript, we let $C(x)$ denote the Kolmogorov complexity of x with respect to U. We also use the concept of conditional Kolmogorov complexity. Here the underlying machine is a Turing machine that in addition to the read/work tape which in the initial state contains the input p, has a second tape containing initially a string y, which is called the conditioning information. Given such a machine M, we define the Kolmogorov complexity of x conditioned by y with respect to M as $C_M(x \mid y) = \min\{|p| \mid M(p,y) = x\}$. There exist universal machines of this type and they satisfy the relation similar to the above, but for conditional complexity. We fix such a universal machine U, and dropping the subscript U, we let $C(x \mid y)$ denote the Kolmogorov complexity of x conditioned by y with respect to U.

For every sufficiently large n and $k \leq n$, for every n-bit string y, $2^{k-2\log n} < |\{x \in \{0,1\}^n \mid C(x \mid y) \leq k\}| < 2^{k+1}$.

The Symmetry of Information Theorem [26] states that for any two strings x and y,

(a) $C(xy) \leq C(y) + C(x \mid y) + 2\log C(y) + O(1)$.

(b) $C(xy) \geq C(x) + C(y \mid x) - 2\log C(xy) - 4\log\log C(xy) - O(1)$.

(c) If $|x| = |y| = n$, $C(y) - C(y \mid x) \geq C(x) - C(x \mid y) - 5\log n$

For integers $m \leq n$, let $b(n,m) = \binom{n}{0} + \binom{n}{1} + \ldots + \binom{n}{m}$. Note that $m(\log n - \log m) < \log b(n,m) < m(\log n - \log m) + m\log e + \log(1+m)$ (since $(n/m)^m < \binom{n}{m} < (en/m)^m$).

All the Kolmogorov extractors will be ensembles of functions $f = (f_n)_{n\in\mathbb{N}}$, of type $f_n : (\{0,1\}^n)^t \rightarrow \{0,1\}^{m(n)}$. The parameter t is constant and indicates the number of sources (in this paper we only consider $t = 1$ and $t = 2$). For readability, we usually drop the subscript and the expression "function

$f : \{0,1\}^n \to \{0,1\}^m$ ..." is a substitute for "ensemble $f = (f_n)_{n\in\mathbb{N}}$, where $f_n : \{0,1\}^n \to \{0,1\}^{m(n)}$, ..."

We say that an ensemble of functions $f = (f_n)$ is computable with advice $k(n)$, if for every n there exists a string p of length at most $k(n)$ such that $U(p, 1^n)$ outputs the table of the function f_n.

We use the following standard version of the Chernoff bounds. Let $X_1, \ldots, X_n$ be independent random variables that take the values 0 and 1, let $X = \sum X_i$ and let μ be the expected value of X. Then, for any $0 < d \leq 1$, $\text{Prob}[X > (1+d)\mu] \leq e^{-d^2\mu/3}$.

2.1 Limited Independence

Definition 1. *(a) The dependency of two strings x and y is* $\text{dep}(x,y) = \max\{C(x) - C(x \mid y), C(y) - C(y \mid x)\}$.

(b) Let $d : \mathbb{N} \to \mathbb{N}$. We say that strings x and y have dependency at most $d(n)$ if $\text{dep}(x,y) \leq d(\max(|x|,|y|))$.

The Symmetry of Information Theorem implies that $|\text{dep}(x,y) - (C(x) - C(x \mid y))| \leq O(\log(C(x)) + \log(C(y)))$. If the strings x and y have length n, then $|\text{dep}(x,y) - (C(x) - C(x \mid y))| \leq 5 \log n$.

3 Limits on Kolmogorov Complexity Extraction

3.1 Limits on Extraction from One Source

We first show that for any single-source function computable with small advice there exists an input with high Kolmogorov complexity whose image has low Kolmogorov complexity.

Proposition 1. *Let $f : \{0,1\}^n \to \{0,1\}^m$ be a function computable with advice $k(n)$. There exists $x \in \{0,1\}^n$ with $C(x) \geq n - m$ and $C(f(x)) \leq k(n) + \log n + 2 \log\log n + O(1)$.*

The following result is, in a sense, a strengthening of the previous proposition. It shows that there exists a string with relatively high Kolmogorov complexity, so that all functions computable with a given amount of advice fail to extract its randomness. We provide two incomparable combinations of parameters. Part (b) is essentially a result of Vereshchagin and Vyugin [22].

Theorem 1. *For every k, every n, any computable function m:*

(a) There exists a string $x \in \{0,1\}^n$ such that for every function $f : \{0,1\}^n \to \{0,1\}^m$ that is computable with advice $k = k(n)$,

(1) $C(x) > n - \log b(M,K) \geq n - K(m-k+O(1))$, where $M = 2^m, K = 2^{k+1}-1$, and

(2) $C(f(x)) < 2k + 2\log k + \log n + 2\log\log n + O(1)$ or $f(x)$ is not defined.

and

(b) There exists a string $x \in \{0,1\}^n$ *such that for every function* $f : \{0,1\}^n \to \{0,1\}^m$ *that is computable with advice* k,

(1) $C(x) > n - K\log(M+1) \approx n - Km$, *where* $M = 2^m, K = 2^{k+1} - 1$, *and*

(2) $C(f(x)) < k + \log n + 2\log\log n + O(1)$ *or* $f(x)$ *is not defined.*

3.2 Limits on Extraction from Two Sources

The following theorem shows that there is no uniform function that from two sources x and y that are α-dependent (i.e., $\mathrm{dep}(x,y) \succeq \alpha$), produces an output whose randomness deficiency is less than $\alpha - \log n - O(\log\alpha)$.

Theorem 2. *Let* $f : \{0,1\}^n \times \{0,1\}^n \to \{0,1\}^m$ *be a computable function and let* $\alpha \in \mathbb{N}$, $\alpha \leq m$. *Then there exists a pair of strings* $x \in \{0,1\}^n, y \in \{0,1\}^n$ *such that the following three properties hold: (1)* $C(x \mid y) \geq n - \alpha - 2\log n$, *(2)* $C(y \mid x) \geq n - \alpha - 2\log n$, *(3)* $C(f(x,y)) \leq m - \alpha + \log n + 2\log\alpha + O(1)$.

The following is the analog of Theorem 2 for distributions.

Theorem 3. *Let* $f : \{0,1\}^n \times \{0,1\}^n \to \{0,1\}^m$ *be a function and let* $\alpha \in \mathbb{N}$, $\alpha \leq m$. *Then there exists two random variables* X *and* Y *taking values in* $\{0,1\}^n$ *such that the following four properties hold: (1)* $H_\infty(X) \geq n - \alpha$, (2) $H_\infty(Y) \geq n - \alpha$, (3) $H_\infty(X,Y) \geq 2n - \alpha$, (4) $H_\infty(f(X,Y)) \leq m - \alpha$.

4 Kolmogorov Complexity Extraction

We construct a Kolmogorov extractor that on input two n-bit strings with Kolmogorov complexity at least $s(n)$ and dependency at most $\alpha(n)$ outputs a string of length $\approx s(n)$ having complexity $\approx s(n) - \alpha(n)$ conditioned by any one of the input strings.

4.1 Construction Overview

We describe the main ideas of the method. We also explain the non-trivial way in which the new construction extends the technique from the earlier works [23] and [25]. For readability, some details are omitted and some estimations are slightly imprecise. Let us fix, for the entire discussion, x and y, two n-bit strings with $C(x) \geq s(n)$ and $C(y) \geq s(n)$ and having dependency at most $\alpha(n)$. We denote $N = 2^n, M = 2^m$ and $S = 2^{s(n)}$. Let $B_x = \{u \mid C(u) \leq C(x)\}$ and $B_y = \{v \mid C(v) \leq C(y)\}$. An N-by-N table colored with M colors is a function $T : [N] \times [N] \to [M]$. If we randomly color such a table T, with parameter $m \preceq 2s(n)$, then, with high probability, no color appears in the $B_x \times B_y$ rectangle more than $2 \cdot (1/M)$ fraction of times (we say that a table that satisfies the above balancing property is balanced in $B_x \times B_y$). Clearly $(x,y) \in B_x \times B_y$ and in a table T balanced in $B_x \times B_y$ there are at most $2 \cdot (1/M) \cdot |B_x| \times$

$|B_y| \approx 2 \cdot (1/M)2^{C(x)}2^{C(y)} = 2^{C(x)+C(y)-m+1}$ entries with the color $z = T(x, y)$. Therefore (x, y) is described by the color $z = T(x, y)$, the rank r of the (x, y) cell in the list of all z-colored cells in $B_x \times B_y$, by the table T, and by $O(\log n)$ additional bits necessary to enumerate the list. Thus, $C(xy) \leq C(z) + \log r + C(\text{table } T) + O(\log n)$. By the above estimation, $\log r \approx C(x) + C(y) - m$. Also $C(xy) \geq C(x)+C(y)-\text{dep}(x, y)$. Suppose that we are able to get a balanced table T with $C(\text{table } T) = O(\log n)$, i.e., a table that can be described with $O(\log n)$ bits. Then we would get that $C(T(x, y)) = C(z) \geq m - \text{dep}(x, y)$, which is our goal. How can we obtain $C(\text{table } T) = O(\log n)$? The normal approach would be to enumerate all possible N-by-N tables with all possible colorings with M colors and pick the first one that satisfies the balancing property. However, B_x and B_y can be enumerated only if $C(x)$ and respectively $C(y)$ are given. Thus we can not check the balancing property in a uniform way. Therefore, instead of restricting to only B_x and B_y, we require that a table T should satisfy the balancing property for *all* rectangles $B_1 \times B_2$ with sizes $|B_1| \geq S$ and $|B_2| \geq S$, where $S = 2^{s(n)}$. The simple probabilistic analysis involves only an additional union bound and carries over showing that such balanced tables exist at the cost that this time we need $m \preceq s(n)$. Now we can pick in an effective way the smallest (in some canonical order) table T having the balancing property, because we can check the balancing property in an exhaustive manner (look at all $S \times S$-sized rectangles, etc.). Therefore this table T can be described with $\log n + O(1)$ bits, as desired. In this way, from any x and y, each having Kolmogorov complexity at least $s(n)$, we obtain $m \approx s(n)$ bits having Kolmogorov complexity $m-\text{dep}(x, y)$. We reobtain $m \approx 2s(n)$ if we change the balancing property and require that for any subset of colors $A \subseteq [M]$ of size M/D, for $D \approx 2^{\alpha(n)}$, for any rectangle $B_1 \times B_2$ with sizes $|B_1| \geq S$ and $|B_2| \geq S$, the fraction of A-colored cells in $B_1 \times B_2$ should be at most $2 \cdot (|A|/M) = 2 \cdot (1/D)$. Such a table can be obtained with $m \approx 2s(n)$, and thus we can extract $\approx 2s(n)$ bits having Kolmogorov complexity $\approx 2s(n) - \text{dep}(x, y)$, which is optimal.

Let us consider next the problem of extracting bits that are random even conditioned by x, and also conditioned by y. Suppose we use tables that satisfy the first balancing property. We focus on $B_x = \{u \mid C(u) \leq C(x)\}$ and we call a column v *bad* for a color $a \in [M]$ if the fraction of a-colored cells in the strip $B_x \times \{v\}$ of the table T is more than $2 \cdot (1/M)$. The number of bad columns is less than S; otherwise the table would have an $S \times S$-sized rectangle that does not have the balancing property. Note that a bad column for a color a can be described by the color a and its rank in an enumeration of the columns that are bad for a plus additional $O(\log n)$ bits. So if v is a bad column, then $C(v) \preceq m + s(n) \approx 2s(n)$. Therefore if $C(y) \succeq 2s(n)$, y is good for any color. An adaptation of the above argument shows that for $z = T(x, y)$ it holds that $C(x \mid y) \preceq C(z \mid y) + C(x) + C(y) - m$, which combined with $C(x \mid y) \succeq C(x) + C(y) - \text{dep}(x, y)$, implies $C(z \mid y) \succeq m - \text{dep}(x, y)$. The above holds only for y with $C(y) \geq 2s(n)$ and since the probabilistic analysis requires m to be less than $s(n)$, it follows that the number of extracted bits (which is m) is less than half the Kolmogorov complexity of y.

The above technique was used in [23] and in [25]. To increase the number of extracted bits, we introduce a new balancing property, which we dub *rainbow balancing*. Fix some parameter D, which eventually will be taken such that $\log D \approx \mathrm{dep}(x, y)$. Let $\mathcal{A}_D$ be the collection of sets of colors $A \subseteq [M]$ with size $|A| \approx M/D$. Let $B_1 \subseteq [N]$ be a set of size a multiple of S, let $\overline{v} = \{v_1 < v_2 \ldots < v_S\}$ be a set of S columns, and let $\overline{A} = (A_1, \ldots, A_S)$ be a tuple with each A_i in $\mathcal{A}_D$. We say that a cell (u, v_i) such that $T(u, v_i) \in A_i$ is properly colored with respect to $\overline{v}$ and $\overline{A}$. Finally we say that a table $T : [N] \times [N] \to [M]$ is (S, D)-rainbow balanced if for every B_1, every $\overline{v}$, and every $\overline{A}$, the fraction of cells in $B_1 \times \overline{v}$ that are properly colored with respect to $\overline{v}$ and $\overline{A}$ is at most $2 \cdot (1/D)$. The probabilistic method shows that such tables exist provided $m \preceq s(n)$ and $\log D \preceq s(n)$. Since the rainbow balancing property can be effectively checked, there is an (S, D)-rainbow balanced table $T : [N] \times [N] \to [M]$ that can be described with $\log n + O(1)$ bits and $m \approx s(n)$ and $\log D \approx s(n)$. Let $z = T(x, y)$ and suppose that $C(z \mid y) < m - t$, where $t = \alpha(n) - c \log n$, for some constant c that will be defined later (in the actual proof we do a tighter analysis and we manage to take $t = \alpha(n) - O(1)$). For each v, let $A_v = \{w \in [M] \mid C(w \mid v) < m - t\}$. For $\log D \approx \alpha(n) + c \log m$, it holds that $A_v \in \mathcal{A}_D$ for all v. Let us call a column v *bad* if the fraction of cells in $B_x \times \{v\}$ that are A_v-colored is larger than $2 \cdot (1/2^t)$. Analogously to our earlier discussion, the number of bad columns is less than S and from here we infer that if v is a bad column, then $C(v) \preceq s(n)$. Since $C(y) \geq s(n)$, it follows that y is a good column. Therefore the fraction of cells in the $B_x \times \{y\}$ strip of the table T that have a color in A_y is at most $2 \cdot (1/2^t)$. Since (x, y) is one of these cells, it follows that, given y, x can be described by the rank r of (x, y) in an enumeration of the A_y-colored cells in the strip $B_x \times \{y\}$, a description of the table T, and by $O(\log n)$ additional bits necessary for doing the enumeration. Note that there are at most $2 \cdot (1/2^t) \cdot |B_x| \approx 2^{-t+1} \cdot 2^{C(x)}$ cells in $B_x \times \{y\}$ that are A_y-colored and, therefore, $\log r \leq C(x) - t + 1$. From here we obtain that $C(x \mid y) \leq C(x) - t + 1 + O(\log n) = C(x) - \alpha(n) - c \log n + O(\log n)$. Since $C(x \mid y) \geq C(x) - \alpha(n)$, we obtain a contradiction for an appropriate choice of the constant c. Consequently $C(z \mid y) \geq m - t = m - \alpha(n) + c \log n$. Similarly, $C(z \mid x) \geq m - \alpha(n) + c \log n$. Thus we have extracted $m \approx s(n)$ bits that have Kolmogorov complexity $\approx m - \alpha(n)$ conditioned by x and also conditioned by y.

4.2 Construction of the Kolmogorov Extractor

For n and m natural numbers, let $N = 2^n$ and $M = 2^m$. Henceforth, we identify $\{0, 1\}^n$ with $[N]$ and $\{0, 1\}^m$ with $[M]$.

We consider functions of the form $T : [N] \times [N] \to [M]$, which we view as N-by-N tables whose cells are colored with colors in $[M]$. Let S and D be parameters with $S \leq N$ and $D \leq M$.

Let $\mathcal{A}_D = \{A \mid A \subseteq [M], (M/D) \leq |A| \leq (M/D)m^2\}$. Thus, the elements of $\mathcal{A}_D$ are those sets of colors having at least M/D colors and not much more than that.

Let $B_2 \subseteq [N]$ be a subset of size S; we name its elements $B_2 = \{v_1 < v_2 < \ldots < v_S\}$. We view B_2 as a set of columns in the table. Let $(A_1, \ldots, A_S) \in (\mathcal{A}_D)^S$. The cell $(u, v_i) \in [N] \times B_2$ is *properly colored* with respect to the columns in B_2 and $(A_1, \ldots, A_S)$ if $T(u, v_i) \in A_i$. A similar notion of a cell being properly colored with respect to *rows* in a set $B_1 \subseteq [N]$ will also be used.

Definition 2. *A table $T : [N] \times [N] \to [M]$ is (S, D)-rainbow balanced if*

(a)
- *for all $B_1 \subseteq [N]$ of size $k \cdot S$ for some positive natural number k,*
- *for all $B_2 \subseteq [N]$ of size S,*
- *for all $(A_1, \ldots, A_S) \in (\mathcal{A}_D)^S$,*

it holds that the number of cells in $B_1 \times B_2$ that are properly colored with respect to columns B_2 and $(A_1, \ldots, A_S)$ is at most $2m^2 \frac{|B_1| \cdot |B_2|}{D}$,

and

(b) if the similar relation holds if we switch the roles of B_1 and B_2.

Lemma 1. *If $S \geq 12D + 3(1 + \ln D)Mm^2 + 6D \ln(N/S)$, there exists a table $T : [N] \times [N] \to [M]$ that is (S, D)-rainbow balanced.*

Theorem 4. *For any computable functions $s(n)$ and $\alpha(n)$ with $n \geq s(n) \geq \alpha(n) + 7 \log n + O(1)$, for every computable function $m(n)$ with $m(n) \leq s(n) - 7 \log n$, there exists a computable function $E : \{0,1\}^n \times \{0,1\}^n \to \{0,1\}^{m(n)}$, such that for all x and y in $\{0,1\}^n$ if*

(i) $C(x) \geq s(n), C(y) \geq s(n)$,

(ii) $C(x) - C(x \mid y) \leq \alpha(n)$ and $C(y) - C(y \mid x) \leq \alpha(n)$,

then

(1) $C(E(x, y) \mid x) \geq m - \alpha(n) - O(1)$,

(2) $C(E(x, y) \mid y) \geq m - \alpha(n) - O(1)$.

5 Impossibility of Independence Amplification

The dependence of strings x and y is given by $\text{dep}(x, y) = C(x) + C(y) - C(xy)$. The smaller $\text{dep}(x, y)$ is, the more independent the strings x and y are. Thus, amplifying independence amounts to reducing dependence. An effective dependence reducer would consist of two computable functions f_1 and f_2 that for two functions $\alpha(n) > \beta(n)$ guarantee that for all x, y of length n,

$$\text{dep}(x, y) \leq \alpha(n) \Rightarrow \text{dep}(f_1(x, y), f_2(x, y)) \leq \beta(n). \quad (1)$$

Note that, since $\text{dep}(u, v) \leq \beta(n)$ whenever $C(u) \leq \beta(n)$ or $C(v) \leq \beta(n)$, dependency reduction would be achieved by two functions that simply output strings with Kolmogorov complexity $\leq \beta(n)$. To avoid this trivial and non-interesting type of dependency reduction, we require that, in addition to requierement (1), $C(f_1(x,y)) \succeq \beta(n)$ and $C(f_2(x,y)) \succeq \beta(n)$. More precisely, we seek two computable functions $f_1 : \{0,1\}^n \times \{0,1\}^n \to \{0,1\}^{l(n)}$ and $f_2 : \{0,1\}^n \times \{0,1\}^n \to \{0,1\}^{l(n)}$ that satisfy the following DEPENDENCY REDUCTION TASK.

DEPENDENCY REDUCTION TASK for parameters $\alpha(n), \beta(n)$, $s(n)$, $l(n)$, and a.

For all $x \in \{0,1\}^n$, $y \in \{0,1\}^n$ with $\text{dep}(x,y) \le \alpha(n)$, $C(x) \ge s(n)$ and $C(y) \ge s(n)$ the following should hold:

1. $\text{dep}(f_1(x,y), f_2(x,y)) \le \beta(n)$,
2. $C(f_1(x,y)) \ge \beta(n) + a \cdot \log n$ and $C(f_2(x,y)) \ge \beta(n) + a \cdot \log n$.

We show that effective independence amplification is essentially impossible.

Theorem 5. *Let $\alpha(n)$ be a function such that $\alpha(n) \le n/2 - 5\log n$ and let $\beta(n) = \alpha(n) - \log n - 3\log\alpha(n)$. Let $s(n)$ be a function such that $s(n) \le n - \alpha(n) - 2\log n - O(1)$ and let $l(n)$ be a function such that $l(n) \ge \beta(n) + 8\log n$.*

There are no computable functions $f_1 : \{0,1\}^n \times \{0,1\}^n \to \{0,1\}^{l(n)}$ and $f_2 : \{0,1\}^n \times \{0,1\}^n \to \{0,1\}^{l(n)}$ satisfying the DEPENDENCY REDUCTION TASK for parameters $\alpha(n)$, $\beta(n)$, $s(n)$, $l(n)$ and $a = 8$.

References

1. Barak, B., Impagliazzo, R., Wigderson, A.: Extracting randomness using few independent sources. In: Proceedings of the 36th ACM Symposium on Theory of Computing, pp. 384–393 (2004)
2. Barak, B., Kindler, G., Shaltiel, R., Sudakov, B., Wigderson, A.: Simulating independence: new constructions of condensers, ramsey graphs, dispersers, and extractors. In: Proceedings of the 37th ACM Symposium on Theory of Computing, pp. 1–10 (2005)
3. Bienvenu, L., Doty, D., Stephan, F.: Constructive dimension and Turing degrees. Theory Comput. Syst. 45(4), 740–755 (2009)
4. Bourgain, J.: More on the sum-product phenomenon in prime fields and its applications. International Journal of Number Theory 1, 1–32 (2005)
5. Calude, C.: Information and Randomness: An Algorithmic Perspective, 2nd edn. Springer, Heidelberg (2002); 1st edn. (1994)
6. Chor, B., Goldreich, O.: Unbiased bits from sources of weak randomness and probabilistic communication complexity. SIAM Journal on Computing 17, 230–261 (1988)
7. Dodis, Y., Elbaz, A., Oliveira, R., Raz, R.: Improved randomness extraction from two independent sources. In: APPROX-RANDOM, pp. 334–344 (2004)
8. Dodis, Y., Oliveira, R.: On extracting private randomness over a public channel. In: Arora, S., Jansen, K., Rolim, J.D.P., Sahai, A. (eds.) RANDOM 2003 and APPROX 2003. LNCS, vol. 2764, pp. 252–263. Springer, Heidelberg (2003)

9. Downey, R., Hirschfeldt, D.: Algorithmic randomness and complexity. Springer, Heidelberg (2010)
10. Fortnow, L., Hitchcock, J., Pavan, A., Vinodchandran, N., Wang, F.: Extracting Kolmogorov complexity with applications to dimension zero-one laws. In: Bugliesi, M., Preneel, B., Sassone, V., Wegener, I. (eds.) ICALP 2006. LNCS, vol. 4051, pp. 335–345. Springer, Heidelberg (2006)
11. Hitchcock, J., Pavan, A., Vinodchandran, N.: Kolmogorov complexity in randomness extraction. In: Electronic Colloquium on Computational Complexity (ECCC) (09-071) (2009)
12. Li, M., Vitanyi, P.: An introduction to Kolmogorov complexity and its applications, 3rd edn. Springer, Heidelberg (2008); 1st edn. (1993)
13. Miller, J.: Extracting information is hard: a Turing degree of non-integral effective Hausdorff dimension. Advances in Mathematics (2008) (to appear)
14. Nies, A., Reimann, J.: A lower cone in the wtt degrees of non-integral effective dimension. In: Proceedings of IMS workshop on Computational Prospects of Infinity, Singapore (2006) (to appear)
15. Rao, A.: Extractors for a constant number of polynomially small min-entropy independent sources. In: Proceedings of the 38th ACM Symposium on Theory of Computing, pp. 497–506 (2006)
16. Rao, A., Zuckerman, D.: Extractors for three uneven-length sources. In: Goel, A., Jansen, K., Rolim, J.D.P., Rubinfeld, R. (eds.) APPROX and RANDOM 2008. LNCS, vol. 5171, pp. 557–570. Springer, Heidelberg (2008)
17. Raz, R.: Extractors with weak random seeds. In: Gabow, H.N., Fagin, R. (eds.) STOC, pp. 11–20. ACM, New York (2005)
18. Reimann, J.: Computability and fractal dimension. Tech. rep., Universität Heidelberg (2004) (ph.D. thesis)
19. Santha, M., Vazirani, U.: Generating quasi-random sequences from semi-random sources. Journal of Computer and System Sciences 33, 75–87 (1986)
20. Shen, A.: Algorithmic information theory and Kolmogorov complexity. Tech. Rep. 2000-034, Uppsala Universitet (December 2000)
21. Trevisan, L., Vadhan, S.: Extracting randomness from samplable distributions. In: Proceedings of the 41st IEEE Symposium on Foundations of Computer Science, pp. 32–42 (2000)
22. Vereshchagin, N.K., Vyugin, M.V.: Independent minimum length programs to translate between given strings. Theor. Comput. Sci. 271(1-2), 131–143 (2002)
23. Zimand, M.: Extracting the Kolmogorov complexity of strings and sequences from sources with limited independence. In: Proceedings 26th STACS, Freiburg, Germany, February 26-29 (2009)
24. Zimand, M.: Two sources are better than one for increasing the Kolmogorov complexity of infinite sequences. In: Hirsch, E.A., Razborov, A.A., Semenov, A., Slissenko, A. (eds.) CSR 2008. LNCS, vol. 5010, pp. 326–338. Springer, Heidelberg (2008)
25. Zimand, M.: On generating independent random strings. In: Ambos-Spies, K., Löwe, B., Merkle, W. (eds.) CiE. LNCS, vol. 5635, pp. 499–508. Springer, Heidelberg (2009)
26. Zvonkin, A., Levin, L.: The complexity of finite objects and the development of the concepts of information and randomness by means of the theory of algorithms. Russian Mathematical Surveys 25(6), 83–124 (1970)

Author Index